ABOUT ISLAND PRESS

Island Press is the only nonprofit organization in the United States whose principal purpose is the publication of books on environmental issues and natural resource management. We provide solutions-oriented information to professionals, public officials, business and community leaders, and concerned citizens who are shaping responses to environmental problems.

In 2006, Island Press celebrates its twenty-first anniversary as the leading provider of timely and practical books that take a multidisciplinary approach to critical environmental concerns. Our growing list of titles reflects our commitment to bringing the best of an expanding body of literature to the environmental community throughout North America and the world.

Support for Island Press is provided by the Agua Fund, The Geraldine R. Dodge Foundation, Doris Duke Charitable Foundation, The William and Flora Hewlett Foundation, Kendeda Sustainability Fund of the Tides Foundation, Forrest C. Lattner Foundation, The Henry Luce Foundation, The John D. and Catherine T. MacArthur Foundation, The Marisla Foundation, The Andrew W. Mellon Foundation, Gordon and Betty Moore Foundation, The Curtis and Edith Munson Foundation, Oak Foundation, The Overbrook Foundation, The David and Lucile Packard Foundation, The Winslow Foundation, and other generous donors.

The opinions expressed in this book are those of the author(s) and do not necessarily reflect the views of these foundations.

FISH CONSERVATION

FISH CONSERVATION

A Guide to Understanding and Restoring Global Aquatic Biodiversity and Fishery Resources

Gene S. Helfman

Washington • Covelo • London

Library of Congress Cataloging-in-Publication Data

Helfman, Gene S.
Fish conservation : a guide to understanding and restoring global aquatic biodiversity and fishery resources / Gene S. Helfman.
p. cm.
ISBN-13: 978-1-55963-595-0 (cloth : alk. paper)
ISBN-10: 1-55963-595-9 (cloth : alk. paper)
ISBN-13: 978-1-55963-596-7 (pbk. : alk. paper)
ISBN-10: 1-55963-596-7 (pbk. : alk. paper)
1. Fishes—Conservation. 2. Biodiversity conservation. 3. Fishery Management. I. Title.
SH327.7.H46 2007
333.95'611—dc22

2007001842

Printed on recycled, acid-free paper 

Manufactured in the United States of America

10 9 8 7 6 5 4 3 2

To Bob Johannes, who cared about fishes and people and made the world a better place for both.

Contents

PART IV: DIRECT CAUSES OF DECLINE: FISHES AS COMMODITIES

PART V: ASKING HARD QUESTIONS, SORTING OUT ANSWERS

Preface

When I first toyed with the idea of writing a general book on fish conservation, beginning in the mid-1990s, the world was a simpler place. I had seldom visited the World Wide Web, and I doubt there were many relevant sites. Coelacanths were known only from the Comoros Islands, a U.S. shark management plan was only a dream, dams were decommissioned primarily as a subplot in ecoterrorism novels written by Edward Abbey, and no one had put together a seafood watch list.

Much has transpired in the world of fish conservation since. More than half of the 2,000-plus references cited in this volume were published after I initiated my writing; new studies appear daily and in growing numbers. Fish conservation has emerged as a topic of intense public interest, fueled by media coverage of scientific findings and policy decisions. What has also grown is a general awareness of our dependence on aquatic environments and their inhabitants, both of which are showing unmistakable signs of human-caused deterioration and loss.

My initial motivation to write this book was to fill a gap, namely the lack of a general treatment of the diversity of topics that fell under the general heading of "fish conservation." I was warned by colleagues early on that the topic was impossibly large and complex but that someone should bring it together in one place. They were right on both accounts. What none of us anticipated was how rapidly the literature would grow, a growth that regularly tempted me to update and expand on subjects I had already covered and add coverage of topics that had newly emerged. This temptation was tempered by the realization that the original need for a general book still existed. Which is my way of saying that I couldn't get everything into this book that I wanted, that I've undoubtedly missed important topics, recent events, and critical references, treated others superficially, and still produced a book that is inordinately—but necessarily—long. If this book provides interested persons with the background needed to explore topics in greater depth I will have been successful. If they then use this knowledge to take action to conserve fishes and their habitats, I will have succeeded beyond my hopes.

THE NATURE AND STRUCTURE OF THE BOOK

Many of the topics and issues dealt with in this book represent general environmental problems in need of scientific solutions. My objective is to focus on the interaction between these issues and fishes, without ignoring the larger-scale nature of the problems. My approach has been to write chapters that deal with generalities, principles, and concepts, followed with specific examples and solutions. Solutions are scattered throughout each chapter—where a specific example is given, suggested solutions are often included with the description. More sweeping solutions are presented toward the end of most chapters in a "Solutions" section. The final chapter attempts to summarize and synthesize some of the more important, recurring themes.

This book purports to be global in perspective. International breadth is most evident in the taxonomic and geopolitical chapters (2, 3) and those on coral reef fishes and the live fish trade (12, 13). In others, most examples are drawn from North America, with some reference to literature from elsewhere. I apologize for this focus—nearly 200 countries are United Nations members, and all have fishes and fish conservation problems—but I can confidently state that the vast majority of problems and their solutions are generalizable, regardless of nationality.

HOW MUCH FISH KNOWLEDGE IS NEEDED TO NAVIGATE THIS BOOK?

I'm assuming a reader has some basic knowledge of ichthyology—if not at the level of an entire course, at least at the level of a college-level general biology class. I've tried to write this book so that the material is usable by anyone concerned with general issues of conservation, minimizing the taxonomic, anatomical, and biogeographic jargon that underlies the science of ichthyology. For those seeking deeper background, a few general treatments can serve as an introduction to ichthyology. The University of Michigan maintains an excellent Web site that presents the basics in readable form, with many luxuriant, cross-referenced illustrations (http://animaldiversity.ummz.umich.edu/site/index.html; see "Bony Fishes"). The best overall Web site for details on just about every fish alive is Froese and Pauly (2006, www.fishbase.org). Another general taxonomic overview with some biological and conservation tidbits is Helfman (2001); the *Encyclopedia of Biodiversity* is a good general reference on taxonomic groups and conservation-related topics. The nicest popularization of the subject I've encountered is Paxton and Eschmeyer (1998). Photos of just about every fish mentioned here can be found at www.A9.com if you click on "Images." For greater coverage, three general ichthyology textbooks are Bond (2006), Helfman et al. (1997), and Moyle and Cech (2004). The definitive treatment of fish taxonomy remains J. S. Nelson (2006).

FISH CONSERVATION IN A LARGER CONTEXT

"Conservation ecology" has always played a critical role in human affairs. Parochial societies conserved resources via land tenureship, taboos, restricted entry, and tradition. Beyond normal social pressures, violators were deterred by the penalties of ostracism, physical punishment, and death. In primitive and traditional cultures, the ecologies of humans, plants, and other animals existed in a continuum of interaction. Humans saw themselves as part of this continuum—if not in harmonious balance, at least in dependency. Western society developed an ecology of domination and alienation, a philosophy of conquest, with nature often the enemy. This philosophy evolved into estrangement between humans and nature as urbanization and industrialization became the norm. In the third world, conservation is often an impractical luxury because critical scarcities of food, shelter, and water are everyday realities, at the same time that people aspire to a Western standard of living. Disharmony with nature, which sits at the base of our present global environmental crisis, is a consequence of the expansion of Western consumptive practices, the collapse of traditional cultures, and the weight of burgeoning human populations.

Aquatic ecosystems have borne the brunt of urbanization, industrialization, agriculture, cultural dissolution, and overpopulation. Approximately 55% of the U.S. population lives and works in counties adjacent to an ocean or one of the Great Lakes; 19 of the 20 largest cities are on a river, lake, or estuary. Similar circumstances apply globally. Few human societies have prospered away from reliable sources of water, and major civilizations have collapsed repeatedly as shifting climate patterns altered the availability of water (e.g., Gore 1992; Diamond 2004). Humans depend on water: for drinking, cooking, hygiene, agriculture, fish and game, transportation, energy, spiritual practices, recreation, and waste disposal. All require more or less healthy aquatic ecosystems. But through carelessness, greed, and lack of understanding, healthy aquatic ecosystems and growing human populations have become increasingly incompatible. The result is degraded waters characterized by low biotic diversity. The flora and the fauna of a stream or river or lake or coastline or reef can tell us how badly we are soiling our nest.

It is the role of ecologists to assess ecosystem health, to warn us when things are in decline, and to offer suggestions for remediation. This role has been well served in the past few decades, as more ecologists have shifted from traditional academic pursuits and instead strive to apply ecological knowledge to the practical problems of environmental degradation and restoration. But surprisingly, given human dependence on water, the emphasis in conservation ecology has been on terrestrial issues. The goals of this book are to help rectify this disparity: to summarize the current state of our knowledge about aquatic ecosystem conservation; to point out areas where work is in progress and where it is needed; and, specifically, to use the world's fish fauna as a bellwether of aquatic ecosystem health.

The tone of this book is one of advocacy. I advocate the ethical conviction that fish diversity deserves to be conserved. Advocacy is a dangerous word in scientific circles, where the Great Fear is that when we become advocates, we lose objectivity. I have tried to maintain objectivity throughout this volume, to explore and report alternative information and interpretations, to advocate for fishes and their ecosystems by reaching logical conclusions based on credible sources. If we as concerned individuals don't do something to halt and reverse planet-wide trends toward environmental degradation—trends that can only have disastrous consequences for all of Earth's life forms—who will?

FISH CONSERVATION IN A PERSONAL CONTEXT

I began work on this book after almost three decades as a behavioral ecologist, engaged in what an unimpressed observer referred to as underwater birdwatching. I had a grand time studying fascinating fishes in beautiful settings, a childhood dream come true (I've been infatuated with fishes since about age nine). I coauthored an ichthyology textbook, which forced me to learn much more about fishes than I could have imagined and made me realize how narrow my knowledge was. Much earlier, I had, with limited success, been a commercial fisherman, working as a deckhand on sport fishing boats, trolling for salmon, jackpoling and seining for tuna. Between my commercial and academic phases, I spent three years in the Peace Corps in Palau, exploring reefs, watching and spearing fish, and partaking in a subsistence (albeit subsidized) lifestyle. As my exploitative and investigative activities progressed, it became increasingly obvious that things were changing. The fishes I knew and their habitats were in alarming decline, a phenomenon that became impossible to ignore.

As fellow behaviorist Hans Fricke so aptly states when asked why he is devoting so much effort trying to save the coelacanth, "It is time to give something back to the fishes." *Fish Conservation* is my attempt to give something back.

ACKNOWLEDGMENTS

Hundreds of people in dozens of countries—academics, professional conservationists, ardent anglers, commercial fishers, underwater photographers, live fish traders, and aquarium keepers—have aided me immeasurably during the writing of this book. Over the past decade, I'm afraid

I've misplaced many names. I apologize to the many people who helped whom I am overlooking. Dozens of students in my ichthyology and conservation biology classes wrote term papers that served as sources of information for many topics. I also apologize to the conscientious reviewers enlisted by Island Press to critique individual chapters. Among their helpful comments were numerous suggestions for additions and updates of material. As useful as these were, my major goal during revisions was to cut rather than add material. The book would have been improved with those suggested additions.

Barbara Dean and Barbara Youngblood of Island Press have shown tremendous patience while encouraging me to finish this project and tremendous faith in its ambition. Jill Mason of Masonedit.com took my final draft from which I could cut no more and cut it another 5% without loss of content but with considerable improvement in style and clarity.

Scores of persons have read all or parts of many chapters. Without ranking by importance or type of contribution, I gratefully acknowledge the help of the following: C. Anderson, L. Anderson, A. Arthington, G. Barlow, J. Beets, W. Bemis, E. Benigno, J. Benstead, B. Best, V. Birstein, N. Bogutskaya, J. Bohnsack, B. Bowen, B. Bruce, A. Bull-Tornøe, N. Burkhead, V. Burnley, B. Carlson, N. Chao, L. Chapman, J. Chavous, A. Clarke, F. Coleman, L. Colin, P. Colin, M. Collares-Pereira, B. Collette, S. Contreras-Balderas, W. Courtenay, I. Cowx, T. Coyle, R. Dirks, P. Doherty, T. Donaldson, R. Edwards, M. Erdmann, P. Esselman, E. Fernandez-Galiano, M. Fitzsimons, B. Freeman, M. Freeman, H. Fricke, G. Galland, D. Haggarty, J. Hamlin, J. Hawkins, P. Heemstra, JB Heiser, D. Helfmeyer, M. Helfmeyer, Z. Hogan, J. Hutchins, C. Jeffrey, R. E. Jenkins, L. Kaufman, C. Koenig, M. Kottelat, S. Kraft, B. Kuhajda, J. Lichatowich, O. Lucanus, R. Lynch, A. MacNeil, P. Maitland, P. Marcinek, K. Martin-Smith, R. Mayden, D. McAllister, F. McCormick, R. M. McDowall, W. McFarland, G. Meffe, W. Minckley, M. Moore, R. Mottice, P. Moyle, J. Nelson, G. Ostrander, P. Pister, D. Policansky, G. Proudlove, B. Pusey, R. Pyle, P. Quong, F. Rahel, P. Rakes, T. Reinert, C. Reynolds, J. Reynolds, C. Roberts, J. Ruiz, Y. Sadovy, C. Safina, C. Scharpf, B. Semmens, N. Sharber, JR Shute, P. Shute, P. Skelton, R. Smith, W. Smith-Vaniz, M. Stiassny, D. Sumang, A. Sutherland, J. Swanagan, L. Taylor, B. Tissot, A. Vincent, M. Warren, J. Williams, and I. Winfield. However, the most thorough and critical reviews and helpful discussions came from members of the University of Georgia graduate-level Ecology 8990 Seminar in Fish Conservation (Fall 2005): D. Elkins, J. Ellis, C. Flaute, M. Hill, D. Homans, G. Loeffler-Peltier, J. Meyer, J. Norman, J. Rogers, J. Skyfield, C. Small, P. Vecsei, and S. Wenger.

I especially thank Ron Carroll and Alan Covich, past directors of the University of Georgia Institute of Ecology, for their moral, financial, and temporal support during the writing of this book. I am also grateful to Dave Coleman and Ted Gragson, University of Georgia, for financial support via the National Science Foundation Coweeta Long Term Ecological Research grant DEB-9632854. Finally, my harshest and most insightful critic and strongest supporter throughout this project has been my wife, Dr. Judy Meyer, distinguished research professor emerita, University of Georgia.

PART I Introduction

1. Fish Biodiversity and Why It Should Matter

To keep every cog and wheel is the first precaution of intelligent tinkering.
—Aldo Leopold, 1953

Why a book on fish conservation? First, because there isn't one. This is not meant to be flippant. The reference list at the end of this volume, which exceeds 2,000 citations and will rapidly become dated, represents only a subset of the available coverage on the topic. Fish conservation is a large and growing field of interest to a surprisingly large audience. To date, no compilation on the diversity of topics associated with the field of "fish conservation" (as opposed to fisheries conservation) has appeared. Someone seeking background information must basically begin from scratch. This volume is designed to at least give interested persons a starting point.

Second, despite widespread interest and abundant research, fishes have been neglected, at least in comparison with more traditionally charismatic taxa: "The word *tuna* still seems conjoined with *sandwich*" (Dallmeyer 2005, p. 414). J. A. Clark and R. M. May (2002) surveyed the more than 2,700 taxonomically focused articles that appeared between 1987 and 2001 in the two leading conservation research journals (*Conservation Biology* and *Biological Conservation*). They found that coverage was exceedingly uneven across taxa, with a strong bias that favored warm-blooded vertebrates (figure 1.1). Fishes make up nearly half of all vertebrate species but were the subject of only 8% of the articles published in these two prestigious journals (similarly, Irish and Norse 1996 reported that less than 10% of the conservation literature focused on freshwater and marine issues, and Lawler et al. 2006 found no trend toward increased coverage of aquatic environments and their organisms across time). Clark and May pointed out an important implication, cause, or perhaps result of this disparity: namely, that the bias toward coverage of birds and mammals reflected an even greater bias in funds devoted to research.

Warm and cuddlies may understandably receive greater attention from the public at large. However, conservationists advocate a holistic approach to the preservation of all biodiversity and cannot hope to "keep all the parts" if they give preferential treatment to a small subset of that diversity (invertebrates, which make up about 80% of all named organisms, fare even worse than fishes, receiving only 11% of coverage). One purpose of this book is to increase the attention that fishes receive among conservation practitioners.

Third, I love fishes.

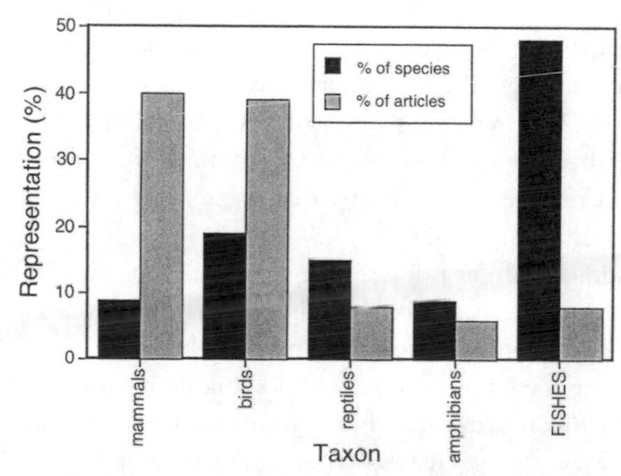

Figure 1.1. Treatment of vertebrate groups in the conservation literature between 1987 and 2001 relative to their actual diversity. Fishes, which make up 48% of all vertebrate species, were the subject of less than 10% of journal articles. After J. A. Clark and R. M. May (2002).

WHERE IS FISH DIVERSITY LOCATED?

Current estimates indicate that the world's ichthyofauna consists of around 27,300 species, with an additional 300–350 new species described annually (Nelson 2006; W. Eschmeyer, pers. comm.; see Helfman 2001 for an overview and Helfman et al. 1997 for details). Taxonomically, fishes can be divided into three main groups: (1) jawless fishes, which include 22 species of lancelets (cephalochordates), 25 hagfishes (myxines), and 35 lampreys (petromyzontiforms); (2) cartilaginous fishes, including 31 chimaeras or ratfishes (holocephalans), 475 sharks, and 725 skates and rays (elasmobranchs); and (3) more than 25,000 bony fishes, which run the gamut from primitive coelacanths, lungfishes, and sturgeons to advanced teleosts, which include most important commercial and recreational fishes, such as bonytongues, eels and tarpons, herrings, minnows, catfishes, salmons and trouts, cods, scorpionfishes, perch-like fishes, tunas, flatfishes, and triggerfishes, to name just a portion.

Fishes occur wherever water of reasonable integrity exists, from deep sea depths exceeding 8,000 m to mountain lakes above 5,000 m altitude. About 58% of all fishes are marine, and 41% live in freshwater, with the remaining 1% designated as diadromous, moving regularly between the ocean and freshwater systems. The proportion of freshwater species is rather striking in light of the availability of freshwater habitats. Approximately 97.5% of Earth's water is oceanic salt water, leaving only 2.5% as fresh. However, 99.7% of the freshwater is frozen in polar ice caps and glaciers, stored as groundwater, or locked up as soil moisture or permafrost (Stiassny 1999). In fact, only about 0.009% of the water on Earth is available as habitat for the more than 10,250 freshwater fish species (and all the other organisms restricted to freshwater habitats). It is for this limited resource of freshwater that we compete with fishes and other organisms.

Additionally, of this available freshwater, about 99% by volume is in lakes and only 1% is in rivers (R. T. Watson et al. 1996). This differential becomes all the more striking given the higher fish diversity of flowing (lotic) as opposed to lake (lentic) habitats. No rigorous analysis of relative diversity in the two habitat types is available, but the difference must be substantial. Diversity is highest in the southern portions of North America, Europe, and Asia, where relatively few natural lakes occur. Similarly, the regions of highest diversity in the tropics—tropical Asia, Africa, and South and Central America—have relatively few large lakes and house faunas dominated by riverine fishes (i.e., in the Mekong, Zaire-Congo, Amazon, Orinoco, Uruguai, and Parana-Paraguay basins). Tim Berra (pers. comm.) estimated that about 1,700 or about 15% of the 10,250 described freshwater fish species could be considered lake endemics, indicating that at least 85% of freshwater fishes live in rivers and streams and less than 1% in caves and springs. The world's rivers have thus been factories of fish evolution. But it is also rivers that are degraded most by pollution, habitat modification, dam building, riparian destruction, siltation, and water withdrawal (see chapters 4–7). Hence, flowing freshwater, especially smaller riverine and stream systems, is home to most freshwater fishes but is also the most heavily affected of major aquatic habitats.

EVIDENCE THAT FISHES ARE DECLINING

Available evidence strongly suggests that fish abundance and diversity are in decline at the same time that human populations and destructive activities, including fishing, are increasing. Bruton (1995) stated that fishes are second only to amphibians among vertebrates in degree of imperilment. Approximately 48% of amphibians are in serious decline (Stuart et al. 2004), and various regional and global estimates on the status of fishes indicate levels of imperilment approaching that of amphibians.

How Many Imperiled Fish Are There?

Estimates abound for the percentages of different faunas that are at risk of extinction. Chapters 2 and 3 chronicle the decline of fishes based on taxonomic and geopolitical criteria. A few examples should suffice to underscore the magnitude of the problem (see Moyle and Leidy 1992; Leidy and Moyle 1997; Stiassny 1996, 1999; Safina 2001a; Allan et al. 2005 for background):

- 66% of the native fishes of western Germany are imperiled;
- 65% of Spain's native fishes are at risk;
- for all of Europe, the number is between 40% and 80%;
- for North America, 27% to 35% are threatened;
- 42% of the 130 native fishes of Nepal are considered severely threatened;

BOX 1.1. IUCN and Red Data Lists of Imperiled Species

The International Union for the Conservation of Nature and Natural Resources (IUCN, now known as the World Conservation Union, or WCU) has been my primary source of comparative information on the conservation status of fishes at national, regional, and global levels (www.iucn.org). IUCN/WCU is an independent, international organization dedicated to natural resource conservation and the protection of endangered species. It is headquartered in Gland, Switzerland, but employs a full-time staff of over 1,000 in 62 countries. As the world's largest environmental knowledge network, it consists of 1,000 member organizations, including 82 states, 111 government agencies, and more than 800 nongovernmental organizations. It is organized around six commissions, which consist of and are advised by more than 4,000 scientists and experts who provide guidance on conservation knowledge, policy, and technical advice. The commission of most relevance to this book is the Species Survival Commission, which maintains and updates the international Red List of threatened and endangered species.

- in Malaysia, where deforestation has been rampant, only 45% of the fish species recorded historically could be found during a concerted four-year collecting effort in the 1980s;
- on the island of Singapore, more than 30% of the species found in the 1930s have apparently disappeared;
- the number of species considered at risk in Mexico rose from 36 in the 1960s to 123 in the 1990s;
- 314 native stocks of Pacific salmon, steelhead, and coastal cutthroat trout are at risk of extinction in the Pacific Northwest of the U.S.;
- since passage of the U.S. Endangered Species Act in 1973, 114 fishes have been listed as federally Endangered or Threatened. None has improved sufficiently to be taken off the list.

Although so-called red lists of threatened fishes have been produced for many countries and regions (chapter 3), estimating the proportion of global fish diversity that is threatened is difficult because the ratio includes ambiguity in both denominator and numerator. The denominator should represent the number of species whose conservation status has been assessed, not the total number of fish species alive today. This number is disappointingly small, although growing. Some estimates give values of 1,813 (IUCN 2004), 2,158 (Stiassny 1996), and 10%–15% of total diversity (= 2,500–3,750) (Greenwood 1992). The assessment of the International Union for the Conservation of Nature (IUCN) is the most recent and is well documented, developed by taxon experts who consult and revise the lists regularly (box 1.1). However, 447 of the 1,813 assessed by IUCN were "Data Deficient," meaning sufficient data are lacking, making it difficult to assign a rank of either threatened or nonthreatened. That leaves 1,366 species with confident assessments for a denominator.

The numerator is a statement, more accurately a hypothesis, of the number of imperiled fishes. It varies because of disagreement over the definition of *imperiled*, as well as over the level of taxonomic differentiation that should be considered (species, subspecies, race, stock, distinct population, evolutionarily significant unit, etc.).

One other basic complication is that our knowledge of both fish diversity and imperilment is heavily biased toward the industrialized world. Many fish species remain to be described. Hence the actual diversity of fishes is unknown, placing a cloud of uncertainty around all estimates of numbers of species at risk. Various extrapolations from rates of discovery and regions of coverage put eventual fish diversity somewhere around 31,500 species (Berra 2001). We know we are underestimating the number of imperiled marine and freshwater species in the tropics because our knowledge relies on the intensity of scientific effort in different countries. Industrialized nations can afford the relative luxury of employing researchers to study fishes, not to mention the luxury of worrying about declining biodiversity. Developing and emerging nations are hard put to deal with poverty, health, and starvation among their citizens. Conservation problems, despite the impact they have on human welfare, take a back seat. Most fish biodiversity resides in the developing world. We know only a subset of what we need to know.

The Number Please

Local and regional estimates of the proportion of a fauna that is threatened must be taken at face value, recognizing the complications discussed above. At the global level, published estimates range as low as 5% for marine species (Leidy and Moyle 1997). For freshwater species, the number ranges from 20% (Moyle and Leidy 1992) to 30%–35% (Stiassny 1999), to 39% (L. R. Brown and Starke 1998). Leidy and Moyle (p. 219) considered their 1992 value of 20% as "probably very conservative."

Another estimate can be derived from the IUCN (2004) assessment. Of 1,366 fishes examined, IUCN considered 965 species (including some subspecies), or 71%, to be either extinct or at a high risk of extinction. If Data Deficient species are included among the nonimperiled, the proportion in trouble is reduced, but only to 53% of assessed species.

IUCN is more likely to assess species already considered to be in trouble, biasing the sample toward imperiled fishes. At the same time, IUCN has stringent, quantitative requirements for assigning ranks. Hence, many species that are in trouble will not be ranked because we lack hard data. IUCN is also demonstrably conservative, in that it typically assigns at-risk status to fewer species than appear on most regional lists. For the 1,366 fishes surveyed, between half and two-thirds are threatened with extinction if current human activities and population trends continue.

A valid number for all endangered fish species cannot yet be given, but an approximation can be made for some fishes based on the country summaries given in chapter 3, IUCN (1996) designations of rank for other countries, and additional sources not included in chapter 3. For 44 countries in North America, South America, Europe, Asia, Africa, and Oceania, an average of between 25% and 41% of freshwater, diadromous, and estuarine fishes were at risk (marine fishes remain too poorly assessed to estimate). The 25% value is conservative, including only species designated by a recognized authority as Critically Endangered, Endangered, or Vulnerable, or their equivalent (it does not include extinct species). The 41% value is more precautionary, as it includes species with any imperiled status except Data Deficient. Recognizing the limitations in the data, we can estimate that at least one-quarter and probably at least one-third of the world's freshwater fishes are imperiled.

EXTINCTION

Numerical estimates of species at risk often include the phrase "extinct or imperiled." At first glance, inclusion of extinct species with those at risk of extinction appears misleading and even counterproductive. Why include what isn't there along with species that we might be able to do something about? Shouldn't we focus our efforts on those in decline?

The answer to this philosophical and pragmatic statement of strategy is unexpectedly complex. For reasons detailed below, it is often difficult (some say impossible) to know whether a fish is extinct. Hence, the cautious option is to allocate resources to both categories, those perhaps gone and those on the way out. How many do we think we have actually lost?

Just How Bad Are Things?

The number of known extinctions is surrounded by uncertainties but for fishes stands somewhere between 95 and 171 (Harrison and Stiassny 1999); unresolved, problematic, and debatable extinctions raise the number to between 210 and 290, depending on how many cichlids from Lake Victoria are included (see appendix). Regionally, the number again varies by source. The greatest extinction number and fastest rate occur on the African continent, where Lake Victoria alone may have lost 200–300 species of cichlids, deforestation is decimating Madagascar stream habitats, and species are being both discovered and lost at high rates. A conservative estimate for Africa, Madagascar included, is 109 species extinct, including 103 from Lake Victoria. Asia may have lost 33 species. Central America and Europe have lost 18–20 species each, South America perhaps 11. For North America, R. R. Miller et al. (1989) estimate 3 genera, 27 species, and 13 subspecies extinguished; Harrison and Stiassny (1999) and IUCN (2004) estimate 17 species. If distinct stocks of Pacific salmonids are included, the number for North America could be as high as 120–130.

Fish extinctions have increased dramatically in the past half century (figure 1.2). Only a few well-documented extinctions, in the neighborhood of a dozen worldwide, were recorded during the first half of the 20th century. The pace accelerated after World War II, although both the number and the rate are underestimated. Our knowledge of many faunas is incomplete, particularly in tropical

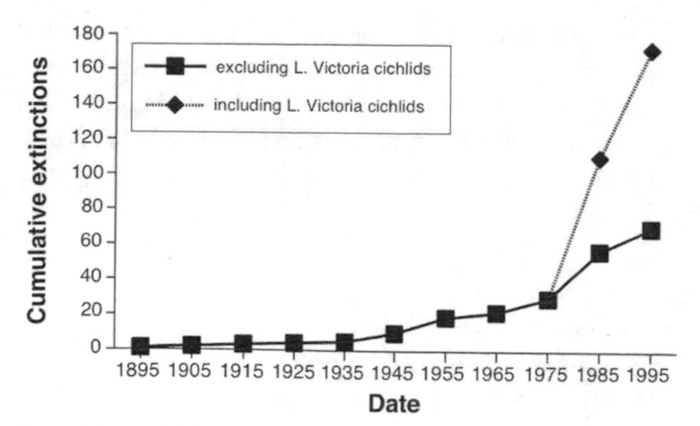

Figure 1.2. Global freshwater fish extinctions over the past century. The cumulative number of known and presumed extinctions is plotted for all extinctions and for all extinctions excluding Lake Victoria cichlids. Data are midpoints of decadal values given in Harrison and Stiassny (1999), figure 1.

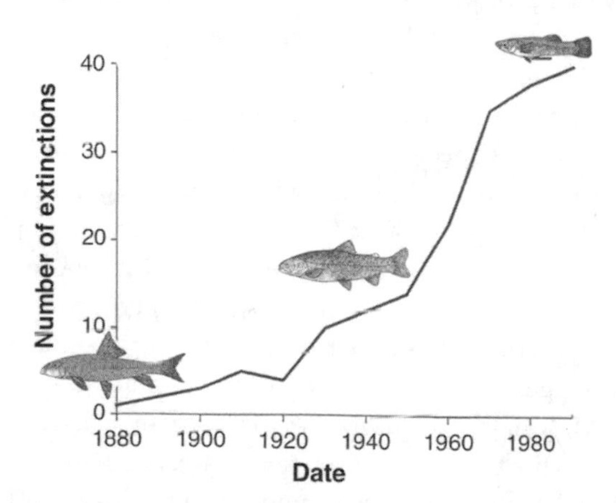

Figure 1.3. Fish extinctions in North America. Extinctions grew steadily over the past century until the latter part, when the rate fell, possibly indicating improved conditions or early elimination of more sensitive forms. Illustrated are the harelip sucker (extinguished ca. 1900), Alvord cutthroat trout (ca. 1930s or 1940s), and San Marcos gambusia (ca. 1980). After Stiassny (1996), based on Williams and Miller (1990); sucker drawing by J. Tomelleri in Boschung and Mayden (2004), trout and gambusia by Sara V. Fink in R. R. Miller et al. (1989); used with permission of the artists.

regions, but even major temperate areas such as Europe have been surprisingly understudied (Kottelat 1997). Given what we know, approximately 85% of all known fish extinctions occurred in the past 50 years, and the number runs closer to 95% if presumed extinctions in Lake Victoria are included (Harrison and Stiassny 1999). Most alarming is the evident increase in the global extinction rate over the past two decades. In North America, where we have relatively good data, the rate accelerated throughout most of the 20th century, with some slowing in recent years (figure 1.3).

Causes and Trajectory of Extinction

The factors responsible for exterminating fishes are the same factors that are initially responsible for population declines (figure 1.4). These are the so-called HIPPO factors taught in conservation biology classes: habitat loss, introduced species, pollution, human population and consumption (the ultimate cause of everything), and overexploitation. To these five, Montgomery (2003) added another H: history—that is, our inability to learn from past mistakes. That extinctions result primarily from human actions should not be taken lightly: "Human impacts have been so profound that not a single case of nonanthropogenic species extinction can be documented in the last 8000 years" (McKinney 1997, p. 496).

For freshwater fishes, the principle cause of declines and extinctions is habitat degradation, including disruption of the bottom, removal of structure, water withdrawal, hydrologic alterations (including impoundments), eutrophication, and sediment deposition. Such alterations have

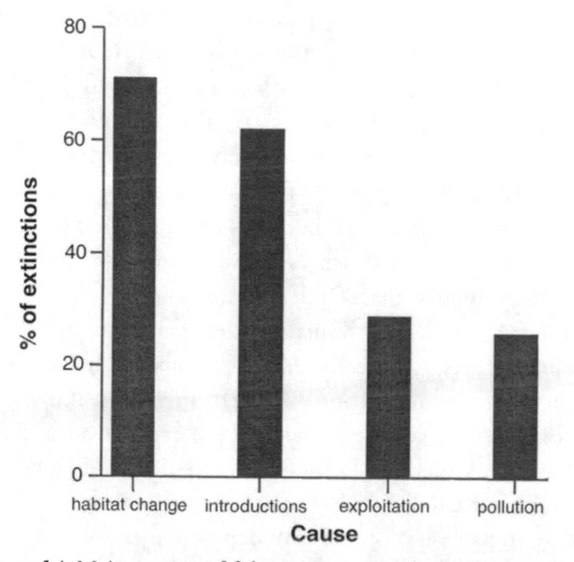

Figure 1.4. Major causes of fish extinctions globally. Habitat alteration, introduced species, overfishing, and pollution are the primary agents, but combined factors cause most extinctions, which is why summed percentages of all columns exceed 100%. Data from Harrison and Stiassny (1999) for 70 extinct freshwater species, excluding Lake Victoria cichlids.

contributed to 71% of extinctions. Not far behind are the various effects of introduced species of animals and plants. Predation (often on eggs and young), competition for food and habitat, and transmission of diseases and parasites accompany introduced species, which have contributed to 62% of extinctions. Aquaculture is an oft-cited source of escapees that establish themselves as introduced species. Overfishing and exploitation, largely for subsistence and commercial use as food but also in the ornamental and aquarium trades, have contributed to 29% of extinctions. Aquaculture again is a major contributor to overexploitation, as fish are captured for seed stock or feed.

Overexploitation, often involving habitat destruction, surpasses all other factors affecting marine fishes (chapters 10–13). Crowder and Norse (2005) categorically declared fisheries to be the greatest threat to marine biodiversity, writing that "the reach of industrialized fishing has gone global and the impact on marine wildlife, whether targeted or not, has been devastating" (p. 184). Chemical pollution and other forms of water quality degradation (acidification, endocrine disrupters) are a particular problem in industrialized nations, where freshwater systems are dumping grounds for organic and inorganic wastes. Pollution has contributed to approximately 26% of extinctions. Other identified factors include diseases and parasites (4%), hybridization (4%), and deliberate eradication (1%). Climate change is an anticipated factor that will affect many fish species, but the actual impacts remain unknown.

Extinction "is a process rather than an event" (Safina 2001a, p. 795); "it is the process . . . that is important, not the recording of the last individual" (Myers and Ottensmeyer 2005, p. 59). The process, or the steps that lead to extirmination of a species, is often referred to as the Extinction Vortex, a downward spiral generally involving progressive stages that feed on one another. The steps include (1) localized population declines brought on by HIPPO factors, (2) localized extirpations, (3) habitat and population fragmentation, (4) interrupted gene flow and loss of genetic diversity, (5) widespread extirpation, and (6) biological extinction. Heavily exploited species take an alternate but parallel path. Initial reductions from overharvesting impair reproduction by depleting spawning aggregations and curtailing seasonal cycles. Ecological extinction occurs, whereby the species becomes too scarce to function in its evolved role. This affects other ecosystem components and is followed by commercial extinction, when profitability plummets to the point that exploitation becomes uneconomical. Biological extinction may result if a species is too commercially valuable not to exploit, or if successful reproduction requires some threshold number of individuals (King 1987; Vincent and Sadovy 1998; Safina 2001a).

A North American Focus

In an in-depth analysis of past and future extinctions among aquatic fauna in North America, Ricciardi and Rasmussen (1999) arrived at estimates of the proportional species loss per decade for different taxonomic groups. They derived a predictive model based on extinction rates over the past century and the assumption that species currently designated as endangered or threatened will not survive for the next century (a conservative assumption given that we are probably unaware of many extinctions and that listings are in themselves conservative). North American freshwater fishes have experienced an extinction rate of 0.4% per decade over the past century, using 40 as the estimate of extinct species (conservative again given that the current numbers for U.S., Canada, and Mexico are closer to 45). Currently, 21.3% (217 of 1,021 species) are imperiled. Ricciardi and Rasmussen estimated a proportional species loss rate of 2.4%, or 24 species per decade for North American freshwater fishes. Paleontological evidence suggests one fish extinction per 3 million species-years (McKinney 1997), whereas the projected current rate is one extinction per 2,600 species-years, about 1,000 times higher than the background rate. The projected rate is also three times greater than the projected rate for terrestrial animals and even falls within the range of values estimated for tropical rain forest biomes, which is 1%–8% of species lost per year. Rain forests are generally considered to be suffering the highest depletion rates in the world. Ricciardi and Rasmussen concluded that "North American freshwater biodiversity is diminishing as rapidly as that of some of the most stressed terrestrial ecosystems on the planet" (p. 1221), and that the situation worldwide is similarly bleak because conditions in North American aquatic ecosystems are likely representative of events on most other continents.

Master (1990) independently arrived at a similar conclusion, that rates of imperilment of North American fishes, crayfishes, and mussels are three to eight times those for birds and mammals (Angermeier 1995). Even if these projections are off by a factor of two, we are still experiencing disturbingly high losses among our aquatic animals, and not just among fishes. But all of the discussions about extinction numbers and rates have to be tempered by a major caveat.

Extinction May Not Always Be Forever

A dictum of conservation biology is that "nothing is quite so final as extinction" (e.g., Ono et al. 1983, p. 209). However, of all natural history phenomena, extinction is probably the most difficult to verify. The number of qualified taxonomists out in the field collecting fishes is not impressively large, whereas the number of river and stream miles and acreage and volume of lakes and ponds (not to mention the ocean) is enormous. Many habitats are difficult to sample, and the likelihood of finding rare animals is low. Add to this the difficulty of telling closely related species apart, assuming the taxonomy of a species leaves little doubt as to its identity. Finally, most trained scientists are reluctant to make definitive public statements that may be wrong, and a pronouncement of extinction is a definitive statement.

Appropriately, guidelines for making such pronouncements are also fairly conservative. IUCN (1994, 1996) considered an organism extinct when "there is no reasonable doubt that the last individual has died" after "exhaustive surveys in known and/or expected habitat, at appropriate times . . . throughout its historic range have failed to record an individual." Harrison and Stiassny (1999), citing precedence in earlier IUCN guidelines, recommended that a species not be unequivocally designated as extinct unless collection efforts had failed to turn up an individual for at least 50 years. Such restrictions would seem to be a fairly good guarantee against making mistakes.

But they aren't. The literature contains abundant examples of species mistakenly thought to have gone extinct. The once abundant Alabama sturgeon, *Scaphirhynchus suttkusi*, is an excellent example (see chapter 2). It was taken out of consideration for endangered species listing because it was considered to be extinct, or at least too rare to bother protecting, with no specimens encountered for at least 15 years. But then at least six fish were collected over a six-year period (see Mayden and Kuhajda 1996). The sturgeon has subsequently been officially designated as Endangered.

Even in countries with a large, active community of researchers, extinctions can be difficult to document. J. E. Williams et al. (1989) compiled the definitive list of North American fish extinctions, with an important caveat in the form of a table of inaccurately reported extinctions. The table lists 25 such species and subspecies, their current conservation status, and the persons responsible for the original extinction designation. Mayden and Kuhajda (1996) analyzed this list and other examples of reportedly extinguished fish species and found that the average time between the extinction declaration and the rediscovery of a species was 62 years (so much for the conservative 50 year rule). The list in Williams et al. of authors who mistakenly declared species extinct is a veritable who's who of North American ichthyologists, who no doubt were reluctant to make public statements that might be wrong.

IUCN has sought an objective and quantitative process by which species are placed in various categories of threat (see chapter 3). Similarly, Harrison and Stiassny (1999), both specialists on tropical freshwater fishes, proposed a thorough and potentially useful protocol that adds objectivity to designations of extinction, as a means of dispelling some of the confusion that often surrounds an otherwise subjective process. These authors recommend that resolving a hypothesized extinction requires evidence along four lines:

- *taxonomic validity*: the species has a valid scientific name and preferably has been treated in a recent taxonomic revision;
- *temporal validity*: an "effective extinction date" (EED) establishes the presumed date of extinction, the EED is based on unsuccessful sampling around that date, and the date occurred more than 50 years in the past but since AD 1500 (presumably, to focus on anthropogenically caused extinctions and to avoid confusion from fossil and subfossil species);
- *subsequent sampling validity*: sampling effort has been expended in the past 50 years in appropriate habitat; and
- *just cause*: evidence exists to suggest that the species was in decline or faced some environmental threat that might have caused its demise, to avoid problems of natural rarity.

These are fairly strict and complicated requirements for determining extinction, as shown by the number of reported extinctions that meet all four criteria. Harrison and Stiassny ran the more than 200 reported fish extinctions (see appendix) through a dichotomous key using evidence from their four criteria. Only 3 species—harelip sucker, New Zealand grayling, and silver trout—met all the qualifications of a "resolved extinction." The remaining fishes fell into nine other categories of likely extinctions: 172 species had insufficient or ambiguous information for at least one criterion, 43 species were unclassifiable, and 30 species were disqualified for other reasons. Some workers may object that recent extinctions will go unrecognized and underappreciated if every species has to pass a litmus test such as this one, but Harrison and Stiassny view their protocol as a practical and informative approach deserving of consideration, and maintain a Web site for updates and comments under the aegis of the Committee on Recently Extinct Organisms (http://creo.amnh.org). At the least, adherence to their protocol would increase scrutiny of declining species, which can only be a positive development.

What lessons can be learned from such a debate? Does it really matter that a species might be mistakenly listed as extinct? First, an inaccurate assessment of a species' existence affects our estimates of extinction rate. One consequence is that concern over the current, unnaturally high, anthropogenically driven extinction crisis could be belittled by those who seek to hamper our efforts. Such complainants could also use inaccuracies to discredit warnings about the perilous status of other endangered species (witness the mileage that opponents of the Endangered Species Act derived from the snail darter debate, chapter 2). Beyond a loss of credibility, conservation efforts are compromised if a species is removed from an endangered species list due to an invalid designation of extinction. The U.S. Endangered Species Act (ESA), although focusing on species, also provides for the protection of the habitats in which designated species live. Once a species is removed from the list, its habitat loses that protection. Many people involved in events surrounding the listing of the Alabama sturgeon feel that a desire to prevent habitat protection motivated efforts to designate the species as extinct. Given that many endangered species co-occur with other rare and vulnerable species, loss of official status affects more than the immediate species. And, of course, if the endangered species is in fact extant but exceedingly rare, it, too, has lost protection once it is declared extinct.

Ecologists and other professionals have a responsibility to inform the public about declining biodiversity and when preservation efforts have been insufficient to save a species. Without this information, complacency will prevail. But caution in declaring a species extinct is justified now more than ever. While it may be precautionary to assume a species is extinct until proven extant (e.g., Diamond 1987), political reality argues that we assume a species is extant until reasonable evidence of its demise can be gathered.

WHY CARE ABOUT BIODIVERSITY, ESPECIALLY OF FISHES?

Convincing the public that we need to manage and conserve fishery resources is not difficult. A much greater challenge is justifying the expenditure of public funds and private effort to protect species with no known or potential human utility. Meeting the challenge requires understanding and appreciating the inherent, nonutilitarian value of diversity.

In its simplest definition, *biodiversity* is the number of species in a defined area or a particular taxon. But such a limited definition misses the richness and interrelatedness of biological phenomena. To avoid such oversight, conservation biologists expand the definition to include processes, habitats, and interactions, scaling up from the gene to the ecosystem and landscape. Biodiversity thus includes the variety of living organisms at genetic, species, and higher levels of taxonomy, as well as the variety of habitats and ecosystems and the processes that occur in them (Meffe and Carroll 1997). This broad inclusiveness is necessary to appreciate how actions and impacts at the smallest scale can influence events at much larger scales. Loss of genetic diversity within a species limits that species' role within an ecosystem, which can affect the ecosystem as a functioning whole. It's the butterfly effect but with real butterflies, or butterflyfishes.

Diversity begets diversity and is in turn dependent on it. Diversity per se is not a good predictor of ecosystem integrity or health, beyond the general observation that systems with intermediate diversity tend to be more resilient and resistant than systems with low diversity. More significant are the actual species present or absent and their roles relative to other species. Lost species and reduced abundances of individual species degrade the integrity of ecosystems (Chapin et al. 2000). *Integrity* is a general term for system health, for sustained capacity to support and maintain naturally functioning, adapting assemblages and processes without human intervention.

One example can serve to hint at the complexity of interactions in healthy, functioning ecosystems. Grunts (Haemulidae) on Caribbean coral reefs depend on diverse coral reef organisms and in turn promote the growth of other organisms (see chapter 12). Grunts first live in grassbeds and mangrove swamps, then among branching corals, and finally amid the structure created by more massive corals. Corals with daytime-resident grunt schools grow faster than corals without grunts because grunt excrement —the metabolic byproduct of fish feeding in other reef habitats at night—stimulates the growth of the symbiotic algae in the corals, thus promoting calcite skeleton production in the corals themselves. Multiply the grunt-coral interactions by the many species of fishes and corals and other habitat-forming invertebrates and plants that make up a "coral reef," and the complexity of relationships that define biodiversity becomes evident.

Appealing to Human Self-Interest

Winter and Hughes (1997, p. 22) characterized loss of biodiversity as "one of the four greatest risks to natural ecology and human well-being." This observation represents the official position statement of the American Fisheries Society (AFS), based on findings of the U.S. Environmental Protection Agency (EPA), neither of which can be dismissed as an environmental do-gooder narrowly focused on nonutilitarian goals. In 1992, the United Nations Convention on Biological Diversity (CBD) recognized the global impact of declining biodiversity, along with acknowledgment of the intrinsic value of biological diversity (www.biodiv.org/convention/articles.asp; as of March 2005, the U.S. was still not among the 168 countries to ratify the CBD, www.biodiv.org/world/parties.asp). This convention has stimulated and facilitated biodiversity protection laws and actions around the globe (chapter 3).

A great deal of the concern over declining biodiversity results from realized and potential impacts on human welfare. Fishes provide for human needs directly and indirectly. They provide food, medicine, entertainment, and jobs. The most obvious use of fishes is for food; they remain the one group of animals that we still exploit primarily in the wild. Human dependence on wild fishes for food is substantial, growing, and unsustainable as currently practiced (McAllister et al. 1997; Parrish 1999; Ormerod 2003; see World Resources Institute 1996 and chapter 10 for details). About one-sixth of Earth's six billion humans rely on fish as their primary source of protein. The proportion is greater in developing regions with high human densities, as in Africa where 30% of total animal protein comes from fishes (Stiassny 1996). Marine-derived protein, mainly in the form of finfishes, accounts for about 16% of humanity's protein consumption, coming from an annual catch of 80–85 million metric tons (MMT) (NRC 1999b). Another 6–12 MMT of protein comes from freshwaters, the large range in the numbers resulting from local subsistence and indigenous fisheries that are largely unreported and may constitute half of the take.

Regardless of source, total catch leveled off at around 100 MMT in the 1990s, while effort increased. The world's fishing fleet doubled between 1970 and 1990, from 600,000 large vessels to 1.2 million large vessels (again, subsistence fishing is unrecorded). Fishing technology has also advanced, increasing the effectiveness of the vessels. Despite increased effort and effectiveness, or because of it, fish catches have fallen in all but 2 of the world's 15 major fishing regions. In 4 regions, the decline exceeded 30% (P. Weber 1995). Given an anticipated human growth rate of 1.6% per year, annual global fish catch will have to increase to 120 MMT by 2010 and to 140 MMT by 2025 to keep up with population growth. The theoretical sustainable marine catch lies between 69 and 96 MMT. Even with increased aquaculture (10–15 MMT/year), supply will not meet and may fall far short of demand. People need fish. To meet this need, fishing must be practiced in a sustainable manner, which will require major modifications to fishery management practices (chapter 11).

Edible fishes are only a small subset of global fish diversity. A major inconvenience, from a utilitarian perspective, is that nature's complexity makes it difficult to know which species or processes are essential to the existence of the organisms we exploit.

Fishes Represent Larger Problems

Although fishes have their greatest instrumental value as commodities, they also serve human needs as repositories of information about the environment. Most important, they are part of larger ecosystems. The trends and conservation issues discussed throughout this book have their parallels in most other aquatic vertebrate and invertebrate groups, all of which are on average more imperiled than terrestrial groups (34% of fish, 48% of amphibians, 75% of unionid mussels, and 65% of crayfish vs. 11% to 14% for birds, mammals, and reptiles [e.g., Allan and Flecker 1993; Karr and Chu 1999]).

"Charismatic or not, in aquatic environments fishes are usually the best-known faunal components and must serve as indicators of the health of many aquatic systems" (Stiassny 1996, p. 8). Assessments of ecosystem health and integrity often rely on the abundance, species composition, and ecological roles of fishes because they are easy to count and identify and are sensitive to degradation. The Index of Biotic Integrity (IBI), widely adopted as a habitat assessment protocol in streams, employs fish numbers and diversity as primary indicators of system health (chapters 5, 7). Butterflyfishes and damselfishes, for example, are closely associated with live coral on reefs, and the diversity and abundance of these families are demonstrated indicators of habitat structure, disturbance, and recovery from stress on coral reefs (Ohman et al. 1998). Because fish sit at the top of many food webs, they bioaccumulate toxins. Hence measurable biological responses among fishes (chemical concentrations, phenotypic responses, anatomical anomalies, disease incidence, altered behavior or metabolism) are commonly used as biomarkers of the condition of aquatic habitats. At the extreme, fish kills and die-offs give us advance warning about water quality conditions that could affect human health.

What Good Is a Darter?

To the biologist, or anyone in awe of the evolutionary process, preserving the products of evolution requires no justification. Destruction of organisms is analogous to destroying spectacular art that required millions of years to create. Degrading habitat is as deplorable as defacing monumental architecture. E. O. Wilson (1984) has even postulated an evolutionary basis for our appreciation of nature, diversity, and evolution, termed *biophilia*. Wilson argues that we have an inherent interest in the biological world because such interest and knowledge are adaptive. But just as some people are indifferent to art and architecture, many are indifferent or even hostile to the notion that organisms have intrinsic value and an inherent right to exist independent of their known or potential utility to humans.

A number of arguments can be constructed in defense of useless fishes. For those swayed more by scriptural guidance than by reference to evolutionary processes, numerous biblical passages refer to humanity's stewardship role toward nature and caring for creation. In the Old Testament (Midrash Ecclesiastes Rabbah 7:28), God says, "Think upon this and do not destroy and desolate My World, For if you corrupt it, there is no one to set it right after you." Although Genesis 1:28 gave humanity "dominion over the fish of the sea," that dominion came with responsibility, and in fact, Rosenzweig (2003, p. 42) insisted that to have dominion "does not mean to ruin; it means to govern." To this effect, an Evangelical Environmental Network has been established in the U.S. and has published an *Evangelical Declaration on the Care of Creation* (www.creationcare.org). The National Association of Evangelicals (NAE), representing potentially 30 million evangelical Christians in the U.S., identified as among "the most pressing environmental questions of our day . . . the negative effects of environmental degradation on . . . God's endangered creatures." This organization concluded that "we must not evade our responsibility to care for God's creation" (National Association of Evangelicals 2004b). In its "Evangelical Call to Civic Responsibility," NAE (2004a) explicitly affirmed that

> God-given dominion is a sacred responsibility to steward the earth and not a license to abuse the creation of which we are a part. We are not the owners of creation, but its steward, summoned by God to "watch over and care for it" (Gen. 2:15). This implies the principle of sustainability: our uses of the Earth must be designed to conserve and renew the Earth rather than to deplete or destroy it.

Reference to the beauty of God's creation and human responsibility in caring for it can also be found in the teachings of Islam, Hinduism, Buddhism, Taoism, and a host of smaller, more "parochial" religions and cultures (see Meffe and Carroll 1997; Rosenzweig 2003). In essence, a conservation ethic appears to be a part of human nature.

In western society, where utilitarianism and dominion without responsibility have characterized attitudes toward the environment, respect for biodiversity grew with the environmental philosophy movement of the mid-1800s. The historical development of appreciation for the intrinsic value of nature can be traced to the preservationist/romantic-transcendental ethic of Emerson, Thoreau, and Muir, up through the evolutionary-ecological land ethic of Leopold and his modern disciples, and to the deep ecology movement (see Meffe and Carroll 1997; Dallmeyer 2005; Groom et al. 2005). At its core, modern conservation biology values biodiversity as good and maintains that destruction of biodiversity is ethically wrong (see chapter 15). One essential component of biodiversity is the evolutionary process; destroying organisms and their habitats arrests the process and discards its building blocks in the genes of exterminated species. Hence, to protect species is to preserve evolutionary potential.

What good is a darter? What good are any of us?

BIODIVERSITY ABOVE AND BELOW THE SPECIES LEVEL

Community and Ecosystem-Level Disruption

Fishes interact directly with predators, prey, competitors, parasites, and symbionts, and indirectly with even more species in food webs and community matrices. Thus, the scale increases from species abundance and presence or absence to the entire fish assemblage, to the biotic community, and to the interacting ecosystem, all embedded in the physical and biotic landscape. Reducing the numbers of a fish species may have little effect on other ecosystem components, but in many instances the effect can be substantial.

Biodiversity losses in this context occur as a result of reductions, deletions, additions, and substitutions. Reductions and deletions often occur because of pollution, overfishing, and habitat degradation. Habitat degradation changes the suitability of an area for native specialists while facilitating the proliferation of introduced species, many of which are generalists and predatory, which then displace natives (chapters 8, 9). The replacement is often accelerated as native, so-called trash species (gars, bowfin, suckers, minnows) are deliberately eradicated to make way for more desirable sport fishes.

One widely heralded result of this process of native loss and substitution is the homogenization of fish faunas (chapter 9). Because the same sport fishes (trout, bass, carp) are introduced in different regions of a country and in different countries, and because a few introduced species displace a larger number of endemic species through predation and competition, the resulting assemblages look more and more alike regardless of where they occur (Rahel 2000, 2002). Such homogenization also occurs on a more local scale within systems as specialized endemics give way to generalized, cosmopolitan species, which may also be native (M. C. Scott and Helfman 2001). The result is proliferation of weedy native and non-native species at the expense of unique endemics.

Human impacts on community- and ecosystem-level characteristics take many other, recently identified forms. Increasingly, we are seeing the result of overexploitation in the form of fished-down food webs (chapter 11). Our tendency to preferentially target the top predators in marine and freshwater systems has implications for the structure of their ecosystems. Bycatch and noncatch discards, disrupted living seafloors, discarded nontarget species, and ejected offal and other forms of "waste" short-circuit the higher links in food webs as we literally race to the bottom. The nutrients and energy previously locked up in and consumed by predators are still there but are now converted into other food web components, such as detritus and plankton, leading to a proliferation of planktivores and detritivores, species humans find less desirable. The ironic result is that our unsustainable fishing practices create ecosystems that can no longer sustain us (see Norse and Crowder 2005).

Ecosystem Services Provided by Fishes

One aspect of biodiversity loss that has direct, indirect, and nonutilitarian impacts on human welfare involves the role that fishes play in ecosystems. Biodiversity is intimately linked to ecosystem function: healthy ecosystems—those that contain natural assemblages of organisms, habitats, interactions, and processes—can sustain exploitation. Disrupted ecosystems collapse.

Organisms in ecosystems provide both goods and services to humans and other members of the ecosystem. Utilitarian goods are obvious: We eat fish, we use them in medicines, we worship them in ceremonies, we buy them as curios, and we derive pleasure from fish-centered recreation. Ecosystem services, in contrast, are the processes that occur as the result of functioning ecosystems, processes that humans (and other organisms) find useful or necessary (Daily 1997; Ecological Society of America 2000; see also the Millennium Ecosystem Assessment, www.maweb.org). Classically, ecosystem services were defined as processes that benefited humans: plant pollination, water and air purification, seed dispersal and germination, drought and flood mitigation, erosion control, nutrient cycling, pest control, and waste decomposition and transformation. These are all products of plant and animal activities. Because of the interconnectedness and coevolution of living things in ecosystems, one organism's output serves as input to another organism. The essential point here is that "ecosystem services are generated by the biodiversity present in natural ecosystems" (Chapin et al. 2000, p. 240).

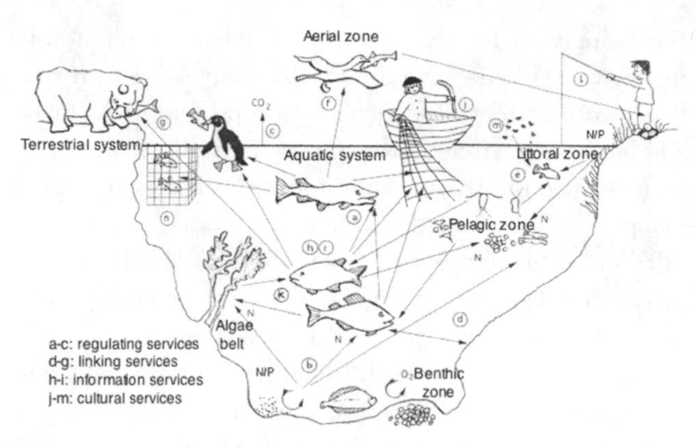

Figure 1.5. A pictorial summary of ecosystem services provided by fishes to humans and other organisms. Services can be described as (a–c) *regulating* populations and processes (e.g., trophic cascades that regulate population dynamics or nutrient cycling, bioturbation of sediments, carbon exchange); (d–g) *linking* different parts of the ecosystem via transport of nutrients and energy (e.g., open water to benthos, littoral zone, birds, and terrestrial mammals); (h–i) *informing* (e.g., indicating and recording past and present ecosystem integrity); and (j–n) *cultural* (e.g., human interactions and direct benefits via exploitation, recreation, water purification, disease abatement, and aquaculture). From Holmlund and Hammer (1999); used with permission.

Functioning ecosystems have their biodiversity largely intact; reduce the abundance of a species or eliminate it from the ecosystem, and the service may be diminished or no longer available. Assuming a lack of "redundant" species that serve the same role, reductions and losses impair ecosystem function.

Fishes provide a number of ecosystem services (Helfman et al. 1997; Holmlund and Hammer 1999; figure 1.5):

- As a result of short- and long-distance movements, fishes transport nutrients between different parts of ecosystems and between different ecosystems. As in the example of grunts mentioned earlier but also in kelp bed fishes, nutrients obtained in one habitat and excreted in another stimulate coral or plant growth. Long-distance migrations of salmonids and other diadromous fishes bring nutrients and energy obtained in ocean regions to distant, upriver habitats (chapter 11). This transport forms the base of the food webs in lakes and rivers, as well as surrounding terrestrial regions. Fishes, birds, mammals, and riparian vegetation are all dependent directly on these fishes, on the invertebrates that feed on the fishes and their offspring, and on the nutrients released as waste products or from decomposing bodies.
- Some fishes are habitat engineers, producing and moving sand and gravel. Parrotfishes generate sand in the process of digesting coral and move sand between different reef areas. Salmonids and minnows redistribute gravel and pebbles in the process of nest building. This redistribution and concentration of bottom types create favorable living conditions for the young, not only of the engineers but also of many other fishes and invertebrates.
- Fishes are at the top of the food web in many habitats, and their feeding activities can cause trophic cascades, affecting species lower in the web. Herbivorous fishes on reefs prevent algae from overgrowing and smothering coral; coral is in turn critical habitat for fishes and invertebrates. Reef-dwelling fishes that eat urchins prevent urchin explosions, which can denude reefs of both algae and coral (chapter 12). Piscivores in lakes eat smaller fishes that eat zooplankton. Zooplankton feed on phytoplankton. Eliminate the piscivores, and lakes experience blooms of algae, some of which are noxious.
- Zooplanktivorous and microcarnivorous fishes feed on larval stages of mosquitoes and biting flies, some of which carry human pathogens. Other fishes feed on snails that are intermediate hosts of human parasites (Stauffer et al. 1997).
- Although no fishes are known to pollinate plants, they do assist in the germination and transportation of seeds, as has been shown for piranha relatives in the riparian forests of the Amazon. In a symbiotic relationship, fish deposit eggs in the mantle cavities of freshwater mussels, whose larvae attach to the gills of the developing fish and are protected and transported until large enough to survive. In areas where host fishes have been reduced in number, bivalve populations crash.

A fully functional ecosystem is therefore dependent on its biodiversity, on the essential parts being present and functioning in their evolved ecological roles. The ecosystem and its constituent biodiversity are inseparable, and protecting diversity requires an ecosystem perspective.

Genes, the Ultimate Measure and Mover of Diversity

Although most concern over loss of diversity is focused at the species level or higher, ultimately, conservation efforts must promote maintenance of genetic variation within species (e.g., Vrijenhoek 1989; Ryman et al. 1995; Thorpe et al. 1995). Ecological theory and observations demonstrate that reduced genetic diversity within a species reduces egg output, hatching success, survival rate, and population size in fishes (Winter and Hughes 1997; Welcomme 2001). Genetic diversity not only affects current survival and reproduction but is fundamental to conservation biology as the raw material of evolutionary change and the insurance that lineages will be maintained over time (Meffe 1990; Bruton 1995; Bowen and Roman 2005).

Definitions of evolutionary and genetic units for conservation include evolutionarily significant units (ESUs) (Ryder 1986; Moritz 1994) and distinct population segments (DPSs) (Waples et al. 2001). These subspecific entities focus on unique genetic characteristics and their role in the evolution of species. If a population's extirpation would represent a significant loss to the ecological and genetic diversity of the species, that population would be an ESU. An ESU is not only isolated from other groups but also "represents an important component of the evolutionary legacy of the species" (Waples 1995, p. 9). Current genetic variability represents the reservoir on which future evolutionary potential depends and on which selection will operate. The variety of genes in today's populations will determine whether a species will be able to adjust to future conditions. Hence we want to preserve and maximize genetic diversity. To do this, we have to protect populations that are genetically distinct, that have a unique genetic make-up. Unique genes are the cogs and wheels that must be saved as we tinker with nature (e.g., Leopold 1953).

The ESU concept found early application as an aid in the identification of Pacific salmon stocks in need of protection (Waples 1991b). More recently, it is being applied to populations of species threatened by habitat fragmentation and other anthropogenic factors that interrupt gene flow and otherwise threaten isolated groups (e.g., gobies, whitefishes, sturgeons; C. T. Smith et al. 2002; Stefanni and Thorley 2003; Turgeon and Bernatchez 2003). Loss of genetic variation—in ESUs, endangered species, exploited populations, and populations subjected to habitat fragmentation—is an observed phenomenon and a growing concern (e.g., P. J. Smith et al. 1991; Ryman et al. 1995; Laroche and Durand 2004).

Understanding population genetics among fishes may give us insight into patterns of imperilment. Freshwater fishes are the most threatened forms, followed by diadromous species (chapter 3). Relatively few marine fishes are genuinely at risk of extinction. Is it purely coincidence that marine fishes show higher levels of genetic heterozygosity within species than do anadromous fishes, and that anadromous heterozygosity is higher than freshwater heterozygosity (Ward et al. 1994; DeWoody and Avise 2000)? From an applied perspective, lower average heterozygosity among freshwater fishes may help explain their higher average vulnerability (Vrijenhoek 1998). Vulnerability appears greatest among riverine species that have suffered habitat disruption and population fragmentation, such as from dam construction. These observations highlight the insights that studies at the level of genes can provide into widespread trends at the larger scale of major habitat types. They also emphasize the importance of maintaining large population sizes and connectivity among populations of freshwater fishes.

Fishing as a Form of Selection

Because genetics ultimately determines biodiversity, the impacts that fishing has on genetics resonate throughout ecosystems (J. B. C. Jackson et al. 2001). Humans are predators, and fishing is a strong form of natural selection (chapter 11). Unfortunately, humans are frequently inefficient and imprudent predators, removing the largest, the strongest, and the most fecund individuals. Such individuals belong to size classes that normally experience the lowest mortality rates (Baum et al. 2003); their global depletion is documented in marine and freshwater systems (e.g., R. A. Myers and Worm 2003). Natural predators prey on juveniles and weakened, post-reproductive individuals, size classes that have evolved to sustain high mortality rates. Because of gear specializations and high market values, humans fish despite drastically reduced prey populations, whereas natural predators switch to alternate prey species when density drops below certain levels.

The consequence of this unnatural predation is selection favoring individuals that are genetically programmed to be smaller; that have shorter life spans; and that breed at smaller sizes and ages, producing fewer eggs (Beacham 1983a, 1983b; Bigler et al. 1996; Hutchings and Reynolds 2004). The outcome is reduced genetic diversity and populations made up of small individuals (P. J. Smith 1994; Hauser et al. 2002), certainly not what we intended.

CONCLUSION

It is tempting to look for an obvious factor, such as overharvesting, to explain the decline or loss of a fish species, but any search for a single or major cause is destined to failure. Unlike dodos, moas, and Steller's sea cows, fishes haven't been hunted into extinction, at least not that we know. The biology of a species is the result of all the selection factors and quirks of history that have operated over millions of years to produce the suite of anatomical, physiological, behavioral, and ecological traits that define that species. Most declines, extirpations, and extinctions, therefore, result from a combination of factors and forces acting on these evolved traits. Forces interact and multiply to drive a population to critically low levels. Unitary causes of decline are a political myth, or evidence of a wish that easy answers will produce easy solutions. Antienvironmentalists are quick to point out the complicated and ambiguous nature of the causes of decline, as if such complexity and ambiguity were justification for inaction. In fact, the conservation of biodiversity requires complex, intact, functioning ecosystems.

It is the job of the conservation scientist then to catalogue and document the causes of decline and extinction, to understand how causes interact, and to devise restorative practices that address each cause, including their interactions. A list of priority actions that corresponds to the degree of threat imposed by each causative factor is desirable, as long as items farther down the list are not ignored when actions are taken near the top. If for no other reason, such "list topping" only moves the major impact to a lower item in the list. Removing a dam does not necessarily cause the trees to grow back in the watershed; curtailing fishing does not necessarily restore physical structure to a system; reducing the inputs of chemical contaminants does not necessarily eliminate introduced species from the river.

Regardless, fishes deserve to be protected for their value as commodities, for their role in ecosystems, for the information they contain that may or may not be useful to humans, and especially for their intrinsic value as the amazing result of the evolutionary process. As 11th Circuit Court of Appeals Judge Ed Carnes wrote in upholding Endangered Species Act protection of the commercially insignificant Alabama sturgeon, "Inside fragile living things, in little flowers or even in ugly fish, may hidden treasures lie" (Eleventh Circuit Court 2007, pp. 1274–75).

PART II Imperiled Fishes

Taxonomy, Geography, and Vulnerability

2. Roll Call I: A Taxonomic Perspective

Fishes need to attain an elevated status in our society so that our flagship species—the coelacanth, totoaba, sturgeon, huchen, asprete, great white shark and arapaima—attract as much attention and empathy as the panda, elephant, whooping crane, gorilla or rhinoceros.

—Michael Bruton, 1995

My major objectives in this chapter are to provide an accounting of families—and, in some instances, species—that are imperiled and to briefly document factors threatening them. The accounts provide background for the topics and concepts presented in later chapters. In the interests of space, many deserving examples are left out here and later—there are too many imperiled fishes to do justice to more than a subset. Thorough, accurate, and readable treatments of many imperiled North American fishes, including some discussed here, can be found in Ono et al. (1983). An excellent source for updated information on fish conservation issues in the U.S. is the Center for Biological Diversity's web page, especially the Species Alert listings for "fish" and "trout" (www.endangeredearth.org/alerts; also see the Programs and Habitats sections). Leidy and Moyle (1997) took a global perspective in their excellent review. Species accounts of many of the world's imperiled fishes have appeared in the series "Threatened Fishes of the World" in the journal *Environmental Biology of Fishes*.

This chapter is divided between freshwater (including anadromous) and marine taxa; groups appear in phylogenetic order from evolutionarily most primitive to most advanced within the two habitat categories. Imperilment ranks and abbreviations used here, explained in detail in chapter 3, are CR (Critically Endangered), EN (Endangered), VU (Vulnerable), and LR (Lower Risk than Vulnerable).

FRESHWATER AND ANADROMOUS FISHES

Lampreys (Petromyzontidae, Mordaciidae, and Geotriidae)

Lampreys receive minimal positive press and are about as low as one can get on the charisma scale for a chordate. However, they offer a unique view of the evolutionary process: 20 nonparasitic brook species, about half of all lampreys, can be linked evolutionarily to extant, ancestral parasitic forms (Helfman et al. 1997). Lampreys consist of 41 temperate, jawless, eel-like, primitive, freshwater and anadromous species. They have a long-lived (up to seven-year), benthic, larval phase that inhabits silty streambeds, feeding on microscopic algae and detritus. Larvae of parasitic species transform into feeding adults and live for one to three years, sucking out the body fluids of other fishes. In nonparasitic brook lampreys, larvae transform into nonfeeding adults, live for six months, spawn, and die. Brook lampreys typically live in headwater streams. Adults undertake a spawning migration that may cover a few kilometers in nonparasitic or landlocked species, or more than 1,000 km in anadromous species. Spawning occurs at predictable locales in headwater areas over gravel and cobble, after which the animals die.

These ecological traits and life history characteristics make lampreys, particularly the nonparasitic brook species, surprisingly vulnerable. Typical habitats for breeding are easily disrupted by human activities. Cobble habitats are degraded by sedimentation, whereas gravel is often extracted for road building and other construction practices. Larvae spend several years buried in bottom sediments, but in Europe stream maintenance programs often involve deliberate flushing of sediment on a shorter time scale (e.g., Kirchhofer 1996). Also, spawning grounds appear to be reused repeatedly, an interesting specialization that usually means limited plasticity in spawning and vulnerability to disruption of spawning areas. One-time spawning and relatively low fecundity (ca. 1,000 eggs) translate into a low replacement rate in populations and drastic consequences if year-classes fail. Migrations and anadromy mean that fishes move through multiple potentially degraded habitats, constituting a gauntlet of environmental insults additive in their impact. Dams are especially deadly, as they block spawning migrations (few fish passage devices have been designed specifically to accommodate lampreys). Repeated evolution of parasitic-nonparasitic species pairs has created isolated species with relatively small ranges and populations, all common characteristics among imperiled species (see chapter 4).

Two species typify the problems that lampreys in Europe and elsewhere have encountered. The river lamprey, *Lampetra fluviatilis*, has undergone range contraction and fragmentation due to dam construction and other factors throughout much of France, where it has Vulnerable rank (Keith and Allardi 1996); it is listed as Near Threatened in IUCN (2004). The European brook lamprey, *L. planeri*, also internationally Near Threatened, was abundant and widely distributed across Europe a century ago. Its populations are now small, occurring in isolated areas of a greatly fragmented range (Lelek 1987). It is classified as Critical in Switzerland and Endangered in the Czech Republic, among other listings (Kirchhofer 1996; Lusk 1996).

Numerous lamprey species have local, regional, and international conservation status. Three appeared in the highest-risk categories of IUCN (2004): Greek brook lamprey, *Eudontomyzon hellenicus* (VU), Lombardy brook lamprey, *L. zanandreai* (EN), and Australian nonparasitic lamprey, *Mordacia praecox* (VU). Four other European lampreys had IUCN-LR (Near Threatened) status, indicating cause for concern. One of these, Vladykov's lamprey, *E. vladykovi*, had Endangered status in Germany (Freyhof 2002). Russia listed three lampreys in its highest-risk ranks, none of which were recognized by IUCN. Chile listed *Geotria australis* as Vulnerable and *M. lapicida* as Data Deficient (Campos et al. 1998), a rank it also held in IUCN (2004).

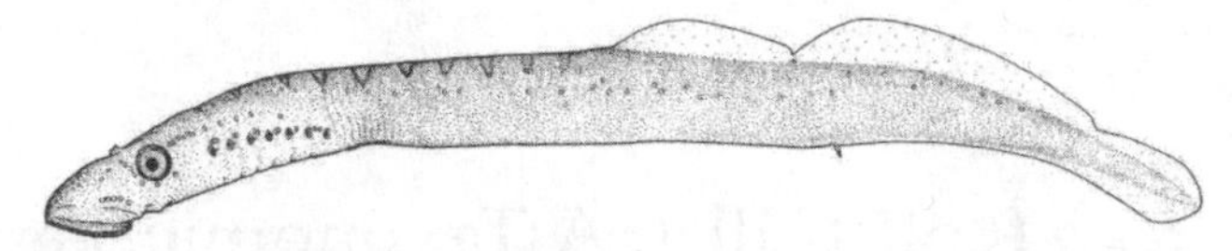

Figure 2.1. The Miller Lake lamprey, North America's smallest lamprey and the world's smallest parasitic lamprey at an adult size of <13 cm. It was exterminated from Miller Lake in the mid-1950s and thought to be extinct for more than 30 years, but small populations were discovered in nearby Oregon streams in 1992. Drawing by Sara Fink. From R. R. Miller et al. (1989); used with permission of the artist and AFS.

In North America, several lamprey species have national conservation status. Three were listed as Vulnerable in Canada (R. R. Campbell 1997), three were Of Special Concern in the U.S. (J. E. Williams et al. 1989), and two were Endangered in Mexico (Secretaría de Solidaridad 1994). The U.S. had the seeming distinction of causing the only "known" lamprey extinction. The Miller Lake lamprey, *L. minima* (figure 2.1), the world's smallest parasitic lamprey, was poisoned into apparent extinction in 1958 because it parasitized introduced trout at its only known locale, Miller Lake, Oregon (Bond and Kan 1973; R. R. Miller et al. 1989). Fortunately, a few remnant populations went unnoticed and were rediscovered in the region in 1992 and thereafter (Lorion et al. 2000). These populations occurred in headwaters of an adjoining river and a tributary to Miller Lake; the lake itself was not recolonized.

The Miller Lake lamprey is part of a species complex of lampreys largely endemic to the northern California and southern Oregon region. In January 2003, 11 conservation organizations petitioned the U.S. Fish and Wildlife Service (USFWS) to grant Endangered Species Act (ESA) protection to four species of western lampreys: Pacific lamprey, *L. tridentata*; river lamprey, *L. ayresi*; western brook lamprey, *L. richardsoni*; and Kern brook lamprey, *L. hubbsi*. Factors cited in the petition included habitat destruction and hydraulic alteration, especially dams that blocked migrating lampreys (the petition was declined in December 2004). Four other species of endemic lampreys in the upper Klamath River basin and upper Sacramento/Pit River system—including the Miller Lake species—are also considered to be at risk (www.biologicaldiversity.org/swcbd/species/lamprey/).

An intriguing and complicated set of circumstances

surrounds the conservation status of a widespread anadromous North Atlantic species, the sea lamprey, *Petromyzon marinus*. Canals constructed during the early 1800s and widened in the early 1900s led to the accidental introduction of this parasitic species into the Great Lakes of North America, where it contributed to the demise of lake trout, whitefishes, and blue pike (Fuller et al. 1999). Extensive and expensive eradication programs (e.g., Hanson and Swink 1989) are practiced throughout the region. In ironic contrast, *P. marinus* is listed in Appendix III (Protected) of Europe's Berne Convention, is one of only 13 species listed by OSPAR (2004), is considered to be under threat of extinction in Russia, and has been extirpated from the Czech Republic. It has Vulnerable status in France, where many runs have been eliminated by dams, weirs, bottom disturbance, chemical pollution, sedimentation of spawning areas, and even overfishing—because it is "highly esteemed for the table" (Keith and Allardi 1996, p. 38). The sea lamprey is one example of the loathed-abroad-but-loved-at-home phenomenon, involving species that are actively sought out and killed as an introduced nuisance in one region but imperiled, desirable, and even protected in their places of origin (chapter 9).

Sturgeons and Paddlefishes (Acipenseriformes)

Sturgeons

Sturgeons figure prominently in this book because of almost universal declines despite enormous abundances a century ago. Causes of declines include migration blockage by dams, habitat destruction from siltation and water withdrawal, pollution, legal and illegal overexploitation, and the politics that often swirl around endangered species issues. Underlying factors include high economic value and a life history characterized by slow maturation, infrequent reproduction, and migratory movements across international boundaries (see Birstein, Waldman, and Bemis 1997; Birstein, Bemis, and Waldman 1997).

The Aral Sea stock of the ship sturgeon, *Acipenser nudiventris*, and the Adriatic Sea stock of beluga, *Huso huso*, are considered extirpated by IUCN (2004). The shortnose sturgeon, *A. brevirostrum*, of North America and the Baltic sturgeon, *A. sturio*, are 2 of the only 9 fish species that appear in Appendix I of the Convention on International Trade in Endangered Species (CITES, www.cites.org; see box 3.1). Internationally, 9 sturgeon stocks or subspecies are Critically Endangered, 25 are Endangered, and 13 are Vulnerable. Collectively, recognized stocks of all 25 sturgeon species have some IUCN imperiled (Extinct through Vulnerable) status. Five U.S. species have ESA protection, the green sturgeon, *A. medirostris*, having been added in April 2006. ESA protection for the Alabama sturgeon, *Scaphirhynchus suttkusi*, the most recent sturgeon to have been described and perhaps "the rarest fish in North America," was delayed for nine years by political battles over the potential impacts that listing might have on barge activity in the Tennessee-Tombigbee waterway. The debate became rancorous at times. Alabama State Representative Johnny Ford declared categorically, "We don't want these ugly fish in the state of Alabama" (Scharpf 2000, p. 6). Only lake and shovelnose sturgeon are not federally listed among U.S. sturgeons, but IUCN assigns them Vulnerable status (see Van Winkle et al. 2002 on North American species and CITES 2001 for a thorough, elegantly illustrated taxonomic guide to all sturgeons and their products).

As with many species, it is difficult to pinpoint a specific human impact that is primarily responsible for sturgeon declines. Most species are subjected to multiple factors: Spawning migrations are blocked by dams, spawning substrates are unusable due to sediment or other debris, industrial and municipal pollutants accumulate at high concentrations in these long-lived (100+ years) fishes, and juveniles are captured as bycatch in net fisheries for other migratory species. Recovery from population depletion is slow because fish do not mature until 6 to 25 years of age, and reproduction in many species occurs at three- to four-year intervals.

Overfishing of adult sturgeon, driven by the high economic value of their eggs for caviar and conducted with nonselective gear, has often delivered a fatal blow. Sturgeon are among the few fishes that are market force-free, intensively fished despite low and declining population sizes. The ovaries of large female sturgeons may constitute 25% of body mass. Sturgeons in excess of 100 kg are not unusual, and historically fish weighing as much as 800 kg (white sturgeon, *A. transmontanus*) and more than 1,000 kg (beluga) were reported. The largest beluga was captured in 1924 from the Tikhaya Sosna River of Russia. It weighed 1,227 kg and yielded 245 kg of caviar (www.guinessworldrecords.com). If its eggs qualified as grade 000 Malossol caviar, which has sold for as much as $165/oz (= $2,640/lb, or $5,280/kg), the fish would have been worth nearly $1,230,000, making it by far the world's most valu-

BOX 2.1. Caviar Crimes and Criminals

Illegal fishing and export of caviar from the Caspian Sea region are frequently estimated to be as much as ten times larger than the legal fishery. This trade is purportedly run by what is widely referred to as the Caviar Mafia (De Meulenaer and Raymakers 1996; Tidwell 2001). In 1994 alone, 11,000 sturgeon poachers were caught in the Astrakhan region of southwestern Russia—the self-proclaimed capital of the Caspian fisheries—leading to the confiscation of 52 *tons* of illegally produced caviar (Kozyreva 1995).

However, poaching and black market trade are not restricted to developing nations and impoverished economies. In 1993, the Hansen Caviar Co. of New Jersey was convicted of selling 1,500 kg of illegally obtained caviar from Columbia River white sturgeon. Hansen sold the caviar as imported beluga and osetra caviar, worth five times what white sturgeon caviar brings. The caviar had been processed in a motel room in Vancouver, Washington, and shipped with fictitious business names, nonexistent return addresses, and various post office boxes. The poachers were paid $247,176 (tax free) for the processed caviar, which may have then retailed for as much as $2,000,000. Fines levied against the poachers and the caviar company amounted to less than $20,000. To obtain 1,500 kg of caviar would require capturing and killing approximately 2,000 white sturgeon, "a significant part of the sturgeon population in the lower Columbia River, where the poachers were operating" (Cohen 1997, p. 425).

able fish. Malossol caviar is considered "the finest food delicacy ever to have been discovered" (www.caviarideas.com; for comparison, American paddlefish and sturgeon caviar had a 2005 price of $12/oz). Although sturgeon fishing is highly regulated nationally and internationally, high prices have promoted rampant poaching and black markets, at the same time that fishery management and enforcement programs have collapsed (box 2.1).

The Baltic (also called European Atlantic) sturgeon, *A. sturio*, exemplifies many of the problems faced by sturgeon species. Now listed in Appendix I of CITES, considered Critically Endangered by IUCN, and protected by a European Union Habitats Directive, Baltic sturgeon historically spawned in rivers along the entire seaboard of western Europe, from the Baltic to the Mediterranean and Black Sea areas. As recently as the mid-20th century, it occurred in southwestern France, the Iberian peninsula, and the Adriatic and Black seas. It is thought to be extirpated from Belgium, Finland, Germany, Iceland, Ireland, Norway, Poland, and Yugoslavia (Birstein, Bemis, and Waldman 1997). Fewer than 10,000 individuals may remain. Two reproducing populations are known, and they are restricted to the Gironde River of France and River Rioni, a tributary to the Black Sea in Georgia, Former Soviet Union (Williot et al. 1997). The Gironde River population may be the only successfully reproducing stock, and reproduction after 1985 may have been limited to the 1988 and 1994 spawning seasons (Pustelnik and Guerri 2002). Dams, polluted and gravel-depleted spawning grounds, and overfishing of both juveniles and adults are considered major causes. Recovery efforts, supported by funding from the European Union, include artificial propagation and restocking.

As with many attempts at culturing endangered sturgeons, these efforts have met with limited success (e.g., Williot et al. 1997). Natural reproduction in sturgeons is an infrequent event, capture of brood fish is stressful, and hormonal induction of spawning frequently fails. Meeting the environmental conditions necessary for maturation of gametes and rearing of young has proven challenging. Meanwhile, fish are taken from the wild-reproducing population for propagation, further depleting the spawning stock. Recovery, if it can occur, will require catchment-wide, ecosystem-oriented management plans and cooperation among international fishing interests to limit bycatch (Pustelnik and Guerri 2002).

In light of widespread publicity over the plight of sturgeons, Web sites and organizations dedicated to sturgeon conservation have proliferated. Notable are the World Sturgeon Conservation Society and the IUCN Sturgeon Specialist Group. The latter group met for the first time in February 2001 and recommended an international course of action focused on reducing poaching and illegal trade while recognizing socioeconomic conditions that drive such activities.

Other groups promoting sturgeon conservation include Caviar Emptor, a coalition of SeaWeb, the Natural Resources Defense Council, and the University of Miami's

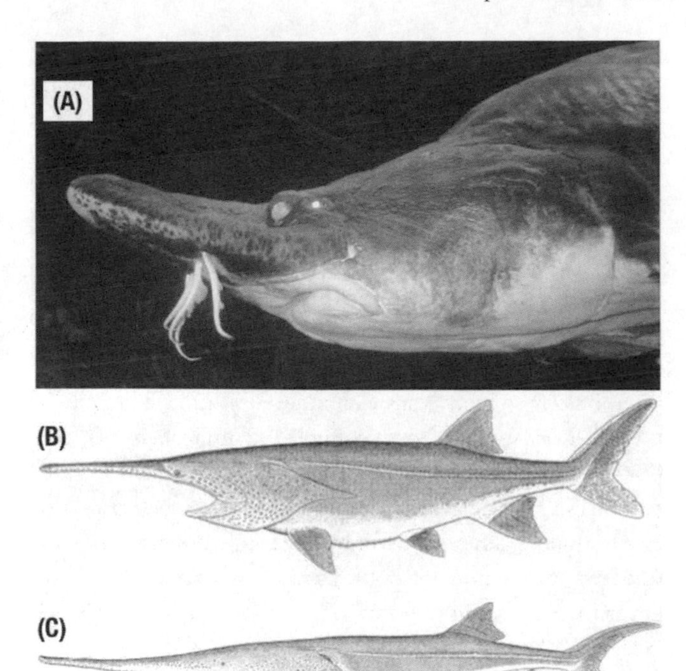

Figure 2.2. (A) The beluga sturgeon, the world's largest sturgeon. Prior to overfishing, this highly endangered species may have reached a length of 6 m and a mass of 1,000 kg. (B) The North American paddlefish. (C) The Critically Endangered Chinese paddlefish. B and C by Paul Vecsei in CITES (2001); used with permission.

Pew Institute for Ocean Science. Its focus has been on obtaining threatened status under the ESA for beluga sturgeon of the Caspian Sea region (figure 2.2A). Beluga populations declined more than 90% in the 1980s and 1990s, accelerated by the breakup of the Soviet Union. A 40% drop in 2001–2002 alone was most likely due to illegal fishing (De Meulenaer and Raymakers 1996; Speer et al. 2000). Full ESA status for beluga would make importation of beluga caviar illegal. Because the U.S. consumes 60%–80% of the beluga caviar produced, such a ban would reduce demand and presumably make illegal fishing less profitable. Opponents of such a ban maintain that caviar from aquacultured fish should not be restricted.

Fortunately for poachers, no practical means exist for distinguishing eggs of the few fish raised in captivity from those caught in the wild. Caviar Emptor launched its campaign in December 2000. Meanwhile, CITES imposed and then lifted a ban on beluga caviar exports and implemented varying catch quotas on the sturgeon. In October 2004, USFWS granted ESA status to beluga, but the official listing did not include a ban on caviar importation (USFWS 2004). Caviar Emptor—and other conservation organizations—have pointed out that many other sources of caviar exist, including products from aquacultured North American sturgeons and paddlefish: "There are many luxuries in life in which we can still indulge. The beluga sturgeon can't afford for us to indulge in this one" (www.caviaremptor.org/press_031502.html).

Paddlefishes

Only two species of paddlefishes exist, each endemic to a major continental river and its tributaries. Both have undergone significant declines, although the North American species, *Polyodon spathula*, appears much more resilient and has a brighter future than its Chinese counterpart, *Psephurus gladius* (figure 2.2B, C). Paddlefishes are distinguished by their paddle or spoonbill snout, which is flat and rounded in the North American species but elongate and almost spearlike in the Chinese species. The smaller North American paddlefish inhabits the Mississippi River system, achieving a length of 2.2 m and a mass of 83 kg. North American paddlefish do not mature until ten years old and then spawn at two- to five-year intervals. These reproductive traits make them vulnerable to overfishing, and the situation is exacerbated by deterioration of clean gravel spawning habitat and backwater nursery habitats, both degraded by dam construction. North American paddlefish have been extirpated from four states on the periphery of their range and are declining in two others. Through active stocking programs, numbers have stabilized or increased in 17 states. Concern about declines has led to prohibitions on sport and commercial fishing in several states. Because of its desirability in the caviar trade and sensitivity to overfishing, the North American paddlefish was placed in Appendix II of CITES in 1992 to control poaching. Although stocking programs are showing promise, preventing future declines requires greater attention to habitat protection and restoration (Gengerke 1986; K. Graham 1997).

The exceedingly rare, critically endangered, and poorly known Chinese paddlefish is larger, reaching perhaps 3 m in length and 500 kg. It may be the most endangered fish in China, at least among the species we know. Declines result in part from overfishing, but more dramatically from habitat destruction brought on by pollution and massive dam construction, the latter blocking spawning migrations in the Yangtze River basin, where the fish lives and was once common. Historically, it also occurred in the Yellow

River. It is probably anadromous, with adults moving up the river to spawn and juveniles moving down to the East China Sea to grow. Completion of the Gezhouba Dam on the Yangtze in 1981 essentially cut the paddlefish's habitat in half and blocked spawning migrations. Since 1988, only three to ten adult paddlefish have been found annually below the dam, and no recruitment to the population is thought to be occurring, despite full protection granted the species in 1983 (Wei et al. 1997). Completion of the massive Three Gorges Dam in 2009 is predicted to drive this already rare species into extinction (chapter 6; Fu et al. 2003). Artificial propagation is the only hope for the Chinese paddlefish, but all attempts so far have failed because the fish cannot be kept in captivity. Nothing is known of its spawning habits, locales, or habitat. Mature fish are becoming increasingly rare. Ironically, North American paddlefish have been introduced into China and are being successfully reared in ponds (Wei et al. 1997).

Bonytongues (Osteoglossomorpha)

The bonytongues are the most primitive group of modern teleost fishes. The group includes the hiodontid mooneye and goldeye of North America, the mormyrid elephantfishes of Africa, and the widely dispersed osteoglossid arowanas (*Osteoglossum, Arapaima, Scleropages*) of South America, Oceania, and Asia (Helfman et al. 1997). *Scleropages formosus*, the Asian bonytongue, Asian arowana, or golden dragon fish, is highly prized as an ornamental species in a specialized aquarium trade that has greatly depleted wild populations. Captive-bred individuals displaying reddish color command prices upward of $5,000. Trade in the species is highly regulated (see chapter 13), and the fish is listed in CITES Appendix I.

The arapaima, *Arapaima gigas*, also known as pirarucu or paiche, occurs in the Amazon basin of Brazil, Guyana, and Peru (figure 2.3). The species is listed in Appendix II of CITES and as Data Deficient in IUCN (2004). It is threatened by overexploitation for food and as a gamefish. Arapaima breed when the Amazon floods, and spawning adults occur predictably in swampy, seasonally flooded floodplain lakes, making them susceptible to both overharvest and any hydrologic alterations that affect seasonal flood cycles. This massive predator grows to 4.5 m and 200 kg, making it among the world's largest freshwater fishes, although fish of this size are rare in the wild. Arapaima are a highly desirable food fish and are sought as a spectacular sport fish because of their size, hard fight, and jumping behavior. Amazonian fishing trips focused on arapaima are still advertised on the Web, and sport fishing also occurs in stocked lakes in Thailand (along with endangered giant Mekong catfish).

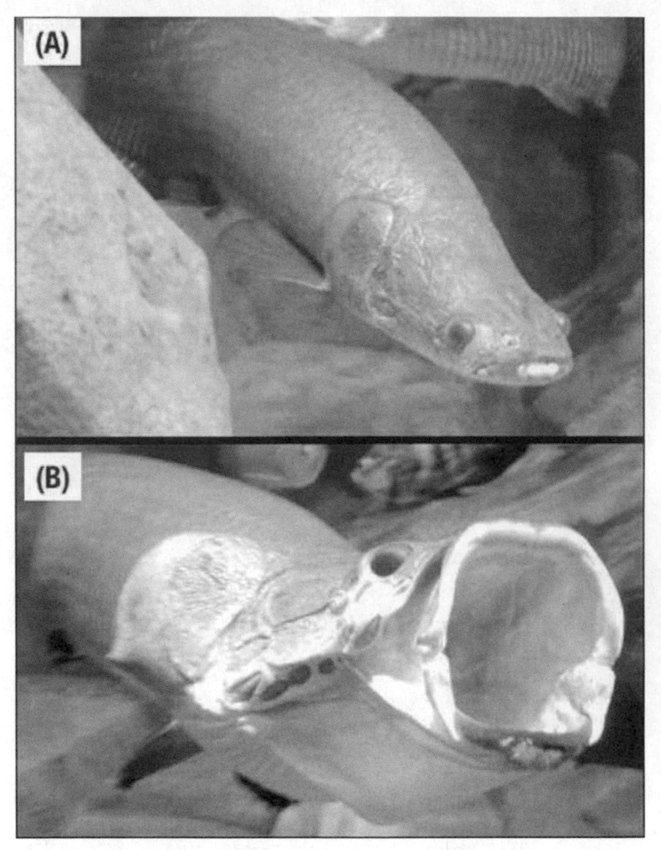

Figure 2.3. Arapaima, South America's largest freshwater fish. This approximately 1-m fish (A) demonstrates how its large mouth (B) allows it to prey on fairly large fishes.

Arapaima have been partially protected in Brazil since 1976 by decree of the Superintendent for the Bureau for the Development of Fishing (SUDEPE), which outlawed killing arapaima during the dry season (October to March), when fish were confined to shallow water and easily captured. A minimum size of 1.5 m was also declared (Melfi 2003). To reduce pressure on wild populations, pond culture of arapaima was initiated in 2000 in Amazonian villages of Peru, with some success (http://pdacrsp.oregonstate.edu/aquanews/fa112002/p13.html).

Minnows and Barbs (Cyprinidae)

The family Cyprinidae contains upward of 2,000 species of carps, minnows, shiners, dace, chubs, and barbs. As the largest freshwater family in the world, cyprinids are abundant and diverse in North America, Europe, Asia, and Africa. Given such diversity, it is understandable that many species appear on national and international red data lists. The Congo blind barb, *Caecobarbus geertsi*, a cave-dwelling cyprinid from the lower Congo basin, is one of only two freshwater teleosts listed in Appendix II of CITES. IUCN (2004) listed 254 cyprinid species with international conservation status, the largest number among fish families. Fifteen species are extinct, and one, the red-tailed shark, *Epalzeorhynchos bicolor*, is extinct in the wild (and is a fish I kept as a child and was the subject of my first behavior study). Forty species have Critically Endangered status, and another 31 are Endangered. Given that 12 of the 15 species known to be extinct were North American and that the greatest diversity of cyprinids is in Asia and Africa, the list again represents the distribution of research and conservation effort as much as actual global status. The extent of this list makes even a cursory overview difficult. A few regions and species deserve detail as representatives of the family and of conservation efforts to protect them.

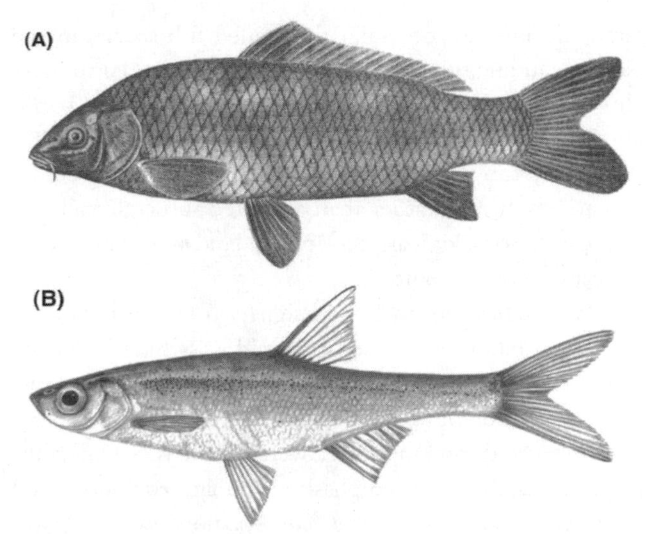

Figure 2.4. Endangered European cyprinids. (A) The wild, and Critically Endangered European carp, the ancestral species for cultured carp throughout the world, including highly invasive populations and the fantastically colored and priced koi or nishikigoi of Japan. From Antipa (1909) in Balon (2006). (B) The saramugo. This small cyprinid embodies conservation problems faced by several imperiled fish species that inhabit rivers in semiarid regions of the Iberian Peninsula. Drawing by Filipa Filipe; used with permission.

European Cyprinids

Europe, with a long history of human activity and aquatic habitat degradation, has a fairly diverse minnow fauna, many of which are known to be in trouble. Even the European carp, *Cyprinus carpio*, one of the most actively cultured, widely introduced, and destructive species globally (see chapter 8), is now designated Critically Endangered by IUCN and may be extinct in its pure, wild, Danube River form (Balon 1995; figure 2.4A). IUCN listed an additional 31 European minnows in high-threat categories. *Leuciscus turskyi* from the Cikola River of Croatia and *Chondrostoma scodrensis*, a cyprinid endemic to Lake Skadar of Albania, Serbia, and Montenegro may be extinct. The petropsaro, *Barbus euboicus*, of Greece and *L. ukliva* of Croatia are the 2 remaining Critically Endangered species. Habitat loss and range contraction and fragmentation have been major contributors to the decline of these species. Five species are considered Endangered, including the saramugo of Portugal, and 22 others are listed as Vulnerable.

The Endangered saramugo, *Anaecypris hispanica*, is representative of fish conservation issues and efforts in Europe (figure 2.4B). It is a 7-cm cyprinid once common across the Guadiana catchment but presently limited to small streams in Portuguese regions of the Guadiana River basin, which forms a major border between Portugal and Spain. The saramugo is considered the most threatened nonmigratory freshwater fish in the Portuguese *Red Data Book* and one of the most endangered fishes in Europe. It also has endangered status in Spain, where it is known as jarabujo.

The saramugo's range has become highly fragmented (Collares-Pereira, Cowx, et al. 2002). Populations are threatened due to habitat degradation and exotic species, but mainly by water resource development schemes. In particular, remaining populations will be affected by the Alqueva Multipurpose Project, which includes construction of Alqueva Dam, a large (96-m-high, 460-m-wide) structure that would create the largest reservoir in Europe, 80 km long and 29,636 ha in area. This massive project, designed to irrigate a semiarid region, includes nine smaller dams, 5,000 km of canals, and water transfer between major river basins.

Investigations of the species' status and plans for its recovery were funded under the auspices of a European Union LIFE-Nature project. A management plan for conservation arose from these studies, focusing not only on the saramugo but also on other threatened fish species in the basin, including six other cyprinids listed in the Portuguese *Red Data Book*, such as the meter-long cumba, or barbo comiza, *Barbus comiza* (Doadrio and Perdices 1998). Admirably, the plan addresses most of the factors affecting the species' health, rather than a limited subset. It includes biological, sociological, economic, land-use, and public education components.

Proposed actions include designation of Guadiana tributaries as candidates for the National List of Sites under the EU Natura 2000 Programme, rehabilitation of degraded systems, integration of fish conservation into the basin management plan, and stocking in recovered habitats. Significantly, project reports have emphasized public education and increased awareness of the conservation value of fish resources in the region. The plan recognizes that short-term incentives and public involvement are essential to overcome market-oriented tendencies among modern consumer societies (Collares-Pereira and Cowx 2004). The effort to save the saramugo can serve as a blueprint for species recovery, especially where multinational rivers are involved.

Asian Cyprinids

One of the best-known and most controversial accounts of massive species extinctions involves the endemic species flock of cyprinids of Lake Lanao, Philippines. There is little doubt that a major cause of the extinctions was introduced predators in the 1960s, but the controversy swirls around just how many fishes were exterminated. Based on the original species descriptions of Albert W. Herre during the 1920s, as many as three endemic genera and 15 to 30 endemic cyprinid species may have existed. These purported endemics evolved in a volcanic lake isolated by an 18-m waterfall at its outflow. All but three cyprinids were extinguished as the result of stocking largemouth bass, a goby, a sleeper, a clariid catfish, gouramies, an eel, and a snakehead, as well as carp, tilapia, and milkfish. Complicating our understanding of this extinction event is the fact that Herre was a notorious "splitter," describing as different species fishes that varied only by a slight degree. Verification of the original species determinations was obscured by destruction of critical museum specimens (holotypes) during World War II (Reid 1980; Kornfield and Carpenter 1984; I. Payne 1987). If the species descriptions are valid, the Lake Lanao extinctions make up 42% of all cypriniform extinctions and 21% of all possibly extinct, noncichlid freshwater fishes (Harrison and Stiassny 1999).

Less controversial but no less alarming is the condition of fishes endemic to the Mekong River, which is the largest river in southeast Asia, with a fish diversity upward of 1,700 species. At least 13 species in the Mekong, of which 6 are cyprinids, have IUCN Red Data List status. Included are

Figure 2.5. Giant, Endangered Mekong fishes. (A) The Mekong giant catfish, probably the world's largest catfish, grows to 3 m and more than 300 kg; (B) the freshwater whipray, 2.4 m and 600 kg; and (C) the giant barb, 3 m and 300 kg. Photos by Zeb Hogan; used with the photographer's permission.

BOX 2.2. The Mekong River and Tonle Sap Great Lake: A Biological Hotspot

The Mekong River is 4,900 km long, the tenth-longest river in the world. More than 60 million people live in its basin, where they have long exploited a fishery that yields almost 2 million tons annually, worth about $1.45 billion. The Mekong is second only to the Amazon in fish diversity (Rainboth 1996; Mattson et al. 2002) and houses numerous endemics, giants such as the Mekong giant catfish, and imperiled species such as the Laotian shad, *Tenualosa thibaudeaui* (once one of the most abundant commercial fishes in the river, now designated Endangered by IUCN; its last large spawning migration was recorded in 1984).

The Mekong giant catfish is only one of several very large, troubled, migratory fish species in the Mekong (figure 2.5). Three other pangasiid catfishes there grow to greater than 80 cm, including the giant or dog-eating catfish, *Pangasius sanitwongsei*, which can be 2.75 m long or longer. Silver-toned catfish, *P. krempfi*, which grow to 80 cm, are unusual among catfishes in being truly anadromous, with a spawning migration that may take them 1,600 km from the Mekong Delta and South China Sea to Thailand and Laos. Other "giants" in the river include the giant barb, *Catlocarpio siamensis* (perhaps 300 cm, 300 kg, and considered depleted by many authorities), and the freshwater whipray, *Himantura chaophraya* (IUCN Vulnerable—240 cm, 600 kg), both actively targeted by sport fishers.

Mekong giant catfish and many other species, including giant barb and Laotian shad, migrate into and within Tonle Sap Great Lake, Cambodia, which houses 500 fish species. The lake expands from 250,000 ha during the dry season to more than 1.25 million ha during the monsoon season. Waters from the Mekong River flow back into the lake, flooding the marginal forest and agricultural lands (at which time the lake covers just less than half the area of Lake Malawi in Africa). This flood cycle contributes to high productivity and supports an annual fish catch estimated at 300,000 to 400,000 tons, valued at $69 to $85 million (Mareth et al. 2001). As the human population around the lake has grown, especially after the Khmer Rouge period in 1979, threats to the lake and its fishes—from overfishing, deforestation, invasion by weed plants, hydrological alterations, and planned development of petroleum reserves—have increased.

To address these threats and their impacts, the Cambodian government issued a Royal Decree in February 2001 establishing the Tonle Sap Biosphere Reserve (TSBR) as part of UNESCO's World Network of Biosphere Reserves. The TSBR plan focuses on "an integrated strategy for the sustainable use of the natural resources of the lake and its watershed" (Mareth et al. 2001). Such efforts are a positive beginning.

four "giant" barbs (70–150 cm long) that undertake long-distance spawning migrations. Representative is the seven-striped barb (or Jullien's golden carp), *Probarbus jullieni*, a species with Endangered status in IUCN (2004) and one of the 9 fish species protected in CITES Appendix I. Large specimens 150 cm long and weighing 70 kg were previously captured, but such fish are now a rarity.

Seven-striped barb have been depleted by overfishing, exacerbated because they congregate at known spawning sites where they are targeted. The species is also threatened by planned hydrologic alterations in the basin, including dams that would block its spawning migrations (box 2.2). Dams are also likely to reduce seasonal flooding of areas into which the fish migrate to feed during the wet season. These fish are known to migrate up the Mekong River between Kompong Cham in Cambodia and Chiang Khong in Thailand. Crossing national borders adds to the fish's problems because its conservation requires international cooperation. Efforts to save threatened Mekong fishes are part of a Basin Development Plan by the Mekong River Commission, an official governing body involving Thailand, Lao PDR, Cambodia, and Vietnam, but unfortunately not the People's Republic of China, where the river arises. The development plan provides for biological studies, monitoring of populations, and captive propagation, with an emphasis on ecosystem-level efforts and adaptive, sustainable management (Mattson et al. 2002).

South African Cyprinids

South Africa has an active conservation community that emphasizes the country's unique biodiversity and biogeography. All 33 temperate freshwater species in South Africa are regional endemics, half occurring in a single river system (Skelton 2002). Cyprinids dominate, constituting 81% of the temperate species. Because of isolation, small average size, and a general lack of native piscivores, the fauna is inordinately at risk. Sixteen of the 27 cyprinids had IUCN (2004) CR, EN, or VU status.

Three of the critically endangered species exemplify problems faced by South African cyprinids. The first, the Twee River redfin, *Barbus erubescens*, is restricted to a tributary of the Olifants River in the Western Cape, occurring only above a 12-m waterfall. Its range has shrunk by 40% due to alien fish introductions and to water extraction and pollution associated with farming. Ironically, two introductions are of imperiled South African fishes, one being the predatory (1-m, 10-kg) Clanwilliam yellowfish, *B. capensis* (VU), itself in decline because of introduced smallmouth and largemouth bass. The second critically endangered cyprinid, the Maloti minnow, *Pseudobarbus quathlambae*, was considered extinct, having disappeared from its single known locale in the Umkomazana River in Natal. It is now known to exist in six isolated, alpine populations in tributaries to the Orange River, Lesotho. Only three of those populations are free of introduced rainbow and brown trout, because of barrier waterfalls, and the largest population is threatened by a planned water development scheme (i.e., dams) that will flood habitat, allow incursions by trout, and promote industry and farming: "The long-term prospect for this critically endangered species is poor" (Skelton 2002, p. 226). The minnow has also been affected by expanding agricultural activities, introduced predators, and blocked spawning migrations due to dam construction (Impson 1997). The third critically endangered cyprinid is in even worse condition. The Clanwilliam sandfish, *Labeo seeberi*, is also an Olifants River, Western Cape, endemic. Once widespread and abundant, it experienced a 90% range reduction in the 1980s and 1990s. Only one healthy population remains, in the Oorlogskloof River.

North American Cyprinids

Cyprinids constitute the largest family of freshwater fishes in North America, making up 230 of the 800 native species. Thirty-nine cyprinids are Threatened or Endangered, and 12 have gone extinct. Dewatering of springs, introduction of nonindigenous species, impoundments, and general habitat loss are major contributors to declines, especially in the arid southwestern U.S. and Mexico.

Losses among the endemic fishes of the Colorado River basin exemplify these problems. Minnows and other

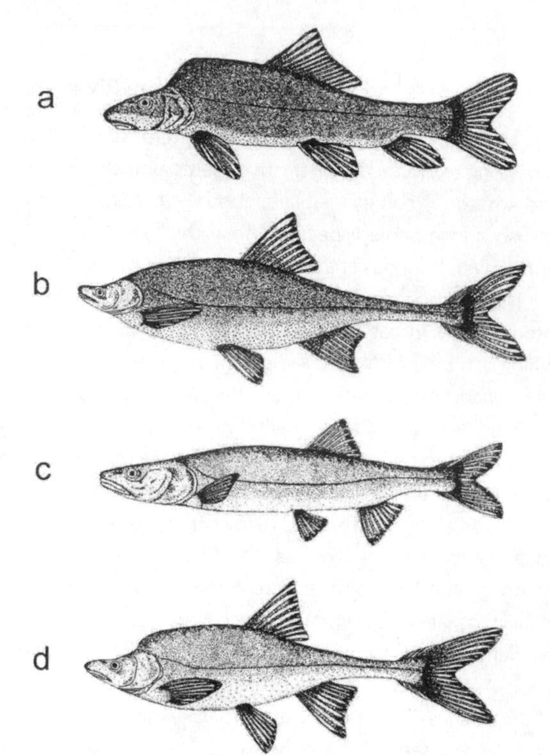

Figure 2.6. Endangered fishes of the upper Colorado River that evolved reproductive habits attuned to its flood cycle: (A) razorback sucker; (B) bonytail chub; (C) Colorado pikeminnow; and (D) humpback chub. Prior to impoundment, the river experienced exceptionally high flows, >9,000 m^3/sec during winter and spring floods, which redistributed sediments critical to spawning and larval rearing. From Portz and Tyus (2004); used with permission.

fishes in the Colorado show adaptations specific to the unique, extreme hydrological cycle of desert regions. Although we tend to focus on lack of water in such locales, when rains occur, they often result in torrential flows that transport sediments and reshape the river. Colorado River endemic cyprinids such as the humpback chub, *Gila cypha*; bonytail, *G. elegans*; Colorado pikeminnow, *Ptychocheilus lucius*; and a catostomid sucker, the razorback, *Xyrauchen texanus*, spawn in response to this flood cycle, which creates seasonal backwater areas that serve as nursery regions (figure 2.6).

During the 20th century, more than 100 dams were built along the Colorado and its tributaries, disrupting flows, temperatures, and sediment distributions. Alterations commonly facilitate invasion. Among the 80 fish species that now reside in the Colorado, only about a third are native, and many of these have federal status. Cyprinids listed as Endangered are humpback chub; bonytail; Colorado

BOX 2.3. Good News, and From a Sucker No Less

A generally happier tale surrounds the robust redhorse, *Moxostoma robustum* (80 cm, 8 kg). Known from only a few, poorly preserved specimens—although at one time apparently abundant in Atlantic slope rivers from North Carolina to Georgia—robust redhorse went unrecorded for over 120 years. It was rediscovered in the Oconee River, Georgia, in 1991. Intensive sampling turned up more fish, but all were adults, and locales were heavily degraded by sedimentation, hydraulic alteration and regulation, and introduced species. Robust redhorse were quickly nominated for federal protection, which would have affected power generation capabilities on southeastern rivers and relicensing of a major power-generating dam. A remarkable cooperative effort among state, federal, corporate, and nongovernmental organizations (14 groups in all) led to formation of the Robust Redhorse Conservation Committee, which reached an agreement that included altered flows, habitat improvement, extensive research, and captive propagation and release. Only minor disruptions in dam operation were needed to protect spawning fish and their habitats, and federal listing has apparently been averted. Ongoing surveys indicate successful establishment and reproduction by propagated fish (www.robustredhorse.com/h/reportpubs.html).

pikeminnow; Virgin River chub, *G. robusta seminuda*; and woundfin, *Plagopterus argentissimus* (and razorback sucker). The Little Colorado spinedace, *Lepidomeda vittata*, is Threatened (Ono et al. 1983; Minckley 1991; Wydoski and Hamill 1991). Recovery plans include periodic experimental releases of high flows from Glen Canyon Dam upstream of the Colorado mainstem to redistribute sediments and restore spawning and nursery habitat, especially for humpback chub (Gorman and Stone 1999; Patten et al. 2001; see chapter 6).

Suckers (Catostomidae)

Suckers, with their public relations nightmare of a name, are one of the most threatened families of predominantly North American fishes. All but 2 species are endemic to North America, and 39 of the approximately 70 described species appear on either the U.S., Canadian, Mexican, or IUCN endangered species lists. Warren et al. (2000) included another 6 species, bringing the number of imperiled suckers to at least 45 species, or more than 60% of the family. Two species are known to be extinct, the harelip sucker, *Moxostoma* (formerly *Lagochila*) *lacerum*, and Snake River sucker, *Chasmistes muriei* (but see box 2.3).

The first bona fide fish extinction in the U.S. was of a catostomid, the harelip sucker (figure 1.3). This 30–45-cm fish occurred in disjunct populations in perhaps 13 U.S. states, mostly east of the Mississippi River in the Cumberland and Tennessee river drainages (Jenkins and Burkhead 1994). It was described in 1877, at which time it had been commonly hunted during its spawning migrations for centuries, as indicated by its occurrence at archaeological sites (Etnier and Starnes 1993). The last specimen was taken in 1893, but the fish was not declared extinct until the 1940s, in hopes that a population existed somewhere. Harelip suckers probably suffered because they were habitat and trophic specialists, occupying deep, rocky pools in medium and large, upland, warm, clear streams, where they fed on small mollusks, predominantly snails. The harelip sucker may have been "the only true mollusk specialist in the family" (Jenkins and Burkhead 1994, p. 461).

We know remarkably little about this once widespread fish. No photos of a live or even a recently dead specimen exist. In fact, only 33 specimens, of which only one was an adult, were placed in museums (D. S. Lee et al. 1980). Extinction of the harelip sucker correlates with deforestation and expansion of agriculture in the eastern-central U.S. in the mid-late 19th century. Logging, farming, and silviculture practices that eventually led to the Dust Bowl conditions of the early 20th century created unprecedented sediment inputs into streams. Sediment settles first in pools in flowing water. Increased silt loads would have covered rocks and created turbid conditions, hampering foraging of this presumed sight feeder. In addition, its disjunct distribution and presumed spawning migrations and aggregations suggest that numerous small dams placed across upland streams and rivers led to separation of spawning populations and eventual fragmentation of range (Jenkins and Burkhead 1994).

Range fragmentation is a recurring contributor to the decline of endangered fishes that live in upland areas, away

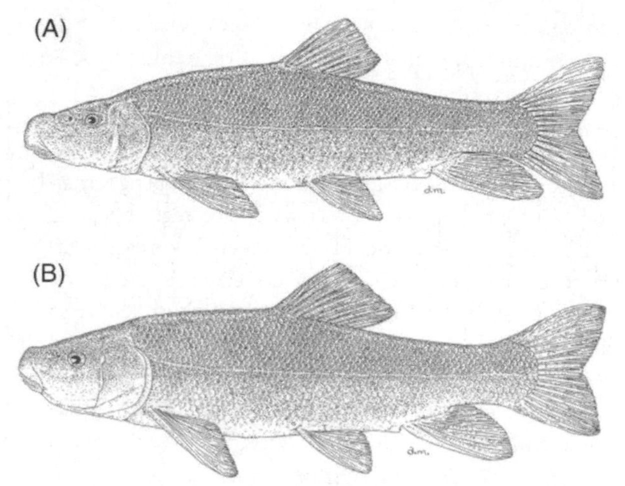

Figure 2.7. Endangered lake suckers. (A) Lost River sucker and (B) shortnose sucker from the Klamath River basin of Oregon and California. Drawn by A. Marciochi from Moyle (2002); used with permission of the author.

from large, lowland river valleys. Notable among extant, threatened suckers are the lake suckers (*Deltistes, Chasmistes*; figure 2.7) of the western U.S. Lake suckers represent remnants of a larger group that evolved in extensive Pleistocene lakes of the now arid Southwest and Great Basin. They are under increasing threat from habitat loss, pollution, and invasive species. Most inhabit remnant lakes and move into inflowing rivers in spring and early summer to spawn. Disruption of these spawning migrations by dams and water diversions contributes to their decline (e.g., see chapter 6).

Lake suckers grow large, many to 50 cm and some to almost a meter, and are long lived, not uncommonly to 40 years. They are surprisingly tolerant of low oxygen, high alkalinity, high temperature, and high ammonia concentrations, apparently having adapted to highly variable, frequently hostile physical conditions (Moyle 2002; NRC 2004b). Large size, long life spans, and physiological tolerance historically allowed them to rely on intermittently successful reproduction to get through prolonged periods of unfavorable climate and water quality. Despite this resilience, several species are at risk or worse. The group includes the now-extinct Snake River sucker (known from a single specimen collected in the Snake River below Jackson Dam in Wyoming) and the endangered June sucker, *Chasmistes liorus*, endemic to Utah Lake, Utah. June suckers were at one time exceedingly abundant, as apparently were all these species. As their numbers decrease, June suckers become increasingly vulnerable to genetic disruption via hybridization with abundant Utah suckers, *Catostomus ardens* (Echelle 1991).

Hybridization also threatens the two endangered lake suckers in the Klamath River basin, the shortnose sucker, *Ch. brevirostris*, and Lost River sucker, *D. luxatus*, both of which previously supported commercial fisheries. Cui-ui, *Ch. cujus*, are an endangered lake sucker endemic to Pyramid Lake, Nevada, where previously abundant spawning runs have been drastically reduced due to water withdrawals from the Truckee River. Withdrawals lowered Pyramid Lake levels and blocked access to spawning habitat (Scoppettone and Vinyard 1991). Cui-ui, which suffered through a period of apparent total reproductive failure between 1950 and 1969, are the only freshwater North American fish listed in Appendix I of CITES. They have been successfully propagated in a hatchery managed by the Paiute Tribe and have been transplanted back into Pyramid Lake, where numbers are increasing.

Catfishes (Siluriformes)

More than 35 families make up the catfish order Siluriformes, which is most diverse in South America, Africa, and Asia. Among the 41 species listed in IUCN (2004), only 1—the trichomycterid *Rhizosomichthys totae* from Lake Tota of Colombia—is considered Extinct. The regions with the largest numbers of imperiled catfish species are sub-Saharan Africa and Southeast Asia, each with 8.

Size matters when it comes to endangered fishes. Large fish are often migratory and thus subjected to the passage-blocking and population-fragmenting impacts of dams. Large fish are often specifically targeted by fishers for both food and sport, adding overexploitation to the factors affecting population success. It is also easier to generate public interest when dealing with superlative species (largest, heaviest, longest, etc.; see Owen 2004). Some of the world's largest freshwater fishes live in the Mekong River and are among southeast Asia's most imperiled species.

The Mekong giant catfish, *Pangasianodon gigas* (Pangasiidae), is probably the world's largest catfish (figure 2.5A). Record individuals have been measured at 300 cm long, weighing more than 300 kg (Hogan et al. 2004). Only the European wells, *Silurus glanis* (Siluridae) may be larger, at a reported 500 cm and 306 kg, although none that large have been seen in years (how many times does that phrase occur in this chapter?). South America's goliath or kumakuma catfish, *Brachyplatystoma filamentosum* (Pimelodidae), attains a length of 360 cm and a mass of 200 kg, whereas North America's largest catfish, the blue catfish, *Ictalurus furcatus* (Ictaluridae), is relatively small at 165 cm and 68 kg.

P. gigas may also be the most threatened fish endemic to the Mekong River basin. It was previously recorded from coastal Vietnam to southern Laos and perhaps as far north as southern China but currently occurs only in Tonle Sap Great Lake, Tonle Sap River, and portions of the mainstem Mekong River. Overall population size has decreased by an estimated 80%–90% since the early 1990s (Hogan et al. 2004; IUCN 2004). At numerous historical fishing sites, catches have dropped from several dozen fish per year to zero in just the last few decades, despite apparent increased fishing pressure (Hogan et al. 2001). These trends have prompted a CITES Appendix I listing and a designation of Critically Endangered by IUCN (2004).

P. gigas undertakes long-distance spawning migrations that carry it across international boundaries from Tonle Sap Great Lake, Cambodia, into Lao PDR or Thailand (Hogan et al. 2001). Juveniles and adults use different habitats at different times of the year, and young fish move into flooded lake margins during the rainy season (Hogan et al. 2004). Hydrologic alterations would affect breeding and feeding movements and access to nursery habitat. The catfish have therefore declined due to overfishing, habitat loss and degradation from impoundments and deforestation, and genetic introgression from released, cultured stocks. Future threats include a plan to construct several migration-blocking dams in the Mekong system (see http://assets.panda.org/downloads/mekong.pdf) and destruction of rapids that may be spawning habitat for the species.

Efforts at conserving *P. gigas* are both national and international in scope. In addition to IUCN and CITES listings, brought about largely by a few concerned biologists, laws prohibiting capture, sale, and transport have been enacted in Cambodia, Thailand, and Laos. Unfortunately, enforcement of the laws is reportedly weak (IUCN 2004). Funds have been obtained for "buy and release" efforts to return captured fish to the wild, and approximately 5,000 individuals of Mekong giant catfish and a dozen other vulnerable species have been tagged and released (Hogan et al. 2004). A captive propagation program was begun by government agencies in 1985. However, researchers have become alarmed at inroads into the wild population caused by capture of brood fish, and by poor husbandry techniques that have resulted in limited parental stock being used in these programs. Genetic analyses suggest that 95% of the progeny from these efforts share the same two parents. This could lead to loss of genetic diversity when captive-propagated fish are released and breed with wild fish.

For many of the other listed catfish species worldwide, overexploitation seems less of a contributor to declines than habitat degradation and migration blockage. Federally and internationally recognized imperiled catfish species in North America include several diminutive madtoms (*Noturus* spp.) of the southeastern U.S. and desert spring and cave species in the southwestern U.S. and Mexico (e.g., Mexican blindcat, *Prietella phreatophila*; widemouth blindcat, *Satan eurystomus*; toothless blindcat, *Trogloglanis pattersoni*; blindwhiskered catfish, *Rhamdia reddelli*). That cave-adapted species are threatened is not surprising. Cave faunas around the world are especially vulnerable because of the sensitivity of their habitats to any disturbance.

Salmons, Trouts, and Chars (Salmonidae)

"The 'salmon problem' is easy to state, hard to analyze, and even more difficult to solve. . . . The salmon problem is the decline of wild salmon runs and the reductions in abundance of salmon even after massive investments in hatcheries" (NRC 1996b, p. 3, 18). Probably no single taxonomic group of fishes is exposed to as extensive a litany of anthropogenic and natural threats as salmons and trouts. The broad array of destructive factors results from their complex life history, in combination with high desirability as a commodity and sport fishing target. Because salmon hatch in freshwater, travel down rivers and through estuaries to the ocean to grow and mature, and then return to freshwater to spawn—specifically to the stream where they hatched—they encounter multiple anthropogenic insults. Because they are fun to catch and taste good, they have high economic value. The salmon's problem then is that they face "a continuous series of threats at nearly every point in their range, throughout their entire life cycle" (Lichatowich 1999, p. 46).

The literature on salmon and their conservation is immense. The University of Washington's library contains over 6,400 articles and books about salmon. A few books are almost universally cited (Netboy 1974; Groot and Margolis 1991; B. Brown 1995), and others stand out, especially NRC (1996b, 2004a), Lichatowich (1999), Stouder et al. (1997), Montgomery (2003), Augerot (2005), and R. N. Williams (2006). The Lichatowich and Montgomery volumes are particularly useful for understanding the historical context of impacts.

The litany of human actions leading to salmon declines, listed below and detailed in several chapters, has been repeated throughout their range to a degree that is almost ludicrous. Overfishing, migration blockage, and habitat degradation due to development, deforestation, and pollution occurred first on the European continent, especially in Germany and France. This cycle was repeated in the British Isles, followed by the U.S. Northeast and eastern Canada, then the U.S. Pacific Northwest and Japan (Lichatowich 1999; Montgomery 2003). At each juncture, spectacularly abundant salmon runs were reduced to remnants and then driven to commercial extinction or extirpation. A problem was detected, the causes identified, predictions of long-term loss were made based on current trends and past experience elsewhere, and rules and regulations were passed to slow the loss. But laws were continually broken, ignored, or circumvented in the name of short-term economic gain, and penalties for infractions were minor if not nonexistent.

In just about every instance of decline, potential problems were recognized early and the need for caution espoused, but repeatedly "the plan was to figure something out later" (Montgomery 2003, p. 195). The major difference across time and space has been the accelerating rate at which declines occurred: It took centuries to wipe out salmon in Europe and the UK, about one century in New England and eastern Canada, and a matter of decades in the Pacific Northwest (Montgomery 2003). History repeats itself and at an accelerating rate. Current salmon management in the Pacific Northwest relies on technological solutions such as hatcheries, which often cause more harm than good because they address symptoms rather than the root causes of decline, namely dams, overfishing, pollution, and habitat loss. Historically, "managing hatcheries as salmon factories promised a painless way to treat the symptom of too few fish without curing the diseases of overfishing and environmental degradation" (Montgomery 2003, p. 150). Hatcheries have been tried but have failed to resurrect degraded salmon runs in France, Scotland, England, eastern Canada, New England, Oregon, and Washington, all places where "the promise of increasing salmon runs without having to reduce fishing was irresistible" (Montgomery 2003, p. 153). Despite repeated failures, managers guarantee success "just around the corner" as technology and hatchery practices purportedly improve (e.g., Brannon et al. 2004; see chapter 14).

In another repeat of bad historical precedence, a November 2004 revised *Biological Opinion*—the federal blueprint for minimizing future impacts on 12 listed salmon ESUs in the Columbia and lower Snake rivers—determined that the 14 sets of dams and reservoirs already in place were historical structures and should be considered "part of the Environmental Baseline" and that "the Action Agencies are not responsible for mitigating the effects of the existence of the dams and reservoirs" (sections 5.2.1 and 9.1, *Final FCRPS BiOp on Remand*, November 30, 2004, www.salmonrecovery.gov). This grandfathering of existing structures while admitting that dams are a major cause of declining salmon runs occurred previously as Europe, Britain, New England, and Canada destroyed their salmon. Dams that blocked or degraded migrations were built in direct violation of existing legislation and then protected by later legislation that outlawed future dams (Montgomery 2003).

Salmon Stocks at Risk

Wild Atlantic salmon, *Salmo salar*, in North America previously spawned as far south as the Hudson River, as far north as arctic regions of Canada, across the North Atlantic, and south to Portugal and Spain. Spawning runs numbered in the hundreds of thousands. Today, Maine is the only U.S. state with wild runs, and its salmon were listed as Endangered in 2000. Only 33 wild fish returned in 2002 to spawn in the eight Maine rivers that house the listed fish (NRC 2004a). Runs have also declined more than 90% throughout Canada, where the inner Bay of Fundy populations have Endangered status (www.sararegistry.gc.ca). European runs have been decimated in every country except Iceland, Ireland, Norway, and Scotland. In those four countries, at least half of runs are still healthy. In the remaining 17 European countries, only about 10% of runs are considered healthy, and runs have been extirpated from more than 250 European rivers (WWF 2001; www.asf.ca).

Assessments of the status of U.S. and Canadian runs, distinct population segments, and ESUs indicate that perhaps 280 Pacific salmon and steelhead stocks (*Oncorhynchus* spp. and ssp.) have been extinguished and another 880 are at high risk of extinction (Nehlsen et al. 1991; T. T. Baker et al. 1996; C. Huntington et al. 1996; Slaney et al. 1996; P. Vecsei, pers. comm.). As of December 2004, USFWS listed two sockeye, two coho, two chum, nine Chinook, and ten steelhead stocks as Threatened or Endangered. The numbers shift periodically, however, in response to legal decisions over whether hatchery fish should be counted along with wild fish. Hatchery fish, reared and released in large numbers, can greatly outnumber wild fish at some locales. Most scientists maintain that wild fish are sufficiently different genetically and behaviorally to warrant separate counting. Hatchery fish also have deleterious effects on wild fish, particularly genetic impacts from crossbreeding and because high numbers of hatchery fish support large numbers of human and nonhuman predators. Wild runs deserve protection under endangered species legislation, and their numbers are artificially inflated by counting co-occurring hatchery fish. Persons inconvenienced or commercially affected by protective actions (or philosophically opposed to strong endangered species legislation) argue that wild and hatchery fish are similar enough to be combined in counts, and that runs are therefore healthy (e.g., Oregon runs of coho salmon, see R. A. Myers et al. 2004).

In the U.S., an additional seven trout species have Threatened or Endangered status, including the Apache trout, *Oncorhynchus apache*; Little Kern golden trout, *O. aguabonita whitei*; Gila trout, *O. gilae*; three subspecies of cutthroat trout, *O. clarkii* (greenback, Lahontan, and Paiute), and bull trout, *Salvelinus confluentus* (ecos.fws.gov). One U.S. species, the silver trout, *S. agassizi*, of New Hampshire, was declared extinct in 1939 as the result of introduced species, overfishing, and hybridization (Harrison and Stiassny 1999); some authorities treat the silver trout as a brook trout subspecies.

A number of other salmonids are recognized by IUCN as globally imperiled. The ala balik, *Salmo platycephalus* (perhaps a subspecies of brown trout), of Turkey has Critically Endangered status. The Danube salmon or huchen, *Hucho hucho*, of eastern Europe is Endangered, as are the satsukimasu salmon, *O. ishikawai*, and Kirikuchi char, *Salvelinus japonicus*, of Japan, and the Adriatic salmon, *Salmothymus obtusirostris*, of Croatia, Serbia, and Montenegro (IUCN 2004). An attractive, small, "morphologically aberrant" char endemic to Lake El'gygytgyn in northeastern Siberia has Vulnerable status (figure 2.8). The longfin Svetovidov's char, *Salvethymus svetovidovi* is threatened by tourism-accelerated pollution and a fishery for the more abundant Boganidsk char, *Salvelinus boganidae* (Chereshnev 1996).

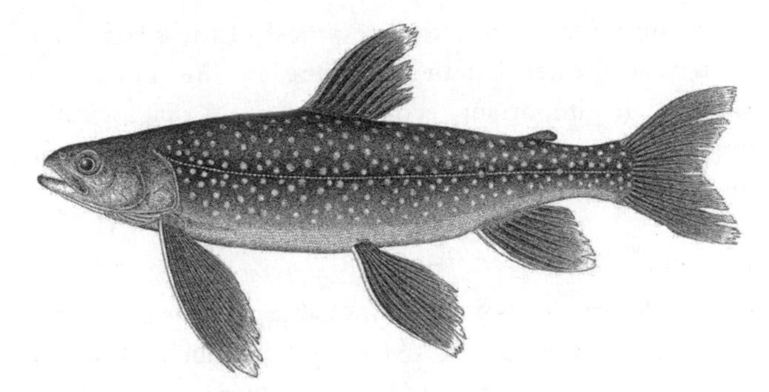

Figure 2.8. The Vulnerable longfin Svetovidov's char of Lake El'gygytgyn, Siberia. Longfin char live at depths greater than 50 m, feed on zooplankton, and take 30 years to reach their maximum length of 30 cm. Drawing by Paul Vecsei; used with permission of the artist.

The brown trout, *Salmo trutta*, although globally secure as a species and often a nuisance as an introduced species, appears on the red lists of Denmark, Estonia, Finland, France, Greece, Portugal, Russia, Slovenia, Spain, and Switzerland. These listings refer to a host of subspecies; *S. trutta* differentiates genetically and has undoubtedly formed many unique races or stocks, the identity of which may be forever lost due to massive stocking programs with nonindigenous strains (e.g., Almodovar et al. 2002). The taxonomy of brown trout is complex and controversial. Among described forms identified as vulnerable are stocks of *S. trutta* with the subspecific designations *S. t. aralensis, carpio, caspius, dentex, exenami, fario, ischchan, labrax, lacustris, letnica, platycephalus, marmoratus, macrostigma, montenegrinus,* and *trutta* (Laikre et al. 1999); some of these forms are given species status. Apart from confusing those unfamiliar with regional faunas, subspecific differentiation within brown trout points out major problems facing those working to preserve biodiversity at the genetic level. Many lists fail to include subspecies and "lower" taxa, and our knowledge of the genetic structure of many species is weak.

Major Impacts on Salmon

Discussions about the decline of salmon focus on the five Hs of habitat, harvest, hydropower, hatcheries, and human inability to learn from history. Contained within each category are multiple subcategories; most of these topics are treated at greater length elsewhere in this book. But salmon are important ecologically and sociologically; impacts on them are representative of the problems facing many other groups and are therefore worth detailing.

Habitat

Direct destruction of instream habitat occurred in the Pacific Northwest due to extensive hydraulic and dredge mining for gold in the mid to late 19th century. Less direct but similarly destructive is deforestation, which degrades spawning and nursery habitats. Logging, road building, land clearing for agriculture, and overgrazing accelerate erosion; silt clogs gravels, smothers eggs, and impedes visual feeding. Sawdust from mills, dumped into rivers by the ton, acts similar to sediment. It clogs and abrades gills, smothers eggs and bottom organisms, and even depletes oxygen when it decomposes (Lichatowich 1999). Streamside logging increases UV penetration into streams, causing eye lesions and vision impairment; it also elevates stream temperatures, which can be lethal. Increased temperature accelerates evaporation, which reduces summer stream flows, further eliminating habitat. Lack of vegetative cover in the spring increases runoff and magnifies peak flows, which wash gravel and woody debris from streams, eliminating habitat if not young salmon directly. Logging removes the source of woody debris that provides crucial refuge from currents for young salmonids. Logging on the land and removal (snagging) of large logs from streams and rivers alter stream morphology, eliminating pools that form around logs and impeding the creation of side channels used as nursery habitat. Splash-damming and log-driving were common practices during the heyday of logging in the late 1800s. These practices used streams and rivers to move massive quantities of giant cedar, fir, hemlock, and spruce logs. Fish were killed outright, while bottom characteristics and even river shape were altered.

Other impacts on habitat include channelization, levees, and diking, which have many similar impacts on fluvial geomorphology. Natural flow regulation can be altered by eradicating beaver, thus eliminating large wood inputs that create backwaters on which salmon rely. Such eradication occurred in the Pacific Northwest during the early 19th century (Lichatowich 1999). Habitat quality is also degraded by urbanization of downstream areas, causing devegetation and an increased percent of impervious surface.

Pollution also affects the suitability of streams. Chemical pollution of rivers occurs from industrial activity such as mining and metal smelters and from municipal and agriculture activity. Some chemical pollutants are endocrine-disrupting compounds (EDCs), which affect reproductive success by reducing sperm count and egg production, increasing intersexuality, and increasing developmental abnormalities (chapter 7). Atlantic salmon are affected by acid deposition (acid rain) from industrial and municipal sources. Thermal pollution results from power plant discharges, as well as reduced instream flows because of water withdrawal for agriculture, river slowing behind dams, and warmed water flowing off impervious surfaces. Thermal pollution is especially deleterious at the southern limits of species' ranges, where populations already live near their critical thermal maxima. These are valuable (evolutionarily significant) stocks because they may possess a genetic tolerance of higher temperatures.

Harvest

Commercial, subsistence, and recreational fishing occur at sea, nearshore as fish approach spawning rivers, and in rivers where natural topographic constrictions (passes, channels, cascades, waterfalls) concentrate fishes, thus facilitating their capture in large numbers. The total take of many runs exceeds 90%—leaving 10% "escapement"—whereas sustainable harvest is generally considered to allow at least 50% escapement. Oceanic fishing with driftnets has been particularly deleterious for Atlantic salmon, especially after their feeding region southwest of Greenland was discovered following World War II.

Natural predators such as fish-eating birds and piscivorous fishes eat large numbers of smolts, whereas otters, sea lions, orcas, and bears consume large numbers of fish returning to spawn. These predators would have minimal impact on healthy populations, but depleted stocks are more vulnerable, especially when predator populations are subsidized by hatchery fish. Predation is maximized at locales that artificially concentrate and disorient migrating salmon, such as dams and locks.

Hydropower

Dams block upstream migrations of spawning adults if no fish ladder is present; poorly designed fish ladders slow migration, exhausting fish. Interdam mortalities can reach 25%. Dams also cause mortality of downstream migrating smolts via entrainment in turbines (more from pressure changes and cavitation than from the blender effect), gas bubble disease from supersaturation of gases injected under pressure from turbines, predation on disoriented smolts, and entrainment of young fish into insufficiently screened intakes for hydropower and irrigation projects.

Water retention behind dams exposes downstream, shallow, shoreline habitats that are important nursery areas for juveniles. At the same time, dams cause heating of upstream water to physiologically stressful levels.

Hatcheries

Hatcheries release large numbers of fish that, although often genetically dissimilar from wild stocks, create an illusion of abundance and rehabilitation. The hatchery environment frequently selects for aggressive fish that grow fast; they outcompete and even cannibalize wild fish. They are often not as well adapted to local conditions, thus undermining the genetic integrity of wild stocks via hybridization. Hatchery fish may harbor pathogens to which they have been immunized, thus transmitting disease to wild fish. "There is no evidence that, economically or ecologically, hatchery-based fisheries can be sustained over the long run" (Montgomery 2003, p. 171), nor are there many examples of success at rebuilding self-sustaining runs with hatchery fish (chapter 14).

Aquaculture, especially salmon farms or ocean ranching, pollutes native habitat with food, fecal waste, and antibiotics and subjects wild fish to increased disease and parasite incidence. High escape rates from aquaculture facilities lead to competition with wild fish for food and spawning sites. Escapees interbreed with wild fish, weakening genetic diversity and local adaptation. Abundant escapees also sustain fisheries that could not exist given low numbers of wild fish, again maintaining fishing pressure on rare, wild genotypes.

History

Salmon runs have been decimated almost everywhere they co-occur with large numbers of humans or with anything more technological than an artisanal fishery. Despite the repetition of this cycle and lessons that should have been learned, we continue our salmon-destroying activities, blame everyone but ourselves for the problem, and predict and then lament the salmon's decline (Montgomery 2003). Dams halt migrations, but we ignore violations of laws prohibiting blockage. Hatcheries have failed everywhere, yet we still build hatcheries rather than deal with the factors that actually cause the populations to decline.

Not only do we fail to accept responsibility for our negative actions, but we also take credit for positive influences over which we have no control. When cyclical oceanographic conditions increase oceanic survival of salmon, we congratulate ourselves for our management success and use the increase as justification for business as usual. Even worse, such periods of abundance are used to justify increased fishing pressure, rather than using temporary population growth to rebuild depleted stocks. The official press release accompanying the 2004 *Biological Opinion* on Snake and Columbia river salmon directly linked increased runs during 2000–2003 to the annual $600 million expenditure on Columbia Basin stocks, a link claimed to have "made possible the first spring Chinook recreational fishery since 1977 and the first summer Chinook fishery since 1973" (NOAA Fisheries 2004). This assessment ignores the findings of the scientific community, which identified favorable oceanic conditions during this period as part of a cycle—the Pacific Decadal Oscillation—that affects marine temperatures, marine productivity, stream flow, and the survival rate of salmon (see Mantua et al. 1997; Hare et al. 1999; Lackey 2003; Beamish et al. 2004).

Other Salmonids

Arctic char and coregonine whitefishes are genetically plastic fishes. They have differentiated repeatedly in lakes in North America and Europe, evolving into distinct phenotypes and genotypes. Unfortunately, they occur in some of the most heavily disrupted lake and riverine ecosystems on earth. Kottelat (1997) documented at least five apparent extinctions and several more extirpations of unique coregonines in Europe. Many more are threatened by hybridization, introgression, and outbreeding depression due to well-intentioned but potentially disastrous stocking programs where the fish had declined (see Freyhof 2002). During the mid-20th century, two coregonines native to the Laurentian Great Lakes—the deepwater cisco, *Coregonus johannae*, and the blackfin cisco, *C. nigripinnis*—disappeared as the result of overexploitation, pollution, siltation, competition with nonindigenous species, and perhaps predation by introduced sea lampreys. Whitefishes exemplify the vulnerability of taxa that develop localized specializations. Ecological specialists seem especially sensitive to anthropogenic disturbance and invasion by nonindigenous species.

An informative Web site maintained by the Wild Salmon Center and Ecotrust (www.stateofthesalmon.org) includes an updated account of all red-listed salmonids, based on information from IUCN's Salmon Specialist Group.

Galaxioids

The galaxioid southern smelts, southern graylings, and galaxiids were dominant among cool-temperate freshwater fishes in the Southern Hemisphere (Nelson 1994). Many have been driven to the edge of extinction and beyond due to introduced species, habitat destruction, and hydrologic alterations. Fully 26 of the 50 species in the superfamily Galaxioidea have some rank in the IUCN listings. Among the galaxiids—often referred to as southern salmonids because of their ecological similarities to that Northern Hemisphere group—16 of the 28 species that occur in Australia and New Zealand are threatened with extinction by IUCN criteria. The superfamily also included the endemic New Zealand grayling, *Prototroctes oxyrhynchus* (Retropinnidae, figure 2.9), driven into extinction by the combined effects of introduced trout and degraded habitat brought on by deforestation (McDowall 1990a; J. Richardson 2005). The fish was abundant in the 1860s, when Europeans were settling New Zealand, but was rare by 1900. The last specimen was collected around 1930, which makes it the first known human-caused fish extinction. The remaining member of the genus, the Australian grayling, *P. maraena*, was designated VU by IUCN (2004).

Although the New Zealand grayling was diadromous (technically amphidromous, its early life history spent in the sea but most of its life spent in freshwater), most other galaxioids can credit their vulnerability in part to their ability to live in marginal, isolated, freshwater habitats. The 6-cm-long salamanderfish, *Lepidogalaxias salamandroides*, inhabits ephemeral freshwater streams and temporary pools as small as 2 m^2 in acidic, black waters among the heathland peat flats in a limited region of extreme southwestern Australia (Berra and Pusey 1997). This area is essentially dry from January to late May each year, and while drying, water temperatures can exceed 34°C. The fish survive via estivation: They burrow into the sediments, as deep as 60 cm, where they remain until rains return. Much of their physiology and anatomy, including urea storage, cutaneous (skin) respiration, exceptional skull thickness and shape, and skeletal flexibility, reflects adaptations to seasonal drought and burrowing in wet sand. Salamanderfish are listed by IUCN (2004) as Near Threatened due to the vulnerability of their habitat to any kind of disturbance and the impact that any introduced species would likely have on this isolated specialist.

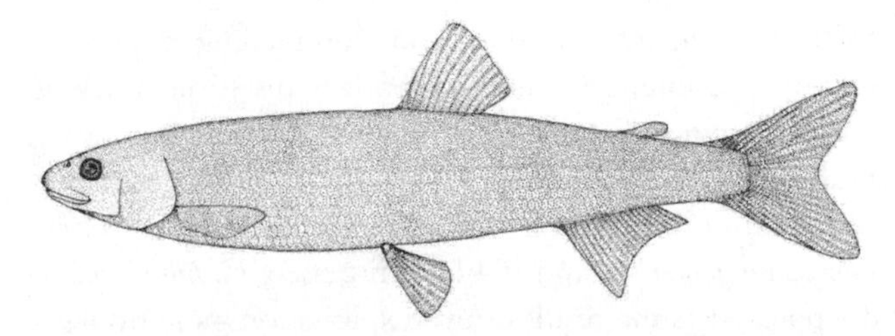

Figure 2.9. The extinct New Zealand grayling. The color pattern is a best guess, as is often the case with extinct fishes, many of which were never recorded properly. From McDowall (1990a); painting by P. M. Morse. Copyright R. M. McDowall; used with permission of the copyright holder.

The roll call of imperiled galaxiids underscores many of the problems of the New Zealand grayling and salamanderfish. At least five New Zealand galaxiids have Vulnerable status in the IUCN (2004) compilation, and another four species received high, at-risk rank in the official New Zealand red list (New Zealand Department of Conservation 2002a). *Galaxias argenteus*, the giant kokopu or Maori trout (VU), is the world's largest galaxiid at 58 cm and 2.7 kg. A coastal, anadromous species that forms landlocked populations, the giant kokopu is featured prominently on the New Zealand $5 coin. It has suffered declining and fragmented populations primarily due to habitat loss and restricted access to the ocean, with suspected impacts from competition with brown trout and bycatch in the whitebait fishery (David 2002). The Canterbury mudfish, *Neochanna burrowsius* (VU), is frequently the only fish species in the spring-fed, swampy wetlands of the Canterbury Plains, South Island, that encompass its entire range. It, too, overcomes drought by estivating in mud. The once extensive wetlands of the region have been converted to agricultural use, with only fragments remaining. Extirpations have been recorded in the locale from which the species was described; continued wetland conversion and water abstraction are major threats (McDowall and Eldon 1996). The dwarf inanga, *Galaxias gracilis* (VU), is known only from 13 dune lakes along an 80-km stretch of the west coast of Northland, New

Zealand. It is abundant in only four lakes, having been reduced in numbers or extirpated where rainbow trout have been released (McDowall and Rowe 1996). Populations of galaxiids throughout New Zealand have been diminished and fragmented due largely to trout introductions. New Zealand is a regular melting pot of salmonids, having welcomed rainbow trout, sockeye salmon, Chinook salmon, Atlantic salmon, brown trout, brook char, and lake trout (McDowall 1990b).

Australian galaxiids have suffered similarly, especially when faced with large predators that are not a part of their evolutionary history. Coldwater species have generally declined due to trout introductions, whereas warmer-water species in southeastern Australia have suffered from habitat loss and introductions of native and exotic predators that can tolerate the extreme climatic conditions in that region. The Pedder galaxias, *G. pedderensis*, has been designated Extinct in the Wild by Australian authorities. Two Tasmanian endemics—the Swan galaxias, *G. fontanus*, of eastern Tasmania and Clarence galaxias, *G. johnstoni*, of central Tasmania—are Critically Endangered. Once widely distributed throughout the Swan and Macquarie River catchments, only three natural populations of Swan galaxias remain along an 8-km stretch of river, with recent extirpations recorded where predatory brown trout and European perch were introduced. Clarence galaxias have been pushed out of the lower part of the Clarence River by introduced brown trout; the species now exists in only six fragmented breeding populations in headwaters that trout cannot invade. Recovery plans have been formulated for both Tasmanian species, and successful translocations into predator-free areas have been achieved or are planned (Crook and Sanger 1998a, 1998b). One other Australian galaxiid is Critically Endangered. The barred galaxias, *G. fuscus*, is currently restricted to 11 small highland streams of the Goulburn River system in Victoria. Its demise is mostly attributed to predation by introduced rainbow and brown trout (Raadik et al. 1996). Four additional Australian galaxiids have Vulnerable status. All things considered, galaxiids are "one of the most threatened fish families on the planet" (R. McDowall, pers. comm.).

Pupfishes and Killifishes (Cyprinodontiforms)

The cyprinodontiforms include rivulines, killifishes, toothcarps, topminnows, livebearers, splitfins, and pupfishes. They are mostly freshwater species, but many show a high tolerance for thermal, alklaline, saline, and even hypersaline conditions. For example, the desert pupfish, *Cyprinodon macularius*, of California and Mexico can tolerate a temperature range of 7°C–45°C, salinities between 0 and 68 ppt (the latter being twice the salinity of seawater), oxygen levels as low as 1–4 mg/l (anything <5 mg/l is considered stressful for most species), and rapid fluctuations in all these factors (Moyle 2002).

Despite this physiological plasticity, or in part because it has allowed them to evolve in habitats that are marginal for most other fishes, many species in the eight or so families are at risk. This is especially the case for ephemeral desert- and spring-dwelling forms in the families Aplocheilidae, Cyprinodontidae, Poeciliidae, Goodeidae, and Valenciidae. Any form of habitat disruption imperils these restricted-distribution species. IUCN considers at least 12 cyprinodontiforms Extinct and another 14 Critically Endangered; both lists are heavily dominated by aplocheilid rivulines, cyprinodontid pupfishes, and goodeid splittails. The goodeids include Mexican species with an evolutionarily unique, complex, placenta-like connection between mother and internally developing young.

Many aplocheilids in South America and Africa are annual fishes. They live in temporary habitats, spawn following rains, and then die, their genes preserved in eggs that wait in bottom muds for the next rainy season (Helfman et al. 1997). All seven of Brazil's IUCN-ranked at-risk freshwater fishes are aplocheilid sabrefins, lyrefins, and killifishes. Several undescribed species of *Pachypanchax* from Madagascar are imperiled by deforestation, dams, and habitat loss, along with many other endemic fishes (see chapter 3). More than a fourth of all 40 goodeid species have IUCN imperiled status, with 3 Mexican species considered Extinct or Extinct in the Wild.

The 100 or so species of cyprinodontid pupfishes, poolfishes, springfishes, perritos, cachorritos, and toothcarps (*Aphanius, Crenichthys, Cualac, Cyprinodon, Empetrichthys, Orestias*) include an alarmingly high number of species in trouble. IUCN (2004) listed 8 as Extinct, 9 as Critically Endangered, 11 as Endangered, and 5 as Vulnerable. USFWS gave threatened or endangered status to an additional 3, and Mexico added 12, bringing the total to at least 48 species, close to half of the described species. Often known as desert pupfishes, species are characteristically restricted to small, isolated habitats in arid regions; both the fishes and their habitats are exceedingly vulnerable

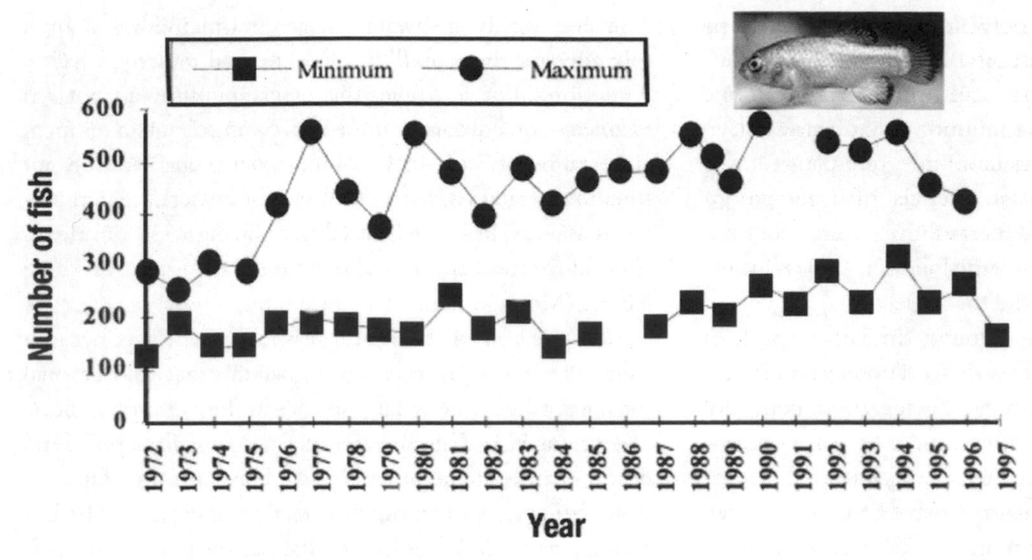

Figure 2.10. Population fluctuations of Endangered Devils Hole pupfish. Annual reproduction creates a natural population cycle that is lowest in spring and highest in autumn (breaks in lines are periods without data). From Andersen and Deacon (2001); used with permission. Inset: Devils Hole pupfish, the smallest pupfish in North America (25 mm); its specializations include a relatively large head, no pelvic fins, and a lack of the vertical barring characteristic of most other pupfishes. Photo by T. M. Baugh; used with permission of the photographer.

to human activity. Isolated habitats such as desert springs and pools can be easily destroyed by dumping of toxic substances, introduction of predators, habitat modification, or ground- and surface-water withdrawal.

Cyprinodontids include the Devils Hole pupfish, *C. diabolis* (figure 2.10), which occupies a small, approximately 3.5 m x 6 m, limestone shelf in a cave-spring system in Ash Meadows, Nevada. This "may be the smallest habitat in the world" containing an entire vertebrate species (Andersen and Deacon 2001, p. 224; see also figure 3.3). *C. diabolis* was among the first organisms protected by the ESA, in 1973. The pupfish, which had existed in Devils Hole for perhaps 60,000 years, was threatened by groundwater pumping for crop irrigation. Pumping lowered the level of the pool beginning in 1968, until water and population reached historic minima in 1972. After a protracted legal battle, the U.S. Supreme Court in July 1976 mandated water levels to protect the species. Population size varies between springtime minima of about 200 fish to autumn maxima of about 550 fish, reflecting recruitment followed by winter mortality in this annual species (Andersen and Deacon 2001). Lowest recorded populations, of only 127–143 fish, occurred between 1972 and 1974. Following the court decision, populations have slowly increased, except for an all-time low of 49 fish in April 2006, caused by an unhappy confluence of flooding and human carelessness. Numbers show a positive annual correlation with water levels in Devils Hole, reflecting primary productivity of diatoms on the submerged shelf where the fish forage and reproduce (Andersen and Deacon 2001; see also http://ecos.fws.gov/species_profile/SpeciesProfile?spcode=E009).

Other notable cyprinodontids include the aptly named La Trinidad pupfish, *C. inmemoriam*. This species became extinct shortly after its discovery in 1983 in Ojo de La Trinidad, an isolated desert spring in Nuevo León, Mexico (Lozano-Vilano and Contreras-Balderas 1993). The spring was already deteriorating, and the authors collected only a single specimen to minimize impact, intending to collect additional fish later (S. Contreras-Balderas, pers. comm.). Upon their return in 1986, the spring was dry due to water extraction. Ojo de La Trinidad had been one of four springs still flowing in 1983; at least 8 others dried up before the valley was adequately surveyed. Springs in the region occur close to but isolated from one another, and they typically house unique pupfish species. All 12 springs are now dry and were never sampled. Professor Contreras-Balderas may have the dubious distinction of discovering and describing more extinct fish species, many of which were pupfishes, than any other living ichthyologist. Of 12 Mexican species he has named, 4 are extinct, 4 are probably extinct, and 3 are endangered. Among 20 species he is currently describing, only 4 are considered safe.

C. inmemoriam is one of six pupfishes in just the Nuevo León area officially recognized by IUCN (2004) as extinct. Due to a 50-years-of-absence criterion for designating extinctions (see chapter 3), IUCN recognizes only eight extinct cyprinodontid species, unquestionably an underestimate. Our knowledge of extinguished species is clearly incomplete, and our estimates of extinction rates are undoubtedly conservative.

A dedicated group of researchers and concerned citizens has formed the Desert Fishes Council to protect pupfishes and other desert forms. Originally organized in 1969 to save the Devils Hole pupfish, the group works actively to educate the public and politicians about the plight of these evolutionarily unique organisms (Pister 1991).

Other threatened cyprinodontiforms include two European toothcarps, the samaruc or Valencia toothcarp, *Valencia hispanica*, of Spain and the Corfu toothcarp, *V. letourneuxi*, of Albania and Greece, both IUCN Endangered. Habitat loss and fragmentation, pollution, hydraulic alteration, and introduced species are major factors contributing to their declines. The Altiplano region of the Andes in Peru, Bolivia, and Chile contains many isolated lakes in addition to Lake Titicaca, the world's highest navigable lake and the largest lake in South America (3,800 m). This region has given rise to complexes of endemic cyprinodontids, totaling perhaps 43 species, largely in the genus *Orestias* (Lussen et al. 2003). Habitat destruction, pollution from mining, and introduced species have led to major declines. The umanto, *O. cuvieri*, the world's largest killifish at 27 cm, is probably extinct (IUCN 2004). The Peruvian government listed 6 other *Orestias* species as Critically Endangered and another 13 as Vulnerable (www.peruecologico.com.pe/extincion.htm); Chile unofficially included several *Orestias* species in its red list (Jara et al. 1995; Campos et al. 1998; chapter 3). Given the remoteness of this region and evidence of impacts from a growing human population, more attention to biodiversity protection is desperately needed.

Pikeperches and Darters (Percidae)

At least 200 species make up the family Percidae, 187 of which occur in North America. Percids include the yellow and Eurasian perch *Perca* spp., walleye and sauger (pikeperches, zander), and about 180 species of small, stream-dwelling, spectacularly colored, and often imperiled darters endemic to North America (see chapter 5 for impacts on darters) (figure 2.11). Darters are disproportionately represented in lists of imperiled fishes because they possess many traits that increase their vulnerability, including small size, small geographic range, headwater habitats, and, especially, benthic breeding and feeding habits. Fully 29 darter species are considered to be Critically Endangered, Endangered, or Vulnerable by IUCN (2004), and one species, the Maryland darter, *Etheostoma sellare*, is considered Extinct. Nineteen darters have official U.S. status as Endangered or Threatened, 6 of which are not on the IUCN list. Another 6 species are classified as candidate species for ESA inclusion (17 IUCN-listed species do not appear in the USFWS list). The Southeastern Fishes Council, which is made up of regional experts, designated an additional 45 species as Endangered, Threatened, or Vulnerable (Warren et al. 2000); 15 of those are subspecies or remain to be described, categories not generally recognized by either USFWS or IUCN. All told, at least 96 described species, or 111 species, subspecies, and undescribed species of darters—somewhere between half and two-thirds of all species—are recognized as being at risk of extinction.

Figure 2.11. The tangerine darter, *Percina aurantiaca*, the world's second-largest darter, at 180 mm. Tangerines are restricted to the Tennessee River system, where they are sometimes caught by trout fishers. Photo by N. Burkhead; used with permission. For more pictures of darters, see www.cnr.vt.edu/efish/families/percidae.html.

The snail darter is representative of the ecological and political obstacles faced by many darters (Etnier 1976; Ono et al. 1983; D. A. Etnier, pers. comm.). First discovered in Tennessee in 1973 and listed as Endangered in 1975, it almost immediately became the focus of controversy. Its only known habitat was directly threatened by a planned hydropower project, Tellico Dam. Promoted by the Tennessee Valley Authority (TVA), local land developers, and the Army Corps of Engineers, the dam was of questionable utility. It was opposed by conservationists, farmers, local landowners, fishers, USFWS, U.S. Supreme Court Justice William O. Douglas, Tennessee Governor Winfield Dunn, and the Cherokee Indian Nation (see Etnier 1976; Helfman et al. 1997). To circumvent the ESA, TVA illegally and surreptitiously transplanted 700 snail darters from the Little Tennessee River to the nearby Hiwassee River and then accelerated work on the dam. Lawsuits followed, and TVA was enjoined against further dam construction in 1977. But TVA appealed on the grounds that dam construction had already made the Little Tennessee unsuitable habitat for the darter by blocking upstream migration of spawning fish. The U.S. Supreme Court denied the appeal but recommended that Congress intervene when economic consequences of species protection were significant. Congress amended the ESA, creating an exemption committee that came to be known as the God Squad or the Extinction Committee.

The committee voted unanimously in favor of the darter in February 1979, but later that year, a special exemption for Tellico Dam was included in general energy legislation and passed without debate. Tellico Dam was completed. Although extirpated from the Little Tennessee, the darter survived the battle. Transplanted populations in the Hiwassee and three other rivers were successful, and additional wild populations were subsequently discovered; the snail darter's status was upgraded from Endangered to Threatened in 1984. Of greater consequence, the battle over the snail darter mobilized conservation forces and educated the public about endangered species issues.

European percids have converged on darter morphology, ecology, and vulnerability. The asprete, *Romanichthys valsanicola*, a small, benthic percid, was described in 1957 and recognized as seriously imperiled by 1971 (Lelek 1987). It inhabits small, upland, fast-flowing streams, where it lives among rocks and feeds on benthic invertebrates such as stoneflies, an ecological description common to most darters. Its original range was quite small, encompassing three tributaries of the upper Arges River, itself a tributary to the Danube River in Romania. This original range, the smallest of any fish in Eurasia, collapsed to a single stretch of the Valsan River, largely due to dam construction and gravel extraction (Banarescu 2002). Aspretes continue to be threatened by dams, deforestation, road building, agricultural runoff, gravel extraction, and introduced species. Breeding populations remain depressed; the asprete was given IUCN CR status in 1994. More recently, governmental regulations were enacted to protect its habitat (Lelek 1987; Banarescu 2002). Attempts at captive breeding are ongoing as part of an EU LIFE-Nature recovery program, and hydrological conditions in the tributaries that make up its entire range appear to be improving (Banarescu 2002).

Approximately half of the remaining European percids are also at risk. The apron or asper, *Zingel asper*, is Critically Endangered throughout its Rhone River, France, range, where it has lost 83% of its former habitat and is now restricted to a few stretches of the river (Mari et al. 2002). Two other *Zingel* species, the streber, *Z. streber*, and the zingel, *Z. zingel*, are Vulnerable. The Vulnerable schraetser, *Gymnocephalus schraetzer*, also called striped ruffe or yellow pope, occurs primarily in the Danube system and has undergone dramatic population declines throughout its range (Leidy and Moyle 1997). Finally, *Percarina demidoffii* is a poorly known Vulnerable eastern European species that occurs in Ukraine and in estuaries of the Black and Azov seas. All of these have suffered due to channelization of rivers, construction of dams, and other types of hydrologic alteration.

Other critically imperiled (and perhaps extinct) species of percids were once widespread. The blue pike, *Sander vitreus glaucus*, a subspecies of the walleye, sustained a large fishery in Lake Erie and Lake Ontario until the mid-1950s; in some years, it made up more than half the commercial catch in those lakes. Pollution, introduced fishes, habitat degradation, overharvesting, and hybridization all contributed to its demise (R. R. Miller et al. 1989). It was officially declared extinct in 1975. Periodically, someone catches a walleye that has a bluish tint, sustaining hope that the fish may still exist (e.g., www.nativefish.org).

Cichlids (Cichlidae)

The sheer number of cichlids, perhaps 1,500 species, makes it likely that many are suffering declines, especially given that their highest diversity occurs in tropical regions with burgeoning human populations. That more haven't been declared extinct or appeared on red data lists is likely a result of incomplete information rather than a true estimate of total threat.

IUCN (2004) listed 160 cichlids as imperiled. Of these, 42 were Critically Endangered and 18 Endangered. An alarming 51 are thought to be extinct, although at least half of these cases are unresolved, mostly because of incomplete collection information. Five are considered extinct in the wild, thanks largely to aquarium aficionados who are helping preserve genetic diversity of this group (assuming overcollecting for the aquarium trade hasn't contributed to problems; see chapter 13). The vast majority of cichlids listed by IUCN are from the Lake Victoria region, where some experts believe that perhaps 200 (some say 300) endemic species have been eliminated due to introduction of predatory Nile perch, *Lates* sp. cf. *niloticus* (see Harrison and Stiassny 1999 for a complete listing; see also chapter 8). Nile perch have also been introduced into small lakes near Victoria such as Kyoga and Nabugabo, with similar impacts on endemic cichlids (Schofield and Chapman 1999).

Many other central African lakes have their own cichlid species flocks and other endemics (figure 2.12). Cameroon contains 28 freshwater species with IUCN (2004) high-risk status, 26 of which are endemic cichlids. Barombi Mbo and Bermin lakes in Cameroon, with respective surface areas of only 4.2 and 0.6 km^2, each possess endemic cichlid species flocks. Barombi Mbo houses 11 such species (plus an endemic clariid catfish), and Bermin has 9, which is spectacular speciation for such diminutive water bodies (Barlow 2000). These locales should be appreciated for the obvious, disastrous impacts that would occur if predators such as Nile perch were introduced (e.g., L. J. Chapman and Chapman 2003).

Six listed cichlid species are from Mexico, four from Madagascar, and others are from Mozambique, Namibia, and South Africa. A Namibian species, the Otjikoto tilapia, *Tilapia guinasana* (Critically Endangered), shows striking ecological similarities to the Devils Hole pupfish discussed earlier. *T. guinasana* occurs as a single population of only 250–400 individuals in a limestone sinkhole. The total habitat for the species is less than 3,000 m^2. Remarkably, the fish occurs in five distinct color morphs, and genetic evidence suggests that individuals mate preferentially with like-colored fish (Barlow 2000). A major threat is disturbance by recreational divers that frequent this clear-water spring. The Mexican species on the IUCN list are also mostly spring dwellers from the states of San Luis Potosí and Coahuila. The Mexican red list includes four additional species in need of special protection. Habitat alteration, dewatering of springs, and introduced species (including tilapia) are major threats to Mexican cichlids. Malagasy cichlids and other fishes face threats due to deforestation and attendant sedimentation, overfishing, and exotic species introductions, especially of tilapia (Reinthal et al. 2003).

Figure 2.12. Lake Malawi cichlids at the Georgia Aquarium; the abundance and diversity exhibited here are not unusual in the wild. Blue, yellow, orange, black, and pink colors are common. For more cichlid photos, see www.cichlidworld.com/photo.html.

Cave Fishes

Approximately 136 freshwater fish species in 19 families and 10 orders are adapted to life in subterranean caves. These unusual fishes—termed variously *hypogean*, *troglobitic*, *phreatic*, and *stygobitic*—occur in scattered locales at tropical and warm-temperate latitudes on all continents except Antarctica (Proudlove 1997b; A. Weber et al. 1998). Cave fishes have delightful and imaginative scientific names that often describe their appearance and their ecology: *Stygichthys typhlops*, *Phreatichthys andruzzii*, *Sinocyclocheilus anophthalmus*, *Nemacheilus troglocataractus*, *Satan eurystomus*, *Lucifuga spelaeotes*, *Schistura oedipus*. They are often taxonomically unique, with 20 species representing monotypic genera; no exclusively hypogean genus contains more than 5 species.

Cave fishes are a classic example of convergent evolution involving multiple anatomical and ecological traits (figure 2.13). These fishes characteristically have reduced eyes and scales and hypertrophied olfactory and acoustic capabilities. They typically live in small populations, grow slowly, have low reproductive output, and are relatively high in the food webs of their respective environments, in which energy is limiting. They are extreme specialists among freshwater fishes. Many are known from only one or a few locations, although obvious sampling difficulties prevent accurate population estimates. But isolation seems to be commonplace: 48 species are known from only their type locality, and many cave species occur on islands in larger countries (e.g., the CR ellinopygósteos *Pungitius hellenicus* and the VU goby *Economidichthys pygmaeus* of Greece; Leidy and Moyle 1997). Many occur in island countries such as Trinidad, Cuba, the Bahamas, Galapagos, Madagascar, and Papua New Guinea (Proudlove 1997b; A. Weber et al. 1998; G. Proudlove, pers. comm.).

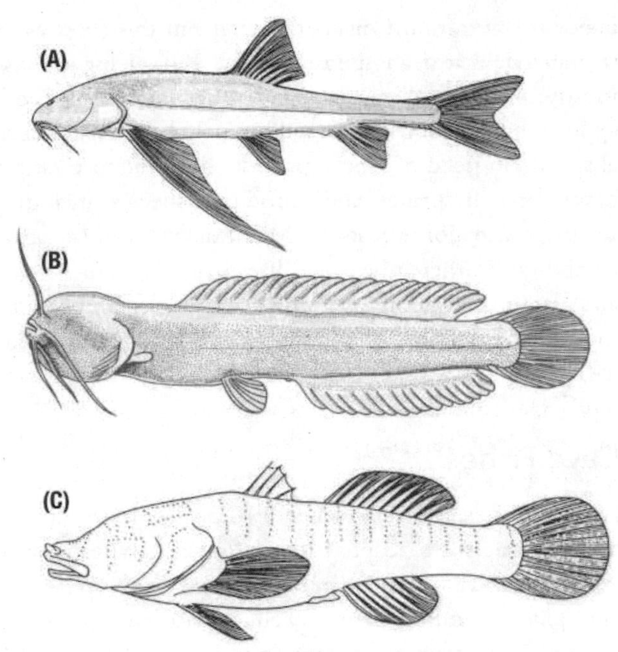

Figure 2.13. Cave fishes from three different orders, showing convergent loss of eyes and other weird features: (A) a balitorid river loach, *Triplophysa xiangxiensis* (Cypriniformes), from China; (B) a clariid catfish, *Horaglanis krishnai* (Siluriformes), from India; and (C) an eleotrid sleeper, *Typheleotris madagascariensis* (Perciformes), from Madagascar. From A. Weber et al. (1998); used with permission. See also Proudlove (2005).

The many specialized traits of cave fishes make them sensitive to habitat degradation via pollutants or water withdrawal, and their isolation in low-diversity assemblages makes them vulnerable to competition, predation, and disease brought in by introduced species. As an ecological grouping, caves must be the habitat type with the proportionately highest rate of imperilment among fishes (and other organisms). Of the 136 known cave fish species, 51 were listed by IUCN (1996) in high-risk ranks (Proudlove 1997a). Proudlove considered 9 species not assessed by IUCN to deserve VU status (IUCN 2004 subsequently acted only on Starostin's loach, *Nemacheilus starostini*, changing its status from Data Deficient to Vulnerable). Several species are sought for the aquarium trade because of their unusual appearance, which is why the Congo blind barb is listed in CITES Appendix II; *Clarias cavernicola* from Namibia has been proposed for listing in CITES Appendix III. In the U.S., the Alabama cave fish, *Speoplatyrhinus poulsoni*, is federally protected as Endangered, and the Ozark cave fish, *Amblyopsis rosae*, is Threatened. Recovery plans that include habitat protection have been implemented for both species. Two other cave fishes that occur in the U.S.—the northern cave fish, *Amblyopsis spelaea*, of Kentucky and Indiana and the southern cave fish, *Typhlichthys subterraneus*, found in five southeastern states, Indiana, and Missouri—are designated as Vulnerable by IUCN (2004) (see Romero 1998; Romero and Bennis 1998).

MARINE FISHES

Approximately 60% of the world's fish species are marine. However—for reasons that are complex, conjectural, and lack consensus—many fewer marine than freshwater species are severely imperiled (see chapter 4). In fact, many imperiled marine species inhabit or depend on estuaries at some stage in their lives. Among purely marine forms, highly valuable commercial species dominate the lists of species in trouble (chapters 10–13). All but a few species fit the estuarine or exploited pattern. However, because of problems of distribution, access, and diversity, we are undoubtedly unaware of many marine species that need protection (e.g., Vincent and Hall 1996; R. A. Myers and Ottensmeyer 2005).

The U.S. currently grants federal status to 2 ostensibly marine species, the tidewater goby, *Eucyclogobius newberryi*, of California (Endangered) and the smalltooth sawfish, *Pristis pectinata*, of Florida (Endangered); both are estuarine inhabitants. The totoaba, *Totoaba macdonaldi*, a Mexican Sea of Cortez species heavily affected by water withdrawal from the Colorado River, also has Endangered status. The Cherry Point herring, once the largest stock of Pacific herring, *Clupea pallasi*, in Washington State, has undergone an 80% loss of habitat and a 90% decline in abundance and has been accepted for consideration as a federally protected species. Other ESA-listed "marine" fish species are also estuarine, anadromous, or freshwater (e.g., sturgeons, smelts, salmonids; TESS 2004). The NOAA Fisheries Office of Protected Resources maintains an active listing of marine and anadromous species proposed for ESA protection and includes another category, Species of Concern, for organisms that meet criteria just short of ESA status. As of January 2005, 14 marine species had such status: 2 sharks, a sawfish, a skate, a topminnow, a hake, a silverside, a rivulin, a rockfish, 4 groupers, and a marlin. Of note are the saltmarsh rivulus, *Kryptolebias marmoratus*, North America's only self-fertilizing hermaphroditic fish; Key silverside, *Menidia conchorum*, North America's smallest silverside; Pacific hake, *Merluccius productus*—a species I discarded by the ton during my collecting and commercial fishing ventures in the 1960s; the boccacio, *Sebastes paucispinis*, a West Coast rockfish that declined by 97% between the 1970s and the late 1990s; and Goliath grouper, *Epinephelus itajara*, perhaps the world's largest serranid, at 400 kg. The American Fisheries Society recognized 82 species and subspecies (or 151 DPSs) of North American marine fishes as endangered, threatened, or vulnerable (see chapter 4). Of these, 22 species are at risk of global extinction given present trends (Musick et al. 2001).

At the international level, CITES Appendix I included only 2 marine species (coelacanth and totoaba). Totoaba are largely endangered because they live in a giant, disrupted estuary. As of the October 2004 Conference of the Parties, Appendix II protected only seahorses, humphead wrasse (*Cheilinus undulatus*), whale sharks, white sharks, and basking sharks. IUCN (2004) was more inclusive. Not counting diadromous or predominantly freshwater species, the tally was 83 species listed, including 13 CR, 14 EN, and 56 VU. The most imperiled families, those with 4 or more species listed, were groupers (17), seahorses (16), toadfishes and butterflyfishes (5 each), and tunas (4). Families with 3 listed species were pegasid seamoths, scorpaenid rockfishes, anthiids, wrasses, damselfishes, and triggerfishes. The remaining species represent approximately 23 families in most of the major tropical and temperate oceans and seas. Leidy and Moyle (1997) emphasized that marine species listed by IUCN (1996) covered fishes that occurred disproportionately in inland seas and nearshore regions, reflecting the influence of human populations and land disturbance. Few other patterns are obvious, other than only 16 of the 83 listed species lacked English common names, suggesting again that our awareness of imperiled fishes is heavily biased toward the industrialized world.

The coverage of marine fishes in the IUCN Red List and elsewhere is still unrepresentative because status assessments of marine organisms have focused on marine mammals, seabirds, marine turtles, and a few other taxa (Vincent and Hall 1996). Those fish groups that have received treatment—namely, relatively charismatic sharks and rays, coral reef fishes, seahorses, and groupers and wrasses—provide evidence of a number of inherent extinction risk factors among marine organisms. These factors include low reproductive potential and restricted range and are compounded by threats such as overexploitation, habitat destruction and degradation, and the effects of disease and invasive species. Efforts to expand red list assessments to more marine species are ongoing—for example, witness addition of white, whale, and basking sharks to CITES Appendix II in 2004. Expanded effort will undoubtedly reveal that extinction risk in the marine environment is greater than currently known and undoubtedly increasing, because marine species share many of the threats that seriously affect terrestrial and freshwater ecosystems (see Baillie 2004).

Sharks, Skates, and Rays (Elasmobranchs)

Upward of 100 million sharks are caught annually, and an equal number may die as bycatch. Shark populations around the world have decreased 50%–80% since the 1970s, the steepest declines occurring in the late 1980s. National regulation of shark fishing has been limited. International management plans, made necessary by the mobility of many sharks, have been largely nonexistent.

Many sharks possess biological and ecological traits that make them especially vulnerable to exploitation (chapters 4, 10), including large size, slow growth, late maturation, low fecundity, long gestation periods, high trophic position, and high commercial value. In these and other respects that affect their conservation, sharks are basically marine mammals with cold blood (and some sharks are noticeably warm-blooded; Helfman et al. 1997).

Several shark species are experiencing population declines because they are the primary target of a fishery (in the case of dogfish, whale sharks, thresher sharks, basking sharks), or they are retained during operations targeting other fishes (e.g., mako and blue sharks in the swordfish fishery of the North Atlantic). These fisheries are at least potentially regulated because fisheries personnel monitor the dynamics of targeted species and make management decisions based on what is happening to them.

In contrast, nontargeted species caught incidentally can suffer high but unnoticed, or at least unregulated, population declines. Bycatch remains the major threat to the U.S. federally listed smalltooth sawfish, once common throughout a range from the Gulf of Mexico (Texas to Florida) and up the East Coast from Florida to Cape Hatteras (Simpfendorfer 2000). The fish now occurs only in peninsular Florida, primarily in the Everglades region at the southern tip of the state; the largetooth sawfish, *P. perotteti*, suffers similarly. Bycatch is also the suspected factor leading to precipitous declines in barndoor skate, *Dipturus* (= *Raja*) *laevis*, which appears in demersal (bottom) fisheries of the North Atlantic (strong evidence exists for a parallel decline in another large eastern Atlantic skate, the common skate *D. batis*). Barndoor skates, which grow to over 1 m wide, were once fairly common in the northwestern Atlantic; catches of 7–21 per tow were not unusual in the first half of the 20th century. Recent surveys indicate that they have fallen to less than 5% of earlier numbers (Musick et al. 2001). Barndoor skates do not mature until 11 years old and produce fewer than 50 very large eggs per year. Meshes commonly used in groundfish trawls are regulated to be no less than 7–14 cm wide; barndoor skates are wider than that at birth and are thus subjected to fishing almost immediately. Continued intense pressure on more rapidly reproducing, targeted species (cod and other bottom fishes) has not been adjusted to accommodate declining numbers of barndoor skates, which may be threatened with extirpation in some regions (Casey and Myers 1998; Musick et al. 2001). They were listed as Vulnerable by AFS and IUCN; common skates were listed by IUCN (2004) as Endangered. Barndoor skates have shown some signs of recovery where groundfishing has been greatly curtailed.

Sharks are harvested for their meat, fins, hides, jaws (worth thousands of dollars in some cases), cartilage, and other body parts for cosmetics and traditional medicines. It has long been believed, erroneously, that sharks do not get cancer (see Ostrander et al. 2004). Shark cartilage pills are sold as anticancer medication, to the great detriment of sharks. Although substances in shark cartilage inhibit the formation of blood vessels often associated with cancerous growths (e.g., A. Lee and Langer 1983), this effect has not been shown in humans. Clinical trials have "unsatisfactory patient outcome[s]," and data are still lacking that show a pharmacological effect from oral shark cartilage (Gonzalez et al. 2001).

IUCN (2004) has greatly expanded its inclusion of sharks in recent years. The Shark Specialist Group assessed about a third of the world's 950 sharks, skates, and rays and assigned high-risk status to 86 species: 14 Critically Endangered, 26 Endangered (including barndoor and common skates), and 46 Vulnerable (including great white and whale sharks). Those numbers represent an almost threefold increase since the 2000 list, when only 3 species were considered Critically Endangered, 17 Endangered, and 10 Vulnerable. CITES listed whale, basking, and white sharks in Appendix II in 2004.

Although the likelihood of biological extinction remains remote, some shark species are on the verge of ecological extinction, with serious implications for marine ecosystems. Sharks are apex predators, and population crashes are likely to have direct and indirect cascading effects in marine food webs. Evidence for such impacts already exists. When a shark fishery in Tasmania experienced a population collapse, octopus, which they had fed on, exploded. Octopi feed on spiny lobsters, so lobsters crashed, and fishermen were put out of work.

Coelacanths (Latimeriidae)

The fossil record of coelacanths extends from the Middle Devonian, about 375 million years ago, to near the end of the Mesozoic, about 80 million years ago; they were thought to have gone extinct with the dinosaurs. Then in 1938, a single specimen was trawled up off South Africa. Named *Latimeria chalumnae*, it was heralded as one of the most significant scientific findings of the 20th century. A second specimen was not discovered until 1952. Between 1952 and 1992, approximately 175 known individuals were captured, primarily at 100 to 500 m along volcanic slopes of the Comoros Islands between Madagascar and Mozambique. The total Comoran population is estimated at 200–600 coelacanths and is thought to be declining (Fricke et al. 1991; Fricke 2001).

Coelacanths grow to 200 cm and 95 kg, living perhaps 40–50 years. More important, they bear only 5–26 live young and reproduce infrequently, with a gestation period of perhaps 3 years, the longest of any known vertebrate. Females do not mature until 15 years old. Replacement rate in the population is therefore slow (Froese and Palomares 2000). Their rarity and limited distribution, limited fecundity and slow replacement rate, late maturation, and an unsustainable incidental capture rate of 5–10 animals per year in local fisheries for unrelated oilfish, *Ruvettus pretiosus*, garnered considerable public sympathy in the 1990s. Major, additional threats included the species' desirability as a curiosity, for scientific specimens, and for live display in public aquaria. During the 1990s, a black market supposedly developed for coelacanth notochord fluid, which was purported to promote human longevity because of the fish's antiquity. Whether any actual trade resulted from this is in question (J. Hamlin, pers. comm.). The scientific and conservation communities (e.g., the Coelacanth Conservation Council, the Society for the Preservation of Gombessa, the Gombessa Association, the African Coelacanth Ecosystem Programme, and the

Figure 2.14. Not only pandas have charisma. Around 1988, the now defunct Coelacanth Conservation Council (CCC) proposed that the World Wildlife Fund (WWF) adopt the coelacanth as a symbol for marine conservation. CCC produced a rather stylized rendition of the fish (A) to serve as the logo. WWF demurred, but in 1999, an ichthyology student heard of the plight of the coelacanth and had the CCC image tattooed on her hip (B). A from Balon et al. (1988); B: Thanks, Gisalene.

Coelacanth Rescue Mission) have mobilized to protect the fish (figure 2.14). The species has been placed in Appendix I of CITES and has Critically Endangered status with IUCN; its capture has been outlawed by the Comoran government (e.g., Fricke 2001). Bycatch has apparently dropped to a few fish per year (J. Hamlin, pers. comm.).

In 1998, another population of coelacanths was discovered at similar depths and habitat in northern Indonesia; it was described as a separate species, *L. manadoensis* (Holder et al. 1999). A small population apparently exists off the coast of KwaZulu Natal, South Africa, based on diver sightings of several juveniles and adults in 2000 and subsequently (Venter et al. 2000), and additional fish have been caught in gill nets off Kenya and the Tanzanian coast (De Vos and Oyugi 2002; www.dinofish.com). Despite the remoteness of its few habitats and international concern and action, coelacanths are threatened by human activities. Alarmingly high levels of DDT and its derivatives have been found in its flesh, raising concerns about effects on reproduction (R. C. Hale et al. 1991; chapter 7). Coelacanths produce the largest egg of any bony fish. Organochlorines such as DDT are highly lipid-soluble, making maternal transfer of these known metabolic poisons likely during embryonic development.

Handfishes (Brachionichthyidae)

The spotted handfish, *Brachionichthys hirsutus*, ironically also known as the common handfish, belongs to an endemic and unusual Australian family (figure 2.15). Handfishes are restricted to southeastern Australia, and five of the eight known species are restricted to Tasmania. The first dorsal spine is modified into a fishing lure (ilicium), and other anterior dorsal spines are often long and webbed, forming a sail-like structure on the head. The pectoral fin is modified into an armlike appendage with an elbow and fingers (Last et al. 1983).

The spotted handfish is restricted to southeastern Tasmania, frequenting shallow (5–10-m) depths, often in estuaries, where it sits on sand or shell bottoms (Bruce et al. 1998). It is small (<12 cm) and colorful (orange or black spots on a white or pink background). Common as recently as the mid-1980s, it has declined precipitously. Exact causes remain unclear, but much evidence implicates an introduced starfish, the northern Pacific seastar, *Asterias amurensis*. The starfish preys directly on handfish eggs, which are few in number (80–250 per clutch) and attached to vertical structures such as stalked ascidians (sea squirts), sponges, or sea grasses. The starfish also eats the ascidians.

Figure 2.15. The Critically Endangered spotted handfish. Handfishes get their name from the unusual arrangement of bones in the pectoral fin. Photo courtesy of David Doubilet, Undersea Images Inc., www.DavidDoubilet.com.

Other suspected causes include siltation of estuaries, which obliterates benthic structure and transports heavy metals, as well as effluent from urban centers. Spotted handfish have Critically Endangered status with IUCN.

Two additional, undescribed species (Ziebell's and Waterfall Bay handfishes, *Sympterichthys* spp.) have been proposed for Vulnerable status with IUCN, and a third, the red handfish, *S. politus*, is Data Deficient (Peterken 1996). Even given uncertainties in taxonomy, it appears that 50%–80% of this geographically restricted taxon should have some IUCN rank, which would be something of a record for a marine family.

Handfishes possess biological traits—restricted geographic range, natural rarity, small body size (<15 cm), benthic habitat, limited mobility, and low fecundity—that are often correlates of vulnerability in fishes. More important, and unlike the vast majority of marine fishes, handfishes lack a pelagic larval stage. Fully formed juveniles emerge from eggs that are guarded by the female for seven to eight weeks. This protracted developmental period exposes eggs to predation, and lack of a larval phase limits dispersal of young. Combined with adult immobility, this contributes to lack of population replenishment or replacement. Interestingly, two marine fishes unrelated to handfishes and once thought to be extinct, Banggai cardinalfish, *Pterapogon cauderni*, and Texas pipefish, *Syngnathus affinis*, also lack planktonic larval stages and are characterized by relatively sedentary adults. The same characteristics apply to other imperiled seahorse relatives such as leafy sea dragons (see chapters 4, 13).

B. hirsutus has been legally protected in Tasmania since 1994 and at the federal level since 1996; it has Critically Endangered status under the Environmental Protection and Biodiversity Conservation Act of 1999. Their size, color, and immobility make handfishes desirable as aquarium animals (Last et al. 1983), although that is not considered a contributing factor to their population declines (B. Barker, pers. comm.). The fish's adaptability to captivity has encouraged attempts at captive breeding, which if successful, could allow future reintroductions, assuming the primary causes of decline have been identified and corrected.

Seahorses (Syngnathidae)

The syngnathid seahorses, pipefishes, and sea dragons are intensively hunted as curios, as live fish for aquaria, and for traditional medicine (Vincent 1996). Because they mate for life and reproduce slowly, seahorses are particularly vulnerable to overcollecting. Fifty-one species have some international protection, with 39 listed in Appendix II of CITES and 16 considered at risk by IUCN (2004). Much of the campaign to understand and publicize threats to seahorses and to promote their sustainable use has been spearheaded by the internationally acclaimed Project Seahorse (see chapter 13).

Seabasses and Groupers

IUCN (2004) listed 17 species in the family Serranidae, which includes groupers, sea basses, hamlets, and sea "trout" (Australia). This number represents a sizable increase over all previous IUCN assessments and was long overdue. Serranids are among the most heavily exploited fishes in nearshore tropical and subtropical waters and are often the first species to disappear from coral reefs when fishing intensifies (chapter 12). Aspects of their biology, especially sex change and predictable spawning aggregations, underlie their vulnerability. Conservation measures that are targeted specifically at protecting serranids and take into account their reproductive traits have proven successful. If serranid biology is ignored, marine protected areas can be successful in protecting other fishes but not groupers and sea basses. On the Caribbean island nation of Saba, establishment of the Saba Marine Reserve resulted in increased biomass of four of five commercially fished families, with some, such as snappers, increasing 220% (Roberts 1995), but serranids did not increase, probably because the reserve served as a refuge for adults but did not protect spawning aggregations per se.

Nassau groupers exemplify the problem. Once widespread and common throughout much of the tropical western North Atlantic, spawning aggregations have been targeted mercilessly. One third of spawning aggregations have been extirpated, and density within some remaining aggregations has declined 90% (Musick et al. 2001). Nassau groupers throughout the tropical western Atlantic were designated as Endangered by IUCN, a Species of Concern by the NOAA Fisheries Office of Protected Resources (2005), and Threatened by AFS.

Many serranids undergo a maturational sex reversal, typically maturing first as female and later changing to male. The largest and most intensively targeted individuals in a population are therefore male. Estimating total numbers but ignoring sex ratios can lead to an overestimation of breeding stock. Too few males in a population because of overfishing has been associated with population declines in some serranid species (e.g., Beets and Friedlander 1999). Commonly employed criteria for placing species on threatened lists frequently overlook the influence that life history complexities—such as maturational sex reversal—can have on a species. Regardless, serranids are disproportionately represented in lists of recognized imperiled fishes, making up more than 20% of all listed marine species, although the family constitutes only 3% of marine fishes.

Croakers and Drums (Sciaenidae)

The totoaba is endemic to the northernmost part of the Sea of Cortez and is the largest member of the widespread croaker family, reaching 2 m in length and a mass of over 100 kg (Ono et al. 1983). It has at least four different ontogenetic stages that migrate through much of the Gulf of California, experiencing different environmental conditions and stresses (Cisneros-Mata et al. 1995, 1997). Populations have been decimated by overfishing on the spawning grounds in the Colorado River estuary, by agricultural pollution and dewatering of the estuary due to water withdrawals upstream in the Colorado River, and by incidental capture of juveniles by shrimp boats. The totoaba was at one time the most important commercial fish in the Gulf of California and the second most valuable fishery, after shrimp, in all of Mexico. Abundance was so high that spawning fish could be speared from small boats. In the 1920s, fish were valued for their large swim bladder, which was dried and exported to make soup, the remainder of the body often discarded, an astonishingly wasteful practice reminiscent of finning for sharkfin soup. Totoaba meat gained popularity in the late 1920s, creating an export market to the U.S., particularly Southern California. Demand was met by large-scale net fishing in the spawning area of the northern Gulf of California.

Catch peaked in 1942 at 2,300 tons and then declined despite increased fishing effort (Ono et al. 1983). The fishery reached an all-time low in 1958, at which point the Mexican government instituted a 45-day fishing moratorium during the spawning season. Mexico also created a sanctuary at the mouth of the Colorado River. Some population recovery occurred over the next 15 years, but in the mid-1970s catches fell to a new low, attributed to water-use practices in the Colorado River. Dams, diversions, and irrigation projects in the U.S. have dried up much of the river's estuary and created hypersaline conditions where water exists. Totoaba spawned only in that estuary, a specialization that could not adjust to changing conditions.

The totoaba fishery in Mexico was officially closed in 1975, although poaching was apparently common into the early 1990s (Cisneros-Mata et al. 1997). USFWS made importation of totoaba into the U.S. illegal in 1978 and declared the fish Endangered in 1979, motivated by the realization that U.S. markets were largely responsible for commercial pressure on the species (Federal Register 44FR29480). The totoaba was listed as In Danger of Extinction in Mexico (Secretaría de Solidaridad 1994) and appears in Appendix I of CITES—it cannot be traded legally. Totoaba were stocked unsuccessfully in California's Salton Sea for sport fishing in the 1950s (see chapter 6). Conservation efforts in Mexico have led to stabilization of the population in recent years, and efforts at captive propagation hold promise (True et al. 1997). However, successful reintroduction will depend on reversing habitat degradation in the lower Colorado River, where competition for and contamination of water remain significant (e.g., M. K. Briggs and Cornelius 1998).

Despite increasingly stringent, national and international protective measures dating back to the mid-1970s, the species has shown minimal recovery at best. Unique biological traits (endemism with a restricted geographic range, specialized spawning requirements, aggregation and vulnerability on the spawning grounds) combined with anthropogenic effects (area-concentrated overfishing; spawning and juvenile habitat degradation; direct mortality of young due to bycatch; and perhaps overfishing of shrimp, which are totoaba prey) have produced one of the few documented near extinctions of a large marine species.

Another, equally large (2-m, 100-kg) sciaenid is equally imperiled (Sadovy and Cheung 2003). The Chinese bahaba or giant yellow croaker, *Bahaba taipingensis*, occurs only along the coast of southern China from the Yangtze River to Hong Kong and has disappeared from most of its range. It enters estuaries seasonally to spawn in large aggregations that are especially vulnerable to overexploitation by fishers because the fish's vocalizations during spawning can be heard by pressing one's ear to a boat hull. It, too, is valued for its swim bladder, which is used in traditional medicine. Estimated landings in the past 40–60 years indicate massive declines: In the 1930s, 50 tons were landed annually, but recently the catch has been negligible (figure 2.16). Swim bladder value has risen in direct relation to scarcity, from a

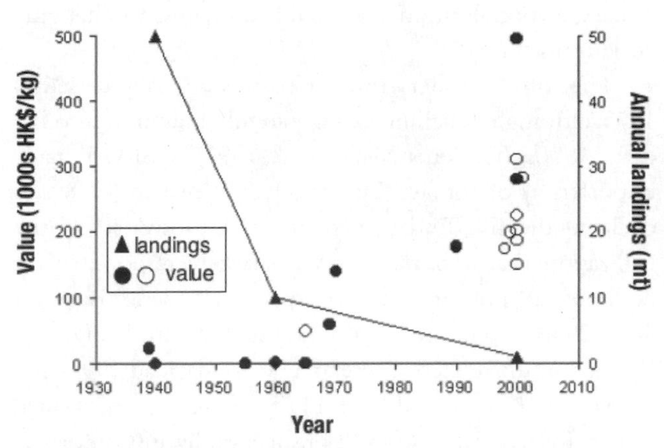

Figure 2.16. Landings and value of the giant yellow croaker. As catches have fallen (solid line and triangles), the price paid for its swim bladder has risen (open and closed circles). The fish would be "commercially extinct" except it is worth too much. HK$ = Hong Kong dollars; MT = metric tons. Adapted from Sadovy and Cheung (2003); used with permission.

few dollars per kilogram of dry weight in the 1930s to as much as $64,000/kg in 2001, seven times more than gold. Despite a greater than 99% reduction in catch, fishing effort remains high, as in the Pearl River (Hong Kong) estuary, where 100–200 boats still pursue the fish on its spawning grounds (and where habitat quality is also declining).

The fishery continues even though the species was listed as a Grade II State Protected Species in the People's Republic of China in 1988 and has long been commercially extinct in Hong Kong (where it is not protected, and where no fishery management is practiced). Giant yellow croaker are "the first marine food fish of extreme conservation concern in Southeast Asia" and may have the distinction of becoming, along with totoaba, one of the first documented examples of the near extinction of a highly fecund marine fish species (Sadovy and Cheung 2003, p. 89).

Life history traits and fishery patterns common to many sciaenids make them especially vulnerable to fishing. Sadovy and Cheung (2003) summarized information on four other heavily fished Asian sciaenids that have undergone major population declines, as have three species in the eastern Pacific, two in the west Atlantic, one in Europe, and another in South Africa. Common to these species, whose large size and concomitant high fecundity should provide some buffering against overfishing, are a suite of characteristics that can be linked to vulnerability. These include large size, which makes them desirable in commercial as well as sport fisheries; limited geographic range centered on heavily exploited coastal regions, leading to reduced availability of refuges from fishing; susceptibility to bycatch because of the multispecies nature of exploitation in the regions; aggregated spawning in estuarine areas that are subject to degradation and excessive fishing pressure (probably exacerbated by spawning stupor, see chapter 11); active vocalization during spawning, which makes them locatable; extremely high value that increases with scarcity; and limited or nonexistent fisheries regulations. The only sciaenid assigned at-risk status by IUCN (2004) was the totoaba.

CONCLUSION

This chapter has a decidedly downbeat tone, which is to be expected from a compilation of species in serious trouble. Describing the causes of decline of fishes results in a depressing redundancy. The same factors, and many of the same mistakes—habitat alteration, introduced species, pollution, and overexploitation—have been repeatedly responsible.

Many programmatic solutions are detailed in chapter 3, which discusses national plans, actions, and regulations designed to determine status and to protect species once they are listed. The first step in solving these problems is recognizing they exist, which occurs as an outgrowth of agency and national listings of species at risk. Official designation of a species as imperiled initiates a recovery plan in many countries, and such plans often include habitat protection (e.g., Australia, Canada, the EU, New Zealand, South Africa, the U.S.). Such listing is also a first step toward restrictions on international trade, as provided by CITES. Several of the species, higher taxa, and habitats presented here are closely monitored by dedicated advocates in international bodies, including the Species Specialist Groups of IUCN, scientific societies, and nonprofessional special interest groups. New groups appear regularly; enter a family or common name and "conservation" into any Web search engine to discover them. The existence of these watchdog entities ensures that people are paying attention and taking action independent of the political process, which can move frustratingly slowly if at all. Much less political leadership would be demonstrated if it weren't for activist citizens. Such groups give anyone concerned about preserving fish diversity an opportunity to join like-minded people and get something done.

3. Roll Call II: A Geopolitical Perspective

The decision of what to conserve, how, and to what extent, is a political decision based on the value given to biodiversity by society.

—Susette Biber-Klemm, 1995

What follows is an overview and synthesis of the world's imperiled fishes at the species level, organized geopolitically. Space considerations forbid listing all known or probably imperiled fishes; coverage here is intended as representative at best. An accurate list of imperiled fishes from different regions would conservatively run to 2,500–4,500 species. A catalog of all internationally recognized, highly threatened fishes is available at the IUCN Web site, www.redlist.org. It lists 67 species of elasmobranchs (sharks, skates, and rays), 3 species of lampreys, and 731 species of bony fishes (95 IUCN-recognized extinct species are included in the appendix to this book). Some discussion of definitions of categories of imperilment and of criteria for listing is unavoidable, because both categories and criteria differ between and even within nations. I have attempted to recognize generalities and patterns of imperilment, both at the taxonomic level and among the various causes, at the risk of ignoring special circumstances. A synthesis of apparent common attributes of vulnerable fish species forms the basis of chapter 4.

Why *species* lists? We can argue about the relative merits of protecting species versus habitats versus ecosystems in reaching an overall goal of conserving fishes, but the reality is that most assessments of declining biodiversity are performed at the species level, because most governing bodies and the public recognize and protect species. States, provinces, regions, and countries maintain species lists, which are updated and revised at varying intervals. Finding these lists can be a challenge, and any such list is soon superceded. Membership on a list depends on widely differing definitions of imperilment, as well as (sometimes arbitrary) political and economic considerations. Consequently, although without them we lack the basic information to monitor trends and assess the status of the world's organisms, official lists—of species or any other category—have to be taken at no more than face value in terms of both factual and temporal accuracy.

One important caveat. I originally compiled this chapter in 1999, comparing regional lists with the 1996 Red Data Lists from IUCN (see box 1.1). I have tried to update my comparisons, and I have also cross-checked most against the 2004 IUCN Red Data Lists. Most lists underwent only minor change, but change did occur. By the time this appears, some will have undergone further revision. Coverage in many countries is improving, and revisions are appearing more frequently—a good sign. Anyone desiring the most current information should consult my sources of information, which are primarily the Web sites cited.

WHAT DOES IMPERILED MEAN?

The chief motivation behind efforts to identify imperiled species is a desire to recognize taxa that are in decline and are at a high probability of extinction given current trends. Categories and lists of imperiled organisms are invariably subdivided or organized by degree of imperilment. Boundaries between categories are necessarily vague, which hasn't deterred well-intentioned efforts to quantify the procedure. At one extreme of generality is the U.S. Endangered Species Act (ESA), which defines an Endangered species simply as one in danger of extinction throughout all or a significant portion of its range. A Threatened species is likely to become Endangered within the foreseeable future. While these definitions appear straightforward and have been implemented widely (more than 1,200 animal and plant species are protected by the ESA), they have been criticized for their vagueness—e.g., what is "significant" and how far off is "the foreseeable future"?

At another extreme, IUCN maintains a Red List of threatened species that are defined by both qualitative and quantitative criteria designed to circumvent or at least minimize criticisms of vagueness and subjectivity. The Red List began in the 1960s; membership on the list and criteria for listing have been updated every few years (most recently in 2006). The need for revision and reassessment points out the difficulty of developing definitive criteria for recognizing imperilment.

IUCN recognizes three major levels of threat—Critically Endangered, Endangered, and Vulnerable (CR, EN, and VU, respectively). Four other categories reflect extremes of population condition: Extinct (EX), Extinct in the Wild (EW), Lower Risk than Vulnerable (LR), and Data Deficient (DD). A species is placed in a threat category when it has crossed specified quantitative thresholds of measurable criteria (G. M. Mace and Hudson 1999). The criteria pertain to decline in population size or distribution, diminished area of occurrence, change in population structure or in certain biological attributes (such as reproductive rate, generation time, and maturation size), and change in the estimated probability of extinction. Thresholds relax progressively in the less threatened risk categories. Glossing over many details, the thresholds can be summarized in the table below.

This attempt at quantifying risk is laudable in that it offers specific targets for classification and specifies the kinds of data needed to assess conservation status. It has been criticized, however, for the arbitrary nature of the numerical cutoffs between categories. Also questioned is application of the cutoff values for disparate taxa (e.g., sedentary freshwater invertebrates vs. highly migratory whales and fishes). To fishery managers, the values appear inappropriate for commercially exploited fishes, which undergo large natural fluctuations in population size. This concern has led to a general critique of applicability of the categories to marine fishes (IUCN 1997; see chapter 4). Regardless of exceptions and complications, the IUCN format represents a useful and carefully crafted effort to establish standardized, objective criteria for determining which taxa are in decline, which are at a high probability of extinction, and which are in need of particular data to allow such determinations.

A summary of fish listed by IUCN (2004) appears in table 3.1. Perusal of the list and comparison with earlier compilations reveal the growth in concern and coverage that fishes have received in recent years. Thanks to IUCN and the people—largely volunteers—who contribute to that program, our awareness of the integrity of global fish biodiversity increased tremendously in the last decade of the 20th century and continues to grow. The 1996 Red List included 695 fish species, subspecies, or stocks in the threatened categories of CR, EN, and VU, plus 91 EX and EW (total 786). Only four nonbony fishes were listed (three lampreys and one shark). Coverage grew in 2000 by 5%, to 734 threatened and 81 extinct (total 815). About 40% of the increase resulted from greater inclusion of sharks and rays, which accounted for 15 added species. Coverage and numbers increased further with publication of the 2004 assessment, with more attention paid to unique stocks and population segments. The three threatened categories contained 870 taxa, with 95 extinctions recorded (total 965). Increased coverage of sharks and rays, with 56 species added, again contributed substantially. Other groups receiving greater attention were sturgeons (+25 taxa), perciforms (+24), and atheriniform silversides (+16). The jump in perciform groups in part reflected greater attention to marine species; seven species were added, five of which were tropical groupers.

Population reduction	*Area of occurrence/occupancy*	*Number mature*	*Extinction probability*
CR: >80–90%/10 yrs/3 generations	100 km^2/10 km^2	<50–250	50%/ 10 yrs/3 generations
EN: >50–70%/10 yrs/3 generations	5,000 km^2/500 km^2	<250–2,500	20%/20 yrs/5 generations
VU: >30–50%/10 yrs /3 generations	20,000 km^2/2,000 km^2	<1,000–10,000	10%/100 yrs

Table 3.1. IUCN (2004) global red list designations of imperiled fishes at higher taxonomic levels

	EX	*EW*	*CR*	*EN*	*VU*
Number of species by order					
Agnathans: lampreys and hagfishes					
Petromyzontiformes: lampreys	0	0	0	1	2
Chondrichthyes: sharks, skates, rays					
Orectolobiformes: nurse sharks	0	0	0	0	7
Lamniformes: mackerel sharks	0	0	1	2	5
Carchariniformes: ground sharks	0	0	5	4	13
Squaliformes: dogfish sharks	0	0	2	3	3
Squatiniformes: angel sharks	0	0	0	2	3
Rajiformes: rays	0	0	6	15	15
Sarcopterygii: fleshy-finned fishes					
Coelacanthiformes: coelacanths[a]	0	0	0	1	0
Actinopterygii: ray-finned fishes					
Acipenseriformes: sturgeons	2	0	10	25	13
Osteoglossiformes: bonytongues	0	0	0	1	0
Clupeiformes: herrings	0	0	0	2	3
Cypriniformes: minnows	17	1	43	37	117
Characiformes: characins	0	0	0	1	1
Siluriformes: catfishes	1	0	10	8	22
Esociformes: mudminnows	0	0	0	0	1
Osmeriformes: galaxiids[b]	1	0	4	0	10
Salmoniformes: salmons	4	0	5	8	12
Ophidiiformes: cusk eels	0	0	0	0	7
Gadiformes: codfishes	0	0	1	0	2
Percopsiformes: troutperches	0	0	1	0	3
Batrachoidiformes: toadfishes	0	0	0	0	5
Lophiiformes: anglerfishes	0	0	1	0	0
Atheriniformes: silversides	1	0	9	7	44
Beloniformes: halfbeaks	0	0	2	3	8
Cyprinodontiformes: killifishes	11	5	20	21	27
Gasterosteiformes: sticklebacks	0	0	1	0	4
Syngnathiformes: seahorses[c]	0	0	1	1	14
Synbranchiformes: swamp eels	0	0	0	1	0
Scorpaeniformes: scorpionfishes	1	0	3	2	5
Perciformes: perchlike fishes	46	6	62	42	145
Pleuronectiformes: flatfishes	0	0	0	1	1
Tetraodontiformes: triggerfishes	0	0	0	0	3
Major taxonomic groups					
Agnatha	0	0	0	1	2
Chondrichthyes	0	0	14	26	46
Sarcopterygii[a]	0	0	0	1	0
Actinopterygii	83	12	173	160	447
TOTAL	83	12	187	188	495

Source. IUCN Red List of Threatened Species, www.redlist.org/search/search-basic.php.

Note: Taxonomic categories follow J. S. Nelson (2006). Categories are EX (Extinct), EW (Extinct in the Wild), CR (Critically Endangered), EN (Endangered), and VU (Vulnerable). Common names are representative of only part of each group. An additional 108 actinopterygians and 5 lampreys fell into two low-risk categories, and 250 actinopterygians, 2 elasmobranchs, and 3 lampreys were listed as Data Deficient, indicating reason for concern but lacking appropriate data on abundance or distribution (= 368 additional imperiled species). Only orders containing listed species are included. Extinct species are named in the appendix.

[a]The IUCN Web site has yet to include the Indonesian coelacanth, *Latimeria manadoensis*, which is in CITES Appendix I.

[b]I follow Nelson (2006) and place retropinnid southern smelts and galaxiids in the Osmeriformes, whereas IUCN includes them with salmoniforms.

[c]IUCN separates seahorses and allies from the Gasterosteiformes, and I have followed that system here.

Increased numbers across the years undoubtedly reflect greater attention to faunas coupled with worsening conditions. Our knowledge remains woefully incomplete both geographically and taxonomically. The IUCN searchable database contains information from 233 countries and territories, but information on fishes is provided for only a fraction of these. The conservation status of the vast majority of fishes remains unknown. In countries with active conservation programs, inclusion is increasingly likely for species facing obvious threats. But because of the bias toward the industrialized world and problem species, estimating the absolute fraction of the world's fish fauna at risk remains a guessing game.

INTERNATIONAL AND NATIONAL PROGRAMS

Lists of imperiled species are generated by governmental as well as nongovernmental organizations and cover regions of diverse scale. Inclusiveness is often a function of the scale of coverage; the more local the list, the more inclusive it tends to be. Local and regional lists are more likely to contain species of problematic or questionable conservation status. These lists are incorporated into national and international lists and in the process distilled to species whose status is more certain. States or provinces likely include such categories as Of Special Concern or Rare, which are seldom included in national rankings, and disagreement on definitions of categories can add to the confusion. At the extreme of exclusivity, the ESA gives legal standing, and disburses funds, only to species officially designated as Threatened or Endangered.

The effectiveness of conservation laws also increases as one proceeds up the governmental hierarchy; penalties increase in severity, and the number of people affected by rulings increases. Arbitration, negotiation, and litigation over inclusion on lists are more likely at national and international levels. For these and other reasons, federal and international lists tend to be more exclusive and conservative; the likelihood that a marginally threatened species will make it onto such a list is also marginal (see box 3.1).

The international standard among catalogs of imperiled organisms is the list maintained by IUCN, whose criteria have been widely adopted by individual countries in creating their own lists. According to the United Nations Environment Programme (Groombridge 1994), by the mid-1990s, red data books or red lists covering threatened and endangered animals had been produced by 62 nations (29 in Europe, 6 in North and Central America, 6 in South America, 4 in Africa, 12 in Asia, and 5 in Oceania). Fish coverage is somewhat spotty, but every major western and southern European nation except Italy lists fishes (a Web search in 2004 did not locate a red list for Italian fishes, although Maitland [1995] maintained that Italy had an equivalent list). Other developed countries or biodiversity hot spots that lacked fish lists were Cambodia, Costa Rica, Cuba, Ecuador, Hong Kong, Indonesia, Israel (reportedly in preparation, 2004), Malaysia, Nicaragua, Pakistan (reportedly in preparation, 2003), and the Philippines. Only Algeria and South Africa among African nations included fishes (based on Groombridge 1994, augmented with a Web search conducted in November 2004). Some countries have simply adopted the IUCN list as their own rather than applying local expertise and personnel to examine regional biodiversity. Given the rarefaction process described above, simple adoption of an IUCN list means we lose much of the information that accumulates when actual, in-country status assessments are conducted, particularly in countries with moderate to high levels of endemism.

GEOPOLITICAL LISTS AND PROGRAMS

Inclusion of a species on a red list is often based on best guesses of abundance or status. Actual data on population size, geographic range, and probability of extinction are lacking for many if not most species. I have selected a few such lists for summary and comment, focusing on countries with accessible records and species that appear to qualify for CR, EN, or VU status. Countries are presented in alphabetical order, and the information is summarized in table 3.2. I provide some information on levels of diversity and endemism within a country, environmental problems, legislative protection where such information is available, efforts at preserving biodiversity, and a comparison with the IUCN (2004) compilation. I have excluded most of the historical comparisons within countries that inevitably show increasing numbers of imperiled species over time. Because efforts to identify and conserve imperiled fishes are increasing almost everywhere, it is difficult to separate accelerating degradation of species and habitats from increasing knowledge about such degradation.

I recognize that this geopolitical approach to imperil-

BOX 3.1. CITES: An International Effort to Protect Endangered Species

The Convention on International Trade in Endangered Species of Wild Fauna and Flora (CITES) is a cooperative, international program designed to protect wildlife from overexploitation and prevent international trade from further threatening imperiled species. CITES involves 167 signatory countries; it bans commercial trade of designated species and regulates and monitors trade in species that may become endangered. Member countries agree to restrict trade in species listed in the appendices of the convention (www.cites.org/eng/app/appendices.pdf). Species in Appendix I are those threatened with extinction and therefore most vulnerable to commercial trading; they cannot be traded commercially between nations, although special permits can be issued for scientific study. Appendix II species are not currently threatened with extinction but could be threatened if trade is not restricted or regulated. Export permits can be issued but only if such trade "will not be detrimental to the survival of the species in the wild." This important "nondetriment" provision basically mandates that exploitation must be sustainable. Appendix III lists species that a particular nation wants to regulate and for which it desires international cooperation to limit trade; export of any listed species requires special permission. Inclusion in Appendix I or II requires agreement among signatory parties after often protracted negotiations. Any country can place a species in Appendix III.

A limited number of fishes are currently (March 2005) listed in CITES. Nine species are protected by Appendix I: both species of coelacanth (*Latimeria chalumnae, L. manadoensis*); shortnose and Baltic sturgeons (*Acipenser brevirostrum, A. sturio*); golden dragonfish, *Scleropages formosus*; Julien's golden carp (or seven-striped barb), *Probarbus jullieni*; cui-ui sucker, *Chasmistes cujus*; Mekong giant catfish, *Pangasianodon gigas*; and totoaba, *Totoaba macdonaldi* (information about most of these is given in chapter 2). All of the remaining 23 sturgeon species have Appendix II status, as do all species of seahorse, *Hippocampus* spp. Only seven other fishes are listed in Appendix II, including the white, whale, and basking sharks; Australian lungfish, *Neoceratodus forsteri*; arapaima, *Arapaima gigas*; the Congo blind barb, *Caecobarbus geertsi*; and the giant or humphead wrasse, *Cheilinus undulatus*. Surprisingly, a number of known endangered species, including many designated as critically endangered by IUCN, have been taken out of the protective appendices in recent years, including paddlefishes, cyprinids, salmonids, aplocheilids, a poeciliid, and a percid. Regardless, CITES remains an invaluable source of information, especially via its relationship with TRAFFIC, publishing reports about taxa that are heavily exploited (e.g., on shark trade; the whale shark fishery in Taiwan; the trade in seahorses, sturgeons and caviar; bluefin tuna fisheries; e.g., D. A. Rose 1996; Vincent 1996; Che-Tsung et al. 1997; see www.traffic.org).

As an international, cooperative agreement, CITES is expectedly complex and open to negotiations, which can be quite convoluted and often play economic interests of countries against jurisdictional disputes over who regulates which species (e.g., CITES vs. FAO; see www.iisd.ca/v0121/enb2145e.html for an example). CITES applies to international trade but not to trade within a country's boundaries, which requires passage of separate legislation in each country. It also focuses on wild organisms; someone rearing an endangered species cannot be prohibited by CITES from selling the offspring. For example, shovelnose sturgeon are protected as an Appendix II species, but young laboratory-reared shovelnose sturgeon are sold on the Internet in the U.S.

ment biases any analysis because fishes and environmental problems do not respect international boundaries (e.g., Ray and Ginsberg 1999). However, preservation efforts and government actions do recognize such boundaries, and the available data on imperiled fishes are organized primarily along political lines. Although the treatment here will likely become rapidly outdated, as mentioned earlier, the data should have at least historical and comparative value and provide a starting point for anyone seeking additional and updated information. (I would be delighted to hear from anyone who knows of additional lists or updates of the lists provided.)

Australia

The Australian freshwater fish fauna consists of more than 200 species, the vast majority of which are endemic, and about 4,100 marine and estuarine fishes, with high endemism in the temperate component (Berra 1998; G. R. Allen et al. 2002; Pogonoski et al. 2002). Protection of this unique fauna was compromised for an inordinate time by jurisdictional disputes between state and federal governments. Following actions by most states and territories, which have their own environmental protection laws, the Australian federal government enacted an

Table 3.2. Summary of the legal conservation status of fish species in countries discussed in the text

		Number listed		*IUCN recognized*		*Country not IUCN*		*IUCN not country*	
Country	*Source*	*FW*	*Marine*	*FW*	*Marine*	*FW*	*Marine*	*FW*	*Marine*
Australia	EPBC 2005	0,14,13	2,3,7	8,5,16	9,10,33	6	4	10	42
Brazil	Fund.Bio. 2004	31,33,70	2,6,14	0,0,7	7,9,23	148	6	0	11
Canada	SARA 2003	2,13,9	0,1,2	3,2,11	2,3,5	22	3	13	15
Chile	Campos 1998	17,12,4	—	0,1,2	0,0,2	28	—	0	2
Europe (EU)	BERNE 1999	18		12		6		21	
	OSPAR 2005	—	13	—	8	—	5	—	—
Japan	Red Book 1991	16,6,17	—[a]	1,6,4	0,4,19	10	—	3	23
Madagascar	Benstead 2003	31,49,24	—[a]	12,13,27	1,3,12	—	—	—	—
Mexico	Conabio 2004	68[b],64	2[b],6	21,29,38	5,5,21	73	3	18	26
New Zealand	DOC 2002	1,2,1	0	0,0,6	1,0,9	4	—	4	10
Russia	Red Book 2001	16[b],18		0,1,2	2,20,14	38	—	2	25
S. Africa	Skelton 2002	7,6,9	—[a]	6,6,9	2,6,19	1	—	0	27
UK	WCA 2002	5	4	0	1,5,8	5	2	—	2
USA	ESA 2005	71,43[c]	16[c]	16,20,88	6,8,27	55	6	20	31
Venezuela	Red List 1995	1,0,3	0,0,2	0,0,0	1,4,14	5	0	—	17

Note: The values given for species protected by a country ("Number Listed") or recognized by IUCN (2004) are the numbers of species with CR, EN, VU, or approximate equivalent ranks in a country (except Canada, which uses Extirpated, Endangered, Threatened). Marine includes estuarine species, except for the U.S. FW = freshwater. Numbers centered between FW and Marine are the total of the two habitats. The "Country not IUCN" and "IUCN not country" columns list the number of species ranked by one entity but not the other. "Sources" given as acronyms refer to country-specific legislation. References are given in the text.

[a]Reference dealt only with freshwater species.

[b]Mexico and Russia basically combine the categories of Critically Endangered and Endangered into one rank.

[c]USFWS recognizes only Endangered and Threatened species, of which none are truly marine; marine species are administered by NOAA Fisheries, which uses the category Species of Concern.

Endangered Species Protection Act (ESPA) in 1992 (Woinarski and Fisher 1999). The ESPA had jurisdiction only over federally managed lands, which accounted for less than 1% of the Australian continent. To its credit, the act went beyond protecting species, subspecies, and populations by recognizing the need to conserve "endangered ecological communities" and to remediate "key threatening processes." An *Action Plan for Australian Freshwater Fishes* (Wager and Jackson 1993) spelled out the status of fishes and proposed solutions to known problems.

Federal authority to protect species and habitats expanded with passage of the Environment Protection and Biodiversity Conservation Act (EPBC) of 1999, which again protected species, ecological communities, and processes, and recognized the value of protected habitats (and may be the only major piece of legislation internationally with the phrase "biodiversity conservation" in its title!). Assignment of ranks is determined via a nomination process, reviewed by a Threatened Species Scientific Committee. The act is administered by the federal Department of the Environment and Heritage and applies to private as well as public lands. Violators "can be liable for a civil penalty of up to $550,000 for an individual and $5.5 million for a body corporate, or for a criminal penalty of seven years imprisonment and/or $46,200" (www.deh.gov.au/epbc/compliance). As of February 2005, if any convictions involving fishes had occurred, they hadn't been widely publicized (B. Pusey, pers. comm.). However, court rulings have led to unanticipated control over actions that occur outside protected areas but affect the habitats and the organisms within. For example, dam construction upstream of the Great Barrier Reef is referable to the act because of its potential impact on the reef, which is a World Heritage Area (D. Burrows, pers. comm.).

Federal protection for marine fishes was hampered until 1998 because they were viewed primarily as commercial species and therefore exempt from monitoring and control under the Wildlife Protection Act. Their management was dependent on fisheries legislation (Vincent and Sadovy 1998). The EPBC list for 2005 identified 27 freshwater and 12 marine species in high, at-risk ranks. The contemporaneous IUCN accounting was more inclusive—particularly for sharks, rays, seahorse relatives, and other marine groups—listing 29 freshwater and 52 marine species.

At the level of expert, nongovernmental identification of fishes at risk, the Australian Society for Fish Biology (ASFB) has catalogued threatened fishes since 1985. Its assessment has been updated almost annually and published in the society's newsletter. Addition of a species to the list involves a nominating process that requires detailed information justifying the designation, which must be ratified by the society. The original 1985 compilation listed 59 species in seven categories; the 2001 list contained 54 taxa in three categories, based on IUCN criteria. Ten ASFB (2001) species were marine, a doubling from 1996; 5 of the 54 listed fish were undescribed species.

Although ASFB cross-checks its list with the IUCN Red List, and in fact Australian species on the IUCN list are compiled by Australian biologists, discrepancies existed, such as those of trout cod, *Maccullochella macquariensis* (ASFB-CR, IUCN-EN); southern bluefin tuna (ASFB-no rank, IUCN-CR), and humphead wrasse (ASFB-no rank, IUCN-EN); a galaxias and a perch ranked EN by ASFB were not included by IUCN. Whether these differences reflect needed coordination between groups or differences in opinion remains in question.

Australia had until recently no documented extinctions among its ichthyofauna. In June 2005, the Pedder galaxias, *Galaxias pedderensis*, was officially designated Extinct in the Wild. The Lake Eacham rainbowfish, *Melanotaenia eachamensis*, was thought to be extinct (Groombridge 1994) but was reevaluated by both ASFB and IUCN and is now listed as Vulnerable due to habitat destruction, flow regulation, hybridization with other rainbowfishes, and introduced species. It had been extirpated from the locale where it was originally described but was subsequently found in at least two other river drainages (Pusey et al. 2004; B. Pusey, pers. comm.; figure 3.1). Of major concern is the CR status of the spotted handfish mentioned in chapter 2.

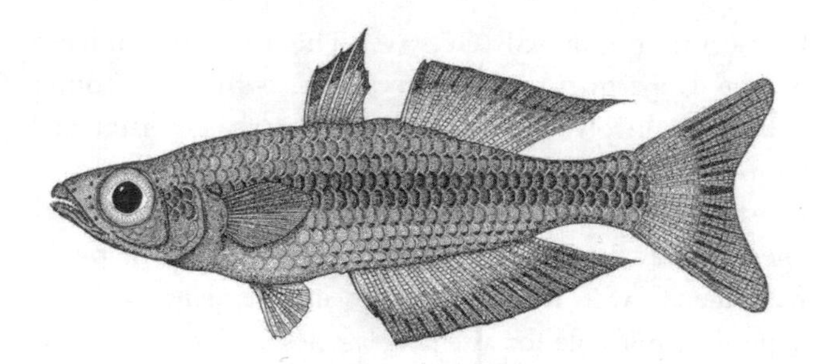

Figure 3.1. The Lake Eacham rainbowfish of Australia. Drawing by B. Pusey in Pusey et al. (2004); used with permission.

Canada

Canada's fish fauna includes 171 native freshwater species (W. B. Scott and Crossman 1973). Endemicity is relatively low, less than 20% (R. R. Campbell 1997). Debate rages over possible subspecific or even specific status for forms of whitefishes and sticklebacks and perhaps other groups in Canada's many isolated lakes. Campbell listed ten distinct, undescribed sticklebacks in British Columbia alone. Ten introduced species have become established (Robins et al. 1991; Fuller et al. 1999).

Despite a large, active, and well-funded fisheries research establishment, financial support and legal protection for imperiled, nongame fishes developed slowly in Canada, relative to other industrialized and British Commonwealth countries. Unlike Australia, where states had protection laws, only seven of Canada's ten provinces and three territories passed endangered species legislation during the 20th century (Scudder 1999). Where lists appeared, they did not necessarily reflect scientific assessments of levels of imperilment: Quebec listed only the copper redhorse, *Moxostoma hubbsi*, on its endangered species list, an optimistic accounting of environmental conditions. No federal act protecting endangered species existed in Canada before 2003, although such legislation had been considered repeatedly in the Parliament (R. R. Campbell 1997, 1998).

In 2003, the legislative picture changed with the passage of the Species at Risk Act (SARA), which resulted from cooperative efforts between federal and provincial authorities. SARA's stated purpose is to prevent Canadian indigenous species, subspecies, and distinct populations from becoming extirpated or extinct. Activities are coordinated through the Canadian Endangered Species

Conservation Council (CESCC). The act is administered by the Department of Environment, with several other authorities involved and consulted. The nongovernmental Committee on the Status of Endangered Wildlife in Canada (COSEWIC), which had been identifying threatened species for decades and working to get their plight recognized, was retained as an independent body of experts responsible for assessing species at risk and recommending them for protection under SARA.

Under SARA, imperiled species are identified, restoration activities are implemented, and habitat is protected via a Habitat Stewardship Program. Fines and jail terms can be imposed, although prosecutions involving fishes remain rare. Protection is given to Extirpated, Endangered, and Threatened species. Protection is a dynamic process; candidate species listed in Schedules 2 and 3 and other nominated species are periodically assessed and, if justified, may be moved into Schedule 1, which contains those species with actual legal status. By protecting extirpated and extinct species, SARA is superior to the ESA, in which a declaration of extinction eliminates a species from protection.

As of August 2005, Schedule 1 contained 27 fish species, subspecies, or populations. Another 9 species were designated Of Special Concern, which is not a protected category. IUCN listed 31 Canadian species, 28 that were not on the SARA 2005 list (IUCN is clearly wrong in listing hogfish, *Lachnolaimus maximus*, as Vulnerable; it is a tropical West Indian reef fish with no business or future in Canada). Only North American paddlefish, *Polyodon spathula*; eastern sand darter, *Etheostoma* (*Ammocrypta*) *pellucida*; and Atlantic whitefish, *Coregonus huntsmani*—all SARA-VU—also had IUCN recognition. Several of the 28 IUCN-only species were Of Special Concern or Schedule 2 or 3 species in SARA, and their Canadian status may be reevaluated soon.

At least two Threatened Schedule 1 fishes represent peripheral species or populations, whose occurrence in Canada is at the northern limit of their range and which would be designated as globally secure if more southerly populations were considered. These are the lake chubsucker, *Erimyzon sucetta*, and spotted gar, *Lepisosteus oculatus*, which are widespread and unprotected in the U.S. Such peripheral Canadian populations may be genetically distinct and worthy of international listing, but their status will remain problematic until additional studies are completed.

SARA represents a major improvement in species protection, but a weakness of the act, one that reflects the continued tension between provincial and national governments, is that its protection and prohibitions apply only to species under federal jurisdiction and on federal lands (R. Campbell, pers. comm.). More inclusive protection remains to be enacted.

Endangered species were not ignored by Canadian scientists before SARA. A comprehensive accounting of imperiled fishes in McAllister et al. (1985) listed 11 species as endangered, 4 as threatened, 3 as extirpated, and 2 as extinct. The final group included the blue pike, *Sander vitreum glaucum*, which is also considered extinct in the U.S. J. E. Williams et al. (1989) counted 22 imperiled fishes in Canada and identified British Columbia and Ontario as the provinces with the greatest numbers of vulnerable species (7 and 8, respectively). More important, and despite a lack of legislative protection, Canada had an effective national, nongovernmental program for identifying species at risk (Scudder 1999). Beginning in 1977, COSEWIC monitored candidate species in jeopardy, assigned ranks, and published updated status reports regularly in the *Canadian Field-Naturalist* (e.g., R. R. Campbell 1997, 1998). Comparison of the COSEWIC lists from the late 1990s with the fishes in Schedules 1–3 of SARA suggests that COSEWIC did a great deal of the legwork that went into identifying candidate species for protection under SARA. COSEWIC continues to assess species suspected to be at risk of extirpation from Canada. As of February 2005, 20 freshwater fish species and 18 marine species were under consideration for assessment (www.cosewic.gc.ca/eng/sct3/sct3_1_e.cfm#3).

All told, Canada's freshwater fishes appear to be in average shape for an industrialized nation, with about 20% clearly imperiled and perhaps 33% under threat (based on species listed in all three SARA schedules). However, only half of Canada's freshwater species had been evaluated by COSEWIC by 2000, which means that large knowledge gaps remain. Also, an alarmingly large proportion of the recognized imperiled fauna is made up of known Canadian endemics. The total amounts to about ten listed species and could be much higher if stickleback and whitefish stocks are as specialized as many believe, suggesting that the unique component of Canada's fish diversity is particularly vulnerable.

European Union

The native freshwater fish fauna of Europe west of the former Soviet Union consists of 358 native species (Kottelat

BOX 3.2. Applying IUCN Criteria to Freshwater Fishes

The IUCN criteria for assignment of categories of risk focus on "extent of occurrence" and "area of occupancy," which correspond roughly to a species' overall geographic range and the actual space where it is found. Quantitative thresholds are then applied in assigning species to ranks, with thresholds of 100 km^2 for occurrence and 10 km^2 for occupancy to qualify a species for Critically Endangered status. What do these criteria imply for the world's freshwater fishes?

Maurice Kottelat (pers. comm., 2000) has analyzed European freshwater fishes in light of the IUCN criteria. Most fishes have relatively restricted ranges, either naturally or because of habitat fragmentation; the habitat of a species that occurs sporadically in a number of rivers is limited to just those river stretches. Extent of occurrence is not all contiguous map space surrounding the lakes or rivers or streams where the species is known to occur; it is the area of the water bodies themselves. In temperate regions in particular, most fishes live in relatively small streams in upland regions or in headwater springs. Many are restricted to one or a few drainage basins. Hence, areas of occurrence and occupancy are no more than tens of meters wide and tens of kilometers long. Kottelat used the example of the Rhine River, which is 1,320 km long by 100 m wide, or 132 km^2. Most species occur over only a portion of that area, and their actual areas of occupancy are even smaller, making all Rhine endemics Critically Endangered by the 100 km^2 criterion. Kottelat concluded that most European freshwater fishes are Endangered or Critically Endangered by IUCN designations, with relatively few falling into the still alarming category of Vulnerable.

A similar analysis of North American freshwater fishes would undoubtedly result in the same conclusions, particularly in the species-rich areas of the Southeast and Southwest, where localized endemism and biodiversity are highest. IUCN categories, with their emphasis on and definitions of area of occupancy, may be appropriate for terrestrial species, but to accurately reflect the status of aquatic organisms, they may need revision.

1997, 1998), a larger number than often cited in older compilations. Several international treaties protect freshwater fishes and their habitats in Europe, and general legal principles, such as International Neighbour Law and the Principle of Shared Natural Resources, give nations a responsibility to ensure that activities within their borders do not damage shared resources or migratory species (Biber-Klemm 1995). Four international conventions contain measures that specifically protect fishes: the Convention on Biological Diversity (Biodiversity Convention) of 1992, the Convention on Wetlands of International Importance (Ramsar Convention) of 1971, the Convention Concerning the Protection of the World Cultural and Natural Heritage (World Heritage Convention) of 1972, and the Convention on the Conservation of European Wildlife and Natural Habitats (Berne Convention) of 1979. Marine species and habitats gained protection from the OSPAR Convention for the Protection of the Marine Environment of the North-East Atlantic of 1992.

The Berne Convention (as modified in 1987) stands out, in that it identified species in need of protection. Its Appendix II listed 18 European fishes as Strictly Protected, prohibiting capture, killing, or habitat destruction (Appendix II was revised May 1999 and had not changed as of January 2005). In Appendix III, 118 species (plus "all species of *Coregonus*") were given Protected status, which emphasized sustainable use and regulated capture, not protection. Prohibitions on exploitation were in effect only until stocks were restored to sustainable levels.

Correspondence between these lists and concurrent IUCN (2004) designations was mixed. All but six Appendix II species were IUCN listed. Appendix III showed minimal concordance with IUCN assessments, with only 26 of 118 Berne-listed species having IUCN ranks. Such inclusiveness by the Berne Convention means more attention will be paid to potentially imperiled species. It is disconcerting, however, that minimal change occurred in either appendix over six years.

Eighty-five species identified by IUCN as at risk (CR, EN,VU, LR, DD) appeared on no Berne list, meaning that the Berne Convention ignores many species that international experts view as in need of protection or assessment. In the opinion of some experts, the Berne list greatly underestimates the proportion of European fishes that are at risk (box 3.2). Also, IUCN regularly updates and generally increases its coverage, whereas Berne appears fairly static. These discrepancies probably reflect a lack of communication between agencies (and perhaps insufficient staffing and resources at the Berne offices). Regardless, disagreement between lists can be misinterpreted as disagreement about the imperiled nature of species and does little to strengthen conservation efforts.

The restrictions and prohibitions of these various conventions may appear toothless because they do not carry the force of legislation; however, signatory states feel obligated to pass national laws that put the recommendations into effect (Biber-Klemm 1995). Biber-Klemm cited legislative action to protect species and habitats in France, Spain, Greece, Italy, and Jordan that grew out of international conventions.

More formally, species and habitat protection has been strengthened with the advent of European Union (EU) jurisdiction over environmental issues. Member states have a legal obligation to "prevent, mitigate, and compensate for" actions that harm protected species and habitat types. Species and habitats are protected under Council Directives, such as Directive 78/659/EEC concerning the quality of fresh waters needing protection or improvement, and Habitats Directive 92/43/EEC that lists in its annexes freshwater and marine species in need of protection. As a result of such directives, habitats and species of EU importance receive protection that carries the force of law, which includes withholding funds for development projects and levying fines against countries that engage in practices detrimental to listed sites and organisms (B. Delpeuch, pers. comm.). Under these directives and the more recent Water Framework Directive, 58 fish species and 15 freshwater habitat types were listed as of November 2000. Species recovery and study programs are funded under the LIFE-Nature program (e.g., see figure 2.4), and habitat types are designated under the Natura 2000 program (see http://europa.eu.int/comm/environment/nature/home.htm).

Finally, international trade in endangered species by member countries is regulated under EC Regulation 338/97 (amended by Regulation 1497/2004), which placed protected species in annexes equivalent to CITES appendices. The sixth edition of this regulation (WCMC 2003) listed nine species in Annex A, matching the nine species in CITES Appendix I; Annex B listed the same species that appeared in Appendix II of CITES. Annexes C and D included Appendix III species and others "imported into the [European] Community in such numbers as to warrant monitoring."

Most of the EU directives and policies deal with freshwater habitats. Marine conservation has lagged behind. This oversight was corrected in part by the OSPAR Convention, which came into force in 1998. OSPAR merged two conventions from the 1970s, the Oslo and Paris Commissions, that had focused on marine pollution, and expanded concern to cover more general issues of biodiversity and habitat protection. OSPAR consists of representatives of 15 governments and the European Commission (the executive body of the EU). In addition to focusing on marine issues, it identifies species and habitats threatened or in decline, develops measures for their protection, and emphasizes an ecosystem approach to the management of human activities (except for issues relating to fisheries management, which must be referred to other administrative authorities such as the International Commission for the Exploration of the Seas [ICES] for further action). OSPAR included provisions for developing marine protected areas (MPAs), but no action had been taken as of February 2005.

Under its Annex V (protection and conservation of the ecosystems and biological diversity), OSPAR developed the Texel/Faial Criteria to create an initial list of threatened and declining species and habitats. These criteria apply to both species and habitats and are more subjective than IUCN guidelines, focusing on global importance, regional importance, rarity, sensitivity, ecological role (e.g., keystone species), and population decline. The species list was expanded in early 2005 to include 13 marine and anadromous fish species, 8 of which had IUCN (2004) status. "The purpose of the list is to guide the OSPAR Commission in setting priorities for its further work on the conservation and protection of marine biodiversity. . . . This list has no other significance" (OSPAR 2004). Although the list appears toothless, as mentioned above, public identification of species at risk can lead to action in individual countries, at least in Europe.

In addition to these regional or continental protective agreements, many EU countries have developed their own red lists and legislation to call attention to local biodiversity issues. A few are summarized elsewhere in this chapter to represent conditions in Europe and surrounding countries; the conservation status of the fish faunas of 14 other European countries were reviewed in Leidy and Moyle (1997). Compilations on endangered species protection and recovery efforts in several European countries can also be found in Kirchhofer and Hefti (1996a) and Collares-Pereira, Coelho, and Cowx (2002).

Japan

Japan has 211 freshwater fish species, 42% of which are endemic. Responsibility for protecting them lies with the Environment Agency of Japan, which publishes a *Red Data Book*. The book lists species, subspecies, and forms of imperiled fishes, including undescribed taxa. An English translation of the 1991 list (Yuma et al. 1998) provided annotations and updates on the status of many freshwater species.

Extant, imperiled fishes in Japan were divided into four categories: Endangered, Vulnerable, Rare, and Declining Local Populations of Widely Distributed Species, with 39 species in the first three categories (Yuma et al. 1998; see also Maruyama and Hiratsuka 1992). Correspondence with IUCN 2004 was fairly good (table 3.2). Interestingly, the Sakhalin sturgeon, *Acipenser mikadoi*, was listed by IUCN as Endangered but was not on the Japan list. A "Sakhalin green sturgeon, *A. medirostris*" appeared in the overall species list for the nation without conservation status. IUCN (2004) gave 34 Japanese species high at-risk status, but 23 of these were marine, a habitat not dealt with in Yuma et al. (1998). Protection under Japanese guidelines appears to apply to organisms but not to habitats or ecosystems.

Three Japanese freshwater fishes, all endemic, are considered extinct. They are the kuni-masu or black kokanee, *Oncorhynchus nerka kawamurai* (Salmonidae), the minami-tomiyo or Kyoto ninespine stickleback, *Pungitius kaibarae* (Gasterosteidae), and the Suwa gudgeon, *Gnathopogon elongatus suwae* (Cyprinidae). The black kokanee was exterminated in 1940 from its only habitat, Lake Tazawa, when strongly acidic water was diverted from a nearby river into the lake as part of a hydropower scheme, killing most organisms in the lake. The Kyoto ninespine stickleback was restricted to streams, springs, and spring-fed parsley paddies near Kyoto. Its primary habitats were destroyed when these areas were turned into land developments. The fish disappeared in the early 1960s despite transplantation efforts. The Suwa gudgeon was restricted to Lake Suwa but no longer occurs there, perhaps due to displacement by a congener. None of these three species appeared in the 2004 IUCN list of extinct fishes.

The Japanese established an additional, commendable category that offers protection to fishes with localized distributions or that have undergone range contraction due to human activity. Species can be registered as national natural monuments. In 1992, 11 fish species had such designation, including 4 that appeared on Japan's red list. Natural monument status can also be granted at the prefectural, city, town, or village level.

Public recognition and concern for freshwater fishes are essential in a country as densely populated as Japan, where aquatic environments have been manipulated for centuries. Only 2 of Japan's 30,000 rivers are undammed or otherwise modified (McAllister et al. 1997). Other threats to Japanese freshwater fishes include industrial and municipal pollution, wetland draining and creation (particularly for rice paddies), and translocation and introduction of alien species. Approximately 10% of the freshwater fish fauna is introduced, mainly from elsewhere in Asia, for food, sport, and the aquarium trade. These introduced species have displaced or replaced natives through predation, hybridization, and perhaps competition (Yuma et al. 1998).

Madagascar

The tropical island nation of Madagascar is a showplace of evolution for numerous taxa; approximately 90% of the flora and fauna of the island are found nowhere else (Benstead et al. 2000). Madagascar is also a prime example of rapid human population growth leading to habitat loss and species decline. Fishes are considered the most endangered vertebrates on the island (Stiassny and Raminosoa 1994).

The fishes of Madagascar are an interesting array of 21 tropical freshwater families (figure 3.2). Two families and 13 genera are endemic to the island, with many species exemplifying *microendemism*—that is, they are restricted to single river basins (Stiassny and Raminosoa 1994; Benstead et al. 2003). Of the 143 recognized native freshwater fishes, 65% are endemic, and new descriptions appear annually while more await description (Sparks and Stiassny 2003). Using Benstead et al.'s assessments, 31 species would qualify as Critically Endangered, 49 as Endangered, and 24 as Vulnerable. Only 16 species would be considered at low risk of extinction. IUCN (2004) took a more conservative approach, listing 52 at-risk species (detailed comparisons seem inappropriate because the Benstead list is unofficial). Benstead et al. (2000) stated that four Malagasy species were extinct, and IUCN recognized all four. Harrison and Stiassny (1999) added a fifth, the atherinid *Teramulus waterloti*, which is ranked EN by IUCN (2004).

Deforestation, often the result of slash-and-burn clear-

Figure 3.2. A pinstripe damba cichlid, *Paretroplus menarambo*, from Madagascar, which is Extinct in the Wild. Malagasy cichlids may be ancestral to the megadiverse species flocks of cichlids of the great and lesser lakes of the African continent. Photo taken at the Tennessee Aquarium.

ing and annual reburning for subsistence agriculture, has caused extensive erosion and "supersedimentation," giving rise to some of the highest erosion rates ever measured (see B. D. Patterson and Goodman 1997). As a result, 75% of Madagascar's native forests are gone (Groom and Schumaker 1993). These and other forms of habitat degradation have extirpated native fish species and facilitated the deliberate and accidental establishment of at least 21 species of introduced fishes (Reinthal and Stiassny 1991; Stiassny and Raminosoa 1994). Some exotics have been stocked to make up for declining catches of native species, the declines probably brought on by supersedimentation as well as overfishing (Benstead et al. 2000). Introduced species are common and dominate in many areas; of 96 widespread sites sampled over a five-year period, only 12 were unpolluted by exotic species. Both natives and exotics were found at 46 sites, and exotics alone at 38. Exotics have displaced natives via the summed effects of predation, competition, reproductive differences, and the oft-observed replacement of native specialists by introduced generalists better adapted to human-altered landscapes (Benstead et al. 2000).

Conservation legislation protecting fishes is lacking in Madagascar. Only one aquatic organism, the Madagascan big-headed turtle, *Erymnochelys madagascariensis*, is protected by legislation. Laws regulating cutting and burning of forests exist, and many natural areas have been set aside as reserves, but programs suffer from lack of funds and enforcement (the 1987 national budget for protected areas was US$1,000). External funding of conservation efforts is necessary and increasing. Recent improvements include creation of national parks and protected areas that encompass watersheds and establishment of national offices for the environment and for protected areas (Benstead et al. 2003). Efforts at captive propagation of Madagascar fishes have been under way, with at least 33 fishes cultured in Europe and North America, including 5 Critically Endangered and 19 Endangered species and a cichlid, *Paretroplus menarambo*, that is considered Extinct in the Wild (Loiselle 2003). Whether any of these fishes can be returned to the wild given levels of disturbance remains to be seen.

Arresting population declines among Malagasy organisms is critical. However, conservation issues in Madagascar take on greater philosophical and strategic importance in light of debates over allocation of conservation efforts and resources. Many Malagasy fishes (e.g., melanotaenioid rainbowfishes, mugilid mullets, and especially African cichlids) are basal or archaic taxa. They diverged early in the evolutionary history of their lineages and represent a type of living fossil that can inform us about ancestral traits in these evolutionarily important groups. Malagasy cichlids may be ancestral to the explosive species flocks that occupy continental African lakes and rivers (see chapters 2, 8). Losing these taxa is like losing the legend to a roadmap or the initial couplets in a taxonomic key. These "unique repositories of phylogenetic information" (Benstead et al. 2003, p. 1110) take on disproportionate importance in advancing our understanding of the evolutionary process in general and the evolution of Southern Hemisphere (Gondwanan) fishes in particular. Basal taxa deserve greater consideration in our decisions about prioritization of conservation efforts (Stiassny and de Pinna 1994).

Mexico

Mexico's continental fish fauna consists of approximately 500 species, including perhaps 30 undescribed forms (Contreras-Balderas and Lozano-Vilano 1994; Contreras-Balderas et al. 2002). The country is rich in habitat types, ranging from wet tropical zones in the south to extremely arid zones in the north. Southern, wet tropical areas are diverse, have fewer local endemics, and contain relatively few threatened fish species; they have also received less study (Almada-Villela 1990).

The vulnerability of fishes in Mexico has long been the subject of study, and the number of vulnerable species appears to be growing. Sources and numbers of species at high risk include Contreras (1969), 48 species and subspecies; Deacon et al. (1979), 59; J. E. Williams et al. (1989), 123; Contreras-Balderas and Lozano-Vilano (1994), 135; Conabio (2001), 140; and Contreras-Balderas et al. (2002), 168. How much this apparent tripling results from deteriorating conditions versus increased knowledge is difficult to evaluate.

Mexico's official list of threatened fishes appears in the Norma Oficial Mexicana (Conabio 2001). The September 2004 revised list contained 185 species in four ranks: P, En peligro de extinción (= CR+EN); A, Amenazada (= VU); and PR, Sujeta a protección especial (subject to special protection, approximating lower-risk ranks of IUCN). Extinct or extirpated species are also listed (the links at the Conabio Web site are especially useful; see www.ine.gob.mx/ueajei/publicaciones/normas/rec_nat/no_059_a2f.html). The 142 species in the two highest-risk ranks (P and A) represented a substantial increase from the previous official number of 119 (59 + 60, Secretaría de Solidaridad 1994).

The Mexican official list included about 33% more freshwater species and 75% fewer marine species than IUCN (2004). IUCN's marine catalog recognized 5 CR and 6 EN species not included in either high at-risk Mexican rank. Concordance was higher for freshwater taxa, with Mexico recognizing 70 of the 88 species listed by IUCN (table 3.2). This agreement improves over the 1994 Mexico list, in which almost 29% of IUCN listed fishes were missing.

Mexico has suffered an inordinate and probably underestimated number of extinctions among its freshwater fishes, with published values ranging between 9 and 25, the larger numbers being more recent (R. R. Miller et al. 1989; Contreras-Balderas and Lozano-Vilano 1994; Contreras-Balderas et al. 2002). Recent losses not mentioned in IUCN's (2004) estimate of 19 extinctions include a cyprinid, 2 atherinids, and at least 1 goodeid (Lyons et al. 1998; Webb and Miller 1998), pushing the number closer to the 25 reported by Contreras-Balderas et al. (2002). Extinctions have resulted largely from water degradation and withdrawal, combined with the effects of introduced species, all of which have increased in recent years.

The fish fauna in the arid north of Mexico is diverse, accounting for approximately 40% of the country's ichthyofauna. This diversity is the result of a high degree of endemism associated with many isolated springs and accompanying small drainage systems in otherwise desert-like conditions. As in the adjoining desert Southwest of the U.S., water use, withdrawal, and pollution pose significant threats to fishes and aquatic habitats. Approximately 90 springs and 2,500 km of river in the northern half of the country have dried up in recent years, extirpating on average 68% of fish species from known locales (Contreras-Balderas and Lozano-Vilano 1994). This trend continues (S. Contreras-Balderas, pers. comm.). Remaining drainages have been degraded as water tables and river discharges lowered, leading to salinization, arsenicism, and other forms of chemical pollution that result from concentrating salts and chemicals in less water. Salt water from the ocean has intruded up major river systems, causing range contraction of freshwater endemics and spread of marine generalists. The lower Rio Bravo del Norte (Rio Grande) drainage contained only 6 marine species in 1950; more recently 32 freshwater fishes native to the area have been replaced by 54 marine or salt-tolerant species (Contreras-Balderas and Lozano-Vilano 1994). Some marine species have penetrated as far as 400 km upstream. Their incursions have been stopped by municipal and industrial pollution coming downstream from growing cities and by the lower salinity. Municipal growth fueled by internationalization of industries (as a result of the North American Free Trade Agreement, NAFTA) is taxing the fragile and ancient water resources of northern Mexico through both competition for and contamination of water.

West-central Mexico is a region of large lakes and rivers and also of high endemism; approximately 70% of its 100 fish species occur nowhere else. Some of this diversity is due to explosive speciation of atherinids in area lakes. Although wetter than the arid north, west-central Mexico has a dry season that runs from October through May, making the effects of impoundments, agricultural diversions, and industrial and municipal pollution particularly severe (Lyons et al. 1998). More than half of localities sampled before 1985 became completely dewatered or were so polluted as to no longer support fishes in the late 1990s. Of 44 native fish species in the Lerma River basin—which includes Mexico's largest lake, Lake Chapala—3 were extinct and 23 were experiencing significant range and population reductions, while 11 exotic species had become established (Lyons et al. 1998).

Mexican law protects listed and unlisted species, their habitats and ecosystems. Penalties include six months to six years imprisonment and fines equivalent to 1,000 to 20,000 times the federal minimum wage (S. Contreras-Balderas, pers. comm.). Although the laws are strong, enforcement is spotty. Environmental protection also requires balancing the water needs of the fauna with the growing population and economy of Mexico. Growth management, if it were to occur, should take into account extreme low rather than average water availability, because such extremes in the desert or in seasonally arid lands can be fatal to aquatic and terrestrial organisms. "If water quality is adequate to support varied and higher life forms, it will generally support human populations as well" (Contreras-Balderas and Lozano-Vilano 1994, p. 384).

New Zealand

New Zealand has 35–40 native freshwater fishes; almost all (86%) are endemic. Several new species have been described in recent years, and molecular techniques are revealing distinct lineages where single species were previously identified. Some undescribed and newly described forms are known to be threatened (R. McDowall, pers. comm.). One species, the New Zealand grayling, *Prototroctes oxyrhynchus*, is known to have gone extinct (see figure 2.9). The Nature Conservation Council published a *Red Data Book of New Zealand* beginning in 1981 and has updated it periodically (e.g., Tisdall 1994). The 1994 edition recognized one High Priority fish, four Second Priority species, and five Third Priority fishes. All were freshwater, and nine of the ten species were in the family Galaxiidae (the tenth was an eleotrid).

New Zealand completely reworked its accounting of imperiled species in 2000. Rather than adopting the IUCN ranking system, it developed its own rather extensive and inclusive system. The New Zealand Biodiversity Strategy (NZBS) was implemented to address the country's obligations under the international Convention on Biological Diversity. NZBS took into account "the relatively small size of New Zealand, the period over which recent declines have occurred, and the large number of taxa with naturally restricted ranges and small population sizes" (Molloy et al. 2002, p. 9). Three general categories were recognized: Acutely Threatened, Chronically Threatened, and At Risk. The Acutely Threatened category contained three subcategories—Nationally Critical, Nationally Endangered, and Nationally Vulnerable—that are considered equivalent to the IUCN CR, EN, and VU ranks (Molloy et al. 2002; the classification system contains seven ranks and 11 qualifiers that can apply to any rank). Under this system, 26 freshwater and 84 marine fish species were assigned some rank (Hitchmough 2002; New Zealand Department of Conservation 2002a, 2002b). These lists are to be reassessed every three years.

In the Department of Conservation (2002) list, four freshwater galaxiids were placed in the three highest ranks. Expanding the list to include Chronically Threatened and imperiled but undescribed species adds four more galaxiids. Recovery plans exist for six listed species plus several other galaxiids (New Zealand Department of Conservation 2004). Although 51 marine fishes had some at-risk designation, none ranked above the fifth-highest category of risk (gradual decline: great white and basking sharks).

New Zealand has adopted an obviously conservative approach to recognizing species at risk and is one of the few countries that listed fewer freshwater species than IUCN (2004) recognized. The actual species in the official New Zealand list and the IUCN (2004) list also differed markedly. Few New Zealand listed species of either freshwater or marine fishes appeared in high IUCN ranks, and vice versa (table 3.2).

Disagreement between national and international authorities here may result in part from the use of different criteria. New Zealand has adopted its own qualitative assessment protocols that are sensitive to special local conditions (Molloy et al. 2002), whereas IUCN applies the quantitative thresholds of G. M. Mace and Hudson (1999). IUCN may also be slow in bringing its list up to date with recent findings and reassessments in New Zealand, which has a long history of biological research and active programs for monitoring its freshwater fauna. Regardless, almost 50% (19/40) of New Zealand's freshwater fishes had some imperiled status in Hitchmough (2002), making this fauna among the most endangered in the world. Such an accounting reflects a high level of both habitat disturbance and knowledge of the fauna and its status.

Although marine species were underemphasized in its threatened taxon list (as is the case almost everywhere), New Zealand has shown leadership in marine reserve creation. It has one of the most active MPA programs in the world. As of 2004, 19 areas had been designated protected, with 15 more proposed and under consideration (www.biodiversity.govt.nz/seas/biodiversity/protected). The government aims to protect 10% of the marine environment by 2010.

As an isolated island with high endemism and low numbers of native browsers and predators, New Zealand has seen its flora and fauna decimated by introduced species (see chapter 8). In a survey of more than 4,200 stream sites throughout the country, Minns (1990) found that native fishes were frequently absent where nonindigenous fishes occurred. The impact of aliens is a direct result of predation on natives by introduced species such as trout but is also a byproduct of trout feeding on the benthic invertebrates that native fishes consume (McDowall 2003). In light of the aggressive stocking of 20 nonnative species, especially salmonids (McDowall 1990b), such displacement is a matter of serious concern. With a half dozen (and perhaps a dozen) endemic fish species at serious risk of extinction, and as much as 50% of the native fish fauna at some risk, strong protective legislation seems critical to ensure the continued existence of what remains. The Conservation Act passed in 1987 expressed concern for "the preservation of indigenous freshwater fisheries (as far as is practicable)" but provided no explicit protection of native freshwater fish (R. M. McDowall, pers. comm.).

Russian Federation and Former Soviet Union

The ichthyofauna of Russia is well known, especially given the upsurge in taxonomic work in the last few decades. The fauna consists of 340 described species and another 8 or 10 that remain to be described (Bogutskaya and Naseka 2002a, 2002b). Endemism is high for a continental fauna, amounting to about 20% (67 species). It is highest in the Baikal region, and regional diversity is highest in the Amur drainage of southeastern Russia, where about one-third of all species occur (Bogutskaya and Naseka 2002a, 2002b). About one dozen introduced species are established, and another 30 species (especially whitefishes; salmons; pikeperch; and the common or carp bream, *Abramis brama*) have been transferred within the country to new locales for fishery purposes.

Information on fish conservation issues in Russia and the former Soviet Union (FSU) is improving after a long period of fragmentary findings and incomplete data. Conservation efforts in the FSU traditionally focused on maximization of commercial production. Fish assemblages were "considered to consist of valuable species and trash (coarse) species, only the former being worthy of protection" (Bogutskaya and Naseka 2002a). This situation gradually improved, although the first edition of *The Red Book of the USSR*, which appeared around 1977, contained no fishes (Pavlov et al. 1985). The second edition, ca. 1985, listed ten species in four categories (I–IV), the first two of which corresponded roughly to CR, EN, and VU.

The 2001 Red Data Book of the Russian Federation listed 43 fish species (51 when all subspecies and local populations were counted) (Bogutskaya and Naseka 2002a). Five categories were recognized: probable extirpations (Baltic sturgeon, *Acipenser sturio*, a species with Critically Endangered status internationally); under threat of extinction (16); declining (18); rare (8); and status undetermined (4). Imperilment is most common among lampreys, sturgeons, herrings, many migratory species of cyprinids, and several endemic salmonids.

Correspondence between the Russian list and IUCN (2004) was minimal (table 3.2). IUCN listed 40 Russian fishes in high at-risk ranks, but 27 were sturgeon subspecies or stocks, and another 4 species were marine, leaving only 3 bony, freshwater fishes as seriously in trouble. This low number contrasts starkly with the well-known degraded condition of freshwaters throughout much of the FSU. Four sturgeon species listed by IUCN did not appear on the Russian list.

Among lampreys, of which Russia has the highest diversity of any nation, 3 species in the *Red Book* were unrecognized by IUCN, whereas IUCN listed 2 European lampreys (as near threatened) that did not appear on the Russian list. IUCN recognized as Endangered only 1 of the 9 freshwater fishes considered highly threatened by Russian authorities, and 6 others were not even listed. Russia gave high rank to only 1 of the 3 so listed by IUCN.

It is once again difficult to decipher why so little agreement exists between national and international authorities. It appears that IUCN needs to update its assessment of the conservation status of fishes in the Russian Federation, and Russian experts should participate actively in the revision of the IUCN document.

Environmental conditions in Russia and the FSU in general are widely regarded as deplorable, with aquatic ecosystems bearing much of the burden (Bogutskaya and Naseka 2002a). Degradation of aquatic habitats in the FSU results primarily from former and current chemical and nutrient pollution, impoundment, introductions, salinization, dewatering, and overfishing and poaching. "There is hardly a big lake or river basin where the fish fauna has not become at least locally impoverished" (Bogutskaya and Naseka 2002a). In addition, and despite the efforts of dedicated biologists, environmental laws and protection for endangered species are weak. The best available data on the condition of the fish fauna comes from the Russian Federation itself. Red data lists from 49 geographic entities existed in 1997, with minimal standardization of categories and criteria for ranking (N. Bogutskaya, pers. comm.). Given the extent of the problems and the level of uncertainty in many FSU regions, estimates of the number of species at risk have to be viewed with caution. They are if anything underestimates of the gravity of the situation, again calling the conservative nature of the IUCN lists into question. More and better information from this region is greatly needed, and publication of authoritative red data books from each geopolitical locale will be welcomed.

South Africa

South Africa's freshwater ichthyofauna is an interesting mixture of widespread tropical (Zambezian) species in the northern portion of the country, and endemic, temperate species in the southern portion. Although two-thirds of the nation's 94 freshwater species are tropical, only 13 (21%) of those are endemic, whereas all 33 temperate species are endemic. Barbine cyprinids, most of them small (<15 cm), dominate the temperate fauna, accounting for 80% of the southern fishes. Many are very restricted in their distributions. Fifteen of the 33 temperate species are microendemics, occurring in only one drainage system, and another 9 are found in only two river drainages. Imperilment appears to correlate strongly with endemism, and temperate species show higher levels of both. Small body size, localized distribution, and limited geographic range characterize many imperiled South African fishes (Skelton 1990), as well as imperiled freshwater fishes worldwide (chapter 4).

Major compilations of threatened South African species began in the late 1970s, with periodic updates (e.g., Skelton 1977, 1987). IUCN coverage also dates back as far as 1976. Agreement between IUCN (2004) and a recent compilation by South African experts (Skelton 2002) is almost total, and this agreement reflects close coordination between local authorities and international agencies (P. Skelton, pers. comm.; table 3.2). The only difference of note is that the Critically Endangered Maloti minnow, *Pseudobarbus quathlambae*, once thought to be extinguished but recently rediscovered, is missing from the IUCN compilation. The number of South African freshwater fishes considered at risk by IUCN has climbed steadily, from 21 in 1987 to 42 in 2004. Although the fauna has received more scrutiny in recent years, Skelton (2002, p. 223) contended that "deterioration in the conservation status of freshwater fishes [had occurred] since 1987."

No official South African accounting of marine and estuarine fishes has been published, aside from the material available from IUCN, although plans for such an assessment exist (P. Skelton, pers. comm.). IUCN (2004) listed 18 elasmobranch species and 11 other marine and estuarine species from South Africa as imperiled.

The two major, direct causes of decline among freshwater fishes are habitat destruction and introduced species (Skelton 1990). Stream channelization, damming, and water diversion and extraction for agriculture are major contributors, particularly in more arid regions. Few of the dams took fish passage into consideration. Diversion projects also led to interbasin water transfer, accompanied by translocations of species. A predatory clariid, the sharptooth catfish, *Clarias gariepinus*, was thus translocated from the Orange River to the Great Fish River, threatening a unique South African anabantid, the endangered Eastern Province rocky, *Sandelia bainsii* (Skelton 1990).

Eighteen foreign species are established in South Africa (Skelton et al. 1995). Introductions of foreign predators such as largemouth and smallmouth bass and rainbow and brown trout constitute a major threat to more than half of the imperiled freshwater fishes (e.g., de Moor and Bruton 1988). Because few endemic fishes are predatory specialists, the fauna evolved largely in the absence of piscivorous fishes. Lack of coevolution with predators, in combination with the small average body size of most endemic fishes, makes them especially vulnerable to introduced predators. In fact, "stream fishes of the Cape are notoriously naive in the presence of predatory species" and have been rapidly eliminated following introductions of piscivores (Skelton 2002, p. 229). The situation has been exacerbated by conflict between the sport fishing and conservation communities, the outcomes affected by policy deficiencies and what Skelton (2002, p. 232) referred to as "a low law-enforcement capacity."

Among other impacts, soil erosion and sedimentation,

salinization due to water withdrawal (some resulting from abundant, water-demanding, invasive plant species), and pollution from agriculture and mining have all been identified as contributors to declines among native fishes (Skelton 2002).

No fish species from South Africa is known to have gone extinct. One estuarine species, the river pipefish, *Syngnathus watermeyeri*, appeared on the 1993 IUCN list as extinct since 1968, but a previously unknown population was discovered in 1996; it was listed as CR in the 2004 IUCN catalog. Extirpations are a major concern, however, because they portend extinction. Known extirpations of populations of threatened fishes include *Pseudobarbus burgi* and *P. quathlambae* (Skelton 2002).

Prior to 2004, South Africa did not have national legislation that protected nongame fish species. Any protection came from provincial regulations, which were only sporadically enforced. Penalties were inadequate to deter violators, resulting in "gaps, overlaps, inconsistencies and loopholes, all of which were exploited by unscrupulous wildlife traders" (TRAFFIC East/southern Africa, www.ewt.org.za). The situation should change following passage in May 2004 of the National Environmental Management: Biodiversity Act 2004 (NEMBA), administered by the Department of Environmental Affairs and Tourism (DEAT). NEMBA provides a framework for enacting regulations that will identify and protect rare and threatened species, including species protected under international agreements (e.g., CITES, IUCN). It contains provisions for the protection of at-risk ecosystems and bioregions rather than focusing solely on species. Ecosystems and species will be ranked as Critically Endangered, Endangered, Vulnerable, and Protected using subjective (i.e., not IUCN's quantitative) criteria. Penalties for violations include jail terms of up to five years and "appropriate" fines that may run as high as three times the commercial value of the violated species. A list of fishes to be protected under NEMBA has been drawn up by local expert groups, informed by the IUCN red lists. National lists were under review as of February 2005 and will be available at the DEAT Web site (www.deat.gov.za).

A unique component of NEMBA is its chapter 5, which provides for control of the introduction and spread of alien and invasive species, including eradication of alien species that are already established. It holds liable anyone responsible for introducing an invasive species, with fines to cover costs incurred in control and eradication. Sport fishing interests in South Africa have reacted with some alarm to passage of NEMBA. A Web site maintained by a bass fishing organization instructed its members in November 2004 to "mobilise . . . to ensure that a catastrophe (from a Bass fishing point of view) is not forced on us" (www.bassfishing.co.za/bassingnews). The devil is always in the details, and implementation and funding had not yet occurred as of February 2005, but the world will take definite notice of the effectiveness of such groundbreaking, inclusive, species protection legislation.

South Africa has an active scientific community monitoring the status of its threatened fauna (e.g., Skelton 2002; see the series "Threatened Fishes of the World" in the journal *Environmental Biology of Fishes*). The country is establishing protected reserves that house many species with conservation status and into which threatened fishes have been transferred (Skelton et al. 1995; Skelton 2002), making it a leader in freshwater protected area development. South Africa was also the first country to ban the capture of white sharks, a species that now has VU status in the IUCN Red List and is included in Appendix II of CITES.

South America

Assessments of threatened fishes in South America vary greatly among countries. Red books and lists covering birds and mammals are more common than treatments of fishes. According to the South American Regional Office of UICN-South (2003), red lists for fishes specifically or included among general vertebrate compilations exist for Argentina (1992), Bolivia (1996), Brazil (2002, updated to 2004), Chile (1993, updated to 1998), Colombia (2002), Peru (1991), and Venezuela (1995). I have selected three accessible (i.e., Web-available) lists to represent conditions on the continent, with supplementation from a published report from Chile.

Brazil has been developing red lists since 1989. The 2004 revised list (www.biodiversitas.org.br), gathered under the auspices of Fundação Biodiversitas, included 13 elasmobranch and 156 bony fishes (148 freshwater, 8 marine) in 30 families (table 3.2). Of 31 CR bony species, 27 occurred in a single state, again showing the influence of limited distribution on risk. Agreement between Brazil's official list and IUCN (2004) was spotty. Among elasmobranchs, 5 of 13 Brazil-listed species did not appear on the IUCN rolls, and 11 of the 20 IUCN-listed species were not recognized by Brazilian authorities. The discordance in bony fishes was even greater. IUCN listed only 7 freshwater species, all aplocheilids (sabrefins, killifishes), ignoring 148 species in 15 other families. Concordance among marine species was only slightly better.

The Venezuela Red List of imperiled fishes was compiled in 1995 (Rodriguez and Rojas-Suarez 1996). It contained 3 catfishes, a herring, a seabass, and a cichlid. IUCN (2004) listed 19 Venezuelan fishes, all marine or estuarine. The only species common to both lists were the Venezuelan grouper, *Mycteroperca cidi*, and the Venezuelan herring, *Jenkinsia parvula*, designated VU in both. The IUCN-listed fishes are widely distributed Caribbean reef species, with the exception of a grunt, a toadfish, and a chaenopsid blenny that are restricted to the southern Caribbean.

Chile houses 46 native freshwater fishes and at least 15 introduced species (Campos et al. 1998). Red lists compiled by the government environmental agency CONAF (Corporación Nacional Forestal) appeared in 1987 and 1993, with an update by academic researchers in 1997 (Campos et al. 1998). CONAF (1993) identified 18 species as Endangered and 23 as Vulnerable. One other fish was Rare and 2 were Data Deficient (*Insuficientemente Conocida*), leaving only 2 species as Outside of Danger (*Fuera de Peligro*). The 1998 reanalysis by Campos et al. was much more thorough and conservative, presenting a species-by-species account of distribution, status in each of the nation's 13 political regions, and justification for ranks, along with probable causes of imperilment. Criteria for rank assignment included 4 characteristics of distribution, 12 parameters of population condition, and 11 parameters of habitat condition.

Campos et al. (1998) assessed 47 species, identifying 30–33 as at risk, 10 as Data Deficient, and 2–3 as Outside of Danger (variability arose from species being assigned different status in different regions). Major identified threats were introduced species (especially trout), alteration of habitat via pollution, changes in hydrology (principally decreased water volume), and alterations in riparian and aquatic vegetation. The most threatened families were cyprinodontid killifishes and trichomycterid catfishes that occurred in the arid northern portion near the border with Bolivia. Water extraction for agriculture and mining appeared to be the major impact, with vegetation loss also affecting cyprinodontids. Riverine species were more strongly affected than lake species, and narrow ranges led to a higher probability of imperilment. Among the 17 Endangered species, 9 occurred in only 1 of the country's 13 regions.

Only 3 of the 31 freshwater fishes in the Chilean list were recognized by IUCN (2004) as seriously at risk. These were 1 diplomystid catfish (EN), 1 trichomycterid catfish (Chile-EN, IUCN-VU), and 1 cyprinodontid (Chile-EN, IUCN-VU). IUCN also listed a butterflyfish and bigeye tuna, but Campos et al. (1998) did not assess marine species.

As of March 2005, Chile did not have enforceable endangered species protection. However, enabling legislation (*La Ley 19.300 de Bases del Medio Ambiente*) was passed in 1993, setting up a legal framework to be implemented by mid-2005. Ley 19.300 calls for the identification of threatened fauna and flora. Currently, the only binding legislation is the Hunting Law, which controls recreational, scientific, and commercial take of some species. Campos et al. (1998) represents the de facto list of the conservation status of Chilean fishes. Prior to 2005, enforcement was largely voluntary, except for the Hunting Law and some protection granted through environmental impact assessments. Implementation of Ley 19.300 will subject listed species to appropriate management plans (C. B. Anderson and J. C. Torres-Mura, pers. comm.).

British Isles

The British Isles have a limited freshwater fish fauna of 41 native species. An additional 15 foreign species have been established. No fish species are endemic only to the British Isles, and fishes identified as imperiled in Britain are apparently globally secure. However, several species (e.g., river and brook lampreys, twaite shad, powan, pollan, vendace) can be anatomically unusual, are restricted to a diminishing number of isolated and probably genetically distinct populations, and may be of conservation value because of genetic uniqueness (Maitland and Lyle 1991). Unique stocks of brown trout and three-spine stickleback also occur. Maitland and Lyle (1991, 1996) summarized the conservation status of these fishes in the UK; Sweetman et al. (1996) and Quigley and Flannery (1996) focused on imperiled fishes of Scotland and Ireland. Analysis of levels of threat focused on range restriction, habitat decline, and population decline. Seven species were listed at high risk and nine at lower risk.

Official government status for endangered species in

Britain is provided by Schedule 5 of the Wildlife and Countryside Act (WCA) of 1981 (revised in 1998). WCA prohibits the killing, injuring, taking, or selling of listed animals and protects refuge and breeding habitat. It also prohibits the importation of harmful exotic species under Schedules 9 and 14. Fish species blacklisted were North American largemouth bass, rock bass, and pumpkinseed sunfish, and European bitterling, wels, and zander. The agency in England responsible for administering WCA is English Nature. The Joint Nature Conservation Committee (JNCC) advises UK agencies (English Nature, Scottish Natural Heritage, and the Countryside Council for Wales) on matters concerning fish and wildlife. JNCC reviews the status of WCA-listed species on a five-year basis. A review was scheduled for completion in 2005. Nine fish species had protection under WCA as of 2002: basking shark, Baltic sturgeon, allis and twait shad, two coregonine whitefishes (vendace and powan), burbot, and Couch's and giant gobies. JNCC in 2002 recommended adding two seahorses (shortsnouted and spiny), four skates (common, black, longnose, and white), and angel shark to the list. IUCN (2004) listed only the Baltic sturgeon, basking shark, common skate, and angel shark—and seven others not in the WCA list—in its highly threatened ranks. All shads and whitefishes were classified Data Deficient and other listed species had no IUCN rank. The Data Deficient designations are puzzling; extensive research has been ongoing, and detailed Species Action Plans have existed for these fishes for several years in the UK. Most discrepancies between lists probably relate to the locally-unique-and-depleted-versus-globally-secure issue alluded to earlier, although the Data Deficient designations remain anomalous.

Fishes are also protected under the Salmon and Freshwater Fisheries Act 1975, a fisheries regulatory document that underwent substantive review in 2000 and February 2005. The 2000 review was 236 pages long and recommended 195 policy and legislative changes, which will take considerable time to accomplish. Significantly, while focusing on fishery species and activities, the review committee adopted an ecosystem perspective for the sustainable management of UK freshwaters. The committee concluded that a principle management objective of government should be "to ensure the conservation and maintain the diversity of freshwater fish, salmon, sea trout and eels and to conserve their aquatic environment" (Department of Environment, Food and Rural Affairs 2005).

Major threats to fishes in the British Isles are loss of habitat, overexploitation, pollution, and introduced species. Migratory species were extirpated from portions of their range due to dams. These problems have existed for centuries, suggesting that the relatively secure condition of most extant fishes masks a history during which sensitive and unique unrecorded stocks likely vanished. However, species have returned to previously degraded areas such as the River Thames as a result of conservation and restoration efforts. Such improvements are a tribute to these efforts and show how resilient some fishes can be (e.g., Colclough et al. 2002; see also Thames Landscape Strategy, www.thames-landscape-strategy.org.uk).

The Nature Conservancy Council assessed the conservation status of 322 marine species that occurred out to the UK continental shelf (Swaby and Potts 1990). Vulnerability ranks were assigned based on then prevalent IUCN terminology, given information on population size and distribution, with caveats about incomplete data and reliance on the qualitative judgment of experts. Of 165 species identified as uncommon, 110 were considered Indeterminate due to a lack of adequate data on abundance. An additional 39 species occurred in normally small populations. Nine species remained in the Uncommon and Scarce categories, reflecting rarity and perhaps declining numbers. Analysis of habitat extent (of restricted or localized populations) indicated that 4 species occurred in single populations. From these lists, 8 species were considered to be threatened or nearly so. Four of these, all brackish water inhabitants (sturgeon, allis shad, twaite shad, and Couch's goby) were granted protection under Schedule 5 of WCA; the only truly marine protected species was the basking shark.

Marine fish receive protection under other general and specific programs, including the Sea Fisheries (Wildlife Conservation) Act 1992 (c. 36). English Nature promotes recovery of depleted populations through its Species Recovery Programme. Species are designated under Species Action Plans and Biodiversity Action Plans, the latter an outgrowth of the UK's participation in the international Rio Convention on Biological Diversity (1992). As of February 2005, Species Action Plans existed for basking shark, twaite and allis shad, vendace, and burbot. Two Biodiversity Action Plans existed for marine fishes, one for 14 species of deep-water fish and another for 9 species of commercial fish. Marine species also receive habitat protection under WCA in the form of designated Sites of

Special Scientific Interest, Marine Nature Reserves, Special Areas of Conservation, and Special Protection Areas.

The Department for Environment, Food and Rural Affairs (DEFRA), through its Wildlife and Countryside program, launched a Biodiversity Strategy for England in 2002. This program focuses on ecosystem management and natural processes as a means of preserving biodiversity and using it sustainably. A goal is for government decisions and individual actions to incorporate sensitivity to biodiversity issues. Species were identified whose status could be used as indicators of the state of biodiversity in the country. Four fish species were designated as such: allis shad, twaite shad, vendace, and burbot. These are the same species protected under Schedule 5 of the Wildlife and Countryside Act.

One note in passing. Although most countries provide legislation on government Web sites, access to the two main UK species protection acts requires purchase of the texts (available from the Stationery Office Limited, https://www.tso.co.uk). Fortunately, nongovernmental organizations have taken the trouble to reproduce the acts and lists on their Web sites (e.g., www.naturenet.net).

United States

A commonly cited number for freshwater fish diversity in the continental U.S. is 790 species (Page and Burr 1991); inshore marine fish diversity to 200-m depth is about 1,600 species (Robins et al. 1991). Climate and habitat types in the continental U.S. (lower 48 states) vary from tropical south Florida to cool, temperate North Dakota and everything between, including temperate rain forests and desert hot springs. The southeastern U.S. has the highest diversity of freshwater fishes and other taxa of any temperate freshwater region in the world (e.g., Lydeard and Mayden 1995; Burkhead et al. 1997). Regional endemism is greatest in the desert Southwest, where isolated springs and their drainages often contain locally restricted species.

The status of fishes in the U.S. is reviewed and updated regularly as a result of activity by numerous fisheries and ichthyological professionals concerned with conservation issues. Several compilations have appeared in reviews of the situation in the American Fisheries Society (AFS) journal *Fisheries*, with periodic updates in other publications (e.g., Deacon et al. 1979; Ono et al. 1983; J. A. Johnson 1987; R. R. Miller et al. 1989; J. E. Williams et al. 1989; Warren and Burr 1994; Warren et al. 1997). Papers with a more regional focus abound, as do overviews of introduced species and other conservation-relevant matters (e.g., Fuller et al. 1999; Warren et al. 2000; J. S. Nelson et al. 2004). Endangered fishes in the U.S. are monitored closely by state natural resource agencies, often in conjunction with natural heritage programs (see www.natureserve.org). Additional accounting and monitoring are provided by numerous nongovernmental agencies with primary interests focused on sport fishing, aquarium keeping, and aquatic ecosystem health (Isaac Walton League, Trout Unlimited, North American Native Fish Association, Native Fish Society, Desert Fishes Council, American Rivers, Sierra Club, The Nature Conservancy, NatureServe, Association for Biodiversity Information, and the Canada-based Aquatic Conservation Network, to name a small subset).

Major threats to U.S. freshwater fishes are similar to those in other industrialized nations, although the ranks of the factors are shuffled. Direct exploitation is a minor threat compared to habitat degradation and species introductions. Indirect effects of fishing have had greater impact than direct removal (although direct impacts of sport fishing may be greatly underestimated; see Coleman et al. 2004, Cooke and Cowx 2004). Significant indirect effects include: bait and predator introductions; habitat alteration via damming and reservoir building to accommodate sport fishes; and "reclamation" of water bodies, whereby native nongame fishes are poisoned, followed by introduction of sport fishes.

Regionally, water withdrawal in the Southwest and water quality in the Northeast are paramount. Southeastern streams and midwestern rivers are more affected by impoundments that block migrations, fragment ranges, and alter hydrologic regimes, and by sediment inputs due to development, mining, logging, and agriculture. Pacific Northwest salmon runs, of which more than 800 distinct stocks are considered threatened, are degraded by all of the above (Allendorf et al. 1997). Levels of species imperilment are highest in the relatively wet, speciose Southeast and the arid, endemic-rich Southwest (Warren and Burr 1994; Warren et al. 1997; Warren et al. 2000).

Introductions of alien species are common and ecologically significant, especially in areas that infrequently experience freezing temperatures. South Florida, southern Texas, Southern California, and Hawaii lead the list, although no state including Alaska is free from introductions (Fuller et al. 1999). Counting foreign and domestic species, hybrids, and unknowns, 536 unique fish taxa have been introduced into the U.S., 52 of which have become established (J. S. Nelson et al. 2004). Most (316) introductions are native transplants that have been moved from one locale to another, the number indicating that nearly half of the country's freshwater fishes have been transplanted outside of their native range. Major sources of introductions include sport fish stocking, escapes from the aquarium trade, and bait and ballast water (Fuller et al. 1999; see chapters 8, 13).

The basis for species protection in the U.S. is the ESA of 1973, as subsequently amended (see Rieser et al. 2005), which is considered by many to be the most complete such legislation enacted anywhere, and which is continually under challenge and threat by special interests affected by its enforcement. Predecessors of the ESA—the Lacey Act of 1900, the Black Bass Act of 1926, the Endangered Species Preservation Act of 1966, and the Endangered Species Conservation Act of 1969—provided some protection, but none were as far reaching (NRC 1995). Enforcement of regulations and updating of freshwater lists are the responsibility of USFWS, which maintains a relevant Web site at http://ecos.fws.gov/tess_public/TESSWebpage. For most anadromous and all marine fishes, responsibility lies with the NMFS-NOAA Fisheries (www.nmfs.noaa.gov/pr/species/fish/).

As of March 1, 2005, USFWS listed 124 fish species in the U.S. (table 3.2). None were truly marine, although 6 salmon and 5 sturgeon species are anadromous, and the tidewater goby, *Eucyclobius newberryi*, and delta smelt, *Hypomesus transpacificus*, are estuarine. An additional 14 species—5 darters, 4 minnows, 2 salmonids, a sucker, a catfish, and a sculpin—languished as Proposed and Candidate Species under consideration for ESA protection. The National Marine Fisheries Service (NMFS) recognized another 16 marine species that fell just short of ESA criteria, placing them in the category Species of Concern.

In contrast, and again showing the broader net cast by nongovernmental agencies and experts, AFS (Musick et al. 2001) identified 22 U.S. marine fishes as at risk of global extinction among 82 marine species and subspecies that deserved Endangered, Threatened, or Vulnerable status. This number far exceeds the totals recognized by the National Oceanic and Atmospheric Administration (NOAA; 16) and IUCN (41).

Of the 124 ESA-protected freshwater fish species, 95 had recovery plans in effect. As of January 2005, presumably as a result of recovery efforts, 5 species (rather ESUs)—the snail darter, *Percina tanasi*, and 4 trout species (Apache trout, *Oncorhynchus apache*; and greenback, Lahontan, and Paiute cutthroat trout, *O. clarki stomias*, *O. c. henshawi*, and *O. c. seleniris*—experienced improved status since their original ESA listings and saw their official designations downgraded from Endangered to Threatened. At the same time, the prognosis for 4 fishes (Alabama cavefish, *Speoplatyrhinus poulsoni*; and three Chinook, *O. tshawytscha* ESUs: fall Snake River, spring/summer Snake River, and winter Sacramento River) deteriorated; their status was upgraded from Threatened to Endangered (http://ecos.fws.gov/tess_public and www.stateofthesalmon.org).

Against this backdrop of protected species are the ones that got away. Although U.S. government agencies do not maintain official lists of extinct species, other groups do keep track. A semiofficial list compiled through AFS included 31 species, subspecies, and undescribed species, half of which were from the arid West (R. R. Miller et al. 1989; see appendix). IUCN (2004) listed only 17 U.S. extinctions. The discrepancy may be due to AFS listing 13 subspecies or unnamed species, whereas IUCN did not include such taxa. Also, IUCN removed the Miller Lake lamprey from its list because the fish was rediscovered in 1992 (chapter 2). IUCN but not AFS included the Maryland darter, *Etheostoma sellare*. This darter was restricted to a single creek in the lower Susquehanna River drainage of Maryland's upper coastal plain and has not been seen since 1989. It is listed as Endangered under the ESA (Fleischman 1996).

Neither list reflected losses among distinctive stocks and populations (ESUs) of salmonids in the West and Northwest. Conservative estimates put the number at 280 salmonid ESUs extirpated. Determining causes of extinctions is a challenge, because most species decline in the face of multiple threats. A single cause can be concluded for only 20% of the instances. Contributing factors in the remaining 80% were, in order of importance, habitat degradation, introduced species, chemical pollution, hybridization, and overharvest (R. R. Miller et al. 1989).

Interestingly, the ESA (Title 50, part 17, subpart B) listed an additional 11 Endangered and 1 Threatened species that did not occur in the U.S. but were native to Mexico, Turkey, Japan, and Southeast Asia. These included the Asian bonytongue, Mekong giant catfish, Mexican blind catfish, and totoaba, all species with IUCN or CITES protection. Their inclusion under the ESA is largely the legacy of pre-ESA legislative action, from a time when protective legislation carried the force of law but lacked enforcement provisions. Protection was needed to discourage international trade in obviously endangered fishes, and eliminating legal markets in the U.S. would remove much of the incentive for commercial exploitation. These efforts were in part apparently successful, as shown by a substantial reduction in fishing pressure on totoaba after it was listed. Because these fishes were still globally threatened, little justification existed for removing them from protection after the ESA passed in 1973, and they were retained on the USFWS list; others, such as beluga sturgeon, have been added (J. Williams, pers. comm.).

Once again, a comparison of regional and international compilations reveals a lack of concordance, which is all the more surprising given the involvement of American authorities in the international arena. IUCN (2004) listed 165 imperiled fish species in the U.S., including 27 marine bony and 14 elasmobranch species. These numbers represent a considerable increase (of more than *40* species) since the IUCN (1996) assessment, especially among marine fishes. IUCN (2004), therefore, recognized about 35 more species than were protected by the ESA, departing from the usual pattern of greater inclusiveness at the national level.

Overlap between the IUCN (2004) and USFWS (2004) lists was only moderate (table 3.2). Ten of the 16 marine fishes in USFWS were recognized by IUCN, including all but 1 of the 6 elasmobranchs, but only a slight majority (56%) of USFWS Threatened and Endangered freshwater fishes appeared on the IUCN list. Among the omissions were 29 of the 81 freshwater species with Endangered status. Admittedly, half of the 29 were ESUs, DPSs, or subspecies, taxonomic ranks that IUCN usually does not consider, at least for fishes other than elasmobranchs and sturgeons. Also puzzling is assignment of different risk levels for the same species. For example, 17 species with Endangered status in the U.S. were listed only as VU by IUCN. This group included the Devils Hole pupfish, *Cyprinodon diabolis*, a closely monitored species that has long been recognized as exceedingly endangered (figure 3.3). Conversely, 51 species ranked by IUCN were omitted from the USFWS listings. Six of these were considered CR (2) or EN (4), a troubling discrepancy between the two programs.

Disagreement was noted previously by Warren et al. (1997) between the 1994 USFWS list and the list created by AFS (J. E. Williams et al. 1989). The AFS list is generally

Figure 3.3. Efforts to save endangered species in the U.S. (A) Devils Hole, Nevada, natural home of the Devils Hole pupfish, the first fish listed under the ESA (see figure 2.10). Visible are water-level monitoring equipment and a platform for people to walk on while counting fish. Photo by J. Barkstedt; used with permission. (B) A breeding tank below Hoover Dam, Nevada, where a refuge population of the pupfish was established in 1972. Photo by J. Heinrich, Nevada Department of Wildlife; used with permission.

considered the compilation that best represents the judgment of ichthyological and conservation professionals in the U.S.; it is relatively free of the political considerations that influence ESA action (e.g., Warren and Burr 1994). Warren et al. (1997) compared the lists for southeastern U.S. fishes and found that about 25% of the species on either list were unrecognized by the other. Another important difference was the level of protection proposed for species, with the USFWS providing lower protection status for many species, especially among Threatened taxa. AFS proposed Threatened rank for almost three times as many species (36 vs. 13). Such disagreement reinforces the observation that, even in industrialized nations with active conservation communities and abundant professionals, coordination within and between programs should be strengthened.

It is regrettable that these three lists (AFS, USFWS, IUCN)—with their high credibility and extensive background research—fail to show greater agreement. Every list assignment likely has a complex story, but the story is undoubtedly more convoluted for species protected by ESA. Lists created by nongovernmental organizations are more inclusive, not because they are more concerned with local than global patterns, but because they are more precautionary. The greater numbers and higher rankings assigned in regional and nongovernmental lists reflect reduced politicization in the decision-making process. Warren and Burr (1994) specifically decried the backlog that occurred in the USFWS list for southeastern states, with perhaps 14 fish species recognized by fisheries professionals as imperiled that were not protected under federal categories.

The take-home messages from these comparisons are that any species receiving legal protection from a governmental agency is undoubtedly imperiled, and that many more species need such protection but have yet to clear the bureaucratic hurdles. We can only hope that the lag time between professional and official recognition of such "candidate" species will diminish.

CONCLUSION

Endangered Species Laws: Good Intentions, Variable Results

Around the beginning of the 21st century, the list of nations with laws protecting species and habitats grew substantially. As noted, Canada, Australia, and South Africa initiated, expanded, or strengthened protection for endangered species and their habitats, and Chile was in the process. Awareness of the problems facing marine species also grew, as evidenced by the OSPAR Convention and EPBC in Australia. Such activity reflects a growing realization of the need to protect biodiversity and is cause for optimism.

However, legislation is no stronger than the ability and willingness of agencies to implement the regulations. The protection provided on paper is too frequently not realized in practice. Sufficient funds and personnel must be allocated for assessment and enforcement, rather than simply adding these duties to the responsibilities of overtaxed agencies within existing, already stretched budgets. Unfunded and underfunded programs are particularly characteristic of many developing nations.

In industrialized nations, the problem of insufficient resources is often secondary to a lack of willingness to exercise the authority provided by the legislation, both in listing species and in exercising the provisions of an act. Few prosecutions have occurred and fewer fines imposed for jeopardizing, damaging, or destroying endangered fish species and their habitats. However, it is not in police actions that endangered species legislation is most effective. Benefits to species are more likely to accrue via provisions for protecting habitats from private and governmental development schemes. Almost all endangered species regulations require that developers perform environmental impact assessments to determine whether sensitive species or habitats would be harmed by a proposed action. If negative impact is likely, plans must be developed to minimize or mitigate it; otherwise, permits are supposed to be denied. Unfortunately, agencies have been known to cave in rather than impede development. Variances have been granted and permits issued despite species vulnerability and obvious detrimental outcomes in numerous countries, but Australia, Canada, Hong Kong, and the U.S. stand out. These failings remain a source of frustration for persons working to protect biodiversity.

Praise for International Efforts . . .

Species protection has advanced immeasurably due to the efforts of international organizations such as IUCN and CITES, which provide an opportunity for experts to assess conditions and publicize problems. Their actions bring otherwise intransigent bureaucracies on board in a concerted and cooperative effort, all backed by good science. Much of the expertise is provided voluntarily by people with full-time-plus jobs doing other things, who contribute as part of various specialist groups, under the auspices of the IUCN Species Survival Commission. Specialist groups exist at present to assess the status of Caribbean fishes, coral reef fish, freshwater fish, grouper and wrasse, salmon, shark (sharks and rays), and sturgeon. Some groups are more active than others, and some fish taxa (e.g., lampreys, killifishes, seahorses, tunas) are obviously well represented in IUCN deliberations despite lacking officially constituted specialist groups.

. . . With the Following Caveats

The programs and practices reviewed in this chapter suggest areas where international organizations could accomplish more to preserve biodiversity.

Too Few Species Are Protected

CITES prohibits international trade of only nine species of fishes, those listed in its Appendix I. European Community Regulation 1497/2004 protects the same nine. For CITES to reach its potential as an international agreement that protects imperiled fishes and maximize its influence on corresponding legislation, Appendix I should be expanded. It should include many more species that are threatened by commercial activity and fall into categories such as Critically Endangered and Endangered and their national and international equivalent ranks. Appendix II species require permitting for export or import, a process dependent on the integrity of the permitting agency in each country. Moving more species from Appendix II to Appendix I would restrict trade in species that clearly should not be traded. Inclusion of a species in Appendix II is a positive first step in raising awareness about species that are vulnerable to trade, especially given the nondetriment proviso discussed earlier (Y. Sadovy, pers. comm.). Within-country restrictions on capture and trade, which are not covered by CITES, require national protective legislation followed by implementation and enforcement. Much has been accomplished in this area, but many more countries need to develop effective legislation.

Disagreement among Lists Is Counterproductive

The most striking pattern in this chapter's compilations, aside from growth in the number of countries working to protect endangered species, is the lack of agreement between national lists and those produced by international organizations. This was discussed by Leidy and Moyle (1997), who noted that IUCN (1996) recognized 5 threatened fish species in Slovenia, whereas a regional authority (Povz 1996) put the number at 31. IUCN (2003, p. 19) suggested that regional decisions, "may *under some circumstances* result in a taxon qualifying for listing at the global but not at the regional level" (emphasis mine). In reality, such disagreement is common.

IUCN and its listing protocol form the recognized benchmarks against which other attempts are measured, and IUCN's extremely useful database allows ease of access and comparison (www.redlist.org/search/search-basic.php). Several country lists lack even 50% accord with IUCN. In fact, only South Africa and to a lesser extent Mexico (for freshwater) come close to agreement with IUCN.

IUCN deserves high praise for its efforts and accomplishments in gathering expertise and data, synthesizing information, establishing standardized criteria for determining levels of imperilment, publicizing the plight of endangered species, and motivating countries around the world to assess their faunas and take action to protect species at risk. The entire "red list" phenomenon is a direct outgrowth of its efforts, and countries without red lists are increasingly in the minority.

However, room for improvement exists. At least, explanations need to be provided—via the IUCN Web site or as appendices to regional lists—for the sometimes extensive differences between lists. Unless we understand the reasons for such discrepancies, our efforts to protect species will be weakened because

- inclusion of a species on one list but not another can be used as an argument against its protection—for example, "IUCN does not recognize this as an endangered species; therefore, we cannot justify allocating resources for its protection"; similarly, "The international experts at IUCN don't agree with our experts, so maybe our experts are not to be believed"; and
- opponents of endangered species protection will be quick to point out that disagreement among international experts calls into question the credibility of all proponents of protection.

Some differences clearly arise due to unavoidable time lags. International programs meet periodically and draw volunteer personnel from far-flung locales; they work more slowly than local offices that deal with new information on a daily basis. IUCN updates the list on its Web site annually. Time lags are therefore more likely to affect the IUCN and similar rankings than locally produced lists. IUCN recognized the potential problems arising from time lags when it created version 3.1 of its Categories and Criteria in February 2000. The revised format mandated new assessments of taxa, and IUCN emphasized that "conversion of all existing assessments on the Red List to the revised system will take time." However, my previous comparison of IUCN (1996) lists with then current national lists showed extensive disagreement similar to that reported in this chapter, suggesting that the recent differences reflect more than simple delays in updating.

Some discrepancies between lists are understandable in light of local versus global perspectives. Regional lists may be used partly to prioritize conservation actions rather than solely to assess extinction risk, which will lead to different results (IUCN 2003). Residents of regions and nations also tend to focus on smaller geographic entities or on regional extirpations, whereas international expertise takes a broader view. International bodies also set more stringent requirements for recognizing imperilment. IUCN has clear guidelines for applying its red list criteria at national and regional levels and for the relationship between regionally derived rankings and those that appear in its red list: "In using these guidelines it is important to note that national or regional assessments are NOT eligible for inclusion on the IUCN Red List of Threatened Species, unless they are for endemics" (www.redlist.org/info/categories_criteria.html). Some inclusions on the current UK and Canadian lists—populations that exist at the periphery of their geographic range—may exemplify this local-global dichotomy.

Less easily dismissed are instances where disagreement existed and force of personality rather than scientific merit prevailed (we are, after all, working with human decision making). Along these lines, Jennings and Kaiser (1998, p. 238) observed that "the allocation of marine fishes to endangered categories has been rather arbitrary and may reflect the personal interests of those who compile the lists rather than the probability that the species will be lost." I have also heard of instances of heavy-handedness on the part of representatives of international agencies, suggesting that additions to some regional lists were based on less than objective criteria. Conflicts of this nature are probably small in number and precautionary in nature, in that they would tend to increase rather than decrease the number of species listed. Regardless, good science and common sense should prevail over arbitrary or authoritarian decisions.

The examples that cause the most alarm are the extremes, where either essentially complete or no agreement exists. Complete agreement could occur where consultation is mutual or where national experts make up both lists; otherwise, it is unlikely in the real world of science and scientists. It may indicate that countries have simply adopted IUCN designations rather than collecting and assessing information themselves, which IUCN's Regional Application Working Group endorsed (IUCN 2003). Lack of resources may proscribe collecting the data necessary to come up with a reasonable and accurate list. A national agency—understaffed and lacking important information about its fauna—might accept the IUCN list and then deal with the more immediate crises of keeping the lights on and the boat operational. By providing a document that can be taken to bureaucrats to initiate action, the international community serves such an agency well. In some instances, local knowledge may exceed external information, but outside authorities often carry more weight with policy formulators and bureaucrats. Local experts may therefore compromise in the name of accomplishing a positive end, but their efforts should influence any final compilations.

Large-scale disagreement also reflects lack of communication or an unwillingness to reach consensus, what Leidy and Moyle (1997, p. 208) referred to as a lack of complete information on the status of certain species, as well as "differences in the interpretation of status categories." It is especially disconcerting where countries have active ichthyological programs, such as in New Zealand and Russia. Disagreement then gives international lists the appearance of having been drawn up without input from local experts. Lack of agreement could also reflect national pride in developing one's own list regardless of outside expertise and opinions, or an unwillingness of local experts to participate in the process. Lack of interest is unlikely, given the dedication of workers in this field and the zeal with which they try to publicize the plight of their animals. The important thing is that conservationists should all be working toward a common goal, one that will be met faster via greater cooperation and maximal transparency in developing criteria and assigning ranks.

Technical and sociological solutions to the problem of discrepancies exist. Some apparent disagreement arises from the way the IUCN Web site delivers results; it does not engage in a country-by-country assessment per se. Enter "elasmobranch" and a country name, and the search function gives the intersection of the different information matrices—that is, any shark, skate, or ray whose range includes that country and which has undergone population depletions in a significant part of its range. The search function does not rely on actual data from the country in question. Perhaps some later version of the IUCN software will be able to note whether a species' status has been assessed for a particular country, whether a Species Specialist Group has been involved and what the results were. IUCN (2003) recognized the potential for differing conclusions and recommended that regional red lists include the IUCN rank, data, and rationale behind each listing; the rationale for assignment of a different rank; and an estimate of the proportion of the global population of the species within the region. Few regional lists include more than the IUCN rank.

IUCN (2003) also presented a protocol for mediating disagreement, but none of the sources I reviewed indicated that such an appeal process had been pursued. It appears that mutual respect and communication between people drawing up lists could be increased. In any event, a precautionary approach mandates that if either party has good evidence of imperilment of a species—due to immediate, continuing, or fluctuating population reductions or habitat loss—the default option should be inclusion.

Quantitative versus Qualitative Assessment: Valuing Local Knowledge

IUCN sets the bar high by requiring that its objective and quantitative (1994) criteria be met (see G. M. Mace and Hudson 1999). It is obvious why IUCN developed quantitative criteria and thresholds: They have a greater probability of being accurate than subjective impressions of decline. A requirement for strict adherence to IUCN guidelines will run into obvious conflict, however, where countries and agencies develop their own criteria for listing, as in the case of New Zealand or the Texel/Faial Criteria of OSPAR.

More important, requiring that specific quantitative criteria be met disregards the wealth of local and traditional ecological knowledge (TEK) available in many regions. Such subjective, often unpublished knowledge can be broad, deep, precise, and biologically valid while lacking quantification or statistical rigor. Quantitative criteria elevate data over knowledge, whereas both have merit. Frequently in the developing world—where most of our biodiversity resides—we have only knowledge to make decisions (e.g., Johannes 1981, 1998).

Sadovy and Cheung (2003) documented the near extinction of giant yellow croaker, a commercially targeted species for which no formal fishery data had been gathered. They used informal, historical, and traditional information to show a near 99% collapse in the fishery and rapid decline toward commercial and eventual biological extinction (see chapter 2). They concluded (pp. 86, 87) that

> careful use of informal, or traditional, information can provide a powerful, sometimes unique, means of identifying and assessing the status and history of species that might be quietly slipping away before we learn anything about them. . . . The judicious use of informal sources of information could, and perhaps should, be much more widely adopted in fishery assessment, especially in regions where resource monitoring is poor or lacking and there is little or no management.

Quantitative, objective, data-rich documentation of the actual status of a species is the ideal, but as with so many aspects of conservation biology, we often have to work under less than ideal conditions. The insights that can be gained from local knowledge about the history, status, and solutions of environmental problems can be immeasurable (Mackinson 2001; Bergmann et al. 2004; Silvano and Begossi 2005). Any assessment of the status of a species should seek out and incorporate informal as well as formal information. We cannot afford to ignore any important source of information in our efforts at protecting biodiversity.

4. Characteristics of Vulnerable Species and Correlates of Imperilment

Some species are extremely vulnerable because they have synergistic combinations of extinction-promoting traits.

—Michael L. McKinney, 1997

Numerous authors have attempted to enumerate and understand the biological traits and circumstances that characterize particularly vulnerable species. This literature has focused primarily on birds and mammals and concluded that extinction-prone species are typified by large bodies, predatory habits (high trophic position), small or slow-growing or highly fluctuating populations, poor dispersal or colonization ability, restricted geographic range (localized endemics), strong migratory habits, group reproduction, low fecundity, late maturation, long generation time, intolerance of human presence, and relative specialization with respect to food or habitat use (e.g., Ehrenfeld 1970; Terborgh 1974; McKinney 1997). Only recently has this analysis been applied to fishes, with mixed success.

Papers on vulnerable fishes have generally approached the topic via characterization or prediction. The former includes descriptive studies that look at local or regional faunas and attempt to derive patterns of vulnerability from biological traits and environmental circumstances (e.g., Maitland 1995; Moyle 1995; Burkhead et al. 1997; Etnier 1997; Johnston 1999; Roberts and Hawkins 1999). Predictive analyses apply statistical methods to data sets derived from descriptive data to determine whether a particular statistical approach or model can be used to predict which species in a region are likely to become imperiled (e.g., Moyle and Williams 1990; Angermeier 1995; Parent and Schriml 1995; Warren et al. 1997).

To begin, it is instructive to ask which taxa have been most affected by anthropogenic activities to the point of extinction. Using all reported extinctions in the appendix, we can assess vulnerability among families as a function of habitat type, habitat or range size, and known natural history traits (tables 4.1, 4.2). Among families with a high extinction rate (5% or more of species), the strongest correlate appears to be restricted distribution or isolated populations, as of species that occurred in a single lake or spring or small drainage basin (e.g., salmonids, bedotiids, adrianichthyids, goodeid splitfins and springfishes, cyprinodontid pupfishes, gasterosteid sticklebacks). This pattern also applies to individual species in families with less than 5% extinctions (e.g., cyprinid minnows, poeciliid livebearers, cichlids). Small body size is a characteristic of many of these fishes, the few exceptions being the anadromous sturgeons (which are also exposed to human exploitation), catostomid suckers, and some of the cichlids. No marine extinctions are known. The solitary elasmobranch on the list, the Ganges shark, *Glyphus gangeticus*, is a riverine species, again underscoring impacts in freshwater habitats.

Most extinct species occurred in lakes (Harrison and Stiassny 1999), which make up 99% of the earth's available surface fresh water. Rivers make up about 1% at any time; caves and springs are negligible in terms of volume (McAllister et al. 1997). A disproportionate number of extinctions have therefore occurred in relatively small habitats such as caves, ponds, and springs, which is not surprising given how easily such habitats are disturbed. Rivers have also suffered extinctions disproportionate to their volume. An alternative comparison focuses on relative diversity in different habitats (Table 4.2). The approximate diversities presented in chapter 1 highlight the relative

Table 4.1. Taxonomic distribution of probable fish extinctions

	Extinct species[a]	*No. species in family*[b]	*Percent extinct*
Petromyzontidae (lampreys)	1	41	2
Carcharhinidae (ground sharks)	1	58	2
Acipenseridae (sturgeons)	**4**	**24**	**17**
Clupeidae (herrings)	1	181	<1
Cyprinidae (minnows)	54	2010	3
Catostomidae (suckers)	3	68	4
Balitoridae (river loaches)	1	522	<1
Characidae (characins)	3	885	<1
Ictaluridae (N. American cats)	1	45	2
Bagridae (bagrid cats)	1	210	<1
Siluridae (sheatfishes)	1	100	1
Schilbeidae (schilbeid cats)	2	45	4
Amblycipitidae (torrent cats)	**2**	**10**	**20**
Trichomycteridae (pencil cats)	1	155	<1
Retropinnidae (New Zealand smelts)	**1**	**5**	**20**
Salmonidae (salmons)	**20**	**66**	**30**
Bedotiidae (bedotiids)	**1**	**9**	**11**
Atherinidae (silversides)	3	165	2
Adrianichthyidae (adrianichthyids)	**3**	**18**	**17**
Fundulidae (killifishes)	1	48	2
Poeciliidae (livebearers)	6	293	2
Goodeidae (splitfins)	**8**	**40**	**20**
Cyprinodontidae (pupfishes)	**16**	**100**	**16**
Gasterosteidae (sticklebacks)	**2**	**7**	**29**
Syngnathidae (seahorses)	1	215	<1
Cottidae (sculpins)	1	300	<1
Percidae (darters)	2	162	1
Apogonidae (cardinalfishes)	1	207	<1
Cichlidae[c] (cichlids)	**64(+75)**	**1300**	**5(11)**
Pomacentridae (damselfishes)	1	315	<1
Eleotridae (sleepers)	1	150	<1
Gobiidae (gobies)	2	1875	<1
Belontiidae (gouramies)	1	46	2
MEAN PER FAMILY	6		7
TOTAL EXTINCTIONS	212 (287[c])		

Note: Families in bold type have experienced a 5% or greater extinction rate. Common names are representative of a family. Cats = catfishes.

[a]Values are based on data in the appendix and in table 3.1, derived primarily from IUCN (2004) and Harrison and Stiassny (1999), with additions from the recent literature.

[b]As given by Nelson (1994).

[c]The larger number in parentheses for cichlids refers to approximately 75 unresolved extinctions in Lake Victoria; see Harrison and Stiassny (1999).

Table 4.2. Distribution of fish extinctions as a function of habitat

	Lakes	*Rivers*	*Caves/springs*
Global freshwater volume (%)	99	1	<1
Global fish diversity (%)	15	85	<1
Number extinct	41 (143)	16	>11[a]
Extinctions (%)	60 (84)	24 (9)	>16 (>7)

Note: Based on 68 extinctions reported in Harrison and Stiassny (1999). Numbers in parentheses include 102 presumed or problematic extinctions from Lake Victoria, per Harrison and Stiassny (1999).

[a]Cave/spring extinction values do not include a number of recent losses from Mexican spring systems; see chapter 2.

vulnerability of small habitats. However, losses relative to diversity reverse the order of lakes versus rivers: Lacustrine extinctions have occurred in greater proportion than would be expected from lacustrine diversity, especially if the situation in Lake Victoria is as bad as many believe. Because lakes refresh their water infrequently compared to flowing water, and given human densities around and intensity of use of lakes, impacts on lacustrine fauna are not surprising.

QUALITATIVE STUDIES OF VULNERABILITY AMONG FRESHWATER FISHES

One of the most thorough studies of the relationship between vulnerability and biological or habitat factors was Burkhead et al. (1997). These authors focused primarily on the speciose Etowah River drainage system of Georgia (91 native species + 15 species extirpated) but reviewed a broader literature and made comparisons with much of the southeastern U.S. ichthyofauna. They sought particular ecological correlates of imperilment in stream fishes, focusing a qualitative analysis on the importance of geographic range size, body size, habitat use, and vertical orientation in the water column (figure 4.1). They found that range size was the trait most strongly associated with imperilment. Fishes with the smallest geographic ranges were most imperiled, especially geographically isolated and local endemic species (figure 4.1A). Among the four darter species with the most restricted distributions, two were federally Endangered and two were candidates for federal listing. Species vulnerability increases as range shrinks because small reductions in available habitat have a disproportionately large impact on the species as a whole.

Other apparent correlates were benthic living habits and small body size. About five times as many benthic as

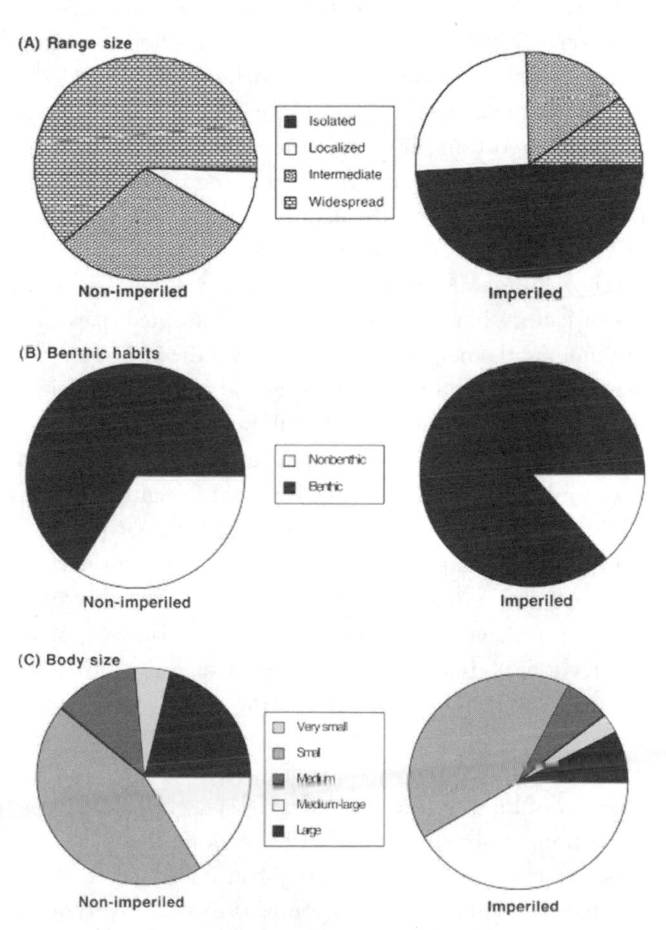

Figure 4.1. Correlates of imperilment among 81 imperiled and 325 nonimperiled southeastern U.S. fishes. (A) Proportions of imperiled and nonimperiled fishes with small (isolated and localized) versus large ranges; (B) proportions of benthic and nonbenthic species; (C) proportions as a function of body size. Redrawn from Burkhead et al. (1997).

nonbenthic fishes made up the imperiled fauna, whereas among nonimperiled groups, only twice as many had benthic habits (figure 4.1B). Many land-disturbing activities occur in the southern Appalachians, most of which lead to increased sediment covering stream bottoms (see chapter 5). Small (<150 mm) fishes made up more than 80% of the imperiled southern Appalachian fauna but less than 60% of the nonimperiled fishes (figure 4.1C). Small body size in stream fishes usually means low dispersal capability, lower fecundity, and shorter lives, all of which could contribute to increased vulnerability. Habitat size was not associated with imperilment, but this could be an artifact of degradation affecting all habitat types sampled (small creeks to large rivers).

Etnier (1997) took a broader look at all southeastern U.S. freshwater fishes, which represent the most diverse temperate freshwater fauna in the world. After reviewing status data for 490 species, he concluded that 91 (19%) were imperiled. He then looked for taxa and habitats that were disproportionately represented and identified the primary causes of decline. Three taxa stood out: sturgeons (83% of species), percid darters (31%), and madtom catfishes (26%) (centrarchid sunfishes contained no imperiled species). Sturgeons are globally jeopardized due to a variety of factors, but blocked migrations and silted spawning grounds are the major impacts in the Southeast. Darters are fishes of creeks and medium rivers and are often endemic to small drainage systems. Small range size, nonpoint-source pollution (siltation of stream bottoms), and impounded (dammed) streams leading to fragmented distributions have all contributed to their declines. Madtoms, many locally endemic, appear particularly sensitive to chemical pollution, perhaps interacting with olfactory-mediated behavior. They have exceptionally large eggs and low fecundities relative to body size: Median clutch size for 14 species treated in Etnier and Starnes (1993) was only 78 eggs, with half producing clutches of 50 or fewer. Minnows of the same size, for example, typically produce clutches of hundreds to thousands of eggs.

Among habitats, Etnier (1997) found that springs, medium rivers, and perhaps caves had higher percentages of imperiled fishes than the Southeast overall (70%, 40%, and 33%, respectively) (table 4.3). Springs often contain isolated endemic species, are easily polluted, and are rapidly altered by impoundment and water withdrawal. Undisturbed medium rivers are naturally clear with silt-free bottoms but are frequently impounded, which affects water flow and leads to silting-in, both detrimental to native fishes. Caves suffer from problems similar to those of springs; additionally, cave fishes are highly specialized for the unique ecological circumstances that typify subterranean habitats. Habitats that have few imperiled members are typically slow flowing or still and naturally sediment rich (lakes, ponds, big rivers), or they are common and generally unimpounded (creeks) (the southeastern U.S. has relatively few natural lakes, making Etnier's assessment of this often altered habitat type unrepresentative). Sunfishes, the group with no imperiled members, are most abundant in slow-flowing or still water. Overall, nonpoint-source pollution (mainly sediment) and altered water flow (impoundments) stood out as major contributors to species declines among southeastern fishes, affecting species with small native ranges the most. Frissell (1993), after analyzing patterns among extinct and endangered western U.S. fishes, similarly concluded that hydrologic alteration (dams, water diversions) and urbanization were major contributors to irreversible losses among both localized and widely distributed taxa in that region.

Johnston (1999) took a taxon-specific, phylogenetic approach to exploring patterns of vulnerability in minnows, the largest fish family in North America and perhaps the world. At least 46 (about 20%) of North America's 231 cyprinid species are imperiled. Johnston explored the interaction between environmental factors and behavioral ecology, specifically "spawning mode." Eight modes were characterized by the substrates over which minnows spawned, efforts at nest construction, and the degree to which eggs were scattered or guarded. Spawning mode was known for only 25% of the imperiled species. Among studied species, Johnston found that (1) all spawning modes except mound building and egg clustering were represented, and (2) the distribution of spawning modes among imperiled and nonthreatened species was roughly equivalent, except that crevice spawners (four *Cyprinella* spp.) and the one pit-ridge builder (sandhills chub, *Semotilus lumbee*) were more likely to be imperiled. Mound builders (*Nocomis*, *Exoglossum*) and egg clusterers (*Pimephales*, *Codoma*, *Opsopoeodus*) either build a raised structure that is kept free of silt, or guard eggs laid on the roof of an egg cavity. Species in these groups and their nest associates are among the most abundant stream fishes in North America, often occurring in areas subjected to heavy sediment inputs (e.g., Berkman and Rabeni 1987; E. B. D. Jones et al. 1999). In contrast, the oft-imperiled

Table 4.3. The distribution of imperiled fishes among habitat types in the southeastern U.S.

Habitat	*No. species*	*% of total*	*No. imperiled*	*% of total imperiled*	*% imperiled in habitat*
Big rivers	45	9	6	7	13
Medium rivers	90	18	36	40	40
Creeks	248	51	32	35	13
Headwaters	16	3	4	4	25
Springs	10	2	7	9	70
Caves	3	1	1	1	33
Lakes, ponds	70	14	3	3	4
Diadromous	8	2	2	2	25
TOTAL	490		91	19	

Source: Based on Etnier 1997 (Table 1, p. 89).
Note: Springs, medium rivers, and caves contained more than their share of species in jeopardy. Although creeks contain many vulnerable species, they are an abundant habitat type and are not disproportionately represented among the totals.

crevice spawners do not clean their nests and would be particularly vulnerable to silting-in of spawning cracks. Johnston's behavioral and phylogenetic approach has relevance to proactive conservation efforts, as it allows us to anticipate environmental problems. By studying the spawning behavior of healthy relatives, we may be able to predict spawning behavior, causes of vulnerability, and appropriate remediation for rare and threatened species about which we know little.

QUANTITATIVE STUDIES OF VULNERABILITY AMONG FRESHWATER FISHES

Moyle and Williams (1990) were among the first to apply a statistical analysis to patterns of vulnerability. They looked at California's freshwater fishes and asked whether declining species differed ecologically from stable taxa. California has 113 native freshwater fishes, 60% of which are endemic. Seven species have been extirpated and 41% are at a high level of risk. Moyle and Williams scored fishes for conservation status, endemism, number and size of drainage basins occupied, number of co-occurring fish species, degree of isolation of habitat type, legal protection afforded a habitat, and ichthyological province occupied. They also rated other influential factors, dividing them into natural factors (biotic agents, natural disturbance) and artificial factors (water diversions, other habitat modification, pollution, introduced species, exploitation). All factors were entered into a principal components analysis (PCA) to separate at-risk from not-at-risk species.

The analysis indicated that vulnerable and healthy species had dissimilar traits. Species in trouble tended to (1) be endemic to California; (2) occupy relatively small areas, often in just one drainage basin; (3) be part of small assemblages (<5 species); and (4) occur in isolated springs, small warm streams, large rivers, or headwater streams in arid areas. Causative factors were largely anthropogenic, including water diversions, introduced species, and other habitat modifications, with less impact from pollution and overexploitation. Populations of healthy species were by contrast often habitat generalists in tributary streams to larger rivers; they could live in reservoirs, and their populations were influenced more by natural biotic and abiotic processes than by human impacts.

Moyle and Williams concluded that the decline of California's fish fauna resulted in large part from "massive water projects that have usurped most of California's water in combination with introductions of fish species better suited to altered habitats" (p. 280). Because particular habitat types and regions had several imperiled species and 75% of imperiled species lived in habitats without legal protection, Moyle and Williams emphasized that an ecosystem-level approach centered on creation of multiple aquatic reserves was needed to protect what was left of California's native freshwater fishes.

Parent and Schriml (1995) looked at 49 life history and other ecological characteristics of 117 species from the Great Lakes–St. Lawrence River region. They compared a nonparametric univariate test with linear logistic regressions to find the model that gave the best separation between at-risk and not-at-risk species. Their goals were to develop a "vulnerability profile" and to test the predictive power of that profile with a different data set from the midwestern U.S.

The nonparametric approach distinguished 14 traits that differed between risk categories. At-risk species were characterized by feeding in relatively high current areas and breeding in deep water. Deep-water breeding characterizes Great Lakes ciscos and the deep-water sculpin, all of which are extinct or appear on IUCN or Canadian imperiled lists. Not-at-risk species had geographic ranges that extended to the north; were piscivorous; fed in lakes, ponds, rivers, and over grasses; bred over gravel in lakes, ponds, and streams; and built nests and engaged in parental care. A logistic regression model permitted calculation of the probability that a species was at risk based on combinations of ecological traits. This probability increased for species that matured late, fed in moderate to strong currents and not over grass, were not piscivorous, did not inhabit lakes, or did not breed in streams or over gravel or pebbles. The species closest to this overall profile of peril was the copper redhorse, *Moxostoma hubbsi*, which was designated Vulnerable by IUCN (2004).

Parent and Schriml's model successfully predicted the risk status of 78% of midwestern species in their alternative data set. The authors were confident that their model had applicability in other areas where human impacts were strong, such as central and northern Europe. It might not apply as well where anthropogenic stress is less and flowing-water habitats dominate (e.g., the southern U.S.) or where effects of invasive species are large, as in isolated habitats of the desert Southwest.

Angermeier (1995) conducted a thorough, quantitative study of extinction vulnerability among southern U.S. fishes, using univariate and multivariate statistical techniques to determine whether extinction-prone fishes were ecologically distinct. Analyzing the well-studied freshwater fishes of Virginia (197 native species), he compared resource use, life history, and geographic distribution of extirpated and extant fishes, focusing on whether species were relative ecological specialists or generalists. Attributes included geographic distribution (occurrence in 12 drainages and 5 physiographic provinces), use of flowing versus still water, benthic versus nonbenthic position, trophic habits (five major food types), maximum length, age at first reproduction, diadromy, five categories of spawning substrate use, and extent of nesting behavior.

Although extirpated and extant species shared many traits, extirpation was linked to the following ecological attributes: (1) restricted geographic distributions, occurring in only one physiographic province; (2) small geographic ranges; (3) narrow range rather than varying with respect to water body size; (4) relatively small body size among minnows; (5) diadromy; and (6) specializations with regard to physiographic province, flow rate, water body size, position in the water column, and trophic habits. Angermeier concluded that four attributes were most strongly correlated with extirpation: diadromy, limited physiographic range, limited range of water body sizes, and ecological specialization. Unlike birds and mammals, large body size and high trophic position were weak predictors of extirpation among Virginia fishes. Angermeier applied these findings to predict which extant species were potentially threatened, and the model identified five species, three of which had been listed as imperiled by various authors. Angermeier's combined univariate and multivariate approach is thus a useful method for predicting problems and taking proactive steps to prevent future declines.

Warren et al. (1997) conducted the most ambitious quantitative evaluation of correlates of imperilment. They investigated southeastern U.S. fishes, using several statistical approaches, particularly analysis of variance and logistic regression models. They detailed the distribution of 530 freshwater species over an 11-state, 33-river-drainage area, seeking statistical relationships between imperilment and range size, family, native taxon richness, endemism, and diversity of stream types. They found that probability of imperilment increased with decreasing range extent. Imperiled fishes were most likely to occur in areas of high endemism and high taxon richness, where stream-type diversity was also high. Thus regions that have historically provided favorable conditions for speciation are also likely to suffer extirpations, particularly of species with restricted ranges. Other data support this conclusion, as in mass extinctions among endemic species flocks, e.g., Lake Victoria cichlids, Lake Lanao cyprinids, Central Mexican atherinids, and Andean cyprinodontids. A political implica-

Table 4.4. Ecological correlates of imperilment in freshwater fishes

Ecological trait	*SE streams*[a]	*Tenn.*[b]	*NA minnows*[c]	*Virginia*[d]	*G. Lakes*[e]	*Calif.*[f]	*SE*[g]	*Europe*[h]	*Europe*[i]
Small range, endemic	x	x		(x)	x	x	x	x	x
Specialized ecology		x	(x)	x	n	x	(x)		
Fragmented range	x	x				x		(x)	
Spring/small habitats		x		(x)		x	x		x
Benthic habits	x		(x)	x			x		
Diadromy/migration				x	n	(x)		x	
Small body size	x			x	n			x	x
Lack parental care			x		x				
Short life	x				n			x	
Not piscivorous				(x)	x				
Dependent on currents				x	x				
Low fecundity	x				n				
Large body size				n	n			x	

Note: Categories do not include generally accepted traits such as crevice spawning, small assemblages, small population size, and late maturation. Several of the traits identified as contributors to imperilment do not occur independently of one another. Most studies are of southeastern U.S. fishes; similar analyses from other geographic regions would be extremely useful. x = trait associated with imperiled fishes; (x) = trait inferred from data; n = trait not correlated with vulnerability; SE = southeastern U.S.; Tenn. = Tennessee; NA = North America.

[a]Burkhead et al. (1997).
[b]Etnier (1997).
[c]Johnston (1999).
[d]Angermeier (1995).
[e]Parent and Schriml (1995).
[f]Moyle and Williams (1990).
[g]Warren et al. (1997).
[h]Maitland (1995).
[i]Kottelat (2000).

tion of Warren et al.'s results is that two-thirds of the identified drainage units they studied cross state boundaries. Hence, solutions to conservation problems in the southeast U.S., as elsewhere, require interagency and interstate cooperation.

VULNERABILITY: AN OVERVIEW AND SYNTHESIS

The lists and accounts summarized in chapters 1–3 combined with the syntheses on extinction-prone fish above suggest a suite of environmental, geographic, habitat, and life history attributes that may predispose and even push fish species toward extinction (table 4.4). Notable in this list are differences between fishes and terrestrial vertebrates and between freshwater and marine species. Also, different authors, perhaps because of biological as well as methodological variation, have arrived at some different conclusions. Overriding any attempt to identify "traits" that may predispose species toward extinction is the realization that biology interacts with anthropogenic change; without human presence, most species would continue to survive. Nonetheless, species with certain traits do appear to be more extinction prone—that is, more sensitive to anthropogenic influences than others.

FRESHWATER FISHES

Large size is a frequently noted conservation liability for birds and mammals, as mentioned. Humans preferentially target large terrestrial animals. Also, a large homeothermic body is energetically expensive to maintain and requires a larger range, which increases encounter rates with humans. In fishes, or at least in the stream fishes analyzed above, smallness—of body, habitat, or geographic range—is associated with risk. Endemic small fishes that live in small, isolated habitats (springs, caves, headwaters) over a limited geographic range are multiply threatened. These traits are also associated with low fecundity and low population density, which contribute additional problems (large size is an unquestioned liability among exploited marine fishes, however; e.g., Dulvy and Reynolds 2002).

South Africa can serve as a test case. Imperilment there is inversely related to body size, with about 70% of listed freshwater species and 90% of listed estuarine species being less than 150 mm long (Skelton 1990). As in other areas, site restriction and limited range also correlate strongly and positively with imperilment. Most indigenous freshwater and estuarine species that are restricted to a single drainage basin or to estuaries along a small section of coast have conservation status (Skelton 1990; Skelton et al. 1995) (this assumes that restricted range is not a criterion for imperiled status).

Benthic habits associated with breeding and feeding predispose these (mostly stream) fishes to extirpation, primarily due to sediment deposition. Freshwater fishes that spawn on the bottom and don't engage in parental care (via sediment removal) are disproportionately imperiled among species considered in the present analysis. Bruton (1995) suggested, however, that a correlation exists globally between parental care and restricted range or specialized habits, which might explain the apparent vulnerability of care-giving species that he found. Given apparent disagreement about the influence of parental care on vulnerability, additional analysis of ecological circumstances and phylogenetic patterns is obviously needed.

Specialized benthic habits also interact with impoundments and a dependence on flowing water. Lotic-adapted, benthic fishes cannot live in the depths and littoral expanses of reservoirs, nor can they tolerate the range fragmentation that often occurs as multiple dams dissect and impound river basins (see chapters 5, 6).

Of the other strong correlates of vulnerability mentioned, it is not surprising that ecological specialists are often vulnerable, given the success with which introduced generalists invade and take over disturbed aquatic habitats. The correlate of smallness probably explains the lack of piscivory among imperiled fishes, unlike imperiled birds and mammals, many of which are high in their food webs. Also, the tendency for diadromous or migratory species to be vulnerable has often been noted in discussions of imperiled fishes. For these and several other vulnerability characteristics, Parent and Schriml's 1995 study of lake fishes (see table 4.4) contrasts with the trends among stream fishes. Parent and Schriml's lacustrine species showed no costs associated with small body, low fecundity, ecological specialization, or migratory habits. These differences suggest that lake fishes (or what remains of the native North American lake fauna) may be under an alternative set of constraints and sounds a precautionary note about generalizing on correlates of vulnerability. Additional data from other lake studies and habitats would be instructive; given multiple extinctions among species flocks in lakes, we desperately need comparative ecological and biological data from these mostly tropical locales.

Many vulnerable fishes use multiple habitats as a normal part of their life history, regardless of migratory behavior (Rosenfeld 2003). The more habitat types required, the more likely individuals will encounter degraded conditions at some time. Thus complex life histories interact with specialized ecologies, something of a double whammy. One example is riverine species that use floodplains, side channels, oxbows, or other backwater regions as a regular part of their biology. Many anthropogenic modifications to large rivers—dams, channelization, levee construction, dredging—disturb natural flood cycles, preventing inundation of floodplains and side channels and causing dewatering of oxbows. Floodplain areas serve as spawning and nursery grounds for many riverine species, including the darters and minnows that we tend to characterize as swift-water inhabitants (Scheidegger and Bain 1995; see chapter 6). Floodplain inundation may occur in the winter or spring or wet season depending on geographic locale. Regardless, flooding is predictable, and species depend on this predictability to cue their activities and to allow them to capitalize on food and habitat availability tied to flood cycles. Human activities—particularly dam building—change, disrupt, and sometimes completely eliminate the cycle. Dry floodplains and oxbows do not support flood-adapted species.

Invulnerable Fishes?

Equally interesting are fishes that prosper despite human impacts. In many studies, complementarity existed between species at risk and not at risk—that is, conditions that jeopardized some species may actually have favored others. Dams, sediment, altered flow regimes, deeper and warmer water, and other anthropogenic factors can create conditions favorable to generalists that are behaviorally, anatomically, or physiologically tolerant of or adapted to the altered conditions. Details are often lacking, but relatively invulnerable species tend to engage in parental care and nest cleaning; are habitat and trophic generalists; are reservoir adaptable and live in the water column; and are widespread, large, and often piscivorous (Moyle and Williams 1990; Angermeier 1995; Parent and Schriml 1995; Burkhead et al. 1997; Johnston 1999). Not coincidentally, these are the kinds of fishes—including large shad and other clupeids; large cyprinids such as carp, goldfish, and golden shiner; brown trout; centrarchid sunfishes, crappies, and black basses; Nile perch; and tilapiine cichlids—that humans actively introduce into altered aquatic systems (chapter 8). It would be interesting to learn more about traits of tolerant species in the locales detailed above. A locale-specific, comparative approach that controlled for phylogeny would contribute to our understanding of vulnerable as well as invulnerable fishes.

A North American-European Comparison

Maitland (1995) summarized literature on primarily European species and identified three characteristics of vulnerable fishes: (1) discrete populations in confined regions, causing genetic distinctness and a high potential for extirpation via habitat degradation; (2) diadromous migrations through multiple, degraded habitats; and (3) large, slow-growing and small, short-lived species, which were both vulnerable to overexploitation. He concluded that highly threatened species typically belonged to small taxonomic units in restricted geographic areas whose populations or ranges had undergone recent contraction largely due to continued anthropogenic disturbance. A substantive difference between Maitland's findings and those from North America is the role of size, which may reflect the longer history of human exploitation experienced by European fishes. However, Kottelat (2000) pointed out that exploitation is more a problem for northern European fishes, many of which live in lakes. In southern Europe, species are often smaller and more restricted to streams and springs. These local endemics are threatened more by habitat degradation and introductions, paralleling conditions among southeastern and southwestern U.S. fish faunas. The trend then is for large, migratory fishes to experience overfishing, as also shown by conditions in southeast Asia, where 8 of the 13 Mekong River fishes listed by IUCN (2004) are large (>70 cm) and tend to undertake spawning migrations (Mattson et al. 2002; Hogan et al. 2004; chapter 2).

A Europe versus North America comparison—with a focus on habitat, body size, and exploitation—may also help explain the apparent discrepancy between northern and southern U.S. areas. Small body size emerges as a strong correlate for imperiled species in the southern U.S. but not in the Great Lakes (e.g., table 4.4, data from Parent and Schriml 1995). Northern species on both continents are more likely to live in lakes because lake-forming glaciers did not advance into the more southern regions of either continent. Lake species are generally larger than stream species and consequently more subject to exploitation. Southern species live in streams and are rarely exploited but are more affected by habitat degradation. Historical geography, then, interacts with current human activities in determining not only the underlying diversity of faunas in a region but also the patterns of vulnerability among the species that evolved there.

MARINE FISHES

Roberts and Hawkins (1999) reviewed literature on marine fishes and developed a list of characteristics that rendered them vulnerable to depletion, based on empirical observations and deductions from ecological theory. They grouped traits into six population categories, with species characteristics relevant to each category. Six population attributes emerged that increased the likelihood of extinction: (1) *population turnover:* long life span, slow growth rate, low natural mortality rate, and low production-biomass ratio; (2) *reproduction:* low reproductive effort, semelparity (single lifetime reproductive episode), old age or large size at sexual maturity, protandrous (male-first) sex reversal, aggregated spawning at predictable locations, and density-dependent reproduction; (3) *capacity for recovery:* poor regeneration from fragments, short dispersal (= low adult mobility, irregular or low larval recruitment), poor colonizing ability, strong density-dependent effects at settlement, and depensatory responses whereby mortality increases as numbers decrease; (4) *range and distribution:* nearshore occurrence, narrow depth range, small geographic range, high degree of population patchiness, high habitat specificity, and high vulnerability to anthropogenic habitat disruption; (5) *small population size*; and (6) *high trophic position*.

Roberts and Hawkins provided numerous examples of fishes that possessed these traits and demonstrated vulnerability. Their examples parallel observations in other taxa: Late maturation, slow growth, low reproductive output, low natural mortality, limited dispersal, and small range characterize vulnerability in general. (Jennings et al. [1998] found that late maturation, large size, and low intrinsic population growth were most important.) With the exceptions of limited dispersal and small range, these are characteristics of rapidly declining shark populations (e.g., Musick 1999a). Some factors may be more specific to fishes than other taxa, especially to marine fishes. Sex reversal is most common in marine fishes. Protandrous reversal would be especially deleterious if large, mature females were disproportionately targeted by fishers, potentially lowering reproductive output of a stock (C. Roberts, pers. comm.). Among imperiled sex-reversing fishes (e.g., groupers), protogyny predominates, however, and these fish are vulnerable nonetheless. Aggregated spawning at predictable locations characterizes many overexploited marine fishes because these fishes (e.g., cods, hakes, croakers, groupers, snappers; chapters 11, 12) become easy targets for exploitation. Density dependence during spawning or recruitment is seldom identified as a risk factor (but see Parrish 1999).

Vulnerability in Reef Fishes: Size Matters

Localized extirpations have been documented for several coral reef species and locales, with serious conservation implications. Extirpations also affect the protein and employment resources of coastal communities (see chapter 12). Few quantitative studies specifically address characteristics of vulnerable reef fishes. Jennings et al. (1999) looked at depletion of Fijian reef families relative to fishery trends. Drawing on data from northeast Atlantic fish stocks (see chapter 11), they postulated that fishes that mature late, attain large size, and have low potential rates of population increase are most likely to suffer from overexploitation. Because detailed life history information is seldom available for reef fishes, Jennings et al. sought a surrogate parameter that would represent many relevant traits. They found that maximum body size correlated well with growth rate, age at maturity, reproductive output, and natural mortality. They then calculated the ability of maximum body size to predict likelihood of overexploitation in 33 species of snappers, groupers, and parrotfishes, families in which abundance declined as exploitation increased. When pairs of closely related fishes were compared (to control for phylogenetic influences), the larger members of pairs declined the most: "For the heavily fished groupers and snappers, size and vulnerability are closely related" (p. 1472). Even without data on fishing mortality, we can therefore determine which species are most likely to need monitoring and protection. In other words, a precautionary approach is practical and possible in reef fisheries, rather than waiting for the fishery to develop, data to accumulate, and depletion to occur.

Risk Factors and Assignment of Conservation Status

The potential risk factors and traits investigated by Roberts and Hawkins (1999), Jennings et al. (1999), and others have gained acceptance among conservation practitioners and are being applied to schemes designed to assess or predict vulnerability. Musick et al. (2001) used similar criteria to identify potentially imperiled North American marine species before they had to be given official government status (see also Musick 1999a). Musick et al. (2001) focused on four risk criteria: rarity, small range or endemism, specialized habitat requirements, and population decline. For population decline, they established quantitative thresholds based on data commonly gathered for commercially exploited species; the thresholds provided a conservative measure of productivity. Species with "very low productivity" would be most threatened and should therefore be considered vulnerable if they had experienced a 70% decline of mature individuals over the longer of ten years or three generations. Species with "high productivity" would not be considered vulnerable unless they had undergone a 99% population decline; intermediate decline levels of 95% and 85% applied respectively to stocks with "medium" and "low" productivity. From this analysis, Musick et al. determined that 82 marine, estuarine, and diadromous stocks (actually DPSs) were at least vulnerable to extirpation in North American waters.

Because their analysis included actual risk factors, Musick et al. (2001) were able to identify probable causative agents of population decline. Fifty-one species and subspecies were vulnerable because of life history constraints such as large size, late maturation, slow growth, and low reproductive output. These fishes—mostly scorpionfishes, groupers, sharks, and sturgeons—were threatened by direct exploitation and by their participation in mixed-species fisheries, which results in continued capture even when their own abundance has declined. Eighteen DPSs, especially anadromous species, were vulnerable because of habitat destruction. Twelve species fell into the endemic/small range/rare category. The most imperiled species, many of which already had official status (e.g., smalltooth and largetooth sawfishes, delta smelt, Atlantic salmon, saltmarsh topminnow, longsnout seahorse, Texas pipefish, totoaba, and tidewater goby), were those threatened by multiple risk factors. Of the 22 species considered most vulnerable to global extinction, 12 were estuarine dependent or anadromous, again reflecting human impacts on freshwater.

The Missing Larval Stage in Marine Fishes

An interesting pattern not specifically identified in the existing compilations is that of imperiled marine fishes departing from the marine norm with respect to larval biology. The vast majority of marine fishes have a planktonic larval phase (Helfman et al. 1997). A disproportionate number of at risk marine fishes appear to have eliminated this early dispersal phase and instead exhibit more direct development and limited larval or juvenile dispersal. The highly imperiled handfishes of Australia, including the Critically Endangered spotted handfish, have a prolonged developmental period of up to two months, after which miniature adults emerge from the egg. Banggai cardinalfish and Texas pipefish, both designated extinct at one time, lack planktonic larval stages. Syngnathid pipefishes and seahorses, which make up about a fifth of the IUCN list of imperiled marine species, have a reduced larval stage compared to most marine fishes. Elasmobranchs lack a larval phase and are disproportionately imperiled; perhaps low dispersal capability adds to their vulnerability.

The link between biological trait and environmental influence here is speculative. Low egg number or fecundity is an obvious contributor, especially if mortality rates increase due to anthropogenic influences. Increased mortality might occur if benthic eggs or retained embryos suffered extended exposure to multiple mortality factors not experienced by planktonic larvae. Alternatively, relatively limited dispersal ability in adult handfishes, Banggai cardinalfishes, and seahorses may also contribute to their problems, as would adult habitat degradation or fragmentation. It will be interesting to see if other vulnerable marine species for which we lack important early life history information follow this apparent trend. Direct comparisons with freshwater fishes are complicated by phylogeny and habitat, but freshwater species seldom possess the long-lived and widely dispersed larvae characteristic of marine groups. Our understanding of vulnerability among freshwater fishes would undoubtedly benefit from greater knowledge and comparisons of dispersal among early life history stages.

Are Marine Fishes Different?

An obvious drawback of identifying common traits among imperiled organisms is the risk of overgeneralizing. Biological entities are by definition unique, and ecological circumstances are by nature complex. Extinctions among freshwater fishes have been unfortunately common, whereas no pelagic marine fish—the most heavily exploited fish group on earth—is known to have gone extinct. In fact, only 3 of the nearly 300 species listed in the appendix are marine: Texas pipefish, Banggai cardinalfish, and Galapagos damselfish. The pipefish and damselfish are arguably extinct; the cardinalfish is bred in captivity and has been reintroduced to the wild (see Roberts and Hawkins 1999 and chapter 13). Although marine fishes account for about 60% of fish biodiversity, they make up less than 20% (151/826) of the species listed by IUCN (2004) as high risk. Among the 30 Critically Endangered marine species, 8 are estuarine, again suggesting a strong freshwater contribution to factors affecting vulnerability.

Is lack of evidence of extinction among marine species evidence that it isn't happening (e.g., Myers and Ottensmeyer 2005)? Although terrestrial extinctions are common, only four known modern extinctions have occurred among marine invertebrates, all of them gastropods. Present-day differences in imperilment also have historical antecedents. The fossil record indicates that marine nonfish taxa have been less extinction prone than terrestrial taxa (McKinney 1997). Does this apparent relative immunity of marine plants, invertebrates, and mammals apply to fishes? If so, what is it about marine fishes that might reduce their risk of extinction?

Two general attributes may explain the differences. First, marine species are more widely distributed than terrestrial species, reducing the likelihood that harsh conditions in one region will affect an entire species. The pelagic larvae that characterize marine fishes account for large geographic ranges. Hence metapopulation size and distribution are large for marine species, reducing the risk factors of small population size and small range observed in many vulnerable freshwater species. However, emerging techniques such as otolith microchemistry suggest that marine fishes may show more genetic distinctness among populations than has been generally assumed (see Conover 2000). Where restricted ranges do exist among marine fishes, as in numerous coral reef species, species are frequently at risk (Hawkins et al. 2000; Roberts et al. 2002).

Second, the marine environment may buffer species against rapid physical changes (McKinney 1997). The sheer volume of the oceans dilutes or at least slows the effects of such factors as pollutants, climatic and weather extremes, and oxygen depletion. This buffering capacity may be particularly important during sensitive early life stages. The distribution of past and present extinctions in the sea supports the dilution hypothesis. Deeper-water species were more likely to have survived the late Devonian and late Permian mass extinctions, and the current extinction crisis appears to be concentrated in shallow, nearshore habitats (McKinney 1997).

The notion that high fecundity protects marine fishes against population declines is increasingly discounted (Jennings et al. 1998; Sadovy 2001b; Myers and Ottensmeyer 2005). The "Millions of Eggs Hypothesis," which argued that marine fishes gained some protection via their high output of eggs and larvae, has been the subject of a growing chorus of criticism: Myers and Ottensmeyer (2005, p. 63) labeled it a "pernicious myth." Empirical findings of depleted but fecund species, combined with theoretical calculations, demonstrate that survival conditions, not initial reproductive output per se, determine later population size.

Differences between Marine and Freshwater Fishes: Conservation Implications

Many aspects of the lives of marine fishes are different from those of terrestrial animals and even from freshwater fishes, indicating that some IUCN criteria may be too conservative and others may be insufficiently precautionary. Marine fishes typically exhibit increased fecundity with age and growth and have larval periods and indeterminate growth, all of which differ from birds and mammals. Life histories can be plastic, including compensating adjustments to exploitation. Reproductive potential is typically high, geographic ranges broad, and recruitment and survival highly variable, producing naturally large fluctuations in population size. Many marine fishes also engage in spawning migrations and aggregations, sex change, and social inhibition of maturation. Some traits are common to taxa that are extinction prone, whereas others are correlated with taxonomic longevity. All of these traits complicate application of the IUCN criteria to determining the conservation status of marine fishes.

It is important to note that many fishes can maintain stable populations even when subjected to the high mortality associated with intense exploitation, greatly complicating efforts to base conservation status on population fluctuations and threshold sizes. Atlantic cod, southern bluefin tuna, and Atlantic halibut were designated Vulnerable, Critically Endangered, and Endangered, respectively, by IUCN (1996) because they met threshold criteria for population declines (IUCN 1994, 1996, 1997; Mace and Hudson 1999). Fisheries managers objected to these designations because their management strategies generally assume population decline. In fact, conventional fisheries management procedures often call for harvesting 80% of the standing stock, but an 80% reduction in population qualifies a species for Critically Endangered status under IUCN criteria. Even a more conservative harvest level of 40%–50%—the level at which surplus production models predict maximum replacement rates (Ross 1997)—would trigger IUCN protection because IUCN identifies a species as Endangered if populations decline 40%–50%.

Obvious conflicts therefore arise between resource managers who aim to maximize sustainable fisheries yields and conservation biologists intent on preserving biodiversity (Vincent and Hall 1996; G. M. Mace and Hudson 1999). Current IUCN guidelines may be too inclusive, causing listing of inappropriate species. Designated thresholds of population decline need to recognize the reality of sustainable commercial exploitation. Many fish species naturally undergo wide fluctuations in abundance and can sustain relatively high levels of exploitation without going extinct (see Hutchings 2000a, 2001a, 2001b for an alternative view of this widely held assumption). Different threshold levels need to be devised for species with a demonstrated ability to recover from intense fishing pressure, as described by Musick (1999a) and Musick et al. (2001) (see also the Texel/Faial Criteria of OSPAR 2004, chapter 3). Declines due to fishing pressure need to be distinguished from those that result from other, presumably less controllable processes such as habitat destruction, pollution, and introduced species.

However, many "managed" fish species have been overexploited to the point of population collapse (see chapter 10). These may be insufficiently protected by IUCN criteria that fail to account for differences in reproductive biology between fishes versus mammals and birds. Fish grow throughout their lives, and bigger fish produce disproportionately more eggs. IUCN criteria focus either on overall population size or on numbers of mature individuals, ignoring the greater reproductive potential of larger fish, which are also more likely to be targeted by a fishery. In addition, heavily exploited populations begin reproducing at younger ages and smaller lengths (see chapter 11). These traits combine to decrease the number of individuals capable of producing the large numbers of eggs needed to replace a population. Focusing solely on population size or even on numbers of mature fish ignores this runaway effect. Attention should be paid instead to altered age or size structure in exploited populations. One suggestion is that "maturity" be defined by some intermediate rather than a minimum size or age. This would provide more individuals with reproductive opportunities while increasing the number of individuals in a population that could produce larger numbers of eggs. Criteria based on generation time should similarly be adjusted upward. Fish should be allowed to grow beyond age at first breeding before being captured, allowing them time to reach the size at which reproductive potential is greatest. Average age and median age of reproductive individuals have been suggested as alternative measures of generation time (G. Mace 1999; see chapter 11).

Other aspects of currently accepted criteria may underestimate the risk of decline. For example, a shrinking range size triggers assignment to a higher rank. Many wide-ranging fishes form spawning aggregations, which makes them extremely vulnerable to overexploitation. Focusing on species' total area of occupancy underestimates this vulnerability. From a population replacement standpoint, the functional range of the species is the area it occupies during reproduction, a much smaller area than total area of occupancy and therefore a more accurate depiction of the species range. Nassau groupers occur throughout the Bahamas area, a region of occupancy of thousands of km^2. However, these groupers came to a limited number of traditional spawning sites, traveling as much as 20 km. Thousands would aggregate to spawn in an area of less than 0.1 km^2 (C. L. Smith 1972), and as a result were decimated. Range size was indicative of vulnerability, but spawning area, not total range of occupancy, should determine conservation status.

Because many marine fishes reverse sex to become males, large males are frequently the most intensively targeted individuals in a population. Population estimates based on total numbers and ignoring sex ratios can overestimate breeding stock (Vincent and Sadovy 1998). Overfishing of males has resulted in general population declines among groupers and sea basses (e.g., Beets and Friedlander 1999), which make up more than 10% of all listed marine species, even though the family constitutes only 3% of marine fishes. Accepted listing criteria overlook the influence of such complex life histories.

A few other relevant points deserve mention. First, our knowledge of the population size and geographic extent of most marine species, especially of nonexploited fishes, is limited (G. Mace 1999 among many). Commercial species are potentially better protected because we know more about them; we are more likely to detect significant changes and apply the benchmarks established by protective criteria. Also, some commercial fisheries have a built-in, self-correcting mechanism that should slow declines brought about by overexploitation. As fish become scarce, fishers generally switch to more easily captured stocks, which often curtails the hunt before a species disappears completely. However, numerous exceptions to this density-dependent relaxation of fishing effort exist. A depleted species that is part of a multispecies fishery will continue to be captured because other species in the fishery remain abundant and profitable. Artisanal fisheries are less influenced by profitability considerations as long as fishers are able to put food on the table. Also, fisheries in which each individual is exceedingly valuable (e.g., bluefin tuna, sturgeon, totoaba, giant yellow croaker, humphead wrasse, whale sharks) continue to be pursued even when the target occurs at very low densities (Vincent and Hall 1996; see chapter 16). Finally, any philosophical differentiation between exploited and nonexploited species becomes irrelevant when they co-occur. Intense exploitation of one species generally inflicts collateral damage in the form of bycatch (see chapter 10). Fisheries personnel monitor the dynamics of targeted species and make management decisions based on target species. Nontargeted species can suffer high but unnoticed population declines, as documented for barndoor skate, *Raja laevis*, captured incidentally by trawls in the North Atlantic. Analogous bycatch scenarios are suspected as causes of population crashes among similarly long-lived sea turtles and sea birds such as albatrosses (Musick 1999b).

Estuaries, Diadromy, and Migration

Roberts and Hawkins (1999) identified dependence on estuarine habitats as a risk factor. Five of eight UK marine fishes designated as imperiled were estuarine, and two others were nearshore, shallow inhabitants, attesting to the impacts of human activities on nearshore environments (Swaby and Potts 1990). The correlation between risk and estuarine dependence reflects the intensity with which estuaries are used and degraded worldwide. Degradation results from concentrated human populations around estuaries and bays and the pollutants and sediments that concentrate in estuaries as rivers slow and release their sediment load and as tides slosh back and forth. "Estuaries have long been recognized for their importance to fishes as migration routes and destinations, spawning grounds and nursery areas. . . . However, they are also areas subjected to considerable anthropogenic pressures through fisheries, urban and industrial development, pollution, agriculture and recreation . . ." (Swaby and Potts 1990, p. 139). Fishes that live in or travel through estuaries are inordinately exposed to human effluent and impacts. The federally threatened delta smelt, *Hypomesus transpacificus*, of California's Sacramento River system is a freshwater-estuarine species that has undergone precipitous declines because of agricultural inputs and urban water diversions throughout the river delta (Moyle and Sweetnam 1994). Estuarine-dependent specialists suffer in the extreme, as in the case of the endangered giant totoaba, which spawned only in the highly modified, polluted, and dewatered Colorado River estuary.

Many estuarine-dependent fishes migrate through estuaries as part of a diadromous life cycle, differentiated by where spawning and growth occur. Anadromous fishes (lampreys, sturgeons, shads, smelts, salmons) spawn in freshwater and grow in the ocean. Catadromous fishes (eels, galaxiids, mullets) spawn at sea and grow in freshwater. Amphidromous fishes (some osmeriforms, gobies) spawn in freshwater and grow in the sea but often spend extra time in estuaries. Hence, diadromy imposes multiple risks: such fishes are migratory and pass through an estuary at least twice in their life cycle. One might assume that diadromous fishes would be generally represented among vulnerable fish species, as they appear to be in Europe (e.g., Maitland 1995; Kirchhofer and Hefti 1996a). However, McDowall (1999) analyzed whether diadromy was a risk factor by reviewing the entire IUCN 1996 list. Among all categories, 45 (18%) of the 250 diadromous species were listed, with 26 (10%) in the high-risk categories. These are relatively low percentages when compared with global numbers, which run somewhere between 20% and 70% (chapter 1). Half of the high-risk "estuarine" fishes were sturgeons, which are compromised because of slow maturation, infrequent reproduction, disrupted spawning, and overexploitation; none of these factors are directly linked to diadromy. Salmon make up a large fraction of the estuarine species at risk, but they also face a multiplicity of compromising factors, only some of which involve estuarine dependence.

McDowall (1999) concluded that, whereas many diadromous species were listed, diadromy alone did not necessarily propel a species into a high-risk category. Diadromous species were probably no more vulnerable than other fishes that undertook riverine migrations or used estuaries as nursery areas. It is noteworthy that among the three major types of diadromy, 32 listed species were anadromous and 13 were amphidromous; both of these life history patterns involve spawning in freshwater habitats and frequent use of estuaries as nurseries. None of the approximately 60 catadromous species appeared in the IUCN lists. This observation suggests that vulnerability and decline in fish species is intimately linked to degradation of conditions during spawning and early life history. Declines likely reflect blocked spawning migrations, compromised spawning habitat, or use of nursery habitat in deteriorated freshwater or estuarine conditions.

DISRUPTED REPRODUCTION AND EARLY LIFE HISTORY

Much literature suggests that reproduction and early life stages are periods of maximal vulnerability in a fish's life:

Among Freshwater Fishes

- The high incidence of imperilment among benthic spawning fishes in southern and midwestern U.S. streams establishes a link between breeding habits and habitat degradation (Berkman and Rabeni 1987; Burkhead et al. 1997; E. B. D. Jones et al. 1999). Crevice spawners are disproportionately vulnerable to imperilment among minnows, North America's largest fish family (Johnston 1999).

- Degraded nursery habitat, such as side channels and floodplains, appears to be strongly linked to declines in riverine fishes. Young of the year (YOY) of some darters occur in temporarily flooded side channels, pools, backwaters, and sandbars; older individuals occur in deeper, swifter water (e.g., endangered Roanoke logperch, *Percina rex*, Rosenberger and Angermeier 2003). Darters constitute at least 25% of North America's imperiled freshwater fishes.
- Many of the negative impacts of introductions occur during early life stages. Young fishes are especially susceptible to a wide range of introduced predators. Competitive interactions may be similarly influenced. In the San Juan River of New Mexico and Utah, habitat overlap was greater between juvenile native and nonnative species than between adults (Gido and Propst 1999).
- Where several factors interact, severe declines can occur. Sacramento perch, *Archoplites interruptus*, are declining throughout their California Central Valley range (Moyle 2002). When bluegill sunfish were introduced into Lake Greenhaven and underwent a population surge, perch recruitment dropped to zero, implying disruption of reproduction or decreased survival of young fish. Sacramento perch have been subsequently extirpated from the lake (Vanicek 1980).

Among Diadromous and Estuarine Fishes

- Anadromous and amphidromous species are more imperiled than catadromous species. All diadromous species move between fresh and salt water, the chief difference being that anadromous and amphidromous species breed, develop, and grow initially in freshwater habitats, which are in much worse shape than marine regions. The impact of freshwater conditions on successful spawning and rearing seems evident.
- Many estuaries are degraded, and many estuarine-dependent species are imperiled. Again, the major use of estuaries for many of these fishes (e.g., anadromous fishes, totoaba, handfishes) is as a nursery, implying that the impact occurs early in life (see Bruton 1995; McDowall 1999).
- Sturgeons are primarily anadromous and are widely endangered. Although overexploitation is a major contributor, even small, relatively unexploited sturgeons (*Scaphirhynchus* and *Pseudoscaphirhynchus* spp.) are imperiled. Adults of these long-lived species can tolerate or at least survive in a broad variety of riverine conditions. Reproductive failure appears to be what limits population growth, resulting in part from blocking of spawning migrations by dams and by siltation of spawning substrate. Less is known about impacts on rearing habitat.

Among Marine Fishes

- Although marine fishes are not generally at risk, fishes are extremely vulnerable to overexploitation during formation of spawning aggregations, which are common among listed species.
- Sex reversal also characterizes imperiled marine fishes. Because large individuals are the most intensively targeted, one sex will be eliminated differentially, leading to gender depletion and population reduction.
- Most marine fishes provide little or no parental care and produce abundant larvae that disperse widely. Many small, imperiled forms (seahorses, handfishes, Banggai cardinalfish) produce few young and either engage in parental care or have limited larval dispersal. A direct link between early life history and risk is evident (J. P. Hawkins et al. 2000; Sadovy 2001b).
- The Critical Period hypothesis states that minor changes in larval mortality rates can have major implications for later year-class strength and population size (see Helfman et al. 1997). Larval survival is dependent on food availability at the onset of exogenous feeding, when yolk resources are exhausted. Hence early life history greatly influences adult population size.

These lists show that many at-risk species—regardless of habitat—experience problems during reproduction and early life stages that are directly linked to anthropogenic activities. Resource managers, conservation biologists, and anyone concerned about or responsible for protecting fishes should look at impacts during these stages for causes and solutions. Larval and juvenile periods, though relatively difficult to investigate, especially deserve greater attention.

PATTERNS AND CORRELATES ACROSS TAXA AND TIME

When we search for patterns, we hope to find specific features or traits that are hallmarks of at-risk species.

However, the patterns that emerge in the present analysis point to multiple factors working in conjunction. At least among freshwater fishes, two traits—specialization and small body size—often correlate with extinction proneness, but even this distillation involves multiple interactions (box 4.1). Specialized species have several extinction-prone traits. They tend to have limited geographic range, live in small populations, and be intolerant of anthropogenic change. Many imperiled freshwater fishes could be characterized thus, particularly spring and cave species, inhabitants of headwater areas, and localized endemics. Fishes in these habitats are often small and are thought to have poor dispersal ability; limited dispersal also characterizes extinct species of mammals and plants (McKinney 1997).

Fishes in small habitats also tend to have small populations. Such rarity has been identified as the single-best predictor of extinction across numerous temporal and spatial scales (McKinney 1997). Some species may be naturally rare (e.g., Sheldon 1987), but others may be artificially rare due to human impacts. Human-induced rarity is more serious because formerly abundant species that have been decimated by human activity may not possess adaptations to rarity. The costs of small population size and the probability that low numbers will lead to extinction are greater in declining populations of formerly abundant species. This complex interaction between rarity and vulnerability highlights the need for population studies of rare species and for careful monitoring of population size in any species experiencing decline.

Rarity is also strongly associated with localized geographic range. Implicit in a small range is poor dispersal, which means fewer source populations to maintain a population or to repopulate an area after extirpation. Widespread species tend to be successful because they are generalists, are capable of exploiting a wide range of resources, and are thus able to withstand human perturbations. They also tend to be strong dispersers (e.g., Albanese et al. 2004) and to occur in locally dense populations, which buffers them against extinction.

Studies have reached different conclusions about the influence of such traits as body size, ecological specialization, and migration (see table 4.4). Some apparent contradictions undoubtedly arise from the diversity of habitat types and locations investigated, which include small streams in high-diversity regions of the southern U.S., lower-diversity lakes in northern and midwestern U.S. locales, and intermediate diversity habitats in California. Water body size, water flow, underlying diversity, and historical geography could all affect the outcomes. We need comparisons that control for these factors, such as investigations of northern streams, or southern streams with lower natural diversity, such as those along the Atlantic and Gulf coastal plains. Some evidence suggests that marine and freshwater fishes differ ecologically in ways that lead to different patterns of vulnerability. Small body size was an identified liability among stream fishes, but large size is an apparent liability among marine fishes. High trophic position correlates with vulnerability in marine systems, but the opposite holds in streams. One outcome of these analyses may be the realization that traits characterizing potentially imperiled species in one region of the U.S. or world will have only partial application to other areas or

BOX 4.1. Some Correlated Risk Factors among Freshwater Fishes

Vulnerability traits are often linked, which undoubtedly increases risk. Whether multiple factors have additive or synergistic impacts and how factors should be weighted remain undetermined.

Small population size is often associated with: small range (and endemicity), small habitat size, small body size (and short life span), limited dispersal capability, low fecundity, and nonpiscivory.

Fragmented range is often associated with: small habitats, small range, specialized ecology, and small population size.

Benthic habits are often associated with: crevice spawning and dependence on currents.

Late maturation is often associated with: large body size (and slow reproductive turnover via low fecundity and long reproductive intervals), slow growth, and specialized ecology (e.g., some cave fishes).

Diadromy implies migration and is additionally associated with: specialized ecology, fragmented range, and large body size.

habitats. Ecological conditions and causes of imperilment are complex; appreciating this complexity and abandoning a search for simple causes and solutions would be a positive development.

APPLYING KNOWLEDGE OF TRENDS TO FISH CONSERVATION

Increased understanding of trends and traits among vulnerable taxa and habitats has proven directly applicable to decisions about resource allocation in fish conservation programs. Successful applications have relied on both quantitative and qualitative approaches, with outcomes directed at populations, species, and habitats.

A major impediment to assessing or predicting extinction risk, and thus mapping strategies for prioritizing conservation effort, is a lack of data on many potentially imperiled taxa, especially nongame fish species. However, commercially important species are better known, and few species are better studied than Pacific salmon. Several assessment schemes have been suggested to determine extinction risk in salmonids. For example, Allendorf et al. (1997) used data and expertise to determine risk and to prioritize conservation efforts among stocks in a region. They used a questionnaire-type approach to assess and rank the genetic, evolutionary, and ecological consequences of extinction of different stocks. Summations of scores from the ranking exercise determined which stocks were in greatest need of immediate remedial action.

Allendorf et al. evaluated criteria concerning population size and change. Population parameters were established either by using population viability analysis (PVA) or by calculating effective or total population size (N_e, N) over time, thus determining rates and extent of population decline. PVA (see Boyce 1992) is an algorithm for estimating probability of extinction over a designated time period. Depending on threshold values and trends, extinction risk of salmon stocks was placed at a Very High, High, or Moderate level, or the analysis revealed where data were inadequate. Priority for action was then determined by ranking the biological consequences of inaction (box 4.2).

Allendorf et al. focused on (1) genetic and evolutionary consequences that affect a species, especially how much of a species' adaptive genetic diversity would be lost if a

BOX 4.2. Worksheet to Establish Priority of Action When Comparing Stocks of Salmonids

Questions focus on the consequences of extinction of a stock to its species and ecosystem. A yes answer to any question receives one point; a no answer or no information available gets zero points. Stocks with the highest point total have the highest priority for conservation action because the consequences of their extinction are greatest. Condensed from Allendorf et al. (1997).

GENETIC AND EVOLUTIONARY FACTORS

1. Does stock diverge greatly genetically from other conspecific stocks using accepted genetic techniques?
2. Does stock occupy unusual habitats compared to conspecific stocks?
3. Does stock possess unusual life history characteristics compared to conspecific stocks?
4. Does stock possess unusual, genetically linked morphological traits?
5. Has stock been reproductively isolated from conspecifics stocks without opportunities for migration?
6. Has stock been free from reproduction with introduced stocks?
7. Has stock avoided breeding bottlenecks (i.e., has N_e remained above 50 for one generation or more)?
8. Does stock occur at the species' range margin or in marginal habitat?

ECOLOGICAL FACTORS

1. Is stock part of a rare or unusual native fish assemblage compared to conspecific stocks?
2. Is this the only biogeographical province or one of a few biogeographical provinces in which the species occurs?
3. Are adjacent or nearby stocks imperiled, extinct, or relictual?
4. Are many other aquatic species in the basin imperiled, extinct, or relictual?
5. Would protecting the habitat of the stock promote recovery of other imperiled populations in the basin for which data are limited?

particular stock were extinguished; and (2) ecological consequences that affect a stock's ecosystem—namely, how much of an ecosystem's structure and function would be lost upon stock extinction. Evolutionary consequences are the basis of the ESU concept, and consideration of ecosystem impacts is a unique strength of this approach. In assessing 20 stocks of four species of Pacific salmonids, Allendorf et al. established a prioritization ranking and identified stocks for which crucial information was lacking.

This outcome is laudable, but the assessment depended on having relatively complete data sets, which forced the authors to rely on current population size for many stocks for which inadequate genetic and historical population trend data existed. This raises questions about the applicability of this approach to less charismatic and well-monitored species and, thus, to its general utility. Wainright and Waples (1998) recommended modifications to the Allendorf approach, including omitting wild-spawning hatchery fish from population estimates because such an augmented population would give a false impression of abundance of the genetically distinct stock. In its defense, the Allendorf approach does provide a prioritization strategy for allocating limited conservation and management efforts. An analogous approach might serve to prioritize conservation efforts among commercially important species, species for which relevant data are likely to be available. If multiple imperiled species or stocks lived in the same habitat or were experiencing a common, correctable threat, efforts directed at habitat or ecosystem conservation would serve all.

Much of what we know about salmon is difficult to quantify but still valuable. Capitalizing on this relative wealth of informal information and available expertise, Wainwright and Kope (1999) used a risk-matrix method to assess extinction risk for 12 Chinook salmon ESUs. They asked a panel of 20 knowledgeable scientists to assign scores of 1 (very low risk, unlikely to contribute to extinction alone or in combination) to 5 (high risk, likely to cause extinction) to a series of qualitative and quantitative risk factors: abundance (population size, distribution, carrying capacity); trends (in abundance, in population variation, in population productivity, in natural vs. artificial production); and genetic integrity (interbreeding, loss of fitness, loss of diversity, anthropogenic selection due to harvest or habitat modification). A fourth factor—recent events that might alter the status of ESUs—was scored from "strong improvement" to "strong decline." This factor would account for the effects of recent changes in management, habitat improvement or decline, and natural events such as floods or volcanic eruptions, all phenomena that could affect the immediate viability of fish in a region. Factors were added to give an overall score for each ESU, which was then assigned to risk categories of Endangered, Threatened, or Not at Risk.

This "linear ranking" approach is meritorious because it combines qualitative and quantitative information from several sources into an overall evaluation of extinction risk. It is also politically expeditious because government authorities and the public at large deal well with distilled and simplified representations. Shortcomings include (1) the question of whether a single score derived from multiple, disparate types of data has real validity; (2) the difficulty of weighting the different factors that constitute the score; (3) inequalities in accuracy and completeness among types of information; and (4) the inability of a single score to accurately represent uncertainty or inevitable disagreement among the panel of experts (see Given and Norton 1993). Given that much of the information we have about imperiled fishes is opinion and best guesses, the procedure mapped out by Wainwright and Kope deserves testing in other contexts.

Prioritization of Habitats

An emphasis throughout this chapter on traits of species distracts from the more general and pervasive threat, namely degraded habitats (chapters 5–7), which fishes share with declining amphibians, mollusks, and crayfishes. We can accomplish more by identifying and protecting vulnerable habitats than by focusing on individual species. Numerous workers have emphasized the need for an ecosystem perspective (see Williams et al. 1989; Warren and Burr 1994), but assessment methods that begin with endangered habitats are relatively uncommon.

Building on an earlier classification system of habitat types used by Moyle and Ellison (1991), Moyle (1996) developed a rating system to determine the conservation status of 66 habitat types in the Sierra Nevada range of California. Values were assigned to habitat rarity, degree of disturbance, and existing protection for each habitat type, and status scores were translated into categories of vulnerability (Gone, Threatened, Of Special Concern, Secure). Patterns included an elevation trend: Threatened and extinct habitats were more common in lowland valley

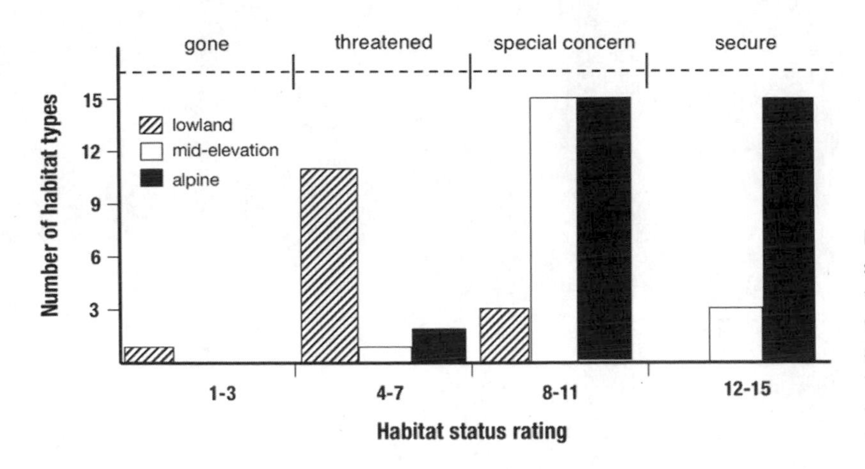

Figure 4.2. Frequency of occurrence, conservation status, and spatial distribution of aquatic habitat types in the Sierra Nevada mountains of California. Status ratings are the sum of values assigned to rarity, disturbance, and legal protection of each type. An altitudinal trend in status is obvious, with lowland habitats under the greatest threat. After Moyle (1996).

areas, mid-mountain and foothill habitat types were in moderate decline, and secure habitats occurred at higher elevations (figure 4.2). Rarity had a strong influence on status calculations, which is appropriate given that unique habitats often contain endemic species and seem to invite disturbance. Moyle (1995) expanded this habitat-based approach as part of a proposal to create a series of Aquatic Diversity Management Areas (ADMAs) throughout California. His approach seems applicable to a wide variety of situations and should be particularly useful in setting priorities for the establishment of protected areas, especially for freshwater habitats that are poorly represented among created reserves.

CONCLUSION

A suite of traits characterize vulnerable fish species. Some identify causes of imperilment; others are admittedly correlates. Some can be generalized across species and habitats; others are more specific to particular locales and taxa. With exceptions, small fishes with limited ranges and mobility, especially endemics in isolated, easily disrupted habitats, are heavily represented among endangered and extinct species. For marine fishes, the impacts of overexploitation fall heaviest on large, commercially valuable species; likelihood of decline increases if fish form predictable spawning aggregations. Marine species with limited larval stages or dispersal are unusually vulnerable. Estuarine dependence, especially early in life, is another identified risk factor. In fresh water, lotic or spring fishes have been more affected than lacustrine species, except for those that evolved in the absence of predators, whose vulnerability to introductions crosses habitats. Among riverine species, small fishes dependent on clean bottoms for feeding and breeding are in greatest decline, suffering primarily from habitat disruption. Although noticeable declines among marine fishes fall heaviest on larger age classes (due to overfishing), for both marine and freshwater groups, human activities that disrupt reproduction and affect younger classes emerge as major causes of declines and deserve greater attention.

The challenge is putting this information into action. Knowing which species or habitats have a high likelihood of declining allows us to take action in anticipation of a problem, rather than after some obvious threshold has passed. Knowledge of relative vulnerability can also guide how we prioritize our conservation efforts, whether we focus on species, on habitats, or, ideally, on ecosystems that are at risk.

PART III Indirect Causes of Decline

Habitat, Water, and Introductions

5. Habitat Modification and Loss

The concept of marine reserves is simple; if protected from human interference, nature will take care of itself.

—James A. Bohnsack, 1993

Fishes lose habitat when it (1) dries up or floods at inappropriate times, (2) fills with sediment, (3) is choked with vegetation or filled with debris, (4) is contaminated by toxicants, (5) becomes unlivable due to hypereutrophication or deoxygenation, or (6) is destroyed or homogenized through structural damage or removal. These effects interact with life history attributes of fishes to produce population losses. This and the next two chapters review types and effects of physical and chemical habitat modification.

WHAT IS FISH HABITAT?

Understanding how our actions alter fish habitat can be facilitated by first envisioning what constitutes habitat to a fish. For simplicity's sake, let's first consider fishes that are strongly dependent on benthic "habitat" for survival and reproduction and then expand and modify observable trends to other habitat types. The example I visualize most easily is a typical darter living in a stream in the southeastern U.S. To varying extents, many other benthic stream fishes—such as brook lampreys, benthic-living or spawning minnows, suckers, madtom catfishes, salmonids, and sculpins in North America and their ecological equivalents on other continents—have the same requirements and activities (this representation will be clearer if you have spent some time underwater; if you haven't, you've cheated yourself).

Habitats as Seen by Darters

Darters are small, often colorful, benthic stream fishes endemic to eastern North America (figure 2.11). They generally prefer fast-flowing, cool, clear stream riffles populated with aquatic insect larvae. Darter habitat is primarily small rocks and large gravel, and to a lesser extent, aquatic plants and woody debris. Darters sit on the bottom behind or between rocks. Rocks, gravel, wood, plants, and leaves are the structures under and in which many aquatic insects live, and darters find their prey by poking around them. Sitting downstream of structures also gives darters access to passing insect larvae and is a means of avoiding predators and strong currents. During the breeding season, males of many species display an array of blue, red, orange, and green. They establish territories around prominent cobbles or flat rocks and repel intruding males or court passing females. Successfully courted females lay their eggs on the sides of or under a rock, and the male guards the eggs until they hatch. Larger rocks and logs create backwater eddies, where darter larvae and juveniles develop until they are physically able to move into faster-flowing adult habitat.

To a darter then, habitat modification is anything that affects the bottom and hence the size, number, availability, stability, and distribution of silt, sand, rocks, gravel, plants, and woody debris. Human activities that negatively affect stream bottoms include filling in spaces between rocks or covering structure with sediment; removing crucial habitat structure, including woody debris; and mechanical disturbance of the stream bottom.

The Role of Riparian Vegetation

Before human intrusion, North America and most other continents had extensive forest cover. Even in 'naturally treeless' areas, such as the arid desert Southwest or Great Plains of North America, streams were typically lined with a corridor of willows, cottonwoods, and other trees and vegetation (Patten 1998). North American fishes, and stream and river fishes in general, evolved in forested landscapes.

Humans, including native peoples, settle preferentially along water courses and typically cut down trees where they settle (e.g., Hudson 1976). Thus, riparian zones—the terrestrial borders immediately adjacent to water courses—have been heavily deforested and converted to agriculture or other human uses. As a result, stream fishes today live in a mosaic of forested and nonforested habitats. Streams are altered, and fishes are affected where the zone closest to a water course has been deforested.

The ecological functions of riparian buffer zones and their role in maintaining a healthy environment for stream fishes have been reviewed extensively (Correll 2000; NRC 2002c; Pusey and Arthington 2003). Although riparian zones are not alone in determining stream health, they play many roles (box 5.1), chiefly keeping the right stuff in and the wrong stuff out of streams. They intercept sediments from upland sources, reduce streambank and channel erosion, process nutrients and contaminants or slow or prevent their movement into streams, control the range and elevation of stream temperatures, and provide inputs of habitat

BOX 5.1. Riparian Trees, Temperature, and Ultraviolet Light

Several surprises have accompanied riparian zone degradation. Logging near a stream increases summer water temperature (Campbell and Doeg 1989) because of the loss of shade and perhaps because increased solar radiation raises the temperature of groundwater. Higher temperatures can make a stream unsuitable as trout habitat. Declining runs of coho salmon and steelhead trout in Washington State have been linked directly to a 4°C increase in stream temperatures following extensive logging (Beechie et al. 1996).

Elevated temperatures of only 1°–3°C following clear-cutting of the Carnation Creek watershed, British Columbia, caused young coho salmon to emerge earlier from the gravel and to grow faster. This led to earlier migration of coho smolts, which then experienced poor ocean survival, probably because they arrived in the ocean out of phase with prey cycles (Holtby 1988).

Logging affects more than temperature. Many animals that live in heavily shaded marine and freshwater habitats lack physiological or anatomical protection against ultraviolet (UV) radiation and can be killed by ambient UV levels characterizing nearby sunlit surfaces (Shick et al. 1996). Deforested stream sections have lower abundances and diversity of invertebrates (fish food), which can be attributed directly to ultraviolet light exposure (Kelly et al. 2003).

Fishes also suffer directly from UV exposure. Fishes can be sunburned (see Blazer et al. 1997); some sun-dwelling fishes are protected by mucous that has a sunscreen function (Zamzow and Losey 2002). Juvenile coho salmon actively seek shade when exposed to UV-rich sunlight (Kelly and Bothwell 2002). In heavily forested streams—the natural habitat of most juvenile salmon—shade is available, and fishes are rarely exposed to intense solar radiation for more than a few hours daily. Deforestation of the riparian zone increases both duration and intensity of UV exposure, including harmful UV-B wavelengths, which penetrate several meters in clear water. Eggs, embryos, and larvae of marine and freshwater species suffer higher mortality when exposed to high but natural levels of solar UV-B (Hakkinen et al. 2002). Excessive exposure to solar radiation induces cataracts in rainbow trout lenses (Cullen and Monteith-McMaster 1993), which diminishes a trout's ability to focus images on the retina. If the lens transmits high levels of UV, damage to the retina could result (Cullen et al. 1993). Impaired vision could influence feeding, predator avoidance, and navigational abilities, among other crucial activities.

UV-B radiation at middle latitudes increased 10% over a 20-year period in the late 20th century, a result of the depletion of stratospheric ozone, which is expected to continue (Kerr and McElroy 1993). Although most studies of UV effects have focused on salmonids, it seems reasonable to expect that other species adapted to life in shaded streams (e.g., minnows, darters) will respond similarly.

and food (large woody debris, leaves, terrestrial insects) to streams. The contribution of woody debris alone has far-reaching implications for streams and fishes. Many species (e.g., catfishes) use the outsides and hollowed insides of logs as spawning sites or as resting sites (Lowe-McConnell 1987). When woody debris inputs are diminished due to deforestation, stream depth, substrate, geomorphology, and current velocity are altered. Streams become homogeneous, wider, shallower, and structurally simpler (e.g., Schlosser 1991; Montgomery 2003).

Riparian zones and seasonal wetlands are mechanical arrestors of and refuges from high flows during floods. In combination with upland forests, riparian areas influence the seasonal delivery of water to streams. They are also permanent or seasonal habitats or corridors of movement for a large variety of terrestrial organisms, many with narrow habitat requirements. Flooded gallery forests along rivers are major spawning sites for fishes that move into the flooded zones during winter or spring floods in temperate locales and during rainy seasons in tropical areas (Goulding 1980). Many of these functions directly or indirectly affect the numbers, kinds, and ecological roles of fishes in a receiving stream, which may in turn influence the types of organisms that live in the riparian zone. The overall biotic integrity of a stream is in large part a function of the proportion of its banks that is covered with natural vegetation (e.g., Steedman 1988).

To a darter, these functions make the riparian zone invaluable. Paramount is the ability of buffer vegetation to arrest the movement of silt and sediment into a stream. Sedimentation is the largest source of contamination in North American streams and rivers (T. E. Waters 1995; USEPA 2000) and is the most important factor limiting the availability of fish habitat. Nearly half of all U.S. streams receive excessive sedimentation due to erosion (Judy et al. 1984; Henley et al. 2000). In what is generally considered the definitive treatise on the subject, T. E. Waters (1995, p. 79) stated that fine sediments constituted "perhaps the principal factor . . . in the degradation of stream fisheries."

Riparian vegetation, be it trees, brush, or grasses, acts as a mechanical trap and filter for sediments exported from upland regions that have been cut, plowed, graded, graveled, or trampled during logging, farming, grazing, mining, or building. Riparian vegetation also stabilizes streambanks, arresting the significant fraction of sediments that otherwise enter a stream from the eroding bank (e.g., 2000 kg sediment per year per meter of unvegetated streambank; Rabeni and Smale 1995). Deeply and extensively rooted trees are much more effective than grasses and shrubs at stabilizing streambanks (Gurnell et al. 1995). Regardless of source, when fine sediments—generally defined as clay, silt, sand, or gravel particles smaller than 2 mm—enter a stream, they cover the bottom and fill the interstitial spaces between larger particles and between small rocks (box 5.2). Especially during periods of high

BOX 5.2. Measuring Sediment, Silt, and Sand

The sediment load of water is commonly measured as total suspended solids (TSS) and nephelometric turbidity units (NTU). TSS describes the mass of sediment suspended in the water (mg l^{-1} or g m^{-3}), whereas NTU measures turbidity, or light penetration as a function of suspended material (see Kirk 1983). Relatively clear water has 10 or less NTU or TSS. Trout populations can be maintained in southeastern U.S. streams at levels below 10–15 NTU (Meyer et al. 1999). During storm events, when sediment load is maximized, suspended sediment levels in streams with disturbed riparian zones commonly exceed 100 NTU or even 1,000 NTU, the maximum level that most turbidometers record.

Unfortunately, sediment measurements are most frequently conducted during normal flow conditions, when it's easiest and safest to do so. This misses the high-flow storm periods when most sediment is flushed into a stream, suspended due to turbulence, and transported and distributed downstream. Although fish behavior is affected by material suspended in the water column, significant biological impacts also result from bedload sediment and sand transported, deposited, and remobilized along the stream bottom (e.g., Wood and Armitage 1997). This shifting, flowing, deposited component is much harder to measure. The infilling process can be measured with a pitfall trap sampler or as *percent embeddedness*, which depicts how clogged a riffle is with sediment; it is measured as the degree to which rocks are eventually overtopped by sediment (Gordon et al. 1992). Embeddedness values in excess of 40% are considered detrimental to benthic fishes (L. D. Zweig, pers. comm.).

flow, much sediment and silt may remain suspended in the water column, reducing light penetration and underwater visibility.

Sediments and turbidity lower fish diversity and abundance through a combination of direct and indirect effects, including habitat loss, avoidance reactions, dispersal and homing, range fragmentation, and physical abrasion affecting respiration and reproduction. Disease resistance, feeding, growth, and survival are also impaired. These effects interact and operate differently depending on life history stage. They generally increase as a function of sediment concentration and exposure duration (Newcombe and MacDonald 1991; Newcombe and Jensen 1996), and they multiply in response to destructive land-use practices (Reice et al. 1990). Fewer trees mean faster runoff because the water retention capability of the surrounding watershed is lost with the trees (Likens et al. 1977; Bosch and Hewlett 1982); faster runoff means a higher silt load flowing into the streams and more resuspension of sediment already deposited there; and increased stream flow means greater transport of sediments by stream water.

Fish diversity in streams is largely a function of habitat diversity (Schiemer and Zalewski 1992). Sediment homogenizes the stream bottom, carpeting it, filling in cracks and crevices, obliterating gravel and cobble, reducing rugosity, turning topographically complex hard bottoms into homogeneous soft bottoms. Total diversity of habitats is therefore decreased by sedimentation (e.g., Berkman and Rabeni 1987; A. B. Sutherland et al. 2002; Mol and Ouboter 2004).

Effects on Reproduction, Development, and Physiology

High sediment loads can directly affect spawning behavior and success. In the tricolor shiner, *Cyprinella trichroistia*, increased sediment concentrations led to decreased spawning frequency, disrupted timing of spawning, and a 93% reduction in number of viable eggs (Burkhead and Jelks 2001). The reductions may have resulted from disrupted visual cues associated with increased turbidity and resulting changes in light transmission. Mate selection in cichlids in Lake Victoria is thought to be impaired by turbidity-caused absorption of long and short wavelengths, leading to altered perception of mating colors and increased hybridization (Seehausen et al. 1997a; see chapter 8). Many colorful minnows and darters are sexually and seasonally dichromatic and could be similarly affected by turbidity. The topic deserves investigation.

As sediment covers the stream bottom, darters and other fishes lose sites for egg deposition and nursery habitat (Burkhead et al. 1997). Sediment-intolerant species (several darters and minnows, sculpins) deposit eggs in cracks and crevices and leave them to fend for themselves. Even darters that occupy lowland habitats and are otherwise unparticular about where they spawn, such as the endangered relict darter, *Etheostoma chienense*, can be limited by the availability of hard, silt-free breeding substrate (Piller and Burr 1999). Salmonids similarly require clean gravel bottoms for spawning. Cobble bottoms are the winter refuge for many salmonid fry, and fry density decreases with increased percent embeddedness. Summer rearing capacity of a stream for salmonids is a function of pool depth; where excessive sedimentation decreases pool depth, salmonid densities suffer (T. E. Waters 1995). Similarly, young-of-the-year darters and sculpin prefer stream pools that have silt-free bottoms with relatively coarse substrates (Greenberg and Stiles 1993).

Sediment can kill eggs and young directly by suffocation and scouring. Smothering, entrapment, and abrasion of fish eggs by silt are primary causes of prelarval mortality in benthic-spawning cold-water and warm-water fishes (D. W. Chapman 1988; T. E. Waters 1995; Montgomery et al. 1996). Very young fishes in particular find their respiration impeded, as gills become clogged and abraded by small particles (A. B. Sutherland 2005; figure 5.1). At high sediment concentrations, oxygen exchange sites in the gills malfunction, and fish die from anoxia and lethal buildup of metabolic wastes such as carbon dioxide and ammonia (Ritchie 1972; Turnpenny and Williams 1980). Fishes that survive may suffer reduced growth. Degradation of juvenile nursery or rearing habitat may be a critical determinant of population size for many stream fishes (Muncy et al. 1979; Elliott 1989). Effects—particularly clogging and abrasion of gills—also apply to older fishes (e.g., Redding et al. 1987; Bergstedt and Bergersen 1997).

Most studies have focused on the effects of fine silt, but sand also has a significant impact on stream fauna because of its shifting and scouring properties (e.g., C. P. Hawkins et al. 1983). After five years of experimental additions of sand to a trout stream in Michigan, sediments increased fourfold, and brook trout declined by 50%. Habitat diversity decreased as the stream channel became a run of shifting sand without pools or riffles. Trout populations did not

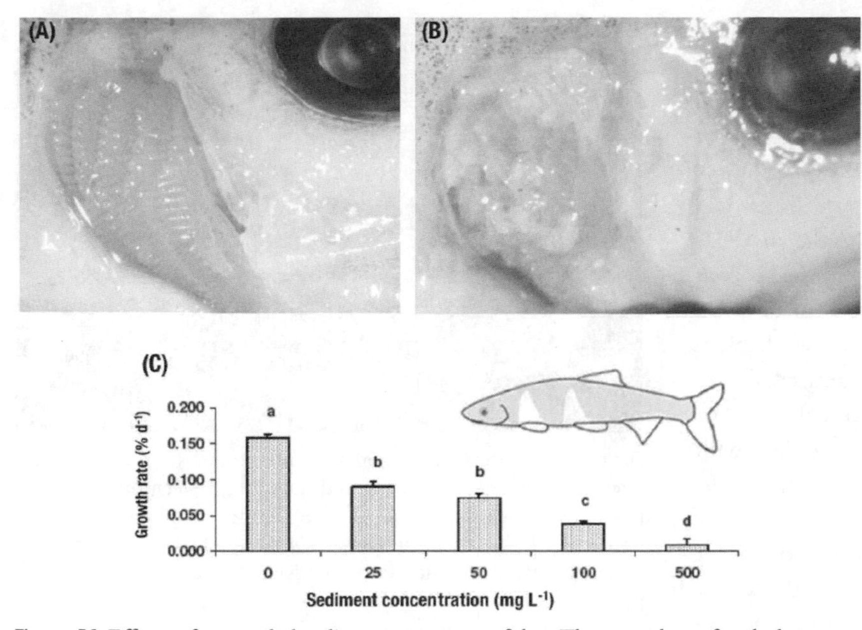

Figure 5.1. Effects of suspended sediments on young fishes. Threatened spotfin chub, *Erimonax monachus*, were raised at various sediment concentrations to study effects on gill morphology and growth. (A) Gill arches and filaments of a young spotfin chub reared for 21 days at low (0 mg/L) sediment concentrations. (B) Gills from similarly aged chub reared at high (500 mg/L) sediment concentrations; note thickening and fusion of filaments and clogging with mucous. (C) Growth rates of young spotfin chub reared for 21 days at various sediment concentrations, showing significantly decreased growth at higher sediment loads; growth rate at highest sediment level was 1/15 that in clean water. High sediment concentrations tested (500 and 100 mg/L) occur for more than one month and six months, respectively, in the wild as a result of watershed development, well within the range of concentrations experienced by these imperiled minnows. From A. B. Sutherland (2005); used with permission. Chub drawing by A. B. Sutherland.

return to pretreatment levels until five years after cessation of sand inputs (Alexander and Hensen 1986).

High sediment loads may also stress fishes indirectly by lowering their resistance to parasites, disease, and contaminants. Studies primarily on salmonids have shown increased incidence of bacterial infection, kidney disease, and protozoan colonization when suspended sediments are at high concentrations. Arctic grayling were more sensitive to pentachlorophenol when the sediment load was high (see T. E. Waters 1995). Increased stress may cause a compromised immune system, indicated by increased plasma cortisol, plasma and serum glucose, and hematocrit (D. S. Lloyd 1987; Redding et al. 1987; Schreck 2000; see chapter 7). Contaminants can be physically adsorbed onto sediment particles, increasing impact; concentrations of pesticides and herbicides tend to be higher in habitats with high sediment loads (Manigold and Schulze 1969).

Effects on Feeding

Sediments affect both food availability and feeding behavior. Infilling of the stream bottom eliminates habitat for insect larvae on which fishes feed (Lenat and Crawford 1994). Fishes that typically decline in numbers are particularly dependent on riffle-dwelling, sediment-intolerant invertebrates, such as the "EPT taxa" of mayflies (Ephemeroptera), stoneflies (Plecoptera), and caddisflies (Trichoptera). These stream insects live under cobbles and among wood and leaves. Riparian deforestation leads to smothering of cobbles and eliminates the source of woody debris and leaves. Decreased food availability affects fish growth and eventually causes fishes to abandon food-poor areas.

The ability of fishes to find prey in the water column diminishes with increasing turbidity (Newcombe and MacDonald 1991; T. E. Waters 1995). Salmonids are heavily dependent on vision for feeding; their food intake often decreases as sediment levels increase. Typically, reaction distances decrease at higher turbidities because small objects must be closer before they can be seen (P. A. Ryan 1991; J. C. Barrett et al. 1992; Sweka and Hartman 2003). A 50% reduction in reaction distance translates into a 90% reduction in the reaction volume hemisphere in which a fish typically views objects, indicating a substantial decrease in an individual's foraging space (O'Scannell 1988). The visibility of small objects is additionally obscured because of poor light penetration; food items blend into the background of silt particles. Feeding is impaired even when food is temporarily abundant, as may happen if the density of drifting invertebrates initially increases with pulses of sediment (e.g., P. A. Ryan 1991). However, invertebrate abundance often declines in sedimented waters because of habitat smothering, so any increase in prey availability is short lived.

Fishes apparently find increased turbidity uncomfortable and unpleasant. Coughing rates increase as they attempt to clear their gills of silt (Servizi and Martens 1992). Divers frequently note that fishes remain closer to structure and are less approachable when underwater visibility is reduced, suggesting increased wariness during turbid conditions (Hobson 1979; E. Hobson, pers. comm.). Such wariness will depress foraging success because fishes forgo feeding in favor of predator avoidance (Lima and Dill 1990), whether or not predators materialize out of the gloom. Some fishes will abandon areas of increased sedimentation (P. A. Ryan 1991; T. E. Waters 1995). Several salmonids actively disperse from such areas, even if the increase is relatively small (>20 NTU). Because fishes typically live in preferred habitat, where feeding, survival, and breeding are maximized, forced emigration may place them in suboptimal habitats. For imperiled species, many of which are food specialists with stringent ecological requirements (see chapter 4), such shifts will likely accelerate population collapse.

In summary, stream fishes that breed on the bottom or feed on benthic invertebrates are most strongly affected by excessive sedimentation. Many species do both, and we find many imperiled species at this intersection of foraging and reproductive habits. "Species that are especially sensitive to high levels of sedimentation are those that evolved in upland streams where historic levels of sedimentation were usually low and water transparency was normally high" (Burkhead et al. 1997, p. 409). Two U.S. federally listed minnow species, the blue shiner, *Cyprinella caerulea*, and the spotfin chub, have suffered range fragmentation and localized extirpations directly linked to sediment impacts on breeding behavior (Burkhead and Jelks 2001). From the standpoint of stream fishes—at the very least, the minnows and darters that normally inhabit riffles in relatively clear water—excess sediment is an unquestioned liability.

Livestock Grazing

Rearing livestock also leads to streamside and instream disturbance (figure 5.2). Cattle grazing, particularly in the arid western U.S., has been especially detrimental, although any activity near water that involves large numbers of cattle, pigs, sheep, poultry, or cultured fishes can have severe impacts (see chapter 7). Livestock grazing occurs on approximately 200 million ha of public and private land in the 11 contiguous states of the western U.S.; about 120 million ha are considered to be in fair or poor condition as a result (Armour et al. 1991).

The impact of cattle grazing is strongest in riparian areas because cattle spend a disproportionate amount of time near water (Kauffman and Krueger 1984; Fleischner 1994).

Figure 5.2. Where riparian zones are not fenced, livestock frequently enter streams, grazing and trampling vegetation and retarding its growth, thus accelerating erosion and inputs of excess nutrients. All these activities have negative consequences for stream habitats and fishes. Photo by Matt McTammany, used with permission.

Several imperiled fishes (e.g., Paiute cutthroat trout, Apache trout) are directly threatened (Kondolf 1994; Clarkson and Wilson 1995). Cattle eat and trample riparian vegetation, trample streambanks, and compact soil. In heavily grazed areas, vegetation disappears, water infiltration is reduced, erosion accelerates, sediment increases, stream temperatures rise, coliform counts go up, and invertebrate densities and fish abundance decline (e.g., Wohl and Carline 1996).

Solutions to the problems are fairly obvious and effective. Fencing cattle out of riparian areas leads to lowered stream temperatures, increased number of pools, channel stabilization, increased riparian vegetation, and increased fish numbers (e.g., Magilligan and McDowell 1997). Economic aid or tax advantages to farmers who keep their cattle out of streams would expedite matters. Armour et al. (1991) emphasized increased public and professional awareness of the value of riparian areas and the impacts cattle have on them. They also recommended increasing grazing fees to discourage overgrazing and to pay for remediation. Such action would prevent gross inequities such as the $12,000 in grazing fees collected versus $260,000 spent to mitigate grazing damage next to a California stream (Kondolf 1994).

Other Sources of Sediment

Although much literature on sediment in streams focuses on riparian function, other sources can be substantial. Any extensive land-disturbing activity can mobilize sediment, which then flows over land as runoff, through storm sewers, and into water courses. Sources include agriculture, forestry operations (especially clear-cutting), home construction, commercial development, golf course construction (recreational clear-cuts), utility and road construction (especially across streams; see chapter 6), and urbanization in general. Worldwide, human activities such as mining, road and house building, and agriculture displace about 45 billion tons of earth annually, much of which winds up in aquatic habitats (Hooke 1999; the U.S., with only 6% of the surface area, accounts for 18% of earth movement). Best management practices (BMPs) have been developed to minimize impacts from all such activities (see International Stormwater BMP Database, www.bmpdatabase.org). BMPs often emphasize installing and maintaining sediment retention devices such as silt fences, riprap, hay bales, and retention ponds.

Significant impacts also arise from sediment already stored in the stream channel or banks. Massive quantities of sediment have been flushed into streams and rivers over decades from agriculture, mining, and road building. These stored sediments are remobilized when streambeds and banks erode, especially during storms. Streambed and bank erosion is particularly prevalent where stormwater runoff is inadequately controlled, as in areas of much impervious (nonporous) surface. Water rushes off the land and into streams directly or via storm drains, bypassing riparian areas, regardless of vegetative cover.

Complications in Understanding and Measuring Impacts

Our understanding of sediment's impacts on stream fishes suffers from taxonomic and habitat bias, incomplete information, flawed metrics, and some impatience. For example, we know the most about the effects of sediments on salmonids (data follow money), and relatively little about the loads at which nonsalmonids such as darters and minnows are affected. Some taxa suffer sediment-induced reductions in diversity and abundance, but not all species respond similarly. Response differences thus complicate the measurement of impacts.

Diversity and Density Measures

Diversity and density are commonly used to measure sediment impacts. Diversity may be a flawed metric of disturbance, however, because diversity indices simply indicate how many species occur at a site, regardless of species replacements (e.g., Fausch et al. 1990; Angermeier and Winston 1997; M. C. Scott and Helfman 2001). Often, riffle-dwelling, benthic-feeding and -breeding fish decline while pool and water-column dwellers increase (Berkman and Rabeni 1987; Rabeni and Smale 1995; E. B. D. Jones et al. 1999).

Fish density may also be a poor indicator of disturbance because numerical abundance may be relatively high in deforested and agricultural areas (e.g., C. P. Hawkins et al. 1983; Schlosser 1991; Harding et al. 1998). Stream invertebrate productivity may increase in response to cattle and otherwise fertilized crops in streamside pastures. These nutrient inputs, combined with increased temperature and solar radiation from removing trees, promote the growth of algae. Algivorous invertebrates and invertebrates that thrive in relatively anoxic pool bottoms (many oligochaete and annelid worms, midge larvae) multiply, and insectivorous fishes that eat them benefit from human disruption of the riparian zone. Habitat diversity in the stream can decline as a by-product of riparian clearing, but the abundance of sediment-laden pool habitats may increase (E. B. D. Jones et al. 1999). Pool-dwelling, schooling minnows, which can attain high densities, often dominate pasture streams. One large school of minnows in a pool can inflate abundance measurements in a sample, overwhelming losses of rarer species.

Sediment-Loving Fishes

Some species and age classes thrive on the sediment that flows naturally into most aquatic systems. Feeding rates of larval Pacific herring are maximal in highly turbid water (as high as 1,000 NTU; Boehlert and Morgan 1985). Increased turbidity apparently favors creek chubs, *Semotilus atromaculatus*, over brook trout, *Salvelinus fontinalis*, chubs actually preferring highly turbid water (Gradall and Swenson 1982). P.W. Smith (1968) found a positive correlation between creek chub abundance and turbidity in Illinois streams. Decreased turbidity in naturally turbid areas can also cause problems. Endangered razorback suckers, *Xyrauchen texanus*, in Colorado River impoundments suffer from increased predation because dams have led to diminished turbidity. Where water is unnaturally clear, native Colorado pikeminnows, *Ptychocheilus lucius*, and introduced green sunfish, *Lepomis cyanellus*, feed on young suckers more than when turbidities are at higher, more natural levels (Johnson and Hines 1999).

Many fishes in a variety of habitats are specifically adapted to finding prey hidden in sediments; these include suckers and cichlids among freshwater families. Some feed by winnowing. They ingest sand and sediment, churn it in their mouth, and sieve out sediment-dwelling invertebrates. Increased sediment is also likely to favor species less reliant on vision for feeding, which may partially explain why visual feeders such as trout, many minnows, and some darters are replaced by species that use olfaction, gustation, or touch, such as catfishes and suckers.

An interaction between sediment and breeding biology also causes species replacements. Fishes such as centrarchid sunfishes and cichlids dig or build nests on the bottom and then sweep sediment away from developing eggs and embryos. Bluegill sunfish use their fins to sweep clear a spawning location in pure sand or silt, exposing small rocks, twigs, or shells on which females will lay eggs (figure 5.3). Sunfishes are also more tolerant of elevated temperatures than darters, benthic minnows, madtoms, and salmonids, which require cool, sediment-free, shaded riffles for breeding. Sunfishes typically increase in diversity and density in streams that flow through pastures and other deforested regions (M. C. Scott 2001; M. C. Scott and Helfman 2001).

Many minnows (e.g., stonerollers, river and creek chubs, cutlips minnows) build pebble pile nests in which they spawn. These small high-rises of clean rock may contain thousands of pebbles, carried several meters and piled up by a male, which then attracts females to spawn. The elevated piles provide structure into which females deposit eggs and which is continually cleaned by the nest-tending actions of the male. Many other minnow species, known as nest associates, spawn in these pebble-pile nests; their young are kept sediment and predator free by the nest-guarding male (Wallin 1992; Johnston 1999). Hence chubs and other nest builders may serve as classical keystone species whose presence and behavior make it possible for other species to exist in areas that sediments would otherwise make unsuitable. River chub have greater relative abundance downstream from deforested riparian patches, as do their nest associates (E. B. D. Jones et al. 1999). Again, diversity may remain constant or even increase as sediment intolerant species are replaced by species that tolerate sediment or associate with species that make sediment tolerable. The displaced species include many that appear on regional and national lists of imperiled species.

Figure 5.3. A male bluegill sunfish over a pit nest on a pure sand bottom in a Florida spring. He is surrounded by abandoned nests from several other males. Bluegills and other sunfishes do well in sediment-rich habitats because they clear sediment from the bottom to create a nest and then continually clean incoming sediment off eggs and larvae. Photo by G. Helfman. Used with permission of the American Fisheries Society.

Most measures of stream quality and fish occurrence decline once a threshold proportion of a watershed is subjected to logging, agriculture, home or road building, or urbanization. This emphasizes the pitfall of focusing solely on riparian zones as opposed to the entire watershed (Newcombe and Jensen 1996); riparian area function cannot make up for all of the damage done to the larger landscape. Work in Michigan and Wisconsin suggests that conversion of 35% or more of wetlands and uplands to agriculture may affect stream fishes as much as or more than riparian clearing (N. Roth et al. 1996; Wang et al. 1997). As bad as agricultural practices can be for streams, they pale in comparison to urbanization. In the Wisconsin study, habitat degradation and biotic integrity did not decline noticeably until more than 50% of the watershed was converted from forest to agriculture, whereas only a 10%–20% conversion to urban land use precipitated a steep decline in biotic integrity (Wang et al. 1997). A key feature of urbanization is the creation of impervious surfaces, a euphemism for paving the landscape. A one-acre parking lot allows 16 times more runoff of water and associated contaminants than a comparable-sized meadow (Schueler 1994). Smoothing, compacting, and converting only 5%–10% of a watershed into streets, sidewalks, parking lots, rooftops, and playgrounds can significantly degrade stream function (Booth and Jackson 1997).

The important point is that deforestation-caused sediment inputs lower habitat diversity, often eliminating unique endemics while favoring widespread, sediment-tolerant generalists. Geographic ranges of sediment-intolerant species contract because they avoid moving through or across sedimented regions, thus the amount of habitat available to them is progressively constricted and fragmented (Burkhead et al. 1997). The biological calamity—regardless of the metrics applied—is that shifts in species distribution and composition are human induced; the rate of change in species composition is accelerated by human activity; and the species that are eliminated from an area tend to be relatively rare, intolerant, and imperiled.

Sediment in Other Settings

Negative impacts of erosion-caused sediment deposition are documented for stream, river, lake, and estuarine fishes in Europe, Australia, New Zealand, South America, Zimbabwe, and South Africa, among others (see Alabaster and Lloyd 1982; Bruton 1985; I. C. Campbell and Doeg 1989; Ryan 1991; Wood and Armitage 1997). Siltation has been directly linked to native fish declines and assemblage disruption in Sri Lankan (Moyle and Leidy 1992) and South American (Mol and Ouboter 2004) streams. Where selective logging rather than clear-cutting occurs, impacts on habitats and fish assemblages can be minimal (e.g., in Malaysia, Martin-Smith 1998); but stream and river fishes everywhere are susceptible to suspended and deposited sediment that results from poorly managed logging, farming, mining, and home and road building (figure 5.4).

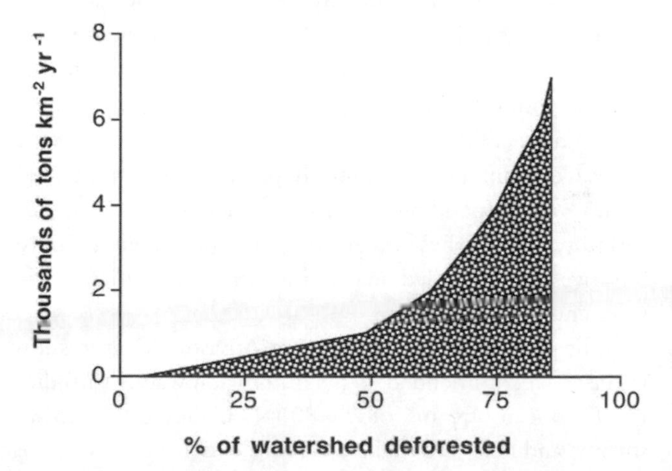

Figure 5.4. Sediment loads in African and Asian rivers as a function of watershed vegetation. The amount of sediment washed into rivers (expressed as tons of sediment deposited per km^2 of watershed per year) is a direct and increasing result of deforestation in the surrounding landscape. After Starmans (1970).

Lakes

Tropical deforestation receives much-deserved attention because cutting and burning tropical trees degrade terrestrial biodiversity and accelerate global climate disruption. However, tropical aquatic systems are also dependent on the complex vegetation of the surrounding forests and jungles, and deforestation has multiple, serious impacts on receiving rivers and lakes (L. J. Chapman and Chapman 2003; C. A. Chapman and Chapman 2003).

Lake Chilwa is a shallow (<3-m), turbid, saline lake in Malawi, in which water levels fluctuate markedly. During wet years, lake surface area is about 675 km^2, with another 1,150 km^2 of swamps, marshes, and seasonally inundated grasslands (Bruton 1985). In dry years, the entire 2,000 km^2 system may go dry. Despite such periodic harsh conditions and high turbidities, the lake is productive and supports large populations of three zooplanktivorous species—a catfish, *Clarias gariepinus*; an endemic cichlid, *Oreochromis shiranus chilwae*; and a cyprinid, *Barbus paludinosus*. High turbidity (5–9 cm visibility) inhibits macrophyte growth, making extensive *Typha* cattails the major structure, as well as providing food in the form of decaying plant material. The cichlid eats zooplankton and plant detritus, and the catfish eats the other fishes, especially the cyprinid.

The normally high turbidity becomes lethal, however, due to agricultural disturbance that increases sediment inputs. When strong winds turn over the increased bottom sediments, the lake goes anoxic, killing off the cyprinid and the endemic cichlid (the catfish breathes air and can survive). Anoxia is aggravated by algal die-offs, which are stimulated by agricultural runoff. A similar pattern of anoxia and fish kills from resuspension of bottom sediments has been reported for other African lakes, including Lake George, Lake Victoria, Mweru-wa-ntipa, and Lake Chad (Bruton 1985). Because many African lakes have high levels of endemicity among both cichlid and noncichlid species, such turbidity-induced mass mortalities are a threat to regional biodiversity. A similar sequence of impacts and results threatens the endemic fishes of Klamath Lake, Oregon (NRC 2004b).

Lake Victoria, the world's second-largest lake, was a showcase of evolution among fishes, with a species flock of perhaps 500 haplochromine cichlids (chapter 8). In the early-1960s, large (2-m, 200-kg), predatory Nile perch were stocked in the lake and came to dominate the fish catch, at the expense of the endemic cichlids on which they fed. The perch also indirectly increased turbidity, which affected the ichthyofauna. Some of the perch's prey were herbivores, whose elimination led to algal blooms, attendant deoxygenation, and subsequent fish kills. In addition, the necessity of smoking Nile perch over wood fires because of their high oil content prompted deforestation of surrounding hillsides, extensive erosion, and nutrient and silt runoff into the lake (I. Payne 1987; D. J. Miller 1989). Silt and algae further reduced water transparency, from 2.5 m to less than 1 m in many areas. Turbid water selectively filters out relatively long (reddish) and short (bluish) wavelengths of light, creating a relatively monochromatic light environment that impaired the ability of females to discriminate among species, thus causing hybridization and a loss of colorful species. Even rock-dwelling cichlids, which were relatively safe from Nile perch, became less colorful and were eliminated. When turbidity "turns the lights off, ecological and species diversity erode rapidly" (Seehausen et al. 1997a, p. 1810).

Estuarine Regions

Estuaries typically and naturally contain much sediment, located as they are at the mouths of rivers. Nutrients are deposited with sediments and detritus, driving productivity and supporting high fish densities, especially of juveniles. Many estuarine fishes are relatively sediment tolerant, and several families contain specialists that feed in sandy regions (e.g., stingrays, mullets, gerreid mojarras, goatfishes). However, many others are surprisingly less so, and their occurrence and distribution are affected by sediment deposition. In the southwest Indian Ocean, sediment tolerance determines the distribution of age classes within species and species within families (Blaber and Blaber 1980; Bruton 1985). For example, in estuaries in Natal, South Africa, juvenile great barracuda, *Sphyraena barracuda*, are sediment intolerant, whereas juvenile *S. jello* are sediment tolerant. Similarly, most juvenile and subadult carangids are relatively silt intolerant, avoiding areas of 20 NTU or greater. But juveniles of two common species, *Caranx ignobilis* and *C. sexfasciatus*, will tolerate turbidities of more than 70 NTU.

Erosion and dredging have caused excessive sand and silt deposition in more than 60% of South African estuaries, diminishing their function as nurseries. Fish kills in South African estuaries have even resulted from sediment carried by flash floods (Whitfield and Paterson 1995). Mass mortalities of juveniles and adults of 16 different species occurred during such floods in January 1995; death resulted from clogged gill filaments and reduced oxygen. And where tropical estuaries adjoin coral reefs, this degradation literally spills over to the reef system (chapter 12).

Solutions to Sedimentation

Sediment inputs can be minimized by employing BMPs that have a sound scientific basis. Similarly, promoting intact riparian zones is an unquestioned way to improve stream water quality. Debate rages, however, over the actual, measurable goals or targets to be sought, the widths of riparian areas to be considered, the types of plants to be used, and the time in which restoration can be reasonably expected.

Sediment Standards

Regulations aimed at curtailing sediment inputs into streams are commonplace, although disagreement exists over the best approach. (See also TMDLs, chapter 7.) Two alternatives are to mandate maximum turbidity levels by restricting activities that increase total suspended sediment above some absolute amount, say 10 NTU; or to restrict activities that increase sediment levels above natural or background conditions—that is, sediment inputs could not raise existing conditions by more than 10 NTU. The former approach is biologically preferable because fish distribution and behavior are influenced by the absolute amount of turbidity in the water column. However, because fishes may acclimate to existing conditions, or because political expediency makes it difficult to halt land-disturbing activities, the latter approach of "additional allowable sediment" is commonly pursued. Recommended values vary depending on natural background levels of turbidity, the fishes involved (salmonids being more sensitive than so-called coarse fish), and the political environment.

The European Inland Fisheries Advisory Commission (EIFAC) recommended a maximum allowable turbidity for European streams of 25–80 NTU, applicable to all streams and taxa (see Alabaster and Lloyd 1982; P. A. Ryan 1991). Although such standards seem appropriate for sediment-tolerant species, they would be inadequate for salmonid streams. Meyer et al. (1999) divided southern Appalachian streams into low- and high-turbidity groups, separated at 15 NTU during base flow conditions. Low-turbidity streams contained rainbow trout, most darters, sculpin, and bottom-feeding or crevice/gravel-breeding minnows. High-turbidity streams were more likely to contain centrarchid sunfishes. Brown trout and nest-building minnows occurred in both types. Meyer et al. concluded that rainbow trout and darters could tolerate turbidities above 10 NTU during only 20% of base flow periods.

Examples of the additional-allowable-sediment approach are more common. Many U.S. states permit activities that increase turbidity by 5–25 NTU above "natural" levels; slightly lower values often apply to salmonid streams (see D. S. Lloyd 1987). Canadian Water Quality Guidelines (see P. A. Ryan 1991) allow up to a 10-NTU increase when existing conditions are less than 100 NTU, and no more than a 10% increase when existing levels exceed 100 NTU. In New Zealand, P. A. Ryan (1991) recommended allowable increases of 2–10 NTU under most conditions, taking into account season, stream size, and ambient turbidity. Implementation and enforcement of these recommendations are another matter.

Riparian Buffer Widths

Because intact riparian zones are critical for stream integrity, riparian corridor protection has gained popularity. The challenge is deciding just how wide a buffer should be to protect stream habitats, water quality, and terrestrial wildlife habitat. Wenger (1999) recommended an adaptive management approach that was responsive to local conditions, resulting in buffers that vary as a function of the steepness and nature of the surrounding landscape. Wenger suggested a minimum buffer width of 30 m, with an additional 0.7 m added per 1% of slope in the riparian zone, thus controlling for accelerated runoff across steeper landscapes. All potentially destructive or disruptive activities should be disallowed in the buffer zone, which should contain native plants (no cutting and replanting with exotics).

Thorough, scientifically based requirements for riparian protection have been developed for Maryland and Virginia watersheds of the Chesapeake Bay. California, Maine, New Hampshire, and Rhode Island also have relatively strong coastal riparian protection laws (see Desbonnet et al. 1994, table 8). Otherwise, comprehensive legislation is spotty, often mandating fixed widths of less than 30 m, despite validated science showing the value of greater widths. For example, Georgia requires 50-ft buffers (reduced from 100 ft in 2001), and only along streams cold enough to sustain (usually introduced) trout; agriculture and forestry practices are exempted, and nontrout streams receive only 25-foot buffers. While public sentiment and political action can be expected to favor trout fishing, warm-water streams are popular fishing areas in Georgia and throughout the South (e.g., Fisher et al. 1998). Additionally, 29 of 55 fishes considered imperiled by Georgia state agencies, including 7 of 8 federally listed fishes, live in streams and rivers too warm to support trout. Most of these fishes are relatively intolerant of sedimentation (B. J. Freeman, pers. comm.).

Riparian Regrowth and Stream Recovery

Riparian fauna and flora are remarkably resilient, because riparian regions are dynamic and subject to periodic, natural disturbance (Kauffman et al. 1997). Where extensive deforestation without clear-cutting has occurred, reforestation is generally successful (e.g., Duff et al. 1995; Wichert and Rapport 1998). Clear-cutting, with its attendant land-destroying activities and effects (road building, skidding, bulldozer and tractor damage, soil compaction, landslides), has the greatest impact and takes longest to repair. The time required for recovery depends on local hydrology and topography, type of disturbance (e.g., cattle grazing, logging extent and methods), and ecosystem function in question. Recovery can be rapid once the agents of degradation are removed, but it can also be slow. Stream flow in an old-growth *Eucalyptus* forest in southeastern Australia doubled in the year following clear-cutting but required 13 years to return to precutting levels (I. C. Campbell and Doeg 1989). In an eastern Oregon river, channel morphology had not returned to the nondegraded condition after 14 years of cattle exclusion from the riparian zone (Magilligan and McDowell 1997).

Simply replanting a riparian zone without regard to the coevolved nature of terrestrial and aquatic components may not alleviate the effects of deforestation. In an Australian forest cleared of native *Eucalyptus* and then replanted in nonnative pine, densities of stream invertebrates that colonized leaf bags declined, presumably because they found foreign pine needles less palatable. Moreover, problems may arise when old growth is replaced with native but pioneer species, as also shown in Australia, where stream insects were unable to process the leaves of

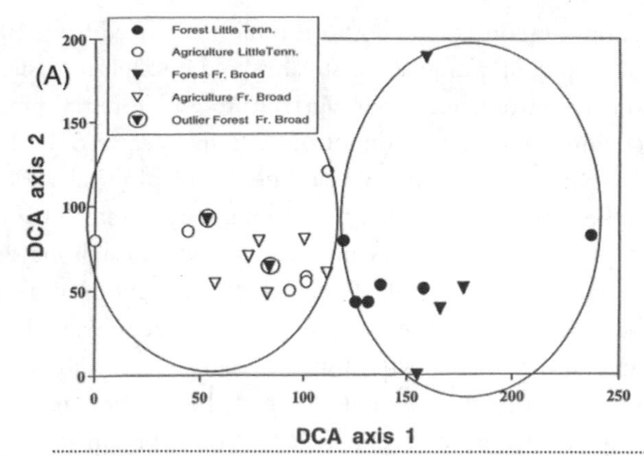

(B)

	Little Laurel Creek	Big Laurel Creek	Other forest sites	Agric. sites
Diversity	15	14	10.2	16.8
Species in common	12	12		
Species shared w/			8	11
Species not shared w/			16	4
Jaccard's Similarity w/			0.16	0.32***
Abundance	1131	1107	577	1772
Indicator species:				
sculpin	"0" (1 fish)	0	16/16	12/24
trout (3 spp.)	0	"0" (1 fish)	3/3, 16/16	1/3, 2/24

Figure 5.5. Stream recovery following intensive agriculture. (A) Detrended correspondence analysis (DCA) shows two groups of stream sites. Each point represents the aquatic insect assemblage of a different stream, sampled in the 1990s. Open triangles and circles in the left oval represent streams flowing through agricultural watersheds; closed triangles and circles in the right oval represent streams flowing through forested watersheds. Two outliers (Little Laurel and Big Laurel creeks, closed triangles inside open circles) were surrounded by >90% forested watersheds in 1990 but were only 55% forested in the 1950s and earlier. Even after five decades, aquatic insects in the two Laurel creeks were more typical of agricultural than forested streams. Table (B) shows similar trends: Laurel Creek fishes were more characteristic of agricultural than forested streams. Little Tenn. = Little Tennessee River system; Fr. Broad = French Broad River system. From Harding et al. (1998); used with permission of the National Academy Press.

an early successional acacia species replanted in a deforested riparian zone (I. C. Campbell and Doeg 1989).

We also cannot conclude that regrowth of a native, mature forest will lead to recovery of stream function and fish composition on the same time scale. Even after extensive logging, mining, and farming cease and a forest recovers, nearby streams may carry the legacy of insults to the land for decades. Southern Appalachian streams flowing through watersheds that were intensively farmed early in the 20th century carried heavy sediment loads more than 50 years later—mainly from resuspension of sediments deposited in the streambed decades earlier—despite forest regrowth, more than 90% forest cover in surrounding watersheds, and intact riparian zones. Both fishes and invertebrates in local streams were sediment-tolerant species, more characteristic of pasture than of forested streams (Harding et al. 1998; figure 5.5).

Channelization: Extreme Disruption and Remediation

River straightening or ditching, usually called *channelization*, frequently results from misguided efforts to move water quickly, often under the guise of flood control. The process is straightforward: Bulldozers straighten the bends, smooth the bottom, cut and "stabilize" the banks so they are angled uniformly, build levees along the tops of the banks to raise their height, and then cover the banks with large stones ("riprap") or concrete. Channelization reduces seasonal inundation of the floodplain, which normally receives overflow during seasonal rains and provides nursery habitat for many riverine fish species (e.g., Bayley 1995; Galat et al. 1998). But because channelization often accompanies deforestation of the floodplain to allow for agriculture and housing development, the entire hydrologic regime of a river is altered, and flooding during abnormally high flows can increase (P. W. Simpson et al. 1982; Moyle and Leidy 1992; Mattingly et al. 1993). The catastrophic flooding of the Mississippi River in 1993 was partly due to decades of channelization (M. A. Myers and White 1993), as was the inundation of New Orleans in 2005.

Yount and Niemi (1990) and Niemi et al. (1990) cataloged ten major categories of riverine disturbance and concluded that recovery was slowest when the actual physical properties of a stream were altered. They identified channelization as "one of the most severe disturbances to which streams are subjected. [Natural] recovery . . . requires at least decades" (Yount and Niemi 1990, p. 558). Channelization alters the natural movement and deposition of various-sized particles among habitats as the stream is diverted from its natural, sinuous, accelerating and decelerating path. Channelized systems consequently have low habitat heterogeneity and flow faster after storms. The stream becomes a ditch or pipe; in fact, many smaller streams, especially in urban areas, are simply diverted through pipes. The biotic community, not having evolved in pipes, responds accordingly. Channelized rivers either lack fishes or are dominated by a few generalist or introduced species. Hardest hit are big river species, species dependent on sandy and shallow areas, and fishes that use backwater areas or floodplains. In North America, affected groups include sturgeons, paddlefish, and *Ammocrypta* sand darters. Channelization-induced loss of the floodplain along the Lower Mississippi River reduced the standing biomass of fishes by an order of magnitude (Rosenberg et al. 1997).

Fish assemblages are greatly altered after channelization. On a relatively small scale, Middle Creek, Tennessee, was channelized annually for flood control, beginning in 1967. As described by Etnier (1972, p. 372), "Bulldozers were used to widen and straighten the channel, smooth the banks, and create a nearly uniform gradient . . . to prevent the development of obstructions to the smooth flow of water." Before channelization, the stream was 5 m wide, contained pools and riffles, and had a bottom of gravel and boulders. After channelization, it was 8 m wide, 15 cm deep, with a medium gravel bottom. Habitat specialists declined: Greenside darters, speckled darters, and banded sculpins disappeared, and snubnose darters decreased more than 80%. Habitat generalists were favored: Northern hogsuckers, stonerollers, and sand shiners remained stable or increased in abundance. Etnier attributed the changes to substrate instability, decreased habitat variability, and decreased invertebrate diversity.

More than half the freshwater fishes in western and central Europe are either extinct or seriously threatened, and channelization is believed largely responsible. Following World War II, the River Melk in Lower Austria, a fifth-order stream, was converted into an "absolutely monotonous" straight ditch with a paved bottom covered mostly by coarse gravel and pebbles (Jungwirth et al. 1995, p. 195). The channelized river had no riffles or pools; it was one long, flat, straight run, 10 m wide, 0.3–0.6 m deep, with a mean current speed greater than 0.3 m s^{-1}. Its "riparian zone" consisted of riprap, with little or no vegetation. Fishes were relatively uncommon, averaging around 150 individuals of ten common salmonid and cyprinid species and less than 20 kg per 100-m reach. Two cyprinids, chub *Leuciscus cephalus*, and gudgeon *Gobio fluviatilis* made up over 90% of the individuals. Age structure was also fairly monotonous, dominated by a few year-classes, with few YOY or juveniles.

In 1988, restoration began. Bottom pavement was removed, the river cross section was enlarged, and groins and bedfalls were added to the bottom. Results were rapid and dramatic. Within three years, depths, flow velocities, and substrate types diversified. Riffles and pools developed, and depth varied from less than 0.1 to 2.0 m. Current speed likewise diversified to include regions of zero current as well as regions flowing at more than 0.9 m s^{-1}. Bottom type included mud, sand, gravel, and woody debris, in addition to the structures placed in the stream. The Melk changed from a monotonous run to "riffle pool sequences, gravel banks, bays and backwaters" (Jungwirth et al. 1995, p. 203). A riparian community of early successional grasses and woody plants developed along the stream's banks (see figure 5.6).

The fish assemblage responded as dramatically. Species number doubled from 10 to 19, and evenness increased from 1.4 to 1.9, indicating less dominance by the two cyprinids. Density and biomass almost tripled, to more than 400 individuals and more than 50 kg/100 m. Age structure diversified to include multiple year-classes, including YOY and juveniles. Abundance of juveniles and small species increased almost ninefold in shallow, slow-current areas, habitats that were essentially nonexistent prior to restoration. New habitats included appropriate spawning habitat in spring, slow-flow shorelines in summer, slow-flow rock and woody debris refuges in winter, and, most important, a riparian zone that slowed water currents during flood periods so young were not flushed downstream during high flows. Fishes occupied and reproduced and recruited within the restored portions of the stream. An increase in spatial heterogeneity resulted in a diversity of habitats appropriate for different life history stages at different seasons and flow conditions.

Ditching and channelizing can be easy and cheap, but restoring a river to something approximating natural conditions is a complicated, costly, and lengthy process. In south Florida, after severe flooding following a 1947 hurricane, the U.S. Army Corps of Engineers channelized the Kissimmee River, transforming a 160-km, meandering, shaded, productive river into a 90-km, straight, 9-m-deep, 75-m-wide series of five stagnant pools (Koebel 1995; Whalen et al. 2002). This massive project required nine years to complete and had many unanticipated consequences. About 20,000 ha of surrounding wetlands were drained. Other effects included increased water pollution and eutrophication, periodic flooding, salt contamination of water sources, water table lowering, land subsidence,

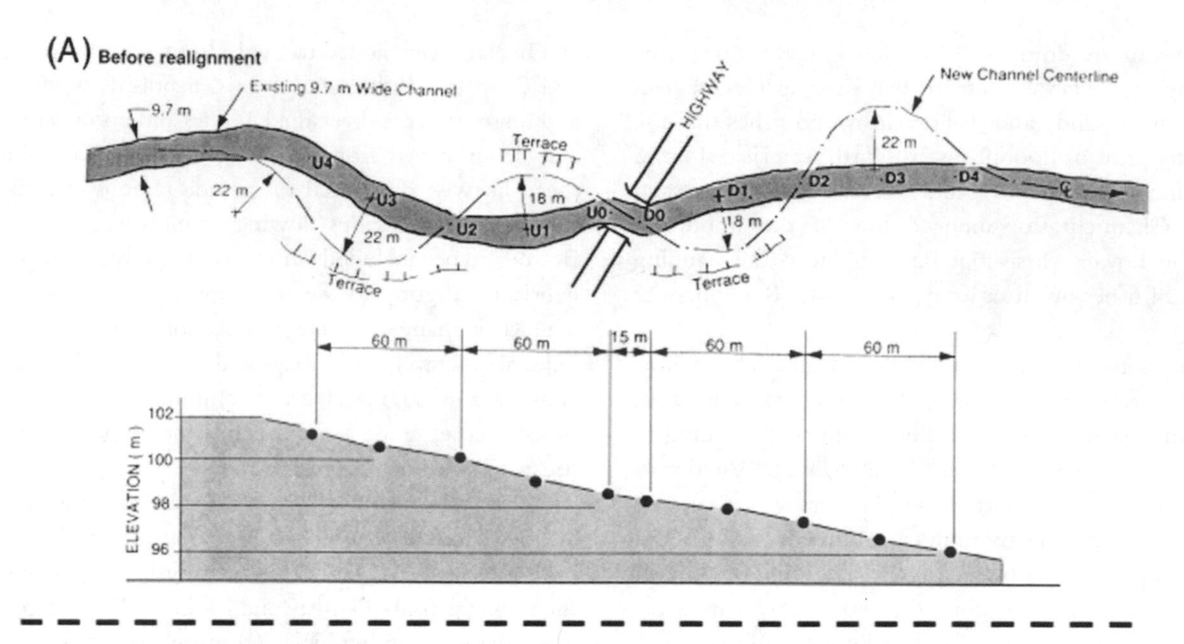

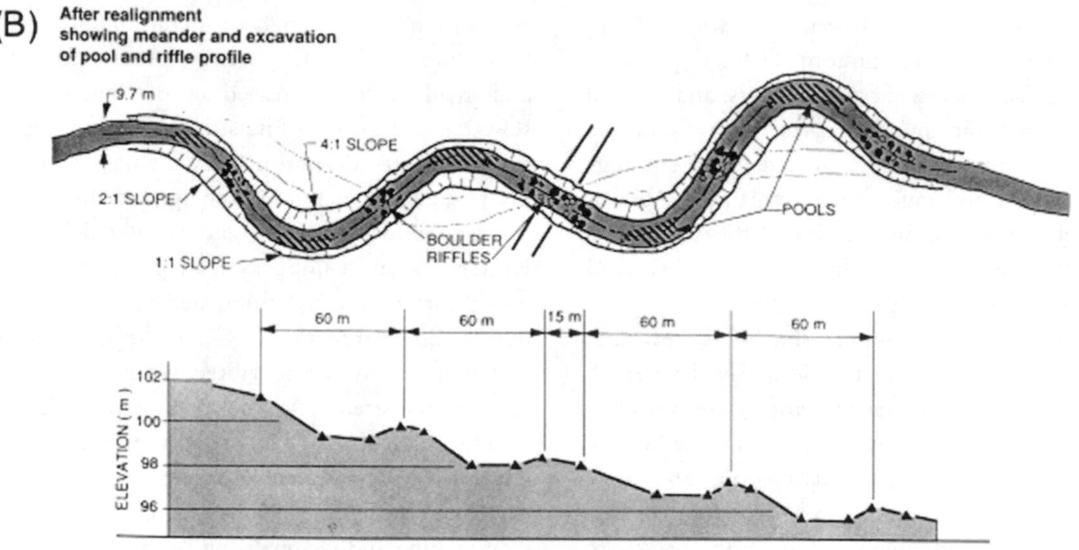

Figure 5.6. Goals in restoring a channelized river (turning a ditch back into a stream). By recreating historical channel structure, a channelized stream can be made to closely approximate natural flow patterns, which in turn alters the deposition of sediments, the occurrence of riffles, and the excavation of pools. Reestablishment of meanders is appropriate where such curves were natural, but forcing a stream to meander unnaturally has failed repeatedly (e.g., Kondolf 1995). Restoration will also be accelerated by replanting of riparian vegetation. From Cowx and Welcomme (1998); used with permission.

oxidation of peat soils, wind erosion, and marsh fires. Wading bird numbers declined 90%, 5 billion fishes and 6 billion shrimp were killed, and six native fish species were extirpated from the Kissimmee River.

In 1976, Florida reconsidered the project. The Corps proposed dechannelization, which began with feasibility studies in 1978–85 and 1990–98. Actual dechannelization is ongoing and will constitute the largest river restoration project ever undertaken in the U.S., requiring more than 13 years and costing over $500 million (this number increases regularly). Restoring the river, which has begun, will take several years more and cost perhaps ten times more than channelization (Koebel 1995, Whalen et al. 2002).

Draining Lakes and Wetlands

As important as lakes and wetlands are for food, water storage, storm mitigation, and recreation, many have been drained to promote agriculture and general development. The impact on lake-dwelling and marsh-dependent fishes is understandably dramatic. "Reclamation" of Lower Klamath Lake, an extensive lake-marsh complex in southern Oregon, in 1927 eliminated perhaps half of the available habitat of two now Endangered sucker species. The agricultural land created proved useless because of the high peat content of the soil, as was predicted by soil geologists (NRC 2004b). Although the practice of draining lakes and wetlands has slowed in most western countries and dams are regularly built to create new lakes in many locales (see chapter 6), lake drainage on a large scale is ongoing in some countries. In China loss of habitat through lake draining has had far-reaching impacts on the fishes of the Yangtze River basin. Between the 1950s and the 1970s, lake surface area in the middle and lower reaches of the basin was reduced from 35,000 km^2 to about 23,000 km^2. Entire lakes were drained. In Hubei Province, the number of lakes decreased from 1,066 in the 1950s to 326 in the 1990s. Fish diversity in the Yangtze, the world's third-longest river and home to many imperiled endemics, has suffered dramatically, while nonindigenous species have flourished (Fu et al. 2003).

STRUCTURE AS HABITAT

Humans regularly alter aquatic habitats by removing or destroying structure. Logs and debris dams are removed to aid navigation and as "habitat improvement," bottoms are dredged for navigation and to obtain construction materials, and bottom structure is damaged during fishing operations.

Woody debris is critical instream habitat and exemplifies the interconnectedness of aquatic and terrestrial habitats (Maser and Sedell 1994; S. Gregory et al. 2003). Logs, branches, sticks, leaves, and other types of natural litter are an essential part of streams and lakes, but they all originate on land. These *allochthonous inputs* create habitat for invertebrates and fishes and constitute a food source for many organisms that fishes eat (Harmon et al. 1986; Wallace et al. 1997). Debris dams in streams and logs ("snags") in rivers and lakes are essential ecosystem components. Debris dams trap silt, organic matter, and nutrients; offer a solid substrate for invertebrate attachment; and are a site for processing organic matter, making it available to invertebrates and fishes. Woody debris slows the flow of the water, making nutrients available longer. In coastal, low-gradient rivers, snags are the most biologically rich habitat. Despite low abundance in such systems, snags can contain 60% of the total invertebrate biomass, provide 80% of the drifting invertebrate biomass, and produce four times more prey than mud or sand habitats. Many fishes obtain more than half of their food directly from snags (Benke et al. 1985), and salmon and trout depend on them. As trees fall into a stream, they change its hydraulics and create pools, a favored habitat of resident trout and a refuge for migrating juveniles and adults. Woody debris, especially large woody debris (LWD) is essential for creating and maintaining salmonid habitat (e.g., Bilby and Bisson 1998; see chapter 2 on the relevance of beaver).

Despite these multiple functions, snags have long been viewed by government agencies as hazards to navigation for both humans and fishes. Logs were removed from navigable rivers of the southeastern U.S. beginning in the early 1800s via a process called "snagging" and from northwestern streams via a process known as "stream cleaning," which included bulldozing, straightening, and smoothing stream channels to aid log transport after massive tree-cutting activities. Rail transportation largely replaced river commerce in the 1850s, but snagging by the U.S. Army Corps of Engineers continued until the 1950s. Some state agencies viewed woody debris removal as habitat improvement that facilitated upstream salmon migrations (Sedell et al. 1997); earlier practices included removal of woody streambank vegetation to "improve trout habitat" (e.g., R. L. Hunt 1979). The more recent realizations that snags are essential fish habitat and that logs contribute to the hydrologic processes that create essential fish habitat have led to dramatic changes in management policies (e.g., Beechie et al. 1996). Riparian protection programs often recommend that vegetation buffers be at least as wide as the trees are tall, to ensure that trees or at least their tops fall into the stream (S.V. Gregory 1997).

Today, somewhat ironically, management agencies often add wood to streams to create better conditions for fishes (e.g., Reeves et al. 1991; Kauffman et al. 1997), because the natural restoration process can be slow. Cutting large trees eliminates the source of logs, which may not be replenished for at least 45 years. Projections suggest that it may take 100–400 years for debris dams to return to normal densities in a logged watershed (Silsbee and Larson 1983; Golladay et al. 1987). Again, careless land-use practices have long-term consequences.

The normal supply of wood in a stream represents a balance between inputs and decay. Disruption of that supply can create problems of both under- and overabundance. Although woody debris is a natural ecosystem component, too much can be harmful. It can block a stream channel and impede normal migratory patterns, and increase scouring during high flows (if a large tree tumbling and churning downstream in a flood constitutes "scouring"). Intensive and careless logging and road building can cause excess inputs. Decaying slash and sawdust piles reduce dissolved oxygen levels, and dense organic debris in the streambed can depress interstitial oxygen concentrations. The deoxygenation effects of slash piles and interstitial organic matter can persist for over six years after logging (Likens et al. 1977; I. C. Campbell and Doeg 1989). Stream fishes evolved in a habitat that is usually oxygen saturated and are easily distressed by reduced oxygen; eggs and embryos developing in stream gravel are particularly sensitive.

Mining and Dredging for Sand and Gravel

Flowing water sorts particles into different fractions, and the resulting deposition by size is exploited by humans for building materials. Sand, gravel, and rock can be obtained in relatively pure deposits, there for the taking. Unfortunately, the taking removes habitat and resuspends the unwanted fraction, namely silt and fine sediment. In Texas' Brazos River, a gravel mining operation resulted in downstream sedimentation and reduced overall abundance of fishes, although species adapted to sand-silt substrates increased (Forshage and Carter 1973). Another common impact occurs as mining lowers and steepens the streambed, creating a nick point that moves progressively upstream in a process known as headcutting. During headcutting, substantial quantities of stored sediment are released from the stream channel and moved downstream (Kondolf 1997).

Kanehl and Lyons (1992) reviewed the impacts of sand and gravel mining and conducted their own study of the Big Rib River, Wisconsin (table 5.1). They showed that mining operations have negative, long-term impacts on streams in general and on fishes in particular. Instream mining and excavation of stream-connected floodplain areas (1) alter the stream channel, disrupting habitats and flow patterns, transporting sediments, and increasing headcutting; (2) degrade water quality; (3) disrupt plant growth; (4) reduce and alter invertebrate populations, including mussels; and (5) alter fish populations. Fishes are eliminated or their assemblages altered. Ecological specialists that frequented a variety of stream habitats, particularly riffles, are replaced by generalists tolerant of homogenized, channelized, sediment-laden, run-dominated, dredged streams (see also chapter 9). Recovery of a stream and its fishes can take months to decades following the cessation of dredging.

Dredging and mining thus constitute a disruption to streams that is often worse than riparian deforestation. Sediment is deposited; the streambed's physical structure is disrupted; and accumulated toxins such as mercury, cyanide, and acid mine waste are released. Kanehl and Lyons (1992, p. 28) recommended "that consideration be given to banning all in-stream mining activities." Short of a ban, they recommended that any site be surveyed before, during, and after mining activities; that mitigation (bank stabilization, erosion control, stream channel rehabilitation, revegetation) be emphasized; and that gravel-washing and other sediment-laden wastewater be recycled rather than dumped into the stream.

Marine habitats are also dredged to create harbors and navigation channels and mined for building materials. During dredging of harbors and channels, the upper layers of sediment are removed along with the resident biota. This "spoil material" is often redeposited in an area that may differ both geologically and biologically. Recolonization of dredged areas can occur relatively rapidly, but the colonizing community is often different from what was present before dredging. Several years may pass before original species are reestablished at a dredged site (Watling and Norse 1998a, 1998b). In addition, harbors are major settling basins for contaminants, which have now been redistributed.

Coral reefs are also subjected to mining and collecting (see chapter 12). Fish biomass, abundance, and diversity decline in mined areas because of the dependence of fishes and their prey on corals for food and shelter. Mining of limestone blocks, used in road and home building and as landfill, can reduce coral cover from 50% to 5% due to trampling and sedimentation as well as direct removal. Recovery can take more than a decade, if it occurs at all.

Table 5.1. Examples of instream and nearstream sand and gravel mining effects

Locale	*Nature of mining*	*Effects on habitat*	*Effects on fishes*	*Recovery*	*Reference*
North America					
Big Rib R, WI	Instream mining	Riffles filled with sand and gravel	IBI scores reduced in most altered sites	10–20 yr	Kanehl & Lyons 1992
4 streams, ID	Suction dredging		Entrainment of cutthroat trout eggs, 100% mortality		Griffith & Andrews 1981
Brazos R, TX	Instream gravel removal, washing	Habitat alteration, food reduction	6 spp disappeared, 8 reduced, 11 increased, 4 unchanged	12 km	Forshage & Carter 1973
3 streams, MO	Gravel mining	Increased sediment in riffles	Benthic feeders and spawners reduced, generalists increased		Berkman & Rabeni 1987
25 streams, AK	Instream mining, riparian excavation	Geomorphological changes, silt, temperature rise	Fish density and diversity decreased, char and grayling replaced by sculpin, whitefish		Woodward Clyde 1980
Kansas R, KS	Instream mining	Geomorphological change, siltation	15 spp declined, 10 increased		Simons & Li 1984
Europe					
R. Loire, France	Gravel removal		28% reduced numbers, 17% reduced biomass		Rivier & Seguier 1985
Fruin Water, Scotland	Gravel washing	Siltation of riffles	Salmon, trout spawning eliminated	6 mo	Hamilton 1961
R. Simojoki, Finland	Dredging for log transport	Spawning, nursery areas reduced	33% reduction in parr production		Jutila 1985

Source: Data mostly from Kanehl and Lyons (1992); original references are cited in table for further details.
Note: Recovery = time to return to pre-mining state after mining activities ceased, or distance downstream where no effect of mining was obvious. R = river; IBI = Index of Biotic Integrity (see chapter 7).

Coral mining and harbor dredging may also stimulate the growth of organisms responsible for ciguatera fish poisoning, which is common in areas around an island that have been subjected to disturbance (chapter 7). The toxin originates in unicellular dinoflagellates that grow on reef macroalgae or newly exposed coral surfaces.

TRAWLING'S IMPACTS ON MARINE HABITATS

Many marine fishes are associated with and obviously adapted to living in sandy and silty areas. Hagfishes, rays, skates, angelsharks, flatfishes, gadoids, tilefishes, eels, grenadiers, and gobies often reach their highest abundances in and over relatively structureless bottom. But overall diversity and often abundance are highest where some type of hard structure protrudes from the bottom. Nearshore marine fishes, whether in tide pools or along a rocky coast, live in and among rocks, kelp and other algae, sponges, tunicates, oyster bars, barnacles, and mussels—basically any hard substrate or living organism attached to hard substrate. Such structure provides refuge from predators, places to rest, vantage and ambush points, habitat for prey and feeding, and breeding sites. Juveniles depend heavily on complex physical structure for hiding places. Nearly every study of the 100-plus fish species associated with kelp beds in California has found a positive relationship between biomass of fishes and kelp density (e.g., DeMartini and Roberts 1990). When we disturb the structure, be it coral, kelp, grass, or other living or dead emergent objects, we disturb a feature of "habitat" for fishes. The most insidious of such disturbance comes from bottom trawling.

"Mobile fishing gear" dragged across the bottom is the most destructive of current fishing practices. Watling and Norse (1998a, p. 1178) went so far as to equate it with

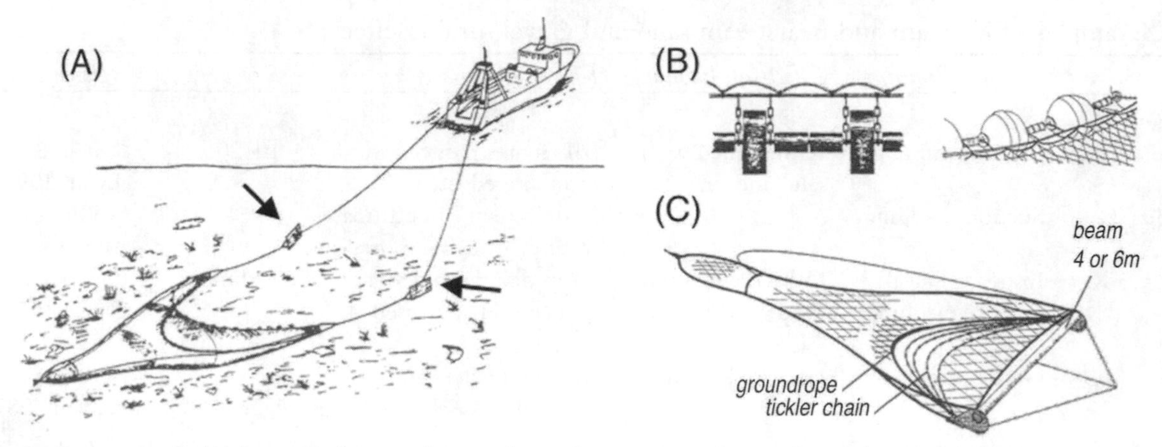

Figure 5.7. Bottom trawls. (A) A trawl being towed across the sea bottom. Note the two "otter doors" (arrows) that precede the net and keep it open. (B) Roller gear attached to the headrope that helps move the trawl over uneven bottom and keep it from hanging up on rocks, coral, and other solid structure. (C) A beam trawl, showing the steel beam at the front, tickler chains just ahead of the net opening, and the actual front of the trawl, called the groundrope. From J. C. Sainsbury (1996); used with permission.

agriculture "as humankind's most important physical disturbance of the biosphere." Mobile gear includes trawls and dredges made of rope, synthetic fiber, chain, chain mail, bars, and metal doors weighing many tons, as well as hydraulic or suction dredges (figure 5.7; Watling and Norse 1998b). It is used from near shore to 200 km offshore at depths as great as 1,800 m. As a trawl is dragged across the sea floor, the seabed is scoured and a plume of suspended material is generated. Severity of disturbance depends on the weight, type, and speed of the trawl; bottom type (particle size and distribution, rugosity); and strength of tides and currents (J. B. Jones 1992). Mobile gear scoops up everything in its path, turning over rocks and destroying attached animals (Raloff 1996; Watling 2005) and bringing to the surface whatever can't out swim the trawl or squeeze through the mesh.

The impact of mobile gear has only recently caught scientific and public attention, in part because science lagged behind the industry. The technology needed to exploit the bottom (sonar, large engines, big nets, improved trawl efficiency) raced ahead of that needed to assess the impacts (submersible and towed cameras, sidescan sonar, LORAN, and global positioning systems). Technological advances increased the weight and speed of trawls, adding to their impact. In 1960, an average trawl weighed 3.5 tons; by 1980, it was 10 tons, with heavier gear penetrating deeper into soft bottoms, increasing disturbance (Gibbs et al. 1980; J. B. Jones 1992).

Studies of the extent and impact of mobile gear indicate staggering damage to the sea floor and to oceanic ecosystems. Few if any trawlable areas on the world's continental shelves remain untrawled or undredged (Pauly and Christensen 1995; Watling and Norse 1998b). Data indicate that in many places, trawls cross every trawlable square meter of ocean bottom annually. Effort is definitely concentrated in certain regions, with specific locales trawled between 2 and 400 times each year (Dayton et al. 1995; Raloff 1996; Watling and Norse 1998b). Each pass removes 5% to 20% of the benthic fauna. Given the slow growth rates of many benthic animals, everything alive is killed or removed in a year or two (Raloff 1996). Biomass and biodiversity decline; the global extent of trawling is so great, however, that few undisturbed areas are available for comparison (M. J. Kaiser 1998). On average, an area the size of the world's continental shelves is trawled every two years. The process and impact of trawling have been likened to clear-cutting under the seas, and the area of the continental shelves trawled annually is 150 times larger than the terrestrial area lost each year to deforestation (table 5.2; Watling and Norse 1998b).

Mobile fishing gear affects the sea floor and in turn the fishes that live there. Living and inert habitat is removed or destroyed, fishes not captured are injured or displaced, sediment is resuspended and redistributed, and nutrient flux and biogeochemical cycles are altered (e.g., Messieh et al. 1991; Jennings and Kaiser 1998; see chapter 11). Fishes are probably harmed most by damage to the benthic invertebrates that provide both habitat and food. Trawling affects benthic organisms by smothering them with sediment; by direct removal; by injury that kills them immediately or

Table 5.2. Similarities between seabed trawling and forest clear-cutting

Effect on:	*Trawling*	*Clear-cutting*
Substrates	Boulders and cobbles overturned, displaced, buried; sediments homogenized, microtopography reduced; long-lasting grooves remain	Soils compressed and exposed to erosion
Infauna and roots	Infauna crushed and buried or exposed to scavengers	Decomposition of roots stimulated then depressed
Emergent biogenic structure and structure builders	Most structure-forming species removed, damaged, or displaced	Above-ground snags, logs, and structure-forming species removed or burned
Associated species	Late-successional species eliminated; pioneer species may persist for decades	Same effect
Biogeochemistry	Accumulated organic matter removed and oxidized, releasing carbon to water column and atmosphere; oxygen demand increased	Similar except carbon released to atmosphere; nitrogen fixation by arboreal lichens eliminated
Recovery to original structure	Years to centuries	Decades to centuries
Interval of repeat	40 days to 10 years	40 to 200 years
Annual global extent	15 million km^2	0.1 million km^2 (net forest and woodland loss)
Latitudinal range	Subpolar to tropical	Same
Ownership of areas	Public	Public and private
Published scientific studies	Few	Many
Public consciousness	Little	Substantial
Legal status	Restricted in some areas	Increasingly regulated and modified to lessen impacts; alternative logging methods and forest preservation promoted

Source: Adapted from Watling and Norse (1998b).

slowly; and by sublethal injury that makes them vulnerable to predators, competitors, and disease and impairs their growth and reproduction (J. B. Jones 1992). In soft-bottom areas, the sea floor is smoothed, obliterating both macro- and microtopographic features, as live and dead structure is knocked over and sediment ripples, accumulations of detritus, and surface traces of infaunal activity (burrow holes and mounds and other signs of "bioturbation") are eliminated (Schwinghamer et al. 1996). Objects not knocked over are moved around: approximately half of all rocks over 5 cm diameter are dislodged and rotated after a single pass (Auster et al. 1996). This smoothing and homogenization reduce habitat complexity, which is again directly related to fish diversity.

Ironically, most faunal decimation is incidental, collateral damage in the assault on the ocean's resources. Emptying a trawl net is a lesson in invertebrate biology and waste because only a fraction of the catch by weight—5% to 10% by some estimates—constitutes target species. The majority is bycatch of small and unwanted fishes and crustaceans mixed with sponges, tunicates, bryozoans, corals, sea fans, sea anemones, sea cucumbers, starfish, sand dollars, clams, worms, and barnacles (e.g., Hutchings 1990). This bycatch—which is the living matrix on which target species feed and in which they shelter, forage, grow, and reproduce—is shoveled back over the side, dead and dying. The decaying mass of dead bycatch settles on the bottom and reduces oxygen levels

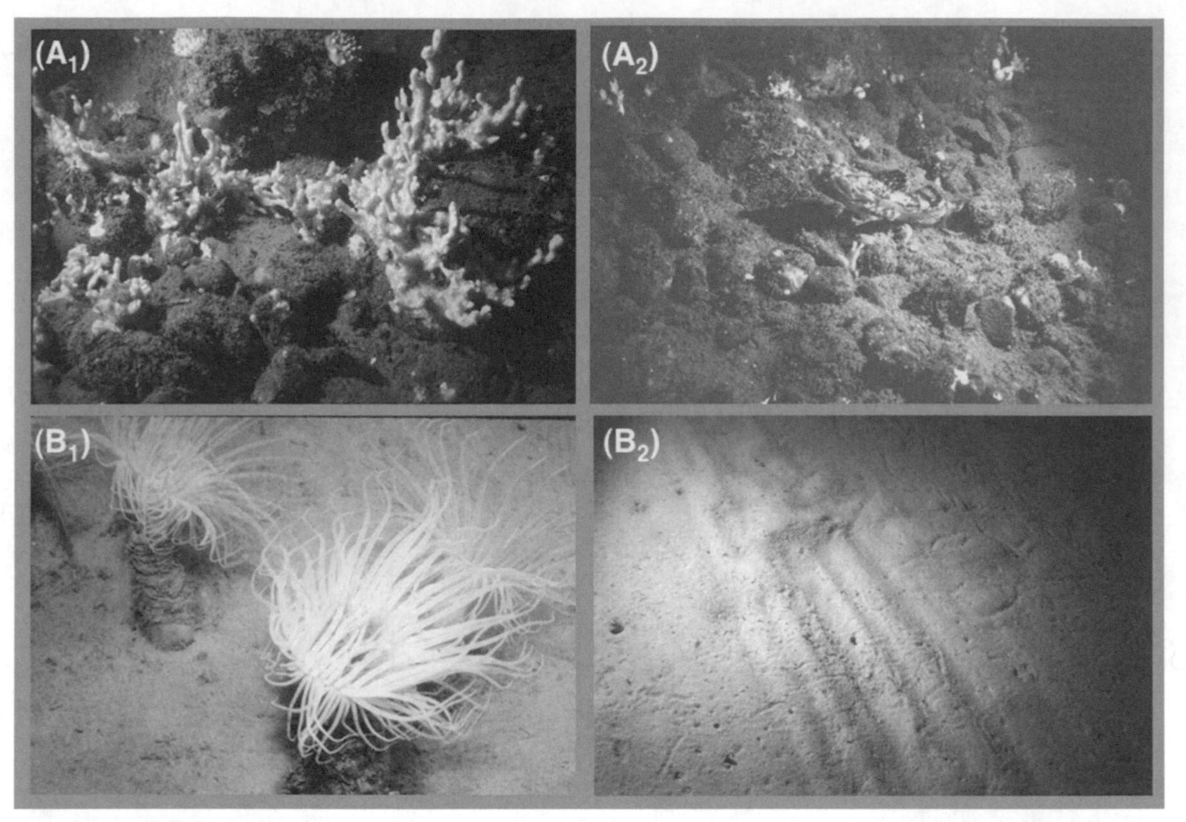

Figure 5.8. Impacts of trawling on gravel and mud habitats, Stellwagen Bank National Marine Sanctuary, Gulf of Maine. Gravel habitats protected from trawling (A_1) contain erect sponges, whereas areas open to fishing (A_2) lack such biogenic structure (a longhorn sculpin, *Myoxocephalus octodecemspinosus*, is visible in the center of photo A_2). Mud habitats often contain biological structure such as burrowing anemones (B_1), but trawled areas (B_2) can be devoid of such structure (note trawl gear tracks). Photos courtesy of P. J. Auster, National Undersea Research Center, University of Connecticut.

by 45%–55%, stressing the benthic community further. Trawling is a self-defeating practice, like harvesting mushrooms with a bulldozer (figure 5.8).

As in freshwater and estuarine ecosystems, too much sediment in the wrong place in the ocean destroys benthic fauna, including fishes. Sediment is resuspended as the otter doors, beam, chain, and trawl dig and plow as deep as 10–30 cm into the bottom, kicking up clouds of silt and detritus that then settle back rapidly (Riemann and Hoffman 1991; J. B. Jones 1992; Brylinsky et al. 1994). Sediment clogs and smothers living things, depressing dissolved oxygen levels and promoting anaerobic microbial activity. Contaminants that accumulated on the ocean floor may also be resuspended. In deep, still waters, trawls may be the major source of sediment suspension because storm events are not felt there (J. B. Jones 1992; Auster and Langton 1998). Sediment resuspension likely contributes to replacement of sediment-intolerant species by more tolerant species. In areas subjected to trawling, polychaete worms increase in numbers, whereas most other invertebrates (sponges, bivalves, coelenterates) decrease (Messieh et al. 1991; Auster and Langton 1998). Sediment deposition is a natural process, but not at the tenfold increased rate that can follow a trawl's passage (Riemann and Hoffman 1991). Again, rate of change is what determines impact and separates natural processes from human-induced, destructive processes.

Juvenile and adult fishes with narrow habitat requirements appear most affected by trawl-related destruction of benthic organisms and habitats. Density of juvenile snapper, *Pagrus auratus*, and tarakihi, *Nemadactylus macropterus*, decreased when the New Zealand bryozoan beds where they occur were subjected to trawling (Bradstock and Gordon 1983). In northwest Australia, commercially valuable snappers and emperors (*Lutjanus* spp., *Lethrinus* spp.) were most abundant in habitats with sponges, alcyonarians,

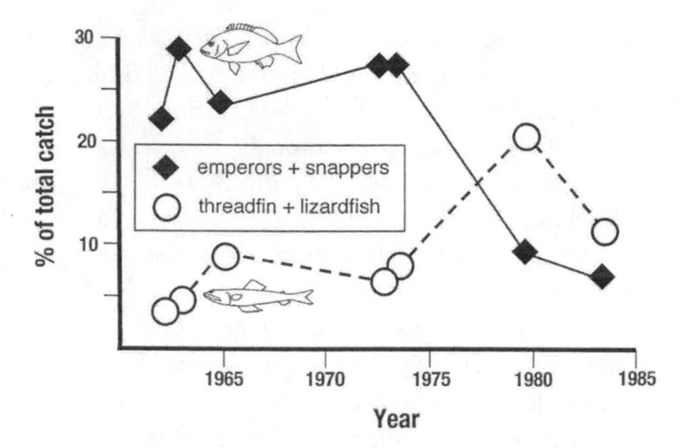

Figure 5.9. Replacement of habitat specialists in a trawl fishery. Trawling in northwest Australia removed large numbers of epibenthic invertebrates (sponges, alcyonarians, gorgonians), resulting in structure-oriented fishes (emperors and snappers) being replaced by sand bottom–dwelling threadfin-bream and lizardfishes. Total catch did not change, but lizardfishes were of low commercial value compared to the other groups. Values are average catches for each species pair. After K. J. Sainsbury (1988); line drawings of fishes by Helen Randall and Loreen Bauman, from Randall (1981); used with permission.

and gorgonians. Trawl-induced damage to these complex benthic areas resulted in replacement by fishes typical of open, sandy bottoms, such as threadfin-bream and less valuable lizardfishes (*Nemipterus* spp., *Saurida* spp.) (figure 5.9; K. J. Sainsbury 1988).

In the North Atlantic, abundance of several commercially important fishes is positively correlated with overall habitat complexity, which is a direct function of the numbers and kinds of benthic invertebrates present. One mechanism is through survivorship of juveniles that find spatial and visual refuge from predators among invertebrates. Juvenile Atlantic cod have higher survivorship in more complex habitats, juvenile silver hake are more abundant where amphipod tubes are more abundant, and Acadian redfish density is higher where cerianthid worm tubes abound (Tupper and Boutilier 1995; Auster et al. 1996, 1997). Losses in the benthos translate directly into losses to the fishery (Gibbs et al. 1980).

Problems caused by bottom trawling can generally be alleviated by three types of actions: fishing reduction, area closure, and gear modification (e.g., NRC 2002b). Reducing effort, the most obvious first step, requires reductions in both trawling intensity and frequency. Area closures, implemented in Australia, New Zealand, Sicily, and off some U.S. states, create trawling-free zones that can serve as refuges for fishes displaced from trawled areas, as source areas to recolonize fish and invertebrate populations decimated by trawling, and as research areas to study recovery rates of both fishes and invertebrates (e.g., the *Oculina* Habitat of Particular Concern off eastern Florida, Reed 2002). One such area in northwest Sicily saw as much as a fivefold increase in some fish groups after four years of closure (Pipitone et al. 1996).

Area closure need not be universal. Trawling appears to have minimal impacts in shallow, sandy areas and in gravel areas with high current velocities (e.g., Harris and Poiner 1991; Auster 1998), but it should be barred from hard, high-relief, and deep-bottom areas, where damage is more extensive and recovery of invertebrates and fishes is slower (K. J. Sainsbury et al. 1993; Safina 2001a). Many deep-water benthic invertebrates are slow growing and long lived. Estimates indicate that sponges live more than 50 years, clams to 145–225 years, and gorgonians to 500–1,500 years. Replacement of mature benthic communities may therefore take a century or longer (Watling and Norse 1998b; Raloff 1996). Area-use rotation, which ensures that some areas will be allowed to recover, could be valuable once we know how much time is needed for recovery. Auster (1998) developed a promising model that predicts habitat destruction as a function of fishing effort for different bottom types. His model can help direct management strategies that lead to sustainable harvests while maintaining biodiversity. Determination of recovery rates, among other important information, is hampered by a lack of data in the Gulf of Mexico, North Pacific, Latin America, Africa, and Asia (Watling and Norse 1998b).

Gear modifications include prohibiting use of "tickler" chains, which drag along the bottom ahead of the net and increase damage to the benthos (while increasing the catch). Increased use of wheels on the groundrope of the net would raise the net off the bottom and reduce damage somewhat—though it would still be considerable; incorporation of wheels into roller and rockhopper trawls allowed trawling in areas of complex topography that were previously inaccessible. The creativity that has gone into improving trawl effectiveness could be usefully redirected toward creating trawls that are less destructive of the habitats they exploit.

The good news is that, although long-lived and slow-growing biogenic substrates may take decades to recover, trawling restrictions can lead to rapid recovery of some benthic habitats and taxa. Where restrictions are in place, increases have occurred in density, species richness, habitat diversity, and productivity of a variety of invertebrates (Bradshaw et al. 2001; NRC 2002b). Recovery among fishes has included flounders, skates, and redfishes (*Sebastes* spp.) (Murawski et al. 2000; J. A. D. Fisher and Frank 2002). Given the dependence of fishes on biogenic habitat, curtailing destructive fishing practices can be expected to aid recovery of other, commercial fishes, including such highly depleted species as cod and haddock.

SOLUTIONS: PROTECTED AREAS AND HABITAT RESTORATION

The importance of intact habitat to fish survival and the deleterious impacts of degraded habitat are recognized throughout the world. For example, the U.S. Congress in 1996 amended the chief legislation that regulates marine and anadromous fisheries, the Magnuson-Stevens Fishery Conservation and Management Act, to acknowledge the need to protect and restore Essential Fish Habitat (EFH). EFH is identified as "waters and substrate necessary to fish for spawning, breeding, feeding, or growth to maturity," which is basically habitat important to all life stages of a species (www.nmfs.noaa.gov/habitat/efh/). Government agencies are required to identify EFH for managed species, must consider EFH in any resource management decision, and must emphasize understanding of EFH in their programs. An extension of the EFH concept involves designation of Habitat Areas of Particular Concern (HAPCs), which are specific types or areas of habitat especially important to a species, exceptionally sensitive to human disturbance, and rare or threatened (see Benaka 1999).

Reversing habitat degradation has proven relatively tractable, although fundamentally different approaches have been applied to marine versus freshwater situations. Success has been generally higher in marine situations.

Protected Areas: Gradients of Protection and Success

Setting aside large areas for protection—a true ecosystem-based approach to management—could be appropriately discussed in the context of overfishing or coral reef decline, but it is too general a solution to limit to one issue, habitat type, or climatic regime. The basic premise and justification behind reserve creation, regardless of locale, is that, historically, exploited species had natural refuges in remote, deep, and inaccessible locales; that such refuges have become increasingly rare due to improved fishing and navigation technologies; and that the situation can only be improved if we designate and enforce refuge areas.

Reserves are created most often to protect habitat; the most effective reserves also eliminate resource exploitation. Some authors distinguish between *protected areas* and *reserves*, restricting the latter term to "no-take" areas. I use the terms interchangeably, as is common. This discussion will address the generally recognized theoretical and actual goals, as well as applications, functions, and limitations of protected areas, with a focus on marine protected areas, because they are more common and better studied. An updated listing and thumbnail descriptions of MPAs globally can be found at the World Conservation Monitoring Center's Web site, http://sea.unep-wcmc.org/wdbpa/ (see also Kelleher et al. 1995); information on MPAs in the U.S. is available at www.mpa.gov and is nicely summarized in Palumbi (2003). Australian MPAs are described at www.deh.gov.au/coasts/mpa/.

Although the concept and utility of protected areas have long been established both on land and in the sea, relatively few freshwater locales have received such protection (e.g., Moyle and Leidy 1992; Doppelt et al. 1993). Saunders et al. (2002) gave examples, discussed shortcomings, and suggested design criteria for freshwater protected areas (FWPAs), with emphasis on ecosystem-based approaches. They focused on the need to limit land disturbance in the surrounding watershed, maintain natural flow regimes, and minimize introductions of nonindigenous species.

Protected areas are intended to maintain, increase, and eventually maximize biodiversity and/or biomass in the area set aside and, ideally, in surrounding, nonprotected areas. They can additionally serve as reference sites for comparisons with exploited regions. Restoration of a degraded area to conditions characteristic of a naturally functioning ecosystem is another general goal. Restoration can result from habitat manipulation or addition or solely through elimination of habitat destruction. Improved conditions result from relaxation of both exploitation (lowered mortality due to fishing) and habitat disturbance, which lead to changes in ecological and life history traits of resident fishes (figure 5.10). Fishes within a protected area may increase in diversity, density, average size and age, and overall biomass. As a result, they disperse from the densely populated, protected area to surrounding areas (the "spillover"

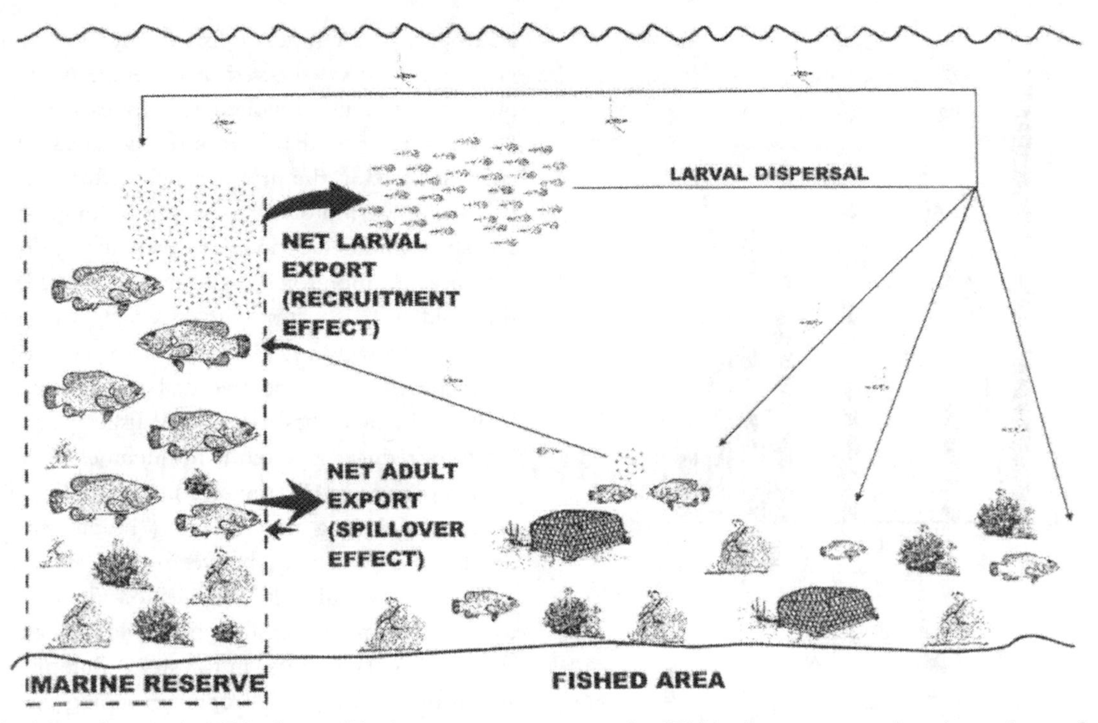

Figure 5.10. Anticipated events in MPAs from a fisheries management perspective. Within the reserve, number, size, and age of residents increase in response to reduced fishing pressure, leading to increased reproduction and net export of adults and larvae into adjacent regions. Diversity can also increase. From Russ (2002); used with permission.

effect). When effective, spillover more than compensates for lost fishing opportunities within the reserve, although that can be slow to occur (e.g., McClanahan and Mangi 2000). At a more regional scale, many large fishes lead to increased reproductive output in the protected area, which should result in increased export of larvae that potentially settle in downstream, often distant areas (the "recruitment" effect). At least in theory.

MPAs have been abundantly and somewhat redundantly reviewed. Russ's (2002) paper, "Yet Another Review of Marine Reserves as Reef Fishery Management Tools," reflects the wealth of literature on the topic. Russ treated 1990s review papers that focused on reef fishery management and had 40 references to discuss. This abundance is also indicative of the breadth and complexity of the issue: MPAs differ greatly in intent, structure, and accomplishments. Other recent, relatively thorough treatments of the topic include NRC (2001), chapters in Norse and Crowder (2005), and papers in *Ecological Applications* in 2003 (especially Halpern 2003); Plan Development Team (1990) is a classic review, and Roberts and Hawkins (2000) provided a thorough and readable piece advocating reserve creation.

How Well Do MPAs Achieve General Goals?

First, marine protected areas work, especially in the tropics (see chapters 11–13). Fish and invertebrate numbers increase quickly when exploitation is restricted, and numbers decline when fishing resumes (figure 5.11). These changes are obvious to local fishers, who are often strong advocates of area closure and take pragmatic advantage of improved conditions by "fishing the line" just outside reserves (Roberts and Hawkins 2000). In contrast, more traditional fishery management practices—such as limitations on effort and fish size, limited entry and seasonal closure, catch quotas and gear restrictions—have failed widely.

Second, again especially in the tropics, MPAs are the only workable solution to fisheries declines because of logistics, enforcement issues, and ethical considerations in subsistence economies where effort and catch restrictions can create unbearable hardships. Third, MPAs diversify economies, providing income from tourists attracted to healthy reefs, while sustaining fisheries in nonprotected areas. An MPA "offers a chance to have our fishes and eat them too" (Russ 2002, p. 421).

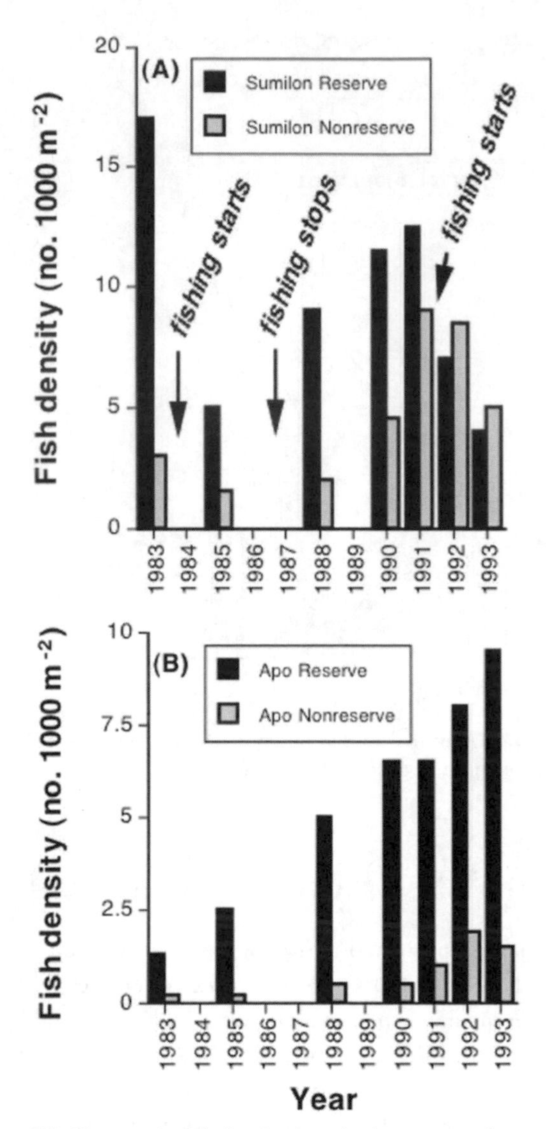

Figure 5.11. Changes in fish density in paired reserve and nonreserve areas in the Philippines. (A) In the Sumilon Island region, fish density increased when fishing was halted, and decreased rapidly when fishing resumed. Arrows indicating onset and cessation of fishing pertain to fishing activity inside the reserve; fishing was continuous in the nonreserve area. (B) In the Apo reserve, fishing ceased in 1982, and density steadily increased; outside the reserve, density remained low. Values are mean total density for jacks, groupers, snappers, and emperors, all heavily targeted fishes. Redrawn from Russ (2002); data from Russ and Alcala (1996).

The local improvements associated with MPAs—increased fish density, size and age, biomass, and diversity inside versus outside reserves or before versus after reserve establishment—are well documented. Roberts and Hawkins (2000) summarized positive results from 16 studies in which fish metrics improved. Russ's (2002) review of 22 tropical and temperate fisheries reserves (mean age 13 yr, range 2–36 yr) showed that average densities of large, predatory, heavily exploited reef fishes were 2.5–3 times higher in reserves than in nonreserve areas. Density regularly increased in the first year or so after area closure, and no obvious asymptote was evident in longer-term studies. Density continued to increase, even after a decade of protection and monitoring. Biomass increased on average threefold. Of 18 sites with relevant data, 16 reported increased average fish size inside reserves. Species targeted by fisheries and nontargeted, large species were most likely to benefit, the nontargeted fishes probably benefiting from reduced bycatch (Mosqueira et al. 2000).

Halpern (2003) reviewed 89 studies in no-take reserves, including nonfisheries ("biodiversity") reserves, divided fairly equally between tropical and temperate examples. His impressive analysis found an average overall twofold increase in density, a threefold increase in biomass, and a 20%–30% increase in fish size and number of species. When broken down by trophic groupings, carnivores and herbivores (i.e., major fisheries targets) did slightly better than planktivores and invertebrate feeders. In general, more sedentary species tended to benefit the most (e.g., Gell and Roberts 2003). Between 60% and 90% of reserves showed increases in all four measures, 0%–30% showed no change, and fewer than 10% showed losses.

Conventional wisdom and conclusions derived from studies of terrestrial reserves strongly promote establishing the largest reserves possible. Small marine reserves (1) are less likely to produce significant numbers of larvae for downstream transport, (2) are less likely to contain the variety of habitats needed by different species at different life history stages, (3) will house fewer species because of classic species-area relationships, and (4) are more likely to be damaged during catastrophic events such as hurricanes and ship groundings. Halpern (2003) assessed effects of reserve size for MPAs between 0.002 km^2 and 846 km^2 (median 4.0 km^2) in area. He found that small reserves were as effective as large reserves on a relative basis; small reserves were as likely as large reserves to improve in all metrics. However, larger reserves had higher absolute abundance, biomass, and diversity values. Determining optimal size is complicated by differences in mobility among species, boundary distinctness, surrounding habitats, and desirability of spillover (e.g., Ratikin and Kramer 1996; Kramer and Chapman 1999). Small reserves will likely encompass the movements of relatively sedentary species, but more mobile species are likely to move beyond reserve boundaries and be exposed to fishing. If feasible, larger is better, but smaller is better than nothing.

A related, unresolved topic concerns how much area should be protected to optimize local and global benefits. Theoretically, maximum benefits accrue when 10%–80% of available area is protected, with a mode between 20% and 40% and a target of at least 20%, the remaining 80% managed for sustainable yields (e.g., United States Coral Reef Task Force 2000; Roberts and Hawkins 2000; Bohnsack et al. 2000). Ideal reserve size and fraction of an area, or of the oceans, requiring protection remain a matter of debate. All agree that the number exceeds the 0.5% of the ocean currently protected. Smaller reserves are less disruptive of people's lives and are an "easier sell"; thus, actual reserve size is often determined more by local politics than by science.

Spillover

Measuring spillover is difficult and complicated by (1) the remarkably sedentary nature of many large, heavily exploited reef fishes, a major reason reef fisheries are quickly depleted and sedentary fishes are among the first to benefit from reserve creation; (2) the decreasing probability of finding a tagged fish as one moves farther from a reserve's borders, due solely to geometry; (3) the need for extensive tag/recapture or resighting efforts both inside and outside a reserve to determine if net emigration occurred; and (4) the need to control for local variation by observing movement from reserve areas paired with movement in similar, nonreserve regions. Many spillover predictions are therefore based on modeling efforts (e.g., DeMartini 1993) rather than empirical studies.

Rigorous data demonstrating spillover are rare. Where it occurs, spillover is generally limited, declines rapidly as one moves away from reserve borders, and varies as a function of species mobility. One oft-cited example from South Africa required tagging of over 11,000 individuals of the coracinid, *Coracinus capensis* (Attwood and Bennett 1994). Only 10% were recaptured, 80% of which were found within 5 km of the tagging locale. However, 200 fish traveled over 25 km from the release point, some up to 1,000 km.

When fish are not individually tagged, spillover is inferred indirectly from increased fishing success in surrounding areas. Creation of Mombasa Marine Park, Kenya, led to more than a 50% increased catch per unit effort and per unit area in adjacent locales. Significant spillover was limited to a few hundred meters on one side and perhaps 2 km on the other side of the reserve (McClanahan and Mangi 2000). Moderately mobile species such as rabbitfishes (Siganidae), emperors (Lethrinidae), and surgeonfishes (Acanthuridae) predominated, as predicted by mathematical models. Spillover in Mombasa was corroborated by decreased fish grazing in sea-grass beds with increasing distance from park boundaries. Attempts at experimental enhancement of spillover have produced minimal effects. Zeller et al. (2003) removed 60%–80% of target species in areas immediately adjacent to an Australian reserve to stimulate movement to regions of lower density. Little movement was found.

Some of the strongest direct and indirect evidence for spillover comes from St. Lucia in the Caribbean and, especially, from Florida's east coast. Five small reef regions in St. Lucia were set aside as reserves in 1995. By 2000–2001, biomass in adjacent areas had doubled, and trap catches had increased 36%–90% (Roberts, Bohnsack, et al. 2001). Merritt Island National Wildlife Refuge, Florida, the oldest fully protected marine reserve in the U.S., produced even more dramatic results. Established as a security zone for the Kennedy Space Center, this 40-km stretch of estuary was closed to fishing in 1962. By the late 1980s, fish abundances there were 2–13 times higher than in adjacent areas (D. R. Johnson et al. 1999). Tagging studies documented export of sport fishes to surrounding areas, and catches of large sport fishes 100 km north and south of the reserve improved in the late 1990s. Significantly, 50%–62% of the world records for red drum, *Sciaenops ocellatus*, black drum, *Pogonias cromis*, and spotted sea trout, *Cynoscion nebulosus*, came out of this area, although only 13% of Florida's coastline was protected (Roberts, Bohnsack, et al. 2001). Spillover of large fish from the reserve is a likely explanation.

Recruitment

Evidence for a recruitment effect is scarce and may remain so given the logistics of tracking very small animals over very large temporal and spatial scales. Recruitment could increase as reproductive output in reserves increases, and greater reproductive output has been documented for reserve invertebrates, including lobsters, abalones, and sea urchins (Dugan and Davis 1993). A recruitment effect would theoretically result from the widely observed increase in numbers and sizes of fishes in reserves, with more and bigger fish expected to produce more eggs. Using known size-fecundity relationships, population size structure, and density of rockfishes (*Sebastes* spp.) in the Hopkins and Point Lobos reserves in California, Paddack and Estes (2000) calculated that egg production could be 2.7–2.9 times higher inside than outside the reserves. Similar calculations were made for the sparid, *Pagrus auratus*, in three northern New Zealand reserves, where higher densities and larger sizes translated into an 18-fold increase in egg production compared to exploited areas (Willis et al. 2003). Gell and Roberts (2003) cited analogous findings for snappers in New Zealand, lingcod and rockfish in Washington State, and several reef species in Kenya. Although encouraging, these are measures of potential output. They say nothing about actual spawning frequency, let alone the fate of the larvae produced and whether larvae contribute to populations outside the reserve.

Direct evidence of a recruitment effect will likely depend on promising technological developments in population genetics and microchemistry of fish otoliths. Fingerprinting using microsatellite DNA markers and amplified fragment length polymorphism (AFLP) procedures may allow assignment of individual larvae to different populations and regions. This would help establish source locales of larvae (e.g., Hellberg et al. 2002). The use of mass spectrometry to detect trace and rare earth elements in specific layers of otoliths can help establish "otolith elemental fingerprints." Such fingerprints help identify habitats and regions from which larvae originate and where individual fish spend time as they grow (Campana 1999; Swearer et al. 2003; Hamer et al. 2003).

How Much Is Taken from No-Take Zones?

An important question concerns the degree to which restrictions are observed and enforced. The data are scarce and disappointing. Kelleher et al. (1995) estimated that "management effectiveness" is known for only about 30% of the world's MPAs, and in only about one-third of known cases can management be considered effective. At best, about 10% of the world's MPAs actually protect their resources.

Compliance and enforcement are problems in both industrialized and developing nations. People violate the boundaries and spirit of protected areas out of need, ignorance, indifference, or greed. California, with its strong reputation for environmental awareness, has experienced widespread unlawful collecting and poaching in some reserves "because enforcement has been virtually nonexistent" (Murray et al. 1999, p. 20). The Great Barrier Reef Marine Park of Australia, where 5% has been set aside as no-take ("green") zones, also experiences violations. After 8–12 years of protection, no discernible difference in density of coral trout, *Plectropomus leopardus*, existed between no-take and legally fished areas in the Cairns section of the park. Poaching was cited as the likely cause: "Thus, the shining example of management success, the largest marine park in the world, in a developed nation and sitting on one of the most sparsely populated coastlines in the world, has probably been rather unsuccessful in reducing [fishing mortality] substantially on coral reefs nominally 'closed' to fishing" (Russ 2002, p. 426).

Permanent versus Temporary or Rotating Closure

Many advocates of MPAs maintain that partial protection achieves only partial success, that any relaxation of totally banned fishing will negate most benefits of protection. In the Sumilon Island reserve in the Philippines, often heralded as a major success story, much of the fish biomass that accumulated during a 5–10-year period of protection was eliminated in 18–24 months when protection broke down (Russ and Alcala 1996; see figure 5.11). On Boult Reef in the Great Barrier Reef, coral trout populations increased substantially during 3.5 years of protection, but fell 25% in the first two weeks when the reserve was reopened to fishing (Roberts and Hawkins 2000). Total closure means that habitat gains protection as well, which is important because of the slow recovery of biogenic habitat. "Permanent no-take" is also easier to define and enforce than alternatives: How much is "some" fishing? Proponents of MPAs consequently emphasize that for benefits to accrue, no-take means no-take, but before this proscription will be widely accepted, additional, quantified verification of the effectiveness of true no-take restrictions will have to be obtained.

Networks of Reserves

Networked reserves rather than isolated protected areas are increasingly emphasized (e.g., Murray et al. 1999; Gell and Roberts 2003; Palumbi 2003). Rationale includes the argument that migratory species frequently use different places and habitats to aggregate and spawn and as nursery regions. They therefore need protection in several habitats during their migratory cycle. Even relatively sedentary fishes utilize multiple habitats, such as mangroves, sea-grass beds, coral habitats, and spawning locales, during ontogeny and also rest and feed in different habitats on a diel basis (Wolff et al. 1999; see chapter 12). A single reserve is unlikely to encompass all relevant and important habitats; hence, an ecosystem-based approach mandates multiple reserves. Another asset of networks is the possibility that they will increase recruitment via connectivity, because networks are "linked by the invisible threads of larval transport" (Roberts, Halpern, et al. 2001, p. 12). Reef fishes produce planktonic larvae with a mean larval duration of about a month. Thus many populations may rely at least in part on upstream larval production for future recruits (e.g., Lipcius et al. 2005). A major conclusion is that a network of small reserves represents a precautionary approach to MPA creation, providing insurance against natural and human catastrophes, as well as human management errors.

Interestingly, some network proponents maintain that several small, networked MPAs are preferential to a few large areas, a contrast with the conventional wisdom surrounding terrestrial reserves. A critical aspect of this unorthodox view is the spacing between network components. The distance between reserves that would balance export of larvae with local retention or self-recruitment, and maximize both within- and between-site recruitment, is thought to be between a few to a few tens of kilometers (Roberts, Halpern, et al. 2001).

Established MPA networks are fairly rare. Examples include the 35-million-ha Great Barrier Reef Marine Park of Australia, the Northwestern Hawaiian Islands Coral Reef Ecosystem Reserve, the Channel Islands National Marine Sanctuary off California, and the Soufriere Marine Management Area on the Caribbean island of St. Lucia. The best-known network in the U.S. is the Florida Keys National Marine Sanctuary (FKNMS), which encompasses 2,800 square nautical miles of diverse reef, grass bed, mangrove, and sand habitat, divided into wildlife management areas, ecological reserves, sanctuary preservation areas, and special-use areas. Less than 1% of the FKNMS has strict, no-take protection, although the protected area includes much of the critical coral reef habitat. Apparently missing from the original FKNMS was a larval source component. A Tortugas Ecological Reserve, which lies up-current from the FKNMS, was added and is expected to partially serve that function. However, studies of dispersing currents, current speeds and directions, and larval duration times suggest that larval sources may also lie along the eastern coast of Central America in a region known as the Mesoamerican Reef (M. Lara, pers. comm.). Cooperative conservation efforts between the U.S. and other nations that established multiple reserve networks may be vital to the health of the United States' largest coral reef ecosystem.

Enthusiasm for networks among conservation biologists is considerable. Gell and Roberts (2003, p. 448) have even suggested that, "by integrating large-scale networks of marine reserves into fishery management, we could reverse global fishery declines." Although the merits of networks have a solid theoretical basis in community ecology, selling the concept is going to be harder than convincing small tropical nations as well as industrialized countries to establish a few isolated, protected areas.

Methodological Complications

Despite widespread evidence of benefits, numerous aspects of the MPA concept have been challenged (e.g., Russ 2002; Gell and Roberts 2003; Halpern 2003; Lipcius et al. 2005). Oft-repeated criticisms focus on methodological shortcomings in reports of MPA results. Accurate population assessments prior to the establishment of a reserve have often been lacking: "before" data have been either unavailable or based on a single assessment before area closure. Apparent trends in numbers post-closure could therefore be unrelated to fishing restrictions. Also, measures of fishing mortality inside and outside reserves are generally unavailable, weakening conclusions about the impacts of relaxed fishing pressure. Studies have also suffered from a general lack of design rigor, such as nesting, subsampling, adequate controls or reference locales, and replication. A lack of methodological rigor, while certainly undesirable, is in part an unavoidable consequence of the complexity of such studies and the physical scale at which they occur. It is difficult to replicate ecosystems.

Finally, a related, serious problem for both analysis and reserve effectiveness concerns knowledge of violation frequency—namely, poaching. Information on this topic is unlikely to improve, given "difficulties in monitoring a clandestine act" (Halpern 2003, p. S129).

Resolution of a number of contentious scientific

points concerning reserves will require statistical verification using criteria and methodologies acceptable to quantitative ecologists. Russ (2002) proposed a long-term, replicated experimental approach to study biomass and spillover, based on the Before-After-Control-Impact-Pairs (BACIP) methodology. His suggested protocol would require 5 to 20 years at replicated sites, duplicated regionally "to ensure the generality of the result" (p. 436). Such a proposal might also be called the Best-of-All-Possible-Worlds approach and appears unrealistic (as Russ freely admits). However, definitive answers to several issues require more rigorous and controlled studies than have been the norm.

Restoration and Rehabilitation of Freshwater Habitat

Unlike most negative impacts on fishes, habitat modification is relatively easy to correct. Hence considerable activity and resources have been expended in rehabilitation and restoration of aquatic habitats, especially in flowing waters. Dams can be breached, bypassed, and removed (see chapter 6); riparian vegetation can be planted, and banks can be stabilized; straightened river channels can be reengineered into more natural configurations; large woody debris and boulders can be placed in streams to provide habitat and alter water flow, creating deep pools and redistributing sediments; and silt can be dredged from water courses and its transport from terrestrial sources arrested.

Improvements may be dramatic and relatively rapid, as in the case of the highly degraded and channelized Austrian River Melk discussed earlier. The desire to "do something," however, sometimes supersedes the need for careful planning, baseline data collection, standardized implementation, monitoring of results, and modifications to the original plan, all valid aspects of a restoration effort. Many restoration projects have unrealistic goals: "Restoration of freshwater habitats towards pristine conditions, which is the objective of most managers, is a utopian view" (Cowx 2003, p. 213). Desired endpoints may not be formally articulated. And finally, a "one size fits all" approach to restoration, which ignores special conditions in the surrounding watershed, the dynamic nature of flowing water, and the hydrologic uniqueness of a stream, often results in more damage than had the stream been allowed to repair itself (e.g., Sear 1994).

Stream Restoration: Successes and Failures

Ultimately, the goal of a rehabilitation, restoration, or improvement project is to return a degraded aquatic habitat to something resembling a naturally functioning ecosystem. It should be free from contaminants and contain self-sustaining populations of many of its original native organisms. A return to original conditions is unlikely, and most attempts focus instead on incremental improvements in conditions over the degraded condition, with emphasis on physical change to instream habitat. Examples of sought-after changes are numerous, including increased habitat diversity; restored riffle-pool sequences; stabilized bottom structure, overhead cover, and banks; removed sediments; and reestablished natural flow variation and connections to adjacent habitat. The kinds of structures employed or actions taken are equally varied and include placement of logs, natural debris, boulders, and overhead structure (sometimes using streamside trees and logs); revegetation of instream, riparian, and floodplain areas; exclusion of livestock; removal of nonnative species; and stocking with native species (Cowx and Welcomme 1998; D. M. Thompson 2002). The goals and actions focus on instream rather than terrestrial conditions.

Stream restoration takes many forms, with varying implied and actual goals (Bernhardt et al. 2005). Such activities as "Adopt-a-Stream" programs involving nongovernmental, volunteer organizations such as Trout Unlimited, Isaac Walton League, and various riverkeeper/riverwatch groups working with businesses, neighborhood volunteers, local governments, and academic or agency scientists are often directed at improving general conditions in urban or suburban streams that have been channelized, deforested, clogged with trash and silt, polluted with municipal and residential runoff, and populated by tolerant, introduced species. Candidate streams are typically situated in a watershed characterized by large amounts of impervious surface and human development, including upstream agricultural activity.

In contrast are habitat improvement and enhancement measures undertaken by government agencies in partially deforested streams that support limited fishing, whose goal is improved angling opportunities. Methodologies range from specific, engineering-focused, "cookbook"-type manuals on instream modifications (e.g., Seehorn 1992; Riley 1998), through more complex but still cookbook fluvial morphology approaches that target water flow and channel characteristics (e.g., Rosgen 1996), to more general, ecosystem approaches that attempt to take into consideration not just the stream but also the surrounding watershed and the ecological principles and sociological impacts that affect restoration programs (e.g., J. E. Williams et al. 1997; Cowx and Welcomme 1998).

The two—"Adopt-a-Stream" versus improved fishing opportunities—contrast sharply in their objectives. Restoration of urban and agricultural streams most often focuses on aesthetic and ecological attributes and the biotic integrity of a stream. Participation by a consortium of organizations is necessary because the work can be complex and time consuming. One example of suburban restoration is the Upper Paint Branch Watershed, a tributary of the Potomac (American Rivers 2003). Heavily impacted by urban development associated with Washington, D.C., suburbs, Upper Paint Branch became the center of a highly integrated protection and restoration program that involved instream manipulations combined with carefully planned development in the watershed, including implementation of best management practices affecting land use.

Efforts resulted in reduced silt and temperature, increased habitat diversity and bank stability, a resurgence of sculpin (which are typically sensitive to sediment and temperature impacts), and, important from the standpoint of public relations, successful reproduction and reappearance of YOY brown trout in tributaries of the system. Although not all stream reaches have shown improvement, and biotic integrity trends have been mixed and complicated by drought and continued land-use problems, Upper Paint Branch serves as an example of coordinated action by scientists, citizens, and more than 30 governmental agencies and nongovernmental organizations (www.eopb.org/publications/restoringpb.pdf).

Systematic assessments of the impacts of such projects in urban, suburban, and agricultural streams, especially on fishes, are rare. The available evidence indicates minimal success, particularly when restoration is limited to instream modifications. Pretty et al. (2003) assessed changes in 13 lowland rivers in Britain that had been rehabilitated with flow deflectors and artificial riffles. They compared abundance, richness, diversity, and equitability of fishes in rehabilitated reaches paired with adjacent unrehabilitated reaches 100 to 1,000 m long. Although some sites showed elevated fish abundance, such increases were not consistent. Overall, Pretty et al. found no significant differences between rehabilitated and reference sections. They concluded that, even though rehabilitation efforts commonly increased heterogeneity of both flow and depth, "physical rehabilitation of a river reach may not lead directly to biological rehabilitation" (p. 260). Lowland river restoration might prove more successful where emphasis is placed on off-channel and stream margin habitats (riparian zones, floodplains, backwaters, off-channel bays) and by reestablishing connectivity between the river and these habitats (Pretty et al. 2003).

The alternative restoration situation, where fishing exists but should or might be better, has been the more popular, relatively well-funded recipient of improvement activities. These efforts are typically administered by state and federal management agencies and focus on the creation or restoration of fish habitat, with emphasis on spawning and rearing habitat for salmonids. Such programs also have a surprisingly dismal track record, at least when success is assessed over time and the real costs and benefits are calculated. If anything, such projects often prove counterproductive because they fail in ways that create new and bigger problems for a stream, presenting "the bleak prospect of restoring many channels that have been damaged by historic stream-restoration efforts themselves . . . [because failed structures create] habitat conditions that are worse than unaltered reaches" (Thompson 2002, pp. 261, 264). Case histories from several western states include numerous instances of rapid failure of structures, neutral or negative impacts on fish production, increased blockage of fish migrations by structures, and even trampling by cattle of structures deployed to overcome grazing-induced erosion (Hamilton 1989, Frissell and Nawa 1992). "Despite evidence of serious inadequacies in many instream-structure designs, many types of structures are still installed" (Thompson 2002, p. 253).

Most recent evaluations of restoration programs, whether targeted at urban rivers or trout streams, recognize the extent of such failed efforts and stress a need for more widespread cooperation, planning, continued monitoring, assessment-based modification of design, and sharing of results on a national and international basis (Bernhardt et al. 2005). One project addressing this need is the National River Restoration Science Synthesis (NRRSS) project designed to link "the practice of ecological restoration and the science of restoration ecology" in the U.S. NRRSS takes an adaptive management approach that emphasizes ecological, engineering, and fluvial geomorphological influences (Palmer et al. 2003; see also http://nrrss.nbii.gov and www.restoringrivers.org). The NRRSS data base was used in Bernhardt et al.'s (2005) analysis of over 37,000 restoration projects in the U.S.

Monitoring and Adaptive Management

The most likely reason for this repeated scenario of manipulation and failure is largely an uncritical acceptance of established practices, as opposed to continued monitoring of outcomes and modification of management plans—that is, adaptive management. "Many opportunities to learn from successes and failures, and thus to improve future practice, are being lost" (Bernhardt et al. 2005, p. 637). Follow-up monitoring, often referred to as post-project appraisal (PPA; Downs and Kondolf 2002) is surprisingly rare. Fewer than 10% of the 37,000-plus restoration projects analyzed by Bernhardt et al. (2005) included an assessment component. Too often, complex and expensive restoration practices have ended with planting trees and placing logs and rocks, "success" defined as completing the physical action of placing objects in a stream. However, engineering success should not be equated with or serve as a surrogate for biological recovery. What the structures do and for how long, and what impacts the restoration activities have on the stream and its inhabitants all need to be assessed, repeatedly and long term. This obvious requirement received surprisingly little attention through the first several decades of restoration practice (Kondolf 1995; Bash and Ryan 2002; Thompson 2002; Pretty et al. 2003).

The product of a failure to assess is acceptance of standard practices whether or not they work, and even if they do more harm than good. As a result, basic techniques and designs developed in the 1930s remain accepted practices in the U.S. and elsewhere (e.g., Seehorn 1992). Some structures fail to accomplish their intended hydraulic modifications, such as creation of riffle and pool sequences; some create problems of unintended flow alteration from the start; and some lead to progressive problems as they degrade. Grade-control structures intended to slow the flow of water—such as various small dams, log sills, or cross logs—may create an upstream pool that later fills in with sediment and allows water to increase in temperature; the structure itself may block upstream fish migration. Current deflectors designed to form downstream pools can instead lead to wider and shallower channels. Structures designed as overhead cover can lead to loss of streamside vegetation and a corresponding reduction in overhead cover when compared with unaltered streams. Revegetation of riparian zones may be impaired because trees cannot grow in riprap, even when it collapses. Widespread restoration efforts may thus eliminate natural cover and prevent its future development. A reliance on engineering and physical manipulation is often counterproductive because "restoration of geomorphic form does not necessarily result in the restoration of the appropriate geomorphic process" (Thompson 2002, p. 250).

Frissell and Nawa (1992) presented a scathing indictment of the uncritical application of stream improvement protocols that dominated U.S. management activities during the 1980s. In 1987 alone, the U.S. Forest Service placed more than 2,400 fish habitat structures in Pacific Northwest streams. Justification was often based on untested assumptions. One economic model used to justify a $1.7 million habitat improvement plan in the Siskiyou National Forest assumed, without documentation, a net gain of 1.4–1.8 kg of anadromous salmonids for every dollar invested in artificial structures. Logging and other land-disturbing activities were assumed to have no negative impacts on fishery value. "The Forest Service assumed that any adverse effects on fish habitat and water quality would be more than compensated by fish habitat created with new artificial structures" (Frissell and Nawa 1992, p. 182).

Frissell and Nawa surveyed 161 fish habitat structures in Oregon and Washington streams. They estimated rates and causes of failure and impacts of failed structures following a fairly average flood event. Failure and impairment characterized about 60% of the structures, almost regardless of physical type, with lowest rates among structures that least modified the preexisting channel. Damage was most common in streams draining disturbed watersheds, especially those carrying high sediment loads and with unstable channels—in essence, the streams most in need of repair. Failure was also common in streams that drained low-gradient alluvial valleys and thus transported sand and gravel into critical spawning and rearing habitats of migratory salmonids.

Although salmonids were the species primarily targeted by stream improvement programs, removing riparian trees and logs to create artificial structures accelerated the deposition of harmful sediments and reduced shade available to salmonids. Additional damages from "improvement" efforts included accelerated bank erosion; damage to streamside and instream areas from heavy equipment; failure of riparian trees to anchor instream structures; creation of sand or silt bars due to altered hydrology, which also led to shallower pools; massive debris flows during floods when structures failed; and release of toxic substances such as epoxy resins and tar paper used in construction. Frissell and Nawa cited earlier literature to support the conclusion that "commonly prescribed structural modifications often are inappropriate and counterproductive" (p. 182), indicating that practices had continued despite knowledge of their limited success. Frissell and Nawa's literature review also implied that successes were widely cited, whereas neutral or negative results were underemphasized.

Frissell and Nawa concluded that common practices failed because restoration practitioners focused on structure design and materials, overlooking more important determinants, namely land-use conditions in the riparian zone and upslope watershed that drive channel dynamics. Instream structural failures resulted from overgrazing and urbanization, and from landslides originating in clear-cuts and road failures, all of which led to massive deposition of bedload material. A preferable approach would involve an integrated, ecosystem-based combination of efforts aimed at (1) preventing slope erosion, channelization, and careless floodplain development; (2) correction of problems associated with failing roads, active landslides, and other sources of sediment; and (3) reforestation of riparian zones, floodplains, and unstable slopes. "Unless these larger-scale concerns are dealt with first, direct structural modifications of channels are unlikely to succeed" (1992, p. 193).

More recently, Roni et al. (2002, table 5) surveyed 11 studies of 29 instream restoration projects targeting Pacific Northwest salmonids, studies that included post-project assessment. Improvement in fish populations occurred in one-third to one-half of projects, complicated by lack of statistical testing; little monitoring after the first few years; and species, age-class, and seasonal differences. Increased density of coho salmon juveniles was the most commonly observed positive result, but other salmonids (cutthroat trout, steelhead, Chinook) were equally likely to decline, show no improvement, or increase. No general pattern related to structure type was detectable. Roni et al. concluded, again, that site-specific, instream modification is the least desirable restoration tactic, and that processes in the overall watershed are more important (figure 5.12). Improving instream integrity requires a watershed perspective focused on "restoring landscape processes that form and sustain habitats" (p. 2). Instream habitat enhancement should be viewed as a tactic of last resort, best restricted to emergency restoration directed at endangered species whose specific habitat needs are understood, with an expectation of only short-term impacts.

One obvious conclusion emerges: Restoration efforts should be viewed as *part* of habitat management, not as separate activities, much less desired endpoints. Increasingly, researchers, policy makers, and resource managers have come to realize that management must be adaptive and flexible, viewed as an experiment in the sense that conditions are incompletely knowable and surprises will occur. An unmodifiable management plan has a minimal likelihood of success. Monitoring's greatest value is in providing information to guide changes to the plan, to tell us which approaches work under which conditions and what adjustments need to be made to improve the effort.

Mandatory Assessment: An Alternative View

Failure to monitor is often more a result of budgetary constraint than of bureaucratic shortsightedness. A restoration plan's success may not be realized for many years, but few governmental agencies can commit funds for agency activity five, ten, or more years in the future. Also, funds allocated to future assessment may be subtracted from the restoration activities themselves. It is therefore unrealistic to insist that all restoration projects include monitoring efforts. An alternative is to monitor those projects and practices most likely to represent conditions at many locales, and apply the lessons learned at the selected sites to other, relevant projects. What remains unclear, however, is whether we presently know which projects are optimal for monitoring.

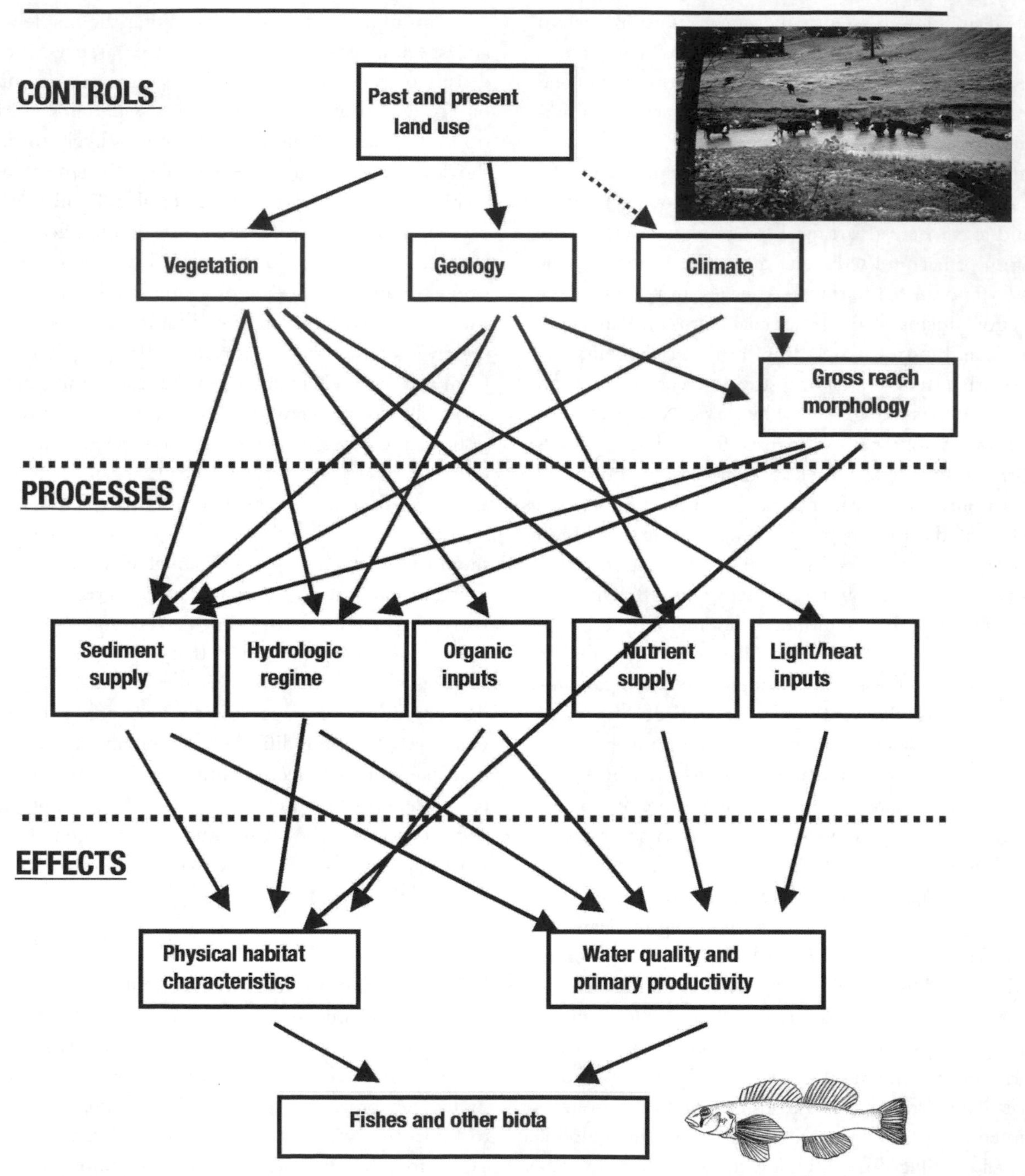

Figure 5.12. Why instream habitat manipulation is unlikely to restore fish populations. Physical habitat and water quality affect fishes, but many other factors, especially land-use practices, ultimately determine population success. Restoration should focus on terrestrial and instream influences that affect, connect, and sustain instream processes, rather than on site-specific modification of physical habitat. Based on Roni et al. (2002).

Until our understanding of the costs and benefits of restoration activities improves, it seems advisable to conduct assessments and practice adaptive management. Bryant (1995) proposed a thorough set of criteria and activities for monitoring watershed and stream restoration activities at different levels of intensity and cost over short-term (3–5 year) and longer-term (10–15 year) intervals (see also Downs and Kondolf 2002).

Restoring Ecosystem Processes via Carcass Supplementation

The selective addition of salmon carcasses to streams is a tactic being applied to stream recovery in the Pacific Northwest that exemplifies the ecosystem approach and recognizes connectivity among habitats. The historically massive numbers of salmon that spawned and died in these streams provided nutrients and energy that influenced both riparian production and food availability for the next generation. Many salmon runs are now less than 10% of historic levels (NRC 1996b). Nutrient levels can be augmented by slow-release addition of chemical fertilizers, but carcass addition has logistic advantages (albeit aesthetic disadvantages) and more accurately mimics natural conditions. Results from carcass supplementation show elevated primary and secondary production and improved growth rates among juvenile salmon (Bilby et al. 1998; Wipfli et al. 1999).

Restoration: Reinventing the Edsel

Efforts at repairing aquatic systems are variously referred to as restoration, rehabilitation, mitigation, augmentation, enhancement, and improvement. This terminology is vaguely reminiscent of the lexicon of fish stocking programs (see chapters 8 and 14), which should immediately raise warning flags. In fact, recent enthusiasm for stream improvement programs among resource management agencies parallels earlier, popular "solutions" to perceived and real problems in aquatic systems. Common themes among these solutions are (1) they addressed symptoms rather than causes associated with habitat destruction, (2) they gained popularity without systematic monitoring of their impacts, and (3) they deemphasized research into best practices and careful monitoring of actual outcomes in favor of popular but largely untested assumptions about methods. Inadequate or declining fishing resources—resulting in no small part from destructive land-use practices that caused hydrologic alterations in surrounding watersheds—were corrected in the late 1800s by massive introductions of exotics, often provided by hatcheries (chapter 8). As native species continued to decline, in part because of uncorrected habitat destruction but now exacerbated by introductions, hatchery rearing and stocking of native species proliferated (Lichatowich 1999; see chapter 14). A lack of regard for genetics and for ecological interactions between hatchery and native fishes, combined with still uncorrected upslope habitat destruction, continued and even accelerated the decline of natives. Habitat issues were finally addressed, but the solution focused on engineering and technological fixes involving manipulation of stream and river courses, with continued disregard for the causative agents in surrounding watersheds. Again, practices that have become widespread not only fail but often exacerbate the problems. In each case, assessment of impacts has lagged behind enthusiasm for action, and solutions have focused on correcting symptoms rather than addressing underlying causes.

CONCLUSION

The different emphases that have typified marine and freshwater restoration efforts—protecting versus actively restoring habitat—reflect logistics, spatial scale, and biology as much as philosophy. Both approaches have merit and reciprocal value. Clearly, more freshwater protected areas are needed. Analogously, insufficient attention has been paid to the impacts of terrestrial activities on marine habitat. Greater emphasis on accurate monitoring and assessment is needed in both. Restoration science follows basic principles that are relevant independent of locale. Practitioners would be wise to borrow and learn more from their counterparts in other systems.

Habitat protection and restoration make sense only in the context of action at the level of the entire aquatic ecosystem, whether salt or freshwater (e.g., Doppelt et al. 1993; Beechie et al. 1996). To preserve or restore habitats, we have to maintain and restore ecosystems, and not just the physical structure but also the processes and functions of the ecosystem. If we can keep ecosystems intact, the constituent habitats and their inhabitants are much more likely to take care of themselves.

6. Dams, Impoundments, and Other Hydrological Alterations

Talented an artist he may be with ten thousand brushes,
Still, he can never paint those peaks in Qutang Gorge.
—Zhang Wentao, 1764–1814

Aswan High, Flaming Gorge, Glen Canyon, Grand Coulee, Hetch Hetchy, Itaipu, Tellico, Three Gorges. This international roll call of famous dams and reservoirs is also a catalogue of world-class ecological disasters. Engineers, economists, and politicians often celebrate these and other dams as monumental human achievements, whereas ecologists frequently vilify them as monuments to human shortsightedness, as "killing fields for native aquatic species" (Gregory et al. 2002, p. 716).

Estimates and criteria vary, but between 39,000 and 45,000 large dams—over 15 m high or impounding more than 3 million m^3 of water—exist worldwide. Of these, 6,575 are in the U.S., 793 in Canada, and another 2,000 in the rest of the Western Hemisphere. Among other large dam-building nations, China has 22,000, India 4,290, Japan 2,675, and Spain 1,200 (Dynesius and Nilsson 1994; Rosenberg et al. 1997; see also WCD 2000 and ICOLD 2003). Dams worldwide impound between 500,000 and 1,500,000 km^2 of water, the latter value equaling the surface area of all natural lakes (St. Louis et al. 2000). Between 75,000 and 2,000,000 dams of all sizes occur on U.S. rivers (Poff and Hart 2002), trapping 600,000 river miles behind them. Fewer than 2% of large U.S. rivers run entirely unimpeded, and two-thirds of the world's rivers are regulated in some way (Vitousek et al. 1997).

Dams are constructed to store water and modify the magnitude and timing of its downstream movement, thus compensating for fluctuations in river flow that otherwise counter human needs and desires (Rosenberg et al. 2000; Poff and Hart 2002). Dams provide control over water resources, which can include hydroelectric power, fisheries, water for domestic and agricultural uses, opportunities for recreation and lakefront development, dilution and storage of waste and toxins, removal of pathogenic bacteria, denitrification of nitrate inputs, entrapment of sediment arising from poor land-use practices, improved navigability, and water storage capacity for various commercial uses as well as for flood control. Benefits to humans, especially in terms of fishery production, are often considerable (Fernando et al. 1998).

However, dams flood upstream wetlands; riparian vegetation; fertile agricultural land; and land with historical, cultural, and archaeological value. Inundated regions are reservoirs for disease vectors, especially in the tropics. Anoxic waters develop in deeper portions of reservoirs, leading to accumulation of toxic substances, conversion of metals to toxic forms, and release of phosphorus from sediments that can initiate harmful algal blooms in the reservoir and in downstream areas. All have negative impacts on fishes. Downstream river, floodplain, and wetland areas are deprived of water and sediment and consequently decrease in productivity and biodiversity. Estuaries are deprived of freshwater inflows, creating marine rather than estuarine conditions or allowing upriver extension of higher-salinity water. Downstream temperatures are unnaturally warm or cold, depending on whether water is released from the surface or bottom of the reservoir. Natural fluctuations in water flow, to which fishes are closely adapted, are overridden by human needs, as water is stored and released at unnatural times. Reservoirs are subject to greater evaporation than the rivers they impound because of their relatively large surface area–to–volume ratio and heat absorption characteristics. The result is a net loss of water up to 33% greater than what was lost from the river and surrounding land prior to inundation (RBSAPC 2001). As much as 6% of river runoff worldwide is evaporated as a result of impoundment (Vitousek et al. 1997). Excess evaporation has salinized the water and surrounding landscape in many areas, decreasing agricultural productivity and stressing fishes not adapted to elevated salinities. Loss and conversion of habitat forces species into new competitive and predatory relationships and promotes hybridization. Interrupted migrations and movement fractionate metapopulations, degrading their structure. The result is biotic impoverishment, which impairs community structure and ecosystem function by threatening endemics and specialists while favoring generalist and invasive species. (For more thorough treatment of dams and regulated rivers, see N. Smith 1972; Baxter 1977; Goldsmith and Hildyard 1992; and McCully 1996.)

LAKES VS. RESERVOIRS AS NATURAL HABITATS

Why can't dams and their attendant reservoirs be viewed as natural phenomena, as analogs to lakes? As with many anthropocentric stressors, dams represent extreme occurrences or extreme rates of change brought about by human actions. Lakes form when flowing water is blocked by landform uplift, glacial retreat, periodic drought, volcanic eruption, or landslide. Uplifts, droughts, and glaciers operate on time scales that allow for natural selection and adaptation, or at least dispersal. Animals that don't adapt or disperse are extirpated. Major showcases of evolution, such as species flocks of cichlids in the African Great Lakes, result from the genetic isolation and rampant speciation that can occur when new water bodies are created by uplift and drought (see Helfman et al. 1997). In contrast, volcanoes and landslides, which we tend to view as ecological disasters (e.g., the defaunation of Spirit Lake following the 1980 eruption of Mount St. Helens), occur on a time scale more akin to that of construction of a dam, which can be viewed as an equivalent disaster. Lotic-adapted species trapped above or below a dam often perish in the still water conditions created by impoundments, or are unable to complete some critical phase in their life history because the dam blocks their movement.

An important difference between natural aquatic systems and reservoirs is the nature of the habitats contained in each. Reservoirs have lower habitat diversity. As natural lakes develop, the surrounding landscape changes accordingly. Lakes tend to be rounded, with a shoreline development ratio (ratio of shoreline length to circumference of a circle) of less than 3. Their gently sloping margins commonly establish an extensive and productive littoral zone filled with fallen logs and abundant vegetation. Reservoirs are constructed faster and assume a very different form. They have irregular shapes (mean shoreline development ratio of 12), with steep-sided ridges and valleys around the periphery. Extreme water fluctuations brought on by dam releases inhibit development of a vegetated littoral zone (Soballe et al. 1992). The cascades, riffles, runs, and pools of the ancestral river are drowned in the reservoir's deep waters. Such habitat simplification reduces ecological opportunities. Impacts of reservoir creation on riverine fauna will likely be greatest where natural lakes are rare. Fish assemblages in such regions—such as much of the southern U.S. and Europe—are chiefly river adapted. Former ecological and habitat specialists, often endemic to a region, become vulnerable to invasion by introduced, ecological generalists, fueling the homogenization of faunas witnessed in so many disturbed habitats (see table 6.1; chapter 9).

Table 6.1. Negative effects of large-scale dams and other water diversion projects on fishes at various locales. Main source Rosenberg et al. (1997) unless noted otherwise.

Area affected	*Project*	*Effects*	*Probable cause*	*Source*
NORTH AMERICA				
Columbia, Snake rivers, U.S.	Multiple dams	10- to 1,000-fold decline in salmon, some extirpations, 1880–1990, sturgeon endangered	Migration blockage, habitat fragmentation and destruction, proliferation of nonnative fishes	

(continues)

Table 6.1. *Continued*

Area affected	*Project*	*Effects*	*Probable cause*	*Source*
Colorado River, U.S.	Multiple dams	Native endemics endangered or replaced by introduced species	Unnatural discharge regime	Minckley 1991
Moose River, Canada	Dams	Declining lake sturgeon populations	Unnatural discharge regime	
R. Grande, R. Lerma, Mexico	Dams, water withdrawal	Saltwater intrusion, replacement of freshwater species with intruding marine species		Contreras and Lozano 1994; Lyons et al. 1998
Multiple rivers, northeastern U.S.	Multiple dams	Migration blockage, habitat fragmentation for anadromous species		
CENTRAL, SOUTH AMERICA & CARIBBEAN				
Colombia, W. Venezuela	Multiple dams	Blocked migrations, reduced productivity among large characins (e.g., *Prochilodus* spp., *Semaprochilodus* spp.) and catfishes		Pringle et al. 2000
Sinnamary River, French Guiana	Petit Saut Dam	Tributary nursery areas for almost half of species degraded due to unpredictable releases from dam		Merigoux and Ponton 1999
EUROPE				
Spain	Multiple dams	Lamprey, sturgeon, shad, eel, mullet range distributions reduced 50%–100%, listed as threatened	Migration blockage	
Rhone, Danube, Rhine, Meuse rivers	River regulation	Reduced species richness and diversity, elimination of specialists from loss of mainstem-floodplain connectivity		Aarts et al. 2004
Volga Delta, Caspian Sea	Multiple dams	>10-fold decrease in fisheries, 1930–70	Reduced downstream flows, increased salinity, reproductive failure	
Azov Sea, Russia	Multiple dams	25-fold reduction in fish catch; loss of 14 of 20 species	Reduced downstream flows, pollution	
Danube River, Slovakia	Dam and canal system	87% decline in fish catches	Reduced downstream flows, migration blockage	Balon and Holcik 1999

BOX 6.1. Reservoir Fish Conservation: Alien Species in Unnatural Habitats

"Nature knows no such organism as a 'reservoir species'" (Noble 1980, p. 139).

I live on 30-ha Lake Oglethorpe in rural northeast Georgia, USA. Georgia has no natural lakes; unlike General Sherman, the Pleistocene glaciers that formed most North American lakes did not make it this far south. My "lake" is really a reservoir, created when private landowners built a dam across Goulding Creek, a second-order stream. Below the dam, Goulding Creek contains an assemblage of mostly minnows and sunfishes. The fish "assemblage" of my lake, in contrast, is a mixture of river species that can tolerate still-water conditions and the subtropical heat of Georgia summers: American eel, brown bullhead catfish, channel catfish, golden shiner, grass carp (introduced for weed control), common carp, largemouth bass, bluegill, black crappie, redbreast sunfish, green sunfish, mosquitofish. Many of these fishes were stocked for sport fishing, some were trapped when the stream was dammed, and others were released from bait buckets and home aquaria. The point is that, with the exception of the exotic grass and common carp, these are all technically native species. But the habitat is artificial, as is the commingling of fishes. This is not an assemblage whose ecological interactions are likely to be the product of millennia of coevolved interactions, as in a kelp bed or coral reef, or a northern U.S. or African lake.

The same can be said of reservoirs throughout the world. Reservoir species come and go at the whim of humans; reservoir "ecology" is consequently something of a non sequitur, because many of the actors lack evolved relationships. Reservoirs are an ecological hodgepodge, a hybrid environment that is neither lake nor stream. A dam is built. Some riverine species can survive the altered environment, but most cannot. Some species that were just marginally adapted to flowing water thrive in the still waters; some species are dumped in the reservoir for human convenience, regardless of origin. Reservoirs are created for fishing, among other purposes, so they are stocked with predator-prey combinations: largemouth bass and bluegill, striped bass and shad, walleye and yellow perch. An initial pulse of productivity (a "trophic upsurge" or "boom") that lasts about five years typically follows construction of a reservoir, as nutrients trapped in the flooded landscape are released (Kimmel and Groeger 1980). Fishing is great, for a while. Then things settle down, literally, as the nutrient pulse is used up and the lake begins to fill with sediments and worse.

Reservoirs may be aesthetically appealing (ignoring for now the aesthetic losses brought about by drowning spectacular natural areas behind dams), but they are biologically ambiguous. As impoundments of rivers, as altered and compromised habitats, they are utilitarian havens for introduced species. Few if any imperiled species occur for long in reservoirs unless we put them there deliberately, as has happened with Australian lungfish and North American paddlefish.

One view is that reservoirs are the aquatic equivalent of city parks: subsets of the natural landscape, highly modified for human needs, populated by an unnatural mixture of native and exotic species. Admittedly, not all reservoirs behave the same, have equally detrimental impacts on native fishes, or are net losers of biomass or biodiversity or fisheries (Courtenay and Moyle 1996). Reservoirs are unquestionably valuable for resource management and research (e.g., Miranda and DeVries 1996) and have strong public support because of their recreational value and agricultural, hydroelectric, and real estate potential. They are worthy of study in their own right because they abound in human-influenced landscapes. The importance of understanding their negative impacts on the landscape and on affected fishes cannot be overemphasized, but working to conserve reservoir fishes distracts us from more important matters.

The lotic-to-lentic conversion represented by a reservoir affects fish diversity because a reservoir duplicates neither the river it replaced nor the lake it attempts to mimic (Soballe et al. 1992; box 6.1). Alterations affecting fishes occur in topography as described, and in at least three other general characteristics of rivers: hydrology, physicochemistry, and continuity.

CHANGES IN WATER FLOW (HYDROLOGY AND HYDROPERIOD)

The natural flow of water in a river is characterized by quantity, timing, and variability, which produce cycles that include flood and drought. These flow parameters determine water quality, energy availability, and physical habitat

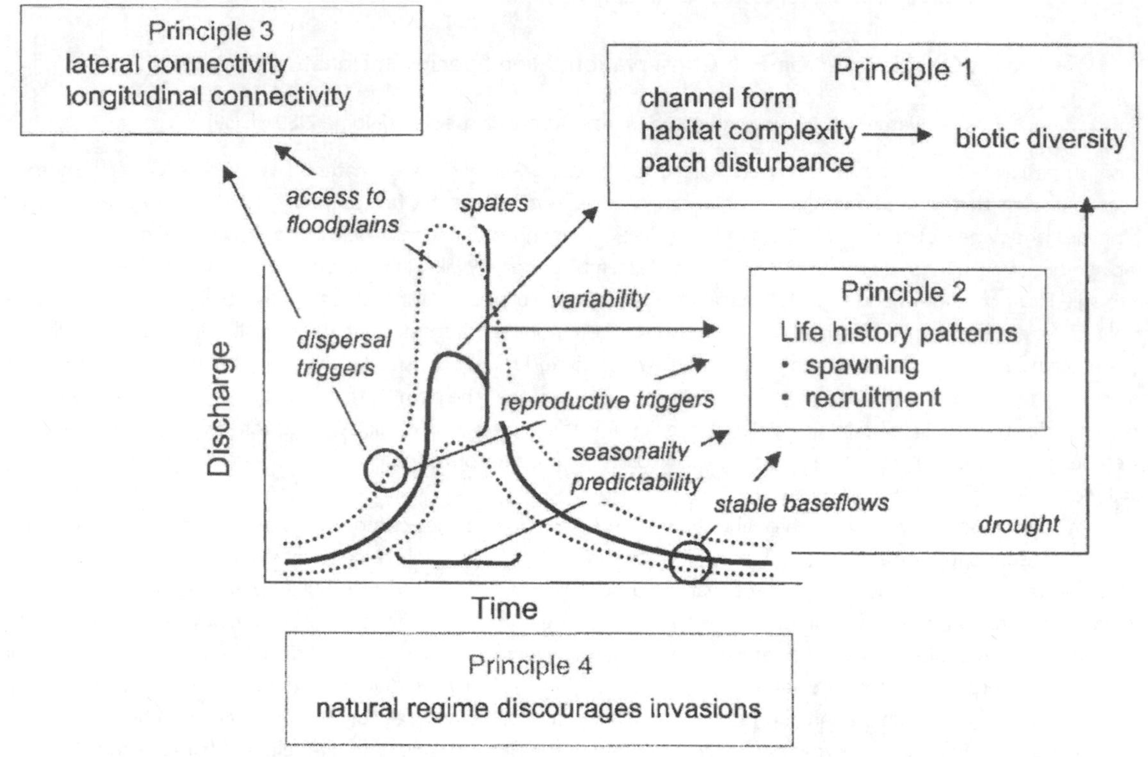

Figure 6.1. A synthesis of major interactions between natural flow regimes and biotic integrity in rivers. Principle 1 relates to large-scale flow events that influence channel form and shape. Principle 2 focuses on temporal patterns in flow that affect life history attributes. Principle 3 involves connectivity among river habitats. Principle 4 relates to the interaction between natural flow and invasive species. Native biota evolve in response to the overall flow regime. From Bunn and Arthington (2002); used with permission.

and serve as selection agents on the anatomy, physiology, behavior, and ecology of fishes and other taxa (Poff et al. 1997). In an interacting hierarchy of actions and effects, alterations in flow influence habitat, fishes, and ecological relationships (Bunn and Arthington 2002; figure 6.1). Flow determines physical habitat, which determines biotic composition and the life history strategies that evolve in response to natural flow regimes. Flow affects longitudinal and lateral connectivity, which is essential to the viability of riverine species. Finally, flow alteration facilitates invasion by nonindigenous species.

Dams are built to regulate flow, and "water levels and releases are controlled to meet specific project objectives," which may not coincide with the needs of the local fauna and flora (Soballe et al. 1992, p. 435). As a result, a major effect of dams is alteration of normal flow and flood cycles upstream and downstream, leading to unnatural and sometimes extreme fluctuations in daily and seasonal water volume and temperature. Upstream, water backs up and inundates former shallow-water, wetland, riparian, and upland terrestrial areas. Water levels then fluctuate in response to dam releases rather than natural hydroperiod. Downstream, a natural seasonal hydroperiod influenced by rainfall and snowmelt is instead regulated by opening and closing gates and valves (figure 6.2). Released water may be diverted through pipes and canals, often bypassing previous downstream areas.

When electricity is not being generated and the water is held behind the dam, flow may cease and downstream regions dry into a series of isolated pools connected by trickles (referred to as *ephemeral flow*, indicating that water flows in the channel only after precipitation). The Tallulah River in north Georgia, which flowed over Tallulah Falls—the "Niagara of the South"—and through spectacular Tallulah Gorge at average volumes of 200 cubic ft/sec (cfs) with peak flows of 15,000 cfs, provides one example. A 40-m-high hydroelectric dam was built across the gorge in 1912. Today, daily flows average only 50 cfs, the rest diverted

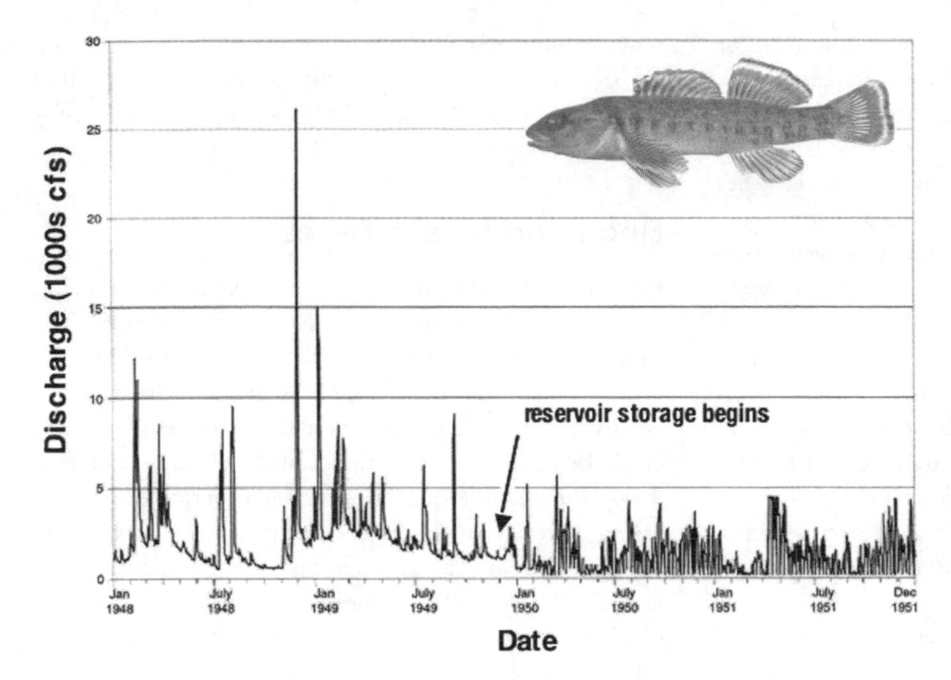

Figure 6.2. Dams alter flow and flood conditions. The Allatoona Dam on the Etowah River of Georgia began operation on the river in 1950, replacing a seasonal cycle of winter high flows and summer low flows with a relatively continuous low-flow state. Endemic fish species reliant on shoal habitats, such as the frecklebelly madtom, *Noturus munitus*; the Etowah darter, *Etheostoma etowahae*; and the amber darter, *Percina antesella*, have been extirpated from regions below the dam. cfs = cubic feet per second. From RBSAPC (2001); used with permission. Inset photo of Etowah darter by N. Burkhead; used with permission.

through pipes to generate electricity. Twenty times a year, 200 cfs flows are released for "aesthetic purposes." Otherwise, the downstream area consists of small, algal-green pools and starkly projecting, dry rocks.

Altered Flows and Fishes

Extremes of flow translate into water level fluctuations that can be detrimental to downstream habitats and fishes. Fish assemblages and juvenile survival below Harris Dam on the Tallapoosa River, Alabama, were dramatically different from those at a comparable site unaffected by dam operation (Freeman et al. 2001). Daily fluctuations below the dam alternately submerged and then dried up shoreline nursery areas, especially in spring when natural water levels should have been consistently high. In summer, low rainfall and mandated minimum flow requirements curtailed dam operation. As a result, natural-type flows—uncommon in spring—were more common during the summer. In spring, the unregulated section contained appropriate, shallow nursery habitats almost four times longer than below the dam (about 2,000 hours versus <500 hours), whereas during summer the two regions were more similar, differing by a factor of about 2.3 (250 vs. 110 hours). This pattern depressed spring-spawning fishes and created a fish assemblage dominated by summer-spawning fishes. Juveniles of five of the six spring-spawning species were less abundant. Analogous problems arise in Pacific Northwest rivers, where different species and runs of salmon migrate, spawn, and exit a river at different times (Lichatowich 1999; Coutant and Whitney 2006). Dam operation should be adjusted to accommodate life history differences among species, rather than simply mandating some minimum flow requirement.

Environmental problems associated with extreme flow regulation are typified by the "pump storage reservoirs" that occur globally (e.g., Snyder 1975). In pump storage, water runs downhill through generating turbines by day and is pumped back up into a reservoir at night. Although it violates basic laws of thermodynamics, this practice makes good economic sense. At night, demand for electricity is low, and its cost to dam operators is consequently cheap. Greater demand during the day drives the value up. Dam operators buy and use cheap electricity by night but produce and sell it at higher prices in the day. Buy low, sell high. Fish and other aquatic organisms suffer in this free-market environment; they are incapacitated as they pass upstream or downstream through pumps and turbines or are thermally stressed by the release of frigid hypolimnetic water during power generation. The unnatural hydrologic regime of daily fluctuating river levels and currents, high by day and low by night, extends above and below the dam, eliminating kilometers of usable habitat and stressing riverine organisms. Where pumped storage routes water to turbines and then releases it into another watershed or coastal area, migrating anadromous species are exposed to false environmental cues, potentially disrupting or delaying their migrations.

Substantial fluctuations below dams can subject fishes to warm, standing water one minute and raging, frigid torrents the next. Water depth, current velocity, oxygen levels, temperature, silt load (from water released from the reservoir and scoured from the river's banks), and chemical composition all change (Burkhead et al. 1997). Few fishes can tolerate such dramatic physicochemical changes; eggs and larvae are most vulnerable. Declines among many dam-impacted fishes are attributed more to recruitment failure than interrupted spawning (e.g., Humphries and Lake 2000). Many species recovery plans prescribe more natural flows from dams upstream of critical habitat (e.g., Poff et al. 1997, table 3).

Such was the impetus behind experimental releases of high, fluctuating flows from Glen Canyon Dam in the Colorado River basin in 1996. Glen Canyon Dam, completed in 1963, trapped virtually all incoming sediment in newly created Lake Powell. Most sandy beaches in the Grand Canyon below the dam vanished. Also lost were shallow, low-velocity eddies behind sandbars that historically provided juvenile habitat for fishes such as humpback chub, *Gila cypha*, designated Endangered in 1973 (Collier et al. 1997). Humpback chub also declined because introduced rainbow trout, a chub predator and competitor, flourished in the cold water released from the dam. Trout numbers increased and chub decreased even more after 1991, when fluctuating flows were diminished as a management tactic mistakenly thought to help chub (Converse et al. 1998).

In early 1996, up to 1,274 m^3/s of fluctuating, beach- and habitat-building flows were released from the dam. These experimental releases were intended to redistribute sediments important for chub spawning and reduce spawning success and larval survival of trout (e.g., Patten et al. 2001). Results were mixed. Abundant sand was deposited above the normal high-water line, rejuvenating backwater habitats important to spawning native fish (Collier et al. 1997); but no significant changes were observed in distribution, abundance, or movement of natives, and density reductions among nonnative trout were short lived (Valdez et al. 2001). Regardless, experimental releases were planned again for early 2004. This time, in addition to benefiting chub, management agencies anticipated that experimental flows would "benefit the trout upstream . . . by reducing the overpopulation of trout, thus increasing the viability and size of the fish" (USBR 2004), via unexplained mechanisms.

Flood Control and Fishes

Flood control downstream is often a goal in dam building that conflicts with fish welfare. Most rivers overtop their banks seasonally, flooding the adjacent landscape. Riverine organisms have not only adapted to seasonal flooding but are dependent on it to complete their life cycles (e.g., Galat et al. 1998; Bunn and Arthington 2002; Lytle and Poff 2004). Riparian, floodplain, backwater, periodic wetland, gallery, oxbow, and varzea habitats provide seasonal feeding, breeding, and nursery regions (e.g., Goulding 1980; Goulding et al. 1996). A massive hydroelectric dam and canal system, the Gabcikovo Water Project, built across the middle Danube River in Slovakia in 1992 dewatered approximately 230 km^2 of downstream wetlands that were formerly spawning, feeding, and overwintering habitats for fishes. This "irreversible destruction of the last inland delta of Europe" caused an 87% decline in fish catches, exacerbated by a lack of functioning fish ladders (Balon and Holcik 1999, p. 13). The decline should not have been surprising. Fish production for human use is generally dependent on seasonal flooding. Welcomme (1995, p. 121) concluded that "fisheries reasonably conducted have proved sustainable with a high rate of catch correlated with the intensity of flooding in the same or preceding years."

Dams in combination with levees and dykes prevent a river from overflowing its natural banks. Dams impede the downstream flow of water, whereas levees and dykes direct water downstream that would normally flow laterally. Such redirection of high flows affects fishes because so many are flood adapted or flood dependent. In fact, "many—if not most—fishes living in the main channel of floodplain rivers use naturally flooded habitats to some degree for feeding and reproduction" (Pringle et al. 2000, p. 810). Flood control efforts affect not only large river and mainstem species; many fishes living in small prairie streams of the midwestern U.S. have undergone significant range and population reductions due to loss of seasonal flooding (e.g., Cross and Moss 1987; Echelle et al. 1995).

The many other species dependent on marginal or lateral flooded habitats, such as oxbow lakes and billabongs, include backwater specialists with adaptations for low oxygen conditions such as gars, bowfin, and catfishes in North America; cyprinids and cobitids in Europe; galaxiids, eleotrids, rainbowfishes, and salamanderfish in Australia; and lungfishes and catfishes in South America and Africa. These habitats are successional, filling slowly and becoming swamps and bogs before drying out. Human disruption of oxbows is much more rapid. Water withdrawals and channelization in the mainstem cut oxbows off from their water source. Channelization also prevents the formation of new oxbow habitats. Impoundments leave downstream side channels and oxbows dry. Reconnecting them to mainstem rivers would provide quiet-water refuges and thus help reduce the impacts of boat wake turbulence, which makes mainstem shorelines uninhabitable for riverine larvae and juveniles (Wolter and Arlinghaus 2003). Oxbows are prime targets for human development via dredging and filling, providing instant "riverfront" property with minimal effort because of their shallow, backwater nature. Approximately 80% of the natural floodplain habitats in Germany have been lost to anthropogenic activities (J. Freyhof, pers. comm.). Protection and restoration efforts that include sediment removal, reconnection of oxbows to river channels, and other restorative measures have been initiated in such areas as Germany's Elbe River drainage (Puhlmann 2000).

Seasonal Inversion of the Hydroperiod

Disruptions of flow often result in unnatural, frequently reversed, seasonal high- and low-water periods downstream from a reservoir. At high latitudes, flow is naturally maximal during spring due to snowmelt and low during summer droughts and winter freezes. Opposite conditions prevail below a hydroelectric dam if water is captured during spring snowmelt and then released to generate electricity for air conditioning in summer and heating in winter. Tropical and subtropical hydroelectric projects also produce an upside-down hydrograph. Rainy-season floods are captured behind a dam and released during hot, dry seasons to produce electricity or provide irrigation water. Dam releases for navigational purposes may capture water during spring floods and release it during normally low summer and fall periods, smoothing out natural seasonal extremes. Animals downstream thus encounter a hydroperiod dramatically different from the one under which they evolved. Life cycles are timed with seasonally dynamic flows; inverted hydroperiods disrupt feeding, growth, migration, spawning, and larval and juvenile development (e.g., Poff et al. 1997; Rosenberg et al. 1997; Bunn and Arthington 2002). In particular, spawning migrations and the spawning act are cued by seasonal flow peaks, bringing spawning adults together and placing eggs in optimal locales for hatching and survival (Nesler et al. 1988; Naesje et al. 1995). Reproductive failure can result if peaks do not occur naturally.

Unnatural periodicities of a different nature result when flows are regulated for irrigation. A constant water supply may be optimal for crops but not for native fishes. High flows maintained for agriculture throughout the summer in the regulated Pecos River, New Mexico, are out of sync with the spawning cycle of the Threatened Pecos bluntnose shiner, *Notropis sinius pecosensis*. The shiner evolved in a desert environment, where summer flows were low, except for an occasional thunderstorm-induced spate. Constant, high summer flows displace the minnow's floating eggs into unfavorable habitat, where mortality is 100%. Reproductive success of this imperiled minnow has been increased by altering summer water releases to duplicate more natural, short-duration flow spikes (Robertson 1997).

Other Downstream Effects

River entrapment and regulated flows deprive downstream habitats of water, sediment, and nutrients, causing disruptions to freshwater, deltaic, estuarine, and marine coastal ecosystems (Poff et al. 1997; Rosenberg et al. 1997; Pringle et al. 2000). Riverine fishes are adapted to normal levels of sediment concentration and deposition. Just as excess sediment has serious consequences (chapter 5), unnaturally low sediment concentrations also create problems. Retention of sediment behind a dam results in scour and bed armoring downstream because sediment is no longer available to replenish downstream regions. *Armoring* describes a hardened and coarsened stream bottom lacking small-size fractions of sediment. It occurs when the water is "sediment hungry"—when a stream can transport more sediment than is available—and it reduces and alters important habitat for fishes and their prey (Donnelly 1993; McCully 1996).

Because some species evolved in conditions of high turbidity, loss of sediment from the water column can cause problems, as in the case of razorback suckers, *Xyrauchen texanus*, in the Colorado River downstream from Glen Canyon Dam. These endangered suckers experience higher predation rates from both natural and introduced predators under conditions of unnaturally reduced turbidity.

Nutrients contained in sediments and trapped in a reservoir deprive downstream areas, lowering their productivity. Floodplains that are normally inundated during floods remain perched above the river year round, which affects not only aquatic organisms but also multifunctional riparian zone vegetation. Fish and fisheries suffer. Fish may achieve up to 75% of their annual growth during the few months they spend in flooded habitats where feeding conditions are favorable (Welcomme and Hagborg 1977). Fish landings downstream from three new dams in northern Nigeria fell 50%; similar results have been reported after dam construction in Zambia, South Africa, Ghana, and Egypt (see table 6.1). The Aswan High Dam in Egypt impounded 50%–80% of the Nile River's flow and is considered largely responsible for a 96% reduction in sardine landings in the eastern Mediterranean. In eastern Europe, dams along the Volga River contributed to a 90% reduction in fish catches in the Caspian Sea (Welcomme 1985; Moyle and Leidy 1992).

Estuaries, regardless of distance downstream, often suffer from upstream dams. Effects include drying of wetlands, increased estuarine and nearshore salinity, intrusion of salt water upstream from the estuary, reduced sediment inputs to coastal regions, and reduced inputs of nutrients and silica to estuaries. The last phenomenon can have substantial, cascading effects on algal blooms, nearshore and oceanic food webs, and global climate (e.g., Friedl and Wuest 2002; Bottom et al. 2006).

Combined, these effects cause alterations in productivity; collapse of fisheries; and biodiversity loss via impacts on mortality, growth and development, and movement of fishes (Gillanders and Kingsford 2002). One well-studied example of estuarine disruption is the San Francisco Bay/Sacramento–San Joaquin estuary area, where reduced freshwater inputs due to river modifications have affected Threatened delta smelt, *Hypomesus transpacificus*, and Sacramento splittail, *Pogonichthys macrolepidotus*. The smelt is federally Threatened due in part to flow reductions; entrainment into irrigation canals; and introduction of the exotic, more salt-tolerant wakasagi, *H. nipponensis* (Moyle et al. 1992; Swanson et al. 2000; Moyle 2002).

In sum, rivers are characterized by natural, dynamic, seasonally recurring flow alterations. High water as well as low are part of this natural cycle and are normal occurrences; "a significant departure from the usual hydrological regimen . . . can be regarded as a disturbance" (Dudgeon 2000b, p. 794).

ALTERED THERMAL AND CHEMICAL REGIMES

The still waters of a reservoir warm as they bask in sunlight. Lotic species intolerant of higher temperatures thus encounter a thermal barrier in the transition zone between river and lake, a barrier that may strengthen as oxygen content declines. Oxygen enters water via solution at the surface and diffuses through the water column, aided by mechanical mixing. Rivers are generally oxygenated throughout the water column because of turbulence and continual mixing, which supports fishes from top to bottom. In reservoirs, the water column stratifies into a relatively thin, warm, oxygenated surface region and a much thicker, deoxygenated, deeper region. In seasonally hot reservoirs, rising surface temperatures in summer force fish into cooler, deeper regions, but they must stop at intermediate depths where sufficient oxygen still exists. The fish are sandwiched between warm water they are trying to avoid and cold water where they cannot breathe. Deoxygenation is especially severe in tropical regions, where organic matter in a drowned river may take centuries to decay, as opposed to perhaps ten years in the temperate zone. Fish kills attributed to release of deoxygenated, hydrogen sulfide–rich water below tropical dams are not unusual. In the case of the Brokopondo Dam in Suriname, South America, oxygen levels remained lethally low as far as 110 km downstream of the dam (Pringle et al. 2000).

Fishes trapped behind a dam or stocked in reservoirs are particularly susceptible to thermal and chemical stratification. Fish kills result. Striped bass, *Morone saxatilis*, have narrow temperature and oxygen tolerances, preferring temperatures of 18°–25°C and oxygen concentrations of more than three parts per million (ppm). During summer thermal stratification of reservoirs, stripers squeeze into an increasingly small region near the thermocline and become emaciated because they will not leave their preferred temperature zone to feed. Mortality results if the fish are forced into the relatively deoxygenated but cooler waters of the hypolimnion (Coutant 1985).

Reservoir construction can also produce unusual water chemistry. The trees, brush, and grasses that are drowned during reservoir filling create a massive substrate for microbial growth. Anaerobic bacteria sequester inorganic mercury already present in vegetation and soil and transform it into toxic methylmercury (Therriault and Schneider 1998), which passes up the food chain and is biomagnified in fishes (Bodaly et al. 1997; Rosenberg et al. 1997). Methylmercury concentrations in predaceous and zooplanktivorous fishes (e.g., suckers, whitefish, lake trout, northern pike, walleye, yellow perch, largemouth bass) can reach six to ten times Canadian allowable limits and remain high for 20 to 30 years. Fishes downstream from dams also become contaminated as they biomagnify mercury via consumption of fish injured while passing through hydropower turbines. In northern Canada, valuable whitefish fisheries collapsed when mercury levels rose because existing lakes were enlarged by damming outlet rivers (Bodaly et al. 1984). In addition to impacts on humans (Cree Indians in northern Canada derive 25%–33% of their wild-caught food from fishing), methylmercury at encountered concentrations delays reproduction and reduces egg production, probably via endocrine disruption (e.g., Drevnick and Sandheinrich 2003; see chapter 7).

Rivers are self-purifying systems, processing nutrients and other material through chemical and biological action; bacteria, algae, and other organisms attached to rocks and wood accomplish much of it. If excess sediment or rapidly changing flows degrade the biota below a dam, a river's ability to remove nutrients and toxins is similarly degraded. Substances stored in one reservoir are then exported downstream and added to those stored in another, leading to rapid eutrophication or accumulation of increasingly concentrated toxins (RBSAPC 2001).

Loss of self-purifying capabilities has additional consequences. Reservoirs both accumulate nutrients and are contaminated by nutrient loading more easily than rivers. A river can assimilate and process phosphorus at concentrations exceeding 100 μg/1, whereas a lake or reservoir is designated nutrient enriched (overloaded) at total phosphorus concentrations over 50 μg/1 (Wetzel 2001). Hence nutrient concentrations that have minimal impact on a river may cause nuisance algal blooms in a reservoir, frequently resulting in major fish kills.

Reservoirs and Climate Change

An environmental surprise literally arising from impoundments—one with long-term and potentially global consequences for humans as well as fishes—concerns the organic material trapped at the bottom of new reservoirs. When landscapes are flooded, dead and decomposing terrestrial plants stop assimilating and sequestering carbon and instead release it as methane (CH_4) and CO_2, two potent greenhouse gases. CO_2 output from reservoirs globally may amount to 4% of other anthropogenic sources, and CH_4 release is estimated at 20% of the total calculated output from all other human activities (St. Louis et al. 2000). CH_4 from reservoirs may exceed the output from biomass burning (forest fires, wood heating and cooking) and from the world's rice paddies. Greenhouse gas emissions from reservoirs can only increase with an anticipated increase in dam building and forest, grassland, and peat bog flooding. And contrary to industry claims that hydropower is "clean" relative to fossil fuel, hydropower reservoirs may emit up to 20 times more greenhouse gas than fossil fuel–burning power plants at the same locales. "Greenhouse gas emissions probably represent the most extensive impact of large-scale hydroelectric development" (Rosenberg et al. 1997, p. 33).

BARRIERS TO FISH PASSAGE AND LOSS OF SYSTEM CONNECTIVITY

Large rivers are relatively continuous habitats for fishes, other biota, and ecosystem processes, an observation formalized in the widely regarded River Continuum Concept (Vannote et al. 1980). Connectivity occurs longitudinally between upstream and downstream areas and laterally between main channel and off-channel regions such as floodplains. Fishes in most riverine systems move seasonally between channels and floodplains; nutrients flow with the water and the fish. Reduced connectivity between the mainstem and seasonal floodplains leads to declines in fish assemblages. Despite improvements in water quality, major rivers in Europe such as the Rhone, Danube, Rhine, and Meuse still suffer reduced species and guild richness because of decreased hydrological connectivity. Generalist species have replaced riverine specialists in these disconnected systems (Aarts et al. 2004), a pattern also observed throughout Asia in degraded rivers that have lost floodplain connectivity (Dudgeon 2000a).

Longitudinal continuity among river habitats arises from the long-distance, seasonal and breeding migrations of many fishes. Such migrations are fairly common everywhere they have been studied. Fishes migrate hundreds or even thousands of kilometers in response to seasonal rainfall patterns and to breed and feed. Long-distance migrations occur on all major continents, undertaken by sturgeons, anguillid eels, catfishes, prochilodontid characins, salmonids, and Australian barramundi, among many others (see Lucas and Baras 2001; Welcomme 2003; Hogan et al. 2004; Pusey et al. 2004). Dams impede migrations, interrupting seasonal cycles and decimating populations, directly by impairing reproduction and indirectly by disrupting the structure and function of metapopulations and compromising their long-term resilience (Liss et al. 2006; table 6.1). Dams also impede movement of colonists back into areas depopulated during droughts, floods, or other disturbances.

Genetic Consequences of Fragmentation

A major result of inhibited migrations is interrupted gene flow between subpopulations of riverine species. Dams fragment the river and isolate reproductive units from one another, leading to loss of genetic diversity and over- or underrepresentation of genotypes (R. N. Williams 2006). In the Rio Grande River of New Mexico, Texas, and northern Mexico, four pelagic-spawning minnows have been extirpated or extinguished (speckled chub, *Macrhybopsis aestivalis*; bluehead shiner, *Notropis jemezanus*; Rio Grande bluntnose shiner, *N. simus simus*; and phantom shiner, *N. orca*), mainly because of habitat fragmentation from dams (Minckley and Deacon 1968; Platania and Altenbach 1998). One remaining species with similar reproductive habits, the Endangered Rio Grande silvery minnow, *Hybognathus amarus*, is now restricted to 5% of its original range, occurring in only 280 km of river that is fragmented by five dams (Bestgen and Platania 1991). Genetic diversity is low and homozygous individuals are overrepresented. Genetic deterioration will likely continue, because dams degrade spawning habitat and prevent movement of spawning adults, drifting eggs, and recruiting juveniles (Alò and Turner 2005).

Gene flow is usually sufficient to ensure genetic similarity among tributary populations in a river basin (e.g., Soulé 1980). Dams fragment this interconnected network, not just because of insurmountable concrete walls but also because reservoirs themselves prevent interchange among populations. Many small stream fishes do not enter lake habitats, thus isolating stream populations (e.g., Herbert and Gelwick 2003). The fragmentation of habitats and isolation of populations caused by dams contribute to the downward spiral in numbers and genetic diversity that characterizes small populations (Burkhead et al. 1997) (figure 6.3).

Small Dams and Small Fishes

Concern over large dams and impoundments is easily justified by the grand scale of disturbance they cause. However, biodiversity is also degraded by small dams. In fact, the summed extent and impact of small dams and impoundments—conservatively estimated to number 800,000 worldwide—may be large. Small dams in the U.S. impound three to four times the reservoir area of large dams (Rosenberg et al. 2000).

For example, the Upper Oconee River watershed in the Piedmont region of northeast Georgia is about 60 km wide and 140 km long; its long axis can be traversed by car in about two hours. The *National Inventory of Dams* identifies 3 large dams and 273 smaller dams in the watershed (USACE 1999). However, the *Inventory* includes only dams higher than six feet. A more detailed geographic analysis revealed over 5,400 impoundments—including farm ponds, small "low-head" hydropower and mill dams, and small reservoirs—there (Merrill et al. 2001). If representative, this analysis suggests that 95% of reservoirs are not tallied, and their effects may be overlooked.

Small dams block fish movement, among other impacts. A low-head dam on a cool-water stream could retain and heat water above the preferred limits of species, affecting fish distribution up- and downstream (Gregory et al. 2002). Tiemann et al. (2004) studied two low-head dams on the Neosho River, Kansas, and found dam-related effects in fishes, invertebrates, and habitats. Fish abundance, especially among benthic species, was lower both downstream and upstream of dams. Tiemann et al. concluded that "the effects of low head dams on fishes, macroinvertebrates, and habitat are similar to those reported for larger dams" (p. 705). For most fishes, a 5-foot-high dam is no less an obstacle than a 25-foot-high dam (the same applies to small barriers created by road crossings; see box 6.2).

Expand the local practice of creating farm ponds, low-head hydropower operations, and small recreational

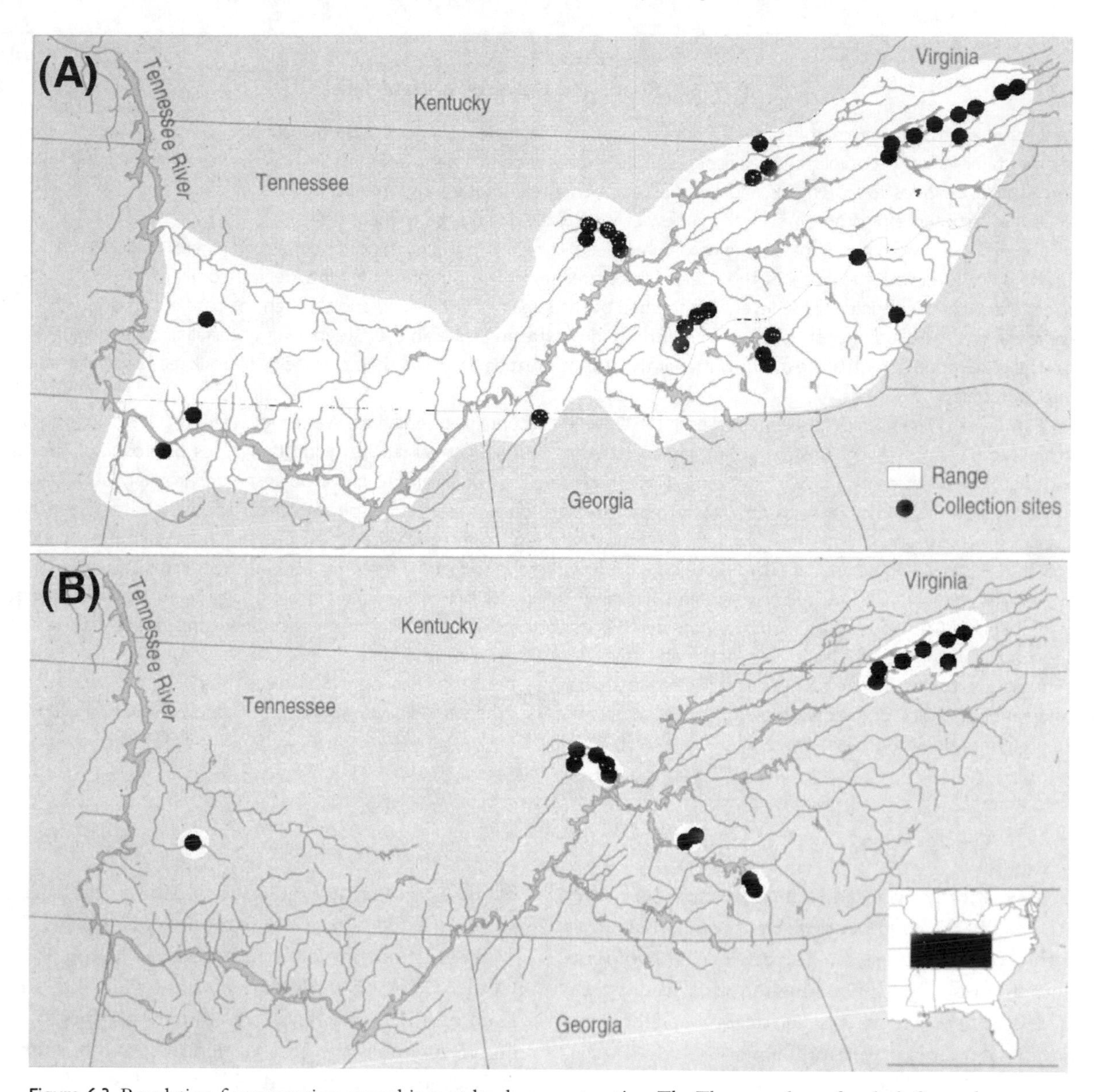

Figure 6.3. Population fragmentation caused in part by dam construction. The Threatened spotfin chub formerly occurred through much of the upland sections of the Tennessee River drainage in seven southeastern states. Its current distribution is limited to a half dozen isolated locales in three states. (A) Probable original range, based on collections made prior to the 1930s; (B) current distribution. The expanded portions of the rivers are areas impounded by dams built mostly after 1933. From Walsh et al. (1995).

and water storage reservoirs to a state, regional, or continental scale, and the impact of small impoundments can easily approach that of more spectacular and more noticed large dams. Many imperiled fishes are small, have limited mobility and geographic ranges, and occur in small habitats (see chapter 4). Small habitats include ephemeral wetlands and headwater regions of river systems. Headwaters are habitats commonly targeted for construction of small dams and road crossings because of the relative ease of construction. Often overlooked is the reality that headwaters are the primary source of water for larger, more noticeable downstream regions. Headwaters function by "mitigating flooding, maintaining water quality and quantity, recycling nutrients, providing an enormous array of habitats for plant, animal and microbial life, and sustaining the biological produc-

BOX 6.2. Roads and Culverts as Waterfalls

Salmon have evolved to negotiate waterfalls during upstream spawning migrations and downstream juvenile passage to the sea. Few other families can accomplish either, especially the leaping waterfalls part. Gobioids such as the waterfall-climbing Hawaiian 'o'opu alamo'o, *Lentipes concolor*, eels; and various torrent-adapted fishes are also exceptional in their ability to move up wet rock faces, but a vertical rise or drop of more than a few centimeters is an impasse to most fishes.

Among the most common impasses are road culverts, typically involving large pipes under a road where it crosses a stream. If a culvert is not level with the grade of the stream, if the stream gradient is more than a few degrees, or if flow velocity exceeds 1 m/sec (even if a culvert initially lies level with the bed of the stream), flow around and through the pipe scours and erodes the streambed. "Hydraulic jump" upstream and "outlet drop" downstream create an ever steepening waterfall and deepening plunge pool. The pipe mouth eventually extends out over the stream, creating a vertical drop that lacks the rock face characteristic of a natural waterfall.

In a study of 21 fish species in seven families in the Ouachita Mountains of Arkansas, Warren and Pardew (1998) found an order of magnitude less movement upstream through culverts than through natural reaches and other types of crossings. They also found that fishes upstream of culverts were less likely than fishes below culverts to move downstream, probably because fishes avoided the accelerated flow associated with culverts. A reluctance to move could isolate upstream populations and contribute to localized extirpations. Culverts can also increase the vulnerability of imperiled species by reducing movement between habitat patches. Federally Threatened leopard darters, *Percina panthera*, in Oklahoma failed to move upstream through culverts, even though water temperatures at downstream sites had risen to undesirable levels and thermal refugia were available upstream (Schaefer et al. 2003). Culvert placement was considered an important contributor to the absence of endangered goldstripe darter, *Etheostoma parvipinne*, from historically occupied locales in Missouri (Winston 2002).

Alternatives to obstructive culverts exist. Designs that minimize impacts on fishes are available (e.g., WDFW 2003) and take into consideration hydrologic, geologic, biologic, and economic factors. Where conditions prohibit fish-friendly culverts, small bridges are a preferred, albeit more expensive, alternative.

tivity of downstream rivers, lakes and estuaries" (Meyer et al. 2001, p. 3). A 2003 survey by the Conservation Committee of the American Society of Ichthyologists and Herpetologists (ASIH) estimated that at least 23 imperiled U.S. fish species were critically dependent on headwater or ephemeral wetland habitats.

Fish Passage around Dams

Fishes cannot swim through concrete walls. Regardless, small and large dams have been built without consideration of fish passage. Approximately 90% of the 1,825 hydroelectric dams regulated by the U.S. Federal Energy Regulatory Commission lack upstream bypass devices such as fish ladders, and 87% lack downstream mitigation structures (FERC 1992). Federal laws, as well as statutes enacted by Washington, Oregon, and Idaho, require fish passage devices at major dams. Section 12 of the Oregon Territorial Constitution (1848) stated, "Rivers and streams . . . in which salmon are found . . . shall not be obstructed by dams or otherwise, unless such dams or obstructions are so constructed as to allow salmon to pass freely."

Legislation notwithstanding, Chief Joseph and Grand Coulee dams on the Columbia River and Hells Canyon Dam on the Snake River are absolute barriers to fish passage. Sometimes the laws are ignored; at other times they are cleverly circumvented. When the Elwha Dam on the Olympic Peninsula of Washington was constructed in 1913, it lacked a fish bypass, in direct violation of state law. To deal with this legal inconvenience, the builders connected a fish hatchery to the dam, making the dam "an official, state-sanctioned fish obstruction for the purpose of supplying the hatchery with eggs" (Montgomery 2003, p. 182); the hatchery was a failure and closed in 1922 (for more on the "hatchery solution" to dam-induced salmon declines, see chapters 2 and 14 and Lichatowich et al. 2006).

Several large dams on the Columbia included fish ladders for spawning adult salmonids migrating upstream but failed to include devices to keep downstream-migrating smolts from entering turbines. Turbines expose smolts to extreme pressure changes and cavitation forces. Without smolt deflectors, mortality is understandably high (NRC 1996b), but mortality also occurs at deflection screens because fish impinged on the screens suffer bruising, scale loss, and stress. Water spilling over a dam into the plunge basin and cold water suddenly mixed with warmer water become supersaturated with dissolved atmospheric gases, particularly nitrogen and oxygen. Fishes swimming in such water develop gas bubbles under their skin; in their fins, gills, and eyes; and in blood vessels. Some bubbles can be 5 mm across. These emphysemas and embolisms can become infected, fish become disoriented, and death can occur within hours to weeks either from anoxia due to obstructed blood flow or from secondary infection. Young fish are particularly susceptible. Large fish kills are not unusual (Weitkamp and Katz 1980; Lutz 1995).

Injured, stressed, and disoriented smolts fall easy prey to piscivores (pikeminnows, gulls, cormorants, terns, etc.) that gather below guidance devices (OHC-NMFS 2000; figure 6.4). Predators also find easy pickings because "many thousands of smolts per hour can be delivered in a small volume of water to the dam tailrace, which provides a concentrated stream of prey for predators" (NRC 1996b, p. 232).

As a result of all these factors, smolt mortalities generally run between 5% and 15% per dam (an average value of 10% is often given), depending on season, passage structures (spillways, turbines, screening, bypasses), fish behavior and depth, predator activity, and turbine operation practices (Muir et al. 2001; Coutant and Whitney 2000). When extrapolated over the length of the Snake and Columbia Rivers—each with eight major dams—the additive, dam-induced mortality for smolts on their way to the sea can exceed 90%. The remaining smolts, having negotiated the dams and expended excess energy swimming rather than drifting downriver because of dam-induced current reductions, encounter the world's largest nesting colony of Caspian terns, approaching 30% of the total North American population of the bird species. The terns and other piscivorous birds occupy six-acre East Sand Island at the mouth of the Columbia River. Terns on this dredge spoil–created island consume an estimated 6.5 million smolts annually (wdfw.wa.gov/factshts/terns2003.htm). Predator control has been proposed, including a USFWS plan to reduce tern numbers by 65%–75% via deliberate habitat destruction or shooting, to the understandable

Figure 6.4. Avian predators downstream of the Dalles Dam on the Columbia River. The Dalles Dam also inundated Celilo Falls, a spectacular, 7-m, cascading artisanal fishing site (Gregory et al. 2002; see also Hockenberry 2001).

consternation of the birding community (ABC 2004; less clamor has been raised over a $5/fish "sport reward" bounty on native, piscivorous northern pikeminnows, *Ptychocheilus oregonensis*). Regardless of cause, high mortality rates are a major reason smolts are transported in barges past the Columbia River dams, again with limited success. Efforts are also ongoing to develop more fish-friendly turbines, juvenile bypass systems, and improved operating procedures (R. N. Williams 2006).

Juvenile salmon are not the only fishes subjected to predation at dams. Faced with a ladderless dam, upstream migrants of many species fall victim to the tailrace effect (Welcomme 2003). When 85-m-high Kainji Dam was constructed across the Niger River, western Nigeria, in 1968, an immediate impact was a 50%–60% reduction in downstream fish catches. Accompanying the declines was a shift in species composition. Small, migratory mormyrid elephantfishes, characins, and clariid catfishes fell prey to large, predatory Nile perch. Similar events likely occur wherever small migrating fishes encounter dams and predators.

Fish ladders may be ill designed, inappropriate, or dysfunctional (figure 6.5). Poorly designed ladders disorient and stress fish and cause them to expend excess energy,

Figure 6.5. Chiloquin Dam on the Sprague-Williamson River, Klamath River drainage of southern Oregon. This dam blocks access of two Endangered suckers, the shortnose sucker, *Chasmistes brevirostris*, and Lost River sucker, *Deltistes luxatus*, to approximately 90% of previously used upstream spawning habitat. (A) The fish ladder on the left side of the dam was based on designs appropriate for salmon but inappropriate for suckers, which do not jump over barriers. (B) Other fish ladders on the right (see arrows) terminate above the water's surface. Removal of Chiloquin Dam has been proposed (NRC 2004b).

increasing mortality and decreasing spawning success. Even well-designed bypass structures elevate stress, delay upstream migration, and reduce spawning success. Mortality rates of adult salmon at the various fish ladders bypassing Columbia River dams vary between 5% and 25% per dam (NRC 1996b), and most ladders were engineered with salmon in mind, ignoring the requirements of other migratory species, such as sturgeons and lampreys.

The international catalog of bad ladders is large. Ten-meter-high Sunbeam Dam on the Snake River was first built without a ladder. Ten years later, the dam was retrofitted with a ladder, which soon collapsed (the dam was removed in 1934). Edwards Dam near Augusta, Maine, blocked migration of ten anadromous species for 160 years because the original ladder washed away within a year and was never replaced (Meadows 2001). Inappropriate "technology transfer" has also occurred, as in the case of the $30 million fish elevators placed at the Yacyreta Dam on the Paraná River of South America. The elevators were designed to lift migrating salmon upstream; however, many Paraná fish species, none of which are salmonids, undergo up- and downstream migrations repeatedly during their life cycle (Pringle et al. 2000).

Dams, because they flood upstream habitats, can also eliminate natural barriers to dispersal. Waterfalls and cataracts are important faunal separators in many parts of the world. When dams flood such barriers, species disperse into areas previously unavailable to them, leading to new conditions of predation, competition, and hybridization. Construction of the Dalles Dam on the Columbia River below Celilo Falls may have facilitated expansion of introduced American shad, *Alosa sapidissima*, a species that, unlike salmon, was unable to negotiate Celilo Falls (Gregory et al. 2002). Shad also move readily through the fish ladders designed for salmon. American shad traveling through fish passage devices on the Columbia River are 1,000 times more abundant than Chinook salmon (M. Moser, pers. comm.). The Guayra waterfall complex in central South America separated two distinct ichthyological provinces. As water backed up behind Itaipu Dam, 18 cataracts taller than 30 m were flooded, allowing dispersal and commingling of previously isolated faunas (Pringle et al. 2000).

COMMUNITY AND ECOSYSTEM IMPACTS OF DAMS

The alteration of upstream and downstream environments brought about by dam construction is reflected in altered fish assemblages. Many North American streams historically rich in native species such as minnows, suckers, trouts, sculpins, and darters have become increasingly populated by sediment-tolerant minnows, catfishes, and sunfishes. Introduced species such as common carp typically thrive in disturbed habitats. Other generalist, lentic-adapted species such as golden shiner, mosquitofish, and yellow perch also come to dominate. Common carp constitute most of the fish biomass in lower reaches of the heavily impounded Rio De La Plata of Argentina and Uruguay (Pringle et al. 2000). Dams on African rivers promote a shift from predominantly riverine species (migratory characins, shilbeid catfishes, riverine cyprinids, mormyrids) to faunas dominated by lacustrine species (clupeids, tilapias, clariid and bagrid catfishes, lacustrine cyprinids, lungfishes, Nile perch). These impacts might be even more marked if natural flood-drought cycles did not create periodic semilacustrine conditions, to which residents are presumably adapted (Welcomme 2003).

As one example, a unique fish fauna of 24 endemics evolved in the high flows, temperatures, and turbidities of the Colorado River. Spawning by many involved migrations of hundreds of kilometers, triggered by and dependent on seasonal changes in water level and temperature. But over 100 dams now interrupt the river's flow for flood control, agriculture, municipal water supply, and recreation, such that less than 1% of the water reaches the river's mouth in the Sea of Cortez. The reservoirs behind many Colorado dams are thermally stratified. Cold, clear, hypolimnetic water chills downstream habitats, leaving them sediment starved, disrupting natural spawning cycles, and killing native fishes. Introduced sport fish and other invaders now make up two-thirds of the ichthyofauna, and most of the remaining native fishes are Threatened or Endangered (figure 2.5). The impacts brought on by disturbed hydrology—habitat modification, altered sediment distribution, altered thermal and hydrologic regimes, blocked migrations, and invasions by exotic fishes—have all contributed to the decline of native Colorado River fishes (Minckley 1991; Wydoski and Hamill 1991).

One other impact of impoundments on fish assemblages is facilitation of hybridization. Species-isolating mechanisms fail when physical habitat is altered to favor nonindigenous and native invasive species that previously found it unfavorable. The remaining, depleted native residents encounter a loss of preferred spawning habitat and a shortage of conspecifics, at the same time that invaders with less specialized spawning requirements abound. This imbalance in numbers and appropriate habitat results in mistakes during courtship and spawning, which leads to hybrid matings, outbreeding depression, and genetic introgression (Hubbs 1955). Hybridization and introgression can eventually extirpate a rare species, as may have happened to endangered June suckers, *Chasmistes liorus liorus*. June suckers hybridize readily with the more abundant Utah sucker, *Catostomus ardens*, to the point that some researchers believe they have been replaced entirely by a hybrid species, *Chasmistes liorus mictus* (Echelle 1991).

Four other endemic western lakesuckers in the Klamath basin of northern California and southern Oregon also engage in hybrid matings. These apparent anomalies may reflect multiple impacts of dam construction, including blocked access to preferred spawning locales and forced approximation with other species during the spawning season. Two species, the shortnose sucker and Lost River sucker, were designated Endangered in 1988 in part because of concern over high rates of hybridization. Hybridization is especially common in reservoirs in the basin and in the area below Chiloquin Dam (figure 6.5), where spawning fish may encounter one another because their spawning migrations terminate unnaturally at this barrier. During the spawning season, Lost River suckers are ten times more common below the dam than shortnose suckers. Intermediate-appearing fish constitute 6%–30% of the spawners present (Moyle 2002; NRC 2004b). Whether hybridization results directly from the multiple influences of dams, habitat disturbance, population decline, and population imbalance is by no means established, but it remains a concern.

Ecosystem-Level Impacts

Reduced abundances and species losses ultimately alter ecosystem function. Migratory fishes—especially diadromous species such as salmon, shad, and eels—link distant ecosystems by transporting energy and nutrients (chapter 11). Interrupted migrations affect ecosystem members, such as predators and scavengers, that depend on migrating species. The best-studied examples involve Pacific salmonids, but effects of interrupted migrations ripple through island ecosystems, western Atlantic coastal rivers, and inland temperate and tropical rivers, among others (Barthem and Goulding 1997; Winemiller and Jepsen 1998; Pringle et al. 2000; March et al. 2003).

Other organisms that are linked intimately to fish species also suffer. Many freshwater mussels and clams utilize fishes as intermediate hosts in a semiparasitic life cycle. Their larvae develop attached to the gills of freshwater fishes. Declines among fish species result in declines in host-dependent mollusks, as evidenced by the elimination of bivalves upstream of dams that block migration by host fishes. Seven unionid mussels in the American Midwest have declined markedly because dams on five rivers block migrations of freshwater drum, *Aplodinotus grunniens*, which host the mussel's larvae. Freshwater mollusks are among the most endangered taxa in North America, with more than 70 native bivalve species listed as federally Endangered or Threatened (ecos.fws.gov/tess_public). This degree of imperilment has been directly attributed to habitat loss and to dams and other obstructions to fish migration (J. D. Williams et al. 1993; Watters 1996).

Another ecosystem-level impact is the effect that a dam can have on downstream food webs. In the Columbia River before dams, the food web was characterized by shallow-water benthic consumers with a food base of macrodetritus. The dams favor pelagic species such as nonnative American shad that feed on microdetritus, at the same time that seasonal alterations to the plume from the river have changed nearshore marine habitats over a large area (Bottom et al. 2006). These circumstances parallel the phenomenon of fishing down food webs that is occurring widely in marine and freshwater habitats (chapter 11).

Dams in the Desert: Catastrophes from Irrigation and Water Withdrawal

Many fish species live in deserts and other seasonally dewatered locales, even though they are marginal habitats. Typical families include pupfishes, killifishes, lungfishes, gobies, and clariid catfishes. Desert-adapted species have specializations for survival through periodic but predictable droughts and attendant high temperatures, concentrated salts, low oxygen tension, and high alkalinity. These adaptations include diapause, annual life cycles, estivation, and accessory breathing structures (Helfman et al. 1997). However, not even desert fishes can tolerate complete or unpredictable water loss. Humans are not particularly drought tolerant. Intense competition for water is almost inevitable when people move into desert areas, and fishes are the usual losers in this battle. Many desert species such as those in the Colorado River are imperiled.

Despite the physics of the situation, humans build dams in deserts and try numerous other means of overcoming naturally arid conditions to establish populations and grow water-demanding plants where they were never meant to grow (U.S. Southwest, much of Australia, Mediterranean lands; e.g., Reisner 1993). Natural precipitation is insufficient to sustain agriculture (or people), so farmers use water from underground sources to irrigate. Desert aquifers and other underground sources recharge slowly, on geologic time scales, whereas water withdrawal occurs daily. The inevitable results are drying of groundwater-fed streams and lowering of water tables. Irrigation leads to salinization of topsoil and water tables as water evaporates in the dry desert air.

The Salton Sea

The Salton Sea, in Southern California, is a textbook example of mismanaged desert agriculture and its aftermath. The largest water body in California, it resulted from an agricultural scheme gone awry. A levy associated with a diversion-irrigation project in California's Imperial Valley (a desert valley) failed in 1905 when the Colorado River broke through and flowed for a year and a half into a geological depression known as the Salton Sink, which sits about 90 m below sea level. When the river finally returned to its natural bed, a 980+ km^2 lake without any outflows had formed where none had existed for several thousand years (the Salton Sink had filled and evaporated at least twice before in geological history, but no humans were around to stock it with fish) (Boyle 1996; J. Kaiser 1999).

Since 1905, 63 freshwater and marine fish species in 27 families have been placed in the lake. About a dozen species remain, with four (orangemouth corvina, sargo, gulf croaker, hybridized tilapia) forming a sport fishery. One Endangered species, the desert pupfish, *Cyprinodon macularius*, occurs naturally in inflowing streams. Salt content in the Salton Sea is continually rising; 1.3 million acre feet of Colorado River water flow annually through the agricultural Imperial, Coachella, and Mexicali valleys and then into the Salton Sea, where the water evaporates and leaves 3 billion kg of salt behind each year.

Introduced fishes that flourished at lower salinities have been eliminated as salinity, which reached 44 parts per thousand (ppt) in 2000, rose. Normal seawater contains 37 ppt salinity. Fish reproduction will be impaired above 45 ppt, and survival is unlikely above 60 ppt. The inflowing water also carries fertilizers, pesticides, selenium, industrial wastes, and sewage; 10%–15% of inflows come from Mexico's New River, designated one of the ten most endangered rivers in North America in 1998 (American Rivers 1998). Algal blooms, including species known to be toxic to fishes (Tang and Au 2004), create oxygen depletion problems in the sea. Bottom waters become anoxic, get mixed by desert winds, and cause massive die-offs of fishes. Fishes are heavily infected with ectoparasites and some with botulism bacteria (Nol et al. 2004). When the latter are consumed by fish-eating birds (cormorants, pelicans, terns, Eared Grebes), the birds die by the thousands (Friend 2002). Fish-eating birds, which are also killed by cholera, salmonellosis, and Newcastle disease, are attracted to the sea because 80% of California's wetlands have been drained, and it is one of the few wetlands remaining. Fish and bird die-offs have convinced many people that the Salton Sea is an attractive nuisance at best, a death trap at worst; others focus on its role as important habitat for birds.

Plans to manage the sea abound and include allowing it to dry up and go away, as has happened geologically; installing evaporation plants to reduce salinity; and diluting it with fresh or seawater from various sources to keep it alive. None of these solutions is without controversy or cost; the dilution solution is estimated at $1 billion over a 30-year period (Kaiser 2000).

The Aral Sea

Whereas the Salton Sea originated from human folly, the Aral Sea (figure 6.6) was natural but is now "perhaps the most notorious ecological catastrophe of human making" (Stone 1999, p. 30; Lake Chad and Three Gorges Dam may compete; see box 6.3). In 1960, the Aral Sea of the Uzbekistan-Kazakhstan region was the fourth-largest lake in the world, at 68,000 km^2. Water came from the Syr Darya and Amu Darya rivers and then evaporated, which kept the Aral at a salinity of 10 ppt. Dams and canals built in the 1950s and 1960s diverted water into vast cotton fields, shrinking the lake's area 55% by 1993. By 1998, its volume was reduced 80%. Lake salinity rose to 50 ppt by the 1990s. Flow in the rivers has been reduced 94%, disrupting the lake, the rivers, and former river delta ecosystems. What little water still entered the lake was highly polluted irrigation return or wastewater.

Because of dams, lowered water level, spawning habitat desiccation, increased salinity and other minerals, and agricultural pollution (petrochemicals, phenols, heavy metals, benzene, organochlorines, DDT—some at 100 times Russian Maximum Allowable Concentrations), an original native fish fauna of 20–24 species has been extirpated. The fish included such widespread species as bream, common carp, wels, northern pike, zander, and Eurasian perch, as well as several endemics, including Aral roach, Aral shemaya, Aral sea trout, Aral stickleback, and four sturgeons. The native fishes were replaced by a progressive series of 15 saline- and pollution-tolerant, introduced species. These aliens have also been eliminated as water quality has deteriorated. Currently, four introduced species—an atherinid, a goby, a flounder, and a stickleback—remain. Analogous changes have occurred in all other faunal and floral groups (Aladin and Potts 1992; Zholdasova 1997).

Four of the extirpated native species were sturgeons. The Syr-Dar shovelnose sturgeon, *Pseudoscaphirhynchus fedtschenkoi*, and possibly the small Amu-Dar shovelnose sturgeon, *P. hermanni*, both regional endemics, are consid-

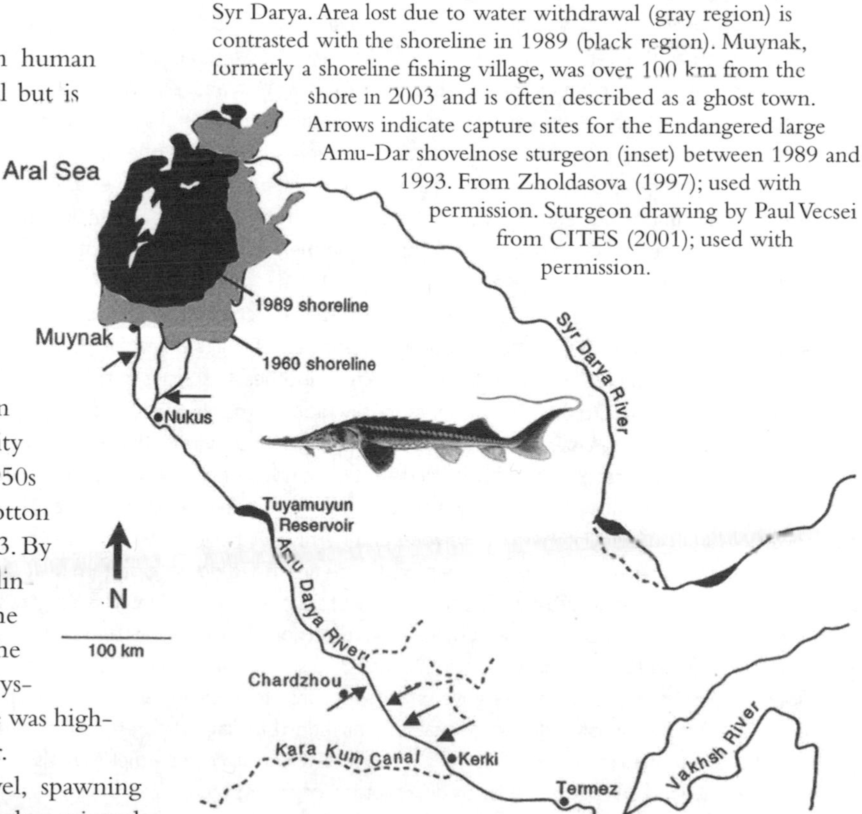

Figure 6.6. The Aral Sea and its inflow rivers, the Amu Darya and Syr Darya. Area lost due to water withdrawal (gray region) is contrasted with the shoreline in 1989 (black region). Muynak, formerly a shoreline fishing village, was over 100 km from the shore in 2003 and is often described as a ghost town. Arrows indicate capture sites for the Endangered large Amu-Dar shovelnose sturgeon (inset) between 1989 and 1993. From Zholdasova (1997); used with permission. Sturgeon drawing by Paul Vecsei from CITES (2001); used with permission.

BOX 6.3. Three Gorges and Growing

The enormity of the Three Gorges Project (TGP), better known as the Three Gorges Dam (TGD), is difficult to comprehend. When completed in 2009, it will be the largest dam in the world at 190 m high and 2 km wide. The lake behind it, which will drown the gorge celebrated by Qing Dynasty poet Zhang Wentao at the opening of this chapter, will extend upstream 600 km, about the distance from Los Angeles to San Francisco, and cover an area of 630 km^2. It will block the flow of the Yangtze River, the third-longest river in the world. The Yangtze, with 3,000 tributaries and 4,000 lakes, has the highest fish diversity of any river in the Palearctic region, with 29 families and at least 361 species and subspecies, including 177 endemic species and subspecies (cyprinid barbs, cobitid loaches, bagrid catfishes, and homalopterid hillstream loaches dominate). The main channel above the dam site alone contains 162 species, 44 of which are endemic.

The better-known Critically Endangered fishes of the Yangtze include the Yangtze or Dabry's sturgeon, *Acipenser dabryanus*, and the Chinese paddlefish, *Psephurus gladius*. The Chinese sturgeon, *A. sinensis*, is considered Endangered by IUCN. Construction of the massive (54-m-high, 2.6-km-wide) Gezhouba Dam in 1988, 38 km downstream of the TGD, created a 1.58-billion-m^3 reservoir. Gezhouba eliminated Yangtze sturgeon from reaches below the dam and heavily impacted the Chinese sturgeon and Chinese paddlefish via blocked migrations (Fu et al. 2003). Gezhouba also affected populations of the Chinese sucker, *Myxocyprinus asiaticus*, the only sucker endemic outside of North America, delivering a near-fatal blow that TGD will probably complete.

An eight-volume environmental impact study (EIS) of TGD was completed by the Chinese government in 1991 and "examined and approved by an expert panel" (SSTC 1991). It concluded that impacts on biota would be minor, specifically stating, "In general, TGP will not damage the survival of any aquatic species" (YWRP 2002). Fewer than three pages focused on fishes and those in a somewhat enigmatic tone, saying, for example, that critically endangered Yangtze sturgeon "will not significantly increase in the reservoir." Separate analyses suggest potentially serious effects on fishes. Park et al. (2003) surveyed the 44 fish species endemic to the main channel above the dam, mostly fluvial or lotic adapted, and calculated their probability of survival as low and dependent on refugial habitat in tributaries. Given habitat preferences, biology, and tributary availability, 6 species are likely to go extinct, 14 have an uncertain future, and 24 "may survive in the tributaries" (Park et al. 2003, p. 1748).

Even these predictions are best-case scenarios. Many tributaries are degraded or slated for additional dam construction, many fishes are already in decline, and important life history information is generally lacking. Little is known about possible interactions with introduced species or with natives when river fish are pushed into new habitats. The greatest impact will befall migratory species such as the Yangtze sturgeon; the cobitid royal clown loach, *Leptobotia elongata*; and the cyprinid *Coreius guichenoti*. Other species likely to go extinct are the cyprinids *Megalobrama elongata* and *Onychostoma brevis* and the goby *Ctenogobius szechuanensis*, which are already restricted to the main channel or a single tributary of the Yangtze. Although Park et al. considered 17 tributaries as potential refuges after the reservoir fills, only 3 (the Min, Tuo, and Chishui rivers) are not already degraded or slated for hydrodevelopment (e.g., 10 dams are planned for the Jinsha River alone). Fisheries will also suffer, because 10 of the 12 commercially important endemics are threatened by the dam. Park et al. concluded (p. 1757), "The TGD will have dramatic impacts on the ecosystems of the Yangtze River" and recommended that the few potentially habitable tributaries be set aside as reserves and protected from further development.

Fu et al. (2003), assessing the entire basin, noted that 25 Yangtze species were designated Endangered in the *China Red Data Book* and that "closure of [TGD] will cause profound ecological changes . . . [but] actions to conserve fish biodiversity in the river are lacking" (p. 1650). Closure of TGD would likely endanger another 40 species in the upper reaches of the river, 19 of which are endemic. Fu et al. identified hydrologic alteration, especially blocked migrations, as the primary cause of declines, even before TGD. Catches of Reeves shad, *Tenualosa reevesii*, averaged 475,000 kg annually until the 1970s. After dam construction, and despite a fishing ban instituted in 1986, research collections fell dramatically; not a single fish was captured in the 1990s. Overfishing is apparent, indicated by declines in diversity and abundance of migratory fishes, decreased catches of large fishes, increased catches of small fish, and fewer old individuals. Catches of piscivores and herbivores have declined, while zooplanktivores and omnivores and detritivores have come to dominate, suggesting that food webs are fished down, a pattern best documented in the world's oceans (chapter 11).

TGD's impact will be substantial, but additional threats loom darkly on the Yangtze's horizon (and only threats to fishes are mentioned here). Several more dams are planned, as is diversion of water from the Yangtze to northern China, "the biggest water diversion project in the world" (Fu et al. 2003, p. 1660). Such diversion threatens riverine, estuarine, and migratory species dependent on Yangtze flows and sediments. The Chinese government's response focuses on captive propagation and aquaculture, which will accomplish little if causative factors of pollution, habitat loss, migration blockage, deforestation and sedimentation, and overfishing are not corrected. Fu et al. emphasized creating reserves, reconnecting lakes to rivers, employing fish passage devices, and increasing research on the many poorly studied species. To its credit, chapter 5 of the EIS recognized the importance of providing variable flows to ensure "suitable aquatic environment for breeding of fishes, such as producing artificial flood peaks by reservoir regulation to promote breeding of the [native] fishes." The EIS also promoted research, especially long-term investigations of rare, endangered, and endemic species.

ered extinct. Two other species survive, but in greatly reduced numbers. The large Amu-Dar shovelnose sturgeon, *P. kaufmanni*, is endemic to the Amu Darya River. It was once common but is now considered Critically Endangered. The ship sturgeon, *Acipenser nudiventris*, is an anadromous species of the Black, Caspian, and Aral seas. It migrated 1,500–1,800 km up both input rivers to spawn. Once commercially abundant (catches up to 4,000 metric tons annually), it is now apparently extirpated from the Aral Sea and the Amu Darya River (Zholdasova 1997). It is Endangered or Critically Endangered throughout its range (Birstein, Waldman, and Bemis 1997b).

Commercial fisheries in the Aral Sea, which landed 48,000 metric tons (MT) in 1957 and once employed 60,000 people, fell to zero by the early 1980s. Fish canneries along the sea's previous shore, which has receded many kilometers, import fish from the Pacific Ocean to provide employment (Gore 1992). Economic losses are estimated at $3.2 billion annually for the region (Micklin 1988). Declining fish and fisheries are a major concern in their own right, but they are also a harbinger of human health problems. With 3.6 million ha of seabed exposed, toxic dust storms visible from space deposit 43 million MT of salt annually over a 200,000 km^2 area. The dust storms also contain fertilizers and pesticides and have led to tuberculosis, anemia, high infant mortality, and a death rate from respiratory ailments that ranks among the highest in the world. High rates of intestinal illnesses and throat cancer also plague human inhabitants around the dwindling sea (Micklin 1988; Stone 1999).

Proposed restoration schemes are ecologically and technologically questionable and would require massive international cooperation and investment (Stone 1999).

SOLUTIONS TO DAMS

Dams and irrigation projects are not going away. They are as much a part of human culture as fire and agriculture. However, much can be done to prevent, minimize, and reverse their negative impacts. The most obvious actions include improving irrigation methods to reduce water loss, conserving water to reduce need, conserving energy and developing alternative sources, and not building dams in the first place. Additional activities to reduce impacts include restoring natural flows to impounded systems, correcting sediment transport and deposition problems, correcting passage and entrainment problems, and, ultimately, removing dams that have outlived their usefulness. Dams modify entire ecosystems, more so than many of the other negative impacts humans visit on aquatic habitats and fishes; therefore, correcting the damage requires an ecosystem perspective and ecosystem-level management actions, a recurring theme throughout this book.

Restoring Natural Flow Regimes

A major advance in understanding the impacts of dams is recognizing that management approaches often overlook the naturally dynamic nature of healthy flowing water systems (e.g., Richter et al. 2006). Implementation of this understanding mandates water releases from dams that mimic or approximate natural flow variation. Equally important is approaching dam operation and flow management as an ongoing experiment, to be adjusted as new information becomes available (e.g., Irwin and Freeman 2002).

Flow has five major attributes: magnitude, frequency, duration, timing, and rate of change. Each attribute affects riverine processes and characteristics and hence the integrity of fish populations. Dams alter all five attributes, which causes declines among native fishes. All five flow components are dynamic, changing hourly with rainfall, daily, and seasonally, and at times unpredictably from storms or droughts. Dams often strive for constant, predictable water output, but riverine organisms are adapted to the dynamics of river flow. Restoring native communities requires reinstituting natural, dynamic flow regimes.

Specific goals have been identified that link the benefits of natural flow regimes to fishes (based chiefly on Stanford et al. 1996; see table 6.2):

- restoring peak flows to reconnect and reconfigure channel and floodplain habitats;
- stabilizing base-flow conditions to revitalize food webs, especially in shallow water;
- reconstituting seasonal temperature patterns and dynamics, such as practicing depth selective withdrawal from dams rather than purely hypolimnetic or surface releases;
- maximizing dam passage in both directions to facilitate migrations, allow recolonization of impacted areas, and allow recovery of genetic structure among metapopulations;
- promoting management that relies on natural habitat restoration and maintenance by improving land-use practices, rather than artificial propagation and overreliance on river engineering and predator control; and
- instituting adaptive, ecosystem management.

Table 6.2. Effects on fishes of different flow levels

Flow component/level	*Impact on fishes*
Low or base[a] flows	
Normal conditions	Provide adequate habitat space for fish and prey taxa; maintain suitable water temperatures, dissolved oxygen, water chemistry; keep eggs suspended; enable movement to feeding and spawning areas
Drought conditions	Purge invasive, introduced species; concentrate prey in limited areas
High pulse flows	Shape physical character of river channel, including pools and riffles; determine size of streambed substrates (cobble, gravel, sand); restore normal water quality after prolonged low flows; flush away waste products and pollutants; aerate eggs in spawning gravels, prevent siltation; maintain normal salinity levels
Large floods	Provide migration and spawning cues; trigger new phase in life cycles (e.g., smoltification); enable spawning on floodplain, provide nursery areas; provide productive, new foraging areas in floodplain; control distribution and abundance of riparian and floodplain plants; shape physical habitat of stream edge; deposit gravel and cobbles in spawning areas flush/redistribute organic material and woody debris; purge invasive species; rework/reconnect side-channel habitats such as oxbow lakes

Source: After Postel and Richter 2003; Richter et al. 2006.

[a]Base flows are average flows maintained primarily by groundwater discharge without recent rain or meltwater.

The challenge is deciding which approach to take. Rivers have been altered for so long and to such a degree that the original flow regime may be unknown. Also, the river may be so physically changed—including containing multiple dams—that return to natural conditions is impractical. Economic and social factors may also militate against radical adjustments. At least 207 different environmental flow methodologies have been applied in 44 different countries, all with the stated or implied purpose of determining the quantities and qualities of water needed to conserve riverine ecosystems and protect aquatic resources (Tharme 2003; see also Morehardt 1986; Arthington and Zalucki 1998; Arthington and Pusey 2003). One commonality among approaches is recognizing that, in determining water allocations, the river is a legitimate user. Consideration should be given to flows "necessary [to] restore ecological processes and the biodiversity of water-dependent ecosystems" (Arthington and Pusey 2003, p. 377). This "adequate environmental flow" philosophy underlies many international efforts and directives focused on river restoration and hydrologic management, including the South African National Water Act (1998), the Water Directive Framework of the European Union, the National Water Initiative in Australia, and the Sustainable Rivers Project of the U.S. Army Corps of Engineers and The Nature Conservancy (see Postel and Richter 2003).

Methodologies differ in approach, complexity, and focus. Among popular, simpler efforts are those based on designated low-flow targets. These keep some minimal amount of water in a stream for specified purposes, such as providing habitat at minimal, good, or optimum amounts (e.g., the "Montana Method" of Tennant 1976; see also Orth and Maughan 1981). Determination of minimal flows is based on how much water a river needs to maintain its integrity and self-sustaining productivity, including maintenance of fish populations. More sophisticated, hydrodynamic habitat modeling approaches include the instream flow incremental methodology (IFIM), which focuses on water allocations that affect habitat. At its simplest, IFIM determines "flows to maximize the amount of microhabitat for a single life stage of a high profile fish species at a few isolated spots in a river" (Bovee et al. 1998, p. 1). Application of IFIM has broadened considerably, but outcomes of IFIM-type models, such as PHABSIM (Milhous et al. 1989), still focus on effects of various water management alternatives on habitat availability for selected species. "Habitat" can include not just microhabitats but also macrohabitats (stream segments, stream networks, and entire drainage basins) and mesohabitats (runs, pools, riffles). IFIM remains fish oriented and, despite its detractors, is widely adopted by fisheries management agencies.

A third approach gaining popularity, especially among aquatic ecologists, is characterized as "holistic." It addresses flow requirements of entire river ecosystems and emphasizes links among flow regime, habitat, and organisms throughout a river and its watershed. Holistic approaches are a reaction to minimal-flow methods, which, in combination with water quality issues, dominated river protection and restoration and dam operation improvements until relatively recently (see Richter et al. 1997a, 2006). Falling below certain minima at critical times of the year can harm fishes, but seasonal differences in flow—including cyclically low/baseline and high flows, periodic drought, and flood conditions—are conditions to which fishes have adapted and on which they depend (table 6.2). As shown in Freeman et al.'s (2001) study of juvenile survival in the Tallapoosa River, persistence of habitat throughout the year is more important than maintaining arbitrary and often low minimum water levels. Flow regime cannot focus solely on minimal flows.

Holistic, dynamic-flow approaches stress conservation of native aquatic biodiversity and protection of natural ecosystem functions. Some, such as the Range ofVariability Approach (RVA) emphasize manipulating flows to approximate natural variation in their five components. RVA has been implemented in the lower Roanoke River, where striped bass recruitment has recovered to the highest levels since impoundment (Richter et al. 1997a). Other, analogous approaches have been used with success, even though background data on natural flow variation are less complete (e.g., the DRIFT methodology for determining flows for fishes in South Africa, Arthington et al. 2003).

The holistic view has been expanded to an adaptive, consensus-based approach for developing environmental flow recommendations for large river systems (Richter et al., 2006). A five-step process includes involvement of all concerned parties from the outset. Literature review is incorporated into development of preliminary flow recommendations that include all five flow components under differing climatic and water availability scenarios. Following implementation on a trial basis, monitoring and reassessment continue. Impacts on fishes are emphasized. This five-step approach has been applied to determining dam operations and flow management as part of the Nature Conservancy's Sustainable Rivers Project. As of late 2005, the approach was being used on 11 rivers in 13 states, involving 26 dams (see http://nature.org/success/dams.html).

Sediment Accumulation

Reservoirs fill rapidly with silt due to river deposition. Silting also results from slumping of steep sides, deforestation of the surrounding land, and erosion of shorelines from fluctuating water levels associated with dam operations. Sediment accumulated behind a dam limits the reservoir's life span (Poff and Hart 2002). Dams across rivers with high sediment loads may fill up in a few decades, after which the reservoir can no longer store water or generate electricity. Lake Powell, created by Glen Canyon Dam, receives up to 140 million tons of sediment annually (Stanford and Ward 1991). It accumulates in the reservoir at a rate of 2–3 m/yr, which means the 51-m-deep lake could fill with sediment in as little as 50 years (estimates vary, depending on definitions of reservoir function). Approximately 1% of the storage capacity of the world's reservoirs is lost annually to sedimentation, necessitating the construction of 300–400 new dams (R. White 2000).

Sediment can be dredged from a reservoir and deposited on land, but that is logistically difficult and costly. Resuspension of sediments during dredging can lead to toxin release, which also occurs when dredge piles erode or become airborne as dust. A simpler and less costly solution is to periodically flush sediments by opening gates in the dam that sit below spillways and turbine gates, assuming they were installed (Atkinson 1996).

But flushing can also affect downstream fishes and invertebrates. Spencer Hydroelectric Dam on the Niobrara River, Nebraska, periodically filled with sand, silt, and organic debris, reducing water storage and clogging turbine intakes. This accumulation necessitated flushing on average four times a year. Each flushing typically lasted a week, during which time water quality standards were exceeded, sometimes as far as 60 km below the dam (Hesse and Newcomb 1982). Dissolved oxygen fell as low as 3.5 mg l^{-1}, 2.9 mg l^{-1} below acceptable levels. Turbidity increased from less than 100 JTU (Jackson turbidity units) to over 1,000 JTU and as high as 3,750 JTU. Suspended solids increased from less than 100 mg l^{-1} to over 10,000 and as high as 21,875 mg l^{-1}. Invertebrate assemblages were likened to those "reported for aquatic environments [occurring] immediately below an effluent discharging raw sewage" (p. 50).

During flushing, fish kills were common. More than 22,400 dead fish among over 30 species were counted on 12 occasions, but deaths were underestimated, because many fish were stranded behind boulders and in depressions and were quickly covered by sediment more than 1 m deep. Death also resulted from high velocities, buffeting by large debris and sediment, low dissolved oxygen, stranding, and stress. Hesse and Newcomb (1982) recommended curtailment of flushing during spawning seasons (mid-April to mid-September) and flushing only once annually to allow faunal recovery. More generally, hydroelectric projects should not be placed on heavily silted rivers. If they are, land-use practices in surrounding watersheds should be improved to reduce the silt load coming off the land (see chapter 5).

Getting Fish Past Dams

The history of dam building has been replete with state-of-the-art engineering while ignoring basic biology. No other explanation accounts for the lack of fish bypass devices at so many dams, although a variety of such structures are available (e.g., Odeh 2000; Larinier et al. 2002).

Upstream passage of fish in dammed waterways is accomplished via pumps, fishways that incorporate ladders or lifts (elevators, locks), and transportation ("trap and truck"). Downstream passage is facilitated via various screens; barrier nets; bar racks and louvers; hydraulic deflection; behavioral guidance via acoustics (ultra, midrange, or infrasound), lights, and electric fields combined with spillways; transportation via truck or barge; or pumping (OTA 1995). The method depends on a variety of factors including species, behavior, injury vulnerability, age class, fish size and swimming ability, river flow (approach and flow-through velocities and turbulence), ice or debris frequency, availability of funds, public attitude, regional preference, and effects on the dam or hydropower facility (OTA 1995).

Public pressure from the fishing and environmental communities has led to inclusion of bypass devices in newly created dams and forced retrofitting of dams built without them. However, many obstacles remain to convince those who construct or operate dams of the need for bypass devices (OTA 1995). No single design applies to all circumstances, attitudes among some agencies and corporations remain entrenched, and motivation to change often requires economic incentives or disincentives. Economics is especially relevant because adding a bypass device after a dam is constructed is much more expensive than incorporating a device in the original design. Not insignificant is the value of the water allocated to bypass systems and that does not enter turbines. For example, in the Columbia Basin, the cost of forgone power production attributed to salmon restoration efforts has been estimated at $1.29 billion since the program began in 1982 (NPCC 2003).

Keeping Fish Away from Dams

Hydropower dams and other water intake structures must allow passage around structures, but they must also keep fish out of the machinery during normal operations, whether the fish are migrating or not (figure 6.7). Economic as well as ethical concerns arise, as in the case of the Richard B.

Figure 6.7. Results of fish passing through an unscreened hydroelectric dam. Remains of (A) an American eel and (B) an alewife after passing through the unscreened turbines of Benton Falls Dam on the Sebasticook River, Maine. Photos by Doug Watts; used with permission of Friends of the Kennebec Salmon.

Russell Dam on Strom Thurmond Reservoir, which impounded the Savannah River. Dam operations and electricity generation at Russell Dam were curtailed by court order for more than 12 years because of potential entrainment of an estimated 13 million fish per year in the pumps and turbines (Viney 2004). Other dams, such as Missouri's Truman Dam, have had similar problems.

To avoid similar injunctions, numerous methods have been developed to keep fish from entering or even coming close to water intake structures. Mechanical barriers and stimuli that frighten fish away have been emphasized (e.g., Popper and Carlson 1998). Physical barriers such as screens of differing mesh sizes, accommodating different species and sizes, are often effective, but impingement on screens remains a problem where strong flows are involved. Flashing lights repel some species. However, lights and screening are counterproductive if some fish are migratory and must negotiate the dam. The challenge then is to chase away the nonmigrators while allowing the migrators through.

Acoustic repulsion shows particular promise where clupeoid fishes such as herring or shad are involved. Most fishes detect sounds no higher than 1 kHz; a few can hear up to 3 kHz. By a quirk of evolution, probably involving detection of predatory dolphins, several clupeoids (alosine shads, menhaden, and herrings; *Alosa* spp.; *Brevoortia* spp.; *Clupea* spp.) are capable of hearing ultrasound frequencies at 180–200 kHz (Mann et al. 2001). They not only detect but are repulsed by ultrasound (e.g., Wilson and Dill 2002; Plachta and Popper 2003). Ultrasonic transmitters have been placed near dams and water intakes with promising results. Frequencies between 122 and 128 kHz projected into the forebay of turbines at an electric generating facility in Nova Scotia resulted in a 42%–49% reduction in passage of American shad, alewife, and blueback herring (*Alosa sapidissima, A. pseudoharengus, A. aestivalis*) but had no discernible effect on eight other, nonalosine fish species (A. J. F. Gibson and Myers 2002). Twaite shad, *Alosa fallax fallax*, a Berne Convention Appendix III species, would not pass a site on the River Wye in Wales when a 200-kHz beam was projected across the river (J. Gregory and Clabburn 2003). Research in this area is expanding to include other species and sound frequencies (e.g., Maes et al. 2004).

Removing Dams

The first dam built by nonnative Americans on a U.S. river was put in place in 1644, shortly after the first wave of European immigration. Enthusiasm for large dam construction in the U.S., which began with the construction of Hoover Dam in 1930, reached record levels in the latter half of the 20th century. Dam building peaked in the mid-1960s, when more than 200 major dams were completed annually. In South America, it accelerated after 1970 as international lending organizations made funds increasingly available (Pringle et al. 2000). Planning and building of new dams continue in Asia (Dudgeon 2000b).

However, many dams no longer function—or never functioned—as intended. Besides their negative environmental consequences, these dysfunctional dams are costly to maintain, generate little or no power, impound basins that have become filled with sediment, and create a liability hazard for owners as construction materials deteriorate. Meanwhile, the concept championed by researchers and environmentalists—that free-flowing rivers contribute to environmental health—is gaining acceptance. So is a need to comply with environmental regulations, some enacted after dams were constructed (e.g., mandated fish bypass devices, passage of the ESA in 1973). In the U.S., assessment of hydropower dams occurs officially on a 30- or 50-year basis as part of a relicensing process administered by the Federal Energy Regulatory Commission (FERC). Profits from electricity generation are balanced against costs of fish bypass structures and upgrading or repairing components. The economics of dam maintenance and operation versus out-and-out removal have frequently led to the conclusion that removal is a viable and desirable alternative. Marginally profitable and counterproductive dams are increasingly being eliminated, a proposition that, until late in the 20th century, was primarily relegated to the sphere of fictional ecoterrorists (e.g., Abbey 1975; Hockenberry 2001).

Dam removal activity is following a time course similar to that of dam building, with the U.S. leading the effort. More than 480 dams have been removed from U.S. waterways in the past 100 years, and another 100 are slated for removal or under consideration (see www.americanrivers.org). Although many are relatively small, they are still injurious to fishes and other riverine organisms. American Rivers maintains a list of actual dam removals, with case histories on 26 and discussion of dams under consideration for decommissioning. Dam removal as a science has advanced to the point that most stages, results, and complications can be anticipated (e.g., Aspen Institute 2002; Heinz Center 2002).

Plans for dam decommissioning are also accelerating in Canada. Almost 20% (400 of 2,200) of British Columbia's small and large dams are considered to have outlived their usefulness and are eligible for removal (see www.recovery.bcit.ca). Similar activities are ongoing in Europe under the aegis of the European Rivers Network, with the International Rivers Network River Revival Project acting as a clearinghouse for information. The 12-m-high Saint-Etienne-du-Vigan dam, which blocked the Allier River, a major tributary of the Loire River in France, was demolished on June 24, 1998 (figure 6.8). Prior to its construction in the late 19th century, approximately 100,000 Atlantic salmon migrated to the headwaters of the Loire to spawn. In 1997, only 389 salmon made the journey. This is the first dam operated by the state-owned electrical utility, Electricité de France, to be decommissioned for restoration of salmon habitat. The Loire salmon stock is considered a vital, remaining genetic resource for anadromous salmon in Europe because salmon runs have disappeared from all large European Atlantic rivers (Epple 2000). Information on decommissioning projects worldwide is available through the World Commission on Dams (WCD 2000).

Funding has also increased for studying the impacts of dam removal, and improved conditions for fishes have been documented. The 160-year-old Edwards Dam on the Kennebec River, Maine, blocked migrations of shortnose and Atlantic sturgeon, American shad, alewives, Atlantic salmon, and striped bass, among others. After ten years of negotiation and collaboration, it was removed in July 1999, at about 40% less cost than installing fish ladders. Within a year, water quality improved dramatically and benthic invertebrate diversity doubled, while abundance increased 40-fold. Ten migratory fishes blocked by the dam began to appear in the 27-km stretch that had previously been unavailable; fish ladders would have aided only three anadromous salmonids. "So far the recovery of the newly undammed stretch of the Kennebec River has exceeded all expectations" (Meadows 2001, p. 32).

Figure 6.8. The Saint-Etienne-du-Vigan dam (A) before, (B) during, and (C) after demolition. Photos courtesy of Roberto Epple/SOS Loire Vivante, European Rivers Network, www.rivernet.org.

Plans to remove 30-m-high Elwha and 85-m-high Glines Canyon dams on the Elwha River of Washington include extensive and expensive pre- and post-removal assessment (Wunderlich et al. 1994). These may be the first high-head dams to be removed in the U.S. They block historically and evolutionarily significant runs of Chinook, steelhead, coho, sockeye, chum, and pink salmon, sea-run cutthroat trout, Dolly Varden, and bull trout, the Elwha being one of the few rivers in the lower 48 states to support all anadromous salmonid species native to the Pacific Northwest. Elwha Chinook were especially noteworthy, commonly exceeding 45 kg (see B. Brown 1995). The dams cut off over 115 km of available fish habitat along the Olympic Peninsula, leaving only 8 km of natural spawning grounds for salmon and trout (see www.nps.gov/olym/elwha/documents.htm). Loss of these and other fishes annually deprives local and downstream regions of almost 400,000 kg of carcasses and the nutrients and energy they contain (Wunderlich et al. 1994). Purchase of the dams cost $30 million; removal and restoration are estimated to cost another $70 million (and could run to $203 million, Wunderlich et al. 1994). But economic analyses indicate that the aggregate benefits to residents of Washington State from dam removal will reach $138 million annually, with a national benefit of $3 to $6 billion (Loomis 1996). The Elwha dam removals will be closely monitored and will potentially provide guidance for other high-head dam decommissioning in the U.S. and internationally.

To acolytes of Edward Abbey, blowing up a dam has substantial appeal; decommissioning is more complicated. Socioeconomic, ecologic, and safety factors come into play. Recreational use of a reservoir will be lost, affecting real estate prices and quality of life. Remnants of dams can become navigational hazards (e.g., Mill Race Rapids on the Saluda River, South Carolina; Heinz Center 2002). A chief consideration is the sediment behind a dam that may be a major factor motivating its removal (e.g., Bednarek 2001; Stanley and Doyle 2002). Massive amounts of sediment suddenly flushed downstream can smother habitats and fishes, clog receiving reservoirs and turbines, and poison downstream ecosystems with contaminants. Airborne dispersal is an additional concern as the mud dries and turns to dust. Nutrients such as phosphorus and nitrogen processed or stored in the sediments or in wetlands created by a dam will be mobilized, potentially increasing concentrations in downstream habitats. This is especially a concern in agriculturally intense regions where years of runoff from fertilized fields are stored behind aging dams and in millponds (e.g., Stanley and Doyle 2002).

Reservoirs can be several kilometers long. Hence erosion and sedimentation problems may persist for years after a dam is removed and unstable sediments are washed downstream. Potential rates and amounts of sediment movement often determine whether a dam will be removed quickly (by dynamiting) or in stages. Dredging and disposing of sediment would seem an alternative but can be prohibitively expensive and time consuming. Such calculations have affected decommissioning plans for Matilija Dam on Matilija Creek, a tributary of the Ventura River of Southern California. This 60-m dam blocks spawning migrations of one of the few remaining runs of federally Endangered southern steelhead trout, *Oncorhynchus mykiss* ssp. It also degrades habitat of state-listed arroyo chub, *Gila orcutti*. Matilija Dam filled rapidly with sediment soon after its completion in 1947, as predicted by U.S. Army Corps of Engineers consultants. It was constructed for water storage and flood control, but these functions disappeared as sediment accumulated. By 2000, water retention was reduced over 90% (Ventura County 2004). In addition, the dam itself deteriorated, the fish ladder proved to be nonfunctional, the sediment-starved Ventura River eroded, critical instream and riparian habitats were destroyed, and economically important downstream beaches disappeared. An estimated 4.5 million m^3 of sediment had to be dealt with. Leaving the dam in place and dredging and trucking out the sediment was one of many solutions considered, but if standard 8-m^3 dump trucks were used eight hours a day, five days a week, at a rate of six per hour, hauling away the sediment was calculated to require almost 50 years (ignoring continued, additional accumulation). After long analysis and consultation, the concerned parties decided on an alternative that entailed full dam removal but staged sediment flushing, at a cost in excess of $100 million (Ventura County 2004).

Probably the most notorious dam removal in the U.S., one that exemplifies how serious the sediment contamination problem can be, involved the Fort Edward Dam. The 10-m-high, 195-m-wide dam was constructed in 1898 on the Hudson River, above Albany, New York, and impounded a small, 80-ha reservoir. Sixty years later, Niagara Mohawk Power Corporation determined that the $4 million needed to repair and upgrade the dam exceeded its value, especially since removal would cost under $500,000. Decommissioning included removing an estimated 3,000 m^3 of sediment from behind the dam. About 40 ha of exposed riverbed was supposed to recover via natural processes.

Apparently unknown, due to "inadequate research and engineering analyses" (www.americanrivers.org), was that capacitor manufacturing plants operated upstream by General Electric (GE) had dumped nearly 600,000 kg of polychlorinated biphenyls (PCBs) into the river over a 30-yr period. This illegal disposal contaminated 80 km of river with known carcinogens, whose downstream dispersal was stopped by the dam. When the dam was removed in 1973, approximately 990,000 m^3 of PCB-laden sediment was mobilized and released, contaminating the river for 320 km, all the way to New York Harbor (see www.darp.noaa.gov/northeast/hudson and www.epa.gov/hudson/background.htm).

In 1976, the Hudson River was closed to commercial fishing, including a $40 million striped bass fishery. Consumption advisories were issued, warning of the dangers of eating fish from the river (see Cronin and Kennedy 1999). Initial cleanup efforts by New York State in 1977 and 1978 removed 138,000 m^3 of contaminated sediments, more than 50 times the original estimated quantity. In 1983, EPA declared a significant stretch of the river a federal Superfund site. Full remediation has yet to be completed, but a dredging project mandated by the Superfund law requires GE to perform the cleanup, estimated to take five years and cost about $460 million, or about 920 times the original cost of dam removal (www.epa.gov/history/topics/pcbs/02.htm).

Fort Edward Dam is, fortunately, an exception in the annals of dam removal. Much has been learned since 1973 about what to anticipate. Thorough sediment testing is now routine; at the time of the Fort Edward removal, PCBs were not considered a pollutant and were not even among the contaminants assayed. Dam removal has progressed from an action to a science, and positive outcomes are overwhelmingly the result (Poff and Hart 2002).

CONCLUSION

John Muir referred to Hetch Hetchy Valley, Yosemite National Park, as "one of nature's rarest and most precious mountain temples," prior to its drowning under a water supply reservoir (see www.sierraclub.org/ca/hetchhetchy). Dams have been constructed despite the aesthetic, cultural, religious, and subsistence fishery value of the locales they destroyed, not to mention the people they have displaced. This repeated scenario is testament to the power that large-scale economic ventures have over other considerations. Biodiversity is another consideration.

That dams hurt fishes has been recognized for centuries, and relevant laws date back at least to 12th-century England. Richard I sought to ensure passage of salmon up rivers by keeping them free of blockage. He established the "King's Gap" around any river obstruction, a space wide enough "that a well-fed three year old pig could stand sideways in the stream without touching either side" (Montgomery 2003, p. 62). Similar prohibitions against impeding migrations were enacted wherever migratory fishes thrived, in Europe as well as eastern and western North America. Under English common law, dams that obstructed fish migrations in early Colonial times could be declared public nuisances and destroyed. The dam builder could appeal, but only after removal, not before (Kulik 1985). It wasn't until passage of various Mill Acts in New England in the early 1700s that the burden of proof changed, but even then subsequent Fish Acts mandated construction of bypass devices. What has been consistently lacking, then and now, is enforcement: Most passage-blocking dams were built despite the laws, laws frequently grandfathered existing structures, and violators have been subjected to minimal penalties (Montgomery 2003).

To prevent, minimize, and reverse the impacts of dams, we need to increase appreciation of rivers and their landscapes as functioning ecosystems and increase understanding of how human activities degrade natural riverine flow regimes (e.g., R. E. Sparks 1995; Fausch et al. 2002). Although our general knowledge of dam impacts is extensive, we have often failed to perform the pre-impoundment studies needed to provide specific knowledge at locales where dams are being built. This is especially true in the developing world, where dam construction is increasing. The many lessons learned in temperate regions may not be directly transferable to the tropics, but we can expect significant negative impacts given the extensive migrations of tropical riverine fish species, the high degree of regional endemism, the importance of seasonal inundation of floodplains to the life cycles of the fishes, and the obvious disruptions of physical and chemical conditions that accompany dam construction (Pringle et al. 2000).

Ultimately, the economics of dams have to be recalculated in light of our increased knowledge and understanding of the value of free-flowing rivers. Benefits to human and nonhuman components of the landscape from intact river and floodplain systems far outweigh the benefits to the few who profit by destroying and degrading instream, estuarine, riparian, and floodplain habitat.

7. Degraded Water Quality

I never drink water because of the disgusting things that fish do in it.
—WC Fields

WC Fields excepted, we all drink water, and we all live downstream. Given our total dependence on water, it seems irrational that degraded water quality is an almost global problem for humans. The vast literature on water quality degradation understandably focuses on human welfare. Fishes, when treated, are primarily viewed as sentinels that detect potential human problems. But fishes not only drink water, they live in it and continually absorb it through their skin and their gills. The impacts of degraded water quality on fish health are direct, sometimes obvious, and unquestionably important. Fishes live downstream from us.

EXTENT, FORMS, AND EFFECTS

The kinds, sources, and impacts of pollutants that affect fishes are frighteningly diverse (table 7.1). Many of the most lethal or insidious are products of technologies and industrial processes developed or advanced in recent times (heavy metals, biocides, endocrine disrupters, acid rain, thermal and acoustic pollution). Other pollutants such as sediments, dissolved solids, and excess nutrients have probably accompanied human sewage disposal and agricultural practices for at least centuries, although not at current levels. The extent of these impacts is a strong indictment of our stewardship of aquatic systems and the organisms that inhabit them. Given our dependence on clean water, it is also evidence of our shortsightedness.

Developed and developing nations alike neglect their water resources. Although forms of pollution vary as a function of available technology and national emphasis on conservation, global commerce and weather patterns carry pollutants far beyond the borders of the countries where they are produced, internationalizing the effects of as well as the need for common solutions. This chapter focuses on impacts in the U.S., but a glance at the international literature suggests that America's problems are fairly representative of the international situation—sometimes better, sometimes worse.

The EPA (USEPA 2000) analyzed the degree to which U.S. lotic and lentic habitats were affected and found 25% to 50% below health standards. Pollution or habitat degradation impaired at least 39% of assessed river and stream kilometers, with 34% impaired for supporting aquatic life, 38% with fish consumption advisories, and 28% impaired for primary contact such as swimming (figure 7.1). Lakes and reservoirs suffered similarly. At least 45% of lakes were impaired in some way, 29% for supporting aquatic life, 35% with fish consumption advisories, and 23% for primary contact (51% of estuarine areas and 78% of Great Lakes shoreline miles were also impaired; groundwater and wetlands lack assessments or characterization methods). An updated survey in 2006 found further degradation (www.epa.gov/owow/streamsurvey).

Although these numbers are sobering, they underestimate conditions. EPA did not include fish consumption advisories for mercury in several northeast and midwest states (see below), nor was a statewide New York consumption advisory for PCB, chlordane, mirex, and DDT included, which would expand the advisory list to all waters in an additional 10 or 11 states. All told then, one-third to

Table 7.1. Overview of water quality issues as they affect fishes

Pollutant class/stressor	*Common forms/sources*	*Impacts on fishes*
Toxins and contaminants		
Heavy metals	Runoff, evaporation, burning, atmospheric deposition	Stress[a], impaired feeding, bioaccumulation, impaired reproduction and development, mortality
Pesticides/herbicides	Runoff from overapplication, overspray, hazardous waste	Bioaccumulation, inhibited smoltification, spawning timing, EDCs
Toxic spills	Petroleum, sewage treatment, industrial accidents, resource extraction	Acute toxicity, impaired development
Dissolved solids/inorganics	"Salts," runoff, evaporation	Osmotic stress
Organics	Industrial discharge, spills, oil refining, hazardous waste	Impaired reproduction/development, bioaccumulation, liver cancer
Endocrine disrupters (EDCs)	Pulp/paper, pharmaceuticals, agricultural/municipal runoff	Impaired reproduction/development, bioaccumulation, chromosome damage, intersexuality
Acidification	Precipitation, dry deposition, runoff	Impaired respiration, reproduction, development; mortality
Nutrient pollution	Agricultural/municipal runoff, sewage	Eutrophication, HABs, deoxygenation
Pathogens	Microorganisms, agricultural runoff	Disease; impaired reproduction, growth
Thermal pollution	Power plants, industry, climate change	Metabolic stress, altered development, intersexuality, distributional change, deoxygenation
Acoustic pollution	Boat engines, military activity	Avoidance, impaired hearing
Sediment/siltation	Deforestation, urbanization, road building	Habitat degradation, impaired reproduction

Note: General types of pollution and representative impacts on fishes are listed, but most water quality problems have multiple causes and effects, occurring simultaneously or sequentially. Studies on impacts unfortunately tend to isolate causes, masking interactive, additive, and synergistic effects. Details in text. EDCs = endocrine disrupting compounds. HABs = harmful algal blooms.
[a]Physiological or behavioral stress can result from any of the pollutant classes or stressors listed.

one-half of U.S. waters are too polluted to support healthy living organisms, are unhealthy for human contact, or present risks from fish consumption. Although EPA categories focus on human health, they also characterize conditions likely to affect fishes.

Describing and Assessing Pollution's Effects on Fishes

Research has focused on acute and lethal effects of contaminants and pollutants because such phenomena are obvious, obviously harmful, and relatively easy to assess. But in the politically charged arena of water quality, terminological debate has crept into the issue of anthropogenic "releases" that kill fishes. Mass deaths are an extreme event with political, economic, and legal ramifications. It has become practice therefore to distinguish between "fish die-offs" and "fish kills." *Die-off* is the general term applied to large numbers of dead fish, especially when the cause is unknown or involves multiple, relatively natural factors such as low oxygen resulting from high summer temperatures or thick ice cover. A *fish kill* results from an identified cause involving a known, "responsible party"—a toxic waste spill, for example. A fish die-off is a matter of concern; a fish kill is a matter for litigation (e.g., K. Gibson 2003; KWUA 2003).

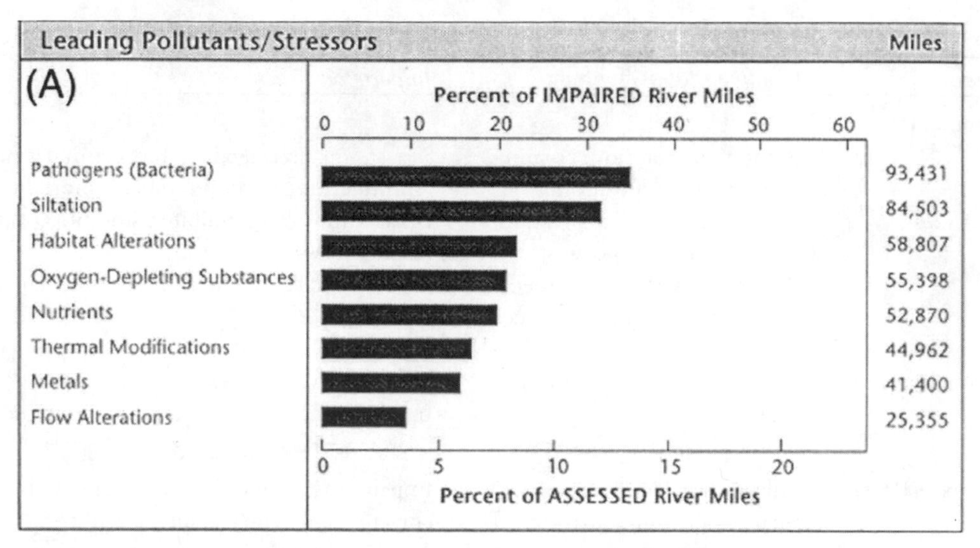

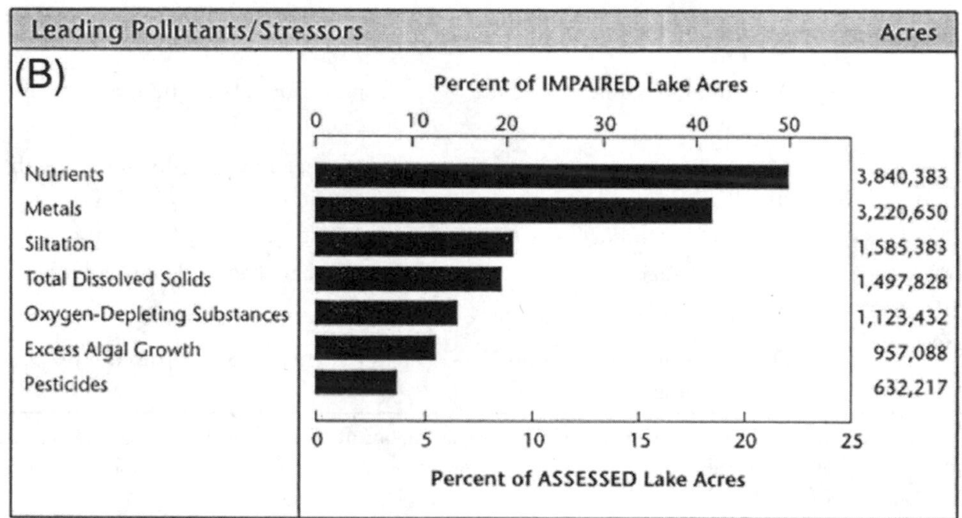

Figure 7.1. Causes of pollution in U.S. aquatic habitats. EPA assessed 700,000 miles of rivers and streams (about 20% of total river miles) and 17 million acres of lakes (about 40% of total acreage). Among assessed systems, 40%–45% suffer some impairment of ability to support biota. The figures show the relative contributions of major pollutants and stressors in impaired (upper axis) and assessed (lower axis) habitats in (A) rivers and streams and (B) lakes. From USEPA (2000).

However, less spectacular, chronic effects may prove more detrimental to fish populations and their ecosystems. Xenobiotics are foreign chemicals or materials not produced in nature and not normally constituting a component of a biological system; they can also be natural compounds present in unnatural concentrations (e.g., Rand and Petrocelli 1985; www.issx.org). Their effects are measured using biomarkers, which are biological responses such as enzyme or hormone concentrations, phenotypic responses, anatomical anomalies, disease incidence, or altered behavioral or metabolic responses. Xenobiotics have become a major focus in studies of chronic pollutants.

TOXINS

Metals, Precious and Otherwise

Metals and metalloids such as aluminum, arsenic, cadmium, chromium, lead, and selenium are toxic or teratogenic (interfering with embryonic development) to fishes at relatively low concentrations, and most contaminated environments contain elevated concentrations of more than one heavy metal (e.g., Beyer et al. 1996; Witters 1998; Clearwater et al. 2002). Fishes take up suspended or dissolved metals via their gills, their skin, or their food (Bury et al. 2003). Most investigations of metal toxicity have involved freshwater fishes; effects on marine fishes are relatively uninvestigated (Clearwater et al. 2002). Saltwater species may be more sensitive to environmental contaminants because they drink seawater to maintain osmotic balance, thus adding a pathway for exposure to environmental contaminants.

Metals such as iron, copper, and zinc are integral components of proteins and play essential roles in many metabolic pathways as micronutrients. But they can also be toxic when concentrations exceed the assimilative and excretory capacity of organisms, causing reduced digestive function, growth, and survival in a variety of species. Copper toxicity occurred at daily dietary doses of 1–45 ppm in channel catfish, Atlantic salmon, and rainbow trout, and zinc toxicity occurred in carp, Nile tilapia, and guppies at doses of 9–12 ppm (Clearwater et al. 2002; 1 ppm = 1 mg/kg body weight of fish or 1 mg/1; 1 ppb = 1 µg/kg body weight; see EPA's water quality criteria for aquatic life at www.epa.gov/waterscience/criteria/aqlife.html).

Sensitivity of fishes to heavy metals differs among species, life history stages, and diet composition. The Clark Fork River in Montana received a century of mine wastes that resulted in elevated concentrations of arsenic, cadmium, copper, lead, and zinc in its waters, sediments, and organisms (D. F. Woodward et al. 1994). The authors calculated that, without metal contamination, each river kilometer could support 1,250 rainbow and brown trout. Actual densities were 21–125 brown trout km^{-1}. Brown trout are the more metal tolerant of the two species. Rainbow trout were essentially restricted to tributary waters that lacked mine wastes. Laboratory exposure of rainbow trout fry to ambient concentrations of metals resulted in poorer growth and survival, associated with reduced feeding and liver degeneration. Brown trout fry tested similarly exhibited multiple abnormalities of the digestive system, such as decreased enzyme production, intestinal mucosal sloughing, and terminal constipation (D. F. Woodward et al. 1995). Elevated metal concentrations from mine wastes could thus explain both the absence of rainbow trout and the reduced standing crop of brown trout in the Clark Fork River, with dietary impairment of very young fish a likely contributor.

Mining as a Source of Contaminants

Toxic metals are released during mining when ore is exposed and as a by-product of ore washing and processing, and are leached from spoil material discarded into water bodies or eroded from mine tailings. Surface mining in particular degrades streams and injures fishes. Fishes are affected via erosion and sedimentation, dewatering of wetlands, diversion and channelization of streams, and contamination of surface and groundwater with acid mine wastes and toxic chemicals (Starnes and Gasper 1995). Sediment released from dredging operations (chapter 5) is relatively localized compared with the chemicals that move continually downstream from a mine site (Yount and Niemi 1990). Mining exposes, accumulates, and mobilizes toxic materials during normal operations and if storage facilities fail.

Coal mining does all of the above. As practiced widely in West Virginia, so-called mountaintop-removal–valley-fill coal mining entails removing the "overburden" that covers large coal deposits by literally bulldozing mountaintops into adjacent stream valleys. This activity buries the stream along with its biota. Fish assemblages downstream from valley fills show reduced diversity and abundance, especially among benthic species. In one study in West Virginia and Kentucky, streams in unmined watersheds contained a median of 17 fish species, whereas "filled" streams had only 8. Benthic species diversity declined from 6 to 1.5. Filled streams also had higher levels of aluminum, iron, and copper, and toxic levels of selenium (Ferreri et al. 2004).

Fishes are also harmed when waste ponds leak or collapse. In October 2000, a 30-ha tailings pond overlying an abandoned mine in Inez, Kentucky, failed, releasing 950,000 m^3 of water and 118,000 m^3 of coal waste into the Big Sandy River and eventually into the Ohio River (MSHA 2001, www.antenna.nl/wise/uranium/mdafin.html). The volume spilled was more than double that from the *Exxon Valdez* oil spill of 1989 (LOE 2003). Rivers were stained black as much as 120 km downstream, contaminating drinking water supplies and causing a fish kill as part of the annihilation of "all aquatic life" (Alford 2002, p. A16). The Inez failure prompted investigations by the Mine Safety and Health Administration and a National Research Council committee, which recommended changes in engineering practices and standards for coal waste impoundments to prevent similar disasters (MSHA 2001; NRC 2002a). Kentucky has 58 such coal-waste impoundments. Nationwide about 635 exist, 240 of which sit over abandoned underground mines (LOE 2003).

Mining's impacts on aquatic systems can be prevented, reduced, and corrected in a number of ways. Starnes and Gasper (1995) suggested that agencies and industries do the following:

- Consider potential impacts on aquatic systems during planning and permitting, before operations commence. Toxicity can be predicted from local chemistry, geology, and hydrology. Areas with a high-impact or low-restoration potential should be designated unsuitable for mining.
- Supervise and modify activities during operations, which should be halted and corrected when excess leakage of contaminants occurs.
- Monitor downstream areas during reclamation, focusing on stream restoration via erosion control, bank stabilization, and riparian revegetation.

Aquatic ecosystems in the U.S. are in theory protected from many mining practices by the Surface Mining Control and Reclamation Act of 1977 (SMCRA), in combination with the Clean Water Act of 1972. SMCRA regulates environmental impacts of surface coal mining and mandates that land be restored to original contours and conditions capable of supporting at least pre-mining uses; this presumably includes maintenance of downstream receiving waters. Enforcement and environmentally focused interpretation of both acts have varied. Extraction of other minerals and gravel is even less controlled than surface coal mining.

Mercury

Elemental mercury (Hg) occurs naturally in rocks, soil, and water. It is mobilized by volcanic activity, forest fires, and volatilization. Anthropogenic contamination occurs via waste incineration, coal burning, chlorine production, and ore extraction. Much of this is atmospheric, and mercury vapor (HgO) returns to earth in rain and snow. Post-industrialization concentrations of mercury in lake sediments are about three to five times higher than in preindustrial sediments (water.usgs.gov/wid/FS_216-95/FS_216-95.html).

In low-oxygen aquatic systems, sulfate-reducing bacteria absorb and convert less toxic inorganic or vaporous mercury to highly toxic methylmercury. Methylmercury production is especially high in warm, acidic systems rich in dissolved organic carbon, including areas subject to acid precipitation. It accumulates in muscle tissue and is biomagnified through the food chain, reaching its highest concentrations in large predators (Morel et al. 1998).

Mercury contamination is widely publicized because of its occurrence in popular food fish species. Human health consequences of consuming small amounts of mercury and of long-term exposure are debated. However, acute dietary exposure to inorganic mercury-contaminated fish causes kidney damage, and direct ingestion causes severe neurological damage, especially during developmental stages (e.g., the Minamata, Japan, poisonings and deformities of the late 1950s). Cognitive impairment, muscle and joint pain, hair loss, and increased coronary heart disease have also been shown to varying degrees (e.g., Hightower and Moore 2003). Recommended allowable levels of consumption vary. The U.S. Food and Drug Administration (FDA) and EPA limit daily intake to a reference dose (RfD) of 1 ppm methylmercury. The World Health Organization (WHO) set its RfD at 2 ppm.

Normal eating patterns often expose people to concentrations that exceed recommended doses. Concentrations of 2 ppm and higher total mercury have been found in sharks, anguillid and conger eels, brown trout, northern pike, Eurasian perch, swordfish, tunas, and marlins. The species list for concentrations of over 1 ppm is much longer (UNEP 2003). Canned tuna is a major concern because it is the most common fish consumed in the U.S. Canned white tuna (albacore, the dolphin-friendly tuna; see chapter 11) averages 0.51 ppm methylmercury, four times greater than light tuna (usually skipjack, yellowfin, or bigeye) at 0.12 ppm methylmercury (Bender 2003). An average woman of child-bearing age weighing 60 kg who ate one 180-gm can of white tuna a week would therefore consume twice the EPA RfD; a 10-kg toddler eating only 60 gm of white tuna per week would consume four times the recommended limit (see www.mercurypolicy.org).

The food chain connection also involves freshwater fishes that live in waters impacted by gold mining. Elemental Hg is used to amalgamate gold ore. The Amazonian gold rush of the 1970s and 1980s led to massive inputs of mercury into Amazon waters and eventually into fishes via the river food web. Many studies reported widespread methylmercury contamination and health impairment from eating fish (e.g., Barbosa et al. 1995). Other reports claimed minimal impact, or emphasized a trade-off between potential risks of mercury poisoning and the benefits of fish as a protein source (e.g., Dorea 2003).

Mercury's Effects on Fishes

The literature on the direct effects of mercury on fishes is limited, dominated by laboratory exposure tests, because more relevant field evaluations are often complicated by multiple contaminants (e.g., Hontela et al. 1995). Mercury can kill fishes, is more toxic to younger fish, and is fatal at lower concentrations than lead or copper (Alam and Maughan 1995). Acute exposure to sublethal concentrations of dissolved inorganic Hg resulted in elevated metabolic rates, probably due to gill epithelium damage (e.g., Jagoe et al. 1996), although prolonged exposure at lower concentrations had no apparent effect on metabolic rate (Hopkins et al. 2003). Chronic exposure to mercury (as mercuric chloride) can lead to impaired feeding ability, manifested as reduced foraging efficiency and capture speed (Grippo and Heath 2003). Dietary ingestion of inorganic mercury appears to have minimal effects at low doses, but methylmercury ingestion causes severe liver damage, even at relatively low doses (Ribeiro et al. 2002). Female mosquitofish exposed to mercury contained fewer young and were less likely to bear young than control females (Tatara et al. 2002). Other laboratory studies have shown impaired gonadal function, reduced sperm motility, and altered sex ratio in a variety of fish species (Friedmann et al. 2002). Among the few field studies able to separate mercury from other factors, Friedmann et al. (2002) found no effects on general health or reproductive parameters in male largemouth bass in a mercury-contaminated drinking water reservoir in New Jersey.

Attention has also been directed at fish as ecosystem components and as prey of higher vertebrates. Studies in the U.S., Canada, and elsewhere have determined that fish-eating birds (e.g., wading birds, loons, kingfishers, eagles) and mammals (mink, otter, whales, pinnipeds) suffer reproductive impairment and even mortality at methylmercury concentrations of 0.077–0.3 ppm (USEPA) and less than 0.033 ppm (Environment Canada; UNEP 2003; Morrissey et al. 2004), levels easily reached in a fish-heavy diet.

Pesticides and Herbicides

Pesticides and herbicides are designed to kill living things, so it is not surprising that fish suffer when exposed to them. Exposure occurs due to poorly timed and directed application to agricultural fields and urban lawns, derailment of transport trains, and illegal disposal, among other sources. Declines among wild runs of Atlantic salmon may in part result from contaminant exposure (NRC 2004a). Salmon smolts often encounter a complex mixture of contaminants during their seaward migration. Atrazine, a commonly used herbicide that inhibits acetycholinesterase activity, widely contaminates surface and groundwater at nonlethal concentrations. Low concentrations of atrazine alone and mixed with the nonionic surfactant 4-nonylphenol impair gill Na+K+ATPase activity, hypo-osmoregulatory capability, and survival in seawater of Atlantic salmon smolts (Moore et al. 2003). Diazinon, a widely used agricultural organophosphate pesticide, occurs in river systems where Atlantic salmon spawn. Mature male Atlantic salmon parr exposed to Diazinon at sublethal concentrations, below those known to occur in rivers, had reduced levels of reproductive hormones in their blood and displayed impaired olfactory-mediated responses (Moore and Waring 1996). Odorants and pheromones in female urine attract males, prime the male reproductive system, induce sperm motility, and aid in synchronization of reproductive behavior. Exposed fish responded less to water-borne prostaglandins and to urine from ovulated females, showing as much as a tenfold reduction in sensitivity. Any of these anomalies could affect spawning success, especially at the low densities that characterize many endangered salmonids.

ENDOCRINE DISRUPTING COMPOUNDS

Although laboratory studies most commonly test for lethal impacts of contaminants (e.g., LD_{50}, or LC_{50}, the lethal dose or concentration at which 50% of test animals die), impaired reproduction can also harm populations. Many pollutants affect reproduction at sublethal concentrations known to occur in aquatic habitats (e.g., Kime 1995). These pollutants include endocrine disrupting compounds (EDCs), which occur in aquatic systems "in sufficient concentrations to elicit responses normally under the control of endocrine hormones" (Arcand-Hoy and Benson 1998, p. 53).

During and after World War II, industrial production of artificial chemicals and their by-products increased dramatically. Among these products—plastics, biocides, cosmetics, solvents, detergents, paints, and so on—were synthetic hormones, such as oral contraceptives and diethylstilbestrol (DES), a synthetic estrogen. DES was widely prescribed to prevent miscarriages and to produce "bigger and stronger babies" (Colborn et al. 1996, p. 48). Its use declined in the 1970s, when children of DES mothers were shown to have greater incidence of reproductive structure abnormalities and cancer of reproductive organs (see www.desaction.org).

DES is an EDC. EDCs mimic the chemical activity of hormones or stimulate the production of hormones, thus altering embryonic development and affecting potentially all bodily functions and processes. Development and function normally are controlled by specific hormones produced at the right times, often in minute quantities. Alterations in type, amount, and timing can have long-lasting and often devastating results. At least 90 chemicals commonly found in food, water, soil, and the air are known EDCs (www.ourstolenfuture.org/Basics/chemlist.htm). A partial list includes persistent organohalogens (e.g., dioxins, furans, PCBs), a food antioxidant (BHA), pesticides (aldrin, atrazine, chlordane, DDT, kepone, lindane, malathion, mirex, toxaphene), phthalates (plasticizers, cosmetics, safety glass, insecticides, printing inks, paper coatings, adhesives, elastomers, explosives, rocket propellant), surfactants, styrenes, and metals (this is only a partial list). These chemicals mimic, induce, or disrupt the actions of reproductive, thyroid, and corticosteroid hormones, among other functions, in most vertebrates.

The classic view was that toxins were metabolic poisons that shut down cellular machinery or overwhelmed the body's defenses. We now know that many environmental contaminants affect organisms by modifying control of embryonic development, often at concentrations much lower than those at which toxicological effects occur (e.g., Krimsky 2000). Metabolic processes interpret these toxins as hormones. Humanity has produced prodigious quantities of them, and they are now found in all the world's regions, at concentrations demonstrated to have disruptive effects on development (www.ourstolenfuture.org; see especially Colborn et al. 1996; NRC 1999a).

What Does This Have to Do with Fishes?

Many reproductive and developmental processes, as well as steroid hormone structure and function, are evolutionarily conserved among vertebrates (Campbell and Hutchinson 1998). The same basic chemicals and pathways that affect development in humans, laboratory mice, and birds occur in lower vertebrates, including fishes. Eggshell proteins produced in the liver of birds are structurally and functionally similar to the liver-produced zona radiata (or vitelline envelope) proteins of teleosts that prevent polyspermy and protect developing embryos from mechanical disturbance (Arukwe and Goksøyr 1998). Synthesis of these proteins is regulated by the primary female reproductive hormone 17_estradiol (Hylland and Haux 1997). Other analogs exist in the pathways involved in gametogenesis and steroidogenesis (Arukwe 2001; see Kime 1993 and Campbell and Hutchinson 1998 for exceptions and differences).

It's important to note that most estrogenic compounds are lipophilic and hydrophobic—that is, they accumulate in fats and do not dissolve in water. Because fish live in water and don't merely drink it, they are particularly susceptible to waterborne contaminants. Lakes, rivers, and the ocean sit downstream from sources of EDC production and release, whether it occurs via precipitation, overland flow, subsurface water, eroded soil, landfill leakage, industrial waste, or sewage discharge (box 7.1).

In fishes, EDCs affect sexual differentiation and reproductive performance, acting early in sex determination as well as late when gonads produce sex products (Devlin and Nagahama 2002). Disruptions include abnormal gonad morphology, reduced rates of sperm and egg production and release, and reduced quality of gametes and can result in unnatural sex reversal or failure to mature, with reproductive failure an ultimate result. Reproductive behavior may also be altered (e.g., J. C. Jones and Reynolds 1997).

The list of examples of EDCs and related chemicals that have caused impaired development and reproduction in fishes is extensive, and impacts on wild populations are becoming increasingly documented (table 7.2; see also Arukwe 2001). Direct effects on population size, growth, and survival are usually implied from biomarker studies. EDCs produce numerous outcomes, but some of the most dramatic involve gender reversals, impaired maturation associated with vitellogenesis, and decreased gamete production.

EDCs and Intersexuality

One of the first demonstrations of widespread EDC-induced disruptions of sexual differentiation involved roach, *Rutilus rutilus*. Jobling et al. (1998) showed that roach throughout rivers in the UK had a high incidence of

BOX 7.1. Fish on Prozac?

Several pharmaceuticals have been detected in water bodies and in fish muscle and organs (e.g., Raloff 1998), including fluoxetine, the active ingredient in the antidepressant Prozac, which at high concentrations reduces survival and impairs reproduction of many species (Brooks, Turner, et al. 2003; Brooks, Foran, et al. 2003). Mosquitofish exposed to more than 60 ppb of fluoxetine exhibit loss of equilibrium and uncoordinated swimming, as well as delayed gonopodium (male intromittent organ) development (Henry and Black 2003; M. Black and T. Henry, pers. comm.). At the lower concentrations thought to exist in the wild, developmental abnormalities and elevated concentrations of female steroid compounds have been found in embryos of Japanese medaka, *Oryzias latipes* (environmental concentrations may reach 0.5 ppb, with measured effects at 0.1 ppb). Delays in developmental processes could be especially harmful where development is normally synchronized with predictable environmental cycles of temperature, hydrology, food availability, or spawning readiness.

The estimated U.S. consumption of Prozac—the most commonly prescribed of a host of similar antidepressants—from over 21,000,000 prescriptions filled annually approaches 19 MT (see www.rxlist.com/top200.htm). More than 1.2×10^{11} liters of Prozac-containing effluent enter public treatment facilities daily, but most facilities do not remove or degrade human and agricultural pharmaceuticals (Prozac release is representative of a *much* larger problem) (Brooks, Foran, et al. 2003). Measurable fluoxetine concentrations have been found downstream from municipal water treatment facilities (Streater 2003). The studies cited here are among the first to look at nonlethal responses to a few of the pharmaceuticals we discharge into water bodies, substances for which we lack guidelines or federal testing standards. These investigations raise a host of questions about the health effects of such substances, alone and in combination. All of the above ignores the issue of what the mood of a bluegill on Prozac might be.

Table 7.2. Examples from *field* studies demonstrating impaired reproduction and development after exposure to xenobiotics

Xenobiotic/source	*Effect*	*Species*
BKME	Masculinization of females	Mosquitofish, eelpout, fathead minnow
Columbia R pollutants, DDT	Phenotypic sex reversal	Chinook salmon, ricefish
Sewage estrogenic compounds	Intersexuality	Roach
PCBs, DDT/sewage effluent, oil spill	Increased egg mortality	Sand goby, arctic charr
Oil spill	Premature hatch, deformities	Pacific herring
North Sea DDE, pollutants	Embryonic deformities	Flatfishes, cod
PCBs, DDT/various discharges	Chromosomal aberrations	Whiting
PCBs, PAHs/urban discharge, landfill leachate	Precocious maturation, decreased gonad development	English sole, Eurasian perch, brook trout
Crude oil/oil spill, BKME	Altered ovarian development	Plaice, white sucker
Alkylphenols/sewage effluent	Altered vitellogenesis	Rainbow trout et al
Pulp mill effluent, oil spill	Reduced plasma steroids, sperm motility	White sucker, Atlantic salmon, flounder
Textile mill, vegetable oil effluent	Retarded/reversed ovarian recrudescence	Airsac catfish, snakehead
EE2 in sewage effluent	Reduced territory acquisition	Fathead minnow

Source: Table expanded from Arukwe and Goksøyr (1998); see that review for references. The literature from lab studies of EDCs, especially effects from injected compounds, is even greater.
Note: BKME = bleached kraft mill effluent; R = river; PAH = polycyclic aromatic hydrocarbons; EE2 = ethynylestradiol; DDE is a metabolic byproduct of DDT.

intersexuality resulting from exposure to estrogenic chemicals discharged from sewage treatment plants. In December 2004, nine male smallmouth bass, *Micropterus dolomieu*, captured in the Potomac River near Sharpsburg, Maryland, had eggs developing in their gonads (Fahrenthold 2004). Intersexual bass were also captured from the South Branch of the Potomac, about 170 miles upstream. The causative agent remains unknown. Sharpsburg is 60 miles above Washington, D.C., on the Potomac, which provides drinking water for about 2.7 million residents of D.C., Maryland, and Virginia.

Among the earliest documented instances of EDC-caused sex change was that of mosquitofish living downstream from paper mills, which were exposed to bleached kraft mill effluent (BKME) and exhibited altered secondary sexual characteristics (Bortone and Davis 1994). Female mosquitofish became externally masculinized, developing a gonopodium (figure 7.2) and engaging in male-typical breeding behaviors, but could still produce normal, fertilizable ova. External sex characters in mosquitofish are regulated by androgens. Microbial degradation of plant sterols released during BKME production was thought to produce androgen-like substances. Similar masculinization has been shown in fathead minnows, *Pimephales promelas*, and eelpout, *Zoarces viviparous*, exposed to BKME in the laboratory or field (see Kovacs et al. 2004).

Laboratory studies indicate that EDC-altered sex determination can be complete and irreversible if it occurs early in development. Eggs of Japanese medaka, which normally produce XX females and XY males, were injected with sublethal concentrations of DDT. Genetic (XY) males developed as functional females, half of which were able to produce fertilized embryos that hatched into viable larvae, a success rate comparable to that of normal XX females. Microinjection of DDT into eggs approximates events that occur if mothers carrying a sublethal body burden of DDT transferred lipophilic xenoestrogens to developing embryos (Stewart et al. 2000).

EDC-altered sex determination may play a role in accelerating declines among endangered species. Many imperiled runs of Pacific salmon spawn in degraded rivers (e.g., Afonso et al. 2002). Approximately 85% of female-appearing Chinook salmon in the Columbia River possessed a genetic marker for the Y chromosome, indicating that they were in fact sex-reversed males (Nagler et al. 2001). When XY females mate with normal XY males, a significant fraction of the F1 generation would be YY males (25% if you do the basic Punnett square calculation), skewing the population sex ratio from a normal 1:1 to a male dominated 3:1. Subsequent matings could increase the proportion of males as YY males mated with normal XX females. This is a common technique used in aquaculture to produce all-male, sterile populations (Pandian and Sheela 1995) but is potentially disastrous for wild runs. Sex ratio disruption would accelerate in subsequent generations, putting already stressed populations in further peril. Problems could be exacerbated by other known EDC-caused abnormalities, such as reduced sperm production and delayed maturation (I. J. Baker et al. 1988). The Columbia River contains at least 92 chemical contaminants found in fish samples, including 14 metals, DDT, chlordane, PCBs, and chlorinated dioxin and furans

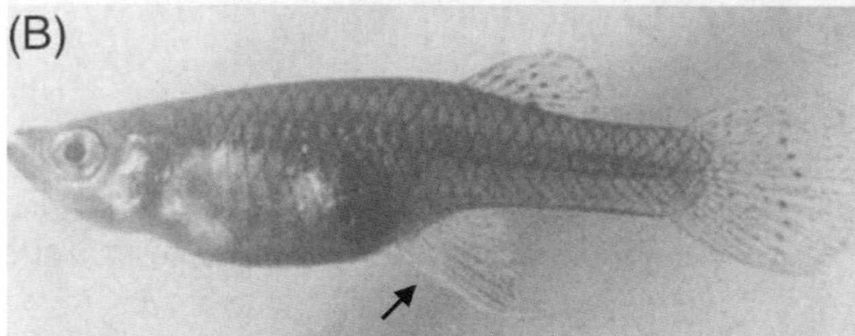

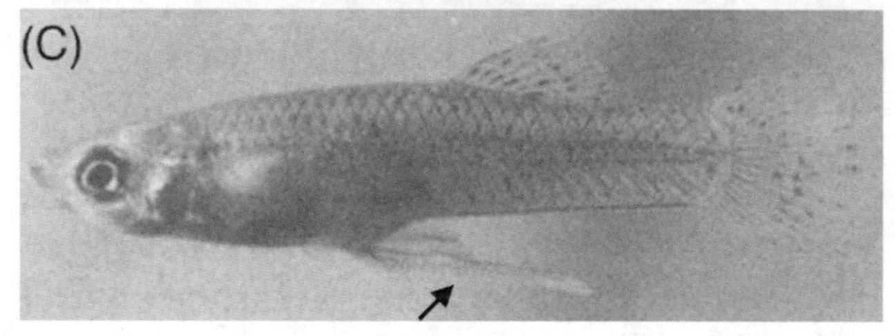

Figure 7.2. Masculinization of mosquitofish females. (A) The anal fin of a normal male *Gambusia* is elongated to form the gonopodium, an intromittent organ used to inseminate females. (B) In normal females, the anal fin is fan shaped. (C) A masculinized female exposed to pulp mill effluent, in which the anal fin has developed into a gonopodium. From Helfman et al. (1997). Photos by S. A. Bortone; used with permission.

(USEPA 2002), some of which are known to disrupt endocrine functions (table 7.2).

Vitellogenin, Maturation, and Gamete Production

Vitellogenin is a glycoprotein precursor to egg yolk that is critical to proper egg and embryo development. It is produced in the liver when that organ is stimulated by estrogens, and it is often used as a biomarker sensitive to estrogen-like compounds (Sumpter and Jobling 1995). Zonagenesis, involving egg shell protein precursors, follows similar pathways. Excess or mistimed production of vitellogenin and zonagenin in females can delay the onset of spawning and reduce the number of eggs produced. Males exposed to estrogens also produce vitellogenin, which can suppress testicular activity and lower fertilization rates.

Even common household products can affect fish reproduction. Sewage treatment plants typically release surfactants that originate in detergents, pesticides, and widely used industrial and domestic cleaning products (including toothpaste!). These surfactants have estrogenic activity in fish, bind to estrogen receptors, induce vitellogenin, and impair gonadal development. Laboratory studies of fish injected with surfactants have shown elevated levels of vitellogenin and inhibited testicular activity in killifish and delayed smoltification in salmon (e.g., Pait and Nelson 2003). Injecting compounds obviously differs from immersion in surfactant-contaminated water, and definitive results require exposure to ambient levels of contaminants (e.g., A. Moore et al. 2003).

EDCs and Behavior, Gamete Production, and Maturation

In many instances, altered physiology can be linked to EDCs, although specific pathways and compounds haven't been determined. Adult female Eurasian perch, *Perca fluviatilis*, in a lake in central Sweden that received runoff from a municipal landfill showed unusually high incidences of ovaries without maturing oocytes (25%–42% mature versus 92%–100% mature in nearby lakes). Female brook trout, *Salvelinus fontinalis*, in the stream that fed the lake (which therefore received relatively undiluted leachate from the landfill) showed even lower frequencies of maturation, with only 17% mature trout compared to 100% in a nearby reference stream (Noaksson et al. 2003).

Sometimes the chemical linkages are more obvious. Human contraceptives released from sewage treatment facilities may act as EDCs at concentrations far below those that cause obvious, immediate health problems. Adult male fathead minnows, a pollution-tolerant species, were exposed for 27 days to low levels of ethynylestradiol (EE2), a component of oral contraceptives found in sewage treatment effluent. Exposed males were unable to acquire and maintain territories because their overall aggression rates were inappropriately high (Majewski et al. 2002). EE2-induced reduction of fertilization success in zebrafish, *Danio rerio*, occurred after exposure to less than 1/50th of the LC_{50}, at around 1.67 part per trillion (Segner et al. 2003). Maturing male rainbow trout exposed to EE2 concentrations 1/100th of the lethal dose produced impaired sperm that resulted in about a 50% reduction in embryo production (Schultz et al. 2003). These findings probably represent the tip of the iceberg with respect to sewage treatment and disposal.

The reach of EDCs extends far beyond the point sources of our pollutants. PCBs and DDT are persistent, bioaccumulated toxins that also act as EDCs. Both impair reproduction through anti-estrogenic and anti-androgenic activity by altering synthesis, metabolism, and reception of hormones (e.g., Kime 1999). The distribution of these chemicals appears to be essentially worldwide, including the ocean's remote regions and depths. Coelacanths, *Latimeria chalumnae*, inhabit nutrient-poor waters at midwater depths of 150–250 m, such as off the Comoro Islands, northwest of Madagascar. Tissue samples from two immature coelacanths contained PCB concentrations as high as 510 ppb and DDT concentrations up to 840 ppb (R. C. Hale et al. 1991). Such concentrations could affect reproductive success in one of the world's best-known and perhaps most endangered marine fish species (see chapter 2). "The remoteness of their habitat has not provided a sufficient buffer to prevent exposure of the coelacanths to PCB and DDT" (p. 366).

EDCs: Conclusions and Solutions

EDCs—whose impacts and our awareness of them are growing—are the products of relatively new technologies. Their effects are nonlethal and appear only in the offspring of exposed individuals. Surveys of the extent and distribution of EDCs were not initiated until general awareness of the problem spread; legislative protection in the U.S. has

(A) EXPOSURE TO POTENTIAL ENVIRONMENTAL CONTAMINANTS
External endocrine disrupters (e.g., xenoestrogens, other xenobiotics).
Natural endocrine disrupters (e.g., environmental stress).
Nonendocrine factors (e.g., nutritional deficiency).

(B) MOLECULAR, CELLULAR, AND BIOCHEMICAL RESPONSES
Changes in blood hormone levels, eggshell or vitellogenin protein synthesis.
Effects on neuroendocrine system (e.g., corticosteroid production).
Impaired synthesis of essential enzymes or hormones.

(C) BEHAVIORAL AND MORPHOLOGICAL EFFECTS
Altered gametogenesis, decreased gamete quality, changes in secondary sex characteristics affecting mating behavior, altered courtship and territoriality; impaired parental care.
Immunosuppression, reduced gamete quality/quantity.
Impaired development and survival, altered sex determination, delayed metamorphosis or maturation.

(D) INDIVIDUAL REPRODUCTIVE EFFECTS
Impaired reproduction: reduced fecundity, fertilization, hatching rate, larval viability; unexpected sex reversal/intersexuality.

(E) POPULATION EFFECTS
Skewed sex ratios affecting mating patterns; impaired mating leads to population declines and eventual extirpations and extinctions.

(F) ECOSYSTEM EFFECTS
Transmittal of contaminants via biomagnification/bioaccumulation; altered role of affected species in food webs or other interactional networks; impaired ecosystem structure and function.

Figure 7.3. Known and hypothesized impacts of EDCs at different levels of biological organization. First cellular biochemistry is affected; then growth, differentiation, and reproduction; and eventually population characteristics. Ecosystem effects are a likely outcome of population change but remain largely speculative. The general sequence outlined here can apply to many other contaminants. Vtg = vitellogenin; Zr = zona radiata (eggshell) protein. Modified from Campbell and Hutchinson (1998) and Arukwe (2001).

been even slower to develop. In 1996, the Safe Drinking Water Act and the Food Quality Protection Act established screening programs for estrogenic substances, administered by the EPA in consultation with the FDA (http://pmep.cce.cornell.edu/fqpa/FQPASlideShow/). However, implementation of these provisions has been hampered by litigation (see Krimsky 2000). In 1999, the EU adopted the Community Strategy for Endocrine Disrupters to "identify the problem of endocrine disruption, its causes and consequences and to identify appropriate policy action" (http://europa.eu.int/comm/environment/docum/99706sm.htm). In 2003, the European Parliament banned the sale and use of two known endocrine disrupters, nonylphenol and nonylphenol ethoxylate (EU 2003). Canada appears to be actively investigating the issue (e.g., *Water Quality Research Journal of Canada*, vol. 36(2), 2001, and www.emcom.ca), but federal legislation remains to be enacted. In all instances, human health considerations can be expected to take precedence, but fishes and other organisms will benefit as a result.

Many major gaps exist in our knowledge of the impacts of these substances on fishes (e.g., Campbell and Hutchinson 1998; Arcand-Hoy and Benson 1998). For the sake of fish health, and to increase the value of fishes as sentinels of human health hazards, our information base needs expansion. The longer-term and larger-scale ecological consequences of EDCs' impacts on individual growth and maturation remain largely unexplored (Arcand-Hoy and Benson 1998; Arukwe 2001; figure 7.3). More species need to be studied because of likely species-specific responses. Such research must be conducted in the field and laboratory and the results compared and linked.

ACID DEPOSITION AND ACIDIFICATION

The hydrogen ion concentration, or pH, of most aquatic habitats is close to neutral (7). Deviation from that value toward a lower pH constitutes acidification, which is primarily a problem in freshwater because the oceans dilute inputs, and seawater is relatively well buffered against pH change. Acidification of aquatic habitats occurs largely as a result of toxic spills, acid precipitation, and runoff from coal and other types of mining (Starnes and Gasper 1995). Spills and runoff have been with us since the Industrial Revolution. Acid precipitation, however, was not recognized as a widespread and biologically disastrous environmental condition until the latter decades of the 20th century (see Likens et al. 1972, 1979).

Acid spills are acutely fatal to fishes. An accidental sulfuric acid spill in the Rio Salinas, northern Mexico, in 1990 killed approximately 2,000 fish per linear meter of river along the upper 2 km of river (= 4,000,000 fish); impacts of the spill were detectable another 20 km downstream (Contreras-Balderas and Lozano-Vilano 1994). Acid mine drainage is highly toxic to fishes as a result of low pH but also because mine wastes typically contain a number of toxic compounds, many of which (e.g., lead, aluminum, copper) increase in toxicity as pH is reduced. Acidification from mining operations tends to be localized to areas downstream from mines, but those areas can be extensive.

Acid Rain

Of much wider impact is chronic, regional acidification resulting from acid rain (aka *acid precipitation* and *acid deposition*; the latter term is more correct, but the three are used interchangeably) (ESA 1999; Galloway 2001). The natural pH of rainwater is slightly acidic at a pH of 5.5–5.6 as a result of CO_2-H_2O reactions; lower pH constitutes acid rain (figure 7.4). Acid rain lower than pH 4.3 is rare in the U.S. (USEPA 1999), but the pH of snow in Sweden and fog in central India can be as low as 4.0 (Aggarwal et al. 2001). Rainfalls of pH 3.8 have been recorded in Ontario, Canada, and as low as pH 2.0 around mining operations in West Virginia (Beamish and Harvey 1972).

Acid rain results chiefly from the burning of fossil fuels in industrial operations (factories, smelters, power plants), emissions of internal combustion engines, fertilized agricultural fields and animal waste, and wood smoke (approximately 10% originates from nonanthropogenic sources such as volcanoes and plant decomposition). These emissions contain oxides of sulfur and nitrogen, such as sulfur dioxide (SO_2) and nitrogen oxides (usually NO_x). When combined with atmospheric moisture, SO_2 and NO_x react to form acids such as sulfuric, nitric, and nitrous (H_2SO_4, HNO_3, and HNO_2), which fall to earth as wet deposition in rain, snow, and fog. Dry deposition falls on its own as ash, soot, and smoke and is flushed into water bodies during rains. The downwind effects of emissions can be felt hundreds or even thousands of kilometers from the source because of high altitude injection of exhaust from tall smokestacks, after which particles and compounds are transported by winds. For example, the smelter stacks at

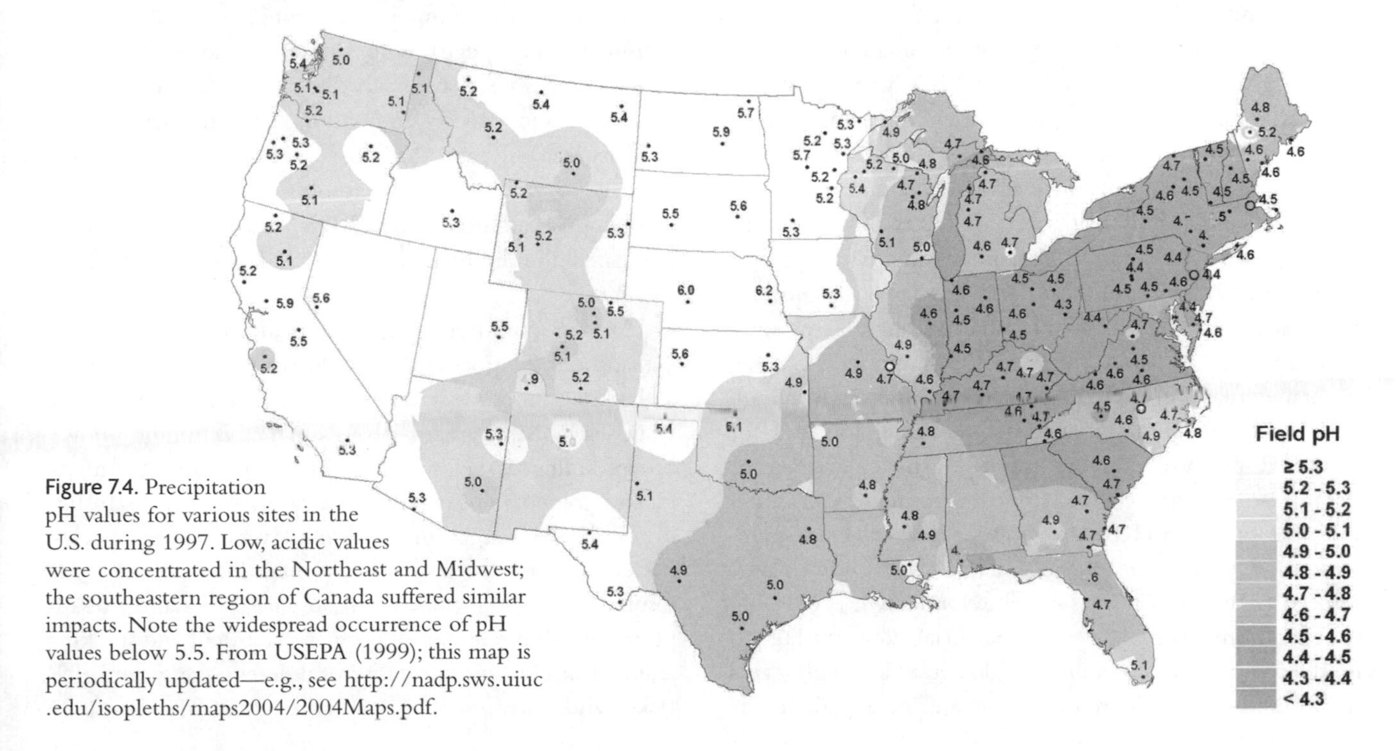

Figure 7.4. Precipitation pH values for various sites in the U.S. during 1997. Low, acidic values were concentrated in the Northeast and Midwest; the southeastern region of Canada suffered similar impacts. Note the widespread occurrence of pH values below 5.5. From USEPA (1999); this map is periodically updated—e.g., see http://nadp.sws.uiuc.edu/isopleths/maps2004/2004Maps.pdf.

Sudbury, Ontario, which pumped out 2.6 million tons of sulfur dioxide annually in the 1940s–1960s, are almost 400 m tall (Beamish and Harvey 1972; Sullivan 2000).

Acid rain is unquestionably caused by industrial activity. Although probably evident during the 19th century, it did not become a scientific and political issue until the 1960s, and actions to curtail it were fairly minimal until the 1970s. Areas most impacted have been the northeastern U.S. and southeastern Canada, western Europe and Scandinavia, central Europe (Poland, Germany, Czech Republic), southern India, eastern China, Korea, and Japan. All are near or downwind from major industrial centers. More moderate but noticeable impacts occur in the western U.S., southeastern Brazil, southeastern Africa, and Indochina (TSLG 2001). Approximately 30% of Sweden's 90,000 lakes and 300,000 km of its rivers and streams are "so acid that only acidification-resistant plant and animal species could survive in them" (www.internat.environ.se). Most of the acid-causing pollutants affecting Sweden originated in central Europe, in the UK, or from oceanic shipping. Massive death of forest trees in North America and Europe (half of the Black Forest in Germany dead or damaged) and wholesale decimation of salmonid populations in the U.S., Canada, and Scandinavia were the major impetus for research into the causes of acid rain and prompted environmental organizations to insist on governmental action.

Low pH in rainfall does not automatically acidify receiving streams and lakes. Final pH is a function of precipitation; inflowing water; and bedrock mineralogy, which determines the buffering capacity of the soil and rocks over which the water sits or flows. Acid neutralizing capacity (ANC) is the ability of a water body to buffer or neutralize strong acid inputs. EPA defines water with less than 50 microequivalents of ANC per liter as acid sensitive. Systems with low ANC are characterized by shallow soils, noncarbonate (nonlimestone) rocks, granitic rock, small watershed or water body size, and high elevation. Although most acid rain–contaminated systems suffer low pH for extended periods, episodic inputs during snowmelt or storms can exacerbate already stressful conditions. Episodes can also tip the pH balance toward stressful conditions, which often include increased acute toxicity from aluminum (Al) and Hg (see Gensemer and Playle 1999). Hg mobilization occurs because bacteria convert Hg to methylmercury more rapidly at lower pH. Spring rainstorms and snowmelt are especially injurious: Acidic compounds accumulate in winter snowpack, flushing occurs when eggs and larvae are most abundant; and early life stages are particularly vulnerable to low pH (e.g., Sullivan 2000).

Impacts on Fishes

Acid-related toxicity to fishes is largely a combined function of the pH of rainfall and the calcium (as Ca^{2+}) and inorganic monomeric aluminum content of the water in which the fish live. Calcium ions reduce aluminum toxicity; fish in Ca-rich waters can tolerate lower pH and higher Al concentrations. Physiological effects result from impaired regulation of body salts mediated via membrane permeability as affected by Ca availability. Respiration is disrupted because gill lamellae become clogged with mucous, also leading to stress and mortality (Sullivan 2000).

In general, areas with low ANC experience the greatest reductions in overall fish diversity and largest changes in species composition (J. P. Baker and Christensen 1991). Although young fish are more susceptible to low pH, older fish (e.g., Atlantic salmon) are more sensitive to aluminum toxicity (Rosseland et al. 2001). Population declines result less often from adult mortality than from recruitment failure, which reflects impaired reproduction that may involve ovarian maturation, inhibited spawning, or actual mortality among fertilized eggs, embryos, or larvae. Decreases in adult populations result from direct mortality and behavioral avoidance of or emigration from impacted areas. Studies have shown that loss of body condition and increased mortality may also result from food chain effects involving invertebrates, which are also sensitive to acid-associated aluminum toxicity (J. P. Baker and Christensen 1991).

Among common lake and stream fishes in eastern Canada and the northeastern U.S., most minnows, darters, and some salmonids (rainbow trout, Atlantic salmon, lake trout) are relatively acid sensitive, whereas mudminnows, brown bullheads, yellow perch, brook trout, most sunfishes, and golden shiner are comparatively insensitive. In figure 7.5 the ranges in pH values vary within species. Differences arise because authors used different metrics of response, including population absence, population disappearance, inhibited reproduction, recruitment and stocking failure, fish kills, and mortality of different age groups (J. P. Baker and Christensen 1991). Differences may also arise pH — Fish species tolerant of that critical pH or higher (range below critical value in parentheses)

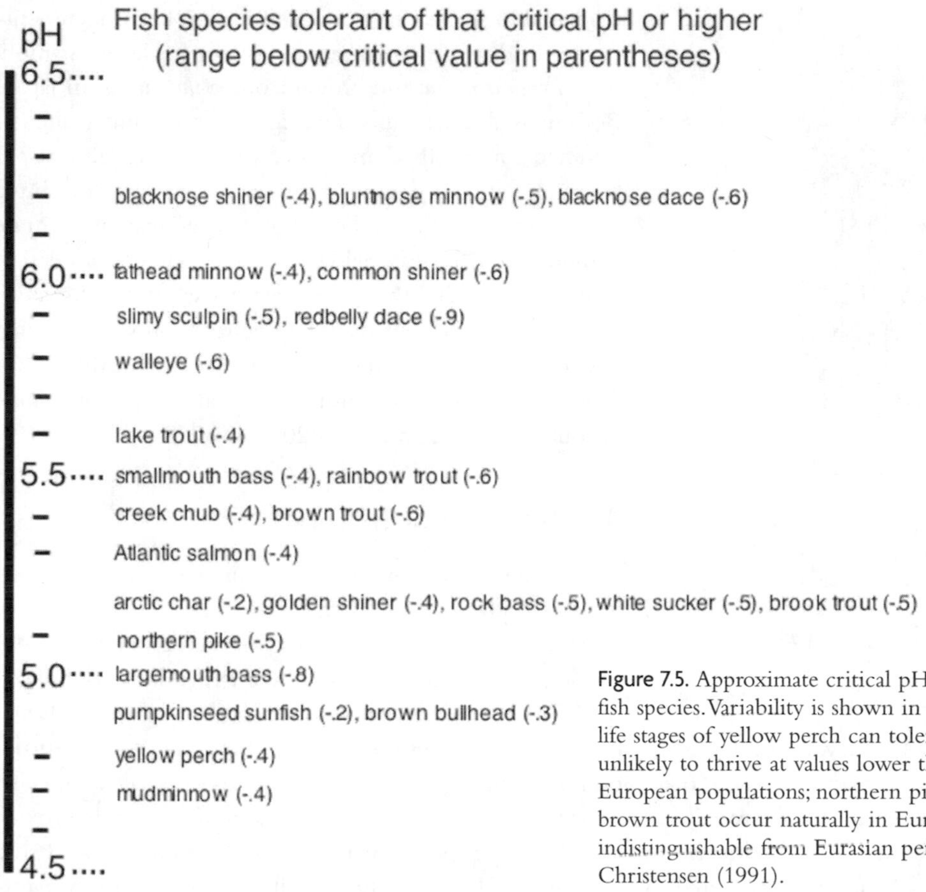

6.5 ....

- blacknose shiner (-.4), bluntnose minnow (-.5), blacknose dace (-.6)

6.0 .... fathead minnow (-.4), common shiner (-.6)

- slimy sculpin (-.5), redbelly dace (-.9)
- walleye (-.6)
- lake trout (-.4)

5.5 .... smallmouth bass (-.4), rainbow trout (-.6)

- creek chub (-.4), brown trout (-.6)
- Atlantic salmon (-.4)

arctic char (-.2), golden shiner (-.4), rock bass (-.5), white sucker (-.5), brook trout (-.5)

- northern pike (-.5)

5.0 .... largemouth bass (-.8)

pumpkinseed sunfish (-.2), brown bullhead (-.3)

- yellow perch (-.4)
- mudminnow (-.4)

4.5 ....

Figure 7.5. Approximate critical pH values for selected North American fish species. Variability is shown in parentheses (e.g., some populations or life stages of yellow perch can tolerate pH as low as 4.4). A species is unlikely to thrive at values lower than the critical pH. Data may apply to European populations; northern pike, arctic char, Atlantic salmon, and brown trout occur naturally in Europe, and yellow perch is essentially indistinguishable from Eurasian perch. Data from J. P. Baker and Christensen (1991).

from methodological inconsistencies and the biological realities of local acclimation.

Salmonids appear to be particularly acid sensitive. Acid deposition is considered a prime contributor to the decline of Atlantic salmon stocks in eastern Canada and will probably prevent their recovery (e.g., Watt et al. 2000). Fish kills in Norway following episodic acidification affected both Atlantic salmon and brown trout (J. P. Baker and Christensen 1991). Norway has lost 18 stocks of Atlantic salmon, with 8 more considered threatened; and brown trout have disappeared from 39% of its lakes, with significant declines in another 17% (Sandøy and Langåker 2001). Brook trout, which are relatively acid tolerant, have disappeared from approximately 11% of the lakes in the Adirondack Mountains of New York due to acidification (J. P. Baker et al. 1993); even more acid sensitive, minnows have disappeared from 19% of surveyed lakes. Approximately half of Virginia's trout streams that do not flow over carbonate bedrock are unsuitable for brook trout because of acidification, even though brook trout are less sensitive to low pH than other salmonids (Bulger et al. 2000).

Shenandoah National Park, Virginia, has the distinction of receiving more atmospheric sulfate than any other U.S. National Park. Comparisons of fish assemblages in low-, intermediate-, and high-ANC streams revealed reductions in species richness, population density, condition factor, age distribution, fish size, and survival during bioassays in streams with low ANC (Bulger et al. 1995). Condition factor of one particularly sensitive species, the blacknose dace, *Rhinichthys atratulus*, was depressed in streams with low ANC and high sulfate concentrations (Dennis et al. 1995). Survival of brook trout embryos and fry was also lower in low-ANC streams (MacAvoy and Bulger 1995).

The Adirondack Mountains are perhaps the most thoroughly studied region in the U.S. with respect to acid precipitation and its effects on organisms. They contain more than 5,000 lakes across a forested landscape that varies in topography, soil, and underlying geology, all impacted by relatively uniform acid precipitation from outside sources.

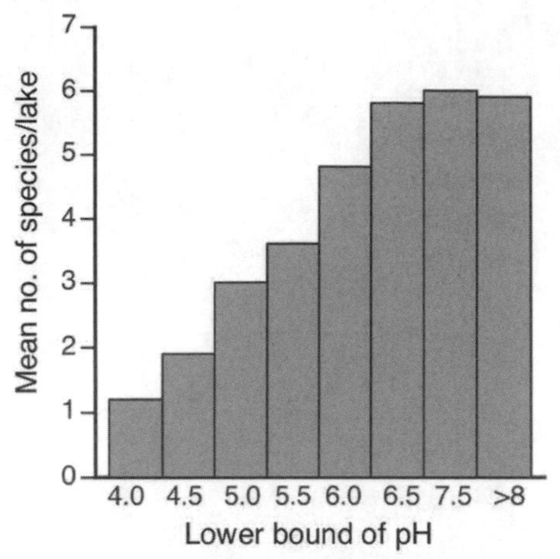

Figure 7.6. Species diversity increases with increasing pH in Adirondack lakes, New York. Bottom axis shows lakes with different pH. Although 55 fish species occur in Adirondack lakes and streams, few lakes have more than 6.0 pH was the most influential variable determining species diversity, although ANC, total aluminum, and oxygen and calcium concentrations contributed. Data from Driscoll et al. (1991), who grouped lakes in the pH range 6.5–7.5.

Fish surveys date back to the late 1800s. Surveys conducted during the 1970s indicated that as many as 24% of Adirondack lakes had become fishless and that diversity decreased with decreasing pH in both lakes and streams (Driscoll et al. 1991; figure 7.6). Most losses occurred between 1930 and 1970 following an 1882–1930 period of relative stability. Lakes with pH values below 5.0 were most likely to experience declines, and lakes with pH of less than 4.5 contained no fish or only acid-tolerant species (e.g., central mudminnow, brown bullhead, golden shiner, or yellow perch). Affected lakes were concentrated in regions of low-ANC, granitic, gneiss bedrock and shallow soils. Despite reduced sulfate emissions and precipitation in recent years, lakes appear slow to recover from an earlier period of intense acidification.

Sublethal Impacts on Development and Behavior

Extremely skewed sex ratios in fishes can cause population declines, and sex ratio in some species responds to relatively minor alterations in pH. Cichlids breeding at pH 5.0–5.8 often produce essentially all-male broods; pH of 6.9–7.1 results in all-female broods. The same occurs in swordtail, *Xiphophorus helleri*, which produce all males at a pH of 6.2 and all females at 7.8 (Rubin 1985). Behaviorally mediated losses due to acidification have been documented in landlocked kokanee or hime (sockeye) salmon, *Oncorhynchus nerka*, and brown trout, *Salmo trutta*, in Japan. Both ceased essential spawning activities under slightly acidified conditions. Females decreased nest digging at pH below 6.4 and all digging below 6.0 (Kitamura and Ikuta 2001). Subsequent studies found that spawning hime salmon avoided areas below pH 6.0 (Ikuta et al. 2001). "Avoidance of slightly acidic water in selection of spawning site or cessation of spawning behavior in weakly acidic environments may be the most potent cause of the reduction of salmonid populations in the early stages of acidification" (Kitamura and Ikuta 2001, p. 875).

Actions and Solutions

The good news is that curtailing sources of acid deposition reverses acidification and can lead to biotic recovery. The "discovery" of acid rain and its impacts resulted in widespread government regulations and expenditure of billions of dollars to decrease production of acid compounds and for remediation, despite decades of resistance and obstruction by polluting industries (see Havas et al. 1984). Specific remedial actions included installation of calcium hydroxide "scrubbers" in power plants to convert sulfuric acid into calcium sulfate and water, installation of catalytic converters in vehicles to transform exhaust gases into CO_2 and N_2, and increased use of low-sulfur coals and other fuels. In 1988, the UN sponsored the Long Range Transboundary Air Pollution Agreement to limit NO_x emissions. In 1990, the U.S. amended the Clean Air Act with an Acidic Deposition Control Program designed to further reduce SO_2 and NO_x emissions, which included a market-based trading and banking system of emission allowances. This landmark legislation resulted in part from the ten-year, $500 million National Acid Precipitation Assessment Program (NAPAP), "the largest environmental research program ever conducted" (Sullivan 2000, p. vii).

As a result, sulfate and nitrate emissions have been reduced, acid deposition has decreased in many regions, and area-specific recovery has been observed (see Gunn and Sandøy 2003; Yan et al. 2003). USEPA (1999) claimed that sulfate levels in precipitation in the northeastern and midwestern U.S. had declined up to 25%. Overall sulfur deposition in Europe and North America has declined up to 50%, although nitrogen deposition has remained relatively constant (Skjelkvåle et al. 2001). However, emissions and acidification in India and especially China remain bad and are expected to get worse (Rodhe et al. 1992).

Improved ANC and pH have also resulted from additions of lime to specific water bodies. Limestone ($CaCO_3$) neutralizes sulfuric acid, converting H_2SO_4 to $CaSO_4$, H_2O, and CO_2 (Olem 1991). Sweden has spread 200,000 tonnes of fine-ground limestone annually in lakes and watercourses, treating more than 7,000 water bodies since the 1970s (www.internat.environ.se). Liming of rivers in Norway has raised pH and is reportedly responsible for increased invertebrate, fry, and parr survival and a more than tenfold increase in salmon catches in some rivers (Walseng et al. 2001). Hg impacts have also been reduced by elevating pH, with positive results at locales distant from sources of contaminants and acidification. A 60% decline in Hg contamination among fishes in Little Rock Lake in remote northern Wisconsin was attributed to deacidification and decreased atmospheric Hg loading (Hrabik and Watras 2002).

Numerous faunal recoveries have been attributed to decreased sulphate deposition. Brown trout have come back in many Norwegian rivers (e.g., Hesthagen et al. 2001), and Eurasian perch have recovered in southern Finland lakes. A 100-fold increase in abundance in one lake over a four-year period corresponded to decreased acid deposition (Rask et al. 2001). Roach, which are more acid sensitive, appear to be slower to recover in Finland (Nyberg et al. 2001). When copper and nickel smelters in Sudbury, Ontario, reduced SO_2 emissions by two-thirds in the early 1970s, water quality in receiving lakes improved substantially. The pH of Whitepine Lake, 90 km north of Sudbury, increased from 5.4 to 5.9 during the 1980s. With improving water quality, sensitive species such as white sucker and lake trout recovered: Young lake trout first (re)appeared in 1982 and became increasingly abundant through the 1980s (Gunn and Keller 1990).

Improvements near Sudbury may, however, represent a localized, worst-to-best-case scenario (Jeffries et al. 2003). The pH and alkalinity of many lakes of southeastern Canada do not appear to be recovering, or at least not as quickly as the extremely degraded lakes near smelters. Many regions in North America and Europe, especially regions with poor ANC or very sensitive fish species, have also been slow to recover. In the UK, national declines in sulfur deposition since 1988 have not led to a general improvement in freshwater chemistry such as increased pH or alkalinity (Monteith et al. 2001). Lakes in the Adirondacks and streams in the Catskill Mountains of the northeastern U.S. have not shown higher pH measurements despite regional reductions in sulfate emissions (e.g., Sullivan 2000). Adirondack and Catskill ecosystems may suffer from continued high inputs of atmospheric nitrogen, from leaching of base cations such as calcium and magnesium that help neutralize acids in soils, and from continued aluminum toxicity associated with low pH (ESA 1999). Improvement may require further reductions in sulfur and nitrogen (USEPA 1999). Because of declines in ANC and accumulation of sulfur and nitrogen in soil, an additional 80% reduction in electric utility emissions of sulfur is needed to change northeastern U.S. streams from acidic to nonacidic in approximately 20–25 years (Driscoll et al. 2001).

Species reintroduction programs may also be necessary, because many defaunated lakes are isolated from sources of recolonization. But the success of such efforts may vary as a function of species sensitivities and unanticipated ecological interactions. Smallmouth bass are as sensitive to low pH as lake trout, *Salvelinus namaycush* (figure 7.5), but smallmouth recover more rapidly when acidity conditions improve. In lakes near Sudbury, reintroduced smallmouth established reproducing populations within five years of water quality recovery, whereas lake trout populations remained below reference levels for ten more years (Snucins and Gunn 2003). Trout recovery was especially slow in relatively species-rich lakes (nine versus two to five species) and in lakes where both bass and trout were introduced. Lake trout exhibited slower growth, lower survival, and delayed recruitment as an apparent result of competition for food that forced them to switch from piscivory to zooplanktivory, as well as from predation on their young by the resurging bass.

DISSOLVED SOLIDS

Most waters carry a natural level of dissolved minerals, organic matter, and nutrients that originate in underlying rocks or soils. These dissolved solids include ionic forms of calcium, chlorine, magnesium, sodium, and potassium, as well as bicarbonate, silica, sulfate, and nitrogenous compounds and phosphates. Excess salts, especially sodium chloride (NaCl), are problematic dissolved solids for most freshwater fishes.

Fish distribution is determined by the natural salt content of water; tolerance of or affinity for salinity forms the basis of traditional biogeographic classifications (see any ichthyology text). Dissolved salts determine the osmotic balance between an organism and its aqueous environment. Hence alterations to natural conditions can create osmotic stress. Salinity can increase due to evaporation, leakage from natural brine deposits or aquaculture impoundments, runoff of road salt used in winter, and saltwater intrusion due to dewatering or reduced stream flow in coastal areas. "Coastal" can extend far inland. In northern Mexico, a diverse, endemic freshwater ichthyofauna has declined as water withdrawal for agricultural and municipal uses has lowered water tables and reduced river discharge, leading to salinization from both evaporation and upstream seawater intrusion (Contreras-Balderas and Lozano-Vilano 1994). Native freshwater fishes have retreated from downstream regions and been replaced by brackish and marine invaders (Contreras-Balderas et al. 2002).

Agriculture in much of the world depends on impoundment and irrigation. Nearly half of total terrestrial food production depends on irrigated lands, and the number will continue to grow (Alexandratos 1995). When rivers are impounded, increased salinity commonly results, especially in arid regions. Fishes suffer to their and our detriment. The Aral Sea and Salton Sea disasters (chapter 6) are just two of the best-known examples of a general pattern. Evaporation rates are high both in a new reservoir and in the irrigated fields that drain into the impoundment. Impoundment, water abstraction, irrigation, and salinization constitute a sequence that culminates in major shifts in fish species composition and abundance. Native assemblages of endemic, lotic specialists often give way to hatchery-produced, introduced, salt-tolerant generalists whose diets must be supplemented with food species tolerant of the altered hydrologic situation. In arid and semiarid nations, drastic declines in species diversity among native fishes have been accompanied by declines in fish yields (Petr 1995).

Lake de Guiers, a natural lake in Senegal, is an example of human-induced salinization decreasing fishery production. It was fed by the Senegal River during floods but dried up during droughts. Lake fisheries produced around 2,500 MT yr^{-1}; river and floodplain fisheries accounted for ten times more, at 23,500 MT yr^{-1}. To reduce flooding and improve lake fisheries, a dam was constructed to retain river water in the lake. Stabilized lake levels did increase production, but only to 3,000 MT yr^{-1}, whereas the dam allowed intrusion of seawater from downstream into the river, leading to a 50% decline in river fisheries. The net change was a loss of 11,250 MT yr^{-1} of fish from the system (Sagua 1997).

Lowered or variable salinity can also affect fishes in places that normally experience high or stable salinity regimes. In Biscayne Bay, Florida, flood control canals periodically release large volumes of freshwater into a coastal area that was historically more stable and saline. Higher fish diversity characterized stable habitats, whereas species abundances were reduced at some sites subjected to variable salinities (Serafy 1997). Fishes were experimentally exposed to rapidly changing salinities (32 ppt salinity to 0 ppt and back to 32 ppt) over a two-hour period, simulating a single, rapid, freshwater pulse such as occurs during thunderstorms. Five of ten local species suffered mortalities of 12%–100% during trials, suggesting that assemblage changes resulted from differential osmoregulatory abilities among species (Serafy 1997). A subsequent study pointed to salinity as a primary determinant of assemblage structure, causing a shift to larger-bodied species and individuals under more variable conditions and thus altering the assemblage and decreasing prey availability for higher trophic levels (Lorenz 1999).

TOXIC SPILLS

Fish kills result from careless handling and transporting of large amounts of toxic substances, often a by-product of large-scale commerce. Sewage and water treatment plants periodically release excess amounts of chlorine that result in fish kills. Many toxic chemicals are released during industrial accidents, be they point-source discharges or truck, train, or ship spills.

Massive oil spills from ruptured tankers, oil refinery mishaps, storms, wartime activities, or failed drilling and storage operations receive considerable and appropriate attention (Burger 1997). Damage is a function of amount and type of oil (sulfur content, volatility, water solubility), temperature, weather, and ocean currents and other forces that disperse or degrade the oil. Attention is most commonly focused on affected birds, mammals, and shoreline algae and invertebrates, especially commercially important shellfish. "Public notice generally ignores those organisms, such as fishes, that cannot be taken ashore, washed off, and photographed with concerned volunteers" (J. McEarchern, pers. comm.). Impacts on fishes tend to be less publicized because they are less charismatic than seabirds and marine mammals; impacts of oil below the water's surface and away from beaches and coastlines are less obvious; and the fraction of oil that sinks is often orders of magnitude less than what washes ashore (e.g., R. F. Lee and Page 1997). Oil is toxic to fishes, especially to eggs and embryos, although most measures of impact revert to pre-spill conditions after a year or two.

The March 1989 *Exxon Valdez* oil spill—so infamous that it has its own acronym, EVOS—had both immediate and persistent impacts on the fishes of Prince William Sound, Alaska (Peterson 2001). The initial spill involved 36,000 MT of crude oil that contaminated an estimated 850 km of shoreline. Many fish species not directly killed exhibited reduced growth, genetic damage, physical deformities, and elevated levels of various petroleum exposure biomarkers. Among important commercial species, Pacific herring, *Clupea pallasi*, suffered egg and embryo mortalities, deformities, genetic damage, reduced growth, hepatic necrosis, viral infections, and lowered reproductive success. Pink salmon, *Oncorhynchus gorbuscha*, suffered reduced growth and survival among juveniles, and adult cutthroat trout, *O. clarki*, and Dolly Varden, *Salvelinus malma*, exhibited lowered growth and survival (Lee and Page 1997; Jewett et al. 2002).

Although most attention focused on commercially important species and the economic consequences to fishing communities, noncommercial and intertidal species were also affected. Rockfishes (Scorpaenidae) were apparently quite vulnerable to the spill (Marty et al. 2003). Overall density and biomass of 21 intertidal species in several families declined as a result of oiling and subsequent cleanup activities (W. E. Barber et al. 1995). Although signs of recovery were evident the second year after the spill, biomarkers indicating continued exposure to petroleum products were present ten years later in benthic fishes such as masked greenling, *Hexagrammos octogrammus*, and crescent gunnel, *Pholis laeta* (Jewett et al. 2002). Leakage from oiled substrates, especially of heavy molecular weight fractions, continued for years and persisted as toxic polycyclic aromatic hydrocarbons (PAHs). PAHs accumulate in the eggs of demersal spawning fishes and cause delayed deleterious health effects arising from damage during embryogenesis (Short et al. 2003; table 7.2).

As bad as EVOS was, it ranks as only the 14th worst in volume spilled. Other major spills causing documented fish mortalities and pathologies include the *Amoco Cadiz* (1978, Brittany coast of France, 233,000 MT of crude oil), the *Torrey Canyon* (1967, Cornwall, UK, 117,000 MT of Kuwait crude oil), and the *Sea Empress* (1996, southwest coast of England, 70,000 MT of North Sea light crude) (Gundlach et al. 1983; Neff 1985; www.swan.ac.uk/empress/empress.htm). On an even larger scale, an offshore drilling platform, *IXTOC-1*, spilled 476,000 MT of crude oil off the east coast of Mexico over a nine-month period in 1979. The Iran-Iraq War, between 1981 and 1987, is estimated to have spilled 260,000 MT of oil into the Gulf of Arabia; the Gulf War of 1991 released an estimated 0.8–2 million MT of oil into the Persian Gulf (not counting an additional 42–126 million MT spilled onto land; see http://multinationalmonitor.org/hyper/issues/1991/ridgeway.html).

Because they are mobile, fishes can often leave affected areas, and not all oil spills result in widespread fish mortalities (e.g., Edgar et al. 2003). However, chronicled incidents where fish kills were minimal generally involved relatively small amounts of lightweight, rapidly evaporating oil and rapid transport of oil away from the spill site via ocean currents.

More exotic lethal events result from discharges of substances not usually regarded as pollutants. In the Fox River of Wisconsin, 58 fish kills occurred during April–October 1988, some involving over 30,000 individuals. The cause was eventually revealed as carbon monoxide (CO) poisoning originating from an outboard motor testing facility (Kempinger et al. 1998). In September 2003, approximately 19,000 fish were killed in the Salt River of Kentucky when 800,000 gallons of bourbon were released from a Jim Beam warehouse following a fire. Species killed included paddlefish, catfishes, freshwater drum, sunfishes, and largemouth bass. Fish died from alcohol poisoning combined with deoxygenation due to respiration by microorganisms attracted to the alcohol. The company was fined $27,000 to help remediation efforts that included restocking of several species. The liquor spill was unfortunately not unique, and the potential for future spills remains considerable. In 2000, fire in a seven-story Wild Turkey warehouse led to a spill that killed hundreds of thousands of fishes along a 100-km stretch of the Kentucky River. The resulting fine ran to $256,000. More than 200 bourbon warehouses, storing about 4,000,000 barrels, exist in Kentucky alone (AJC 2003).

Solutions: Remedation, Regulation, Reduced Demand

Oil spills may be bad for fishes and other organisms, but the "cures" employed during remediation are often worse. Beaches contaminated by the *Exxon Valdez* spill were washed with hot water from pressurized hoses. Limited data indicate that some fish that survived the spill died from the cleanup; saddleback gunnel, *Pholis ornata*, present in the intertidal region prior to the cleanup were absent afterward (W. E. Barber et al. 1995). The detergents and other surfactants used to disperse spilled oil cause widespread damage. Fishes exposed to oil dispersed via chemical methods can exhibit higher levels of stress and mortality than those exposed to the untreated oil alone (NRC 2005).

Many of the institutional impediments to solving oil spill disasters were made evident after the *Prestige* spill of 2002 off the coast of Galicia, Spain. A single-hulled tanker carrying 60,000–70,000 MT of heavy fuel oil, the *Prestige* was built in Japan, registered in the Bahamas, managed by a Liberian-listed Greek company, and chartered by a Russian-owned Swiss company with British directors. It was carrying Russian oil from Latvia to Singapore. The ship had been recently inspected and surveyed in St. Petersburg, Dubai, and China. Repairs performed in China later failed (www.foe.co.uk/pubsinfo/briefings/html/20021126160752.html). This potpourri of international involvement complicated the issue of liability.

Preventing future disasters requires national and international regulations that start with clear identification of responsible parties. Restricted movement of single-hulled petroleum carriers and use of double-hulled vessels are increasingly common and prevent some types of spills. In addition, Marine Exclusion Zones containing sensitive areas should be declared off limits to the movement of oil tankers and other large-scale shipping. By 2003, only six such areas, referred to as Particularly Sensitive Sea Areas (PSSAs), were recognized: the Florida Keys, the Great Barrier Reef of Australia, the Malpelo Islands of Colombia, the Sabana-Camaguey Archipelago of Cuba, the Wadden Sea in northern Europe, and the Paracas National Reserve of Peru (IMA 2003). Ultimately, a reduction in the demand for oil and accelerated development of alternative, renewable, clean sources of energy are the only reliable solutions.

NUTRIENT POLLUTION

A major by-product of human activity, especially in the industrialized world, is production of excess nutrients that are released into waterways. A fertilizer effect results, stimulating growth of plant species that had been limited by lack of an essential nutrient or kept in check by grazers. The resultant algal blooms can be fatal to fishes because of toxicity, habitat destruction (smothering, overshading), and deoxygenation of the water column.

Nutrient pollution usually results from excessive nitrogen or phosphorus compounds. Sources include runoff and discharge from sewage, agriculture, aquaculture, lawns, golf courses, ball fields, and other recreational clear-cuts; atmospheric deposition; and groundwater flow. In general, phosphorus limits plant growth in freshwater, whereas nitrogen limits plants in estuarine and marine systems (D. M. Anderson et al. 2002). The result is a eutrophic, or overly productive, system. Eutrophication is a natural aging process in aquatic systems, exemplified by the classic lake-to-pond-to-marsh successional sequence. Cultural or anthropogenic eutrophication accelerates the process manyfold.

Eutrophication and resulting harmful algal blooms (HABs) can occur in almost any aquatic system. Upper Klamath Lake in southern Oregon contains two endemic Endangered suckers, the Lost River sucker, *Deltistes luxatus*, and the shortnose sucker, *Chasmistes brevirostris*. Both were abundant but have been severely depleted as a result of multiple factors, including hypereutrophication influenced by phosphorus and nitrogen inputs (Moyle 2002; NRC 2004b). Upper Klamath Lake is naturally eutrophic because of phosphorus-rich volcanic soils, but increased agriculture and large-scale losses of riparian and wetland vegetation in the mid-20th century apparently tipped the balance from a lake dominated by diatoms to one favoring bluegreen algae (cyanobacteria). The cyanobacterium, *Aphanizomenon flos-aquae* (AFA), is a nitrogen fixer (converts atmospheric N_2 to NH_3) and hence can take advantage of naturally abundant phosphorus. In late summer, AFA achieves monoculture densities that turn the water pea-soup green and approach the theoretical maximum possible biomass. Growth is limited only by light availability due to self-shading, rather than by nutrient availability. High AFA biomass at high summer temperatures leads to higher pH, depleted oxygen, and increased concentrations of un-ionized ammonia, all of which impair growth, reproductive success, and disease resistance of the fish. These conditions culminate in periodic massive fish die-offs that kill thousands of the already depleted species (NRC 2004b).

Noxious Blooms and Pathogenic Algae

Algal blooms are a natural phenomenon that has been recorded for centuries; Cabeza de Vaca reported in 1542 that fish kills occurred annually in Florida (Landsberg 2002). However, the frequency and global range of HABs appear to have increased in the past three decades, some involving novel algal species and affecting regions where blooms were not previously observed (Burkholder 1998; D. M. Anderson et al. 2002). Of additional concern is the documented dispersal of harmful plankton species in the 10 billion tons of ballast water transported by ships each year (Carlton and Geller 1993; Hallegraeff 1998; see chapter 9). Also troubling is the observation that excess nutrients have brought about "an increase in the number of formerly benign species that have become toxic" (Landsberg 2002, p. 115). This increase in toxic blooms has serious ecosystem, economic, and human health consequences.

Massive phytoplankton blooms involving toxic algal species (marine microalgae and heterotrophic dinoflagellates) have occurred in the South China Sea, the Black Sea, Hong Kong Harbor, Chesapeake Bay, and the northern Gulf of Mexico, to name a few locales. The Gulf of Mexico, which receives nitrate-enriched outflows from the Mississippi and Atchafalaya rivers, encompasses an 18,000-km^2 hypoxic "dead zone" off the Louisiana coast that develops each summer in the midst of one of the most important commercial and recreational fisheries in the U.S. (Rabalais et al. 2002). One harmful alga whose bloom and later decomposition in summer contributes to the dead zone is *Pseudo-nitzschia*, a diatom that produces neurotoxins implicated in deaths of seabirds and marine mammals (Landsberg 2002).

Excess inputs of industrial, agricultural, and municipal wastes create favorable growing conditions for HAB species. Inputs of phosphorus to the oceans are three times greater than during preindustrial and preagricultural periods. More important, nitrogen inputs into the oceans in just the past 40 years have increased four- to tenfold, depending on location (NRC 2000a). Given the nitrogen-limited growth of many marine phytoplankton species, increases in the frequency and extent of noxious blooms are to be expected. Between 1970 and 1995 the annual tonnage of fertilizer applied in China increased from around 4×10^6 to over 20×10^6, and the number of red tides (toxic dinoflagellate blooms) reported in Chinese coastal waters rose 50-fold (Anderson et al. 2002). Concern over HABs has stimulated organization of international symposia, with resulting publications. A journal, *Harmful Algae*, has been created to deal with the topic (see Burkholder 1998 and particularly Landsberg 2002).

Anthropogenic sources and pathways of nutrients are many. Sewage outfalls and other point ("pipe end") sources contribute variously but are relatively easy to monitor and (potentially) control. More problematic are diffuse, non-point sources such as runoff from agriculture fields, animal farms, and stockyards (concentrated animal feeding operations, or CAFOs); septic leakage; and atmospheric deposition. Worldwide use of nitrogen in fertilizers has increased eightfold, and of phosphorus threefold, since the 1960s (Anderson et al. 2002). A growing portion of the N (6% in 1960, 38% in 1990) is applied as urea, a form taken up readily by many harmful algal species.

Livestock in CAFOs and similar operations deposit their organic wastes on the ground, and they wind up in rivers and estuaries. USEPA (1998) concluded that animal feedlots impaired more river miles in the U.S. than combined sewer overflows, storm sewers, and industrial sources. Livestock on the North Carolina coastal plain alone generated 124,000 tons of nitrogen and 29,000 tons of phosphorus annually (Mallin and Cahoon 2003). These wastes are commonly held in "lagoons" before being sprayed on land. Massive failure of such lagoons during rainstorms has released millions of gallons of pig and chicken feces and urine into rivers, causing numerous fish kills and algal blooms (Burkholder et al. 1997; Mallin 2000). Livestock operations may contribute to a majority of fish kills in the U.S. (e.g., Stubbs and Cathey 1999; www.nrdc.org).

On June 21, 1995, the waste-holding lagoon for 12,000-pig Oceanview Farm, Onslow County, North Carolina, ruptured, flowing 0.5 km over neighbors' fields and lawns and spilling 98,000 m^3 (26×10^6 gallons) of pig feces and urine into the New River and its estuary. Toxic algal blooms persisted for three months, killing over 4,000 fish in the river and over 10,000 in the estuary. Dissolved oxygen in the river dropped to a lethal 0–1 ppm (concentrations below 5 ppm stress most fishes); fecal coliform concentrations were 300–1,500 times greater than "safe for human contact" levels; and ammonium was measured as high as 46 ppm, which is 150–4,600 times greater than normal levels; the recognized lethal-to-fish concentration is 40 ppm. In the New River estuary, algal densities were 2.5 to 7.5 times higher than the "nuisance" benchmark of 40 ppb, with high abundances of the ichthyotoxic prymnesiophyte

Phaeocystis globosa and the predatory and toxic dinoflagellate *Pfiesteria piscicida* (Burkholder et al. 1997; Mallin 2000).

Many nutrients from agriculture also enter aquatic systems as wet and dry deposition, either as phosphate adsorbed onto fine particles or as NO_x transformed into nitrate and falling in rain or dust. The nitrogen component of atmospheric deposition is a major contributor to acid rain, but its fertilizing effect may also be significant. Disposal of animal wastes is a constant problem in intensive agriculture. A common solution is to spread dry litter or spray liquified wastes ("liquid swine manure") on nearby fields. Although plants take up some of the nutrients, the unabsorbed fraction leaches into groundwater or flows overland during rains. Also, nitrogenous compounds evaporate from the ground and plants, and the spraying operation itself volatilizes the waste, injecting it into the atmosphere. Between 7% and 16% of the nitrogen from swine wastewater applied to experimental wetlands escapes via NH_3 volatilization (Poach et al. 2002). Wetlands are generally more effective at nutrient uptake than the crop plants on which such wastes are commonly applied. Sampson County, North Carolina, is representative of agricultural regions on the U.S. coastal plain. The county's human population in 1997 was 60,000; the swine population was nearly 2,000,000, up from less than 500,000 ten years earlier. Each hog produces 3.5 times the solid waste, 3 times the total nitrogen, and 5 times the phosphorus of a person. Atmospheric ammonium (NH4+) over those ten years increased from around 0.1 to 0.4 ppm, largely due to spray application of hog wastes, and much of it fell back into local rivers and streams (Mallin 2000). North Carolina, by no means unique in regard to weak regulations, exempts farm development from zoning laws and mandatory inspection programs (Burkholder et al. 1997).

HAB Taxa, Agents, and Effects

Harmful blooms involve a variety of algal types, most commonly dinoflagellates, but also diatoms, cyanobacteria, raphidophytes, prymnesiophytes, pelagophytes, and silicoflagellates (Landsberg 2002). Cyanobacteria (especially *Anabaena* and *Aphanizomenon* species) cause the majority of HABs in freshwater, whereas dinoflagellates (*Alexandrium*, *Cochlodinium*, *Gambierdiscus*, *Gonyaulax*, *Gymnodinium*, *Karenia*, *Noctiluca*, *Pfiesteria*, *Prorocentrum*) and some diatoms (*Chaetoceros*, *Pseudo-nitzschia*, *Skeletenoma*) are the chief culprits in estuarine and marine habitats. Although the usual depiction of an HAB involves massive phytoplankton densities of billions of cells ml^{-1} that cause water discoloration (e.g., foams, scums, mats, red tides, brown tides), some toxic algal species can cause fish kills at densities of only a few hundred cells l^{-1} (figure 7.7).

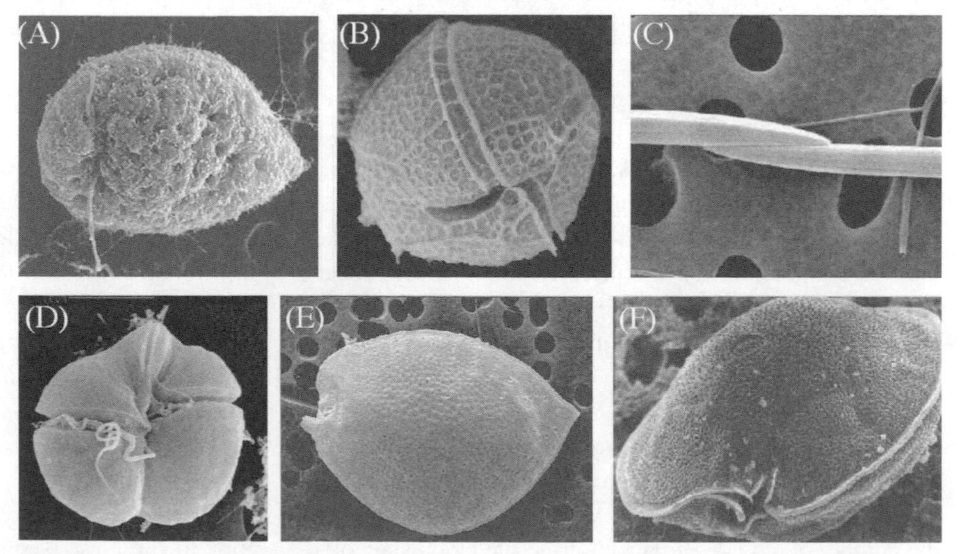

Figure 7.7. Scanning electron micrographs of algal species involved or suspected in toxic and harmful algal blooms. (A) *Chattonella subsalsa*, a raphidophyte. *Chattonella* spp. produce brevetoxin-like neurotoxins; brevetoxins cause neurotoxic shellfish poisoning. (B) *Gonyaulax grindleyi* (~50 µm, aka *Protoceratium reticulatum*), a red tide–causing dinoflagellate; relatives of *Gonyaulax* spp. in the genus *Alexandrium* produce saxitoxin, a cause of paralytic shellfish poisoning and reportedly used as a biological weapon (www.mbari.org). (C) *Pseudo-nitzschia pseudodelicatissima* (~200 µm) a chain-forming, widely distributed toxic diatom responsible for amnesic shellfish poisoning. (D) *Karenia brevis*, a red tide–causing dinoflagellate that produces brevetoxins. (E) *Prorocentrum micans* (~50 µm), a dinoflagellate implicated in diarrhetic shellfish poisoning; other *Prorocentrum* spp. are suspected as part of the ciguatera complex. (F) *Gambierdiscus toxicus*, the main causative agent of ciguatera fish poisoning. Photos courtesy of the Florida Marine Research Institute (www.floridamarine.org).

Familiar HABs include the causative agents of amnesic, diarrhetic, neurotoxic, and paralytic shellfish poisoning; ciguatera fish poisoning; red and brown tides; *Pfiesteria* toxicity; and putative estuary associated syndrome. Identified toxins include domoic and okadaic acid, brevetoxins, and ciguatoxins, produced by diatoms, dinoflagellates, and raphidophytes, to mention just a few (some species produce multiple toxins, several species produce similar toxins, and the taxonomy of HAB species is under revision). These toxins cause acute and chronic ailments in humans, other mammals, birds, turtles, fishes, and numerous invertebrates that consume prey contaminated via bioaccumulation, bioconversion, and biomagnification. Direct contact with HAB water or inhalation of aerosolized toxin(s) can also cause reactions (Van Dolah et al. 2001). Toxic by-products of HABs include ammonia, nitrite, and hydrogen sulfide. Oxygen depletion is common when blooms collapse and bacterial populations explode and consume oxygen; it can also result directly from nutrient stimulation of bacterial or microbial growth. Some HAB species excrete exotoxins, others produce mucus that clogs gills, and some have sharp extensions that cause capillary hemorrhage of gill structures or induce mucus production and suffocation (Burkholder 1998).

The 55-plus species of toxic dinoflagellates form an important subset of harmful algae. Dinoflagellates are unicellular, flagellated, photosynthesizing organisms that, under the right temperature and nutrient conditions, divide rapidly, creating the red tides that plague many coastal regions. Red tides, usually caused by *Alexandrium*, *Gymnodinium*, *Pyrodinium*, and *Karenia* (also *Ceratium* and *Gonyaulax*), have been linked to fish kills involving millions of individuals, as well as mass die-offs of other organisms (figure 7.8). Red tides are also associated with paralytic shellfish poisoning (PSP) and neurotoxic shellfish poisoning (NSP), named because they affect people who eat contaminated shellfish; mollusks, fishes, and nonhuman mammals are also sensitive to dinoflagellate toxins. Fourteen humpback whales died in Cape Cod Bay in 1987 after consuming Atlantic mackerel, *Scomber scombrus*, that had accumulated PSP toxins. As an apparent result of eating contaminated fishes, 117 Endangered Mediterranean monk seals died in 1997 from PSP. Brevetoxins from *Karenia*-caused red tides are toxic to both fishes and marine mammals. *Karenia* red tides have been linked to mass strandings and deaths of dolphins that fed on brevetoxin-contaminated fishes. In 1996, 150 Endangered manatees in Florida died from brevetoxicosis, which is caused by inhaling aerosolized brevetoxin and includes a variety of chilling respiratory pathologies (Landsberg 2002).

HAB-fish-mammal interactions can involve other phytoplankton groups. Some diatoms produce neurotoxins that move up the food chain. In May-June 1998, over 400 California sea lions died along the central California coast. The deaths were eventually linked to domoic acid poisoning resulting from a bloom of the diatom *Pseudo-nitzschia australis*. Northern anchovies, *Engraulis mordax*, ate the diatoms, and sea lions ate the anchovies (Scholin et al. 2000).

The public is aware of acute toxicity and fatalities among highly visible, charismatic marine mammals (and humans), but nonlethal, chronic, and ecosystem effects of HABs are relatively poorly studied and potentially important (Burkholder 1998; Landsberg 2002). Many "emerging" toxic algae proliferate in estuaries, which are preferred sites for aquaculture operations, resulting in spectacular fish

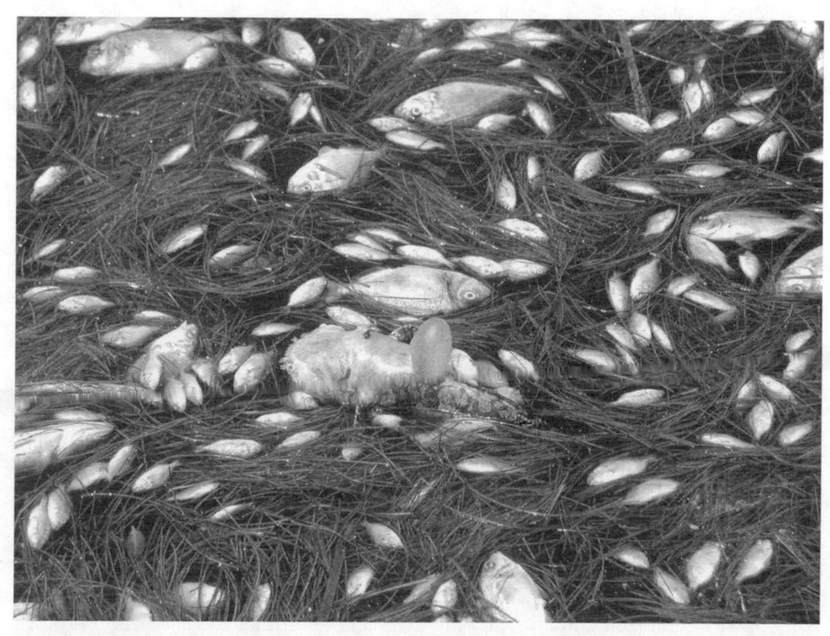

Figure 7.8. Fish killed by a red tide in Coquina Key, west-central Florida, 2005. Photo courtesy Florida Fish and Wildlife Conservation Commission.

kills. Estuaries are also nursery grounds for commercially important fishes. Estuarine algal blooms can cause recruitment failure, year-class collapse, and eventual species losses or declines, which affect food web and ecosystem structure and function. Toxin accumulation can produce sublethal impacts including depressed feeding, impaired reproduction, and increased susceptibility to cancers and other diseases. Habitat impacts, including decimation of important plants such as sea grasses, result from long-term shading by persistent algal blooms. Toxic species can grow selectively on damaged surfaces, as can ciguatera-causing agents (see below). Also, HAB species tend to grow optimally under warm conditions, raising the specter of increased bloom frequency and range expansion in response to climate warming. The list goes on.

Ciguatera: Humans in Coral Reef Food Chains

Among the best-studied phenomena involving toxic algae and fishes is the tropical ailment known as ciguatera fish poisoning (figure 7.9, chapter 13). People who eat certain reef fish from extensive areas in the Caribbean, tropical Pacific, and western Indian oceans can suffer a variety of gastrointestinal, neurological, and cardiovascular symptoms, including nausea, diarrhea, abdominal pain, perspiration, headache, numbing, tickling, burning, sensation reversal (e.g., ice cream feeling hot), coma, and occasional death from respiratory failure. Symptoms differ slightly by locale, with Caribbean incidents characterized more by gastrointestinal ailments, Pacific poisonings more neurological, and Indian Ocean incidents (the most lethal) by hallucinatory poisoning (R. J. Lewis 2001). Drying, freezing, and cooking fail to denature the toxins, which internationalizes the effects. Twenty people at a dinner in Calgary, Alberta, suffered ciguatera poisoning from eating thawed and cooked reef fishes imported from Fiji (Daley 2002); dried barracuda caused poisoning in Montreal (CCDR 1997).

Ciguatera originates in dinoflagellates, primarily *Gambierdiscus toxicus*. These dinoflagellates grow on common reef macroalgae or newly exposed coral surfaces (caused by, e.g., storms, bleaching, dredging, blasting, or dragging anchors). Some of the macroalgal hosts are also pioneer species that grow on disturbed surfaces (Kohler and Kohler 1992; R. J. Lewis and Ruff 1993). Outbreaks of macroalgae and *Gambierdiscus* have been linked to sedimentation pulses and excess nutrient inputs, as have other harmful algal blooms (Burkholder 1998). Herbivorous fishes and invertebrates—including planktivores—ingest the dinoflagellates directly or incidentally when feeding

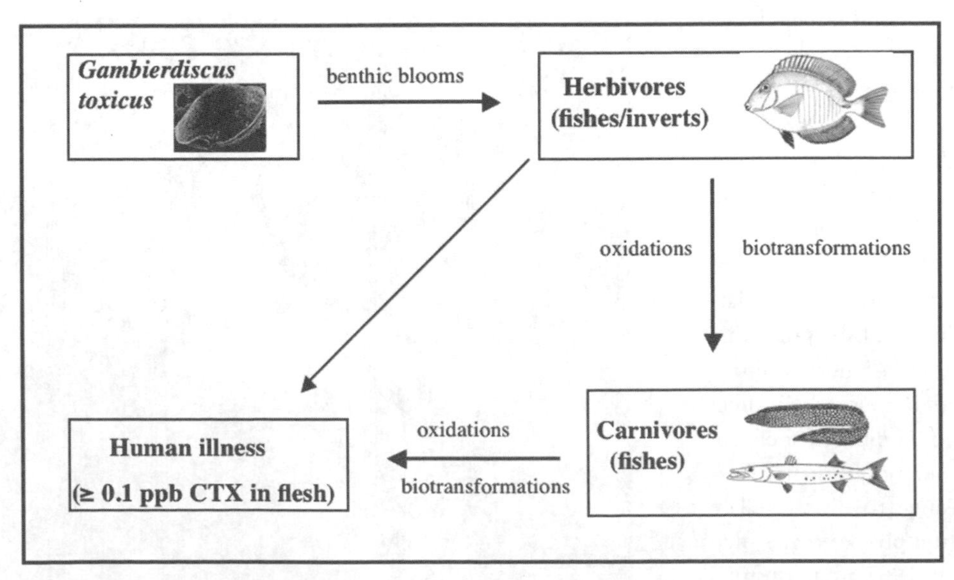

Figure 7.9. Events associated with ciguatera fish poisoning on a reef. Top predators, including humans, are affected because a single meal can contain significant amounts of a highly potent neurotoxin. *Gambierdiscus toxicus* is the dinoflagellate implicated in most ciguatera poisonings. CTX = ciguatera toxin; ppb = parts per billion. After Lewis (2001). Inset photo courtesy of the Florida Marine Research Institute (www.floridamarine.org).

on other algae or detritus (Kelley et al. 1992). The toxin bioaccumulates because it does not degrade, remaining stable in fish for up to 30 months (Banner et al. 1966). Predatory fishes can, in a single meal, acquire the entire body burden of ciguatoxins accumulated over a prey individual's lifetime.

The ecological impacts of ciguatera on fishes and other reef organisms are poorly understood but potentially significant. Ciguatera is toxic to fishes when ingested or even when dissolved in water. Fish fed extracts of ciguatoxins or flesh of ciguatoxic fish exhibit symptoms similar to those from exposure to paralytic shellfish toxins; they stop feeding, swim erratically, leap from the water, become immobile and disoriented on the bottom, display uncoordinated escape and avoidance movements when chased with a net, and may die (Davin et al. 1988). Any fish on the reef exhibiting such behavior would be easy prey, thus facilitating transmission of the toxin to higher-level predators. When ciguatoxins are injected into developing fish embryos, cardiovascular, muscular, and skeletal abnormalities occur, and hatching success declines. Maternal transfer of ciguatoxins to eggs "may represent an unrecognized threat to the reproductive success of reef fish and a previously undetected ecological consequence of proliferation of ciguatoxin-producing algae in reef systems" (Edmunds et al. 1999, p. 1827). Some major fish die-offs in southern Florida may have been a secondary result of ciguatera-induced stress (Landsberg 1995).

More than 400 species of tropical and subtropical fishes have been implicated as ciguatoxic, but the most toxic species are large predators such as moray eels, groupers, snappers, pompanos, mackerels, and barracudas. Herbivores such as parrotfishes and surgeonfishes can also be ciguatoxic. A suite of ciguatera toxins is suspected, and toxicity varies depending on locale, species, and presumably dinoflagellate strain. Besides ciguatoxins, maitotoxins, gambierols, gambieric acids, and scaritoxin are implicated (Landsberg 2002). Because the most toxic fishes are large predators, it has long been concluded that the toxin is biotransformed and biomagnified in the food chain.

Preventing ciguatera outbreaks has been a largely after-the-fact response to the human health implications. People are warned to avoid areas and species known to have caused illness. Prior to development of chemical assays, the only easy method of determining whether a fish was toxic was to feed it to a cat, chicken, or mongoose, an activity with economic and ethical ramifications. Recently, ToxiTec, Inc., developed the Cigua-Check Fish Poison Test Kit (available at www.cigua.com, which also contains information on ciguatera). Outbreaks may be preventable by minimizing conditions that promote the growth of ciguatera-associated dinoflagellates. Although the environmental factors that promote ciguatera blooms are incompletely understood, minimizing damage to corals by controlling nutrient pollution and sediment inputs is an obvious measure, as is restricting coral mining, blasting, dredging, and careless anchoring. Once again, the welfare of reef fish assemblages can be directly linked to the integrity of the reef ecosystem itself (e.g., chapter 12). Where ciguatera is involved, human health is an added consideration.

Pfiesteria: A Special Case

Most fish deaths that result from HABs can be considered incidental, via secondary plant chemicals or deoxygenation. However, some dinoflagellates apparently produce toxins as part of a predatory lifestyle. The most alarming and publicized example of toxic predatory behavior involves the recently discovered "phantom dinoflagellate-cell from hell," *Pfiesteria piscicida*, and related dinoflagellates that make up the toxic *Pfiesteria* complex (TPC) (figure 7.10).

When fish kept in brackish-water laboratory aquaria in North Carolina began to die for unknown reasons in 1988, the responsible organism was eventually identified as a new family, genus, and species of predatory, toxic dinoflagellate, named *Pfiesteria piscicida* (Burkholder and Glasgow 2001). Two species of *Pfiesteria* are currently recognized, *P. piscicida* and *P. shumwayae* (Burkholder 2002). TPC species possess two main attributes: strong attraction to live fish; and production of toxins, triggered by the presence of live fish or fresh fish tissues and excreta, that cause stress, narcosis, disease, or death in fish. The two species are thought to be native to the southeastern U.S. between New York and Mobile Bay but have been detected in estuaries in Scandinavia and New Zealand (ballast water introductions are a possible and frightening agent of dispersal, see chapter 9). Technically, TPC species are not photosynthetic; they consume algae and utilize photosynthetic products while the prey's chloroplasts continue to function for a few days. A number of similar, toxic dinoflagellates exist but lack one or more of the nasty attributes of the TPC.

Among the unique characteristics of the TPC is a life cycle of 20 or more flagellated, amoeboid, and cyst-like

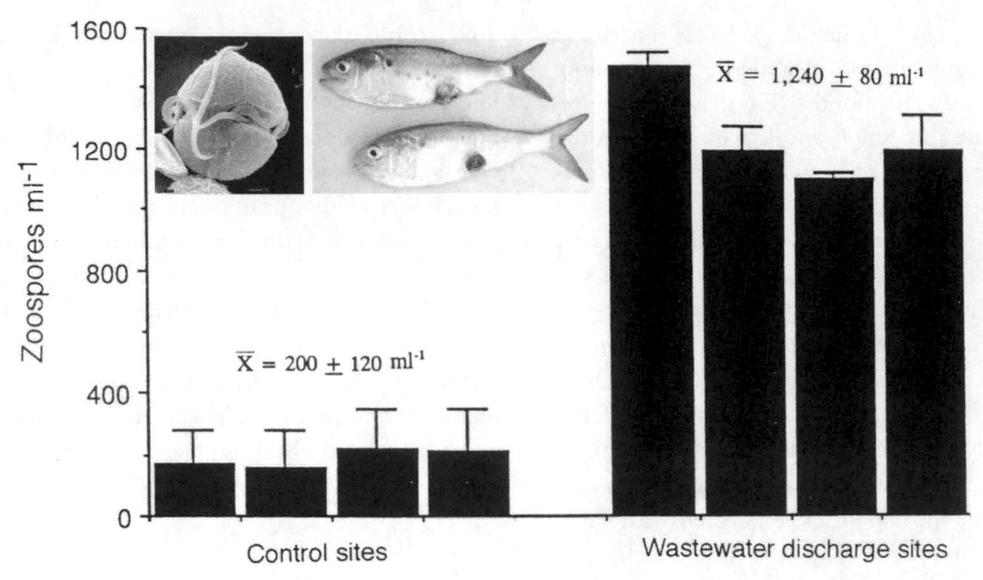

Figure 7.10. *Pfiesteria* growth relative to nutrient conditions. Water samples from regions near and distant from sewage outfalls in the New River estuary, North Carolina, show the dramatic increase in *Pfiesteria* zoospores within 100 m of wastewater discharge sites. *Pfiesteria* responds most to phosphorus inputs, but both phosphorus and nitrogen compound concentrations ≥100 ppb stimulate cell growth. From Burkholder and Glasgow (1997); used with permission. Left inset, a *Pfiesteria* zoospore; right inset, deep focal lesions on menhaden taken from a *Pfiesteria*-induced fish kill, Pamlico estuary, North Carolina. Insets courtesy of North Carolina State University Center for Applied Aquatic Ecology, www.waterquality.ncsu.edu.

stages. Transformation between stages is controlled by the kind and abundance of prey types, such as algae, bacteria, and finfish. The presence of fish causes sexual reproduction and production of toxic flagellated zoospores. The toxin induces immobility and death in fish; later-produced amoeboid cells and zoospores feed on moribund and dead fish. During fish kills, zoospore density typically reaches over 300 zoospores ml^{-1}.

Many dead and dying fish associated with *Pfiesteria* have ulcerated lesions (figure 7.10). In the laboratory, fish attacked by *Pfiesteria* exhibit epithelial and skeletal destruction, including characteristic deep lesions that bleed and often expose viscera. Gills, eyes, kidney, liver, and brain can be damaged. Lethal strains of *Pfiesteria* can narcotize finfish in minutes, causing them to become sluggish and swim erratically (Burkholder and Glasgow 1997). Death occurs in a matter of minutes or hours, depending on fish sensitivity, TPC strain lethality, and cell density (Burkholder 1998).

Pfiesteria outbreaks are often associated with nutrient pollution from sewage and agricultural activities. Zoospore production is maximal at temperatures above 26°C, salinities of 10–15 ppt, and concentrations of N and P at above 100 ppb, when both algae and finfish prey are available. A variety of estuarine fishes have experienced fish kills, but juvenile Atlantic menhaden, *Brevoortia tyrannus*, under quiet, warm, poorly flushed conditions have been the primary victims; "in addition, waters that have been degraded by nutrient pollution from human sewage, animal wastes, cropland and lawn fertilizer runoff, urban runoff, and atmospheric sources appear to be preferred habitat" of *Pfiesteria* (Burkholder 2002, p. 2439).

At least 75% of *Pfiesteria* outbreaks in North Carolina were associated with high anthropogenic nutrient concentrations (Burkholder and Glasgow 1997). In the estuaries of the southeastern U.S., large-scale hog (North Carolina) and poultry (Chesapeake Bay) facilities upriver from estuaries are prime suspects. One fish kill in the Neuse River Estuary, North Carolina, in 1995 followed shortly on the heels of a major swine effluent spill, during which approximately 1×10^8 liters of raw hog sewage were discharged from a ruptured, upstream, sewage lagoon (Burkholder and Glasgow 2001).

Confirmed *Pfiesteria*-related fish kills have to date been restricted to the two largest estuaries on the Atlantic seaboard. The Albemarle-Pamlico Sound in North Carolina—a large, shallow, poorly flushed, culturally eutrophied estuary—has experienced four to nine major

Pfiesteria-related fish kills. Massive die-offs were frequent during summer and fall in the 1980s and 1990s, involving over a billion fish, mostly menhaden. Four events have been documented in the Chesapeake Bay in Maryland. A strict set of requirements must be met to verify that *Pfiesteria* is responsible for a fish kill (see Burkholder 2002). Additional episodes at these and other locales (e.g., the Oceanview Farm incident detailed earlier) may have been *Pfiesteria* induced. Even more likely is the possibility that fish die-offs involving *Pfiesteria* result from a combination of "*Pfiesteria*, co-occurring microbial pathogens, periodic or sudden exposure to water with low dissolved oxygen, a sudden salinity shift, and other factors" (Burkholder and Glasgow 2001, p. 834).

Of consequence to humans is the unique nature of the *Pfiesteria* toxin, which can be aerosolized and inhaled and cause respiratory, visual, and neurocognitive impairment in humans. Symptoms include asthma-like respiratory problems, skin lesions that do not respond to antibiotics, severe headaches, joint and muscle pain, nausea, blurred vision, and autoimmune system dysfunction. Laboratory workers and fishers have reported symptoms when working around *Pfiesteria*-related fish kills, some of the most frightening involving Alzheimer's-like suppression of learning ability and short-term memory, including "severe cognitive impairment involving learning disabilities . . . in some subjects who were evaluated in the bottom 2% of the U.S. population in ability to learn or remember" (Burkholder 2002, p. 2445–46). Recovery to normal abilities required three to six months. Research on *Pfiesteria* is now considered hazardous and requires Biohazard Biosafety Laboratory 3 (BSL 3) facilities and precautions.

Despite such health and welfare issues, information about *Pfiesteria* and its effects was initially downplayed, underreported, and even actively suppressed by agriculture, aquaculture, tourism, seafood, and coastal development interests. These groups "viewed publicized information about *Pfiesteria* as an economic threat" (Burkholder and Glasgow 2001, p. 837). The threat is not illusory: A 1997 *Pfiesteria* episode in the Chesapeake killed an estimated 50,000 menhaden, a relatively inconsequential number economically, but the ripple effects from depressed seafood sales in Maryland alone resulted in approximately $48 million in lost revenue (Hoagland et al. 2002). Growing concern and confirmation of cause and effect prompted legislation in North Carolina and the Chesapeake region aimed at reducing nonpoint pollution from CAFOs. Funding was increased for study and detection of *Pfiesteria* outbreaks and promotion of public education about threats from toxic algal blooms (e.g., Magnien 2001).

Pfiesteria is not without scientific controversy. Some researchers contend that it has a typical haplontic dinoflagellate life cycle involving cysts and zoospores or dinospores, not multiple life cycle stages, and that amoeboid cells could represent contamination of culture media with other organisms. Others maintain that it kills via physical attack by zoospores or via secondary infection, not with a toxin (e.g., Vogelbein et al. 2002). The actual toxin(s) involved has only been partially purified (Burkholder et al. 2005); its structure suggests that it "mimics an ATP neurotransmitter, and increases intracellular calcium membrane permeability" that apparently affects the immune response of brain cells (Burkholder 2002, p. 2445). Such activity would explain lesion, inflammatory, and neural responses to *Pfiesteria* attack. Complicating the picture is the possibility that endosymbiotic bacteria may be the source of the toxin (Burkholder and Glasgow 2001). Several bacterial and fungal pathogens in estuarine waters besides *Pfiesteria* can cause lesions, perhaps secondary to *Pfiesteria* attack (e.g., Blazer et al. 1999), although fungal lesions can occur without any apparent prior stress (Kiryu et al. 2002). Also, other toxic algae, such as the raphidophyte *Chattonella* cf. *verruculosa*, may be present during presumed *Pfiesteria*-caused fish kills and could cause neurological pathologies in humans (Bourdelais et al. 2002).

The publications of the opposing sides in this discussion have fueled a public and sometimes ugly discourse in both the scientific and the popular literature (e.g., Barker 1997; J. Kaiser 2002). There can be little doubt, however, that *Pfiesteria* causes lesions and kills fish, regardless of the mechanism (Burkholder et al. 2005). Also, blooms of *Pfiesteria* and other HABs are clearly related to nutrient-rich conditions that are probably caused by agricultural or municipal inputs.

Economic Impacts

The U.S. Harmful Algal Bloom and Hypoxia Research and Control Act of 1998 estimated that economic losses resulting from HABs in the U.S. during the 1990s approached $1 billion and involved commercial fisheries, tourism and recreation, medical expenses, beach cleanups, lost labor, and monitoring. A more thorough analysis by Hoagland et al. (2002) for the period 1987–92 estimated the annual average costs at $46 million ($24–$83 million), or about half the federal decadal estimate. The largest portion, $20 million, went toward public health costs, primarily for ciguatera fish poisoning. Commercial fisheries costs were $18 million, mostly in shellfish losses. An average loss of $46 million is only 0.4% of the $11 billion annual value of commercial and recreational marine fisheries in the U.S. However, impacts at the local level, where costs of an individual event can be thousands or millions of dollars, are unquestionably significant and serious.

Reversing Toxic Blooms

The good news is that cultural eutrophication is often reversible. Toxic and noxious algae decrease when nutrient loading is reduced. Examples can be cited from small and large bodies, both fresh and salt water. Lake Washington, which received sewage inputs from metropolitan Seattle, was the site of multiple noxious cyanobacteria blooms until sewage inputs were diverted to Puget Sound in 1968, after which the blooms largely ceased. The western basin of Lake Erie was choked with the green macroalga *Cladophora* until improvements to wastewater treatment were made and bans on phosphate-containing detergents were instituted. Diversion of sewage outfalls in Tolo Harbor, Hong Kong, reversed a 13-year trend of increasing red tide frequency. The frequency of fish-killing red tides in the Seto Inland Sea of Japan increased sevenfold between 1965 and 1976, in conjunction with increased domestic and industrial waste production. Legislated reductions in nutrient loadings in 1973 led to bloom reductions, including a 50% drop within three years of implementation. Blooms leveled off in the 1990s at less than one-third of the peak bloom frequency. These and other examples (Black Sea, Kaneohe Bay in Hawaii, Mumford Cove in Connecticut, Lake Michigan) indicate that algal communities are largely regulated by nutrient inputs and that the types, densities, and impacts of blooms are determined by the nutrients that limit their growth (with exceptions and complications noted by D. M. Anderson et al. 2002).

NONCHEMICAL POLLUTANTS

Thermal Pollution

The metabolic processes of fishes are evolved responses to long-term thermal regimes characteristic of different climatic regions. Alterations in thermal regime can affect the kinetics of such processes. Human activities affect local water temperatures in the short term, and regional temperature alterations are likely to result from global climate change. The greatest threats from thermal pollution result from elevated temperature because fishes often live close to their critical thermal maxima (e.g., Magnuson and Destasio 1997), oxygen solubility is reduced at higher temperatures at the same time that metabolic requirements increase, and many pollutants are more toxic at higher temperatures.

Local temperatures can increase when heated water is discharged from power plants; when inflows of cool water are diverted to agriculture and municipal use and returned as irrigation return water; when heat budgets of water bodies are affected due to eutrophication or ice cover; and when riparian vegetation is removed and its shade-producing function degraded. Fish respond to temperature changes by altering metabolic processes, reproduction, behavior, and distribution, creating a potential for problems where unnatural temperature regimes become established. Fish kills have resulted when fish attracted to thermal plumes acclimate to the higher temperatures and then the heat source is turned off. Intermediate responses include distributional shifts to avoid areas with unfavorable temperatures. Ultimately, species' ranges can be altered via extensive dispersal or population collapse where lethal conditions cannot be avoided.

The impacts of changing temperature on fish reproduction have been well studied. Sex determination in fishes can be sensitive to thermal alteration (Devlin and Nagahama 2002). Experimental studies generally find masculinization of individuals or male-skewed sex ratios when eggs or larvae of species of minnows, gobies, silversides, loaches, rockfishes, cichlids, and flounders are reared at higher temperatures, with the effect increasing as temperature rises. Femininization or female-biased sex ratios have resulted at higher temperatures in lampreys, salmon, livebearers, sticklebacks, and sea basses. Gonadal development and germ cell viability are also temperature sensitive (Strüssmann et al. 1998).

The mechanisms underlying these effects appear to involve either altered enzyme activity or endocrine disruption (hormone synthesis or impaired steroid receptor function). Aromatase is an ovarian enzyme that converts testosterone to estradiol, a process vital to oocyte growth. In Nile tilapia, *Oreochromis niloticus*, and Japanese flounder, *Pleuronectes olivaceus*, elevated temperatures resulted in masculinization associated with reduced aromatase activity (Devlin and Nagahama 2002). Estradiol production in carp was maximal at 24°C and fell 95% with just a 5°C difference in temperature (Manning and Kime 1984). Testosterone production, which is essential for spermatogenesis, was reduced at elevated temperatures in male rainbow trout (Manning and Kime 1984).

The latitudinal and altitudinal distribution of many fish species is determined by water temperature. Elevated temperatures often prevent cold-water species from occurring at lower latitudes and elevations. Land-clearing activities that reduce stream shading result in elevated temperatures, degrading or eliminating cool-water fish habitat. The temperature dependence of some species squeezes them into seasonally reduced habitat space (Power et al. 1999). In southern portions of their range, striped bass are dependent on cool-water springs and seeps when ambient river temperatures exceed 25°C. Large aggregations of stripers crowd into places where cool groundwater emerges, growing more emaciated as the summer progresses because they are unwilling to leave the cool water (Coutant 1985). Water withdrawals that reduce spring flows and reduce their influence on river temperature are lethal to such populations (Power et al. 1999).

Climate Disruption

Since 1750, atmospheric concentrations of greenhouse gases—mostly carbon dioxide, methane, and nitrous oxide—have increased 17%–151% due to fossil fuel and wood burning, deforestation, cattle grazing, rice growing, and industrial pollution (IPCC 2001). The 20th century was the warmest in the past 1,000 years, with average temperatures increasing about 0.6°C. The decade of the 1990s was the warmest. Arctic sea ice thickness declined 40% and sea levels rose 0.1–0.2 m because of ice melting and thermal expansion of the oceans. Drought frequency and intensity in Africa and Asia increased. Temperatures are predicted to continue to increase perhaps another 1.4°–5.8°C over the next century, depending on which climate model is applied. Sea level will rise 0.09–0.9 m, and warming will be most evident in northern North America and northern and central Asia (IPCC 2001).

The postulated consequences for fishes of predicted warming in some regions and cooling elsewhere, of drier or wetter conditions, and of sea level rise are potentially dramatic. Shifts in distribution of commercial and noncommercial marine species have been observed in the North Atlantic, where bottom temperatures increased 1°C between 1977 and 2001 (Perry et al. 2005). Among 36 species assessed, two-thirds moved northward or deeper toward cooler waters over that period. Such large-scale changes can have multiple serious impacts on community structure, ecosystem function, and recovery of depleted fisheries.

Most other consequences of climate disruption remain hypothetical albeit plausible (see McGinn 2002). Sea level rise will flood coastal marshes, affecting many species that feed, breed, or grow there. Coral reefs, already stressed by periods of slight temperature elevation, will be devastated. Altered oceanic currents could affect the distribution and production of pelagic species that make up 70% of the world's fisheries. Timing of reproduction, particularly in migratory fish, would undoubtedly be disrupted. Migrations of anadromous salmonids are timed to take advantage of increased flows and cold water temperatures associated with snowmelt. Genetically determined migration times would be decoupled from altered melt cycles. Drought will likely escalate human impacts on stream fishes by reducing stream flow, elevating temperatures, and increasing pollutant concentrations. Drought would also affect water-stressed areas such as deserts and their already imperiled fish species.

Significant warming would aridify areas that now have intermittent rainfall and lead to contraction of the habitat space available for many species. Increased evaporation or decreased rainfall would decrease river flows and lake levels, causing wetlands to disappear and water tables to decline. The volume of cool water in many lakes would shrink, especially in summer. This would affect fishes such as striped bass and lake trout that use deep, cool, oxygenated water as a summer refuge. Such fish will be squeezed into anoxic areas by the decrease in cool, oxygenated water. Cool-water species whose ranges extend into warmer regions, such as brook trout, would be excluded from lower portions of streams during the summer. A few degrees of warming could be catastrophic for fishes that live near their critical thermal maximum, because groundwater temperature is strongly dependent on air temperature (Power et al. 1999). Many stream fishes in the southwestern U.S. find temperatures above 38°–40°C lethal. When temperatures in southern rivers exceed these limits, heat-related deaths occur, as they do with salmonids on the West Coast at even lower temperatures (NRC 2004b). A

3°C temperature rise would potentially exterminate 20 species of fishes endemic to the Southwest (Matthews and Zimmerman 1990). Warming would contract the geographic ranges of arctic species, pushing the southern edge of their ranges northward (e.g., IPCC 2001).

Not all species would suffer from general climate warming. Some warm-water species would benefit from an increase in available habitat space at northerly latitudes, especially where the ice-free period was lengthened (Magnuson et al. 1990; Magnuson 2002). Some cool-water species would gain access to higher altitudes and latitudes that are currently too cold to inhabit. It has been estimated that 27 species currently confined to the lower Laurentian Great Lakes would gain access to lakes Huron and Superior, but these shifts would dramatically alter assemblage relationships, with unknown consequences (Mandrak 1989). Smallmouth bass, for example, would expand their range northward into lakes where overwinter mortality among young of the year currently limits colonization (D. A. Jackson and Mandrak 2002). If elevated temperatures led to longer growing seasons, some increases in productivity might be realized.

But any "gains" would be offset by an overall loss of genetic and species diversity, especially because climate appears to be changing too quickly for genetic change to keep pace. Cold-water species will probably be both replaced and displaced by warm-water species, especially invasive generalists, accelerating the process of faunal homogenization. By the year 2100, the northward expansion of predatory smallmouth bass would likely extirpate 25,000 populations of at least four cyprinid prey species in Ontario alone (D. A. Jackson and Mandrak 2002). New species will not have time to evolve to take the place of those that cannot adapt (IPCC 2001). A likely reduction in biodiversity is a serious potential negative impact of climate warming.

Acoustic Pollution

Much recent attention has been paid to the impacts of anthropogenic sounds, especially ship noise, sonic pulses used in mapping and mining, and military applications of sonic technologies on marine mammals (NRC 2003). Cetaceans and pinnipeds depend on hearing and sound for navigation and communication, but mammals are not the only aquatic organisms affected by noise. Evidence indicates that fishes, too, are harmed, that "short- or long-term exposure to loud sounds may alter behavior, and also result in temporary or permanent loss of hearing" (Popper 2003, p. 24).

Fishes communicate with sounds and listen passively to biotic and abiotic sounds to gain information of biological relevance (Popper and Fay 1999; Popper et al. 2003). Reception occurs in the ear and via lateral line receptors (Helfman et al. 1997). Fishes are most sensitive to sounds in the 100 to 400 Hz range, although so-called hearing specialists such as ostariophysans (minnows, catfishes, loaches, suckers, characins) detect sounds up to 3 kHz, and some clupeoids can detect dolphin-emitted sounds at frequencies up to 200 kHz. Large diesel engines and small outboard motors produce loud sounds at frequencies below 1 kHz, well within the hearing range of most fish.

Fish can be affected by noise that masks or interferes with natural sounds or causes short- or long-term hearing loss. Impairment due to masking remains largely a subject of speculation, although evidence for such interference is growing. Fishes in the croaker and drum family Sciaenidae produce courtship sounds by vibrating their gas bladders. In eastern Florida, silver perch, *Bairdiella chrysoura*, and spotted sea trout, *Cynoscion nebulosus*, ceased calling when approached by boats and did not resume calling until the boat motor had been shut off for 30–60 secs. Sea trout stopped calling for up to 30 min when trains passed by. Oyster toadfish, *Opsanus tau*, similarly ceased calling when cars passed over a nearby bridge. In all cases call frequency was masked by the frequency and intensity of the human sound source (R. G. Gilmore, pers. comm.). It is not difficult to imagine situations where such traffic could be close by and relatively incessant. Another intriguing possibility concerning sound masking is the observation that larvae of at least six families of coral reef fishes are attracted to traps that broadcast nocturnal reef sounds, chiefly snapping shrimp and fish calls. Masking of such sounds by boat noise or drilling could affect recruitment by eliminating the directional cue that planktonic larvae use to find reefs (S. D. Simpson et al. 2005).

Hearing loss in fish due to sound pollution is well documented. When fathead minnows were subjected to the sound of a 55-horsepower outboard motor idling in neutral for 2 h, they suffered significant hearing loss at their most sensitive frequencies (Scholik and Yan 2002; figure 7.11). The frequencies and intensity of the sound were well within the range of what fish in the wild might experience if they lived near a small boat marina or along a shoreline

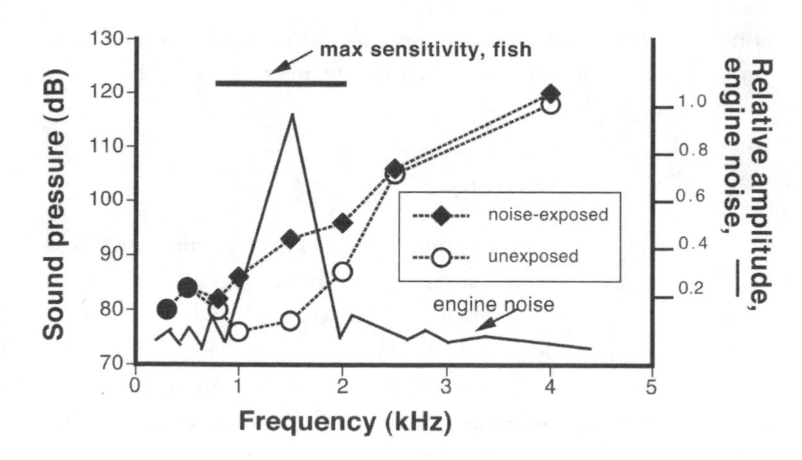

Figure 7.11. Fish suffer hearing loss when exposed to outboard motor noise, as shown by the decreased hearing sensitivity of fish exposed to outboard motor noise. Fathead minnows exposed to such noise lost hearing at 1 and 2 kHz, frequencies to which they are most sensitive (indicated by solid bar at top). Redrawn from Scholik and Yan (2002).

frequented by powerboats and personal watercraft. The noise levels tested were, in fact, less intense than outboard motor sounds recorded in the field. In other studies using 24 hr of white noise, hearing did not recover after 14 days (Scholik and Yan 2001). Irreversible anatomical damage to the sensory epithelium of the inner ear of cod, cichlids, and porgies occurred after exposure to loud pure tones (>180 dB) and pulses from an air gun used in seismic exploration (see Popper 2003).

These studies demonstrate that common intense anthropogenic sounds experienced in nature impair hearing and damage the sensory hair cells that make up the ears and lateral lines of fishes. "Even if the sounds do not kill the fish directly, permanent (or even temporary) loss of hearing will clearly affect the chances of survival of exposed fish" (Popper 2003, p. 29). The topic of acoustic pollution and its effects on fishes deserves greater attention.

INDIVIDUAL AND ECOSYSTEM STRESS

Stress Due to Impaired Water Quality

Stress, broadly explained, results from exposure of an organism to any of a suite of environmental factors that upset normal metabolic homeostasis (e.g., Adams 1990, 2002). Although we are most interested in effects of pollutant-induced stress at the population level, those effects are usually expressed long after exposure to stressors. They are also laborious to measure. More immediate and easily assessed indicators are usually sought, even though this compromises our ability to link initial and later responses, and to understand the interactions among environmental causes, internal responses, and population effects.

Three levels of stress response are recognized in fishes (e.g., Schreck 1990; Mommsen et al. 1999; B. A. Barton 2002). *Primary* responses include increased endocrine production, chiefly of catecholamines (epinephrine, norepinephrine) and corticosteroids (cortisol), which are fairly easy to measure in the bloodstream with standardized assays (e.g., radioimmunoassays, RIA; enzyme-linked immunosorbent assays, ELISA; Morgan and Iwama 1997). Primary responses lead to *secondary* responses, which include changes in metabolism (measured as plasma levels of glucose and lactate, muscle levels of glycogen and ADP:ATP); hydromineral balance (plasma ions, electrolytes, and proteins); blood components (hematocrit, blood cell characters); and cardiovascular, respiratory, and immune function. *Tertiary* responses indicate fitness-associated alterations in growth (absolute or as physical condition indices), reproduction, disease resistance, and behavior.

Among primary responses, the most commonly measured reaction involves the hypothalamic-hypophyseal-interrenal (HPI) axis that produces elevated levels of plasma cortisol. Cortisol is a corticosteroid hormone released by interrenal (kidney) tissues in response to adrenocorticotropic hormone (ACTH) secreted from the adenohypophysis of the brain (see Helfman et al. 1997). Cortisol levels rise dramatically and quickly, increasing by 10- to 100-fold in a matter of minutes, in response to stressful water quality conditions. Stressors include chemical pollutants (including EDCs), temperature and ionic shock, oxygen depletion, pH change, and sediment inputs. Stress can also be induced via inadequate nutrition, capture and handling (analogous to predatory attack), and crowding and negative social interactions (Sumpter 1997). Cortisol levels often remain elevated for a week but can stay high for weeks or months, as in salmon exposed to sublethal concentrations of heavy metals (Schreck 2000). Cortisol production causes a cascade of metabolic, physiological, and behavioral changes in the stressed individual (Mommsen et al. 1999). It is adaptive in that cortisol affects the metabolism of carbohydrates, proteins, and lipids and fuels elevated metabolic rates necessary to combat various stressors, as in classic "fight or flight" situations (e.g., Sumpter 1997).

Exposure to general and specific stressors may elevate concentrations and occurrences of additional substances or cell types in the blood or tissues. Exposure to high levels of certain heavy metals, especially cadmium, mercury, zinc, and copper, leads to the production of metallothioneins, which are proteins and polypeptides that control the concentrations of these elements and neutralize their harmful effects (Kägi and Schäffer 1988). Macrophages are white blood cells that phagocytize foreign cells and cellular debris. They aggregate in the spleen, liver, and head kidney, where they store cellular debris and antigens, thus keeping foreign materials isolated from other tissues (A. E. Ellis 1989). Increased numbers of macrophages indicate exposure to environmental contaminants and stress (Helfman et al. 1997).

Behavioral Indicators of Stress

Laboratory-assay methods for measuring primary and secondary responses are invaluable but have drawbacks of cost and training (Morgan and Iwama 1997). Measuring tertiary responses, especially behavior, is considerably cheaper and deals with responses that are often as immediate as primary responses, are sustained for long enough to allow accurate measurement, and can be especially informative because of the direct connection between behavior and fitness (Schreck et al. 1997; Weis et al. 1999).

Behavior types that are altered in response to environmental stressors include avoidance, sensory behavior, activity levels and cycles, predator-prey interactions, foraging, shelter seeking, courtship and mating, aggression and territoriality, and learning and conditioning (figure 7.12; see

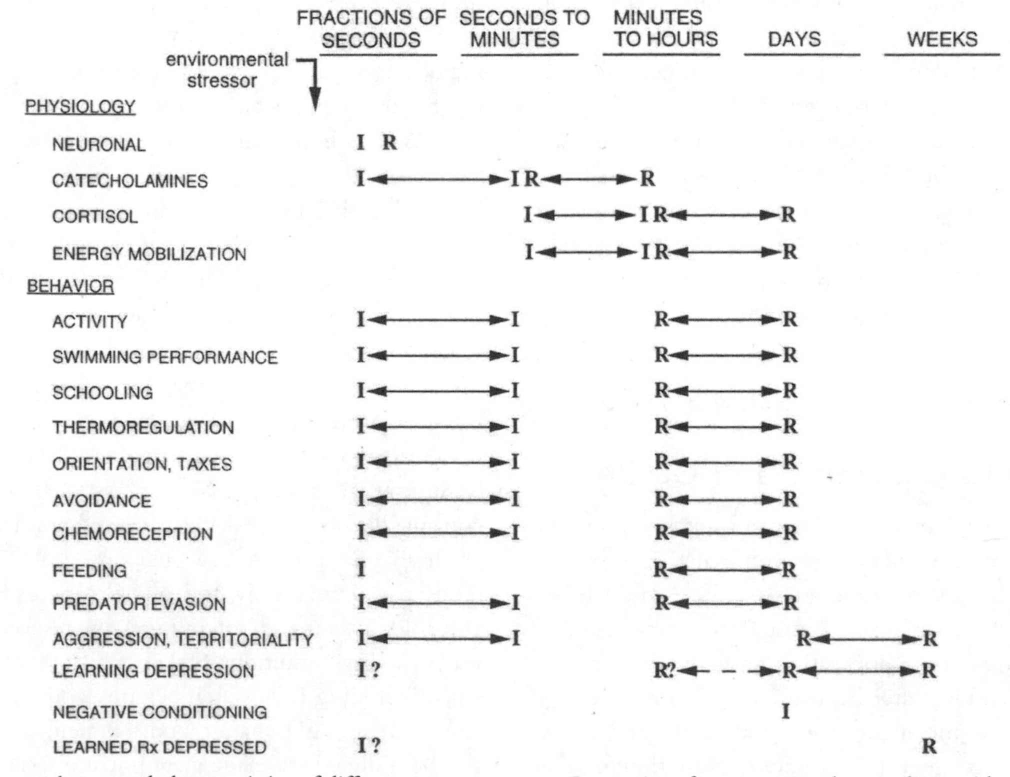

Figure 7.12. Extent and temporal characteristics of different stress responses. Categories of response are shown, along with time to initiation after exposure to a stressor (I) and time course for responses to return to pre-stress levels (R). Arrows indicate temporal variation in response; dashed arrow and question marks indicate a degree of uncertainty; Rx = reaction or response. From Schreck et al. (1997); used with permission.

Newman and Unger 2002). Chemical pollutants that elicit measurable behavioral stress responses include organochlorines, heavy metals, asbestos, petroleum products, ammonia, chlorine, and acids. Responses have been demonstrated in salmonids, livebearers, killifishes, and centrarchid sunfishes (Schreck et al. 1997), a short list that is more a reflection of limited testing than of the actual universe of likely stressors, responses, and responding taxa.

Weis et al. (1999) looked at feeding ability and predator avoidance of various life history stages of mummichogs, *Fundulus heteroclitus*, living in a New Jersey estuary contaminated with organic chemicals, mercury, and lead. They also exposed laboratory fish to contaminated estuarine water or sediments. Foraging (strike number and accuracy, success rate) and antipredation responses (activity, swimming ability, predator avoidance behavior) were impaired, which resulted in reduced growth and longevity and reduced population density. Altered behavior in mummichogs was associated with increased density and altered size of prey in the estuary, suggesting that behavioral impacts eventually led to ecosystem-level modifications.

Despite increasing emphasis on nonlethal effects of pollutants such as EDCs, research on behavioral impacts, especially on reproduction, remains relatively uncommon. J. C. Jones and Reynolds (1997) reviewed the relevant literature and found that only 19 studies—0.1% of 20,000 published studies on fishes—assessed impacts of pollution on reproductive behavior. Eleven of the 19 found effects due to low pH, herbicides, and thermal pollution. Alterations occurred in courtship (changes in frequency of displays, increased duration of courtship, behavioral masculinization of females) and parental care (altered nest building, decreased defense of young, altered division of labor). Surprisingly, no studies measured effects on actual reproductive success or extrapolated to population-level effects, two ultimate measures of impact.

Chronic Stress

If a fish is subjected to stressful conditions for extended periods, short-term physiological reactions can have long-term negative consequences. Cortisol affects osmoregulation, growth, and reproduction, all of which can be impaired if levels remain elevated. Low-level chronic stress can arrest growth and development and impair ability to mate, avoid predators, and resist disease (Schreck et al. 1997; Sumpter 1997; Schreck 2000). Chronic stress can also suppress the immune system, causing a decrease in number of white blood cells and other cells that attack pathogens, thus decreasing ability to combat infections, diseases, and parasites. Red drum, *Sciaenops ocellatus*, exposed to a variety of stressful conditions (handling, transport, crowding, disease) displayed elevated cortisol levels and suppressed immune function (P. Thomas 1990).

Knowledge Gaps

Stress responses in fishes have been fairly well studied, in no small part because of real-world relevance to aquaculture operations. However, much remains to be learned. As usual, we know little about physiological or behavioral stress reactions relative to population and ecosystem phenomena. Of particular relevance are additive impacts involving multiple and sequential stressors characteristic of the situations encountered by wild populations of fishes (Schreck 2000). Multiple and sequential stressors are most likely to affect species with complex life histories or extensive migrations, which describes vulnerable taxonomic groups such as salmonids and sturgeons.

Ecosystem-Level Effects

Individual impacts summed become population problems, significant population consequences affect interactions between species, and changes in species interactions can alter ecosystem processes such as cycling of elements and nutrients. Fish may link parts of ecosystems or may benefit from human actions directed at preserving ecosystem functions, both in the context of impaired water quality. "The full economic impacts of [degraded water quality] have often been underestimated because the resulting loss of ecosystem services has been overlooked" (Postel and Carpenter 1997, p. 210).

Wild Salmon Transport Contaminants

Considerable attention has been paid to the role of migratory anadromous fishes in the nutrient dynamics of ecosystems far removed from the ocean. Much of the literature focuses on spawning salmon and fish-based food webs, how decaying salmon carcasses feed not only the next generation of salmon but also many terrestrial animals, which in turn stimulate growth of riparian trees (e.g., Wipfli et al. 2003; see chapter 11).

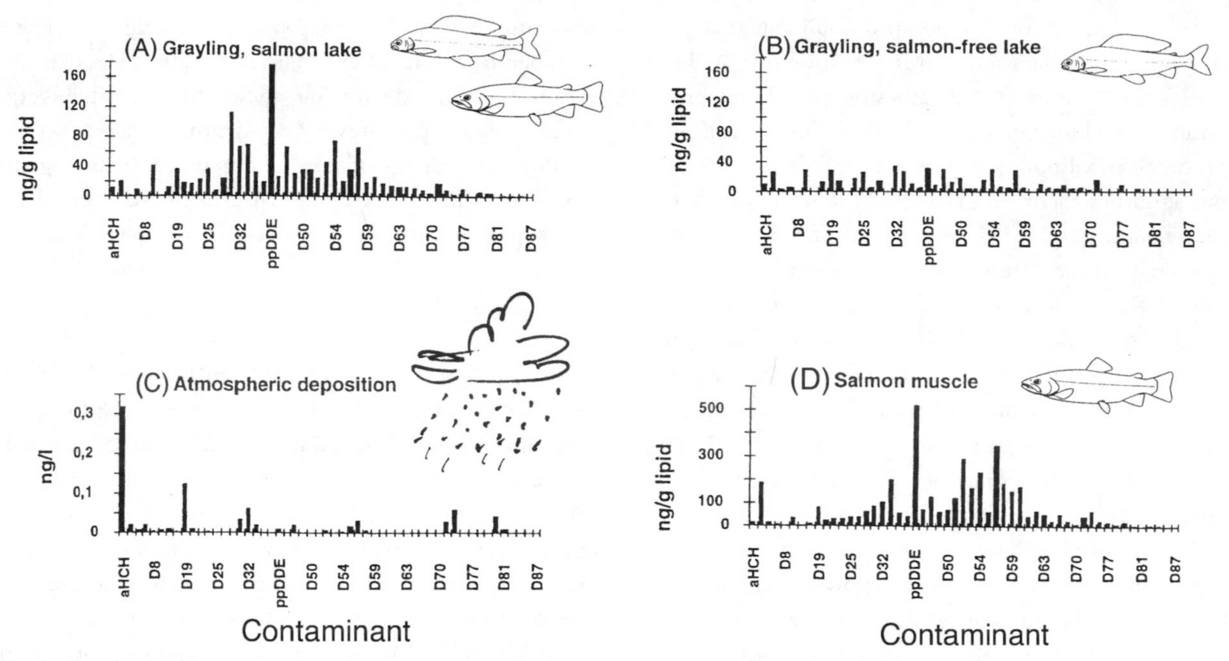

Figure 7.13. Contaminant concentration in Alaskan lakes from different sources. Salmon can be a significant vector of organic pollutants in distant, otherwise uncontaminated lakes. Grayling—a nonmigratory, lake-dwelling species—consume fish eggs and invertebrates that feed on fish carcasses. (A) Grayling living in lakes into which sea-run sockeye salmon migrate have higher concentrations of PCBs and DDT than (B) grayling in salmon-free lakes that receive only (C) atmospherically derived contaminants. (D) The types and concentrations of pollutants in salmon-lake grayling match closely those found in the migrating salmon. Pollutants are shown on the x-axis. From Ewald et al. (1998); used with permission.

Such complex interconnectedness also carries unanticipated liabilities. Inland arctic habitats are far removed from industrial and agricultural pollution and are justifiably viewed as relatively uncontaminated. Organic pollutants, if they occur, come from atmospheric deposition (rain, snow, drift) that originates from distant sources. It is now known that migrating salmon are an even more important source of contaminants. A study of contamination in nonmigratory arctic grayling, *Thymallus arcticus*, revealed that PCB and DDT levels differed greatly in grayling taken from lakes in the Copper River watershed of Alaska due to the presence of sea-run sockeye salmon (Ewald et al. 1998). Grayling in a lake where sockeye migrated and spawned had concentrations of persistent organic pollutants two or more times higher than grayling in a salmon-free lake that received pollutants only from atmospheric deposition (1,024 ppm vs. 548 ppm for PCBs; 239 ppm vs. 41 ppm for DDT; figure 7.13). The pollutant composition in the salmon-spawning lake matched that found in the salmon themselves. The salmon accumulated the pollutants during their three to five years of feeding and growth in the ocean, concentrated them in lipids, and then "biotransported" them to pristine lakes 400 km inland. Grayling fed on salmon eggs as well as on invertebrates that consumed salmon carcasses.

Ironically, transport by salmon concentrates pollutants and makes them more biologically available than atmospherically deposited organic contaminants. Storage of PCBs and DDT in muscle and gonad lipids of migrating salmon protects these organic pollutants from several oxidative processes that break down atmospherically deposited organics. During migration, lipids and the contaminants they contained were mobilized and transferred to the maturing ovaries, ovarian concentrations of PCBs and DDT more than doubling as fish moved upstream. That Copper River sockeye are a potential source of persistent organic contaminants—with muscle PCB and DDT concentrations approaching 2,800 ppm and 1,400 ppm respectively—is in itself ironic. Copper River sockeye are reputed "throughout the world" for their quality as a smoked product. The spring and early summer runs (May-July) have achieved a status akin to the first bottles of fine Beaujolais Nouveau among wine aficionados. Cold-smoked Copper River sockeye, when available, was selling on the Web for $34–$42/pound in spring 2003 (see Krümmel et al. 2003).

Ecosystem Services

Among the important ecosystem services provided by lakes, rivers, streams, wetlands, and riparian buffers is water purification. Intact ecosystems are excellent processors of nutrients, pollutants, and sediments; human engineering solutions to water purification are, by comparison, exceedingly expensive. Even after costs of habitat acquisition, restoration, and protection are factored in, purchasing and maintaining intact watersheds can be many times cheaper than building water treatment facilities. New York City's drinking water comes from the Catskill Mountains over 150 km away, where deforestation and agriculture were causing water supplies to deteriorate. The city opted for natural water purification by buying up forested land and supporting less harmful agricultural practices, rather than building treatment plants. It saved an estimated $5–$8 billion in initial costs and another $300 million in annual operating costs compared with the technological solution (NRC 2000b). Inadvertently, by addressing the problem at its source through ecosystem restoration and protection, New York also improved water quality for all the organisms that live between the Catskills and the city. Had the city opted for the water treatment solution, aquatic habitats above the treatment facility—which would likely have been placed as close to urban areas as possible—would have continued to degrade. Fishes were therefore direct beneficiaries of a decision made primarily on economic grounds. New York City's positive experience has encouraged other agencies and municipalities to promote riparian and wetland protection, restoration, and even construction as solutions to pollution problems.

POLLUTED FISH IN THE FOOD CHAIN

Fish Consumption Advisories

People are concerned about fishes that accumulate pollutants because people eat fish. However, monitoring of contaminated fishes tells us just how pervasive our pollutants have become, which can inform us about fish health, too. In the U.S., EPA maintains a National Listing of Fish and Wildlife Advisories (NLFWA), which tracks fish consumption advisories issued at federal and state levels (USEPA 2003, updated at www.epa.gov/ost/fish/advisories). Advisories are issued primarily by state agencies, whose criteria and thresholds for taking action frequently differ. Advisories provide no-consumption and restricted-consumption guidelines with respect to meal size and meal frequency for the general public, pregnant and nursing mothers, and children. During the 1990s, the number of advisories increased 125%, to a total of 2,800, identifying 33% of the nation's total lake acreage and 15% of the nation's total river miles as contaminated (figure 7.14; this excluded the Great Lakes and connecting waters, although all five lakes have advisories). In addition, 71% of the U.S. coastline was under advisory (92% of the Atlantic coast, 100% of the Gulf coast). The 125% increase reflected higher water quality standards, improved monitoring and detection capabilities, and deteriorating conditions.

The chemical contaminants that have made consumption advisories necessary are primarily mercury, polychlorinated biphenyls (PCBs), chlordane, dioxins, and dichlorodiphenyltrichloroethane (DDT and its by-products). However, 34 other chemicals were also responsible (e.g., arsenic, cadmium, chromium, lead, kepone, dieldrin, lindane, mirex, toxaphene, creosote, pentachlorophenol, vinyl chloride, polycyclic aromatic hydrocarbons). After 1993, EPA also issued "safe eating guidelines." These are "clean lists" that states maintain to advise the public about water bodies and species known to contain low levels of contaminants. That number also grew, from 20 in 1993 to 164 in 2002 , encompassing less than 0.01% of the nation's river miles and about 10% of lake acreage. The increase reflected greater participation in the program as much as decreased contamination.

Although consumption advisories are most common in freshwaters, marine systems under advisories are no less contaminated. For example, common sport fishes in San Francisco Bay contained contaminant levels that exceeded EPA safe consumption guidelines (Fairey et al. 1997). The contaminants included PCBs, pesticides (dieldrin, chlordane, DDT), trace elements (mercury, arsenic), and dioxin/furans. Muscle concentrations of PCBs exceeded "screening levels" by as much as 200 times, and mercury levels were up to 9 times greater than established standards. Advisories limited consumption of fishes from San Francisco Bay, especially larger striped bass and sharks, to two meals per month, with greater restrictions for young children and pregnant women.

Fish consumption advisories in Canada are adminis-

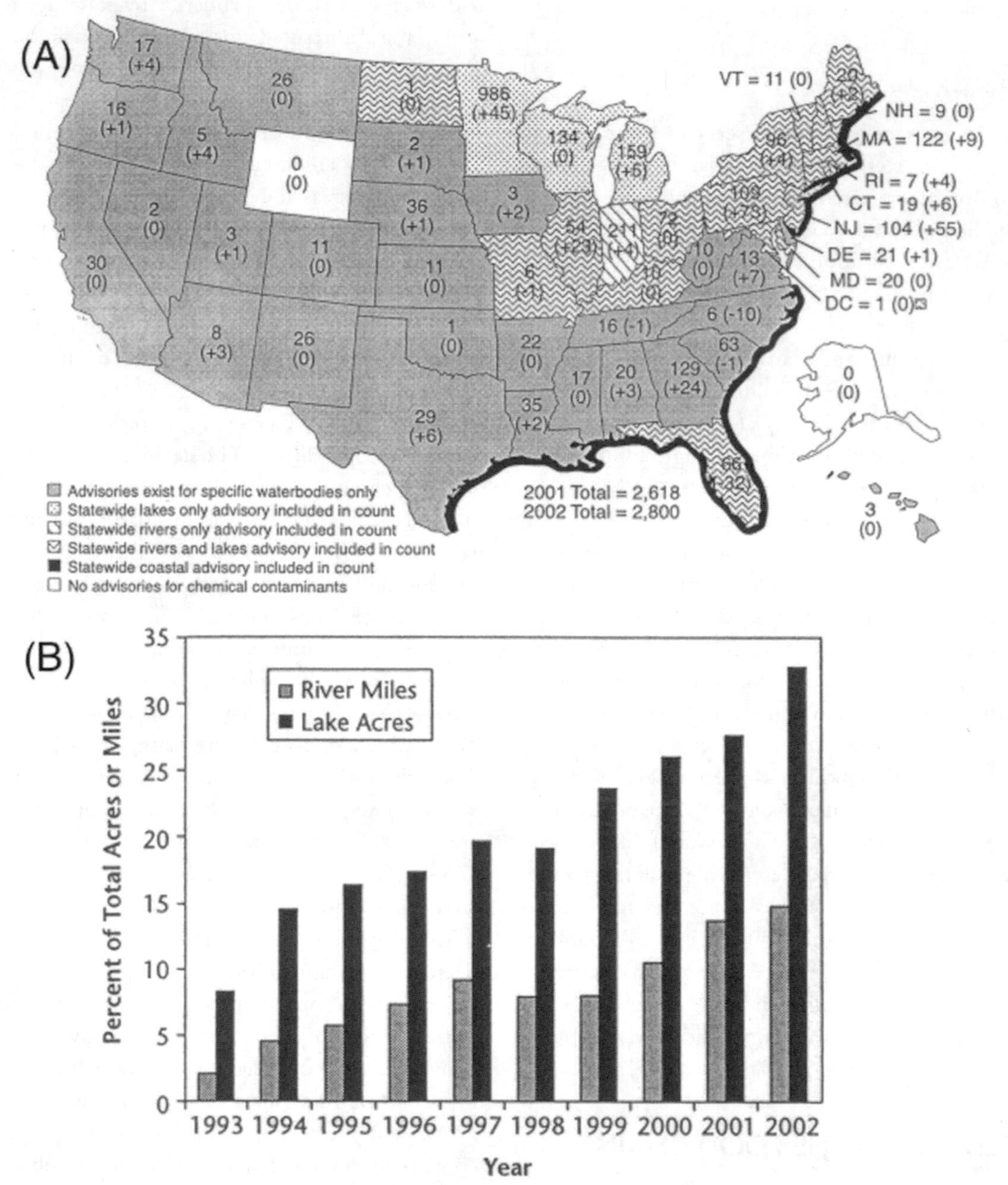

Figure 7.14. (A) The total number of fish consumption advisories by state in 2002, the change in numbers since 2001 shown in parentheses. Only Alaska and Wyoming lacked advisories. (B) Trends in number of advisories issued, 1993–2002. The percentage of total lake acreage under advisory increased from around 8% in 1993 to 33% in 2002, while river miles increased from 2% to 15%. The Great Lakes are not included; their inclusion would increase numbers substantially. From USEPA (2003).

tered more by individual provinces than by federal agencies. The focus of advisories is on mercury contamination in large predators such as walleye and northern pike, although dioxin and furan advisories also exist. Provinces on the Great Lakes share all the concerns identified by the EPA for those waters (www.ec.gc.ca/mercury/en/index.cfm).

Although fishes have no escape from this anthropogenically degraded landscape, advisories carry a slight benefit for them. Reduced fishing pressure—or at least increased catch and release—may aid some species because of their known high body burdens of toxins (e.g., striped bass and American eels in the Hudson River). "Although the consequences of pollution are all unintended, surely protective inedibility of game fish would be among the least expected" (J. Gorman 2003, p. 3).

Wild-Caught versus Farm-Raised Fish

Eating fish is supposed to be good for you. However, popular species may contain contaminants, and cultured fish may be more contaminated than wild-caught fish. In a preliminary study, farmed salmon had 10 times more PCBs, 15 times more polybrominated diphenylethers (PBDEs), and 3.5 times more organochlorine pesticides than wild fish (Easton et al. 2002). The apparent source of contamination was the commercial feeds, which are known to contain elevated contaminant levels. Easton et al. (p. 1054) concluded that "the use of fish oil and fish meal in feed serves not only to produce high energy feeds with good growth performance qualities but also to act as a pipeline for contaminants into the human food chain." At the PCB concentrations found, consumption of more than one to three portions of those fish per week would exceed tolerable daily intake (TDI) recommendations established by WHO. A similar study of Scottish farmed salmon found higher levels of dioxins, furans, and PCBs in farmed fish; a single meal exceeded TDI levels (Jacobs et al. 2002).

These studies were based on relatively small sample sizes. A later study by Hites et al. (2004), analyzing over two tons of farmed and wild salmon from around the world, showed that contaminant concentrations were significantly higher in farmed than in wild salmon. The authors concluded that "consumption of farmed Atlantic salmon may pose health risks that detract from the beneficial effects of fish consumption" (p. 226; see chapter 14 for other reasons to avoid farmed salmon).

Formulated feeds used in aquaculture often contain high concentrations of organochlorine compounds that are biomagnified and accumulated by cultured fish (e.g., Serrano et al. 2003). The exact origins and pathways of contamination remain open to speculation. Commercial feed given to salmon contains significant amounts of fish meal, much of it from processed herring, which is high in lipids and hence a potential accumulator of lipid-soluble organochlorine compounds (Jacobs et al. 2002). Farmed salmon tend to be higher in fat content than wild fish, perhaps because of the high lipid content of their diet, as well as their relatively sedentary living conditions. Jacobs et al. recommended changing from feed formulations based on marine fish oils to formulations using more vegetable oils. Vegetable oil–based diets would be economically competitive, meet the salmon's requirement for a high-energy feed, reduce organochlorine contamination, and thus produce "a nutritious food using environmentally friendly methods" (p. 190), thereby addressing consumer concerns about both human and environmental health.

WATER QUALITY AND FISH SPECIES IN PERIL

Water quality generally ranks below habitat destruction, overexploitation, and introduced species as a contributor to declines among fishes. However, we may be underestimating its impacts because pollution often acts indirectly, and impacts can be delayed. Aside from their role in relatively spectacular and widely publicized die-offs of listed species—such as those of the Endangered sucker species in Upper Klamath Lake in the mid-1990s and a Threatened run of coho salmon in the Klamath River in summer 2002 (NRC 2004b)—many pollutants occur at sublethal concentrations. Their impacts may take years to be expressed and are therefore often suspected rather than demonstrated. Such suspicion is usually based on valid science, however, and the effects are potentially severe, as in the following examples:

- In the highly polluted Columbia River, white sturgeon, *Acipenser transmontanus*, contained the highest concentrations of several metals, pesticides, EDCs, and other chemical contaminants found in fishes. Recorded concentrations were 1,400 ppb DDE, 200 ppb PCBs, 0.05 ppb chlorinated dioxins and furans, 29 ppb chlordane, 1,000 ppb chromium, and 2,700 ppb selenium (USEPA 2002). White sturgeon attain large size and great age and are the top predator in the river, which may explain their tendency to accumulate toxins. They appear to be recovering from a 90% population decline brought on by overharvesting in the late 19th century, but reproduction could be affected by chemicals such as PCBs and selenium (e.g., Moyle 2002).

- American and European eels, *Anguilla rostrata* and *A. anguilla*, have undergone precipitous population declines throughout their large geographic ranges since the early 1980s. American eels were proposed for ESA protection in 2005. The causes of the declines remain a matter of intense speculation (see Richkus and Whalen 2000). One possibility is that during the reproductive migration, adult eels metabolize their substantial lipid stores, thus mobilizing persistent pollutants accumulated during the long freshwater life stage (Robinet and Feunteun 2002). Xenobiotics are then concentrated in the gonads during gametogenesis, with all their attendant reproductive implications.
- The plight of Pacific salmon on the west coast of North America and especially in the Columbia, Snake, Fraser, and Klamath river basins is well known. Chemical pollutants, when discussed, are usually considered secondary to the effects of dams and overfishing. In otherwise comprehensive treatments of northwest salmon, NRC (1996b) and Lichatowich (1999) give scant coverage to nonacute chemical pollutants. We can now add to the list of potentially significant factors the insidious actions of EDCs, as evidenced by EDC-induced sex reversal in Columbia River Chinook, discussed earlier.
- To make matters worse for northwest salmonids, fluoride contamination from municipal water supplies and industrial wastes may pose an additional threat. Fluoride is an anesthetic as well as an enzymatic poison. Salmon not killed by relatively low concentrations of fluoride become lethargic and reluctant to ascend fish ladders. At the John Day Dam on the Columbia River during the mid 1980s, Chinook and coho salmon took more than 150 hours to traverse the dam, and 55% died. Fluoride levels at the dam were 0.3 to 0.5 ppm due to waste discharges from an upstream aluminum smelter that released on average 384 kg fluoride per day. When the smelter reduced fluoride releases, concentrations dropped to under 0.2 ppm, mortality fell to 1%–5%, and salmon negotiated the dam in 28 hours (Damkaer and Dey 1989). In addition to industrial discharges, at least 11 municipalities release artificially fluoridated water (for tooth decay prevention) into the Columbia and Snake rivers (Foulkes and Anderson 1994). Fluoride is not significantly degraded by primary sewage treatment, and municipal release can contain relatively high concentrations of 1.2 ppm, which is six times above the British Columbia allowable standard for salmon protection. Aqueous fluoride concentrations in rivers and streams of the U.S. Northwest of 0.1 to 0.5 ppm are not uncommon, and concentrations as low as 0.2 ppm have been shown to be lethal, with toxicity to salmon increasing at higher temperatures and lower pH (Foulkes and Anderson 1994).

SOLUTIONS: ASSESSMENT, REMEDIATION, REGULATION

Solutions to water quality problems involve scientific inquiry, citizen involvement, and legislative action. Discussed here are some popular methods for assessing water quality; corrective measures based on realistic recovery goals subjected to accurate monitoring; and national and international regulations designed to address, correct, and eventually prevent water quality problems.

Assessment: Alternatives to Aquatic Toxicology

The scientific discipline usually charged with assessing and solving water quality problems is aquatic toxicology (for fish-focused toxicology, start with Kennedy and MacKinlay 2000). Some common criticisms of toxicological approaches to aquatic ecosystem management are that they emphasize laboratory over field measurements; assess single compounds rather than contaminant mixtures; and focus on chemical and physical parameters to the neglect of biological factors, which reflects a chemical and physical emphasis in most regulations. However, emphases do little to promote the conservation of biodiversity. The toxicological approach also reflects shortcomings in the legal protection afforded aquatic resources. For example, the U.S. Clean Water Act focuses on point-source chemical pollutants. Diffuse, nonpoint pollution and factors such as introduced species, flow alterations, habitat degradation and loss, and thermal alterations can also degrade water quality (figure 7.15). Habitats can meet physical or chemical standards and still be incapable of sustaining function-

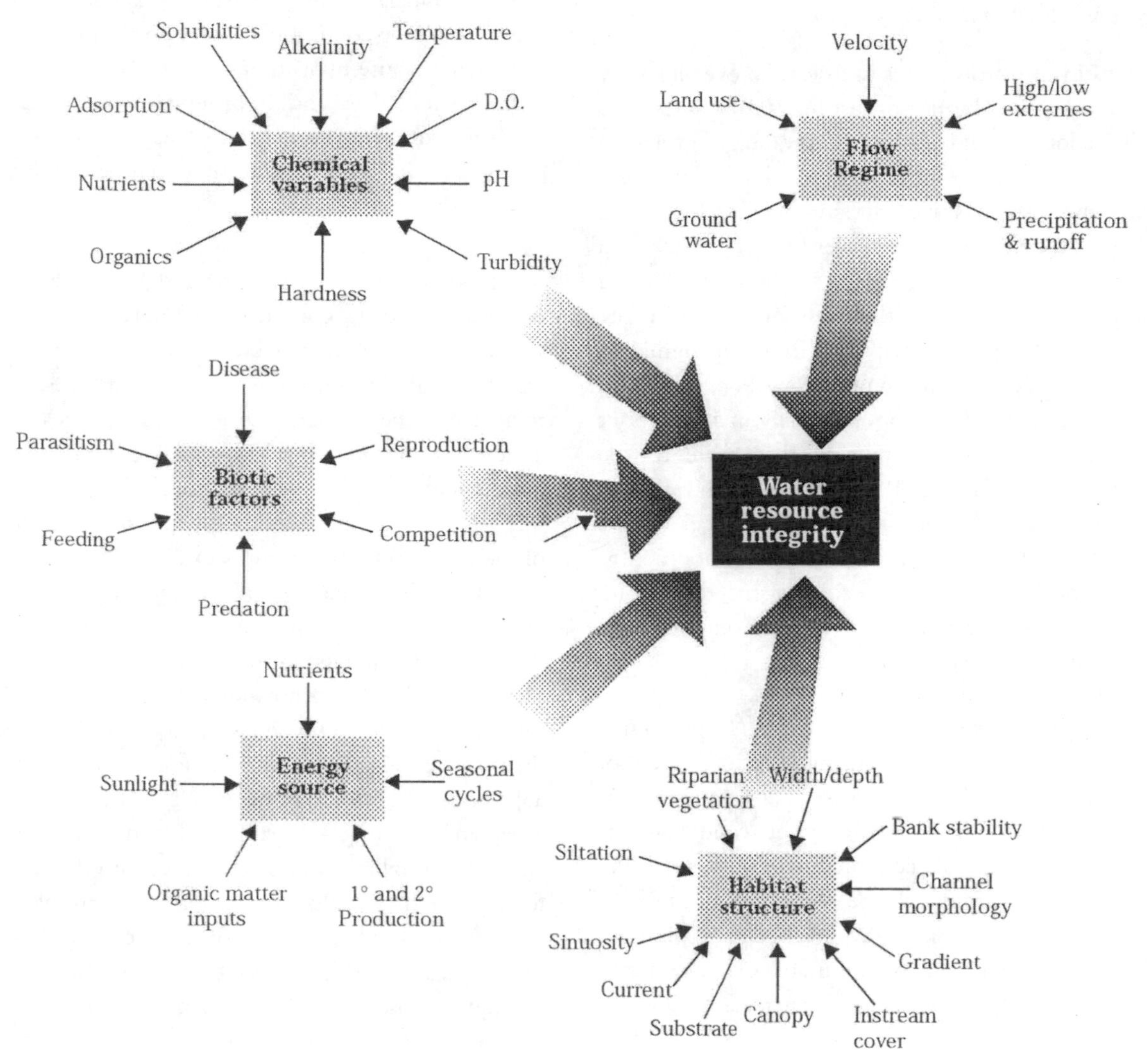

Figure 7.15. Some factors that influence aquatic ecosystem integrity. The integrity of an aquatic ecosystem—defined broadly as its ability to maintain its natural function and composition in the face of environmental change—depends on much more than chemical inputs. From NRCS (1998).

ing, natural ecosystems (Karr 1991). Any accurate measure of the health or integrity of an ecosystem must account for biological components as well as physical and chemical ones because "many environmental impacts that degrade ecosystems are simply too diverse and too complex to be detected and understood by chemical assays" (Van Dyke 2003, p. 243).

To correct for this perceived underemphasis, bioassessment and biomonitoring metrics have been developed that include both biotic and physical parameters. Others focus primarily on the biotic components themselves as integrators of physicochemical conditions. Simpler protocols, such as the Index of Well-Being, measure abundance, density, catch rate, biomass, diversity, or evenness (e.g., Barbour et al. 1999). Such measures are relatively easy to obtain but insensitive to ecological attributes and roles of species. They also suffer from the assumption that high diversity and abundance equate with high integrity, whereas many disturbed systems have higher diversity and abundance than undisturbed systems (e.g., Fausch et al. 1990; Lyons et al. 1996; M. C. Scott and Helfman 2001).

The Index of Biotic Integrity

The object of most protocols is to assess the overall ability of a stream to support biota, not just fishes. But fishes have a number of logistic and ecological advantages over taxa such as invertebrates as indicators of overall ecosystem integrity. Fishes are sensitive to water quality, long-lived and relatively sedentary, and relatively easy to catch and identify (see Moyle 1994; Simon and Lyons 1995). Among protocols that focus on fishes, the most widely used is the Index of Biotic Integrity (IBI). The IBI was designed to detect gradients of disturbance among a series of streams in a given geographic locale. It conventionally includes about a dozen measured criteria (metrics) that relate to the "compositional, structural, and functional" organization of fish assemblages. IBIs generally follow an overall format presented in Karr et al. (1986) but should be modified to account for local biogeography, ecology, stream size, and temperature regime (Karr and Chu 1999). Common metrics include the following:

- *Species richness and composition*, measured as (1) total number of species or native species; (2) number or percent of riffle-benthic species such as darters and sculpins; (3) number of species sensitive to pool conditions or instream cover such as centrarchid sunfishes in the Midwest and salmonids in the West; (4) number of species sensitive to or generally intolerant of siltation, low flow, low oxygen concentrations, or toxic chemicals; (5) proportion of individuals of particularly tolerant species, such as green sunfish, carp, goldfish, and black bullheads. Alternative metrics, depending on locale, may include a category for water column–dwelling species or percent introduced species.
- *Trophic composition*, where alterations in food type or availability reflect an impaired energy base. Metrics include (1) proportion of omnivorous individuals, because trophic generalists often increase as food types become less reliable; (2) proportion of insectivorous individuals, sometimes separated into water column insectivores or pool-benthic insectivores, whose numbers decrease in degraded streams as insect numbers decrease; (3) proportion of top carnivores or piscivorous individuals, where viable predators usually indicate a healthy, trophically diverse community.
- *Fish abundance and condition*, to account for overall density and subpopulation characteristics, including (1) number of individuals in the site, assuming larger numbers indicate healthier conditions; (2) proportion of hybrid individuals, because hybridization is often more common in disturbed habitats; and (3) proportion of individuals with disease, tumors, parasites, fin damage, or deformities, all problems that tend to increase in frequency with increasing degradation.

In an IBI, each metric is scored 1, 3, or 5, depending on its similarity to conditions at a relatively undisturbed, geographically relevant reference site. The various metrics are added to calculate a single index score for each site. The reference or benchmark site represents the least impaired conditions, and sites with lower total scores represent varying degrees of human perturbation.

IBIs were developed first for fishes in shallow streams of the north-central ("midwestern") U.S. (Karr 1981; Karr et al. 1986). They have been subsequently refined (Simon 1999) and widely applied and modified for use in other regions and habitats (Simon and Lyons 1995; R. M. Hughes et al. 1998; Karr and Chu 1999). Because of its multimetric nature and demonstrated ability to detect gradients of water quality degradation—not to mention its apparent "cookbook" simplicity—the IBI has gained wide acceptance among governmental and nongovernmental agencies. Simplicity is, however, a liability because an IBI focuses on the results of degradation without exploring causes and possible solutions for low scores (e.g., Attrill and Depledge 1997). When possible, workers also develop IBIs for aquatic insects and amphibians and then combine all IBIs with other water quality and habitat measures (e.g., Lyons et al. 1995).

Other criticisms of the IBI approach include a lack of hypothesis and statistical testing associated with the results; circularity in the designation of the endpoints of human impact; unavailability of "unimpacted" reference sites in many regions; and redundancy, autocorrelation, or contradiction among different metrics in an IBI. These and other criticisms are discussed and largely rebutted in Karr and Chu (1999).

Remediation and Recovery

Assessment is the first step toward correcting water quality problems. Ultimately, however, pollution must be stopped at its source. Once inputs cease, efforts can begin to attempt to repair ecosystems.

Restoration of polluted habitats is in many ways simpler and faster than restoring habitats that have been physically destroyed. Except where a pollutant is long-lived, as in persistent pesticides or heavy metals, removal of the source of contamination often leads to fairly rapid biotic recovery, often in less than three years (Niemi et al. 1990; Detenbeck et al. 1992). For persistent pollutants, dredging may be necessary. But dredging can be exceedingly costly, present new problems of contaminant disposal, and require years before anything resembling preimpact conditions return. PCB contamination of Hudson River sediments and fishes will require removing approximately 2.65 million m^3 of sediment, which must then be processed and dewatered, followed by offsite disposal. Dredging alone is estimated to take six years and cost $460 million (www.epa.gov/hudson).

Where fishes or their prey bioaccumulate toxins, returning to uncontaminated conditions may require a generation, decades, or longer (Yount and Niemi 1990; Lappalainen and Pesonen 2000). After the pollutant source has been eliminated, recovery depends on natural repopulation and recolonization from surrounding areas. If those processes are impaired, active stocking may be necessary.

Variation in time to recovery also reflects the definition of recovery applied to a site, an issue subject to considerable debate. Some authors require merely the first reappearance of a formerly absent species, whereas others require return of a predisturbance assemblage containing actively reproducing individuals leading to subsequent recruitment (Niemi et al. 1990). Regardless of definition, realistic goals and proven methods for measuring attainment are crucial, often involving comparison of a recovering site with conditions at one or more physically comparable, relatively undisturbed sites in the same region (Hughes et al. 1990; Ensign et al. 1997; Yan et al. 2003).

Water Protection Legislation

Aquatic resources cross state and international boundaries and receive pollutants from local and distant sources. Humanity has used lakes and streams as public sewers for too long, and the costs of degrading this common resource have traditionally been externalized (not borne by the polluter, private or commercial). As a result, ensuring high water quality requires government protection and regulation. Such actions have been taken by government agencies in most states, provinces, and countries, as well as some individual municipalities and tribes, and include guidelines and legislative mandates that establish standards for water quality parameters. As expected, standards focus on human health issues, but fishes benefit indirectly from human-oriented regulations, and many regulations actually target biotic integrity.

In the U.S., protection of aquatic habitats is legislated at the national level, with implementation by federal and state agencies. Particularly pristine or attractive streams and rivers are candidates for protection under the Wild and Scenic Rivers Act of 1968, designed to preserve such systems in their free-flowing condition, although water quality per se is not at issue. Only a few percent of the nation's waterways qualify for and have received such protection (www.nps.gov/rivers). A variety of laws, acts, and executive orders prevent additional loss of wetlands (see M. Williams 1990). The primary legislation protecting water, aquatic habitats, and aquatic biota is the Clean Water Act (CWA) of 1977 (as amended from 1972), which regulates discharges of pollutants and sets water quality standards for all contaminants in surface waters. CWA's goal is to restore and maintain the chemical, physical, and biological integrity of the nation's waters to support "the protection and propagation of fish, shellfish, and wildlife and recreation in and on the water" (www.epa.gov/watertrain/cwa/). The CWA initially focused on individual pollutants originating from point-source polluters, as detailed in the National Pollutant Discharge Elimination System (NPDES) but has progressively incorporated a more ecosystem-oriented perspective. It has also included nonpoint sources with provisions such as the Total Maximum Daily Load (TMDL) approach. TMDLs are a means of calculating the maximum amount of any pollutant (chemical, nutrient, sediment) that a water body can receive and still meet water quality standards; it is the sum of the allowable loads of a single pollutant from all contributing point and nonpoint sources. Fulfillment of CWA objectives is the responsibility of the EPA, which also handles permitting to achieve NPDES and TMDL targets. This responsibility is transferred to states, although their standards must be at least as rigorous as those of EPA.

Australia, New Zealand, and Canada lack federal clean water protection acts. Water quality issues are treated individually by provinces or states. Australia and New Zealand in 2002 finalized Guidelines for Fresh and Marine Water Quality; in Australia, these guidelines are used to formulate water quality objectives, which are interpreted, adjusted, and applied regionally by each state and territory (A. Arthington, pers. comm.). In September 2003, New Zealand was developing national drinking water standards based on the 2002 guidelines. Canada also developed guidelines (as opposed to regulations) for water quality, as detailed in the Canadian Water Quality Guidelines for the Protection of Aquatic Life (CCME 1999). Provinces then determine standards based on the guidelines, but noncompliance is rarely punished (W. Donahue, pers. comm.). Canada also has a highly credible national wetlands policy instituted by each province (D. A. Rubec 1994). Mexico passed a Ley de Aguas Nacionales (National Water Law) in 1992 that protects both surface and subsurface water (the CWA deals only with surface waters).

The European Union views water as "not a commercial product like any other but, rather, a heritage which must be protected, defended and treated as such" (EC Water Framework Directive 2000/60/EC). The EU has issued a number of directives that deal with drinking water, bathing water, nitrates and dangerous substances, and river basin management (europa.eu.int/comm/environment/water/index.html). The 1978 EC Freshwater Fish Directive (78/659/EEC), linked to the more general Water Framework Directive (WFD, 2000/60/EC), protects freshwater bodies suitable for sustaining fish populations. It sets physical and chemical water quality objectives for salmonid waters and cyprinid waters. The WFD, enacted in 2000, covers inland and coastal waters managed on a river basin basis and applies to both surface and groundwater quantity and quality. Member states had until 2003 to enact national laws enforcing provisions of the directives, with total implementation by 2015. The UK, as a member nation, is meeting WFD requirements via a variety of acts (DEFRA 2003), although additional implementing legislation will be needed to meet objectives with regard to diffuse nonpoint pollution (C. Mainstone, pers. comm.). In England and Wales, existing legislation includes the Control of Pollution (Discharges by the National Rivers Authority) Regulations and Water Act of 1989, which was replaced by the Water Industry Act 1991 and the Water Resources Act 1991. These acts include provisions for classifying surface waters that contain fish life, and they address pollution control and water resource management (DEFRA 1997). South Africa's National Water Act 36 of 1998 recognizes that water is a scarce and unevenly distributed national resource; the act has the goal of achieving "the sustainable use of water for the benefit of all users" (www.elaw.org/resources/text.asp?id=1153).

Even the best-intentioned legislative attempts to regulate and protect water resources are ineffective if they lack significant penalties for violations and resources for enforcement. Unfortunately, few sources include information about enforcement. A danger here is that the existence of a law may encourage the assumption that the problems it was created to address have been fixed. This is often false: "In many respects, society has been lulled into believing that our individual and collective interests in water resources are protected by national, state, and local laws and regulations . . . when, in fact, our habits as a society and the way we have implemented our laws have progressively compromised our fresh waters" (Karr and Chu 1999, p. 9).

CONCLUSION

The acute and chronic impacts of water pollution—the disgusting things that humans do to water—include many truly frightening known, presumed, and unknown causes and outcomes. Emerging technologies present newer concerns, because enthusiasm for novel products usually races ahead of consideration of potential environmental impacts. One example is nanotechnology, "another watershed in scientific innovation" (Owen and Depledge 2005, p. 609). Nanoparticles of carbon (buckyballs, fullerenes) under 100 nm in diameter are produced by the ton annually and incorporated into a wide variety of products (cosmetics, car bodies, tennis rackets), and are proposed for use for administering drugs. No standards for monitoring, acceptable levels of release, regulations on production and use, labeling requirements, or methods for removing them from water supplies have been presented. Yet preliminary tests indicate that nanoballs are lipid soluble, cause oxidative damage, are rapidly taken up by cells, and can even cross the blood-brain barrier of vertebrates. Largemouth bass exposed to water containing uncoated nanoparticles at concentrations of 500 ppb exhibited brain tissue damage via lipid peroxidation (Oberdörster 2004). Perhaps there is nothing to worry about, but assumptions of minimal effect have proven inaccurate in the past.

In many respects, water pollution issues leave greater room for optimism than other factors that contribute to population declines and biodiversity losses among fishes.

Perhaps because humans are directly affected by impaired water quality, and because we can easily and directly associate impacts on fishes with our own welfare, corrective measures by governmental and nongovernmental entities are common and politically popular. The relative reversibility of compromised water quality, especially compared with habitat destruction, is another cause for optimism. That the disciplines of chemistry and biochemistry contribute to discovering solutions is an added benefit. Conservation biologists and fisheries scientists have little such aid when dealing with issues of habitat destruction, overexploitation, and introduced species. It's encouraging to be able to end this discussion on a relatively positive note.

8. Alien Species I: Case Histories, Mechanisms, and Levels of Impact

What havoc the introduction of any new beast of prey must cause in a country, before the instincts of the indigenous inhabitants have become adapted to the stranger's craft or power.

—Charles Robert Darwin, 1871

Under natural conditions, population growth of a species is constrained by coevolved parasites, predators, prey, and competitors. In response to negative ecological interactions, as a normal part of life history, or by accident, animals disperse; movement of species into new areas is therefore a natural zoogeographic process. Natural dispersal is normally limited by species' mobility and by physical and biological barriers. Establishment of species in new areas is limited by climate; habitat and food availability; and predators, competitors, and parasites. When individuals enter an alien environment, they are likely to encounter new, inhospitable, and even lethal physical and biological conditions.

Nevertheless, "the human species seems beset with an irresistible urge to transport live animals and plants from one part of planet Earth and liberate them in another" (Randall 1987b, p. 490). When range extensions occur as a result of human actions, they constitute introductions (see box 8.1). Most introductions are doomed to failure for the reasons listed, but, in many instances, and especially when the habitat has been altered or degraded, human-transported species are exceptionally successful. Residents often lack adequate defense mechanisms to cope with new competitors, predators, and pathogens. Success also emerges from a suite of biological characteristics that typify many introduced species. As weedy generalists, we humans often carry or introduce other weedy generalists that can tolerate a variety of climatic conditions. Liberation from natural biotic control may remove most checks on population growth of the introduced species, which then not only survives but becomes a pest. Ultimately, anthropogenic actions accelerate the rate at which species are transported and established around the world; native residents cannot adapt fast enough to the tide of introductions that overwhelms them.

This chapter addresses general issues, particular alien species, and the ecological impacts brought about by an increased rate of introduction. Chapter 9 focuses on trends in why and how introductions occur, attributes of invaders and invasible (and noninvasible) communities and habitats, beneficial introductions, and solutions for reducing the frequency and impacts of introductions.

The section of the Global Invasive Species Database of IUCN's Invasive Species Specialist Group (www.issg.org/database) titled "100 of the World's Worst Invasive Alien Species" includes eight fish species: walking catfish (*Clarias batrachus*), common carp, rainbow trout, brown trout, mosquitofish, largemouth bass, Nile perch, and Mozambique tilapia (see table 8.1 for details on most of these). Many of these species occur almost globally; their ecological impact varies but is often substantial. They are held in low regard in conservation circles, despite the economic activity that surrounds their establishment (tilapia admittedly supply significant protein to humans in need, although alternative edible native species exist, or existed, in many locales). These fishes are listed by IUCN alongside such bad actors as chestnut blight, water hyacinth, prickly pear cactus, kudzu, fire ants, malaria mosquitoes, gypsy moths, zebra mussels, cane toads, brown tree snakes, starlings, rats, rabbits, pigs, and goats. You are known by the company you keep.

BOX 8.1. Definitions.

Nonnative, nonindigenous, introduced, alien, exotic, transplanted, translocated, allochthonous, invasive, feral, biological pollutant: The literature on introductions is crammed with more synonyms and disagreement than just about any other topic in conservation. The U.S. Congress attempted to dispel the confusion with a standardized definition, which I will use here. Under the Nonindigenous Aquatic Nuisance Prevention and Control Act of 1990, a "nonindigenous species" (NIS) is one that has been "moved beyond its natural range or natural zone of potential dispersal." This definition focuses on human transport and avoids political boundaries, which are meaningless in a biological context. It recognizes that dispersal, range expansion, and biological invasions are natural events; human-induced introductions are not.

Earlier definitions discriminated between exotics that originated from another country and transplants that were moved within a political zone such as a state or country. However, the impacts of an exotic or transplant on receiving systems and species are unlikely to depend on its politics of origin. Cichlids translocated from northwestern Lake Malawi and released in the lake's southern region have undergone ecological release; they grow faster and produce more young than conspecifics in their original locales (Barlow 2000). Ecological release might also occur if they were moved a shorter distance to a lake in neighboring Tanzania. As Fuller et al. (1999, p. 12) point out, we should not assume that "introduction of a species into an adjacent watershed poses a less serious threat than establishment of a foreign species in the same system."

By this definition, *indigenous* and *native* mean the same thing. *Nonindigenous, nonnative, exotic, foreign*, and *alien* are synonyms for species that have been transplanted, translocated, or introduced. Sometimes, "successful" introductions are referred to as *established, acclimatized*, or *naturalized*. The third term in particular, used frequently by Lever (1996), carries an almost positive connotation, as in "naturalized citizens." The "acclimatized" designation derives from the practices of so-called acclimatization societies, which were groups of (often displaced or expatriated) Europeans who took it upon themselves to distribute sport fishes from Europe and North America into places, such as Australia, New Zealand, and South Africa, that lacked such species (e.g., Lever 1992; McDowall 1994).

Regardless, these species are still alien. There may be ecological merit in distinguishing *invasives*, a term botanists have traditionally applied to introduced species that are highly noxious because they take over an area once they become established (S. Pearson, pers. comm.). Kolar and Lodge (2001) defined *invasive* as an alien that spreads and becomes abundant. IUCN refers to an invasive as an agent of change that threatens biological diversity (www.issg.org). Contrast this with Lever's (1996, p. ix) definition of an invasive as "an introduced species (not necessarily one that has had a negative ecological impact)." One's viewpoint can affect one's definition.

MECHANISMS BEHIND SUCCESSFUL INTRODUCTIONS

Successful invaders can affect individuals, populations, communities, and ecosystems (J. N. Taylor et al. 1984). Because of multiple causes and levels of interaction and effect, it is usually difficult to pinpoint why and how an introduced species causes declines in a native species. Adults of natives may experience accelerated mortality, or older individuals may not be replaced by young fish because of impaired reproduction or larval mortality. Reproductive failure can result from competition for breeding sites or food necessary to produce eggs, destruction of spawning habitat, disrupted spawning, and hybridization. Larvae may face increased predation or decreased feeding success.

The mechanisms that cause native declines are consequently often implied through correlation rather than demonstrated by manipulation or observation. Among the 31 studies reviewed by S. T. Ross (1991), in only 5 were experimental tests of interactions performed. Three reductions of natives resulted from predation by introduced fishes: brown trout and green sunfish preying on several native species (Garman and Nielsen 1982; Lemly 1985); and mosquitofish preying on Gila topminnow, *Poeciliopsis occidentalis* (Meffe 1985). In two systems, superior competitive ability of introduced species was demonstrated: brown trout (Fausch and White 1981), and rainbow trout (Hearn and Kynard 1986). Since Ross's (1991) pioneering synthesis, additional studies have identified actual mechanisms, but definitive explanations remain elusive.

Replacement versus Displacement

Alien species invade and become established in new habitats via two hypothesized processes: replacement and

Table 8.1. Ten commonly introduced but controversial fish species

Species	*No. countries where established*	*No. states where introduced*	*Native to*	*Original purpose of introduction*
Cyprinus carpio, common carp	49	49	Eurasia	Food, ornamental
Carassius auratus, goldfish	>40	49	E. Asia	Ornamental
Ctenopharyngodon idella, grass carp	9	45	E. Asia	Vegetation control
Oncorhynchus mykiss, rainbow trout[a]	56	48	W. North America	Game fish, aquaculture
Gambusia spp., mosquitofish	67	35	E. North America	Mosquito control
Poecilia reticulata, guppy	34	15	N. South America	Mosquito control
Micropterus salmoides, largemouth bass[b]	53	43	E. North America	Game fish
Lates niloticus, Nile perch[c]	3	1	E. Africa	Food
Tilapiine cichlids[d]	94	13	E., C. South Africa	Aquaculture, vegetation control
Cichla ocellaris, peacock cichlid	6	2	Amazon basin	Game fish

Sources: Based on Welcomme (1984, 1988), Courtenay et al. (1984), Lever (1996), and Fuller et al. (1999). Number of countries and island groups where established is from Lever (1996), number of U.S. states where introduced is from Fuller et al. (1999). Modified from Helfman et al. (1997), presented in roughly phylogenetic order.

Note: Species listed have often been effective in the purpose for which they were introduced but have subsequently posed serious ecological problems.

[a]Related species: Eurasian brown trout, *Salmo trutta* (28 countries, 47 states), and North American brook trout, *Salvelinus fontinalis* (31 countries, 38 states).

[b]Related species: smallmouth bass, *M. dolomieu*.

[c]*L. longispinus* or *L. macrophthalmus* may also have been introduced into Lake Victoria (see Ribbink 1987; Witte et al. 1992).

[d]Numbers are for *Oreochromis mossambicus* and *O. niloticus*; other widely introduced tilapiines include *O. aureus*, *O. macrochir*, *O. urolepis* ssp., *Tilapia rendalli*, *T. zilli*, and others, including hybrids.

displacement (e.g., Douglas et al. 1994). *Replacement* occurs when habitat degradation creates conditions that are no longer favorable for natives but are acceptable to aliens. Replacement requires little interaction between native and nonnative but instead implies that physiological tolerances and habitat preferences favor some species over others. Replacement may explain many major faunal changes in highly degraded and modified aquatic habitats in western North America (e.g., Herbold and Moyle 1986). The replacement hypothesis implies that native species are relative habitat specialists, whereas invaders are generalists, an idea supported by many entries in table 4.4 and box 9.1.

A replacement scenario would explain observed shifts in assemblage structure among stream fishes affected by increased sediment loads (chapter 5). As benthic habitats fill with sediment, riffle-feeding or crevice-spawning fishes would suffer, whereas water column feeders or fishes less dependent on clean bottoms would be favored (e.g., Berkman and Rabeni 1987; E. B. D. Jones et al. 1999). Other examples of habitat disturbance that might result in replacement include removal of kelp beds through harvesting or storms, loss of macrophyte beds in lakes from herbicides, destruction of live corals by dredging or climate change, and loss of woody debris due to snag removal or flash floods.

Displacement involves biotic interactions in the form of predation, behavioral inhibition of reproduction, competition, or hybridization. Predation can be documented by watching one species eating another or through stomach contents analysis. Inhibition of reproduction, such as interference with spawning pairs or harassment of nest-guarding parents, has been observed occasionally, as in the case of male introduced western mosquitofish sexually harassing female endangered Gila topminnow (Schoenherr 1981).

Competitive interactions are seldom directly observable. Unequivocal evidence of competitive displacement is rare, although its existence is often strongly implicated (e.g., Schoenherr 1981). It can be inferred from shifts in resource use (food, habitat) when species with similar habits are brought together. Planktivorous bloaters, *Coregonus hoyi* (designated Vulnerable by IUCN), in Lake

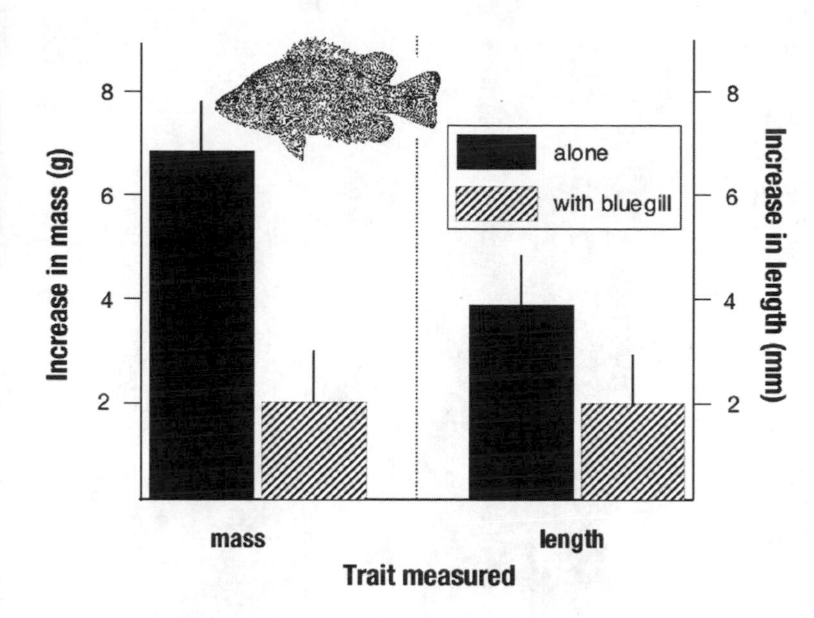

Figure 8.1. Evidence of competitive displacement of Sacramento perch, an imperiled native sunfish of California, by introduced bluegill sunfish. Sacramento perch when alone in aquaria (dark bars) grew significantly more in mass and showed a trend toward greater increase in length than when kept with bluegill (crosshatch). After Marchetti (1999); fish illustration by A. Marciochi in Moyle (2002); used with permission.

Michigan were replaced by introduced, planktivorous alewives, *Alosa pseudoharengus* in the 1960s. Bloaters declined in numbers, shifted to feeding on benthic invertebrates at an earlier age, and apparently evolved a reduced number and length of gill rakers in response to increasing numbers of alewives (Crowder 1984). Five other coregonid species were extirpated from Lake Michigan during the same period.

Strong inference implicates competitive displacement in other fish assemblages. Douglas et al. (1994) postulated that interference competition via spatial displacement caused population declines in spikedace, *Meda fulgida*, a federally Threatened minnow in the lower Colorado River in Arizona and New Mexico. Spikedace disappeared simultaneously and progressively as red shiners, *Cyprinella lutrensis*, a well-known invasive, populated the area while dams and water withdrawals led to degraded habitat. When alone, both species occupied slow-current regions. When they co-occurred, red shiners occupied slow-current regions, but spikedace were displaced into regions of swifter current. The authors concluded that the introduced generalist competitively displaced the native specialist via aggressive behavior.

Introduced trout are often thought to competitively displace native trout (e.g., Gatz et al. 1987; Fausch 1988). In a Michigan stream, adult introduced brown trout displaced adult native brook trout from the best foraging habitats, forcing them into faster water, where the energetic costs of maintaining position were likely to be higher and they were more likely to be caught by anglers (Fausch and White 1986; see also Waters 1983). Rainbow trout displaced two native Japanese salmonids (Dolly Varden, *Salvelinus malma*, and white-spotted char, *S. leucomaenis*) because of timing differences in spawning. Rainbows spawned in spring at a time when embryos of the fall-spawning natives were developing in the gravel. The digging and spawning activities of the introduced trout disturbed the redds of the natives (Taniguchi et al. 2000).

Ideally, demonstrating competitive displacement entails manipulation of species and resources (Fausch 1988; Ross 1991). Marchetti (1999) investigated competitive interactions as a possible cause of population declines and local extirpations in a native California centrarchid, the Sacramento perch, *Archoplites interruptus*. Sacramento perch are least numerous where introduced centrarchid sunfishes are most abundant. Marchetti found that Sacramento perch, when placed with bluegill sunfish, *Lepomis macrochirus*, in lab aquaria, grew less and shifted their habitat use to less natural habitats. Bluegill fed more actively than perch and chased and harassed the perch (figure 8.1). Marchetti postulated that flow regime might also influence interactions between the species via replacement. Native perch would be more likely to persist under the natural, fairly extreme water fluctuations that characterize the arid Southwest, conditions to which the introduced species was not adapted. Stabilized hydrologic regimes would likely favor the introduced species. Several studies of introduced species in arid regions have reached similar conclusions. Dams and water diversion schemes often stabilize flow regimes, apparently tipping the balance in favor of alien invaders (e.g., Moyle and Light 1996a; Allibone 1999); restoration of natural flows can facilitate the return of natives (Marchetti and Moyle 2001). Again, mechanisms seldom operate alone when invaders increase at the expense of natives.

Impacts of Introduced Predators

The evidence for the impacts of introduced predators is fairly unequivocal. Predators have played a well-documented role in decreasing native fish abundance and diversity, often in the name of improved commercial, sport, and subsistence fishing. Introduced predators—removed from assemblages where they have coevolved and coexisted with native prey—can decimate naive prey that lack evolved escape and avoidance mechanisms.

Nile Perch in Lake Victoria: Biomass versus Biodiversity

East Africa's Lake Victoria, the world's largest tropical lake, contained 350–500 species of endemic haplochromine cichlids and 50 other fishes, in what may well have been the richest lake fish fauna in the world. Although the East African Fisheries Research Organization opposed the action based on observations of negative ecological impacts in nearby lakes, centropomid Nile perch, *Lates* cf. *niloticus*, were stocked into Lake Victoria in the early 1960s to provide sport fishing and "to feed on 'trash' haplochromines . . . [and convert the cichlids] . . . into more desirable table fish" (Ribbink 1987, p. 9; I. Payne 1987; haplochromines are cichlids in the genus *Haplochromis*). The stocking has affected all levels of ecosystem structure and function (Kaufman 1992; Ogutu-Ohwayo et al. 1997; Lowe-McConnell 1997).

Nile perch, a predator growing to 2 m long and 200 kg (figure 8.2), was naturally excluded from the lake by 43-m-high Murchison Falls. Once introduced, it spread through Victoria, slowly at first but increasing exponentially in the 1970s and 1980s. It fed preferentially on abundant species, then shifted to rarer species as initial prey declined. Declines were reflected in commercial landings of cichlids, which, between 1977 and the early 1980s, went from 27 to 0 kg/hr and fell from 32% to 1% of the catch. Nile perch landings meanwhile increased to 169 kg/hr. Although overfishing affected endemic cichlids, native species also disappeared where fishing was light, implicating the perch as a cause of native declines (Ogutu-Ohwayo 1990; Witte et al. 1992; Seehausen et al. 1997b).

Figure 8.2. A large Nile perch. Photo courtesy of L. and C. Chapman.

Comparative sampling between 1978 and 1990 indicated that perhaps 200 species of Lake Victoria haplochromines were driven extinct or nearly so (Witte et al. 1992; Seehausen et al. 1997b). Exactly how many were exterminated remains unknown because sampling was inadequate, rarity complicates capture, and extinction is almost impossible to prove (see chapter 1). Some undescribed species may have been eliminated, given that two-thirds of 163 species found in the southern lake in the early

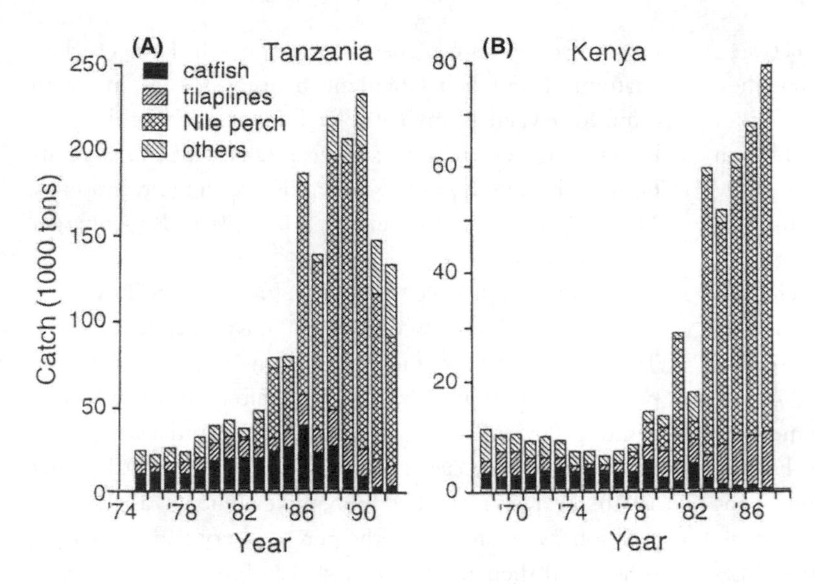

Figure 8.3. Impact of Nile perch on commercial Lake Victoria catfishes. Catch data are for the (A) Tanzanian and (B) Kenyan portions of the lake. From Goudswaard and Witte (1997); used with permission.

1990s were new to science (Seehausen et al. 1997b). Based on apparent trends, "probably more vertebrate species are at imminent risk of extinction in the African lakes than anywhere else in the world" (Ribbink 1987, p. 22).

This once incredibly diverse fish fauna is now dominated by three species: introduced Nile perch; introduced Nile tilapia; and a native zooplanktivorous cyprinid, the dagaa, *Rastrineobola argentea*. Nile tilapia evolved with Nile perch and can coexist with it in large numbers (e.g., Payne 1987), perhaps because the tilapia grows quickly to an adult size larger than most haplochromines (>30 cm), making it too large for many Nile perch. The minnow is a small (5 cm) pelagic species that forms evasive schools; its density is perhaps four to five times greater than before Nile perch invaded (Kitchell et al. 1997).

More than species abundances have changed. Food webs have been simplified. Before *Lates*, the lake's top predators included piscivorous catfishes and haplochromine cichlids, which fed on lungfishes, mormyrids, characins, cyprinids, catfishes, and haplochromine and tilapiine cichlids, which ate invertebrates and algae. After *Lates* decimated most groups, the food web consisted largely of *Lates* feeding on juvenile *Lates*, *Rastrineobola*, and the introduced tilapia (Ligtvoet and Witte 1991). Increased nutrient inputs from deforested hillsides promoted eutrophication. Bluegreen algae replaced diatoms as a dominant primary producer, and ensuing algal blooms caused oxygen depletion in deep water and periodic fish kills (e.g., Kaufman 1992; Hecky 1993). Algal blooms and die-offs produced more detritus, which fed a superabundant prawn, *Caridina nilotica*, the major prey of juvenile Nile perch. More recently, water hyacinth, *Eichornia crassipes*, invaded the lake, forming vast mats 5 km long in bays and along shoreline areas. These mats foul fishing nets and cause anoxic nearshore conditions, exacerbating the problem of anoxic bottom waters. Water hyacinths have also compromised human health because they provide habitat for schistosomiasis-causing snails (Kitchell et al. 1997).

Turbidity increased because Nile perch has to be smoked, not air dried; hillsides were deforested as firewood was collected. Cichlids moved to clearer areas, even though this increased their encounters with Nile perch (Seehausen et al. 1997a). Colorful species and overall diversity declined via hybridization brought about by altered light transmission that interfered with sexual selection (Seehausen et al. 1997a; chapter 5).

Losses also occurred among noncichlids. Ten catfishes, including three top predators, constituted one-third to one-half of fisheries landings prior to introduction of Nile perch (Goudswaard and Witte 1997; figure 8.3). Catfishes disappeared first from deep water, where Nile perch densities were highest, with juveniles disappearing fastest, implicating perch predation as a cause. Overfishing was ruled out as a cause of catfish declines, although competition for prey and deoxygenation of deep water may have contributed. The endemic *Xenoclarias eupogon*, which inhabited deep water, may be extinct, and other catfish species moved into shallow, rocky areas; inflowing streams; or marginal wetlands that Nile perch seldom frequent. Catfishes are no longer of economic importance in the lake.

Nile perch populations and catches peaked between 1985 and 1990 at around 150–170 kg/ha and then declined, despite increased fishing effort (Reidmiller 1994). The number and types of haplochromine cichlids in perch stomachs also declined, constituting only 10% of the diet of adult (3- to 11-yr-old) perch. The diet of Nile perch became dominated by cannibalism (40%–60% of stomach contents) and introduced tilapiine cichlids (25%–30%) (Kitchell et al. 1997). Condition factors of Nile perch declined to the lowest values known for the species anywhere (Ogutu-Ohwayo 1999).

With declines in the Nile perch population, some haplochromines may be recovering, at least partially. Habitat preference or plasticity may also protect some haplochromines and allow others to recover from earlier population crashes. Detritivorous and zooplanktivorous fishes in deep water (13–15 m) have shown some increase in numbers compared to the late 1980s (figure 8.4). Deep rocky habitats, which are avoided by Nile perch, appear to be a refuge both for rock-dwelling species and for species that began using such habitats after Nile perch were introduced (Seehausen et al. 1997b). This recovery is not without new problems. Fishers now target rock-dwelling cichlids for bait in the long-line fishery for Nile perch. Other haplochromines continue to disappear from the lake; 13 species that survived the 1980s disappeared from shallow-water habitats in the 1990s (Seehausen et al. 1997b).

Ironically, oxygen depletion may actually favor recovery of some native species. Nile perch lack efficient anatomical and behavioral mechanisms for coping with reduced oxygen. Many native cichlids and other fishes deal better with oxygen deficits and can take refuge in oxygen-poor environments such as wetlands, avoiding predation by Nile perch (Schofield and Chapman 2000; Rosenberger and Chapman 2000).

Nile perch in Lake Victoria are not universally viewed as a liability (Pitcher and Hart 1995; Kitchell et al. 1997). Total fisheries landings increased from 100,000 MT before perch introduction to 500,000 MT after, with perch constituting 90% of the catch (ignoring an initial lag period when overall fisheries decreased by about 80%; D. J. Miller 1989). Fishery-related employment doubled as 150,000 new jobs were created by the perch fishery, and many local fishers and their families regarded Nile perch as "the saviour" (Reynolds and Greboval 1988).

However, the boom times have increased immigration, and recent declines in perch catches, combined with losses among other edible species, are a major cause for concern (Nile perch landings declined 35% from a 1990 peak of 372,000 MT to 241,000 MT in 2002; FAO 2004). Artisanal methods of fishing from small boats for haplochromines are often supplanted by mechanized, large vessels and larger nets needed for Nile perch. Dagaa are turned into inedible fish meal, and local consumption of haplochromines has given way to export of Nile perch, especially to Europe (except when embargoed due to bac-

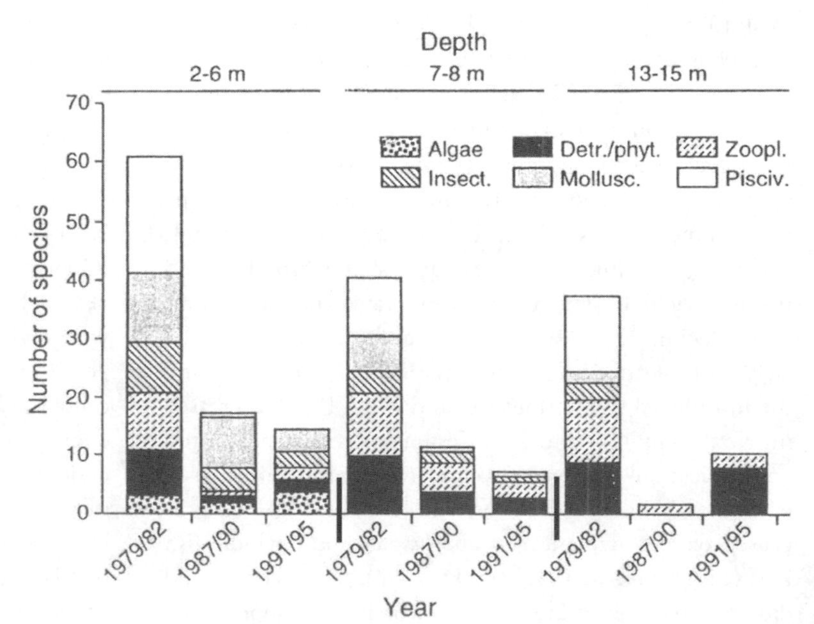

Figure 8.4. Temporal patterns in cichlid distribution as a function of feeding guild, Lake Victoria. Data are from the southern, Mwanza Gulf region, based on trawl and gillnet fishing. Shadings indicate feeding guilds. From Seehausen et al. (1997b); used with per-

terial contamination): "The present fishery is dominated by national and international capital . . . [and] the price of fish . . . has risen beyond what most consumers can afford" (Abila and Jansen 1997, p. 4). The 35 million people living around the lake will increase in number and will need to eat (for a truly sobering depiction of the human misery brought on by this export industry, see the movie *Darwin's Nightmare*).

A major challenge then is how to manage lake fisheries to maximize protein production while minimizing the ecological impact of the introduced predator. Kitchell et al. (1997) developed a bioenergetics model to determine the impact of different fishing methods on Nile perch populations. They concluded that a closely regulated perch fishery that emphasized large-mesh gillnets to catch large adults and beach seines to catch smaller fish would be sustainable and would increase availability of dagaa and tilapiine cichlids. Such a fishery would reduce predation on haplochromines to about 10%, promoting their recovery and continuing a trend observed in lake regions where perch fishing was intense (Kitchell et al. 1997; see also Schindler et al. 1998).

Whether emphasizing biomass production or biodiversity protection, the lesson from Lake Victoria is one of largely irreversible changes due to the introduction of a single species. Similar events transpired when Nile perch were introduced into nearby Lake Kyoga. These events challenge the wisdom of recent proposals to introduce Nile perch into lakes Malawi and Tanganyika and underscore the inadvisability of introducing large predators into any aquatic system (Witte et al. 1992).

Trout, Unlimited

Brown trout and rainbow trout have been introduced into temperate waters around the globe, usually to improve sport fishing. These introductions have been wildly successful, to the detriment of native trout and other fishes. They outcompete and hybridize with natives and carry exotic diseases, but their chief impact has been as predators, eating natives and displacing natives that avoid predation.

Brown trout are native to Europe and northern Africa, occurring naturally south to the Atlas Mountains of Africa and east to the Ural Mountains and Caspian Sea (Elliott 1994). The species diversifies into distinct races and subspecies, many of which are themselves severely threatened (see chapter 9). But because of its large size (to 1 m and 16 kg); qualities as a game fish; and tolerance of warm, silty conditions relative to other salmonids, it is among the most widely introduced temperate species. Brown trout have been established in at least 28 nations in Asia; Africa; North, Central, and South America; and Australia (W. B. Scott and Crossman 1973; Lever 1996; Fuller et al. 1999). The only continent on which they do not occur is Antarctica.

Brown trout were first successfully introduced into the U.S. from Germany in 1883, after unsuccessful attempts at least 30 years earlier (Behnke 1992). They have been stocked in every U.S. state except Louisiana, Mississippi, and Alaska (yes, Hawaii has been stocked). Many introductions are maintained through continued stocking, but reproducing populations are widespread (Fuller et al. 1999). The species inhabits streams from clear headwaters to turbid rivers, where summer temperatures do not exceed 25°–30°C (MacCrimmon and Marshall 1968). Other stream salmonids such as brook trout seldom survive in water warmer than 20°C.

Brown trout populations persist even in heavily fished areas because they are harder to catch than other salmonids (Behnke 1992). Standardized daytime fly-fishing in an Arizona stream yielded 55 Apache trout, *Oncorhynchus gilae apache*, but only four brown trout, although unbiased electrofishing indicated that brown trout outnumbered Apache trout by as much as 20:1. In Palisades Reservoir on the Snake River, brown trout supplanted native cutthroat trout, *O. clarki* ssp., through differential susceptibility to angling. Cutthroat constituted only 28% of the trout stock but 84% of the angling catch. Stocking brown trout may stimulate fishing in an area, but that may affect native salmonids disproportionately and lead to their depletion.

Brown trout compete with native fishes for food when small and feed on natives when large. Their impact on native species as a predator has been well documented (see Fuller et al. 1999). Impacts have been particularly acute in Australia and New Zealand, where they were stocked in the 1860s (Crowl et al. 1992; Townsend 1996). Perhaps 60 million brown trout were released in New Zealand by 1921, largely through the efforts of acclimatization societies and fish and game councils (e.g., McDowall 1994). They were "stocked in every conceivable lake, river, or stream"; "few species can have had so much effort devoted towards successful invasion outcomes" (MacCrimmon and Marshall 1968, p. 2530; Townsend 1996, p. 14, respective-

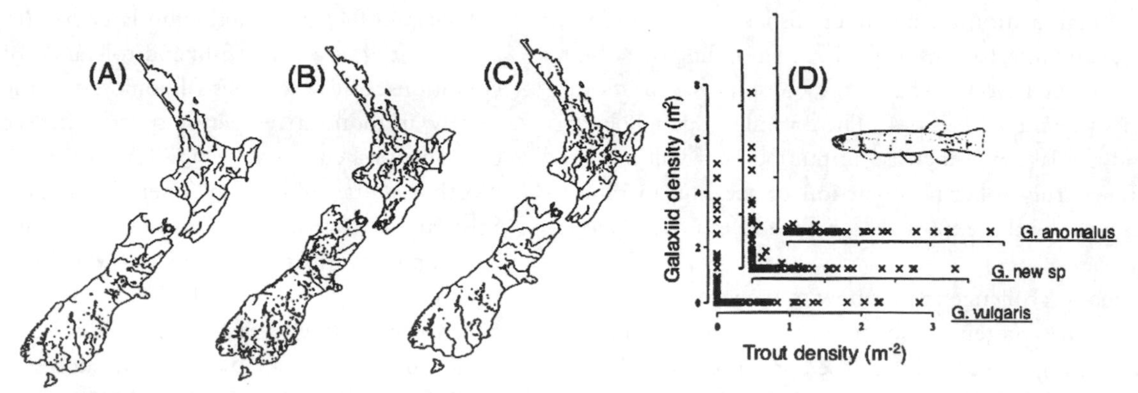

Figure 8.5. Distribution of native and introduced fishes in New Zealand. (A) common river galaxias, *Galaxias vulgaris*; (B) brown trout; (C) rainbow trout. (D) Although distributions of river galaxias and brown trout appear to overlap, native species are often relegated to river regions that are inaccessible to or unoccupied by trout. The figure shows that galaxiid density is high only where trout are rare or absent. A–C from McDowall (1990b), D from Townsend (1996); used with permission.

ly). As a result, they occur in almost all North and South Island habitats where they can survive (figure 8.5).

This wholesale stocking effort has led to the extinction, extirpation, redistribution, and range fragmentation of native fishes, probably because New Zealand freshwater fishes evolved in the absence of large piscivores. The once common New Zealand grayling, *Prototroctes oxyrhynchus*, is thought to have been extinguished in the 1920s in part due to trout predation. Other native species, especially galaxiids, co-occur with brown trout but not in the same river sections. Natives are typically found above waterfalls that brown trout cannot ascend, or in streams with unstable rocky and bouldery bottoms subject to seasonal floods, conditions that brown trout find inhospitable (McDowall 1990b; Townsend 1996; McIntosh 2000; figure 8.5). Such redistribution of natives, although not lethal, can cause metapopulation fragmentation and have serious genetic consequences for species in decline (e.g., Townsend and Crowl 1991). A similar interaction between a native galaxiid and brown trout was observed in Australia (Closs and Lake 1996). Brown trout actively feed on galaxiids, eating as many as 135 per day in laboratory settings, again suggesting that galaxiids lack effective adaptations for avoiding predators (Townsend 1996; McIntosh 2000).

Brown trout are also credited with depleting, extirpating, and even extinguishing several native U.S. fishes, including Endangered Modoc sucker, *Catostomus microps*, and golden trout, *O. aguabonita* (Fuller et al. 1999). For many years, California fish and game personnel actively stocked brown trout (J. N. Taylor et al. 1984; Rahel 1997). California agencies now actively eradicate brown trout from the habitat of native golden trout, the California state fish. Brown trout are sufficiently effective predators to have even been employed in biological control. They were introduced into the notorious Flaming Gorge Reservoir on the Green River, Wyoming (see chapter 15) to eliminate Utah chub, *Gila atraria*, thereby reducing chub competition with introduced kokanee salmon, *O. nerka* (Teuscher and Luecke 1996).

Rainbow trout approach brown trout in extent and impact of introductions. They are native to western North America but have been actively and enthusiastically stocked in perhaps more places than any other fish species. They are easier to raise in hatcheries than brown trout and have consequently been established around the world via hatchery propagation (Behnke 1992; figure 8.6). Lever (1996) listed 56 countries and island groups where rainbow trout have become established and an additional 39 countries where introductions have been attempted (see table 8.1).

Native to the Pacific Slope of western North America (including northern Baja California), rainbow trout have been transplanted into every Canadian province and every state in the U.S., including Hawaii and Alaska (Lever 1996; Fuller et al. 1999). Stocking began in the U.S. in the 1870s with fish from the San Francisco Bay area. International transportation of rainbow trout saw fish first shipped to Japan in 1877; France in 1879; Germany in 1882 (proba-

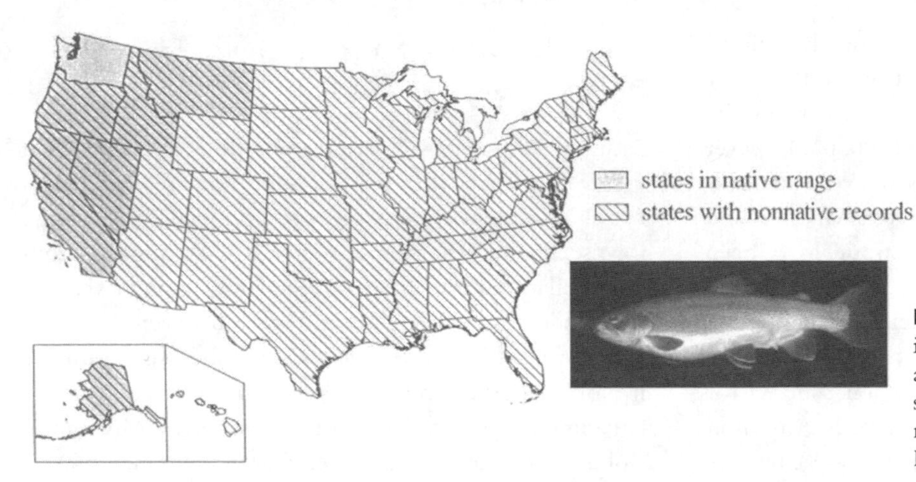

Figure 8.6. Native distributional regions of introduction of rainbow trout. Gray areas are native range; diagonal shadings are stocked fish. Six northwestern states in native range also contain stocked fish. From Fuller et al. (1999); used with permission.

bly in exchange for brown trout); New Zealand in 1883 (and then on to Australia); England in 1884; and throughout Europe, largely from the German stock (Behnke 1992; Lever 1996). Populations are maintained even where water is too warm for reproduction (>13°C) via constant restocking, thus creating "put-and-take" fisheries.

Rainbow trout (and its anadromous sea-run variant, steelhead trout) grow to 114 cm long and 20 kg. Their predatory impact is considerable, although unlike brown trout, they also imperil native salmonids because of a greater tendency to hybridize. Predation has led to declines among native stream fishes in Australia and New Zealand, causing displacement and range fragmentation. Rainbow trout predation has been specifically implicated in the decline of barred galaxias, *Galaxias fuscus*, a once widespread but currently endangered endemic of the central highlands of Victoria, Australia (Raadik et al. 1996). In New Zealand, rainbow trout are implicated in the demise of lake-dwelling koaro, *Galaxias brevipennis*, and dwarf inanga, *G. gracilis* (Lever 1996). Trout prey directly on galaxiids and also dramatically deplete the forage base of native fishes (McDowall 2003). Rivers containing rainbow trout are characterized by trout downstream and galaxiids upstream above barriers that trout have difficulty surmounting, a distribution that parallels that of brown trout. When rainbow trout are removed from reaches of a river and prohibited from reinvading, galaxiids quickly redistribute and establish breeding populations. Impacts of introduced predators may therefore be reversible in some circumstances (Lintermans 2000).

In the U.S., transplanted rainbow trout have caused problems outside their native range, particularly where large salmonid predators were not a natural assemblage component. In northern Arizona, Little Colorado spinedace, *Lepidomeda vittata*, were formerly abundant and occupied a wide variety of habitats through much of the Little Colorado River (Blinn et al. 1993). They are now a federally Threatened species, occurring in disjunct populations that are distributed inversely to rainbow trout. In the same creeks with rainbow trout, spinedace occur in downstream, warmer areas, and rainbow trout occur in upstream, cooler areas, with minimal overlap. Blinn et al. (1993) found that trout fed actively on spinedace and that when trout were present, spinedace shifted habitats from preferred undercut banks to open water, making it more vulnerable to trout predation. Hence instream and interstream distribution of the threatened cyprinid may reflect predator avoidance as well as reduction in numbers due to predation.

Also in the Colorado basin, Marsh and Douglas (1997) assessed stomach contents and predator density and concluded that rainbow trout could easily consume more than 1,400 juvenile and adult humpback chub, *Gila cypha*, annually from an adult chub population that numbered only 5,000–10,000 individuals. Nonnatives such as rainbow trout and other predators could therefore "exert a major negative effect on the [chub] population there" (p. 345).

In the interest of improved sport fishing, introduced rainbow trout have been provided opportunities to depress native fishes in far-flung regions. In Lake Titicaca in the Andes Mountains, a species flock of endemic cyprinodontids has been decimated by rainbow trout due to direct predation and competition for food. This introduction also undermined the protein base for native Indian populations (e.g., Villwock 1993). Rainbow trout are similarly indicted for having "greatly depleted the populations of those few indigenous fish species with which they occur" in the Argentine Patagonian cordillera (Lever 1996, p. 24). Lever also documented negative interactions between rainbow trout and native species in Denmark, Yugoslavia, Chile, and Colombia. In South Africa, rainbow trout have been directly linked to declines, displacements, extirpations, and extinctions of at least 11 native fish species (Skelton 1987; de Moor and Bruton 1988).

Brook trout (or char) often suffer when rainbows or browns are introduced into eastern North American streams (e.g., Elwood and Waters 1969; chapter 9), but the tables turn when brookies are moved outside their native range (J. B. Dunham et al. 2002). Although smaller and less tolerant of warmth than browns or rainbows, brook trout have been successfully introduced into 31 countries and 38 U.S. states (table 8.1). They have caused documented declines among several imperiled salmonids, including Chinook salmon in the Columbia River basin, various subspecies of cutthroat trout in the western U.S., golden trout in California, and bull trout throughout a large part of that species' range (Fuller et al. 1999; Levin et al. 2002; Moyle 2002). Vulnerability to invasion doesn't mean a species can't become an invader when introduced to new locales.

Flathead Catfish: More Problems from Transplanting Top Predators

Catfishes are popular food and game fishes and have been widely introduced for sport, despite documented warnings about potential negative effects on native fishes (e.g., Townsend and Winterbourn 1992). The impacts from moving flathead catfish, *Pylodictis olivaris*, among river systems underscores why discriminating between "exotics" and "transplants" is meaningless. One of the largest catfishes in North America and the top predator in many parts of its natural range, the flathead is native to the Mississippi, Mobile, and Rio Grande river drainages (figure 8.7). It grows rapidly, reaches lengths of 1.5 m and masses of over 55 kg, and can exceed 30 years of age (D. C. Jackson 1999).

The fishes of the Atlantic coastal plain rivers of the southeastern United States did not evolve with giant catfish as a selective force. In the mid-20th century, flathead catfish were introduced into those rivers from Florida to Virginia by fishery agency personnel and by private individuals seeking better fishing opportunities (Fuller et al. 1999). The fish dispersed throughout most of the systems that were stocked. From a 1966 introduction of 11 individuals in the Cape Fear River, North Carolina, they dispersed over 200 km. By 1984, they accounted for 10% of

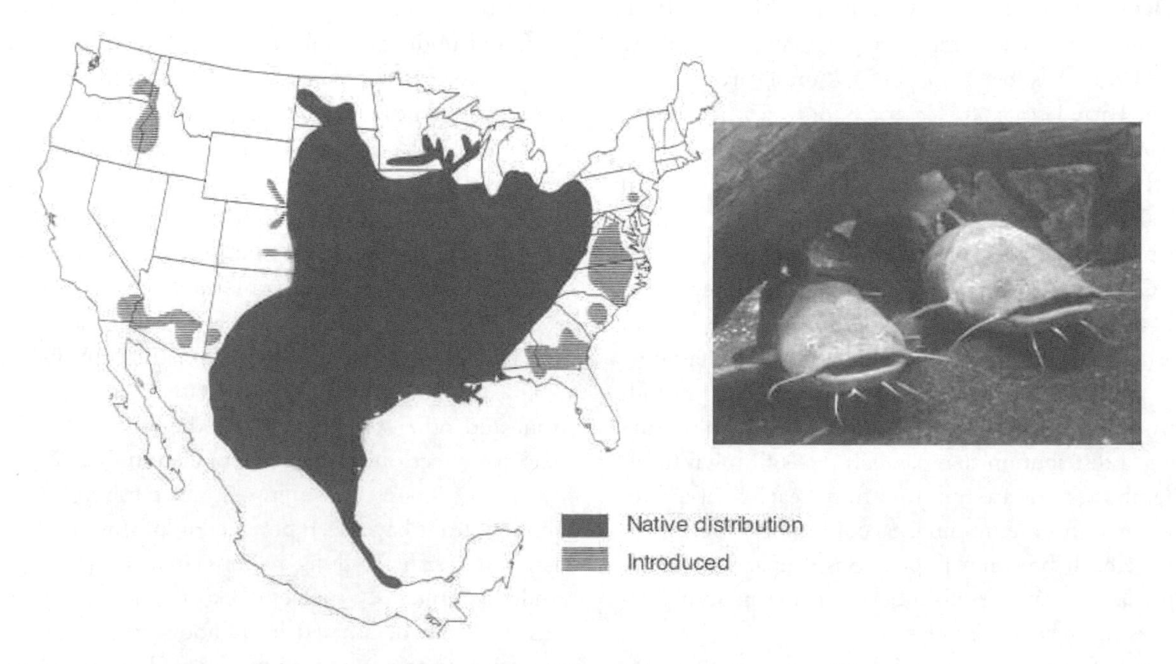

Figure 8.7. Distribution of native and introduced flathead catfish. Where it has been introduced into Atlantic coastal plain and Colorado River drainages, it has contributed to declines among native species. From D. C. Jackson (1999); used with permission.

total fish numbers and 65% of total fish biomass (Guier et al. 1984). At the same time, populations of many native fishes declined. Flathead catfish feed heavily on other catfishes, sunfishes, suckers, and shad, first consuming abundant prey species and then shifting to rarer species (Ashley and Buff 1988; M. E. Thomas 1995). Think Nile perch.

Flatheads were released by an unknown person into the Ocmulgee River, Georgia, in 1973 and spread downstream into the much larger Altamaha River in the early 1980s (M. E. Thomas 1995). Bullhead catfishes, *Ameiurus* spp., disappeared from systematic samples in the Altamaha between 1988 and 1992. Redbreast sunfish, *Lepomis auritus*, which had been the most commonly caught species in the river, also declined, catch rates falling from 0.59/hr to 0.26/hr. Meanwhile, flathead occurrence in electrofishing samples rose progressively from 21/hr to 84/hr.

Small bullhead catfishes appear particularly susceptible to flathead predation (Dobbins et al. 1999). A total of 111 flatheads were stocked in a 31-ha reservoir in Virginia that contained abundant stunted brown bullheads, *A. nebulosus*. After about four years, catch rates fell from a high of 2,285 bullheads (1.43 fish/hr/ha) to 25 (0.03 fish/hr/ha), a 48-fold decrease (Odenkirk et al. 1999). Flatheads have even been recommended as a biological control agent for nuisance populations of bullheads, with caveats about stocking flatheads "outside their native range" (p. 477). Efforts at removing flatheads via electroshock have generally met with limited success (Moser and Roberts 1999). Once the catfish is out of the bag...

Unauthorized stocking by fishers remains a problem. Flathead catfish grow very large, fight hard against fishing tackle, and are good eating. D. C. Jackson (1999, p. 23) praised the fish as "a cultural icon among catfish fishers . . . they are considered the principal big game fish, and anglers willingly accept low catch rates for the opportunity to catch exceptionally large, aggressive fish." The species' impact goes beyond otherwise abundant and recreationally popular fishes. Reintroductions of Endangered juvenile razorback suckers, *Xyrauchen texanus*, into the Colorado River have been impeded due to intensive predation by introduced flathead and channel catfish (Marsh and Brooks 1989). Carter Gilbert, an authority on North American freshwater fishes, has stated that flathead catfish are "probably the most biologically harmful of all fish introductions in North America" (quoted in Fuller et al. 1999, p. 208). Think Nile perch.

Basses

Large predatory fishes called basses are introduced widely, including the cichlid peacock bass, *Cichla ocellaris*, and moronid striped bass, *Morone saxatilis*. The two most extensively introduced predatory basses are the centrarchid black basses, largemouth bass, *Micropterus salmoides*, and smallmouth bass, *M. dolomieu*. Black bass fishing is an approximately $14 billion annual industry in the U.S., fueling commerce in fishing gear, boats and motors, fishing guides, travel, magazine sales, aquaculture, and fisheries management employment (Pullis and Laughland 1999).

Largemouth bass are unquestionably the most popular freshwater sport fish in warmer regions of the U.S. Fisheries biologists and fishers have consequently exported largemouth widely and enthusiastically to improve sport fishing. Authorized and unauthorized stocking has happened in over 50 countries and 40 U.S. states. U.S. stockings involve transport from eastern states where the species is native and transplantation into systems within the native range that lacked it (table 8.1). Introduced largemouth bass are directly implicated in declines and extirpations of numerous endemic species, including at least five federally Endangered western U.S. species (Clover Valley speckled dace, *Rhinichthys osculus oligoporus*; White River springfish, *Crenichthys baileyi*; White River spinedace, *Lepidomeda albivallis*; Pahranagat roundtail chub, *Gila robusta jordani*; and Owens pupfish, *Cyprinodon radiosus*) (Minckley et al. 1991; Fuller et al. 1999). Largemouth are cited as contributing to declines of 21 of the 48 federally listed U.S. fish species for which introduced species have been a factor (Lassuy 1995).

Internationally, largemouth bass have been established in temperate and subtropical lakes, reservoirs, and rivers. Imported into Japan in 1925, it spread extensively via dispersal and secondary introduction to the detriment of native fishes, leading to a ban on additional translocations. "Its popularity as a sport fish, however, resulted in numerous clandestine and illegal movements of the species into many waters throughout Japan, where it has had the impact on indigenous species that had been anticipated" (Lever 1996, p. 219). In Lake Lanao, Philippines, largemouth and other introduced species eliminated all but three species in a unique flock of perhaps 30 cyprinids (Kornfield and Carpenter 1984; I. Payne 1987). In Madagascar, where over 90% of the native freshwater fishes may be endemic, largemouth (in combination with extensive deforestation and numerous introduced species) have contributed to perhaps 5 extinctions and the inclusion of another 35 species in various imperiled lists (Reinthal and Stiassny 1991; Benstead et al. 2000).

In temperate South Africa, largemouth and smallmouth have been widely introduced, largemouth into warmer regions and smallmouth into cooler areas. Largemouth occur in all major river systems (Weyl and Hecht 1999) and have "arguably had a greater negative impact on native species in South African waters than any other naturalized piscivore" (Lever 1996, p. 225), although other authors bestow that distinction on smallmouth (e.g., Skelton et al. 1995; P. Skelton, pers. comm.). The two basses have extirpated the Cape kurper, *Sandelia capensis*, from some lakes, caused declines of ten native catfishes and minnows in the Olifants River, and displaced other imperiled cyprinids in the eastern Cape region (Lever 1996). South Africa exemplifies the multiple and cascading effects of predatory introductions. To provide forage for introduced largemouth bass, other species were introduced, including mosquitofish; bluegill sunfish; four species of tilapia; and redtail barb, *Barbus gurneyi*. Mosquitofish are a significant predator in their own right, and both bluegill and Mozambique tilapia have had significant effects on the South African ichthyofauna (de Moor and Bruton 1988).

Largemouth bass are held responsible for similar community disruptions in Kenya, Zimbabwe, northern Italy, Mexico, Cuba, Guatemala, and Honduras, most commonly in systems that lacked large natural predators (Welcomme 1988; Lever 1996). In Cuba, increased outbreaks of malaria were linked to largemouth bass predation on a native species of *Gambusia* that fed on a malaria-carrying mosquito (Robbins and MacCrimmon 1974).

Smallmouth bass have been introduced successfully into only 11 countries, with failed attempts in at least 19 (Lever 1996). Smallmouth require cooler, clearer waters than largemouth. They were exported from the U.S. to South Africa in 1937 to fill a vacant sport fish niche between the cool headwater regions occupied by introduced trout and the warmer lowlands occupied by introduced largemouth bass (Lever 1996). The region contained no large native piscivores (Skelton 2002). Hatchery programs and stocking distributed them throughout the Cape region; they were inhibited only by dams and waterfalls. The Cape region is a biodiversity hot spot, with 33 endemic Gondwanan relict species (chapter 3). Many native species that survived the earlier introductions of trout and largemouth declined with the arrival of smallmouth. Remnant populations of natives were fragmented, being confined to tributaries or to areas above waterfalls. Such isolation raises the specter of lost genetic diversity, some of which is already being detected (E. R. Swartz, pers. comm.). At least nine endemic, nationally listed cyprinids and two catfishes are directly threatened by smallmouth (Skelton et al. 1995; Cambray 1999; Impson 2001).

South Africa does not lack native sport fishes. The now rare Clanwilliam yellowfish, *Barbus capensis*, of the Olifants River grew to 98 cm and 10 kg. Cambray (1999, p. 150) lamented, "This once formerly abundant species was known as one of the best freshwater angling species in the world. It is ironic that the introduction of a much smaller, more common alien angling species [smallmouth bass] . . . is now one of the main negative impacts on the remaining *B. capensis* populations!" Plans to eradicate invasive predators are being developed in South Africa, although such efforts have often met with limited success (P. Skelton, D. Impson, pers. comm.).

The tropical peacock bass (also peacock cichlid, butterfly peacock, tucanare, or pavon) and its close relatives have had similar impacts on native species. Peacock bass are 75-cm, 7-kg, predatory cichlids native to northern South America. They have been established in south Florida and in Hawaiian reservoirs to improve sport fishing and to control other introduced cichlids (Fuller et al. 1999). They are best known among ecologists for their effects on fishes and food webs in Gatun Lake, Panama, a water body created by the construction of the canal. They were introduced into an impoundment but escaped into the Chagres River during floods and eventually invaded Gatun Lake. After two years, they had eliminated six common fish species (four characins and two poeciliids) and reduced an abundant atherinid by 50%. Lake Gatun already contained two native predators, a sleeper (*Dormitor*) and a predatory characin (*Hoplias*), but both were crepuscular, ambush, and solitary, whereas the peacock bass is a diurnal, schooling predator that pursues its prey. The native fishes apparently lacked evolved avoidance and escape tactics appropriate to the novel threat it imposed (Zaret and Paine 1973; Zaret 1974; Swartzmann and Zaret 1983). Ironically, Tom Zaret, writing in the early 1970s, was surprised that *Cichla* affected Gatun Lake fishes dramatically whereas Nile perch recently introduced into Lake Victoria appeared to be having minimal impact. Had Zaret lived into the late 1980s, he would have seen that

Lake Gatun was a harbinger of the disaster that eventually befell Lake Victoria.

Some Gatun natives have persisted by taking refuge in the turbid Chagres River, from which they recolonize the lake each wet season. Predation pressure has also been redistributed with the introduction of common carp and Nile tilapia as prey species (Swartzmann and Zaret 1983; D. J. Miller 1989). Apparent partial recovery of natives via refuging and alternative, introduced prey parallel some events in Lake Victoria. Whether generalizations about management actions can be drawn from these parallels remains to be determined.

Pike and Pikeperch in the Northern Mediterranean

Introductions in southern European countries also show the vulnerability of predator-naive fish assemblages. Lake Egridir in Turkey was predator free, with nine native species—six endemic to the region and two (*Phoxinellus handlirschi* and *P. egridiri*) endemic to that lake. In 1955, the percid zander, *Zander lucioperca*, was introduced to improve commercial fishing. Zander catches reached as high as 480 MT/yr, but endemics disappeared. Five species were extirpated, including both lake endemics, and other native species declined in number. The catch of natives dropped from 850 MT in 1960 to 140 MT in 1981 (Crivelli 1995); overall catch in the lake thus declined. Northern pike, *Esox lucius*, introduced into the River Esla in Spain had similar impacts. The Esla contained seven native and three introduced species but no predators. After pike introduction, native minnows declined sharply, whereas the introduced gudgeon, *Gobio gobio*, which is sympatric with northern pike over much of its range, remained relatively stable (Rincon et al. 1990).

Gambusia: Preying on the Young

Introduced predators generally evoke an image of something large and menacing—Nile perch, largemouth bass, flathead catfish—swallowing up native adult fishes. But the impact of many introduced predators may fall heaviest on young and small natives, especially eggs, larvae, and juveniles. The mosquitofishes, *Gambusia affinis* and *G. holbrooki*, have had that impact on a broad geographic scale. Mosquitofish subspecies are native to eastern Mexico and the eastern U.S. up to New Jersey but are among the most widely introduced fish species (Lever 1996; table 8.1). A small (<6-cm), water-column livebearer, mosquitofish were introduced to control disease-bearing mosquitoes early in the 20th century, with repeated efforts through the 1960s (Lever 1996). They can survive environmental extremes, including stagnant freshwater pools and moderately saline conditions, although they do poorly in fast-flowing water (Haas and Pal 1984; L. N. Lloyd et al. 1986). Their popularity arose from a reputation for eating insect larvae, broad physiological tolerances, and high reproductive output.

Whether mosquitofish ever really controlled mosquitoes is disputed. They are effective and have minimal ecological impact when stocked in ditches, rice fields, and other highly artificial environments that lack native mosquito predators. However, most accounts summarized in Lever (1996) indicate that (1) they are no better at controlling mosquitoes than the native fishes they displace; (2) mosquito larvae are a minor diet component because mosquitoes often lay eggs in dense vegetation, a habitat *Gambusia* generally avoid; (3) mosquitofish may actually benefit mosquito larvae by preying on invertebrate predators; and (4) *Gambusia* can make up more than 90% of the fish biomass in some systems because they feed on fish eggs and fry, nip fins of larger fish, and compete with small fishes for invertebrate prey.

As a result, mosquitofish often eradicate smaller native fishes, including some that are also predators on mosquito larvae. For example, S. M. Nelson and L. C. Keenan (1992) showed experimentally that plains killifish, *Fundulus zebrinus*, were as effective as introduced mosquitofish in controlling mosquito larvae. They concluded (p. 303) that, "in the haste to disseminate mosquitofish for use in mosquito control, the native fish populations that were acting as larvivores, at less environmental and economical costs, have been decimated."

Gambusia's impact, as predator or competitor, has been felt widely. Schoenherr (1981) credits mosquitofish with reducing or eliminating at least 20 fish species and numerous invertebrates. McKay (1984, p. 185) went so far as to declare that mosquitofish in Australia "may be the most harmful of our introduced fishes." In southern Europe, mosquitofish contributed to the displacement of *Valencia* and *Aphania* cyprinodontids, several of which are imperiled (Crivelli 1995; Economidis et al. 2000). In the American Southwest, mosquitofish are directly implicated in declines among endangered pupfishes, springfishes, and

Gila topminnows. Gila topminnow, once the most abundant fish in southern Arizona and an effective mosquito larva predator, is now restricted to a dozen populations (Haas and Pal 1984; Williams and Meffe 1998). In summary, "mosquitofish do not appear to perform any useful roles," are "a pest or ecological weed," and become so abundant as to constitute an "infestation" (Lever 1996).

Sea Lampreys, Whether Predators or Parasites

Most introduced predators arrive after deliberate efforts by government agencies, acclimatization societies, or zealous anglers. However, some introductions are inadvertent, as in the case of the sea lamprey, *Petromyzon marinus*. Sea lampreys entered the western North American Great Lakes and adjoining waterways when the Welland Canal was built and later enlarged (1914–32), providing an easily surmounted connection between Lake Ontario and Lake Erie that bypassed Niagara Falls. Exactly how lampreys arrived in Lake Erie and beyond is a matter of conjecture. They may have moved under their own power, been attached to ships or the bodies of migrating host fishes, or been dumped from bait buckets (Morman et al. 1980; Fuller et al. 1999; Daniels 2001).

The sea lamprey is native to Atlantic coasts and rivers of both North America and Europe and may have been a landlocked native in Lake Ontario, Lake Champlain, and the New York Finger Lakes (Daniels 2001). As anadromous fish, sea lampreys spend most of their lives at sea and then migrate up rivers to spawn and die. Parasitic activities usually occur during the marine feeding stage. There is little evidence that parasitism of usual hosts depressed populations when the hosts were at normal population levels, even in lakes where landlocked lampreys and hosts naturally co-occurred (e.g., Hardisty and Potter 1971). However, in the Great Lakes that lacked natural sea lamprey populations, new hosts apparently also lacked defenses or were vulnerable because of pollution and overfishing. Perhaps the lack of natural lamprey predators such as harbor seals allowed lamprey populations to explode, thus causing population crashes among host species.

The lamprey's spread, which coincided with increasing pollution and overfishing in the Great Lakes, helped drive the loss of large native fishes. Commercially important species such as whitefishes, ciscoes, lake trout, and walleye plummeted. Annual lake trout landings in lakes Huron and Michigan fell from 1.4–2.2 million kg to less than 225 kg between the 1930s and 1950s. The greatest decline came within five years after introduction of the sea lamprey (W. B. Scott and Crossman 1973; Courtenay 1993). Restoration efforts directed at lake trout alone cost an estimated $12 million annually. The blue pike, *Zander vitreum glaucum*, supported an annual fishery of 5–10 million kg in Lake Erie until the early 1950s (C. L. Smith 1985). The fish is probably now extinct, with sea lampreys a contributing factor (R. R. Miller et al. 1989). Three endemic coregonids (*Coregonus alpenae*, *C. johannae*, and *C. nigripinnis*) were exterminated. With large predators and many zooplanktivorous competitors in decline or missing, alewives underwent periodic population booms and crashes. Alewives may or may not be native to Lake Ontario (Daniels 2001), but their subsequent spread to other Great Lakes in the 1930s and later corresponds with the Welland Canal connection (Fuller et al. 1999). Their massive die-offs and rotting corpses created a significant aesthetic nuisance and health hazard along Great Lakes shorelines.

Considerable effort and cost have gone into controlling sea lampreys in the Great Lakes, including passage of the Great Lakes Fish and Wildlife Restoration Act of 1990. Tactics in the war on lampreys include electric fences, mechanical traps, and, most successfully and controversially, chemical lampricides. The chemicals, chiefly 3-trifluoromethyl-4-nitrophenol (TFM), kill larval sea lampreys but also native parasitic and nonparasitic lampreys (as well as walleye, brown bullhead, margined madtom, white sucker, cutlips minnow, river chub, creek chub, blacknose and longnose dace, introduced rainbow trout, chain pickerel, banded killifish, pumpkinseed, and tessellated darter; R. Daniels, pers. comm.). TFM has reduced some sea lamprey populations by up to 90%, but application must be repeated and costs reach $8 million a year (B.R. Smith and Tibbles 1980; Hanson and Swink 1989; Fuller et al. 1999). Ironically, sea lampreys in many European countries receive protection because they are popular food fishes and their populations are declining due to interrupted migrations and habitat degradation (see chapter 2, 9).

Introduced Native Predators, Too

One general observation emerges from the preceding examples. Alien predators are particularly deleterious when introduced into systems depauperate in native predatory fishes, but significant impacts can also occur whenever predators are introduced into predator-free environments, regardless of a coevolutionary history. In a survey of temperate Adirondack lakes, native minnow diversity in lakes with large piscivores was about one-third that of lakes that lacked piscivores (Findlay et al. 2000). The predators were mostly northern pike, largemouth bass, and smallmouth bass that had been introduced to improve fishing. At least 4, maybe 5, minnow species were essentially lacking from lakes with piscivores, and of 13 minnow species in the region, only 2 nonnative species (bluntnose minnow, *Pimephales notatus*, and golden shiner, *Notemigonus crysoleucas*) were unaffected by introduced piscivores. The authors concluded that "introduction of top piscivores to small temperate lakes puts native minnow communities at high extinction risk" (p. 570). Any proposal for introducing a predator should be approached with extreme caution.

And Even Introduced Prey

It is uncertain whether alewives are native to Lake Ontario or were introduced in the mid-1800s. Regardless, their numbers were very low before 1860, then increased dramatically. The impacts of this population increase have been substantial, operating via unanticipated mechanisms. A body of experimental work implicates alewives as directly responsible for declines among native and introduced salmonids in the Great Lakes and their tributaries (Ketola et al. 2000). Alewives and other clupeids contain high concentrations of thiaminase, an enzyme that catalyzes the destruction of thiamine. Thiamine deficiency leads to high mortality rates among young salmonids, a phenomenon known generically as Early Mortality Syndrome. Adult salmon that feed on alewives produce eggs with significantly low thiamine concentrations. Fry from such eggs may suffer 98% mortality rates, whereas fry from females injected with thiamine experience only 2% mortality (Ketola et al. 2000). Atlantic salmon declined in Lake Ontario and the New York Finger Lakes in the mid-1800s, caused by a combination of blocked spawning migrations, pollution, habitat destruction, and overfishing (S. H. Smith 1995). However, abundant salmon and other predators may have kept alewives in check at that time. Alewives at low densities were probably not a dominant food item. With salmon extirpated, alewives erupted in the late 1800s and became major prey for stocked fish (as did rainbow smelt, *Osmerus mordax*, also high in thiaminase and introduced in the 1930s). Hence declines among predators liberated prey that, at high densities, continue to depress predator numbers (e.g., S. B. Brown et al. 2005).

Hybridization

Hybridization is relatively rare in nature because it is subjected to powerful negative selection (it may be more common in fishes than other vertebrates; Epifanio and Nielsen 2000; Scribner et al. 2000). An individual's genome is made up of coadapted genes that work well together. Hybridization disrupts these gene complexes, often producing individuals less adapted to local conditions. Hybridization between species may produce F1 offspring that are so maladapted they die. But if the F1 survives, it may breed with a parental species. This backcrossing results in introgression, the injection of alien genes into the genome of the original species (Echelle 1991; Leary et al. 1995). Backcrossed individuals are even more capable of breeding with parental species, accelerating dilution of the original genome.

Human activities accelerate hybridization rates. Human-caused hybridization can tip the balance toward extirpation or extinction in relatively few generations, perhaps even faster than competition or predation, especially when a species is relatively rare (e.g., Rhymer and Simberloff 1996). Several human activities facilitate hybridization (see Scribner et al. 2000). Habitat alteration removes physical barriers that keep species apart. A species' preferred spawning habitat may be destroyed, forcing it to spawn where other species breed. Human-induced reductions in population can make it difficult to find conspecific mating partners. When a rare species breeds in the same place and time as an abundant species, interspecific matings are more likely.

Such rarity-induced hybridization has likely caused species declines in North American fishes. In Monterrey, Mexico, three introduced *Xiphophorus* livebearer species hybridized with endangered endemic *X. couchianus*, with massive introgression. Introduced *Cichlasoma cyanoguttatum* hybridized with the endangered endemic Cuatros Cienegas cichlid, *Cichlasoma minckleyi* (Contreras-Balderas and Escalante 1984). Guadalupe bass, *Micropterus treculi*, a Texas endemic designated Vulnerable by Warren et al. (2000), is threatened by hybridization with introduced smallmouth bass (Morizot et al. 1991). In the southwestern U.S., threatened Clear Creek gambusia, *Gambusia heterochir*, have hybridized with introduced mosquitofish, and endangered Mohave tui chub, *Gila bicolor mohavensis*, have hybridized with introduced arroyo chub, *G. orcutti* (Echelle and Echelle 1997). The federally Endangered Leon Springs pupfish, *Cyprinodon bovinus*, endemic to a small, spring-fed system in West Texas and thought to be extinct for more than 100 years, was rediscovered in a spring 15 km downstream from the original site. Recovery efforts began, but a widespread invasive, the sheepshead minnow, *C. variegatus*, whose native range extends from the Yucatán Peninsula to coastal Massachusetts, invaded from a nearby reservoir and within five years had interbred with the pupfish. Alleles unique to *C. variegatus* now occur in all known *C. bovinus* populations, indicating extensive introgression.

C. variegatus has introgressed into the genomes of other imperiled fishes. It invaded the Pecos River of New Mexico and Texas in the early 1980s, probably via a single bait bucket introduction involving a few individuals. In less than five years, hybrids between sheepshead minnows and Pecos River pupfish, *C. pecosensis*, had completely replaced the endemic pupfish along 500 km of the Pecos River; *C. variegatus* genes occurred at frequencies as high as 80% at some locales (Echelle and Echelle 1997). Hybrids have been found beyond the natural southern limit of *C. pecosensis* (Echelle and Connor 1989). Pecos River pupfish suffered a population crash in the 1980s, along with a 60% range reduction. It is now restricted to two sinkhole habitats and has been designated Critically Endangered by IUCN (2004).

Sheepshead minnows apparently replaced Pecos pupfish in part because sheepshead males are more aggressive and because female *C. pecosensis*, at least in the lab, preferred male *C. variegatus* to conspecific males. F1 hybrid males then mated equally with females of both species (Rosenfield and Kodric-Brown 2003). Hybrids and backcrossed individuals also exhibited greater swimming endurance and more rapid growth than *C. pecosensis* (Rosenfield et al. 2004). Such "hybrid vigor," combined with sexual selection favoring the invader, undoubtedly promoted genetic introgression. The hybrid's multiple advantages suggest that introgression will continue until the endemic is genetically exterminated, and that "eradication of hybrids and restoration of *C. pecosensis* to its native range is unlikely" (Rosenfield et al. 2004, p. 1590).

Introduced salmonids have caused numerous problems because of hybridization. Rainbow trout, in addition to eating and competing with natives, also hybridize with a number of imperiled salmonids, including federally Threatened Lahontan cutthroat trout, *Oncorhynchus clarki henshawi*; golden trout, *O. mykiss aquabonita*; Gila trout, *O. gilae*; and Apache trout, *O. apache* (Dowling and Childs 1992; Fuller et al. 1999). Hybridization with rainbows has led to the virtual extinction of the Alvord cutthroat, *O. c. alvordensis*. Brook trout hybridize with brown trout in Germany to form "tiger trout" and with threatened bull trout, *Salvelinus confluentus*, throughout their U.S. range (Lever 1996; Fuller et al. 1999). Bull trout x rainbow trout hybrids are sterile, causing further declines in bull trout populations. Hybridization may help explain why brook trout introductions have caused declines and even extirpations of threatened greenback cutthroat trout, *O. clarki stomias*; westslope cutthroat trout, *O. c. lewisi*; Lahontan cutthroat trout; and golden trout (Behnke 1992).

In Europe, hybridizing brown trout have compromised endemic salmonids. The marbled trout, *Salmo marmoratus* (sometimes referred to as *S. trutta marmoratus*), endemic to the Adriatic River basin, hybridizes readily with European brown trout, *S. t. trutta*, which are not native to this region. Brown trout were introduced into Slovenia throughout the 20th century. As economic development accelerated in the 1960s, hybridization became common. Frequency of marbled trout in Slovenia fell to 13% in 1965, to 2.5% in 1970, and disappeared from all but one watershed by 1985, the few existing trout being hybrids and brown trout. Similar events occurred in Italy, where hybrids can make up 75% of trout (Crivelli 1995). A genetic rehabilitation project initiated in the Soca River of Slovenia in the early 1990s focused on reintroduction of large numbers of genetically pure marbled trout, rather than attempting chemical eradication of brown trout. The project is meeting with apparent success (Crivelli et al. 2000).

Hybridization raises fears of possible escapes of searanched Atlantic salmon, especially in its native range in North America. Genetically distinct Atlantic salmon runs have been greatly depleted and could easily be swamped by the genes of aquacultured fish, including genetically engineered strains (NASCO 1991; see chapter 14).

In Europe, nonsalmonid endemics impacted by hybridization include the Italian bleak, *Alburnus albidus* (IUCN Vulnerable), which hybridizes with an introduced cyprinid, *Leuciscus cephalus cabeda*, in southern Italy; and another endemic cyprinid, *Chondrostoma toxostoma arrigonis*, which hybridizes with introduced *C. polylepis polylepis* in Spain (Crivelli 1995).

Parasites and Diseases

Diseases and parasites lower an individual's fitness. In roundtail chub, *Gila robusta*, two subspecies of which are federally Endangered, higher parasite loads were significantly correlated with reduced growth in length and mass (Brouder 1999). Introduced species can carry novel diseases and parasites for which native fishes lack evolved immunity (Lafferty and Kuris 1999), making introduced species particularly potent as vectors of disease (see Langdon 1990; Lever 1996). Imperiled species in degraded habitats have increased susceptibility to infection, elevating the threat of introduced fishes with their parasite and disease burdens. Furunculosis, a fatal infection linked to introductions and caused by the bacterium *Aeromonas salmonicida*, is endemic to western North American rainbow trout. When rainbows were introduced into Europe, the disease spread to brown trout and now occurs wherever salmonids are cultured (Bernoth et al. 1997).

Although most known introduced parasites are protozoans, some metazoans have been linked to major declines. Widely transported as a bait species, red shiners have become infected with a variety of pathogens. After red shiners were introduced into Utah, the Asian tapeworm, *Bothriocephalus acheilognathi*, entered the Virgin River and infested endangered woundfin, *Plagopterus argentissimus*. The tapeworm has been singled out as a major cause of woundfin declines during the 1980s (Deacon 1988). Declines among American eel, *Anguilla rostrata*, stocks (Haro et al. 2000) may be a partial result of a recently introduced nematode parasite, *Anguillicola crassus*. *Anguillicola* apparently arrived with infected European eels, *Anguilla anguilla*, imported from Japan for aquaculture; young eels of several species are commonly exchanged and cultured. *Anguillicola crassus* is endemic to Japanese eels, *Anguilla japonica*, where it is relatively nonpathogenic (Barse and Secor 1999). However, in European eels, infection rates approach 100%; and infections cause enlarged abdomens, swim bladder rupture, skin ulcers, and bacterial infections that can be fatal. American eels cultured in Taiwan have suffered high mortality, some linked to *Anguillicola*. Survey data suggest that *A. crassus* has become widely dispersed among North American Atlantic coastal rivers and estuaries (Barse and Secor 1999).

Several protist parasites have been spread via introduced fish species. Attempts at introducing a sturgeon species into the Aral Sea between 1927 and 1934 failed, but in the process, *Nitzschia sturionus*, a protist gill parasite of the native ship sturgeon, *Acipenser nudiventris*, was introduced. The parasite is considered to have contributed substantially to the decline of the native sturgeon, which is now Critically Endangered (Rosenthal 1980; D. J. Miller 1989). More common is ich, or white spot disease, a non-host-specific, debilitating gill and skin parasite caused by the ciliated protozoan *Ichthyophthirius multifiliis*. Ich originated in Asia and spread throughout temperate regions via introductions largely linked with the aquarium and aquaculture trades (Hoffman and Schubert 1984; Welcomme 1984). Ich is usually fatal and causes substantial economic losses in mass cultured species such as salmon and catfish (Dickerson and Clark 1998). It is transmitted to native fishes living near or downstream from aquaculture facilities, or via cross-infection from aquaculture escapees. Aquaculture is the suspected source of ich in wild Pacific salmon. It was first reported from four sockeye salmon runs in the Skeena River watershed in British Columbia in 1994 and 1995 (Traxler et al. 1998). This outbreak caused high adult mortality and reduced fry output by an estimated 154 million fish.

Perhaps the most widely publicized instance of introduction-caused infection involves whirling disease, caused by the protozoan *Myxosoma cerebralis*. Particularly virulent in young salmonids, especially rainbow trout—although brook and imperiled cutthroat and bull trout are not immune (e.g., Hedrick et al. 1999)—whirling disease is considered the single greatest threat to many U.S. wild trout populations (MWDTF 1996). Swimming in tight circles is followed by postural collapse and immobility. The parasite inflames cerebrospinal fluid, deforms the brain stem, and causes degeneration of nerves connecting the medulla and spinal cord (J. D. Rose et al. 2000). This fish analog of mad cow disease has decimated entire year-classes of trout in many locales and is of great concern because of its potential impact on imperiled salmonids.

Whirling disease is native to European salmonids, in which it was originally nonpathogenic. It was first detected in North America in 1956 in Pennsylvania and spread progressively through eastern and midwestern, then western states. By the mid-1990s it was a recognized problem in at least 21 states (Bergersen and Anderson 1997). In the Madison River of Montana, formerly a world-class trout fishery, whirling disease has reduced rainbow trout populations by 90%. Ironically, the exportation of infected hatchery fish and even frozen North American salmonids has spread the disease back to Europe, where it is now pathogenic (Moyle et al. 1986). The aquarium trade may also contribute to the disease's spread because the only known intermediate host for the parasite is the commonly sold oligochaete worm *Tubifex tubifex* (Fuller et al. 1999; see www.whirling-disease.org).

Darwinian Management of Parasitism

One explanation for why alien species are often successful colonizers is that their native parasites are absent in new habitats (e.g., Lafferty and Kuris 1999). In New Zealand, brown trout harbor a parasite load of only 17 species versus the more typical 63 species in their native United Kingdom (Townsend 1996). Introduced fish may not have carried all the pathogens because they arrived in New Zealand as eggs. Susceptibility to disease can also work against an introduced species. Exotics may fail when they encounter pathogens for which they have not developed defenses, as when an unidentified *Tilapia* species was introduced into the Balsas basin in Mexico to create a new fishery. A native cichlid, *Cichlasoma istlanum*, harbored an apparently nonpathogenic nematode, which infested the introduced cichlid, essentially wiping it out (the native cichlid was also eliminated, for reasons that are unknown) (Contreras-Balderas and Escalante 1984).

Susceptibility differences may also arise from ecological differences other than evolved immunity. In Palisades Reservoir on the Snake River, nematode parasites caused greater mortality in native cutthroat than in introduced brown trout, allowing brown trout to increase while native cutthroat trout decreased. The species may have been differentially susceptible to infection because cutthroat fed more heavily than brown trout on zooplankton that were intermediate hosts for the parasite (Behnke 1992). The role of differential susceptibility to parasitism as a determinant of invasion success is an area for additional research.

Differential susceptibility can also evolve within species. Young Chinook salmon that originated in streams where the whirling disease parasite was endemic suffered lower mortality rates than salmon from streams where the parasite did not occur naturally (Bartholomew 1998). These results highlight the importance of monitoring the sources of stocked individuals and the danger of escape from pen-rearing operations. If infected but immune fishes are introduced or escape into areas where specific diseases do not occur naturally, they can survive and transmit the disease, with disastrous consequences for native populations.

Intraspecific differences in susceptibility have formed the basis of management decisions. One proposed solution to whirling disease entails stocking disease-resistant trout strains to breed with native strains and transmit immunity (MWDTF 1996). Such applied introgression might solve short-term problems of susceptibility but could decrease genetic diversity, lessen immunity to other pathogens, and reduce overall adaptation to local conditions (Allendorf et al. 2001; see chapter 14). Because whirling disease is most prevalent in disturbed habitats, Allendorf et al. proposed a Darwinian management approach that emphasized habitat protection while allowing populations to adapt to a changing environment that included the *Myxosoma* pathogen.

ASSEMBLAGE, COMMUNITY, AND ECOSYSTEM EFFECTS

Introduced fishes affect individuals and populations, which can affect assemblages and the ecosystems where they occur. One assemblage-level impact commonly observed is a substantial shift in relative abundances, with introduced species assuming numerical dominance. North American mummichog, *Fundulus heteroclitus*, and mosquitofish have been widely introduced to Atlantic coastal systems of southwestern Spain, where they have become the first and second most frequently captured fishes (Gutierrez-Estrada et al. 1998). In streams around metropolitan Atlanta, Georgia, introduced species can make up 90% of the individuals at a site (DeVivo 1996). Welcomme's (1988) analysis of over 1,300 introductions indicated that about 25% involved an introduced species becoming a significant or dominant element in the host assemblage. If failed introductions were eliminated from the analysis, dominance occurred in about one-third of all introductions.

Introduced nonfish species also cause declines and losses among fishes. Water hyacinths created anoxic zones in Lake Victoria, zebra mussels competed for planktonic prey and caused distributional shifts in the Great Lakes and Hudson River, crayfish caused habitat shifts and reduced growth in sculpins in western U.S. streams, ctenophores competed with and ate fish larvae in the Black and Azov seas (see chapter 9). Many others could be cited, but the large list highlights the impacts that introduced species, regardless of taxon, can have when released into novel environments.

Ecosystem Processes and Properties

Conservation biologists are increasingly realizing that programs targeted solely at the species level are shortsighted and simplistic. To save species, ecosystems and the processes and interactions that occur within them must be protected. "Ecosystem management" has become a watchword of conservation efforts at all levels. Unfortunately, comparatively little attention has been directed at the effects that introduced fish species may have on ecosystem processes and properties. Where studied, alterations have been shown to have direct and indirect impacts on many ecosystem components, including human health and the economy (e.g., J. D. Williams and Meffe 1998).

Food Webs and Trophic Cascades

Ecosystems often contain cascading feeding interactions, which respond in sometimes unpredictable ways to introductions. Trout introduced to New Zealand reduced grazing insect biomass, which led to an increase in algal standing crop; native galaxiids did not depress insect populations as greatly (Flecker and Townsend 1994). The events surrounding the proliferation of peacock bass in Gatun Lake, Panama, included elimination of native prey fish species, which cascaded throughout the lake ecosystem, affecting all levels of the food web, including humans. Seven native freshwater fish species were decimated, eliminating prey for migratory tarpon, which also declined. Piscivorous birds such as kingfishers and herons declined due to elimination of their food base, and black terns because they relied on tarpon to chase zooplanktivorous atherinids across the lake, the terns feeding on atherinids that were eluding tarpon. Elimination of zooplanktivores led to increased zooplankton and decreased phytoplankton abundance. Elimination of insectivorous fishes (characins, livebearers, cichlids) resulted in a resurgence of malaria-bearing mosquitoes and a shift in the dominant form of malaria in the Panama Canal Zone from relatively benign *Plasmodium vivax* to much more lethal *P. falciparum* (Lever 1996).

In Flathead Lake, Montana, a trophic cascade involving fishes has become a classic example of unexpected ecosystem consequences initiated by an introduced species (figure 8.8). Opossum shrimp (Mysidae) were introduced into lakes upstream of Flathead Lake to stimulate production of introduced kokanee salmon (kokanee were introduced in 1916, displacing native westslope cutthroat trout; unfortunately, mysids live too deep during the kokanee's daytime feeding period to be a significant food source). The shrimp drifted downstream and entered Flathead Lake via inlet streams around 1981. They multiplied and fed heavily on zooplankton, particularly cladocerans and copepods, depleting food resources for the planktivorous kokanee. Kokanee had formed an annual spawning migration approaching

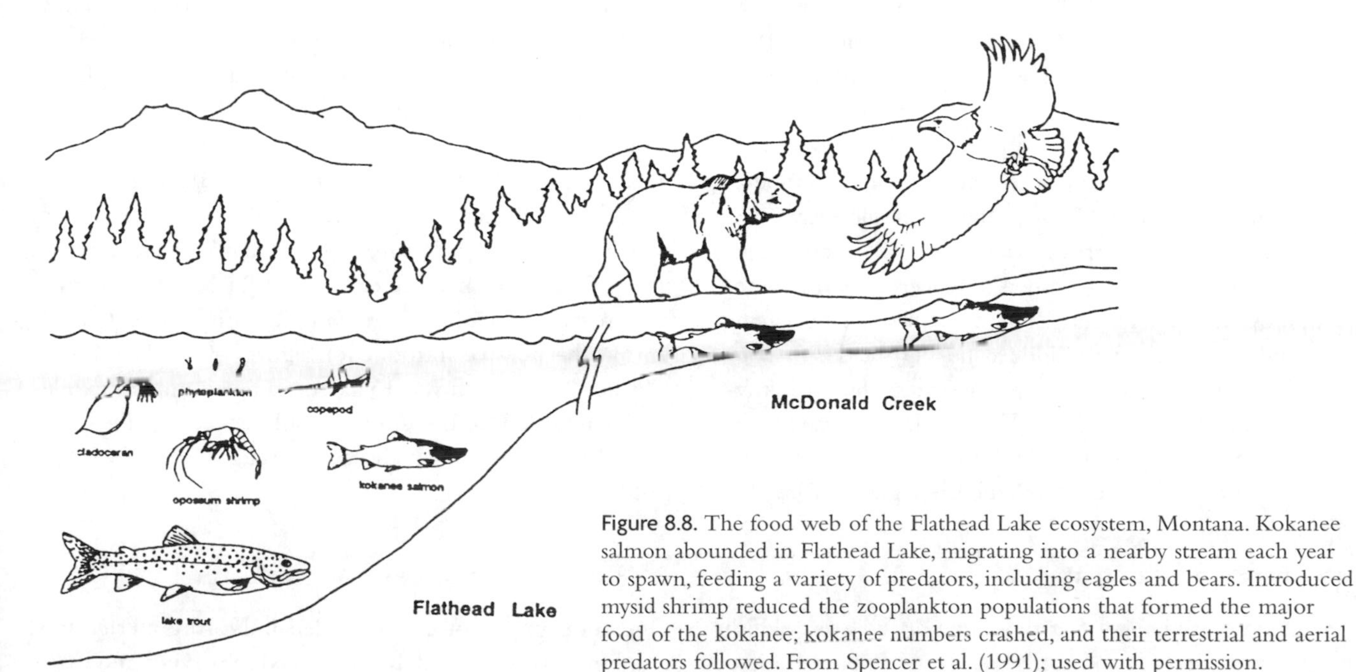

Figure 8.8. The food web of the Flathead Lake ecosystem, Montana. Kokanee salmon abounded in Flathead Lake, migrating into a nearby stream each year to spawn, feeding a variety of predators, including eagles and bears. Introduced mysid shrimp reduced the zooplankton populations that formed the major food of the kokanee; kokanee numbers crashed, and their terrestrial and aerial predators followed. From Spencer et al. (1991); used with permission.

118,000 fish, feeding predators and scavengers including river otter, mink, brown bear, coyote, and the densest concentration of bald eagles south of Canada, sometimes numbering over 600 birds. Mysid density peaked in the mid-1980s. Within two years, the kokanee population collapsed almost completely, and angler catches fell from 100,000 fish in 1985 to no fish in 1988. Elimination of kokanee depressed populations of all the terrestrial carnivores. In 1989, when 50 spawning kokanee migrated, only 25 eagles appeared in the region. Tourism, which focused largely on eagle watching, declined from 46,000 people in 1983 to 1,000 in 1989 (Spencer et al. 1991).

Capitalizing on trophic cascades within ecosystems, management schemes have employed species introductions to improve water clarity in some northern U.S. lakes. In a typical cascade, piscivores eat zooplanktivores that feed on herbivorous zooplankton that eat phytoplankton. Increasing piscivore numbers should reduce zooplanktivores, increasing zooplankton and leading to phytoplankton removal and greater water clarity (Carpenter and Kitchell 1993). A biomanipulation exploiting this cascade was carried out in Lake Mendota, Wisconsin, which had undergone eutrophication due to excessive fertilizer inputs that caused phytoplankton blooms. When abundant large piscivores such as northern pike and walleye were stocked, both primary production and turbidity were reduced. The piscivores ate fishes that ate zooplanktivores, leaving more zooplankton to feed on phytoplankton (Kitchell 1992; Carpenter and Kitchell 1993). An inadvertent biomanipulation occurred in Lake Michigan. Periodic undesirable "whiting events" occurred when limestone crystals precipitated and turned the water milky white. Whitings result from increased photosynthetic activity by phytoplankton at elevated summer temperatures, which removes CO_2 from the water, increasing pH. Limestone solubility decreases at higher pH, leading to precipitate formation and suspension. Large numbers of salmonids such as coho and Chinook salmon and lake, rainbow, and brown trout introduced during the 1970s consumed zooplanktivorous alewives, allowing phytoplanktivorous zooplankton to multiply. The zooplankters ate phytoplankton, which kept the pH lower and prevented the occurrence of whiting events (Vanderploeg et al. 1987).

Habitat Modification

Few fish modify habitat, but those that do, such as goldfish and various carp species, can significantly impact habitats and species. Goldfish are benthic herbivores that uproot vegetation, resuspending sediment from mud or soft bottoms while feeding. When introduced into lakes, they reproduce rapidly, achieving densities as high as 17,000 fish/ha. Their foraging activities may completely defoliate the bottom, stir up the sediment, and increase water turbidity (M. J. Richardson et al. 1995). Herbivorous carp also destroy rooted aquatic vegetation and resuspend sediments, increasing turbidity, decreasing light penetration, and impeding macrophyte regrowth. Plankton blooms occur because nutrients are released into the water column from resuspended sediments and consumed or dying macrophytes, further impairing macrophyte growth. Increased turbidity creates a refuge for small goldfish and carp, defeating attempts to control them by adding predators. Rooting in the bottom also destroys the feeding and refuge habitats of other small fishes and homogenizes complex substrates that house food organisms (e.g., Moyle et al. 1986). J. D. Williams and G. K. Meffe (1998, p. 122) equated introduced common and grass carp to feral hogs that "trample, uproot, and destroy native vegetation."

In Lake Mikri Prespa, northwestern Greece, goldfish were introduced in the late 1970s as a fishery species. Fisheries goals were met, goldfish constituting 33% of the catch in 1990; but as their numbers climbed, so did turbidity (Crivelli 1995). In India, the bottom-disturbing habits of introduced carp were directly linked to elimination of commercially valuable native cyprinids in the genus *Schizothorax* (Jingran and Sehgal 1978). Grass carp, *Ctenopharyngodon idella*, were introduced into Lake Oubeira, Algeria, resulting in complete disappearance of submerged vegetation and reed beds, with negative consequences for local fish and duck populations (Crivelli 1995).

When introduced carp are removed from a water body, which is no small task, turbidity frequently decreases and native fish abundance increases (J. N. Taylor et al. 1984). The reverence with which common carp were treated during the massive introduction efforts of the late 1800s (chapter 9) has given way to a general revulsion; this exotic is now regarded "as the ultimate trash fish, to be killed and discarded in disgust if inadvertently caught" (Moyle 1984, p. 48).

Ecosystem Processes and Function

One of the more notorious invaders in North America, the red shiner, has been widely introduced in western and eastern states and can achieve densities of more than 50 individuals/m^2 (Gido and Propst 1999). This native of the Mississippi River basin feeds on water column prey, primarily terrestrial insects and zooplankters. When stocked in experimental streams at high but realistic densities of 27 fish/m^2, red shiners caused algal productivity to increase threefold, probably due to nutrient enrichment as their feeding and excretion transferred nutrients from the water column to benthic producers (Gido and Matthews 2001). If similar processes occurred in the wild, impacts on stream processes could be significant. Headwater and upland streams are typically low in nutrients, dependent on terrestrial sources for energy, and low in algal productivity. They often contain regional endemics, which are displaced by invaders in conjunction with habitat disturbance. If red shiners and other introduced, water column species increase primary production, the harmful, positive feedback loop so often associated with introductions could be accelerated.

The complexity of interactions between natives and nonindigenes underscores the difficulties of predicting potential impacts, sounding an additional cautionary note about any proposed introduction. One species introduced into adjacent systems can have different effects, mediated through differing interactions, resulting in different responses at population, community, and ecosystem levels. Sparkling Lake and Crystal Lake are two small lakes in northern Wisconsin with similar species composition, separated by only a few km (Hrabik et al. 1998, 2001). Zooplanktivorous rainbow smelt were introduced into both lakes, probably as a result of bait bucket releases. The outcomes have differed dramatically. In Sparkling Lake, smelt preyed on young cisco, *Coregonus artedi*, to the extent that they were extirpated from the lake. In Crystal Lake, smelt competed with both juvenile and adult yellow perch, *Perca flavescens*, for limited plankton resources. Growth rates of larval perch declined, and yellow perch are expected to be eliminated from the lake. In both lakes, the affected native species were previously quite abundant. Introductions affect others besides rare and imperiled species, and impacting a common species is more likely to reverberate through an ecosystem, affecting entire lake systems rather than just a few component species.

CONCLUSION

Disturbed systems tend to be particularly susceptible to invasion by alien species (Fausch et al. 1990; Moyle and Leidy 1992). This can be generalized across habitats and taxa (Williams and Meffe 1998; see chapter 9). In fishes, success of invaders is often directly proportional to the degree of disturbance. For example, in watersheds around the city of Atlanta, Georgia, the percent of nonnative fishes was positively correlated with pesticide and heavy metal concentrations in streams, as well as with the intensity of land use (figure 8.9). In the most disturbed watersheds, alien fishes exceeded 90%, dominated by red shiner, an invasive that is exceptionally tolerant of environmental extremes (Douglas et al. 1994; DeVivo 1996). The widespread nature of the relationship between disturbance and invasion has prompted using "percent nonnative fishes" as a direct measure of degree of disturbance (e.g., in the Index of Biotic Integrity, chapter 7).

Habitat degradation may also interact with the impacts or extent of altered competition, predation, or hybridiza-

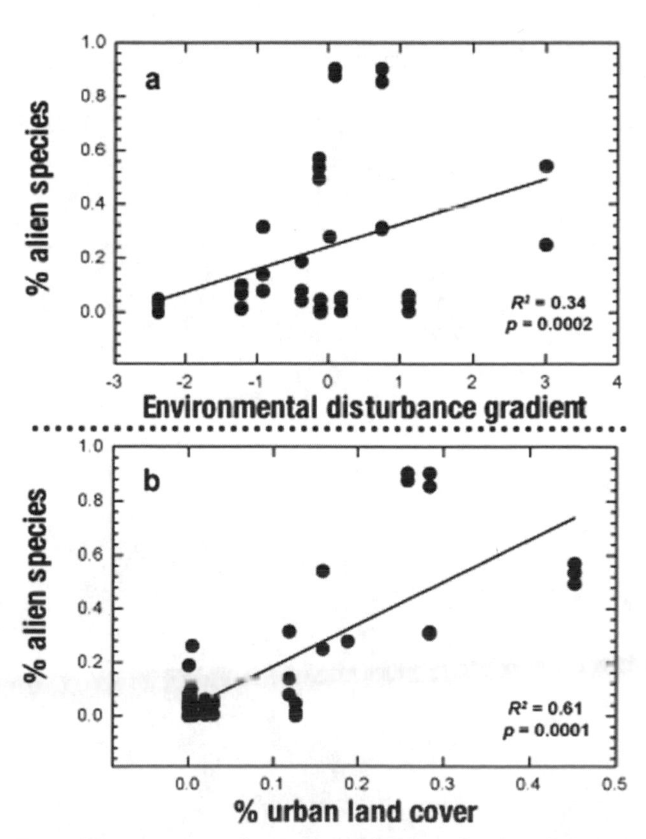

Figure 8.9. Percentage of nonnative fishes correlated with (A) an environmental disturbance gradient (mostly pesticides and heavy metals) upstream from sites, and (B) percent of the landscape converted to urban land use upstream from sites. Data from Atlanta, Georgia. Most of the nonnatives were red shiners. From DeVivo (1996); used with permission.

tion. Changes in habitat that alter food availability could easily shift the balance to favor one species over another in an otherwise symmetrical competitive contest. Habitat alteration could also eliminate essential refuges from predators, and hybridization has long been known to occur more readily in disturbed habitats (e.g., Hubbs 1955). Competition could combine with predation, particularly during stressful periods such as drought when fishes are forced together in shrinking habitats. Laboratory experiments that test for just one of these factors may miss the potential relevance of combined and interacting impacts. Competition may not be evident when explored alone, but conditions of increasing predator threat brought on by shrinking habitat could force a species into an unusual and unfavorable competitive situation. It is widely accepted that multiple factors combine to cause the demise of imperiled species; interactions among competition, predation, and habitat degradation only make sense. These interactions once again point out that to understand the processes that affect the outcomes of introductions, we must look not just at the introduced species and a few potentially impacted species but at the ecosystem as a whole.

9. Alien Species II: Understanding the Process, Minimizing the Impact

The U.S. and Germany swapped fish the way collectors swapped baseball cards. . . . Everyone could use a better fish.

—Kim Todd, 2000

Regardless of source or intent, established alien species too often cause declines among natives (see Welcomme 1984, 1988; Lever 1996). Analyzing 31 studies in 26 different systems that included preintroduction data, S. T. Ross (1991) showed elimination or decline in native fish abundance in 77% of the studies. Two-thirds of North American extinctions and over 60% of global extinctions can be linked to introductions (R. R. Miller et al. 1989; Harrison and Stiassny 1999). About 70% of imperiled U.S. fish species are threatened due in part to introduced species (Lassuy 1995).

The sequence of events and extent of effect may vary but can be rapid and large. Within one year of a largemouth bass introduction to Monkey Spring, Arizona, an undescribed endemic pupfish became extinct (Minckley 1973). Vendace, *Coregonus albula*, are now restricted in the UK to only two lakes. Within a decade of introduction of roach, *Rutilus rutilus*, and ruffe, *Gymnocephalus cornuus* (probably from released baitfish), vendace relative abundance fell from 16% and 41% to 1% and 5%, respectively, while total fish abundance rose only slightly (Winfield et al. 2002). In Lake Atitlan, Panama, introduced largemouth bass eliminated a livebearer and a cichlid that were central to subsistence and commercial fisheries (Zaret and Paine 1973). In Mexico, the introduction of potential food fishes has led to an average 80% decline in number of native, often endemic species (Contreras-Balderas and Escalante 1984).

WHY AND HOW ARE SPECIES INTRODUCED?

Humans have always transported and introduced animals and plants they found useful (e.g., Diamond 1999). Fishes are no exception. Fish introductions occur as a deliberate result of human endeavor and as a by-product of human activity. People have numerous reasons for importing and introducing species, some more admirable than others. The most common reasons behind planned introductions are to enhance aquaculture, improve sport or commercial fishing, augment and improve imperiled wild stocks, promote ornamental species, and practice biological control (Welcomme 1988; Lever 1996).

Alien species introductions are often categorized as intentional introductions, escaped intentional introductions, and unintentional introductions. *Intentional introductions* describe sport fishes that are deliberately stocked to improve fishing. *Escaped intentional introductions* include aquarium and aquaculture species (e.g., piranhas, swamp eels, Atlantic salmon, tilapia), as well as unauthorized stockings, referred to as secondary introductions and leapfrogging (Rahel 1997; Fuller et al. 1999). Unauthorized stockings are perpetrated by private citizens who take fishes from one water body and put them in another. In California, unauthorized stockings have grown increasingly common, even as fisheries managers lose enthusiasm for stocking programs (figure 9.1). Unauthorized stocking of

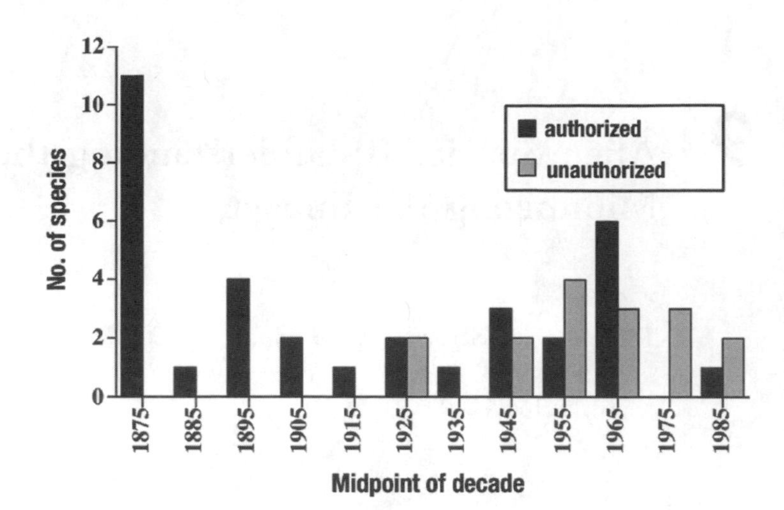

Figure 9.1. Fish introductions in California. The increasing incidence of unauthorized introductions performed by private citizens occurs across the U.S. Redrawn from Courtenay and Moyle (1996).

white bass, *Morone chrysops*, and northern pike, *Esox lucius*, in central California reservoirs costs the state millions of dollars annually. Fish and Game personnel are continually attempting to eradicate these foreign predators on imperiled salmonids and other fishes (Courtenay and Moyle 1996). The third category, *unintentional introductions*, describes hitchhikers (round goby, ruffe, sea lamprey) that are captured and released as by-products of human activity, such as through ballast water and canal introductions. Among U.S. fish introductions, species have proven equally harmful regardless of source (OTA 1993).

Stocking of aliens has historically been justified because the introduced species was "deemed to be of more immediate benefit to humans than what persisted of the native community in the . . . ecosystem" (Lassuy 1995, p. 395). Such activity often initiates a sequence, with one introduced species necessitating the introduction of another (Welcomme 1988). Predators placed in predator-free assemblages can decimate native prey species, requiring the introduction of new nonnative prey (e.g., largemouth bass followed by sunfish, peacock cichlid followed by tilapia).

This sequence can also be reversed, as when introduced species form dense populations of stunted individuals, necessitating the stocking of predators to control their numbers (e.g., tilapias requiring peacock bass, sunfishes requiring largemouth bass). Macrophyte-eating fishes often excrete or free up nutrients that lead to noxious plankton blooms, necessitating the introduction of phytoplanktivores. For example, herbivorous Chinese carps are followed by phytoplanktivorous tilapias; silver carp, *Hypophthalmichthys molitrix*; or threadfin shad, *Dorosoma petenense*. Such serial introductions in closed systems amount to polyculture, which is reasonable in an aquaculture setting. Unfortunately, this cocktail of species is likely to escape into adjoining natural habitats.

Serial introductions and other sport fishing–related efforts have often been prompted by a perceived "empty niche," such as a missing detritivore or herbivore, followed by a piscivore to feed on the introduced species. People assume that resources in a water body are not being used efficiently, due to a lack of some suitable species (Welcomme 1988). The empty-niche justification has been amply criticized from both ecological and management perspectives; it views ecosystems as simple entities open to tinkering. In reality, ecosystems are complex and likely to respond to species additions in unexpected ways (Moyle et al. 1986 and subsequent papers; OTA 1993). Introductions are often undertaken to solve immediate human problems, such as insufficient sport fishing opportunities, weeds or pests, or plankton blooms. The solution is seen in a single dimension, but its impact is likely to be multidimensional because of ecosystem interconnectedness. Moyle et al. (1986) referred to this as the "Frankenstein Effect," after Dr. Frankenstein's attempt to create a better human being without considering how it would fit into society; the attempt resulted in tragedy, as do many well-intentioned but short-sighted species introductions. "If the broad-scale consequences of each introduction are not considered, it may ultimately cause more problems than it solves" (Moyle et al. 1986, p. 422).

Welcomme's extensive (1988) analysis showed that, among intentional introductions worldwide, aquaculture-related introductions were by far the most numerous, followed by those intended to improve sport fishing, improve wild stocks for capture fisheries, and supply ornamentals. A breakdown by category for the U.S. showed 44% stocked for sport, 26% aquarium releases, 16% bait releases, 4% for conservation purposes, 2% for biocontrol, 2% from ballast water, and 6% unknown or miscellaneous (Fuller et al. 1999; chapters 13, 14). Aquaculture species, many maintained in ponds but often released or escaping into the wild, have undergone shifting popularities; salmonids came into vogue in the late 19th century, followed by common carp and goldfish (early 20th century), tilapias (mid-century), and Chinese carps (1960–80). Each has precipitated its own impacts. Salmonids preyed on and competed with natives, whereas goldfish and tilapia became overabundant pests because of their impacts on indigenous fishes and their tendency to stunt (stunting describes dense populations that breed at small sizes, becoming recreationally and commercially useless). Concern over the ecological impacts of grass carp and molluscivorous black carp is growing.

Well-intentioned introductions commonly follow a sad path of unintended consequences and "rarely meet their stated objectives" (Courtenay and Moyle 1996, p. 244). Deliberate, monitored agency efforts with plans to minimize environmental impact devolve into uncontrolled, unassessed introductions with major ecological impacts. Extensive and repeated stockings by management agencies are followed by capture and unauthorized secondary releases into additional water bodies by fishing enthusiasts. Few introductions of any type remain confined, and few successful introductions are reversible (e.g., Cooper 1987; Tyus and Saunders 2000).

At least 10% of introductions are accidental or inadvertent (Welcomme 1988). Fishes can be carried along during water transport (via canals, interbasin water transfers, ballast water introductions, diffusion through natural waterways), escape during live bait fishing, hitchhike during sport and commercial fish introductions, and be carried or pumped from aquaculture ponds during flooding or cleaning. Even Lever (1996, p. xix) concluded that accidental escapes and releases "have almost invariably proved ecologically and/or economically detrimental," although he cited as exceptions three commercially viable populations that originated as aquarium introductions (giant gourami, *Osphronemus gouramy*, in Madagascar; tawes, *Barbus javanicus*, in Sulawesi; and oriental weatherfish, *Misgurnus anguillicaudatus*, in the Philippines).

Until recently, fisheries agencies commonly felt an obligation to transport and help establish useful (to humans) species wherever and whenever possible. Ecological impacts received little attention or concern. Introductions were seen as "a welcome enrichment of the [native] biota" (Williams and Meffe 1998, p. 117), because little distinction was made between native and introduced species. Stocking fish was "the epitome of natural resource management" (Rahel 1997, p. 8). But official attitudes toward introductions are undergoing a dramatic change. Although hatcheries still abound and fisheries agencies often view their major responsibility as providing catchable fishes for the angling public, the wholesale stocking of game species regardless of ecological impact has lost favor. Awareness of the impacts of stocking on natural biodiversity is widespread and growing. A biocentric view that values the intrinsic qualities of nature is challenging the traditional anthropocentric view that nature exists primarily for human benefit (Rahel 1997).

CASES IN POINT: INTRODUCTIONS IN THE UNITED STATES

Historical Trends

The earliest documented Western fish introductions occurred in the first century AD, when Romans transported common carp, *Cyprinus carpio*, from the Danube River to western Roman provinces and maintained them in reservoirs called *piscinae* (Balon 1974, 1995). The Chinese may have domesticated *C. c. haematopterus* as much as five centuries earlier (Balon 1974).

Carp were also among the first fish introduced into the U.S. The U.S. Fish Commission's initial task was to determine which species could be introduced to bolster the nation's food supply (Fritz 1987; Rahel 1997). In 1874, it turned to common carp because, it reported, "It is quite safe to say there is no other species that promises so great a return in limited waters" (U.S. Fish Commission 1874 in Fritz 1987). Carp were lauded as having "as fine a flavor as any fish . . . [being] more palatable than even the salmon or halibut. . . . [As a game fish] there is nothing in our waters that equals them" (Smiley 1884 in Rahel 1997). The hype reached dizzying heights: "We trust the day is not far off when the carp pond, shaded with big trees and willows, and decorated with rose bushes and flowers, will be the possession of every farmer who aspired to thrift, taste, and good living" (*Texas Farm and Ranch Magazine*, as quoted in Fritz 1987, p. 19).

The Fish Commission began importing carp from Germany in 1877, starting with 345 fish. These were raised and bred in a hatchery on grounds now occupied by the Washington Monument. By 1879, 6,000 fingerlings were shipped to 24 states; in 1882, 260,000 carp were shipped to 9,872 applicants in 298 of the 301 U.S. congressional districts; between 1879 and 1896, the Commission distributed 2.4 million carp across the U.S. and to Canada, Costa Rica, Ecuador, and Mexico; and in 1899, 6.3 million pounds of carp were harvested from the Illinois River alone. Fritz (1987, p. 19) concluded that "the U.S. Fish Commission had indeed accomplished what it had set out to do" because carp today may be the most abundant (and least regarded) freshwater fish in North America (Moyle 1984).

The magnitude and pace of introductions in the U.S. accelerated after the late 1800s. Approximately 536 alien fish taxa (species, hybrids, and unidentified forms) have been introduced, 35% imported from foreign countries and 61% translocated within the nation (Nico and Fuller 1999; Fuller et al. 1999; http://nas.er.usgs.gov; table 9.1). Half of the foreign exotics established breeding populations. Before 1850, carp and goldfish (*Carassius auratus*) were probably the only introduced fishes. Over the next 50 years, 71 mostly native fish species were moved around, primarily from species-rich eastern states to relatively depauperate western states, capitalizing on the newly constructed transcontinental railway system. Some species moved into new watersheds via newly constructed canals. Between 1900 and 1950, 110 additional species were introduced into and around the country. Although food and sport fishes were most popular, many small mosquito-eating fishes were transported to combat malaria. About 18 exotic species, some associated with the nascent aquarium trade, were also introduced at this time.

In the past 50 years, the pace has quickened, the number almost tripling between 1950 and 2000, to 536. Food, sport (including bait usage), and insect and weed control are reasons, but ornamental species contributed increasingly to the growing number of nonnative introductions, estimated at 150 since 1950. Transcontinental trucking, airlines, and ballast water supplemented earlier modes of transport. The current distribution of introduced fishes among states (figure 9.2) shows that warmer and western states received most species from outside or across the country, whereas introductions in eastern states are likely due to "native" species. Florida's large number (ca. 90) of exotic species largely reflects escapees from ornamental farm operations (Nico and Fuller 1999).

Table 9.1. The thirteen most commonly introduced fish species in U.S. waters

Species	*No. states where introduced*[a]	*Native to*	*Primary purpose of introduction*
Cyprinus carpio, common carp	49	Eurasia	Food, ornamental
Carassius auratus, goldfish	49	E. Asia	Ornamental
Oncorhynchus mykiss, rainbow trout	48	W. North America	Game fish, aquaculture
Salmo trutta, brown trout	47	Eurasia	Game fish
Ctenopharyngodon idella, grass carp	45	E. Asia	Vegetation control
Micropterus salmoides, largemouth bass	43	E. North America	Game fish
Zander vitreum, walleye	43	E. North America	Game fish
Micropterus dolomieu, smallmouth bass	41	E. North America	Game fish
Salmo salar, Atlantic salmon	40	E. North America, W. Europe	Game fish, aquaculture
Pomoxis annularis, white crappie	38	E. North America	Game fish
Tinca tinca, tench	38	Europe, W. Asia	Food, game fish
Morone saxatilis, striped bass	37	E. North America	Game fish
Esox lucius, northern pike	36	N. North America, Europe, Asia	Game fish

Source: Number of states where fish were introduced is from Fuller et al. (1999).

Note: Note overlap between this list and the bad actors identified in table 8.1.

[a]Includes translocations within states into areas where the fish did not naturally occur.

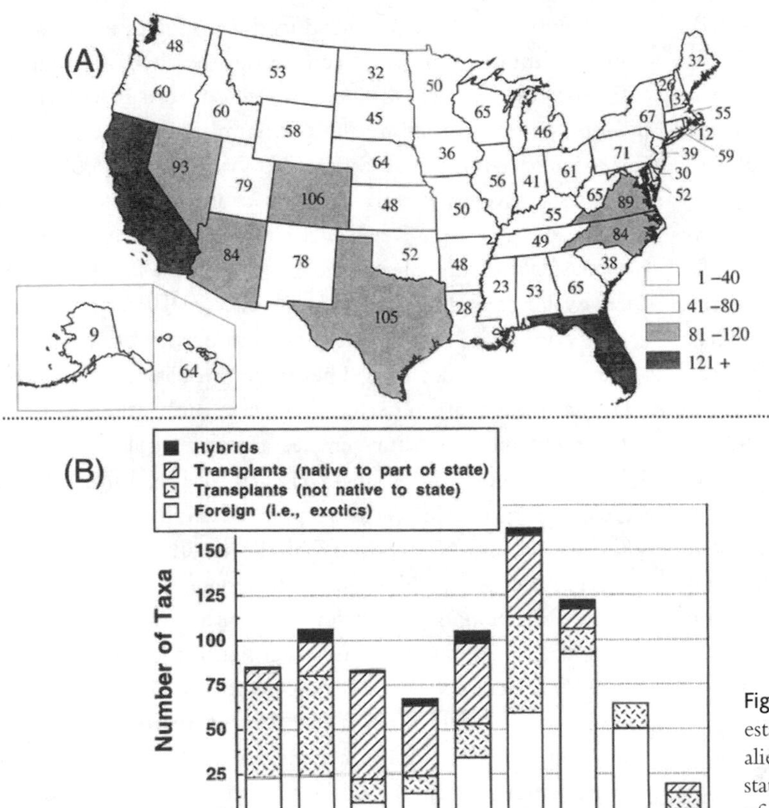

Figure 9.2. Introduced fishes in the United States, including established and nonestablished species. (A) Number of alien fish taxa by state (Fuller et al. 1999). (B) Selected states and their introduced fish faunas, showing the sources of introductions (Nico and Fuller 1999).Used with permission.

Aliens Across the Landscape

The geographic extent of introductions and translocations is impressive. At least 32 species have been introduced into more than half of the 50 U.S. states (Fuller et al. 1999; Rahel 2000). Of these, 28 are native to the U.S. In fact, 41% (327 of 790) of U.S. native freshwater fishes have been transported to at least one nonnative locale. Such active transport is understandable for sport species, but the extent to which small, seemingly nonutilitarian species have turned up in new, often distant locales is mystifying. More than 100 species of cyprinid minnows and 20 species of percid darters (plus 27 suckers, 3 mudminnows, 6 silversides, 13 topminnows, 24 livebearers, 4 sculpins, and 9 gobies, among other mostly small fishes) have been moved outside their native ranges; most did not become established (Fuller et al. 1999).

Some imperiled species have been translocated as part of recovery plans, especially when native habitats were degraded but nearby habitats could support fish (e.g., minnows, madtom catfishes, pupfishes, livebearers). Some populations became established in new areas as the result of scientific experiments, such as in the 1930s to test the effects of changed environments on countable and measurable traits (e.g., tui chub, *Gila bicolor*; arroyo chub, *G. orcutti*; speckled dace, *Rhinichthys osculus*) (R. R. Miller 1968). An unknown but presumably small number of introduced natives represent aquarium releases, but this number will increase as the practice of keeping native fishes in aquaria grows (see chapter 13).

Many small fishes have apparently hitchhiked along with stocked game fishes. Tadpole madtoms, *Noturus gyrinus*, are native to eastern drainages of the Mississippi River but have contaminated shipments of channel or bullhead catfish sent to New Hampshire, Massachusetts, Idaho, Washington, and Oregon. Similarly, river carpsucker, *Carpiodes carpio*, entered Lake Erie in a shipment of buffalofishes, *Ictiobus* spp. Several species of suckers have accompanied trout introductions; and minnows such as the sand shiner, *Notropis ludibundus*, are thought to have accompanied (introduced) largemouth bass into Arizona. Bigscale logperch, *Percina macrocephala*, a darter native to Texas and adjacent states, found its way into California and Colorado in shipments of largemouth bass and bluegill sunfish (Fuller et al. 1999).

Interbasin water transfers explain the appearance of some species in new areas. Among others, the Owens sucker, *Catostomus fumeiventris*, an Owens River, California, endemic, was transported to June Lake and the Santa Clara River when water was transferred from the Owens River valley via the notorious Los Angeles Aqueduct of *Chinatown* and *Cadillac Desert* fame (Fuller et al. 1999; see Reisner 1986). Becker (1983) listed at least one gar, two minnows, a topminnow, and two darters native to the Mississippi River that appeared in the Lake Michigan basin after the two were connected by the Wisconsin-Fox Canal. Federally endangered tidewater gobies, *Eucyclogobius newberryi*, are potentially threatened by introduced shimofuri goby, *Tridentiger bifasciatus*, that moved from San Francisco Bay to other estuarine regions via canals and reservoirs (Matern 2001).

Ballast Water Introductions

When large ships traverse oceans or coastlines without full holds, they often take on water as ballast in special tanks. This water and any included organisms are pumped out when cargo is loaded later. Ballast water sampled from five vessels in Hong Kong Harbor contained 81 species in eight animal and five protist phyla (Chu et al. 1997). Extrapolating from the international extent of shipping, Carlton (1999) estimated that over 7,000 species are transported daily in ballast water, including serious human pathogens. In Peru in 1991, 1 million people were infected and over 10,000 died from cholera by drinking water contaminated with *Vibrio cholerae* transported in ballast water (Bright 1998). Ballast water brought to the U.S. zebra mussel, *Dreissena polymorpha*, and predatory spiny water flea, *Bythotrephes cederstroemi*, both of which have drastically altered lake ecosystems and food resources for fishes. An American export, the ctenophore, *Mnemiopsis leidyi*, was introduced via ballast water into the Black and Azov seas of Asia, where it reached densities of 180 individuals m^3. It competed with and ate native fish larvae and has been linked to loss of a $250 million anchovy fishery (Ruiz et al. 1997; Shiganova and Bulgakova 2000).

Globally, at least 32 introduced fish species in 11 families are thought to have been transported by ballast tanks; 31 species have actually been collected from ballast tanks. Gobies and blennies are the two families most commonly associated with ballast water (Wonham et al. 2000). Four gobies native to eastern Asia (shimofuri goby; yellowfin goby, *Acanthogobius flavimanus*; chameleon goby, *Tridentiger trigonocephalus*; and shokihaze goby, *T. barbulatus*) were ballast water introductions to San Francisco Bay, where they often dominate fish catches at the same time that pelagic fish density has declined. Their invasion has been aided by habitat modification because concrete rubble used to reinforce channel shorelines creates abundant goby spawning habitat (Meng et al. 1994; Matern and Fleming 1995). Yellowfin gobies have also appeared in San Diego Bay, California, and New South Wales, Australia.

Round goby, *Neogobius melanostomus*, and tubenose goby, *Proterorhinus marmoratus*, both native to the Black and Caspian seas, arrived in the Laurentian Great Lakes around 1990 via ballast water and quickly spread through all five lakes. Round gobies reach densities of up to $133/m^2$ and are egg predators. At high densities, they could compete with native sculpin and affect benthic spawners such as lake sturgeon, *Acipenser fulvescens*, and lake trout, *Salvelinus namaycush*, thus compromising expensive rehabilitation efforts (e.g., Dubs and Corkum 1996; Chotkowski and Marsden 1999; K. M. Jones, pers. comm.). Electric barriers are being tested on the Chicago Waterway as a means of preventing round gobies from entering the Mississippi River from Lake Michigan. These barriers may also keep Asian carps in the Mississippi from moving in the other direction (Savino et al. 2001).

Another ballast water transportee is the Eurasian ruffe, a small percid pumped into Duluth Harbor, Minnesota, in the 1980s. Ruffe are highly adaptable generalists and proven invaders in European locales (Devine et al. 2000). Ruffe spread along the south shore of Lake Superior and are now the most abundant fish in the St. Louis River estuary of the lake. Continued spread and negative interactions with native yellow perch, *Perca flavescens*, and whitefish, *Coregonus* spp., are major concerns, prompting considerable effort and expense to arrest their movement and minimize impacts.

Proposals for preventing future ballast water introductions range from the practical to the practically impossible (see NRC 1996a) and include ultraviolet irradiation, filtration, salinization, heating, chilling, chemical additives, water exchange, nitrogen purging to eliminate oxygen, and shoreside retention facilities for ballast water (Rigby et al. 1999; Tamburri et al. 2002). The quantities of water involved are prodigious—an average commercial ship holds over 30,000 MT of ballast water (L. D. Smith et al. 1999)—disqualifying several solutions outright. Ships traveling at sea between freshwater ports could replace freshwater with seawater, which would presumably kill most stowaways. Water exchange far from shore in the open ocean (mid-ocean exchange) or the Great Lakes should eliminate most nearshore-adapted species and slow their transmission.

Ballast introductions will continue because international shipping is so prevalent, and because the future bodes "more ships with larger volumes of ballast water arriving from more regions in less time (due to faster speeds)" (Ruiz et al. 1997, p. 623). However, care in where water is loaded, unloaded, and exchanged should at least slow the spread (e.g., Moyle 1991; G. J. Ryan 1996). Implementation of such procedures, especially mid-ocean exchanges, has been greatly facilitated by provisions and penalties established in 2002 and 2004 in accordance with the U.S. Nonindigenous Aquatic Nuisance Prevention and Control Act and the National Invasive Species Act. Enforcement of these regulations is written into cooperative agreements between U.S. and Canadian governments, and in industry guidelines for control and management of ballast water "to minimize the transfer of harmful aquatic organisms and pathogens" (UN International Maritime Organization, http://globallast.imo.org; U.S. Department of Transportation, 33CFR Part 401, SLSDC 2002-11358-3).

Forage Fish and Bait Bucket Introductions

Small fishes are also introduced via stocking as forage species for (often introduced) predatory sport fishes and by bait buckets. About one-third of U.S. established introductions have followed these routes (Courtenay and Taylor 1984). The stocked-as-forage category includes lake chubsucker, *Erimyzon sucetta*, introduced into Nebraska; clupeids such as alewife, *Alosa pseudoharengus*, and various shad species stocked as forage for (often stocked) largemouth bass and striped bass; silversides (e.g., inland silverside, *Menidia beryllina*); and numerous cyprinids (Fuller et al. 1999).

Bait buckets are the most common means of moving small fishes. Bait is collected by individuals using seines and traps and by commercial operations that culture species in tanks and ponds. The commercial scale of the bait industry is valued at more than $1 billion annually in the U.S. and Canada. In Ontario and Minnesota, more than 930 million baitfish are harvested annually (Litvak and Mandrak 1993).

Bait buckets have led to many unexpected introductions. The only characin native to the U.S., the Mexican tetra, *Astyanax mexicanus*, occurs naturally in the Texas–New Mexico region but has been introduced via probable bait bucket dumping (or aquarium releases) in Arizona, New Mexico, Oklahoma, Louisiana, and Ohio. The margined madtom, *Noturus insignis*, native to Atlantic slope drainages from New York to Georgia, is a popular bait for black basses and has been introduced into Michigan, Tennessee, New Hampshire, and Massachusetts, plus two sites in Ontario, Canada, at least 250 km north of the known native populations (Litvak and Mandrak 1993; Jenkins and Burkhead 1994). Blacktail shiner, *Cyprinella venusta*, native to Gulf of Mexico drainages, has established populations in Georgia, Tennessee, Missouri, Nevada, and possibly Utah, which is a long way to haul a bait bucket. Minnows are especially popular. Fuller et al. (1999) concluded that populations of at least 53 U.S. cyprinids, including several federally listed forms, are known or probable bait introductions. It is estimated that Montana alone has received 210 illegal fish introductions, largely from bait bucket dumping (Vashro 1995).

Both harvesting and introducing bait affect ecosystems (Litvak and Mandrak 1993). Impacts on source regions include bycatch, alteration of food webs, physical damage to habitat due to capture methods, and reduction of forage fish populations (including imperiled species). In receiving systems, negative effects include habitat alteration (e.g., from goldfish and other rooting herbivores), predation on eggs and larvae of natives, competitive displacement of natives, gene pool alteration in the case of translocations, introduction of disease, and incidental transport of nontargeted species such as zebra mussels and spiny water fleas.

Among the 45 U.S. states and 10 Canadian provinces surveyed by Litvak and Mandrak (1993), few prohibited release or limited transportation of baitfish. Where regulations existed, definitions of bait and restrictions on production, harvest, transport, and use varied. Restrictions often applied only to trout waters, where the concern was with trout capture during bait harvesting and with deep-hooking and other injuries to trout during bait use. The purpose of most regulations appears to be more of regulating commerce and promoting local over nonresident exploitation of bait resources, with minimal consideration given to protecting diversity or encouraging sustainable use (Meronek et al. 1995).

Public education and enforcement are also secondary. Most users of baitfish are either ignorant of regulations and their importance or violate them knowingly. About 40% of anglers surveyed released unused live bait when they were done fishing, and most "thought they were doing something beneficial for the ecosystem" (Litvak and Mandrak 1993, p. 10). In Ontario, 15 legal bait species were considered Vulnerable or Threatened in Canada. At four bait dealerships in Toronto, 6 of 28 species in holding tanks were illegal. Anglers intended to use 18 of the 28 outside of the species' known geographic ranges. Meronek et al. (1995) found that more than half of the commerce of the bait industry was likely to involve interstate transport of fishes. These real-world events mean that more species are going to be spread to more places. H. R. Ludwig and J. A. Leitch (1996) calculated that the probability of translocating alien fishes from the Mississippi River to the Hudson Bay drainage is only 1.2% for any one angler on any one day, but when extrapolated to an expected 19 million angler days per year, the probability of 10,000 introductions per year approaches 100%.

The combination of ineffective regulation and minimal public awareness prompted Litvak and Mandrak (1993) to propose greater limits on the species and habitats that could be used for bait production; limits on the number of bait collecting trips allowable at any given site; emphasis on cultured bait production, ideally through production of sterile bait fishes to minimize reproduction by escapees; and promotion of preserved baitfish, which would curtail both reproduction and disease transmission (but would probably prove politically unpopular). The most promising and practical of their recommendations entails restrictions on transport of live bait coupled with allowable harvest by individual anglers or commercial operators only for use at the harvest site. Such proscriptions are unlikely to be well received by commercial bait producers. Meronek et al. (1995), focusing on the contradictory regulations among states, recommended cooperative arrangements that led to legislation that at least attempted "to achieve more uniformity among states consistent with protection and wise use of aquatic resources" (p. 22). Bait shops should be provided with posters to educate customers about the impacts of bait introductions. At the level of effective individual action is the simple act of tossing unused bait into the woods or flushing it down a toilet (ethical issues of humane disposal are treated in chapter 13).

As an aside, in my home state of Georgia, a recreational fishing license gives me the right to use a small seine to capture, keep, and transport live bait fishes as long as they are not a game or state-listed species. I cannot, however, keep native fishes in a home aquarium as pets. As a consequence, I maintain a 200-liter glass bait bucket in my living room stocked with local, common minnows. I just don't get out to fish with them very often.

MARINE INTRODUCTIONS: THE EXCEPTION AGAIN

Numerous marine invertebrates and algae have been introduced, with an ecological impact that is truly astounding, particularly in nearshore environments (Carlton 1999, 2000); given the extent of international shipping, the number can only grow. Introduced invertebrates affect fishes by disrupting the mix of predators, competitors, and prey in an area, altering ecosystem function. One of the few truly imperiled marine species, the Critically Endangered spotted handfish of Tasmania, is in part threatened by an introduced sea star that eats both handfish eggs and the ascidians to which the handfish attaches its eggs (Bruce et al. 1998). Marine fish species are also being transported and established, chiefly through deliberate introduction to enhance fisheries, as hitchhikers in fisheries shipments, via ballast water release, on ship bottoms, and as a result of canal construction and water transfer (Baltz 1991; Wonham et al. 2000). Relatively few fully marine fish introductions have been documented, and their ecological impacts appear relatively minor or poorly understood. Most occur in inland seas and estuarine habitats, or involve euryhaline species. This distribution may reflect high human population densities, extensive habitat disturbance, and the relatively closed nature of inland seas. Estuaries and bays are also major ballast water sources, which could explain the predominance of estuarine organisms among introduced species (Ruiz et al. 1997).

Even successful marine introductions reflect a freshwater influence. The only established marine exotics on the U.S. Pacific coast are anadromous species (striped bass and American shad; Baltz 1991). Freshwater's influence is also evident in the distribution of introductions into the San

Francisco Bay area. At the estuary's upper, fresher end, 20 of 42 fish species are introduced and dominate the assemblage, 2 native species are extirpated, and a third is Endangered. In the more saline bay, only 5 of 57 species are introduced. Along the open ocean coast, striped bass and American shad are the only introduced species among 400 native species (Baltz 1991).

The lack of successful marine introductions doesn't reflect lack of effort. The former Soviet Union attempted repeatedly to introduce 29 fishery species, only 11 of which were deemed successful (Baltz 1991), 10 in the relatively closed Aral and Caspian seas. In the more open Baltic and Barentz seas, only anadromous pink salmon, *Oncorhynchus gorbuscha*, succeeded among 9 introductions. Several "successful" Aral Sea introductions are thought to have contributed to the demise of all fisheries in that water body (see chapter 6). Stock enhancement programs, in which larvae of local marine species are reared to replenish depleted wild fisheries, have also been generally unsuccessful and have been shown "to have little or no impact on the enhancement of marine fish stocks" (W. J. Richards and Edwards 1986, p. 78; see chapter 14).

Hawaiian Introductions

Hawaii is the only area in Oceania known to contain introduced marine fishes (Baltz 1991), but most attempts failed (Randall 1987b; Planes and Lecaillon 1998; see chapter 13). Groupers and snappers are naturally scarce in the geographically isolated Hawaiian archipelago. To fill the "empty" large-predator niche, management agencies introduced 11 grouper and snapper species. Only 3—bluespotted grouper, *Cephalopholis argus*; blacktail snapper, *Lutjanus fulvus*; and bluestripe snapper, *L. kasmira*—were established, and only bluestripe snapper are abundant. In fact, they are now one of the more common species one encounters diving in moderately deep Hawaiian waters (Friedlander et al. 2002). Fishers consider it a nuisance because of its small size, abundance, and bait stealing habits. In all, 16 marine fish species have been introduced into Hawaiian waters. Thirteen were deliberate; the other 3 probably arrived as contaminants in shipments of intended introductions. Interestingly, the 3 inadvertent releases (a mullet, a goatfish, and a herring) have been successful, although inadvertent introductions that failed would probably go unrecorded. Only 4 of the deliberate releases have survived. Seven brackish water species were also transported to Hawaii to provide live bait in the tuna fishery. Individuals escaped, and 3 species are now well established, including 2 tilapia that have assumed pest status in nearshore areas. "None of these introduced fishes have fully attained what was expected of them, and all have been criticized for one negative attribute or another" (Randall 1987b, p. 500).

Why have Hawaiian or for that matter most deliberate marine introductions failed? In Hawaii, some introductions were just stupid, e.g., temperate, anadromous species such as Chinook salmon and striped bass. Failure could have been similarly anticipated for Soviet introductions involving inappropriate temperature, salinity, and spawning habitat (Baltz 1991). But most Hawaiian introductions were of Pacific reef fishes that presumably would find Hawaiian waters hospitable. The blame may lie less on biology and more on chance. Failed introductions typically involved small numbers of individuals—dozens or a few hundred—whereas successful introductions often involved larger numbers. Given the astronomically high mortality rates of planktonic larvae, lack of familiarity of mature fish with appropriate spawning locales, and probable insufficient numbers of adults to form active spawning aggregations, among other factors, reproductive failure should have been anticipated if not predicted. Introductions into tropical environments seem ill advised given the diversity and abundance of native fishes. Fishing can be promoted and sustained through habitat protection and sustainable fisheries practices, without resorting to importation of exotics, with all their proven and unpredictable liabilities.

Transfer via Canals

More marine fishes have been introduced through canals than any other human-influenced means (Baltz 1991). In canal transfers, most climatic and ecological barriers that normally inhibit establishment are absent. Fish are transferred repeatedly to a similar latitude in uncontrolled numbers between faunas occupying similar climatic zones. Two marine examples with different outcomes show the range of possible impacts. The Suez Canal, a sea-level canal that opened in 1869, has been a major conduit for introduced species moving between the Red Sea and the Mediterranean basin, with additional species transiting annually. At least 59 species have traversed the canal from south to north, and as of 2001, 41 Indo-Pacific species in 35 families were established in the Mediterranean (Golani 1999; Golani et al. 2002). Only 3 species—2 gobies and a serranid—are thought to have moved the other way, perhaps as ballast water transports (Baltz 1991). Some migrants have undergone population explosions and range expansions in the invaded region. The brushtooth lizardfish, *Saurida undosquamis*, was first captured in the Mediterranean in 1952. Only three years later, landings reached 266 MT and the species accounted for more than 20% of the trawler catch off the Israeli coast. Brushtooth lizardfish and another successful invader, the Red Sea goatfish, *Upeneus moluccensis*, have apparently displaced two Mediterranean natives (another goatfish and a hake) into deeper, cooler water (J. C. Briggs 1995). Not surprisingly, many of the successful migrants are widely distributed generalists (Golani 1993).

The Panama Canal connects the Caribbean Sea with the Pacific Ocean but involves a marine-to-freshwater-to-marine transition. Five or so marine species (all gobies and blennies) have made the 80-km trip across the Isthmus of Panama, perhaps as ballast water introductions (Wonham et al. 2000). A sea-level canal was proposed by the U.S. in the 1960s to overcome political and economic instability in the Panama Canal Zone (one early plan called for excavation using nuclear devices!). The scientific community opposed a sea-level canal because of anticipated long-term ecological impacts, in part because of known effects of the Suez Canal (J. C. Briggs 1969a, 1969b; Aron and Smith 1971). Such a canal would mix the long-separated faunas of the Pacific and Atlantic, potentially causing the extinction of many species. Scientific concerns held sway over economic and political forces, and the project was abandoned.

A BIOLOGICAL IRONY: PESTS ABROAD, IMPERILED AT HOME

Some widely reviled introduced species face an uncertain future in their native habitats. Wild stocks of sea lamprey, brown trout, rainbow trout, brook trout, and common carp suffer from the usual litany of insults including habitat degradation, overfishing, fragmented populations, and disrupted migrations. Some are also victims of their success as introduced species; their genetic integrity is continually compromised due to widespread culturing efforts, followed by release of cultured stocks into native habitats. Some are threatened by introduced species.

Few introduced species are as maligned in North America as the sea lamprey. Given its depredations on sport and commercial fishes in the Laurentian Great Lakes and its limited aesthetic appeal, the sea lamprey has few North American advocates. Yet, as mentioned in chapter 8, it is a desirable, commercially exploited food species throughout Europe, requiring protection in several regions (Elvira 1996; Keith and Allardi 1996; *Red Book of Russia* 2000). It is red-listed in several European countries (e.g., Portugal, Spain, France, Russia) and declared "in need of designation of special areas of conservation" in the 1992 *European Union Habitat Directory*. Sea lampreys are threatened by a number of human endeavors, including dams that block spawning migrations, pollution (particularly from olive oil production in Portugal), spawning habitat decline associated with water extraction for agriculture and tourism, overfishing, and illegal poaching. A UK conservation program for sea lamprey has been funded by the LIFE-Nature Programme under the European Community Habitats and Species Directive. Measures are being instituted in Portugal to reverse population declines (R. Araujo, pers. comm.).

Brown trout, *Salmo trutta*, are large, adaptable, generalized predators that have driven numerous native fishes into decline (chapter 8). Major impediments to further spread of introduced populations are warm temperatures and natural physical barriers. The trout's relative inability to surmount geomorphological barriers may help explain its tendency to form genetically and ecologically distinct populations, stocks, and management units in its native Europe (e.g., Sanz et al. 2000; Carlsson and Nilsson 2001). In many areas, strains have become genetically homogenized because of introductions of hatchery-reared brown trout, often a result of well-intentioned augmentation programs. In central Spain, seven of eight previously unique brown trout populations were found to contain genes derived from introduced, hatchery strains (Machordom et al. 1999). Brown trout also form genetically distinct stocks in Irish lakes (e.g., Ferguson and Mason 1981), some of which have been depleted by introductions of northern pike, *Esox lucius*. Controlling pike has become a major task of fisheries managers in those regions. Brown trout, identified as *Salmo trutta* or one of its recognized subspecies or stocks, have conservation status in several European countries, including Austria, Germany, Portugal, Russia/FSU, Spain, Switzerland, and Turkey.

Common carp—the "ultimate trash fish" by U.S. standards—have been introduced into almost 50 different countries, where they are often implicated in the demise of native species. Carp are probably the most widely cultured and domesticated fish in the world, accounting for close to

10% of the world's total production of aquacultured freshwater fish. Yet in their native range, genetically pure strains of wild carp are rare and have Endangered status in such large river systems as the Danube. Their diminished natural occurrence is a function of destruction of floodplain spawning habitat and introgression with escaped or stocked, domesticated forms (Balon 1995).

A similar loss of genetically pure, wild populations has happened to *Oncorhynchus mykiss*, another widely introduced species. *O. mykiss* occurs as coastal and inland varieties of rainbow trout, inland strains of redband trout, and sea-run steelhead salmon (Behnke 1992). Steelhead have been extirpated from 45% of their historic range in the western U.S. Healthy runs are rare in California, and remaining runs are commonly contaminated with genes from hatchery fish, complicating efforts at enumerating native stocks (C. Huntington et al. 1996). Approximately 38% of steelhead stocks in British Columbia and Yukon, Canada, are extinct or threatened (Slaney et al. 1996). Nehlsen et al. (1991) listed 23 extinct stocks of steelhead from the Pacific Northwest region of the U.S. and an additional 74 that were threatened with extinction. Nine steelhead taxa have federal Endangered or Threatened status. Eight unnamed subspecies of rainbow trout are considered extinct (R. R. Miller et al. 1989), and another eight subspecies are listed as imperiled by J. E. Williams et al. (1989). Among the few remaining pure strains are rare Mexican subspecies at the southern fringe of the range, such as the Baja California rainbow trout, *O. m. nelsoni* (Nielsen 1998b).

Several other species have been widely introduced but have undergone serious population depletions in their native habitats. Atlantic salmon, *Salmo salar*, are imperiled on both sides of the Atlantic, with millions of dollars spent annually on habitat improvement and species rehabilitation. One major threat to recovery of Atlantic salmon is the inevitable escape of and probable crossbreeding with abundant sea-ranched salmon, some of which may be genetically modified. Striped bass, *Morone saxatilis*, became so depleted in Chesapeake Bay from overfishing, pollution, and habitat destruction that a complete moratorium was declared on sport and commercial fishing (NMFS 1993; M. R. Ross 1997; chapter 10); meanwhile, they were introduced into California in the late 1800s and now occur from British Columbia to Baja California (Baltz 1991). In San Francisco Bay, striped bass may prey on young, Endangered salmon (Safina 1997).

In Italy, interactions among cyprinids show that even relatively short-range transfers can produce similar results. *Rutilus rubilio* has been extirpated from native locales in central Italy due to transfers of native *R. aula*, among others, while *R. rubilio* itself is an introduced pest in rivers in southern Italy, where it is thought to have eliminated the endemic cyprinid, *Alburnus albidus* (Leidy and Moyle 1997). Vendace, *Coregonus albula*, are imperiled in the UK. Predation on vendace eggs by introduced ruffe (a probable bait bucket introduction) has caused declines in one of the two lakes where vendace still occur (Winfield et al. 1998). Vendace were introduced into Finland and spread to the Pasvik River on the Norway-Russia border; after six years, they occupied the entire 120-km watercourse, where they dominate the pelagic zone and displace commercially more desirable whitefish, *Coregonus lavaretus*. This invasion "represents a threat for both biodiversity conservation and the commercial fishery in the watercourse" (Amundsen et al. 1999, p. 405).

The phenomenon of loathed-abroad-but-loved-at-home shows that species can be both noxious invaders and beleaguered natives. The most likely explanation for ecological success in foreign lands probably results from the phenomenon of ecological release. Species normally live in coevolved communities in which natural interactions keep populations in check. When released from these constraints, introduced species undergo population explosions (e.g., the Pasvik River is north of the natural range of vendace and probably lacks many community members with which vendace evolved).

INVASIBILITY: ATTRIBUTES OF INVADERS AND INVASIBLE COMMUNITIES

How concerned should we be that Asian swamp eels have been found in the Everglades, that piranhas have been caught in a Texas reservoir, that snakeheads are in the Potomac River in Maryland, or that Atlantic salmon have escaped from aquaculture facilities in British Columbia? From a conservation standpoint, it is important that we understand the likelihoods that an introduced species will successfully establish itself in a new habitat and, if established, will seriously affect native fishes and their ecosystems. If we feel confident that a species has a low probability of surviving and reproducing, or that its impact will be minimal, resources can be directed at greater threats. It is also important to understand what conditions might prevent invasions. What regions and species seem immune or

BOX 9.1. Why Invasions Succeed

Certain characteristics are often regarded as improving the chances of successful invasion by fishes. The following lists are based on data and syntheses primarily in J. D. Williams and Meffe (1998), supplemented with information in J. N. Taylor et al. (1984), Welcomme (1988), Moyle and Light (1996a, 1996b), Nico and Fuller (1999), Gido and Brown (1999), McIntosh (2000), and Kolar and Lodge (2001, 2002). In the first list, "Characteristics of Successful Invaders," asterisked traits are the opposite of those that appear to characterize vulnerable fish species (see table 4.4). Different traits may affect success at different stages in the invasion process (e.g., Kolar and Lodge 2001, 2002).

CHARACTERISTICS OF SUCCESSFUL INVADERS

- *High reproductive rate, including high fecundity and short interbreeding interval
- *Short generation time with rapid maturation, particularly if species form dense populations of stunted individuals
- Long-lived but with rapid maturation (In fishes, long life often accompanies long generation time and late onset of maturation, which increase vulnerability.)
- *High dispersal rate
- *"Pioneer species," e.g., good colonist (encompasses a number of "r-selected" life history traits)
- Biparental care unnecessary and possibly a liability (but see Annett et al. 1999)
- *Broad native range
- *Abundant in native range
- *High genetic variability and phenotypic plasticity
- *Tolerant of wide range of water quality criteria
- *Ecological generalist with respect to habitat and trophic requirements (Piscivores, detritivores, and herbivores are most successful; zooplanktivores may be successful.)
- Previous history of successful invasions by the species itself or close relatives
- Gregarious
- Ability to breathe air
- Medium body size (Large size is advantageous for predators.)
- Human commensal

CHARACTERISTICS OF HIGHLY INVASIBLE HABITATS AND COMMUNITIES

- Human-modified habitat, such as those with altered flow regime, impoundment, channelization, sedimentation, deforestation of riparian and uplands, urbanization, pollution
- Decreased natural variability in hydrology or geomorphology (e.g., impounded, channelized, regulated)
- Relatively benign and thermally or chemically stable habitats, as opposed to those exposed to extreme seasonal fluctuation (although extirpation of invaders may be more likely in extreme habitats)
- Assemblages with trophic vacancies, such as few or no predators or zooplanktivores
- Low-diversity assemblages (but this may be a sampling artifact; see text)
- Insular streams and lakes
- Highland lakes, and rivers in the tropics

at least resistant to invasion, and what mechanisms promote this immunity?

Many researchers interested in introduced species have addressed the topic of invasibility. Discussions have tended to be qualitative in nature, and fishes in particular have not been subjected to much comparative, quantitative analysis (but see Kolar and Lodge 2001, 2002; Ruesink 2005). The resulting literature is diverse and sometimes contradictory, in part because of taxonomic differences and because site-specific characteristics often do not generate valid offsite predictions. Also, success of an invader is a complex, combined product of interactions involving the characteristics of the introduced species, the determination of humans promoting the introduction, the receiving habitat, and the ecology of the fishes and other species in the receiving habitat. The topic can, however, be clarified by discussing attributes of invading species versus attributes of invasible assemblages and communities (box 9.1). This approach has

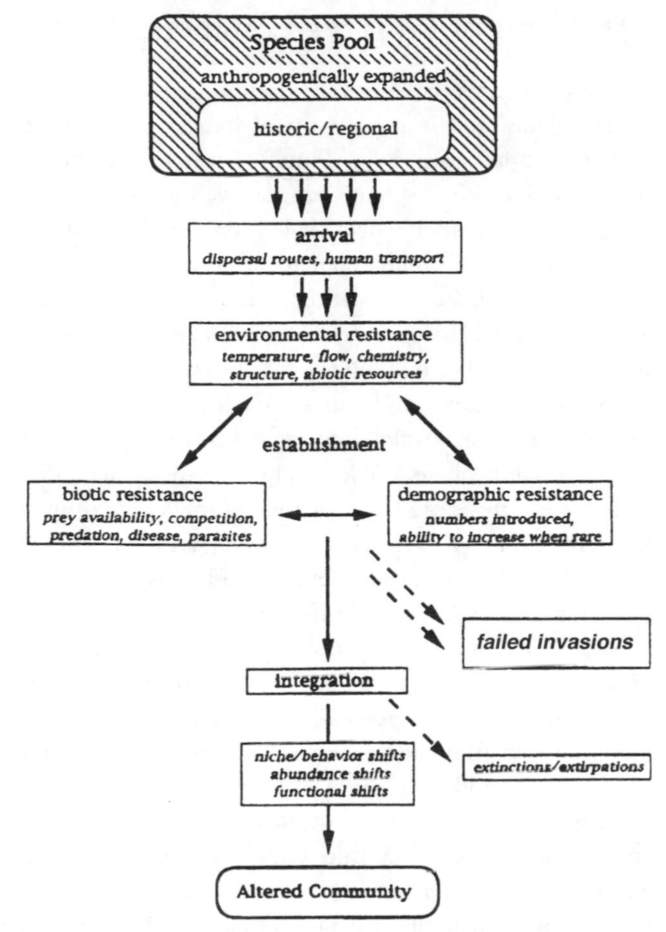

Figure 9.3. One conceptual model of the events that occur during invasions and the factors that determine success. Emphasis is placed on three critical phases—arrival, establishment, and integration—that may favor different traits in invading species. From Moyle and Light (1996a); used with permission.

been synthesized in a model developed by Moyle and Light (1996a; see figure 9.3).

Attributes of Invaders

Numerous authors have proposed a list of traits that characterize successful invaders; all but the last three traits in box 9.1 are commonly observed. Overall, generalist species with predatory habits are among the most successful introductions (except apparently among birds; Kolar and Lodge 2001). Among fishes, this observation holds, especially where habitat disruption is severe and where humans actively work at introducing predators. Omnivores and detritivores also often do well because their food resources are abundant and widely distributed. Hence tilapiine cichlids, Asian carps, and some clupeids are successful invaders in a variety of habitats. Dams and the habitats they create are especially favorable for the kinds of species that become established, including ecological generalists that can live in or over deep water, such as the many species placed in reservoirs to promote recreational fisheries.

Although not generally mentioned, introduced zooplanktivores do well in many lacustrine, impounded, and big river systems, where zooplankton usually abound, suggesting that an abundant food source promotes invasion. Examples include alewives in the Laurentian Great Lakes; the clupeid, *Stolothrissa tanganyikae*, in lakes Kivu, Kariba, and Cahora Bassa in southern Africa; and rainbow smelt, *Osmerus mordax*, in Lake Michigan and Sparkling and Crystal lakes in Wisconsin. A great deal of concern has been expressed over introduced bighead and silver carp, *Hypophthalmichthys nobilis* and *H. molitrix*, which escaped from aquaculture facilities in the Mississippi River; experienced exponential population growth; spread into the Missouri, Ohio, and Illinois rivers; and appear headed for the Great Lakes (figure 9.4). On the positive side, zooplanktivores may be relatively resilient in the face of introduced species. Zooplanktivorous haplochromine cichlids are among the few natives that appear to be staging a recovery in Lake Victoria.

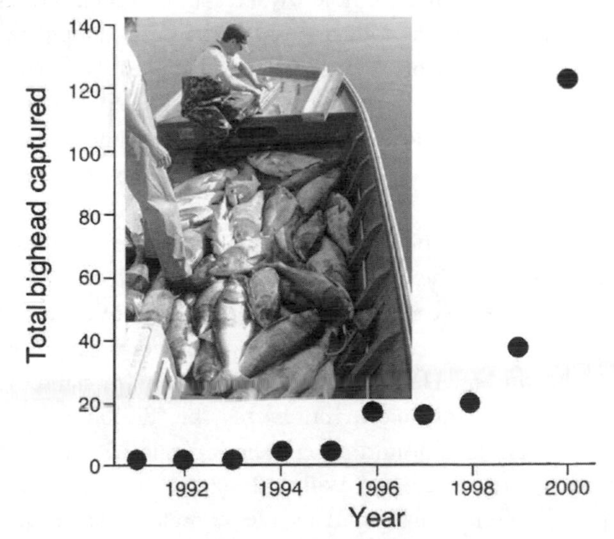

Figure 9.4. Increase in bighead carp introduced into the Mississippi River, as shown by catch statistics from Navigation Pool 26. After Chick and Pegg 2001. Inset: Illinois Natural History Survey personnel haul bigheads out of a lake at the confluence of the Illinois and Mississippi rivers. Photo by E. Gittinger. Used with permission.

An overriding influence on the success of invaders is human preference. Moyle et al. (1986) analyzed introduced species in the western U.S., where 35%–59% of fish species are introduced. They observed that a search for common characteristics among successful introductions had to take into account angler dissatisfaction with native fishes and desirable sport traits of introduced species. Angler preferences often determined which species were introduced and how much repeated effort (propagule pressure) went into establishing them. It appears that, given sufficient persistence, most species can become established outside their native range given suitable habitat.

Different phases of the invasion process may favor different traits within and among species. In ballast water introductions, success is affected by what happens during transportation, release, establishment, and spread (Kolar and Lodge 2001). Parallel events occur in streams (figure 9.3). Ecological traits of an organism that favor it during one phase may not guarantee its success in the next phase; traits that promote invasion during one phase may even work against establishment in a later phase (Kolar and Lodge 2001, 2002). Gobies, herring, and sticklebacks were the families most commonly found in ballast tanks; gobies, blennies, and flatfishes dominated the list of species that appeared in new habitats, but gobies and blennies were the most commonly established families (Wonham et al. 2000). Gobies and blennies are hole dwellers and breeders, favorable traits for survival during ship transport though not identified as traits of successful invasive species. Nocturnal fishes may do better as ballast water hitchhikers because of the lighting conditions that typify ship holds; nocturnality as a general trait of introduced species has not been explored.

The different phases involved are therefore filters that slow the otherwise unlimited onslaught of invaders. Traits that promote success may differ as a function of means of transport, ballast water differing from bait buckets, canals, or other means. This variability as a function of transport method probably contributes to the difficulty that researchers have found in predicting success rates of invaders (see Kolar and Lodge 2002 for an exception). Again, conservation problems are complex and require intimate knowledge of a host of conditions, traits, and interactions. Generally accepted conclusions suggest that anything we do to complicate the process for invaders (e.g., increasing the number of filters via water exchange, quarantine periods) will decrease their success.

Attributes of Invasible Communities

The most frequently observed condition associated with successful invasion is some form of habitat degradation; few exceptions have been found (but see "Synthesis" below). Undisturbed systems are relatively impervious to invasion, whereas highly degraded systems are habitable by only a small set of highly tolerant species, native or introduced. Intermediate degradation seems to be most favorable to invaders. Basically, diversity increases with increasing disturbance, but only up to some point.

Invasions often follow quickly on the heels of disturbance, such as a major flood during a drought or following use of an ichthyocide. Habitats subjected to repeated disturbance are therefore likely candidates for repeated invasions, which increases the chances of establishment (Moyle and Light 1996a). After four severe floods over a two-year period in Valley Creek, Minnesota, Elwood and Waters (1969) found that native brook trout had not recovered because of habitat degradation and lack of a source for recolonization. Brook trout were replaced by introduced rainbow trout from downstream ponds, which were later replaced by introduced brown trout.

Impoundment is one of the most common forms of aquatic habitat alteration. Impoundments often flood and eliminate shallow-water structures, such as the snags or riffles on which many invertebrate species live. This reduces the food base for specialized feeders, which facilitates invasion by food type and feeding habitat generalists, or at least promotes establishment of a suite of fishes not dependent on clean bottoms. Impoundments also eliminate refuges (cracks and crevices, specific structures for structural specialists), which facilitates invasion by piscivores.

Disturbance is sufficient but not necessary for invasion. Aliens can invade and affect native populations without habitat disruption. Levin et al. (2002) found that, even in relatively pristine regions of the Columbia River basin, introduced brook trout led to a 12% reduction in survival and a 2.5% reduction in population growth rate of threatened Chinook salmon. After analyzing R. R. Miller et al.'s (1989) findings on fish extinctions in the U.S., Lassuy (1995) concluded that habitat alteration did not always precede severe impacts by introduced species but did make a species and its supporting ecosystem more vulnerable to invasions and their impacts.

The next most frequently discussed trend among invasible systems is that invaders often have difficulty establishing themselves in naturally variable environments (e.g.,

Meffe 1991; Baltz and Moyle 1993; Moyle and Light 1996a, 1996b). Factors that vary often involve hydrology, in combination with climate and habitat degradation. Highly seasonal environments such as arid regions and regions with Mediterranean climates are characterized by winter floods and summer droughts. Human modifications often reduce variation, making habitats more susceptible to invasion. Dams and irrigation projects reduce seasonal variation. Even such widespread invaders as bluegill sunfish, *Lepomis macrochirus*, have difficulty establishing themselves in habitats where natural flow is unimpeded and unstabilized, although they successfully invade adjacent systems that are impounded and stabilized (e.g., Meffe 1991; Marchetti 1999).

Natural environmental variation interacts with behavioral, physiological, and life history traits, thus often favoring species adapted to local conditions. Threatened Gila topminnow, *Poeciliopsis occidentalis*, suffer in part from predation by widely introduced mosquitofish, *Gambusia affinis*. In streams with natural hydroperiods, topminnows persist because they exhibit adaptive behavioral responses to periodic flash floods, including orientation to strong currents and rapid movement to shoreline areas as waters rise. Mosquitofish, which evolved in southeastern U.S. regions in habitats that seldom experience flash floods, behave inappropriately and are flushed downstream (Meffe 1984; Galat and Robertson 1992). Exotics annually invaded unregulated Deer Creek, California, from both downstream and upstream sources but failed to become established in undisturbed reaches where variable flows eliminated them (Baltz and Moyle 1993). When the hydroperiod is stabilized through dam building, the balance is tipped in favor of the invader. A similar advantage of natives over exotic poeciliids, mediated by flow extremes, has been observed in Hawaiian streams, with the added twist that maintenance of high, natural flows can reduce transmission of parasites carried by invasive species (Fitzsimons et al. 1997).

Adaptation to drought may also favor a native over an invader if hydroperiod remains natural. Brown trout introduced into Australia prey heavily on a native galaxiid, *Galaxias olidus*, but the interaction is mitigated by periodic drought. In intermittent Lerderberg River of central Victoria, Closs and Lake (1996) followed the distribution of both species over a seven-year period that altered between wet and dry years. During wet years, water flowed throughout the system, and trout moved progressively upstream, displacing the natives and fragmenting their population. During dry years, many upstream sections dried into a series of unconnected pools; trout suffered high mortality upstream (100% vs. 60%–80% for the galaxiid in the summer of 1988) and were restricted to downstream regions (figure 9.5), which allowed recolonization of upper sections by the galaxiid (*G. olidus* can survive in wet mud; brown

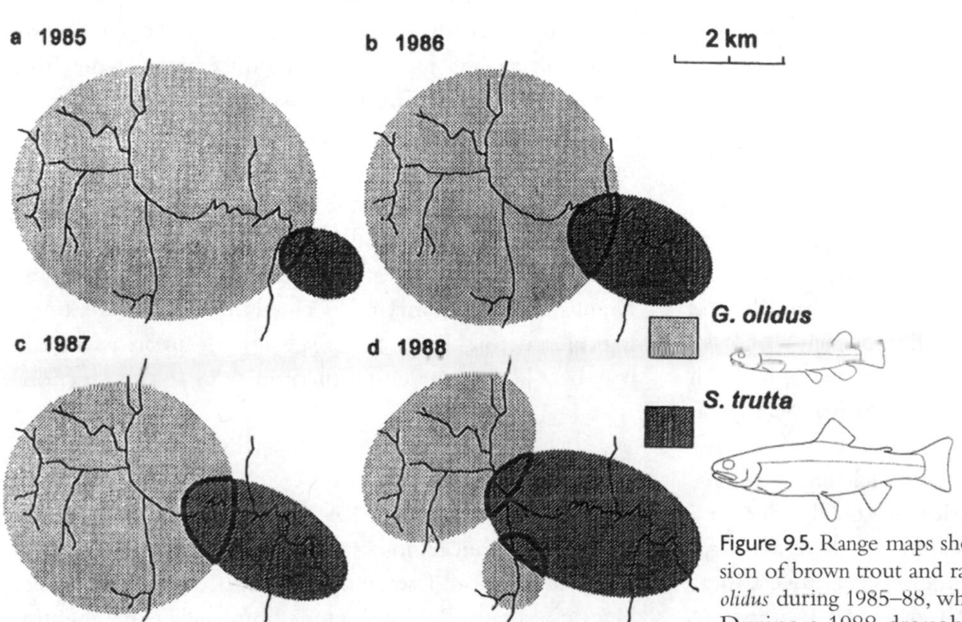

Figure 9.5. Range maps showing the progressive upstream expansion of brown trout and range contraction of Australian *Galaxias olidus* during 1985–88, when rainfall was relatively abundant. During a 1988 drought, trout advance upstream was reversed. From Closs and Lake (1996); used with permission.

trout cannot). This study shows the interplay between local adaptation and incursion of introduced predators. One could predict that flow regulation of the Lerderberg River that minimized extreme low-water periods would lead to eventual extirpation of the native species. Maintaining natural flow regimes is crucial to resisting invasions and minimizing impacts of introduced species.

Invasion also occurs in naturally invariant systems that are perturbed. Fishes endemic to springs are adapted to the remarkable constancy of physical and chemical conditions that typify spring habitats. Habitats downstream from springs are not as constant and contain more widespread species. Many spring species are imperiled because of modifications to this constancy, such as reduced flow brought on by water withdrawal or drought, which alter temperature and pH. Loss of constancy turns the springs into hostile environments for the endemics and appropriate habitats for more widespread stream species. Faunal change involves endemic versus widespread cyprinids, poeciliids, centrarchids, and percids. As springs are homogenized into commonplace habitats, their inhabitants also become commonplace (Hubbs 2001).

Noninvasible Species and Habitats

Truly noninvasible habitats probably do not exist. As mentioned earlier, species can be forced into habitat if humans are sufficiently determined, but some habitats appear more resistant than others. What can we learn from such locales that might help us identify harmful invaders and minimize their impacts?

The Lake Victoria story (chapter 8) is one of unparalleled ecological disaster. However, Nile perch, while devastating most of the lake, have been unsuccessful at invading its edges and some nearby lakes. They were introduced into Lake Nabugabo, which drains into Lake Victoria through a papyrus swamp, without decimating the local ichthyofauna, though the two lakes are climatically and geologically similar. But much of the periphery of Lake Nabugabo is swampy and oxygen deficient, and Nile perch are highly susceptible to deoxygenation. Native Nabugabo fishes possess behavioral and physiological traits that allow them to live in low oxygen conditions, including extensive use of aquatic surface respiration (Rosenberger and Chapman 2000; Schofield and Chapman 2000). Low-oxygen regions therefore serve as refuges from a proven, invasive predator. Maintaining habitat diversity in impacted locales is a means of reducing the effects of invaders.

It is frequently stated that high-diversity habitats are more resistant to invasion. Gido and Brown (1999) analyzed data from 125 drainages across temperate North America to look for patterns of invasion susceptibility. They found only a weak, indirect correlation between native species richness and number of introduced species. Depauperate assemblages may be more vulnerable to invasion, but the relationship is complicated because drainages with low native species richness have also faced a greater pool of potential invaders and have longer histories of disturbance. A weak inverse relationship between diversity and invasion implies that diverse assemblages are not immune to invasion, suggesting that there is no such thing as a community so ecologically saturated that it can't be invaded by alien species.

The relevant question may not be which habitats are noninvasible, but which are least affected when invaded. Although few are absolutely noninvasible, it appears that most assemblages can withstand invasion without collapsing. Moyle et al. (1986) maintained that most adjust fairly rapidly to the addition of new species. Some species may be lost, but most adapt via shifts in their feeding, refuging, and breeding habits. Native species that are likely to be extirpated are often extreme specialists that cannot adjust to competition or predation from an invader. By making adjustments—in food types, feeding locales, refuge sites, breeding times and places—species minimize interactions with the introduced species.

Successful invaders also make adjustments. Welcomme (1988) observed that some slow-growing and slow-reproducing specialists grow and reproduce faster when they invade new habitats. This helps explain why relative specialists seem most vulnerable and relative generalists survive both as introduced species and as hosts to introductions. Generalists adjust to the invader, or if invading, to the new mix of species and conditions in which they find themselves. Hence the new assemblage contains not only a different number and mix of species, but also a different ecological structure. Plasticity of species within an assemblage may be a key to predicting the impact of an introduction.

Synthesis

Moyle and Light (1996a, 1996b) presented a list of general predictions concerning species and habitats that were prone to invasion. They emphasized the interplay between "biotic resistance" (biotic interactions such as competition and predation) that deterred invasions, and "abiotic suitability" (physical suitability of the habitat) that promoted them. In summary, they predicted the following:

- The most successful invaders were those adapted to the local hydrologic regime. Nonnatives that lack adaptations to high flows frequently lose juveniles to springtime washouts, impeding their establishment. Stabilized flows, such as produced by dam operations, reduce invader mortality.
- A wider range of species will invade greatly disturbed systems than will invade less disturbed systems, and invasions often closely follow disruption. The influence of disturbance holds regardless of species diversity (e.g., R. M. Ross et al. 2001).
- In unmodified systems, the most successful invaders will be predators, omnivores, and detritivores. Predators have the greatest overall biotic impact, especially where they were not historically present. Feeding conditions in altered systems also often favor omnivores and detritivores, although these trophic groups should have less impact on natives than predators do, except in the case of bottom-disturbing species such as carps.
- Most fishes can act as invaders and adjust to invaders, with extinction a rare result. Few species have been extinguished as a direct result of invasions alone (discounting Lake Victoria and some desert springs), although many are negatively affected.
- All fish assemblages will be invasible, regardless of diversity; low-diversity, simple assemblages are not more invasible. Some low-diversity systems such as Deer Creek, California, can repel invasions, whereas some high-diversity systems such as megadiverse Lake Victoria cannot. However, extinction of natives does seem more likely and the overall effects of invaders more obvious in low-diversity systems, both possibly a simple matter of math (Moyle and Light 1996a; Lake Victoria may remain an exception here).

Moyle and Light (1996b) concluded that a match between invader characteristics and hydrologic regime played the most important role in determining which species and assemblages would take part in invasion events. Biotic resistance plays a minor role, "except in the early stages of invasion when numbers of the invader are small" (p. 1669). Habitat suitability is heavily influenced by hydrologic regime type and hence invasion success. Moyle and Light maintained that if abiotic conditions in a habitat were appropriate for an exotic species, successful invasion was likely. Failure to establish was probably due to an inability to adapt to abiotic conditions rather than biotic resistance by the receiving community (see also Fausch et al. 2001).

Conclusions Concerning Invasibility

S. T. Ross (1991, table 1) reviewed more than 350 fish introductions from North America, Europe, and New Zealand and found that 38%–77% led to successful establishment depending on locale and taxa, with an overall mean of 55%. His average value suggests that there is an even chance or better that any introduction will become established. Because three-quarters of all introductions result in a decline among native fishes, it is precautionary to conclude that introductions should be approached with trepidation; benefits should be obvious and unequivocal, and risks should be carefully assessed.

Although a few traits appear to consistently apply to many successful invaders, the high degree of variability among outcomes suggests that chance and local conditions have an overriding influence. This conclusion has been articulated by Williamson (1996, p. 63): "The only predictors of invasion success that appear repeatedly in statistical studies are propagule pressure, suitability of the habitat, and previous success in other invasions." Kolar and Lodge (2001) similarly emphasized propagule pressure ("the magnitude of introduction effort") and a previous history of invasion success. Propagule pressure has multiple forms, including the repeated opportunities of ballast water and canal-transported invaders, as well as deliberate, repeated introductions of sport fishes. Habitat suitability represents a match between the physiology of the invader and the physical attributes of the receiving area, e.g., tropical fishes cannot live in temperate climates. Previous success is a good indicator that, all things being equal, some species have a high likelihood of success regardless of propagule pressure and habitat suitability. Such species belong on a black list of suspect species.

"POSITIVE" INTRODUCTIONS: THE OTHER EDGE OF THE SWORD

Is there such a thing as a good introduced species? The answer is largely a matter of perspective and emphasis. In fact, "unbiased evaluation of introductions is virtually impossible because of the many different viewpoints that are brought to bear" (Welcomme 1988, p. 26). The same introduction may be claimed as a success story or reviled as a disaster (e.g., Welcomme 1984). The general perspective taken here is that introductions are neutral at best and negative on the whole. But given that 25%–50% of the freshwater fishes caught by anglers in the continental U.S. come from introduced populations (Moyle et al. 1986), most anglers would provide a more positive appraisal.

It is difficult to argue with introductions that increase protein yields in the developing world. A pelagic clupeid introduced into lakes in southern Africa escaped downstream into Lake Cahora Bassa and created a fishery with an annual yield of 20,000 tons (Welcomme 1988). Largemouth bass introduced into a Mozambique lake experienced ecological release, grew two to three times faster than in the U.S. (reaching 40 cm in two years), and came to constitute 9% of a subsistence fishery that employed over 500 fishers (Weyl and Hecht 1999). Forty years after tilapia and a cyprinid, *Barbus gonionotus*, were introduced into Lake Ranu Lamongan, Indonesia, they support a thriving fishing industry that appears to have had little impact on the native assemblage (I. Payne 1987). "It must be recognized that the welfare of people and not fish is the raison d'être for a management program" (Cooley 1963 in Dill and Cordone 1997, p. 15).

Data can be used to support differing conclusions. The U.S. Office of Technology Assessment analyzed the relative impacts of introduced species in terms of harm or benefit (OTA 1993). Of 111 species of alien fishes introduced into the U.S., 28% had primarily beneficial effects, 17% had beneficial and harmful effects, 30% were primarily harmful, and the remaining 25% had neutral or unknown effects. From this assessment, one can draw half-full or half-empty conclusions. More than two-thirds (70%) of all introduced species had beneficial or neutral impacts, indicating that introductions generally served intended purposes or caused no problem. Alternatively, nearly half (47%) had some harmful impacts, which means there's a 50% probability that harm will result from an introduction. How these odds are incorporated into management decisions probably depends on how risk averse a manager or program decides to be.

Welcomme (1988) successfully maintained a relatively neutral approach to the topic, in contrast to my skeptical view and to the Panglossian statements and rationalizations scattered through Lever's (1996) otherwise impressive compilation. Welcomme surveyed international fisheries authorities and summarized relative costs and benefits of introductions. He divided species into those that are generally viewed as pests, those that have created ecological problems but are also perceived as useful, and those that are generally viewed as beneficial (box 9.2). He found that only about a third of introduced species warranted comment from his informants. Among those that drew comment, species were equally divided among three categories: generally negative, mixed reaction, and generally beneficial. Species viewed negatively were often (1) predatory species that contributed to the eradication of endemic

BOX 9.2. Assessment of introduced species based on their perceived costs and benefits. Modified from Welcomme's (1988) tables 16–18 to reflect changes in taxonomic names. Species are listed in roughly alphabetical order:

- *Species viewed generally as pests where introduced:* black and brown bullheads; bleak; bluegill, green, pumpkinseed, and redbreast sunfishes; Chinese false gudgeon; Chinese sleeper; *Micropercops swinhonis*; Mozambique tilapia; Nile perch; pirambeba; *Rhinogobius similis*; roach; rudd; sharpbelly; spotted steed; stone moroko; three-lips; walking catfish; white bream.

- *Species viewed with mixed feelings (i.e., that have had both negative and positive impacts):* brown and rainbow trouts; chum, coho, and pink salmon; European and Nile perch; goldfish; grass and common carps; guppy; jaguar guapote; largemouth and peacock bass; molly; mosquitofish; northern pike; pejerrey; redbelly, redbreast, and rufigi tilapia; zander.

- *Species viewed generally as beneficial where introduced:* Alluaud's haplo; bighead, black, and silver carps; blue, Nile, and singida tilapia; catla; cha cham; channel catfish; Chinook salmon; European and Japanese eels; goldfish; heterotis; houting; Java barb; Lake Tanganyika sardine; rohu; snakeskin gourami; tambaqui; threadfin shad.

fishes, often in assemblages that previously lacked predators, and (2) species that tended to form dense, stunted populations. Positive introductions created useful commercial or sport fisheries, served as effective biological control agents, or restored imperiled species via captive breeding and translocation.

Reanalysis of Welcomme's categories today might move some names around. Why goldfish and Nile tilapia appear in both the mixed and the overall beneficial groups is puzzling. Most recent discussions of goldfish would place it in the generally negative list, whereas Nile tilapia's tendency to stunt would probably place it only in the mixed category. Mosquitofish and possibly common carp (except for the highly inbred sport varieties) would move from the mixed to the negative list, except where mosquitofish are placed in small or temporary, otherwise fishless, bodies; black carp, *Mylopharyngodon piceus*, would likely be moved from positive to mixed. A few species could be added to the list, such as Asian swamp eel (negative), ruffe (negative), round goby (negative), smallmouth bass (mixed), and Atlantic salmon (mixed). Overall, Welcomme's treatment appears accurate, evenhanded, and representative. His numbers, which reflect global events, are not greatly different from those compiled for U.S. alien species. The proportions of generally harmful and generally beneficial species are about equal, although relatively few U.S. introductions are viewed as having had no effect.

Many successful and otherwise useful introductions suffer from stunting. Stunting occurs in a variety of fish but is common in lepomine sunfishes, *Perca*, cyprinids including goldfish and common carp, and tilapiine cichlids. In tilapia, particularly *O. mossambicus*, the tendency to stunt "has gained them pest status" (Welcomme 1988, p. 192).

Fishes Introduced for Biological Control

The category of positive introductions might be expanded to include fishes introduced to control pest organisms, although many such efforts produce debatable outcomes (e.g., mosquitofish, chapter 8). Biological control is laudable because it reduces the need for chemicals such as insecticides, herbicides, and ichthyocides. Fishes have been used to control disease vectors such as mosquitoes and snails, to control aquatic vegetation (often alien), and to eliminate unwanted fish species (often introductions). Use of molluscivorous fishes to combat snail vectors of such human diseases as schistosomiasis has met with limited success (Lever 1996). Sterile, snail-eating black carp have proven effective in controlling trematode parasites in aquaculture facilities, but escape of nonsterilized fish and their inevitable predation on endangered native mollusks remains a major cause for concern (Nico and Williams 1996; see chapter 8).

Grass carp, *Ctenopharyngodon idella*, have been widely and successfully used for vegetation control, and production of sterile, triploid grass carp has reduced problems of escape and spread (S. K. Allen and Wattendorf 1987; Cassani 1995). On the downside, grass carp prefer certain tender-leaved plant species, thus encouraging the growth of tougher plants, "which can prove an even greater problem" (Lever 1996, p. xix). More often, however, they remove all rooted macrophytes, essentially denuding a water body of vegetation. This creates new problems, including release of nutrients into the water column and promotion of phytoplankton growth, increased turbidity, and elimination of essential cover for young fishes and habitat for many invertebrates (fish food) (Bain 1993).

Biocontrol agents deal with symptoms more often than causes. The problem of algal blooms resulting from eutrophication is ultimately solved by reducing nutrient inputs, not by introducing phytoplankton feeders. Outbreaks of mosquito- and snail-borne diseases often occur where natural flowing water is impounded, creating shallow, stagnant habitats where mosquitoes breed and snails flourish. Habitat restoration, not introduced species, is the best solution. But if causes are intractable, eliminating symptoms can be a partial solution. The secret to successful control agents seems to lie in understanding the evolutionary histories of the organisms involved, using host-specific controls, and employing insects and microbes rather than vertebrates. Fishes do not reproduce quickly enough nor eat enough to keep pace with short-generation pathogens. Also, the host specificity desired in most biological control programs is unlikely in fishes. Few fishes eat only one prey or plant species, which almost ensures that an introduced control species will affect, if not focus its activities on, something other than the intended species (e.g., Courtenay and Williams 1992).

HOMOGENIZATION: A RESULT AND MEASURE OF INTRODUCTIONS

Because certain species have been favored and widely distributed regionally, nationally, and internationally (table 8.1), strikingly similar fish faunas now occur in very different places. The similarity, known as biotic homogenization, applies broadly to taxonomic groups. Homogenization results when a short list of alien species is introduced into many regions at the same time that native species are lost (e.g., Lockwood and McKinney 2001). The process is so widespread, its outcome so ubiquitous, and the disruption so thorough that J. D. Williams and G. K. Meffe (1998, p. 118) declared, "The continued homogenization of the world's flora and fauna . . . is an ecological holocaust of major proportions." Gordon Orians of the University of Washington has suggested we refer to the era in which we live as the Homogocene.

Homogenization of U.S. fish faunas has happened on a grand scale (Fuller et al. 1999; Rahel 2000, 2002). Of the 76 species introduced into 10 or more states, 32 have been placed in 25 or more states, and 13 have gone into more than 35 states (table 9.1). Eight of the latter 13 are relatively large piscivores that continue to be stocked as game species in many places. Three (common carp, goldfish, and tench) stocked extensively during the late 19th century are now generally regarded as nuisance species.

What effect has this enthusiasm for transplanting mostly sport fishes had on native fish distribution and diversity? Rahel (2000) compared historical lists with current species lists and found that state faunas have grown significantly more similar over time (figure 9.6). Subsequent to European occupation of North America, the similarity of fish faunas of states has increased by an average of 15 species, with almost 20% of states sharing 25 or more additional species. Only about 30 (3%) of paired comparisons indicate a decrease in number of shared species. Over half of the fish faunas of Nevada, Utah, and Arizona are non-

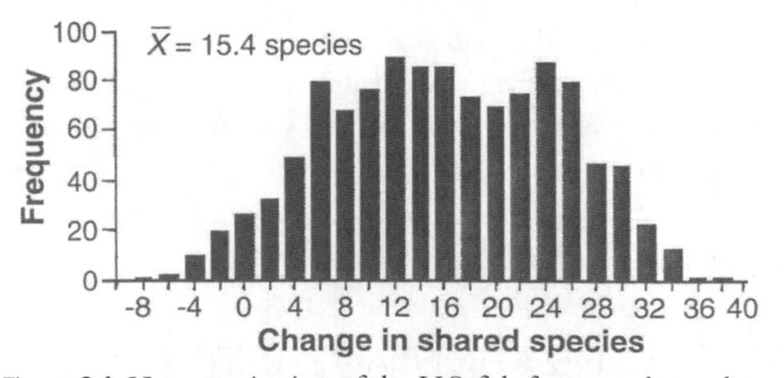

Figure 9.6. Homogenization of the U.S. fish fauna, as shown by change in the number of species shared between states. Plotted is the frequency distribution of change in shared species for 1,128 pairwise comparisons of states, calculated from the current number of shared species minus the historical number of shared species. From Rahel (2000); used with permission.

native, containing species brought in primarily from the eastern part of the country to improve angling. Construction of large impoundments provided appropriate lacustrine habitat in a region that was naturally lacking in large lakes, aiding the establishment of eastern lake species. The only species moved extensively eastward was the rainbow trout.

Faunal convergence has also occurred within regions because distinctive habitats and their associated, endemic faunas have been replaced by common, widespread species in repetitively modified habitats. In California, the freshwater fish fauna increased from a pre-European 67 species to 114 species, a 70% increase. Fifty nonnatives have invaded, while 5 native species have been extirpated (Dill and Cordone 1997; Marchetti et al. 2001). Urbanization and other human impacts (dams, channelization, logging, agriculture) have reduced the hydrologic diversity of the state, homogenizing aquatic habitats by increasing physical homogeneity as development progressed (Marchetti et al. 2001). Habitat homogenization preceded biotic homogenization. Generalist alien species are tolerant of the altered and simplified conditions that spread across the state, such as the proliferated reservoirs, which are more similar in species composition than the streams and rivers they replaced. The more anthropogenic influence in a watershed, the less the fish assemblage resembles the pre-European fish assemblage (figure 9.7).

The homogenization that occurs as a direct and indirect result of sport fish stocking has been particularly evident in central Minnesota. Walleye were actively stocked, but sunfishes and bullhead catfishes were inadvertently transferred with walleye, a fairly commonplace mistake in stocking programs (Radomski and Goeman 1995). Unstocked lakes showed no change in diversity and maintained uniqueness, whereas stocked lakes converged toward similar species lists. Walleye fishing improved, but only 4% of total walleye harvest could be attributed to stocking, an insignificant improvement that occurred at the expense of unique fish assemblages and regional distinctiveness of the fauna. Similar gains could often be realized through habitat protection and restoration in lieu of stocking. Radomski and Goeman concluded that the value of natural diversity should be considered in decisions of what and where to stock.

For the U.S. fish fauna as a whole, introductions have exceeded extirpations by almost 5:1, indicating that increased homogenization is due more to introductions

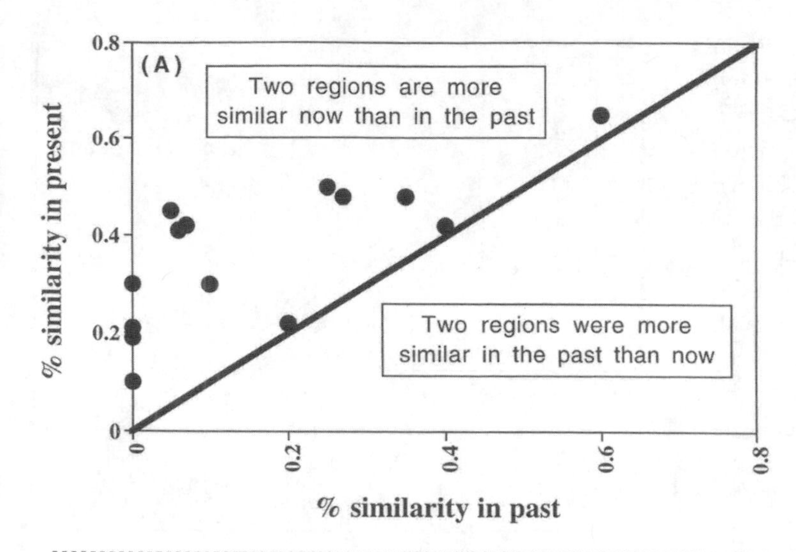

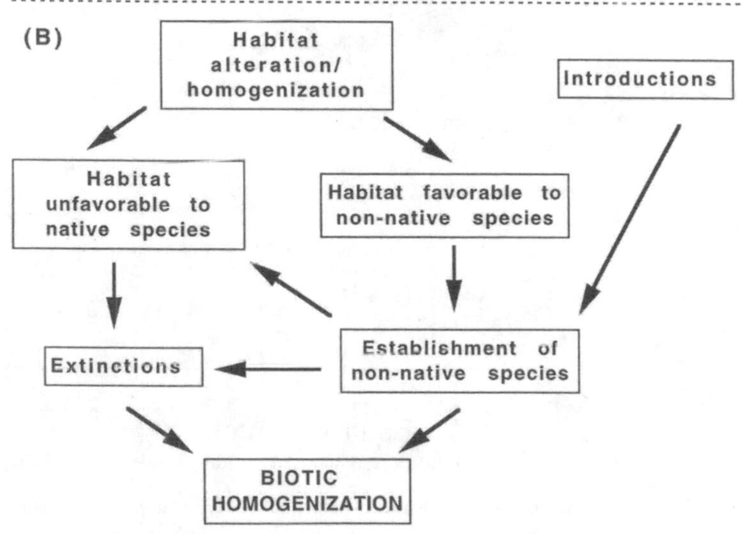

Figure 9.7. Homogenization of California's fish fauna. (A) Comparisons indicate that zoogeographic provinces are more similar today than they were 150 years ago; scores above the diagonal indicate greater similarity in the present; scores below the diagonal would indicate greater similarity in the past. (B) Homogenization involves an interaction between habitat alteration, species introductions, and the habitat requirements or tolerances of natives vs. introduced species. Based on Marchetti et al. (2001).

than to losses (Rahel 2000). This result can be viewed with some optimism: All introductions have not necessarily led to wholesale destruction of the native fish fauna. However, the extirpation of nearly 200 species amounts to serious disruption of native faunas and may reflect the impact of introductions.

Homogenization of European Fish Faunas

Europeans have been at least as active as Americans in introducing fishes. The history of environmental degradation and fish transportation in Europe has produced similar faunas in countries that previously housed distinct assemblages, with alien fishes dominating in many regions. Cowx (1997) produced a list of 166 species introduced or translocated, amounting to about 46% of Europe's 358 native species (Kottelat 1998), which Cowx believed was an underestimate.

The fraction of a nation's ichthyofauna constituted by alien species is one indicator of the impact of introductions and the potential for homogenization. Bianco (1990) proposed that the ratio of native species to total number of species amounted to a zoogeographic integrity index that could be used to compare the health of ichthyofaunas. Values close to 1.00 indicated high integrity (no introductions), and lower values indicated lower integrity and many alien species. Values for Europe ranged from 0.56 for Italy to 0.88 for Greece, with many countries falling close to the overall European (and American) value of 0.65 (e.g., Spain 0.63, Portugal 0.65, Czech Republic 0.66, France 0.69) (Bianco 1990; Elvira 1995; Lusk et al. 2002).

The actual list of shared species confirms the suspicion that Europe's ichthyofauna is highly homogenized. Fifteen species have been introduced into 10 or more of the 30 countries covered in Cowx's treatment, including carp (24 countries), brook trout (23), grass carp (20), pumpkinseed sunfish (19), and rainbow trout (9). Other countries with low native diversity or high numbers of introduced species—where extensive homogenization would be expected and where its impacts should be monitored—are New Zealand, Australia, and South Africa.

Homogenization via Native Invasion: Diversity Is Bad

The emphasis in most discussions of homogenization has been on fairly large geographic scales and species introductions. However, in individual streams, native species affecting other native species can produce homogenization.

In small streams of the southern Appalachian region, M. C. Scott and Helfman (2001) found a pattern of species distributions involving (A) narrowly distributed, endemic, upland, mostly benthic fish species (e.g., darters, sculpin) specialized for life in clear, cool, low-productivity streams with high habitat diversity, and (B) widely distributed (cosmopolitan), generalist, often water column–dwelling species (e.g. minnows, sunfishes) typical of downstream regions that are more turbid, warmer, and higher in productivity but with fewer definable habitat types. The two groups include upland versus lowland members of the same river fauna, and almost all are native to the area. Few exotics or even transplants have invaded this region (see also D. M. Walters et al. 2003).

Homogenization occurs when land-disturbing activities increase silt and sediment inputs to streams, which affects the availability and diversity of instream habitats; overall habitat diversity declines, deeper pools become more common, and riffles become covered with sediment (E. B. D. Jones et al. 1999). Upland endemic species dependent on riffles and clear, cool water decline, whereas downstream species that are more tolerant of sediment and prefer slower-moving, warmer water find the altered habitats more hospitable. Basically, homogenization of habitat leads to homogenization of the fauna, as cosmopolitan species from downstream invade the habitat of narrowly distributed, upland fishes.

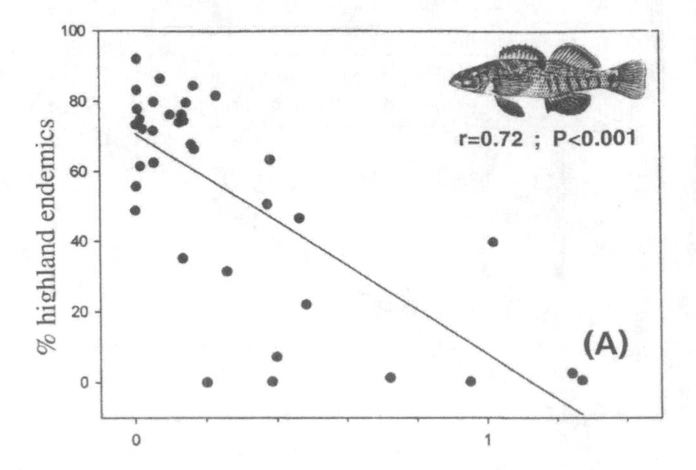

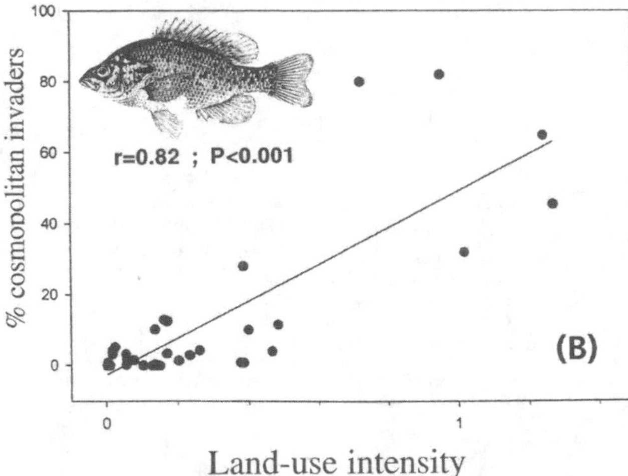

Figure 9.8. The relationship between relative abundance of fishes and land-use intensity in southern Appalachian streams. (A) Relative abundance of upland endemic species and (B) of cosmopolitan downstream fish species as a function of intensity of land use. Regressing relative abundance of all fishes (i.e., lumping highland endemics and cosmopolitan species together) showed no relationship between richness or diversity and land-use intensity. From M. C. Scott and Helfman (2001); used with permission.

Homogenization is not obvious if a river's fish fauna is viewed as a whole, which is the usual approach. We tend to think of all fishes in a river as part of an interacting ecosystem. When all species are considered, no statistical relationship exists between diversity and intensity of land use. If instead the fauna is divided into highland endemics and downstream cosmopolitan species, a different pattern emerges (figure 9.8). The representation of highland endemics decreases, and the frequency of cosmopolitan species increases as a function of increasing human impact. In essence, two very different native fish faunas are involved. Homogenization of the regional fauna occurs as watershed disturbance and native invasions are repeated across river basins, as unique elements in different drainages are progressively replaced by regionally common species.

Three generalizations about local homogenization emerge. First, habitat homogenization promotes biotic homogenization (e.g., Boet et al. 1999; Marchetti et al. 2001; Rahel 2002). Second, upland regions are invaded by native—not alien—species that normally occur in downstream regions. Third, conditions in upland streams can change sufficiently to facilitate establishment of native invaders while still housing reduced numbers of upland endemics (E. B. D. Jones et al. 1999). Native invasions may initially produce no change or even increased diversity; increased species richness has been found in a variety of aquatic systems subject to invasions (Radomski and Goeman 1995; Courtenay and Moyle 1996; Gido and Brown 1999; Rahel 2000). Diversity alone is therefore an unreliable indicator of the conservation status of a fish assemblage. It is high initially due to a strong contribution from regional, specialist, endemic species. After initial disturbance, such as moderate deforestation, losses among endemic species are balanced or may be outweighed by invasion of native, generalized, cosmopolitan species from lowland regions into the increasingly homogenized highland habitats (figure 9.9). With continued disturbance, upland streams become unsuitable for highland endemics and even for some cosmopolitan species, but all may eventually be replaced by widespread, introduced, tolerant exotics. Deliberate introduction of exotic predators at any point can accelerate the loss of native species.

What may be missed by emphasizing numerical measures of diversity and homogenization is changes in ecological interactions and disruptions in actual species composition. At the least, traditional metrics such as species richness and evenness should not include invasive species, native or exotic. Measuring changes in diversity alone will give an incomplete and inaccurate picture. Greater attention should be paid to the actual species present and their ecological requirements and roles. Measuring species richness alone tells us little about the conservation status of assemblages and regions (Radomski and Goeman 1995; Duncan and McKinney 2001). Biodiversity can degrade without declining.

Conclusions about Homogenization

Homogenization, in which habitats and their species composition converge toward a general form or lowest ecological common denominator, results from interactions between habitat disturbance and introduced species and reflects alterations at the level of species, communities, landscapes, and regions (e.g., Lodge 1993). At the local scale it involves interactions between native species that differ ecologically, whereas at the regional scale it can give an international appearance to previously unique, indigenous faunas and the habitats they formerly occupied. Homogenization has significant relevance to the general issue of biodiversity loss. Specialist endemics that lose out in the process of homogenization are likely to be species already imperiled because they are characterized by small natural ranges,

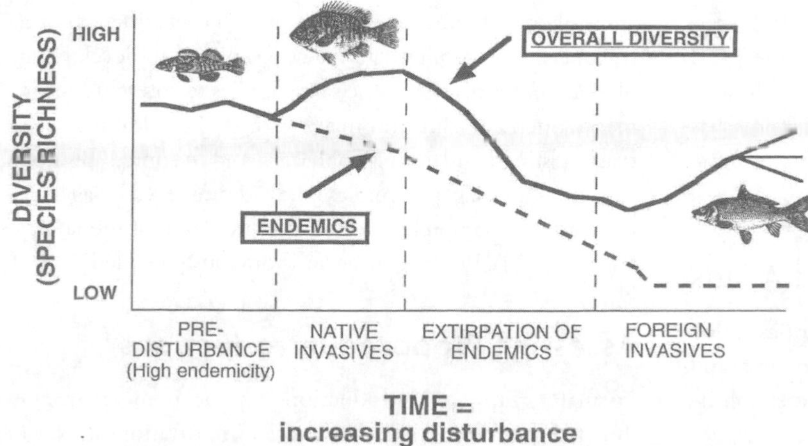

Figure 9.9. Postulated time course over which homogenization occurs in southern Appalachian highland streams as a function of increasing watershed disturbance. From M. C. Scott and Helfman (2001); used with permission.

sensitivity to habitat disturbance, and vulnerability to replacement by introduced generalists (chapter 4). Homogenization can be considered an early warning sign of impending, significant degradation of regional faunas, degradation that ultimately leads to extinctions. To detect homogenization, it is critical that we understand natural assemblages, particularly natural and historical distributions.

SOLUTIONS

Introductions—deliberate in the past and increasingly becoming by-products of trade and other human activities (see figure 9.1)—invariably cause unanticipated and often disastrous ecological and economic problems. The most obvious solution is to avoid them. This may appear a facile and even flippant dismissal of accepted practices, but experience seems to bear out the generalization that native species can serve most human needs, especially where habitats remain intact. In degraded systems, rehabilitation and restoration appear less intrusive and more productive than seeking exogenous species that tolerate the altered environmental conditions. But a proscription against introductions is probably naive because "it is unrealistic, and likely to be impossible under international laws . . . to ban all movements of live fish" (McVicar 1997, p. 1097). Given that introductions will continue, what tactics should be adopted to minimize their negative impacts and maximize their benefits to humans and the environment?

Solving introduced species problems entails (1) minimizing or reversing the impacts of already established alien species; (2) minimizing the likelihood of accidental introductions; (3) assessing the relative benefits and costs of proposed deliberate introductions and making those responsible for most introductions today (shipping, aquarium, bait, and aquaculture industries) financially responsible for introductions that are by-products of their activities; and (4) increasing public awareness of the problem. Solutions abound in the first three areas, and public awareness has increased greatly as the global extent and impacts of invasions have been publicized.

Minimizing Impacts of Established Aliens

Prevention is always easier than eradication or control. The suite of biological characteristics that describe pest and invasive species (box 9.1) makes their survival and establishment likely and their eradication unlikely. Getting rid of carp or red shiners or tilapia may require such drastic means as ichthyocides, which also harm natives (e.g., M. M. Hale et al. 1995). One approach is to accept the inevitability of established introductions and turn them into some socioeconomic good, as has happened in the artificial canals of south Florida ("a biological cesspool of introduced life"; Lachner et al. 1970, p. 1). Peacock cichlid, *Cichla ocellaris*, were stocked to feed on abundant, introduced spotted tilapia, *Tilapia mariae*, and to provide a sport fish for urban anglers. Stocking accomplished these goals, contributing an estimated $1 million annually to the south Florida economy, with little detectable impact on the remnant, natural, catchable centrarchids, according to Shafland (1995). One can only hope that the fishing public will remain satisfied fishing for cichlids where they already exist and not transplant their catch to supplement fisheries in other waters.

Minimizing Accidental Introductions

Solutions to ballast water and bait bucket introductions were discussed earlier. Because international shipping is regulated and subject to some scrutiny, public pressure on that industry could facilitate implementation of recommended actions. Bait bucket introductions, however, are so diffuse and commonplace that public education is the only likely means for achieving any significant reduction. Fishing with live bait obtained offsite should be discouraged or at least highly regulated.

Dissemination and retransportation of introduced species remain a problem because well-intentioned resource users commonly move introduced species beyond their original point of entry (e.g., Manchester and Bullock 2000). Welcomme (1988) emphasized this spillover effect, whereby fisheries introductions in one country invariably migrate or are deliberately transported into waterways of adjacent countries, necessitating international coordination to develop regulations and procedures (the same applies to interstate movement). Such coordination may be somewhat idealistic, given that "existing legislation is haphazard and in some cases nonexistent . . . [full of] loopholes [and] administered by agencies having scant comprehension of the problems of the aquatic sector" (p. 34). Public education is obviously needed.

Assessing Proposed Introductions

Formal scrutiny of introductions has been most strongly directed at accidental or commercial importation of exotics and deliberate introductions promoted by agencies. In the U.S., agricultural pests (plants and insects) have been regulated since 1900, but little protective aquatic legislation existed before the late 1970s. In 1977, a good start was made with Executive Order 11987, which restricted the introduction of exotic species into natural ecosystems. The Great Lakes Fish and Wildlife Restoration Act of 1990 targeted the sea lamprey, and the Nonindigenous Aquatic Nuisance Prevention and Control Act of 1990 addressed other aquatic pest species. Comprehensive proposals for investigating and regulating potential fisheries introductions have focused on protocols for deciding which species should and should not be imported into a country or stocked into a new system. C. C. Kohler and J. C. Stanley (1984) developed a step-by-step, five-level decision model for evaluating the potential biological and sociological costs and benefits of proposed introductions. Unfortunately, such protocols have received minimal application (J. D. Williams and Meffe 1998). Escapes from aquaculture and aquarium-related activities are classified as agricultural, which means they have considerable legal protection that legislators are hesitant to weaken or change.

It is generally acknowledged that much can be gained by thorough, scientific study of the biology and history of introductions of a candidate species. The International Council for the Exploration of the Seas (ICES) developed a Code of Practice for assessing the behavioral, ecological, genetic, and pathological risks of proposed introductions, which was adopted by the FAO European Inland Fisheries Advisory Commission (Welcomme 1988). It emphasizes using the best available science in assessing costs and benefits of introductions (Sindermann 1986). Another common recommendation in recent protocols is that imported species be subject to quarantine to minimize disease and parasite introduction. Disease problems can also be reduced by importing and releasing species only in the egg or larval stage, although this is no guarantee of pathogen-free introductions. Switzerland has passed legislation worth emulating. Rainbow trout can be introduced only to lakes and reservoirs lacking a free connection to a river. In waters connected to river systems, restocking programs must use native species, preferably of similar genetic make-up (Kirchhofer and Hefti 1996b).

One practice advocated increasingly is the development of a "white list" or "clean list" approach prior to approval of release, or where release is likely to happen (e.g., Lodge et al. 2000; Van Driesche and Van Driesche 2000). The clean-list approach reverses the older practice of viewing introductions as positive until proven otherwise, as innocent until proven guilty. Agencies develop black lists or dirty lists of known harmful species, but these are ineffective because determinations occur after harm is done and because little is known about most proposed introduced species. Past practices have resulted in an abundance of unanticipated negative consequences.

The white-list approach views potential introductions as guilty until proven innocent. A species should have a proven track record of minimal impact under known conditions before it can be imported, thus eliminating frivolous introductions. This approach is proactive and precautionary, making prohibition the default situation and permission the exception. This shifts the burden of proof from those attempting to reduce unregulated introductions to those who see introducing species as some sort of intrinsic right. Australia and New Zealand are pioneers in this approach. The aquarium industry is, not surprisingly, unenthusiastic about clean lists (e.g., Meyers 2001).

Another means of reducing the frequency and impact of introduced species is to create a mitigation fund paid into by those who trade in exotic species (Van Driesche and Van Driesche 2000). The fund, which would require "that businesses and individuals trading in risky organisms take collective responsibility for resolving the problems that arise from their activities," would support eradication and mitigation of invasives and their impacts (Van Driesche and Van Driesche 2001, p. 16). Currently, costs associated with invasions are covered by society at large. We no longer tolerate subsidizing the costs of most forms of pollution—why should it be different for alien species (P. B. Moyle, pers. comm.)?

Stocking to supplement fisheries does, however, remain a major goal and activity of fishery management agencies. Welcomme (1988, p. 34) articulated the traditional view that "further introductions will continue to be needed for sound development and management of inland fisheries and aquaculture." This view has been actively debated in the literature. Courtenay and Moyle (1996) were particularly critical of the quick-fix nature of many introductions, initiated as the fastest and cheapest solution to a perceived problem, one that may have arisen from past management debacles. Philipp (1992) similarly argued against a management ethic that justifies compromising the genetic integrity of native stocks and maintained that many introductions have a remote chance of improving a fishery, improvement "being a short-term boost to the creel or the appearance in the creel of a few large individuals." Philipp concluded that such an ethic "fails to recognize the consequences that may result from that practice and abrogates responsibility for the long-term stewardship of those populations" (p. 690).

In the past, stocking was often encouraged and accomplished with little consideration for natural biodiversity or the genetic integrity of species. This has become increasingly rare, and greater attention is given to minimizing negative impacts. Ham and Pearsons (2001) developed a stepwise decision framework for assessing potential risks of fish stocking programs. Their framework takes a cost-benefit approach and focuses on risk assessment, risk minimization, and uncertainty resolution. An important aspect of their framework is the recognition that uncertainty about actual risks must be taken into account in deciding whether to stock a species. Their approach is also adaptive; it promotes repeated monitoring and reevaluation of impacts and uncertainty after stocking to improve the balance between fisheries benefits and ecological costs whenever possible. Management agencies following such a protocol would accomplish much toward preventing the problems and reducing the criticisms brought about by an unquestioned reliance on and enthusiasm for stocking.

Increasing Public Awareness

Much is being done to educate the public about the ecological and economic costs of introductions. Considerable publicity surrounded the unauthorized release of northern snakeheads, *Channa argus*, into public waters of Maryland in 2002 and the efforts of management agencies to eradicate it (figure 9.10). Such an uproar would have been unheard of a few years earlier. The recent accelerated pace of legislation, study group formation, and international cooperation all bode well.

Much can still be done at the level of citizen action. Bait and aquarium stores need to display stickers and signs, hand out pamphlets describing threats and providing guidelines for disposal, print warnings on the plastic bags in which live fishes are placed, take back unwanted animals for disposal, and display hotline or other phone numbers or Web addresses for information (see chapter 13). Basically, commercial establishments need to be encouraged to actively participate in solving the problem. Government agencies responsible for managing nuisance introductions are probably best situated for conducting public outreach campaigns, although private conservation organizations are contributing substantially.

CONCLUSION

In recent years, an upsurge in scientific and public concern over invasive species has led to the creation of agencies and programs dealing with the problems at international and national levels. IUCN has an Invasive Species Specialist Group, which drafted guidelines for the prevention of biodiversity loss due to biological invasions. The multinational, multiagency Global Invasive Species Program (GISP), created to provide new tools for understanding and dealing with invasive species (Mooney 1999), focuses on synthesis of new information, incorporation of societal values and economic consequences, public education, global change, risk analysis, and best management practices. A journal, *Biological Invasions*, was launched in 1999 to address scientific and policy issues associated with introduced organisms.

In the U.S., the Invasive Species Executive Order of 1999 coordinates the activities of several federal agencies under the National Invasive Species Council to prevent the introduction of invasive species; provide for their control; and minimize the economic, ecological, and human health impacts of invasive species. The Nonindigenous Aquatic Species program of the U.S. Geological Survey maintains a Web site on introduced fishes and other aquatic organisms (http://nas.er.usgs.gov); a more general search function is provided at www.nisbase.org.

Among nongovernmental organizations, the most active is the Introduced Fish Section of AFS, which is involved in research, education, and policy formation on introduced species. AFS also periodically publishes informational and position papers and recommended procedures relevant to alien species in the journal *Fisheries*. At the level of citizen action, the Native Fish Conservancy, otherwise focused on aquarium keeping, has instituted an Exotics Removal campaign that encourages people to fish for introduced species and adopt a "once caught never returned policy. . . . If you catch [exotics], aquarium keep them, grill them, feed them to your pets, turn them into fertilizer, do anything except return them to their former homes" (www.nativefish.org).

Finally, many deliberate introductions, including those responsible for some of the worst ecological disasters, have occurred in developing nations, where human economic

Northern Snakehead

Distinguishing Features
Long dorsal fin • small head • large mouth • big teeth
length up to 40 inches • weight up to 15 pounds

HAVE YOU SEEN THIS FISH?

Northern snakehead. (USGS)

The northern snakehead from China is not native to Maryland waters and could cause serious problems if introduced into our ecosystem.

If you come across this fish,
PLEASE DO NOT RELEASE.
Please KILL this fish by cutting/bleeding or freezing and REPORT all catches to Maryland Department of Natural Resources, Fisheries Service. Thank you.

Phone:	**410 260 8320**
TTY:	**711 (within MD) (800) 735-2258 (Out of State)**
Toll Free:	**1 877 620 8DNR (8367) Ext 8320**
E-mail:	**customerservice@dnr.state.md.us**

Figure 9.10. Flier distributed by the Maryland Department of Natural Resources following the discovery of snakeheads in Crofton Pond, July 2002. Used with permission of Maryland DNR, www.dnr.state.md.us/fisheries/education/snakehead/snakeheadflier.pdf.

and nutritional problems are paramount. In addition to negative ecological consequences, many introductions require new harvesting technologies, which eliminate traditional fishing methods or replace local artisanal fishers with commercial or sport fishers. "The long-term results of an introduction may be less protein available to local people, not more" (Courtenay and Moyle 1996, p. 245). Developing nations are in desperate need of both capital and animal protein, but better attention to native species distributions and local fishing techniques could minimize the impacts of introductions that destroy both local biota and local culture.

PART IV Direct Causes of Decline

Fishes as Commodities

10. Fishes versus Fisheries I: Overfishing

. . . all the great sea-fisheries, are inexhaustible.
—Thomas Henry Huxley, 1884

The literature on fisheries conservation is voluminous. S. J. Hall (1999) counted over 125,000 articles and 1,200 textbooks in the scientific literature on fisheries-related subjects, and that number has grown considerably. Some recent treatments of major issues, controversies, and solutions at greater length than given here include Pitcher et al. (1998), Iudicello et al. (1999), NRC (1999b), G. Moore and Jennings (2000), Welcomme (2001), and P. J. B. Hart and Reynolds (2002); see also www.worldfish.org.

This chapter focuses on (1) evidence that fisheries resources are overexploited, (2) causal mechanisms that lead to overexploitation, (3) examples of specific stocks and species that have suffered (and some that are exploited sustainably), and (4) solutions to overfishing. Because most literature concerns marine fishes, freshwater fisheries will be underemphasized, recognizing that they are subject to the same impacts, with the same results (see box 10.1). Several of the terms used throughout the chapter are defined in box 10.2.

EVIDENCE THAT FISHERIES ARE OVEREXPLOITED

An accumulating body of evidence indicates that the world's marine fisheries have been and continue to be exploited to an extent that can only lead to serious depletion of many if not most individual stocks. "The history of fisheries is one littered with examples of the demise of a species following its exploitation" (Levin et al. 2001, p. 1153). This generalization emerges from statistics on the number of stocks fished at or beyond sustainability, declining catch numbers despite increased fishing effort, and theoretical calculations of the productive capacity of the world's oceanic ecosystems. A widely held conclusion from the accumulated evidence is that "fisheries globally are really in deep trouble . . . far more so than is admitted to the public, to whom technical fixes are still being sold" (Pauly et al. 1998b, p. 410).

Beyond Sustainability?

Global fisheries expanded substantially in the last half of the 20th century, the "industrialized fishing period." Marine fish production rose annually by about 4%, increasing fivefold from 15–20 MMT in the early 1950s to around 85 MMT in the late 1990s (figure 10.3). However, the rate of change slowed at the end of the century, declining from 6% per year in the 1950s and 1960s, to 2% in the 1970s and 1980s, and finally falling to zero growth in the 1990s. This global leveling-off reflects trends in most regional fisheries, "which have apparently reached their maximum potential for capture fisheries production, with the majority of stocks being fully exploited" (FAO 2000b). Recognizing that China apparently reported inflated numbers during the 1990s (see box 10.3), one can only conclude that marine fisheries production has in fact been declining, despite increasing effort.

In the 1990s, the actual worldwide annual marine fish catch hovered around 80–85 MMT; the sustainable catch is theoretically somewhere between 69 and 96 MMT (FAO 1995b, 2000b; NRC 1999b), again suggesting we are near or beyond what is sustainable. Many fisheries may already

BOX 10.1. Freshwater Fisheries at Risk

Fishing is one of humanity's oldest protein-acquiring activities that requires tools. The oldest known fishing implements are bone harpoons used to catch a very large, now extinct, freshwater catfish (Pauly and Froese 2001), recovered from a 90,000-year-old site in the Democratic Republic of Congo. Ignoring aquaculture (30 MMT) and recreational fishing (2 MMT)—and recognizing that inland fisheries are poorly monitored and perhaps underestimated by a factor of three or four (e.g., Dudgeon 2000a)—annual total freshwater catch is about 9 MMT, or 10% of marine catch (FAO 2003, 2004).

Although overfishing is relatively more influential in marine systems, freshwater fishes have suffered widely from unsustainable practices. In the U.S. over the last century, commercial fish harvests declined 80% in the Missouri and Delaware rivers, 95% in the Columbia River, and 100% in the Illinois River (Karr and Chu 1999). The collapse of fisheries in the world's great lakes is well documented (see Allan et al. 2005). The impacts of overfishing in northern European freshwaters, especially on larger species, have been profound (see chapter 4). Overfishing of large species has depleted several catfish and barb species in the Mekong River basin of southeast Asia (Mattson et al. 2002). Lake Malawi, an African great lake, has experienced a typical cycle of high initial catches, then declining catches, prompting the use of larger trawls with finer mesh. As ever smaller fishes were taken, large species declined and some disappeared. Nearby, smaller Lake Malombe also saw a decrease in catch, a shift to smaller sizes within species, and disappearance of at least nine large haplochromine cichlids (Barlow 2000).

Welcomme's (2001) *Inland Fisheries* provides a thorough review of freshwater fisheries issues and solutions, with special emphasis on developing nations (summarized and supplemented in Welcomme 2003), and shows that fresh and saltwater fisheries have much in common. For example, the concept of fishing down food webs—whereby the average trophic level and individual size of exploited individuals decrease over time—was popularized by Pauly et al. (1998a) for marine food webs, but the phenomenon has also occurred in large rivers, such as the Ouémé River of West Africa. Large predatory centropomids and catfishes were replaced by large bonytongues and characins and eventually by small catfishes, cichlids, and cyprinids (figure 10.1). Landings data from large-scale fisheries on the Yangtze River indicate decreased abundance of large individuals and species and increased abundance of small fishes during the latter half of the 20th century (Fu et al. 2003).

Overfishing, intensive management, recovery efforts, and complicated results also occur in freshwater fisheries. Lake trout, *Salvelinus namaycush*, fisheries crashed throughout the Laurentian Great Lakes from a combination of factors, especially overfishing and sea lamprey predation (Link 2002a). Extensive recovery efforts have included reduced fishing effort and extensive stocking, but self-sustaining populations have redeveloped only in Lake Superior (Hansen 1999).

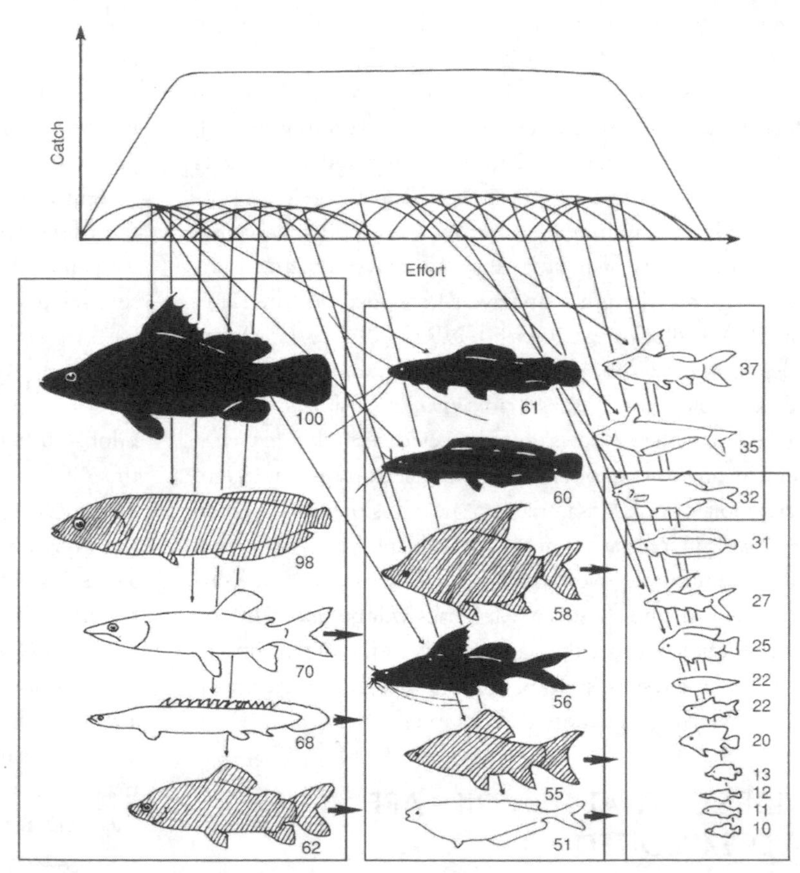

Figure 10.1. Progressive changes in catch and effort, Ouémé River, Africa. Fishing down of food webs occurs in freshwater situations as well as marine. As fishing effort increased, large fish disappeared and the fishery shifted to progressively smaller species, while remaining individuals of larger species were on average smaller. Species in black disappeared from the fishery before 1965; hatched species' numbers were seriously depleted; short arrows to the right of a drawing indicate species whose average breeding size was reduced. Numbers are mean body lengths (cm). From Welcomme (2001); used with permission.

BOX 10.1. Continued

Recreational stocks can suffer similarly from overexploitation. Post et al. (2002) surveyed 30 high-profile recreational fisheries for 14 species in Pacific, Arctic, Hudson Bay, and Atlantic drainages of Canada. Twenty-two fisheries had declined historically, and 19 declines could be attributed in part to overfishing (justifying statements by recreational fishers such as "It ain't as good as it used to be!"). Post et al. attributed declines to a variety of causes, notably a general depensatory population response (higher mortality rates with decreasing population size) arising from additive interactions among fish life history traits, angler behavior, mismanagement, and food web alterations. Combined factors reduced numbers to an unstable equilibrium, culminating in population collapse. Classical fisheries models often ignore depensatory processes and tend to project optimistic levels of sustainable yields, further exacerbating the situation. Recreational fisheries have frequently been portrayed as resilient and self-sustaining because they are less subject to the social and economic forces of the open market that drive commercial fisheries to collapse. Post et al. show otherwise, and they concluded that "many recreational fisheries are headed in the same direction as are the world's commercial fisheries" (p. 15).

BOX 10.2. Fishery Terminology

Much classic fisheries terminology has undergone analysis and revision to reflect criticisms of past practices, not to mention some degree of political correctness. Terms have fallen into disuse, or been modified, corrected, and divided. Some commonly used terms relevant to this discussion are defined below (for a more thorough treatment, see Hilborn and Walters [1992], Caddy and Mahon [1995], and www.nefsc.noaa.gov/techniques/tech_terms.html):

- *Biological, ecological, and commercial extinction* distinguish among impacts of a species' disappearance. *Biological extinction* is the traditional concept of a species vanishing entirely. *Ecological extinction* occurs when a population falls below the level at which the species can function normally within an ecosystem as a predator, prey, competitor, symbiont, nutrient processor, etc. It is difficult to document because of our incomplete knowledge of ecosystem function, and generalizations about the population size at which it occurs appear to be rare. *Commercial extinction* occurs when a species declines to a level at which it is no longer worthwhile for fishers to pursue it. Because it usually occurs at population sizes much higher than biological extinction, it may save a species from further exploitation, except for exceedingly valuable species such as bluefin tuna and sturgeon, which are too valuable not to pursue every individual.
- *Harvest* implies a crop in which the user has made some investment, as a farmer plants seeds and reaps the result. Wild animals and plants are extracted, captured, gathered, fished, landed, collected, exploited, or taken, but not harvested (NRC 1999b; Safina 2001a). Stocked and put-and-take fisheries are exceptions.
- *Highgrading* is the practice of retaining only the largest or most valuable size classes or species of a catch. It often results from size limits or quota systems; fishers highgrade at sea so that dockside landings do not exceed the quota, even though the actual number of fish caught exceeds the limits (Speer et al. 1997).
- *Overfishing* covers a host of sins and has been divided into several categories (fishery management plans in the U.S. alone use 117 fishery-specific definitions, according to G. M. Mace and Hudson [1999]):
  - *Demographic overfishing* refers to changes in age structure brought on by exploitation. Depleting specific year-classes reduces the number of age classes and makes a population more vulnerable to natural sources of mortality and reproductive failure and therefore more likely to crash.
  - *Recruitment overfishing* depletes breeding individuals, resulting in insufficient recruitment to sustain the population.
  - *Genetic overfishing* targets specific genotypes or reduces population size to the point that genetic variation is lost. Depletion of unique stocks of Pacific salmon is one example (see chapter 11).
  - *Growth overfishing* occurs when too many small or young fish are caught before they reach maturity. The average-size fish remaining in the population is small, which affects the number of available spawners. A fishery suffering growth overfishing is characterized by many small fish in the landings. In sex-changing species, overfishing of smaller individuals can deplete one sex differentially.
  - *Malthusian overfishing* describes too many fishers exploiting a stock and results from a need to feed too many hungry humans. Such fishers often turn to destructive fishing practices out of desperation, accelerating collapse of the resource.
  - *Serial or sequential overfishing* occurs when first one species and then another is depleted, as frequently occurs in overcapitalized fisheries.
  - *Ecosystem overfishing* results when the composition of a community is altered via overfishing of species that played important functional roles in the ecosystem. It includes serial overfishing, loss of large individuals and species, declines among piscivores and in average trophic level, and greater population sensitivity to environmental fluctuations (Murawski 2000).
- *Sustainable fishing* is fishing that causes no undesirable changes in biological or economic productivity, diversity, or ecosystem structure or function (NRC 1999b). Sustainable yield, or sustainability, is an extraction level that does not exceed the growth rate of the stock (or the entire fishery where multiple species are concerned). Implicit in most definitions is the concept of intergenerational equity (e.g., Hilborn 2005).
- *Stock* refers to a population with characteristic life history attributes that shows spatial or temporal integrity and perhaps genetic uniqueness and can be treated as a separate, self-perpetuating management entity.

BOX 10.2. Continued

- *Underutilized* has been abandoned in favor of *not overexploited, moderately exploited*, or *lightly fished* (NRC 1999b).
- "Yield" from a fishery has multiple meanings. *Maximum sustainable yield* (MSY) recognizes biological limits; it is the largest average catch that can be taken from a stock under existing environmental conditions, or the largest long-term average yield or catch that can be taken from a species' stock without depressing the species' ability to reproduce (Iudicello et al. 1999; NRC 1999b). MSY is based on the classical ecological concept of log-phase population growth at intermediate population sizes (the logistic growth equation). Fishing at intermediate levels in populations of intermediate size maximizes productivity by yielding a surplus of extractable biomass. Theoretically, a moderately exploited population of intermediate size should produce more individuals with faster individual growth rates than smaller, larger, overexploited, or underexploited populations (see M. R. Ross 1997; P. J. B. Hart and Reynolds 2002). *Maximum economic yield* (MEY) maximizes profits relative to fishing costs rather than maximizing catch; landings targeted at MEY are usually lower than landings at MSY and require less fishing effort (figure 10.2).

In the U.S., *long-term potential yield* (LTPY) has replaced MSY as a goal, with the added consideration that such a catch should be maintainable in the future (NMFS 1999). U.S. fisheries statistics also record *recent average yield* (RAY) and *current potential yield* (CPY). RAY is the reported landings averaged over the three most recent years. CPY is the immediately available catch given the existing population size and current production rate. LTPY is the most conservative estimate, because it takes future production into account. Unfortunately, when NMFS lacks sufficient information to calculate LTPY, it may substitute CPY or even RAY (NMFS 1999), which is like saying that even though we lack vital information, our current actions are appropriate. A more conservative measure of MSY promoted by NMFS is B_{MSY}, the average spawning stock biomass that produces or maintains MSY; B_{MSY} may be ten times MSY (e.g., Sissenwine et al. 1998).

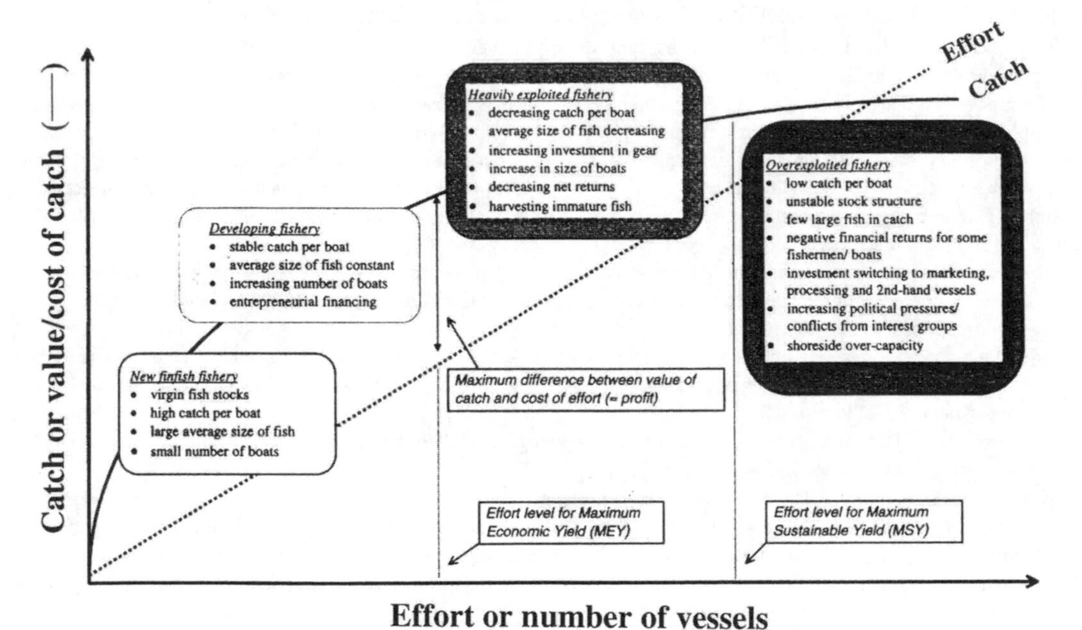

Figure 10.2. Interactions among effort, catch, and economics at different levels of exploitation of an open-access fishery. The curve depicts the classical "catch relative to effort" models of fisheries; rounded boxes show where stocks exploited at different levels reside along the curves and characteristics of fisheries at these different levels. Modified from Iudicello et al. (1999); used with permission.

be exploited at unsustainable rates. Globally, FAO (2000b) presented data on 441 stocks, which it classified as underexploited (mostly in the Indian Ocean), moderately exploited, fully exploited, overexploited, depleted, or recovering (figure 10.4). The classifications were based on MSY, a concept often criticized as insufficiently precautionary. Regardless, by FAO's estimation, stocks designated as overexploited and depleted were fished unsustainably, fully exploited stocks were fished at levels close to MSY, and stocks designated underexploited and moderately exploited may be able to sustain greater fishing pressure. Of the world's assessed fishery stocks, 28% were fished unsustain-

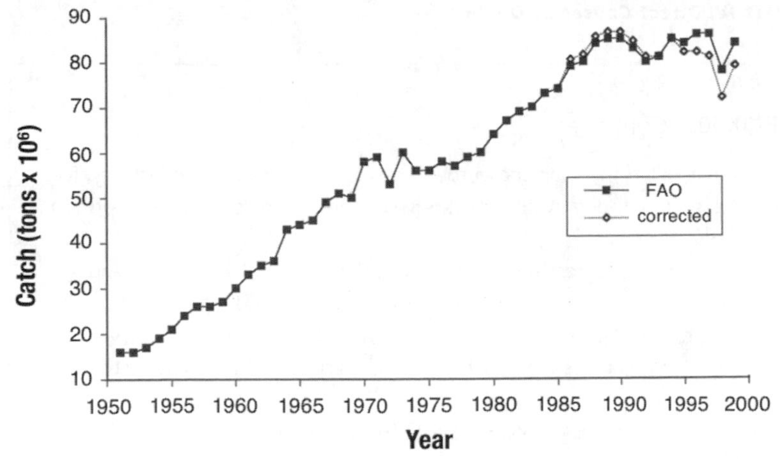

Figure 10.3. Global marine fish production, 1950–2000. Marine capture fisheries increased through much of the last half century, peaking in the late 1980s. The dotted line with open diamonds between 1985 and 2000 is corrected for apparent overreporting by the Chinese government; it shows overall declines in catches in the 1990s. Decreased catches around 1972, 1993, and 1998 are attributable to El Niño events, when stocks of small pelagics were lower. Data are global capture numbers from FAO (2000b), with freshwater capture values of 5–7 MMT (from FAO [2000b] and Safina [1995]) subtracted. Chinese data include invertebrates, which make up 25% of that nation's catch. Corrections for Chinese overreporting are from R. Watson and Pauly (2001).

BOX 10.3. FAO Statistics, Chinese Duplicity, and the Real State of Global Fishery Resources

> The statistics of the world's fisheries . . . are still incomplete and riddled with guesses, inadvertent errors, omissions, and even, perhaps, some perjuries.
>
> —Peter Larkin, 1977

Everyone cites statistics from FAO. This important UN organization is the recognized authority on and clearinghouse for information on international fishery production. FAO's periodic report, *The State of World Fisheries and Aquaculture* (SOFIA) is the definitive source for monitoring levels and trends in production.

However, FAO relies on individual countries to provide landings and production records. Data are submitted on a voluntary basis, and no independent verification occurs. Despite the efforts and honesty of most fisheries statisticians, fudging, book cooking, missing values, and guesstimating are not uncommon. (I knew a fishery program director in a South Pacific nation who encountered her statisticians filling in monthly catch records in a local pub: "What should we say for snapper?" "Oh, 3,500 pounds sounds believable." "What's that in kilograms?")

Small nations have a minor effect on global figures, but when large nations engage in questionable reporting practices, the impacts can be substantial. China accounts for an estimated 15% of global landings, but the exact magnitude of that number and its apparent misrepresentation may have distorted the picture considerably.

Fisheries landings for the rest of the world leveled off by the late 1980s (see figure 10.3), but officially reported Chinese landings grew exponentially from 1985 to 1998. R. Watson and Pauly (2001) and R. Watson et al. (2001) listed three reasons for skepticism about Chinese landings data: (1) major fish stocks along the Chinese coast had been classified as overexploited for decades and fishing effort had increased, which usually corresponds to greater depletion; (2) officially reported CPUE remained constant when effort and catch were claimed to have increased, in contrast to stock survey data indicating declining fish abundances; and (3) when expressed on a per-area basis, officially reported catches were far higher than catches from geographic areas that were similar in latitude, depth, and primary production.

R. Watson and Pauly (2001) specifically tested the third observation by comparing Chinese landings with FAO statistics for 32 other regions on the basis of oceanographic features such as depth, primary production, ice cover, surface temperature, distance from shore, and occurrence of regions of upwelling. Most high-catch areas were predicted correctly by their model and corresponded to areas of upwellings (e.g., coastal Peru). The one outlier in the model was China, which apparently lacks such upwellings and other oceanographic features that contribute to high productivity. China's exclusive economic zone can be expected to produce no more than 5.5 MMT per year; China's reported catch for 1999 was 10.1 MMT.

Why would China consistently report an increasing catch that may have been as much as double the actual catch over a 15-year period? The answer may lie in rewards to officials, who are promoted for increased production from the areas under their administration: "The state entities that monitor the economy are also given the task of increasing its output" (R. Watson and Pauly 2001, p. 535). The data are even more suspect given subsequent events. In 1998, the Chinese government, in an apparent effort to

BOX 10.3. Continued

prevent overreporting, instituted a "zero-growth policy" that called for constant output. Presto, reported fisheries landings for 1999 and 2000 were the same as those for 1998. "Such measures, although well motivated, do not inspire confidence in official statistics, past or present" (R. Watson and Pauly 2001, p. 536).

When global fisheries statistics are corrected for this apparently progressive overreporting, the stabilized worldwide landings of the 1990s turn into declining catches and resources, with an annual rate of decline of 0.36 MMT, which is a greater than 10% decline between 1988 and 1998 (figure 10.3, broken line). Progressive declines make sense given the documented overall total and steadily increasing number of overfished stocks. Reporting stable catches masked this overall decline and probably justified a degree of global complacency about world fish stocks.

Interestingly, FAO may have suspected inaccuracies in Chinese numbers for some time. SOFIA 2000 calculated total world production with and without Chinese numbers; no other country received such treatment.

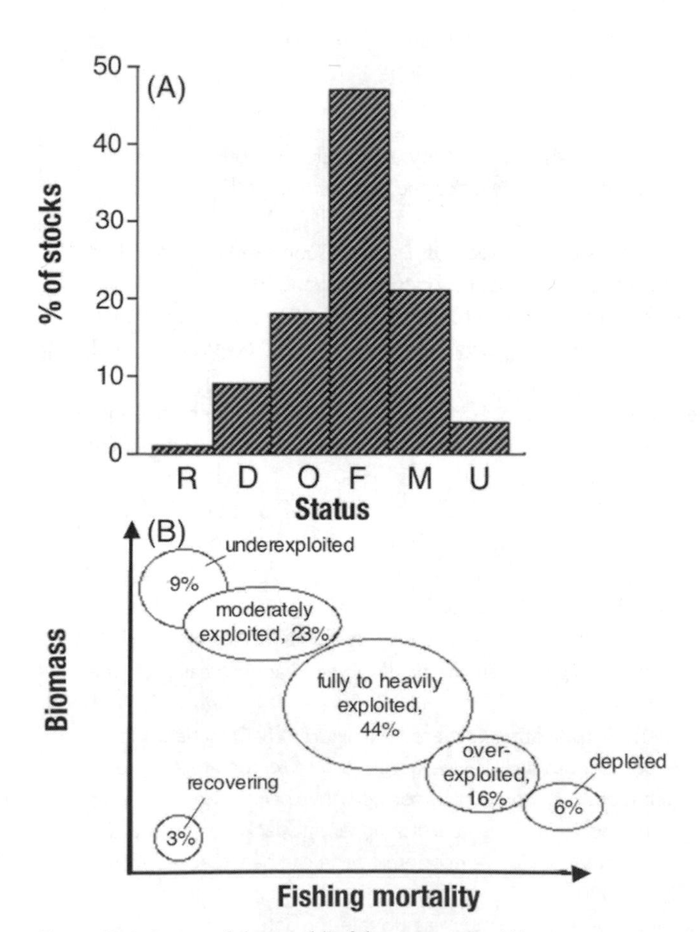

Figure 10.4. Status of the world's fisheries stocks. (A) Underexploited (U), moderately exploited (M), fully exploited (F), overexploited (O), depleted (D), and recovering (R). Data from FAO (2000b). (B) Status viewed from the standpoint of the relationship between biomass and fishing mortality for different exploitation levels. Redrawn from Botsford et al. (1997).

ably and will crash if fishing continues at anything close to current rates, 47% were fished at or near capacity, and only 25% may be able to sustain greater fishing pressure.

Hidden among these statistics are alarming shifts in species composition from high-value demersal species to lower-value pelagic species: "The world fish supply is increasingly relying on variable, small pelagic and other low-value species, thereby concealing the slow but steady degradation of the demersal, high-value resources" (Garcia and Newton 1997, p. 23). In the last three decades of the 20th century, total world fleet and technology increased, total catch stabilized, but landing rates of valuable species fell by 25%.

Many observers would contend that inaccuracies in stock assessments and the overly optimistic MSY concept mean that "fully exploited" stocks are probably overfished rather than sustainably fished. Even FAO (2000b) recognized that "these stocks are in need of (and in some cases already have) effective measures to control fishing capacity." A half-full versus half-empty contrast would view 73% of stocks as capable of sustaining fishing at current or greater rates (F+M+U), or 75% of stocks as fully or overexploited (F+O+D+R) with no room for growth and in need of "stringent management of fishing capacity" (FAO 2000b). FAO (2000b) concluded that most areas are being fished at their maximum.

Of added concern is the progressive change in these values over the past 25 years. Overfished stocks (O+D+R) tripled, from under 10% of available stocks in the early 1970s to almost 30% in the late 1990s. Correspondingly, "underutilized" stocks (U+M) were cut almost in half during the same period, from about 40% to 25%. If these trends continue, overfished stocks will outnumber underfished stocks early in the 21st century (FAO 2000b; see

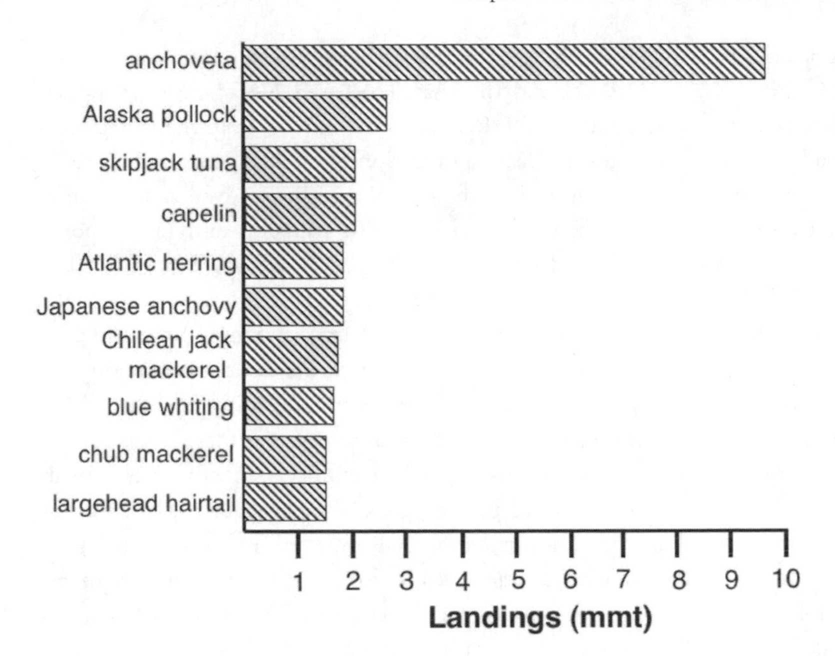

Figure 10.5. The top 10 fisheries species in 2002 have consistently been among the top 15 or so for the past two decades, with minor changes due to a combination of overfishing and climatic shifts. Data from FAO (2004).

Dulvy et al. 2005 for similar conclusions about northeast Atlantic stocks and Worm et al. 2006 for global trends).

Are there major unexploited stocks? Although new fishable stocks are discovered from time to time, a small number of species make up most of the commercial catch. Roughly 10 species consistently show annual landings in excess of 1 MMT and constitute a third of the biomass of annual landings; about 20 species make up half of the catch (figure 10.5). Of biological significance, 8 of the top 10 species are open-ocean, pelagic fishes, and all are schooling species. With the exception of skipjack tuna, walleye pollock, and the mackerel species, the most productive fisheries are primarily rendered into fish meal products rather than consumed directly by humans, often at a net loss of edible protein because fish meal is used in the culture of predatory fishes (chapter 14).

U.S. Stocks

In the U.S., which ranked fifth among nations for fisheries landings, the overall picture resembles the international situation. During the late 1990s, of the 283 stocks under the jurisdiction of NMFS or state agencies and for which population data were available, 30%–40% were below, 30%–40% were near, and 12% were above the population levels considered sustainably fishable (i.e., having long-term potential yield, LTPY; NMFS 1997, 1999); an additional 85 stocks were at unknown levels (figure 10.6). In an

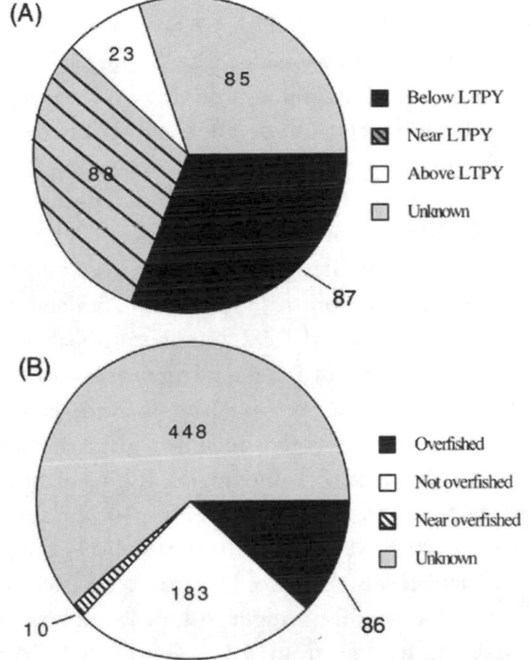

Figure 10.6. The status of exploited marine fisheries stocks in U.S. waters. (A) Among the 283 stocks for which a national or state agency was responsible in the late 1990s (including approximately 30 invertebrate species), about a third were below and another third were near a population level that is considered sustainable; only about 10% existed in numbers sufficiently large to sustain additional exploitation. Data from NMFS (1999). (B) Of all 727 species commercially exploited in U.S. waters, 25% were fished at sustainable levels, and 13% were overfished or soon will be; for the vast majority, 62%, data were insufficient to determine whether species were exploited in a sustainable matter. Data from NMFS (1997). LTPY = long-term potential yield.

alternative analysis of all U.S. marine species subjected to fishing (both analyses included invertebrates), NMFS (1997) reported that 86 species were overfished, 10 were approaching an overfished condition (were likely to be overfished within two years), and 183 were not overexploited. These values indicate that 34% were overexploited or nearly so, and that 70%–80% of U.S. stocks for which we have data were fully or overutilized. It is alarming that at least one-third of known stocks are probably overfished. It is even more disconcerting that for 448 species, or more than 60% of exploited stocks, status with respect to overfishing is unknown in a country that keeps relatively complete fisheries statistics.

Increasing Effort, Declining Catch

Some of the strongest evidence indicating that global fisheries practices are not sustainable comes from decreasing catch per unit of effort (CPUE). As catches have declined, more effort has been expended. But "more fishing [has not produced] more fish, and almost always seemed to mean fewer fish per unit of fishing effort" (T. D. Smith 1994, p. 213). During the late 20th century, decreasing fish abundance led to more and bigger boats with bigger engines, all fishing farther offshore (Pauly and Froese 2001). The size (gross tonnage) of the world's fishing fleet increased by about 3% per year, from 13.5 million to 25.5 million gross tons, between 1970 and 1992. Number of boats also doubled, from 585,000 to 1.2 million large (industrialized) vessels (FAO 1995b). Many new boats became part of offshore and distant fleets that included giant factory ships, extending the reach of the major fishing nations. But world catch increased by only 1.8% per year, for a total increase of 20 MMT (from 60 to 80 MMT) during this time, which means that effort far outpaced catch. In fact, the ratio of tons of fish caught to tonnage of boats fishing has been decreasing, from 4.3 tons/ton in 1970 to 3.0 tons/ton in 1992 (NRC 1999b).

Advances in technology accompanied increases in size and number of boats, magnifying "fishing power," a truer measure of the ability to catch fish than tonnage ratios. Fishing power may have increased 4.4% per year over the last 30 years (Fitzpatrick 1995), even though catches increased by less than 2% per year. Military technologies developed during and after World War II were quickly applied to fishing. Radar allowed navigation in fog, storm, and darkness; sonar made once invisible schools and bottom topography graphically obvious; electronic navigation, first LORAN and then satellite-based global positioning systems, allowed fishers to locate and relocate the exact coordinates of fish aggregations; satellite-generated weather maps allowed prediction of weather, as well as location of water masses associated with fish concentrations; spotter planes and helicopters allowed aerial tracking of fish that previously could not be seen over the horizon. Long-liners can put thousands of baited hooks along 100 km of line; drift nets can be 60 km long; trawls can be large enough to engulf 12 jumbo jetliners (Safina 2001a). The fish had few innovations at their disposal to counter these advances.

Increased fishing effort, ability, and power eventually result in overcapitalization and overcapacity. Capitalization is the amount of capital invested in vessels and gear. Overcapitalization leads to overcapacity, which means more money has been invested in fishing capacity than the resource can return (too many boats chasing too few fish). Garcia and Newton (1997) put world overcapacity at around 30%. In 1987, the U.S. had 23,000 commercial fishing vessels, more capacity than required to achieve a sustainable, long-term potential yield; off Alaska, capacity was about 2.5 times that necessary to catch available resources. The collapse of the New England groundfish fisheries (detailed later) resulted in part from overfishing made possible by excess capacity. If catches decline despite increased effort, additional increase in effort is likely to decrease rather than increase production (NRC 1999b).

One would expect that overcapitalization-driven overfishing would lead to reduction of effort. Unfortunately, as CPUE declines, fishing effort tends to increase, because investors try to get their money back. Such unsustainable fishing inevitably leads to local fishery collapse. However, with so much money invested and so many boats in the water, the response to local fisheries collapse is often to move the boats to another fishery, displacing the problem. "What is fascinating—and also tragic—about the fishing industry is that it so actively participates in its own annihilation" (McGoodwin 1990 in Newton and Dillingham 2002, p. 142).

The overcapacity-overfishing cycle has created the great irony of fishery economics, namely that global commercial fishing is not economically justifiable. The value of the annual global catch is around $70 billion, but the costs of procuring that catch run between $91 and $116 billion, for a net loss conservatively placed at $21–$46 billion annually (Garcia and Newton 1997; these numbers do not include losses associated with debt servicing or the opportunity costs of alternative investment, or, of course, losses related to environmental degradation or declining biodiversity). The red ink in the balance sheet was pointed out by FAO at least as early as 1993. In addition to balance sheet costs are government subsidies of as much as $27 billion annually via fuel tax exemptions, price controls, low-interest loans, support facilities (ports, ice plants) construction, research and gear trials, and outright grants (Bailey 1988; NRC 1999b). When stocks in developed countries collapse, government-subsidized industries often export their excess capacity to developing nations, internationalizing the problem. All these interventions only postpone the day when real-market events and transactions control investment practices.

Declining Ocean Productivity

Fish production in the oceans is determined by the primary productivity of marine plants, just as terrestrial animal production depends on terrestrial plant output. The chief primary producer in the ocean is phytoplankton, which is eaten by zooplankton (and a few fishes). Zooplankton is eaten by fishes, which are eaten by larger fishes. Phytoplankton productivity is determined by nutrients, sunlight, and water temperature. The upper limit of fisheries productivity—including potential sustainable global yield—can therefore be calculated given the amount and location of phytoplankton productivity, the trophic levels of the fishes involved, and the general observation that about 90% of the energy is lost at each trophic level due to metabolic inefficiency (Pauly and Christensen 1995). Depending on how the calculations are performed, the theoretical fisheries productivity ranges from 80 to 700 MMT (NRC 1999b); the higher values are admittedly upper limits. More important, the larger numbers were calculated "using unrealistic assumptions about food web structure . . . , effects of bycatch, feedback effects of fishing on other fish populations and marine ecosystems [e.g., imperfect management and multispecies interactions], and the technical and economic feasibility of new fisheries" (NRC 1999b, p. 24). When corrected (especially for bycatch, discards, and unaccounted mortality), calculations indicate lower values, of around 80–100 MMT. An obvious conclusion is that recent catch levels of around 80–85 MMT are close to what the ocean can produce (Buckworth 1998). We are near or past what is sustainable.

At the same time that we may have exceeded oceanic capacity to produce food for humanity (not to mention other species), natural or human-induced factors may be causing oceanic productivity to decline. Evidence comes from patterns of climate change, oceanic regime shifts, and plankton production best known in the northeast Pacific Ocean, where long-term (decadal-length) reductions in zooplankton production correspond with declines in predatory species dependent on zooplankton. In the latter half of the 20th century, zooplankton biomass off Southern California decreased 80%, and zooplankton-dependent foragers showed similar declines (e.g., Roemmich and McGowan 1995; Veit et al. 1997). Significant reductions in deep-sea benthic productivity have also been linked to these changes (Druffel and Robison 1999). Similar events have been observed or suspected in northeast Atlantic, Baltic, and sub-Antarctic regions (Flinkman et al. 1998; B. P. V. Hunt et al. 2001; Sims and Reid 2002), although plankton and fish production in the south Atlantic has increased (Verheye 2000).

Whether decreased plankton productivity is a permanent or short-term phenomenon is for practical purposes immaterial. Oceanic productivity undergoes cycles with yearly and decadal periodicities, reducing and increasing food availability. When these cycles occur in combination with anthropogenic degradations in marine food webs (e.g., J. B. C. Jackson et al. 2001), food availability for humans and other top predators is reduced. Expecting the oceans to feed humanity at current and even past levels of consumption is unjustifiable, especially if our actions affect long-term climate change in additional, negative ways.

Analysis of the trophic levels subjected to fishing provides additional evidence that we have reached the theoretical (and real) limits of oceanic productivity. The phenomenon known as fishing down marine food webs (FDFW) characterizes global and regional catches in the most heavily fished regions of the world (Pauly et al. 1998a). Pauly et al. compared diets and average trophic level of 220 fishery species in FAO fisheries regions between 1950 and 1994. They calculated that the average trophic level exploited has decreased about 0.1 level per decade (figure 10.7), from a mean of 3.3 in the 1950s to 3.1 in 1994. In the heavily exploited northwest Atlantic, this number fell from 3.2 to 2.9. The fishery has changed from one dominated by long-lived, high-trophic-level, piscivorous bottom fishes (flatfishes and gadoid cod, haddock, hake) to shorter-lived, lower-trophic-level, planktivorous

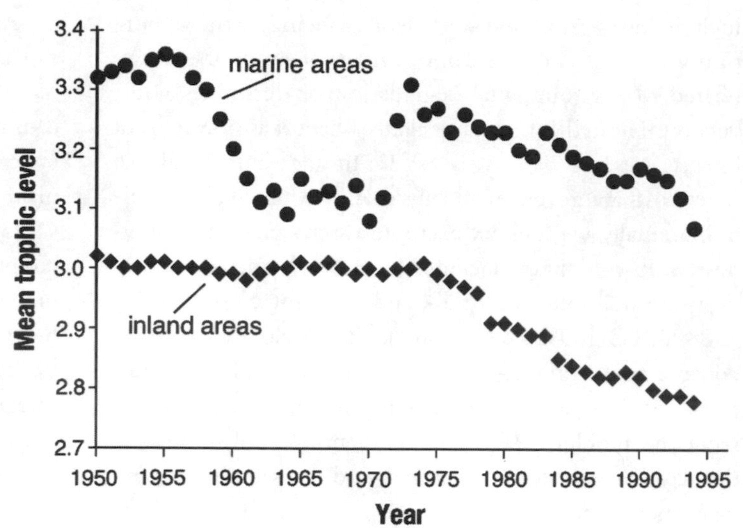

Figure 10.7. Fishing down food webs. Over the past half century, most of the world's fisheries have been taking species progressively lower in food webs, in the oceans and in freshwater. (A) Mean annual trophic levels of fisheries landings for all marine areas, calculated as total landings times the fractional trophic level of species groups (primary producers and detritus = 1, top predators = 4; species that feed at more than one level are assigned fractional values). The dip in mean trophic levels in the 1960s represents extremely large catches of planktivorous Peruvian anchoveta, a fishery that collapsed in the early 1970s. (B) Inland areas. A parallel trend exists in freshwater fisheries. The plateau region between 1950 and 1975 is thought to reflect incomplete information more than stability in trophic level captured. Redrawn from Pauly et al. (1998).

pelagic species (small, planktivorous mackerels, herrings, and anchovies). Large pelagic predators (e.g., billfishes, tunas, sharks) declined globally by more than 90%, contributing to the trends (R. A. Myers and Worm 2003).

Events other than a fishing-down mechanism could explain observed changes in the composition of fisheries landings. Caddy and Garibaldi (2000) focused on the ratio of piscivores to zooplanktivores (PS:ZP) in different regions. Reduced mean trophic levels would produce lowered PS:ZP ratios. Ratios did decline over time (with substantial variance) in regions with long histories of extensive exploitation, such as the northwest Atlantic, northeast Atlantic, and northwest Pacific. However, ratios remained relatively stable or increased (with substantial variance) in the northeast Pacific, most tropical regions, and the southwest and southeast Pacific. Fishing down may be occurring in certain intensively exploited regions but not worldwide.

FDFW is also a descriptive statement about the relative abundance of different trophic groups in the landings. A decreased PS:ZP ratio could reflect simple increases in zooplanktivore numbers, independent of fishing. Zooplanktivores might increase in relative abundance if increased runoff or upwelling brought in nutrients that stimulated zooplankton productivity, a bottom-up control process. Changes in fishing practices that resulted in greater emphasis on catching zooplanktivores would also produce higher PS:ZP values. Technological advances, especially development of synthetic net materials for use in high-speed midwater trawls and large purse seines, favored greater effort directed at small pelagic fishes. Development of distant-water fleets and changing market patterns (increased fleet sizes, increased fishing power and mobility, increased global trade) also increased the capture of previously unexploited and abundant planktivore stocks in the developing world. Hence FDFW may have occurred where piscivores have been depleted or where short-lived invertebrates have increased in numbers, but it should be used sparingly as a general explanation while alternative factors have yet to be accounted for. "Any attempt at a definitive statement as to prime causes [of changing trophic levels or ratios may be] presumptuous" (Caddy and Garibaldi 2000, p. 646).

Fishing at lower trophic levels may not produce higher or sustainable catch rates. In fact, landings have tended to decrease as mean trophic level fished has decreased (Pauly et al. 1998a). Production may increase initially but has often been followed by declines, as in the huge, failed fisheries for planktivorous Peruvian anchoveta and menhaden, and in overall landings in the northwest Atlantic, northeast Atlantic, and Mediterranean. Eventually, after fisheries for larger, edible species vanish, we will increasingly target very small, inedible fishes or zooplankton. These trends, projected into the future, foretell widespread fisheries collapse.

A SAMPLING OF STOCKS AT RISK

Sharks

Shark populations around the world have decreased 50%–80% since the 1970s, the steepest declines occurring in the late 1980s, when international fishing for sharks accelerated (e.g., D. A. Rose 1996). In the northwest Atlantic, where historical reductions had already occurred, substantial additional declines were documented between 1985 and 2000 (Baum et al. 2003). Conservative estimates indicated that hammerheads were reduced another 89%, thresher sharks 80%, white sharks 79% (over 6,000 captured), oceanic whitetips 70%, tiger sharks 65%, the coastal species complex 61% (range 49% to 83%), and blue sharks 60%. Only mako sharks had not declined by more than 50%. These numbers are likely representative of trends in other oceans, indicating that "overfishing is threatening large coastal and oceanic sharks . . . [and] several sharks may also now be at risk of large-scale extirpation" (Baum et al. 2003, p. 390) (not all researchers concur on all points here; see Burgess et al. 2005).

Accelerated targeting of sharks occurred because of (1) a decline in worldwide production of more traditionally targeted finfish species; and (2) increased buying power among Asian nations, especially China, where there was a high demand for shark fins (D. A. Rose 1998; Fong and Anderson 2002). Meat is useful, but fins build fortunes. Shark meat in the U.S. may bring $1.10–$1.30/kg to the fisher; steaks and fillets run $7.70–$22.00/kg depending on species (mercury contamination depressed prices of some long-lived species such as tope, *Galeorhinus galeus*; Vannuccini 1999). Shark fins—one of the world's most valuable fishery products—can be worth $55/kg wet. When dried, processed, and sold in Asia, fins can retail as high as $1,650/kg (Rose 1996, Vannuccini 1999).

Processing and storing shark meat is time and space consuming, whereas fins can be removed quickly and take up less room. The economic and logistic disparities between meat and fins help explain the practice of finning, whereby the fins are cut off a live shark, which is then referred to as a "log" and discarded to starve on the bottom while being slowly eaten by crabs (figure 10.8). Fins are dried with skin removed but fin rays in place, or rays may be further processed into "noodles, needles, or nets." Fin products are primarily used for shark fin soup, which in Hong Kong is expected on the menu for wedding banquets (Kuang 1999).

In its final form, shark fin soup may actually contain little shark, the fin material having been minimized to keep the price affordable (a bowl may sell for as much as $150). Or the fins may be used primarily as a thickener, the flavor coming largely from chicken or other meat stock. The soup

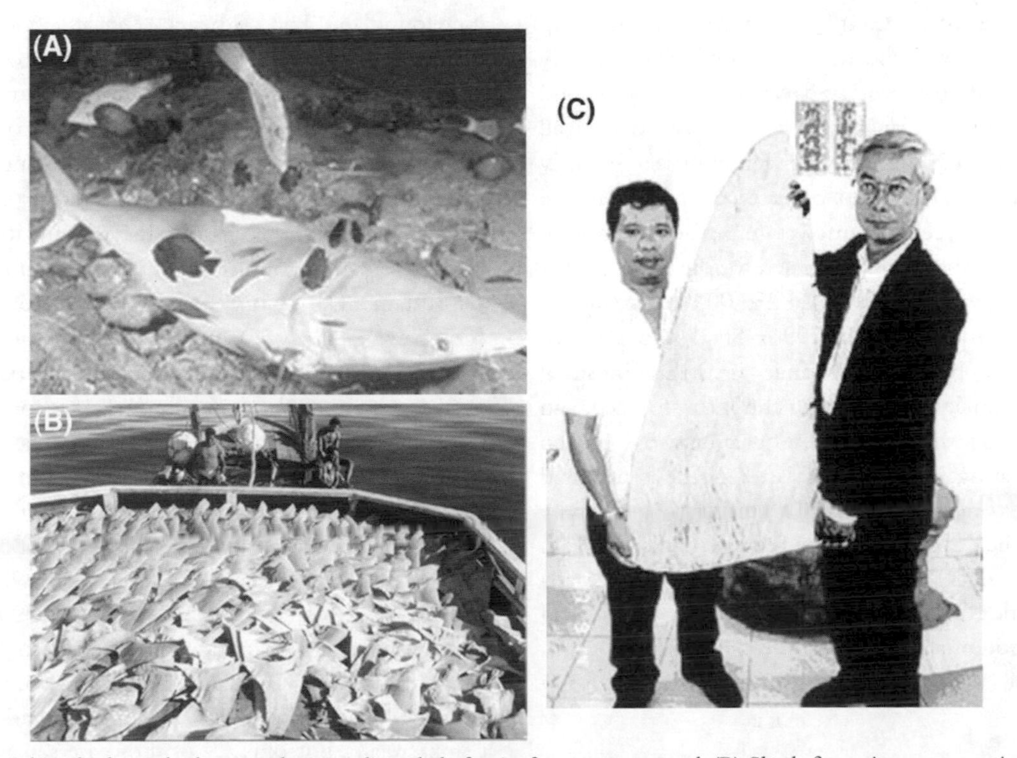

Figure 10.8. (A) A live shark on the bottom that was discarded after its fins were removed. (B) Shark fins prior to processing. (C) Processed whale shark fin. A by Mark Strickland, www.oceanic-impressions.com; B from www.closerfree.org; C from Kuang (1999) in Vannuccini (1999); used with permission.

may even be filtered before serving and the solid fin parts discarded. Hong Kong imported 7,800 MT of shark fins worth US$250 million in 1996: "Stall after store after shop, block after block, on both sides of the street. At any given moment, the fractional remains of hundreds of thousands of sharks must be moving through [Hong Kong] markets" (Safina 1997, p. 400). By comparison, the entire 1996 U.S. shark catch, exclusive of dogfish, weighed 7,000 MT and was worth only $10 million (Branstetter 1999; Kuang 1999).

Most notably of late, besides fins and meat, sharks are captured for their cartilage as a cancer cure. Because cancer in sharks is supposedly rare, shark cartilage has been proposed as a cancer preventive in humans. Sharks actually do get cancer, including carcinomas of cartilage (Borucinska et al. 2004; Ostrander et al. 2004; G. K. Ostrander, pers. comm.), but that fact hasn't deterred entrepreneurs from capitalizing on the need for a cure. Nor have they been deterred by the U.S. Food and Drug Administration and Federal Trade Commission, which filed injunctions against companies making anticancer claims for shark cartilage products (S. Barrett 2000). Shark cartilage can sell from $0.25 to $145/gm (Vannuccini 1999).

The anticancer properties of shark cartilage remain in dispute (Ostrander et al. 2004). The purported relationship is that cancers grow because of increased blood supply (angiogenesis) and that processed cartilage inhibits blood vessel proliferation (antiangiogenesis, e.g., Berbari et al. 1999; Cho and Kim 2002). However, clinical trials generally show that oral administration of cartilage pills has little effect on cancerous cells themselves (e.g., Gonzalez et al. 2001). Regardless, sharks are processed for their cartilage. A single plant in Costa Rica processed 235,000 sharks/month to make cartilage pills (Camhi 1996). Sharks do produce anticancer drugs, but the substance is the aminosterol squalamine, originally derived from the liver, stomach, and gall bladder rather than cartilage. Squalamine has proven effective in treating lung and ovarian cancers, is patented and produced by reputable drug manufacturers, and can be synthesized, which obviates the need to kill sharks to obtain it (Bhargava et al. 2001; Zhang et al. 2003; see www.nugen.com). Shark conservation can therefore be justified on the basis of a potential for useful anticancer drugs.

Levels of Exploitation

Sharks cannot sustain intense fisheries. As slow-growing, slow-maturing, apex predators with long gestation periods and low fecundities, most sharks have too low a reproductive potential to overcome the high mortality rates common to commercial exploitation. Most shark populations cannot tolerate annual fishing mortalities as low as 5% of standing stock (Pratt and Castro 1991), whereas many commercial fisheries for teleosts use as a target 50% removal of virgin biomass to achieve MSY (e.g., Hilborn and Walters 1992; virgin biomass is the size of a previously unexploited population). Sharks also show a surprisingly strong, direct relationship between current and future stock levels (between current spawning biomass and future recruitment). This denies them "the variability which enables most teleost fish populations to be boosted by better-than-average year-classes," meaning that sharks are less likely to rebound from intense exploitation (Pawson 2001).

Given these traits, it is not surprising that commercial shark fishing has a long history of collapse (Pratt et al. 1990; Musick et al. 2000). Examples include thresher and angel sharks in California, school sharks in Australia, spiny dogfish in the North Atlantic, basking sharks off Ireland, bull sharks and sawfish in Lake Nicaragua, soupfin sharks off the U.S. Pacific coast, and porbeagle sharks off Newfoundland. The porbeagle, *Lamna nasus*, fishery of the western North Atlantic collapsed after only six years (Campana et al. 2002). Exploitation began in 1961, and landings averaged 4,500 MT/year prior to the collapse. Slow recovery occurred during the 1970s and 1980s, when landings were around 350 MT, but catches of 1,000–2,000 MT during the 1990s again reduced population. The population remains around 10%–20% of virgin biomass and can sustain a fishery only at recent quotas of 200–250 MT (Campana et al. 2002). In California before 1937, about 270 MT or 6,000 individual soupfin sharks, *Galeorhinus zyopterus*, were landed annually. A market for shark liver oil stimulated the fishery, which expanded rapidly to 4,000 MT in 1939, fell to 2,300 MT in 1941, and by 1944 was back down to 270 MT, despite intensive effort (catch rates fell from 60 sharks per set to 1 shark per set between 1939 and 1944). Thirty years later, populations were still below 1937 levels (Moss 1984; E. D. Anderson 1990).

The most active shark fishing and finning countries are India, Indonesia, Pakistan, USA, Taiwan, Mexico, and Japan. Globally, about 100 species of sharks are sought commercially, but only 29 of those are separated in statistical reports (Vannuccini 1999; figure 10.9). Total shark landings tripled in the latter half of the 20th century, from

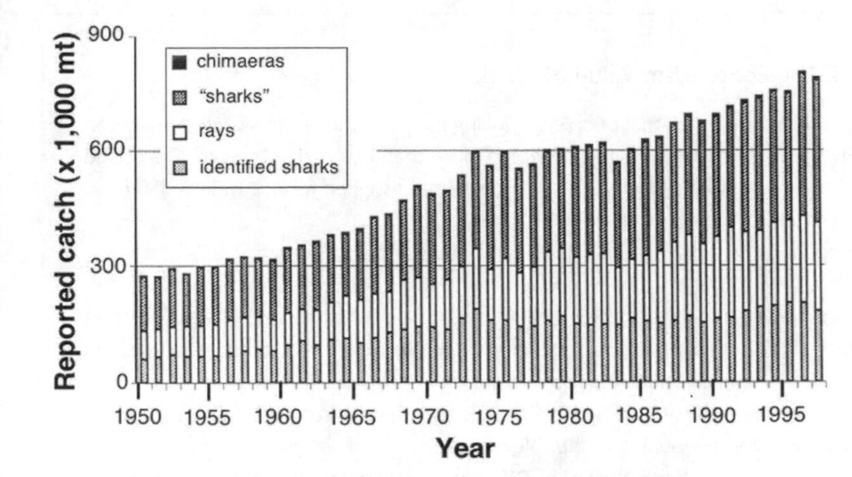

Figure 10.9. Reported world catch of cartilaginous fishes, according to FAO data. Numbers are considered to be underestimated by about 50%. "Sharks" are complexes not separated into species. From Vannuccini (1999); used with permission.

around 272,000 MT in 1950 to 804,000 MT in 1996. Catches are dominated by spiny dogfish, *Squalus acanthias*, a species that has shown signs of overfishing in several areas. Other heavily utilized shark species include silky (*Carcharhinus falciformis*), tope, narrownose smoothhound (*Mustelus schmitti*), porbeagle, basking shark (*Cetorhinus maximus*), and blue shark (*Prionace glauca*). Heavily utilized batoids include cuckoo ray (*Raja naevus*), thornback ray (*R. clavata*), and whip stingray (*Dasyatis akajei*). Holocephalans (chimaeras, ratfishes) account for about 6,000 MT, or less than 1%, of the annual chondrichthyan catch. Unfortunately, most sharks caught are recorded only as genera, families, or orders, adding to the challenge of managing shark fisheries (Vannuccini 1999).

Between 30 million and 100 million sharks, skates, and rays are caught annually, and an equal number may die as bycatch (Vannuccini 1999). These numbers do not include subsistence and recreational fishing, the impacts of which have only recently gained the attention they deserve (Coleman et al. 2004; Cooke and Cowx 2004). In an attempt to estimate actual catch, Vannuccini (1999) combined FAO statistics with calculated bycatch, additional data provided by Bonfil (1994), and presumed artisanal and recreational catches and discards and arrived at a landings total of 1.6 MMT annually. This total is about twice the FAO reported catch, which indicates serious underreporting and higher exploitation rates than have gone into fisheries models.

Population reductions of sharks correspond to periods of increased fishing, underscoring sharks' sensitivity to exploitation. In the U.S., shark populations were already in decline during the 1970s and 1980s and were driven to even lower numbers when Atlantic shark landings increased by 250%–300% between 1985 and 1994 (D.A. Rose 1998; Poffenberger 1999). Florida, the major shark fishing state in the eastern U.S. and Gulf of Mexico (Camhi 1998), saw an order of magnitude increase in landings over the decade of the 1980s, from 132,000 kg in 1980 to 3.3 million kg in 1990 (S.T. Brown 1999). U.S. Atlantic yields exceeded UN-calculated MSY by an average of 170%.

The fishery for whale shark, *Rhincodon typus*, exemplifies trends. Whale sharks, the world's largest fish, are harvested in Taiwan at the rate of around 250 per year, despite declining CPUE (Che-tsung et al. 1997; S. Fowler and Cavanagh 2001). Fish are caught opportunistically—whenever encountered—via either set nets or harpoons (in Philippine artisanal fisheries, fishers jump onto the backs of whale sharks and hook them through the eye by hand; Alava et al. 1997). The popularity of whale shark meat—marketed as tofu shark because of its soft, white texture—is increasing, as is the price. Before 1985, a single fish might bring $300 in Taiwanese markets. The price skyrocketed to $7/kg, making it the most expensive of shark meats. A 9,000-kg shark could now be worth $70,000, a value that can only work against efforts to restrict hunting. Whale sharks are nonetheless fully protected in Australia, the Philippines, Honduras, Malaysia, USA, and the Maldives, mostly to preserve revenues from shark-watching ecotourism (see box 10.4). Such bans apparently stimulated a targeted fishery for whale sharks along the Gujarat coast of India in the late 1990s to meet the demand in Taiwan, prompting the Indian government to protect the fish in 2001 (Dutta 2001). Whale sharks are listed as a Vulnerable species by IUCN. Growing concern over high effort and declining captures and sightings in several locales led, after

BOX 10.4. The Ecotourism Value of Sharks

Shark conservation makes economic sense because sharks are fishes that people are willing to spend money to watch. A shark can be captured only once, but it can be viewed many times; "one live shark may be worth more than 10,000 dead sharks" (Daves and Nammack 1998, p. 228). An informal calculation of the value of individual sharks in the Bahamas assigned returns of $40–$50 dead compared to $750,000 live (S. Gruber, pers. comm. in Daves and Nammack 1998).

A better-substantiated comparison from the Republic of Maldives produced a less spectacular but still impressive 100- to 1,000-fold difference in value (R. C. Anderson and Waheed 2001). The two major economic activities in the Maldives are fishing and tourism. Half of the tourists come to dive, and many dive at specific shark-watching sites. Shark watching in 1992 was valued at US$2.3 million, making a single gray reef shark worth $3,300 to $33,500/year depending on locale, whereas a captured, dead shark had a one-time value of $32. Whale shark and ray watching in the Maldives has also grown in popularity. Willingness-to-pay calculations indicate that shark watching in 1997 had a direct annual value of $6.6 million and manta ray watching, $7.8 million (disregarding multipliers involving travel and accommodations, which would raise the value further).

Shark fishing at some Maldive locales in the 1990s reduced shark populations by 95%, with concomitant loss in tourist revenues. Recognizing the tourist value of live sharks and rays, the Maldives passed regulations designating specific dive sites as protected areas, banning shark fishing in most dive areas, and prohibiting whale shark fishing and the export of ray products (Anderson and Waheed 2001; see also chapter 12). Whale shark watching is a growing commercial activity in the Maldives, western Australia, the Philippines, and Mexico, with spotter airplanes coordinating the diving activities of fleets of boats.

Ecotourism involving sharks is not without controversy. An oft-raised concern is that fishers denied income from shark fishing may not benefit from the large profits associated with shark watching (international tourists are seldom willing to dive from the average fishing boat, and fishers can seldom afford the boats and ancillary gear associated with tourist diving operations). Another controversy surrounds interactive diving, encounter diving, and in-water shark feeding, in which dive leaders or even tourist divers hand-feed sharks and other predators (groupers, moray eels, barracuda) deliberately attracted to certain areas. In 2001, these activities occurred at an estimated 300 locales in 40 countries (www.namibian.com.na). A media campaign against shark feeding cited diver and swimmer safety as major concerns: Sharks will come to associate humans with food. Other issues include acclimation of sharks to humans (eventually endangering the sharks), and larger monetary losses to vulnerable local economies that might result if attacks occurred (see www.marinesafetygroup.net). Citing similar concerns, the Florida Fish and Wildlife Conservation Commission banned the feeding of marine life by divers; a similar ban has been considered in Hawaii. Dive tour operators, among others, opposed the ban largely because hard data demonstrating danger were scarce. A lack of data aside, anyone who doubts that fishes can become conditioned to regular feedings at specific locales and are likely to bite humans when not fed has never kept a goldfish. Allow me an aside that demonstrates fish learning. At the Aquarius Undersea Habitat in St. Croix U.S. Virgin Islands, a diver fed yellowtail snappers, *Ocyurus chrysurus*, small Vienna® sausages whenever he went out to use the "latrine" area near the habitat. After a few days, he ran out of sausages. On his next latrine visit, snappers attacked that anatomical part which bore the closest resemblances to the previously available food. The wound required several sutures to close. This incident was recounted to me by the attending physician, William Shane.

a number of failures, to listing them as a CITES Appendix II species in 2004. Ironically, the Taiwanese market would be little affected, however, because Taiwan is not recognized as a country by the UN and, therefore, cannot join or be bound by CITES (Daves and Nammack 1998).

Shark Fisheries Management

Few countries have effectively regulated their shark fishing. New Zealand instituted shark management in 1986, followed by Australia in 1988; South Africa prohibited capture of white sharks in 1991; Brazil, Canada, Costa Rica, Israel, the Maldives, and the Philippines have also instituted regulations, with many other countries considering legislation (Vannuccini 1999). Despite evidence of decline and collapse, and warnings and specific proposals for management (e.g., Manire and Gruber 1990), the U.S. lacked regulations before 1993. At that time, NMFS passed a shark fishery management plan (FMP) that limited the catch of 39 species in the Atlantic and Gulf of Mexico and banned finning. The FMP did not include dogfish, the primary commercial species. More significantly, NMFS regulations apply only to federal waters, which extend from 3 to 200 miles offshore. Most sharks spend part or most of their lives nearshore and are affected by inshore fisheries, over which only states have jurisdiction; states have been even slower in regulating shark fishing. The FMP also did not regulate Pacific states, including Hawaii, which has the highest shark bycatch rate and where 61,000 sharks were finned in 1998 alone (Camhi 1999).

The 1993 FMP limited fishing to 2,436 MT, but it was too little too late. Populations continued to decline because sharks were still being fished unsustainably. In 1997, the quota was halved to 1,285 MT, additional species were added, recreational fishing was reduced from four to two sharks/day/boat, and complete bans were instituted for whale, basking, white, sand tiger, and bigeye sand tiger sharks. The fishery was divided into three management groups: large coastal species (e.g., tiger, lemon, hammerhead, blacktip, bull), small coastal sharks (sharpnose, finetooth, bonnethead, angel), and pelagic sharks (mako, blue, porbeagle, thresher, oceanic whitetip). Large coastal sharks were considered overutilized (recent average yield is 120% of potential yield), small coastal sharks were fully utilized, and pelagic sharks were an unknown.

In truth, among the metrics that NMFS normally applies to fish stocks, potential yield estimates were available only for large coastal sharks, and long-term potential yield was unknown for all three groups. Sharks were being managed out of ignorance, but at least they were being managed. NMFS recognized these shortcomings and proposed numerous improvements in data collection and management goals, particularly among pelagics and large coastal species (Poffenberger 1999). Meanwhile, six elasmobranch species—barndoor skate (*Raja laevis*), smalltooth and largetooth sawfishes (*Pristis pectinata* and *P. perottetti*), sand tiger shark (*Odontaspis taurus*), dusky shark (*Carcharhinus obscurus*), and night shark (*C. signatus*)—were added to the federal Candidate List of Threatened and Endangered Species, which was a step toward greater protection (Musick et al. 2000). Only smalltooth sawfish received ESA status, although others are Species of Concern and are protected under the shark management plan (skates and rays are especially vulnerable to overfishing and bycatch). Overall, the FMP has served as a model because of its relative thoroughness, and U.S. fishery scientists have been active advocates of improved international monitoring, management, and cooperation in regulating shark fisheries.

Prospects for International Regulation

The issues of sharks at risk and global overharvesting have received considerable attention. Castro et al. (1999) assessed vulnerability among the world's exploited sharks and identified 22 species as having life history characteristics that made them especially vulnerable; 12 additional species had experienced historical declines or local extirpations. IUCN (2004) assessed the conservation status of 400 elasmobranch species and designated 14 as Critically Endangered, 26 as Endangered, and 46 as Vulnerable. A disturbing aspect of the IUCN analysis is that only 93 species, or less than 25% of those assessed, could be considered Low Risk/Least Concern.

Despite evidence that many shark species are in decline, and despite the efforts of dedicated biologists and others, management of shark fisheries at the international level remains elusive or nonexistent. International governing bodies have failed repeatedly to institute needed restrictions (e.g., Daves and Nammack 1998). Opponents from shark fishing nations have pointed out the incompleteness of population data and argued that local (vs. international) regulation would suffice. CITES studied shark fisheries beginning in 1994. Prohibitions on the trade of sawfish products date back to 1997, and other prohibitions followed but failed to pass prior to 2004, when whale, white, and basking sharks were finally listed, whale and basking sharks in Appendix II and white sharks in Appendix III. Some regulatory action is likely from the UN Agreement on the Conservation and Management of Straddling Fish Stocks and Highly Migratory Fish Stocks, which was formalized in 1995 and is slowly being ratified by member nations. This agreement sets guidelines for cooperative management actions among signatory nations, recognizes problems of bycatch, and emphasizes a precautionary approach to ensure long-term sustainability (Daves and Nammack 1998).

In 1999, FAO produced an International Plan of Action for the Conservation and Management of Sharks, another positive step (www.fao.org/waicent/faoinfo/fishery). Member states were encouraged to develop national plans of action that included regular assessments of shark stocks and data sharing among nations, particularly for migratory species. Action by member nations has been slow at best (Pawson 2001). After two years, the U.S. was the only shark-fishing nation among 125 with a national plan completed and available; Australia had completed fishery assessments and was developing a plan. Otherwise, perhaps 14 nations appeared to be actually working on plans. European nations were developing stock assessments under the Common Fisheries Policy program, with responsibility delegated to the ICES Elasmobranch Study Group and completion scheduled for late 2002.

In the U.S., migratory pelagic fishes (tunas, billfishes, sharks) in the Atlantic are managed through the NOAA Office of Sustainable Fisheries, Highly Migratory Species Division. This program interacts closely with the International Commission for the Conservation of Atlantic Tunas (ICCAT), whose success in managing tunas and billfishes is spotty to a degree that the official policy statement of the American Fisheries Society on management of sharks says, "ICCAT's management effectiveness is questionable" (Musick et al. 2000, p. 11).

Obstacles and Solutions

Shark conservation is justifiable on scientific, ethical, recreational, and commercial grounds. More so than most other fisheries, sharks have to be managed at local, national, and international levels, which greatly complicates efforts.

Some of the largest impediments to meaningful shark management are geopolitical, resulting directly from sharks' high mobility. Sharks travel daily, seasonally, annually, and throughout their lives. Females give birth in different areas from where they spend the rest of their lives (in part to minimize predation on pups by other adult sharks); pups move seasonally among nursery grounds; males and females typically occupy different places but congregate (and move) during breeding periods; some species undertake seasonal longshore migrations, whereas other species move regularly across entire oceans (Helfman et al. 1997). All these peregrinations translate into management difficulties because sharks cross jurisdictional boundaries and often enter international waters where no one has jurisdiction and no restrictions apply.

Management is also hampered by the non-species-specific nature of landings data. India, the world's most active shark fishing nation, accounting for about 15% of global catch, reports solely as "elasmobranchs" a catch including 20 species of sharks and 45 species of skates and rays (Vannuccini 1999). Sharks differ in abundance, life history characteristics, habitat, feeding, behavior, and, consequently, vulnerability to exploitation. Managing "elasmobranchs" as a group makes little sense.

Specific solutions and recommended actions to slow the decline of exploited shark stocks abound (Camhi 1998, 1999; Daves and Nammack 1998; D. A. Rose 1998; Vannuccini 1999; Musick et al. 2000). Few proposals call for outright bans on shark fishing; most emphasize the need for better data collection and monitoring prior to considering restrictions. Solutions and recommendations include the following:

- collect data that focus on species rather than groups, particularly with regard to population status, life history information, and landings information; easily used shark identification tools are essential
- greater attention to long-lived species, which appear to be especially vulnerable to overexploitation
- emphasis on management of species exposed to mixed-species fisheries. Species with high production rates can support a fishery, but the fishery may also catch species with lower reproductive potential, which can be extirpated while the overall fishery remains active and profitable. For example, U.S. federally listed sand tiger and dusky sharks continue to be caught along with more numerous sandbar sharks, *Carcharhinus plumbeus*. "Thus the traditional paradigm that fisheries will become commercially extinct before the targets of those fisheries become biologically extirpated may be false" (Musick et al. 2000, p. 10.).
- recognizing that most large sharks are migratory and cross national and international jurisdictional boundaries. Landings statistics from different areas need to be combined to gain a true picture of exploitation levels, and quotas should take into account such movement
- including recreational, subsistence, and artisanal landings in fisheries statistics to gain a more accurate estimate of exploitation levels
- better data on bycatch discards to estimate total fishing mortality; bycatch reduction should be emphasized when developing gear regulations; live release of bycatch should be encouraged
- enacting area and time closures during the pupping season; restricting fisheries on juvenile sharks
- prohibiting finning and requiring that sharks be landed with fins and head intact
- developing regulations and quotas that are precautionary even before assessments are completed and comprehensive management plans are developed

- encouraging nonconsumptive, sustainable use of elasmobranchs such as ecotourism, while ensuring that profits from such activities return to local communities that might otherwise be hurt by restrictions on traditional shark fishing (box 10.4).
- increasing effort and investment in research, monitoring, and enforcement

Sharks fortunately have a large and dedicated constituency with a strong international voice, including the American Elasmobranch Society and the Shark Specialist Group of IUCN, the Pelagic Shark Research Foundation, the Shark Trust in the UK, and the European Elasmobranch Association (see Dudley and Gribble 1999 for more).

New England Groundfish

The North Atlantic Ocean was one of the most biologically productive regions of the world, supporting immense numbers of invertebrates, pelagic and benthic fishes, marine and aquatic mammals, and sea birds. "It is probably impossible for anyone now alive to comprehend the magnitude of fish life in the waters of the New World when the European invasion began" (Mowat 1996, p. 165). Beginning in the 16th century, European merchant mariners ventured across the Atlantic and systematically decimated most large marine forms, mainly to meet European demands for oil, fur, and leather (Mowat 1996). The last flightless great auk or spearbill, a goose-sized bird that once numbered in the millions, was killed on June 3, 1844. Bears, wolves, foxes, musk bearers, seals, and whales experienced extinctions, extirpations, population declines, and range contractions. Overexploitation of fishes occurred relatively late in this sequence but with similar results (e.g., Pauly and Maclean 2003).

More than 25 groundfish or bottom species are actively fished in the 1,200,000-km^2 North Atlantic region of banks, inner basins, channels, and saddles that runs from Cape Hatteras, North Carolina, to the Arctic Circle. The main fishery area is the "New England" continental shelf region from Cape Cod to the Grand Banks off Newfoundland. The largest landings are of gadoids (Atlantic cod; pollock; haddock; silver, red, and white hake), flatfishes (summer, winter, yellowtail, and witch flounder; American plaice; and formerly Atlantic halibut), Acadian redfish, goosefish, spiny dogfish, and skates (NMFS 1999; figure 10.10). Each of these stocks

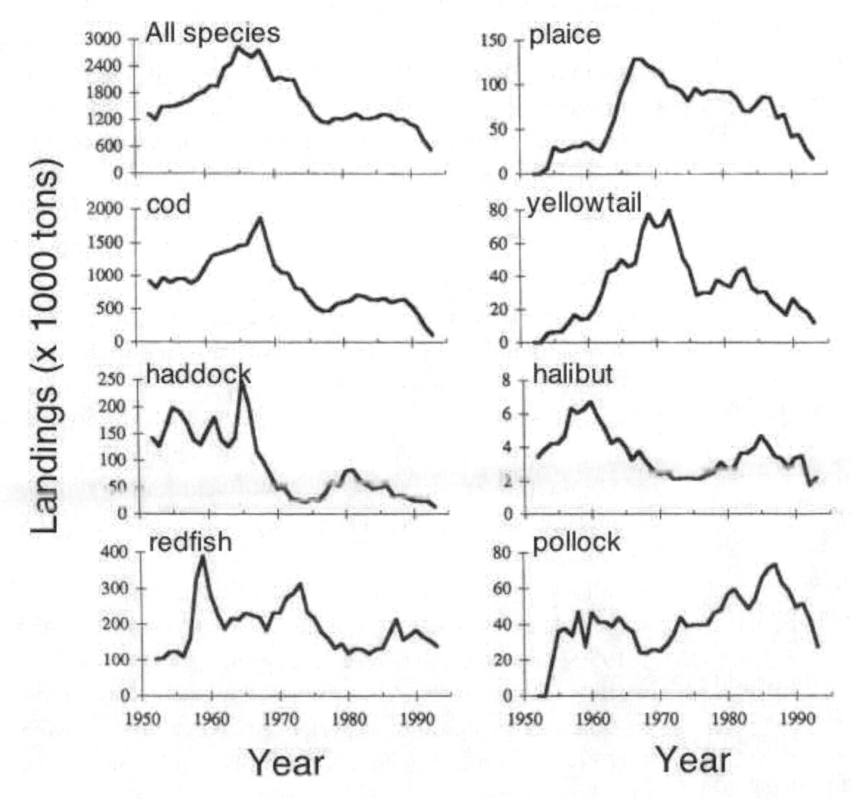

Figure 10.10. Landings trends among North Atlantic groundfishes. The major groundfish of the western North Atlantic region have all experienced rapid or continual population declines in the past 30–40 years and entered the 21st century at or close to all-time low levels. From Sinclair and Murawski (1999); used with permission.

except pollock exists below the population level necessary to provide its long-term potential yield: ten are overfished, six are fully fished, and none are underfished (NMFS 1999).

Each species is also entangled in a history of anthropogenically caused population fluctuations. They have suffered directly from overexploitation, have replaced an overfished species, and have undergone population increases as competing species were overfished. Overfishing has often followed a timeline of technological and political changes that includes conversion from sail to steam to diesel and progressively larger ships, shifts from hook-and-line to trawl fishing, exploitation by local then foreign distant-water then local distant-water fleets, shifting administrative entities and management plans (55 geographic regions are contained in the 36 different subareas, divisions, and subdivisions recognized by the Northwest Atlantic Fisheries Organization, NAFO, for statistical purposes), and climatic vagaries (Murawski et al. 1997; Murawski et al. 1999). Multispecies fisheries often involve bycatch and discard of nontarget fishes, contributing to depletion. Most major fisheries have unhappy histories:

- Flounder, or members of the "flatfish complex," are primarily caught by trawl. Although initially caught by hook and line incidental to species such as cod, flatfish grew in significance when distant-water, foreign fleets from Europe and Asia expanded in the late 1950s and early 1960s. Landings climbed dramatically from 60,000 tons in 1955 to 315,000 in 1968, followed by declines. Most flatfish stocks in the 1990s were at record low population levels. According to Murawski et al. (1997), "All plaice stocks are considered to be in poor to moderate condition. Recent yields are among the lowest recorded." Witch flounder stocks in different regions are "near the lowest observed" or "the lowest observed." Yellowtail flounder landings on the Grand Banks fell almost 20-fold, from 39,000 tons in 1972 to 2,000 tons in 1994, after which the fishery was closed. Biomass in 1996 was 5%–10% of 1960s levels because fishing mortality had exceeded sustainable rates by two- to fourfold. Even before overfishing by distant-water fleets, stock depletions led to serial exploitation. For example, the U.S. fishery for yellowtail began in the 1930s after preferred winter flounder declined (Murawski et al. 1997).
- Atlantic halibut, *Hippoglossus hippoglossus*, are the world's second-largest flatfish (to 3 m and 320 kg). "The history of the halibut . . . must be written largely in the past tense, for their numbers have been sadly depleted" (Bigelow and Schroeder 1953, p. 254). Halibut were fished first by hook and line, later by trawlers. Early catch records indicated abundant fish that were considered something of a nuisance until the 1820s, when a market developed in the Boston area. Catches were at first phenomenal: 250 in three hours, a single boat landing 9,000 kg in a day, large vessels filled in a couple of days (Bigelow and Schroeder 1953). Fishing intensity grew, and numbers declined, with depletion first noted in 1839 near Cape Cod, probably as a result of slow replacement due to late maturation at 8–12 years old (Cargnelli et al. 1999). Progressive depletion followed northward and offshore, and many fisheries were exhausted by the 1850s. By the 1890s, most halibut sold in the east came from Iceland or the West Coast. North Atlantic landings more recently have cycled up and down (6,700 MT in 1960, 4,000 MT in 1985), stabilizing below 2,000 MT in the 1990s (Sinclair and Murawski 1999). The total catch of Atlantic halibut in all of New England in 1999 was 15 MT (NMFS 1999).
- Haddock, *Melanogrammus aeglefinus*, initially a bycatch species in the cod fishery with little commercial value, became the principle targeted species in the 1920s with the large-scale advent of trawling (Murawski et al. 1999). The New England catch jumped from 20,000 MT to 100,000 MT between 1900 and the late 1920s. Landings plummeted in the 1930s, attributed to overfishing, much of which involved bycatch of juveniles. The bycatch-to-fish-retained ratio was two or three to one, with 70–90 million juveniles discarded at sea (mesh regulations to protect small haddock were not implemented until 1953). Some recovery occurred when the fishery moved to Nova Scotia, and catches rose to 250,000 MT in 1965 as foreign fleets took part, but this apparently precipitated another crash, from which stocks have yet to recover (figure 10.10). Recovery of the Georges Bank stock was further hindered when strong year-classes were used to justify increased catches instead of to rebuild the stock (Sissenwine et al. 1998). Recent yields in the New England fishery have been only 3,800 MT (NMFS 1999).
- Pollock, *Pollachius virens*, are a relatively good-news story. Pollock became a major fisheries species only in the 1950s; steady catches rose to a peak of 73,000 MT in 1987, followed by the severe downturn that most exploited groundfish experienced. Regardless, NMFS (1999) recognized pollock as the only major commercial

species to be fully exploited but still near, rather than below, a stock size capable of providing a long-term yield.

- Redfish, or Acadian redfish or ocean perch, is the group name for *Sebastes fasciatus*, *S. marinus*, and *S. mentella* (with some taxonomic confusion). These deep-water, demersal scorpaenids are long-lived, slow growing, and easily overfished, similar to many eastern Pacific scorpaenids. Redfish became targeted when deep-water trawling proliferated in the 1930s. Catches approached 300,000–400,000 MT, second only to cod. Stocks were discovered, fished, and exhausted, forcing fleets to move to more distant areas in a pattern familiar to North Atlantic fisheries (Murawski et al. 1997; Murawski et al. 1999). The overall landings reflect this exploitation-exhaustion cycle, which shows a series of peaks and declines over the last 50 years. U.S. landings peaked in 1942 at 60,000 tons and have declined continually since. "The current U.S. fishery for redfish is virtually nonexistent" (Murawski et al. 1997, p. 41); 1999 landings were only 400 tons, less than 1% of the 1942 maximum. Much of the 1990s North Atlantic catch was as bycatch in Greenland halibut, *Reinhardtius hippoglossoides*, and shrimp fisheries, and included many juveniles (www.nafo.int). AFS considered Acadian redfish to be conservation dependent, with some recovery evident under a current fisheries management plan (Musick et al. 2001).
- Goosefish, *Lophius americanus*, also called monkfish or poor man's lobster (for obvious marketing reasons if you've ever seen one), grew in importance as other groundfish declined. U.S. landings averaged only 300 MT annually in the 1960s and 1970s, rose to 8,800 MT in the 1980s, and peaked at 28,800 MT in 1997. The 1997 ex-vessel value was $35 million, making it the top-ranked demersal species in both poundage and monetary value. "As a consequence, goosefish abundance has dropped to low levels and the stock is now overexploited" (E. D. Anderson, Kocik, and Shepherd 1999; E. D. Anderson, Mayo, et al. 1999, p. 4).

Cod

Atlantic cod, *Gadus morhua*, deserve special treatment as the quintessential commercial species, with one of the saddest histories. The former abundance of Atlantic cod is legendary, and its demise shameful. Martin (1995, p. 6) observed, "The rape of the northern cod stocks ranks with the greatest of human insults against this Earth." Newfoundland was formerly known by its Portuguese name, *Baccalaos*, which means "land of cod." Some authors feel that the discovery, occupation, and political development of North America largely resulted from its abundant cod resources (e.g., Kurlansky 1997). Over the centuries, travelers attempted to describe the bounty, declaring the fish so numerous that at times "they even stayed the passage" of ships (1516), "so thick by the shore that we hardly have been able to row a boat through them" (1620), and "a kind of inexhaustible manna" (1670) (Mowat 1996, p. 166–69). Cod was "a species of fish too well known to require any description . . . [occurring in numbers] that will baffle all the efforts of man to exterminate" (Homans and Homans 1858 in Kurlansky 1997, p. 109).

Cod have deservedly received the most attention of any North Atlantic groundfish, and perhaps of any fishery species (e.g., Hannesson 1996; Kurlansky 1997). They dominated North Atlantic landings for most of the latter half of the 20th century, accounting for 54% of annual groundfish landings. Catches of northwest Atlantic stocks were relatively stable at around 1 MMT until the 1960s and the arrival of foreign fleets, chiefly from eastern Europe. Landings increased to a peak of nearly 2 MMT in 1968, after which they began a steady decline, to about 0.5 MMT in the 1980s and even lower in the 1990s. Average size also decreased. Based on archaeological records, it was 80–100 cm for the first 3,500 years of exploitation and closer to 30 cm in the last 50 years (J. B. C. Jackson et al. 2001). This reduction has significant repercussions for the likelihood of recovery (chapter 11).

Overfishing of cod in the Canadian North Atlantic occurred despite official recognition of depleted stocks and even though targeted fishing mortalities were relatively conservative, at least in later years (Martin 1995). Overfishing ultimately caused the collapse of Canadian cod stocks, but it resulted from multiple mistakes, including systematic errors in assessment of stock size and fishing mortality rates (involving lack of data, improper use of available data, and unjustified confidence in scientific assessment methods), unreported and illegal discards of small fish, unreported catches, and finally a slow response by management once stock assessments were corrected (NRC 1999b). The annual catch of Canadian (northern) cod fluctuated around 200,000 MT (±50,000 MT) between 1850 and 1950, rose to 240,000 MT in the mid-

1950s, then to 800,000 MT in 1968 "when European stern trawlers flocked to the Grand Banks" (Hannesson 1996, p. 90). From 1970 onward, it declined steadily.

In response to declines, in the late 1970s Canadian management agencies set total allowable catch (TAC) targets above 300,000 MT, or 20% of assumed standing biomass, to sustain the fishery. Actual landings never again exceeded 260,000 MT, indicating fewer fish than estimated, particularly in inshore regions. In 1988, TAC was reduced to 266,000 MT, and assessment methods were reviewed. The review found that stocks were substantially smaller than previously estimated and fishing mortality was twice what had been calculated (probably due to unreported catches and discards, especially of small fish; R. A. Myers et al. 1997a). An advisory committee recommended that TAC be lowered to 125,000 MT, but officials feared excessive economic hardships in an overcapitalized fishery and reduced it only to 235,000 MT (Hannesson 1996; NRC 1999b).

Catches continued to decline. The cod fishery off Newfoundland, among the world's largest for nearly 400 years, plummeted 99% by the early 1990s. It became evident that (1) spawning stock biomass of Canadian cod had declined 16-fold between 1962 and 1992, from 1,600,000 MT to 100,000 MT; (2) fishable biomass had been overestimated by as much as 220% going back to 1977; and (3) through much of the 1960s, 1970s, and 1980s, total catch had approached or even exceeded spawning stock biomass, a worst-case scenario. In 1992, a two-year moratorium on cod fishing was enacted. Stocks continued to deteriorate, and the moratorium was extended indefinitely, putting entire fishing communities out of work. Despite this, the northern cod stock in 1995 was at a historic low, and as of 1999, there were no significant signs of recovery (NRC 1999b).

Canadian cod are only slightly worse off than more southern and eastern stocks. U.S. cod landings in 1997 were the lowest on record (NMFS 1999). Stocks in the North and Irish seas declined 90% after the 1960s (Schiermeier 2003). Northeast Atlantic catches declined from a high of 2 MMT in 1969 to a low of 1 MMT in 1990. Spawning stocks of Arcto-Norwegian cod declined 94% between 1946 and 1989, from 3.8 MMT to 250,000 MT; between 1978 and 1994, Norway's catch exceeded TAC in every year but one, sometimes by as much as 130% (Hannesson 1996).

Will cod recover? The evidence is not promising. Despite fishing moratoria on western Atlantic stocks instituted in 1992 (and mandated fisheries reductions in the eastern Atlantic), little evidence of recovery exists (Hutchings 2000a). Cod biology almost proscribes recovery (Bigelow and Schroeder 1953; Knutsen and Tillseth 1985; R. A. Myers et al. 1996; R. A. Myers, Hutchings, and Barrowman 1997; R. A. Myers, Mertz, and Fowlow 1997). Cod mature slowly, taking three to eight years; five to six years are required for 50% of a year-class to mature, and eight to nine years are probably needed for full recruitment into the spawning population. These are first-time spawners, which produce fewer, smaller, less energy-rich eggs than larger females. These differences translate into fewer, smaller, slower-growing, and less competent larvae (see Trippel 1995). Larger females also spawn repeatedly over a season, increasing the probability that some larvae will encounter favorable growth conditions. Hence, fish spawned at the beginning of the moratorium achieved reproductive status only in the late 1990s and probably did not contribute large numbers of successful young for several more years. Strong year-classes may not occur until well into the first decade of the 21st century, assuming favorable oceanographic conditions. Data indicate little or no rebound among North Atlantic cod stocks 15 years after recovery "should have" begun (Hutchings 2000a).

Additional real and postulated factors work against recovery, independent of spawning biomass, including continued bycatch of larvae, juveniles, and adults due to the nonselectivity of bottom-trawling gear; long-term destruction of benthic juvenile habitat by bottom trawlers; declining cod prey because of heavy exploitation (e.g., variously of capelin, mackerel, herring, smelt, shad, alewives; FAO 1997b); predation by pinnipeds (an oft-used rationalization in many depressed fisheries that is just as often refuted); destructive oil exploration activities; overall decreases in productivity of the North Atlantic due to climatic changes; and shifts in the productivity regime that now favor shrimp, crabs, dogfish, and skates, species previously eaten and outcompeted by groundfish but that may now find their ecological roles exchanged.

The dogfish and skate trends are particularly interesting because they are well documented and arguably irreversible (see chapter 11). Abundance indices and catch rates for (desirable) groundfish versus (less desirable) dogfish and skates in the northwest Atlantic produce almost mirror images (figure 10.11). As groundfish decreased through the latter 20th century, skate (mostly winter, little, and thorny)

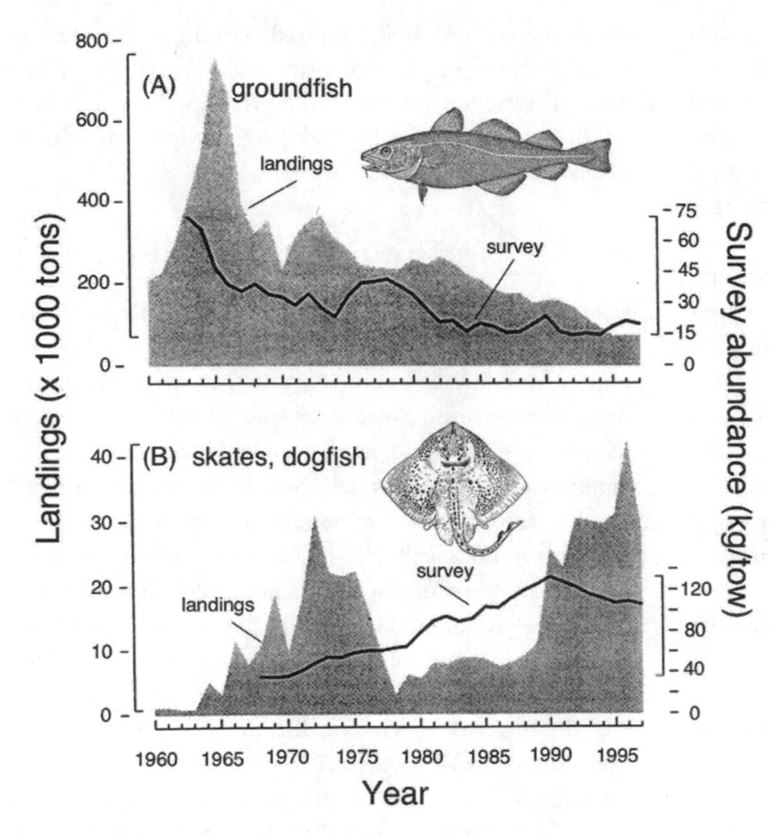

Figure 10.11. Cod wars involve more than boats and people; exploitation may tip the balance between competing species. (A) Landings and abundance index (based on NMFS survey tows) for principal groundfish and flounders off the northeastern U.S.; (B) landings and abundance index for skates and spiny dogfish, showing increases in these less desirable species while groundfish declined. From E. D. Anderson, Mayo, et al. (1999); fish illustrations from Bigelow and Schroeder (1953).

and dogfish (mostly spiny) increased. Skate and dogfish abundances were not even surveyed until the late 1960s but were considered to be low; they rose progressively, as did landings, from 2,900 MT in 1978 to 31,500 MT in 1992 to 42,500 MT in 1996. Discards equaled landings or may have exceeded them by a factor of 25 (depending on source; e.g., NMFS 1999 vs. Sherman et al. 1996). But even these species can't withstand continual high fishing pressure. Dogfish, targeted in recent years as "grayfish" for fish and chips, are responding as would any species that matures at 20 years old and has a 2-year gestation period. The biomass of mature female dogfish decreased by more than 50% between 1989 and 1997 and is now considered overexploited.

New England Groundfish: Causes, Effects, and Prospects

Different stocks followed a repeated pattern of serial overfishing: boom, bust, redirected activity, and more boom and bust, involving first cod, then haddock and redfish, flatfish, goosefish and dogfish (Murawski et al. 1997). Evidence of overfishing existed long before the obvious stock reductions of the 1990s, as both small boats inshore and large boats offshore experienced reduced CPUE (Doeringer et al. 1986). In 1965, 4,000 fishers worked 700 large vessels (over five gross tons) and 5,400 fishers worked 10,500 smaller boats; by 1979, the numbers had increased to 5,000 fishers on 1,300 large vessels and 10,000 fishers on 21,000 small boats. These numbers represent an almost doubling of effort while landings declined from 291,000 MT to 280,000 MT (1979 was actually a good year; in 9 of the 14 years between 1965 and 1979, landings were below 250,000 MT).

Warning signs may have been missed, because the economic value of the resource did not parallel stock levels. It instead reflected a more classical supply-and-demand relationship. As fish became scarcer, their value increased. The catch in 1965 was worth $39 million, or $134/MT, but in 1979, the smaller catch was worth $144 million, or $514/MT. Ecology eventually catches up with and supersedes economics, however, as decreasing supply overshoots price compensation.

When a stock crashes, everyone is hurt, regardless of

price per fish (Doeringer et al. 1986). Overfishing thus devastates local economies along with fish populations. The big question is whether improved management and reduced exploitation will permit stocks to recover sufficiently to allow renewed fishing. For New England groundfish, not just cod, available evidence suggests that recovery will be too slow to allow the current generation of fishers to reenter the fishery. Hutchings (2000a) analyzed recovery rates of 90 marine fish stocks that had experienced 15-year declines in reproductive biomass of 13% to 99%, tracking many for 15 additional years. For North Atlantic gadids, redfish, and flatfishes, recovery was not evident 5 years after the declines, and for gadids, no significant recovery was evident 15 years later. These findings indicate that "the time required for population recovery in many marine fishes appears to be considerably longer than previously believed" (Hutchings 2000a, p. 885).

These findings also emphasize the need for a precautionary approach to managing marine fisheries. The quantitative risk criteria used by organizations such as IUCN have been criticized as too conservative to apply to exploited marine fishes. However, natural population variability and high reproductive potential may not necessarily allow commercial species to recover readily from population collapse (e.g., Sadovy 2001b). "Creation of a productive and sustainable ground fish fishery for the future will require a long period of rebuilding to restore depleted stocks" (Murawski et al. 1997, p. 60).

High Seas, Large Pelagic Fishes

Large pelagic fishes—the tunas and billfishes—sit near the pinnacle of teleostean evolution (Helfman et al. 1997). All aspects of their anatomy, physiology, behavior, and life history are remarkably adapted to and coupled with an open-water, migratory, predatory existence. Large size means high reproductive output, and large geographic range provides refuge from overexploitation. As a result, until recently, most large pelagics withstood commercial exploitation. Unfortunately, their high individual value and large size, coupled with recent advances in fishing technology and increased international fishing effort, have caused steady and alarming declines and serial depletion in stocks of large pelagics. Since the 1960s, the top species harvested in U.S. Atlantic waters have shifted progressively from bluefin tuna to swordfish to yellowfin tuna as species have been overfished. Yellowfin tuna in both the Atlantic and the eastern Pacific are now "fully" utilized, meaning that additional fishing pressure or reduced spawning or recruitment success will produce yet another overfished species (NMFS 1999). Regulations have developed slowly because these species migrate across oceans.

Bluefin Tuna: Charismatic Megafish

Many tunas (Scombridae: skipjack, yellowfin, albacore, bigeye, and bluefin tuna; chub and Atlantic mackerels) form huge schools; wahoo and dogtooth tuna, however, and king and Spanish mackerels are relatively solitary. Schooling species are, understandably, most subject to commercial fishing. Although these species are exceedingly valuable, information on population structure, landings, mortality, and sustainable yield is lacking for many of them, as are management plans (see Safina 2001b and papers cited there). Management often requires treaties drawn up by an alphabet soup of international agencies, particularly the International Commission for the Conservation of Atlantic Tunas (ICCAT, North and South Atlantic), the Inter-American Tropical Tuna Commission (IATTC, eastern tropical Pacific), the Commission for the Conservation of Southern Bluefin Tuna (CCSBT), and the Indian Ocean Tuna Commission (IOTC). Of the 20 or so stocks of the five most heavily fished species (bluefin, yellowfin, albacore, bigeye, and skipjack tuna), approximately 16 are overfished, skipjack and yellowfin being the apparent exceptions (Garcia 1994; Safina 2001b). Many management problems arise from unreported catches by unflagged vessels, inadequate data on landings and stock structure, voluntary compliance and minimal enforcement, heavy fishing of juveniles (60%–80% of catch among some yellowfin and bigeye stocks), and excessive bycatch (Safina 2001b).

Among tunas, the most valuable and most depleted are the bluefin, *Thunnus thynnus*; Pacific bluefin, *T. orientalis*; and southern bluefin, *T. maccoyii* (e.g., Maggio 2000). Bluefin are prime candidates for overfishing because of their extraordinary commercial value, size that can exceed 3 m and 650 kg, delayed maturation until they are 8–12 years old, and restricted spawning areas. Between the 1970s and 1990s, the west Atlantic breeding population may have plummeted 90%, from 250,000 to 22,000 animals, while eastern Atlantic stocks fell by 50% (Safina 1993, 2001b). Fishing pressure did not decrease: Fish were worth too much to not hunt. Added to their restaurant value is their recreational value as one of the world's premier game fish;

a bluefin tuna can reach speeds in excess of 70 kph on its initial run after being hooked. "Recreational" fishing is a minor threat for most fish (except sharks), but relentless recreational pursuit becomes a risk factor for a depleted species.

Interestingly, bluefin tuna, particularly the so-called giants of over 150 kg, were of little commercial value as recently as the early 1970s. They were captured chiefly by purse seines for the pet food trade, bringing only \$0.11/kg. Juvenile fish were easier to process, leaving the giants for the sport fishers, who often discarded large fish after being photographed with them at dockside scaffolds (T. Williams 1992). This changed in the late 1970s, as western Atlantic fishers began airfreighting giants to supply the sushi trade in Japan. The value went up to \$26/kg in 1986 and has continued to climb. Now even sport fishers sell their catch to commercial buyers for export to Japan (W. K. Stevens 1997). In 2001, a 201-kg bluefin tuna caught off northern Japan was sold at the Tsukiji Central Fish Market in Tokyo for a record \$173,600 (\$863/kg, \$391/lb) (AP 2001). A serving of bluefin sushi containing a few thin slices of fish can cost \$75 in Japan.

Atlantic Bluefin

"Resource assessments and management of western Atlantic bluefin tuna are subjects of severe controversy" (Sissenwine et al. 1998, p. 838). Western Atlantic populations have been declining for at least 30 years, while fishing pressure may have risen 2,200% (figure 10.12; T. Williams 1992; Magnuson et al. 2001; Block et al. 2001). Western Atlantic landings peaked in the 1960s at 20,000 MT, then fell in the 1970s to around 5,000 MT and lower. Japanese boats fishing near Brazil dominated 1960s landings, but this fishery was quickly exhausted and should have sounded a warning (a 1960s fishery off Sweden also

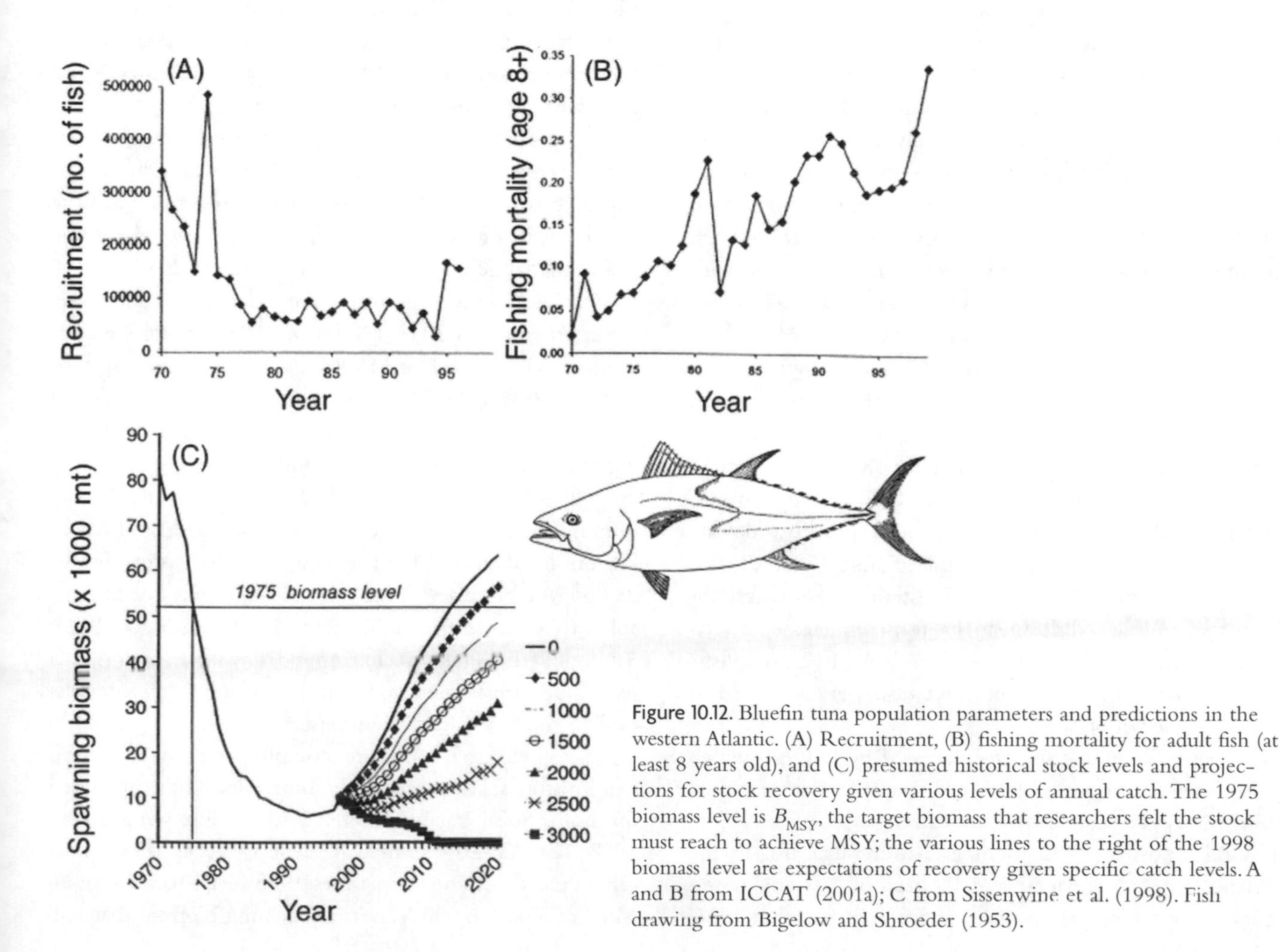

Figure 10.12. Bluefin tuna population parameters and predictions in the western Atlantic. (A) Recruitment, (B) fishing mortality for adult fish (at least 8 years old), and (C) presumed historical stock levels and projections for stock recovery given various levels of annual catch. The 1975 biomass level is B_{MSY}, the target biomass that researchers felt the stock must reach to achieve MSY; the various lines to the right of the 1998 biomass level are expectations of recovery given specific catch levels. A and B from ICCAT (2001a); C from Sissenwine et al. (1998). Fish drawing from Bigelow and Shroeder (1953).

disappeared). ICCAT responded in 1982 with catch quotas for the western Atlantic of 1,160 MT, a number that was increased (against scientific advice) and has varied between 1,995 and 2,560 MT annually. The initial quotas represented a compromise between what environmental groups and fishery scientists felt was necessary (as near zero as possible) and what industrial proponents considered tolerable. Capture of juvenile fish under 6.4 kg was restricted, and this minimum was later raised to 30 kg or 115 cm. Landings approached targeted limits between 1982 and 1987 but rose to and exceeded limits from 1988 to 1997. Stocks were further reduced by unknown amounts due to illegal, unreported, and unregulated (IUU) fishing. As many as 23 fishing nations not bound by ICCAT regulations catch thousands of metric tons and fish during ICCAT-closed seasons (ICCAT 2001a). By some estimates, IUU catches in the Atlantic may exceed members' catches by 80% (Safina 1993) (similar circumstances apply to southern bluefin ; see Hayes 1997; Safina 2001b).

ICCAT instituted a recovery plan centered on catch quotas to restore spawning stock biomass to mid-1970s levels, using a 1975 spawning biomass of 55,000 MT as a benchmark. That benchmark probably underestimated natural stock biomass because extensive long-lining and seining in the 1960s had already reduced stocks (see figure 10.12C). The overall result was too little too late, and fish numbers continued to decline for at least a decade after restrictions were imposed. Spawning biomass finally stabilized in the early 1990s, but only at about 10%–20% of what it had been in the early 1970s (Sissenwine et al. 1998). Given continual declines, IUU catch, and apparent miscalculations of stock history and growth trends, a coalition of conservation groups in the early 1990s proposed that Atlantic bluefin be listed in CITES' Appendix I, restricting all international trade. This proposal was defeated by commercial fishing interests (Safina 1993, 2001b), which maintained that they had restrained themselves for decades and that stabilized stocks justified increased catches. Uncertainty in the data and disagreements among fisheries workers about data interpretation were also used to justify higher quotas.

Atlantic bluefin tuna are managed by ICCAT, which some conservation advocates feel stands for International Conspiracy to Catch All the Tunas (e.g., Safina 1997). ICCAT consists of 22 member nations that fish in the Atlantic, including Japan and China (for Taiwan). It sets quotas for bluefin allowable catches and sizes, but those quotas often exceed the scientific findings and recommendations of its Standing Committee on Research and Statistics (SCRS): "ICCAT's science has generally been of high quality. . . . Its management, however, has often responded inadequately to the advice of its scientists" (Safina 2001b, p. 445).

For management purposes, Atlantic bluefin are divided into independent western and eastern stocks, based chiefly on different ages at maturity (eight years in the west, four or five years in the east) and spawning grounds (Gulf of Mexico vs. Mediterranean Sea). Eastern fish are thought to be more abundant than western. However, recent movement data indicate that stocks may breed in different locales but come together when feeding (genetic differences are not apparent). Mixing, thought previously to involve 1%–3% of fish crossing the Atlantic from west to east, may involve as many as 30% (mixing from east to west has yet to be thoroughly researched) (Lutcavage et al. 1999; Block et al. 2001). The new mixing data, based on electronic tagging, indicate that the western, depleted population is much smaller than ICCAT's earlier estimates. More important, fish from the western Atlantic, where fishing is heavily regulated, are exposed to intensive fishing pressure in the largely unregulated eastern fishery, where no quotas existed before 1994. As a consequence, "overfishing in the eastern Atlantic jeopardizes recovery of bluefin in the western Atlantic" (Magnuson et al. 2001, p. 1268).

Data for eastern fisheries are relatively incomplete. Landings averaged 30,000 MT from 1950 to 1965, rose as high as 53,000 MT in 1996, and have fallen since to 32,000–38,000 MT. These numbers plus IUU catches add up to declining numbers of spawning fish, increased fishing mortality rates, reduced recruitment, growth overfishing (40%–60% of overall catches are below the 6.4-kg minimum), and exceedance of recommended total catch. Recommendations call for reducing TAC of eastern fish to 25,000 MT because "the current catch level is not sustainable" (ICCAT 2001a, p. 60; see also P. M. Mace 1997; Safina 2001b). This conclusion and the new mixing data paint a pessimistic picture for both stocks and call for additional reanalysis and reassessment.

Although data are more complete for western than eastern stocks, conclusions are controversial or convoluted for both. Some analyses (NRC 1994; Sissenwine et al. 1998; ICCAT 2001a) indicate that western stocks need rebuilding to three to ten times 1998 levels to achieve an MSY of 5,000–6,000 MT, or a spawning biomass that will maintain MSY (= B_{MSY}) of 55,000 MT. ICCAT (2001a) concluded, however, that the TAC of 2,500 MT in place at that time had a 50% probability of building stocks back to 1975 levels by 2018. ICCAT's numbers were challenged because they were calculated after ignoring stock-recruitment relationships based on larger, 1970s-level stocks. The ICCAT (2001a) calculations produce a smaller B_{MSY} and, correspondingly, higher allowable catches because the target is lower. Reanalysis using all available data suggested that catch should be closer to zero to achieve B_{MSY} in 20 years (Safina 2001b). Regardless of the model used, the data included or ignored, or the assumptions employed, it was obvious that the western stock was overfished (well below MSY) and that current fishing mortality exceeded what would achieve MSY. Two approaches were consequently suggested to bring stocks up to the targeted goal: (1) an "active" management approach that would reduce fishing pressure and mortality until stocks rebuilt, or (2) a "passive" approach that would permit fishing at current levels while "waiting for natural variability in recruitment to bring a fortuitously strong year-class" (Sissenwine et al. 1998, p. 838).

Bluefin Tuna in a Larger Management Context

The two approaches epitomize a debate that rages across valuable but declining fisheries. Managers promote an active approach entailing conservative fishing quotas to allow stocks to rebuild, although the outcome is uncertain; this tactic carries a greater risk of short-term social and economic costs. The fishing industry prefers a passive approach, despite a higher risk of further decline, invoking uncertainty in data and pointing out socioeconomic hardships that would result if fishing were reduced, perhaps unnecessarily (Sissenwine et al. 1998). From a precautionary viewpoint, the passive approach seems like madness, and it has failed in the past. Georges Bank haddock were depleted by the 1930s, and fishing was curtailed. When several strong year classes appeared, they were used to justify increasing allowable catch instead of protecting the future spawning stock. Stocks subsequently decreased, and landings have remained down 98% from their 1965 peak (Sissenwine et al. 1998). Despite this example, a 1987 strong year-class of southern bluefin was fished hard rather than conserved. Stocks continued to decline, and the species is now listed by IUCN as Threatened.

More generally, Hutchings (2000a) found that stocks of depleted, long-lived marine species can require more time to rebuild than would be expected were natural population variability and high reproductive potential the major determinants. There is, therefore, little reason to anticipate fortuitous, salvation year-classes anytime soon among Atlantic bluefin. Recruitment in the late 1990s was lower than expected given spawning stock size, suggesting that other (environmental? ecological?) factors were slowing recovery. Bluefin appear to be confirming the generalizations drawn by Hutchings (2000a).

Even given optimism and a passive approach, SCRS data suggest that if western bluefin are to have a 50% chance of achieving a targeted spawning biomass of 55,000 MT, catches should be reduced to under 500 MT annually, which would lower allowable catches by 20%. Uncertainty in the data and calculations further complicate the management task (Sissenwine et al. 1998). Fishing industry advocates claim that stocks are more abundant than SCRS assessments, based on sighting frequencies of tuna schools using airplanes. However, outside reviews indicate that aerial counts are in basic agreement with other assessments and should be revised downward (Sissenwine et al. 1998). The controversy goes on and is unlikely to abate soon. The fish are just too valuable. To everyone.

Billfishes, Especially Swordfish

Blue marlin, white marlin, and sailfish (*Makaira nigricans*, *Tetrapturus albidus*, *Istiophorus platypterus*) are overfished in the Atlantic, as are blue marlin in the Pacific (Turner 1999). Catch statistics imply a pattern of serial overfishing by long-liners, targeting first blue marlin, then sailfish, then swordfish (R. A. Myers and Worm 2003). Commercial retention of these fish is banned in U.S. waters, a policy driven largely by recreational fisheries interests. However, marlin and sailfish are a significant and often unreported bycatch component in longline fisheries for tuna and swordfish, bringing those commercial fisheries into direct conflict with the powerful recreational fishery for billfishes.

Swordfish, *Xiphias gladius*, are large (to 545 kg) and tasty, and each individual is valuable; thus normal restraints imposed by diminishing returns from a depleted fishery are once again insufficient. Swordfish are the most popularized as well as the most controversial billfish, starring in print and on screen in Sebastian Junger's *The Perfect Storm*, and in Linda Greenlaw's book on swordfishing (Greenlaw 1999; see also C. D. Gibson 1998). Notwithstanding Greenlaw's contention that good fishers still catch fish and ought to be left alone, Junger's portrayal of dangerous conditions and difficult-to-find fish concurs more strongly with stock assessments and historical aspects of swordfishing in the North Atlantic. Swordfish politics are complex, largely due to the species' commercial importance. For starters, consensus is lacking within scientific circles on the status of swordfish stocks. IUCN (Hilton-Taylor 2000) listed North Atlantic swordfish as Endangered and other Atlantic, Mediterranean, and Pacific stocks as Data Deficient. However, AFS (Musick et al. 2001) did not list swordfish among the 41 marine species considered at risk.

Prior to the 1960s, swordfish in the North Atlantic were taken primarily by harpoon. This highly selective and relatively low-tech (but skill-demanding) method placed inherent limits on landings. More important, it targeted large adults that would have already spawned several times. In the early 1960s, the fishery expanded geographically and technologically as long-lining came to dominate. Catches increased and then inevitably decreased, and average size of landed fish plummeted.

Historically, swordfish were abundant. In the Gulf of Maine, 25 or more fish could be seen at the surface daily; one skipper counted 47 in sight at one time (Bigelow and Schroeder 1953). Now fishers will go days without seeing a swordfish, and harpoon boat catches may be less than 15 fish in a season (Safina 1997). Such numbers, although often anecdotal, are representative of recent declines and reinforce fishery data. Total Atlantic swordfish landings rose through the 1980s to a 1995 peak of 38,500 MT but have fallen since to about 25,000 MT. North Atlantic catches peaked in 1988 at 5,000 MT and fell to 3,100 MT in 1995; south Atlantic catches rose fourfold to peak in 1995 before declining. Catch statistics are revealing but can reflect effort as much as population size; population estimates indicate that biomass in the North Atlantic fell by 68% between 1960 and 1995 (NMFS 1997/NRDC 1998).

Much of the debate over swordfishing focuses on average fish size, which has decreased over the years. North Atlantic swordfish apparently mature at five to six years of age, 21 kg for males and 74 kg for females (Nakamura 1985). Prior to 1960, harpooned fish averaged 135–180 kg, but average size declined to 121 kg in 1963 and to 40 kg in 1995. The average fish caught recently on longlines is therefore immature (NMFS 1997/NRDC 1998; ICCAT 2001b).

ICCAT is also responsible for managing Atlantic swordfish stocks and again has a somewhat blemished record on sustainability. Despite evident population declines through the 1980s, no quotas were set until 1991 for the northern Atlantic and until 1994 for the southern Atlantic. MSY in the North Atlantic was estimated at between 7,600 and 15,900, while catches during the 1990s averaged 15,000 MT, meaning MSY was exceeded in most years, perhaps by as much as 100%. In only four years were catches lower than the average MSY of 13,400. Calculated biomass by the late 1990s was only 65% of B_{MSY} with a fishing rate 34% above what the stock could sustain (South Atlantic landings exceeded MSY in all years but one, sometimes by as much as 60%, despite mandated catch reductions) (NMFS 1998; Turner 1999; ICCAT 2001b). To slow the decline of North Atlantic swordfish, ICCAT established country-specific quotas for 1994, but the 1996 assessment showed increasing harvest levels (NMFS 1998). More specifically, ICCAT estimated that recovery could occur by 2012 if landings were reduced to 8,000 MT. Quotas were set at 11,300 MT in 1997, 11,000 MT for 1998, and 10,700 MT in 1999. Meanwhile, average yield remained around 15,000 MT, 33%–43% above mandated quotas and 88% above the 8,000 MT recovery value. This does not include IUU catches, many of which are dead, discarded juveniles.

ICCAT's response to overfishing of juveniles has been equally ambiguous. In 1991, it set a minimum size of 19 kg, considerably below the average maturation weight for females; in 1996, in the face of continued declines, it *reduced* the minimum to 15 kg. Regardless, such restrictions are ineffective because long-lining is nonselective with respect to size, and smaller fish die faster on longlines, hence the high number of dead, discarded (and underreported) juveniles. In known nursery areas off South Carolina, Florida, and the Gulf of Mexico, 60% of the swordfish caught in 1994 were discarded, presumably because they were undersize. Overall, the U.S. fishery in 1995 and 1996 discarded approximately 40% of the fish caught on longlines, which amounts to 35,000 to 40,000 fish per year. Including small but legal fish, approximately 58% of fish (80% of females, 35% of males) caught by U.S. long-liners may be immature (NMFS 1997/NRDC 1998).

Despite apparent systematic overfishing, ICCAT claimed that catch reductions and reproductive parameters portended stock recovery, "*if the recent year-classes are not heavily harvested*" (ICCAT 2001b, p. 95; emphasis mine). If catch quotas were kept low (10,700 MT in the North Atlantic) and enforced, prospects for recovery to B_{MSY} over 15 years were 50%. By ICCAT's calculations, a catch quota only 1,000 MT higher was unlikely to lead to recovery. Less reliable South Atlantic data suggested sustainability at catch quotas of 13,500 MT, but not at actual fishing levels of 14,600 MT. The margin for error in both fisheries seems rather small, about 10%. Alternatively, NMFS (1997/NRDC 1998) concluded that continued decline of the North Atlantic stock along the 1970s–1990s trajectory doomed the fishery early in the 21st century (NRDC 1998).

The status of the Pacific swordfishery remains open to debate because of considerable data uncertainties. NMFS Southwest (1999) estimated that recent average yield exceeded long-term potential yield by about 20% and then declared the stocks as only "nearly fully fished." The Indian Ocean is, as usual, a large unknown. Generalizing from other fishery stocks, its swordfish are not yet overutilized; however, fishing intensity there grows annually, and it seems prudent to collect better landings and population structure information, if only to prevent the depletions that have occurred elsewhere. Mediterranean data are incomplete and assessments impossible, but stocks appear to be overfished; 64% of the Mediterranean catch consists of immature fish under 120 cm. EU and national regulations are apparently being developed.

Swordfish Solutions

Swordfish populations *can* recover from overfishing if pressure is reduced. When high mercury levels were discovered in swordfish in the early 1970s, the U.S. banned swordfish imports. The commercial fishery in the Atlantic experienced a respite, and after about six years, populations recovered almost to pre-long-lining levels. When the U.S. relaxed mercury contamination standards, fishing resumed and populations again declined (Safina 2001b). Regardless, six years is rapid recovery, suggesting that sustainable management is achievable with minimal economic hardship. NRDC (1998) consequently offered recommendations for (1) reduced fishing of juveniles (time and area closures in known nursery regions), (2) increased minimum sizes of retained or imported fish to 68 kg to reduce fishing pressure on immature females, (3) shorter "soak times" (intervals between setting and pulling longlines) to decrease hooking mortality, (4) inclusion of dead discards in any quotas, and, most controversial, (5) "Give Swordfish a Break," a moratorium on buying swordfish until other measures were implemented. Over 700 U.S. chefs, several restaurants, hotels, cruise lines, airlines, and grocery chains agreed not to serve or sell swordfish. As a result, despite much hue and cry from commercial fishing interests, NMFS in 2000 instituted a time and area closure to long-lining encompassing 133,000 square miles of nursery areas, a closure estimated to reduce juvenile mortality by 30%–40% (see seaweb.org).

Debate over the swordfish boycott centered on the issues of overfishing in the Atlantic versus other areas, and whether Pacific and Indian ocean fish should be tarred with the brush applied to Atlantic fisheries. At the turn of the 21st century, Atlantic swordfish were overfished, and Pacific swordfish were fully or almost fully fished, whereas Indian Ocean swordfish were probably not overexploited. A concerned consumer should have selectively avoided Atlantic swordfish but accepted Pacific or Indian Ocean swordfish. Unfortunately, labeling requirements that would inform such a decision do not exist, meaning the consumer cannot know the origin of the swordfish offered in a restaurant or fish market (the 2002 U.S. Farm Bill included a provision for country of origin labeling, COOL, but fish caught on either U.S. coast or in a U.S. territory would be labeled "U.S."; plus, restaurants were exempted). Consumer action and pressure can, however, motivate a fishery to engage in informative labeling, as well as sustainable fishing; fisheries policy and practices can be modified, with positive results. Without such pressure, history suggests that little progress will be made. As of late 2005, changes in fishing practices in U.S. fisheries had allowed stocks to rebuild once more, prompting the Monterey Bay Aquarium's Seafood Watch program to reclassify American-caught swordfish as a Good Choice (www.mbayaq.org).

Finally, the mobilized, effective public reaction to the swordfish boycott now needs to be directed at more seriously depleted and less easily recoverable, large pelagic species such as bluefin tuna and sharks. The Give Swordfish a Break effort is a positive beginning.

Small Pelagic Fishes

The world's most productive, variable, sometimes vulnerable, and perhaps resilient fisheries involve small pelagic species, including clupeoids (anchovies, herrings, pilchards, shads, menhadens), carangid scads and jacks, scombrid mackerels, and a few other anatomically convergent groups such as osmerids (smelts, capelin, and eulachon), ammodytid sand lances, scomberesocid sauries, and the oddball trichiurid cutlassfishes or hairtails. Small pelagic fishes in these and other groups—whitefishes, smelts, herrings, silversides, cyprinids—also constitute a significant fraction of lake fisheries (see box 10.1). When taken as a group, small pelagics may account for 50% of the world's capture fisheries (see figure 10.5).

Regardless of taxon, small pelagics have converged on a number of anatomical, behavioral, physiological, life history, and distributional traits that affect their productivity. Many traits are obvious adaptations to variable but productive environmental conditions (Helfman et al. 1997). Small pelagics generally are less than 30 cm long, silvery, streamlined, and lipid rich; live in schools; migrate to take advantage of spatially and temporally dispersed food sources; swim near the surface in continental waters; and live near regions of upwelling, where they feed relatively low in the food chain. Phytoplankton production in upwelling areas is quickly transformed into fish biomass because these fishes grow and mature quickly. Fish production of 8 tons/km^2/yr^1 is not uncommon among pelagic fishes; demersal fishes may produce less than half that. However, productivity can vary greatly in response to changing food availability, as when climatic alterations such as El Niños disrupt upwellings, reduce primary production, and precipitate population crashes. The distribution of fish scales in sediment cores tells a tale of repeated swings in population size (Jennings and Kaiser 1998). These species boom and bust, as do their fisheries.

Many characteristics that contribute to the success of small pelagics in nearshore, open waters make them subject to human exploitation. They are fished heavily not because they are individually valuable but because their numbers and behavior make them easy and economical to catch (Pitcher 1995). The tendency to aggregate, largely an antipredator response, is disadvantageous when the predators are people with large nets (Parrish 1999). By exploiting schooling behavior and predictable times and places of aggregation, or by inducing these with fish attraction devices (FADs), fishers increase efficiency. Importantly, small pelagics remain vulnerable to fishing regardless of stock density (G. I. Murphy 1980), meaning that fishing mortality in small pelagics is depensatory rather than compensatory. Pelagic schooling species continuously reaggregate when under attack, which keeps them concentrated in a small area and vulnerable to capture by nets, particularly purse seines. When individual schools are depleted through fishing, remaining individuals aggregate with individuals from other depleted schools. Hence, the absolute numbers of a stock may shrink, but continuously reconstituted schools in an area can be repeatedly located and exploited by fishers. Fish captured per haul may not decline even when total population has been reduced, until the fishery crashes. Pitcher (1995) referred to this as catchability-led stock collapse, or CALSC (also referred to as hyperaggregation).

Some of the world's most spectacular stock collapses have involved small pelagics responding to the combined effects of climate change and overfishing. World fishery landings grew progressively between 1950 and 2000 except in the 1970s and early 1990s (see figure 10.3). The first stagnant period corresponded to collapse of the Peruvian anchoveta, *Engraulis ringens*, fishery, and the second occurred when Japanese and South American pilchard (*Sardinops melanosticta, S. sagax*) stocks collapsed (Garcia and Newton 1997). These and other boom-bust cycles are interwoven in a history of overcapitalized pelagic fisheries (Ueber and MacCall 1990; Botsford et al. 1997).

Pacific sardine, *Sardinops sagax caeruleus*, occur from northern Mexico to the Bering Sea. By 1925, California landings of Pacific sardine reached 175,000 MT, creating a sardine culture artfully depicted in Steinbeck's *Cannery Row* (1945). Sardines were canned for human consumption, and waste was "reduced" into poultry food and fertilizer. However, reduced sardines became more valuable than the canned product, prompting canneries to reduce whole fish. Floating reduction plants proliferated, anchored outside the three-mile limit to bypass regulations. In the 1930s and 1940s, Pacific sardines supplied the largest fishery in the Western hemisphere (T.D. Smith 1994; Speer et al. 1997). Catches peaked at 790,000 MT in 1937, then leveled off at around 600,000 MT per year for the next decade. Fishery calculations indicated, however, that MSY was closer to 250,000 MT. Catches began to decline, to 230,000 MT in 1946–52; 55,000 MT in 1953–62; and finally 24,000 MT in 1963–68 (figure 10.13). By 1970, spawning biomass had declined from 3.8 MMT to only 5,000 MT, a 760-fold decrease (Leet et al. 1992). The fishery ceased. Boats and processing equipment were sold below cost, or costs were subsidized by international agencies.

The collapse of the California sardine fishery, and a

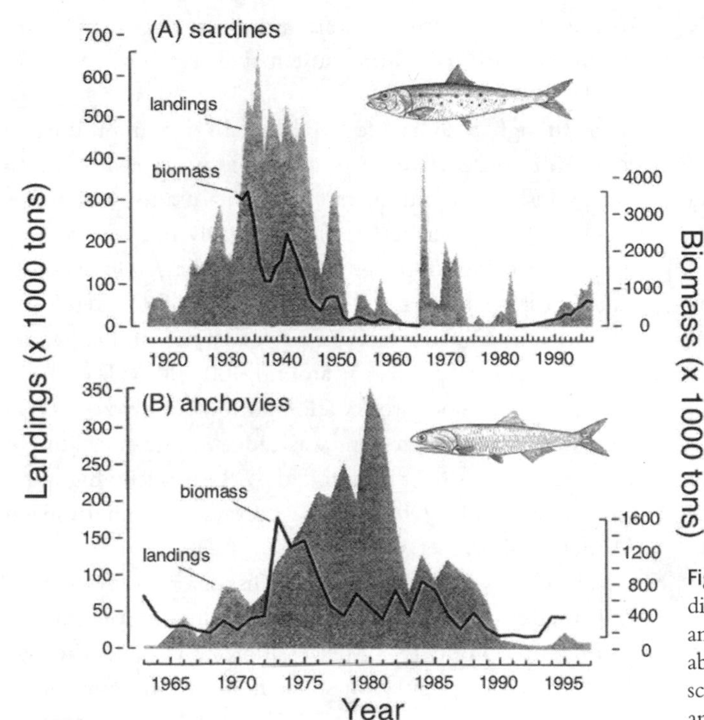

Figure 10.13. Boom-and-bust cycles in biomass and landings of sardines and anchovies, U.S. and Mexico Pacific coast. (A) Sardines and (B) anchovies and their fisheries. No biomass data are available for sardines from 1966 to 1982. Note the different time scales. From Jacobson et al. (1999); fish illustrations from Bigelow and Schroeder (1953).

similar fishery off Japan, likely resulted from "intense fishing pressure on the resource . . . [that] probably accelerated a long-term pattern of natural decline" correlated with cooler ocean temperatures (Jacobson et al. 1999, p. 1). The demise contributed to later development, overexploitation, and eventual collapse of the Peruvian anchoveta fishery, which began in the 1950s primarily for human consumption. Again, reduction plants were built and boats were added to the fleet, many from the defunct California sardine fishery. By 1969, Peru was the world's most productive fishing nation, anchoveta constituting 98% of the catch. Landings were largely unregulated, a 12.4-MMT catch exceeding MSY by 5 MMT. The fishery collapsed to below 1 MMT in the mid-1970s, probably from overfishing combined with unfavorable climatic shifts as upwellings were disrupted by El Niño events in 1972–73 (Caviedes and Fik 1990; Sharp and McLain 1995).

California saw its sardine fishery replaced 20 years later by northern anchovy, *Engraulis mordax*, which boomed in the mid-1970s and busted in the late 1980s (figure 10.13). Sardines then staged a comeback, with biomass exceeding 600,000 MT in 1997 and catches approaching 100,000 MT (Jacobson et al. 1999). This time around, managers based quotas not only on standing stock biomass and productivity, but also on climate, with cooler ocean temperatures justifying lower allowable catches (Speer et al. 1997). The sardine-anchovy story in California was repeated with surprising similarity in South Africa. A fishery for sardine, or pilchard, *Sardinops sagax*, boomed and busted between 1950 and 1965, followed almost immediately by a fishery for anchovy, *E. capensis*, that expanded manyfold between 1962 and 1990. The anchovy fishery in South Africa, however, appears to be more carefully managed (Butterworth et al. 1997).

In the face of such repeated, large-scale fluctuations, controversy now surrounds the health and future of menhaden fisheries of the U.S. Atlantic and Gulf regions. At 860,000 MT annually, Atlantic and Gulf menhaden (chiefly *Brevoortia tyrannus* and *B. patronus*) together constitute the second most productive fishery in U.S. waters, behind only eastern Bering Sea–Aleutian groundfish fisheries (the Pacific fisheries involve 27 stocks vs. 2 for menhaden).

Menhaden are small (to 15 cm), bony, oily clupeids that are seldom used for human consumption, most going into animal feeds. They have been fished commercially since the mid-1800s (and were used as fertilizer by Native Americans long before that; the name apparently comes from the Wampanoag *munnawhatteaug*, meaning "fertilizer"). The fishery was first pursued in more northerly

Atlantic states but collapsed in the 1930s–60s, apparently due to overfishing (ASMFC 2001). Effort shifted south to North Carolina and Virginia, where 2 reduction plants remain, down from 23 during the heyday of menhaden exploitation. Total landings climbed steadily after World War II, as spotter planes were employed for locating schools. Catches peaked in 1956 at 712,000 MT, dropped as low as 162,000 MT in 1969, and then cycled around an average yield of 300,000 MT. More important, spawning biomass peaked in the 1950s–60s at 350,000 MT, fell dramatically to around 10,000 MT in 1966, and has fluctuated around 50,000 MT since. During the 1990s, recruitment fell steadily. Atlantic menhaden are considered fully utilized and currently healthy by NMFS, but low recruitment in recent years has led to concern over future stock levels (Vaughan and Smith 1999; ASMFC 2001; Chesapeake Bay Program 2001). Garcia and Newton (1997), using FAO data, came to the less optimistic conclusion that stocks overall were more than fully fished.

Disagreement centers on past versus present abundance, ecosystem-level effects of a presumed menhaden decline, and the relative importance of overfishing versus climatic variation, the latter determining estuarine conditions affecting juvenile fish. Huge schools a kilometer or more across previously occurred along the Atlantic seaboard but are now rare (Franklin 2001). Prereproductive fish are captured in the highly efficient purse seine fishery, and spotter planes make it difficult for schools to avoid detection. CALSC-like conditions lead to overfishing, also affecting predators such as striped bass, bluefish, and marine mammals historically dependent on abundant menhaden. As filter feeders that previously numbered in the billions, menhaden may have filtered 6%–9% of annual phytoplankton production and 100% of daily production, thus controlling phytoplankton blooms and maintaining water quality in East Coast estuaries (ASMFC 2001). Recent blooms of noxious and toxic phytoplankton may be linked to reduced numbers of menhaden, while pollution and altered food webs in estuaries favoring jellyfish may contribute to poor juvenile survival and recruitment of menhaden. Overfishing, climatic fluctuation, and habitat degradation may once again interact with natural population variation to drive a commercially and ecologically important marine species to dangerously low levels. Area and seasonal closures, mesh size restrictions, catch quotas, and especially adaptive management that responds to population fluctuations are among the remedial steps recommended to ensure sustainable catches (ASMFC 2001).

Although a short life cycle means that reproductive failure in consecutive years can rapidly drive a stock to dangerously low numbers, rapid maturation facilitates recovery. Small pelagics are demonstrably resilient, repeatedly recovering from population crashes, regardless of cause. During the early 1970s, landings of mackerel and herring in the northwest Atlantic, heavily exploited by distant-water fleets, peaked at around 400,000 MT each. In the late 1970s, both stocks fell eightfold to below 50,000 MT. When fishing pressure was reduced and recruitment improved, both stocks rebounded to "historically high levels" in the mid-1990s and are considered underutilized (Anderson, Mayo, et al. 1999).

Hutchings (2000a) found that fishery records for herring, menhaden, sardines, and sprats, including some of the world's most productive fisheries, indicated frequent recovery to previous population sizes after extreme population crashes (see also Hilborn 2005). Hutchings attributed recovery to life history traits as well as fishing methods. Clupeids mature relatively quickly, giving them a higher intrinsic rate of population increase and a greater rebound capability. Also, clupeids swim in relatively species- and size-specific schools that are usually caught in purse seines or midwater trawls. These gear types can target individual schools. Greater gear selectivity produces one to two orders of magnitude less bycatch and unintended fishing mortality. All together, these conditions are significantly better than less selective bottom trawl and longline fisheries for slow to recover demersal and large pelagic species.

Old-Growth Fisheries: Rockfish, Roughy, and Toothfish

In contrast to small pelagics are fishes that are slow growing, long lived, and late to mature. Slow growth and reproduction are common traits among overfished species, which is not surprising given the general vulnerability of fishes with such life histories, exploited or not (chapter 4). But some valuable, recently discovered commercial stocks, several at relatively high latitudes, exhibit these traits in the extreme. Their fisheries show a repeated pattern of rapid and large-scale exploitation, rapid collapse, and lack of recovery. Among these species are the rockfishes of the U.S. West Coast, orange roughies of mostly New Zealand

and Australia, and Patagonian toothfish (Chilean sea bass) of southern oceans.

Long Live Rockfish!

Approximately 70 of the 102 species of scorpaenid rockfishes occur in the eastern Pacific, most in the genus *Sebastes* (Love et al. 2002). Rockfishes are second only to sablefish, *Anoplopoma fimbria*, in value among groundfish on the U.S. West Coast (sablefish, another species discarded as trash when I collected from trawlers in the 1960s, have been reduced to 41% of original biomass despite intensive management; Koslow et al. 2000). Many rockfish populations have declined due to serial overfishing and adverse climatic conditions; TACs have followed, dropping by more than 77% in 1998.

Rockfish biology causes their vulnerability to overfishing (PMCC 1999; Rogers and Builder 1999):

- Rockfishes are ovoviviparous (livebearing from internally hatching eggs) and hence have low fecundity, which decreases the likelihood of fortuitous, large, salvation year-classes once populations are diminished. Rockfish are also long lived, many species reaching 40–100 years old. Rougheye rockfish, *Sebastes aleutianus*, have been aged to 147 years, meaning a large rougheye caught in 2005 could have been swimming in Alaskan waters when Abraham Lincoln was president and the region still belonged to Russia. Rockfish mature slowly, many species not maturing until 5–8 years old (Lea et al. 1999). Late maturation means that intense fisheries will catch many juveniles before they are able to reproduce.
- Rockfishes have low adult mobility. Eight of 13 species showed essentially no mobility over periods of up to 3.5 years; 3 species moved no more than 2.1 km, and only 2 species showed more extensive movement (Lea et al. 1999). Strong residency means overfished populations are unlikely to rebuild via colonization.
- Rockfishes normally experience low natural mortality rates of less than 15% and are thus unlikely to possess built-in mechanisms for rapid replacement. Low adult mortality is expected, given the large size reached by several commercially desirable species (e.g., shortraker, *S. borealis*, to 108 cm; rougheye to 97 cm; bocaccio, *S. paucispinis*, to 91 cm; yelloweye, *S. ruberrimus*, to 91 cm; at least 12 others to >60 cm; D. E. Kramer and O'Connell 1995; Lea et al. 1999).
- Rockfishes form multispecies aggregations, making depleted species vulnerable even when not targeted if they co-occur with abundant, targeted species. Many rockfishes are hence managed as part of a "*Sebastes* complex" of 10 species, with a TAC defined for the complex that exceeds what any single species can sustain. This builds overfishing into the targets, especially given that 80% of the fish are caught by notoriously nonselective bottom trawling.
- Rockfishes show variable recruitment, even for a marine species. From one year to the next, 20-fold differences in juvenile survival occur. Coupled with extreme longevity, variable recruitment produces stocks that consist largely of a single year-class spawned many years earlier. In 1999, California populations of chilipepper rockfish, *S. goodie*, were heavily dependent on the 1984 year-class, and imperiled bocaccio were dominated by the 1977 year-class.
- Rockfishes are subjected to significant climatic vagaries. At the warm sea temperatures associated with an El Niño, reproductive success and recruitment decline.

Accurate population data are available for only 15% of species, but many of these showed declines of 77%–98% at the end of the 20th century. For others, obvious declines have been assumed, based on the better-monitored species. AFS listed 13 species of Pacific rockfish as Vulnerable to regional extinction because of life history characteristics (Musick et al. 2001); 2 of these—cowcod, *S. levis*, and bocaccio—were globally Vulnerable. Four species were listed by NMFS as overfished (bocaccio; cowcod; Pacific Ocean perch, *S. alutus*; and canary rockfish, *S. pinniger*). Pacific Ocean perch, previously the most important commercial scorpaenid and exploited heavily by Japanese and Soviet Union fleets in the 1960s and 1970s, declined as much as 87% from its 1965 peak of 450,000 MT (Koslow et al. 2000). Prefishery population structure in the Gulf of Alaska included fish between 0 and 90 years old, but by 1990 only 12% of fish were older than 15 years (Goñi 1998). Recent average yield—which theoretically results only from bycatch—exceeded current potential yield by a factor of 400 (PMCC 1999; Rogers and Builder 1999). IUCN has listed bocaccio as Critically Endangered beginning in 2000 due to a reduction of over 80% in populations. In January 2000, the U.S. Department of Commerce officially recognized the crisis in rockfish stocks and declared the groundfish fishery on the West Coast a fishery disaster, facilitating economic relief to fishers and additional funding for improved management.

Rockfishes declined despite 20 years of active management. Fishery targets are known to have exceeded sustainable yield, often by 10%–15%, because targets assumed reproduction typical of species with higher reproductive output. Authorities maintain that fishing mortality should be reduced further, a difficult goal given the emerging fishery for the live fish restaurant trade (see chapter 13). Other recommendations include creating refuges that contain habitats used by habitat-specialist species, focusing on individual species rather than lumping all into a "complex," catch moratoria, limited entry permits, vessel buyback programs to reduce overcapacity, individually transferable quotas (ITQs), increased fishery observer coverage, and adaptive management on a precautionary rather than a crisis basis.

Long-Lived, Southern Hemisphere Fishes

Some of the best (most egregious) examples of rapidly depleted fish species are from deep, cold, Southern Hemisphere waters. Valuable resources combine with lack of biological information, rapidly expanding fisheries, and difficult enforcement conditions to produce boom-and-bust fisheries for species biologically incapable of withstanding intense exploitation. The long-term potential of these fisheries is poor because they occur in deep (>600 m), cold, dark, polar waters, "a system of low energy and low productivity" (M. Clark 2001, p. 123). Fishing is expensive and dangerous and began only late in the 20th century. Many fishing nations, having overfished their coastal and shelf fisheries, were forced to send distant-water fleets to less accessible, less desirable, and more marginal fishing grounds (Koslow et al. 2000). The fishes were locally abundant but slow growing, slow maturing, and long lived, with low fecundity and episodic recruitment leading to low replacement rates. Their fisheries have typically gone from unexploited to commercially extinct within five to ten years (Koslow et al. 2000; M. Clark 2001). The two best-known examples are orange roughy, *Hoplostethis atlanticus*, and Patagonian toothfish, *Dissostichus eleginoides*. Future poster children of overexploitation include oreosomatid oreos (*Pseudocyttus*, *Allocyttus*), which live over 100 years; macrourid grenadiers (*Coryphaenoides*), which live 60 years but had CPUE declines of 50% in 5 years; lotine cuskfishes and ling (*Molva*, *Brosme*); pentacerotid pelagic armorhead (*Pseudopentaceros wheeleri*), from 200,000 MT to commercially extinct in 6 years; and deep-water dogfish sharks (Merritt and Haedrich 1997; Koslow et al. 2000).

Orange roughy (figure 10.14) occur in the southern Atlantic, Pacific, and Indian oceans (and northeast Atlantic and Mediterranean) at depths of 500–1,500 m, often in

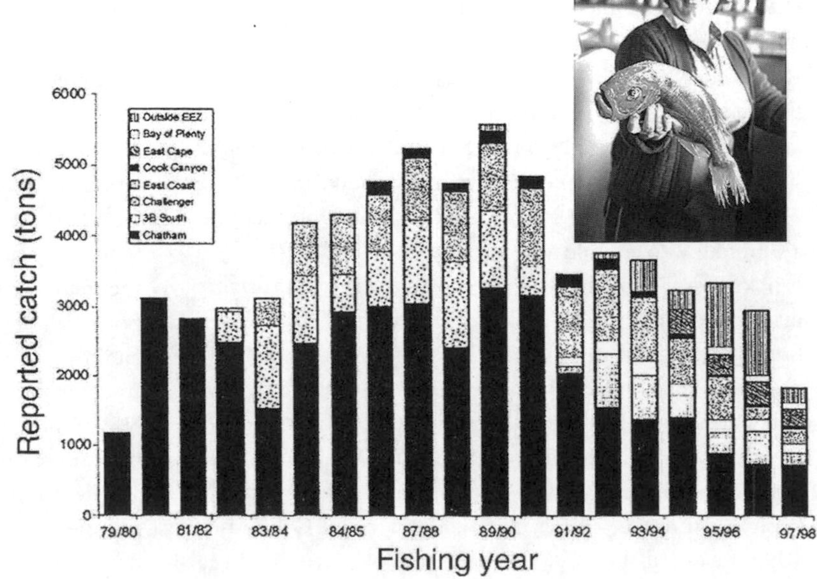

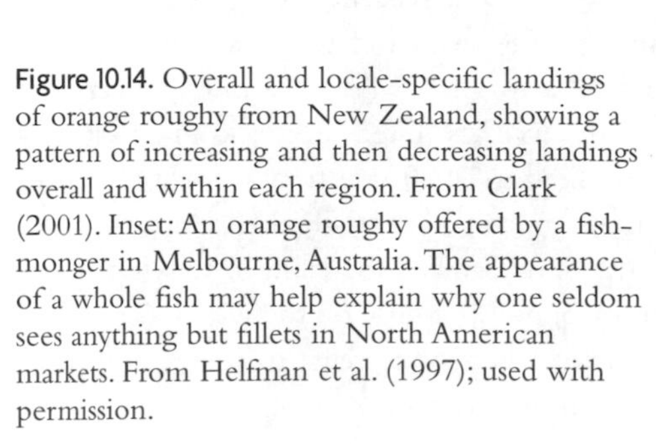

Figure 10.14. Overall and locale-specific landings of orange roughy from New Zealand, showing a pattern of increasing and then decreasing landings overall and within each region. From Clark (2001). Inset: An orange roughy offered by a fishmonger in Melbourne, Australia. The appearance of a whole fish may help explain why one seldom sees anything but fillets in North American markets. From Helfman et al. (1997); used with permission.

association with seamounts. Their attraction to seamounts facilitates location of schools. Roughy grow slowly to 50 cm and 3 kg, may reach ages of 120–130 years, and do not mature until they are 23–31 years old (Tracey and Horn 1999). Natural mortality is exceedingly low, estimated at less than 5% a year for New Zealand populations. Fishable stocks were not discovered until the 1970s, first off New Zealand and then Australia. They are caught in trawls, often as part of winter spawning aggregations (CSIRO 2000).

Orange roughy can withstand little more than moderate fishing pressure. In New Zealand waters, roughy fishing began in 1979, with new fishing grounds sought as old grounds became unproductive. This serial overfishing was typified by initially large catches, contractions in stock distribution and abundance, and reduced catch levels as stocks became overexploited. Little evidence of stock rebuilding existed after fishing pressure was reduced (Clark 2001). The chief roughy fishery (Chatham Rise area) went from a virgin biomass of about 300,000 MT to a depleted biomass of 50,000 MT in only ten years, a decline of 80%. A nearby area (the Spawning Box) shrank from multiple fishable aggregations spread over a 135-km area to a single area 30 km wide, suggesting that individual populations were wiped out (or perhaps coalesced in the familiar CALSC pattern). Meanwhile, overall CPUE plummeted. Effort doubled in the 1980s, but catches fell from 7–9 MT/tow to 2–4 MT/tow by the end of the 1990s. Despite quota reductions and increased fishing effort, TACs were not met.

Most populations have shown few if any signs of recovery. Size distributions in catches remained similar after ten years of fishing, indicating that the same year-classes were being fished with little new recruitment. No indications of biological compensation, such as smaller size or younger age at maturation, arose. After the first 6 years of fishing at three New Zealand spawning sites where stock biomass was reduced by 70%, genetic diversity declined (P. J. Smith et al. 1991), although other populations did not show similar losses (P. J. Smith and Benson 1997). The overall picture is that New Zealand and Australia orange roughy stocks are being depleted and not replenished. Sustainable yield for orange roughy may be only 1%–2% of virgin biomass, compared to 50% for many shallow-water species (Koslow et al. 2000). Traditional fisheries models do not apply.

The more recent fishery for Patagonian toothfish, or Chilean sea bass, has followed similar patterns (figure 10.15). It is even less well regulated because it is located in more remote areas and falls under the jurisdiction of developing nations that are ill equipped to control legal and IUU fishing. Toothfish are in the nototheniid cod icefish family and occur at depths of 700–2,500 m off the coasts of Chile and Argentina and around subantarctic islands (e.g., the Falklands, South Georgia Island). They are large (to 2.4 m, 130 kg) benthopelagic predators that live more than 50 years. They do not mature until 6–12 years old and produce relatively small numbers of large (4.5-mm) eggs (Kock 2000). The fishery primarily employs longlines during the winter: difficult, dangerous work in the Southern Ocean. Densities are relatively low, on the order of 0.4 to 1.3 toothfish/km^2 (Yau et al. 2001), making profitable catches even harder to achieve.

Intensive fishing for toothfish began in 1994 when deep-sea longline fisheries for hake and cusk-eels off Chile collapsed (Lack and Sant 2001) and fishing spread eastward to several Southern Ocean locales. Chief markets are Japanese and U.S. restaurants, Canada and the EU also contributing demand. Legal fishing is practiced by the 24 member nations of the Commission for the Conservation of Antarctic Marine Living Resources (CCAMLR), but vessels flagged in these countries and several other "convenient" countries (e.g., Belize, Argentina, Panama, Cayman Islands) engage in illegal fishing. IUU catch and importation are practiced by at least 11 countries, notably Spain, Denmark, Norway, Mauritius, Japan, Chile, Namibia, and South Africa, which is to say IUU fishing and marketing are international and extensive.

Three major, controversial; and linked issues surround the toothfish fishery: stock depletion, IUU fishing, and seabird bycatch. Toothfish biology is poorly studied, and findings are published in limited-circulation journals such as *CCAMLR Science*. We lack knowledge about interstock

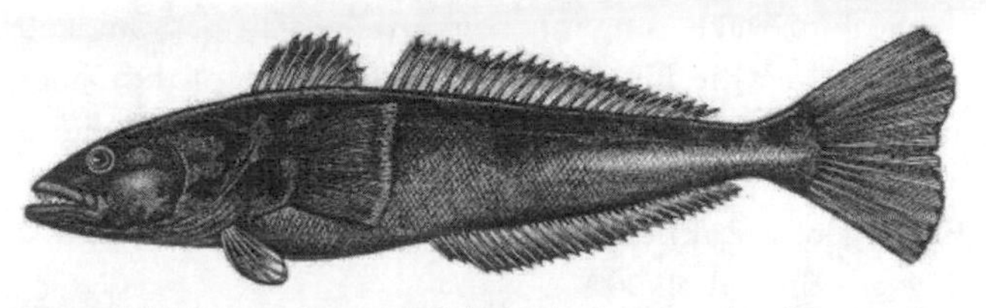

Figure 10.15. Patagonian toothfish. A classic example of a recently discovered fishery that has expanded rapidly, has become just as rapidly overfished, and is largely unregulated. Drawn by Bruce Mahalski from Lack and Sant (2001); used with permission of www.traffic.org.

movement, growth, natural mortality, and even stock size. Stock-recruitment relationships are a mystery; projections of future stocks are mainly guesswork. But evidence of decline is obvious. In just the first few years of the fishery, "overfishing has occurred and stocks have diminished in many areas . . . to the point of imminent collapse" (Bray 2000, p. 42). Landings in South African regions fell 90%, and biomass around the Crozet Islands declined 70%–75%, while number of hooks set increased fourfold and hours fished increased 40% (Lack and Sant 2001). Average size dropped 30% in the first year of fishing at several locales (ISOFISH 2002).

The likelihood of meaningful regulation of a fishery is inversely proportional to the fraction of the catch taken by IUU fisheries. For Patagonian toothfish, which is vulnerable given its biology, prospects for meaningful regulation are low because a large and largely unknown IUU component exists. IUU fishing is calculated either from reported sightings of unlicensed vessels multiplied by likely days at sea and catch rates, or by taking the difference between legal catch and total imports. ISOFISH (formerly www.isofish.org, now www.colto.org), an NGO made up of conservation groups and licensed fishing companies, claimed that IUU catch in 1997–98 was 9.5 times the legal catch, 100,000 MT versus 10,500 MT. Greenpeace estimated IUU catch at 100,000 MT in 1997 and 130,000 MT in 1998 versus respective TACs of 33,000 MT and 18,000 MT. These amount to a three- to sevenfold greater IUU than legal catch (Bray 2000). Using the legal catch-import difference method as a conservative estimator, Lack and Sant (2001) estimated IUU catch at around 50% of legal catch, still four times greater than the value calculated by CCAMLR. Imports of 48,000 MT to Japan and the U.S. in 1999–2000 contrasted with a 25,000 MT reported legal catch, indicating IUU fishing added over 80% to the legal catch. The differences in the estimates reflect the uncertainty surrounding IUU statistics.

Seabirds take longline baits as they are set out and drown when the lines sink, killing hundreds of thousands annually (read Safina 2002). Among the mortalities are 3 petrels and 19 albatrosses, including 2 Critically Endangered, 2 Endangered, and 12 Vulnerable albatross species (Tuck et al. 2001).

Numerous measures have been proposed to reduce IUU catches as well as seabird bycatch. CCAMLR, with its strong economic interest in conserving toothfish stocks, is developing satellite-linked vessel monitoring to identify IUU vessels. It suggested that identified illegal vessels should be blacklisted and prevented from selling catches (some vessels have apparently changed their names to circumvent this). A Catch Documentation Scheme (CDS), implemented in 2000, will monitor catches and international trade and eventually institute export and import permits that limit trade and processing, particularly of catches from "open register" (flag-of-convenience) vessels.

Seabird mortality is a potent ethical issue. Numerous bycatch reduction actions and devices exist, including flying streamers behind a boat to repel birds during longline setting, attaching weights to make lines sink before birds can grab baits, setting at night when most birds are inactive, and limiting offal release from boats while setting lines to reduce attraction of birds (Melvin and Parrish 2001; see www.ccamlr.org, Conservation Measure 29/XIX).

Orange roughy, toothfish, and biologically similar fishes create a true dilemma in promoting sustainable exploitation. They can sustain light fishing at the most, but their value and the high overhead associated with capture make it uneconomical to fish them at low rates. Such fisheries seldom cease before a stock goes commercially extinct. Hence, management options are largely limited to the extremes of (1) tightly controlling fisheries to save the species from commercial extinction but with minimal profits, or (2) abandoning constraints and opening up the fisheries to intentional overfishing. The latter has been advocated for roughy: "The fish stocks should intentionally be 'mined' or overfished, rather than attempting to manage the fisheries toward low volumes in the long term" (Clark 2001, p. 124). The argument for intentional overfishing turns on the assumption that a species that rapidly achieved commercial extinction would be quickly abandoned and therefore unlikely to suffer biological extinction. The task of resource managers in such circumstances would be simplified if demand for the resource were reduced, as when consumers refuse to accept a product because its exploitation is not sustainable (as of most old-growth fishes) or is ethically repugnant (as of longline and gill net fisheries that kill seabirds). Such actions will also reduce illegal fishing because "IUU fishing is only worthwhile if a market exists for the product" (Lack and Sant 2001, p. 11). When the market evaporates, pursuing the quarry is bad business. Until conditions improve, take a pass on Chilean seabass. Orange roughy are too ugly to eat.

Success Stories: Striped Bass and Pacific Halibut

All fisheries are not necessarily doomed. Successful management is possible. Case histories exist that are worth detailing because they provide reasons for optimism. More important, positive outcomes highlight conditions where success can be anticipated as well as specific actions that are likely to lead to success. In addition to the striped bass and Pacific halibut examples, Alaskan fisheries for Pacific salmon could be featured (see Hilborn 2005).

Chesapeake Bay Striped Bass

The striped bass, *Morone saxatilis*, also called rockfish on the U.S. East Coast, is a popular sport and commercial species that is pursued with great zeal and has achieved iconic status for many people (R. A. Richards and Deuel 1987). This enthusiasm and the economic activity it stimulates underlie both the demise and the recovery of Atlantic migratory striped bass. Its well-documented resurrection represents a unique and instructive cooperative venture among often disparate agencies, individuals, organizations, and administrative entities.

Striped bass are anadromous, grow to 60 kg, and are fished from boats and beaches in the ocean and rivers. Four different migratory strains occur along the U.S. Atlantic seaboard, linked to spawning grounds in the Chesapeake Bay, Hudson River, Delaware Bay, and Roanoke River of North Carolina (E. D. Anderson, Kocik, and Shepherd 1999); they are also widely introduced into reservoirs and along other coastlines. Commercial fishing dates to the 1600s. Landings increased from the 1930s until the mid-1970s, when populations crashed due to overfishing, pollution, and spawning habitat destruction. Combined impacts resulted in reproductive failure and severe year-class disruption that worsened through the 1970s (figure 10.16).

The focus of management was the Chesapeake Bay because fish produced there migrated up and down the Atlantic coast and historically contributed up to 90% of total striper catches (R. A. Richards and Deuel 1987). Roanoke and Delaware river populations also experienced

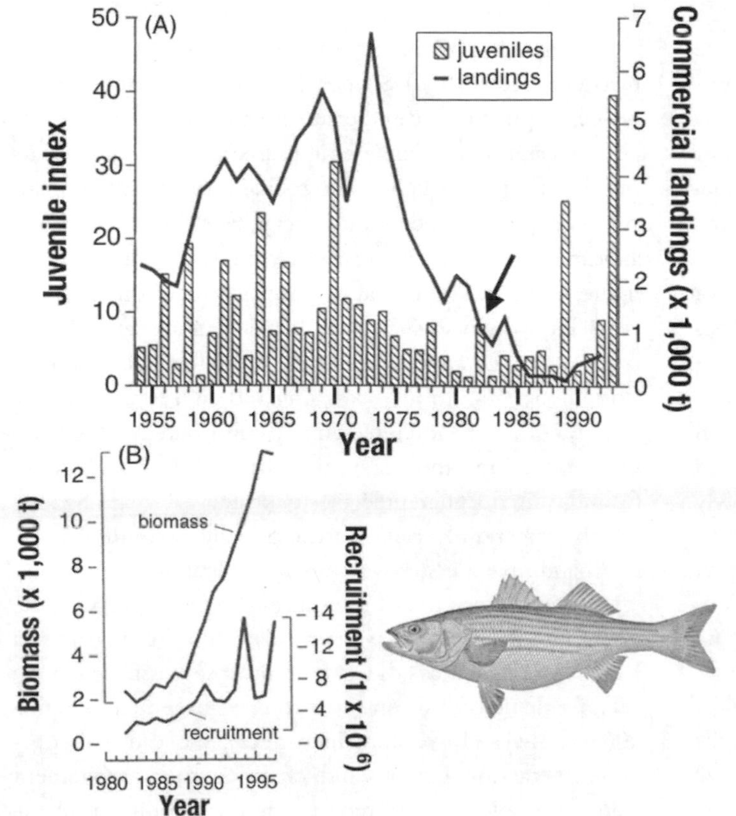

Figure 10.16. Striped bass recovery through effective management. (A) Commercial landings of striped bass in Maryland waters of the Chesapeake Bay peaked in 1973, then plummeted. Protection of the 1982 year-class (arrow) and subsequent cohorts, along with other fishing restrictions, contributed to resurgence and recovery of the stocks. From NMFS 1993. (B) Recruitment (abundance of 1-year-old fish) and spawning biomass (female standing stock), 1980–95. From E. D. Anderson, Kocik, and Shepherd 1999; fish drawing by H. L. Todd, from NMFS (www.photolib.noaa.gov).

population declines, although the Hudson River stock did not, in part because PCB contamination led to relaxed fishing pressure. A greater than 90% decline in Chesapeake landings over ten years coupled with a 99.7% fall in recruitment prompted the Atlantic States Marine Fisheries Commission (ASMFC) to institute an interstate fisheries management plan, including size restrictions and spawning area closures, beginning in 1981 (ASMFC 1981). Before 1981, regulations were left to individual states and were "diverse, inconsistent, and intermittent" (nmfs.noaa.gov/irf/asbca.html).

Continued declines led to greater ASMFC restrictions, including greatly curtailed commercial harvest. Multiple year-classes likely failed as the combined result of high fishing mortality rates (30%–60% on coastal stocks, 90% in Chesapeake Bay) and minimum size limits that allowed capture of prereproductive fish (Secor 2000). By steadily increasing the minimum capture sizes, the expanded ASMFC restrictions focused on protecting the 1982 year-class until 95% of those females had matured. "Shepherding the [1982] class into adulthood to stimulate more egg production was the linchpin of the recovery " (Safina 2001b, p. 441). The U.S. Congress assisted by passing the Atlantic Striped Bass Conservation Act of 1984 (Washington, D.C.'s location on the Chesapeake may have contributed to the willingness of Congress to get involved; T. Reinert, pers. comm.; *all* politics are local). The act put federal teeth into the restrictions and specified that all 15 Atlantic states must comply with the management plan or be subject to a total fishing moratorium. Cooperation earlier was voluntary. Without this stipulation, "few states would have fully complied given the political and economic unpalatability" of the measures (NMFS 1993, p. 27). Their hearts and minds followed.

The fish cooperated: 1989 saw spectacular reproductive success from the 1982 cohort (see figure 10.16B), which was repeated in 1993 and 1996. The fishery was reopened on a limited basis in 1989, but restrictions allowed newly recruited fish to survive. Recreational fishing supplanted and eventually surpassed commercial landings, in part because the management plan and related state actions favored recreational activity (M. R. Ross 1997). By the late 1990s, recreational fishing accounted for 75% of landings, which totaled 8,300 MT and exceeded the peak catches of the mid-1970s, showing once again that a healthy, well-managed stock produces more fish than an overexploited stock.

The plan succeeded in large part because it departed from tradition (NMFS 1993; M. R. Ross 1997) with the novel premise that populations should be managed more to restore and maintain spawning stocks than to produce fishery yields. Federal oversight and control encouraged state-federal cooperation, overcoming local noncompliance. The promotion of recreational interests helped marshal public support. Careful monitoring by a small army of biologists, made possible by emergency funding provided in the Conservation Act, facilitated tracking of populations. The plan was adaptive, allowing for flexible regulations such as adjusting allowable catch in response to changes in the stock. Sport fishers, rather than increasing fishing when restrictions were relaxed, showed their commitment to the plan by requesting that fishing restrictions be maintained *despite* improving numbers. At the insistence of recreational fishers, the Massachusetts Marine Fisheries Commission in 1995 instituted more stringent size, daily bag, and commercial catch limits than were required under ASMFC guidelines (M. R. Ross 1997). The recovery continued, and the species returned to mid-1970s, pre-decline levels. The Hudson and Delaware stocks also improved (NMFS 1993). The fishery was declared fully recovered in 1994.

Striped bass biology undoubtedly contributed to recovery (Secor 2000). Stripers live for more than 30 years; hence, populations can persist despite multiple years of poor recruitment. One good year-class every few years may be enough to maintain a stock. Historically, dominant year-classes occurred about every 6–8 years, but none appeared for almost 20 years between 1970 and 1990 (see figure 10.16A). The management plan's focus on the 1982 year-class, which wasn't much larger than average, capitalized on a single cohort's potential to influence later population dynamics. Annual spawning and high fecundity, particularly in older individuals, likely contributed to high larval production and later recruitment. Although high fecundity is no guarantee of stock recovery (e.g., Sadovy 2001b), it certainly can't hurt, especially if conditions for larval and juvenile survivorship are favorable.

Few observers doubt the success of the recovery or the importance of the management plan, and the turnaround has been heralded as "the world's most successful recovery yet of a depleted fish due to strong management" (Safina 2001b, p. 441). However, stripers declined because of complex, interacting factors, and crediting the management plan as the sole cause of recovery is as misleading as blaming commercial fishing as the sole cause of their demise. The Chesapeake Bay is a notoriously polluted, disrupted ecosystem that has experienced catastrophic ecological change (J. B. C. Jackson et al. 2001). Water quality during the 1980s was periodically so poor that all striped bass larvae died in certain areas (Richards and Deuel 1987; NMFS 1993). Low oxygen; high temperatures; pH stress; sedimentation; and heavy metal, pesticide, and nutrient pollution affected survivors (Goodyear 1985; Price et al. 1985). Richards and Deuel (1987) emphasized that reproduction and recruitment hinged on reduced fishing mortality "and on water quality and environmental conditions in the spawning and nursery areas" (p. 64).

At about the time the management plan was being implemented, a consortium of agencies, citizen groups, businesses, schools, and institutions initiated a multipronged attack on water quality and habitat problems throughout the bay. This massive Chesapeake Bay Program involved more than 300 watershed organizations and led to substantial reductions in pollutant inputs and improvements in other water quality parameters, as did parallel efforts in the Delaware and Hudson systems. These efforts and their impacts can be neither ignored nor discounted. Improved reproduction undoubtedly contributed to and accelerated recovery, but when serious water quality problems remain, focusing solely on reproduction is misleading, just as captive propagation and stocking programs for endangered species are ineffective if causative agents of decline remain uncorrected.

Pacific Halibut: A Sustainable Fishery

Serial overfishing commonly follows on the heels of resource collapse. An exception is the fishery for Pacific halibut, *Hippoglossus stenolepis*, one of the first species to be carefully managed and regulated. The transition from unexploited to full exploitation was not glitch free, but the story shows what can happen when fishers and governments cooperate, use science to advantage, and develop and hold to a management plan that responds to changes in the fishery and its scientific basis.

Pacific halibut (figure 10.17) occur from central California to the Bering Sea and over to Russia and Japan; they frequent banks at depths of 10 to 1,200 m but are most common between 30 and 300 m. Heavier than even Atlantic halibut, they grow to over 2.7 m, 360 kg; males live 55 years, females 42 years. Almost all animals over 45 kg are female. Maturation occurs after approximately 7–8 years for males, 8–12 years for females (F. H. Bell 1978; IPHC 1998).

A commercial fishery for Pacific halibut developed in the 1880s when Atlantic halibut stocks were depleted.

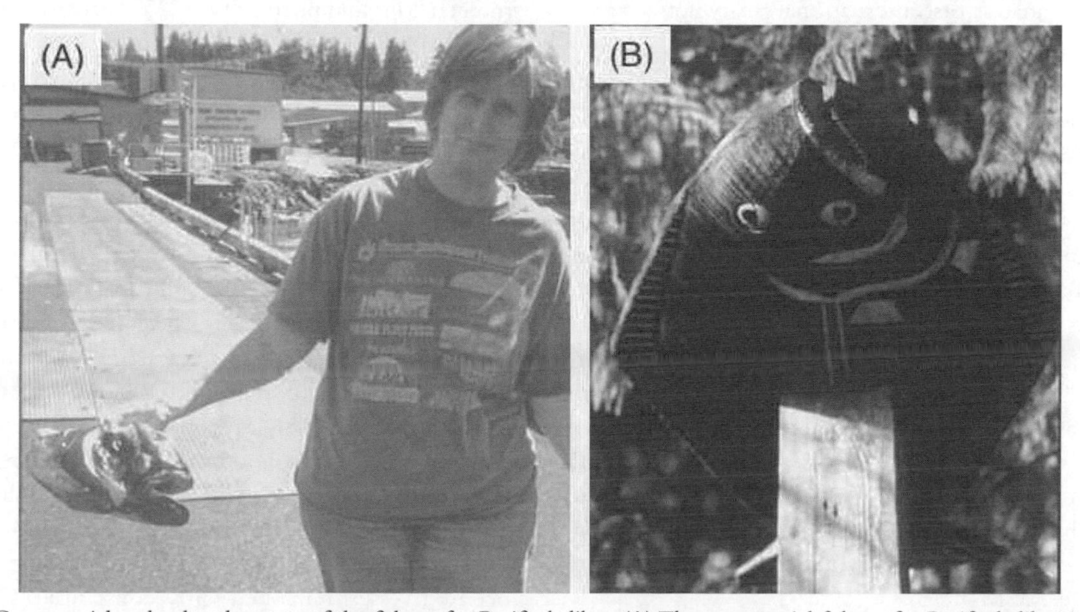

Figure 10.17. Commercial and cultural aspects of the fishery for Pacific halibut. (A) The commercial fishery for Pacific halibut is successful and well managed. High-value products include fillets, steaks, and "cheeks," the latter involving removal of the jaw muscles, as shown here at a fish processing facility in Bellingham, Washington. (B) Pacific halibut figured importantly in the culture of native peoples long before commercial fisheries came upon the scene. This large carving of a halibut sits atop a totem pole near Ketchikan, Alaska.

Completion of the transcontinental railway made a market for another, similar species feasible. But lessons learned from the overfished Atlantic stocks were applied relatively early in the new fishery. Pacific halibut were sought by both American and Canadian boats, fishing at least initially in nearshore regions along the Washington and southern British Columbia coasts. Local stocks were quickly depleted, forcing the fishery into offshore, nonterritorial waters, from northern California to the Bering Sea. Despite local depletions, the continual discovery of more distant stocks kept supply high, and the expanding fishery apparently glutted the market with fish, driving prices down. Controls were first instituted in 1914, including a closed winter season because of dangerous fishing conditions and the low quality of spawning fish. Restrictions were less management and more market oriented, created chiefly to stabilize prices.

The prominent biologist W. F. Thompson wrote three classic fisheries reports aided by accurate logs kept by the fishers (e.g., W. F. Thompson 1916; W. F. Thompson and Bell 1934). He was thus able to document the progressive decline of nearshore populations and the need for conservation measures. After some inevitable haggling, Canada and the U.S. in 1923 signed the Convention for the Preservation of the Halibut Fishery of the Northern Pacific Ocean. The Halibut Convention was the first international treaty focused on the conservation of a deep-sea fishery and Canada's first international treaty not cosigned by Great Britain.

The Convention provided for a three-month winter closure, regulated incidental catch during the closed season, and creation of the International Pacific Halibut Commission (IPHC), which it charged with collecting fishery-relevant data for developing management plans. The evolving plan first focused on CPUE as the metric of resource abundance, then on catch plus age data, to set catch limits that would permit the stock to build. Later, adjustments were made to allow for changes in the size structure of the populations and for bycatch mortality. Improved assessment methods and modified regulations resulted in a stable if not growing resource (T.D. Smith 1994; IPHC 1998). That the management approach was adaptive, changing in response to changing populations and increased knowledge, contributed in no small part to its success (e.g., SBI 1998).

Initially, the major approach was to limit catch by controlling season length. But as the fleet grew and technology improved, fishing "seasons" became shorter, creating a derby or Olympic fishery of less and less duration. The U.S. season was reduced from nine months to five months to two months to 24–48 hours to 12 hours to 10 hours. Management goals were achieved but at tremendous cost because the season or derby opened regardless of weather patterns. Sea states in the North Pacific frequently achieve horrendous conditions, with gale-force winds and 10-m seas not uncommon. Licensed boats had no choice but to fish when the fishery opened, and boats and lives were lost as a consequence (Emerson and Cox 2000). Finally, in 1995 (1991 in Canada), the U.S. implemented an individually transferable quota (ITQ) system for the Alaska fishery. The total allowable catch was divided by the number of licenses sold, and quotas were allocated to boats based on previous catches. Under ITQs, skippers could catch their allotment while fishing throughout the March-November open season, or they could sell (transfer) their quota to another licensed fisher. This stabilized and even drove up prices because supply became less concentrated. ITQs potentially reduce incentives to overinvest in boats and can lead to a reduction in fleet size as quotas are bought and sold. A drawback to the ITQ system is that it encourages highgrading and underreporting of catch (M. R. Ross 1997; IPHC 1998).

Results indicate that the plan is working. Landings in 1997 were at near record levels of 54,000 MT. The commercial, longline portion was 73%; recreational catch was 9%. Bycatch in other fisheries (mostly trawls or longlines for sablefish) accounted for 15%, a remarkable number if for no other reason than it entered into quota calculations. Personal-use fishing (subsistence, tribal treaty fishing) was about 1%, and mortality due to lost gear and discards was estimated at 2%. All four regional fisheries (Bering Sea, Gulf of Alaska, U.S. Pacific coast, Canadian Pacific coast) were fished at full capacity, but stock levels remained near their long-term potential yield. Exploitable biomass remained fairly constant over the decade of the 1990s at around 300,000 MT, down from a maximum in 1988 of 360,000 MT. The resource is considered to be in good condition (Low et al. 1999). Relatively constant landings from a healthy resource pursued in a less dangerous way occurred despite an overall decline in the size of the fishing fleet. Canadian-licensed vessels remained fairly constant at 435, whereas the U.S. fleet in 1997 was 2,000 vessels, down from 3,400 four years earlier (Low et al. 1999). Overcapitalization has been avoided because the cycle of local depletion and serial overfishing was prevented through good management.

(WHY) HAS FISHERIES MANAGEMENT FAILED?

Yes.

The history of fishing is a history of overexploitation, often despite abundant warning signs and predictions of imminent collapse. Even though MSY values, calculated from fisheries yields and population structure data, were frequently overestimated (because of unaccounted mortality), they indicated that current capture levels were unsustainable. Why then were targets frequently set higher than what a stock was thought to be able to sustain, and even these inflated targets exceeded?

Let me quickly acknowledge that I am unqualified to critique the methods and models of fisheries management. As an active, complex endeavor, fisheries science attracts capable researchers who are continually developing, assessing, and revising their field. I do not think that fishery science per se is to blame for the failures of fisheries management. The application and misapplication of that science, the political and socioeconomic pressures that affect its application, the unpredictability of marine ecosystems and complex biological entities, and the need or temptation to use incomplete or inaccurate data and ignore accurate data in formulating management plans all contribute to bad management decisions and failed management efforts. What's flawed isn't the science—it's the sociology. But it is often said that fisheries science is less about managing fishes than about managing people (e.g., Jentoft 1998).

Botsford et al. (1997) eloquently distilled the failure down to a conflict between fisheries science and sociology, complicated by environmental unpredictability: "The root cause of the lack of sustainability is the sociopolitically biased response of management to intrinsic uncertainty" (p. 514). They concluded that underlying management failure was (1) incessant sociopolitical pressure to maintain catches at high levels, and (2) the intrinsic uncertainty associated with predicting the fishing pressure at which stock collapse would occur (see also Buckworth 1998). As a result, fisheries are managed close to their collapse point instead of at some lower level of exploitation that provides insurance against collapse, a collapse made more likely by the inherent and unpredictable variability of oceanic systems and processes.

Exacerbating the overriding influences of sociopolitical pressure and unpredictability are recurrent problems associated with underlying assumptions, time lags between obvious trends and actions, and fatal shortcomings in mortality data.

Sustainability

Probably the most widely criticized assumption of fisheries science (and perhaps of conservation biology) is the concept of sustainability and sustainable yields. At first glance, managing fisheries sustainably appears complicated but not insurmountable. Data are needed on population size and age structure, individual growth parameters, maturation and reproductive output, stock distribution, landings relative to effort, natural mortality, and fishing mortality. From these traits, limits can be placed on seasons, size and age, sex, mesh, and catch that would produce a sustainable yield: "No one can deny that hypothetical animal populations can produce hypothetical maximum sustained yield . . . the dogma was this: any species each year produces a harvestable surplus, and if you take that much, and no more, you can go on getting it forever and ever (Amen)" (Larkin 1977, pp. 1, 3).

In practice, however, sustainability has seldom been achieved. Maximum sustainable yield—with its roots in the logistic growth equation and surplus production (see box 10.1)—has been criticized for a number of reasons. It puts fisheries at risk in part because the recommended harvest level is based on historical conditions that no longer exist. It also fails to account for spatial variability, differential responses in multispecies fisheries, climatic unpredictability, genetic variation, and depensatory population responses. The criticisms are not so much of the concept of sustainability, but of its inapplicability to complex ecological problems subject to social and political influence: "Fishermen vote; and once a person has become a fisherman, he can almost be counted on to vote against anyone who doesn't help him continue to be a fisherman and ensure him a decent standard of living" (Larkin 1977, p. 6).

History demonstrates that fisheries management decisions are subject to a one-way (upward) ratchet effect, whereby fishery *managers* succumb to political pressure and allow harvest to increase because fishery *scientists* "cannot specify with certainty that the next increase will lead to overfishing and collapse" (Botsford et al. 1997, p. 512). Political pressures seldom promote reduced yields, at least not from those who benefit economically from a fishery. The ratchet effect arises from accelerated investment in the fishery during good periods, often spurred by government subsidies, which increases pressure on the fishery and leads to overcapitalization. The effect continues because of little incentive to divest during poor periods—spurred by continued government subsidies—with resulting fisheries collapse. Thus, the ratchet effect leads to overfishing, which is by definition not sustainable (D. Ludwig et al. 1993; Botsford et al. 1997).

However, MSY is still used in fisheries management and remains informative in light of declining marine catches. A major criticism is that it is unsustainable and insufficiently precautionary, that it typically overestimates the catchable portion of a stock. Predictions that use MSY as a benchmark will therefore provide an overly optimistic portrayal of real conditions, and political pressure inevitably drives estimates of "maximum" higher. In this light, evidence that stocks have fallen below MSY means that they are overfished and MSY should have been lower. As figure 10.18 shows, all six stocks of managed pelagic Atlantic species declined, four to levels below the calculated MSY, suggesting miscalculation of what was a sustainable yield. Because of the frequent overestimations of abundance that have characterized MSY calculations, FAO—as part of its Agreement on Straddling and Highly Migratory Fish Stocks—proposed that MSY be used as a minimum standard rather than as a target for setting catch levels (Garcia and Newton 1997).

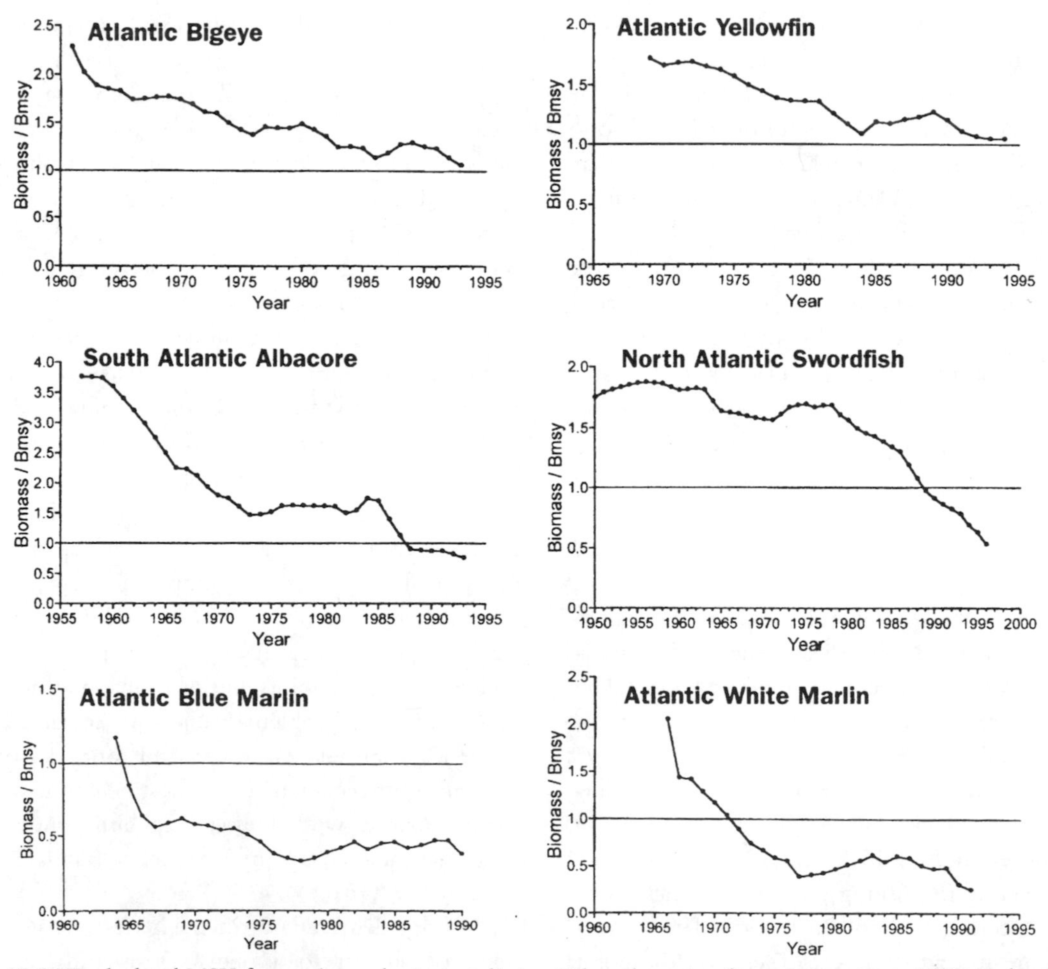

Figure 10.18. ICCAT-calculated MSY for species under its jurisdiction. When the ratio of population size to MSY is plotted over time, it is evident that all six stocks declined and four were fished beyond what was thought to be a sustainable biomass (1.0 would be MSY). From Speer et al. (1997); used with permission.

Stock-Recruitment Relationships, Fecundity, and Recovery

Controversy swirls around the issue of the stock-recruitment relationship, or what influence existing parental biomass has on future recruitment. MSY assumes that present and future stocks are related. However, four possible hypotheses apply: (1) an inverse relationship, resulting in compensatory recruitment; (2) a direct relationship, resulting in depensatory recruitment; (3) no relationship, resulting in random recruitment relative to stock; and (4) constant recruitment, a special case of no relationship (e.g., Frank and Leggett 1994). Although MSY assumes compensatory stock growth (hypothesis #1), the long-held view in many fisheries is #3, that little relationship exists and one cannot predict recruitment at time *t* based on how large the parental biomass was at time *t*–1. The constant-recruitment alternative, #4, has been applied even more commonly, a practice that Frank and Leggett (1994, p. 405) considered "dangerous because the underlying assumption . . . is that a stock capable of producing constant recruitment at any stock level cannot be collapsed by fishing." A lack of relationship has been used historically to argue against recruitment overfishing being a serious problem (e.g., Laevastu 1993); as justification for optimism concerning the resilience of exploited stocks (i.e., even if we drive a fishery down to relatively low levels, its chances for recovery remain at least statistically reasonable); and as justification for managing stocks on a regional rather than a local basis (e.g., Rothschild 1986; Wootton 1990).

However, recent analyses suggest that a positive relationship does exist between stock and recruitment in many fisheries (hypothesis #2). R. A. Myers and N. J. Barrowman (1996) analyzed spawner-recruit relationships for 364 data sets from 83 mostly marine species in nine families. They found an overall and case-by-case positive relationship, indicating that (1) higher recruitment occurred when spawner abundance was high, (2) lower recruitment occurred when abundance was low, and (3) populations below the median abundance level had lower recruitment than populations above it. Specific examples included such well-known depleted fisheries as bluefin tuna, Atlantic cod, and large sharks.

If hypothesis #2 holds in a significant proportion of fisheries, driving a fishery down is likely to keep it down. Hutchings's (2001b) analysis of 90 marine populations confirms that overfished stocks have a low likelihood of recovery. Following stock collapse, many populations do not recover, even if fishing mortality is greatly reduced, indicating that other factors, including climatic uncertainty and ecosystem integrity, can determine or at least influence recovery. A direct stock-recruitment relationship also undermines the notion that high individual fecundity can overcome low spawner abundance. Sadovy (2001b) showed that high fecundity was a poor guarantor of recovery. Highly fecund species (e.g., Atlantic cod) were just as much at risk as low-fecundity species. Fecundity evolves along with other ecological traits to accomplish replacement: Low-fecundity species such as lamnid sharks (with a total lifetime reproductive output of maybe 50 young) and high-fecundity species such as Atlantic cod (with hundreds of millions of eggs over a lifetime) have both evolved to replace, on average, one male and one female (Hutchings 2000b). In American shad, *Alosa sapidissima*, very different, genetically based life histories in northern populations (late maturation, multiple spawnings, low fecundity) versus southern populations (early maturation, one lifetime spawning, high fecundity) result in similar mean lifetime fecundities (Leggett and Carscadden 1978).

These converging phenomena put the lie to the myth of fecundity. Long life and slow maturation make large, valuable species vulnerable to overfishing regardless of fecundity, especially if they are part of intense, mixed-species fisheries. Rebuilding an overfished stock requires allowing biomass to build up—through reduced fishing, protected habitat, and ecosystem management—rather than hoping passively for high fecundity to produce fortuitous increases in larvae and an occasional big year-class.

Maybe we didn't get it wrong originally, but conditions have changed, making the stock-recruitment relationship stronger. Depensatory population growth may have become more common as more stocks have been driven down. That highly depleted bluefin tuna, cod, and sharks show depensatory growth may reflect the effects of lowered population size. Earlier studies described large populations before overfishing was so widespread. Our sampling universe consists now of stocks that have become so small that a direct relationship has emerged (see R. A. Myers and Barrowman 1996 for alternative explanations). Regardless, behavior of populations at low abundance is one of the major sources of uncertainty affecting our ability to predict what constitutes a sustainable harvest (Botsford et al. 1997).

Warning Signs Ignored

Failures to manage stocks sustainably despite mounting evidence can be attributed to bad data, a lag between data collection and application of management, and economic and political pressure to minimize fishing restrictions. However, fisheries managers often had sufficient information but failed to curtail or were prevented from curtailing fishing. M. R. Ross (1997, his table 12.1) gave a telling account of relevant chronologies for two overfished New England species, haddock and yellowtail flounder, based on annual reports from the Northeast Fisheries Science Center:

- *Haddock. 1983*, "stock biomass will continue to decline given anticipated levels of fishing mortality"; *1985*, "stock biomass dropped to very low levels . . . and is expected to decline even further"; *1987*, "the stock will probably decline further in 1987–1988"; *1989*, "The 1988 autumn survey index . . . is only 20 percent of the 1979 value"; *1991*, "The low abundance of incoming year-classes suggests that . . . [abundance] will remain at record or near record low levels"; *1992*, "abundance and biomass are at all-time lows."
- *Yellowtail flounder. 1983*, "stock biomass should continue to decline . . . if fishing mortality remains high"; *1985*, "indices of abundance have declined sharply"; *1987*, "fishing mortality remains substantially above [the maximum fishing mortality that can be sustained]"; *1989*, "the population is severely overexploited"; *1991*, "rebuilding . . . will require a major reduction in fishing mortality and several good years of improved recruitment"; *1992*, "the stock is still at a very low level and is comprised of few age groups."

Ross's analysis indicates that, when fishing finally came under control, it often involved too little management applied too late. Catch limitations, whether they involved time, size, area, or numerical restrictions, were imposed in reaction to obvious declines but represented compromises between the necessary and the politically expedient. By the time such regulations were imposed, stocks had fallen well below sustainable levels, and allowable fishing rates were often set higher than what would allow recovery (particularly because a significant portion of the catch was illegal and unreported). This is not just a recent management problem. Belknap (1792), characterizing fisheries regulations of the late 18th century, observed, "After the mischief was *done*, a law was made against it."

Ironically, had stocks been allowed to rebuild earlier to a spawning biomass capable of producing the maximum sustainable yield (B_{MSY}), permissible catches would actually exceed the quotas that had to be imposed when stocks were depleted but still being fished. It is, quite simply, a matter of principle and interest. A higher principle means a higher return. B_{MSY} represents the principle at which a higher amount of interest can be withdrawn, and that interest is more than what can be withdrawn when stocks have stabilized at a depleted level. Drawing on the principle is a bad business practice, yet the exploiters of many stocks insist on dealing only with short-term gains. If resource users recognized that they actually had a long-term investment in the stock and acted accordingly, they would ultimately reap larger benefits. It is a repeated, classic case of reactive rather than precautionary management of an abused common resource (e.g., Hardin 1968).

Unaccounted Mortality

A major cause of failure in fisheries management is unaccounted mortality. Management practices depend on age-specific mortality data that are accurate and complete. Fishing mortality occurs in addition to natural mortality, and both must be known to determine allowable future catch. Reported landings are often misleading because they deal only with "nominal catch," which represents fish brought to the dock for sale or distribution (Pitcher 1995). Nominal catch excludes fish discarded due to highgrading, bycatch in targeted and nontargeted fisheries, fish injured by gear but not captured (noncatch waste), losses from ghost traps and nets, and illegal or misrepresented landings and poaching.

Bycatch—the incidental capture of nontargeted organisms—usually goes unreported (Pacific halibut being an exception). It often involves juveniles and small species discarded because they are illegal or unprofitable (see chapter 11). Bycatch is generally shoveled overboard dead or dying (Carl Safina prefers the term "bykill"), although in many developing countries bycatch is a bonus utilized for human or animal consumption (Britton and Morton 1994; Kennelly 1995). Highgrading, sometimes called "economic discards," is a special form of bycatch whereby less profitable but legal fish are discarded to make room for more valuable

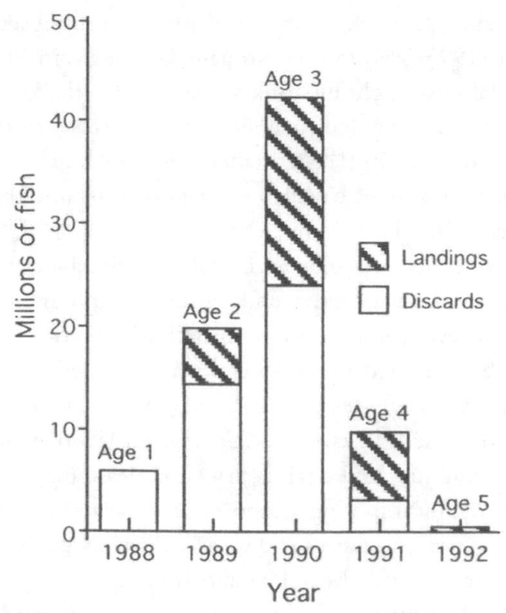

Figure 10.19. Bycatch of southern New England yellowtail flounder. Between 1988 and 1992, the total catch of the 1987 year-class amounted to 77 million fish, but 46 million (60%) were discarded as undersized bycatch. From Alverson (1997); used with permission.

species and sizes. In U.S. fisheries, the most common bycatch discard is so-called regulatory bycatch, which means undersized fish, prohibited species, or quota exceedance catches (S. K. Brown et al. 1998). Juveniles of commercial species, "which, if left to mature, would most likely produce significantly higher weight yields" predominate among discarded catch (Alverson 1997, p. 120; figure 10.19). These reductions of potential yield can be commercially significant, as in the case of plaice in the North Sea (Berghahn and Purps 1998). Capturing juveniles can also threaten survival of species, such as imperiled barndoor skates in the North Atlantic (Casey and Myers 1998). By some estimates, bycatch has contributed to declines among 42%–48% of ESA- and IUCN-listed marine and diadromous species (Kappel 2005).

Gear-induced mortality affects fish that escape capture but die from injuries, stress (chapter 7), acute temperature change, or exhaustion from fighting or avoiding the gear (e.g., Olla et al. 1997, 1998). Stress and exhaustion can reduce growth and reproduction, which become particularly significant in intensive fisheries that involve repeat fishing effort in an area. Another gear-related factor is habitat mortality, exposing fishes to predation because the gear (especially trawls) destroys important habitat (see chapter 5). Such habitat degradation can lead to reduced ecosystem productivity and again result in reduced growth and reproduction of target and nontarget fishes.

Ghost gear mortality describes the well-documented problems associated with lost fishing gear. Pots and gill nets are often made of long-lasting plastic and nylon and keep on fishing when lost. Fish killed by such gear are not counted. Depending on the fishery, annual gear loss rates vary between 2% and 30% for traps and about 20% per year for high-seas gill nets (e.g., Dayton et al. 1995). The time course of ghost fishing also varies. Gear lost in storms or near shore is likely to be destroyed quickly, and nets lost in clear, shallow water will become overgrown with fouling organisms, making them more visible to fishes, weighing them down, and generally decreasing their potential to continue fishing. However, in favorable conditions, pots and traps can fish 2 to 15 years after being lost; gill nets are known to fish for up to 7 years and maybe for a decade (S. K. Brown et al. 1998; Goñi 1998). More than 2,200 lost nets were estimated to occur throughout a 220-km^2 area of New England, and calculations put the average number of lost crab pots near Kodiak, Alaska, at 42 pots/km^2. Approximately 20%–30% of the 30,000 km of drift nets set in one North Pacific fishery were lost each year, and a grapnel survey of the bottom of Georges Bank yielded 1.2 ghost nets per tow, totaling 341 actively fishing ghost nets (Jennings and Kaiser 1998). More than 111 MT of derelict gear and debris were removed from the remote Northwest Hawaiian Islands in 2003 (NMFS 2003).

High-seas fisheries deploy millions of kilometers of longlines and drift nets annually, meaning untold thousands of kilometers of ghost gear take a staggering, unaccounted toll. Nets continue to fish regardless of maintenance, and traps fish without rebaiting because fish trapped become the new bait in a continued cycle of capture, death, and attraction. Ghost trap fishing may have added 30% to the fishing mortality for sablefish off British Columbia (S. K. Brown et al. 1998); sablefish are fully utilized, and stocks are below levels necessary to produce long-term sustainable yields. Many fisheries require that pots be constructed with break-away panels that will decay and create an escape hole large enough to accommodate anything that can enter the trap.

An additional contributor to unaccounted mortality is IUU fishing. Nations left out of or unwilling to join international fisheries organizations, commissions, and agreements are not subject to agreed-upon catch restrictions and quotas. For example, Taiwan—with *reported* fisheries production of 1.35 MMT in 2000 (www.fa.gov.tw)—is not officially bound by UN regulations, including FAO-sponsored agreements. ICCAT includes 20 member nations but regulates a bluefin tuna fishery exploited by at least 43 (known) nations (ICCAT 2001a). As detailed above, IUU landings of Patagonian toothfish may be 0.1 or as much as 9.5 times larger than the reported catch (again depending on source). FAO (2000b) gives a general number of up to 30% IUU fishing in "some important fisheries." Most important, "when IUU fishing is unchecked, the system on which fisheries management decisions are based becomes fundamentally flawed" (FAO 2000b). Attention to the issue of IUU fishing is increasing, including formulation of an FAO International Plan of Action (see www.fao.org and www.oceanlaw.net).

Calculating and Reducing Unaccounted Morality

Estimating how all these unknown agents contribute to and affect world fisheries catch remains a challenge. FAO (1995b) calculated that discarded biomass in the late-1980s was 25%–30% of nominal catch. Alverson et al. (1994) and Alverson (1997), the acknowledged authorities on this subject, estimated that discards ranged conservatively between 18 and 40 MMT annually, which is 15%–50% of nominal catch (several sources take a median number and report annual discards at 27 MMT). Shrimp and prawn trawling account for about one-third of all discards (Alverson 1997). Northern Australia's prawn fisheries catch 411 fish species. Southeastern U.S. shrimp fisheries take 105 finfish species and numerous sharks and skates; shrimp trawlers in the Gulf of Mexico alone discard nearly 10 billion individuals and 180 million kg of incidentally captured fishes annually (Nichols et al. 1990; Stobutzki et al. 2001). S. K. Brown et al. (1998) estimated that bycatch discards in Atlantic and Gulf of Mexico fisheries have negative effects on 40%–63% of overfished stocks.

Regardless of data uncertainties, significant numbers of fish that are not landed die as a result of fishing. Mortality rates vary widely, depending on bycatch species, life history stage, gear type, gear deployment methods (soak and trawl times, haul rates), temperature, and season, among other factors (e.g., Berghahn et al. 1992; M. J. Kaiser and Spencer 1995). Mortality estimates range from 0% to 100%: Halibut caught by hook and line suffered 33% mortality, whereas trawled halibut suffered 78% mortality (Davis and Olla 2001). Mortality rates are high enough that stock assessment calculations generally assume that all discarded fishes die (Mesnil 1996).

The need to reduce bycatch is obvious. Reductions can be achieved with incentive and disincentive programs that ultimately reward efforts at minimization; by gear modification such as colored mesh panels, increased mesh size, and bycatch reduction devices (BRDs) with built-in escape holes and funnels (Isaksen and Valdemarsen 1994); by use of revival techniques and devices prior to release (e.g., Farrell et al. 2001); by imposing seasonal and areal restrictions to minimize bycatch of sizes and sexes known to aggregate at particular times and places; by establishing quotas that close fisheries when bycatch reaches a certain level, independent of landable target quotas; and by increased use of independent observers on ships (see S. J. Hall 1999). Norway's "no discards" policy requires all catch to be brought ashore, thus providing incentive to avoid fishing practices that fill the boat with low-value bycatch species (M. A. Hall et al. 2000).

Bycatch has high visibility, becomes more important as more fisheries approach their upper limit, causes conflicts among fishers because one person's bycatch is someone else's targeted catch, and will receive increasing attention and action, if for no other reason than "it gives fishers a bad public image" (M. A. Hall et al. 2000, p. 204). The U.S. Magnuson-Stevens Fishery Conservation and Management Act, reauthorized in 2006, requires that all federal fishery management plans include provisions for minimizing bycatch (www.nmfs.noaa.gov/sfa/magact/).

If bycatch adds 33% and IUU landings 30%–200% to the reported catch (ICES 1995; Alverson et al. 1997), 60%–230% more fish are killed globally than are "officially" caught and reported. And these numbers do not include recreational and subsistence catch and discards. By some estimates, 90% of the world's fishers are engaged in small-scale artisanal and subsistence activities that go largely unrecorded by FAO (Bailey 1988). When all this unaccounted mortality is left out of fisheries statistics, any management scheme based on MSY or any measure of "sustainability" is doomed to failure. This is the fisheries equivalent of calculating a budget without including a substantial proportion of expenses; it is a flawed way to manage resources. These additional, largely unknown con-

Box 10.5. Promoting Sustainability, Conserving Biodiversity, and Reducing the Probability and Impacts of Overfishing

GENERAL APPROACHES

- Ecosystem-based management—focus on sustainable ecosystems rather than sustainable fisheries (NRC 1999b; Pitcher and Pauly 1998)
- Emphasis on multispecies rather than single-species management (Roberts 1997)
- Adaptive management that is responsive to changing ecological, climatic, sociological, and economic conditions (Bundy 1998)
- Proactive, precautionary management in the face of uncertain data on stock size, population trends, exploitation rates, and the likely impacts of management actions (Botsford et al. 1997)
- Emphasis on sound science, best available science, and improvement of that science to reduce uncertainty in collection and interpretation of statistics, particularly on IUU and unaccounted mortality

ACTIONS (SEE M. R. ROSS 1997 FOR A GENERAL REVIEW)

- Identify critical habitat and rehabilitate degraded critical habitat.
- Establish permanent reserves, no-take zones, and MPAs to create safe havens where fish can grow large and spawn (Sumaila 1998; NRC 2001; see chapters 5, 12).
- Develop time and area closures to protect spawning aggregations, migratory phenomena, and predictable concentrations in nursery areas. Enact closures and moratoria in advance of collapse, as a precautionary rather than a reactive tactic. Reduce bycatch and discards via development of BRDs for imperiled species and sensitive life history stages.
- Enforce mesh size restrictions, recognizing their limitations.
- Establish rational minimum size restrictions to reduce growth overfishing and maximize reproductive output; set size limits above minimum size at reproduction to reduce recruitment overfishing.
- Promote ownership of the resource rather than exploitation of common resources; employ ITQs, licenses, and limited entry to reduce overcapitalization.
- Initiate and fund buyout programs to reduce fishing capacity (Buckworth 1998).
- Promote enforcement through observer programs while encouraging self-policing within fisheries to reduce IUU fishing and unaccounted mortality.

MECHANISMS TO FACILITATE APPROACHES AND ACTIONS

- Promote public education to develop widespread support and community participation in management actions (Starnes et al. 1996; M. Williams 1998).
- Emphasize stakeholder involvement, cooperation, and inclusion—of fishers, processors, conservationists, NGOs, governmental agencies, academics—in planning, monitoring, and enforcement to counter political pressure with political pressure (see www.pmcc.org for a good example).
- Enlist local and artisanal fisheries and communities to capitalize on empirical knowledge about local resources (Ruddle and Johannes 1985; Johannes et al. 1993).
- Promote green fisheries by applying consumer power and enlisting citizen action through organizations such as Seafood Watch, the Marine Stewardship Council, and product certification programs (Peterman 2002).
- Legislate regulations rather than relying on voluntary compliance (Welcomme 2001).
- Maximize participation in international agreements for straddling stocks, migratory fishes, and boycotts on IUU fishing (Welcomme 2001; www.fao.org).

tributors to mortality are particularly troubling if estimates for the maximum productivity of the world's oceans—placed at 80–100 MMT—are accurate. We are already taking 80–100 MMT annually. Adding bycatch and IUU landings means that fisheries have exceeded world sustainable harvest levels for at least two decades. Inaccurate landings data are inconvenient and frustrating if all we are trying to do is assess productivity. They could be tragic if they allow us to outstrip the planet's capability to produce food for us and other species.

SOLUTIONS TO OVERFISHING, STEPS TO SUSTAINABILITY

Every failed fishery teaches specific lessons, whose sum has been incorporated into a body of knowledge applicable to new situations. The ultimate goals in fisheries should be prevention of overfishing through sustainable exploitation, and conservation of biodiversity. The latter goal, widely ignored in the past, is receiving increased emphasis as workers recognize that diminishing stocks are embedded in and dependent on functioning ecosystems (e.g., Bohnsack and Ault 1996; Ducroty 1999). Sustainability accomplishes both economic and environmental objectives and can be achieved via a general process incorporating approaches, actions, and mechanisms, most of which are well known (box 10.5). These components are widely discussed in the fisheries literature cited here and in the next chapter (see especially Pikitch et al. 1997; Pauly et al. 1998b; NRC 1999b).

Past Management Actions in Response to Overfishing

Overfishing on a large scale occurred first in the North Atlantic during the first half of the 20th century. A short respite during World War II was followed by fisheries expansions and overfishing in the North Pacific (1950s), eastern Atlantic and eastern Pacific (1960s), Indian Ocean and Antarctic (1970s), and South Pacific and southwest Atlantic (1980s) (Garcia and Newton 1997). Solutions have been well known, requiring management and regulation. Implementing and enforcing such controls have proven difficult because of the international nature and mixed-species composition of many fisheries, extensive movements of species, the common nature of the resources, the inexact nature of fisheries data, climatic uncertainty, social and economic constraints, and resistance to regulation (including of unlawful behavior).

Because these intervening forces mitigate against proactive management, fisheries are and always have been managed in reaction to declines, with limited success. Effort has not been lacking. The science was in place long before it was put in practice. As a reaction to declining stocks, major national and international management agencies were created by the early part of the 20th century (e.g., U.S. Fish Commission, 1871; International Council for the Exploration of the Seas, 1902), and the pioneers of fishery science—S. F. Baird, W. Garstang, C. H. Gilbert, W. C. Herrington, J. Hjort, C. G. J. Petersen, O. E. Sette, H. Thompson, W. F. Thompson—were actively studying the world's major fisheries (T. D. Smith 1994). By the early 1940s, most of the fundamentals were known, including Graham's Great Law of Fishing, which posited that catches do not necessarily increase in proportion to effort: "As the fishing power increases the stock falls, but the yield at first rises. Later . . . it ceases to rise, and that creates the main problem of fishing. . . . After a certain point the total yield of a fishery fails to increase any more, whatever the fishermen do. This is the key to the history of fishing, all over the world" (Graham 1943 in T. D. Smith 1994, pp. 230–31).

Despite this early recognition and understanding of declining landings regardless of increased effort, concerted management programs directed at marine fisheries came into existence only in the mid-20th century (NRC 1999b); even then most important stocks continued to decline.

It is important to remember that fisheries can and have recovered. Recovery has occurred less from the actions of management than from larger sociopolitical forces. When naval activities in the North Atlantic during both world wars caused understandable reductions in fishing effort, the numbers and sizes of many North Sea stocks increased (NRC 1999b). Total cessation of fishing, however, is not necessary for recovery. Large numbers, broad geographic ranges, and high fecundity (combined with favorable growing conditions) promote resilience in marine stocks, meaning that reduced overfishing and protection of habitat can lead to stock recovery for many species, although a decade or two may be necessary (Safina 2001a; see Sadovy 2001b on problems with high fecundity, Hutchings 2000a on time lags). Knowledge is necessary but not sufficient for recovery; patience is also required. Government intervention in the form of alternative income for displaced fishers, boat and license buyouts, and other subsidies would help make patience a more realistic virtue.

General Approaches

Although most previous efforts at managing fisheries sustainably have failed (e.g., D. Ludwig et al. 1993), abandoning such efforts leaves few alternatives besides free-market exploitation. Few (aside from some economists) would argue that uncontrolled exploitation is a better solution to the crisis facing global fisheries: "The track record of leaving natural resource management up to 'market forces' is far worse than attempts at sustainability" (Meyer and Helfman 1993, p. 569). At national and international levels, emphasis on sustainability is, if anything, growing (see OECD 1997; FAO 1999a; Charles 2001), although Pitcher and Pauly (1998) emphasized caution in applying a goal of sustainability if only ecosystems in a depleted condition far removed from their primal, pre-exploited state were sustained.

Achieving sustainability requires new thinking and new approaches. The majority view is that the ideal (if not idealistic) solution lies in ecosystem management, that "sustaining fishery yields will require sustaining the ecosystems that produce the fish" (NRC 1999b, p. 14; see also Jennings 2005, Rice 2005). The simplified argument for ecosystem management (expanded in chapter 11) is that (1) fishes occur as part of multispecies complexes, embedded in a web of trophic and nontrophic interactions; (2) overfishing affects not only the target species but other components of an ecosystem, including nontarget species and habitats; (3) ecosystems provide recreational, subsistence, commercial, and aesthetic goods and services beyond landings of specific stocks; (4) disruptions of ecosystem structure and function feed back on the dynamics of target and nontarget species; and, therefore, (5) by focusing on management that deals with an ecosystem's structure and function, rather than on a target species, we are more likely to achieve sustainability of multiple components including the embedded target species. Ignoring the ecosystem, as we have traditionally done, leads to resource depletion and system collapse.

NRC (1999b) discussed specific elements of an ecosystem management approach, including monitoring ecosystem (rather than just individual species) properties, monitoring the effects of management on sociological phenomena including economic and bureaucratic interactions, application of ecosystems science (e.g., food web dynamics, functions of MPAs), and application of the precautionary principle. NRC cited recent work involving the Large Marine Ecosystem Approach and the UN Agreement on Straddling and Highly Migratory Fish Stocks as models. Reference to the precautionary principle—which is basically the risk-averse notion that doing something about an apparent problem is less likely to lead to disaster than not doing something—appears repeatedly in discussions of ecosystem and adaptive management (e.g., Botsford et al. 1997). A precautionary approach with its risk-averse credo is also at the heart of the FAO Code of Conduct for Responsible Fisheries (see L. J. Richards and Maguire 1998). Adaptive management recognizes that fisheries are dynamic and complex and that management plans have to change with changing conditions and information (Botsford et al. 1997).

Actions

> If engineers adopted the same approach [as fishery managers] by building bridges strong enough for just the average flow of traffic, they would soon share the reputation that fishery managers enjoy.
>
> —Callum M. Roberts, 1997

MPAs, reserves, and no-take zones, discussed in more detail in chapters 5 and 12, are the most effective, proven actions available for rebuilding depleted stocks or preserving existing stocks (NRC 2001). Reserves provide a place where larvae can settle, grow, achieve reproductive condition, spawn repeatedly, and seed or spill over into nonreserve areas (Roberts, Halpern, et al. 2001). In combination with reduced fishing effort, they provide insurance against the many uncertainties that surround fisheries management. They address underlying causes of decline, whereas reduced fishing attacks the problem after biology and the environment have acted on stock and recruitment (e.g., Roberts 1997). Not all reserves work equally well or all the time (Boersma and Parrish 1999; NRC 2001). When MPAs don't incorporate fishing reductions, fishing effort in nearby areas may even increase; if local resource users are not included in MPA planning, success is unlikely (Baum et al. 2003; Christie et al. 2003). Nonetheless, MPAs provide a general management tool that operates at the ecosystem level and has been recommended as a solution to local, regional, and global problems (e.g., Pauly et al. 1998; Lipcius et al. 2005).

As MPAs grow in popularity, so does opposition from people who view setting aside such areas as a denial of access to resources. In 2004, a Freedom-to-Fish Act was introduced into the U.S. Senate by recreational fishing interests; it would limit the authority of agencies to create no-take zones, with specific amendments to the Magnuson-Stevens Conservation and Management Act and the National Marine Sanctuaries Act (see www.joincca.org). The Freedom-to-Fish Act includes requirements to make closed areas "as small as is scientifically necessary" to achieve conservation goals, to reopen closed areas once recovery goals are met, and to demonstrate that recreational fishing is a cause of the problem and prove that more traditional conservation measures are inadequate before recreational fishing can be curtailed. The "small as necessary" provision stands in direct conflict with a general dictum of conservation biology that reserves should be as large as possible (e.g., Meffe and Carroll 1997).

Proving that recreational fishing is directly responsible for the demise of a fishery is irrelevant: If a stock is depleted and needs to be closed, capture of individuals from that stock should be limited until the stock recovers. Dispensation for recreational fishing can only delay recovery. Insisting that closed areas reopen quickly seems reasonable, although "recovery" lacks a universally accepted definition. Efforts at ecosystem management have already been affected by legal actions arising from the freedom-to-fish movement. An area closure to protect imperiled gag grouper, *Mycteroperca microlepis*, habitat off Florida was rescinded in June 2001 on the grounds that it unfairly curtailed fishing for other species (closure would have restricted trolling in surface waters above grouper habitat because grouper, which usually live near the bottom, will take surface lures; www.ccatexas.org). Support for the Freedom to Fish Act is not unanimous among sport fishing groups (i.e., Bacher 2000).

Other widely applied actions are less effective than they would first appear. Minimum fish size and minimum mesh size regulations are examples. Almost all management schemes include some minimum size below which retention is prohibited. This size is usually either the smallest size at which the species is known to reproduce or, slightly better, the average size at which reproduction occurs, with or without regard to gender. For females, "minimum size" applies to only a fraction of the population; many more fish do not mature until they are older and larger. Importantly, as shown for cod and striped bass (NMFS 1993) and assumed for many species, larger fish produce not only more eggs but also better eggs. The contribution of older fish to recruitment is disproportionate to their numbers but directly (and perhaps exponentially) proportionate to their size or age. Also, different size females may spawn at different times, distributing spawning over a longer period and increasing the chances that some larvae will encounter favorable growing conditions.

Success of the striped bass management plan resulted in part from a stipulation that fishing be restricted until 95% of the females in the 1982 cohort had reached maturity. This created variation in the size of spawning fish, allowed for spawning of older and larger fish, extended the spawning season, and ensured more reproduction than a traditional minimum size designation would have. Size restrictions that take these biological complications into account and allow most fish in a population an opportunity to spawn are preferable. No-take refugia automatically protect large individuals and allow for their disproportionate contribution to the population. Recreational fisheries address this issue by imposing slot limits that protect both smaller and larger fish (see chapter 11, M. R. Ross 1997).

Net mesh size restrictions would seem guaranteed to reduce bycatch of immature fish: Larger mesh sizes allow larger fish to escape and increase the likelihood that small fish will slip through. However, this filtering effect works only in the initial fishing period before a net fills with fish. Many small fish entering a net later become trapped among the larger fish and invertebrates and never encounter the netting. Mandating shorter trawling times and more frequent net hauling reduces this problem (e.g., Farrell et al. 2001), but such regulations are unpopular among fishers and difficult to enforce without an extensive observer program.

Agency and Individual Approaches and Actions

Public education, stakeholder involvement, market forces, and citizen action all come together in such issues as certification programs that promote so-called eco-friendly, certified, and green fisheries. The best-known efforts in the U.S. are the dolphin-safe tuna and Give Swordfish a Break campaigns and their larger relative, the Seafood Watch Program. In the UK, a Thames blackwater herring fishery is ecocertified, as are Alaskan commercial salmon fisheries (Peterman 2002). The swordfish boycott and related lawsuits to reduce bycatch have led to ICCAT and governmental actions that reduced catch quotas, bought out and retired fishing licenses, and created nursery reserves along the Atlantic and Gulf coasts. Limited long-lining for swordfish in nursery reserves will also reduce pressure on tuna stocks (Safina 2001b).

The dolphin-safe tuna campaign grew out of concern over bycatch of dolphin in the eastern Pacific fishery for yellowfin tuna (see chapter 11). Pressure from consumers prompted tuna canners to certify that their fish were not captured under dolphin schools. Subsequently, the U.S. Congress passed legislation prohibiting importation of tuna caught with dolphins, but this governmental action "merely followed where the marketplace had already gone" (Safina 2001b, p. 458). Importation restrictions of the dolphin-safe program conflict with provisions of World Trade Organization rules, a problem that voluntary campaigns such as the swordfish boycott avoid (Peterman 2002).

Tuna and swordfish boycotts have inspired more general consumer action programs that assess the environmental impacts of fisheries in terms of sustainable catch rates, bycatch, socioeconomic justice, and the broader impacts of capture and culture methods. The Monterey Bay Aquarium, among others, produces a downloadable Seafood Watch chart with regional and national lists of species to favor or avoid when consuming seafood (www.mbayaq.org; the Marine Conservation Society in the UK, www.mcsuk.org, produces an expanded "Good Fish Guide," but it costs £10). The ultimate goal in an eco-certification effort is to label products that have met rigorous criteria, thus ensuring both sustainable fishing practices and economic viability for the fishers (Peterman 2002). The UK-based Marine Stewardship Council has developed standards for sustainable and well-managed fisheries. It certifies and labels products that have been produced via environmentally responsible fishery management and practices.

Other forms of citizen action include buying up but not using fishing rights, an outgrowth of ITQs and other licensing schemes (see chapter 14). In fisheries for individually valuable species, such as bluefin tuna and Patagonian toothfish, documentation and tracking of each fish discourages poaching and illegal trade. It is widely felt (at least among conservationists) that ICCAT did little to reduce overfishing of bluefin until pressure was brought by NGOs and conservation groups (Safina 2001b).

National and international regulations are too numerous and complicated to summarize responsibly (see M. R. Ross 1997; Welcomme 2001; Rieser et al. 2005). In the U.S., a series of laws, most notably the Magnuson-Stevens Fishery Conservation and Management Act of 1976 and the Sustainable Fisheries Act of 1996, contain provisions for defining and reversing overfishing. When a stock drops below 25% of virgin biomass, it is classified as overfished and is to be rebuilt. Provisions also exist for setting up regional fisheries management councils; identification of essential fish habitat for spawning, nursery, and migration routes for all species that have management plans; and bycatch minimization requirements. The bycatch provisions prompted lawsuits to reduce the catch of undersized swordfish and helped create the protected nursery region on the U.S. East and Gulf coasts (Safina 2001a).

Among the many important international agreements, treaties, and conventions, the most far-reaching are

Table 10.1. Comparing industrial and artisanal marine fishing

	Industrial/corporate fishing	*Artisanal fishing*
No. fishers employed	ca. 450,000	>12,000,000
Fish caught for human consumption (MMT)	24	20
Capital cost/job/vessel	\$10,000–\$100,000	\$100–\$1,000
Bycatch discarded (MMT)	20	1
Fish reduced to meal and oil (MMT)	19	ca. 0
Fuel oil consumption (MMT)	10–14	1–2
Fish landed/MT fuel used	2–5	10–20
Fishers employed/\$1 million invested in vessels	10–100	1,000–10,000

Source: Modified from P. J. B. Hart and J. D. Reynolds (2002).
Note: MT = metric tonnes, MMT = million metric tonnes.

the UN Convention on the Law of the Sea (UNCLOS) of 1994, the UN Agreement on the Conservation and Management of Straddling Fish Stocks and Highly Migratory Fish Stocks of 1995, and the FAO Code of Conduct for Responsible Fisheries of 1995 (see Awaluddin 1999). General goals include commitments by signatory nations "to sustain their national fisheries, cooperate to sustain international fisheries, address the problems of overcapacity and bycatch, [and] base management on sound scientific information" (NRC 1999b, p. 93). The FAO Code of Conduct begins with the premise that a precautionary approach should be taken in fisheries management (i.e., that negative impact can be assumed until science shows otherwise), thereby shifting the burden of proof from managers to fishers. These agreements, of course, depend on participation, cooperation, and compliance, all often voluntary. Most significantly, they reflect a widespread (although not universal) recognition that fisheries resources have been overexploited, that old practices of unfettered exploitation are unsustainable, and that new, more cautious management approaches are required

CONCLUSION

Reversing the trends of the past half century—switching from overexploitation to sustainable use—requires a dauntingly large set of specific and general actions. Specific actions can focus on economic return, an approach the industrialized world should find easy to grasp. Garcia and Newton (1997) offered suggestions for decreasing net economic losses to commercial fishing, especially losses resulting from overcapitalization and overcapacity, which create an annual bottom-line imbalance of between $20 and $45 billion. Profitability, or at least economic sustainability, can be achieved through reducing capacity by 25%–53%, reducing operating costs by about 43%, and increasing fish prices by 71%. Cost reductions would come from replacing high-tech, distant-water fleets with local and artisanal fishing, which also maximizes employment. (see table 10.1). In nearshore, populous regions—which encompass 90%–95% of all fishers and perhaps 40% of marine fish caught for human consumption—a switch to local fishing would increase efficiency, although, admittedly, it would take little pressure off the resource (Bailey 1988, 1997b). The socioeconomic and ecological impacts of large-scale and small-scale fishing differ in many significant ways; with respect to providing protein and employment and reducing environmental costs, small-scale fishing improves many conditions. Absolute numbers in table 10.1 may change over time, but the relative values should hold.

Overexploitation is self-defeating because it leads to massive net economic losses. Reducing fleet capacity would lessen overfishing, allow stocks to rebuild, and actually increase total catches by 20% to 60%. Priority should be given to replacing the practice of promoting fishing fleet growth with fleet reduction via reduced subsidies and direction of funds toward reduced effort, such as buyback plans and license retirement. Such efforts would return the resource to safer and more economic levels (Garcia and Newton 1997). But this goal runs smack into the hard reality that "for a variety of understandable political, social, and economic reasons, the Hilborn-Walters Principle holds: 'The hardest thing to do in fisheries management is reduce fishing pressure'" (Reynolds and Jennings 2000, p. 242).

Our ability to manage sustainably depends on (1) changing our approach from single-species to flexible, multiple-species and ecosystem management; (2) recognizing that influential large-scale physical factors such as climate cycles, climate change, and regime shifts may act in a density-independent manner; (3) shifting from common resource use to resource ownership and from competition-driven, short-term gain favoring vested interests to long-term cooperation; and (4) reducing the ratchet effect of increasing effort despite decreasing yield by resisting political pressure (Botsford et al. 1997). Achieving these goals internationally could be accelerated through debt relief or relaxation, thus promoting resource conservation rather than forcing countries to exploit their resources to reduce external debt. "The best hope for greater sustainability of marine ecosystems is to insulate management from pressure for greater harvest while attempting to reduce uncertainty through a comprehensive ecosystem view" (Botsford et al. 1997, p. 514). Pauly and Maclean (2003), among others, offered a related set of solutions that emphasized market-oriented changes and suggested we identify "allowable fishing areas" in addition to creating reserves.

Regardless of the specific solutions followed, we know that dramatic changes need to be tried and soon. If we stay the course set in the past, the future is bleak. Huxley's famous statement about the inexhaustibility of the great sea fisheries, quoted at the beginning of the chapter, referred specifically to cod, herring, pilchard, and mackerel, all fisheries that have suffered exhaustion. We know considerably more now. On an altered course, the ultimate resiliency of the great sea fisheries and their seemingly inexhaustible capacity may yet return.

11. Fishes versus Fisheries II: Behavior, Life History Evolution, and Ecosystems

The ecologist has never been asked before how to harvest an ecosystem optimally.
—Martin Hall, Dayton Alverson, and Kaija Metuzals, 2000

Fishing and overfishing affect more than population size and fluctuation; they also affect behavior and evolution of life history traits, food webs, and ecosystem interactions. Numerous authors have explored these additional impacts (two thorough examples are M. J. Kaiser and S. J. deGroot 2000 and Jennings et al. 2001). A common organizing approach has been to divide impacts into direct and indirect effects (e.g., Goñi 1998; Jennings and Kaiser 1998; Vincent and Sadovy 1998; S. J. Hall 1999). Direct effects include population responses in target and nontarget populations, such as demographic changes, habitat destruction, and bycatch issues. Indirect effects commonly include impacts on species interactions, ecosystem processes, and bycatch.

This organization makes initial sense but can devolve into confusion because authors disagree on what constitutes a direct versus an indirect effect (e.g., Goñi 1998 vs. S. J. Hall 1999 on effects of discarded bycatch). Also, some impacts, such as behavioral and evolutionary changes in organisms, fit poorly in either category. An alternative approach, taken here, is to treat fishing's effects in a traditional ecological framework, viewing humans as predators in the ecosystem and focusing on how such predation affects levels of ecological organization. Predation causes both immediate and evolved adjustments in behavior and life history and influences community interactions and ecosystem function, especially trophic relationships and food web dynamics.

HUMANS AS PREDATORS

Many researchers have concluded that viewing humans as predators will improve our understanding of the impacts of fisheries (e.g., Policansky 1993b; Kitchell et al. 1997; C. W. Fowler 1999; Hutchings 2000b; Stergiou 2002). This approach is appealing because it immediately places us in context as part of the natural world of fishes, and because the responses that fishes show to our actions are likely to depend on their already evolved adaptations to predation. C. W. Fowler (1999) applied the idea to help us understand our impacts on fishes, but also as a means of achieving sustainability in fisheries: By mimicking natural predator-prey interactions in terms of what and how much we consume, we can engage in a relationship in which the prey have evolved defensive mechanisms and consequently have a greater likelihood of continued existence. Granted, as Jennings and Kaiser (1998) aptly pointed out, the analogy can be carried too far. Most fish predators are generalists, readily switching to alternative prey when one prey species is driven to low numbers, or to cannibalism if prey become scarce; humans tend to be imprudent, voracious predators that drive their prey to unnaturally low numbers (ignoring for now the cannibalism issue). An alternative view is that human activity represents extreme environmental change to which species may or may not be able to adapt (e.g., Conover 2000). In either case, the emphasis is on human actions as selection agents and on the responses of fishes as adjustments. Behavioral and ecological adjustments occur in the short term, and evolutionary adjustments occur in the long term.

FISHING AND FISH BEHAVIOR

"Behavioural interactions lie at the heart of all predator-prey relationships, and the capture of marine species by humans is no exception" (J. D. Reynolds and Jennings 2000, p. 255). Human predation can affect the behavior of individual fish and the behavior of interacting, social groups of fish. Because of the summed behavioral responses of many individuals, fishing can affect social structure in large and small groups. Excessive fishing depletes large portions of schools. The remaining members then coalesce into new groups, masking population declines and eventually resulting in catchability-led stock collapse (CALSC, or hyperaggregation; chapter 10). Many of the failings of traditional management approaches can be linked to a tendency to ignore the effects of fishing on behavior.

Learning

An individual fish subjected to fishing will likely respond adaptively via learning in ways that reduce its future catchability. These behavioral changes can influence growth and reproduction, which determine individual fitness, which when summed across a population may alter fisheries yields and ultimately affect assemblage composition and biodiversity.

At the simplest, individual level, fish can learn, and their learning abilities have direct conservation implications. Fish learn to recognize humans as predators. In areas where they are subjected to spearfishing, they become wary of divers. In places such as underwater parks, where they are not exposed to spearfishing and may be fed by divers, they become quite tame. An example is Hanauma Bay Nature Park, Oahu, one of Hawaii's major tourist attractions, drawing over 1 million visitors per year (www.soest.hawaii.edu). Much of the park's appeal is the abundance and tameness of the fishes, a direct consequence of years of freedom from fishing of any sort. The conservation value of these situations cannot be overstated: Snorkelers come to appreciate the beauty of the habitat and its inhabitants through direct observation, making people more likely to appreciate and value fish as something other than a meal. No video, virtual tourism, or public aquarium can provide as meaningful an experience.

Fishing and Fish Reproduction

Fishing influences reproductive behavior. It can disrupt spawning activity, influence mating patterns through differential removal of sexes, and influence life history characteristics of species that adopt alternative reproductive tactics in response to altered abundance (Vincent and Sadovy 1998).

Fisheries management practices that fail to account for behaviorally mediated alterations in life history traits are less likely to succeed. For example, dynamic pool models assume equal sex ratios and ignore sex-specific growth and mortality alterations that may occur as a result of exploitation (Vincent and Sadovy 1998). Such sex ratio- and sex-specific factors typify many exploited populations, particularly in hermaphroditic, sex-reversing serranid groupers, many of which appear on U.S. and IUCN lists. In sex-reversing species, sex change often occurs naturally in response to depletion of one gender (see Helfman et al. 1997). For example, groupers are protogynous (female-first) sex changers: Females change to male as they grow and when males are removed. By targeting larger individuals, fishing disproportionately eliminates males from a population and skews sex ratios heavily toward females. Females then change sex at smaller sizes, which means that remaining, reproducing females will be smaller and hence less fecund. The physiological processes involved in sex change occur at a slower rate (weeks to months) than the rate at which individuals are removed from the population. Thus males become rare, potentially leading to sperm limitation and reproductive failure.

Just such a scenario played out in the depleted fishery for red hind, *Epinephelus guttatus*, in the U.S. Virgin Islands (Beets and Friedlander 1999). The red hind is a popular food fish throughout the Caribbean. Red hind form spawning aggregations during full-moon periods between December and March, with a peak in January. Their aggregations are therefore predictable and easily exploited. Fish first mature as females at around 250 mm and become males at a median length of 300 mm. Red hind were the second most common grouper in the Virgin Islands fishery,

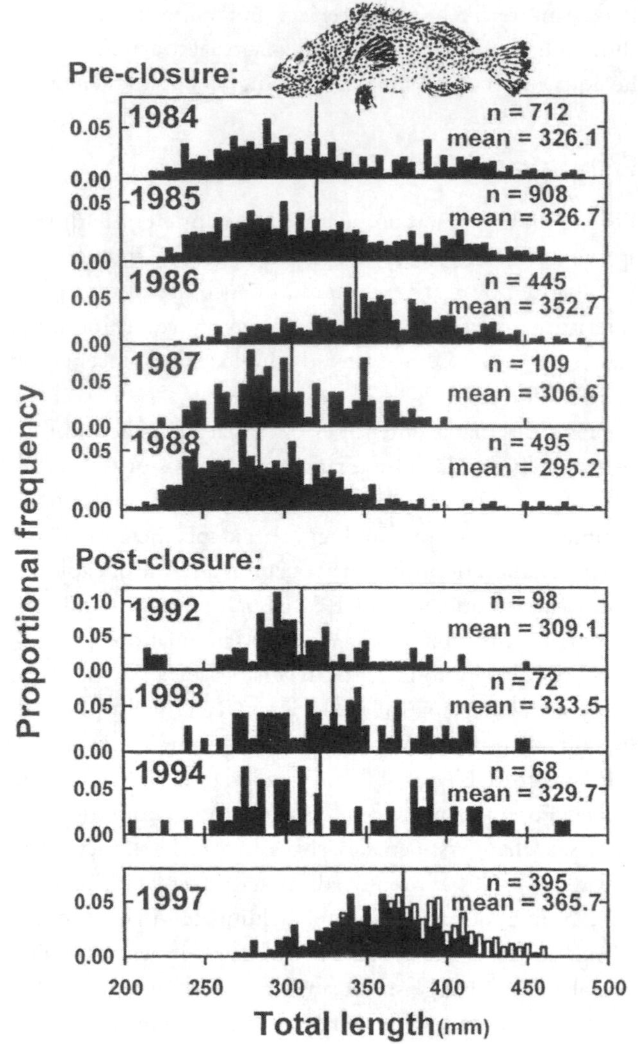

Figure 11.1. Vulnerability of sex-changing fishes to size-selective fishing. Larger size classes of red hind in the U.S. Virgin Islands were differentially eliminated in a fishery that exploited predictable spawning aggregations. Size-selective fishing reduced males to the point that managers feared sperm limitation would occur. Closure of the spawning grounds to fishing in 1990 led to overall recovery, increase in average size, and a progressive increase in relative abundance of larger males. Black vertical line is overall mean size; open bars are males. 1997 data are from the spawning aggregation. From Beets and Friedlander (1999), inset illustration of red hind by W. N. McFarland; used with permission.

after Nassau grouper, *E. striatus*, whose stocks were wiped out in the 1970s when their traditional spawning locales were discovered, displacing pressure onto red hind. Red hind catches declined between 1984 and 1988. More significantly, mean length declined from 340 mm in 1975 to 326 mm in 1984 to 295 mm in 1988 (figure 11.1). A highly skewed sex ratio of 12 females to 1 male in 1988 reflected this altered size distribution. In 1989, divers saw no males in the spawning aggregation. Beets and Friedlander proposed that the spawning aggregation site be declared a no-take reserve to rebuild male abundance and eventually replenish the stock.

The prohibition went into effect in 1990, creating what is now known as the Red Hind Bank Marine Conservation District, the first no-take fishery reserve in the U.S. Virgin Islands. Catch data from the 1992–94 fishery outside the spawning site showed increasing mean size and, importantly, increased frequency of large fish. In 1997, average size at the spawning aggregation had increased to 366 mm, with many fish over 400 mm. The sex ratio improved to a more normal 4 females to 1 male. The red hind fishery itself was not closed, and fishers were still catching fish all year; but protecting the spawning site appears to have been critical in arresting further declines.

These results hold promise for resurrecting depleted stocks of other grouper species, as in greatly depleted gag grouper, *Mycteroperca microlepis*. Gag males declined from 17% to 1% at Gulf of Mexico locales and from 20% to 6% in Atlantic locales (Coleman et al. 1996; Musick et al. 2001). Sperm limitation may have contributed to these declines (Vincent and Sadovy 1998). Restoring such fisheries requires addressing causative factors associated with reproductive behavior, such as size- and gender-selective fishing, and not just regulating the number and size captured. IUCN, in its 1994 and 1996 revised guidelines for determining marine species at risk, emphasized the importance of assessing sex change, relative abundance of the limiting sex, and effects of density on reproductive output (Vincent and Sadovy 1998).

Spawning Stupor

Many fishes form predictable spawning aggregations that make them vulnerable to overfishing. This is common among coral reef fishes, especially large groupers. Vulnerability increases because fishes that are normally wary become remarkably oblivious on the spawning grounds, even to disturbances such as spearfishing. Johannes (1981) called this phenomenon "spawning stupor." Aggregating on spawning grounds makes groupers easy to find, and spawning stupor makes them sitting ducks. "Unusually large numbers and high densities of [docile] fish converging at predictable times and location . . . provide exceptional fishing. So fishermen flock to the spawning grounds with lines, nets, spears, and sometimes, dynamite" (Johannes 1981, p. 36). Spawning stupor has undoubtedly influenced declining grouper stocks.

Alternative Tactics

A final category of interaction between fishing and fish reproductive behavior involves alternative life history tactics within species. Such tactics exist in both sex-changing and non-sex-changing families (see Helfman et al. 1997; Vincent and Sadovy 1998). Alternative tactics interact with overfishing in Pacific Northwest salmonids (e.g., Groot and Margolis 1991), including coho salmon, *Oncorhynchus kisutch*, half of which are Threatened or Endangered. Male coho salmon differ in mating tactics (Gross 1984, 1991). Larger, older, more colorful, "hooknose" males (ca. 52 cm long, 2.5 years old) go to sea as smolts, feed and grow for about 16–18 months, then return and court females that have dug nests (termed *redds*) in the gravel bottom near the middle of a spawning stream (figure 11.2). Male mating success depends on size and proximity to females. Males fight to be closest to females, larger males winning fights and spawning more frequently. The alternative tactic involves smaller, younger males called "jacks" (ca. 34 cm long, 1.5 yrs old), which return after only 4–6 months at sea or travel only to the estuary before returning to their natal river to spawn. Jacks use debris, rocks, and shallow water as refuges that allow them to gain proximity to spawning fish; they dash onto the redd to deposit sperm when hooknoses are spawning with females.

The preferential targeting of larger fish creates an interaction between fishing and mating tactics. Although the "decision" to become a jack or a hooknose is influenced by both genetics and environment, human activities appear to favor jacks. Young fish that develop under conditions that promote faster growth are more likely to become jacks. Clear-cutting riparian forests along salmon streams raises water temperatures and increases debris inputs, promoting faster growth and habitat that conceals jacks. Hatchery conditions also promote fast growth and a higher proportion of jacks, which increases jack genes in a population when hatchery fish breed with wild fish (chapter 14). Finally, fishing targets larger fish, meaning hooknose males, both when at sea and when they return to spawn. Hence human activity favors the smaller, "alternative" males that humans find less desirable (Gross 1991). Reduction in the proportion of hooknoses is also likely to lower genetic diversity and, consequently, population fitness. Increased frequency of jacks has also been found among sockeye salmon in the Columbia River (Ricker 1972). Through many of our actions, "we remove from the breeding population those individuals with the traits we like" (Safina 2001a, p. 793).

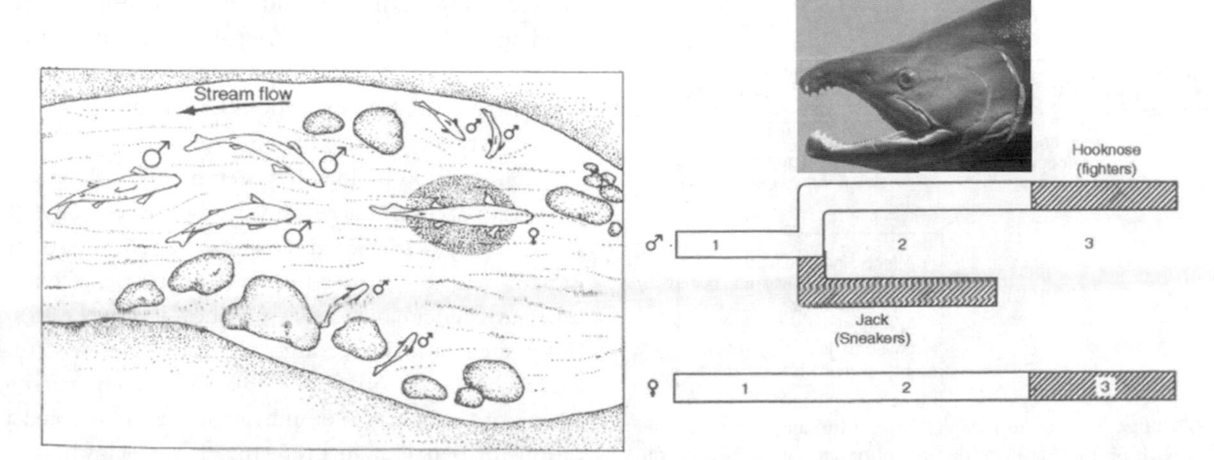

Figure 11.2. Alternative male forms in coho salmon. Larger, older, territorial, "hooknose" males grow in the ocean and then return to fight other males and court females. Smaller, younger, "jack" males hide along stream edges and interpose themselves during spawnings by larger males. All females go to sea and mature at a relatively large size and old age. Fishing pressure targets larger fish, which reduces the number of hooknose males and favors production of jacks. Numbers are age in years; cross hatching indicates spawning. From Helfman et al. (1997) based on Gross (1984, 1991); used with permission.

Behavior, Exploitation, and Future Population Traits

Vincent and Sadovy (1998) explored the larger context of fishery impacts, emphasizing the importance of understanding behavioral responses to fishing in determining future abundance, density, genetic diversity, community interactions, and ecosystem processes (figure 11.3). They stressed that exploitation acts on target and nontarget species, habitats, and ecosystem-level outcomes. The size, gender, and genotypic and phenotypic traits of individuals removed will depend on the fishing gear used. Such human predation immediately affects stock abundance via depletion (catch composition) but also affects future population size by altering reproductive traits that influence the fecundity and longevity of remaining individuals. Traditional management viewed fisheries as a simple relation between fishing effort and fish yield—that is, fishing had "numeric effects" on exploited stocks. Later, dynamic pool management models incorporated the effects of fishing on recruitment, growth, natural mortality, and fishing mortality but neglected behavioral aspects of reproductive biology that produced differences in reproduction and mortality. Vincent and Sadovy (1998) concluded that management's inability to account for these behavioral factors contributed heavily to unsustainable fishing outcomes.

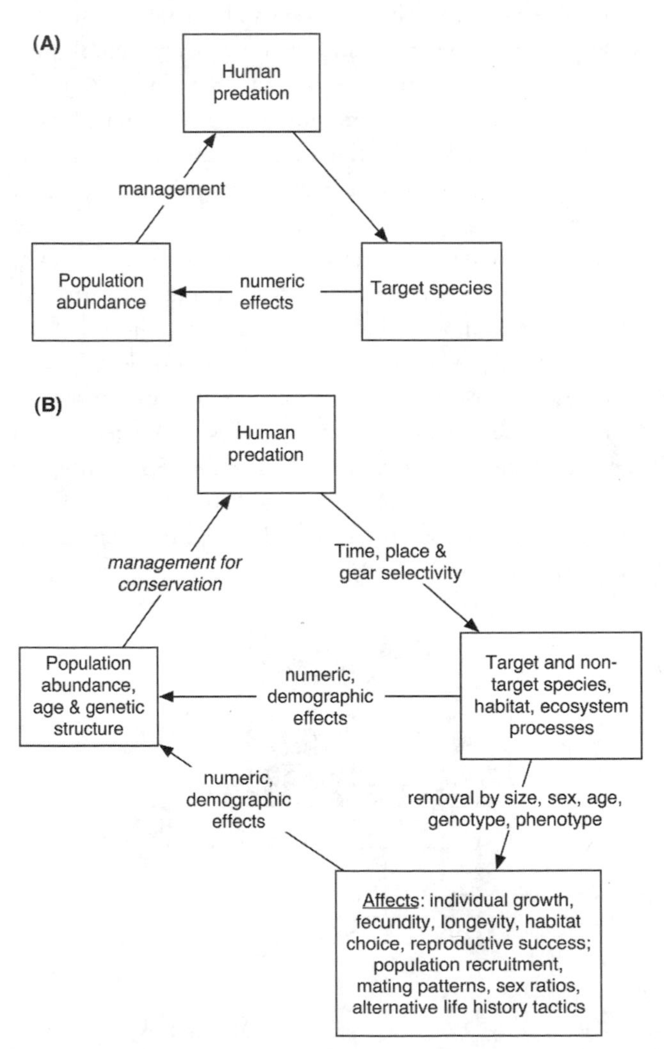

Figure 11.3. How exploitation affects fishes. Human predation interacts with biology and ecology to influence future population size via various actions and mechanisms. (A) Numeric effects are easily and traditionally measured, but (B) demographic effects, which have often been overlooked or underemphasized, arise from complex behavioral adjustments to selective removal of different ages, sizes, and sexes. Based on Vincent and Sadovy (1998).

Selective removal of particular size or age individuals alters life history, such as reducing size at sexual maturity, which influences individual and stock fecundity. By failing to account for the effects of such selective removal, management often overestimated potential yield and set fishing targets too high (see also Sadovy 2001b).

Studying Behavior to Strengthen Management and Conservation

The behavioral interactions and phenomena discussed above point out how much more we can understand about the biology of exploited and imperiled fishes by understanding the behavior of individuals and groups. If the benefits of incorporating knowledge of behavior into conservation and management actions are so obvious, why has the subject received so little attention?

Conservation biology has been traditionally wedded to population, community, and landscape ecology (Soulé 1985). Fisheries science and management have also focused most heavily on population-level phenomena. The "new fisheries science" recognizes the shortcomings of a population focus and increasingly emphasizes ecosystem-level approaches (e.g., NRC 1999b). Still largely missing are studies of the behavior of individuals, despite known applications of behavior in protecting biodiversity (e.g., J. M. Reed and Dobson 1993; Harcourt 1995). Behaviorists, ethologists, and behavioral ecologists have been surprisingly (embarrassingly?) slow to join forces with conservation biologists, especially in the aquatic sciences and fisheries

(Helfman 1999b). The complexity of environmental problems mandates the inclusion of all potentially contributing disciplines.

Exclusion of behavior from conservation biology is also counterproductive. Without knowledge of behavioral interactions between individuals, or behavior relative to a species' habitat, we have little hope of being able to reassemble degraded communities and ecosystems. Behavior also has political and economic relevance. To many nonscientists, conservation biology is largely synonymous with behavior and natural history. Public knowledge of and enthusiasm for protection of endangered species has grown in no small part from natural history programs on television, programs that highlight behavioral research. Without such programs, the conservation movement would have a weaker scientific base and diminished public support.

Participation of behaviorists in conservation is increasing, as evidenced by compilations that show the direct application of behavioral knowledge to solving conservation problems (e.g., Clemmons and Buchholz 1997; Caro 1998; Helfman 1999a; Gosling and Sutherland 2000). Papers on fishes and fishery science, while in the minority, have been included in these compilations (e.g., Vincent and Sadovy 1998; Helfman 1999a; J. D. Reynolds and Jennings 2000; Levitan and McGovern 2005; J. K. Parrish 2005). From these overviews, certain topics emerge as most amenable to and most able to profit from behavioral analysis, forming a subdiscipline referred to as ethoconservation (Helfman 1999b; Shumway 1999). In addition to traditional areas such as captive breeding and toxicological bioassays, promising topics include

- how the disruption of reproductive behavior endangers fishes, and how knowledge of reproduction can aid in recovery efforts
- species-habitat interactions, and interrelations among habitat loss, habitat choice, and habitat protection
- the value of protected areas as a management and conservation tool, how the behavior of fishes within reserves ultimately determines reserve success, and what gaps exist in our knowledge of how such reserves function
- predator-prey interactions and the complexity and diversity of behavioral interactions between predators (including humans) and prey in causing and mitigating biodiversity loss
- ethnography, or human behavior, including the realization that human behavior, not fish behavior, is ultimately responsible for fishery declines, and what can be done to change our behavior patterns as we become aware of their destructive effects

Recent modeling efforts highlight the insights into fishery issues that are available via behavioral studies. C. Walters (2000) began with the recognition that predation risk has been an overriding selective force in fish evolution and that fitness-enhancing activities such as feeding and reproduction are often traded off against predator avoidance. Avoidance of predation affects spatial and temporal patterns of fish aggregation, food resource depletion, supply rates of prey to predators, and coexistence among competitors. These factors, in turn, influence the population and community dynamics that underlie much of fisheries management, including stock-recruitment relationships and compensatory mortality patterns.

C. Walters (2000) and C. J. Walters et al. (2000) applied this approach in developing Ecosim and Ecopath models of fisheries dynamics, ecosystem interactions, and protected area function. The models depict fishes as risk-averse foragers. The modeled outcomes mimic natural systems more faithfully than traditional fisheries models that treat organisms "as though they were chemicals in well-mixed reaction vats" (C. Walters 2000, p. 311). Walters concluded that "many of the most striking and puzzling features of how marine ecosystems respond to fisheries arise very directly from risk-sensitive foraging behaviors" (2000, p. 309). By incorporating behavior into such models, fisheries science progresses from traditional, population-based analysis to ecosystem-based management, with behavioral interactions and outcomes as a bridge.

FISHING AS A SELECTIVE AGENT AFFECTING EVOLUTION

Exploitation often produces predictable shifts in population abundance and structure, age and size at first reproduction, and other life history traits. Some changes are apparently immediate and largely statistical. Other changes reflect elicited, inherited responses that evolved due to previous, presumably nonhuman, selection regimes. Most interesting but difficult to document are alterations that represent recently evolved and evolving adaptive responses to human predation.

Change in Life History Characteristics

Exploitation-caused changes in population structure have long been documented in the fisheries literature. In fact, growth adjustments underlie the basic theory of fishing: Growth compensation by surviving individuals will accompany moderate exploitation, and maximal harvest rates (maximum sustainable yields of surplus production) are achieved by maintaining a population at some intermediate level through fishing (e.g., M. R. Ross 1997; P. J. B. Hart and Reynolds 2002; or any basic fisheries text; see box 10.2). Typically, reductions in abundance cause increased body growth in remaining individuals, and the ensuing faster growth rates generally result in maturation at smaller sizes and younger ages. The proximate stimulus of this generalized compensatory change is assumed to be reduced competition for limiting resources. The ultimate cause, according to life history theory, is that heavy predation pressure favors fish that are capable of initiating reproduction sooner, before they are eaten (see Trippel 1995).

The fisheries literature contains many examples of exploited stocks that show changes in weight-at-age, length-at-age, length-at-maturation, and age-at-maturation, with most species showing reduced weights, lengths, and ages, as well as accelerated growth. Law (2000, table 1) reviewed findings on 16 species of flatfishes, gadoids, and salmons, most of which showed decreases in life history traits (exceptions were generally explained by special conditions). Data on Atlantic cod are complete and telling: Among eight exploited stocks monitored for 7–53 years, median age at maturity declined between 16% and 56%, representing a 0.9–3.6-year reduction in age at maturity; longer periods of exploitation produced greater changes in age. Natural variation in age at maturation was small or negligible when fishing pressure was light (Trippel 1995). Data on flatfishes, gadoids, and salmon and on commercially fished western Atlantic sharks, herrings, scorpionfishes, snappers, drums, groupers, eel pouts, butterfishes, and mackerels similarly revealed reduced lengths and ages at maturation (Upton 1992; L.. O'Brien et al. 1993; Lessa et al. 1999; Vannuccini 1999). The mean length of captured Patagonian toothfish declined 30% in just the first few years of fishing (ISOFISH 2002; chapter 10). Among other examples are freshwater species subjected to intensive commercial and recreational fishing (bluegill, walleye, whitefish, yellow perch, northern pike, brown trout, and Arctic char; Trippel 1995; Drake et al. 1997). Again, decreased length and age at maturity are the rule, with exceptions noted by Trippel (1995) and Rochet (1998).

Such adjustments can create equal-size populations with different age, size, and growth structures, which presents a dilemma to the fisheries manager. If fishmeal is the ultimate product and the objective is only to maximize individual growth rate, with little premium placed on larger fish, the fishery can be treated analogously to pine trees being grown for pulpwood. Fast-growing, small trees with low-quality wood are just fine. However, many fisheries also have a recreational component, for which larger fish are preferred (also, larger salmon, groupers, and eels are fatter and thus more valuable). More important, pine trees are harvested and replanted, whereas most marine fishes are captured without human replacement. The future of the fishery depends on successful natural reproduction. Age and size distribution and growth characteristics take on greater significance as stocks decline in abundance. That significance increases as we become more aware of the influence of current stock on future recruitment, the greater reproductive potential and value of large fish, and the difficulty with which many marine stocks recover from overexploitation (see chapter 10; Hutchings 2001a, 2001b; Sadovy 2001b).

The critical issue becomes whether reduced mean size and younger age at maturation are simple phenotypic responses that are quickly reversible, versus an evolved response that is harder to change and therefore likely to lead to continued population decline. As the evidence mounts of an evolutionary influence, "it is proving increasingly difficult to account for the observed patterns of change in life histories on the basis of environmental factors alone" (Law and Stokes 2005).

Merely Statistical Differences?

Dramatic changes in size distributions can result from fishing, without natural selection. Such differences arise from gear selectivity. Some gear types remove larger individuals; others remove smaller fish. Trap nets and pots tend to selectively capture larger individuals, unless they are too large to enter the pot. Longlines catch all but the smallest individuals. Gill nets capture fish within a particular size range but usually take larger fish. Larger fish can outswim trawls and some large seines. Recreational fishers and highgraders let the small ones swim free. Removing large individuals from a population, the most usual result of exploitation, produces a size distribution with a smaller mean, without any compensation or heritability. When such gear or fisher preferences operate, altered population attributes are largely statistical descriptions summed across individuals.

Alternatively, smaller fish can migrate into an area and replace removed individuals. If growth compensation occurs, it can result from individuals responding to increased availability of food, or might represent a physiological reaction to encountering fewer school members, or fewer large conspecifics. Again, an evolutionary adjustment can't be assumed beyond an already evolved ability to alter feeding or growth rate in response to changing environmental conditions (much like slower growth in colder water, a normal physiological response). Also, removing larger fish from a population eliminates fish with slower average growth rates; the remaining portion of the population is thus characterized by faster growth only because they are at an age when they grow faster. Over time, the population "shifts" from one characterized by slow growth rates to one characterized by faster growth; this shift is referred to as Baranov's fishing-up effect (Policansky 1993b). No heritability in growth rate need be invoked. That comes next.

Evolution in Response to Human Actions

Palumbi (2001) argued that humans have become the world's greatest evolutionary force, exerting strong natural selection in such areas as disease resistant viruses and bacteria, pesticide resistant insects and plants, artificial selection via domestication (Japanese koi, farmed catfish, hatchery salmon), and altered characteristics in introduced species. However, before the early 1980s, the idea that "fish populations can undergo genetically based changes in response to fishing" was disputed (Policansky 1993b, p. 2). Support for that idea came from a few field studies or was inferred from laboratory manipulations using small, short-generation, nonexploited species (Hutchings 2000b). Geneticists had little doubt that fishing could cause evolution. Allendorf et al. (1987, p.141) unequivocally stated, "All populations of fish that are included in a sport or commercial fishery will inevitably be genetically changed by harvesting." However, the accepted view among many fishery scientists was that exploitation did not cause heritable change in commercially exploited populations (e.g., R. R. Miller 1957). Managers paid little attention to the potential impact of evolution on exploited fish populations and did not incorporate evolution into management models and plans (Policansky 1993a).

Reluctance to accept fishing as a selective agent predominated because it was difficult to disentangle genetic effects from the plastic, compensatory growth responses that fishes commonly showed to changing environmental conditions (e.g., Reznick 1993). An added complication involved incomplete knowledge of how long human predation might have influenced compensatory growth. Theory suggested that adaptive responses were most likely early in the history of a fishery. Most fish stocks had been harvested and overharvested long before studies began, meaning evolutionary changes could have taken place long ago, and little was revealed by recent trends in life history traits (Policansky 1993b). Until these complications could be factored out, and because plastic phenotypic responses were likely to occur more quickly and be more noticeable than evolved changes (Jennings and Kaiser 1998), fisheries managers showed "a continuing reluctance . . . to take seriously the threat of genetic change brought about through fishing" (Law and Stokes 2005, p. 241). Not all fisheries researchers discounted the influence of evolved responses. Handford et al. (1977, p. 960) accused management of being "seriously deficient in [its failure] to take into account the possibility of adaptive genetic change in exploited stocks of fish."

In retrospect, little reason existed to doubt fishing's evolutionary potential, especially given (1) the nonrandom, size- and locale-selective nature of most fisheries; (2) the strength of fishing as a source of mortality; and (3) the known genetic basis of many life history traits that influence fisheries yields (relatively high heritability of growth rate, fecundity, age and size at maturation) (Policansky 1993a, 1993b; Law 2000; Palumbi 2001). These observations fit general requirements for recognizing the operation of natural selection on a species' traits. Also, life history traits display sufficient variation to be changed by evolution, including (1) variation upon which natural selection can act (the "additive genetic variance" of quantitative genetics), and (2) total phenotypic variation, which includes plasticity in response to changing environmental conditions and is the basis of density-dependent compensatory responses (Trippel 1995).

Time has also been sufficient, in terms of number of generations needed for significant genetic change to occur. Across a range of taxa, evolution has occurred in less than 10 generations, sometimes in as few as 2 or 3 (e.g.,

Falconer and Mackay 1996). Field studies of salmonids have shown detectable divergence among populations in 8 to 13 generations (Haugen and Vollestad 2001; Hendry et al. 2000; Hendry 2001). Laboratory and theoretical studies suggest that selection forces with strong coefficients (i.e., that cause significant mortality) can exert a detectable evolutionary effect in a single generation, as shown in platyfish, *Xiphophorus maculatus*, where one allelic substitution can alter age and size at maturation (McKenzie et al. 1983).

Most convincingly, Conover and Munch (2002) demonstrated rapid evolution in an exploited species based on the kind of selection imposed by fishing. They simulated size-selective fishing by rearing fast- and slow-growing Atlantic silversides, *Menidia menidia*, in the lab, using randomly sampled fish from founder populations obtained in the wild. After only 4 generations of directional selection for growth rate, large-harvested fish (study groups from which fish with the fastest growth rates were removed) and small-harvested fish (groups from which slow-growing individuals were removed) reversed their growth rate characteristics. The previous fast growers (large-harvested fish) had mean weights nearly half those of the small-harvested lineage that previously possessed slow growth characteristics. The growth differences had a demonstrated genetic basis. Important differences also occurred in egg size and biomass yield, indicating that continued harvest of the largest members of a stock reduced biomass and egg production. "Selection on adult size caused the evolution of a suite of traits likely to influence population growth rate and productivity" (Conover and Munch 2002, p. 95). The traits evolved are largely the opposite of what the fishing industry and society would prefer.

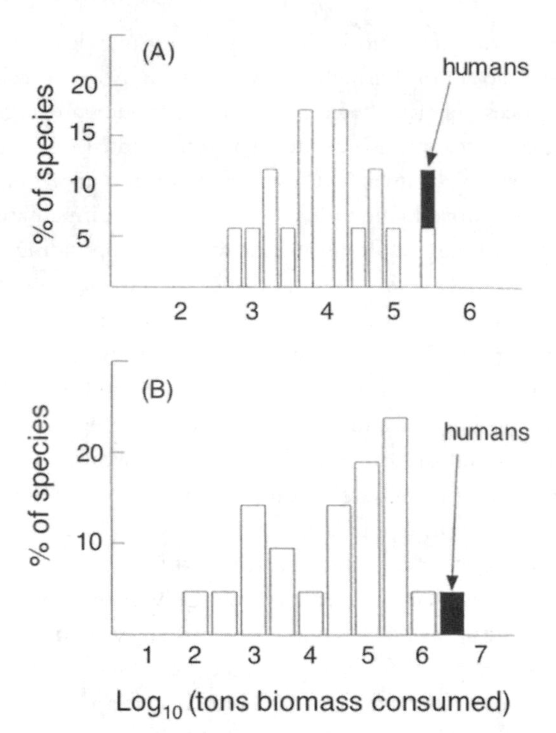

Figure 11.4. Humans as significant predators on exploited fishes compared to other major predators. (A) Annual mortality of sand eels, mackerel, herring, and silver hake in the northwest Atlantic caused by 16 species of predatory fishes, marine mammals, and birds, versus humans (only spiny dogfish consume at a rate similar to ours). (B) Mortality of all commercially harvested bony fishes in the eastern Bering Sea that are also eaten by marine mammals. Vertical axis is proportion of predator species responsible for a particular level of mortality; mortality is expressed in log units of biomass consumed annually. Note that consumption rates do not include additional mortality due to bycatch, escape, or ghost gear or other anthropogenic factors (habitat destruction, pollution, etc.). From C. W. Fowler (1999); used with permission.

Given the rapidity with which evolution can proceed, it is reasonable to conclude that human predation has been strong enough for long enough to have an evolutionary impact (figure 11.4). At least 40 generations of North Atlantic cod, herring, plaice, and sole have experienced intensive fishing, sufficient to alter genetic makeup (Policansky 1993b). Fishing mortality exceeds natural mortality in many if not most heavily exploited species, often by a factor of two or three (e.g., T. K. Stokes et al. 1993; K. Stokes and Law 2000). Natural, post-larval and post-recruitment mortality rates in many finfish species run less than 10%, whereas mortality rates targeted to achieve maximum sustainable yields are around 50%, and actual fishing mortality often ranges between 70% and 90%. Given the history and intensity of most commercial fisheries, fishing can be considered a long experiment with "more than enough time for selection to produce substantial genetic changes on almost every quantitative character that has been examined" (Policansky 1993b, p. 6).

Evolutionary changes in growth rate, size, and body shape were, in fact, documented decades ago. Data from five Pacific salmon species subjected to commercial and recreational trolling, gillnetting, and seining showed responses in different species that could be linked directly to exploitation patterns (Ricker 1981, summarizing 13 papers published between 1957 and 1981). Overall, average size in all species declined. Chinook salmon over a 60-year period matured on average 2 years earlier at half the original size. In 1950, when coho and pink salmon fishers began to be paid according to size and not number of fish caught, larger-mesh gill nets were employed to catch larger fish, accelerating the shift in stocks to smaller, younger fish. Ricker (1981) attributed these shifts to cumulative genetic effects of removing fish of larger than average size (subsequent analysis of pink salmon runs have shown reduced mean body weight of as much as 34%, McCallister et al. 1992). Chum and sockeye salmon decreased less in size than Chinook, coho, and pink. Chum and sockeye were fished after they completed their ocean-growth phase, gill nets tended to selectively take smaller chum, and sockeye increased growth rate when exposed to cooler ocean temperatures. These exceptions strengthened the conclusion that selection and not immediate environment caused the differences in age, size, and growth patterns.

A gill net fishery in Lesser Slave Lake, Alberta, for lake whitefish, *Coregonus clupeaformis*, produced size differences that could best be explained by natural selection (Handford et al. 1977). Gill nets removed large, heavy, fast-growing fish, leading to declines in growth rate and condition factor and an increase in mean age but little change in mean length-at-age. Declining growth rate and increasing mean age contrast with what normally occurs when density-dependent, compensatory (phenotypic) responses operate. The progressive changes in condition factor were particularly informative. Condition factor is calculated by dividing body mass by length: Fat fish have higher condition factors. Fat fish are also more likely to become trapped in gill nets. Condition factor declined dramatically with time in all age groups, to the extent that fish of a given age and length in the 1970s often weighed less than half of what fish of similar age and length weighed in the 1940s (figure 11.5). Gill nets have subsequently been shown to exert size-dependent directional selection on age and length at maturation in grayling, *Thymallus thymallus* (Haugen and Vollestad 2001) and are considered determinants of body size and shape in sockeye salmon (Hamon et al. 2000).

Handford et al. (1977) considered and eliminated factors other than an evolved response that might explain these changes, including classic compensatory responses to reduced competition. They argued that compensatory changes and natural selection acted in opposite directions and could mask one another, hiding the impacts of overexploitation. Because strong compensatory responses are common, "selection is only rarely capable of producing an effect larger than and opposed to that associated with density-dependent compensation," which is why selection's effects normally went unnoticed (p. 960). But natural selection was probably occurring commonly in fisheries, influencing population dynamics and parameters. Handford et al. concluded that, until fisheries models incorporate evolutionary change, "it is unlikely that the behaviour of fish populations under exploitation can be fully understood" (p. 961).

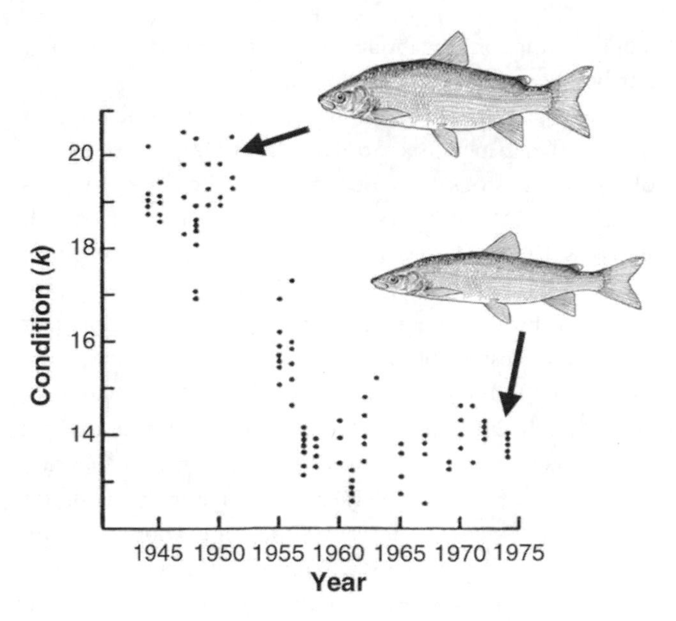

Figure 11.5. Evolution of body shape in exploited whitefish. Lake whitefish in Lesser Slave Lake became skinnier over time as a result of gillnetting. Fat fish were more likely to be captured. Data plotted are condition factors k ($k = 10^5$ Weight/Length^3) for male whitefish, 1945–75; females showed similar patterns but were not used because of weight changes caused by egg bearing. From Handford et al. (1977); used with permission.

One additional example underscores how genotype and phenotype interact to produce life history alterations. In studying 20th-century fisheries data for plaice, *Pleuronectes platessa*, in the North Sea, Rijnsdorp (1993) found smaller sizes and younger ages at maturation as the century progressed, with some respite and trait reversal during World War II and the 1960s. The reversals were strong evidence that character changes resulted from compensatory growth—that is, were phenotypic in nature. However, statistical analysis indicated that compensatory growth accounted for only 3.4 cm of the 5.8-cm reduction in average length-at-maturation for four-year-old fish. A 1°C increase in water temperature could account for an additional 0.7 cm of the change, leaving 1.7 cm of the length reduction, or about 29% of the length difference, attributable to genetic change (Law 2000).

An expanding literature has confirmed the findings of these earlier papers (see Stokes et al. 1993), and evolution of exploited stocks has become an acknowledged phenomenon in the fisheries literature (e.g., Royce 1996; M. R. Ross 1997). Matter-of-fact statements about evolution can now be found, such as, "Simply through the action of fishing, fishers generate selection, causing evolution that changes the sustainable yield" (Law 2000, p. 659) and, more succinctly, "Fishing mortality is highly selective" (Conover and Munch 2002, p. 94). Fisheries management has not rushed to incorporate evolved responses into models and management plans, but workers are increasingly calling for such an approach, arguing that "the application of theory from evolutionary ecology will improve the success of fishery resource management in the long term" (Browman 2000, p. 299; see also Trippel 1995; Conover 2000; Law 2000).

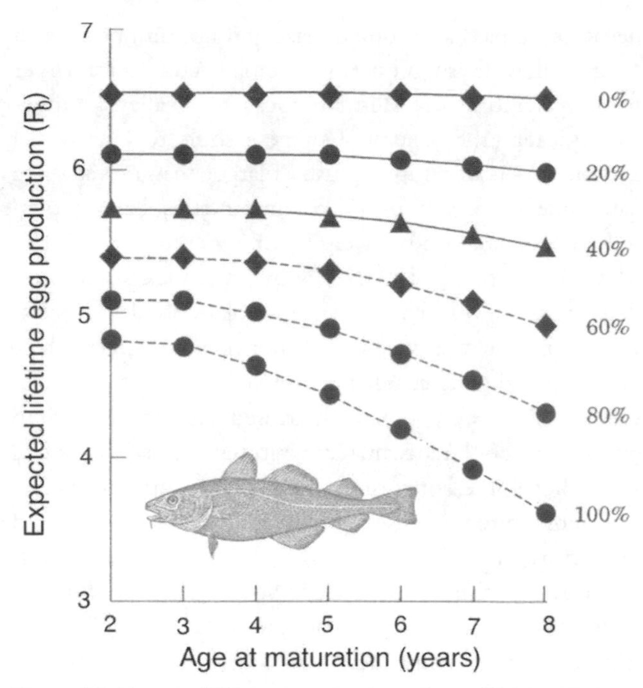

Figure 11.6. Expected lifetime production of eggs (R_o) in relation to age-at-maturation at varying levels of fishing pressure (0%–100% fishing mortality). Fish experiencing intense fishing mature earlier (bottom curve). Under no fishing pressure, no penalty is incurred by delayed maturation (top curve); these fish are more likely to mature later. Relaxation of fishing moves the stock toward the top curve, where, because expected reproductive success does not change greatly from one year to the next, an increase in maturation age is unlikely (see text). Data are for North Sea cod; from Law (2000); used with permission. Cod drawing from Bigelow and Schroder (1953).

Solutions to Evolution-Based Problems, Evolution as Solution

Can reversing fishing pressure also reverse evolution? Put another way, can we manage for evolution to improve growth and reproductive traits? Where stock recovery is concerned, the answer appears to be negative. If reproductive traits were phenotypically plastic, relaxed fishing pressure should result in reversals in traits, such as reversion to older age at maturation. Such a pattern of rapid change and reversion was seen in plaice and in Atlantic herring stocks when populations were allowed to recover in the 1980s (Trippel 1995).

But age at maturation involves both phenotypic and genotypic responses, and it appears that the evolved component requires longer to reverse, if it can reverse at all. Law (2000) analyzed trends in North Sea cod, using data collected by Rowell (1993). Rowell calculated the impact of fishing-induced mortality on age at maturation, using expected lifetime output of eggs as a function of age at maturity as a measure of reproductive success. A family of curves depicting lifetime reproductive success as it relates to fishing regime (selection) resulted (figure 11.6). Two important and reasonable assumptions were that (1) reproduction and growth compete for an individual's energy resources; and (2) bigger, older fish produce more eggs. The "decision" is whether to produce a few eggs while young or to wait and produce more eggs when larger with the risk of being killed first. When no fishing occurs ("0%" top curve in figure 11.6), natural mortality is relatively light and has a constant probability, as does expected lifetime egg production. Few advantages accrue to individuals that reproduce early at a small size or age; delaying maturation allows production of more and better eggs. In contrast, under heavy fishing pressure ("100%" bottom curve in figure), fish are increasingly likely to die as they grow older, especially if the fishery targets bigger fish. It becomes advantageous to reproduce as early as possible. Intermediate fishing pressure produces intermediate curves, but all deflect downward, indicating some advantage of earlier maturation.

Two important conclusions emerge. First, visual inspection of the curves reveals why intense fishing pressure results in earlier age-at-maturation. The costs of delayed maturation increase with increasing fishing pressure, increasing the likelihood of death and selecting for individuals that mature progressively earlier as fishing intensifies. Overfishing thus generates strong directional selection that favors genetic changes that lead to early maturation (Law 2000).

Second, of relevance to managing for evolutionary reversals in life history traits, reverting from heavy to light fishing is unlikely to result in later, larger sizes-at-maturation because relaxed fishing pressure moves fish from the bottom curve toward the top curve of the graph. The flatness of the top curve means little difference exists in reproductive output (i.e., the selection differential is weak) from one year to the next, and selection pressures are too weak to cause maturation to diverge from the newly established, younger condition. "In other words, it could be hard to undo the effects of inadvertent selection caused by fishing [because it is] much easier to select for early maturation than for late maturation" (Law 2000, pp. 664, 666). An added factor is that a population exposed for many generations to intense fishing is likely to evolve life history and other adaptations to counter fishing pressure. These adaptations may occur at the cost of adaptations appropriate to the original, more "natural," lower mortality conditions that previously existed. Reversion to light fishing may not mean reversion to original genetic conditions because "the surviving genotypes in the stock may be those with reduced fitness with respect to natural selective forces" (Conover 2000, p. 304). This loss of original adaptations brought about by directional selection—which invariably reduces genetic variation—is likely to impede recovery.

Because evolutionary change is difficult to reverse, precautionary measures are needed that reduce or eliminate the strong directional selection inherent in size-selective or spatially selective fishing. Foremost, total allowable catch in the fishery should be limited. Catch limits on individual fishers do not achieve this, nor do minimum size limits. Size limits are often misguided because they fail to recognize the ecological and evolutionary importance of large reproducers, resulting in smaller, younger fish, with implications for the future of a stock. A simple comparison of reproductive output of small versus large individuals drives this point home. In red snapper, *Lutjanus campechanus*, which are heavily fished and depleted throughout their North American range, a single ten-year-old, 12-kg, 61-cm female can contain 9 million eggs, which is equivalent to the total egg output of 212 females that are three or four years old, 1 kg, and 42 cm (a 12-fold increase in mass results in a 200-fold increase in egg production). Allowing females to grow, or protecting larger size classes within a stock, can have a disproportionate impact on egg production and probably on future recruitment (Bohnsack 1994).

Some recreational fisheries overcome these problems by imposing maximum size limits or by instituting slot or window limits. Slot limits protect fish within an intermediate size range; window limits protect fish smaller and larger than the size range. Commercial fishing could accomplish similar ends by taking the "window" approach and developing gear that selectively captured intermediate-size individuals. Policansky (1993a) analyzed why slot and window limits are often ineffective in practice and determined that it is mostly because "many factors have variable and unpredictable effects on fish populations, and compliance with fishing regulations is less than perfect" (p. 657). Reducing spatially selective fishing, such as fishing on spawning grounds, is another common and effective remedy.

Ecologically based limits are certainly preferable to passive or traditional management practices, especially counterproductive practices that ignore selection and result in diminished long-term fisheries yields (e.g., W. J. Sutherland 1990). Managing proactively with evolution in mind can at least theoretically increase yields. In reality, fisheries managers must deal first with short-term goals, such as this year's catch limits and perhaps next year's stock and catch. Evolutionary thinking becomes something of a luxury "because [evolution] is slow in comparison with the periods in which managers, operating under contemporary socioeconomic constraints, have to act" (Jennings and Kaiser 1998, p. 237). However, taking an evolution-based, precautionary approach—one that prevents a worldwide fishery landscape dominated by small, prolific reproducers—is our best assurance that "future generations of humankind are afforded the same opportunities for fishing as this generation" (Stokes and Law 2000, p. 308).

FISHING'S IMPACTS ON COMMUNITIES AND ECOSYSTEMS

Fishery science has traditionally focused on population dynamics. Population-level phenomena, such as abundance, numerical change, life history, and age structure, can be characterized and translated into management actions such as regulating catch levels and mesh size, releasing small fish, avoiding concentrations of young fish or spawning fish, and delaying fishery opening dates. By comparison, interactions among species and between species and habitats are more complex and subtle, making curative measures less obvious. Fishery-induced habitat destruction can alter community composition, which affects food web interactions, which may affect habitat and change community composition even more. Traditional fisheries management downplayed the importance of understanding such ecosystem-level effects. Although the pioneers of fishery science investigated ecosystem processes, their findings were seldom incorporated into management actions (T. D. Smith 1994; Jennings and Kaiser 1998; Link 2002a). But an ecosystem approach gained popularity in the late 1980s, in part because classical population-based fishery science failed to manage fisheries sustainably. New approaches were sought. A larger-scale view such as ecosystem-based management is an obvious alternative.

A shift in focus stimulated investigations into fishing's impacts on communities and especially on ecosystems (see Jennings and Kaiser 1998; S. J. Hall 1999). Impacts include the effects of mortality among target and nontarget species on other species' abundances and on community structure, the destruction of habitat by active and passive fishing gear, and impacts that alter food web interactions. Fishing affects all these processes and events. Successful management of fisheries therefore requires an ecosystem-level perspective, rather than a focus solely on individual species.

Community Composition Shifts

Ecosystems are connected subsets of biological and physical phenomena; a functioning ecosystem has, at some time, all or most elements in place. Eliminating components or their connections results in a fractionated and eventually nonfunctional or unrecognizable ecosystem, analogous to progressive removal of pieces of a jigsaw puzzle. The disarticulated ecosystem is no longer capable of providing goods and services to its constituent parts, including humans. Theoretically, the best way to maintain naturally functioning ecosystems is to maintain ecosystem structure by avoiding practices such as overfishing that deplete species and destroy habitat. Still debated is whether maintaining the original species composition of an ecosystem is a practical or even necessary goal, the debate hinging largely on the issues of functional redundancy and replaceability of system components.

Some ecosystems, such as offshore-pelagic and temperate-benthic oceanic systems, are apparently more resilient than others in the interchangeability of their parts. Removal of one or several species has been followed by increases in other species, with fairly constant production of protein. This characterizes cycles of abundance of small pelagic zooplanktivores, such as occurred between anchovies and sardines off California, South Africa, and Peru. The presence of nutrient upwellings in these examples may influence events as much as the species involved.

Among benthic systems, dramatic shifts in species composition have occurred over large areas of the North Atlantic, but total fish biomass appears to have remained relatively constant. Groundfish (cod, haddock, winter flounder, and yellowtail flounder) along Georges Bank in the northwest Atlantic declined markedly but were replaced by skates, spiny dogfish, and other small sharks (figure 10.11). Groundfish constituted 70% and skates and sharks 30% of the catch in the early 1960s, but groundfish fell to 35% and skates and sharks rose to 65% of the catch in the late 1980s. Abundant skates and flatfishes also appear to have replaced overfished gadoids in the eastern Bering Sea and Gulf of Alaska (Goñi 1998). Between 1960 and 1990, total biomass in survey trawls changed little, averaging 121 kg per tow in the early 1960s and 141 kg per tow in the late 1980s (J. D. Reynolds and Jennings 2000). When production is the measure of stability, low-diversity pelagic and high-diversity benthic systems appear stable, despite shifts in relative abundances.

However, stable production and biomass can be misleading, at least from a conservation perspective. Catch statistics tend to be aggregated in many benthic fisheries, ignoring individual species. "Sharks" and "rays" are often treated as composite groups, regardless of the different species involved. The danger in aggregate reporting has become evident for skate fisheries off the British Isles. When viewed collectively, CPUE of Irish Sea and Bristol Channel skates remained remarkably constant from the mid-1960s to the early 1990s, as did aggregate surveys of abundance and biomass from 1988 to 1997 (Dulvy et al. 2000). Exploited species there included eight skate species (Rajidae) that range in adult size from 50 to 250 cm. Hidden in the apparently stable catch numbers were declines among large species, while small species increased. The three largest species—common, long-nose, and white skates (*Dipturus batis*, *D. oxyrhinchus*, *Rostroraja alba*)—disappeared locally, and three other large (>85-cm) species declined. Meanwhile, the two smallest (85-cm) species increased in abundance. When broken into taxonomically related pairs, the larger member of each of two pairs declined, and the smaller species increased in both abundance and biomass.

Similar size-related changes in abundance have occurred among six skate species in the North Sea, where an additional small species has invaded (Walker and Heessen 1996). Dulvy et al. (2000) concluded that smaller species experienced competitive release for commonly exploited benthic prey resources. These results are all the more interesting because strong evidence that fishing leads to competition-mediated changes in abundance is rare. These observations are also important because of the demonstrated vulnerability of large skates to extreme depletion via nontargeted as well as targeted fishing. Shifts in competitive balances could contribute to apparent skate extirpations (e.g., Casey and Myers 1998).

In highly diverse coral reef and kelp bed systems, the impact of species loss appears to depend greatly on the ecological role of those species. Removal of keystone species, of so-called strong interactors, or of some predators can have a pronounced impact. Hence overfishing of sea urchin predators creates areas denuded by urchin grazing, which affects the abundance of fishes dependent on plants for food and habitat. Ecological theory predicts that predatory species removed from most systems are least likely to be replaced naturally. The abundance and diversity of predators are constrained by energetics because they sit at the top of food webs (see Kappel 2005); replacement species are unlikely to be locally available. Some pieces of the jigsaw puzzle that is an ecosystem are more interchangeable than others.

Trawling's Effects on Ecosystem Structure and Function

Much attention has been focused on the extensive physical alteration that trawling causes to benthic habitats, damage that affects fishes through the removal of food resources as well as physical habitat (chapter 5). Trawling also affects sediment and nutrient distribution and alters food webs from the bottom of the sea to the surrounding landscape. Disruptions to energy and nutrient transfer can last for generations.

Trawling suspends and redistributes sediments among habitats, particularly in deeper water where currents are weak. Sediment budgets for regions of the mid-Atlantic continental shelf indicate that in deeper water (100–140 m), trawling contributes significantly to offshore transport of sediments (Churchill 1989). Various nutrients and elements that stimulate plankton productivity—most notably forms of nitrogen and silicon—are suspended with, adsorbed onto, or dissolved in these sediments and accompanying pore water. The amounts involved and their potential importance in food web dynamics are substantial. Redistribution of these materials can influence biogeochemical cycles, including the balance between naturally aerobic and anaerobic sediments, rates and pathways of nutrient regeneration, and nitrification and denitrification processes. Pilskahln et al. (1998) calculated that 9.1 kg sediment m^{-2} were resuspended annually by trawling activity in the Gulf of Maine, releasing 306 x 10^6 uM (= 4.29 kg) nitrogen. These inputs constitute an unrecognized contribution to the annual nitrogen budget of the region that could shift the balance between diatom and noxious phytoplankton production. Because phytoplankton is a major determinant of oceanic food web dynamics, intense trawling could alter ecosystem processes.

Bycatch as an Ecosystem Subsidy

Bycatch, whose impact is greatest in trawl fisheries, depletes nontarget species and size classes (chapter 10), but not all populations suffer from extensive bycatch. "The scavengers of the world's oceans have been receiving enormous carrion windfalls" (Britton and Morton 1994, p. 405). The ca. 30 million tons of catch discarded annually by commercial fishing vessels does not lie unused on the bottom of the sea. Much of it never even has a chance to sink. Fishing vessels are followed by active flocks of surface-feeding, scavenging seabirds (usually gulls and terns but also Bald Eagles and albatrosses); unseen scavengers lurk underwater. These foragers anticipate the discarded fish heads and body parts; dead and dying fish and invertebrates shoveled overboard; and injured, exhausted, and confused prey escaping from a trawl or dredge. Much of what is discarded consists of benthic species otherwise unavailable even to diving birds.

The tonnage of waste and abundance of scavengers is impressive. Jennings and Kaiser (1998) estimated that in North Sea fisheries, where the best data are available, 475,000 MT of bycatch (fishes, invertebrates, and offal) were discarded annually. Garthe et al. (1996) put the number at 790,000 MT. Seabirds consumed approximately half

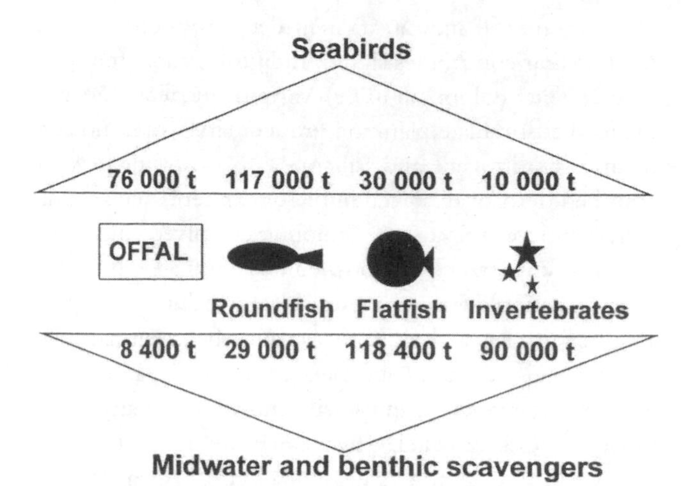

Figure 11.7. Distribution of bycatch discards to the scavenging biota. Based on data in Camphuysen (1993); recalculations in Garthe et al. (1996) put the distribution at 40% seabird consumption, with 60% going to other scavengers and decomposers. From Jennings and Kaiser 1998; used with permission.

of that (figure 11.7), sufficient to support about 2.2–5.9 million seabirds, which exceeds the population of scavenging seabirds in the North Sea. That population increased tenfold during the 20th century, which could easily reflect increased fishing intensity. Similar seabird increases have occurred in Australia, Canada, and the Falkland Islands, all areas of extensive high-seas fisheries. Discards in other fisheries are comparable to those in the North Sea, meaning analogous events likely occur wherever major fisheries exist.

The large number of birds supported by this waste has conservation implications beyond losses of fish and invertebrate species. In the absence of fishing, oceanic productivity may be insufficient to sustain such large seabird populations. Regulations that reduced bycatch or curtailed fishing would remove the discard subsidy. A trawling moratorium in Spanish waters affected foraging success and reproduction in Threatened Audouin's Gull. When trawling ceased, a mainland gull population switched to alternative feeding sites and obtained food easily in nearshore and inshore habitats, but an offshore, island population was forced to fish actively at apparently greater energy expense and suffered a 48% reduction in breeding success in one year (Oro et al. 1996). Before changes in fishing practices are instituted, their ecosystem-level impacts have to be assessed. Modifications—such as introducing changes gradually to minimize the rate of negative impacts on seabirds, especially imperiled species—may be necessary. Such impacts could range from loss of food to an increase in predation by species such as skuas that now feed on discards (Camphuysen 1993; Tasker et al. 2000).

The 40%–60% of the North Sea bycatch not eaten by seabirds sinks and becomes available to midwater and benthic scavengers. Little is known about who eats what parts and how much of this component over how long, what impacts the discarded biota has on different trophic groups, and what if any alterations in oxygen availability, energy flow, and food web structure occur as a result of trawling. Groenewold and Fonds (2000) trawled repeatedly over an area and deployed traps and baited lines in trawled and untrawled areas. They identified scavengers, calculated consumption rates, and assessed the relative importance of scavenged food to different components of the benthic ecosystem (figure 11.8). Scavengers included gadoids, flatfishes, dragonets, gurnards, hermit crabs, swimming crabs, sea stars, brittle stars, isopods, and amphipods. Two-thirds of discarded fish, twice the marketable catch, fell to the sea floor. One pass of a beam trawl produced 0.2 g ash-free dry weight (AFDW) m^{-2} scavengeable discards, to which was added 1.1 g AFDW m^{-2} noncatch scavengeable material left in the wake of the trawl (an important component in fisheries discards that is generally ignored). Most of the scavengers were opportunistic generalists that normally fed on a variety of prey; one isopod and two amphipods were the only species considered to be specialized scavengers. Most discards were thoroughly consumed within a few days to a few weeks, depending on size and water temperature.

Groenewold and Fonds (2000) compared the amount of food mobilized by trawling with values of secondary production and benthic biomass in untrawled areas to determine impact on food web properties and dynamics. They calculated that each pass of a beam trawl makes available to scavengers and the detrital food chain 6%–13% of annual secondary production and 11%–32% of standing stock of macrobenthos. This accelerates release of organic material, which is consumed by the scavengers, transferring material into the benthic food web and making energy available to foragers that is normally tied up in benthic, suprabenthic, midwater, and pelagic species. Under undisturbed circumstances, carrion is produced infrequently and dispersed widely (e.g., Britton and Morton 1994); transfer of energy to the benthos would take many years via excretion and natural mortality rates. "Theoretically, the short-

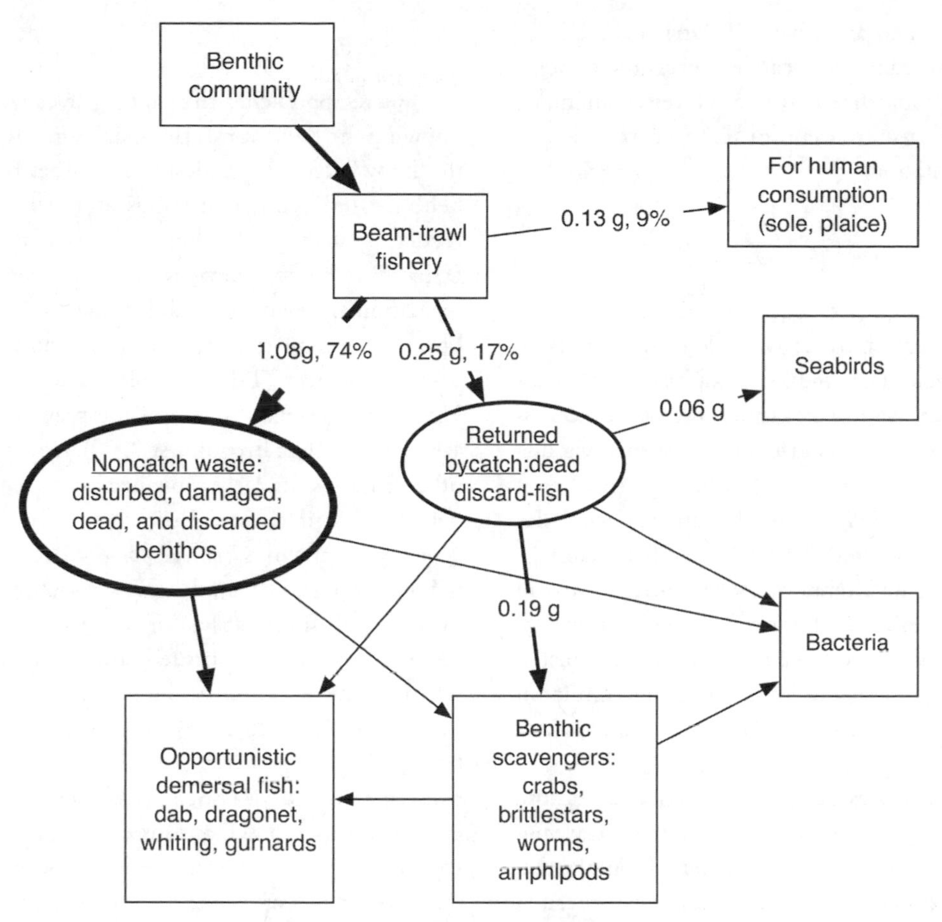

Figure 11.8. Food webs in the wake of trawls. Of the 1.5-g ash-free dry weight m^{-2} "mobilized" by the single passage of a beam trawl, 9% is used for human consumption, 17% is discarded (returned bycatch), and 74% is noncatch waste that is never hauled to the surface but is made available to scavengers. After Groenewold and Fonds (2000); used with permission.

cut in the trophic relationships may lead to a faster transfer of organic matter, higher ecological efficiency, and therefore, higher secondary production" (Groenewold and Fonds 2000, p. 1404).

Left unresolved is whether faster transfer, higher efficiency, and higher production come at the cost of food web disruption. This accelerated transfer of energy to lower trophic levels is another example of humans shifting energy from higher to lower trophic components of the ecosystem. Discards may, in fact, be one causative factor, along with overfishing, in the fishing down of food webs. Fishing removes large and predatory species, and discards and noncatch waste rapidly mobilize energy normally sequestered in higher trophic levels, fueling the growth of scavengers and detritivores while inhibiting recovery of predatory stocks (e.g., turning cod into isopods). Overfishing and overdiscarding may thus drive the fishing down–microbialization syndrome that eliminates apex predators and large species while transforming the ocean into a simplified system increasingly dominated by microbes, jellyfish, benthic invertebrates, plankton, and planktivores (e.g., Pauly et al. 1998a; J. B. C. Jackson et al. 2001; see chapter 10).

Given the amount of organic material discarded worldwide, the amount turned into carrion by ghost fishing, and the importance of detritus-based food webs in the ocean, our incomplete understanding of consumption and processing by scavengers represents a major gap in our knowledge of the impact of discards on ocean ecosystems. Increased abundance of lesser-spotted dogfish, *Scyliorhinus canicula*, off Spain and thorny skate, *Raja radiata*, off Greenland have been linked to finfish discards (Goñi 1998). Are fisheries for scavengers such as sharks, eels, hagfishes, shrimps, and crabs heavily dependent on energy that comes from discards and an enhanced detrital food web? Would curtailed discarding of bycatch depress these lucrative fisheries? The topic needs to be addressed in many more regions, with greater attention to food web disruption and supplementation.

Noncatch Biomass: An Attractive Nuisance

Noncatch biomass is the fraction of catch that is not hauled to the surface but is made available to scavengers by the action of trawls and dredges. Groenewold and Fonds (2000) estimated that five times as much noncatch biomass remained in the wake of a trawl than was returned via discards; Lindeboom and de Groot (1998) put the quantity at about the same level as that discarded at the surface. S. R. Jenkins et al. (2001) concluded that most of the damage done to large benthic invertebrates "occurs unobserved on the seabed, rather than in the bycatch" (p. 297). The magnitude of the numbers and discrepancies among calculated values indicate the level of our ignorance concerning an alarming component of what we "consume" of the ocean's productivity.

Because injured fish escape from trawls and because trawls and dredges disturb biota on the sea floor, trawling has another, unexpected effect on marine food webs. Trawl and dredge zones are attractive nuisances, recruiting and concentrating scavenging and predatory taxa from distant locales. Fish shoals can be 35 times more dense over recently trawled regions than over nearby unfished regions, and the density of predatory fishes in recently dredged areas can be 3–30 times higher than in nearby undredged areas. Gadoids aggregate around newly disturbed pits in sandy bottoms, and flatfishes and dragonets are recruited from areas 4 to 5 times the size of the area affected by a trawl. The attraction phenomenon underscores how trawling's impacts exceed the immediate trawled area (Jennings and Kaiser 1998; S. J. Hall 1999).

This attraction phenomenon has implications for fisheries and benthic ecosystems. Fishers know about and exploit the attraction of fishes to trawl paths and tracks by lining up along the same trawl paths or returning to retrawl areas repeatedly. Retrawling, while obviously convenient to fishers, intensifies the degree of destruction done to a region and concentrates fish species that might otherwise be spread over a region, making stocks and areas even more susceptible to overfishing.

Solutions to Bycatch Problems

Innovative approaches to reducing bycatch rely heavily on knowing the behavior of the species involved and applying this knowledge to gear design and fisher behavior. Applied behavior (ethoconservation) is at the heart of efforts to reduce bycatch of dolphins in tuna nets (see box 11.1). Bycatch reduction devices (BRDs) rely on mechanical separation and behavioral differences and have been developed for several species. The best known are the turtle excluder devices (TEDs) employed in shrimp fisheries in U.S. and Australian shrimp and prawn fisheries (NRC 1990; Perra 1992; Broadhurst 2000). Devices developed for other fisheries similarly capitalize on passive separation and behavioral differences. Bycatch of small fishes in the Norwegian prawn, *Nephrops norvegicus*, trawl fishery can be reduced using large-mesh escape panels in upper portions of nets, with minimal loss of prawns. The young of many benthic fish species "pursued" by a trawl rise over 70 cm when exhausted, whereas the prawns seldom rise that high (Main and Sangster 1985). Applying behavioral information gathered by divers or video cameras can lead to bycatch reduction in such notoriously nonselective fisheries as bottom trawls for groundfish. For example, saith, haddock, and cod separate vertically when escaping a trawl, the first two species moving upward and the cod remaining closer to the bottom. Separator panels can then segregate as much as 90% of the haddock, 60%–70% of saith, and 65%–70% of cod (Engas et al. 1998). Norway's "no discards allowed" policy (chapter 10) promotes use of BRDs via the disincentive of having to take to the dock any fish caught, regardless of value.

Trophic Interactions and Food Web Disruption

Fishing and Trophic Cascades

Evidence of top-down, predator-controlled ecosystems is strongest from temperate lakes, where predator removal and augmentation have become a management tactic for manipulating water clarity (Carpenter and Kitchell 1993). A lake trophic cascade might involve piscivores such as salmonids, pike, and basses that feed on zooplanktivores such as herrings and minnows, which feed on herbivorous zooplankters, which eat phytoplankton. Experimental increases in the numbers of piscivores, via additions or

BOX 11.1. Is a Tuna Sandwich Worth Dying For?

Probably the best-known bycatch problem involves the capture of spinner, spotted, and common dolphins (*Stenella* spp., *Delphinus delphis*) that associate with yellowfin tuna in the eastern Pacific. Dolphins began dying during tuna fishing operations in the 1950s and 1960s, when the fishery changed from a relatively low-tech, pole-and-line activity to one involving very large, fast-sinking purse seines. I spent a month in 1970 working on one of the purse seiners.

Tuna aggregate under dolphin. In the heyday of purse seine fishing, tuna fishers searched for dolphin, which were easier to spot than the tuna that swam below them (why this association occurs remains a matter of conjecture). The dolphin were then herded together by small chaser boats directed by an observer positioned in the crow's nest of the tuna boat, who communicated with the chaser boats via radio headsets. The tuna remained below the dolphin, and the net was set around and under the dolphin-tuna aggregation (termed "encirclement fishing"). When the net was pursed and brought aboard the catcher boat, the tuna were kept and the dolphin were released, in theory. In fact, many dolphin became entangled in the net and drowned, were crushed as the net passed through winches and blocks, or died falling onto the deck of the ship from the boom used to winch the net aboard. Between 1959 and 1972, an estimated 4.9 million dolphin, or about 350,000 per year, were killed in the fishery, representing 20%–72% of the dolphin populations, depending on species and stock. At those capture rates, and given a conservative (low) estimate of dolphin recruitment of 2% per year, some dolphin stocks were in distinct danger of extirpation (T. D. Smith 1983; P. R. Wade 1995).

A public outcry led to changes in the fishery, not to mention stimulating passage of the U.S. Marine Mammal Protection Act of 1972. Fishers developed improved nets and a method of backing down the net, which submerged the float line and allowed many dolphin to escape, often aided by fishers who jumped into the melee, at considerable personal risk (NRC 1992a). More effective was a 1990 U.S.-mandated certification program for labeling tuna as "dolphin safe," ensuring (again in theory) that no dolphin had been captured in the process of netting the fish. Observers placed on some tuna boats authenticated the dolphin-safe designation, but setting on dolphin was otherwise a matter of self-policing and self-reporting. Most U.S. fishers switched from setting on dolphin to setting around logs and other objects at the surface ("log sets") or around mobile tuna schools that were not associated with dolphin or other objects ("school sets"; fishers also set around turtles, whale sharks, and any immobile or slow-moving object at the surface under which tuna aggregated). The (reported) kill of dolphin dropped more than 98%, to 133,000 in 1986 and to 2,600 in 1996. The latter figure represents an annual mortality rate of 0.1%, which is well below the 2% recruitment rate and is, from a purely statistical standpoint, sustainable (M. A. Hall 1998). Some concern remained about the nonmortality stress caused by entrapment in the net and its effects on reproduction, plus the orphaning of nursing dolphins separated from entrapped mothers.

Emotions overruled statistics. Many people feel that there is no excuse for killing dolphin to catch tuna, and that no dolphin sets should be tolerated. Public opinion led to a series of U.S. laws focused on the dolphin-safe label. In addition to the Marine Mammal Protection Act, Congress progressively passed the Dolphin Protection Consumer Information Act (1990); the International Dolphin Conservation Act (1992), which made it unlawful to import tuna that was not dolphin safe; and the International Dolphin Conservation Program Act (1997)—all of which led to reduced U.S. participation in the fishery, which was supplanted by a less regulated international fishery. International law, particularly the General Agreement on Tariffs and Trade (GATT), found in 1994 that U.S. restrictions against importing dolphin-unsafe tuna violated international treaty agreements (www.oceanlaw.net). The legal basis for the finding was that countries could control products but not processes, and encirclement fishing was a process. Amid these and other international agreements and lawsuits, "dolphin safe" has undergone a series of modified definitions designed to encourage stronger international participation. The modifications have met with mixed reactions from the environmental community because they in effect relax restrictions on encirclement fishing.

Further complicating the picture from economic, ecological, and ethical standpoints are the impacts of alternative fishing methods on other ecosystem components. Dolphin sets have very low bycatch rates for other fishes and capture tuna of uniformly large sizes, with a discard rate of only 1%. Log sets in contrast often take depleted turtles, sharks, billfishes, and other non-target species at a rate 10 to 100 times greater. Log sets are also inefficient, catching many small, unmarketable tunas, which make up 20%–25% of the catch before being discarded dead. Dolphin sets, given improvements in net design and color and capture methods, result in bycatch of a small number of dolphin at a rate that dolphin populations can apparently sustain. From an ecosystem management perspective—emotions aside—dolphin sets are the optimal fishing method (M. A. Hall 1998). One alternative solution is to return to pole-and-line fishing, which entails minimal bycatch and would increase employment but would probably lead to higher-priced tuna sandwiches.

controls on fishing, lead to a reduction in zooplanktivores. Fewer zooplanktivores mean more zooplankton, which reduce phytoplankton biomass. The chain of events can be reversed if piscivore abundance is reduced experimentally or through overfishing; fewer piscivores mean that zooplanktivorous fish increase in numbers and reduce zooplankton, allowing phytoplankton to bloom. The present-day, fished-down structure of many exploited systems in both freshwater and marine habitats undoubtedly reflects historic overfishing of apex predators.

Other examples of exploitation affecting ecosystem structure and function come from coral reefs, where top-down effects are well documented. The classic example involves overfishing of predators on reef urchins, leading to well-known "urchin barrens," or removal of large predators and large herbivores, thus enhancing growth of algae and small herbivores (detailed in chapter 12).

Humans as Competitors

The humans-as-predator analogy explains much about the way we deplete fish stocks. But as seemingly insatiable predators, humans are also likely to deplete resources necessary to sustain other species in an ecosystem. Are we the ultimate competitor in addition to being the ultimate predator? The evidence for human impacts on competing species is abundant if not definitive. Many of the world's largest fisheries involve small pelagic species. Given that these abundant prey fishes are likely to have many predators that potentially compete with one another for prey, it is not surprising that precipitous declines in prey fish abundance due to human exploitation affect many components of a trophic web.

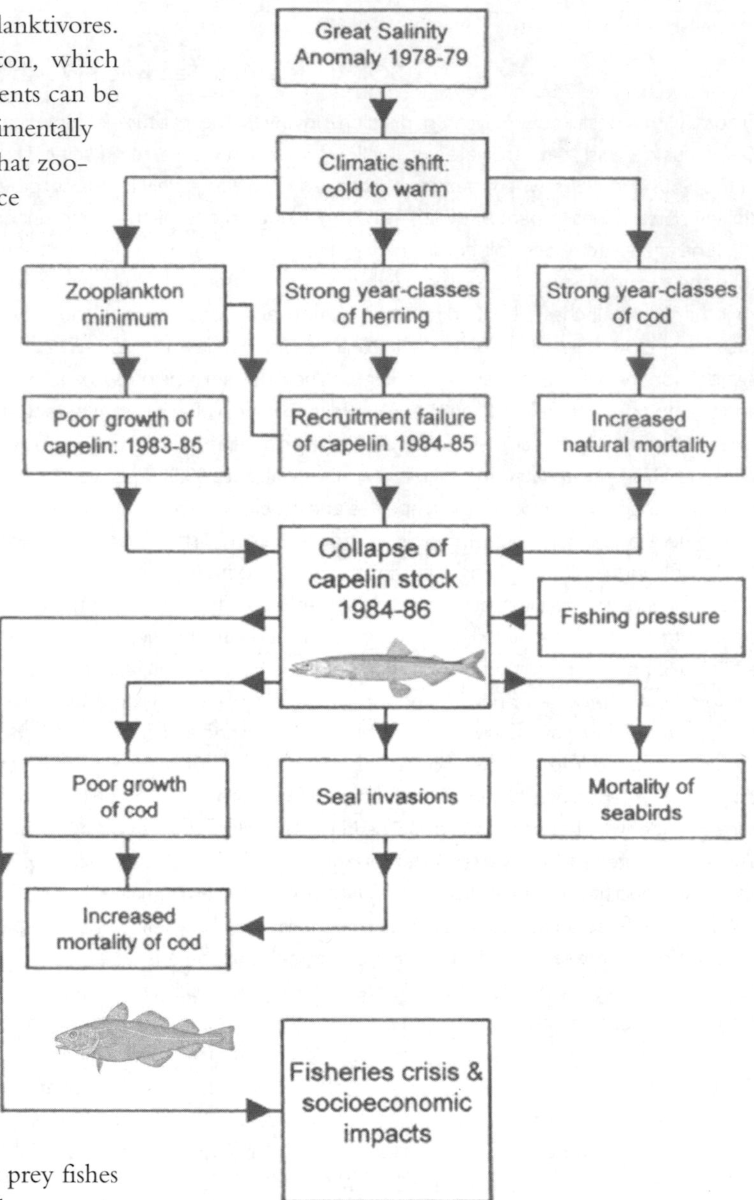

Figure 11.9. Ecological events in the Barents Sea ecosystem. Interactions among environmental factors, fishing pressure, and competition for capelin eventually led to declines among the competing predatory species of cod, seals, seabirds, and humans. From Jennings and Kaiser (1998), redrawn and modified from Blindheim and Skjodal (1993); used with permission.

An oft-cited and -interpreted example of this kind of interaction involves the decline of capelin, *Mallotus villosus*, in the Norwegian Sea–Barents Sea ecosystem in the 1970s and 1980s (figure 11.9). The players in this system included herring and capelin (and zooplankton and juvenile cod) as a food source; human fishers, cod, seabirds, and seals as competing predators; and climatic shifts involving temperature and salinity fluctuations as environmental drivers (see Jennings and Kaiser 1998; J. D. Reynolds and Jennings 2000). Briefly, warmer water in the late 1970s resulted in increased year-classes of herring and cod but decreased zooplankton abundance, which contributed to poor recruitment of capelin in the early 1980s. Poor recruitment combined with heavy fishing led to collapse of the capelin stock in the mid-1980s. Because cod, seabirds, seals (and humans) relied on capelin as prey, these populations crashed or suffered economic decline.

The Barents Sea capelin story underscores how trophic interactions associated with overfishing can be mediated through competition and affect seabird abundance. Seabird populations rise and fall in direct relation to the availability of prey fishes (Furness and Camphuysen 1997; Tasker et al. 2000). Guillemot populations fell 80% in 1985–87 with the crash of capelin in the Barents Sea; puffins off Norway experienced low or no breeding success associated with overfishing of herring; guillemots, murrelets, puffins, and arctic terns all showed substantial population declines in response to possible overfishing of sand eel stocks in Prince William Sound, Alaska (where the *Exxon Valdez* ran aground); and overfishing of anchovy species off Peru and southwest Africa have been accompanied by major reductions in guano-producing birds (boobies, cormorants, pelicans) that fed on the fishes. Flightless seabirds also suffer. Magellanic penguins off Argentina and imperiled African penguins off southern Africa have shown reductions in breeding pair numbers, chick starvation, and even extirpation of breeding colonies because of apparent competition with commercial fisheries for prey fishes (Boersma 1997; Crawford 2001). Because of the tight coupling between fish abundance and seabird ecology in oceanic ecosystems, some authors have proposed that seabirds be used as indirect measures of abundance for pelagic fish species that are otherwise difficult to assess (e.g., Furness and Camphuysen 1997), assuming that the diets of these birds are not heavily supplemented by fisheries discards (Britton and Morton 1994).

Marine mammals depend on the same prey species as seabirds and humans, but because of their ability to store fat, they are often less dependent on specific, local food sources. However, declines in the abundance of pinniped species have been linked to periods of overfishing. In the Bering Sea, where marine mammals and seabirds consumed about 3 MMT of fish annually, expansion of the walleye pollock fishery occurred concurrently with declines of 75% in Steller sea lions, 60% in northern fur seals, and 85% in harbor seals (Goñi 1998; analogous declines have been observed among fur seals off Peru and gray seals in the North Atlantic). Events in the North Pacific have been interpreted as a classic trophic cascade, where declining seal and sea lion populations have been linked to increased predation of orcas on sea otters as alternative prey, thus releasing sea urchins from predation by the otters, and culminating in a loss of kelp forests because of overgrazing by the urchins (Estes et al. 1998). In other mammalian interactions, changes in whale distributions have been linked to the abundance of fisheries resources, as when humpback and fin whales appeared in the southern Gulf of Maine to feed on a proliferation of sand eels, but northern right and sei whales left the area in apparent response to decline of the copepods, which are the mutual prey of the sand eels and the whales. P. M. Payne et al. (1990) described these changes in distribution as likely responses to anthropogenically induced alterations of planktivorous fish abundance.

It thus appears that impacts of overfishing on competing fish species are often difficult to demonstrate, whereas impacts on seabirds and mammals are clearer. These differences can be partially explained by energetic phenomena. As "cold-blooded" vertebrates with relatively low metabolic rates and requirements, most fishes can survive extended periods on relatively little food, once they are past the earliest life history stages. In contrast, birds and mammals that co-occur with and depend on fishes for food have high metabolic rates and are much more vulnerable to starvation. Hence overfishing can have dramatic direct and indirect impacts on these higher vertebrates (for more on human-whale interactions, see chapter 15).

Selective Fishing and the Fishing-Down Phenomenon

The fishing-down phenomenon (chapter 10) has strong implications for the trophic structure of the world's oceans. Fishing-down practices have altered the structure of marine food webs because eliminating top carnivores can cascade throughout an ecosystem. In the North Sea, Norway pout, *Trisopterus esmarkii*, were eaten by large gadids such as cod and saithe. After cod and saith were depleted, pout were targeted by the fishery. When pout numbers then declined, recovery of cod stocks became less likely. Fishing pressure on pout also reduced their predatory impact on krill. Krill eat copepods, which are a more important food source for young gadids than krill, again arresting the recovery of the most important commercial species (NRC 1999b).

Fishing at lower levels in the food web could be viewed as ecologically more efficient. If herbivores or detritivores rather than predators were targeted, less energy would be lost from the multiple energy transfers that occur moving up the food web. Such fishing would permit harvesting more biomass with less impact (e.g., M. A. Hall et al. 2000). Although proactive application of fishing down is theoretically feasible, it would require reworking current practices and customs. At present, fishing down occurs primarily because more desirable, higher-level predators have been overfished. Planktivores are not pursued as human food but are used extensively for animal feed, as in the production of farmed salmon, which adds steps to the food web and is inherently inefficient (chapter 14). Convincing consumers to forgo tuna and salmon in favor of menhaden and anchovies would require some dramatic shifts in attitude. Regardless, Pauly et al. (1998a) provided evidence that fishing down food webs, a well-documented phenomenon, has not created more food from the sea. Small pelagic fishes are being overexploited and depleted not unlike the way we overexploited and depleted the benthic and predatory trophic groups they appear to be replacing (see chapter 10).

Palatability aside, an important drawback of shifting fishing effort to lower-value species is their unreliability as a food source. Small pelagics fluctuate wildly in numbers because their populations are tightly linked to vagaries of climate. Historical shifts in major climatic forces have contributed to spectacular crashes in these fish species (e.g., California sardines and Peruvian anchoveta). Future scenarios of climate change could make them even less reliable as a food source (Garcia and Newton 1997).

What a Bear Does in the Woods May Be Important

Distant and seemingly unrelated ecosystems can be connected via fishes, Pacific salmon being a prime example. Salmon life histories are complex because the fish migrate between and depend on different habitats at different stages. Larval salmon depend on intact headwater streams embedded in old-growth forests, and the health of these forest ecosystems may be greatly influenced by the spawning migrations of adult fish.

The intricacies and interdependencies of salmon and their habitats have been demonstrated in a series of studies, summarized by Willson et al. (1998) and Schindler et al. (2003). Vertebrate and invertebrate predators feed directly on migrating salmon adults, eggs, and fry, while scavengers feed on the carcasses of spawned-out adults. Willson and Halupka (1995) identified 40 species of inland mammals and birds that feed on salmon, including ducks, geese, gulls, dippers, and robins on eggs; loons, mergansers, herons, terns, kingfishers, and crows on juveniles; and eagles, hawks, magpies, ravens, and jays on adults and carcasses. Mammalian piscivores and scavengers include bears, mink, otters, wolverines, wolves, foxes, seals, mice, squirrels, and deer (this doesn't include eight to ten riverine fishes that eat eggs and juveniles, numerous stream invertebrates that scavenge salmon carcasses, and a rich community of bacterial and fungal decomposers that recycle salmon-derived nutrients). The life histories of some species—including hibernation and lactation of bears, birthing of mink, reproduction of megansers, overwintering success of eagles, and survival of juvenile salmonids that feed on the next year's eggs—are clearly adjusted to the timing of salmon spawning (Willson et al. 1998).

A direct trophic linkage between salmon and their predators and scavengers is interesting, but the feedback loop that travels through salmon, streams, and surrounding terrestrial habitats is stunning (figure 11.10). Growth of juvenile salmon is closely linked to the availability of salmon carcasses in a stream (Bilby et al. 1996), in part because young salmon feed on invertebrates that feed on salmon carcasses. But juvenile salmon also benefit because streams with well-developed riparian vegetation provide better salmon habitat for spawning adults and developing fry via temperature-regulating shade, insect inputs, and woody debris inputs that control flow, sediment filtration, and bank stabilization.

Importantly, the salmon may also contribute to the well-being of the riparian trees. Bears and eagles feed on spawning salmon and salmon carcasses. They carry carcasses into the woods and defecate there. Carcasses and feces release phosphorus, nitrogen, carbon, and micronutrients that originated in the ocean. By some calculations, phosphorus deposition by bears approximates 7 kg/ha^{-1}/yr^{-1}, which is similar to the recommended commercial rate of application of phosphorus fertilizer for evergreens and other trees (Willson et al. 1998). Also, marine-derived nitrogen in carcasses and feces may provide 20%–25% of the nitrogen taken up by riparian trees as far as 100 m from a stream (Helfield and Naiman 2001; the percentages have been debated, but some nitrogen enrichment remains plausible e.g., Helfield and Naiman 2003). Sitka spruce, an old-growth species, grows three times faster along salmon-bearing streams, producing the 50-cm-diameter trees that

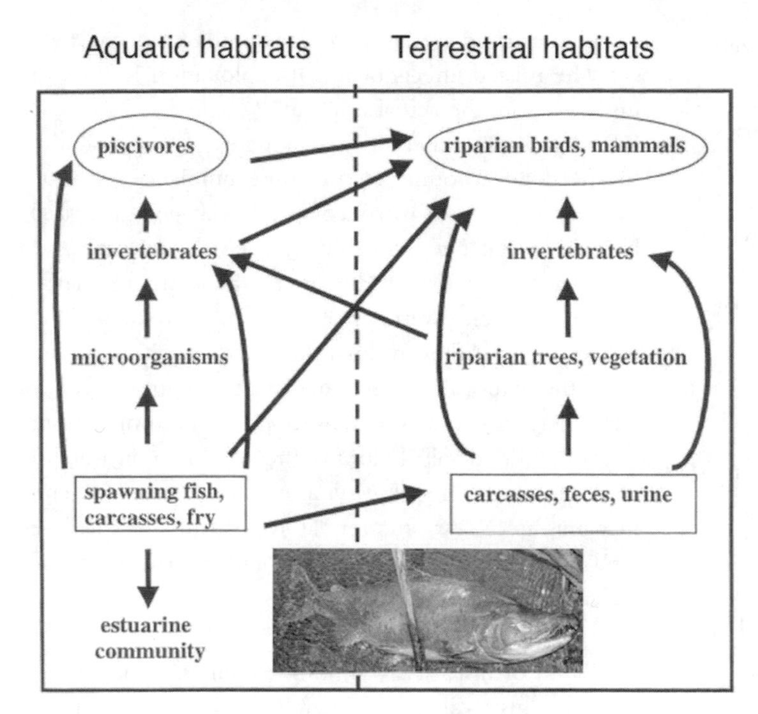

Figure 11.10. Salmon-derived nutrients and food webs based on anadromous fishes. The Pacific salmon ecosystem includes oceanic, estuarine, riverine, and old-growth forest components; aquatic and terrestrial habitats are linked via predators and scavengers that feed on different life history stages. Modified from Willson et al. (1998); used with permission. Inset: A decomposing sockeye salmon carcass.

are most useful as woody debris inputs. Trees growing along salmon streams may achieve this size about 200 years faster than trees along streams without spawning salmon (86 years vs. 307 years). Via a feedback loop that includes salmon, bears, and trees, "spawning salmon help to enhance the survivorship of subsequent salmonid generations" (Helfield and Naiman 2001, p. 2407).

Because of the role of salmon and other migratory species (e.g., lampreys, shad, smelt, char) as connectors and enhancers of ecosystems, Willson et al. (1998) referred to them as "cornerstone" species, species that provide a resource base that supports coastal and inland ecosystems (they may also transport contaminants from the ocean to inland waters; see Ewald et al. 1998 and chapter 7). Overfishing erodes and eventually removes this cornerstone, which slows tree growth, produces shorter trees, reduces woody inputs, degrades riverine habitat, and eventually accelerates the decline of salmon production. Recruitment in salmon fisheries is intricately linked to spawning stock in ways few managers or ecosystem ecologists could have imagined, reinforcing the conclusion that managing salmon demands managing the entire ecosystem, from the North Pacific to headwater streams. "The links between ocean and land mean that management of an ocean fishery can have far-reaching effects on distant ecosystems, and vice versa" (Gende et al. 2002, p. 924). Defining sustainable yield or allowable catch of salmon is much more complicated than traditional fisheries models would predict. A significant fraction of the catch must be allocated to the streams and forests, not just to the fishery.

ECOSYSTEM CAVEATS

At least four important topics complicate our ability to understand and predict the impacts of fishing on ecosystems.

Shifting Baselines and Fisheries Science

Shifting baselines (Pauly 1995) refer less to ecosystem traits and more to our perceptions and interpretations of the effects of fishing, especially its historical impacts. The ecosystems we study today, and by extension the interactions among species within these ecosystems, do not necessarily represent the real, evolved set of species or interactions. Most fisheries experienced significant exploitation long before we began to measure populations and interactions. With many components missing or greatly depleted, we have a distorted idea of what should be there, what might be happening, and what actions we should take to restore ecosystems to something approaching pre-impact, functioning conditions. Different authors have emphasized different aspects of this legacy of past actions:

- Pauly (1995) focused on the unwillingness of fisheries scientists to place credence in historical, anecdotal literature that described the incredible former abundance of many fisheries resources. This widely held "objective" view biases researchers into using as their baseline for comparison population parameters and conditions that existed when they began their work. Because this baseline shrinks with each generation of workers, expectations about natural population levels and desirable recovery targets are progressively lowered. A major consequence is that the true extent of the crisis facing the world's fisheries is unappreciated.
- Mowat (1996) discussed the defaunation of the North Atlantic, citing numerous historical references in making the point that the diversity and abundance of organisms we observe in the Atlantic Ocean today pale in comparison to what that ocean contained before Europeans began their assault on its biota. MacIntyre et al. (1995) used Mowat's historical documentation to arrive at an estimate of just how much had been lost. They concluded that the Atlantic coast of Canada contained perhaps 10% of its original biomass of fish and other exploitable organisms (whales, seals, mink, otter, bear, large seabirds).
- Larkin (1977) was among the first to emphasize the management implications of anthropogenic change in exploited systems. He stated (p. 6) that "there is little doubt that in many parts of the world the species assemblages of fishes that we observe today must be profoundly different in their composition and interrelationships from the assemblages of a century ago, and so are the communities of organisms on which the fish feed."
- Dayton et al. (1995) referred to the shifting baseline problem in an ecosystem context. They concluded that efforts to evaluate the impacts of bycatch, habitat destruction, and other environmental effects of intense fisheries are destined to fail because "most sensitive species have long been impacted, leaving no concept of natural relationships or patterns" (p. 206).
- Jennings and Kaiser (1998) pointed out that fishing has its greatest effects on habitat, genetic diversity, age structure, life history traits, and stock size when unfished areas or stocks are first exploited (see also R. A. Myers and Worm 2003). Areas and resources that have been under exploitation may appear resilient and unresponsive to additional fishing because the greatest changes have already occurred and only the most resilient species persist. The missed effects of initial exploitation lead us to underestimate the actual impacts.
- J. B. C. Jackson et al. (2001) gave increased respectability to anecdotal accounts of previous abundance by augmenting them with paleoecological, archaeological, and historical data. They concluded that such retrospective data demonstrated that "historical abundances of large consumer species were fantastically large" (p. 629). The major impact of overfishing in coastal ecosystems has been the removal of apex predators, leaving us today with altered, degraded systems subject to eutrophication, disease, and toxic algal blooms. Present-day indicators of degradation are in reality symptoms of past actions and may misdirect our attempts at restoration because we tend to focus on recent problems rather than their historical causes.

Because of historically shifting baselines, what we *can* measure is misleading. It is, therefore, important that we create protected areas large and diverse enough to allow systems to recover to something resembling naturally functioning ecosystems, thus giving us more realistic baselines for comparison. Pitcher and Pauly (1998) emphasized that shifting baselines can undermine the focus on sustainability that characterizes much ecosystem-based management. Fished-down systems with abundant planktivores and other species low in food webs are probably easier to sustain than fisheries containing large predators, but what we should do instead is (1) redouble efforts at setting aside and protecting fully those few remaining areas that have been least impacted, and (2) wherever possible, rebuild ecosystems toward their primal condition, before they were devastated by industrial fisheries. Merely sustaining ecosystems in a depleted condition achieves neither ecologic nor economic goals.

Multiple Causation versus Correlation

Many of the observations and conclusions presented here are built on correlations between species trends and other, potentially causative factors. Correlations can be confused for causation, a problem that grows when dealing with ecosystem-level phenomena, which are complex and respond to multiple, interacting factors. Jennings and Kaiser (1998) emphasized the pitfalls of concluding that fishing has ecological effects based only on correlative or inferential data that do not also account for alternative explanations such as climate change. Their major concern is that scientists "are now claiming that fishing has many wider and deleterious impacts on ecosystem function because they intuitively believe this to be the case and not because they can convincingly demonstrate an effect" (p. 293).

The existence of alternative explanations should be acknowledged. Seabird and pelagic fisheries fluctuations may result from fishing intensity but also from simultaneous climatic shifts that affect currents and upwellings and lead to plankton blooms that encourage population growth of prey species, independent of fishing. Debate rages over which factors are most influential (e.g., see Jennings and Kaiser 1998 vs. Tasker et al. 2000).

If claims that overfishing has caused damage to ecosystems were shown to be in error, the credibility of anyone offering management advice based on ecosystem solutions would be compromised. The strength of the evidence of significant and deleterious ecosystem-level impacts of fishing vary considerably across habitat types, being clearest in benthic habitats and organisms and in coral reef systems in general. Here an ecosystem-level rather than a population-level focus for management is justified. However, large oceanic systems involving heavily exploited, multispecies fisheries, including some large pelagics, provide less clear-cut examples of overfishing; and disentangling ecosystem-level effects from other correlated factors is therefore difficult. Small pelagic fishes provide the weakest case because their populations are known to have fluctuated in response to climate, in the absence of exploitation.

Definitive answers are therefore difficult to come by in this arena. Although I have tried to avoid examples that appear to have equivocal explanations, almost all represent conclusions drawn from correlations because ecosystem and evolutionary phenomena are difficult to manipulate experimentally. Conclusions drawn from correlations should be viewed with a degree of caution.

Regime Shifts and Alternative Stable States

In a related vein, increased attention is being given to major oceanographic climatic cycles as determinants of population dynamics and ecosystem structure. Changes in species composition induced by overfishing interact with such "regime shifts" to alter food webs. These long-term, wide-scale climatic shifts occur across entire ocean basins on a periodicity of several years to a few decades (Steele 1998). The shifts can involve major water masses and affect chlorophyll concentrations and productivity, or alter wind patterns and influence the distribution of planktonic organisms, including fish larvae, which may ultimately influence recruitment. Changing physical conditions interact with fishing patterns to determine population and ecosystem properties (see Botsford et al. 1997; Goñi 1998; Jennings and Kaiser 1998). The relative importance of climate versus overfishing can be debated, but "uncertainty about effects of environmental variability should not be used to excuse continued overfishing" (NRC 1999b, p. 5).

The biological analog of climatic regime shifts is the concept of alternative stable states within ecosystems. The "state" is a dominant species or mix of species, and the implied interaction is that conditions that lead to replacement of a previous dominant (including climatic regime shifts) may favor the continued dominance of the newly abundant taxa, sometimes abetted by predatory or competitive actions of the new taxa. Hence overfishing of gadoid groundfish in the northwest Atlantic, coupled with habitat destruction from trawling, has permitted skates and small sharks to increase in abundance. These elasmobranchs may prey on or outcompete depleted gadoids and flatfishes, impeding their recovery (recovery being further slowed by continued fishing pressure and perhaps by water temperature changes that also affect cod negatively) (Fogarty and Murawski 1998). The Baltic Sea may be fluctuating between cod and herring dominance (Rudstam et al. 1994). In the unfished Baltic Sea, adult cod fed on adult herring and young cod fed on larval herring, leading to cod as the dominant species. Overfishing of cod released adult herring from predation, and abundant herring feed on larval cod, depressing future cod recruitment (eutrophication and pollution may additionally retard cod recovery).

Other proposed examples of alternation between stable states include (1) cephalopods and various groundfish alternating as dominants off northwest Africa, the Gulf of Thailand, and the Adriatic Sea in response to overfishing, sometimes combined with upwelling regimes (Caddy and Rodhouse 1998; S. J. Hall 1999); (2) increasing dominance of flatfishes in trawl fisheries in the North Pacific (S. J. Hall 1999), another region characterized by changes in major climate patterns; and (3) the change from a cichlid-dominated to a Nile perch–dominated assemblage in Lake Victoria, where cultural eutrophication is likely to inhibit return of remaining cichlids (Kitchell et al. 1997). In all instances, it is unlikely that altered fishery regulations will restore anything resembling the previous ecosystem structure and species composition, unless we are willing to drastically curtail effort, perhaps for decades. Lake Victoria is pretty much a lost cause from a biodiversity standpoint.

Nonprecautionary Statistics

Impacts of fishing are commonly analyzed with statistics that test the null hypothesis of no effect from fishing. This approach represents an explicit avoidance of claiming incorrectly that an effect exists (see chapter 16 for details). In scientific circles, it is a greater sin to accept a falsehood than to fail to accept a truth: "Getting it wrong is worse than not getting it right" (S. J. Hall 1999, p. 217). Dayton et al. (1995) and Jennings and Kaiser (1998) emphasized that indirect effects and ecosystem impacts, because of their inherent complexity and long-term nature, are particularly susceptible to mistaken analyses that begin with an assumption of no effect. Alternatively, mistaken rejection of the null hypothesis of no effect can result in unnecessary curtailment of fishing activities, with considerable economic and political ramifications.

On the other hand, skimpy data, high variance, and inappropriate statistical tests with insufficient statistical power are inadequate to prevent the alternative sin of proclaiming no effect where one exists. The statistics commonly applied are likely to lead to a failure to reject the null hypothesis and a conclusion of no effect, when in reality the outcome is an inability to detect an effect of fishing. This result translates into a lack of action, the classic passive management approach.

From a precautionary, conservation standpoint, we should be most concerned about not finding an effect when one exists. Peterman (1990) maintained that this tendency was a major cause of problems in fishery science. Peterman and others since (Peterman and M'Gonigle 1992; Dayton et al. 1995; NRC 1995; Jennings and Kaiser 1998) recommended using statistical power analysis, which provides information on the likelihood of committing either type of error. Power analysis and a greater concern over failing to detect real effects would shift the burden of proof from those concerned with preserving the resource to those who are exploiting the resource.

Other statistical problems that arise in analysis of fishery data include the inability of most sampling programs to detect transient but important interactions, such as short-term co-occurrence of species or life history stages, short-lived but significant temperature changes, and recruitment pulses (Jennings and Kaiser 1998). Additional inaccuracies arise from the use of correlation analysis to identify causation, a flaw that is exaggerated when hypothesis testing is replaced with statistical testing, basically permitting the tests to drive the conclusions. Such occurs when researchers take a large, previously collected data set; plug it into a series of sophisticated, usually multivariate, analyses; and attempt to interpret fisheries-relevant biological patterns from the output. Results are often equivocal: The ensuing discussion includes extensive elaboration of sources of error or assumptions that are irrelevant, oversimplifying, or unmet. The conclusions are compromised by data inadequacies, requiring mental gymnastics to disregard outliers, anomalies, and counterintuitive results. None of these outcomes is surprising given that the data were collected for other purposes and with other analyses in mind. Large, previously collected data sets are invaluable sources of information with considerable potential for insight, but it is always preferable to plan data collection in concert with statistical testing, or at least fit the tests to the data rather than vice versa.

WHAT ABOUT BIODIVERSITY LOSSES?

Little has been said in this chapter about the threat of fishing to biodiversity. Given the species losses chronicled throughout this book, concern over potential losses of biodiversity would seem justified. However, fisheries problems are seldom biodiversity problems, in the sense of extinction of taxa. Overfishing causes commercial extinction, which usually occurs long before a species is threatened with biological extinction. Evidence of exploitation-induced extirpations and extinctions is therefore limited. Coral reef systems appear to have suffered most in this regard (see chapter 12). Otherwise, fishing has been less a cause of biodiversity loss than of other anthropogenic impacts for two major reasons.

First, fishing usually functions in the realm of the economics of return per unit effort. As a result, pressure on species tends to relax as commercial extinction approaches, with noted exceptions among inordinately valuable species and marginal subsistence fisheries. Economic factors thus allow species to persist but at low population sizes, no longer functioning in their natural, evolved role within a community but still appearing on species lists for the area. Second, shifting baselines affect issues of diversity as well as production; it's late in the game to be assessing the full extent of fishing's impacts on biodiversity. Impacts of exploitation before anyone started monitoring trends include community composition changes as well as underlying habitat disruption. We don't know what we may have lost because we don't know what was originally there.

A relevant area that has only recently received attention concerns genetic overfishing—loss of genetic diversity within species—and the need to manage with genetics in mind. Attention to genetics has increased with the development of such concepts as ESUs and DPSs, which have genetics at heart. That such issues were neglected is not surprising given historical resistance among fisheries agencies to the idea that fishes could evolve in response to fishing pressure. Lost genetic diversity precedes and undoubtedly contributes to the likelihood of extinction. Emphasis on evolutionary processes and recent advances in molecular techniques for assessing genetic characteristics have stimulated research on these topics (S. Mustafa 1999a). The focus was initially on salmonids, but the 1990s witnessed a growth in studies using molecular techniques to identify genetic variation and population differentiation in such exploited species as mako sharks, herring, anchovy, cod, pollock, orange roughy, snappers, barramundi, snook, billfishes, and tunas (Graves 1996; W. S. Grant et al. 1999; S. Mustafa 1999b). Genetic considerations are increasingly incorporated into notions of sustainable management (S. Mustafa 1999a). Still, a major gap in our understanding of the impacts of fishing on biodiversity is a lack of information about the genetic structure of most populations.

SOLUTIONS

Fisheries management traditionally focused on individual species, a focus that has proven myopic. Managing with ecosystems in mind is the most promising alternative being considered (see S. J. Hall 1999; NRC 1999b). NRC (1999b, p. 15) preferred the term *ecosystem-based management* to recognize that we cannot manage ecosystems per se; "instead, it is human activities that are managed." Management solutions that take an ecosystem perspective abound. With respect to overfishing, three aspects of these solutions deserve particular attention.

Marine Protected Areas: Ecosystem-Level Thinking Applied

Most treatments of ecosystem impacts conclude with a discussion of the design and effectiveness of MPAs, often combined with emphasis on the need for flexible, ecosystem-level management and a precautionary approach to management (e.g., Dayton et al. 1995; Jennings and Kaiser 1998; S. J. Hall 1999; NRC 1999b, 2001; Agardy 2000; Shipley 2004; see chapter 5 for details). GoXi (1998, p. 57) stated, "Areas closed to fishing may be the only way to evaluate impacts of fishing." Roberts (2005, p. 275) declared that "marine reserves are the most powerful tool we have for marine conservation." Murawski (2000) even advocated protected areas as a more plausible and realistic goal than the often-ill defined concept of an "ecosystems approach."

Why have MPAs become a recent, necessary, and general solution to the impacts of overfishing? Put differently, why have we been unaware of the value of creating reserves if such a solution has such widespread success and appeal? The answer probably lies in the increased pace and scale of habitat destruction and overfishing that characterized the late 20th century. Earlier, many nearshore fisheries could withstand intense exploitation because their fish populations were maintained or subsidized by natural refuges, large tracts of undisturbed offshore habitat inaccessible to fisheries. Only after World War II, with the growth of distant-water fleets and the use of military technologies, did these areas become accessible and exploitable (to appreciate historical contrasts in technology, go to www.photolib.noaa.gov/historic/nmfs).

Jennings and Kaiser (1998) pointed out a related asset of MPAs, their role as unfished reference sites where structure and variability are more natural. Having such reference sites allows us to pursue management actions as controlled experiments. The rarity of unfished systems and natural refuges has also eliminated the comparative baseline we require to assess the success or failure of new management approaches.

Intervention and Population Control

Ecosystem-based management can and should be adaptive and experimental whenever possible. Trophic interactions are relatively easy to manipulate, especially predator-prey dynamics involving top-down, bottom-up, and competitive effects. Experiments have been performed with success in temperate lakes, where the trophic cascade that starts with predatory fish and ends in phytoplankton has been manipulated to improve water clarity. Applications in large marine systems have apparently met with mixed results, as in management schemes focused on promoting certain prey species or tipping the balance between competing target species. Theoretically, if prey populations are controlled by predators or competition among predators limits population size, actions taken against predators and competitors should promote growth of "desirable" stocks.

Active management efforts such as these raise practical, empirical, and ethical questions. Practicality is an issue because abundant and successful marine species, desirable or not, are difficult if not impossible to remove regardless of time, effort, and resources invested. This has been demonstrated by our inability to eradicate harmful nonindigenous predators (chapter 9). The empirical question revolves around whether we have convincingly demonstrated that particular predators or competitors decisively influence the abundance of other fish species. With the exception of some nonindigenous species, few instances can be pointed to where such interactions have been convincingly demonstrated, a point that Jennings and Kaiser (1998) emphasized repeatedly. Large marine ecosystems contain too many species that play redundant or substitutable ecological roles and that are capable of shifting their use of resources. Direct, top-down, bottom-up, or competitive control is uncommon. Even when strong correlations exist between the population dynamics of interacting species, it has proven difficult to rule out the influence of climatic, anthropogenic, and random factors. Such uncertainty calls into question the wisdom—and ethics—of attempting to change the population size of one species by limiting the numbers of another (e.g., chapter 15).

Recognizing that few ecosystems are clearly controlled by a single species, Jennings and Kaiser (1998), following Christensen (1996) and Jennings and Polunin (1996), proposed that restoring damaged fisheries and preventing adverse shifts in existing fisheries required an ecosystem-level approach that emphasized (1) minimizing activities that damage habitat, and (2) distributing fishing effort across several species. Habitat destruction is known to have a greater effect on fish populations and fisheries than simple species removals (e.g., impacts of trawling). Targeting many species rather than focusing effort on a single species extends the humans-as-predators analogy into the realm of ecosystem function, requiring humans to act as natural predators with diverse and flexible diet preferences. Few true specialists exist among fish predators, particularly in large marine ecosystems. Predators switch as prey populations rise and fall. "The human species will have to give ecological principles a higher priority when choosing a foraging strategy" (M. A. Hall et al. 2000, p. 204).

If humans adopted the switching strategy employed by most nonhuman piscivores—changing targets and effort in response to year-class strength or species abundance—commercial extinction would be much less likely. More desirable still is a proactive management plan that mandates prey switching in *anticipation* of fluctuations in prey numbers (Jennings and Kaiser 1998). Single-species management approaches have failed repeatedly; little rationale exists to contend that intervention focused on control of one or a few predators or competitors would be any more successful.

Ecosystem-Based Management in Theory and Practice

The NRC (1999b) view of ecosystem-based management can serve as a general guideline for permutations on the concept of managing fisheries with ecosystems in mind. NRC emphasized the need to take all major ecosystem components (including humans) and services into account, focused on understanding larger ecosystem processes, and maintained a goal of sustainability. Ecosystem-based management using this approach has become a commonplace practice in terrestrial conservation (Meffe and Carroll 1997); application to fisheries issues is much more recent. Interest in and concern for ecosystem-level consequences of fishing has become a regular part of planning discussions in many agency deliberations. The International Council for the Exploration of the Seas has formed a Working Group on Ecosystem Effects of Fishing (WGECO), from which useful annual, regional, and general reports arise. FAO also is slowly coming around to accepting an ecosystem approach (e.g., FAO 2000b, part 2) .

The U.S. has embraced an ecosystem perspective in marine fisheries management (see NMFS 1999), as exemplified by provisions for habitat protection. These provisions were incorporated into the Magnuson Fishery Conservation and Management Act in 1996 (now the Magnuson-Stevens Act) by the inclusion of concern over *essential fish habitat* (EFH). EFH includes "habitat required to support a sustainable fishery and the managed species' contribution to a *healthy ecosystem*" (emphasis mine). The revised act mandates that EFH be considered in fisheries management plans. Implementation is the responsibility of NMFS and the various regional fisheries management councils created by the act. Unfortunately, jurisdiction over EFH is limited to the direct effects of fishing and does not provide for control over habitat degradation that results from coastal zone development, pollution, or other anthropogenic factors and forces that affect habitat. But EFH considerations are a positive step toward ecosystem-based management.

Ecosystem-based management emphasizes production, diversity, and variability, all in the context of sustainability. Importantly, "the solution is not to shut down fisheries but rather to modify the type of management" (Agardy 2000, p. 764). Issues requiring attention thus include quantification of resource status, with an emphasis on maintenance and sustainability of biodiversity, structure, function, and yields across wide temporal and spatial scales; understanding feedback among components; protection from pollution and habitat degradation; maintenance of resilience to normal and abnormal stressors, including major regime shifts producing alternative stable states; and provision of social and economic benefits consistent with biological objectives (NRC 1999b; Murawski 2000).

The major difference between ecosystem-based management and more traditional management is the focus on sustainability and interaction among components, rather than production of specific stocks. Both approaches focus on stresses, responses, and benefits; but in the traditional view fishing is the only stress, the response is defined by its effect on a target species, and the outcome (benefit) is the catch. As pointed out by several authors, ecosystem-based management does not ignore traditional approaches; it incorporates and extends them to include all human and environmental stresses, direct and indirect responses of the ecosystem to the stresses, and the various monetary and nonmonetary benefits to human and nonhuman users or components of the ecosystem (NRC 1999b).

Practical Considerations

It is all well and good to advocate for an ecosystem perspective in managing fisheries, but beyond protecting entire ecosystems (as in designating MPAs), the concept grows fuzzy. Managers charged with implementing an ecosystem-based approach are faced with a host of vaguely defined tasks based on vaguely defined terms, including *sustainability*, *ecosystem health*, *ecosystem integrity*, and *ecosystem overfishing* (e.g., Link 2002b). Managers are asked to determine the objectives of an ecosystem-focused strategy, develop quantitative metrics of ecosystem attributes, monitor the effects of any actions, and determine the "success" or "failure" of their activities (Murawski 2000). What does one measure? Several ecosystem properties and metrics (as opposed to population attributes) have been identified that describe diversity, trophic structure, and productivity at the ecosystem level. These include statistical distributions and their attributes, such as the slope of the multispecies size spectrum, diversity-at-size distributions (diversity spectra), the slope of the biomass spectrum, proportions of different trophic groups, and landings composition indices (FAO 2000b, part 2; Murawski 2000).

Although these measures have statistical appeal, they suffer from a lack of objective criteria concerning desirable thresholds: What is a critical threshold for the slope of the size spectrum? More important, "it is not possible to predict how specific management measures would affect these metrics" (Murawski 2000, p. 653). In other words, a manager working with traditional models of resource exploitation can make predictions about yield given certain levels of harvest, but not about how the ecosystem will respond given a particular harvest strategy targeted at maximizing, minimizing, or optimizing the slope of the multispecies size spectrum.

No single metric will deal with all the elements of managing with ecosystems in mind. Murawski (2000) suggested that emphasis be given to measures that could detect significant alteration in a number of components that individually or in combination indicated ecosystem overfishing, such as

- reduced biomass of important species or components beyond biologically defined and meaningful thresholds (with respect to MSY, extirpation, recruitment, and recovery)
- significant declines in various biodiversity measures
- extraordinary variation in populations or landing rates
- altered species composition or demographics that compromise resilience
- economic and social benefits reduced from what was provided under less intense exploitation practices
- impaired long-term viability of ecologically important nonresource species such as birds, mammals, and reptiles

Any of these elements could be used to justify changes in current fishing practices. Murawski (2000) and Agardy (2000) emphasized that monitoring these factors does not require managers to abandon traditional programs and methods that rely on fishery-dependent and -independent information. Ecosystem-based management begins with traditional measures and expands that information base to include a wider array of species, interactions, components, and processes. Fisheries science is still developing those expanded metrics, with the help of ecosystem scientists.

Management Based on Examples of Sustainability

While working out the nuts and bolts of ecosystem-based management, a complementary approach has been proposed. This approach is simpler and especially appealing from the standpoint of ecosystems-as-evolved-entities and the idea that impacts result from our actions as predators. C.W. Fowler (1999) suggested that nature be used to guide our levels of resource exploitation: Humans should mimic consumption rates shown by natural predators, targeting the average consumption rate as a goal (see figure 11.4). Such predators have had their consumption tactics and rates honed to a sustainable level by natural selection and extinction. Interactions among species over time have served as "experiments in the trial-and-error process of natural selection" (p. 930); we would be capitalizing on the outcome of these large-scale experiments.

Fowler's ideas can be applied to determining quotas for individual species or multispecies fisheries, assuming we know for starters the mix of predator and prey species, their size-specific feeding habits, and their abundances. For example, mackerel consumption by nonhuman predators in the northwest Atlantic is around 226,000 MT annually, 47% of the total consumption of small pelagics by nonhuman piscivorous fishes, mammals, and birds (Fowler 1999). Average annual consumption by each of 16 predator species is around 14,000 MT. Humans consume 33,500 MT of mackerel per year (E. D. Anderson, Kocik, and Shepherd 1999), more than 2.3 times the average. Halving our mackerel take would put us closer to the mean consumption rates of other, evolved predatory components of the ecosystem. This lowered target contrasts markedly with NMFS-calculated potential mackerel fisheries yield of 317,000 MT, which ignores the needs of all other predators in the system. Fowler's proposal solves another basic dilemma. S. J. Hall (1999) concluded that ecosystem problems brought about by overfishing would be addressed if we "reduce effort, reduce effort, reduce effort" (p. 247), long before the details of ecosystem-based management have been worked out and applied. A major obstacle to Hall's simple and obvious solution is deciding how much we should reduce effort. Fowler's approach provides a starting point for such decisions that has at least some ecological basis. It is an idea worth exploring.

Fowler's approach also allows direct application of knowledge gained from the legions of descriptive studies of feeding habits that have occupied researchers for so many decades, or as Larkin (1979, p. 13) put it, "the food of the blank in blank lake." All those fish that were sacrificed for stomach contents analysis need not have died in vain.

CONCLUSION

From all the foregoing, one emerging conclusion is that we need to view fisheries in an ecosystem context, manage them based on preserving the functions of entire ecosystems, and include ourselves as an ecosystem component that operates according to ecological principles. To someone more interested in conserving biodiversity than in maximizing fisheries yields (recall the subtitle of this book), focusing on ecosystem impacts is easy to justify. The conclusion of Jennings and Kaiser (1998, p. 313) has special appeal: "Conventional fisheries management [should] be part of an ecosystem-based approach, where the benefits of fishing would be balanced against the need to maintain diversity, seabird and mammal populations, ecosystem service functions and the intrinsic conservation value of the ecosystem."

But why should a fisher whose livelihood depends on exploiting a resource, or a fisheries manager charged with finding a way to maximize short-term yield from a resource, take an ecosystem perspective to exploitation and management? The argument of biodiversity loss is not persuasive to those not inclined to be concerned about such issues, especially given that we lack clear-cut evidence of such losses in exploited marine systems, where population reductions are common but extirpations rare. Proving extirpation, like extinction, is exceedingly difficult.

Arguments that appeal to self-interest will always have greatest impact. Activities that lead to environmental degradation reduce the capacity of degraded systems to support sustainable fisheries. Because of our dependence on exploited systems for goods and services, and because our activities affect those systems, humans are a part of them. It is therefore in our best interest to be concerned about their long-term welfare. Agardy (2000) emphasized that the common goal and agenda of all groups concerned with marine resources should be to prevent activities that would cause irreversible damage to the ecosystem, that would reduce both biodiversity and production.

It is tempting to proclaim that all fisheries problems can be solved by adopting an ecosystem approach, but many managers have not reached that conclusion and may never. Fisheries tend to target specific fishes in specific areas, and the sociology and politics of management have historically mandated that targeted fisheries in specific locales be managed as individual entities. Any movement toward incorporating an ecosystem perspective—with its emphasis on habitat conservation, species interactions, and humans as prudent predators—will represent an improvement over historical practices such as MSY. If we manage right, we get more fish: Where an ecosystem perspective has been added to standard fisheries practices, especially in concert with local control over resources (as demonstrated in the next chapter), conditions and catches have improved. If the growing number of authorities advocating ecosystem-based approaches are correct, additional case histories with demonstrated positive results will be the best testimonials for promoting change.

Because of the repeated, dramatic failures of past management practices, society is turning to ecology for new solutions. It's encouraging that ecologists are helping develop the tools needed to perform the task, in the form of increased understanding of ecosystem processes and functions. The new solutions will incorporate pragmatic successes of the past with promising ideas developed from a broader perspective of fishes, fisheries, and the ecosystems in which they and we are contained.

12. Coral Reefs, Fishes, and Fisheries: Exploitation in Fragile Ecosystems

So here is the paradox. Although coral reefs are the most productive communities in the sea, the fisheries of coral reefs are among the most vulnerable to overexploitation.

—Charles E. Birkeland, 1997c

Coral reef ecosystems and their fishes provide clear-cut evidence of the interdependency of fishes and their habitats; the effects of overfishing and destructive fishing practices; human reliance on intact ecosystems; management strategies and foibles; and practical, often successful, solutions. Loosely defined as concentrations of hermatypic (reef-building) live coral in clear, shallow (<50 m) water, coral reefs today cover about a third of the coastlines of tropical regions, occupying between 255,000 km^2 and 1,500,000 km^2, depending on author and methodology (Spalding and Grenfell 1997 vs. Copper 1994). About 9%–12% of the fish consumed by humans comes from reef areas (S. V. Smith 1978), and because coral reefs occur along tropical coastlines populated by developing nations, tens of millions of people depend on them for food and livelihood (e.g., Salvat 1992).

This dependence results in large part from the exceptional biodiversity of reefs. Fishes and many invertebrate taxa reach their highest regional diversities on coral reefs (see J. C. Briggs 1995; Paulay 1997), and an astounding array of these organisms are consumed and otherwise used by humans. Because diversity often begets diversity, the trophic relationships, symbiotic associations, and ecosystem connectedness among reef taxa achieve exceptional levels of variety and complexity (see Birkeland 1997c). This diversity also supports a multiplicity of human uses: Animals are extracted for consumption and the aquarium trade, corals are mined for building materials or turned into jewelry and curios, ecotourism provides income; all parts and aspects of reefs have incalculable sociocultural uses and value (see Moberg and Folke 1999).

Coral reefs house the greatest diversity of fish species of any of the world's aquatic habitats, with roughly 9,000 of 27,000-plus known species (if we include deep-reef habitats, this may be an underestimate; see box 12.1). Thus, 33% of the world's fish species live in only 0.17% of the world's ocean area or occupy 15% of sea floor at depths less than 30 m (S. V. Smith 1978; Helfman et al. 1997). The four major zoogeographic regions that house coral reef fishes are the Indo-West Pacific (>3,000 species), western Atlantic–Caribbean (1,200 species), eastern Pacific (800 species), and eastern Atlantic (500 species). The number of species regularly captured and consumed runs to about 180 in the Caribbean and 250–300 in the Indo-West Pacific, numbers that are almost an order of magnitude greater than those for fisheries in most other marine habitats. The difference applies to tropical soft-bottom fishes, as well: Demersal trawl fisheries in southeast Asia typically catch between 175 and 350 species, whereas North Atlantic bottom trawls capture only around 50–60 species (Pauly 1979).

THE STATUS OF CORAL REEFS AND THEIR FISHES

Mounting evidence suggests that coral reefs worldwide are experiencing unnaturally high rates of stress, and that reef degradation is occurring at unprecedented levels (e.g., Sebens 1994; R. T. Barber et al. 2001; J. B. C. Jackson et al. 2001). Reefs worldwide have been damaged by pollution, oil spills, sediment from runoff and dredging, ocean warming/climate change, debilitating diseases, poison and dyna-

BOX 12.1. The Deep Reef

Although fishes reach their highest diversity on coral reefs, estimates of diversity focus on species found in water less than 50 m deep, depths accessible to collectors using scuba gear. Although significant coral growth is limited by light, "reef" fishes go deeper. Largely unappreciated and barely explored is the deeper (50–150-m) "twilight zone" portion of the reef face (e.g., Thresher and Colin 1986), which only became accessible with the advent of specialized, mixed-gas rebreathing equipment (Pyle 1996a, 1996b, 2000, in press). Most fishes in this zone seldom occur above 75 m. New species, and even genera, are now being discovered at a rate of four to seven species per hour dive time, with no decline when additional dives have been made at a locale (Pyle 2000, in press). Most species belong to 20 families common in shallower water, the top 6, in descending order of diversity, being gobies, sea basses, wrasses, damselfishes, cardinalfishes, and angelfishes. Although collections have been largely limited to Indo-Polynesia (e.g., Rarotonga, Palau, Hawaii, American Samoa, Fiji, New Guinea), already more than 100 new fish species have been found (R. Pyle, pers. comm.; figure 12.1).

Figure 12.1. The "Dr. Seuss fish," *Belonoperca pylei*. This 6-cm, brilliantly colored new species of sea bass from the deep-reef twilight zone of Rarotonga has a bright yellow head and back, pink body, and orange spots. Photo by Richard Pyle. Reprinted from Helfman (2001), courtesy of Academic Press.

The discoveries suggest that endemism is higher among deep-reef than shallow-reef taxa and that unexplored areas may house many additional new species (Pyle 2000). Extrapolating to unsampled Indo-Pacific reef areas, Pyle (2000, in press) conservatively estimated that between 750 and 2,250 fish species remain to be described from deep-reef habitats, constituting a 25%–67% increase in overall regional fish diversity. Coral reef regions are even more diverse than we thought.

mite fishing, trawling, set nets, ship groundings, anchors, collecting, careless divers, storms, and unnaturally abundant native enemies. Approximately 60% of the world's coral reefs are at high or medium risk of degradation, including more than 80% of southeast Asia's reefs and 60%–70% of Indonesian and Philippine reefs. In Japanese waters, 400 coral species are threatened with extirpation. Eastern Pacific reefs are eroding faster than they can grow. The incidence of coral bleaching, disease, and mass mortalities was maximal in the last decades of the 20th century, at the same time that recovery rates appear to have slowed (Birkeland 1997b; B. Brown 1997; FAO 2002).

Fishes, actively exploited or not, have suffered as a result. A multiyear (1997–2001), comprehensive diver survey at over 1,500 reefs in the Atlantic and Indo-Pacific found significantly decreased abundance of butterflyfishes. Groupers over 30 cm and parrotfishes over 20 cm were absent from 55% and 48% of sites, respectively, and Nassau grouper (*Epinephelus striatus*), bumphead parrotfish (*Bolbometopon muricatum*), and humphead or Napoleon wrasse (*Cheilinus undulatus*) were absent from 82%–89% of sites where they were expected (Hodgson and Liebeler 2002). Several of these indicator species are extremely valuable in the food and ornamental trades and continue to be targeted despite declining abundance (see chapter 13). Although no coral reef fish species is known to have gone extinct, a widespread, repeated pattern of population loss is often a harbinger of extermination. Commercial and noncommercial, widespread and restricted-range reef species will experience continued and even accelerated declines as human populations grow and reefs are further degraded by anthropogenic impacts (e.g., J. P. Hawkins et al. 2000; Roberts et al. 2002).

Fishing harms reefs and reef species whether or not they are targeted. Declines in nontarget species result from bycatch and destructive fishing practices that degrade the living reef. Many nonfishing impacts also cause loss of the reef structure on which the fishes depend. Diverse, commercially unimportant species often decline because reef fisheries use nonspecific gear that has a multispecies bycatch component (e.g., Birkeland 1997a). In those few reef areas where management has been attempted, the approaches used were often developed in temperate locales, and were appropriate for monospecific stocks captured by relatively few gear types; these efforts were largely unsuccessful on reefs.

All problems are exacerbated by lack of information. Far-flung, isolated, small tropical islands are hardly centers of scientific inquiry, assessment, and collection. A steady stream of new species described from coral reefs attests to how little we know about tropical biodiversity, an ignorance compounded by biodiversity in habitats beyond the reach of conventional collecting methods. Also lacking is information relevant to fisheries biology, including stock sizes, mortality rates, landings numbers, distribution of effort, and important biological characteristics of exploited species (e.g., Jennings et al. 1999).

PATTERNS OF EXPLOITATION AND DECLINE IN REEF FISHERIES

Reef fisheries have sustained coastal and island societies for at least 35,000–40,000 years (J. B. C. Jackson et al. 2001). Yields of fishes and invertebrates may reach 44 MT $km^{-2}yr^{-1}$ (e.g., Wass 1982 reported a value of 18.3 MT $km^{-2}yr^{-1}$ from a Samoan reef with a depth of 8 m; various authors "expanded" the depth to 20 or 40 m and calculated the 44 MT value from those extrapolations). Sustained harvests of 20–30 MT km^{-2} have been reported from several Indo-Pacific regions; Caribbean areas are generally less productive (Medley et al. 1993; Munro 1996). Russ (1991) reviewed yield values from 18 Pacific locales, mostly from the Philippines, and calculated an average yield of about 14.5 MT $km^{-2}yr^{-1}$. These numbers approach the production of trawl fisheries on temperate shelves (Munro 1996); standing stocks of coral reef fishes may be 30 to 40 times greater than demersal stocks of southeast Asian, Mediterranean, and other temperate fisheries (Russ 1984).

Fisheries on reefs exploit many species with a variety of gear types (Munro and Smith 1984). Although 100 to 300 fish species are commonly targeted, about 20 species constitute 75% of the catch by weight (Munro 1996). Gear types include traps, barrier and gill nets, cast nets, throwing spears, spearguns, extraction poles, trolled lines, and static hook-and-lines (plus chemicals and explosives). Few boat-pulled nets, such as trawls and seines, are used because the high topographic complexity of reefs causes fouling, damage, and loss of gear.

At issue is not whether reefs can sustain fishing and fishing-dependent cultures (they can), but whether they can be fished intensively for long periods and participate in export economies. Birkeland (1997a, 1997d) reviewed reef fisheries that had been subjected to unsustainable fishing and found the following:

- CPUE declined 70%, and fish density decreased 75% over a 15-year period, with dramatic losses among preferred species in American Samoa, where the world's highest reef yields were reported
- an 82% decline over 15 years in CPUE of trap-caught fish on Southeast Pedro Bank in Jamaica
- an 80% decline in numbers among reef fish populations in Bolinao, Philippines, over only 4 years, a 33% decline in number of species reaching adulthood, strong declines among preferred species, and a shift to targeting juveniles and employing destructive fishing methods, which reduced living coral cover by 60%
- a loss in productivity throughout the Philippines with increasing exploitation; lightly exploited and well-managed reefs produced 32 to 37 MT $km^{-2}yr^{-1}$, whereas degraded reefs produced only 5 to 6 MT $km^{-2}yr^{-1}$
- declining catches in Bermuda over 4 years, from 678,000 kg yr^{-1} to 380,000 kg yr^{-1}, while species composition changed from desirable groupers to less desired parrotfishes
- Rapid deletion of reefs in Kosrae (Federated States in Micronesia). Despite warnings from the local Marine Resource Division, fishing on fringing reefs intensified following donation of 70 fishing boats, motors, and gear from Japan; reefs became seriously depleted within months, and remained depleted for at least a decade.

By any definition of overfishing, reefs worldwide have been overexploited. Most locales that have kept reasonably good catch statistics have experienced, singly or in combination, reductions in biodiversity, biomass, abundance, CPUE, overall catch, and average size of individuals; shifts to less desirable species; altered sex ratios; and loss of predatory groups. Documentation comes from at least 26 countries and locales representing all tropical oceans.

EVIDENCE AND CAUSES OF MISMANAGEMENT IN REEF FISHERIES

Munro (1996, p. 11) summed up the history of reef fisheries management, concluding, "Reef fisheries have a dismal management record [and] . . . there are still singularly few examples of effective management." Why have reef fisheries been so poorly managed, assuming any management has been attempted? Bohnsack (1996) credited three primary factors: (1) inadequate data on the biology of exploited species and on the fisheries directed at them, (2) poor regulatory compliance and enforcement, and (3) a lack of models appropriate to coral reef situations.

Inadequacy of data has been cited repeatedly. Classical fisheries management requires extensive data on reproductive cycles, age and growth, population size and age structure, maturation and fecundity schedules, as well as accurate catch and effort statistics. Ideally, data are obtained using fishery-dependent and -independent sampling. Reef fisheries occur in relatively poor countries with limited resources for gathering the detailed, time-series data needed to apply a fisheries model. Conventional management depends on top-down development, implementation, and enforcement that originate in bureaucratic administrative entities far from fishers and their villages. In addition, national governments tend to focus more resources on large-scale, industrialized, commercial fisheries than on small-scale, local artisanal and subsistence fisheries (Y. Sadovy, pers. comm.).

Fisheries models have been largely developed for the world's most heavily exploited and managed fishes, pelagic species in temperate and tropical regions, and demersal temperate and polar species. These fishes tend to aggregate in huge schools, which makes them susceptible to location and capture with advanced electronic technology, species-specific gear types, and large-scale fishing practices. Scale in tropical reef fisheries is smaller and more localized, gear types are multispecific, and fishes are more dispersed. Few of the conditions incorporated into conventional models apply to reefs.

The result of Bohnsack's 1996 identified triad of inadequate data, lax enforcement, and inappropriate models is a recognizable, recurring pattern of events as reef fisheries proceed from unexploited to overexploited conditions (Koslow et al. 1988; Munro 1996; Jennings and Lock 1996; Jennings and Kaiser 1998). Species composition in the catch and on the reef changes, and the gear used changes to reflect altered availability of species. Light fishing of previously unexploited regions affects species composition, with piscivores rapidly declining and herbivores rapidly rising in relative abundance. This initial phase often focuses on large individuals, with herbivores such as surgeonfishes targeted by spear and net fishers and large predators (groupers, snappers) dominating hook-and-line catches. As fishing pressure intensifies to moderate, sustained levels, progressive changes occur in catch characteristics, species composition, and fishing methods (figure 12.2). Declines occur in catches, size of individuals, biomass, and species richness. Initially targeted groups decline, large fishes become rarer in the catch, and netting dominates as a gear type, with less hand-lining. On Indo-Pacific reefs, small emperors (Lethrinidae) dominate catches regardless of gear type. In the intense phase, hand-lining is abandoned in favor of netting, and small herbivorous fishes such as rabbitfishes (Siganidae) dominate the catch.

Such cyclical overfishing occurs despite the lack of technological advances that have contributed so heavily to the decline of fisheries in temperate locales. Few coral reef systems are subjected to industrial-scale fishing, and yet depletion is commonplace. The biology of many reef fishes makes them vulnerable to overexploitation via fishing methods developed centuries ago. Groupers are heavily targeted and disappear quickly because their breeding habits, involving aggregations and spawning stupor, make them available to commonly used, simple gear. Grouper aggregations and fisheries have vanished from numerous Indo-Pacific, Gulf of Mexico, and Caribbean locales (Coleman et al. 1996; Johannes et al. 1999; Sala et al. 2001). Nassau grouper provide the best-known example; they migrated to breeding aggregations of over 100,000 individuals, attracting fish from as far away as 110 km (C. L. Smith 1972). Fishers capitalized on and decimated these aggregations. At one site in Honduras, two fishers hooked 17,000 kg and speared another 5,400 kg of fish in only two weeks. A spawning aggregation in the U.S. Virgin Islands declined from 100,000 to 13 fish over a six-year period, and landings of Nassau grouper in Bermuda fell 33-fold between 1975 and 1984 (Olsen and LaPlace 1978; Fine 1990; Butler et al. 1993).

Other spawning reef fishes use predictable, accessible locales (e.g., Domeier and Colin 1997). In Palau, at least 57 reef species in 12 families spawned predictably at three well-known sites (Johannes 1981; Johannes et al. 1999). Knowledge rather than technology determines the efficiency of reef fisheries.

WHY ARE CORAL REEFS AND THEIR FISHES (SO) VULNERABLE?

Snorkeling on an unfished coral reef, one is immediately struck by both the diversity and abundance. But biomass should not be equated with production or exploitability (e.g., Hatcher 1997). The numbers and kinds of fishes and invertebrates amassed on a coral reef result from many

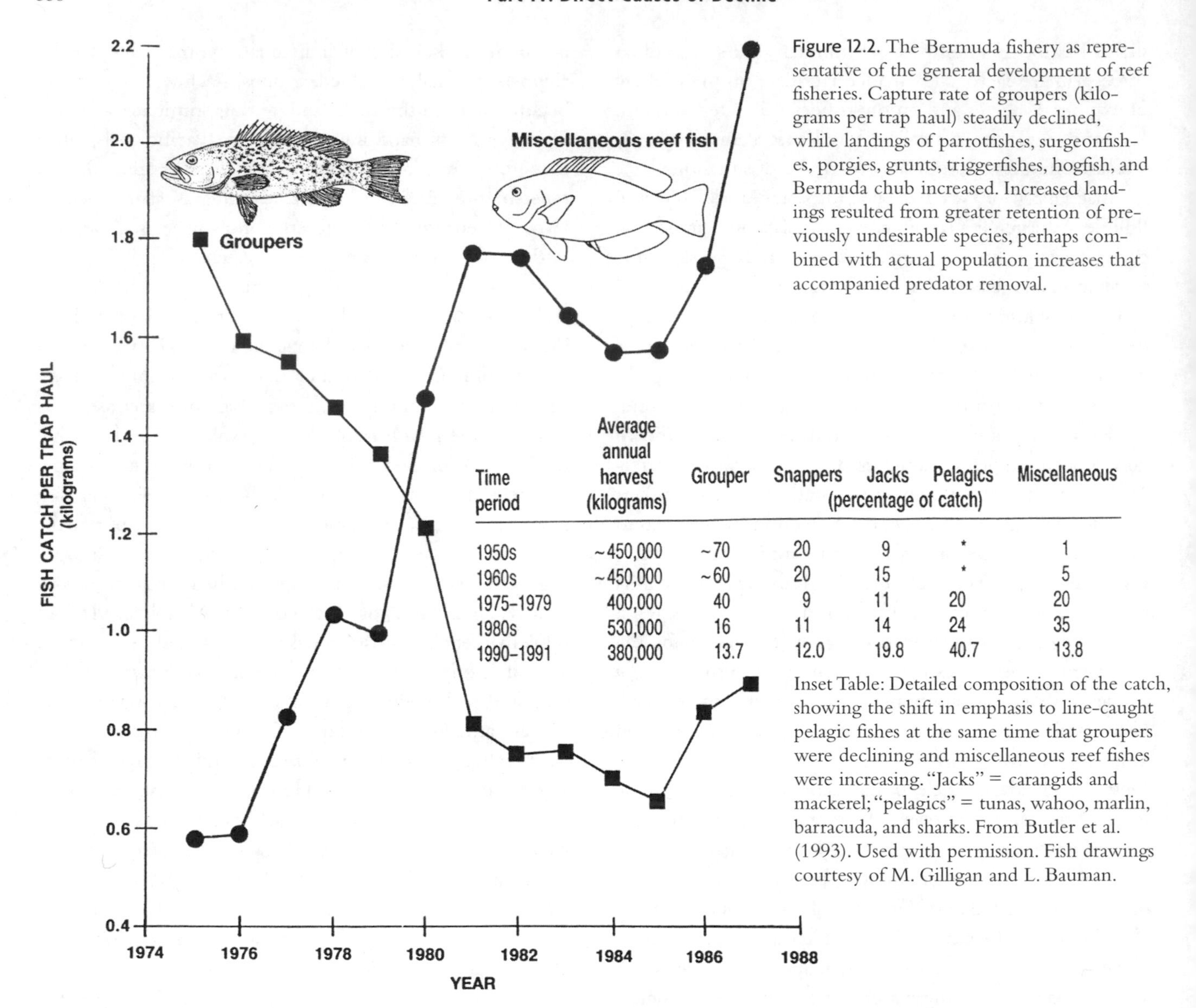

Figure 12.2. The Bermuda fishery as representative of the general development of reef fisheries. Capture rate of groupers (kilograms per trap haul) steadily declined, while landings of parrotfishes, surgeonfishes, porgies, grunts, triggerfishes, hogfish, and Bermuda chub increased. Increased landings resulted from greater retention of previously undesirable species, perhaps combined with actual population increases that accompanied predator removal.

Time period	Average annual harvest (kilograms)	Grouper	Snappers	Jacks	Pelagics	Miscellaneous
			(percentage of catch)			
1950s	~450,000	~70	20	9	*	1
1960s	~450,000	~60	20	15	*	5
1975–1979	400,000	40	9	11	20	20
1980s	530,000	16	11	14	24	35
1990–1991	380,000	13.7	12.0	19.8	40.7	13.8

Inset Table: Detailed composition of the catch, showing the shift in emphasis to line-caught pelagic fishes at the same time that groupers were declining and miscellaneous reef fishes were increasing. "Jacks" = carangids and mackerel; "pelagics" = tunas, wahoo, marlin, barracuda, and sharks. From Butler et al. (1993). Used with permission. Fish drawings courtesy of M. Gilligan and L. Bauman.

years' accumulation of long-lived individuals. Remove them, and the system changes rapidly and recovers slowly. The life histories and ecological interactions of coral reef organisms make them poorly adapted to exploitation.

Setting aside for the moment the issue of human overpopulation in tropical regions, what aspects of reef biology and reef fisheries contribute to a pattern of rampant overexploitation? Commonly cited factors focus on the multispecies nature of reef fisheries, gear types and their impacts on habitat, the closed nature of reef ecosystems, fish sizes and life history traits, aggregated spawning, complex food webs, and the socioeconomics of reef fisheries.

Multispecific Catches, Nonspecific Gears, and Rare or Small Fishes

The multispecificity of both assemblage and gear makes it economically advantageous to continue fishing even after highly prized species are depleted. Many common reef species are acceptable, if not preferred, as food. This jeopardizes the welfare of both rare but desirable and vulnerable species because fishers can continue to fish profitably as long as they are catching something; occasional encounters with prize species become a bonus. Pressure is kept on rare species. Common species therefore cause overfishing of uncommon species, and economic overfishing turns into biological overfishing (Birkeland 1997a). No less important is the overall intensity of exploitation: "Virtually every component of the reef community has value as a commodity . . . there is apparently no such thing as a trash fish in many impoverished developing countries" (Munro 1996, pp. 8, 10; see box 12.2).

Temperate and offshore fisheries commonly suffer from either recruitment overfishing or growth overfishing. Reef fisheries, because of the multispecific composition of the catch and the body sizes attained and targeted, suffer from both. Many targeted reef fishes are relatively small.

BOX 12.2. Nonfood Uses of Small Reef Fishes

The unsustainable capture of reef species for traditional medicines and curios affects both the ecology of the reef and the local economy. Most heavily targeted are the syngnathid seahorses, pipehorses, sea dragons, and pipefishes (figure 12.3).

Seahorses are evolutionary marvels and ecological novelties. Males become "pregnant" and give birth to live young, show active mate selection (a female prerogative in most animals), and are highly monogamous. Seahorses are captured for the aquarium trade but seldom live long because they require live food and are susceptible to disease (Vincent and Sadler 1995; Vincent 1996; see chapter 13). Many more seahorses are netted, dried, and sold as curios and keychains in gift shops or, most commonly, processed into traditional medicines in China, Indonesia, Japan, and Korea. Ailments treated with seahorse (dried whole, chopped, ground, and powdered) include fatigue, throat infections, respiratory disorders, injuries from falls, heart disease, circulatory problems, kidney and liver disease, immune function, difficult childbirth, incontinence, skin disease, mental disorders, and sexual dysfunction (Vincent 1996).

Figure 12.3. Dried seahorses, seamoths, and pipefishes for sale in Hong Kong. Photo courtesy of Project Seahorse/A.C.J.Vincent.

Seahorses are actively sought by individual fishers and captured extensively as bycatch in shrimp trawls. Major export nations include India, the Philippines, Thailand, and Vietnam. Annual trade in the mid-1990s amounted to at least 45 *tons* imported to Asian countries alone, which amounts to about 16 million fish (not including domestic capture and consumption, non-Asian imports, or the aquarium trade). Prices vary widely, but seahorses in 1995 retailed for $1–$17 each, depending on location, size, and species, which at around 500 dried seahorses to the kilo equals $500–$8,500 (Vincent 1996, table discussion 4). Fishers received $0.02–$3.00 per fish.

The seahorse trade increased about tenfold in the late 1980s, reflecting the rapid growth of Asian economies. By the mid-1990s, many seahorse populations had declined by 50%, catches dwindled, and large and desirable species became increasingly rare. Seahorses are susceptible to exploitation because of lengthy parental care and small brood size, reluctance to remate when a mate is lost (a cost of monogamy), small population sizes, naturally low mortality rates of adults, restricted ranges and mobility leading to low recolonization rates, and high valuation that rewards exploitation even as populations dwindle. IUCN (2002) red-listed 25 seahorses and pipefishes as Vulnerable, one as Endangered, and one as Critically Endangered.

Solutions are many and varied, with potentially far-reaching conservation benefits and implications (A. Vincent, pers. comm.). Seahorse fishing communities in the central Philippines are setting up no-take marine reserves. CITES in 2002 agreed to limit exports to levels that would not damage wild populations; these regulations were implemented in 2004. Chinese medicine traders and practitioners in Hong Kong have collaborated on trade research and developed voluntary codes of conduct that reduce pressures on seahorses. Public aquariums have participated by establishing live seahorse conservation displays that reach more than 10 million visitors a year. Cautious aquaculture for the live trade may be possible if long-term economic and ecologic viability can be secured. Such programs have been advanced by an international organization, Project Seahorse, that is widely recognized for its success in educating the public, preserving biodiversity, and promoting local economies (see www.projectseahorse.org).

Leiognathid slipmouths or ponyfishes and gerreid mojarras, which constitute the bulk of many demersal reef catches, have an average maximum length of 12 cm, below the size usually retained in bottom trawl fisheries in temperate regions (Pauly 1979; S. J. Hall 1999). Targeting small adults with nonspecific gear types means also catching many juveniles of valuable, larger species. Overfishing of adults of some species can result in overfishing of juveniles of other species.

Nonselective gear combines with species vulnerability to depress several reef groups (Munro 1996). Large predators are susceptible to a wide variety of gear types, including line fishing. Groupers are especially vulnerable because of their reproductive traits (chapter 11) and because they are recruited to the fishery (are desirable and catchable) at relatively small sizes, before they begin reproducing (Sadovy 1994; Levin and Grimes 2002). Depletions for similar reasons characterize other groups, such as sparids in South Africa (Buxton 1993). Deeper-bodied fishes are more vulnerable to trap fishing because they are less likely to escape once in a trap. Koslow et al. (1988) found that deep-bodied groupers, snappers, parrotfishes, triggerfishes, angelfishes, and surgeonfishes declined rapidly relative to less deep-bodied groups.

Habitat-Dependent Species versus Habitat-Destructive Fishing

Coral reefs are often thought of as oases in an aquatic desert resting on a thin veneer of life. The idea refers to the low nutrient content and low productivity of open tropical oceans and the few millimeters of polyp and symbiotic algal cells that constitute the living surface of a healthy reef. Any disruption of the veneer disrupts the system.

Coral reef fishes depend on the hard and soft corals and other living organisms (anemones, sponges, bryozoans, alcyonarians, ascidians, tube-building worms, coralline algae, etc.) that make up the physical structure of a reef (Birkeland 1997d; Hixon 1997; Ogden 1997). Abundance and especially diversity of reef fishes are positively correlated with habitat complexity: Dead reefs have fewer numbers and kinds of fishes, although algal-covered reefs can support high numbers of herbivores (e.g., Sheppard et al. 2002). Because the reef is living structure, this dependence is coevolved and is undoubtedly more complex than fish-habitat interactions where nonliving structure is involved (e.g., Meyer et al. 1983). Destructive fishing practices such as use of explosives, chemicals, and netting damage the reef and affect reef-dependent fishes. Fish removed from the reef can be replaced through recruitment and colonization, but damaged reefs afford newcomers few places to live. In addition to habitat loss, destructive fishing kills many non-target species, disrupts species interactions, affects human health, and is an overall catastrophe with long-term consequences. Fishing is equivalent to cutting down trees; destructive fishing is removing the mountain.

Blast Fishing

Dynamite or blast fishing, although certainly not restricted to tropical areas, is most common there, reported from at least 40 tropical nations and outlawed in most (e.g., Ruddle 1996). World War II produced "leftover ordnance" in the form of unexploded bombs and mines around Pacific Islands. These provided the resources for blast fishing for decades. Where military munitions are unavailable, explosives include dynamite, TNT from mining operations, sodium azide, potassium and sodium chlorate, and ammonium nitrate fertilizer mixtures. The opportunities are hardly limited.

The impacts of blast fishing on fish and habitats are dramatic (figure 12.4). Reef bombs range in size from 0.5-kg bottles to 10-kg drums. Charges set near the bottom excavate a crater surrounded by pulverized coral. Each bottle bomb shatters corals over a 1.15-m radius; a gallon drum bomb destroys corals 5 m away. P. J. Rubec (1988, pp. 144–45) described the physical impacts of blast fishing in the Danajon Bank region of the Philippines, which had been among that country's most productive fishing areas before dynamiting became commonplace: "From an underwater perspective, it is a disaster area. In many places, hectares of coral have been toppled and shattered by explosions. . . . Most of these prostrate corals, the standing fragments and the remnants of structurally intact reef, are dead and covered with algae. Coral fragments wreak additional havoc when moved by storm waves. . . . Food fish desert a dynamited reef."

Fish killed near the blast are characteristically limp. A blast-killed mullet can be draped over your forearm much like a wet, sand-filled sock, because bones and their connections are broken. Fish more distant from the blast are usually killed due to ruptured swim bladders, "which literally explode when fish are within the lethal zone of an explosion" (Saila et al. 1993, p. 54). Such destruction

Figure 12.4. Damage to reef habitat from dynamite or blast fishing. (A) A healthy Indo-Pacific reef; (B) a recently dynamited reef showing the characteristic circular blast crater; the reef framework prior to the blast was about 1 m deep; (C–D) extensive blasting over time reduces a reef to scattered, dead rubble that takes decades or longer to recover. A ©2001 www.ecoreefs.com; B and D ©2001 Michael D. Moore; C ©2001 Mark V. Erdmann; used with permission.

extends well beyond the zone of obvious physical damage: Fishes and invertebrates subjected to a "typical" charge are killed nonselectively within a radius of about 77 m, an area of almost 19 ha (Alcala and Gomez 1987). Fish species diversity and abundance are reduced on dynamited reefs (e.g., Riegl and Luke 1998), which may take 25 to 160 years to recover (e.g., Saila et al. 1993; Riegl 2001). Only a fraction of the dead and dying are used for human consumption because small fish are more susceptible than large fish to a blast.

In addition, blast fishers are frequently maimed or killed in the process. Despite the obvious health and ecological hazards, blast fishing has been popular. Riegl and Luke (1998) found that 65% of reefs surveyed in the Egyptian Red Sea showed dynamite damage. P. J. Rubec (1988) estimated that perhaps 50% of the 700,000 small-scale fishers in the Philippines used explosives at least part of the year, even though blast fishing was outlawed in 1972. The Philippine military reportedly purchased, transported, and marketed fish caught by blast fishing. Hand grenades appear ready-made for such activities.

Fishers engage in blast fishing because it is quick; easy (albeit dangerous); usable in rough weather; and profitable, providing economic returns "comparable to the highest incomes in the conventional coastal fisheries" (Pet-Soede et al. 1999, p. 83). It is also short-sighted and selfish. Blast fishing often permanently destroys the reef, creating areas with shifting, unstable rubble that inhibit colonization by both fish and corals (Fox and Erdmann 2000). Pet-Soede et al. (1999) showed that, in Indonesian areas where ecotourism was likely, the economic costs to society of blast fishing were about four times greater than the net benefits to individuals. Over 20 years, losses amounted to about \$307,000 per km^2 of reef where tourism potential was high, and \$34,000 per km^2 even in areas with low tourism potential. Blast fishing is inherently nonsustainable.

Poisons

Fish poisons, or so-called stupefacients, have a long history of traditional use in many cultures. Natural products, such as *Derris* root (the basis of rotenone) and extracts of tobacco, *Barringtonia*, *Tephrosia*, and *Wikstroemia* plants have been used to stun and kill fishes in both freshwater and marine habitats. Because their use was relatively restricted by availability and by the effort involved in extraction and use, their impact through most of human history was probably minimal and local. Recently, more "modern," industrially produced and distributed synthetic chemicals such as chlorine bleach, quinaldine (a known carcinogen), formalin (also carcinogenic), and especially cyanide have caused widespread and wholesale destruction of reef faunas, not to mention human suffering and death. The chief use of cyanide is to capture fish for the live-food fish and aquarium trades, detailed in chapter 13.

Net-Based Fishing

Trawling is uncommon around reefs because of potential gear loss. Where used, its impacts are severe. In northwest Australia, trawling destroyed biogenic habitats and caused commercially valuable snapper and emperor populations to decline (K. J. Sainsbury et al. 1997). Where trawling was prohibited and habitats were allowed to recover, species rebounded.

Other netting methods also have substantial negative impacts. Gill nets are employed almost universally in fisheries, leading to habitat destruction and species depletions when they are lost and enter the ghost fishery. A netting technique more specific to reefs is drive netting, often called *muro-ami* fishing. Muro-ami damages coral more from the behavior of the fishers than from the net. It involves a large set net with wings and a large bag into which fish are driven by divers holding weighted scare lines. Nets may be 36 m wide, with 100 m wings. If poles replace scare lines, the technique is called *kayakas* fishing. The process can involve a handful or as many as 400 divers, often young boys. The scare lines end in 3–5-kg stones, which are repeatedly dropped on the coral to scare fishes into the net. In shallow water, coral is deliberately broken to chase groupers, snappers, and other large fishes from holes (Ruddle 1996; Jennings and Kaiser 1998). A single muro-ami set in a 1-ha reef area with 50% coral cover and involving 50 divers damaged 300 m^2, or about 6% of the coral (K. E. Carpenter and Alcala 1977). Many areas are fished repeatedly.

Muro-ami has also been criticized on human welfare grounds. Commercial operations often employ 7- to 15-year-old boys, who earn less than \$1/day, live in relative squalor, are undernourished, and work in dangerous conditions; deaths from drowning and other causes are apparently not rare (McManus 1996). Additionally, lost reef productivity affects local communities more than the muro-ami overseers. Managers of large-scale fishing methods are mostly outsiders who depart an area when its productivity plummets. This impoverishes coastal communities that had been dependent on local resources; their residents commonly constitute the poorest socioeconomic, and politically weakest, sectors in many tropical nations (e.g., P. J. Rubec 1988). Changes in capture methods, gear, and social conditions have been recommended to improve both environmental and human welfare issues (Jennings and Lock 1996; see www.howardhall.com/stories/muro_ami.html and www.asiaobserver.com/Phillippines-story2.htm for contrasting views).

Fish Traps

Fish pots and traps can be sizable and heavy (see Dalzell 1996). In Bermuda, before they were outlawed, traps measured over 4 m^2 and weighed over 100 kg, requiring hydraulic gear to raise them. They damage reefs when they or their anchors are dropped on live coral, retrieved in anything other than a vertical direction, or moved around by waves and currents during storms, especially after a trap is lost and enters the ghost fishery (Butler et al. 1993). Quantitative data on this issue are rare (e.g., Eno et al. 2001). A review of three unpublished reports from the Virgin Islands and Puerto Rico (K. A. Johnson 2002) indicated that traps were set on living substrates 40%–80% of the time and that damage during setting and hauling was common (5%–50% of organisms under traps were damaged). This topic deserves more study because decisions about regulating trap fishing should consider ecosystem as well as direct fishing impacts.

Are Coral Reef Ecosystems Open or Closed?

Many reefs sustained indigenous subsistence fisheries for centuries but are no longer capable of doing so. What has changed and why? What might seem like an esoteric ecological argument has important implications for fisheries management and reef conservation. The debate concerns whether coral reefs are primarily closed or open systems. Closed systems depend heavily on their own primary productivity. They characteristically retain and recycle nutrients and energy, depending little on import while being unable to tolerate much export. The management options for a closed system are restrictive. A closed reef is less able to recruit larvae or colonists to replace fish removed, and fish exported from a reef area constitute a net loss of energy and nutrients that will not be subsidized from the outside. Open systems receive energy and nutrients from external sources, are replenished from external sources, and can tolerate greater levels of fishing and export. High-seas, pelagic and demersal fisheries are classic open systems.

Coral reefs were among the first ecosystems to be intensively studied with regard to ecosystem processes (e.g., Odum and Odum 1955; Johannes et al. 1972), and the investigations are true classics in the ecological literature (see Hatcher 1997). Early studies depicted coral reefs as relatively closed systems, biological oases in a sterile oceanic desert, where inputs were minimal and nutrients were limiting. The high diversity and biomass on reefs resulted from capture, retention, and tight cycling of nutrients and energy. Later work revealed abundant zooplankton from presumably external sources and nitrogen fixation by reef cyanobacteria (blue-green algae) (e.g., Wiebe et al. 1975), meaning reefs were less dependent on retention and recycling than originally thought. Subsequent analyses indicate that much of the zooplankton, including fish larvae, is generated and retained by the reef itself. Remove the resident fishes and invertebrates, and you remove the source of the zooplankton. And although abundant blue-green algae fix nitrogen, they are unavailable as a food source to most reef organisms. The emerging view remains one of a closed, tightly cycled system: "Coral reefs abound with adaptations that enhance recycling at all levels of organization" (Hatcher 1997, p. 151).

Another relevant debate focused on whether reefs were recruitment (i.e., settlement) limited—that is, appropriate habitat is abundant but too few planktonic larvae arrive at a reef to occupy the available habitat. The alternative view is that appropriate habitat is limiting or that post-settlement biological interactions (predation, competition) determine the kinds and abundances of fishes on a reef, regardless of larval abundance. Recruitment-limited reefs would be below carrying capacity and capable of producing even more fish and invertebrates than fisheries remove. Evidence of recruitment limitation includes variability in species composition and density among unexploited reefs, differences in year-class strength, and rarity of larvae of some species. The habitat-limited or interactive scenario depicts a reef at carrying capacity, one that is less able to replace fish removed through fishing. Evidence includes superabundant larvae around reefs, reefs packed with recruits, and rates of predation that exceed 99% during the first year post-settlement (e.g., Shulman and Ogden 1987; reviewed in Roberts 1996; Hixon 1998; Levin and Grimes 2002).

Munro (1996) argued that if reefs were in fact relatively open and settlement limited, they should be able to sustain high levels of exploitation. Munro cited average yield values of 20–30 MT km^{-2} and maximal values of up to 44 MT km^{-2} for Samoa (from Wass 1982) as indicative of the exploitation potential of Pacific reef systems. Caribbean reefs would produce less because of smaller reef flats and fewer planktivores. Munro contended that healthy Pacific reefs could produce harvests "of the same order of magnitude as those once taken in trawl fisheries in northern Atlantic and Pacific shelves," constrained more by transportation and marketing problems than by production capacity (p. 9). Birkeland (1997a, 1997d) cited empirical findings of rapidly diminishing yields as evidence against the resilience of reefs to exploitation. Initial yields of 20–30 T $km^{-2}yr^{-1}$ have been reported, but CPUE reportedly falls 70% and number of fish per hectare declines 75%, with major shifts in relative abundances among species and disappearance of preferred species.

Can reefs sustain fishing and other extractive activities, especially activities that lead to commercial export of reef resources? Or will such activities cause deterioration of the fishery and degradation of the reef ecosystem? All reefs are obviously not equal and span a continuum of closure, larval delivery, and nutrient input and recycling (e.g., Hatcher 1997). The debate with regard to settlement limitation is certainly not resolved to the level that we can adopt widespread management policies that assume that fish we remove by fishing will be quickly replaced through reproduction. But compared to pelagic tropical, nearshore temperate, and polar demersal fisheries, coral reefs are more closed and less able to sustain high levels of exploitation. Intensive and extensive fishing in relatively closed systems is risky, especially given the proven inability of other, more resilient regions to sustain fisheries (e.g., D. Ludwig et al. 1993).

Life Histories, Ecosystem Processes, and Sustainability

Birkeland (1997a, 1997d) cited differences in long-term productivity among different ecosystems in building the case that reefs cannot sustain intense fishing and export economies. Coral reefs show high rates of primary productivity and high standing-stock biomasses that should theoretically lead to sustainable high yields. Actual yields fall below expectations, however, whereas tropical pelagic systems conform to fisheries models that predict sustainable output. The different outcomes may result from differences in life history traits and ecosystem processes.

Life History Traits

Large pelagic species are often fast growing, early maturing fishes that forage over large ocean areas in search of patchily distributed but concentrated food. They have growth rates of 1–4.5 kg yr^{-1} and higher. In their first two years of life, skipjack tuna (*Katsuwonis pelamis*) can grow to 5 kg, mahi mahi (*Coryphaena hippurus*) to 9 kg, and yellowfin tuna (*Thunnus albacares*) to 14 kg. These species and their relatives seldom live longer than four or five years, which means the population and the biomass it contains turn over relatively rapidly (Birkeland 1997d).

Large coral reef fishes are sedentary by comparison and depend on reef resources rather than dispersed or imported food. Home ranges of many large species such as groupers and snappers may not exceed 0.1 km^2 (Pittman and McAlpine 2003). Large reef fishes are also relatively slow growing (<0.5 kg yr^{-1}) and slow maturing and may live for decades (Goliath grouper to 37 years, other groupers 10–25 years, snappers 5–25 years, grunts 5–15 years; even some surgeonfishes may live several decades; see Roberts 1996; Choat and Axe 1996; Birkeland 1997d). Many groupers, wrasses, snappers, and barracuda reach large sizes but much more slowly than large pelagic species.

Coral reef systems also contain relatively large numbers of predators relative to prey (Hixon 1991). Unexploited and lightly exploited reef systems have a remarkable density and diversity of large piscivorous predators, as well as predators on small fishes and invertebrates. Competition within the predator guild must be intense. But because of gape limitation in all except barracuda and sharks, susceptibility of prey to predation declines rapidly once prey reach adult size and perhaps much sooner. Many reef species are also habitat specialists that change habitat preferences as they grow. Collectively, these are traits and conditions that favor large size, late maturation, relative longevity, and multiple reproductive episodes, which characterize imperiled fish species in many habitats (see chapter 4). Slow turnover of biomass is another characteristic of less productive fisheries systems. Collectively, reef fishes are likely to be locally vulnerable to exploitation, in marked contrast in almost all regards to pelagic fishes. On these grounds alone one would predict that tropical pelagic fisheries could support higher levels of exploitation without collapse than coral reef assemblages.

Characteristic Ecosystem Processes

Ecosystem properties exacerbate the impacts of fishing on coral reefs. Pelagic fisheries yield more fish biomass per unit of gross primary productivity than reef fisheries, which produce 20–40 kg $ha^{-1}yr^{-1}$ supported by 500–8,000 g C $m^{-2}yr^{-1}$ primary production. Pelagic systems such as the upwelling regions off Peru have ratios of yield to primary production 10–60 times greater than coral reefs (ca. 300 kg $ha^{-1}yr^{-1}$) (Nixon 1982). Oceanographic factors, energy transfer characteristics, and food web levels are probably responsible. Upwelling systems receive continual inputs of nutrients and have short food webs, with only one or two steps between phytoplankton and primary fishery species such as anchovies, sardines, and herrings, and just one more step to predatory tuna or mahi-mahi (and another to sharks, which may help explain why sharks are easily overfished) (e.g., Ryther 1969). On coral reefs, incoming nutrients are limited, and the major target species in the fishery are piscivores and other large predators that may be five or six trophic levels above the producers (Grigg et al. 1984). At each step, on average 90% of the energy is lost to metabolic processes, leaving less for the human consumer. On reefs, "far less than 1% of the gross primary productivity is converted to production that is meaningful for human consumption" (Birkeland 1997a, p. 414).

As a result, fishing's impacts on coral reefs are greater because the fishes removed represent such a large proportion of the nutrients and energy accumulated there. Each capture and export from a reef eliminates a higher proportion of the accumulated energy than occurs when a net is pulled through a school of pelagic anchovies. Also, in the pelagic realm, fishers remove only the fish and do not affect the major pathway for nutrient and energy inputs, which is upwelling water. On reefs, destructive fishing practices destroy the reef itself, which is both the source of and pathway for nutrients and energy.

At least one additional factor contributes to differences in productivity. The pelagic ecosystem responds rapidly to pulsed availability of nutrients. Unicellular phytoplankton, the production base in open water, is capable of rapid uptake of pulsed nutrients and rapid population growth. Phytoplankton-eating zooplankters increase feeding and breeding rates when phytoplankton increase in availability, which expands the energy base for higher trophic levels. Coral reefs are also dependent on algae, but the dependence involves slower-growing benthic algae and slow exchange of intercellular products between coral polyps or other invertebrates and their endosymbiotic zooxanthellate algae. Endosymbiosis places physical limits on growth, to the point that if algae in corals divide too rapidly, corals may expel them. "Hence regulatory processes within the complex physiological adaptations of reef animals tend to dampen the responses of intracellular algae to pulses of nutrient input" (Birkeland 1997d, p. 365). Thus, secondary production responds relatively slowly to increased nutrient levels or losses due to fishing. All the evidence indicates that coral reefs are intricately designed to retain and recycle matter and energy but poorly designed to sustain excessive exploitation and removal.

Vulnerability to Export Markets

Even as relatively closed systems, coral reefs have demonstrated an ability to sustain some degree of fishing pressure. Island cultures persisted at relatively high population densities, sustained by the surrounding reefs. If people removed the predators through fishing, reef herbivores should have been able to sustain levels of exploitation equal to what the predator fish removed, especially if the nutrients in herbivore bodies were returned to the reef system. Nutrient return was likely because reef-dependent human societies generally discharged wastes back into the lagoons, reef flats, and mangroves located nearest human habitation. Given such practices, humans are great recyclers of energy and nutrients. However, the system breaks down when toxic pollutants are involved, or when human populations aggregate in district centers and discharge sewage into harbors and bays with their characteristic poor circulation patterns.

Added to these disruptions in energy recycling are irreversible losses that occur when export markets encourage overfishing. When fish and invertebrates are overfished, the source of zooplankton prey and future recruits is lost. Export also permanently removes nutrients, and the surrounding open ocean is slow to resupply the food web with the materials necessary to sustain the reef ecosystem. The most serious environmental problems come from outside the reef, including distant-water fleets supporting export markets, ship groundings, coral bleaching from El Niño, and silt deposition due to poor land-use practices. The impacts are additive. Fishing becomes an outside influence when reef products supply global markets. Reefs can feed small, indigenous cultures, but they cannot feed the world.

ASSEMBLAGE, COMMUNITY, AND ECOSYSTEM EFFECTS OF HUMAN ACTIVITIES

Fishing and nonfishing activities degrade a reef's ability to sustain its human and nonhuman components. Coral reefs provide some of the best evidence of the impacts of fishing on ecosystem structure and function, impacts manifested in alterations to fish assemblages, broader reef communities, and ultimately ecosystem properties.

Food Web Disruptions

The effects of fishing on reef ecosystem components are fairly well understood, with similar events and outcomes observed in the Caribbean, Kenya, selected Indo-Pacific locales, and the Mediterranean (see Jennings and Kaiser 1998). Under relatively undisturbed (unfished or lightly fished) conditions, both top-down and bottom-up trophic cascades operate. Corals proliferate, algal turfs (dense mats

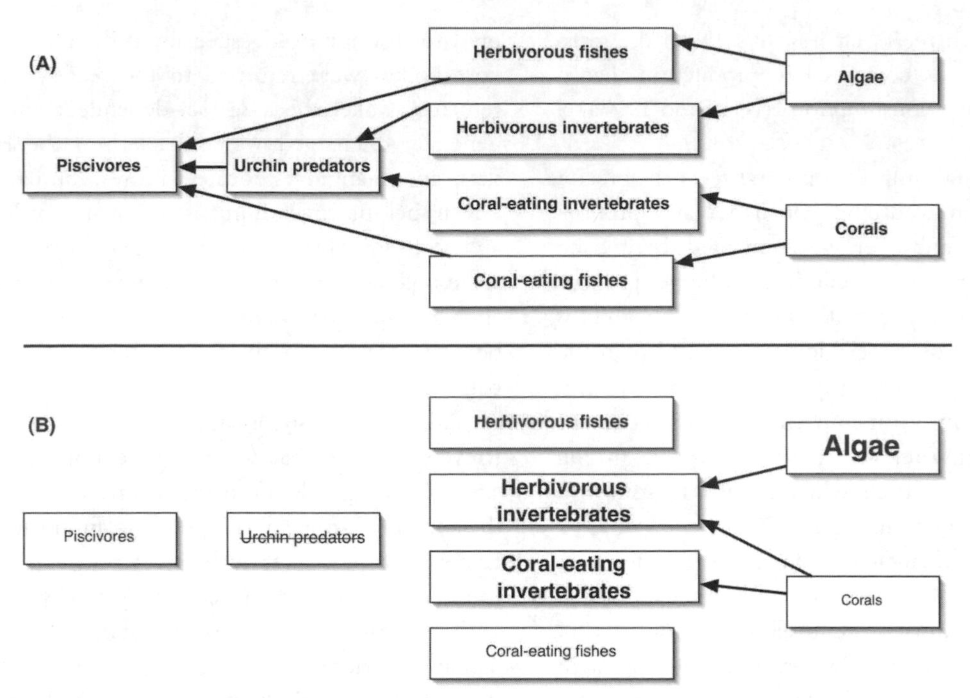

Figure 12.5. Typical effects of overfishing on coral reefs. Overfishing frequently leads to dominance of invertebrates and algae via a series of interrelated pathways. (A) Unfished or lightly fished reefs house a diversity of fishes in several trophic guilds, as well as abundant coral cover. (B) Removal of large piscivores initiates a sequence of overfishing that results in elimination of fishes that feed both on sea urchins and on invertebrates that feed on corals. Herbivorous and coral-eating invertebrates proliferate, which leads to overconsumption of corals or permits algae to overgrow coral. In either situation, overall fish diversity declines because of strong direct and indirect dependence of many fish taxa on live coral cover. Size of lettering denotes relative abundance.

of algae with short filaments) coexist with algal lawns (longer filaments), biological interactions inhibit domination by any one trophic group, and diversity among fish and invertebrate groups remains high (figure 12.5). When overfishing occurs, regardless of the trophic level targeted, either invertebrate herbivores (usually sea urchins) or algae proliferate, shifting the reef away from a state in dynamic equilibrium without obvious dominance, at the expense of fishes and corals. The new conditions appear to be difficult to reverse (i.e., an "alternative stable state" has been created; see Done 1992), and even light fishing can have detectable impacts (see R. C. Carpenter 1997; Hixon 1997).

The resource targeted can influence the outcome. Initial exploitation usually focuses on piscivores, which can make up more than half of the more visible species on the reef (Hixon 1991). Because of the diversity of predators and prey, removal of piscivores has variable and relatively unpredictable results. When predators are removed, herbivorous fishes that were their prey become more numerous. Herbivorous invertebrates (urchins, snails) might be expected to decline because of increased competition, but they commonly persist. Algae may keep up with the added grazing and browsing pressure, but turfs typically prevail over filaments. Herbivorous fishes may clear reef surfaces of algae, facilitating settlement and growth of corals. Reefs subjected to predator-targeted fishing appear to retain recognizable structure and function, although overall fish diversity declines (figure 12.5A).

The problem is that fishing seldom stops after piscivore removal. Urchin predators and fishes that feed on coral-eating invertebrates are next to be hunted (figure 12.5B). This trophic group includes large triggerfishes and wrasses, including the highly prized humphead wrasse, one of the few species that eats crown-of-thorns starfish. Depletion of this group apparently tips the competitive balance between fishes and invertebrates to favor invertebrates. Starfish and especially sea urchins become abundant because sea urchins outcompete herbivorous fishes under conditions of intense fishing on sea urchin predators (e.g., McClanahan and Muthiga 1988; Levitan 1992).

Unlike most herbivorous fishes, grazing urchins actually scrape the substrate. Overabundant urchins accelerate bioerosion, reduce coral cover, reduce topographic complexity, and produce a reef surface dominated by algal turf (the omnivorous urchins in the Indo-Pacific cause relatively more biocrosion of coral substrates than the chiefly herbivorous urchins of the Caribbean; e.g., Hay 1984; McClanahan and Muthiga 1988). Ultimately, fish diversity and fisheries productivity decline (Glynn 1997).

Similar endpoints result when coral-eating invertebrates such as crown-of-thorns starfish or the gastropod *Drupella* become abundant (the role of fishes in controlling starfish and snail abundance remains debated; see Sweatman 1995; R. C. Carpenter 1997; J. B. C. Jackson et al. 2001). Coral-feeding fishes such as butterflyfishes and a few parrotfishes are apparently incapable of causing coral death of the magnitude caused by starfish or snails. Because many reef fishes are coral dependent, urchin-dominated or coralivore-dominated systems house less diverse fish assemblages.

Alternative pathways can lead to coral destruction and loss of coral-dependent species. If large, grazing herbivorous fishes (parrotfishes, surgeonfishes) and other large herbivores (sea turtles, sirenians) are overfished, and if sea urchins cannot proliferate because of low initial abundance or mass mortality, filamentous algae overrun the reef and apparently either overgrow corals or use up available space that would otherwise receive settling coral larvae (Sebens 1994; J. B. C. Jackson et al. 2001; figure 12.6). Large algae also dominate in situations involving sewage and wastewater effluent, and these algae have superior chemical and physical defenses against herbivores (Hatcher 1997). Heavily overfished reefs in the Caribbean experienced "luxurious" algal growth following the mass mortality of *Diadema* sea urchins in the early 1980s (e.g., T. P. Hughes 1994; J. B. C. Jackson 1997); small, unfished herbivores such as damselfishes often became abundant in such areas. Between 1977 and the early 1990s, coral cover in Jamaica declined from 52% to 3%, while algal cover increased from 4% to 92% under these conditions (T. P. Hughes 1994). Rapid algal growth and reduced settling success of coral larvae apparently reduce the ability of reefs to recover from hurricane damage (e.g., Liddell and Ohlhorst 1993). Loss

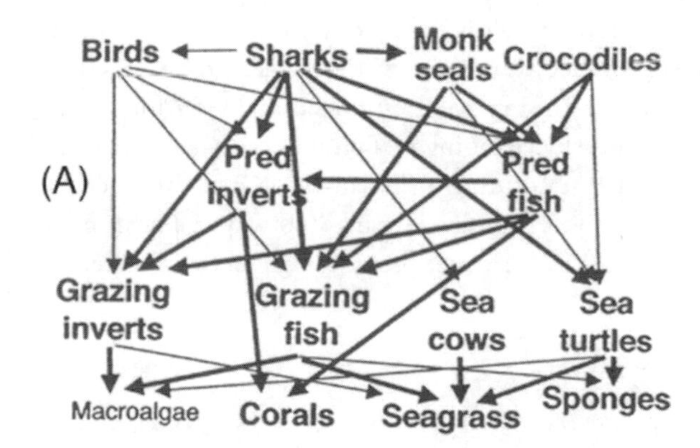

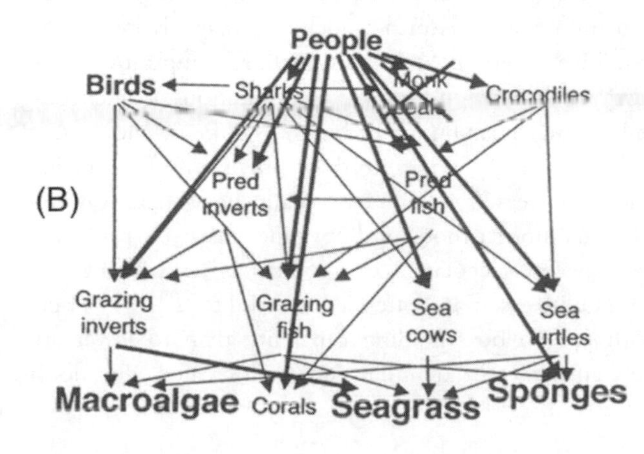

Figure 12.6. Widespread changes on coral reefs due to historical overfishing and overhunting of large predators and herbivores. The food web consequences of such historical practices can be generalized as a shift to dominance by lower trophic levels that includes reduced abundance of corals and increased abundance of algae; sea grasses may have increased in abundance except where sea-grass wasting disease has occurred. (A) Before and (B) after overfishing. Bold fonts represent abundant taxa, normal fonts represent rare taxa; thickness of arrows is relative to strength of interaction. Monk seals have been exterminated in the Caribbean and Mediterranean. From J. B. C. Jackson et al. (2001); used with permission.

of coral again typically leads to declines in reef fish diversity (see Carpenter 1997; Jennings and Kaiser 1998; J. B. C. Jackson et al. 2001).

Understanding the web of interactions on reefs has led to some surprising conclusions and recommendations to prevent repetition of biodiversity loss and fisheries collapse. Birkeland (1997a) suggested that reef fisheries were inherently unsustainable because the abundance of predatory species and complexity of food web interactions already placed them at maximum sustainable yield. Adding human predation quickly pushes the system beyond sustainability. Birkeland (1997a) predicted that CPUE would decline rapidly with even minimal exploitation, a prediction supported by Jennings and Kaiser (1998), who found that initial, light fishing in reef systems has the strongest impacts on abundance and dominance (the classic fishing-down phenomenon, see chapter 11). McClanahan (1995) suggested that fisheries management programs should specifically protect urchin-eating fishes such as large triggerfishes and wrasses, which would prevent the negative diversity and abundance consequences of urchin proliferation. At the same time, protecting urchin predators would lower urchin numbers and direct benthic production into abundant herbivores such as parrotfishes, surgeonfishes, rudderfishes, and rabbitfishes.

Effects of Nonfishing Activities on Reef Fishes and Ecosystems

Many human activities that do not target fishes themselves have direct and indirect, deleterious effects on reef fish populations because of the interdependence between reef organisms and their habitat.

Coral Mining and Collecting

Coral mining is widely regarded as the most destructive and least sustainable use of a reef. Coral rock and sand are extracted for house and road building material, landfill, cement production, and lime production. Live corals are collected for the aquarium, jewelry, and curio trades (see chapter 13). Coral mining is widespread in the Maldives, Indonesia, Sri Lanka, Tanzania, and the Philippines (B. Brown 1997). It is especially common in areas where alternative natural hard building materials are lacking or prohibitively expensive, as is often the case on coral atolls and islands.

Coral mining is nonrenewable because of the disturbance that occurs when blocks of limestone, rubble, and sand are removed from the reef and because of the slow rate at which corals recover (e.g., Moberg and Folke 1999), particularly the massive, slow-growing coral species preferred for construction materials. In places where coral mining has been heavily practiced, such as the Maldives and Tanzania, coral cover has declined 67%–88% because of removal and because trampling, sediment production, and scouring by wave surge and currents inhibit recovery (B. E. Brown and Dunne 1988; Dulvy et al. 1995). Fish biomass, abundance, and often diversity decline on mined reef flats, affecting all trophic groups, especially species dependent on live coral cover and structurally complex habitats (J. D. Bell and Galzin 1984; Shepherd et al. 1992; Dulvy et al. 1995). The Maldives instituted a partial ban on coral mining in 1990 and created incentives for using alternative building materials such as imported cement. Sri Lanka has also banned coral mining in its coastal zone (Berg et al. 1998).

Coral mining yields a net economic loss, affecting multiple sectors of a country's economy beyond the obvious effects on fisheries. In Sri Lanka and southeast Asia, net losses to tourism alone were calculated at \$2–\$3 million per km^2 of reef over 20 years (Berg et al. 1998). Losses due to increased erosion were estimated at \$1–\$4 million per km^2, because live fringing reef provides anti-erosional wave protection. Substituting a human-made structure such as a breakwater is expensive, about \$9,000 per linear meter (Hameed 2002). The reef provides this service for free.

Anchoring

Damage to reefs by anchors is large and underappreciated. Saila et al. (1993) concluded that, in the Philippines, where dynamiting is widespread, anchor damage is even more extensive. Every boat has an anchor; comparatively few carry a bomb. Solutions include prohibitions on anchoring in live coral areas, but enforcement is difficult. More useful and increasingly widespread is provision of anchoring buoys in areas frequented by many boats (e.g., Quirolo 1994). Although mooring buoys incur costs of placement and upkeep, acceptance among boaters tends to be high because it's easier and often less stressful to tie up to a buoy with proven boat-holding capability than to lower and raise an anchor in unfamiliar waters. Increased education is a must.

Ship Groundings

Ships run aground everywhere, but damage in noncoral areas is comparatively small, assuming oil or other toxins are not released as the ship breaks up. Because a coral reef is alive, a ship grounding there can have extensive, long-term consequences. The initial strike on the reef is destructive, but events that follow are often more lethal. Efforts to back ships off reefs frequently cause more damage than if the ship remained in place until the tide rose or professional rescue personnel arrived. Ships hard and fast on reefs tend to be pushed around by storm waves, grinding down the reef, reducing it to rubble similar to that caused by blast fishing. Reef material broken off causes additional damage as it is pushed around by surge and waves long after a ship is removed. Recovery times from ship groundings have been estimated at as long as 100–160 years, depending on location, coral type, and damage extent. Under some conditions, natural regeneration may be almost impossible (Precht 1998; Riegl 2001). Not surprisingly, fish abundance is lowered in grounding sites (Riegl 2001).

It's not just oil tankers and freighters flying flags of convenience that collide with reefs. Late one August night in 1994, the University of Miami's 155-foot research vessel, *Columbus Iselin*, went hard aground on Looe Key National Sanctuary in the Florida Keys. During the 38 hours the *Iselin* spent on the reef, it scraped coral and actually cracked the underlying fossil limestone, killing about 345 m^2 of live coral and destroying another 340 m^2 of reef framework. The damage was doubled over the next five years as loose rubble moved around during storms. The university was assessed $3.76 million in fines for resource damage (including a $200,000 civil penalty). The money has gone toward reef restoration and acquisition of equivalent habitat (www.sanctuaries.nos.noaa.gov/special/columbus/columbus.html).

Terrestrial Inputs

Fringing and barrier reefs along continental margins or associated with islands large enough to capture rainfall have an intimate relationship with the land, receiving the products of terrestrial activities. Land-reef interactions create a naturally more diverse reef system under undisturbed conditions. Nutrients and sediments flow out to the reef, and the shallow water at the sea-land interface makes grassbeds and especially mangroves possible.

Major human impacts come from sewage and agricultural pollution, and from poor land-use practices that include deforestation and result in excess sediment runoff. Sedimentation rates on undisturbed reef systems range from less than 1 to 10 $mg\,cm^{-2}day^{-1}$; corals are stressed above 10 $mg\,cm^{-2}day^{-1}$. Human-disturbed systems may experience rates in excess of 50 $mg\,cm^{-2}day^{-1}$ (B. Brown 1997). Sediment is harmful because it (1) is deposited at rates that exceed the self-cleaning capacity of the coral; (2) reduces light levels needed for photosynthesis by zooxanthellae; and (3) carries adsorbed pollutants such as nutrients, toxic pesticides, and heavy metals. In Asia, soil erosion rates as a result of careless agricultural practices are 1,000 to 10,000 times greater than soil erosion from undisturbed forests (Pimentel et al. 1995). Asia is not alone in exporting terrestrial pollutants to reefs. Agricultural practices along the Queensland coast of Australia exported four times more sediment, nitrogen, and phosphorus to the Great Barrier Reef than occurred prior to western agricultural practices (Brodie 1995).

Reefs unfortunate enough to be located downstream from abundant human populations suffer from sewage, industrial and agricultural pollutants, and sediment from land-disturbing activities (B. Brown 1997). Secondary sewage from a treatment plant was discharged into Kaneohe Bay on the windward side of Oahu, Hawaii, throughout the 1960s and 1970s at rates of about 1,900 m^3day^{-1}. Water clarity in the southern bay, where the discharge occurred and where water exchange was least, was terrible due to high phytoplankton and zooplankton density—I dove there repeatedly in the early 1970s as a grad student at the University of Hawaii. The reefs progressively declined as they were overgrown and eroded by the green bubble alga *Dictyosphaeria cavernosa*. Corals were replaced by particle feeders such as zoanthids, sponges, and barnacles. In 1979, the sewage outfall was relocated to discharge into the open ocean. By 1982, filter feeders had decreased in number, and dead reef areas became covered with a red macrophytic algae. By 1985, coral colonies were visible throughout the bay with some areas of high coral coverage. I dove the same reefs in 1990, and the contrast in water clarity, coral coverage, fish diversity and density, and absence of *Dictyosphaeria* were striking. Coral growth has continued, with a minor setback caused by freshwater runoff due to a storm in 1988. Significantly, corals recovered from the freshwater pulse much more rapidly in 1988 than following a similar incident in 1965, when under the added stress of sewage pollution (B. Brown 1997).

Other examples abound. In Barbados, nutrients, sediments, and toxic pollution created a pollution gradient along the west coast of the island. Coral diversity, skeletal growth, and reproductive effort were lower and algal abundance higher in more polluted sites (see Tomascik 1991). White pox, which has killed more than 70% of the elkhorn coral in the Florida Keys, is apparently caused by a fecal bacterium that lives in the intestines of humans and other animals (K. L. Patterson et al. 2002). Its source is undetermined, but high human populations in the Florida Keys and south Florida are suspect. Along the Great Barrier Reef of Australia, agricultural runoff from cattle operations and sugar cane production creates elevated levels of sediment, nutrients (especially nitrogen), pesticides (diuron, dioxins, dieldrin), and heavy metals (mercury and cadmium), which accumulate in sediments and organisms and degrade inshore reefs and sea grass beds (Haynes et al. 2000; Brodie et al. 2001).

Solutions to terrestrial runoff problems lie in recognizing the connectivity of coral reefs to the surrounding landscape, the responsibility that users of the land have for the welfare of the reefs, and the necessity of managing the entire land-reef ecosystem complex. Once the relationship between land and reef is appreciated, a landscape-seascape perspective can emerge (see Moberg and Folke 1999). Australia, with its ongoing Representative Areas Program along the Great Barrier Reef, is a leader in promoting such ecosystem-level management.

Mangrove Habitats and Reef Ecosystems

The importance of management at the ecosystem level, including terrestrial connections, is particularly evident in the role that mangroves play in reef environments. Coral reefs are complex habitats, of which coral-dominated areas are only a fraction. Sand flats, sea grass meadows, gorgonian fields, mud flats, algal plains, fossil terraces and benches, passes and channels, and mangroves all contribute to the productivity and complexity of reef ecosystems.

The seaward portion of a reef system is the outer reef, where live corals grow to the surface and dissipate the energy of breaking waves, creating lower-energy, depositional, lagoonal and inshore habitats (e.g., sand and grassbeds). The farthest inshore, lowest-energy habitat is the land-fringing mangrove forest. Mangroves buffer the land from any remnants of wave action, and also trap sediments and nutrients running off the land that might otherwise smother or eutrophy offshore reefs. The living reef promotes growing conditions for mangroves and sea grasses, and the living mangroves keep the reef areas free of sediment (Ogden 1997). These physical-chemical functions and services are seldom disputed. More controversial is the role of mangroves as a transitional habitat in the lives of fishes.

Snorkel in a mangrove, and you will be struck by the variety of fishes, especially juvenile fishes. Mangroves are clearly a preferred habitat for adult fishes such as mangrove snapper, mangrove jack, gobies, blennies, killifishes, gerreids, cardinalfishes, scats, archerfishes, pipefishes, and halfbeaks. More abundant are juveniles of species whose adults occur in other, often coral-dominated habitats (including squirrelfishes, groupers, snappers, grunts, angelfishes, butterflyfishes, parrotfishes, surgeonfishes, jacks, and barracuda). An important question is whether the juveniles are mangrove dependent: How many of the juveniles that occupy the inshore mangroves wind up out on the reef, or using ecological jargon, are mangroves sources of reef fish adults or merely sinks for reef fish larvae? If mangroves are critical nursery areas, protecting reef species requires protecting mangroves.

It has been dogma that mangroves and nearby sea grass beds are nursery grounds for coral reef species. Ecology textbooks and general reviews contain statements without equivocation (or reference) such as "Mangroves function as nurseries for a variety of fishes and invertebrates that spend their adult life on coral reefs" (Ogden 1997, p. 289). Although juveniles occur in mangroves and adults on reefs (Laegdsgaard and Johnson 2001), the issue of mangrove dependence has lacked rigorous testing (e.g., Beck et al. 2003).

An unequivocal demonstration of ontogenetic connections between mangrove and reef would require direct evidence such as tagging and following individual fish (practically impossible) or chemical analysis of otoliths of adult fish resident on reefs that showed the isotopic signature of mangroves in early-deposited growth rings (e.g., Thorrold et al. 2002). The closest to a definitive test was performed in Curaçao, where Cocheret de la Moriniere et al. (2002) recorded the distribution of fish species over five months. They found a size-related temporal progression indicative of initial settlement in mangrove and sea grass habitats and eventual movement to coral habitats (figure 12.7). Several

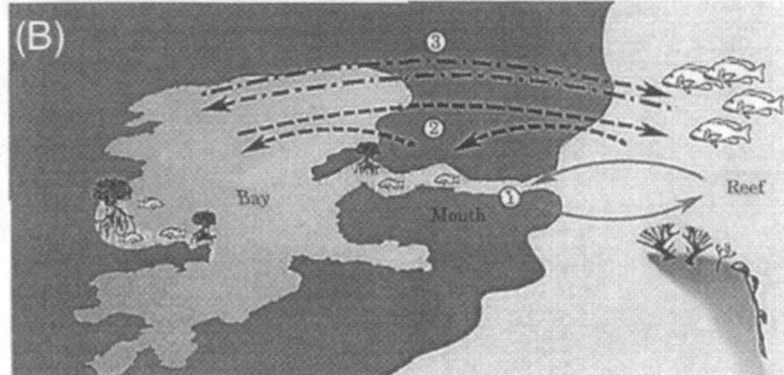

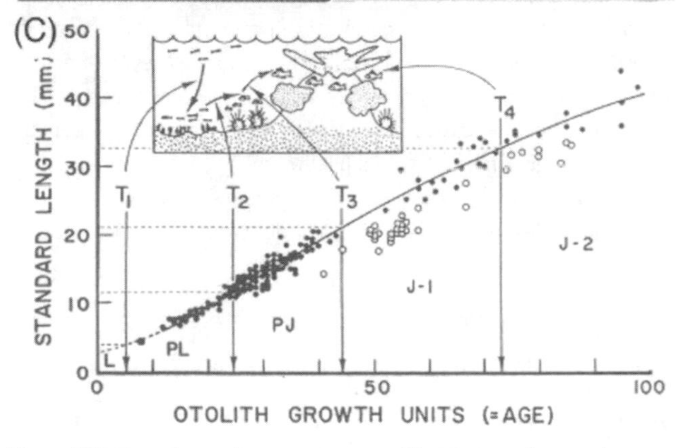

Figure 12.7. Interdependence among reef habitats and organisms. (A) Juvenile French and white grunts migrate at dusk between a coral head and the nearby grassbed, transporting nutrients from grass to coral daily. (B) Grunts, among many species, use separate nursery areas prior to moving to hard-bottom reef areas. Three such patterns of post-settlement movement exist: (1) growth in shallow reef habitats and migration to deeper reef habitats; (2) settlement in intermediate zones, movement to mangrove and sea grass, and eventual occupancy of the reef; and (3) settlement in sea grass and mangroves and progressive migration to reef habitats. From Cocheret de la Moriniere et al. (2002); used with permission. (C) French grunts have four juvenile phases and occupy different habitats before becoming subadults (PL = postlarvae; PJ = pre-juvenile; J-1 = juvenile 1; J-2 = juvenile 2). From McFarland (1979); used with permission.

species were clearly "nursery species" typified by high densities of juveniles in mangroves or sea-grass beds and older fish predominantly on reefs. Other species displayed reverse or more random patterns. The most logical explanation for the patterns is that nursery species moved between habitats as they developed. Fish surveys in Belize also showed strong relationships between densities of fishes and whether the reefs they occupied were associated with mangroves (Mumby et al. 2004). Some grunts and snappers had up to 25 times higher biomasses on mangrove-rich than mangrove-scarce reefs. The Caribbean's largest herbivorous fish, the rainbow parrotfish, *Scarus guacamaia* (IUCN Threatened), is so heavily dependent on mangroves as juvenile habitat that it has disappeared from a number of locales that have suffered mangrove destruction.

Most important, mangroves are an essential and productive part of island ecosystems (box 12.3) and are under attack globally (Valiela et al. 2001). They have been destroyed to create shrimp farms, milkfish ponds, and salt ponds and cut for lumber, firewood, and charcoal production. They are regularly fragmented by roads and cut and filled for shoreline real estate development and as landfills. Globally, less than half of the world's mangrove forests remain, and more than half of the remaining forests are considered degraded. Mangrove forests are being lost in the Americas at a rate of 2,250 km^2yr^{-1}, a destruction rate that exceeds that of tropical forests. By 1984, 90% of the mangrove areas of the Philippines had been destroyed; 55% of Thailand's mangroves were destroyed between 1961 and 1993; only 50% of Bangladesh's original mangrove habitat remains, primarily in one forest preserve and some offshore islands (P. J. Rubec 1988). The nursery function of mangroves underscores their ecological as well as their economic importance. For information on mangrove conservation, see Earth Island Institute's Mangrove Action Project (www.earthisland.org/map).

Global Climate Change: The Ultimate Overexploitation

"It's a great time to be an herbivore."

My wife's comment in August 2000 when we dove in Palau on algal-covered reefs killed by the 1998 El Niño.

Among humanity's most serious and long-term environmental impacts are global climate disruptions resulting

BOX 12.3. Everything's Connected

The interdependence and connectedness of reef habitats and species become apparent when the biology of an individual species is viewed in detail (figure 12.7 and 12.8).

French grunts, *Haemulon flavolineatum*, in the Caribbean initially settle from the plankton and inhabit sea grass and mangrove areas (Cocheret de la Moriniere et al. 2002). These structured habitats provide food and especially protection from predators for small fish. As they grow, the grunts move progressively to reef areas, where large juveniles and adults form daytime resting schools. At night they forage on small invertebrates in sand and grass areas (McFarland et al. 1979). Hence the health of a French grunt population is heavily dependent on the health of at least four distinct reef habitats: mangrove, sea grass, sand flat, and live coral.

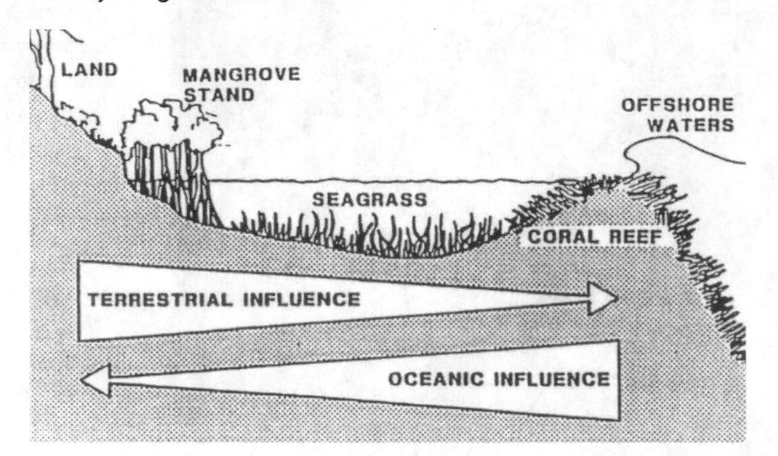

Figure 12.8. Interconnectedness of various "reef" habitats. Offshore reefs buffer mangroves from wave action while mangroves buffer reefs from sediment load from the land. From Ogden (1997); used with permission.

French grunts apparently contribute to the welfare of their daytime reef refuges. When grunts return from their nightly foraging in the grassbeds, they reestablish schools over the same coral heads they used previously, occupying the same sites used by previous generations of grunts (Helfman and Schultz 1984). As they metabolize their food, their excretory products are taken up by the symbiotic algae living in the coral. Coral heads with grunt schools grow faster than heads without schools (Meyer et al. 1983). Hence corals protect grassbeds and mangroves, mangroves protect corals, and grunts connect grassbeds to coral heads.

French grunts are just a well-studied example; their biology is probably no more or less complex than that of many other reef species. Ontogenetic habitat shifts and daily intrareef migrations are common among reef fishes (Helfman 1993; Helfman et al. 1997), implying that connections and interdependence among reef organisms are much greater than we suspect. The diversity of reefs begets the diversity of reefs in a positive feedback loop of immensely complex interactions. Our wonder would only increase if we knew more.

from continued overproduction of greenhouse gases (Walther et al. 2002). Climate change directly affects coral reefs via increased sea surface temperature (SST), which is exacerbated during El Niño–Southern Oscillation (ENSO) events. Strong El Niños occurred in 1992–93 and 1997–98. Tropical SSTs in 1998 were the highest on record, concluding a 50-year trend of increased temperatures in many tropical oceans. Also in 1998, coral reefs globally "suffered the most extensive and severe bleaching (loss of symbiotic algae) and subsequent mortality on record" (Reaser et al. 2000, p. 1500; see Wilkinson et al. 1999). The 1997–98 bleaching event apparently reduced the world's live coral cover by about 10%, with many local areas suffering greater losses (Hodgson and Liebeler 2002).

SST during the 1997–98 event was elevated 1°–4°C for two to five months. Although that increase might not seem catastrophic, it was for the world's coral reefs. Corals live close to an upper thermal limit of about 30°C (e.g., Sebens 1994). Slightly higher temperatures cause corals to lose their symbiotic zooxanthellae algae and take on a light, "bleached" appearance. They may die within months. Bleaching events were recorded as early as the beginning of the 20th century, but their frequency and intensity increased in the mid-1980s. During 1998, coral colonies died that were over 1,000 years old and had survived hundreds of El Niños. Average increase in temperature in many of the hardest-hit areas during the 1997–98 warming event was no more than 1°C (Goreau et al. 2000).

Significant impacts occurred throughout tropical oceans:

- Across the Indian Ocean, water temperatures in 1998 peaked at only 1°–2°C above long-term averages, yet coral mortalities ranged between 70% and 99%, many areas experiencing 90% mortality (Goreau et al. 2000). Economic losses from the 1998 bleaching in the Indian Ocean were estimated at between $700 million and $8.2 billion (Wilkinson et al. 1999).

- On Rangiroa Atoll in the South Pacific, where fast-growing *Pocillopora* corals experienced over 99% mortality, large colonies of massive, slow-growing *Porites* also died, "a phenomenon not previously observed in French Polynesia and virtually unprecedented world-wide" (Mumby et al. 2001, p. 183).
- In the Arabian Gulf, *Acropora*-dominated regions were turned from "previously lush coral gardens into a dead framework that was increasingly bioeroded" (Riegl 2002, p. 29). Fish diversity decreased from 95 to 64 species, with increases among herbivores and planktivores and decreases among invertebrate feeders. Herbivores commonly increase after major bleaching events, probably in response to proliferating algae that grow on the dead coral (Goreau et al. 2000; Sheppard et al. 2002).
- Belize, home to the world's second-largest barrier reef and heavily dependent on reef tourism, experienced average temperature increases of 2.2°C during late September 1998. Catastrophic coral die-offs occurred in the central shelf lagoon, with near eradication of dominant corals on reef slopes. Reef cores indicated this die-off to be "the first bleaching-induced mass coral mortality . . . in at least the last 3,000 years" (Aronson et al 2002, p. 435). Recovery may be impeded by increased growth of an encrusting sponge.
- Coral disease outbreaks (epizootics) in the Caribbean proliferated in the 1990s and included black band, white band, white pox, and white plague diseases. Other pathogens have attacked coralline algae, gorgonians, sponges, and echinoderms (Goreau et al. 1998). Diseases spread faster and cause higher mortalities during periods of elevated temperature. Penetration of harmful UV radiation and bleaching of UV-absorbing chemicals increase at elevated temperatures, further stressing corals (Reaser et al. 2000; S. Anderson et al. 2001).
- I dove Palau's spectacular reefs from 1967 to 1970. When I first took my family to the same reefs in 2000 to show them their beauty, the level of devastation brought me to tears.

Sebens (1994), discussing the extent and probable causes of widespread bleaching events in the Caribbean in the late 1980s–early 1990s, wrote almost presciently, "Extensive events comparable in geographic extent to those in the Caribbean have not been observed in the Indo-Pacific region to date" (p. 126).

Bleaching affects some coral species more than others, with branching corals commonly more susceptible than massive, head-forming corals (e.g., T. P. Hughes et al. 2003). Among Indo-Pacific species, the corals most affected were the most abundant and fastest growing, branching and table-top *Acropora* species, which "are the most important coral species in terms of providing shelter for fish and in breaking waves and protecting the coastline" (Goreau et al. 2000, p. 10). Coral-dependent fish species, especially coral-eating species, are among the first to disappear from reefs after a bleaching event (e.g., Sheppard et al. 2002).

Prospects for recovery from the 1998 mass bleaching are not good. Heavy bleaching was extensive, raising the question of whether upstream sources of coral larvae would be sufficient to provide recruits to downstream, affected areas. Coral diseases appear to be proliferating, jeopardizing corals that escaped previous warming events and increasing the susceptibility of regions exposed to elevated temperatures. Elevated atmospheric CO_2 reduces a coral's ability to produce carbonate skeletons, and recent average carbonate accretion rates may already be insufficient to keep up with predicted rates of sea-level rise (e.g., Ogden 1997). Storm-generated waves, predicted to increase with global warming, further break down reefs that are being eroded by sponges, algae, and algal grazers. Projections under most climate models call for elevated surface temperatures "to become more frequent, more intense, and more widespread in the decades ahead. . . . If bleaching events recur in the coming years, even the cleanest and most remote reef areas may fail to recover" (Goreau et al. 2000, p. 12).

The present trajectory of human activity, climate change, and coral reef response suggests that declines will continue. Global reef surveys indicate that 2002 was second only to 1998 in severity of bleaching events, without the occurrence of El Niño conditions. The Great Barrier Reef was especially hard hit in 2002 (ReefBase 2002; Oliver and Noordeloos 2002). Even reefs protected as marine sanctuaries and managed for sustainable use cannot escape climate disruption, which weakens their potential to help recolonize more-disturbed areas (T. P. Hughes et al. 2003). Significant action requires curtailment of greenhouse gas emissions. "Conservation goals can no longer be achieved without taking global climate change into account . . . without immediate action, the massive world decline of coral reefs will continue" (Reaser et al. 2000, p. 1501). (This quote represented an official resolution of the U.S. Department of State, recorded at www.coralreef.gov. In 1998, President Clinton established the U.S. Coral Reef Task Force, charged with identifying "the major causes and consequences of degradation of coral reef ecosystems." By January 2003, under President George W. Bush, the Web site made no reference to global warming or climate disruption among the identified and proposed causes, and by July 2005 discussion of causes of reef decline had ceased, although a linked NOAA site mentioned "global change" as an issue.)

SOLUTIONS: PATHWAYS TO SUSTAINABILITY

Reefs can be conserved, and reef fishes can be used sustainably. Global warming aside, the outlook for reefs is relatively positive compared to imperiled fish species and fisheries management in general. Some optimism can be justified given the biology of reefs and reef fishes and the sociology of the people who exploit them.

Biology, Socioeconomics, and Reef Fisheries Management

Although standard fisheries management schemes have failed on coral reefs, coral reef fisheries can be managed, can recover from overfishing, and can be fished sustainably, assuming destructive methods have been minimized.

Spawning patterns among reef fishes have many attributes that lend themselves well to management. Aggregated spawning and larval dispersal potentially replenish depleted areas, promoting recovery, as long as fishing pressure on aggregations is controlled in upstream areas where habitat is appropriate to maintain reproducing populations. In addition, local areas can be set aside and protected as sources of recruits and colonists. No-take zones and MPAs—especially those where spawning aggregations occur—have been demonstrably successful on coral reefs and have the added benefit of attracting reef tourists.

The nature and pattern of exploitation that typically exist on reefs also facilitate sustainable fishing. By targeting large predators, human predation can replace fish predation. Because of energetic considerations and the abundance of piscivores in most relatively undisturbed reef systems, large predators consume substantial quantities of prey fishes, more than many fisheries take from the reef (Jennings and Lock 1996). Carangids alone (giant and blue trevally, *Caranx ignobilis* and *C. melampygus*) at French Frigate Shoals in Hawaii consumed about 50 MT km^{-2} annually, exceeding the yields of many multispecies reef fisheries (calculated from Sudekum et al. 1991). Theoretically, removing top predators could lead to an increase in prey fish biomass, although the supporting evidence on this matter is equivocal (see chapter 15). Alternatively, Jennings and Lock (1996) pointed out that cropping prey fishes directly is a better way to maximize yields than relying on top-down food web effects, because prey fishes are lower in the food chain, more abundant, and often faster growing and quicker to mature.

A wealth of relevant information is available on reef fish biology and can be used in place of the "complete data sets" required by standard fisheries models. In reef locales fishers also have an intimate knowledge of the biology of the species, gathered over centuries. Fishing ability is a matter of pride and is woven intimately into the culture, giving it more than subsistence or economic value. "Reef fishermen are vastly more numerous than reef biologists, and have been plying their trade and passing on their accumulated knowledge for many more generations than the latter" (Johannes 1997, p. 381). The data aren't there, but the knowledge is. When that knowledge is applied and combined with local concern for the resource, management can and does succeed (see Cordell 1991; www.culturalsurvival.org).

The first and most obvious step is for reef fishing nations to commit resources to study their fisheries, beginning with compilation and verification of local knowledge. Only then can decisions be made concerning which management schemes seem most appropriate. Instead of trying to predict yields or regulate catches via classical fisheries models, it may be wiser to take a pragmatic approach, examine analogous reef fisheries that have produced sustained yields, determine which if any local fisheries have strong similarities, and use information available from ecologically similar areas.

Geography

Island nations are scattered throughout the tropics, often separated by hundreds or thousands of kilometers. Many islands are too small to accommodate jet airline traffic or so widely spaced as to make airfreight prohibitively expensive, necessitating transportation of goods by small ships. In a world economy, this makes commerce slow, costly, and unprofitable. Multiple examples exist of failed efforts at exporting reef fishes (e.g., Johannes and Riepen 1995), either frozen (expensive) or live (logistically complicated). The failure of commercial fisheries does not mean that reefs can't support subsistence fishing. As Birkeland (1997a, p. 429) pointed out, "It takes at least an order of magnitude of more fish to support a few commercial fishermen than it does to support a population by subsistence because most of the proceeds from the sale of fishes go to supporting the marketing and transportation infrastructure" (see table 10.1). Geography works against entry of small island nations into the global economy, at least in extracting and exporting reef goods.

Tradition

A positive trend of changing attitudes and practices emerged at the end of the 20th century in many tropical regions, especially among small Indo-Pacific island nations. In the late 1970s, the outlook for reef conservation was relatively bleak due to increasing consumerism, intrusive technology, developing cash economies, and burgeoning export markets (Johannes 1978b). However, many of these trends slowed, reversed, and were replaced by local control and implementation of traditional conservation practices, practices that reflected "a marine conservation ethic, that is, an awareness of the finite nature of their marine resources and a commitment to prevent their overexploitation" (Johannes 1997, p. 383). These traditional practices include limited entry, MPAs, closed areas and seasons (the *sasi* of Indonesian cultures, e.g., Soselisa 1998), and restrictions on destructive and extremely efficient fishing methods (Johannes 1977). Several nations have demonstrated that traditional non-Western attitudes toward nature can "provide a sound foundation for contemporary natural resource management" (Johannes 2002, p. 317).

SOLUTIONS IN PRACTICE

Three approaches have experienced repeated if not universal success in developing sustainable reef fisheries: the transfer of responsibility for reef conservation and management from distant to local authorities, the replacement of extractive exploitation with ecotourism-based economies, and the designation of protected areas.

Community Control

Island nations managed their fisheries for centuries before the arrival of westerners, sometimes at human population densities greater than today's: "All the basic conservation measures for marine fisheries that textbooks suggest were invented by Europeans at the turn of the last century, were, in fact, in operation in Oceania . . . centuries earlier" (Johannes and Lam 1999, p. 10). Artisanal, local management was seldom democratic, often immensely punitive, but generally successful. A recurring theme in recent examples of reef management successes is initiation from the inside rather than imposition from the outside—reef conservation works best where local jurisdictions have or take control over their own resources, especially when control includes the ability to limit access of outsiders to reef resources. Where open access exists and catches can be sold and exported, a tragedy-of-the-commons situation (Hardin 1968) develops, because it makes sense for every fisher to catch all he can. However, when the resource is controlled by a village, clan, or chief, it is in the best interest of those in control to promote sustainable use of the resource (Birkeland 1997a). When local control is exerted, both yield and CPUE increase substantially (e.g., Alcala and Russ 1990).

Johannes (2002) reviewed events in six Pacific nations where community-based coastal resource management (CBCRM; sometimes called Territorial Use Rights Fisheries, TURFS) had been successfully applied as well as events that led up to its application and the lessons to be learned. At the heart of most examples was the institutionalization of customary marine tenure, which is the right to control both access to and activities in nearshore fishing grounds traditionally under local control.

In Vanuatu, measures applied widely included fishing ground and seasonal closures, outright bans on capture of certain species, and restrictions on night spearfishing and gill net use. All measures required short- or medium-term sacrifices justified by the expectation of long-term benefits. Compliance was voluntary, but demonstrated successes in one village led to adoption of similar management practices in other villages. During the 1990s, 21 villages implemented 51 new management measures and only 5 measures were dropped. Success in protecting trochus (a shell-bearing mollusk) stocks led to voluntary seasonal and area closures that protected finfishes, lobsters, clams, sea cucumbers, and crabs. All were initiated and enforced at the local level, and "this locally funded shoestring operation has enjoyed greater success than a foreign-aid-funded fisheries development project in Vanuatu costing tens of millions of dollars" (Johannes 2002, p. 319).

In Samoa (formerly Western Samoa), a strong tradition of local marine tenure was replaced by state control during colonial rule, allowing outsiders to fish in waters previously controlled by local villages. In the late 1980s, partially in response to depleted marine resources, laws were passed that recognized chiefly authority, the primacy of village rights, and the right to manage nearshore fisheries. These laws returned to villagers the incentive to manage their traditional fishing grounds. Villages developed marine resource management plans at a rate almost twice that anticipated by Fisheries Division projections. Common measures included creation of no-take reserves, bans on destructive fishing practices, mesh size limits, bans on rubbish dumping, and minimum size limits. Again, "because [the programs] are being managed by communities with direct interest in their success, compliance with bans on fishing is high and there are not the enforcement costs associated with national reserves" (M. King and Faasili 1999).

Fiji also saw increased adoption of traditional customs for managing marine resources. Seasonal and area closures, bans on commercial fishing, bans on nighttime and scuba-based spearfishing, control of fishing methods, and species restrictions were implemented. Fishers reported improved fishing, including increased CPUE and increased numbers and sizes of fish, impressions that were later substantiated by systematic sampling. Initial efforts were copied elsewhere (Johannes 2002, p. 325): "Word of the success spread to other villages in the area, and seven of them implemented their own tabooed fishing areas. By 2000 the total protected area in these waters had increased [from 24 ha] to 7 km^2. Following local media coverage of the Ueunivanua project, similar efforts began in four other sites across Fiji, and the Ueunivanua monitoring team was in high demand to serve as trainers."

Johannes (2002) discussed similar examples and outcomes in Cook Islands, Palau, Tuvalu, and Hawaii. No region is without attendant problems and setbacks, and each nation presents a unique set of conditions and complications. Regardless, efforts at conserving reef resources are increasing because of reinstitution of cultural restraints and practices focused on limited entry, enforced through local control. Analogous conclusions have been reached in tropical freshwater fisheries (e.g., McGrath et al. 1993; Chao et al. 2001).

Two important advantages of tradition-based management programs are (1) that culture and locale specificity make them appropriate to the needs and customs of individual villages, and (2) that traditional law can adapt to changing circumstances within a country because the local people make the laws. Neither of these assets characterize national, Western-based fisheries regulations and practices, whose "uniformity and slowness to change is based more on the need to grease the wheels of commerce rather than for community harmony or equity" (Johannes 2002, p. 331; see also A. T. White et al. 1994; Pomeroy and Carlos 1997; and the *Traditional Marine Resource Management and Knowledge Information Bulletin* at www.spc.org.nc/coastfish/News/Trad/trad.htm).

Ecotourism as an Alternative to Extractive Exploitation

The greatest sustained economic value of coral reefs comes from nonextractive, locally controlled activities managed for ecotourism. Few people will pay to see a school of anchovies or herrings, but legions will spend large sums to have a diving experience that includes seeing luxuriant, colorful corals and large reef fish (see box 10.4). Nondestructive subsistence fishing can occur without significant depletion and can exist side by side with a thriving tourist trade. In addition, ecotourism succeeds most when economic returns go directly to the local communities. This vested interest motivates fishers to forgo capture of species that attract tourists and makes resource protection and compliance with regulations a local responsibility. Such community involvement, often voluntary, greatly reduces the logistic difficulties and expense of law enforcement.

Small nations can derive a significant fraction of their national budget from ecotourism (Birkeland 1997a). Reef tourism contributes about $13 million annually to Palau's gross domestic product of about $170 million. Bonaire derives about $21 million annually, or about half its gross domestic product, from scuba-related tourism. Reef and beach-directed ecotourism in the Caribbean in 1990 was valued at $8.9 billion, employing over 350,000 people (Dixon et al. 1993). Even industrialized nations can benefit substantially from reef tourism. Over two million tourists visit Florida's John Pennecamp Coral Reef State Park and Key Largo National Marine Sanctuary each year, bringing in approximately $1.2 billion (English et al. 1996).

Tourism allows small island nations to participate in the world economy, because tourism in general, and ecotourism in particular, involves people with abundant financial resources, especially when compared with participants in extractive, commercial fisheries. Transportation costs are borne by the tourists themselves rather than being subtracted from the profits derived from the sale of exported fish. Pelagic fish are worth much more when pursued by recreational "big game" fishers than when caught commercially and sold abroad. After all, charter boat operators get paid whether or not fish are caught, which is not the case in commercial fisheries.

Ecotourism is by no means without liabilities (see Mieczkowski 1995; Honey 1999). Negative impacts include pollution from engines, damage from anchoring, coral breakage by inexperienced divers' flippers and hands, and collecting and spearfishing by divers (e.g., J. P. Hawkins and Roberts 1993; Plathong et al. 2000; Tratalos and Austin 2001). All of these liabilities are manageable, however. Four-cycle outboard engines are more efficient than two-cycle engines at combusting fuel and oil; permanent mooring buoys at popular dive sites greatly reduce anchor damage, as does drift diving rather than static diving. Instruction in buoyancy compensation can reduce inadvertent coral breakage, and prohibitions enforced by dive operators—in combination with information sessions on reef conservation—can discourage collecting. Spearfishing has been regulated and banned outright in many countries that derive income from dive tourism. One clever means of reducing breakage and collecting of corals is a prohibition on wearing gloves, which is a self-enforcing way of minimizing contact with the bottom. Many localities also limit the number of divers allowed to visit an area each day (Schleyer and Tomalin 2000).

Marine Protected Areas: A Popular and Debated Solution

MPAs were treated in detail in chapter 5 because of their general relevance, but they have had some of their greatest successes on coral reefs, and many classic examples come from tropical countries (see NRC 2001; Russ 2002). Sebens (1994, p. 130) summarized experiences with reef MPAs and concluded that "the establishment of [MPAs] is probably the quickest and best first step any country can take to ensure the continuation of its fishery and the preservation of its reefs."

MPAs provide an attractive solution to reef conservation problems because they encourage ecotourism while creating a framework for restoring reefs and fish populations, thereby conserving biodiversity while enhancing biomass (e.g., Russ et al. 2004). Examples abound of reserves of varying sizes being established and quickly showing positive results that are obvious to and subsequently promoted by artisanal fishers. Not all reserves work equally well or at all; arguments rage over designs, sizes, and locations; and many biological and sociological trade-offs arise in establishing and maintaining reserves with or without ecotourism (e.g., Christie et al. 2003). However, among the solutions prescribed by those most familiar with reef conservation and fishery management problems, establishing reserves—especially no-take reserves—is as close as we've come to a consensus approach to reef management.

Doing It Right: Bermuda, Palau, Australia, Maldives

Different countries and reef systems present unique combinations of ecological, geological, geographic, economic, and social factors, affecting the kind of reserve that is most appropriate. As Butler et al. pointed out, "there are no established formulae for the management of coral reef fishery resources" (1993, p. 32). However, a few countries stand out as exemplary in their efforts and provide outlines of approaches and attitudes that work. Other countries can turn to these successful experiences for guidance if not blueprints.

Bermuda

Bermuda experienced declining catches, populations, and relative abundances of important commercially fished reef species. The situation came to a head in the late 1980s as economic conflicts increased between a small, commercial, reef-fishing sector worth about $2 million annually and a growing, $9 million sector focused on reef tourism (e.g., hotel owners, dive- and tour-boat operators, charter-boat fishers) (Butler et al. 1993). Economic as well as ecological arguments were used in 1990 to justify a decision to discourage reef fishing via a ban on the use of fish pots, the chief technique. The stated concern was that overfishing portended "a collapse [that] would affect not only the future of the fishing industry but also Bermuda's tourism industry, the quality of life for residents, and the very fabric of the islands" (Butler et al. 1993, p. 27). Fishers were offered as much as $75,000 to cease pot fishing, although fishing for lobster and hook-and-lining for pelagic species were allowed to continue. Many factors contributed to making a ban possible, but critically important was a well-educated public willing to accept ecological findings about reef trophic dynamics, especially that removal of herbivorous reef fishes would encourage algal growth and discourage regrowth of corals. Fishers were in the middle rather than the lower socioeconomic class and less dependent on continued fishing to survive (not that the decisions were popular with them). Good historical catch statistics contributed substantially to the ban, as did the realization that Bermuda was on a collision course with reef disaster akin to what had happened in Jamaica.

Reef surveys have shown gradual increases in most parrotfish and surgeonfish populations (R. Smith, pers. comm.), which are the major herbivorous groups that the pot ban was implemented to protect. Long-term results from ongoing studies will prove invaluable in testing the effectiveness of Bermuda's pioneering conservation strategy and determining its applicability to other regions.

Palau

Bermuda's experience is encouraging, but Bermuda is a small, wealthy, Western country (53 km^2, 64,000 people, per capita income $35,000) with centralized control of fisheries. Its fisheries primarily supply the tourist trade rather than providing subsistence protein or being exported out of country. Can Bermuda's resolve and results be repeated where the government is less centralized, subsistence fishing plays a greater role, and substantial impetus exists for export of reef products?

The Republic of Belau (Palau) exemplifies progressive, sound management practices focused on conservation despite pressure for export markets, in a country where subsistence fishing is important. Palau is an island nation of 19,000 people on a land mass of 500 km^2, stretched across 350 islands along a 100-km north-south axis. It is located in the western Pacific Ocean, about 600 km north of New Guinea and 900 km east of the Philippines. Per capita income is in the range of $5,000–$9,000, depending on whose estimators are used (www.worldbank.org or www.cia.gov/cia/publications/factbook/index.html). Fishing is central to the lives of Palauans (see Johannes 1981), and fish are a part of almost every meal. In 1989 Palau was declared one of the Seven Underwater Wonders of the World by CEDAM International, an organization dedicated to conservation, education, diving, and marine research (www.wonderclub.com/WorldWonders/UnderWaterWonders.html).

When I was in Palau in the late 1960s, most fish landed were frozen and exported to Guam and Saipan (reef fish exports averaged about 1,500 MT per year; Johannes et al. 1999). Palau was governed as part of the U.S. Trust Territory of the Pacific. Marine resource decisions came down from the High Commissioner's office in Saipan, Mariana Islands, 1,500 km away. Economic development plans encouraged export markets. The Palauan family with whom I lived as a Peace Corps volunteer seldom ate fresh fish, even though the male head of our household was employed full-time at the Palau Fishermen's Cooperative in the district center of Koror. Reefs around Koror were depleted, and reef fish at the co-op were too expensive. We ate canned tuna and, more commonly and cheaper, canned mackerel, the potted meat product of fish cuisine.

Palau became independent in 1981 and assumed control of its own economy. External and internal pressures existed to export reef products. However, a growing tourism trade and equally growing national concern over conserving Palau's natural heritage caused the leadership to reconsider unrestricted export of fish and other organisms. Export competed directly with recreational fishing, dive tourism, restaurant and hotel needs, and subsistence fisheries. Palauans witnessed growing defaunation of their reefs. Grouper populations were depleted because of intense fishing on spawning aggregations to supply the live-fish food trade in the 1980s (Safina 1997; Johannes et al. 1999). Palauans also bemoaned their economically constrained access to fresh fish. A fisherman had to catch and export 5 kg of reef fish to afford a 0.2-kg can of tuna after all the expenses of marketing, processing, and transporting (Birkeland 1997a).

Economics was a key factor in the decision to discourage export. Hotels and restaurants needed fresh reef fish for their menus. Dive tourism in the early 1990s brought in $12–$13 million annually, not including revenue from hotels, meals, and transportation. One popular dive site off Peleliu accounted for daily revenues of $14,000 per day in peak season, with no extraction of resources (Birkeland 1997a). People were willing to pay to see fish. Through a series of regulations focused on the Marine Protection Act of 1994, Palau

- banned export of humphead (Napolean) wrasse, bumphead parrotfish, giant clam, lobster, mangrove crab, coconut crab, sea cucumbers, and corals
- restricted spearfishing to snorkel diving
- set closed seasons on trochus and black-lipped pearl shell collecting
- restricted the number of aquarium fish licenses issued annually to 20
- banned fishing on rabbitfish and groupers to protect spawning aggregations
- instituted a reporting system for everyone engaged in export of any marine resource

The Palau Division of Marine Resources' (1998) brochure on fishing laws contains a candid explanation of the ban on export of Napoleon wrasse (p. 13): "A big Napoleon wrasse, alive in the wild, is a sight that divers from all over the world would like to encounter. So, as a diving attraction, a live Napoleon wrasse can bring a lot more money to the Palauan people than if it is caught and sold for food consumption. Palau's overall population of Napoleon wrasse is small and fragile. To protect it further, export is forbidden."

When I revisited Palau in 2000, dive tourism was booming. Blue Corner, a famous, now protected dive site, abounded with sharks, snappers, groupers, jacks, mackerels, and tuna. I seldom saw large fish there in the 1960s. Several active fish markets operated in Koror, prices for fresh fish were reasonable, and fresh fish was once again a regular part of Palauan meals. Conservation of reefs and natural areas is a national priority, spearheaded by governmental action and by an NGO, the Palau Conservation Society. Although not without problems and complications (e.g., T. Graham and Idechong 1998), Palau serves as an example of a small island nation that is trying to do things right.

Australia

Australia was an early leader in reef conservation. Its Great Barrier Reef is the largest in the world, extending along more than 2,300 km of coastline and encompassing 347,800 km^2. During the late 1960s and early 1970s, the reef came under increasing threat from proposed coral mining and oil drilling. At the same time, crown-of-thorns starfish outbreaks, fishing by Taiwanese vessels, and unregulated tourism increased. Parliament passed the Great Barrier Reef Marine Park Act of 1975 (see Kenchington 1990). Although the act emphasized multiple uses and attempted to establish a balance between economic development and conservation (but banned all mineral extraction), its major objectives were protection of large areas for public appreciation and enjoyment and preservation of areas in their natural state undisturbed by humans.

Since original passage of the act, protection has increased, in no small part because public support has been strong. The Great Barrier Reef Marine Park Authority reviewed the initial zone designations to implement changes that reflected increased understanding of the extent of species and habitats on the reef. No-take ("green") zones in 2003 accounted for about 5% of the park, over 16,000 km^2, but a developing Representative Areas Program (RAP) proposed to expand the no-take zones to include examples of all habitat types, as well as to recognize connectivity among aquatic and terrestrial habitats. RAP hoped to increase green zones to eventually encompass 25%–33% of the park's area (seewww.gbrmpa.gov.au/corp_site/key_issues/conservation/rep_areas). The Great Barrier Reef was declared a UN World Heritage Area in 1981, adding to its conservation value and protection. The annual value of tourism on the Great Barrier Reef to the state of Queensland has been estimated at $1.5 billion (Reaser et al. 2000).

Maldives

The Republic of Maldives is an island nation (maximum elevation 2.5 m) of 320,000 people located about 500 km southwest of India. Tourism is the Maldives' largest industry, accounting for 45% of its gross national product (Reaser et al. 2000). Fishing is the second most important economic activity, creating a potential conflict between extractive and nonextractive uses. However, half of all tourists come to dive, and the Maldives government recognized the value of live fish and healthy reefs. Tourism has been developed with considerable government control and planning (Kenchington 1990). Several programs and laws were formulated to protect marine resources (R. C. Anderson and Waheed 2001; see box 10.4). Many environmental laws were spelled out in a carefully crafted report to the UN Environmental Program's Convention on Biodiversity. The report (Hameed 2002) listed national goals and objectives with respect to conservation and sustainable use of biodiversity, authorized inventories of resources and assessment of threats, and outlined a National Biodiversity Strategy and Action Plan (NBSAP). The NBSAP included creation of 25 MPAs, with a strong emphasis on local involvement in the decision process; environmental education centers throughout the country; protection of valuable species, including humphead wrasses and whale sharks; prohibition of export of several fish species of value to the tourist trade; limits on export of aquarium species; and provisions for monitoring of species, habitats, and resources. The latter is often lacking from conservation programs in developed and developing nations alike.

And Doing It Wrong: Galapagos

The Galapagos Islands—despite tremendous ecological value and a historical connection to Charles Darwin—show how marine conservation plans can fail if they don't begin with local concern and control. The Galapagos, a territory of Ecuador 1,000 km to the east, were the first designated UN World Heritage Site, in 1978. In 1984, they were declared a UN Biosphere Reserve. The value of tourism and conflicts with extractive fisheries provided impetus for the creation of the Galapagos Marine Resources Reserve in 1986, which included use plans and zones modeled after the Great Barrier Reef Marine Park (Kenchington 1990; see *Oceanus*, 1987, vol. 30, no. 2). Despite creation of no-take zones and expansion of the marine reserve to 40 miles offshore in 1998 as part of a Special Law for Galapagos, illegal fishing was extensive and destructive (N. Boyce 2000). Shark and other populations declined (Burns 1995), but the greatest controversy surrounded the sea cucumber fishery, which began in 1994. Originally, the fishery took about 550,000 sea cucumbers annually. It quickly grew to involve 800 fishermen, who reportedly collected 6–10 million sea cucumbers (plus seahorses, snails, sea urchins, black coral, and sea lions, the latter for their penises for the Asian medicinal trade). The Ecuadorian government closed both shark and sea cucumber fisheries in 1995, and the fishermen responded by taking over the Park Service office and the Charles Darwin Research Station, threatening to set fires on the islands. Ecuadorian troops were sent in, and the "sea cucumber crisis" ended but not before an estimated 30–40 Galapagos tortoises were reportedly killed, apparently out of spite (Boyce 2000; Bremner and Perez 2002; see also TRAFFIC South America 2000)

This situation exemplifies the conflicts that can arise when well-intentioned efforts run afoul of socioeconomic realities. In the Galapagos, valuable ecological and ecotourism resources ran up against a growing migratory human population, extractive enterprises, and weak regulatory mechanisms. Although international concern was great and conservation organizations such as the Charles Darwin Foundation were involved, no native tribes, clans, or communities that inhabited the islands were included in the decision making. When conservation is created and enforced from afar, it faces an uphill battle.

Lessons Learned

Among the patterns that emerge from these examples are (1) the elements required for effective reef protection fall into place best where ecotourism contributes substantially to local economies; (2) tension between fishing and tourism creates potential for reef exploitation in sustainable, nonextractive ways; and (3) without local involvement at all levels, even thriving ecotourism is no guarantee that reef conservation plans will be implemented and enforced (Christie et al. 2003).

Other Solutions

Management plans focused on local control, ecotourism, and MPAs regulate human behavior as much as anything. More active, interventionist techniques and practices have also been tried, with mixed success.

Rehabilitation through Hatcheries

Hatchery restocking, and for that matter introduction programs in general, have a weak and spotty history when it comes to rehabilitating reef fish populations (Bohnsack 1996). Few examples exist, most have failed, and the practice is inordinately expensive and time consuming (see chapters 9, 14). One example is an experimental effort to introduce hatchery-raised Nassau grouper to the U.S. Virgin Islands (Roberts et al. 1995). Twenty-nine fish were raised in Florida to about 35 cm, flown to St. Thomas, acclimated in cages for two weeks, and released. Sightings on the reef dropped from 17 to 5 in the first month, after which only 2 fish were observed. The fish behaved differently from wild fish in many respects (they hovered in the open and approached divers), and their behavior likely subjected them to human or other forms of predation.

The authors were unenthusiastic about the value of such propagation efforts, concluding that "restocking depleted populations of marine fishes is an expensive exercise rife with uncertainties" (p. 163). They recommended simpler management methods focused on the causes of depletion—protection of spawning areas, size restrictions and catch quotas, fishery closures, and creation of reserves—rather than technological fixes such as propagation and release. Given the high cost and intense effort of propagation, such programs should be regarded as last-ditch actions to replenish rare or endangered species, rather than stock enhancements or substitutes for rational management. Empirical evidence notwithstanding, enthusiasm for propagating "high-value" recreational and commercial species seems undiminished (e.g., C. S. Lee and Ostrowski 2001).

Artificial Reefs

Artificial substrates are regularly placed in aquatic habitats to improve fishing (Pickering and Whitmarsh 1997). Increased catches are common around such fish attraction devices, and they are popular with the sport fishing public. However, the conservation value of such structures remains a matter of heated debate. Disagreement hinges on the "attraction versus production" issue: Do artificial reefs increase the carrying capacity of an area and augment production of new biomass, or do they just serve to concentrate fishes in an area via redistribution? If new biomass is produced, they can serve a conservation function. If concentration is the primary action—one that may be desirable from a sport fishing standpoint—they can impede recovery of depleted or imperiled species by making them more vulnerable to exploitation (Lindberg 1997; figure 12.9). For conservation, artificial reefs should be more than "just another fishing method in which habitat is used as an attractant instead of, or in addition to, bait" (Bohnsack 1996, p. 295).

The issue seems straightforward and amenable to testing, but definitive answers have remained elusive due to methodological complications, incomplete data, and misdirected effort (Bohnsack 1996). Increased production has been shown for spiny lobster and octopus and for some fishes when appropriate habitat is lacking for reef-dependent larvae (e.g., reefs or oil platforms placed where

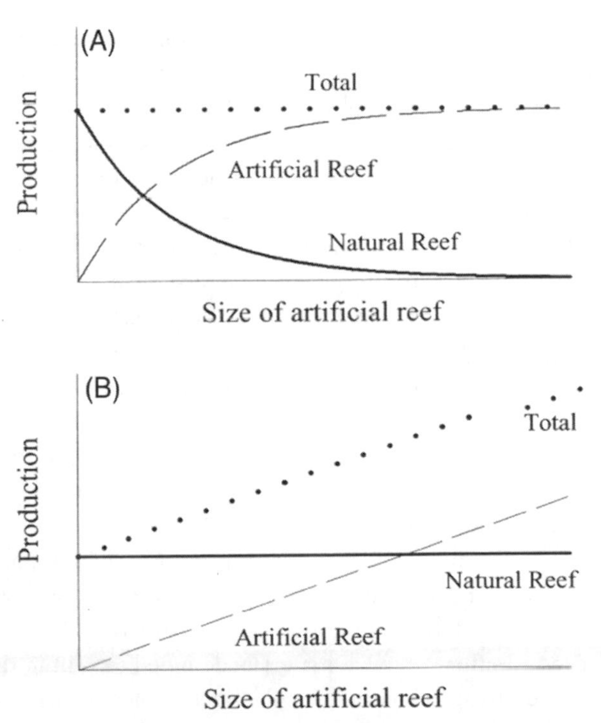

Figure 12.9. The attraction vs. production debate with respect to artificial habitats. (A) Artificial reefs merely attract and concentrate fishes. *Attraction*: As reef size increases, artificial reef production will increase, while total fish production remains constant; hence natural reef production decreases. (B) *Production*: Artificial reefs increase biomass. Total production and artificial reef production increase, while natural reef production remains relatively constant. From J. Wilson et al. (2001); used with permission.

Figure 12.10. The EcoReef. Live coral branch tips wired in place grow rapidly and encrust the ceramic substrate. Structures are also colonized by small fishes, whose feeding helps control algal growth. Photo courtesy of EcoReefs, ©2001, www.ecoreefs.com; used with permission.

hard substrate is lacking; Carr and Hixon 1997). Artificial reefs have proved additionally beneficial if they function as guard structures that physically protect natural habitats from trawling (Polovina 1991; Sanchez-Jerez and Ramos-Espla 2002). They may be able to help in production of ornamental species and thereby reduce pressure on natural habitats and populations (J. Wilson et al. 2001; see chapter 13). But on the whole, and recognizing that reefs probably serve both production and concentration functions, artificial reefs appear to be net attractors rather than producers of valuable species (Bohnsack et al. 1997; Pickering and Whitmarsh 1997).

Artificial substrates are finding some direct application in programs designed to rehabilitate degraded coral reefs (e.g., S. Clark and Edwards 1994; M. Moore and Erdmann 2002; Richmond 2005). Such restoration could require 150 years, depending on cause and circumstances, but may be accelerated through provision of artificial substrates. Successful efforts involve materials designed to mimic the size and shape of natural corals. Such materials are also proving effective in providing habitat for coral reef fishes and solid substrates on which coral larvae settle or to which transplanted coral colonies can be attached. One promising design uses unglazed ceramic stoneware shaped like acroporine (staghorn or antler) coral; the ceramic material is chemically inert, nontoxic, and porous (figure 12.10; www.ecoreefs.com).

The "EcoReef" approach is meeting with some success in accelerating the early stages of reef succession. Attraction of small herbivorous fishes to the interstices of these natural mimics keeps algae in check, thus facilitating coral settlement and growth. Natural mimics are proving more effective than traditionally employed materials (e.g., old tires, cars, sunken ships, and concrete blocks and chunks). Such "junkyard reefs" (M. Moore and Erdmann 2002) are aesthetically unattractive to tourist divers and leach toxic materials that may actually discourage larval establishment. Although promising, the EcoReef concept lacks long-term results, especially of significant coral growth. Any restoration method must convince skeptics educated by the long list of failed costly attempts at reef manipulation (e.g., Ogden 2002). Restoration approached as a solution to reef degradation, or any environmental problem, must first address the causes of degradation.

Is Science the Answer? Little Fish Ecologists versus Big Fish Ecologists

We know a great deal about reef fish biology and management issues, but we don't know enough about the biology of fishes that have to be managed (e.g., Russ 1991). The population, community, and behavioral ecology of small, coral reef species have been intensively studied (e.g., Thresher 1980, 1984; Sale 1991, 2002a). However, fisheries science has focused on large temperate species, which differ biologically from coral reef species in many management-relevant ways. As a result, our knowledge about reef fishes differs greatly with respect to "ecologic" versus "economic" species (DeMartini 1998).

An additional obstacle is that the two groups of researchers don't communicate with each other. Jennings and Lock (1996, p. 213) bemoaned the fact that "there seem to have been few positive exchanges between tropical fishery biologists and reef fish ecologists." Sale (2002b) felt that a gulf between the two groups resulted from the different missions and publication venues of academic ecologists compared to those working for management agencies. Agency ecologists focus on commercially important species, solve pragmatic problems, and publish reports in the so-called gray literature, with variable if any peer review. Academic ecologists, who are more interested in building general scientific principles and careers, publish "their exciting, but often esoteric, new ideas about economically quite unimportant organisms in the peer-reviewed literature" (p. 363). Overstating things only slightly—to academics, applied research is routine and unimaginative; to managers, traditional ecology is largely irrelevant.

An additional complication is that reef ecologists have studied at spatial scales that are probably too small to influence most fisheries. Patch reefs, small sections of reef fronts, or grassbed patches may be manageable and can be manipulated, but can the information obtained be applied to fisheries that cover entire reef systems or national jurisdictions?

One could conclude that this historical estrangement between groups that should have common goals is regrettable, institutionalized but correctable. Sale (2002b) concluded that ecologists have to show how ecology can help improve management and educate themselves better about the management process. Fishery ecologists could make better use of the models and findings of traditional reef ecologists because small reef species share many distributional and biological similarities with larger species.

One area of obvious consensus concerns the importance of understanding larval and juvenile ecology (e.g., Jennings and Lock 1996; DeMartini 1998; Sale 2002b). Small changes in survival of superabundant larvae, whose mortality rates are as astronomical as their numbers, can have dramatic impacts on numbers of fish settling, growing, and eventually recruiting into adult populations (Houde 1997). Improved understanding of mechanisms affecting larval and juvenile production, dispersal, survival, and settling or recruitment has potentially large practical and theoretical value. Standard fisheries production and yield models are dependent on such information (Appeldoorn 1996), and application of those models to reef fisheries is hampered by a lack of it. Reef fish ecologists have studied, experimented with, and speculated extensively about larval dynamics (Leis and McCormick 2002; Cowen 2002), emphasizing economically unimportant (except as ornamentals) families such as gobies, wrasses, and damselfishes (Doherty 2002) and only secondarily fishery taxa such as grunts and groupers (e.g., Shulman and Ogden 1987; Shenker et al. 1993; Levin and Grimes 2002). However, many of the findings appear to be generally relevant (DeMartini 1998).

An exciting development in this area is the recognition that larval retention and self-recruitment contribute substantially to the population dynamics of reef fishes. The traditional, albeit not universally accepted, view was that reef fish spawning and larval passivity maximized dispersal away from natal areas (e.g., Johannes 1978a vs. Barlow 1981). In recent years, oceanographic, genetic, and behavioral observations have forced reanalysis of this view (Planes 2002). It is now known that large numbers of larvae return to the reefs from which they were produced (see Doherty 2002; M. S. Taylor and Hellberg 2003).

Larval retention and self-recruitment, best known for commercially unimportant species, have direct and substantial implications to management and conservation issues, including source-sink dynamics, the value and siting of reserves, and identification of metapopulations. Importantly, if future recruitment depends heavily on local reproduction, local jurisdictions must take responsibility for local conservation. Action cannot be relegated to protection of upstream habitats or delegated to external agencies (see Warner et al. 2000). Biodiversity conservation depends not only on protection of local habitats to serve as hospitable settling sites for larvae produced elsewhere, but also on protection of local populations as sources of such larvae. This realization—the intersection where studies of little fishes have application to big fishes—greatly expands the spatial scale at which management strategies should be applied.

CONCLUSION

The picture that emerges of the current state and likely future of the world's coral reefs is far from promising. At the same time, the number of people concerned about reef conservation and the growing awareness of the economic, ecological, and cultural value of coral reefs, especially in countries where the reefs occur, provide reason for hope (see www.reefbase.org and www.reefcheck.org). From a biological perspective, coral reefs are fascinating and imperiled because of the intimate, evolved linkages that exist among the myriad species that constitute them. Reefs are unquestionably the most complex of marine ecosystems and perhaps the most complex on earth. Whether they are absolutely or relatively more dependent on biological interactions for stability than rain forests is immaterial, as is the issue of whether their biological complexity is a source of stability or vulnerability. What we do know about reefs indicates that perturbations to a few components cascade throughout the ecosystem, with often detrimental and largely irreversible effects on humans as well as other members of the ecosystem. These impacts can be local in scale and extent, such as overfishing and destructive fishing; or regional or global, such as climate change, ocean warming, sea level rise, and coral bleaching. The causes are anthropogenic, as are the solutions.

13. The Trade in Live Fishes

> Only a relatively small number of countries have comprehensive regulations that are rigorously enforced.
>
> —Elizabeth M. Wood, 2001

Most commerce in fishes involves dead animals, consumed for food. The exceptions are trade sectors that target and transport live fishes for table fare in restaurants or as display animals and pets in the aquarium trade. Both sectors focus on tropical fishes, with marine species dominating the live food trade and freshwater species dominating aquarium sales. Because coral reef habitats are especially vulnerable to overfishing, the ecological impacts of the live fish trade are often direct and obvious, in no small part due to habitat-destructive fishing practices specific to the live fish trade. Local and widespread species depletions have been documented, and several species listed by IUCN are marketed in the live fish trade. Socioeconomic impacts are also substantial because tropical fishers and collectors reside chiefly in the developing world, where poverty and human exploitation are not uncommon.

THE LIVE-FOOD TRADE: AN EASTERN LUXURY MARKET

The live-food fish trade shares many similarities with the trade in aquarium species, and it differs from conventional "dead fish" fisheries in several ways that make its impacts stronger and its management more difficult. The differences revolve around the remoteness of locales, the species and habitats exploited, and the often destructive methods used to catch fishes.

The trade in live reef fishes, mainly to supply restaurants in Hong Kong and China, has undergone recent and substantial growth. All indicators suggest that this fishery as currently practiced is nonsustainable. Although not restricted to coral reef fishes, the greatest impacts involve reef species (figure 13.1). And although focused on Asia, the seafood markets and restaurants in almost any "Chinatown" district in North America feature live fish in holding tanks.

Typical Scenario

Live-food fish are captured throughout the tropical Indo-Pacific by fishers using hook-and-line, traps, drive nets, and poisons. Small-scale fishers hold fish in net pens or ponds and wait for large foreign companies, which place fish in live-fish transport vessels (LFTVs). LFTVs may be 40 m long and worth \$1–\$3 million; they can hold up to 20 tons, which amounts to 20,000–24,000 fish. Fish are transferred to airfreight facilities or carried in LFTVs directly to Hong Kong, the major importer, with some trans-shipment to mainland China. Live-food fish are typically retailed in seafood restaurants that cater to a well-heeled clientele. The fish are kept in large display tanks in the restaurant, and the customer can select menu items directly. The fish is killed (but may be cleaned while still alive) and cooked immediately. Freshness is therefore guaranteed, as are sweet flavor and soft texture, which are said to decline quickly if rigor mortis is allowed to set in (Johannes and Riepen 1995).

Fish are cooked "Cantonese-style," which usually involves steaming. Price is determined by freshness, flavor, texture, size (plate-sized fish are preferred), color (red is

Figure 13.1. Live Indo-Pacific reef fishes for sale at a fish market, Hong Kong. (A) Juvenile humphead wrasses and groupers; (B) the bulk of the trade consists of groupers, with some other reef fishes such as snappers. Photos courtesy of IUCN Groupers and Wrasses Specialist Group/Yvonne Sadovy.

"auspicious"), and wildness (vs. farm raised); even known natural or economic rarity can drive up price and demand (Johannes and Riepen 1995). Social status accrues from being able to offer guests or business clients expensive species. A meal of top-quality fish, such as humphead or Napoleon wrasse, *Cheilinus undulatus*, can run as much as $180 per kg, wrasse lips ("an exceptional delicacy") selling for up to $225 per serving.

The export and import of live-food fish are minimally regulated. Hong Kong, the major importer, does not require declaration of live imported fishes or define live-food fish as "food," thus bypassing inconvenient health checks and monitoring. Hence information on amounts and types of fish traded is confused and unreliable, based on "informal and unstandardized data collection methods" (Sadovy and Vincent 2002, p. 399).

Economic Extent and Impact

The live-food trade is a high value-to-volume fishery that supplies luxury Asian markets. Major source nations are Indonesia, Philippines, Malaysia, Thailand, and Australia. The fishery began in the South China Sea, then moved to the Philippines as stocks were reduced. Indonesia replaced the Philippines as the major supplier because of depleted Philippine stocks, but live-food operations now occur as far east as Fiji and Tonga and as far west as the Seychelles in the Indian Ocean. Major destinations are Hong Kong, China, Singapore, and Taiwan. Hong Kong is responsible for about 60% of the total trade, and live fish account for 10% of Hong Kong's $6 billion annual food imports (Sadovy 2001a). Live-food fish are worth more than Hong Kong's annual seafood production from capture fisheries (approximately $345 million vs. $270 million) (C. Lee and Sadovy 1998; Lau and Parry-Jones 1999; Sadovy 2001a).

In 1997, nine families and 56 species made up the bulk of approximately 22,000 MT marketed in Hong Kong—30 species were in the sea bass family Serranidae (Cesar et al. 2000, Sadovy and Vincent 2002). Groupers, snappers, and wrasses are preferred, with humphead wrasse, high-finned or panther grouper (*Cromileptes altivelis*), coral trouts (*Plectropomus* spp.), giant grouper (*Epinephelus lanceolatus*), and stonefish (*Synanceia* spp., a recent, somewhat surprising addition) most favored (figure 13.2). The panther grouper, with its unusually small head and black polka dots, is also targeted by the aquarium trade when a juvenile. Humphead wrasse, high-finned grouper, and giant grouper can bring as much as $30–$100/kg wholesale and $170–$180/kg retail, depending on size. Fish in the 1-kg range are most valued because of aesthetics (they just fit on an average serving plate) and because larger fish are more likely to be ciguatoxic (see "Ciguatera" below). Single large fish have sold for $7,500–$10,000 (Lee and Sadovy 1998). Given an average wholesale price of $17–$22/kg, live fish are typically worth four to eight and as much as ten times what dead, fresh-chilled fish of the same species bring in Hong Kong. Live fish fetch prices beyond the means of locals who live where the fish are captured, thus reducing the availability of seafood to local

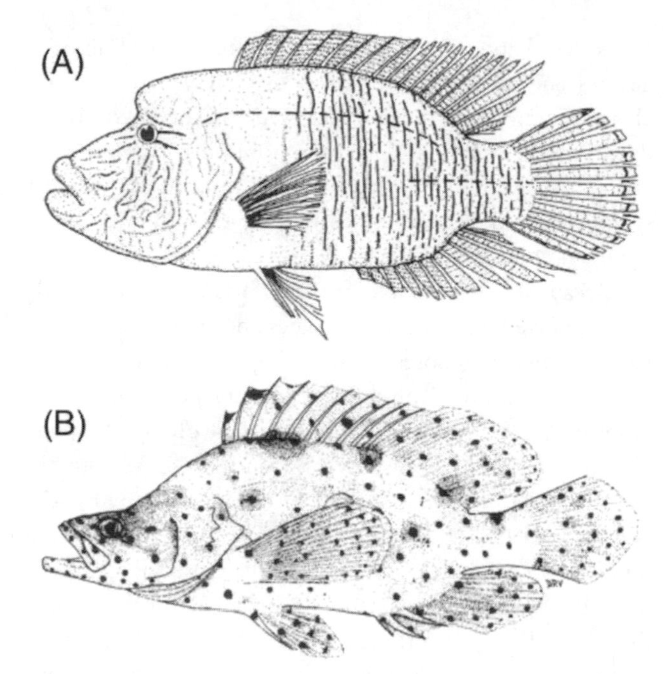

Figure 13.2. Among the most widely sought and highly valued reef species sold as live-food fish are the (A) humphead or Napoleon wrasse, the largest member of its family at 2.5 m and 200 kg; and (B) the panther or high-finned grouper. Illustrations courtesy of the Secretariat of the Pacific Community.

communities. Fishers are typically paid a few dollars per kg for the fish they catch.

Live-fish landings are largely unmonitored and unregulated nationally and internationally and are not at present included in FAO fishery statistics. This makes tracking the trade something of a guessing game. Estimates imply that the trade expanded rapidly in the 1990s and is now considerable. Hong Kong imported 1,000–2,000 MT in the late 1980s, 20,000–25,000 MT by the early 1990s, and 30,000–35,000 MT by the end of the 1990s (Johannes and Riepen 1995; Cesar et al. 2000). Global annual trade in live-food fish may now approach 50,000–54,000 MT and over 50 million individual fishes, worth between $830 million and $1 billion and focused on a small number of species (C. V. Barber and Pratt 1998; Cesar et al. 2000; Sadovy and Vincent 2002). An oft-quoted average estimated mortality of 50% between capture and consumer makes total catch conservatively 100,000 MT (Sadovy and Vincent 2002). Cesar et al. (2000) felt that official statistics provided by Hong Kong underestimated actual import values by 50%. Underestimation is prevalent because underreporting reduces tax burdens and because a substantial portion of the catch results from poaching, often using illegal methods (Johannes and Riepen 1995).

These and other economic factors create a fishery even more conducive to overexploitation than conventional reef fisheries (Sadovy and Vincent 2002). Minimal alternative sources of income for fishers and a tendency for species to become more valuable as they become rarer keep the fishers fishing even as fishing approaches unprofitability. Live-food fisheries are especially attractive because they offer relatively high economic returns in regions where few employment opportunities exist, and where fishers are often among the poorest of the poor. Indonesian live-food fishers earned $150–$500/month in 1996, which was 3 to 30 times the average salary of artisanal fishers (and one to three times the salary of a university lecturer); comparable disparities were calculated for Philippine fishers (Sadovy and Vincent 2002). Overexploitation also occurs because some sought-after reef species are not locally desirable as table fare, as in the Solomon Islands and parts of Indonesia where pelagic tunas are preferred to reef fishes. Such devaluing of reef species results in unrestricted fishing using destructive fishing practices such as cyanide fishing, which affects all reef resources and human welfare. The live-fish food trade thus spirals out of economic and ecological control.

Ecological Issues

Because of the species targeted and the fishing methods employed, the live-food trade affects human and nonhuman components of reef systems.

Overfishing of Predator Species

The live-food trade has depleted stocks of several species that were not otherwise overfished, including several large predators listed by IUCN (see Donaldson and Sadovy 2001). Much of the depletion results from "aggregation fishing" on predictable spawning aggregations of predators such as groupers and wrasses, species that are especially vulnerable because of slow growth, late maturation, and sex reversal (Johannes et al. 1999). Aggregation fishing has become so widespread that the Society for the Conservation of Reef Fish Aggregations was organized to study and minimize the effects. Bans on fishing in spawning aggregations are increasing, spearheaded by efforts in Palau, Pohnpei, and the Solomon Islands. Predator depletion frequently leads to fishing-down-food-web effects and removes animals such as humphead wrasse that play a keystone role in controlling reef-disturbing invertebrates (sea urchins, crown-of-thorns starfish, coral-eating snails; Sadovy et al. 2003).

The trade frequently results in sequential exhaustion of reef resources. LFTVs move into a reef area, fish it intensively with some help from local villagers, and move on to new areas when catches decline, leaving the villagers with greatly depleted stocks and damaged reefs. One visit from "the Asian fishing boats provides the islanders with quick cash, but eliminates large spawning aggregations of fishes that have previously sustained coastal villagers for centuries" (Birkeland 1997a, p. 423).

This scenario was repeated as the live-fish fishery moved farther from Hong Kong, where it originated. Important target species became increasingly rare, and average size of many targeted species declined (Sadovy and Vincent 2002). In the case of stonefish, whose dorsal fin contains deadly venom, the live-fish trade competes directly with a capture industry involved in the production of antivenin (Johannes and Riepen 1995; Fewings and Squire 1999). Stonefish disappeared from areas of Papua New Guinea, representing the first documented, overfishing-caused depletion of a marine fish in that country (Fewings and Squire 1999).

Bycatch

Live-food fishing tends to be highly selective, targeting a limited number of species. The bycatch is nevertheless high because of the methods used. Hook-and-line fishing, although preferable to destructive methods such as blast fishing and cyanide, is multispecific, with bycatch rates of 10:1 reported in Papua New Guinea (Johannes and Riepen 1995). Because much live-food fishing occurs in remote areas that lack refrigeration, excess catch quickly exceeds what a fisher can eat or sell and is often discarded, accelerating depletion of reef resources.

Ciguatera

International trade creates economic opportunities because products previously restricted to certain geographic areas become available in new areas. But these economic transactions move the liabilities of a product as well. One such liability is ciguatera, a toxin usually restricted to tropical fishes (see chapter 7). Ciguatoxins originate in reef dinoflagellates and are biomagnified as they move up the food chain, accumulating in large reef predators. The toxin is not denatured by cooking. Ciguatera poisoning causes a variety of gastrointestinal, neurological, and cardiovascular symptoms, and occasional death (see N. D. Lewis 1986; Helfman et al. 1997).

Because most desirable live reef food species are predators exported from known ciguatera hotspots in the Indo-Pacific, poisoning poses an unanticipated but predictable problem. In Hong Kong, few ciguatera poisonings were reported before 1984, but 23 instances involving 182 people were reported between 1984 and 1988, 31 cases in 1991–92, 244 in 1993–94, and 117 instances involving 420 people in 1998 (A. Stewart 1999; Sadovy 2001a; these numbers do not reflect instances misdiagnosed as food poisoning). Despite these well-publicized and dramatic events, the Hong Kong government took little official action to protect the public. Warning posters were circulated, but their effectiveness was limited. Sadovy (2001a, p. 188) reported: "I have never seen one posted in any major retail area, with the exception of one on Lamma Island, western Hong Kong, which had had the warning removed, leaving attractive pictures of suspect species!"

Ciguatera in the live-fish trade injects a note of irony and perhaps symmetry into what is otherwise a one-sided exploitation of subsistence economies by industrialized nations. Unscrupulous Asian importers supply a luxury market that depletes both the reefs and the human capital of tropical island nations. However, the fishers sell ciguatoxic fish to foreign buyers with full knowledge that they are providing a tainted and even deadly product. Ciguatera hot spots are well known to inhabitants of coral reef areas and are assiduously avoided when they catch fish for their own consumption. Thus an otherwise "useless" resource from a local standpoint becomes a source of external income. Although some Hong Kong buyers have recalled shipments that were identified as originating from ciguatera-contaminated areas, other dealers have proven less scrupulous. One importer knowingly shipped fishes to Hong Kong that had tested positive for ciguatoxin. When Hong Kong government officials requested that the fish not be landed, the importer diverted the shipment to mainland China (Sadovy 2001a). Regulations are needed that require accurate species identification, licensing and bonding of buyers, testing for ciguatera, and careful monitoring of points of origin and import.

Cyanide Fishing: Extent, Impacts, Alternatives

The prevalence of cyanide fishing in both live-food and aquarium fisheries is undeniable. Its use, despite being generally illegal, is often the norm, and even fishers who despise the practice are forced into it because nondestructive capture is less efficient.

Cyanide, chiefly in the form of sodium cyanide (NaCN), comes in powder or tablets; it is dissolved in seawater in small bottles and squirted by divers at fish or into crevices where fish hide, often at night (P. J. Rubec 1986, 1988; figure 13.3). NaCN in water disassociates into hydrocyanic acid (HCN), which in fish tissues is converted to thiocyanate (SCN^-). Fish are stunned due to impaired oxygen uptake at the gills and can be easily netted. They are then placed in clean seawater to recover. Larger fish are less susceptible to cyanide and more likely to recover. Cyanide can be purchased from pharmacies and other sources, its availability often justified because of its use in gold and silver mining (cyanide binds to gold and silver and is recovered as a precipitant). Cyanide is also a reactant and by-product in electroplating processes in jewelry making. One major U.S. distributor quoted me a price of $1 per pound ($2/kg) for 98% pure granular sodium cyanide, available in 100-pound drums. In the Philippines and Indonesia, cyanide powder costs about $3.50–$5/kg, but live-fish exporters often provide divers with cyanide, referring to it as magic powder or medicine (Johannes and Riepen 1995).

Figure 13.3. A young boy with hookah breathing gear and home-made wooden flippers uses sodium cyanide and a hand net to catch small reef fish in the Philippines. Photo by Jeff Jeffords, www.divegallery.com; used with permission.

Because the object of most cyanide fishing is to obtain live fish, one would expect fishers to go to great lengths to minimize mortality. However, because cyanide is highly poisonous and concentrations are essentially uncontrollable when applied on the reef, mortalities are commonplace (good data are elusive). Available numbers, derived mostly from the aquarium trade, indicate high mortality at each step in the capture-transport-holding-wholesale-retail chain. P. J. Rubec (1986, 1988) estimated the following rates (which do not include incidental and bycatch mortality of nontargeted fishes): during collecting, commonly 50% and as high as 75%; during transport and holding prior to first sale, 25%–50%; during export and wholesale, 15%–75% with 30% a probable average; during holding prior to retail sale, 30% or more. These rates suggest that for every 100 fish captured on the reef with cyanide, between 2 and 15 wind up in a tank in a restaurant.

Cyanide is also used to catch as many fish as possible, dead or alive. Chemical concentrations are largely ignored, and large quantities may be used—a 55-gallon drum may be poured onto a reef from the back of a circling boat, with divers then deployed to pick up dead and dying fish. This practice is especially useful "when the weather is rough, conventional fishing is difficult, and the fishermen need money" (Johannes and Riepen 1995, p. 8). Chum or bait may be laced with cyanide and spread on the water or placed in traps. Trolled, cyanide-laced baits are used to catch large pelagics because poisoned predators are weakened and much easier to land. Large fish that ingest cyanide bait often regurgitate it, after which it sinks to the bottom and kills corals and invertebrates.

Physiological Effects

The causes of death among cyanide-affected fish are varied, unspecified, and unsurprising. Fishes exposed to sublethal doses exhibit no apparent pathologies until one to two weeks after exposure, when they may die due to accumulated physiological complications from the poisoning. Cyanide interferes with enzyme systems such as cytochrome oxidase that drive respiratory metabolism, which among other actions reduces the oxygen-carrying capacity of hemoglobin. Cyanide also interferes with enzymatic pathways in the liver. Fish that die later exhibit histological abnormalities in the liver, kidneys, spleen, brain, and gastric tract (K. C. Hall and Bellwood 1995); sloughing of the gastric mucosa followed by cell degeneration characterize fishes that die from "starvation syndrome." Delayed mortality constrains the use of cyanide somewhat in the aquarium industry, but in the live-food trade, deaths that occur days after fish have been purchased and mixed with other fishes in an LRTV are largely untraceable.

Extent of Use

The occurrence and scale of cyanide fishing have been the object of considerable discussion, although the numbers are only estimates. Because fishing with cyanide is usually illegal, "the market is inherently not transparent, as participants are reluctant to convey any information" (Cesar et al. 2000, p. 137). In the mid-1990s, 90% of Indonesian live-fish capture vessels had cyanide on board (Sadovy and Vincent 2002). P. J. Rubec (1988) reported that large ships in the live-fish food trade in the Philippines carried as many as 50 divers and used up to 1,250 kg of cyanide over 10–20 days in a fishery that, by the mid-1980s, had been in operation for at least 15 years. McAllister (1988) estimated that approximately 150 tons of cyanide were applied to Philippine reefs each year just for the aquarium trade. Johannes and Riepen (1995) put the number at 150–400 tons per year, with an even higher application rate in Indonesia. Mark Erdmann (pers. comm. in Rubec 1986) estimated that each cyanide user collected from 50 coral heads per day, working 225 days per year. Given 3,000 collectors (half catching aquarium fish, half catching live-food fish) in the Philippines in the mid-1980s, extrapolation indicates that 33 million coral heads were doused with cyanide each year (Rubec 1988; the number of heads may be smaller because fishers frequently return to the same head). C. V. Barber and V. R. Pratt (1997) put the number closer to 4,000 hard-core users squirting an estimated 150,000 kg (165 tons) annually.

Collateral Damage

Evidence from lab and field studies suggests that cyanide is often fatal to corals, but that death may not occur until days, weeks, or months after exposure, which may explain some earlier negative findings. Mortality is increased if corals receive repeat doses, even months apart (P. J. Rubec 1986).

When reef-building corals are stressed, they become bleached from loss of their symbiotic zooxanthellae. Death often ensues. Bleaching can occur after ten minutes' exposure to a cyanide concentration of four parts per thousand (ppt), with 90% mortality within four days. At 0.1 ppt, bleaching occurs within three to four days and death after nine days; at 0.0001 ppt, corals begin to die after three weeks (R. Richmond, pers. comm. in Johannes and Riepen 1995). Typical squirt bottle concentrations approach 20 ppt, "or two hundred thousand times more concentrated than the lowest concentration eliciting coral mortality in [Richmond's] experiments" (Johannes and Riepen 1995, p. 26; see also R. J. Jones and Steven 1997; R. J. Jones and Hoegh-Guldberg 1999).

How these findings apply to coral loss on the reef remains a matter of conjecture. Coral death from cyanide fishing pales in comparison to the destruction wreaked by blast fishing and climate change. Coral loss due to cyanide fishing in Indonesia, including physical damage caused while extracting anesthetized fish from holes, may not exceed regenerative capacity, according to Mous et al. (2000). However, their calculations may have underestimated long-term damage because revisitation of sites was not considered, nor was impaired regenerative capacity due to multiple stressors such as elevated temperature, disease, dynamite fishing, and other anthropogenic impacts (Y. Sadovy, pers. comm.). P. J. Rubec et al. (2000), chronicling cyanide use among fishers on Olango Island, Philippines, found that many of the 300 collectors there were third-generation cyanide users, dating back to the early 1960s, and said, as a result, "they have destroyed the coral reefs for over 300 miles in every direction" (p. 29).

Cyanide fishing also affects human health. Cyanide is toxic to humans when inhaled, ingested, or absorbed across the skin, and cyanide in gaseous form has been used as an execution method in many societies. Accidental cyanide poisoning in fishing communities has occurred but apparently not as frequently as might be thought (P. J. Rubec 1986). Deaths have resulted when fish were carried in the same plastic bags used to transport the cyanide tablets and when people ate fish that contained cyanide-laced baits.

Cyanide probably leads to more deaths and debilitating injuries indirectly because its use requires scuba or hookah (air line) diving. When distant-water fleets move into a remote area, local men are recruited to participate in the fishing. They are handed scuba or hookah gear but given minimal training in its use, resulting in diving-related accidents, many because divers work as deep as 200 feet. Such depths must be fished because these large-scale operations quickly deplete fishes in shallower water. Johannes and

Riepen (1995) documented diving-related incidents among live-food trade divers in the Philippines and Indonesia; they found 15% of divers with the bends and a mortality rate of 5%. Best (2002) found similarly high rates (20%–40% with diving-related injuries, 4% mortality per year). It is likely that thousands of cases of bends-related paralysis and hundreds of deaths have occurred from cyanide fishing. Compensation is minimal. According to Johannes and Riepen (1995, p. 21), "The widow of someone killed in a diving accident typically receives a single payment of the equivalent of about $50—less than what a pound of humphead wrasse can fetch in a Hong Kong restaurant."

Alternatives

Cyanide fishing is illegal in virtually every country in southeast Asia and the Pacific (C. V. Barber and Pratt 1997), but cyanide is readily available because of its legal applications in mining and electroplating. However, alternative methods and chemicals exist for both industries (see USEPA 1994; www.earthworksaction.org/pubs/CyanideFactSheet.pdf), prompting bans against cyanide use even in these trades.

For fishing, nondestructive practices and less toxic alternatives to cyanide and other toxicants abound. In the mid-1980s, the nonprofit International Marinelife Alliance (IMA) was formed to protect coral reefs. One of the organization's first goals was the replacement of cyanide fishing with nondestructive capture methods as part of its Philippines Destructive Fishing Reform Program. The Philippines in particular have targeted cyanide fishing and have changed from being among the worst offenders in southeast Asia to something of a role model for nondestructive live-fish fisheries (Cesar et al. 2000). Central to these efforts were development and implementation, by IMA and the Philippine Bureau of Fisheries and Aquatic Resources, of a cyanide detection test (CDT) and testing facilities, as a means of determining whether cyanide has been used in the capture of live fish. Facilities have been established throughout the Philippines and other countries. The CDT procedure can detect cyanide residue in fish tissues at very low concentrations. Other parts of the Destructive Fishing Reform Initiative include public education, promotion of local control, legislation, monitoring, inspection, training in alternative fishing methods, and biological research on targeted species (http://eapei.home.att.net).

IMA and other groups, most notably the Secretariat of the Pacific Community (SPC, formerly the South Pacific Commission), have now established similar programs that include training collectors in netting techniques without cyanide use. SPC publicizes its efforts and publishes relevant research articles in the *Live Reef Fish Information Bulletin* (see www.spc.int/coastfish).

Many efforts at reducing cyanide use have been successful, especially with regard to the aquarium trade, where nets are effective at capturing small fishes and where mortality at any step is a commercial liability. For the live-fish trade, however, incentives to the fisher are weak because of the mobility and unlikely return of the buyers: "Many of the fishermen are not concerned if corals are killed or the collected fish die later in an aquarium or cage—they have already been paid" (Erdmann 1999, p. 5).

One recent positive development is the promotion of clove oil (eugenol) as a fish anesthetic of low toxicity (see Erdmann 1999). It has been used as a topical anesthetic for centuries and is known to be an inexpensive, ethanol-soluble fish anesthetic that reduces handling stress (e.g., Munday and Wilson 1997; Keene et al. 1998). It has many advantages over such anesthetics as quinaldine and tricaine methane sulfonate (MS-222). Although the toxicity of clove oil to corals apparently has yet to be tested, published results on efficacy, cost, and toxicity to fishes are promising.

Clove oil could also have positive legal implications (e.g., C.V. Barber and Pratt 1997, 1998). Although cyanide fishing is widely illegal, cyanide can be used legally as an anesthetic for holding and transporting live-food fish. This complicates prosecution of vessels with cyanide or even with cyanide-tainted fish on board. Clove oil appears to provide a cheaper, effective, relatively ecologically friendly alternative to cyanide and would facilitate all-out bans on cyanide use.

Toward a Sustainable Live-Food Fish Trade

It is unrealistic to think that the live-food trade is likely to disappear, given its extent, the strength of the demand from increasingly affluent Asian societies, and the short-term economic returns to local fishers. However, the status quo cannot be justified on ecological grounds. Nor can it be justified on economic grounds at a national level. An analysis of Indonesian live-fishing involving large-scale cyanide use quantified annual losses of $280 million against profits of only $234 million (sustainable, nondestructive practices netted $322 million) (Cesar et al. 2000). Sadovy and Vincent (2002) analyzed fishing levels for live-caught groupers in southeast Asia. They calculated that 68,000 MT could be harvested sustainably, but capture and total trade in 1999 were closer to 100,000 MT per year. These calculations lack confidence intervals and are open to interpretation, but if they are anywhere close to accurate, they portend present and future problems.

Many authors agree that the trade can be promoted where stocks are not depleted, fishing practices are not destructive, and monitoring and regulations are enacted. To achieve these goals, Cesar et al. (2000) called for a "marine market transformation," based on models provided by the Forest Stewardship Council and the Marine Stewardship Council. At the heart of a market transformation are two components: (1) sustainable wild catch of marketable-size individuals; and (2) mariculture, either as closed-cycle operations that use juveniles produced on site, or based on grow-out operations that rely on sustainably caught juveniles. Market forces, especially consumer demand for eco-labeled, certified products, would accelerate the transformation to sustainably produced live fish.

Achieving the goal of a sustainable wild catch requires improved monitoring and regulation via national and international laws. Although existing regulations are well intentioned, enforcement remains an issue because of the limited resources of tropical nations and because bribing is apparently commonplace. National and international oversight and control may be more realistic in industrialized nations and international fisheries. For small-scale coral reef fisheries in tropical countries, local and community-based efforts that incorporate traditional conservation practices have a better record of success.

Local Control and Attributes to the Trade

Local stakeholder involvement typifies successfully managed reef fisheries because local people have a vested interest in the long-term welfare of the resource. After surveying the live-food trade in several nations, Johannes and Riepen (1995) concluded that sustainable fishing practices resulted most from reinstating traditional patterns of limited access to the resource. They observed, "The most effective weapon for combating the problem has proven to be providing villagers with the incentive to protect their marine resources through government-supported marine tenure rights" (p. 67). Where such traditional rights are granted legal standing and force, villagers have been able to prevent or curtail unacceptable live-fish operations, as witnessed in Palau, Yap, and some locations in Papua New Guinea (see Johannes 2002). Such a "co-management" arrangement incorporating local communities and centralized governments is proving increasingly successful for artisanal fisheries (e.g., Pomeroy and Rivera-Guieb 2006). The lack of alternative sources of income can serve to involve locals in monitoring and maintenance of reefs, as long as they see some economic gain for their efforts. Social stigma and peer pressure brought against those who engage in destructive fishing practices have also been effective, as demonstrated in the Philippines (C. V. Barber and Pratt 1997).

When locals are presented with facts about the value of large reef fishes in nonconsumptive uses such as tourism, the attraction of one-time, consumptive capture diminishes. Reef fishes and intact reefs provide repeated income from glass-bottom boat and sport diving operations (e.g., Spurgeon 1992). The value of reef tourism has led countries such as Palau, Palawan (Philippines), Australia, Pohnpei, and the Maldives to ban trade in humphead wrasses because these spectacular fish are highly sought by tourist divers and photographers.

In addition to promoting local control of resources, Johannes and Riepen (1995) made several other recommendations concerning the live-fish trade (see also Barber and Pratt 1997, 1998):

- Create binding contractual obligations between local communities and foreign fishing companies that include live-fish–specific licensing.
- Monitor catch and export, keeping accurate records on size, numbers, species, methods, and locations of capture.
- Prohibit all fishing on spawning aggregations because of their inherent vulnerability.
- Curtail use of compressed air (scuba, hookah) diving because cyanide fishing depends on it, and promote testing for cyanide.
- Educate fishers and the public about problems and solutions.
- Ban export of fingerlings.

Australia has banned the use of LFTVs in favor of land-based recirculating saltwater systems and air shipment. This ban curtails fishing on spawning aggregations, reduces mortality on board LFTVs, limits the spread of disease among wild stocks brought about by dumping dead and diseased fish, and prevents mixing of genetic stocks that result from escapement (Johannes and Riepen 1995). Australia, Palau, Pohnpei, and the Maldives are among the nations that have clamped down on cyanide fishing, prohibited capture from spawning aggregations, and restricted take of certain vulnerable species such as humphead wrasse.

Most early proposals restricting and regulating the live-fish trade focused on actions taken by exporting nations. However, targeting the demand side of the equation has had demonstrated success in achieving conservation goals, as evidenced by effective boycotts of tuna, swordfish, orange roughy, and Patagonian toothfish. Importing countries can also lessen the negative impacts of the live-fish trade by eco-labeling and certifying sustainably captured and cyanide-free shipments, prohibiting importation of vulnerable species and size groups, keeping better records of all aspects of arriving shipments (including cyanide and ciguatera testing), installing electronic Fishing Vessel Monitoring Systems (VMSs) aboard distant-water vessels, and financially supporting international agencies that are working to improve practices in the live-fish trade (see C. V. Barber and Pratt 1997; Sadovy 1998).

Aquaculture, often promulgated as a means of supplying the live-fish trade while stabilizing the fishery and reducing its impacts (e.g., Cesar et al. 2000), has spread and experienced success in several southeast Asian nations (Pomeroy et al. 2002). Currently, about 25% of the live-fish trade consists of cultured fish (Cesar et al. 2000). The four most commonly cultured species are green grouper, *Epinephelus coioides*; greasy cod, *E. tauvina*; Malabar cod, *E. malabaricus*; and the centropomid "sea bass" *Lates calcarifer*. Beginning in the early 1980s, green grouper and Malabar cod were produced in commercial quantities in Taiwan via closed-cycle culture, but, according to Cesar et al. (2000, p. 146), "It is likely that several decades will be required to establish a thriving grouper aquaculture industry capable of supplying [more than a few species to] the live reef fish trade" (see also chapter 14).

Aquaculture is also often promoted on the grounds that it reduces pressure on wild populations. This impact is largely a function of whether farming uses an open- or closed-cycle culture. Only 10%–15% of culture systems, mostly involving groupers, are closed-cycle operations that produce their own juveniles (Pomeroy et al. 2002). Open-cycle aquaculture involves live capture of breeding adults and, particularly, juveniles ("seed") and growing them to marketable size in ponds or floating net pens. This places pressure on wild stocks of juveniles. In the first half of 1995, more than 13 MT of grouper fingerlings were exported from Manila to fish farmers in Taiwan and other locales (Johannes and Riepen 1995). Such extensive capture fisheries for juveniles add growth overfishing to fisheries already suffering from recruitment overfishing. Depletion of recruitment stocks can result because juveniles have passed through younger, larval periods of mass mortality and therefore have low natural mortality rates (e.g., Beets 1997; Sadovy and Pet 1998; Koenig and Colin 1999). Concern over the impacts that depleting juveniles could have on future stocks has led Vietnam, Malaysia, and China to ban export of small fishes, "although significant illegal export persists" (Sadovy 2001a, p. 189). However, Johannes and Ogburn (1999) saw no evidence of depleted wild stocks in Philippine areas subjected to intense juvenile harvest. Standardized monitoring of these fisheries is needed to determine their actual impact.

Another concern over seed collection is the up to 75:1 ratios of bycatch that result when artificial substrates are deployed to attract juveniles, with almost 100% mortality in nontargeted species (Sadovy and Vincent 2002). Also, most live-food species are carnivores, necessitating the capture of small fish to feed them. Feed conversion ratios are poor, perhaps 10:1 (Johannes and Riepen 1995). Formulated feeds are becoming increasingly available but are used in only a small fraction of grow-out programs (Pomeroy et al. 2002). Food loss and feces production have caused local pollution problems, exacerbated by already poor water quality conditions in such places as heavily populated Hong Kong. Disease and deoxygenation add to the costs of culture. Trans-shipment of juveniles to grow-out facilities carries a threat of introductions through escape or deliberate release, as happened with tiger grouper juveniles that escaped from aquaculture facilities in Hong Kong (Sadovy and Cornish 2000). Another criticism is that captive propagation often takes the place of addressing underlying causes of declining reef resources. Basically, aquaculture of live food fish is subject to the drawbacks inherent in aquaculture programs in general (see chapter 14). Finally, aquaculture creates counterintuitive, increased pressure on wild stocks because "whatever species remain uncultured will hold special appeal for many Chinese consumers, for whom rarity and 'wildness' are major gastronomic virtues" (Johannes and Riepen 1995, p. 650).

On the positive side is the incompatibility of seed collection with cyanide fishing, which has converted many former cyanide fishers to juvenile gatherers policing the reef against cyanide users (Johannes and Ogburn 1999). Another advantage is that cultured fish are ciguatera free, an increasingly important consumer issue in Hong Kong. However, closed-cycle operations would put seed collectors out of business.

An obvious answer to criticisms leveled against the industry would be closed-cycle aquaculture that produced acceptable species in commercially viable numbers. But without regulatory reform combined with active and adaptive management, aquaculture will likely add to problems of unsustainable fishing and reef degradation. The aquaculture industry is aware of the many negative perceptions and actual problems, and of the potential it has to reduce pressure on overfished stocks and alleviate poverty. Several organizations are taking steps to promote responsible aquaculture practices and expand consumer awareness of the possible advantages of sustainably cultured species over the wild-caught product (see Network of Aquaculture Centres in Asia-Pacific, www.enaca.org/Grouper/; see also World Bank 2000).

THE AQUARIUM TRADE: NORTHERN RECREATION, SOUTHERN SURVIVAL

Ornamental fishes! I find the term itself offensive, implying that spectacularly colored, complexly behaving, intricately evolved reef and rainforest species are trifling baubles. Newspaper articles referring to tropical fishes as "colorful, mellow eye-candy" (e.g., Watson 2003) do little to lessen my misgivings. Nor am I mollified by my personal experience keeping aquarium fishes, especially marine fishes, that has often involved killing beautiful creatures. Admittedly I'm somewhat cavalier about testing and changing water, but from conversations with fish aficionados and pet store owners, so are many other people. One recognized leader in the American aquarium trade admitted to me that some retailers depended on hobbyist incompetence. Capture and transportation mortalities among wild caught fishes are shocking, as is the short life span of most marine tropical fish once they arrive in someone's home. Knowing all this, walking through the marine section of a pet store for me is a visit to death row.

However, my personal feelings overlook the realities of a complex situation. Aquarium keeping is a $15–$30 billion international industry with strong appeal in the U.S., Europe, and several developed Asian countries. Most wild-caught aquarium fishes come from developing nations in Asia, the Pacific, the Caribbean, South America, and Africa. Exploitation of third world resources and people typifies the aquarium trade, often with insufficient attention paid to sustainability: Habitats are destroyed; bycatch is substantial; most captured fishes die in weeks or months; collecting competes with ecotourism; and released exotics invade distant ecosystems, spreading pathogens in the process. At the same time, aquariums, especially conscientious development of public aquariums, contribute substantially to public awareness of biodiversity issues. Aquarium collecting, when performed in a sustainable manner, can have significant positive impacts on the economies of rural communities. Negative impacts can be reduced via aquaculture of some (mostly freshwater) species, managed harvests, certification of sustainably captured or produced organisms, promoting the protection of intact ecosystems by establishing protected areas.

Secondly, and more generally, many people make a living from the aquarium trade. Viewed thus, it is just another commercial fishery, no more or less reprehensible than any other commercial fishery. In fact, because many small fishes are strongly habitat dependent, sustaining an ornamental fishery is likely to promote conservation of habitats—more so than commercial food fishing with its dependence on habitat-destructive practices. If ornamental fisheries provide sustenance, a fisher's best interests will be served by protecting the ecosystems on which the fisheries depend.

Third, the aquarium trade provides lasting aesthetic pleasure to its practitioners, as opposed to the one time benefit of fish captured for consumption.

Fourth, aquarium collecting typically removes a few individuals, often juveniles, relative to the total number of each species in an area. On a global scale, ornamental fisheries take perhaps 100 tons of fish annually, whereas food fisheries remove 100 million tons, not including bycatch (Dawes 1999). Except where destructive collecting occurs, the ecological impact is probably relatively small.

Finally, as aquarium keeping is an established, international, economic enterprise, efforts to eliminate the trade are unlikely to succeed. While it persists, a common goal for conservationists, consumers, and merchants should be developing sustainable practices that minimize its liabilities and maximize its benefits.

Overview of the Aquarium Trade

Historical, Economic, and Geographic Extent

Freshwater aquaria were popular through most of the 20th century limited only by the availability of electric pumps, lights, and heaters. International trade in marine tropicals developed in the 1950s, with major growth in the 1970s, as jet transport and plastic bags became readily available. Data on numbers of fishes involved are scattered, incomplete, often contradictory, seldom documented, plagued by misidentifications, and generally out of date. Nevertheless, estimates put the global export value of all marine and freshwater fishes and invertebrates in the late 1990s at around $200 million (FAO 1999b; figure 13.4). Import values are roughly twice that number, and the retail price doubles it again (E. M. Wood 2001). The global retail trade is therefore worth roughly $800 million, close to the $963 million value calculated independently by Tomey (1996). Freshwater fishes dominate the trade, accounting for 80%–90% of the estimated 350 million fishes traded annually (Andrews 1990; Young 1997). Annual import value of marine tropical fishes is between $28 and $44 million, involving 10–40 million fish (Wood 2001), or about 10% of the total trade (90% freshwater species, 9% marine, and 1% estuarine; 98% tropical and 2% cold water; Young 1997).

These rough estimates are derived largely from statistics reported to FAO by member countries. Independent estimates from individual regions portray a larger enterprise. Ng and Tan (1997) reported that Singapore alone exported $80–$90 million worth of freshwater fishes, and that exports of freshwater species from southeast Asia (Thailand, Malaysia, Indonesia, Singapore, Brunei, and the Philippines) were officially valued at $150–$200 million but were probably closer to $300–$400 million, or 50% of FAO's global estimate. In 2001, annual exports of wild-caught freshwater fishes from Barcelos, Brazil, alone amounted to 58 million fish (Chao and Prang 2002; see Monteiro-Neto et al. 2003 on underestimation of Brazilian marine exports). Only 10% of the freshwater fishes traded are wild caught, making the official number of 350 million individuals traded annually a gross underestimate. The trade is much larger than available statistics indicate.

Total expenditures for aquarium keeping include the value of tanks, pumps, filtration, lighting, chemicals, plants, foods, transportation, and packaging. A 1970s estimate put the total retail value at $4 billion (Pyle 1993), which climbed to $15 or even $30 billion by the mid-1990s (Tomey 1996; Bartley 2000). Although 60% of income from the sale of fishes goes to developing, tropical countries, nonfish components are produced more widely, and much of those profits go to developed nations.

Major exporting countries, in order of value, are Singapore, Hong Kong, Thailand, Indonesia, U.S., Philippines, Netherlands, Brazil, Germany, and Japan (Bassleer 1994). Both wild-caught and cultured fish come from many locales. Freshwater ornamental species are biogeographically tied to southeast Asia, South America, and Africa, although wild-caught species most recently are taken from the Amazon and the major river systems of southeast Asia (C. A. Watson and Shireman 1996). African ornamentals originally came from Nigerian rivers, the Congo (Zaire) river basin near Kinshasa, Lake Malawi, and Lake Tanganyika (S. M. Grant 1995; L. DeMason, www.cichlidnews.com, pers. comm.). Centers of production of farm-raised freshwater species in the Far East are Thailand, Singapore, Indonesia, Hong Kong, and Malaysia; in the U.S., Florida, Texas, California, and Hawaii rear and distribute species native to a variety of countries.

The U.S. is the largest importer, with an estimated

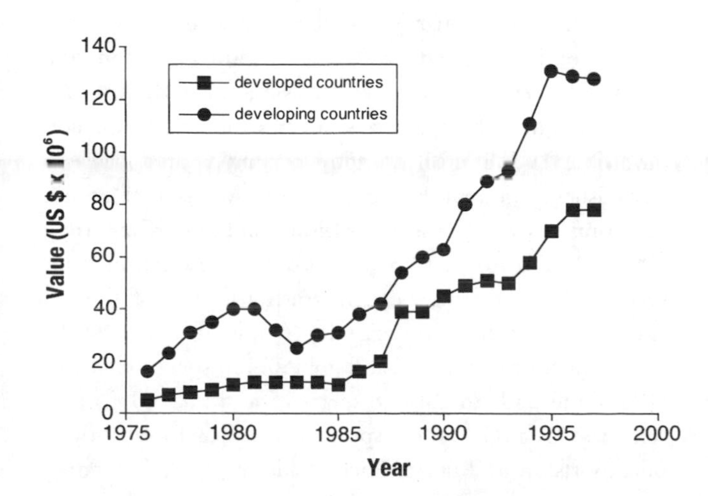

Figure 13.4. Export value of fishes and invertebrates in the global aquarium trade. The global value of exports increased about 15% per year between 1985 and 2000, to around $200 million, developing countries receiving more than 60% of the money. Redrawn from FAO (1999b).

700,000 to 1 million hobbyists, followed by western Europe (0.5 million homes, in primarily the UK, France, Germany, Italy, and Belgium), and Asia, especially Japan and Taiwan (F. A. Chapman et al. 1997; Dufour 1997; WRI 2002). The U.S. may account for 49% of world aquarium expenditures, followed by western Europe (21%) and Japan (13%) (Young 1997). If import numbers alone are used, Europe accounted for 42%, North America 36%, and Asia 22% of the trade (Davenport 1996, using FAO data). With the exception of exports from Hawaii (in 1995, 423,000 fish, or 1%–4% of the global total), a north-south dichotomy between importer and exporter is noticeable.

Tracking the extent of the trade is central to assessing its ecological impact. Unfortunately, record keeping is poor in most countries (see Chapman et al. 1997). Few data are kept on catch, export, or import of individual species, and calculations are based heavily on unpublished reports, informal records, and personal communications (E. M. Wood 2001). Records are even less complete on trade within countries, which involves many millions of fish captured or reared in ponds and distributed domestically. Between 50% and 95% of the U.S. market comes from freshwater culture ponds in Florida, at an estimated 1995 value of $80 million (Young 1997), twice the value of imported fish.

Export numbers do not include fish deaths during capture and holding. After-harvest, pre-export mortality is about 10%–20% for reef species, which means that losses from the wild are underestimated by several million more fish. A major improvement in the aquarium trade would be better record keeping at all points in a chain of supply that involves collection, holding, transport, wholesaling, and retailing. Ongoing efforts to create a global marine aquarium database should help rectify many of these problems for marine tropicals; similar efforts focused on freshwater fishes have yet to be developed.

Water Gardens and Temperate Fishes

Water gardens are indoor or outdoor closed-system ponds designed to hold temperate freshwater fishes and aquatic plants. They make attractive, sometimes expensive additions to a home and are a fast-growing segment of the ornamental fish trade.

A 1997 survey of European fish importers indicated that "cold-water" fishes kept in water gardens were second to tropical freshwater fishes and more popular than marine fishes in the trade (www.ofish.org). One-third by value of ornamental fish imports into the UK were cold-water fishes, with carp, *Cyprinus carpio* (especially the ornamental form known as nishikigoi or koi), and goldfish, *Carassius auratus*, dominating among 20 commonly imported species (Davenport 1996). Native to Eurasia, farm-produced ornamental carp are exported from Israel, Thailand, Japan, and the U.S. (Andrews 1990).

Because the fish and plants kept in water gardens are cold tolerant, the activity creates potential environmental risks. One threat is the potential for escape. Carp and goldfish have been widely introduced worldwide and have caused considerable economic and ecological damage (see chapter 8). Whether highly domesticated escapees pose a direct threat is debatable because of their behavioral and coloration-related vulnerability to predators. However, goldfish revert to a drab, evasive, wild type after only a few generations and often dominate systems into which they are introduced. Fortunately, most water gardens are closed systems that recirculate their water, with minimal connection to surface waters. Open gardens pose a risk of disease transmission to downstream ecosystems. Koi are susceptible to a variety of external and internal diseases, any of which might be transmissible to native species, especially other cyprinids (Waddington 1997; Noga 2000). Chemical treatment of diseases poses an additional pollution hazard.

Species Involved

As many as 1,000 species from 50 families of reef fishes are traded for marine aquarium use (E. M. Wood 2001). Desirable characteristics are "small size, bright or gaudy colouration, non-restrictive diets, and overall adaptability to captive environments" (Pyle 1993, p. 137). Among the most commonly traded marine families (those in which >20 species are traded) are damselfishes (including anemonefishes), angelfishes, butterflyfishes, wrasses, and gobies. Other commonly traded groups are seahorses and pipefishes, scorpionfishes and lionfishes, groupers (and related basslets), hawkfishes, cardinalfishes, grunts and sweetlips, blennies, surgeonfishes, triggerfishes, cowfishes, and pufferfishes (Pyle 1993; Wood 2001). At any one locale, a few families and species dominate the trade. In Australia, 60% of the fish exported came from five families (damselfishes, butterflyfishes, angelfishes, wrasses, and gobies). Ten species made up 60% of the trade in Palau, 8 species constituted 81% of the trade in Costa Rica, and 5 species accounted for two-thirds of exports from Puerto Rico (Sadovy 1992; T. Graham

1996; Wood 2001). In the 1980s, the flame angelfish, *Centropyge loriculus*, accounted for up to 70% of annual earnings of exporting companies from several Pacific islands (Pyle 1993). In Hawaii, 90% of fish exported were from 11 species, with yellow tang, *Zebrasoma flavescens*, accounting for 52% (Tissot and Hallacher 2003).

Most marine tropicals sell for between $2 and $25, but price can reach thousands of dollars for endemic, deep-water, colorful angelfishes. Value is based largely on color, gender (males are often more colorful), juvenile traits (smaller and often more colorful), and availability, with naturally rare species or difficult-to-capture species commanding the highest prices. Endemicity, which is rare in most reef fish families, also drives up demand. For example, clarion angelfish, *Holacanthus clarionensis*, have sold on the Internet for more than $2,000 each (www.thepetstop.com), even though the fish is endemic to a "no-take, no-touch" national marine park in the remote Revillagigedo Archipelago off Mexico's Pacific coast (Almenara-Roldan and Ketchum 1994; Almenara-Roldan 2001). Intense pressure on endemic species is an obvious conservation concern, because overcollecting can affect the species as a whole.

Although more than 730 freshwater species are commonly imported to the U.S.; 32 species dominate. Guppies, *Poecilia reticulata*, and neon tetras, *Paracheirodon innesi*, account for 37% of the total number of fish imported. These two and four other species (platy, *Xiphophorus maculatus*; betta, *Betta splendens*; Chinese algae eater, *Gyrinocheilus aymonieri*; and goldfish) constituted 50% of total imports in 1992 (Chapman et al. 1997). The trade divides freshwater species into two general categories, livebearers and egg layers. Poeciliid livebearers include the familiar guppies, mollies, platys, and swordtails. Egg layers make up the rest: Major groups include barbs, "sharks," koi, and danios (Cyprinidae); loaches (Cobitidae); tetras (Characidae), catfishes (32 families); cichlids (Cichlidae); gouramis (anabantoids); and a host of others. Import values of popular species ranged from $0.06 (Chinese algae eater) to $7.23 (arowana, *Osteoglossum bicirrhosum*). Although popularity shifts with time, 18 of the 20 species that were most popular in 1971 remained in the top 20 in 1992.

Most freshwater tropicals are artificially propagated, often in nonnative locales. Among species imported to the U.S., many of those cultured in southeast Asia originated in South America and Africa: 54% of guppies (native to northern South America) came from Singapore; 89% of oscars, *Astronotus ocellatus* (South American) came from Thailand; 86% of neon tetras (Amazonian) from Hong Kong; 75% of platys (Middle American) from Singapore (Chapman et al. 1997). Large numbers of Lake Malawi and other African Rift lake cichlids are bred and exported from the U.S., Singapore, Malaysia, Hong Kong, Indonesia, Sri Lanka, Israel, Taiwan, the Czech Republic and other former Eastern bloc countries (S. Grant, www.lakemalawi.com, pers. comm.). This globalization phenomenon has its good and bad points. About 80% of U.S. imports of high-priced discus, *Symphysodon discus*, which may be seriously depleted in their native Amazonian habitats, come from culture facilities in Thailand, which reduces pressure on wild stocks; such global swapping and rearing of species, however, will inevitably lead to wide-scale introductions.

In the U.S., production of ornamentals is the fourth-largest component of the aquaculture industry, after catfish, trout, and salmon farming (Tlusty 2002). Freshwater fish farms occur in southern Texas, Southern California, and Hawaii but are most abundant in central and south Florida (figure 13.5). Florida houses about 300 tropical fish farms that contain 20,000 ponds, an average pond covering 0.2 ha to a depth of 2 m. Florida produces 80%–95% of the tropical fish sold in North America, amounting to 700 varieties of ornamental fish (many species come in multiple "varieties") (C. A. Watson and Shireman 1996). Approximately one- to two-thirds of varieties are completely farm raised.

Figure 13.5. Aquaculture ponds for ornamental freshwater fishes in Florida, the major producer of such fishes in the U.S. Typical ponds in the Tampa region are earthen and filled with well water. During winter, they are covered with plastic to retain heat (upper right). From Florida Tropical Fish Farms Association (www.ftffa.com); used with permission.

The remaining species are cultured or wild-caught fishes that originate chiefly in South America and Africa, are imported, and then are trans-shipped (Young 1997). Most farm-raised individuals take two months to one year to reach marketable size. The highest density of farms is in the Tampa region of central Florida because of favorable water chemistry (neutral to high pH) and proximity to a major airport. Tampa International Airport ships 15,000–20,000 boxes of fishes and plants per week (Young 1997). The 1995 value of fish and plants shipped from Florida to North American destinations was $60 million (PIJAC 1996).

Collecting Methods

Capture methods differ in marine and freshwater, reflecting characteristics of the habitat and the fishes. For reef fishes, divers target individuals and sometimes small schools. Hand nets are employed two at a time for individuals, along with a "tickler stick" to coax fishes out of coral refuges (no one uses a slurp gun anymore). In relatively open areas, divers herd fish along natural reef features toward a small barrier net on the bottom, capitalizing on the hesitancy of small reef fishes to flee up into the water column. Tiny, baited, barbless hooks are also employed. Many species can be coaxed into clear containers by placing broken sea urchins inside. Narcotic chemicals such as the coal tar derivative quinaldine and clove oil are effective but are also relatively expensive compared to nets; quinaldine is also toxic to humans. Cyanide remains in use in the aquarium trade despite its almost universal illegality. Other chemicals include diesel fuel, chlorine bleach, and insecticides (Dufour 1997). In freshwater, collectors usually use seines, although divers with hand nets pursue nonschooling, deeper-water species such as some catfishes.

An "average" reef collector employing these methods can collect 25–50 fishes per day, based on observations in the Cook Islands (24–36), Australia (20–45), and Sri Lanka (30–50) (E. M. Wood 2001). Unfortunately, CPUE data are lacking from most of the 50-plus other countries with significant ornamental marine fisheries, making sound management difficult.

Costs, Concerns, and Problems

Destructive Fishing Practices

Fish captured with chemicals suffer high mortality rates. P. J. Rubec's (1988, p. 145) account of cyanide fishing is both descriptive and chilling:

> As an aquarium fish collector approaches, the fish seek refuge in the coral. The diver seals off all the exits by squirting clouds of HCN onto the coral head. . . . Not being able to control the concentration from the squirt bottle, about 50% of the exposed fish die entombed in the coral. The remaining fish flee; but many are stunned by the HCN. The aquarium fish collector selects about 10% of the exposed fish, taking the colourful species of interest to aquarists. The remaining fish lying on the bottom may be collected for the dinner table or left for scavenging predators.

As discussed earlier, mortality rates of 80% for cyanide-captured fishes are not unusual. In home aquaria, cyanide-caught fishes may die when fed, due to irreversible, progressive liver damage. Such delayed mortality may occur several weeks after capture. Another cyanide-related problem is so-called sudden death syndrome (SDS), which occurs when aquarium-held specimens are stressed for a variety of reasons, including abrupt changes in photoperiod (such as when someone turns on the lights in the aquarium). SDS is linked to thiocyanate activity, either as a toxin directly affecting neuromuscular function or because stressed fish convert SCN^- back to HCN due to a drop in blood pH. All these problems would also affect reef fish that had been exposed to cyanide during the fishing operation but not captured; their impaired health and reproduction and delayed deaths are not included in estimates of the impacts of cyanide fishing.

The prevalence of cyanide use among aquarium collectors is again difficult to quantify. In the 1980s, 80%–90% of fish caught in the Philippines may have been collected using cyanide (P. J. Rubec 1986, 1988); similar numbers likely characterized other countries (C. V. Barber and Pratt 1997; Sadovy and Vincent 2002). Collectors who used cyanide were often able to catch three times more fish than collectors using alternative methods (Rubec et al. 2000), placing more conscientious collectors at a competitive disadvantage.

High fish mortality rates and impaired collector well-being gave the Philippines a much deserved bad reputation, precipitating a chain of events in 1986 that included cyanide testing, collector retraining, and certification of cyanide-free fish. The Marine Aquarium Council's certification program is particularly laudable because it promotes responsible action at the import-retail end of the supply chain rather than placing all responsibility on collectors and exporters.

Although cyanide is still used, it has been legislated against, alternatives have been provided, and economic incentives against the practice are increasing (see S. Simpson 2001). Fish with detectable cyanide residues in the Philippines fell from over 80% in 1993 to 47% in 1996 and 20% in 1998 (P. J. Rubec et al. 2000). However, despite being taught alternatives, approximately 30% of collectors reverted to some cyanide use. This apparent backsliding reflects economics as much as attitude. Net capture is slower, net-caught fish do not yet bring higher prices, and some middlemen will not purchase net-caught fish unless collectors buy cyanide tablets (Baquero 2001). Importantly, the campaign is only beginning in Indonesia, the other major producer of reef fishes.

Habitat Depletion through Collecting

Although corals can be killed inadvertently when fish are collected, a greater source of loss of coral, invertebrates, and reef habitat is direct collecting for the aquarium trade. The marine minireefs and dynamic aquaria that have become popular often include so-called live rock, coralline substrate such as dead coral or noncoral limestone covered with live algae and a variety of sessile and even mobile invertebrates. This assemblage of organisms photosynthesizes, filters, and otherwise processes and produces organic material, creating a self-contained ecosystem. An aquarium filled with live rock (recommended density is 0.25 kg live rock per liter of water) more closely simulates the reef ecosystem than one with sand, dead coral, and traditional filtration (Adey and Loveland 1998; Falls et al. 2000). Many reef fishes do better in or even require live rock to survive. However, collecting live rock from the reef degrades the ecosystem on which fish in the wild depend. The upsurge in marine aquarium keeping during the 1980s led to extensive removal of live rock from many reefs, often via dynamite, crow bars, and other reef-unfriendly methods. Until the practice was prohibited in Florida in 1989, 300 tons of live rock were removed from the Florida Keys annually (Derr 1992; Falls et al. 2000). A U.S. federal ban on live rock collection was imposed in 1997.

The natural production of live rock involves deposition of limestone followed by growth of encrusting organisms. The underlying substrate takes 4,000 to 7,000 years to be deposited and cannot be considered a renewable resource. However, the plants and invertebrates that constitute the living component of live rock take only two to four years to become established, which has led to more sustainable, commercial farming or aquaculture of live rock. Businesses quarry limestone from land-based deposits and then "plant" the rock in offshore areas where marine growth is favorable and fast. In 2000, Florida issued 36 aquaculture permits for live rock production (Falls et al. 2000), and several companies were selling such eco-friendly live rock over the Web. Unfortunately, a number of countries still allow the sale of live rock quarried from the wild.

Direct collection of live corals also occurs. Whereas an aquarist can maintain live rock indefinitely with proper lighting and filtration, live corals have stringent requirements: 98% of live corals die within 18 months of collection (Derr 1992). Coral is also collected and sold purely for ornamental purposes, placed in tanks where it becomes little more than a growth surface for algae. The Philippines provided over 90% of the world trade in such ornamental corals during the 1970s, exports exceeding 2,000 MT in 1976 (P. J. Rubec 1988). Exporting coral was banned in 1977 and additionally in 1980, which led to a temporary reduction in trade. Lax enforcement led to a resumption of trade, however. Coral gathering was a major source of income for about 30,000 people, who gathered around 48,000 MT of stony corals annually from the Philippines (Rubec 1988). Many other countries (e.g., Bahamas, Maldives, Palau, Virgin Islands) have banned coral, shell, and sand collecting because of concerns over reef welfare and conflicts with tourist industries.

Collecting live coral destroys reefs because replacement of anything other than small pieces requires sexual reproduction and larval settlement. This slow process is additionally hindered by the variety of contributors to the overall decline of coral reefs worldwide. Aquaculture of live hard and soft corals is rapidly replacing wild-collected corals as a source for responsible aquarists (S. Ellis 1999; Moe 2002). Because keeping corals alive in captivity is difficult, most large, public aquaria use realistic, artificial corals molded from fiberglass. Home aquarists should do the same.

The volume of live rock and coral removed from coral reefs for the aquarium trade admittedly pales in comparison to that removed in coral mining for construction, but both are nonrenewable activities. Both need to be regulated, restricted to limited areas, and monitored, as is happening in the Maldives, among other places. Live rock culture, along with culture of various reef corals, should be encouraged for the aquarium trade.

Fish Mortality

The aquarium trade suffers from three principle kinds of mortality: instantaneous or initial, bycatch or incidental, and delayed (e.g., Sadovy 2002). Targeted fish that die during collecting due to careless or destructive fishing practices, incompetent handling, and bad luck are examples of *instantaneous mortality*, which is money lost from the pockets of the collector. It is also wasteful, because fish killed during collecting must be "replaced" by further collecting, thus increasing pressure on the resource.

Bycatch mortality describes incidental deaths of nontargeted organisms and occurs in all fisheries to some degree. Because aquarium collecting tends to target individual fish, bycatch is relatively minimal, especially when destructive fishing practices are avoided.

Delayed mortality describes deaths that occur post-capture, usually during storage and transport but also after purchase by the final buyer. Delayed mortality occurs because of stress imposed during capture, storage, handling, and transport; starvation; or bad water quality at any point. Using chemicals, especially cyanide, during collecting causes delayed mortality. Some delayed mortality occurs because fish caught below 10 m are insufficiently decompressed (Pyle 1993). Species particularly prone to delayed mortality should be exempted from collecting. Delayed mortality can often be reduced through education of collectors, handlers, and shippers (e.g., recommended storage densities and times; information about incompatible species, water quality testing, temperature control). Market forces tend to minimize delayed mortality because post-collection deaths decrease profits and damage the reputation of collectors and sellers.

Putting numbers on the different kinds of mortality is difficult, in no small part because standardized reporting does not exist. For noncyanide-caught reef fishes, values from Sri Lanka, Puerto Rico, the Philippines, and various Pacific Island nations indicate 10%–40% mortality during holding prior to export, 5%–10% during initial transport, and 5%–60% during holding after import (E. Wood 1985; Sadovy 1992; Pyle 1993; Vallejo 1997; P. J. Rubec et al. 2000). For wild-caught South American and African freshwater fishes, pre-export mortality has been placed at 50%–70%, with as much as 80% additional loss for cardinal and neon tetras shipped from South America to the U.S. (Waichman et al. 2001). High-priced, wild-caught African cichlids fare considerably better. Only nets are used in capture, holding facilities are well maintained, and crowding is minimized. A well-known exporter from Lake Malawi, Red Zebra Tours (www.lakemalawi.com), estimated their shipping mortality at 5%–8%, depending on distance, with actual mortalities of under 1% common.

In sum, mortality can be high when fish are mistreated but minimal when care is taken. Aquarium trade proponents maintain that irresponsible collectors and merchants are progressively replaced by reputable suppliers who can guarantee live delivery of healthy fish. It will be the responsibility of those concerned with the welfare of fishes to test the validity of these assertions.

The Ethics of Aquarium Mortalities

A fourth, seldom-treated category, *ultimate mortality*, refers to the eventual death of fishes kept in aquaria, independent of capture and handling stress. Marine aquaria are much harder to maintain than freshwater setups because marine fish generally have more stringent environmental requirements. This makes evolutionary sense: Freshwater fishes have evolved in a variable environment with respect to many physical and chemical factors, including pH, temperature, seasonal water quality, and oxygen availability (aerial respiration and "lungs" have repeatedly evolved among freshwater fishes but are relatively rare among marine species).

Seawater is relatively well buffered, and marine fishes, except for tide pool species, live in the ultimate diluting environment, which protects them from most extremes of rapid temperature and chemical change, and deoxygenation. Most marine fishes experience environmentally extreme conditions infrequently and lack relevant coping mechanisms. As J. B. Heiser aptly put it, "Freshwater fishes evolved to live periodically in their own [waste products]; marine species were never afforded that luxury" (pers. comm.). Freshwater fish are preadapted to survive if not thrive in a glass bowl.

Marine species are therefore relatively incapable of living long in aquaria because of environmental intolerance or specialized feeding habits. They are "impossible or difficult to keep, even when maintained under ideal conditions by experienced aquarists" (E. M. Wood 2001, p. 31). Sadovy and Vincent (2002) estimated that perhaps 40% of frequently traded ornamental marine species were unsuitable for the average aquarist. Hard-to-keep species include those dependent on live coral and other live organisms for food, such as some butterflyfishes and angelfishes. These are often colorful species and hence desirable, but their capture and sale are unjustifiable.

Data on aquarium longevity are largely anecdotal and subject to unknown biases but far from encouraging, nevertheless. E. Wood (1985), surveying UK hobbyists, reported that 50% of marine fish died within six months of purchase and nearly 70% within a year. The Marine Aquarium Council, in establishing core standards for the trade, states that more than a few percent deaths per species per year is unacceptable (www.aquariumcouncil.org).

Is mortality among aquarium fishes a valid reason for concern? On the one hand, aquarium deaths could be viewed as a nonissue: Fisheries kill fishes. If the same fishes that were collected in the aquarium trade were instead simply killed and eaten, who would complain? Mortality is an economic liability to collectors and traders and an inconvenience to hobbyists.

On the other hand, aquarium mortalities provide several, largely ethical, reasons for concern. First, aquarium deaths are wasteful. We do not normally eat fish of the size and kinds kept in aquaria; there is no good reason to kill them. Second, reef fish commonly sought in the aquarium trade are also a mainstay of the tourist diving experience; if they are going to be removed from the reef and no longer be viewable by divers, they should at least serve their intended function of thriving in aquaria. Third, if the stated purpose of collecting is to provide live fish for public and private aquaria, it is hypocritical to accept a high degree of mortality in the collection and transport processes. Finally, in many commercial food fisheries, fish die quickly because slow death would use up fish nutrients from which we could benefit (hence the use of billy clubs and flash freezing). In the aquarium trade, fish often die slowly, from bad handling or water conditions or starvation. It is difficult to justify subjecting to slow death animals that were captured to provide educational or aesthetic experiences.

These and other ethical issues surrounding aquarium keeping are being addressed by the industry in an effort to promote a sustainable trade and cast the practice in a positive light. The American Marinelife Dealers' Association (among others) has produced an "Ecolist" that classifies aquarium fishes according to their likelihood of surviving in captivity. Details are provided under "Solutions to Trade Problems" below (and see box 13.1).

BOX 13.1. The Ethics of Catching and Keeping Fishes

The high mortality that occurs in the aquarium trade is a direct result of inappropriate collecting, transportation, and holding practices. Suffering occurs between capture and death. This suffering is often avoidable because it results from curable ignorance or callous disregard. If catch-and-release fishing can be considered ethically questionable (chapter 15), what about catching fishes and keeping them in captivity? What are the rights and responsibilities of individuals who keep aquaria in their homes? Do the same issues apply to public aquaria?

Abuses are not confined to tropical fish collected on another continent and purchased in a business. Many individuals in temperate, industrialized nations collect and keep native fishes. These fish keepers should question their motives, abilities, and actions because in the past, practitioners have acted irresponsibly. They have rationalized their collecting, keeping, and breeding activities in the name of advancing science, but their results have too seldom been subjected to critical scientific review or communicated outside a circle of like-minded aficionados. They have also claimed to promote the welfare of imperiled species via captive propagation. Although knowledgeable and conscientious home aquarists have a role to play in species and habitat protection, conservation, and education, collecting fish from the wild and keeping wild-caught fishes in captivity should be treated as a responsibility. Keepers' actions should follow a code of ethics, such as that promoted by the North American Native Fishes Association (www.nanfa.org/admin/NANFAmission.htm).

Home aquaria present an ethical challenge because of issues of individual rights and responsibilities. Public aquaria come under greater scrutiny because they are supported by the public, sometimes with tax dollars, and claim to be educational and therefore tax-exempt. As with zoos and herbaria, it is their capacity to educate that puts them on the strongest footing. The case for public aquaria has been nicely articulated by L. Taylor (1993, p. xxi):

> How do aquariums justify their actions [regarding] the ethics of removing animals from their natural environment and displaying them to the public? In the best of all possible worlds we would not *need* aquariums. . . . Nowadays, the universal justification . . . is that it helps people understand and appreciate the natural world so that their political and economic decisions will act to protect it. Aquarium proponents rely on their institutions to inspire stewardship for the natural world in their visitors.

Stress and Suffering

Fish suffer in the aquarium trade, whether or not they die. Many if not most wild-caught tropical fish are collected far from centers of international commerce, leading to a long, slow, stressful transport (figure 13.6). Although tender care could be expected from people dealing with live animals—at least where profits are dependent on the animals' welfare—Baquero (1996) found minimal awareness among Philippine collectors. Specifically, he discovered that "the concept of water quality was new to all" (p. 28).

Baquero found that fish were mishandled, held in stressful conditions, and transferred rapidly and frequently to new conditions of temperature, pH, and salinity without acclimation. Fish placed in bags in the field were regularly dumped into buckets with as many as 30 other fish, then transferred into additional bags. They were typically held in these bags for three to five days with only one or two daily abrupt water changes, if water was changed at all. A single bag may have contained 10 lionfish, 15 butterflyfish, or 70 damselfish. Death from ammonia buildup was common. Fish were next transported to inland holding facilities, where they were dumped into large tanks, again without acclimation. Many fish were not fed for weeks (perhaps to reduce waste buildup), and filtration was primitive or nonexistent. Shipping arrangements were no better, with high ammonia, low pH (which exacerbates ammonia toxicity), and low oxygen concentrations prevalent. Baquero's findings have been repeated by other observers (e.g., Meyers 2001).

Handling stress also characterizes many freshwater collecting operations. Waichman et al. (2001) monitored capture, handling, and transport practices for fish captured in the Barcelos region of Amazonas, Brazil. They measured water quality and chemistry and found (1) hypoxic oxygen levels in holding facilities averaging 2–4 mg/1, sometimes as low as 1 mg/1; (2) ammonia concentrations nine times greater than ambient levels and reaching toxic values; and (3) pH varying over 2 full units, which is known to be stressful to many fishes. Water was renewed only once during the 24–30-hour downriver journey, and that water was taken from the river, sometimes via "the same pump and hose used to refuel the boat" (p. 284). Waichman et al. concluded that the cumulative stresses encountered during capture, handling, and transport "could be responsible for the high fish losses reported by the importers and hobbyists" (p. 295).

Population Depletion and Biodiversity Loss

Reefs are large, most fish populations are often immense, and only a fraction of the individuals of any one species are collected. However, some species that are naturally low in abundance may experience minimal natural predation and

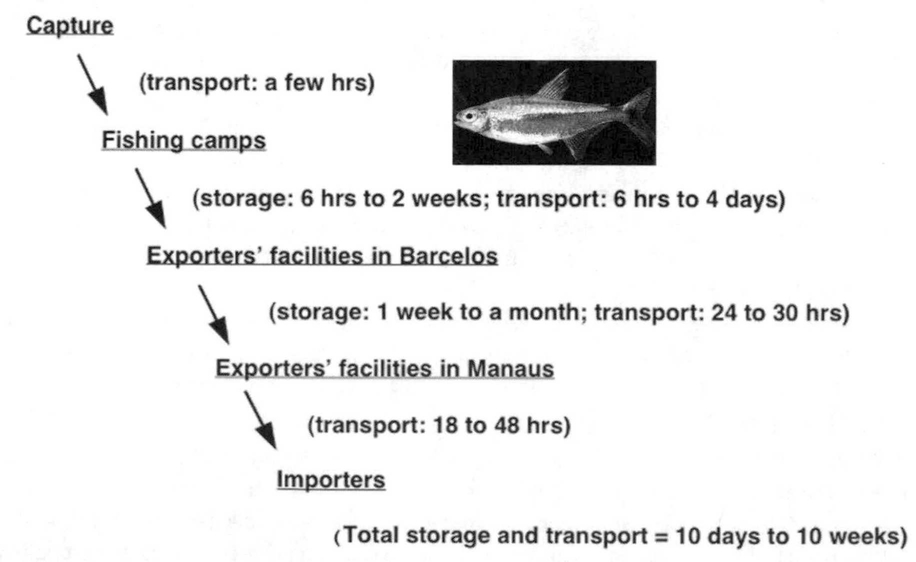

Figure 13.6. The movement of marine and freshwater fish from capture to retailer. The diagram shows the general pattern and time intervals involved in capture, storage, and transport of fishes such as cardinal tetras (inset) collected in the Rio Negro basin of Amazonas, but it is also descriptive of coral reef ornamentals. All steps and activities stress fish. Based on Waichman et al. (2001).

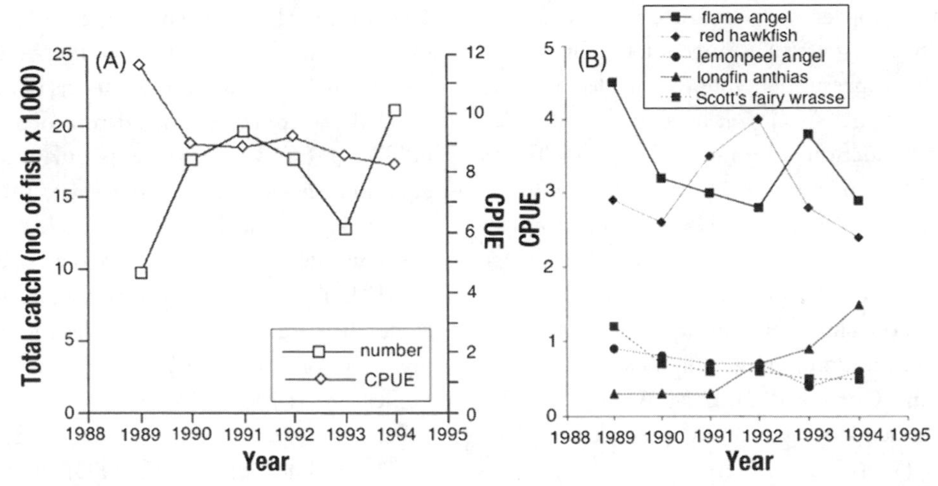

Figure 13.7. Impacts of collecting on reef fish abundance in the Cook Islands, 1989–94. (A) Overall catch numbers for nine collectors using scuba gear, showing total number captured and CPUE for five species combined (CPUE = no. of fish/no. of scuba tanks). From Bertram (1996); used with permission of the Secretariat of the Pacific Community. (B) Data separated by species, suggesting declines in two species, increase in one, and variability in two others.

be replaced infrequently, rendering them vulnerable to overcollecting. For example, lionfish (*Pterois* spp.) are never abundant, are well protected against predators by their venomous spines, and are almost always targeted by collectors because of their popularity in the trade (they may also prove undesirable in home aquaria, given the frequency with which released specimens are encountered by divers). Anemonefishes and cleanerfishes may also be locally vulnerable because of popularity. Anemonefishes are additionally at risk because they depend on anemones, which are also targeted by collectors. Both anemonefishes and cleaner wrasses change sex, a process that can require weeks. Heavy collecting of large individuals would significantly deplete numbers of female anemonefishes and male cleanerfishes, potentially affecting future recovery if populations are self-recruiting (see chapter 12).

Although "just another fishery," a troubling aspect of the aquarium trade is the overall intensity of exploitation, which occurs across age groups of fishes. Tropical reef fisheries differ from temperate fisheries in that both large and small reef fishes are sought in the former. Juveniles are the frequent target of aquarium collectors: Chan and Sadovy (1998) estimated that 56% of the marine ornamentals for sale in Hong Kong were juveniles. Given that coral reef fisheries have collapsed repeatedly, this across-the-board exploitation, to which aquarium collecting contributes, may exacerbate population declines and ecosystem collapse.

Although accurate assessments of the impacts of aquarium collecting on reef fish populations have been rare, "the public perception is that the continual collection . . . is causing populations [of reef fishes] to reach critically low numbers" (Young 1997, p. 147). Various studies have addressed the issue, but the results are often complicated by design flaws (Pyle 1993). For example, Bertram (1996) claimed that CPUE data from the Cook Islands during the initial six years of exploitation of five species "remained consistent . . . suggest[ing] that resources are sustainable with the current levels of exploitation" (p. 10). Ignoring the ambiguity of the term "consistent," a glance at Bertram's accompanying figure (figure 13.7A) suggests highest catches during the first year, a decline the next year, and a gradual downward trend in subsequent years. The initial drop is reminiscent of similar declines in traditional reef fisheries (e.g., Jennings and Kaiser 1998). Bertram's conclusion was based on catch of all species combined, masking any trends among individual species, especially rare ones. Replotting Bertram's data for the five monitored species (figure 13.7B) shows a downward trend in CPUE in at least two species and an upward trend in a third. At best, the data can't be used as evidence for or against the impact of collecting. Given that only six full-time and three part-time fishers were engaged in collecting in the Cook Islands, a conclusion of minimal impact appears qualitatively sound, albeit less than reassuring.

Despite assurances by knowledgeable authors that collecting has minimal impact on populations of reef fishes (e.g., Randall 1987a; Pyle 1993), anecdotal information from several locales suggests that intensive collecting can lead to depletion of target species, at least on a local scale. Collecting-caused reductions in abundance of angelfishes have been reported from the Great Barrier Reef, Florida Keys, Philippines, and Mexico; declines among popular butterflyfish species have been reported from Sri Lanka and the Philippines; and previously common species of triggerfish and surgeonfish targeted by aquarium collectors have become rare in the Philippines (Lubbock and Polunin 1975; Albaladejo and Corpuz 1984; E. M. Wood 2001). In Indonesia, "intensive collecting . . . [has] selectively depleted several species" (Wood 2001, p. 26).

Quantitative studies of impacts are unfortunately rare. Edwards and Shepherd (1992) used visual censuses to estimate population densities and sustainable yields for 65 aquarium species in the Maldives. They concluded that 12 species were exploited at or above sustainable levels, and an additional 12 species would be overexploited if exports tripled from the 1989 levels of 54,000 fish annually. E. M. Wood (2001) calculated that during 1997–99, 167,000 to 262,000 fish were exported annually from the Maldives, representing a three- to fivefold increase over 1989 levels. Given Edwards and Shepherd's calculations, at least a third of Maldivian aquarium species may be overfished.

Disagreement about the impacts of collecting in Hawaii has led to controlled, standardized assessments. Randall (1987a, p. 30) stated, "The populations of [the top ten aquarium species] are enormous and the take by aquarium fish collectors negligible. . . . [Collectors] can return to previous sites in a few months and see the reefs replenished." L. Taylor (1993, p. 63) concluded similarly that "controlled harvesting has very little lasting effect on the standing crop of [Hawaiian] reef communities." In contrast, Pfeffer and Tribble (1985) maintained that the intensity of collecting in Hawaii was producing substantial depletions.

Tissot and Hallacher (2003) pointed out flaws in Taylor's statistical design, which prompted them to conduct the first rigorous analysis at any locale. They monitored 23 sites along the Kona coast of Hawaii, comparing areas subjected to collecting with matched reference locales where collecting was prohibited and presumably minimal. They found that seven of the ten targeted, relatively common species were significantly depleted at collection sites, whereas only two of nine ecologically similar but nontargeted species showed reduced numbers at collection sites (figure 13.8). Declines among aquarium

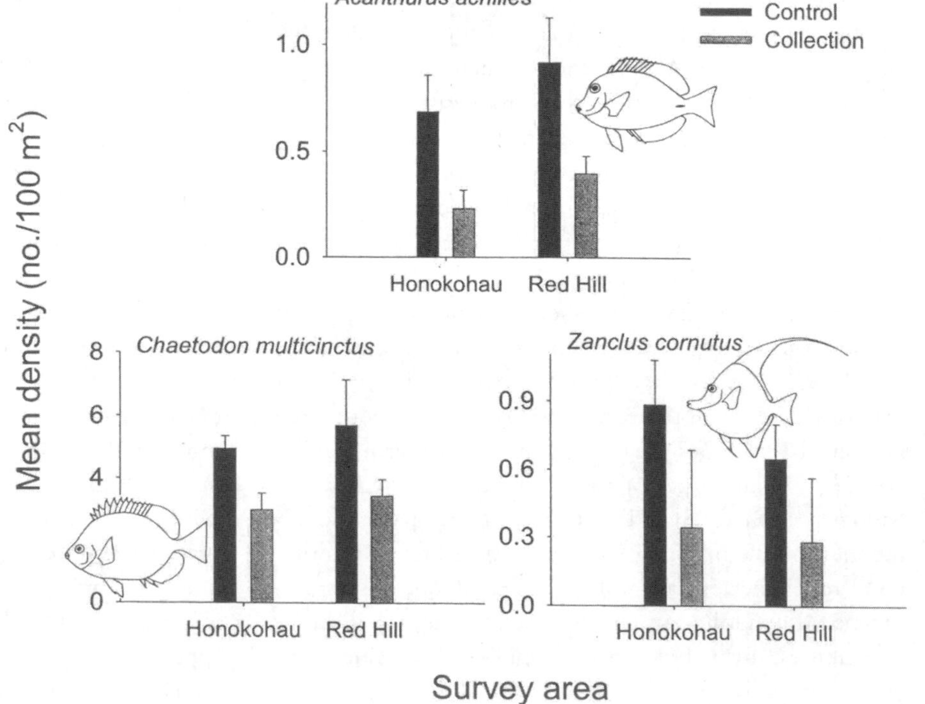

Figure 13.8. Impacts of aquarium collecting on reef fishes along the Kona coast of Hawaii. Replicated counts at areas subjected to collecting ("Collection") and protected from collecting ("Control") showed significant declines among targeted species at collection locales. Data shown are for three aquarium species: Achilles tang (pakuikui, *Acanthurus achilles*), multiband butterflyfish (kikakapu, *Chaetodon multicinctus*), Moorish idol (kihikihi, *Zanclus cornutus*). Nontarget species at the same locales showed little impact from collecting. From Tissot and Hallacher (2003); used with permission. Line drawings of fishes by Helen Randall and Loreen Bauman from Randall (1981); used with permission.

species ranged from 38% to 75%, including significant reductions in seven of the ten species Randall (1987a) described qualitatively as negligibly affected. This discrepancy is most likely the result of a pronounced increase in collecting in Hawaii. In 1973, only 90,000 fish were exported; by the early 1980s, exports reached 150,000 fish, and the number grew to 423,000 in 1995 (Randall 1987a; E. M. Wood 2001).

With Tissot and Hallacher's (2003) study, the burden of proof shifts to those who maintain that collecting has minimal impact. Anecdotal evidence should no longer suffice to support either side in this discussion. In all likelihood, moderate levels of collecting have minimal impact, especially for abundant species, as suggested by the Hawaii data. The challenge is determining the level at which moderate becomes excessive.

Freshwater Depletions

Instances also exist of depletion of freshwater species due to collecting, but the subject is again complicated by poor or contradictory data. South American examples include the reported commercial extinction of discus (Bayley and Petrere 1989), for which no supporting data were presented beyond statements by collectors that stocks had diminished; Prang (1996) countered that fishers in the mid-1990s were able to collect more discus than they could sell.

Bayley and Petrere (1989) also reported that cardinal tetra, *Paracheirodon axelrodi*, were commercially extinct from middle reaches of the Rio Negro of the Amazon, but Prang (1996) responded that 20 million cardinal tetras were exported from the Manaus region of Brazil annually throughout the 1980s and 1990s, rising to more than 50 million fish in 2000. Admittedly, increased export numbers masked an apparent drop in CPUE, with collectors complaining, "It now takes a week to capture the same number of fish that once took one day about thirty years ago" (Prang 1996, p. 2). Post-capture mortalities approaching 100% increased pressure to catch more fish. A two-year fishery closure was proposed but never enacted (Prang 1996), although collecting was restricted during the May-July breeding season in the middle Rio Negro.

Prang (1996, p. 2) went so far as to say, "The ornamental fishery of the middle Rio Negro has been productive without any evidence of *devastating* environmental destruction for the last fifty years" (emphasis mine). Other researchers studying *piabas* or ornamental fishes in Amazonas maintain that the fishery is sustainable despite intensive collecting. Sustainability is possible because of natural hydrological constraints on fishing (inaccessibility of spawning fishes during high water in winter), combined with short life spans, rapid maturation, and high reproductive output (Chao, Prang et al. 2001; Chao and Prang 2002; Norris and Chao 2002).

Valuable species have been heavily depleted, however, in Sri Lanka and Malaysia due to overcollecting (Banister 1989). Well-known examples include the Asian bonytongue, *Scleropages formosus*; bala shark, *Balantiocheilos melanopterus*; and cherry barb, *Puntius titteya*. Captive breeding appears to have taken pressure off these species and is contributing to recovery of previously depleted species, such as several *Barbus* species from Sri Lanka (Tlusty 2002).

The extent and impact of the freshwater aquarium trade in most of Africa are something of a mystery. Documentation is not easily obtained. Many cichlids, killifishes, catfishes, cyprinids, and characins are captured and exported from tropical African water bodies, but little information exists on ecological or even socioeconomic impacts. Long term ecological impacts are reportedly minor. S. M. Grant (1995) claimed that, after more than 20 years of nondestructive collecting on Lake Malawi and 70,000–300,000 fish exported annually, "there is no fish that was available years ago . . . that is not available at the present time" (p. 6). However, escapes from holding facilities and deliberate releases introduced at least 15 native Malawi cichlids into Lake Malawi National Park, among other locales (Bull-Tornøe 1992). Socioeconomic data indicate positive effects, again from Lake Malawi. The sole exporter from Lake Malawi in the early 1990s employed over 100 people and benefited an additional 350 people (Bull-Tornøe 1992). Workers made about 2.2 times the minimum wage. Approximately 50,000 individuals of 193 cichlid species were shipped annually, down from 250,000 individuals in the 1970s. Abundant species were wild caught; rarer species were propagated and contributed about a third of annual income.

One view is that impacts of the ornamental trade in Africa are minor relative to species depletions caused by commercial and subsistence fishing, habitat destruction, chemical and nutrient pollution, and species introductions (e.g., www.lakemalawi.com/faq.htm). For example, in Malawi, 40,000 traditional fishermen supply fish as food for 10 million people, who derive 70% of their animal protein from fish. The impact on aquatic systems of a few dozen collectors of ornamentals is likely to be comparatively small (S. M. Grant 1995).

Introduced Species

The aquarium trade is a major source of species introductions in freshwater environments due to pet releases by hobbyists and escapees from fish farms. However, it is difficult to determine the exact route or the relative contribution of the two sources. Hobbyists are probably responsible for the spread of a variety of illegal species such as piranhas and other predators that often outgrow both living room tanks and the willingness of owners to meet their dietary needs.

Some aquarium species whose import into the U.S. is illegal because of concern over escape have nevertheless escaped and proliferated due to the combined actions of importers, culture facilities, and other commercial dealers. Cox et al. (1997) reported that aquarium industry agents intentionally mislabeled shipping containers to move illegal species past overworked inspectors; for example, electric eels (Electrophoridae) were labeled as "common catfish," and 3,000 parasitic candiru catfish (Trichomycteridae) as "kuhli loaches." R. W. Ward (1998), while working at a tropical fish wholesaler in Minneapolis, observed shipments coming from Florida that contained illegal species such as red piranha, *Pygocentrus nattereri*. Detection at airports is unlikely given the volume of the trade. During the mid-1990s, Miami's sole inspector was responsible for monitoring more than 7,800 fish shipments that entered annually, each involving up to 500 containers. Cox et al. (1997) estimated that as many as 2,400 illegal shipments were passing annually through the Miami airport.

The contribution of fish farms to the problem is well established (but see Shafland 1996a, 1996b). Where farming of ornamentals is common, escapes are commonplace. It is widely felt that culturing nonindigenous species "nearly always leads to escapes of fishes into open water" (Courtenay 1997, p. 120). Courtenay et al. (1974) surveyed locales in central and southern Florida and found 38 exotic fish species, 20 of which were reproducing. Of the 38 species, 36 were held or propagated in fish farms (goldfish and carp were the exceptions). R. W. Ward (1998) conducted stream surveys in an outflow ditch and a nearby creek adjacent to fish farms in Hillsborough County, Florida, which has the highest density of tropical fish culture operations in the country. Ward collected six tropical exotic species in the drainage ditch leading away from the farms and another in the creek. The presence of small juveniles of some species indicated that populations were reproducing. The abundance of exotics decreased with distance from farms. Ward concluded that four of the exotic species (three livebearers and a cichlid) probably originated from fish farms.

Although Florida state law prohibits the release or escape of exotic fishes from a fish farm (FGFC 1997–98), fishes get out of farms and into surrounding waterways due to a variety of reasons:

- *Unscreened drainage pipes pumping out ponds to dispose of uneaten feed and metabolic wastes and turning over stock.* Outflows should be screened, and effluent water should pass through sand filters (Courtenay and Stauffer 1990).
- *Deliberate dumping of unwanted fish because of undesirable qualities (usually color) or overproduction.* Breeding ponds are periodically cleaned out and less colorful or desirable fishes discarded, discards sometimes exceeding half of total production (C. A. Watson and Shireman 1996; R. W. Ward 1998). Black acara, *Cichlasoma bimaculatum*, a South American cichlid, was a popular aquarium fish cultured extensively in Florida until the 1950s (Courtenay and Stauffer 1990). The advent of cargo jets brought more colorful species into the market, and relatively dull species declined in popularity. Farmers apparently discarded black acara, which spread through south Florida's canals and may have been the first aquarium fish established in Florida's open waters; it was until recently the most abundant cichlid in the state (Fuller et al. 1999).
- *Pond overflow during heavy rains.* If outflows were screened and farms were situated in dry areas distant from flowing surface water, both accidental and deliberate releases would occur less frequently, but such locations would negate the economic advantage of siting operations near flowing water. Florida regulations require that levees or banks around ponds be at least one foot higher than the regional 100-year flood elevation (FGFC 1997–98).
- *Purposeful dumping of fishes whose legal status has changed.* Importation and possession of walking catfish, *Clarias batrachus*, a southeast Asian predator, was made illegal in 1968 when they began dispersing from holding ponds in Broward County, near Miami; they soon appeared in the Tampa Bay area, 325 km to the north. *Clarias* can "walk" over land but not that far that fast. Farmers apparently released catfish to avoid being caught with illegal fish.

The fish spread throughout southern Florida and is now considered a pest species, threatening especially the ornamental fish industry and forcing fish farmers to fence their ponds against this predator (Courtenay and Stauffer 1990). One farmer informed Ron Ward that *Clarias* was such a problem that it was necessary "to pour bleach directly into drainage ditches adjacent to his pools in order to control its numbers" (R. W. Ward 1998, p. 14).

- *Transport by predators such as mammals or birds.* Bird predation is an oft-cited explanation for the existence of exotic fishes in waterways around fish farms, at least as far as fish farmers are concerned (R. W. Ward 1998). Newspaper articles validate these contentions, documenting how various species are dropped by eagles and ospreys (e.g., "Fish Falls from Sky, Smashes Windshield," Richmond, Virginia, July 6, 2001; "Big Fish Drops from the Skies into Alberta Woman's Raspberry Patch," Waskatenau, Alberta, August 27, 2001). However, falling from the skies seems an unlikely means of establishing populations of nonindigenous species.

That fish farmers consider predatory birds a problem worthy of remedial action is evident from an incident that occurred near Tampa in October 2002. State and federal agencies charged the owner of Terraqua Aquatics with six violations of the Migratory Bird Act and other laws for shotgunning 4,000 birds, including wood storks, blackneck stilts, cattle egrets, eastern meadowlarks, and herons (and one alligator) that were perceived as threats to her aquaculture operation. Placing nets over ponds, though expensive, would reduce losses to birds, curtail real or presumed export of exotics, and prevent such public relations disasters.

Problems created by fish farming are exemplified by the statistics on nonindigenous species in Florida, where "rates of introduction largely parallel the growth of the aquarium-fish-culture industry" (Courtenay 1997, p. 121). Before 1950, only 3 nonindigenous freshwater species were recorded in the state: carp (from Europe); green sunfish, *Lepomis cyanellus*; and orangespot sunfish, *L. humilis* (both native to the southern U.S.) (Courtenay 1997). With the increasing popularity of aquarium keeping and the proliferation of fish farming, more than 100 nonindigenous fish species have been recorded from the state (Fuller et al. 1999). Approximately 65%–75% of the introductions are thought to be the result of the aquarium trade, either deliberate fish tank releases or fish farm escapees (Courtenay and Stauffer 1990; Courtenay 1997). Because novelty within the ornamental trade tends to reap economic rewards, the fish farming industry is continually breeding or importing new and interesting fishes.

Although laws exist in Florida and elsewhere that prohibit release of exotics by private citizens and fish farmers, better efforts at educating the public are needed. Courtenay and Stauffer (1990) suggested that retailers display educational posters informing fish buyers about the ecological and legal dangers of releasing aquarium fishes and print a simple message on the plastic bags in which buyers take their purchases home. Even Florida authorities haven't gotten the message; they have sanctioned the use of exotic platys, *Xiphophorus maculatus*, for use as a bait fish (FGFC 1997–98; R. W. Ward 1998). Platys are desirable because they are colorful. Official sanctioning of tropical fishes as bait could open the door "to other exotics to be used as bait fishes, leading to introductions of exotic fishes with more potential to affect native fishes and the environment" (Ward 1998, p. 65).

Releases from culture facilities are not restricted to Florida. In the south-central Mexican state of Morelos, 64% of the 22 species in major rivers were introduced, and ornamental species accounted for more than one-third of the ichthyofauna (Contreras-MacBeath et al. 1998). Three ornamentals—two livebearers (*Xiphophorus* spp.) and the convict cichlid, *Cichlasoma nigrofasciatum*—were especially abundant. The convict cichlid dominated within three years of its introduction, accounting for 50% of total fish biomass and abundance. It escaped when a large tank in an ornamental fish farm spilled into a nearby river. Convict cichlids are implicated in the displacement of a native cichlid, *C. istlanum*, and the eradication of an endemic catfish, *Ictalurus balsanus*.

Research scientists are also to blame in releasing unwanted fishes. The Central American pike killifish, *Belonesox belizanus*, was first recorded in south Florida in 1957. Its introduction there apparently resulted from releases following termination of a medical research project (Courtenay 1997; Fuller et al. 1999). It spread widely and accounts for up to 20% of total fish biomass in some areas (Courtenay et al. 1974). This piscivorous livebearer, the largest member of the family Poeciliidae, is an effective predator capable of depleting populations of small native fishes such as mosquitofish, other poeciliids, and cyprinodontids (Fuller et al. 1999).

Marine Introductions

Aquarium escapes or discards have caused relatively few marine introductions until recently (Baltz 1991), but between August 2000 and January 2002, at least 17 lionfish, *Pterois volitans*, were photographed or collected at 11 locations along a 1,400-km stretch of the Atlantic coast of the U.S. between north Florida and Long Island (P. E. Whitfield et al. 2002; figure 13.9). Two fish were also reported from Bermuda. Five records involved groups of two to six fish. Many additional sightings have been reported but not verified, including sightings of juvenile fish 2.5 cm long that probably indicate successful reproduction. These observations represent the first successful invasion of Atlantic coastal waters by a marine fish native to the Indo-Pacific.

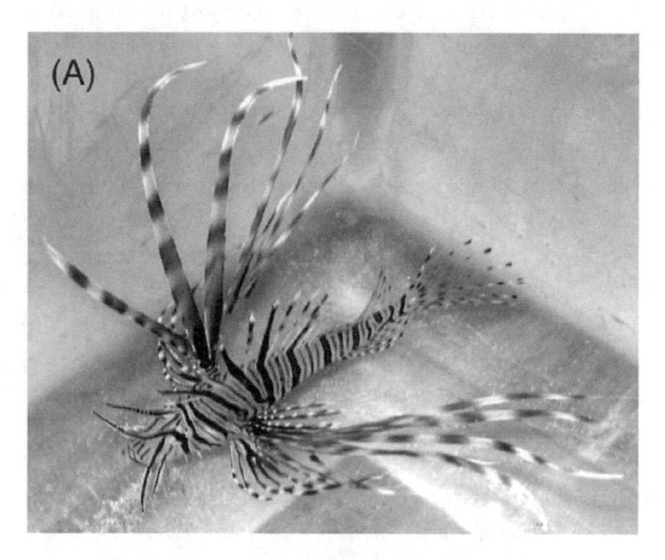

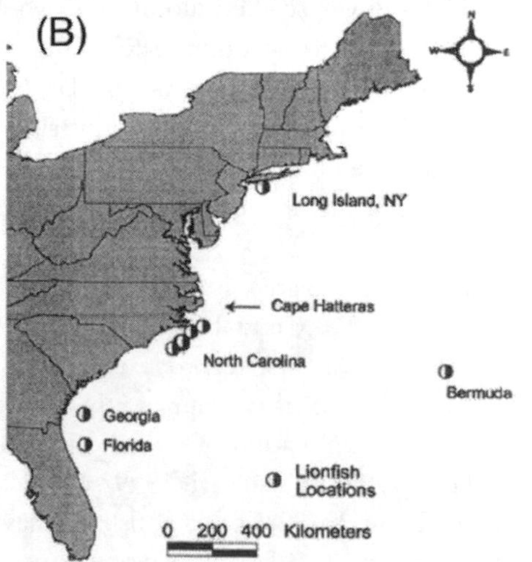

Figure 13.9. Introduction of an Indo-Pacific fish to Atlantic waters. (A) A lionfish captured off Andros Island, Bahamas. Photo courtesy of S. Kraft. (B) Records from Atlantic Florida to New York and Bermuda indicate that the species is reproducing successfully. From Whitfield et al. (2002); used with permission.

A much publicized effort to capture three Indo-Pacific batfish, *Platax orbicularis*, from Molasses Reef in the Florida Keys in 2000 (FKNMS 2000) represents the tip of the iceberg. The Reef Environmental Education Foundation (REEF) monitors sightings of exotic fishes in south Florida. As of February 2003, REEF divers had identified at least 15 Indo-Pacific species in the region, all of which likely resulted (at least initially) from aquarium releases. Sightings included lionfish, panther grouper, raccoon butterflyfish (*Chaetodon lunula*), five angelfish species, batfish, Moorish idol (*Zanclus cornutus*), and six surgeonfish species (see www.reef.org/exotic; Semmens et al. 2004; B. Semmens, pers. comm.). Two anemonefish were seen hovering around an anemone near Key West in 1998, but that sighting has not been repeated (REEF 2000). Undoubtedly, the list will grow. One irony of this phenomenon, pointed out by coral fish ecologist Shane Paterson, is that the offensive, biogeographically incorrect T-shirts sold in the Florida Keys depicting the "wrong" fish species are slowly but surely proving accurate portrayals of the south Florida ichthyofauna: "Dive Ft. Lauderdale and see the world."

Introduced Pathogens

Diseases and parasites can be transmitted from cultured to native fishes. Many documented instances have resulted from importation of eggs, larvae, and older stages of fishes (see Hoffman and Schubert 1984; Hedrick 1996). In a survey of introduced helminth parasites of fishes in the British Isles, Kennedy (1993) concluded that more than two-thirds could be linked to ornamental fishes. New, lethal pathogens are periodically discovered in both marine and freshwater ornamental species (e.g., Paperna et al. 2001), with substantial economic consequences arising from infection of cultured ornamental and food species. Concern over potential ecological and economic impacts led to incorporation in the U.S. of the Pet Industry Joint Advisory Council (PIJAC), which in the early 1970s monitored imported ornamental fishes for pathogens (Meyers 2001). Complicating the issue is a lack of regulatory protection relevant to nonfood items such as ornamental fish. For example, U.S. health regulations do not require inspection of ornamental fishes for possible diseases (Hedrick 1996).

Imported ornamental fishes can carry a variety of parasites and pathogens. Surveys of five freshwater species commonly imported into Australia revealed ten transmissible parasite species, three of which were considered high-risk pathogens (Evans and Lester 2001). The fish were sampled after they had completed a mandated quarantine period. A study in South Africa focused on the actual transmission of parasites via the freshwater ornamental trade and demonstrated how presumably regulated fishes transmitted pathogens (Mouton et al. 2001). Ornamental fish imported into South Africa must be accompanied by a certificate from the country of origin verifying that both the fish and the water they are in are parasite free. However, responsibility for monitoring, quarantine, and treatment of fish and transport water rest with the importer. A survey of four batches of imported goldfish, koi, guppies, and cardinal tetras revealed the presence of one pathogenic bacterium and several parasites, including the ciliated protozoan that causes ich, a ciliophoran, a trichodinid, and monogenean flatworms. Bacteria and parasites were found on the goldfish, koi, and guppies, which were all captive reared, whereas the cardinal tetras had been wild caught and were disease free (whether this is an additional liability of captive propagation is worth considering).

Conservation concerns arise because of the imperiled status of South Africa's temperate freshwater fish fauna, a fauna dominated by cyprinids and therefore potentially susceptible to generalist pathogens hitchhiking on goldfish and koi. Ornamental imports are thought to have been the source of several exotic parasites now affecting South African fishes, including *Argulus japonicus*, *Ichthyobodo necator*, *Chilodonella piscicola*, *C. hexasticha*, *Trichodina acuta*, *T. reticulata*, and ich. A related cause for concern is the widespread, prophylactic use of chemicals and antimicrobials in culture operations, which raises the specter of surviving strains of various pathogens that are resistant to commonly available treatments and hence even more virulent (e.g., Alderman and Hastings 1998).

Hawaii's native freshwater fish fauna is limited to six species: four gobies, an eleotrid sleeper, and a kuhliid flagtail. Most of these are Hawaiian endemics (Yamamoto and Tagawa 2000). At least one, the 'o'opu alamo'o, *Lentipes concolor*, has been proposed for federal protection (Devick et al. 1995). In contrast, the introduced fish fauna is large and growing, with more than 50 fish species documented and many established. Introduced parasites of native fishes have spread because of the aquarium trade and mosquito control efforts, seven introduced livebearers serving as primary vectors (Font 1998). Six of the 13 helminth parasites of native fishes were introduced, as were a tapeworm, a leech, and a roundworm. Also, the most abundant and widely distributed parasites infecting native fishes were exotics. Native parasites of native fishes occurred more sporadically and showed lower prevalence and abundance.

Diseases resulting from escaped or released exotics could at least be curtailed if more aggressive inspection and quarantine procedures existed. Quarantine could occur either at the point of origin or at import centers but obviously requires governmental monitoring. Chlorination of transport water prior to disposal should be universal. Sophisticated testing protocols involving molecular assays are ultimately needed, because many pathogens exist as "subclinical infections," meaning that affected fishes show no outward signs of disease (Hedrick 1996). It is hard to imagine that either export or import dealers would greet such requirements with enthusiasm, given costs and inconvenience. Expecting voluntary compliance is wishful thinking.

Ecosystem-Level Impacts

Impacts on ecosystem processes due to aquarium collecting have not been demonstrated. Despite a 32% reduction in dominant herbivores found by Tissot and Hallacher (2003) in Hawaii, macroalgae abundance did not differ significantly between collected and reference sites, indicating no effect due to removal of herbivores.

A potential problem could arise from wide-scale collecting of cleanerfishes and shrimps that pick parasites and necrotic tissue off reef fishes (e.g., Edwards and Shepherd 1992). Cleaner wrasses such as *Labroides dimidiatus* and *L. bicolor* are heavily targeted by collectors. E. M. Wood and A. Rajasuriya (1999) reported that 20,000 *Labroides* were exported from Sri Lanka annually, and even the tiny Maldives exported over 3,700 *L. dimidiatus* in 2000 (http://www.unep-wcmc.org/marine/GMAD/index.html). Removal of such cleaners could affect the host species that visit the cleaners, but before-after studies of the impacts of widespread removal by aquarium collectors have apparently not been performed.

Economic Conflicts

"Fish watching and fish collecting are generally not compatible" (E. M. Wood 2001, p. 41). Recreational divers want to see large fish, but they also want to see plentiful colorful, small fish and the corals with which the fishes live. Commercial food fisheries can exist side by side with dive tourism when they target something other than what tourists want to see. However, overlap in target species between the aquarium trade and dive tourism is about 100%. Active collecting, as well as spearfishing, makes fish wary of and unapproachable by divers, whereas in no-take reserves, fish are relatively oblivious to divers. Given the profits, added value, and multipliers associated with the tourist trade and the reusable versus consumptive nature of the activities, potential for conflict between the industries is substantial. Exploitation will probably lose out to fish watching in most instances. Because of its incompatibility with ecotourism, collecting has been regulated or restricted in Australia, Florida Keys, Hawaii, the Maldives, Mozambique, and Palau (e.g., Dybas 2002). Conflicts also arise between the aquarium trade and sport fishing interests. The economically important fishery for cardinal tetras in the Rio Negro of Brazil is apparently running into opposition from a developing sport fishery for large cichlids (Bartley 2000).

Conflict between fisheries for food species and collection of ornamental fishes should be minimal because the species targeted are dissimilar (Dufour 1997). However, Palau's restrictions on aquarium collecting arose in part because fishermen "saw the aquarium industry as unwanted competition for foodfish" (T. Graham 1996, p. 13). Conflict can be reduced by restricting the capture of juveniles of popular food fishes for the aquarium trade. The list cannot be very long, but it would include some intensively targeted food species such as the panther grouper.

Human Welfare Issues

The aquarium trade has not been as widely condemned as the live-food fish trade with regard to diving-related problems, probably because most aquarium fishes are caught in relatively shallow water (e.g., Pyle 1993). Collector safety issues may be growing, however. Descriptive reports from the Philippines and Indonesia indicate that, as shallow stocks become depleted, divers must work longer and deeper to obtain specimens in sufficient numbers to make a living (e.g., Baquero 2001). Anecdotal reports suggest that similar problems may be arising in freshwater situations, although relatively little comparative evidence is available on the environmental and socioeconomic impacts of freshwater collecting. Cichlid collectors in Lake Tanganyika collect at depths of over 25 m (S. M. Grant 1995), raising the specter of the bends. Divers there are shot at and robbed for equipment and face crocodiles, hippos, cobras, malaria, bilharzia, and HIV-AIDS, as well as political strife (O. Lucanus, pers. comm.).

An article in the November 5, 2001, *New York Times* (Rohter 2001) described apparent abuses associated with collecting the zebra pleco catfish, *Hypancistrus zebra*, in the Rio Xingú of Pará, Brazil (figure 13.10). The article refers variously to "a clandestine international market" and "a network of black-market dealers" who make "obscene" profits, while divers "receive no more than $5 . . . [although] a single rare albino specimen can fetch as much as $600 . . . in Japan." Divers are "plunging to ever deeper and more dangerous depths, gasping for air as they surface. . . . Almost everybody gets hearing damage as a result of their eardrums bursting from the pressure they have to deal with . . . in order to find the fish that buyers want. . . . Almost none of the divers use scuba tanks, and many do not even have masks."

On closer examination however, the story does not stand up (thanks to Mr. Oliver Lucanus, a collector and importer with several decades' experience, and Dr. N. Labbish Chao of Project Piaba):

- Export of zebra plecos has been legal since 1996 under Brazilian law. Fishermen receive around $6–$10 apiece

Figure 13.10. A spectacular albeit diminutive (64 mm maximum) loricariid armored catfish, the zebra pleco. This and other catfishes support an active ornamental fishery in the Rio Xingú of Pará, Brazil. The zebra pleco has been called "the most high-profile aquarium catfish of our time" and is the logo species of www.planetcatfish.com. Endemic to the Rio Xingú, its actual color is white with black lines. Photo by Oliver Lucanus, www.belowwater.com; used with permission.

for the fish. Retail prices on the Web in May 2003 ranged between $20 and $50, a two- to fivefold markup from collector to retailer, hardly an "obscene" profit. Albino *zebra* plecos do not exist, although other "albino plecos" are captively propagated, selling for $8–$15, with "rare" wild-caught albino Adonis plecos, *Acanthicus adonis* bringing $150 (www.geocities.com/rainforest/jungle/4327/pricelist.html).

- No one is gasping for air. Collecting involves use of hookah rigs connected to surface compressors. No one free dives for zebra plecos in the 18–30-m depths where they occur. All divers have masks (have you ever tried to see underwater without one?). Ear damage may occur, but as any diver knows, the greatest pressure differential, and the one that causes most ear damage, is in the first few meters.

Web and newspaper articles are not scientific publications. They are not refereed. The aquarium fish trade may attract unscrupulous individuals, some of whom exploit collectors, bribe officials, deal in black markets, and circumvent environmental laws; but efforts to ensure collector safety, correct injustices, and promote fair trade—as exemplified by Project Piaba, Project Seahorse, and the Marine Aquarium Council—deserve more attention and emulation.

Benefits of the Aquarium Trade

Income to Tropical Nations

Trade in ornamental fishes—when practiced sustainably, free from destructive methods, and not directed at food species—can provide benefits to tropical nations, especially small or remote villages. It can provide employment, promote conservation, and reduce pressure on fishes as well as other resources. Successful commerce can reduce outmigration to urban centers, thus retaining rural youth, keeping cultural practices intact, and slowing overpopulation of urban areas (S. M. Grant 1995; Prang 1996; Tlusty 2002). The live-fish trade also improves balance-of-trade conditions.

Impacts on subsistence economies can be considerable. Dufour (1997) calculated that every 100,000 fish exported in the trade created 10–20 jobs, generating $200,000 income annually. Extrapolating to the 15–40 million fish exported each year, trade could be worth $30–$60 million to local economies, employing 1,500–3,000 people. A collecting operation on Lake Malawi employed 25 divers and 75 additional workers and supported another 1,000 people directly and perhaps 500 indirectly (S. M. Grant 1995).

Economic benefits can be substantial. Among Pacific island nations, collectors' annual earnings equal or exceed average salaries (Pyle 1993). Along the Rio Negro, ornamental fishes, especially the cardinal tetra, accounted for about 60% of local income, involving up to 80% of households (Prang 1996, 2001). In the Peruvian Amazon, ornamental collecting was found to be more lucrative than agriculture or day labor (Kvist et al. 2001). Unfortunately, most of this trade involved illegal collecting from the Pacaya-Samira National Reserve.

However, collectors (and fishers in general) typically see only a fraction of eventual retail value. When middlemen are minimized, fishers receive 7%–9% of the retail price (E. M. Wood 2001). When middlemen proliferate, as in the Philippines and Indonesia, collectors realize perhaps 1% of retail value (P. J. Rubec et al. 2000; Baquero 2001). Markups at each step in the supply chain typically double the previous price. Freshwater fisheries provide similar numbers. In 1999, cardinal tetra collectors received up to $5 per 1,000 fish depending on size. Intermediaries received $10, exporters $100, importers $260, wholesalers $650, and retailers $2,000.

Admittedly, each step experiences overhead costs (freight and handling, customs fees and taxes, medicines, feed, utilities, packaging and marketing, labor, and losses). Regardless, the markup between collector and retailer is about 400-fold (Prang 2001). Large profits are made, but collectors usually earn only a subsistence return (e.g., Goulding et al. 1996). Improvements include minimizing the number of steps (e.g., Axelrod 2001; Baquero 2001), although initial buyers (intermediaries) in the Rio Negro apparently play important traditional, social roles (Prang 2001).

The potential value of some ornamental fishes helps promote sustainable collecting practices and thereby healthy reefs, rivers, and rain forests. Aquarium fish on Indian Ocean reefs are worth almost 85 times more per kilogram exported than as food fish (FAO 1999b). Tlusty (2002) argued that freshwater collecting inhibited destructive agriculture and mining practices because collectors knew that flooded, intact forests were essential habitat for fishes. By providing stable income, collecting in Amazonia diverts people from more dangerous, ecologically destructive, extractive activities such as gold mining (which causes sedimentation and mercury pollution problems), cattle raising (requiring large-scale deforestation), uncontrolled lumbering, and exploitation of endangered manatees and turtles (Prang 1996; Norris and Chao 2002). Ornamental fishes are "a sustainable by-product of an intact forest. . . . Buy a fish and save a tree" (Norris and Chao 2002, p. 30).

Education on Biodiversity Issues

The ornamental fish industry actively participates in educating the public. Specialty and trade journals such as *Tropical Fish Hobbyist* and *Freshwater and Marine Aquarium* frequently contain articles on biodiversity issues. Web sites devoted to marine tropicals seem to be particularly active in this area, as exemplified by the Marine Aquarium Council's campaign to develop an ecologically responsible chain of supply. Project Piaba's efforts in the Amazon (see "Solutions to Trade Problems" below) are a rare but encouraging departure from the overall neglect of environmental issues in the freshwater ornamental industry. The pet trade industry could do much more to promote conservation through retail stores and chains. Perhaps such retailers don't want the buying public to think about the origin of their fishes or the ecological and social impacts that collecting and culturing fishes might have. When hobbyists become aware of the people who catch the fish and the trade's economic importance, they may be willing to pay more to promote sustainable practices (e.g., Dowd 2001).

The Role of Public Aquaria in Conservation Ecology and Education

The intersection between ornamental fishes and conservation education is most obvious in large public aquaria. Whereas zoos and particularly aquaria traditionally focused on passive viewing and entertainment (and making money), justification and goals have changed as public awareness of environmental issues has increased. Circus-style performances involving marine mammals have progressively given way to habitat-, issue-, concept-, and taxon-focused exhibits. The economic bottom line remains an overarching consideration, but promoting conservation is not incompatible with free enterprise. If anything, the credibility of a public aquarium is raised by visible efforts in conservation education and research (L. Taylor 1998; Nightingale et al. 2001). Ideally, public aquaria "should not limit themselves to only zoological information. Conservation issues, even if controversial, must be presented" (Swanagan 2000, p. 30).

Many seaside communities and even large sporting goods stores have display aquaria (for a list, see www.infoplanets.com/worldaq.html). The focus and impact of these creatures-of-the-deep facilities are usually local (figure 13.11). Proprietors are to be commended for any efforts at conservation education, but their financial and staffing resources are often limited. It is in the larger public aquaria that we should expect to find a strong education and conservation ethic. Witness, for example, the excerpted mission statement of the Waikiki Aquarium (www.waquarium.org): "to understand, love, care for, and work to protect the life of the ocean, through a commitment to excellence in educational and entertaining experiences, research and conservation."

Most public aquaria espouse similar missions. Among many possible examples, the Two Oceans Aquarium, Cape Town, South Africa "promotes effective stewardship of the natural world by . . . applying and advancing conservation and environmental awareness" (www.aquarium.co.za). The Bermuda Aquarium, Museum, and Zoo aims "to inspire appreciation and care of island environments" and spearheads a Bermuda Biodiversity Project (www.bamz.org). The Monterey Bay Aquarium's mission is simply "to inspire conservation of the oceans" (www.montereybayaquarium.org). The Vancouver Aquarium proclaims itself "Canada's leading aquatic conservation organization, with a mission statement that promotes "direct action"

Figure 13.11. Small local aquaria showcase their most charismatic display animals. Charisma is obviously relative.

(www.vanaqua.org). Web sites maintained by these facilities include links to conservation and educational experiences, teacher's guides, lesson plans, and classroom-focused activities, as well as descriptions of conservation-related research projects. Such an advocacy role appears to be increasing (L. Taylor 1998).

Many facilities have joined networks of national and international organizations. The World Association of Zoos and Aquariums ("United for Conservation") coordinates activities with a focus on conservation goals. In North America, the American Zoo and Aquarium Association (AZA) accredits and certifies facilities that keep, display, and conduct research on live organisms. Members "have as their primary business the exhibition, conservation, and preservation of the earth's fauna in an educational and scientific manner." In 2003, AZA accredited 46 aquaria among international institutions and certified related facilities. The potential educational value of this coordinated effort is considerable: L. Taylor (1993, 1998) estimated that total annual ticket sales at American aquariums approached 10 million visitors in the mid-1990s, exceeding that of many sports. It's nice to think that oscars are more popular than NASCAR (NASCAR 1998 total attendance was 9,000,000+, www.hhmotorsports.com/demographics.htm).

In addition to passive exhibits, textual messages, videos, and interactive displays, many facilities publicize their success at maintaining animals for extended periods, which minimizes the need for continued collecting from the wild. Whereas 70% of marine aquarium fish in private homes are dead within a year, public aquaria may keep their animals for decades. Such longevity has multiple values. Collecting, acclimating, and introducing new organisms into existing exhibits are risky and costly in time and money. Long quarantine periods are the norm for new individuals. The Tennessee Aquarium places new fish in exhibits only after they have spent at least 30–45 days in an isolated quarantine tank, where each fish undergoes a series of decontamination procedures involving expensive antibiotics and antiparasitics. The aquarium experiences less than 1% annual mortality among its fishes (much of which results from predation in large community tanks), and many of the fishes on display have been in residence since it opened in 1992 (Rob Mottice, Tennessee Aquarium, pers. comm.).

Achieving a stated conservation mission is not without impediments. Aquarium staff typically have scientific and academic backgrounds, whereas boards of directors are more likely to contain business people focused on the financial realities of maintaining an expensive operation. Boards of directors can be politically, personally, and financially conservative and less enlightened or concerned about environmental problems. I have been informed by more than one aquarium administrator that their board discouraged efforts at conservation, education, or discussions of evolution because controversial topics make visitors (or board members) uncomfortable. Also, educational displays require stopping and reading, which slows the movement of visitors and ultimately decreases admissions.

An additional important consideration is the need to avoid overwhelming the public with "gloom and doom" messages about environmental decline. Surveys indicate that the most successful environmental efforts emphasize a positive message, workable solutions, and success stories: "People need to feel that effective solutions are available, and, if followed, will actually make a difference" (Immerwahr 1999). As a result of these complications, even aquaria with a strong conservation theme focus on environmental issues in only 20% of exhibits. The remaining 80% are generally dedicated to more traditional animal identification and biology (Falk and Adelman 2003).

However, according to a 1995 Roper survey, 90% of people interviewed considered the most important function of an aquarium to be educating the public about animals (Roper Starch Worldwide 1995; see also Reade and Waran 1996). Similarly, an AZA-sponsored study found that although visitors do not necessarily view zoos and aquariums as conservation organizations, they believe that such facilities could assume a leadership role in educating the public about environmental issues (Dierking et al. 2001). These findings suggest that the public is confused about where facilities should and actually do stand on environmental issues and advocacy. If mission statements are being put into action, many facilities are failing to adequately publicize their efforts.

Most attendees pass through a facility only once, but a subset return and become involved in focused programs, field trips, and other activities, all often containing a strong conservation message. Aquariums also have internship programs and train volunteers to become docents, enriching the experience and understanding of program participants, as well as the aquarium visitors with whom they interact (B. Carlson, Georgia Aquarium, pers. comm.). Such activities go beyond the facilities themselves. In the U.S., the Coastal Ecosystem Learning Centers program links numerous governmental agencies with aquaria and marine education centers. The Wildlife Conservation Society and the New York Aquarium administer the Glovers Reef Marine Research Station in Belize. The Monterey Bay Aquarium maintains an ocean science program through its Research Institute and participates in programs such as the Monterey Bay National Marine Sanctuary and the Communication Partnership for Science and the Sea. The Ocean Project, an international network of aquariums, zoos, museums, and conservation organizations focused on ocean conservation, provides information on over 100 aquaria and marine science centers.

Assessing the Message

It is somewhat unclear how well aquaria are achieving their conservation goals because data are scant. Informal questioning of visitors exiting Underwater World in San Francisco indicated that 21% "reported learning a little more about marine life and feeling more positive toward marine animals" (one visitor "felt more fearful of sharks than before") (Kidd and Kidd 1997, p. 1086). In Falk and Adelman's (2003) pre-entry and post-exit interviews of visitors to the National Aquarium in Baltimore, visitors ranged from "cognitive novice" to "expert" in background and from "uninterested" to "zealous" in initial interest (I would have qualified as a "zealous expert" by their criteria, as would most readers of this book; see figure 13.12). Falk and Adelman found significant increases in both conservation knowledge and attitudes and concern as a function of the visit and interaction with exhibits. In general, the proportion of people whose knowledge and interest were "minimal" decreased slightly, whereas the proportion of visitors whose knowledge and interest could be considered "extensive" increased dramatically. The amount of change did vary as a function of background: The message got through best to those in most need of education, which is the audience generally targeted by most museums, zoos, and aquaria (Falk and Adelman 2003). My fellow zealuos experts are already on board and are unlikely to be lost.

Figure 13.12. Two "zealous experts" identify fishes in the 160,000-gallon Pacific Barrier Reef tank at the Georgia Aquarium, which replicates reef drop-off habitats in the Solomon Islands. It's the next best thing to being there.

What remains unknown is whether positive, short-term effects carry over into the longer term. Does visiting an aquarium lead to actual involvement and action, such as donating to environmental organizations, signing petitions, consuming wisely, and writing or phoning legislators (e.g., Swanagan 2000)? Personnel responsible for carrying out the conservation goals of public aquaria recognize the challenge they face in accurately assessing long-term change in visitor understanding and behavior (Dierking et al. 2001; Carol Hopper, Waikiki Aquarium, pers. comm.).

Health Benefits of Private Aquarium Keeping

The pet industry frequently refers to the health benefits of aquarium keeping and fish watching, citing evidence such as a survey showing that 94 of 100 aquarium store customers felt they benefited by keeping aquaria, with "relaxing" and "stress reduction" reported by most (6 saw aquarium fish as nothing more than room decoration) (Kidd and Kidd 1999). The American Pet Products Manufacturing Association reported that 71% of fish owners agreed that their fish offered relaxation or relieved stress, and 23% felt that fish were good for their health or their family's health (APPMA 2002).

Many medical and dental offices have fish tanks in their waiting rooms, partly in response to reports that patients who watched fish closely for 15–20 minutes were more relaxed and less anxious about their visit than patients who only glanced briefly at the tank (Katcher et al. 1983). Katcher et al. found greater reductions in blood pressure in normal and hypertense subjects when they were asked to view fish in an aquarium that contained plants and an airstone as opposed to staring at a blank wall or at a fish tank minus the fish. Doctors periodically prescribe keeping aquaria and watching fish to patients suffering from high blood pressure (Kidd and Kidd 1999).

The stress reduction benefits of aquarium watching are fairly evident to any aquarium keeper, making these largely qualitative reports believable. On a purely editorial note, if the intent of doctors and dentists is stress alleviation, it could be achieved using freshwater aquaria that contain readily cultured species rather than rare, short-lived, and difficult-to-maintain marine tropicals.

Aquaria as Research Tools

It is admittedly self-serving to advocate keeping fish in tanks to conduct research. That caveat notwithstanding, much valuable research is conducted using large and small aquaria, expanding our understanding of the biology of many species (e.g., Reebs 2001). Results have direct application to environmental problems and are interesting to an informed citizenry. Public aquaria often have large research programs increasingly directed at habitat requirements and captive propagation of species in trouble.

A short-lived journal, *Aquarium Sciences and Conservation*, was published by Kluwer Academic Publishers between 1997 and 2001 (ISSN 1357-5325; see www.kluweronline.com). It focused on applying studies of aquarium-kept organisms to conservation issues. In the mid-1990s, the Canadian-based Aquatic Conservation Network published a bulletin, *Aquatic Survival*, which contained information on many tropical fish species and efforts to rear them in captivity (many of the articles remain available at http://www.peter.unmack.net/acn/as/). Perhaps as aquarium-keeping and environmental issues grow in importance, such efforts will be repeated and sustained.

Some aquarium hobbyists emphasize keeping and studying native fishes. In North America, in addition to local and state organizations, the North American Native Fishes Association (NANFA) publishes a quarterly journal, *American Currents*, that frequently includes articles on imperiled species and their propagation, as well as updates on additions to federal and state endangered species lists and actions and threats affecting aquatic diversity. NANFA also funds research and education activities through a small grants program. Its members include both amateur and professional aquarists and ichthyologists.

Case Histories in the Aquarium Trade

Case histories of three high-profile species show how ecologic, economic, and social costs and benefits interact in the worldwide trade in ornamental fishes.

Banggai Cardinalfish

Exemplifying much of the aquarium trade's actual and purported effects, good intentions, and often unpredicted impacts is the Banggai or Buddha cardinalfish, *Pterapogon kauderni*, one of the few marine species ever thought to have been driven to extinction. Roberts and Hawkins (1999), citing a personal communication from Gerry Allen (a recognized authority on Indo-Pacific fishes), declared that overcollecting of Banggai cardinalfish for the aquarium trade "has possibly caused its extinction in the wild" (p. 242). Hence my surprise when I walked into a pet shop in Athens, Georgia, in July 2001 and found two medium-sized Banggai cardinalfish, priced at $26.99 each.

Everything about the Banggai cardinalfish is spectacular and unique—its appearance, breeding biology, biogeography, and involvement in the pet trade. Originally described in 1933 from the Banggai Islands of Indonesia, the fish remained relatively unnoticed until "rediscovered" in 1994 (see Allen and Steene 1995). The reported range of the fish was a small, 700-km^2 region around four islands (Vagelli and Erdmann 2002), among the smallest known ranges for a marine fish anywhere (figure 13.13). Small range size is commonly associated with vulnerability to overexploitation (see J. P. Hawkins et al. 2000). Banggai cardinalfish produce only about 50 large eggs at a time, the lowest number of any cardinalfish and another risk factor. Males brood eggs orally like other cardinalfishes, but Banggai hold their young for an additional week after they hatch; the young then disperse a short distance at most (Vagelli 1999), another extinction risk factor.

Banggai cardinalfish became popular with aquarists because they were attractive, survived well in tanks, accepted dried foods, and bred readily in captivity. Demand skyrocketed in the late 1990s, but limited geographic range, low fecundity, and intense collecting led to dire predictions about the species' future (e.g., Michael 1996). Roberts and Hawkins (1999) announced their possible extinction in the wild, and G. R. Allen (2000) called for their inclusion on the IUCN Red List. However, collecting continued. Independent estimates of capture numbers in the Banggai

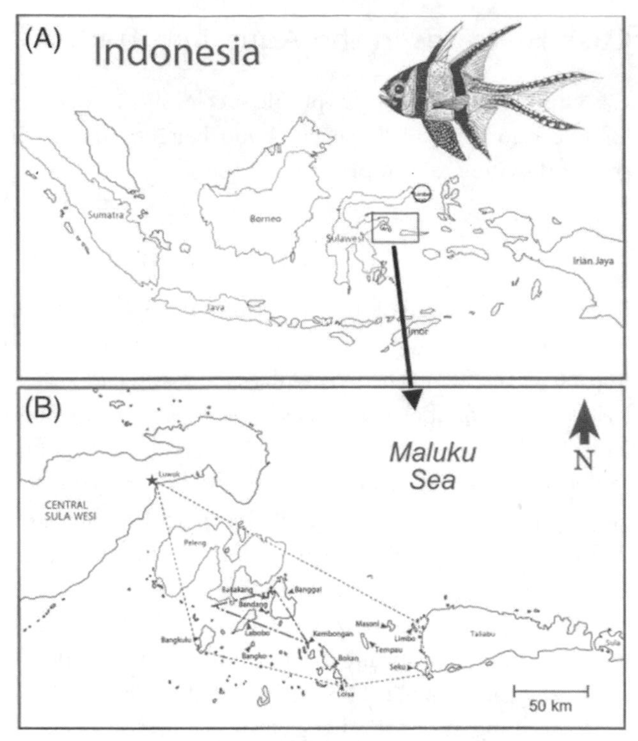

Figure 13.13. Range of Banggai cardinalfish. (A) The Indonesian Archipelago. Rectangle shows the area of the lower map, circle denotes a translocated population in Lembeh Strait. (B) Natural range of Banggai cardinalfish in the Banggai Archipelago. Known range prior to 2001 is denoted by dash-dot triangle; expanded range as determined by Vagelli and Erdmann (2002) is denoted by the larger dashed polygon. From Vagelli and Erdmann (2002) with inset fish drawing from G. R. Allen (2000), drawn by Susan Morrison; all used with permission.

Archipelago by Lunn and Moreau (2002) and Vagelli and Erdmann (2002) indicated 600,000–700,000 individuals taken annually. Suspicions about high shipping mortality (Vagelli and Erdmann 2002) were confirmed by global trade figures, which accounted for at most only 15,000 of the 600,000+ fish collected (http://www..unep-wcmc.org/marine/GMAD/index.html).

Despite this collecting pressure, pessimistic predictions were not realized. The species has if anything increased its range, perhaps as a combined function of collecting and captive breeding. Erdmann and Vagelli (2001) found that the range was perhaps 13 times larger than originally reported, including the Sula Spur region to the east and central Sulawesi to the west, with densities of 0.03–1.2 fish m^{-2} in several shallow habitats (Lunn and Moreau 2002; Vagelli and Erdmann 2002). In addition, a growing population of several hundred individuals was discovered in the Lembeh Strait region of North Sulawesi, approximately 400 km from the Banggai Islands. This population initially occurred a few hundred meters from the holding pens of an aquarium fish exporter and likely represented escapees and their offspring (Erdmann and Vagelli 2001). Sightings at locales where the fish had not previously been seen could result from similar escapes (e.g., Tackett and Tackett 2001).

Banggai cardinalfish are one of the easiest marine tropicals to breed in captivity (Marini 1996; Moe 2002), and captive-bred individuals are widely available. Banggai will likely provide a test of the ability of certified, environmentally friendly products to displace wild-caught fish in the marine ornamental trade, despite the greater cost of cultured fish. Captive-bred Banggai individuals wholesale for $12.50–$15.00, whereas wild-caught fish wholesale for around $6–$8 (Vince Rado, www.orafarm.com, pers. comm.).

Seahorses and Their Relatives

International commerce of seahorses and their relatives is regulated under CITES. No aquarium-traded marine species were listed by CITES before 2002, when *Hippocampus* seahorses were placed in Appendix II. CITES regulation was long overdue, given that 1 species was classified as Endangered and 20 were considered Vulnerable by IUCN. Seahorses are predictably vulnerable to overcollecting because of monogamous mating habits and strong site fidelity. The impact of inclusion in CITES will be interesting to follow, assuming enforcement and record keeping are up to the task.

Some regulation of trade in seahorse relatives existed before 2002. The Australian leafy sea dragon, *Phycodurus eques*, endemic to southern Australia, was "fully protected" in South Australia, Victoria, and Western Australia (www.ccsa.asn.au; figure 13.14). The sea dragon, seahorses, and their relatives also received federal protection under the Environment Protection and Biodiversity Conservation Act of 1999. Capture of leafy sea dragons requires a permit, only one of which has been issued annually in South Australia since 1997 (Jeremy Gramp, Dragon Search, pers. comm.). The permittee collects one pregnant male, raises the young from the 30–160 eggs carried by the male, releases the male back into the wild, and sells three-month-old young primarily to public aquaria. Export numbers for 1998–2000 averaged 52 individuals per year, with more than 90% coming from Victoria (data from Environment Australia via K.

Figure 13.14. The leafy sea dragon, a seahorse relative endemic to southern Australia. The light-colored vertical lines on the body and light patches on the head and snout can be almost electric blue; the body varies from green through brown to orange and rust red. Photo by Jeff Jeffords, www.divegallery.com/Leafy_Sea_Dragon.htm, used with permission.

Martin-Smith, Project Seahorse, pers. comm.). Export numbers climbed to 150 "leafies" in 2001–2 (P. Quong, pers. comm.). Individual leafy sea dragons have apparently sold for up to $10,000, but juveniles generally range between $52 and $200 and may run as high as $520 to $1,600 (K. Martin-Smith, Dragon Search, pers. comm.; R. Mottice, Tennessee Aquarium, pers. comm.).

A multigroup conservation organization, Dragon Search, coordinates activities aimed at protecting both leafy and weedy (*Phyllopteryx taeniolatus*) sea dragons and promotes research on the biology of these spectacular but poorly understood fishes, including producing a manual for scuba divers on minimizing impacts on the fish (see www.dragonsearch.asn.au).

Piranhas in the Pond

Periodically, public concern focuses on whether piranhas can invade U.S. waters. Ichthyologists receive phone calls from fishers, reservoir managers, fisheries personnel, park rangers, mayors, and police officers about piranhas captured in local lakes, ponds, and reservoirs. I received such a call about two fish caught by anglers in Stone Mountain Lake, a popular fishing and swimming reservoir just outside of Atlanta, Georgia (34° N latitude). Lake operators were contemplating closing the lake, draining it, poisoning it, doing something before panic set in. They sent me the fish, and, with the help of some knowledgeable taxonomists, I determined that they were not flesh-eating piranhas (e.g., red or red-bellied piranha and redeye piranha, *Serrasalmus rhombeus*), but a close relative, the herbivorous pacu, *Colossoma* sp. (see Gery 1977; molar-like teeth in the lower jaw identified them [W. L. Fink, pers. comm.]). Pacu reach 90 cm and 20 kg, quickly outgrowing most home aquaria. A common solution—the same solution reached for plecostomus catfishes, oscar cichlids, piranhas, and a number of other fast-growing, large aquarium fishes—is to dump them in a nearby lake.

Had they really been piranhas, how justified would agency personnel have been in taking drastic action to rid the lake of them? Hollywood images notwithstanding, little scientific evidence suggests that piranhas individually or in large numbers constitute a threat to human health. Documented reports of piranhas attacking live humans in the water are largely anecdotal; upon forensic examination, many turn out to involve piranhas scavenging on already dead victims (e.g., Sazima and de Guimaraes 1987). Piranha "attacks" on healthy humans usually involve bites during handling of captured fish. Other than tissue damage, a major threat from such bites is infection (e.g., Revord et al. 1988). Piranha pose a potential minor threat to net fisheries (Agostinho et al. 1997), but their greatest impact would probably be as predators on native fishes, either through whole consumption or fin nipping. Fish that have coevolved with piranhas have coloration and behavior that apparently reduce the piranha's predatory impact (Winemiller 1990). The most likely impact of introduced piranhas would be their effects as a nonindigenous piscivore.

How likely are piranhas to establish reproducing populations in the U.S.? Piranhas are native to tropical South America and enter U.S. waters via releases by home aquarists (Fuller et al. 1999). This pathway has been verified in Hawaii. By applying otolith aging technology, Radtke (1995) clearly demonstrated that a 21-cm red-bellied piranha angled from a Hawaiian reservoir had spent 151–164 days in the reservoir following initial existence in an aquarium environment. Uniform otolith growth rings had been laid down during the relatively constant captive phase, whereas variable growth rings were laid down in the reservoir. These results additionally suggested that the individual was not the offspring of a wild-spawning population. Piranhas have also been captured in Florida, Massachusetts, Michigan, Minnesota, Ohio, Oklahoma, Pennsylvania, Texas, and Virginia. Only Florida, Hawaii, and Ohio locales involved more than a single fish; and reproduction is known only from Florida and Hawaii (Bennett et al. 1997; Fuller et al. 1999).

Bennett et al. (1997) looked at the temperatures at which red-bellied piranhas could presumably survive in the wild, given U.S. latitudes. Their laboratory results indicated that piranhas could not overwinter where average winter temperatures fell below 12°–14°C, eliminating every place except Hawaii; most of Florida; and southern portions of Texas, Arizona, and California. Farther north, fish could live through the summer and might occasionally overwinter but would be unlikely to establish breeding populations (figure 13.15). Most northern reports of piranhas involve fish caught in late summer or found dead in fall or winter (Fuller et al. 1999).

Piranhas might survive habitats fed by warm spring systems, such as those in north Florida, much of the deep southern U.S., the desert Southwest, or any area where flows above 14°C were maintained. Tropical neon tetras have been found in a geothermal spring in Colorado (Fuller et al. 1999). Nico and Fuller (1999) reported small populations of tropical fishes persisting in geothermal springs in Colorado, Idaho, Montana, Nevada, and Wyoming. Hot springs near Banff National Park in Alberta, Canada, have apparently contained a diverse tropical fauna (McAllister 1969). Heated effluent waters from power plants are another potential habitat. Introduced piranhas would pose little threat to humans but could

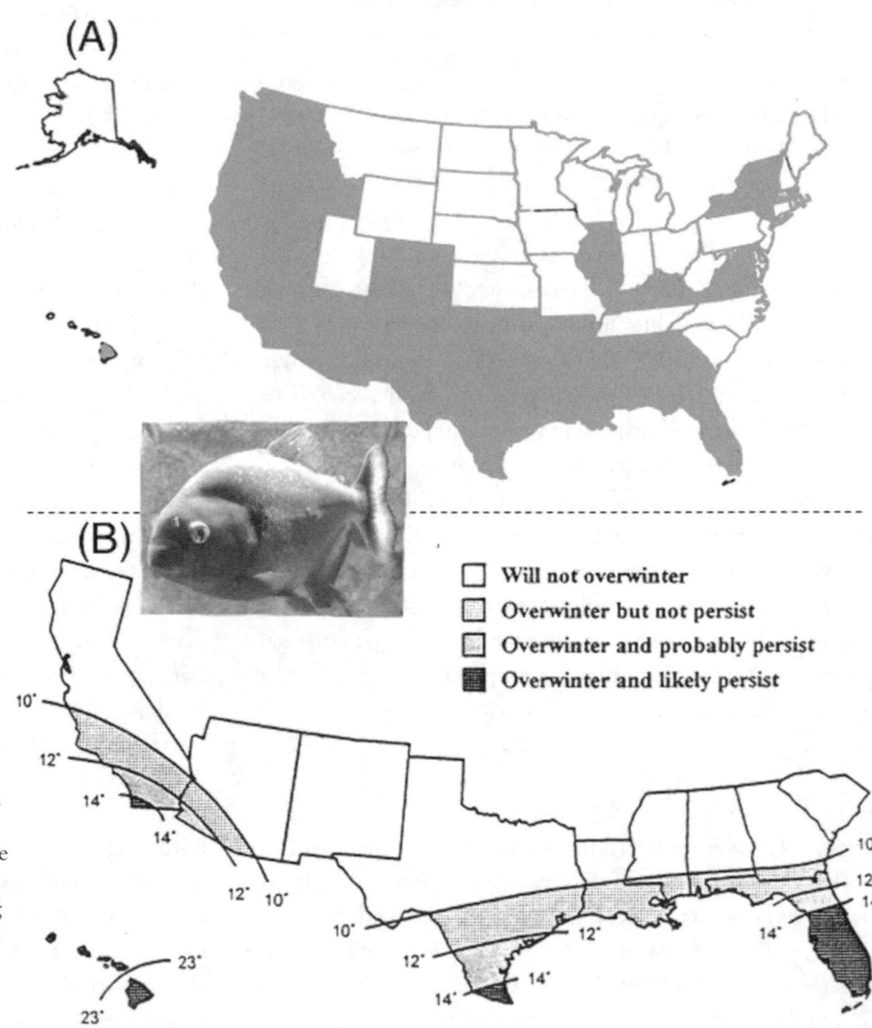

Figure 13.15. Legal restrictions on importation and climatic restrictions on survival of piranha in the U.S. (A) Shaded states prohibit the sale, possession, or transport of piranhas. (B) Piranhas are unlikely to develop reproducing populations where winter temperatures drop below 12°–14°C. Numbers denote winter temperature isotherms. From Bennett et al (1997); used with permission.

affect native fishes because of predatory habits. Concern is justified, given the endemicity common among spring fishes, the appearance of many spring endemics on state and national imperiled species lists, and the general vulnerability of spring systems to invasion.

What should be done? Eradication with poison is effective, albeit nonselective (e.g., Shafland and Foote 1979). Bans on import and possession have preventive value and exist in 24 states, although they are largely irrelevant, given the actual threat to humans (Bennett et al. 1997). Such laws restrict the number of entry and distribution points for piranhas, which should reduce their spread to places where they might survive. It would seem precautionary and laudable to consider a federal ban on piranha importation given (1) the number of places where they have been released; (2) the likelihood of people to continue to release them into the wild; (3) the marginal hysteria that surrounds their release; and (4) the potential ecological threat they pose as introduced predators. No industry is in any way dependent on piranha importation for legal livelihood, and it seems unlikely that the public would mount much opposition to such a law. Public aquaria and other educational facilities could be exempted via a permitting provision. Any black marketeering stimulated by such a law would probably be no more active than what goes on now.

Finally, responsible pet ownership should be emphasized for aquarium keepers who find themselves with piranhas or any fish that has become inconvenient to keep. Such unwanted pets can be donated to a public facility or even returned to a pet store. If neither can be found, the animal should be euthanized. A quick blow to the head is probably best (see Robb and Kestin 2002), although aesthetically unappealing to many people. Placing a tropical fish in a water-filled container and putting it in a freezer is a reasonably humane method of killing, after which the dead fish can be thawed and disposed of by conventional means (see www.habitattitude.net).

As far as Stone Mountain Lake was concerned, waiting until winter would have solved all problems, imagined or real.

SOLUTIONS TO TRADE PROBLEMS

Regulation, Licensing, and Certification

Management plans and regulations are even less common for the aquarium trade than for reef fisheries (E. M. Wood 2001). Where comprehensive regulations do exist, compliance and enforcement tend to be problematic. In Hawaii, only 13% of licensed collectors filed all required reports, and provisions for assessing the accuracy of reports do not exist (Moffie 2002).

E. M. Wood (2001) generalized that marine ornamental fisheries should be managed to (1) ensure sustainability and be integrated into other resource uses (other fisheries, ecotourism); (2) minimize harvesting and post-harvesting mortalities, including bans on the collection of species unsuited to aquarium life; and (3) address socioeconomic issues of fair and equitable trade. Among specific actions, Wood recommended that countries

- collect accurate data on life history; population sizes and dynamics; CPUE for individual species, not all species combined; and exports, imports, and domestic trade
- implement nontransferable licensing of all sectors of the industry and limit the number of collecting permits issued (see also Pyle 1993); make license renewal contingent on accurate record keeping and reporting of numbers captured and discards due to injury and mortality
- ban destructive collecting and train collectors in nondestructive techniques; observe collectors and inspect holding facilities
- establish data-dependent quotas for rare and endemic species and restrict capture of species with low aquarium survivability, including juveniles of many species
- establish no-take areas as reference sites for determining impacts of collecting
- promote aquaculture of native species where it is logistically realistic
- promote international cooperation and establish a certification program that brings higher prices for certified fish
- provide funding sufficient to enforce all regulations
- distribute the burden of responsibility equally between producers and consumers
- educate all trade sectors and the public about recommendations and actions

Wood (2001) described how Oman, Jamaica, and Puerto Rico have, for economic and ecological reasons, instituted restrictions and out-and-out bans on the aquarium trade. To the list can be added Hawaii and the Florida Keys, Palau, Sri Lanka, Maldives, and Mozambique (T. Graham 1996; Dufour 1997; Sadovy and Vincent 2002). Brazil maintains a "white list" of about 180 species that can be legally exported; species not on the list are illegal to trade, although enforcement is variable (O. Lucanus, pers. comm.).

Protected Areas

Establishing reserves protects both species and habitats. Although most reserves focus on larger food fishes, prohibiting all extraction activities would benefit smaller ornamental species. Tissot and Hallacher's (2003) findings that aquarium collecting in Hawaii caused species depletions led to the establishment of nine Fish Replenishment Areas on the west coast of Hawaii, placing 30% of the coastline off limits to aquarium collectors.

Captive Propagation

Perhaps as much as 90% of freshwater aquarium species are captive bred (Andrews 1990; Tlusty 2002), in part because methods developed for culture of commercial freshwater species can be applied to the propagation of other freshwater fishes (e.g., transfer of koi and grass carp methodologies to rearing of other cyprinids; C. A. Watson and Shireman 1996). Many programs are so successful that taking fish from the wild is only occasionally necessary to "revitalize" the genetics of breeding stocks (R. W. Ward 1998).

Many aquarists' organizations are actively involved in the culture of known threatened species, breeding fishes for gene banking and reintroduction purposes in so-called species survival programs. Greatest successes have been with African cichlids, Central American cyprinodontids, and Asian *Barbus* species (e.g., Hemdal 1995; Tlusty 2002; Loiselle 2003).

Because wild harvests can deplete populations, captive propagation can reduce pressure on wild stocks. Species that have been heavily depleted in the wild but continue to be traded because of captive breeding include the Asian bonytongue and the bala shark.

The Asian bonytongue, also known as the Asian arowana or golden dragon fish, was designated Endangered by IUCN and the U.S. in the mid-1970s and placed in CITES' Appendix I, prohibiting all trade (Bartley 2000; figure 13.16). Native to Indonesia, Malaysia, Thailand, Cambodia, and Vietnam, it was heavily overfished, largely for the ornamental trade. Populations in Malaysia and Sumatra were extirpated due to overcollecting (Ng and Tan 1997). Unfortunately for the fish, it suffers from status as a good luck symbol and is often referred to as the feng shui fish: "Many well-known Chinese businessmen swear that their good fortune is due to keeping live arowanas in their offices" (www.wofs.com, search "arowana"). Golden dragon fish are now bred under close government supervision in several countries; every individual is tagged with a microchip and issued a Certificate of Identity specifying its color variety (red, red-tail, green, gold). Six-month-old fish 15–20 cm long can sell for $3,800 apiece, with rare AAA-grade red fish costing more than $5,500. Red fish with particular color patterns in their scales bring the highest prices, and unscrupulous breeders have been known to immerse fish in hormone or color baths, raise them in bright sunlight, starve them, and even blind them to artificially bring out red color (www.dragonfish.com, www.arofanatics.com). As a result of successful breeding, the Indonesian population was transferred from Appendix I to Appendix II of CITES, which allows permitted trade of captive-bred individuals, although the species as a whole remains in Appendix I (Andrews 1990; IUCN 2004).

The silver or bala shark is native to Thailand, Borneo, and Sumatra. This 25-cm cyprinid was formerly abundant but was overfished for the aquarium trade and suffered losses due to deforestation. Juveniles and adults were targeted, and populations were decimated by the discovery and exploitation of breeding aggregations. Many local populations were apparently extirpated (Ng and Tan 1997). High collection mortalities occurred because the fish is relatively easily stressed ("Avoid sudden movements near the tank"). Bala sharks were listed as Endangered by

Figure 13.16. The Asian bonytongue. Photo by Marcel Burkhard, http://en.wikipedia.org/wiki/Image:Arowanacele4.jpg.

IUCN in 1996. Subsequently, aquarists successfully bred the fish; almost all animals now in the aquarium trade are captive bred (Ng and Tan 1997).

Selective breeding to create varieties and "sports" of some species may also reduce pressure on wild stocks. Goldfish, koi, guppies, platys, bettas, cichlids, and dragon fish among others are actively bred for desirable traits promoted by fish-keeping organizations. The International Fancy Guppy Association recognizes 76 varieties based on fin and body color and conformation. The Goldfish Society of America promotes competitions that assess body, fin, and scale shape and size, eye type, body color, and general condition and deportment. Aquaculture of ornamental carp (koi, nishikigoi) dates back over 2,000 years (Balon 1995). About 13–15 koi color groups are officially recognized. Grand champions selected at the annual Zen Nippon Arinkai (ZNA) All-Japan Nishikigoi Show have sold for as much as $1.2 million, with juvenile offspring of such champions selling for $10,000–$20,000 (V. Burnley, University of Georgia, pers. comm.). Aquarium keepers who devote time and money to producing and maintaining such highly domesticated stocks may be less likely to purchase wild-caught species.

Breeding Marine Fishes

Success in breeding marine ornamentals has, by comparison, been limited. Only 1% to 5% of marine species imported into the U.S. are successfully cultured (Moe 1999; Tlusty 2002), although the basis of estimates remains obscure. Regardless, "people prefer and will pay extra for the cultured fish" (Vince Rado, pers. comm.) because (1) culturing reduces capture of depleted wild stocks, and (2) captive-bred individuals are often more adaptable to life in a fish tank (Sadovy and Vincent 2002).

Regardless of desirability, propagation of marine tropicals is hampered by logistics, inadequate technology, and bad history. Successes remain limited despite repeated predictions of imminent breakthroughs. Approximately 100 species of marine tropicals have been bred in captivity (E. M. Wood 2001; Tlusty 2002), but only 21–25 species meet the three major criteria for commercial success of high value, in high demand, and relatively easy to culture (Moe 1999; Tlusty 2002). Ease of culture appears to be the major impediment. Water-column spawning reef fishes have proven difficult to breed in tanks, and pelagic larvae have proven difficult to feed.

Among marine success stories are anemonefishes (*Amphiprion* and *Premnas*), gobies (*Gobiosoma*, *Gobiodon*, *Amblygobius*), pseudochromid dottybacks (*Pseudochromis*), jawfish (*Opisthognathus*), basslets (*Gramma*), and Banggai cardinalfish. This group is dominated by species that spawn benthically, engage in some parental care, and produce relatively large larvae or lack larvae (cardinalfish), all traits uncommon among reef fishes (see Thresher 1984). Availability of even commonly bred species is apparently limited. For example, saltwaterfish.com, a large mail-order aquarium business in Miami, listed 43 species of reef fish for sale in late 2002. Only 8 were denoted "aquacultured," and 5 of those were anemonefishes, a group that has been cultured by aquarists for decades.

Other assessments reveal similarly limited availability. The American Marinelife Dealers' Association's *Captive Bred Livestock Directory* in 2003 listed 33 species (1 angelfish, 1 cardinalfish, 17 anemonefishes, 9 dottybacks, 1 goby, and 4 seahorses) as constituting "all known captive-bred [fish species] available to the marine aquarium trade to date" (www.amdareef.com/captive_main.htm). Moe (2002) included 32 species that commercial U.S. breeders felt could be bred with economic success, adding 5 species (2 sharks, 1 pipefish, 1 threadfin, and 1 carangid) to the AMDA list. If market forces rather than logistics determined the popularity of captive-bred species, many more would be cultivated.

Captive breeding of ornamentals lacks many of the environmental drawbacks that plague the culture of commercial food fishes (chapter 14), and has also been subjected to less criticism than grow-out programs for the live-fish food trade with its dependence on "seed" collection of larval fish. Aquarium aquaculture almost always involves closed-cycle techniques that use captive breeding rather than seed collection.

However, captive breeding has socioeconomic consequences. Wild-caught specimens benefit small villages and local fishers. Mariculture, especially of reef species on a commercial scale, involves substantial capital investment and fairly elaborate technologies; it is out of the reach of small villages in island nations. Not surprisingly, commercial culture of aquarium species is concentrated in the U.S., Europe, and Taiwan, rather than the countries from which breeding stocks originate (E. M. Wood 2001; Tlusty 2002). Such expropriation of genetic resources is a form of bioprospecting and contravenes fairness provisions of the Convention on Biological Diversity (UNEP 1992). These issues trouble countries such as Brazil, where many small communities depend on freshwater fish collecting for income. Proponents of indigenous rights and practices, such as Tlusty (2002, p. 215), have advocated discouraging captive cultivation "when the wild harvest maintains habitat and a cultural or economic benefit would disappear if collecting was stopped."

One could argue, however, that aquaculture of common or easily cultured species would create opportunities for local fishers able to collect rare or difficult-to-breed species that demand high prices. Such free-market competition could theoretically eliminate poorly trained and inefficient collectors that supplied low-quality but common specimens, favoring collectors able to produce a better, higher-value product. Who will make these decisions and promote their implementation remains to be resolved.

Certification, Self-Policing, and Grassroots Organizing

In recent years, trade organizations and NGOs have joined forces to ensure that ornamental species are captured, transported, held, and sold in a responsible manner that results in a sustainable industry. Project Piaba, a community-based organization centered in Amazonas, promotes a sustainable fishery by implementing many of the actions recommended by E. M. Wood (2001). It emphasizes diversification of the fishery, rotation among collecting areas, and ecosystem protection, with a goal of developing a "green certification" process that would identify sustainably caught fish (Chao, Prang et al. 2001; Chao and Prang 2002; see http://opefe.com/piaba.html).

Industry-based organizations include Ornamental Fish International and the UK-based Ornamental Aquatic Trade Association. Both have a code of ethics, which in the latter case includes specific guidelines on water quality. The American Marinelife Dealers Association (AMDA) has detailed Standards of Practice "to promote environmentally responsible marine aquarium keeping" that also mandate water chemistry criteria. The UK Marine Conservation Society is preparing a guide called *The Responsible Marine Aquarist* with the aim of encouraging marine ornamental keepers to choose species that are appropriate to their tank conditions and originate from well-managed fisheries.

AMDA publishes a useful and revealing "Ecolist" that classifies aquarium fishes by the skill required to keep them in captivity (www.amdareef.com), focusing on the approximately 650 species frequently imported into the U.S. (figure 13.17). Categories are *beginner, intermediate, reef only* (should be kept in a tank with live rock; R. Goodlett, pers. comm.), *advanced*, and *requirements unknown*. To its credit, AMDA recommends that fish dealers should avoid carrying difficult-to-keep species and should sell such species only to customers with advanced skills, for research.

Categorizing reef families by the AMDA criteria indicates where different groups generally fall. For example, some representative categories and families are

- 1, beginner: anemonefishes, humbugs and other damselfishes, groupers, hawkfishes, triggerfishes, basslets, dottybacks
- 2, intermediate skill: *Heniochus*, snappers, *Bodianus* and some other wrasses, puffers, squirrelfishes, *Zebrasoma*, many angelfishes, some hogfishes, butterflyfishes, hamlets, some surgeonfishes
- 2–3: blennies, gobies, dragonettes, some wrasses
- 3, reef tank/live rock only: parrotfishes, some blennies
- 3–4: *Anthias*, some wrasses
- 4, advanced skill: lionfishes, sharks, rays, morays, many butterflyfishes, many surgeonfishes, many angelfishes,

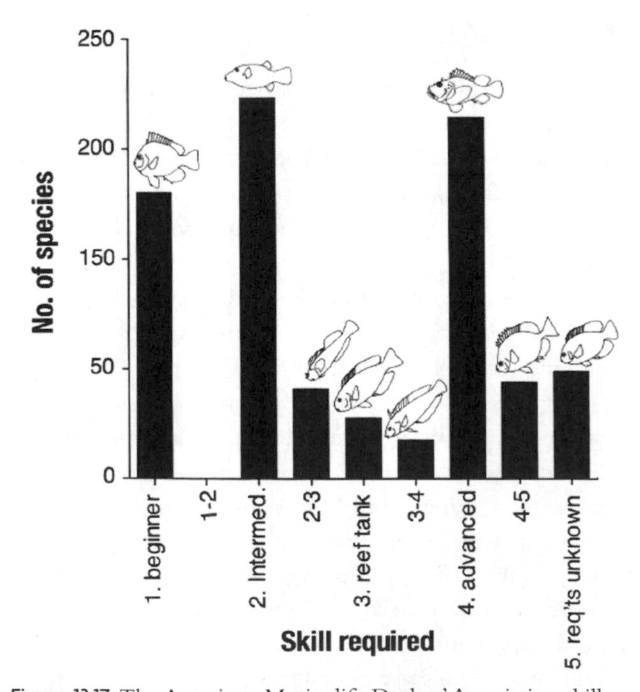

Figure 13.17. The American Marinelife Dealers' Association skill levels required for keeping fish species commonly found at U.S. pet stores. Data from www.amdareef.com/ecolist_main.htm. Drawings from Randall (1981); used with permission.

anglerfishes, scorpionfishes, jawfishes, most hogfishes, trunkfishes
- 4–5: cleaner wrasses, some gobies, some butterflyfishes, rays, parrotfishes
- 5, requirements unknown: tilefish, many angelfishes, some morays, some butterflyfishes, sweetlips

Using categories 4 and 4–5 as a metric, 35%, or 208, of the 598 species for which requirements are known are inappropriate for most aquarists. If category 3 (live-reef tanks) and above are included, the number climbs to 47% (303/647), indicating that about half of all imported fish species should not be kept by anyone not willing to devote considerable time and resources to the activity. Even with AMDA's guidelines, hundreds of reef species are suitable for most aquarists. Given that aquarists are understandably unhappy when their fish die, AMDA's Ecolist creates a win-win situation for fishes and fish keepers.

Fast becoming an umbrella organization for efforts directed at a responsible, sustainable aquarium trade is the not-for-profit Marine Aquarium Council (MAC), which focuses on certification of collectors, exporters, importers, and retailers, e.g., "From Reef to Retail" (www.aquarium-council.org). MAC certification (figure 13.18) assures purchasers that marine fishes and invertebrates are of high quality and were obtained via sustainable practices, with accurate data collection, ecosystem management, and promotion of reef conservation and socioeconomic equitability (Holthus 2001). MAC certification guidelines are detailed in a booklet, *Information for Marine Aquarium Hobbyists,* and organizations first gained certification in 2002. MAC's efforts include cooperation with a variety of NGOs, industry groups, and government agencies in training collectors in nondestructive collecting practices. MAC has also partnered with the UN's World Conservation Monitoring Center to create the Global Marine Aquarium Database, which compiles export and import records by species and country (http://www.unep-wcmc.org/marine/GMAD/index.html).

The MAC certification program has developed slowly and is not without controversy. Its effectiveness ultimately depends on (1) consumer willingness to pay more for sustainably produced aquarium products, and (2) collectors benefiting from the increased prices. Willingness to pay more, referred to as "green equity," has become a mantra among marine aquarium aficionados and is widespread among U.S. aquarists (e.g. Dowd 2001; Moe 2002), who

Figure 13.18. The Marine Aquarium Council certification logo. On an aquarium store, this logo indicates that the fish sold there were caught sustainably and transported using best practices, and that the store promotes reef conservation and economic benefits for rural, coastal communities. Used with permission of the Marine Aquarium Council.

make up 60% of the market. Additional benefits to collectors are necessary because people living a subsistence existence can't be expected to voluntarily restrain their livelihoods to conserve coral reef ecosystems (Baquero 2001).

Public Education

The public needs to be made aware that their purchases can affect natural habitats, imperiled species, and socioeconomic justice. Ecolabeling and certification programs consequently include activities focused at educating collectors, transporters, agents and merchants, and the buying public. In particular, the public needs to appreciate the ecological damage done by introduced species, whether stocked to improve sport fishing or simply released to be gotten rid of. Much of the problem lies in the emotional attachment aquarium keepers develop for fish they have kept for a long time and the revulsion they feel toward euthanizing pets. When all else fails, informing the public that releasing fish is illegal should be persuasive (figure 13.19).

Why Not Keep Natives?

A small number of hobbyists in the industrialized world keep temperate, native fishes. Organizations such as the North American Native Fishes Association encourage this through their activities, publications, and Web sites. NANFA's mission statement focuses on responsible captive husbandry and "the appreciation, study and conservation

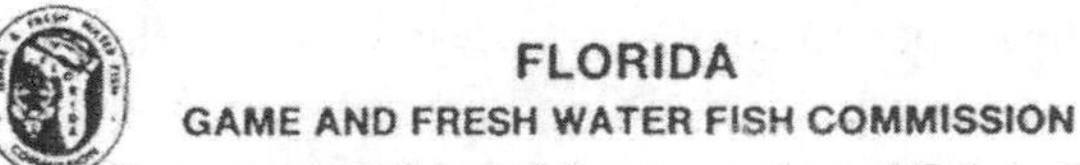

Figure 13.19. A poster from the Florida Game and Fresh Water Fish Commission informing the public about the ecological and legal ramifications of releasing nonindigenous species. Note inclusion of spotted tilapia, *Tilapia mariae*. From Cox et al. (1997).

of the continent's native fishes" (www.nanfa.org/mission.shtml). NANFA argues that collecting and keeping native fishes in aquaria via environmentally responsible practices constitutes "a valid use of a natural resource," as long as all environmental and habitat protection laws are observed. NANFA promotes a Code of Ethics emphasizing strict adherence to legal proscriptions concerning capture, transfer, and maintenance of fishes; habitat protection; private property rights; and avoidance of species unsuitable for captive husbandry. NANFA's code is especially adamant on the sins of releasing aquarium-kept fish into the wild, "even to places where they were originally collected . . . [because of] complicated and often unknown ecological processes at work in aquatic systems which may make fish introductions detrimental to the system."

Keeping native fishes, in home aquaria or classrooms and nature centers, can be justified on the following grounds:

- It increases knowledge and fosters appreciation of local biodiversity, thereby promoting its protection (Scharpf 2001).
- Through collecting, one directly experiences local natural surroundings (Bock 2002; Denkhaus 2002), something that is lost when one buys from a pet shop. Collecting provides satisfaction similar to that derived from catching one's food or collecting wild, edible plants. Keeping a wild fish rewards an aquarium hobbyist with a sense of accomplishment not available with store-bought fish.
- The activity is open to all, rather than to an economic elite. Many native fishes are abundant, colorful, behaviorally interesting, and easy to catch and keep. Any child can catch a small bluegill or minnow from a local pond and keep it in a ten-gallon aquarium with occasional water changes and minimal filtration, lighting, heating, and aeration.
- Pressure on distant, tropical habitats is reduced, and many of the environmental, ecological, and social problems detailed in this chapter are decreased.

However, these factors argue against keeping natives:

- Problems from releasing fish into the wild, including introducing diseases acquired in captivity. If moved even short distances, captives can alter the genetic makeup of local populations. When moved outside their native range, all the problems associated with introduced species arise. Native fish keepers enjoy trading species. If no fish were ever returned to the wild, fish swapping would be harmless. But released, temperate native species pose a strong threat because they are likely to (1) survive in the wild, (2) transmit diseases, and (3) disrupt ecosystems. Many pet stores will take back tropical fish that people have grown tired of, but this does not apply to native species, in part because commercial trade in native fishes is often illegal. A related, condemnable complication involves statements by authors of relevant books who advocate release of wild-caught fishes back to their points of capture (e.g., Quinn 1994; Schlesser 1998).
- Collecting natives threatens species that would otherwise be left alone. Many rare species, especially those with limited geographic ranges, are particularly vulnerable to habitat loss and introduced species. These fishes gain some degree of protection because of their limited economic value, small size, out-of-the-way locales, and relative invisibility to the general public. Their only direct threats come from collectors, both scientific and hobbyist. No one else wants them. Also, many are difficult to identify, and accurate species determinations are often not made until they have been removed from the wild. This places a particular responsibility on ichthyologists and fish keepers to question their own motives, abilities, and actions. Members of both groups have abused their rights and responsibilities in the past, rationalized collecting in the name of advancing science, cited the inconsequential impact of taking "just a few," or even claimed to promote the welfare of the species via captive propagation. All such justifications should be carefully scrutinized and weighed against the stresses placed on the individuals and populations under exploitation.

Overall, and when practiced responsibly, keeping native fishes should cause little damage and can ultimately sensitize a larger segment of the public to aquatic conservation issues. Keeping natives can even be viewed in terms of national pride: "Isn't it also patriotic to learn about our *natural* history, and to help preserve our *natural* heritage?" (Scharpf 2003, p. 74).

Sources of information on collecting and keeping North American native fishes can be found at the NANFA Web site and in such books as Quinn (1994) and Goldstein et al. (2000), although both authors are misinformed about the hazards of releasing captive animals into wild habitats. Native fish aficionados in Australia are supported by the Australia New Guinea Fishes Association.

Can Science Provide Answers?

Although most fishes studied by coral reef ecologists are of minimal commercial value, ornamentals may be an exception. Findings by reef fish ecologists may have direct application to managing ornamental fisheries, because study subjects and exploitation targets are small, often colorful, reef fishes. Bolker et al. (2002) developed a management model for ornamental fisheries based on field studies of the small, tropical (and ornamental) humbug damselfish, *Dascyllus trimaculatus*. They found that the presence of humbug individuals in preferred settling habitat (sea anemones) could increase the probability of additional settlement—that is, anemones could be seeded with catchable fish and sustain harvest. Their model predicted that yield increased with increasing catch rate up to 80% exploitation. Although simplifying assumptions and natural variability complicated the outcome, their approach holds promise for developing management strategies at a scale appropriate to marine ornamental species.

Also of direct relevance is the possible application of findings from small artificial reefs used to attract ornamental species. Theoretically, such reefs could be placed at appropriate depths in areas lacking structure attractive to settling larvae. They could increase the number of larvae recruited to an area without depleting supplies at natural structures, assuming they intercepted only larvae that had missed natural areas (e.g., J. Wilson et al. 2001). The possible applications of artificial reef technology take the issue of recruitment limitation beyond the realm of pure theory (chapter 12). If replenishment of reef populations is not limited by the delivery of new larval individuals to reefs, artificial reefs become a viable means of increasing the production of commercially valuable aquarium species.

CONCLUSION

Fisheries that target live fishes and sell them as such differ ecologically and economically from traditional food fisheries, and these differences complicate their management. Significantly, both the live-food trade and the ornamental trade are "luxury" fisheries that originate in developing economies but whose products are intended for relatively wealthy consumers. Because most such fisheries occur in tropical countries, their additive impact on global biodiversity can be substantial. Destructive fishing practices in the live-fish food trade have the greatest and best-documented impacts on fishes and on the people who exploit them. The aquarium trade operates on a smaller scale, targets a wider variety of small fishes, and, because fish mortality is a greater liability, can benefit from improvements at all stages. Its overall ecological impact is small compared to the live-food trade. In both sectors, similar solutions to environmental and socioeconomic problems can promote sustainability, including regulation, enforcement, certification, captive propagation, local control, directed research, and protected areas. Improving sustainability in the live-food trade is more challenging because profits are higher and the ultimate consumers have demonstrated little concern about sustainability. Changes will require government regulation affecting collection practices, and increased outcry from the conservation community. The chain of supply for aquarium fishes can be improved more readily because consumers are educated and increasingly concerned about the ecological and social implications of their hobby. Hobbyists can insist on sustainable practices, which can be instituted through voluntary action and cooperation, driven by market forces.

14. The Promise of Aquaculture and Hatcheries

Aquaculture is a possible solution, but also a contributing factor, to the collapse of fisheries stocks worldwide.

—Rosamond Lee Naylor et al., 2000

Aquaculture has a large and vocal group of advocates in commercial, academic, and government circles (Parker 2002; Stickney 2005; www.was.org; www.aquaculturemag.com) who have championed the postulated and realized benefits of farmed fish. My focus is more on the impacts of aquaculture on wild fishes and their ecosystems, with some discussion of captive propagation of imperiled species, realizing that what affects fishes and their ecosystems eventually affects humans. My treatment concentrates on two, often-overlapping types of aquaculture: (1) grow-out aquaculture for direct human consumption, and (2) hatchery and stocking operations that release cultured fishes into the wild to supplement exploited or declining stocks.

Aquaculture, or fish farming, involves growing aquatic organisms under controlled conditions (Bardach et al. 1972). Culturing marine organisms is sometimes called mariculture. Aquaculture usually entails artificial propagation and rearing in a confined space, often at relatively high densities. It can start with (sometimes genetically modified) eggs and sperm or with captured young organisms that are grown out to marketable or releasable size. Products can be harvested as part of a sport or commercial fishery or can be imperiled species never intended for recapture (the term *captive propagation* is often reserved for efforts involving imperiled species). Facilities range in size from small laboratories and small, rural paddies and ponds to large, land-based and sea ranch or net pen operations. Facilities can be closed (e.g., net pens, catfish ponds), where fish spend their entire lives; or open, where fish spend initial phases in a hatchery and are then released (e.g., salmon and other grow-out hatcheries for stocking of commercial and sport fishes or reintroduction of imperiled species). *Extensive* aquaculture usually contains species in some sort of pen but relies on natural food supplies; *semi-intensive* aquaculture involves feeding supplements or adding fertilizer to encourage feed production; and *intensive* aquaculture is highly subsidized, involving large inputs of often-formulated feed and other chemicals, with the highest yields and the greatest ecological impact (Kautsky et al. 2001). This chapter focuses on intensively cultured fishes, although many other aquatic organisms are also produced (see NRC 1992b; Pillay 1992, Stickney 1994, Beveridge 1996).

DOES AQUACULTURE MEET HUMAN NEEDS?

Fisheries provide 16%–19% of human animal protein consumption. Approximately one billion people rely on fish for most of their protein, especially in developing nations (Botsford et al. 1997; FAO 2000b). Pressure on aquatic resources and wild fish stocks can only increase as the human population grows. Wild supply and CPUE decreased as capture fisheries leveled off at 90 MMT in the late 1980s, but in recent years, total fisheries yields increased to 120 MMT, due largely to aquaculture, which expanded from 5 MMT in 1950 to around 30 MMT in the late 1990s. Given past and current trends in marine productivity and freshwater degradation, the increasing demand for fish protein will have to be met by aquaculture. The primary environmental issue is whether it can meet those needs sustainably, without unacceptable costs to the environment that will eventually be borne by humans.

Most aquaculture today utilizes freshwater organisms and is land-based in Asia, particularly China and India. On a live-weight basis, 59% consists of freshwater, 35% marine, and 6% estuarine species. The marine values are inflated by the high water content of seaweeds and the shell weight of mollusks (FAO 2000b), meaning the relative freshwater contribution is even higher. Among fishes, the majority of cultured freshwater species are carps (silver, grass, bighead, and common); Nile and related tilapia; *Labeo* cyprinids; and channel catfish, *Ictalurus punctatus* (Naylor et al. 2000). Brackish water species are dominated by milkfish, *Chanos chanos*. The primary, cultivated, marine species is Atlantic salmon (*Salmo salar*), others being Pacific salmons, striped bass (*Morone saxatilis*), serranid sea basses, and sparid and pagrid sea breams. Carp and tilapia production generally goes toward meeting local protein needs, whereas salmon (and shrimp) mostly support export markets.

Capture versus Culture Fisheries

Naylor et al. (2000) addressed the question of whether aquaculture contributes to or subtracts from the available fish supply. They concluded that some aquaculture practices could compensate for declining capture fisheries, but pressure on capture fisheries would continue because of the growing demand for fish products. Aquaculture as currently practiced may create additional pressure on wild stocks because of competition for (1) larvae and other fishes that are fed to cultured stocks, (2) coastal ecosystems and their services, and (3) world markets where products are sold. Also, capture fisheries and marine ecosystems will likely suffer due to problems of waste production, chemical pollution, exotic species invasions, and pathogen transmission (figure 14.1).

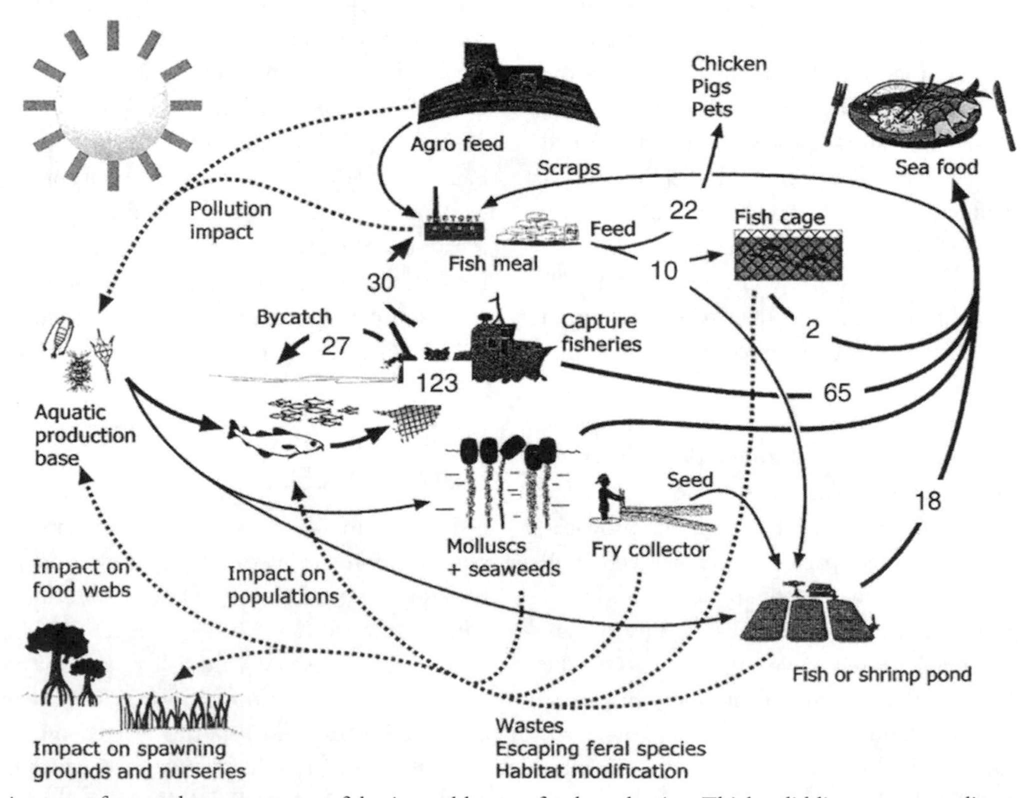

Figure 14.1. The impacts of aquaculture on capture fisheries and human food production. Thick solid lines represent direct energy sources made available to humans; thin solid lines are indirect and minor flows. Broken lines represent negative feedbacks between human activities and aquatic production. Numeric values are MMT for 1997. Starting with the "aquatic production base," energy flows directly via capture fisheries to human consumption ("sea food"), or indirectly via aquaculture ("fish meal" to "feed" to "fish cage," or "fish or shrimp pond" to "sea food"). Marine fishes ("fish cage") and shrimp aquaculture produce a net loss of production because of inefficient protein conversion. Modified from Naylor et al. (2001); used with permission.

The New Alchemy: Turning Sardines into Shrimp

"The appropriation of aquatic productivity for fish feeds reduces supplies of wild fish that could potentially be consumed directly" (Naylor et al. 2000, p. 1020). This loss occurs because of the ecologies of many cultured species. Most cultured marine and diadromous fishes are carnivores reared in floating pens or net cages in nearshore areas. They are fed feeds formulated from wild-captured fishes. As aquaculture has expanded, species previously used directly for human consumption have gone increasingly into feeds for cultured fishes. One-third of fisheries landings (about 30 MMT) is converted into fish meal. Between 1986 and 1997, 8 of the top 20 capture species, including 4 of the top 5, were converted into feed for aquaculture and livestock production (figure 10.5). The proportion of fish meal used in aquaculture rose from 10% to 33% over that period (Naylor et al. 2000). The 8 fish species include various sardines, anchovies, and mackerels, species that are also eaten by humans in many regions.

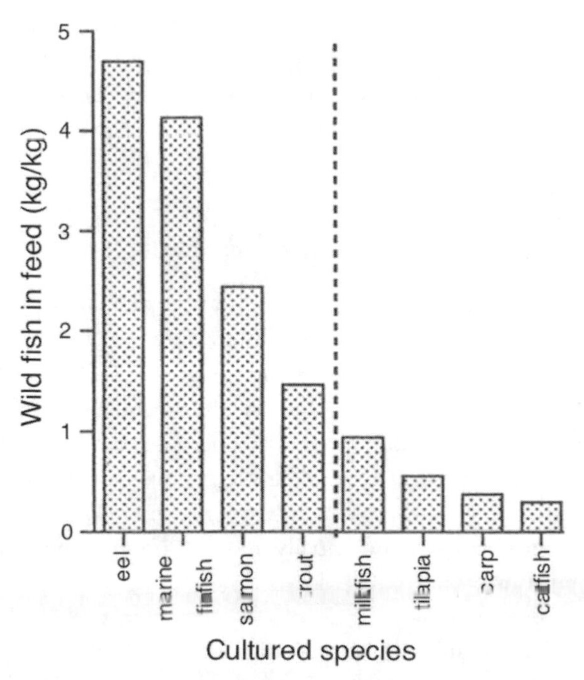

Figure 14.2. Biomass ratios of wild fish used in formulated feeds to farmed species produced for eight of the most commonly cultured fishes. Species left of the vertical dashed line are carnivores that require more protein inputs than they produce. Omnivores (right of the line) can be produced using less animal protein than they convert into fish biomass. "Marine finfish" include flatfishes, gadoids, drums, sea basses, and tunas. Carp are nonfilter-feeding species fed compound feeds. Data from Naylor et al. (2001).

Farming marine finfishes results in a net loss of fish protein for humans. Because of conversion inefficiencies, 2 to 5 kg (avg = 3.4 kg) of fish protein are required to produce 1 kg of farmed carnivorous fish (Tacon 1996; figure 14.2). For salmon and shrimp, the ratio is about 2.5:1. When the weight of captured ocean fishes fed to farmed fish is subtracted from aquaculture production (30 MMT – 10 MMT), aquaculture's contribution to human protein supplies drops to 20 MMT. Capturing marine fishes to feed cultured predators is an inefficient means of producing food.

Aquaculture does not have to cause protein losses. Carp and tilapia are net contributors to protein supplies because they are omnivores or herbivores. They are grown in ponds, and their production is often integrated into agriculture production (= polyculture) by using waste products from domesticated animals to fertilize water and produce fish food (e.g., Mathias et al. 1997). Considerable effort is also going into reformulating feeds and replacing animal protein and fish meals with seed oils.

ECOLOGICAL IMPACTS OF AQUACULTURE

Proponents argue that aquaculture helps meet human protein needs while reducing demand for overexploited wild species. For example, it is widely held, although difficult to prove, that exploitation of wild Atlantic salmon declined because the species was so successfully farmed (C. M. Cunningham, pers. comm.). Detractors maintain that fish farming and hatchery or stocking programs have a number of negative effects on surrounding fishes, biotic communities (including humans), habitats, and ecosystems. Safina (2001a, p. 791) stated, "Aquaculture will do no more to save wild fish than poultry farms do to save wild birds." The bulk of the evidence supports his contention.

Competition for Food and Spawning Sites

Competitive impacts of hatchery-grown, released fish on wild fish were detailed by Levin et al. (2001), who analyzed 25 years of abundance data for migrating smolts and returning adult Chinook salmon, *Oncorhynchus tshawytscha*,

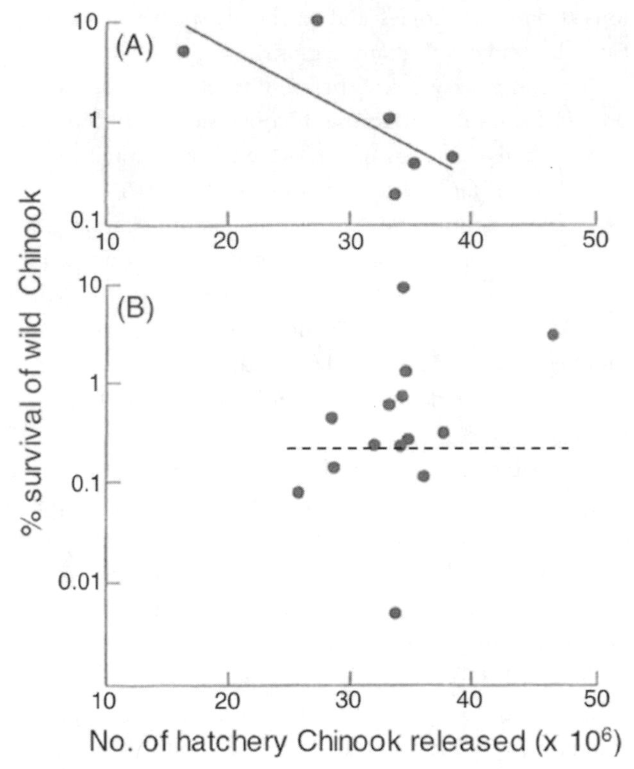

Figure 14.3. Survival of wild Chinook relative to abundance of hatchery fish under different conditions of oceanic productivity. Hatchery fish depress wild salmon populations when oceanic conditions are worst for wild fish. (A) During years of poor ocean conditions, survival of wild Chinook salmon declined as number of hatchery fish released into the Snake River increased (r^2 = 0.73, p = 0.03). (B) During years of average oceanic productivity, no relationship existed, indicating no impact of hatchery fish on wild fish (r^2 = 0.06, p = 0.40). From Levin et al. (2001); Chinook drawing inset by K. Spencer, www.katespencer.com, in Augerot (2005); both used with permission.

in the Snake River, a tributary of the Columbia. Chinook are reared and released to supplement wild populations, an activity that is often promoted as boosting wild fish productivity. Over 130 million Chinook juveniles were released from Snake River hatcheries every year; hatchery fish constituted about 80% of the Columbia's returning adults. When oceanic productivity, measured mainly as plankton production, was accounted for, hatchery fish appeared to have a negative impact on wild fish when they were most vulnerable (figure 14.3). During years when plankton productivity was average, no relationship existed between wild fish survival and number of hatchery fish released. However, during years of low oceanic productivity—for example, during strong El Niño–Southern Oscillation (ENSO) events, wild fish survival was inversely proportional to hatchery output.

The mechanism driving the declines appears to be a density-dependent interaction determined by food availability for young fish shortly after they enter the ocean as smolts. When food is plentiful, no effect is evident. But when food is scarce, abundant hatchery fish compete with rarer wild fish and overwhelm them, driving them to even lower densities. Supplementation programs, therefore, have an effect opposite of their purported goals, reducing wild fish abundance when wild fish can least afford additional insults and populations are at historic lows. Climate models suggest that climate change will likely result in conditions more like those during ENSO events, indicating that poor oceanic conditions for juvenile salmon will become more prevalent. Hatchery activities and population supplementation cannot be justified on the grounds of conserving wild populations, now or in the future.

Competitive impacts on native fishes have been shown elsewhere. In Spain, aquacultured mussels filter phytoplankton out of the water, depressing zooplankton abundance and apparently contributing to declines in local sardine, *Sardinella pilchardus*, fisheries (Kautsky et al. 2001). In North America, escaped blue tilapia, *Oreochromis aureus*, compete for food with native shads and for spawning habitat with centrarchid sunfishes.

Competition for Larval Resources

Grow-out programs capture wild larvae or juveniles and raise them to marketable size. Fishes grown include milkfish, anguillid eels, reef fishes such as groupers for the live fish trade (Johannes and Ogburn 1999), cod, and bluefin tuna. Grow-out is often necessary because adults will not spawn in captivity, nor do larvae survive initial growth phases. Grow-out bypasses a period of extremely high natural mortality and potentially reduces pressure on adult populations.

However, collecting so-called seed stock may deplete recruits for wild populations, although little evidence exists to suggest serious impacts. Theoretically, a rare species with a limited spawning period, low egg production, and narrowly defined and localized larval settlement requirements could be depleted, but the arithmetic of natural larval production and mortality makes that unlikely. High bycatch mortality is, however, a recognized liability. Few marine fish larvae occur in isolation from other fish larvae. Naylor et al. (2000), extrapolating from milkfish fry data in the Philippines, estimated that collecting 1.7 billion milkfish fry annually killed 10 billion fry of other fishes via discards. The seine fishery for larvae of giant tiger shrimp, *Penaeus monodon*, in India and Bangladesh kills 160 fish and shrimp fry for each tiger shrimp larva captured. Whether such bycatch or bykill is directly responsible for declines in populations of nontargeted species remains to be determined.

Competition for Ecosystem Goods and Services

Culturing milkfish and shrimp often entails modifying mangroves or salt marshes (Kautsky et al. 2001), ecosystems that provide numerous important services, including erosion and flood control, trapping sediments, and processing wastes. They are also nursery habitats for many coastal and offshore fishes. Shrimp farming in particular causes mangrove destruction and productivity loss, especially in southeast Asia and Central and South America (Primavera 1998). In Thailand, the world's largest producer of farmed shrimp, every kilogram of shrimp produced results in 0.4 kg of lost fisheries biomass, mainly wild fish and shrimp. Mangrove conversion also has social impacts, because farming displaces wild-capture fisheries that provide employment and food for local communities. Farmed shrimp use less human power and are exported. Artisanal fishery practices disappear with the lost fisheries, and local fisheries are converted from common to private property (Naylor et al. 2000).

Aquaculture is by nature a water-demanding activity, although water use varies greatly among culture species and practices. Catfish farming as generally practiced is relatively benign. Catfish live in turbid, warm, and relatively nutrient-rich water and can tolerate lower water quality than species such as trout or striped bass. Catfish farming is most intense in the southeastern U.S., where rainfall is relatively high and soils are heavy in clay and relatively impervious to leakage. Water replacement is infrequent, and evaporation seldom exceeds rainfall. Because catfish ponds are drained every three to ten years, demand on water resources is relatively minimal (Boyd and Tucker 1995). Catfish ponds require only about 6,500 m^3 water per MT of fish produced.

In contrast, trout raised in raceways can require 210,000 m^3/MT of clear, cool, oxygen-rich water that must be replaced continually (Phillips et al. 1991). Trout production in the U.S. is concentrated along the Snake River of Idaho, extracting water from the Eastern Snake River Plain Aquifer. Large farms can use as much as 64 million gallons of water per day. Trout aquaculture diverts approximately 1.5 million acre-feet of water annually, placing it in direct competition with other forms of agriculture, particularly during drought years. Water use by and effluent pollution from trout farms in the middle Snake River have prompted the EPA to declare the region "water quality limited." Expansion of the trout industry was prohibited until pollutant releases such as phosphorus were reduced 40% over five years (Goldburg and Triplett 1997). Not all salmonid culture is as profligate. Spring Creek National Fish Hatchery on the Columbia River produces 15 million Chinook salmon annually and recycles 90% of the water used (www.fws.gov/gorgefish/springcreek).

Feed production also affects ecosystem-level interactions. In the North Atlantic, small pelagic fishes such as capelin, sand eel, and Norway pout are turned into fish meal. Their depletion may have contributed to declining cod fisheries and depressed seal and seabird populations; similar impacts are suspected in the Peruvian upwelling system and in the North Pacific (figure 11.9).

Increased Predation on Wild Fish

Aquaculture operations can have indirect predatory effects. A variety of piscivorous birds (terns, cormorants, pelicans, gulls, egrets, herons, kingfishers, etc.) are commonly attracted to net pen and pond areas, particularly when salmonid smolts are released. Aggregations of such predators could affect wild fish moving through an area (Bayer 2000). A coho salmon ranching operation in the Yaquina River, Oregon, apparently led to declines in native coho stocks. Large numbers of predators attracted to abundant cultured and released fish also fed on wild fish (see Utter 1998). This type of indirect effect parallels one impact of salmon enhancement programs. Releasing large numbers of farmed fish produces a mixed-stock fishery and increases harvest pressure on wild populations. Legal harvest rates are established to capitalize on abundant hatchery fish, with rarer wild fish suffering collateral damage as bycatch in the fishery. Approximately half of the Pacific salmon stocks considered at risk suffer from excessive fishing pressure due to such mixed-stock fisheries (Nehlsen et al. 1991).

Aquaculture facilities can also be "attractive nuisances,"

creating conflict between the public, especially birders, and farm operators because aquaculture operations find it necessary to control unwanted predators. Reported *authorized* activities in the U.S. between 1989 and 1993 involved killing more than 51,000 predatory birds of 38 species, including over 12,000 herons, over 5,000 egrets, and over 200 pelicans; many authors feel the actual number is considerably larger (T. Williams 1994; Goldburg and Triplett 1997). Although seals and sea lions are protected under the U.S. Marine Mammal Protection Act of 1973, nuisance animals can be killed under a subsequent Marine Mammal Exemption Program. Canadian laws are similarly lenient. The number of seals killed around salmon operations in British Columbia was estimated at about 500 per year between 1990 and 1994 (Goldburg and Triplett 1997). From both an ethical and a PR standpoint, nonlethal predator control would seem advisable, although economics may overrule ethics when expensive nets and repulsion devices are the alternative.

Because many cultured fishes are large predators, their ecological impacts after release could be significant. In ecosystems where such species evolved, such as salmon-growing regions in the Northern Hemisphere, released fish should have relatively minimal impacts, assuming predator biomass is within natural limits. Evidence to the contrary does exist, however, as in the effects of released brook trout on wild Chinook juveniles in the Snake River (Levin et al. 2002). Of potentially substantial impact are truly alien introductions, such as the large, sea-run salmonids farmed in South America, New Zealand, and Australia. It would be interesting to know if coastal assemblages have suffered ecological disruption from escapees and released fishes.

Competition for World Markets

Aquacultured products compete successfully with wild-caught species on the world market because cultured fish can often be sold at lower prices (e.g., Zimmermann 1996). If wild stocks decline in value, commercial fishers are driven out of business, an unfortunate social cost of aquaculture practices. Large-scale aquaculture can increase exports and foreign exchange but again with social consequences. In the 1980s and 1990s, pressure from national governments, international development agencies, and the private sector caused Asian countries to convert from traditional aquaculture systems—largely involving herbivorous fish species consumed locally—to modern, large-scale production of exportable, carnivorous species such as shrimp (Muluk and Bailey 1996).

Ironically, declining profit margins and fewer commercial fishers allow conservation groups to purchase commercial fishing rights, which they never implement. Because of depressed fish prices, several North Atlantic conservation organizations were able to purchase fishing rights and then not fill the Greenland salmon fishery quotas for 1993 and 1994 (Speer et al. 1997). Conservation and sport fishing organizations bought out but did not use coastal netting stations around Britain, which may have contributed to a 30-year high of wild salmon returns in Scotland in 1995 (Gross 1998). Via such actions, a growing international aquaculture industry can reduce pressure on capture fisheries.

Pollution

Aquaculture entails rearing large numbers of fish in confined quarters, which means large amounts of feed, large amounts of waste, and subsequent disease problems requiring addition of chemicals. Organic wastes such as feces and excess feed accumulate in sediments, creating anaerobic conditions that produce toxic gases and reduce benthic diversity. Feed and feces also enrich the water column, stimulating growth of noxious algae and dinoflagellates, which are treated with herbicides (e.g., Heil et al. 2001). For every kilogram of salmon produced in Norway net pens in the late 1980s, about 1.1 kg of waste was released (0.6 kg unused feed, 0.3 kg feces, 0.1 kg slaughter waste, and 0.1 kg dead fish; Seymour and Bergheim 1991). The efficiency of feeding operations was subsequently improved, but two-thirds of the nutrients added through feed still "escaped" during intensive fish farm operations (Wu 1995; Kautsky et al. 2001). For every ton of salmon produced, 33 kg of N and 7 kg of P were released. Thus, 40 salmon produce as much organic sewage as one person (Naylor et al. 2001). Salmon farms along the British Columbia coast discharged as much organic sewage as 500,000 people (D. Ellis and Associates 1996). From a pollution perspective, aquaculture is the second-largest city in the province. The 150–175 million 3-kg Atlantic salmon held worldwide in aquaculture facilities annually produce around 400 million kg of organic pollutants, which are dumped into relatively still waters within 1 km of net pen operations (e.g., Wu 1995).

Crowding of fish promotes disease and parasites, which must be treated with medicines. Because rapid growth increases profits, chemical supplements (growth hormones, vitamins) may be added to feed as growth accelerators (e.g., McLean et al. 1997). Cage materials can also leach chemicals, and cages are frequently treated with antifouling agents and preservatives, which can be mobilized during periodic scrubbing of pens. The chemicals used in salmon farms for all these conditions include (1) food components, such as herring and shark oils, fish meal, blood, poultry, dyes, and antioxidants, but also organochlorine pesticides, PCBs, and other persistent chemicals; (2) medicinals, such as pesticides for parasites, disinfectants for viruses, and antibiotics for bacteria (including pyrethrins, cypremethrin, dichlorvos, azamethiphos added to water and ivermectin, emamectin benzoate, diflubenzuron, teflubenzuron, oxytetracycline, oxolinic acid, amoxycillin, sulfadimethoxine, erythromycin, romet 30, sulfamerizine, furazolidone, chloramphenicol and ormetoprim added to feed); and (3) construction materials, such as wood, plastics, paints, metal, antifoulants (often copper, zinc, diquat, malachite green, and chloramine T), and wood preservatives (Herwig et al. 1997; Haya et al. 2001).

Antifoulants, although applied to pen surfaces, find their way into the fish. Tin is used as an antifoulant in Philippine milkfish aquaculture. Tin concentrations of 0.3–0.7 mg/kg body weight were measured in fish 6–12 months after tin application had ceased (Kautsky et al. 2001). These concentrations are about 1,000 times greater than WHO acceptable intake levels of 0.0005 mg/kg of fish. Anything not incorporated into the fish or degraded in the water column can be exported to sediments or transported downstream by currents.

Pesticides applied in salmon operations can spread and contaminate surrounding water. They are toxic to invertebrates such as lobsters, an obvious concern in New England waters (Scholz 1999; Ernst et al. 2001). An organophosphate insecticide, azamethiphos, is commonly used for treating sea lice on salmon. Infested fish are isolated in a tarp and bathed in the pesticide solution for 30–60 minutes; the solution is then released into the environment. Several treatments per day are not unusual. Lobster die after 15 to 30 minutes of exposure to azamethiphos at 10% of the recommended bath concentration of 100 μg l^{-1}, with sublethal effects such as reproductive failure occurring at 5% to 10% doses; some effects are observed at 1% concentrations. Such impacts are not surprising given that sea lice and lobsters are both crustaceans. Ivermectin, another pesticide used to control sea lice, is added directly to salmon feed. Ivermectin-medicated feed is lethal to sand shrimp at concentrations below the recommended dosage (Haya et al. 2001), meaning that excess feed sinking below net pens is likely to affect shrimp and perhaps other benthic crustaceans attracted to the area (ironically, green crab, *Carcinus maenas*, a European invasive affecting substantial portions of the New England coastline, survives concentrations of azamethiphos of up to 500 μg l^{-1}). Azamethiphos and other pesticides were developed for use on terrestrial mammals and approved only for emergency use on fishes; "information on their effects on the marine ecosystem is extremely limited" (Haya et al. 2001, p. 494).

Substantial pollution can also occur when waste and chemicals are released from land-based aquaculture facilities such as flow-through raceways, or during the cleaning of holding ponds (Schwartz and Boyd 1994). Gilthead sea bream, *Sparus aurata*, are intensively cultured in saltwater ponds in coastal Spain. One such facility, located along the tidal saltwater San Pedro River, flowed into a protected area, the Natural Park of the Bay of Cadiz (Tovar et al. 2000). The facility produced 1,000 tons of fish/year, withdrawing and releasing 240,000 m^3 of water daily. Discharge from the ponds lowered the pH of receiving river water, probably because of the acidic nature of feces and fish food. Suspended solids and ammonium, nitrite, nitrate, and phosphate levels in river water were elevated where fish farms were located; ammonium concentrations exceeded 2.5 mg l^{-1}. Dissolved oxygen concentrations in affected river regions were as low as 2 mg l^{-1}, which is lethal for many fishes. Hence a 4-km section of river in a protected natural area received effluents from three fish farms, lowering water quality to substandard levels (Tovar et al. 2000).

Large, active fish farms also cause visual pollution, a political issue where property values are high, such as in Puget Sound, Washington, or coastal southern British Columbia. Property owners who invest in ocean-view real estate are unlikely to welcome fish farms in the foreground. Devaluation of property because of nearby sea ranch pens is difficult to document, but potential conflicts are of such importance that fish farm organizations incorporate aesthetic issues in their Best Management Practices (e.g., www.wfga.net).

Escape: Biological Pollution

"Escape" refers to the unintentional release of fishes into the wild from aquaculture operations; its impacts on native species and ecosystems constitute a major environmental concern. Many impacts are admittedly hypothetical but are grounded in the best available science.

Despite assurances to the contrary, almost all culture programs produce escapees. "In nearly every instance where an exotic aquatic species has been the subject of culture, escape into open waters has occurred. . . . Escape or release is likely, if not inevitable" (Courtenay and Williams 1992, pp. 52, 74–75). Probability of escape varies as a function of operation type and locale. Closed, recirculating, land-based systems holding fish at temperatures or salinities very different from ambient conditions have relatively low probabilities of escape. Open-water, net pen operations have the highest probabilities. Escapes occur even though the aquaculture industry's economic and PR interests suffer as a result. When a facility fails as an economic venture, which many do, it is easier and cheaper to allow remaining animals to escape than to dispose of them in a more responsible manner (Courtenay and Williams 1992).

Species with the longest records of culture are among those best known to escape. Common carp have escaped from ponds in Europe and Asia for two millenia, with serious genetic impacts on wild genotypes (chapter 9). Mozambique tilapia, *Sarotherodon mossambicus*, began escaping and affecting native species in Java in the 1930s (Bardach et al. 1972). Blue tilapia escaped from south Florida facilities and spread through the extensive canal system into the Everglades, where they became a major pest (Loftus 1989). Blue tilapia are also established in North Carolina, Oklahoma, Pennsylvania, Colorado, and Texas, among other states (Courtenay and Williams 1992).

Similar histories characterize other cultured species. In the U.S., ide (*Leuciscus idus*) and tench (*Tinca tinca*) were among the first species brought from Europe. They were grown in federal hatchery ponds in Washington, D.C.; both species escaped into the Potomac River when ponds overflowed during a flood (Fuller et al. 1999). Asian carps, especially black carp (*Mylopharyngodon piceus*), bighead carp (*Hypophthalmichthys nobilis*), and silver carp (*H. molitrix*), escaped from aquaculture facilities into the Mississippi and Missouri river systems; black carp have undergone exponential population growth, threatening imperiled mollusks.

Escaped Sea Ranch Salmon: Numbers

Most of the literature and concern over escapes has focused on salmon that are ranched at sea in large net pens anchored to the bottom. The most commonly ranched species is Atlantic salmon, with 150–175 million fish cultured annually in at least 14 countries, notably Chile, Australia, Canada, the U.S. (Maine, Washington), and northern Europe. The species' popularity results from its general hardiness, disease resistance, fast growth, reduced space requirements, good flavor, and, most important, domesticated nature. Norway is the largest producer of sea ranched Atlantic salmon, worth $1 billion annually. A large net pen operation can cover hectares and house millions of fish (Gross 1998).

Net pen–sea ranch facilities lose fish due to extreme weather, strong tidal currents, normal wear and tear, and the predatory activities of pinnipeds and river otters, not to mention vandalism. The largest documented release on the U.S. East Coast occurred during a December 2000 storm that struck Machias Bay, Maine, where 33 net pen operations housed 773 cages. Steel frames buckled and moorings broke, releasing 100,000–170,000 year-old fish. The loss cost one fish farm approximately $1 million. Ecological concern focused on three nearby rivers that contained federally listed runs of Atlantic salmon. Wild-type Atlantic salmon spawning in Maine waters are rare, with only 20–83 fish returning to Gulf of Maine rivers between 1995 and 1999. The total adult population may be only 500–1,000 fish. Escaped farm fish accounted for 22%–100% of the runs in various rivers (Colligan et al. 1999; ASF 2001; NRC 2004a).

In Norway, 700,000 Atlantic salmon reportedly escaped during a single storm in 1988, and an average of 1.3 million Atlantics escaped from farms annually during the early 1990s. In Scotland, depending on the source, between 353,500 and 745,000 fish escaped during as many as 31 incidents from 1997 to 2000, with 395,000 escaping during the first five months of 2000. Between 1987 and 1996, 255,540 Atlantics escaped from Washington and British Columbia fish farms, of which about 4% were subsequently captured by fishers. From a farm in Washington State approximately 100,000 Atlantics escaped in July 1996, 300,000 in 1997, and 100,000 in 1999. On average, 60,000 Atlantics escape annually from aquaculture facilities in British Columbia (Gausen and Moen 1991; McKinnell and Thomson 1997; Steelhead Society of B.C. 1999).

The July 1996 release in Puget Sound, Washington, was well documented and reliably researched (McKinnell and Thomson 1997). Strong currents during high tides destroyed seven of ten net cages near Cypress Island, releasing 101,000 fish, most of which weighed 2.7 to 3.7 kg. The fish dispersed rapidly and began to appear two days later in a test fishery for sockeye and pink salmon, 167 km to the north. In 1996, escaped Atlantics were reported from 29 different streams in British Columbia. Farther north in Alaska, the catch of Atlantics peaked in 1996 at 135 fish, more than the sum of all catches in all previous years. No Atlantic salmon were cultured in Alaska at that time, making escapees from Vancouver Island and Washington farms at least 500 km to the south the likely source (McKinnell and Thomson 1997).

Impacts of Escaped Salmon

How justified is concern about these escaped salmon? Many escapees are young and ill equipped for life in the wild, but their numbers ensure that some will survive to interact with native salmon and eventually spawn. Real and potential impacts include competition for food and spawning sites, disease transmission, and hybridization and other genetic consequences. Impacts are a function of relative numbers of wild versus farmed fish, and of genetic and ecological differences between farmed and wild species.

For Atlantic salmon in Europe and eastern North America, escapees often outnumber wild fish. The 15 farms in Loch Roag, Scotland, where 20,000 fish escaped in one incident in 1999, held about 5 million fish (potential escapees) per year; wild salmon returning to nearby rivers numbered only 300 in 1998 (Staniford 2002). Escapees in Scottish rivers may outnumber wild fish by as much as 4:1 (based on a 91,000 wild fish catch in 1998 and a reported 395,000 escapees in 2000; NASCO 2000). Farmed fish make up 80% of salmon runs in Norway and 90% of the catch in the Baltic Sea (NASCO 2000; figure 14.4).

Increasing numbers of escapees are spawning in the wild, and sexually mature, cultured fish dominate many runs. Between 20% and 55% of the eggs in salmon redds in the Magaguadavic River of New Brunswick, Canada, were from cultured fish (J. W. Carr et al. 1997). Genetic tests confirmed hybridization between wild and aquacultured fish, which is alarming given that Atlantic salmon in North America are endangered and salmon farmed in North America are from European genetic strains that were selected for different attributes than wild fish (Gross 1998; Colligan et al. 1999).

Cultured Atlantic salmon behave differently than wild fish, many differences being genetically determined.

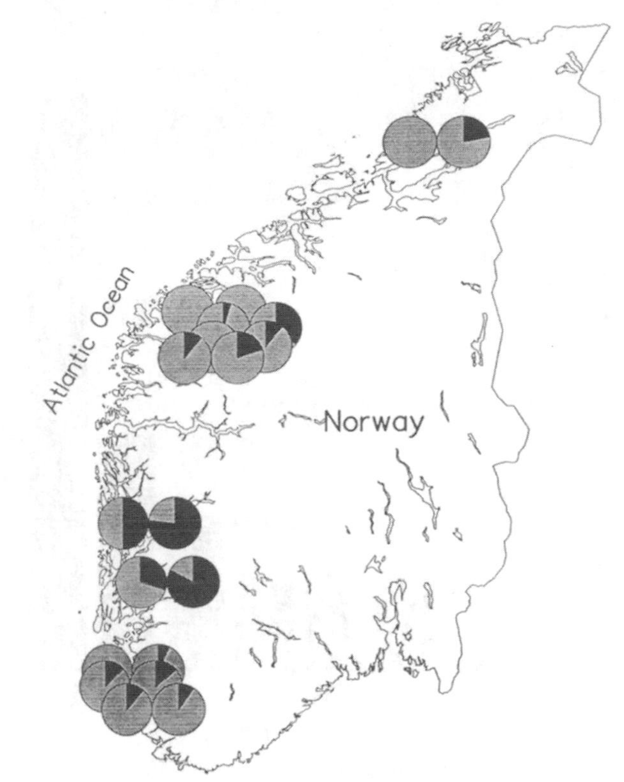

Figure 14.4. The proportion of escapees among Atlantic salmon spawning in Norway. Data are for ten rivers in southern Norway during the summers of 1987 (left circle of each pair) and 1988 (right circle). The dark portion of each circle represents the percent of net pen escapees among spawning fish. In eight rivers, the proportion of escapees increased from 1987 to 1988. Numbers and proportions of spawning escapees have continued to rise since these data were collected. From Hindar et al. (1991); used with permission.

Cultured females dig fewer nests and cover them poorly. Cultured males are generally poorer competitors at spawning sites, making cultured females likely mates for wild males, particularly where cultured females outnumber wild females. The offspring of cultured salmon are more risk prone than wild offspring, which lowers their survival rate and probably the survival rate of hybrid offspring (B. Jonsson 1997). Lifetime reproductive success of farmed fish is only 16% that of native salmon; hybridization would therefore lower the fitness of any offspring and any hybrid population, the progressive loss in fitness increasing with each wave of escapees (Crozier 2000; Fleming et al. 2000).

Even without hybridization, behavioral differences can threaten wild stocks. Escapees tend to arrive later in spawning rivers, perhaps lacking olfactory cues to guide them to appropriate sites, and consequently spawn later

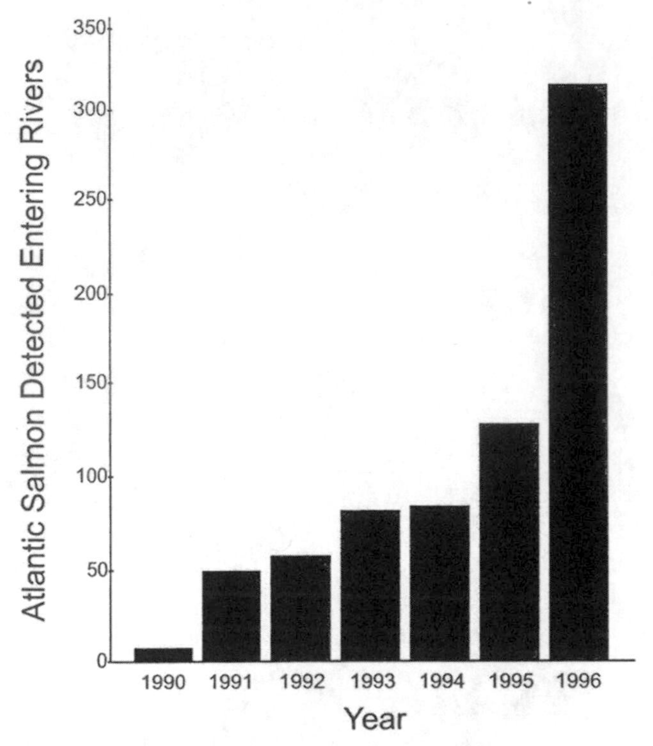

Figure 14.5. Escaped Atlantic salmon in rivers of British Columbia and Washington State. As numbers grow, potential negative interactions with imperiled native stocks of Pacific salmon become more likely. From Gross (1998); used with permission.

than wild fish but often at the same sites. Their courting and digging activities displace the eggs of wild fish that spawned before them (J. H. Webb et al. 1991).

Along the Pacific coast of North America, escaped Atlantic salmon are relatively rare, accounting for only 0.004% of the British Columbia salmon catch in 1996. However, they are increasing numerically, are known to have reached spawning condition, and could compete with native salmon for spawning sites (figure 14.5). Spawning Atlantics have been caught in the Elwha River of Washington (Fleischman 1997), and juvenile Atlantics that resulted from spawning between escapees were reported for the Tsitika River, British Columbia, by Volpe et al. (2000).

Intergeneric hybridization between Atlantic *Salmo* and Pacific *Oncorhynchus* species is unlikely (Noakes et al. 2000). Of more concern are ecological interactions and disease transmission, particularly affecting federally listed Pacific salmonids in Washington and Oregon, where millions of caged Atlantics and hundreds of thousands of escaped Atlantics could have an impact. Resident fish won encounters during behavioral interactions between Atlantic salmon and steelhead rainbow trout, *Oncorhynchus mykiss*, regardless of species. If Atlantics enter a stream and establish territories, they may exclude migrating native fishes from preferred breeding and feeding sites (Volpe et al. 2001). Because Pacific salmon spawn only once and die, a female that spawned with an Atlantic male would be wasting her entire lifetime output of eggs (Gross 1998). Such egg losses can only exacerbate the plight of imperiled stocks.

Atlantic salmon represent an exotic species with considerable invasive potential. As thousands of escapees are pumped into northwest ecosystems, it becomes increasingly likely that they will become established, the region being historically suitable to salmonids. As the number of exotics increases, so does the probability of interspecific interactions over food and spawning sites. This competition is bound to take a toll on native Pacific salmon, further depressing population size and increasing the likelihood of inbreeding depression, among the other perils experienced by small populations.

Parasites and Disease

High fish densities in aquaculture facilities promote the spread of pathogens (Hastein and Lindstad 1991). Cages placed in waters where wild fish reside or migrate become potential sources of disease transmission, as are effluent water and escapees. High densities of caged fish also increase the potential for pathogen transmission from wild to captive fish, creating a positive feedback loop of reinfection. Disease transmission interacts with genetic problems when genetically resistant stocks are transplanted into an area and introduce an opportunistic pathogen to which local stocks lack immunity.

Salmon Pathogens and Impacts

Among the many diseases that afflict salmon, most attention has focused on exotic organisms and diseases to which native fishes are unlikely to possess immunity or defenses. These include infectious salmon anemia (ISA), sea lice, furunculosis, the monogenean parasite *Gyrodactylus salaris*, and cold-water vibriosis (see Hastein and Lindstad 1991). ISA is particularly lethal. It was transferred from Europe to North America, where it appeared first in New Brunswick, Canada, fish farms in 1996 and then in Maine in 2001. ISA is viral, causes haemorrhagic kidney syndrome (HKS), is often fatal, and can be transmitted via sea lice and even fish carcasses (Lovely et al. 1999; Bouchard et al. 2001). In Maine, any infected fish and any fish they have come in contact with are supposed to be destroyed. Boats associated with infected pens are barred from leaving the area. Between March and September 2001, an ISA outbreak in Cobscook Bay, Maine, forced the destruction of more than 900,000 fish (Revkin 2001). If ISA spread to imperiled Maine Atlantics, results could be devastating; salmon populations are particularly vulnerable to new pathogenic exotics (McVicar 1997).

Sea lice are the ectoparasitic copepods *Lepeophtheirus salmonis* and *Caligus elongates*. They browse on skin and cause lesions that lower marketability; reduce growth; and can be fatal through ulceration, osmotic imbalance, stress, immune suppression, and secondary infections. According to McVicar (1997), they "are now considered to be the greatest threat to farmed salmon in all parts of the North Atlantic area" (p. 1099). Sea lice increase production costs by at least 9%, a significant figure in an industry with tight margins (A. Mustafa et al. 2001). Sea lice on pen fish can be transmitted to wild fish that swim near the pens (McVicar 1997), and lice lead to pollution because infestations are often treated with pesticides and other toxins. Furunculosis, now treated with a vaccine, was introduced into 20 Norwegian rivers in 1985 via Scottish hatchery fish, causing an epidemic that cost $85 million to eradicate (Hastein and Lindstad 1991; Fleischman 1997); it can be carried by water movement as far as 10 km from cage sites (McVicar 1997). *Gyrodactylus* appeared widely in western Norwegian rivers beginning around 1975, the likely vector infected fish stocked from hatcheries in the Baltic Sea. Baltic salmon stocks show resistance to the parasite (Bakke et al. 1990). The parasite first reduced the number of Norwegian parr, with runs of returning adults declining in later years. Salmon populations in at least 30 Norwegian rivers were decimated following introduction of the parasite (Johnsen and Jensen 1991; Scholz 1999).

Concern over pathogens in western North America varies. British Columbia has an eggs-only importation policy to reduce the likelihood of introducing exotic diseases and parasites. Pathogens found in salmon farmed there generally already exist in wild fish (Noakes et al. 2000). Of greater concern in British Columbia, and of general concern worldwide, is the overuse of chemicals to combat salmon pathogens, to which bacteria and other pathogens may develop resistance. These chemicals are added directly to the water and food or can enter the food chain when salmon excrete, escape, or are released (Noakes et al. 2000).

The overapplication issue appears to be lessening as the aquaculture industry gains greater understanding of effective dosages, although this may come too late to prevent some ecological problems. In Norway, antibiotic application went from 900 g per ton of salmon produced in 1987 to 6 g per ton in 1996. The difference is indicative of a poor level of understanding and a tendency to overmedicate during salmon ranching's early years. In the 1980s, concurrent with the high levels of antibiotic application, antibiotic-resistant strains of *Vibrio salmonicida* (the cause of cold-water vibriosis) and *Aeromonas salmonicida* (furunculosis) became widespread in Norway and elsewhere (Wu 1995; McVicar 1997), often in the sediments below net pens. Wild trout near trout farms also accumulated antibiotics, and antibiotic-resistant fecal bacteria increased after nearby trout farms were treated with antibiotics (Ervik et al. 1994).

Commonly used salmon antibiotics persist in the environment for over a year, increasing the opportunity for bacterial resistance (Herwig et al. 1997; Schmidt et al. 2000). The American Society for Microbiology (1995) identified the evolution of antibiotic-resistant bacteria as a major concern, citing as causes (1) the use in aquaculture of antibiotics that are also used to treat human infections, and (2) aquaculture application methods that allow wide dissemination of antibiotics into the open environment.

Vaccines are increasingly used for combating salmonid diseases, lessening the need for chemical additives in feed or water (Alderman and Hastings 1998; Noakes et al. 2000). Overuse of antibiotics still occurs in tropical and developing countries, which have limited access to vaccines and therefore turn to more widely available antibiotics (Alderman and Hastings 1998). Sea lice can be controlled by placing corkwing wrasse, *Crenilabrus melops*, in net pens; the wrasse serve as a cleaner fish, picking copepods off the bodies of infected salmon (Sayer et al. 1996). Unfortunately, wrasse are also a known host of furunculosis (Treasurer and Laidler 1994).

Diseases in Other Cultured Species

Pathogens have been transmitted from cultured to wild fish of many other species besides Atlantic salmon. Whirling disease, imported from Europe and spread from hatcheries to wild North American trout and back to Europe, has devastated North American trout stocks. Along equally convoluted and international pathways, aquaculture of anguillid eels has allowed interspecific transmission of several parasites, such as the nematode *Anguillicola crassus*. Shipments of eels from northeastern Asia to North America brought two other lethal pathogens native to the Japanese eel, the monogenean gill parasites *Pseudodactylogyrus bini* and *P. anguillae*. These have appeared in wild American eels in South Carolina, the Chesapeake Bay, and Nova Scotia (Hayward et al. 2001). Exotic parasites may contribute to observed declines among American eel stocks.

Tropical fish diseases transmitted via aquaculture include mycobacteriosis, a bacterially caused ailment characterized by splenic lesions (splenomegaly) and other visceral and integumental lesions. Mycobacteriosis was introduced into the Red Sea region around 1990, apparently from infected, cultured European sea bass, *Dicentrarchus labrax* (Diamant et al. 2000). Wild fishes are attracted to net pen operations because of the structure and the continuous supply of food. Small wild fishes pass easily through the mesh, moving between the cage and the surrounding area. Mycobacteriosis thus spread from cultured serranids and sparids to wild rabbitfish, *Siganus rivulatus*. Infection rates among wild rabbitfish were 21% to 66% as far as 10 km from the net cages. Nine other reef species living near the cages were also infected. The causative agent of mycobacteriosis, *Mycobacterium marinum*, can be transmitted to humans, resulting in subcutaneous and other lesions.

Comparatively few pathogens have spread to fishes from the extensive, low-density, pond culturing of carps, tilapia, and milkfish (Kautsky et al. 2001). However, human pathogens can be transmitted during the handling of such freshwater species. People handling raw, farmed tilapia in Canada have become infected with the fish pathogen *Streptococcus iniae* (Weinstein et al. 1997). *S. iniae* infection rates in farmed tilapia in the U.S. can range as high as 27% (Shoemaker et al. 2001). Developing countries with less technological capacity could likely have higher rates. Wounds received while handling the St. Peter's fish, *Tilapia zilli*, frequently lead to serious infections caused by *Vibrio vulnificus* and can necessitate amputation (Said et al. 1998). At least 12 bacterial species found in fishes—including strains of *Salmonella*, *Vibrio*, *Campylobacter*, *Aeromonas*, *Edwardsiella*, *Pseudomonas*, *Erysipelothrix*, *Mycobacterium*, and *Leptospira*—have been shown to cause food poisoning, diarrhea, wound infection, and leptospirosis in humans (P. Smith et al. 1994; Lehane and Rawlin 2000). Treatment of these bacterial ailments is becoming increasingly difficult because of growing antibiotic resistance in pathogens (Durborow 1999; Lehane and Rawlin 2000).

GENETIC CONSEQUENCES OF AQUACULTURE

In the long term, which means evolutionarily, a major concern about culture facilities and hatcheries involves their inability to manage genotypes or the selection environment in a way that approximates nature (figure 14.6; Allendorf and Ryman 1987; Meffe 1992; Utter 1998). Pollution, competition, and disease are potentially curable. Genetic alterations, particularly hybridization and subsequent introgression, are essentially irreversible and likely to result in extinction of wild genotypes (e.g., Leary et al. 1995). In cases where genetic effects have been documented, "they always appear to be negative in comparison with the unaffected native populations" (Hindar et al. 1991, p. 945). The literature focuses on salmonids because they are best known, but the evidence is applicable to any farmed species.

Indirect Genetic Effects

Negative reproductive interactions don't necessarily require that fish breed successfully. Wild populations can be depressed by *nonintrogressive hybridization*, which means that spawning between wild and hatchery fish produce nonviable or incompetent offspring. Brown trout in Europe cross increasingly with Atlantic salmon as more Atlantics escape from net pens. Resulting offspring have low viability and present a significant threat to dwindling brown trout populations in many northern European regions (Galbreath and Thorgaard 1995).

Direct Effects

Lost Genetic Variation Due to Outbreeding Depression

Atlantic salmon farmed throughout Europe and exported elsewhere come from a small number of hatchery strains. Also, farmed fish are genetically less variable than wild fish (McGinnity et al. 1997), perhaps because fish farms use standardized growing conditions and practices, creating similar selection pressures. Because aquaculturists started with a

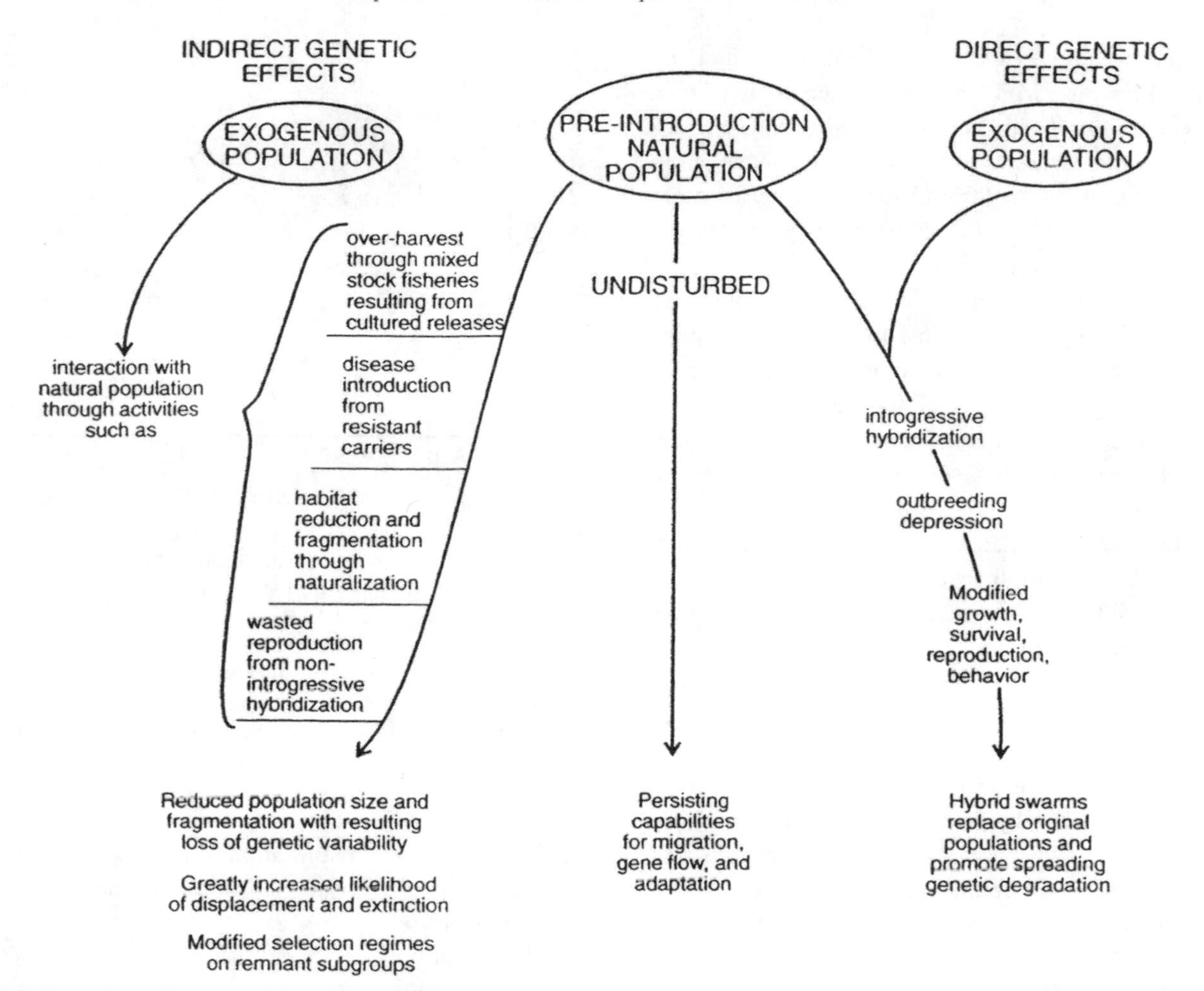

Figure 14.6. Indirect and direct genetic effects of farming, releasing, and escape of cultured fish on natural populations. Hatchery fish can cause genetic changes, termed *indirect genetic effects*, whether or not they breed with wild fish. *Direct effects* occur from introgressive hybridization. From Utter, F. 1998. Genetic problems of hatchery-reared progeny released into the wild, and how to deal with them. *Bulletin of Marine Sciences* 62:623–640; used with permission.

limited sample of the salmon gene pool and further reduced variation via artificial selection, genotypes of cultured fish appear to be converging toward a homogenized, relatively invariant genome best suited for culture conditions.

Before 1960, few hatchery-reared salmon of any species survived to spawn (Lichatowich 1999). Improved handling and rearing methods sensitive to the biology and natural history of different species changed that. Hatcheries are now more successful at producing young fish that survive to the adult stage. When large numbers of hatchery fish breed with wild strains, *outbreeding depression*—the overwhelming of the wild genotype with foreign alleles—can result. Offspring from matings between wild and farmed fish contain a mishmash of genes but a limited subset of the possible genetic variation of a species. Repeated breeding episodes move the resulting offspring further and further from wild, locally adapted genotypes. As genetic diversity is lost and genetic differences among wild populations are homogenized, the capacity for wild fish to adapt to changing environments declines (Meffe and Carroll 1997; Gross 1998). This irreversibility occurs in no small part because, once genetic variability is lost, evolution has less material with which to work (see chapter 11).

A simple model of the rate at which introgression of hatchery genes can occur predicts that a stocking rate of 10% can cause a halving of wild-type genes in the wild population after only eight generations (see Masuda and Tsukamoto 1998). Such rapid introgression is particularly troublesome because escaped, farmed Atlantic salmon far outnumber returning wild spawners in many locales. Stock supplementation of Pacific salmon has led to similar problems. In northern Japan, where two billion juvenile chum salmon, *O. keta*, are released from hatcheries annually, more than 90% of captured fish are from hatchery releases (Masuda and Tsukamoto 1998).

Two fundamental concepts in conservation biology are effective population size (EPS) and minimum viable population (MVP). EPS is the actual number of breeding individuals in the population. MVP is the population size below which loss of genetic variation becomes highly probable due to inbreeding, genetic drift, and mutation rate (Meffe and Carroll 1997). As EPS decreases, the many negative outcomes associated with small population size, especially increased probability of extinction, increase. MVP values differ in theory and reality, but somewhere between 50 and 500, or perhaps 5,000, individuals is considered a threshold (the so-called 50–500 rule). Many wild salmon runs in both the Pacific and Atlantic are below 50 and 500, and considerably more are below 5,000 individuals. Problems of genetic variation loss due to small numbers are compounded by introgression with escapees or stocked fish. As wild populations decrease in size and escapees become more abundant, the impact of all these negative factors increases (see McElhany et al. 2000).

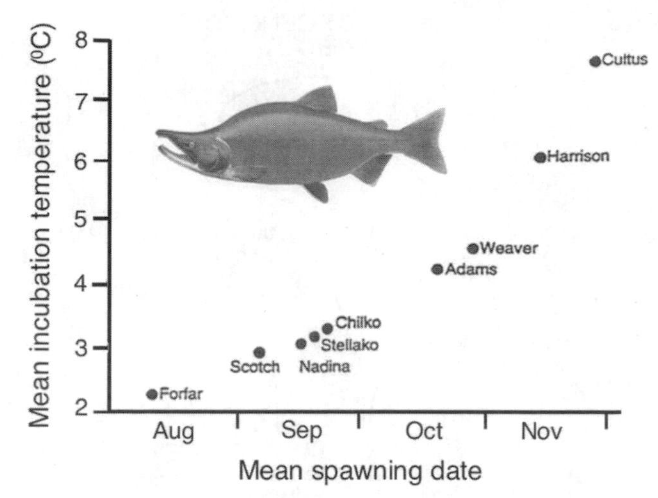

Figure 14.7. Genetically based, local adaptation in Fraser River sockeye salmon. Among stocks (named dots), fish spawn on different dates, and eggs are incubated at different mean temperatures. These differences lead to emergence dates favorable to juvenile feeding. Different spawning dates also help to coordinate migrations among smolts originating at different distances from the sea. From Brannon (1987); sockeye drawing inset by Kate Spencer, www.katespencer.com, in Augerot et al. (2005); both used with permission.

Genetic Distinctness and Local Adaptations

Local salmonid populations are often morphologically and genetically distinct from one another, which generally (but not universally) reflects adaptation to local conditions (figure 14.7; Ricker 1972; Hindar et al. 1991; E. B. Taylor 1991; see Philippart 1995 on European salmonids, cyprinids, and cyprinodontids; Meffe and Snelson 1989 on poeciliids; M. A. Bell and Foster 1994 on gasterosteids; Conover 1998 on atherinids). Hatchery fish differ genetically from wild fish because brood fish are often brought to hatcheries from other regions. Also, genomes change over time in hatcheries due to low numbers of brood fish (founder effects, genetic drift) and the selective regime that characterizes hatcheries (e.g., Philippart 1995; Flagg et al. 2000). Hatchery steelhead trout planted throughout Columbia River drainages until at least the mid-1980s originated almost exclusively from the Skamania Hatchery, 45 km below Bonneville Dam, the closest dam to the ocean (Chilcote et al. 1986). Managers are increasing their use of native broodstock in hatchery operations, although "translocation of hatchery fish outside their natal systems . . . is still part of management practice" (Brannon et al. 2004, p. 25).

The genetics of origin and domestication are reflected in behavior and fitness. Among steelhead trout returning to spawn in the Kalama River, Washington, hatchery fish were only 28% as successful as wild fish in producing smolts (Chilcote et al. 1986). Lower reproductive success in hatchery fish was apparently the result of 30 years of artificial selection for earlier spawning, which subjected eggs to washout because of the timing of snowmelt, flooding, and streambed scouring; wild fish usually spawned after these events. However, 62% of the next generation of smolts were the offspring of former hatchery fish because they outnumbered wild fish 4.5 to 1.

McGinnity et al. (1997) crossed an established lineage of farmed Atlantic salmon with wild fish and looked at the characteristics of pure and hybrid parr and smolts reared under natural conditions in Ireland. They found that hatchery fish grew faster and were more aggressive than wild-type fish, which would be expected given the traits favored by hatchery conditions. Survival of hatchery fish to the smolt stage, however, was lower by almost 50%. Hybrid fish were intermediate in many characteristics, again attesting to the heritable nature of traits. These findings suggest that hatchery and hybrid fish will displace wild juveniles, and other data indicate generally lower survival at sea and lower return rates of spawning hatchery and hybrid fish. All these findings point to reduced viability of salmon runs, particularly as repeated cross-breeding and introgression further replace wild genes with hatchery genes in a population, "potentially resulting in an extinction vortex" (McGinnity et al. 1997, p. 1006).

The literature on genetic differences indicates that cultured Atlantic salmon showed genetically based increases in growth rate, weight, disease resistance, juvenile aggression, and tameness, and decreases in stress response, genetic diversity, predator evasion, and survival in the wild (Gross 1998). They also mature later, produce more and smaller eggs, and have poorer stamina. Cultured fish differ enough in genetically based traits that Gross (1998) proposed they be recognized as a different, domesticated animal and renamed *Salmo domesticus* (figure 14.8).

The Selective Regime That Produces Hatchery Fish

Hatcheries produce a hatchery phenotype and genotype. Hatchery stocks are more similar to each other in morphology and life history than to wild stocks (Reisenbichler and Rubin 1999), reflecting the uniformity that exists in hatchery operations. The strength of selection in hatcheries is evident from genetically based differences in hatchery fish that have appeared after as little as one generation in the hatchery (Waples 1999).

Genetic change is not necessarily detrimental, but the changes shown in hatchery strains represent adaptations to the hatchery environment and may be liabilities in the wild (e.g., Utter 1998; Waples 1999). Hybrids between hatchery and wild fish can thus be expected to perform at intermediate (i.e., poorer) levels under wild conditions than pure wild fish, and data confirm these expectations (Reisenbichler and Rubin 1999). Hatchery fish typically

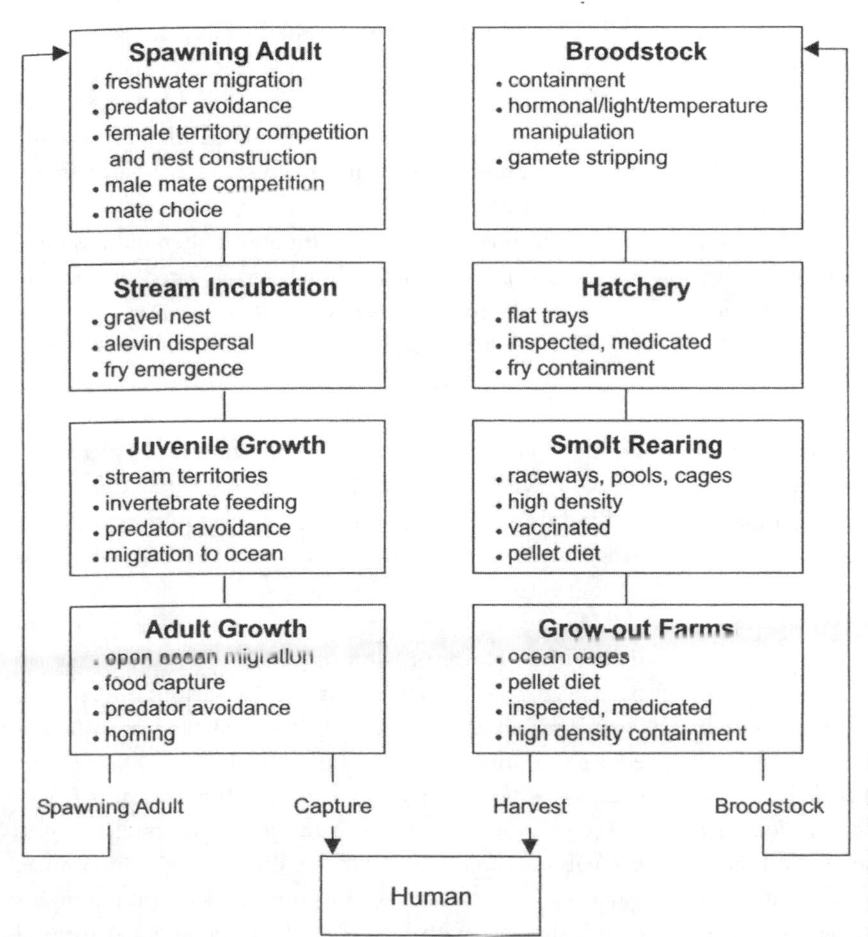

Figure 14.8. Environmental factors and selective pressures encountered by wild versus hatchery-reared Atlantic salmon. The resulting organisms can be said to occupy different niches and can be regarded functionally and ecologically as different species. From Gross (1998); used with permission.

grow and develop quickly as a result of higher temperatures and greater food availability, and perhaps active selection for faster growth by hatchery personnel (Chilcote et al. 1986). Additional growth selection occurs during downstream migration by smolts, where survival during passage through dams and turbines and past predators is size dependent. Fish growing faster than average were 30 times more likely to survive dam passage than fish growing slower than average (Reisenbichler and Rubin 1999). Relatively large size confers a general fitness advantage in all fishes but particularly in younger life history stages (see Helfman et al. 1997). Dams and their bypass structures, among their other negative impacts on fishes (chapter 6), select for hatchery fish.

The conditions that minimize mortality in the hatchery therefore combine with characteristics that reduce mortality during seaward migration to produce large year-classes of hatchery fish entering the sea. This artificial abundance justifies sustained fishing pressure, which further eliminates wild fish and their genes from the population (e.g., Flagg et al. 1995). Abundant adult hatchery fish eventually return to rivers and spawn with wild fish. Advantages conferred by hatchery selection can be later liabilities among the offspring of these fish. Rapid growth demands high food availability, high aggression leads to more fights and injuries, and poor predator avoidance is a direct source of mortality. Productivity of the wild-hybrid population declines (Reisenbichler and Rubin 1999). Low fitness in now-wild fish further reduces productivity each year until the "wild" stock cannot exist without supplementation from hatcheries. This continually subsidized system provides catchable salmon in the short term, but the stock is destined to crash once supplementation ceases and may crash even with supplementation if a new pathogen appears to which cultured stocks lack resistance (Hindar et al. 1991).

The Overall Impact of Hatcheries on Wild Salmon

Reisenbichler and Rubin (1999), after reviewing studies on the fitness consequences of artificial propagation in Atlantic salmon, brown trout, chum salmon, coho salmon, steelhead salmon, and Chinook salmon, concluded that supplementation programs degraded wild strains and "artificial propagation poses a genetic threat to conservation of naturally spawning populations" because "fitness [in terms of] natural spawning and rearing can be rapidly and substantially reduced by artificial propagation" (p. 459). Most supplementation programs target declining natural populations, with the expressed purpose of producing hatchery fish that will interbreed with wild fish. Supplementation hatcheries are in essence net pens that build escape into their design. In the context of declining wild salmon populations worldwide, aquaculture is a potential third strike against imperiled stocks, compounding the problems caused by overexploitation and habitat destruction (Gross 1998). Hatcheries accelerate extinction.

AQUACULTURE, STOCK ENHANCEMENT, AND SUPPLEMENTATION IN MARINE FISHES

Marine aquaculture most often entails stock enhancement and supplementation. "The appeal of stock enhancement rests in its simple premise and its bold promise" (Travis et al. 1998, p. 305). The premise is that stock sizes of depleted marine species can be increased by rearing and releasing large numbers of larvae or juveniles, thus reducing the over 99% mortality associated with early life history. The promise is that our activities will compensate for anthropogenic sources of mortality (overfishing, habitat destruction) that have caused depletion in the first place, which "appeals to our reluctance to impose harsher and less popular conservation measures" such as regulated fishing and habitat restoration (Travis et al. 1998, p. 305).

Among marine species frequently grown for enhancement purposes are striped mullet, *Mugil cephalus*, and Pacific threadfin, *Polydactylus sexfilis*, in Hawaii; red sea bream, *Pagrus major*, in Japan; Atlantic cod, *Gadus morhua*, in Norway; white sea bass, *Atractoscion nobilis*, in California; and red drum, *Sciaenops ocellatus*, in the southern U.S. Are issues surrounding the culture of salmon applicable to these species?

The answer at present suggests less overall impact from these fishes. However, although marine hatchery releases have occurred for over a century, few serious efforts have been directed at assessing either impact or success (Leber et al. 1995). In fact, "in the United States and other countries, stocking to augment depleted [marine] fish stocks was commonly practiced without demonstrable success from about 1880–1950" (Grimes 1995, p. 593). In essence, "success" in stocking programs was predicated on numbers of fish stocked, with little regard to documented survival (e.g., W. J. Richards and Edwards 1986; Wahl et al. 1995). "Any increase in stock abundance . . . was taken as evidence that the stocking was working. Failure of stocks to increase was taken as evidence of the need for even more enhancement" (Hilborn 1998, p. 662). Assessment was hampered by problems associated with marking large numbers of small, fragile fishes. Improvements in microtags and development of genetic markers have overcome most such difficulties. However, massive, repeated hatchery releases without assessment are a strong indictment of past hatchery practices: Fisheries personnel and the public accepted that hatcheries were beneficial without knowing whether they caused damage to natural ecosystems or even if they worked at all.

Recent assessments have produced conclusions that remain open to interpretation. Leber et al. (1995) tested for displacement of wild striped mullet by hatchery fish in Hawaii. They found no evidence of displacement or greater mortality among wild fish and concluded that hatchery enhancement was a viable approach to replenishing overfished mullet resources, once carrying capacity was known. Carrying capacity is notoriously difficult to measure, however, particularly if one defines it as numbers of a single species. Given (1) the natural population variation and large geographic ranges of marine populations, and (2) considerable overlap in food and habitat resource requirements of many nearshore fishes, any estimates of carrying capacity would have to be surrounded by uncertainty (e.g., Grimes 1995). Also left untested were such issues as genetic effects resulting from inbreeding or domestication in the hatchery, followed by disease transmission and outbreeding depression in the wild (e.g., Blankenship and Leber 1995).

Genetic problems have occurred in marine aquaculture programs when basic ecological rules of hatchery genetics have been violated. Small numbers of brood fish can be expected to result in inbreeding depression and lowered overall heterozygosity in offspring, as found in Japanese aquaculture of red sea bream. When only 8 to 26 parents were used as brood stock (in violation of the 50–500 rule), heterozygosity in offspring was reduced 18%–23% compared with wild stocks. After release, these cultured fish interbred with wild fish, leading to altered allele frequencies in the offspring (Utter 1998).

Unlike that from salmonids, evidence from marine species of adaptive, genetic distinctness in local populations is weak. Marine fishes are apparently "population poor" (Conover 1998), meaning that distinct populations are uncommon. General surveys of total gene diversity, a measure of genetic uniqueness, produce low values in marine fishes: 2% in marine fishes, 7% in anadromous fishes, and 30% in freshwater fishes (Gyllensten 1985). The average degree of genetic subpopulation differentiation is 0.22 in freshwater fishes and 0.06 in marine fishes, anadromous species again showing intermediate levels (R. D. Ward et al. 1994). A relative lack of genetic distinctness among marine fishes probably reflects the more continuous nature of marine populations, resulting from large effective population sizes, considerable gene flow, and high dispersal, particularly of larvae. Exceptions can be explained by unique features of life history, especially a lack of planktonically dispersed larvae.

For example, the atherinid *Craterocephalus capreoli*; the plotosid catfish, *Cnidoglanis macrocephalus*; and the damselfish, *Acanthochromis polycacanthus*, show considerable subpopulation differentiation over distances of only a few kilometers (Shaklee and Bentzen 1998). Each has no or a relatively short planktonic phase. *Acanthochromis* is the only member of its family that lacks a planktonic larval stage. Although localized differences in adaptive traits have been shown for a few nearshore marine species, such as silversides, striped bass, and killifish (see Conover 1998), the general pattern is one of relatively low population distinctiveness, as long as comparisons are between areas without significant geological or ocean current barriers.

One potential genetic concern arises from the high fecundity of marine fishes, which often produce millions of eggs per individual (Utter 1998). High fecundity, high survival in hatcheries, and small numbers of brood fish mean that any genetic deficiencies in hatchery stocks are likely to insert themselves into the genome of wild fishes, particularly where native stocks are in decline and the effective population size is low.

GMOS: FRANKENFISH, CODZILLA, PHARM ANIMALS, OR PROTEIN FOR THE MASSES?

The controversy over genetically engineered fish will not be solved here. This book focuses on the application of science to environmental issues related to fish (and ultimately to humans); treatment of genetic engineering will be similarly restricted. The view of genetically modified organisms, or GMOs, subscribed to here parallels that of FAO Director-General J. Diouf, who in Wijkstrom et al. (2000) stated, "We have no problem with . . . GMOs as long as they are proved to be safe to human beings and have no negative impact on the environment."

Genetically engineered, genetically modified, recombinant, or transgenic fish fall under the general class of GMOs. The biotechnology of producing GMOs usually involves microinjection of a gene and a switch that controls expression of the gene into newly fertilized eggs or into sperm of the species to be modified. The switch is a so-called fusion gene or promoter that may contain retrovirus sequences that direct the integration process (Fyt 1997; Wijkstrom et al. 2000). More than two dozen species of transgenic fish have been produced, and the number grows annually. Engineered species with aquaculture potential include common and Crucian carp; African catfish (*Clarias gariepinus*) and channel catfish; mud loach (*Misgurnus mizolepis*); Arctic char (*Salvelinus alpinus*); cutthroat trout (*Oncorhynchus clarki*) and rainbow trout; Atlantic salmon, Chinook salmon, and coho salmon (*O. kisutch*); striped bass; tilapia (*Oreochromis niloticus*, *O. hornorum*); gilthead bream; and red sea bream (for information on genetically modified aquarium fish, see www.ornamentalfish.org/aquanautstatement/gmfish.php).

Figure 14.9. Transgenic and nontransgenic Atlantic salmon. Transgenic lines show accelerated growth. The large fish is a second-generation transgenic that contains a salmon growth hormone gene and an antifreeze protein promoter. The smaller siblings are nontransgenic controls of the same age. From Hew and Fletcher 1997; used with permission.

Who Benefits, Who Will Be Hurt?

Two opposing attitudes about GMOs exist among interested parties (Wijkstrom et al. 2000). The first, which is more precautionary, views transgenic technologies as posing unknown risks that need to be carefully monitored and regulated to ensure protection of both the environment and human health. The alternative attitude views genetic engineering as little different from other technologies involving genetic improvement or domestication, and GMOs as highly domesticated and therefore unlikely to survive in the wild if they escape; the process, therefore, needs little additional testing or oversight.

The major appeal of genetic engineering to aquaculture practitioners is its potential to improve fish growth characteristics and feeding efficiency, environmental tolerance, and disease resistance. Demonstrated, intended effects include faster growth to marketable size, larger maximum size, higher feeding efficiency, and reduced feed intake while attaining marketable size (figure 14.9). Environmental tolerance has focused on temperature effects, including improved cold tolerance via splicing of antifreeze protein genes into species that lack them (e.g., from winter flounder, *Pleuronectes americanus*, to Atlantic salmon). Such cold tolerance would extend the growing season of a cultured species by producing growth hormone year-round, as well as allowing fish to be grown in otherwise lethally cold conditions (Hew and Fletcher 1997; Fletcher et al. 1999). Disease resistance involves identifying genetic factors that promote immunity or resistance and transferring them to cultured species (Hew et al. 1995; Hew and Fletcher 1992, 1997).

Actual changes in genetically modified fishes can be quite dramatic. Transgenic tilapia, *Oreochromis hornorum urolepis*, displayed 60%–80% faster growth, consumed 3.6 times less food, and converted feed 2.9 times more efficiently than wild-type fish (de la Fuente et al. 1999; Martinez et al. 2000). Such results could cut by two-thirds the six-month period required to produce marketable 250-g fish. Incidental benefits claimed include better taste and lower plasma cholesterol concentrations (de la Fuente et al. 1999). Chinook salmon with growth hormone and antifreeze protein genes showed 10- to 30-fold increases in growth rate; common carp with salmon and human growth hormone genes grew faster, had increased disease resistance, and showed greater tolerance to low oxygen concentrations (Hew and Fletcher 1997; Wijkstrom et al. 2000). Mud loach spliced with their own growth hormone gene and a mud loach promoter showed "extraordinary gigantism," achieving 35-fold increases in growth rate and body masses four times greater than observed in wild fish, reaching marketable size in one-sixth the time of nontransgenics (Nam et al. 2001).

Environmental concerns primarily focus on ecological

interactions between escaped or released GM fish and natives. Will faster growing transgenic fish outcompete native conspecifics and contraspecifics for food and habitat? Will larger transgenics prey on smaller natives, particularly where predatory species are concerned (e.g., very large salmon)? Because transgenics are deliberately bred for wider environmental tolerances (e.g., antifreeze proteins, disease resistance), will they be able to invade new habitats and displace local populations? If transgenics interbreed with natives, will transgenes be transferred to nonengineered organisms, disrupting evolved gene complexes and lowering genetic diversity? Concerns basically pertain to escape issues, with the added—to some people, sinister—complication that the new genetic combinations have been produced in test tubes rather than emerging in the wild as a result of natural selection.

The evidence supporting or refuting claims and concerns is limited and mixed. Although some transgenic lines show no difference from nontransgenic individuals, anatomical and behavioral abnormalities are not uncommon:

- Transgenic coho salmon showed increased somatic muscle hyperplasia; had 2.2 times more intestinal surface area; exhibited cranial deformities, opercular overgrowth, and distorted body shapes; and swam significantly more slowly than nontransgenics (Farrell et al. 1997; Ostenfeld et al. 1998; Hill et al. 2000; E. D. Stevens and Devlin 2000b). Deformed individuals had lower viability (Devlin et al. 1995).
- Transgenic Atlantic salmon had about 25% more gill area than nontransgenics (E. D. Stevens and Devlin 2000a).
- Transgenic tilapia, *O. hornorum*, had lower feeding motivation and lower dominance status than control wild-type tilapia (Guillen et al. 1999). No behavioral differences were observed between transgenic carp that contained a trout growth hormone gene and nontransgenic control carp (Dunham 1996 in Hew and Fletcher 1997). However, transgenic coho salmon showed higher feeding motivation than nontransgenics, consuming 2.9 times more food during a feeding trial, which could lead to resource depletion and food competition if large numbers of transgenics were released or escaped (Devlin et al. 1999; see Levin et al. 2001).
- Accelerated growth in transgenic fish usually results from excess growth hormone production by the fish. Behavioral evidence suggests that growth rates and feeding behavior of wild fish are, in fact, adaptations to natural conditions, and that changes induced by genetic engineering are nonadaptive. Transgenic Atlantic salmon spent significantly more time feeding in the presence of predators, whereas control fish avoided feeding when predators were present, suggesting that the normal growth rates of wild fish have been optimized to reduce predation risk (Abrahams and Sutterlin 1999). Similarly, transgenic rainbow trout exposed to attacking predators foraged closer to the water surface and resumed feeding more quickly after predatory attacks than nontransgenics. Predation apparently selects against high production of growth hormone in wild fish (E. Jonsson et al. 1996). Similar results were found with growth hormone–transgenic channel catfish, which were poorer at predator avoidance than normal fish (R. A. Dunham et al. 1999).

The major ecological impact in all these examples would arise from release or escape of large numbers of transgenics. Aquacultured fishes escape in large numbers, breed with wild fish, and pass their genes along. Little reassurance is given by statements such as "Any ecological effect [of escapes] would be unlikely because the increased susceptibility of transgenic [fish] to predators would most likely decrease or eliminate the transgenic genotype" (R. A. Dunham et al. 1999, p. 545).

Concerns related to human health and other impacts run from clinically testable pathologies to ethical questions about manipulating genomes ("playing God") and beyond (see Newton and Dillingham 2002). Public concern over GMOs in general has focused on modified plants that produce pesticides, herbicides, and antibiotics that may be toxic to humans, cause allergic reactions, or promote antibiotic resistance in bacteria. Fish transgenics have not yet involved toxin and antibiotic production, which should minimize such concerns. Theoretically, a disease-resistant transgenic fish could harbor pathogens without succumbing to them and therefore become a vector for fish and human diseases. Alternatively, wild populations could lose immunity after mating with introduced transgenics that lacked genetic protection against locally evolved pathogens. No documented instances of either have been found.

Public health concerns are specifically addressed in recommended protocols for approval of transgenics for human consumption (Scientists' Working Group on Biosafety 1998; FAO 2001). Clinical tests in this arena have also occurred. Before incorporating transgenic tilapia into aquaculture programs in Cuba, researchers administered tilapia growth hormone to rhesus monkeys and found that it had no biological activity (Guillen et al. 1999). Researchers also conducted a food safety test on 22 volunteers, who ate transgenic and nontransgenic fish twice daily over a five-day period. Blood samples after five days revealed no differences in several blood components between test groups, including antibody production against tilapia growth hormone (de la Fuente et al. 1999). These tests satisfied national safety requirements, and the fish are now produced and consumed in Cuba with apparent enthusiasm (Intrafish 2001). Whether this sample size and length of study are sufficient to satisfy concerns in other countries remains to be seen.

A problem arises when attempting to objectively assess characteristics and impacts of transgenic fishes. A significant proportion of the literature showing no or little impact of genetic modification is the work of a small number of authors employed or funded by biotechnology firms. Their papers appear in symposium proceedings; edited volumes; trade journals; and other secondary, gray, captive, or advocacy publications. Anyone familiar with publishing standards in science will attest to the greater rigor brought about by the critical review process, which is often lacking in the secondary literature. The intimate economic link between aquaculture and transgenics is probable cause for skepticism when reading non-peer-reviewed publications.

Surprising Results and Unintended Consequences

Early in the development of transgenic methodologies for fishes, workers did not know whether fishes with gene constructs for faster growth would need more or less feed (e.g., Kapuscinski and Hallerman 1990b). The now oft-repeated result of improved food conversion efficiency must have come as a pleasant surprise. Similarly, the researchers who spliced tilapia growth hormone genes into *O. hornorum* were unlikely to have suspected that their fish would contain lower plasma cholesterol levels or that volunteer human subjects being tested for health effects would find that the GM fish tasted better (de la Fuente et al. 1999). Genetic engineering is a complicated technology performed on complicated biological systems; results will produce complications. It is reasonable and precautionary to assume that "traits other than those targeted by gene transfer are likely to be affected" (Kapuscinski and Hallerman 1990a, p. 2). It is the unanticipated and unintended consequences that fuel the fires of concern and public discourse and justify a precautionary approach to transgenics (e.g., box 14.1). Such concerns affect purchasing decisions, which will ultimately determine economic success in a free-market environment.

Solutions to Genetic Engineering Problems

Workers developing GM fishes are demonstrably aware of real, potential, and perceived environmental and health risks. Published papers on the topic commonly include admissions of concern and recommendations for proceeding cautiously in applying the technology (e.g., R. A. Dunham 1999; Perez et al. 2001). Two frequently proposed precautions are to restrict the culture of transgenics to land-based, closed-circulation (but expensive) operations (figure 14.10), and to limit production to sterile individuals. Such operations may only be economically feasible if genetic engineering is employed to boost growth and feed conversion efficiency (C. M. Cunningham, pers. comm.). Sterility can be achieved through established procedures (heat or pressure shock to produce monosex lines, chemical immersion such as androgen treatment of juvenile fish, induced triploidy; see Johnstone 1996) or even by transgenic manipulations involving insertion of a "loss of function" (e.g., "terminator") gene. Monosex and sterile lines are the result, often with over 99% efficiency; the variance around that number, however, is cause for concern and debate. Also, sterility is often accompanied by unanticipated alterations in behavior and physiology, and sterile fishes still have potential as predators and competitors (Kapuscinski and Hallerman 1990a).

Other recommendations include avoiding genetic complications via a general proscription against cultivating transgenics in the vicinity of conspecifics or congenerics, which underlies the controversy over GM Atlantic salmon. More specific is the suggestion that manipulations of temperature and salinity tolerance be abandoned because escaping transgenics would have considerable invasive potential.

Approaches that would boost consumer confidence include avoiding incorporation of genes with potential human health risks. This seems obvious, but precedents exist to indicate otherwise, including a Brazil nut allergen

BOX 14.1. GM Salmon and the Environment

At present, the greatest controversy surrounding GM fish swirls around salmon. Salmon are among the world's most extensively aquacultured fishes, and it is not surprising that salmonids would be candidate species for transgenic efforts, as is evident from the salmon-heavy literature reviewed here. But because salmon are both widely imperiled and popular sport and food fishes in industrialized nations, it could be expected that manipulating their genes is going to evoke controversy.

A good source of updated information on GM science can be found at a Web site maintained by one of the main actors in this drama, A/F Protein/Aqua Bounty Farms of St. John's, Newfoundland, and Waltham, Massachusetts (now Aqua Bounty Technologies, Inc.). This business came under focused attack in the GM salmon debate because of its development and proposed farming of salmon that contained a growth hormone transgene from Chinook salmon and an antifreeze gene promoter from winter flounder or ocean pout that stimulated growth hormone production in the pituitary and liver (www.aquabounty.com). The process produced salmon that grew four to six times faster, had a 20% improvement in feed conversion efficiency, and reached a marketable size of 3 kg in 14–18 months instead of three years (whether fish grow larger than normal remains unknown) (Fletcher et al. 1999).

Hew and Fletcher (1997, p. 313) maintained that "there is no evidence that transgenics disrupt the ecological balance." The statement is general enough to be innocuous, particularly if one adds "per se" after "transgenics." However, aquacultured fish have caused significant, recurring ecological disruptions, and transgenics are primarily intended for aquaculture. There is little reason to suspect that transgenic salmon would be less disruptive of wild genomes than massively released or escaped domesticated, farmed salmon.

In fact, the ecological impacts of free-swimming, interbreeding transgenics are potentially serious, and laboratory findings contradict the generality of "no disruption." Muir and Howard (1999, 2001) ran lab experiments and constructed a model to predict the possible impacts of a transgene that reduced juvenile viability but increased male mating success. A critical determinant of the spread of the gene was that larger males had a mating advantage; such a size advantage is commonly observed in fishes, including salmonids (Fleming 1996). Muir and Howard's model predicted that transgenic males that produced the least fit offspring would tend to obtain a disproportionate share of matings. As a result, any wild population invaded by such a transgenic would be reduced by 50% in about 6 generations and be destined for extinction in 40–150 generations, depending on variation in offspring viability and male mating success. The researchers called this phenomenon the Trojan gene effect because "the mating advantage provides a mechanism for the transgene to enter and spread in a population, and the viability reduction eventually results in population extinction" (Muir and Howard 1999, p. 13855). This theoretical scenario is strikingly familiar: Genetic engineers produce transgenic fish with augmented growth characteristics (e.g., "extraordinary gigantism"), while aquaculture conditions are known to select for fish with reduced viability when released into the wild. The evidence at present is theoretical, but the process and its environmental impacts are not difficult to imagine.

Concern over environmental and human health impacts of GM salmon have been expressed by a litany of environmental, scientific, and even commercial organizations (e.g., American Fisheries Society, Atlantic Salmon Federation, Conservation Law Foundation, Ecological Society of America, International Salmon Farmers Association, North Atlantic Salmon Conservation Association, Trout Unlimited; see respective Web sites). Representative of this concern is a resolution passed by the American Society of Ichthyologists and Herpetologists (ASIH 2000), a deliberative body not given to environmental hysteria. ASIH concluded that transgenic salmonids were neither an effective nor ecologically safe solution to problems of food shortages. It proposed, among other recommendations, an international moratorium on creation or marketing of transgenic salmonids until it is firmly established that such fish will not gain access to natural waters, by accident or intent. Efforts would be better spent achieving sustainability of existing aquatic ecosystems.

spliced into soybeans (Nordlee et al. 1996). Similar PR disasters in cultured fishes can be avoided by bypassing certain opportunities for gene splicing; for example, bivalves may have useful genes that affect metabolism or longevity, but many people are allergic to shellfish. Proponents of GM technology downplay public concern over transgenic fish because, to date, only fish genes have been used to produce transgenic fishes, and because "there are no toxin genes in the fish genome" (Hew and Fletcher 1997, p. 313). Fin and spine toxins, egg toxins, and skin toxins in a wide variety of species (Helfman et al. 1997) argue against this simplistic view of fish antipredator defenses.

Another concern is domestication selection in cultured fishes that may escape and breed with wild fish.

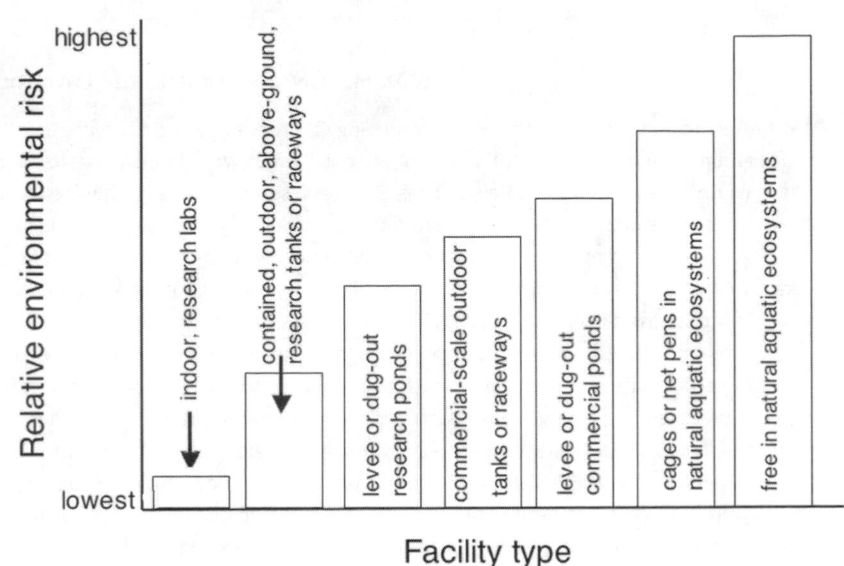

Figure 14.10. The potential for environmental disruptions by transgenic fishes depending on holding facility. Land-based, indoor, recirculating systems pose the lowest risk of escape and impact; fish kept in net pens under ambient conditions have a high likelihood of both escape and impact. Some aquaculture operations involve free-swimming fish fed at floating stations (e.g., in Japan; Masuda and Tsukamoto 1998). After Kapuscinski and Hallerman (1990b); used with permission.

Establishing lineages of broodstock would allow quality control of the product, but such strong selection increases the likelihood of domestication. Aqua Bounty Technologies, Inc., is apparently producing just such broodstock fish for commercial aquaculture facilities in Canada, New Zealand, Chile, and the U.S. (Fletcher et al. 1999). The company is neither ignorant of nor insensitive to the concerns; it recommends that farmers grow the fish in closed, terrestrial systems, and that fish grown in net pens should be sterile (see Hew and Fletcher 1997). Whether eggs sold by Aqua Bounty are in fact sterile is unclear.

SOLUTIONS

Demand for seafood will increase, and aquaculture will help meet the demand. "The real challenge for aquaculture is to develop farming practices that are in tune with ecosystem processes and functions in a fashion that enhances seafood production" (Kautsky et al. 2001, p. 190).

General Recommendations

Numerous conferences have produced recommendations for improving the efficiency and reducing the environmental impact of aquaculture facilities and related programs (e.g., UNEP 1990; FAO 1995a; GESAMP 1996, 1997). A bounty of reasonable, general recommendations exists (Goldburg and Triplett 1997; Goldburg et al. 2001; Kautsky et al. 2001). Many are incorporated into increasingly common Codes of Conduct and Codes of Practice adopted by national and international industries, agencies, and trade organizations (e.g., FAO 1997a; PIRSA 1999; FEAP 2000; SEAFDEC 2001). Six of the more general recommendations follow:

1. *Improving formulated feeds.* The fraction of animal protein in fish feeds should be reduced to the minimum necessary for adequate nutrition. This is happening by replacing animal protein with seed oils. Economic benefits result because plant protein is cheaper to acquire and easier to store than animal-based feeds. For example, catfish feed contains only 3%–5% fish meal (Boyd and Tucker 1995). Unfortunately, salmon farming, the largest user of prepared feeds, requires 50%–70% fish meal because oilseed proteins do not contain essential amino acids and unsaturated fatty acids that salmon require (Kautsky et al. 2001). Culturing species lower on the food chain, as is done in China and Africa, would reduce the depletion of fish stocks for feed production.
2. *Reducing escapes.* Open-water net pens should be replaced with closed, recirculating, land-based systems that, if connected to natural waters, have screens over all outflow pipes. Species held at conditions different from surrounding habitats have relatively low probabilities of establishment. To minimize impacts of escaped exotics, native fishes could be cultured

(Welcomme 1988)—for example, instead of importing tilapia to Central and South America, a native cichlid with appropriate ecological and growth characteristics could be farmed there. A laudable example is the farming of the characid tambaqui, *Colossoma macropomum*, in the Amazon (Araujo-Lima and Goulding 1997). Finally, to obtain a license, owners should be required to post bond against the costs of escapes, including guaranteed proper disposal of discarded stock, with fines levied in the event of escape (e.g., Courtenay and Williams 1992).

3. *Minimizing genetic impacts of stocking and escapes.* Genetic pollution (loss of variation, introgression of foreign alleles and genes) can be reduced by identifying genetic variation in captive and wild stocks (Ryman and Utter 1987; Hindar et al. 1991; Utter 1998; Waples 1999). Tracing parentage, keeping broodstock size large, and frequently replacing brood fish with new, wild fish are increasingly common practices in hatcheries. Ideally, emphasis will shift from managing fitness to maximizing genetic diversity (Busack and Currens 1995; Philippart 1995).
4. *Eliminating pollution.* Pollution comes largely from waste, which ultimately erodes the bottom line. Reduced overfeeding and optimized use of chemicals are obvious, positive steps. Closed, recirculating, land-based systems, again are a major improvement. Diking ponds reduces the possibility of overflow and spills. In ocean-based systems fish can be kept in bags instead of nets; ocean water is pumped into the bag and solid waste removed for composting (www.bucksuzuki.org; this approach is obviously constrained by currents). Ecological engineering, integrated fish farming, and polyculture have been successfully developed and involve biofiltration of nutrients by seaweeds, marsh plants, lettuce, and bivalves (Mathias et al. 1997; Troell et al. 1999). If seaweed is used to recycle nutrients inside pens or cages, the final discharge of N is reduced by 20%–27%, and P by 39%–47% (Krom et al. 1995).
5. *Periodic decommissioning.* Fallowing of sea ranch areas benefits both the ranch and the environment. New Brunswick encourages salmon ranchers to allow an area to remain fallow (without active net pens) for up to six months before reusing pens, to reduce the spread of infectious diseases (Haya et al. 2001). Fallowing would also allow pollutants in sediments to degrade and oxygen concentrations to build back up, although heavy-metal contamination from antifoulants apparently takes longer to reverse (e.g., Morrisey et al. 2000).
6. *Culturing species appropriate to conditions.* Species differ in their environmental impact as a function of their traits and the ability of a region to assimilate and process wastes and chemicals. The "ecological footprint" of a species can be calculated (Folke et al. 1998) to determine how much surrounding area is needed to provide energy and nutrients in open systems and to process the waste produced. If the area required is excessive, the venture may be neither economically nor ecologically viable without substantial investments (e.g., of pipelines, water treatment, integrated farming). Chinook and Atlantic salmon have the largest footprint among salmonids, at 16 ha/MT and 13 ha/MT respectively. Commercial capture of salmon is less energy intensive, with a smaller footprint (11 ha/MT for commercially caught Chinook and only 5 ha/MT for pink salmon, *O. gorbuscha*) (Tyedmers 2000). Intensive tilapia cage farming has a footprint one-fifth that of salmon culture, whereas pond culture of tilapia fertilized with by-products from other activities has a negligible footprint (Kautsky et al. 1997). Footprint calculations are not without criticism. E. Roth et al. (2000) viewed the concept as too ecologically and economically simplified for direct management application.

A shortcoming of this subset of recommended actions is a focus on the "supply" end of economic production. Addressing the "demand" component can also stimulate (force) change in production methods. Goldburg and Triplett (1997) correctly recommended that an organic or ecocertification program for responsibly produced aquaculture products would provide aquaculturists with incentives to engage in environmentally sound practices. Such green products can bring higher prices, at least in the industrialized world. The Marine Stewardship Council's guidelines for certification of capture fisheries could serve as a framework (www.msc.org; see chapter 10).

Improving Stock Enhancement Practices

Culturing fishes to augment wild stocks creates a set of environmental challenges beyond those involving production and consumption of fishes without release. Tactics to maximize success and minimize negative consequences of stock enhancement programs were summarized in Travis et al. (1998) and subsequent papers in the *Bulletin of Marine Science* (vol. 62, no. 2). Travis et al. focused on life history characteristics of species with aquaculture promise; positive genetic and numerical characteristics of brood fish; optimal hatchery and release conditions; and optimal (rather than maximal) sizes, numbers, and timing at release. The authors recommended that ecological costs be factored into any cost-benefit calculations.

Improvements are also being made in developing, evaluating, and managing enhancement and stocking programs, particularly for depleted marine fishes. Some proponents have advocated a "responsible approach" that follows a ten-step evaluation protocol (e.g., Blankenship and Leber 1995). The elements include an adaptive management plan that prioritizes target species and incorporates life history knowledge, stated management objectives, defined success and assessment criteria and methods, emphasis on minimization of negative genetic effects and disease introduction, and public education. The responsible approach concept is a vast improvement over older dump-and-run tactics but still focuses on production as the major metric of success. Interactions with and impacts on native and nongame, nonprey species are underemphasized. Obtaining knowledge about ecosystem dynamics, if considered at all, is sometimes viewed as a costly inconvenience. Overviews of recommended stocking strategies include such statements as "Unfortunately, research on competition is costly and time consuming. At the least, a literature search should precede species introductions" (Murphy and Kelso 1986, p. 307); and "It does not seem practical to hold off on stock enhancement research until the ecological mechanisms are completely understood" (Blankenship and Leber 1995, p. 171). Complete understanding is seldom achieved. Opponents of enhancement programs could (just as reasonably) insist that culture efforts be curtailed until understanding of the life history of the species was complete.

Literature searches cannot provide crucial information if no one has conducted relevant ecological research. Some understanding seems preferable to introducing hatchery fish and seeing what happens. As presented, the responsible approach appears well intentioned but still focuses on single-species solutions to problems that exist in a multi-species and ecosystem context, diverting attention and resources away from habitat restoration and overfishing. A preferable approach, advocated by Wahl et al. (1995), attempts to take into account biotic and abiotic factors and interactions to determine what, when, where, and how many fish, of what size, should be stocked (figure 14.11). The Position Statement of the American Fisheries Society, while recognizing considerable potential benefits from stock enhancement efforts, spells out the concerns and the best approach: "We strongly emphasize that enhancement must be a part of an integrated program that focuses primarily on conservation of natural stocks and the ecosystems that produce them" (Grimes 1995, p. 594). In line with such emphasis, the U.S. federal approach to managing marine fisheries has focused more on regulating fishing effort and restoring essential fish habitat and less on hatcheries and stock enhancement (Benaka 1999).

Genetic Restoration Pros and Cons

Conservation programs have frequently relied on stock enhancement that depends on captive propagation. Because of limited availability of brood animals, this approach has often produced introgressed or otherwise genetically polluted wild stocks. Such genetic alterations are essentially irreversible and often result in extinction of wild genotypes. To solve emerging genetic problems, a lake or stream may be poisoned, killing off all hybrids and probably all other fishes. The water body is then restocked with native fishes of the original genotype (assuming some remain). For logistic as well as ethical reasons, chemical eradication is a less than desirable course of conservation action.

Less radical is genetic restoration via reversal of introgression (Leary et al. 1995). If large numbers of genetically pure wild-type individuals are repeatedly introduced, they will breed with hybrids and slowly convert an introgressed population to one more similar to the original genotype. Theoretically, and assuming (among other things) a closed population of equal numbers of each genotype breeding randomly, the proportion of foreign genes can be halved with each generation. Leary et al. (1995) recommended as a target that a restored population contain 1% or fewer foreign genes. For fish maturing at age three, 24 stockings would be needed to change the proportion of foreign genes from 50% to 1%. Leary et al. acknowledged that few fisheries programs had the resources (appropriate brood stock, genetic analysis facilities, humanpower, hatchery and hold-

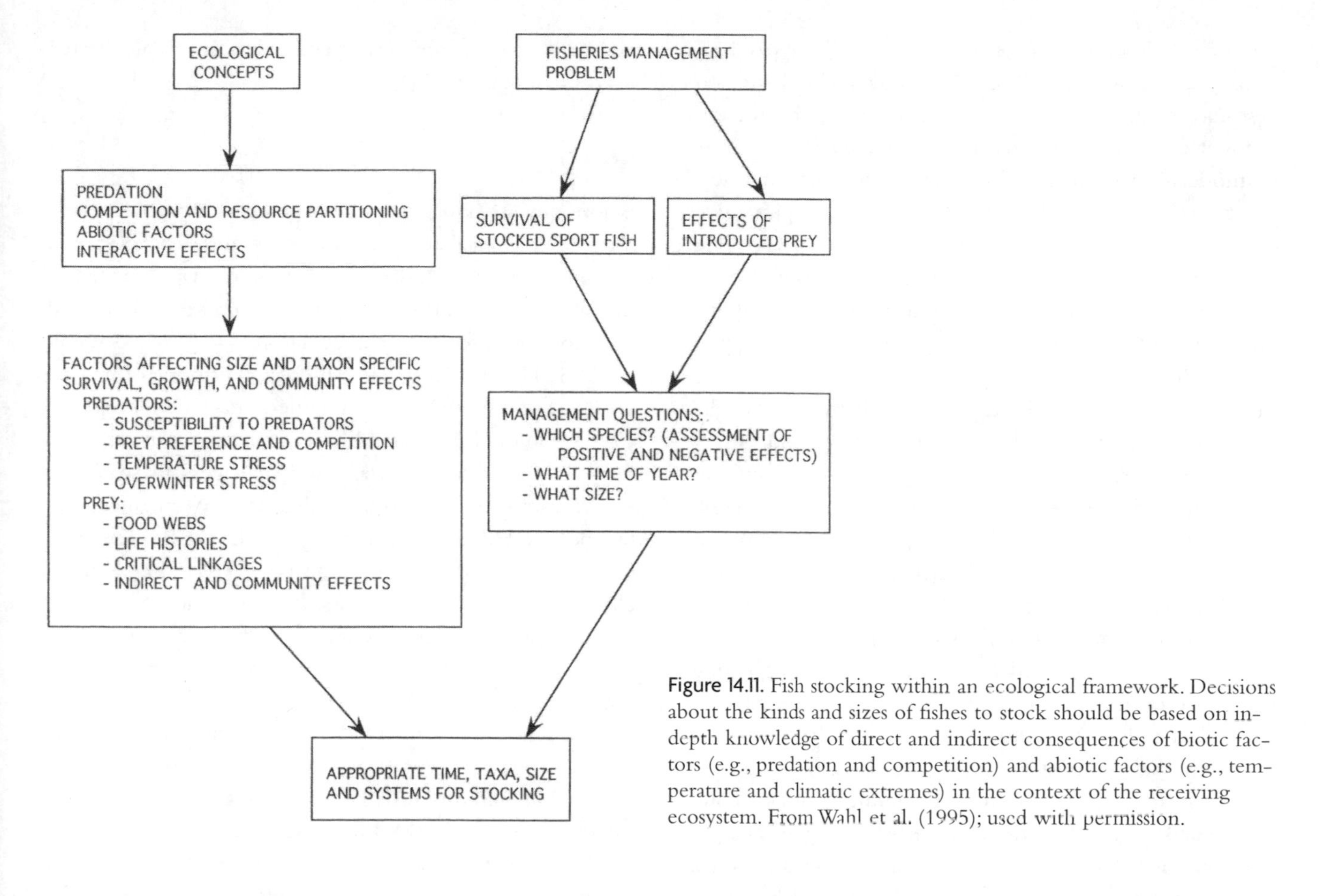

Figure 14.11. Fish stocking within an ecological framework. Decisions about the kinds and sizes of fishes to stock should be based on in-depth knowledge of direct and indirect consequences of biotic factors (e.g., predation and competition) and abiotic factors (e.g., temperature and climatic extremes) in the context of the receiving ecosystem. From Wahl et al. (1995); used with permission.

ing space) or time to perform such an operation. They concluded that genetic restoration is probably only practical for introgressed populations that contain 10% or fewer foreign genes. Whether such restoration has been attempted and succeeded remains unknown.

Genetic restoration may also have a dark side. Genetic enhancement of depressed or imperiled stocks has been proposed previously. The basic idea is to infuse wild lineages with desirable traits that should overcome identified genetic inadequacies. But such programs could be disastrous because they are on a collision course with fundamental evolutionary processes. For example, whirling disease has been introduced into 21 U.S. states, where it infects at least four salmonids (see chapter 8). Genetically resistant strains of trout exist that could be bred with native strains to produce immunity in wild fish (Incerpi 1996; MWDTF 1996; future steps could involve gene isolation and recombinant DNA splicing). Allendorf et al. (2001) viewed this approach as a quick but misguided technological fix that divorces the disease from its ecological and evolutionary contexts. Their major concerns were (1) that disease resistance is usually controlled by multiple genes; (2) that strains resistant to one pathogen are often susceptible to others; (3) that introducing exogenous strains generally leads to homogenization of genotypes; and (4) that regions most heavily affected by whirling disease have typically experienced significant habitat alteration. Allendorf et al. concluded that focusing on a genetic solution to whirling disease will make trout populations even more susceptible to future epizootics, thereby trapping us "in a cycle of repeated actions to combat the failure of our previous attempts. . . . Stocking trout that are clinically resistant to whirling disease perpetuates a cycle of treating the symptoms of a problem but not the underlying causes" (p. 28). The pattern is hauntingly familiar.

Hatchery Bashing?

Hatcheries and related programs are also criticized as technologies that treat symptoms instead of causes. Hatcheries cannot solve habitat problems. Meffe (1992) applied the terms "technoarrogance" and "halfway technologies" to describe technological solutions (hatcheries) to problems brought about through technology (depleted runs caused by modern agriculture and forestry, hydropower development, overfishing, pollution). Such technological fixes commonly fail to deal with the habitat unsuitability and overfishing that are the root causes of a population decline. Producing more fish to cure population declines is like providing a blood transfusion to a patient without stopping the bleeding. The solution ignores the real problem, gives the impression of solving it, goes only halfway to a solution, and can eventually do more harm than good.

Is this merely hatchery bashing, a "biased, narrow, and unrealistic attitude towards fish culture [that] is certainly not substantiated by science" (Incerpi 1996, p. 28)? Waples (1999) systematically explored criticisms and myths surrounding hatchery programs and concluded that hatcheries have many adverse effects on wild fishes and many criticisms were supported by the science, although some risks were overstated. Incerpi's defense is representative of aquaculture advocates who admit that mistakes have been made, say that the mistakes are being corrected, and insist that the future holds great promise. Opinions differ according to perspective. Traditional fisheries managers, charged with producing fishes for commercial and recreational use and focused on objectives "largely determined by public demand" (Radonski and Loftus 1995, p. 1), are likely to view things differently than people concerned primarily with conservation of biodiversity. Progress will come only from mutual respect, communication, and compromise (see R. J. White et al. 1995 and Mobrand et al. 2005 for middle ground).

Many critics of hatcheries see a valuable role for culture programs in conserving genes. Depleted or genetically degraded stocks can be rejuvenated via hatchery programs that follow "strict genetic operational guidelines" (Meffe 1992, p. 353; see also Meffe 1986). Redirecting some hatcheries from production to research would help answer genetic questions central to the debate over the utility of hatchery programs, such as those concerning domestication selection, inbreeding and outbreeding depression, and behavioral anomalies (Waples 1999). Such selective redefining of hatchery goals could, however, be manipulated into justification for continued, large-scale hatchery activity, analogous to the debate over "scientific whaling," whereby commercial whaling activities are justified on the basis of the meager and redundant data collected from the carcasses.

Just Don't Stock!

Before any stocking program begins, an objective assessment of ecological and economic costs and benefits should be completed. Such assessments are seldom performed. If they were, many existing programs would be discontinued and most proposed programs abandoned (e.g., Moksness et al. 1998). Hilborn (1998), in a thorough critique of marine enhancement programs, maintained that such programs were promoted by a coalition of resource users who desired more fish, politicians and managers in search of easy answers, and uncritical advocates of enhancement technology. Hilborn maintained that among the few programs that calculated economic costs, only the Japanese chum salmon program appeared to be a clear economic success (with possible, uncalculated losses to North American fishers who depend on North American chum that compete in the North Pacific with chum that were produced in and returned to Japan). Among economic failures are some of the largest stocking programs in the world. Pink salmon in Alaska return 50 cents on the dollar (to the tune of a $100 million subsidized debt). Salmon hatcheries in Canada had a net value of –$142 million (Canadian dollars) for coho and –$252 million for Chinook; the benefit-to-cost ratio never broke even for either program, and both were often in the 0.4 or 0.5:1 range.

Similar analyses are largely unavailable for U.S. hatcheries, but survival rates for U.S. salmon have been lower than for Canadian salmon, suggesting even poorer economic performance. Oregon's hatchery system in the early 1990s cost about $15 million annually. Production costs ran from $10 to $25 to over $100 per fish (Montgomery 2003). Japanese flounder programs in Hokkaido operated at benefit-to-cost ratios of roughly 0.4:1. Norwegian cod stocking, ongoing since the late 1800s, shows no evidence of having increased cod abundance. Economic analyses are lacking for other oft-cited programs such as those of red drum in Texas and striped mullet in Hawaii. "Economic costs and benefits are often not evaluated, perhaps because the supporters of the programs recognize that they are not economically justifiable" (Hilborn 1998, p. 669). Existing analyses generally fail to account for adverse effects on wild fish, such as population reduction or loss of genetic variability, which would drive the bottom lines down further. These results led Montgomery (2003, p. 170) to conclude, "Hatchery fish are neither an ecological nor an economic bargain."

Why do enhancement programs regularly fail to increase stocks? First, a long-established lack of correlation exists between adult spawning stock and later recruitment for many species (chapter 10). Programs increase spawning stock numbers in hatcheries while ignoring the likelihood that this will not affect subsequent recruitment. Second, hatchery rearing-and-release may bypass the earliest phases of high mortality, but mortality continues at high levels in the open ocean because of possible density dependent interactions such as competition for food. Third, as stated earlier, stock enhancement is a response to symptoms, not a solution of problems. Population declines that result from overfishing or habitat destruction can be solved only by reduced fishing or restored habitat, not by pumping more fish into the ocean. "It is more cost effective to rely on natural production by wild stocks than to rehabilitate, enhance, or replace depleted stocks to support continued exploitation" (Shaklee and Bentzen 1998, p. 589).

Then why do we keep stocking despite evidence suggesting that the programs are failures? The answer is largely political. Stock enhancement programs exist because the real problem of dwindling returns forces fishery managers to respond to public demand for visible action (e.g., Radonski and Loftus 1995). Politicians and resource managers are likely to be concerned less with economics or ecology and more with "whether the program relieves political pressure to do something" (Hilborn 1998, p. 662). Scientists are not blameless in this cycle of unrecognized or unadmitted failure because research funding to study enhancement often depends on a perception of success. Too many "evaluations" of programs appear in conference proceedings or the gray literature and are conducted by advocates of enhancement whose research funding comes from the enhancement program under study (Hilborn 1998). Peer-reviewed, primary literature on the subject is greatly needed.

Hilborn's (1998) analysis indicates that enhancement programs for depleted fisheries are likely to achieve economic success only when wild stocks are essentially nonexistent, as in the case of Japanese chum salmon. But such a scenario is ecologically unsound because of the genetic damage to any remaining wild stocks caused by outbreeding depression. Regardless, Hilborn concluded that most enhancement programs are a waste of public funds and that decisions on whether to stock should "begin with the assumption that fish stocking programs are less likely to succeed than improved enforcement or habitat protection" (p. 670).

Conservation Aquaculture: Farming Imperiled Fishes

Despite its shortcomings, aquaculture has taught us much that has direct application to the restoration of imperiled fishes (see Schramm and Piper 1995). Although the basic goals of production aquaculture and conservation aquaculture are very different (numbers maximization versus genetic and ecologic sufficiency), methodologies are largely interchangeable. In conservation aquaculture the emphasis is on genetic diversity, on keeping small numbers of fish at low densities, and on simulating natural conditions; hence, problems of domestication selection, disease, and behavioral aberrance are avoided (Anders 1998; but see box 14.2).

Government agencies, private organizations, native peoples, and private individuals culture and reintroduce imperiled fishes, into either restored native habitat or acceptable, alternative habitat (e.g., Wikramanayake 1990; Minckley 1995). A private, nonprofit organization, Conservation Fisheries, Inc., of Knoxville, Tennessee, has successfully propagated and reintroduced several imperiled minnows, madtom catfishes, and darters to streams of the southeastern U.S., particularly in and near Great Smoky Mountains National Park (Rakes et al. 1999). Public aquaria have obvious potential, as well as facilities for engaging in rearing and restoration, such as the Southeast Aquatic Research Institute (SARI), the research arm of the Tennessee Aquarium in Chattanooga. At the aquarium and at a converted state-run trout hatchery, SARI personnel have reared and released lake sturgeon, *Acipenser fulvescens*; spotfin chub, *Erimonax monachus*; and Barrens topminnow, *Fundulus julisia*.

The Dexter National Fish Hatchery and Technology Center in Dexter, New Mexico, is a federally run facility dedicated to holding, culturing, and distributing endangered fishes (Rinne et al. 1986; J. E. Johnson and Jensen 1991). Its activities focus on imperiled species of the southwestern U.S. and northern Mexico, and it has aided recovery of over 60 species of minnows, suckers, catfishes, livebearers, pupfishes, and darters, including Colorado

BOX 14.2. A Brave New World for Fish?

How far should we go with captive propagation to aid our attempts at conserving biodiversity? Ideally, for species in apparent trouble, we can anticipate problems and be proactive. If we maintain captive stocks as insurance against loss of genetic variation, we can periodically infuse them with genetic material from wild populations (Allendorf and Ryman 1987; Echelle and Echelle 1997). However, given limited funding and the usual triage, reactive nature of most conservation programs, this proactive approach may be unrealistic.

Another, more complex solution is the frozen zoo approach, in which stocks of frozen gametes or DNA banks are maintained for some aquacultured species and are applied to conservation. Such cryopreservation methods could serve as genetic reserves to ensure brood stock diversity of imperiled fishes, but at present, methods do not exist to freeze fish eggs and embryos because of their large size, large amounts of yolk, and thick chorion (Linhart et al. 2000; N. H. Chao and Liao 2001). It would seem regrettable if funds were diverted from conventional captive propagation and species rehabilitation programs and directed instead toward such technologically demanding solutions with their heavy reliance on energy, personnel, funds, and political stability.

Along these lines, and one step short of frozen zoos, are proposals to create intensive, closed, recirculating aquaculture facilities for endangered fishes. One such proposed program envisioned a 600-m^2 (basketball court–size) facility capable of housing 1,000 individuals each of 200 species of small endangered fishes, at densities of 320 fish/m^2 (Kaplan 1999). Initial cost was estimated at $2 million. The objective of such a program would be to use the technology of intensive aquaculture "to concentrate populations of endangered species for extended periods of time until they could be repatriated into refugia" (Kaplan 1999, p. 5). Left unresolved are, among other issues, problems of artificial and domestication selection in such factory farms. I hope it doesn't come to this.

pikeminnow, *Ptychocheilus lucius*; Guzman beautiful shiner, *Cyprinella formosa*; bonytail chub, *Gila elegans*; razorback sucker, *Xyrauchen texanus*; and Yaqui catfish, *Ictalurus pricei* (for other U.S. facilities, see www.fws.gov/southwest/fishery and www.fws.gov/fisheries). Philippart (1995) provided a review of global efforts at captive propagation, emphasizing activities in Europe and details on the Tihange Fish Breeding Center in Belgium, where the focus has been on imperiled European cyprinids (see also J. Fish Biol. 37, Suppl. A).

Native peoples were dependent on fishes for sustenance, and many formerly abundant species were celebrated in rituals and traditions (e.g., Fobes 1994). Tribes in the U.S. are active in restoration activities that frequently involve propagation and release. The Paiute tribe of Pyramid Lake, Nevada, has maintained a hatchery for Endangered cui-ui, *Chasmistes cujus*, suckers since 1973, producing more than 5 million four-inch fingerlings between 1995 and 1999. The Paiute people traditionally call themselves Cuyui Ticutta, which means "cui-ui eaters," documenting the historical importance of the species to the tribe (Pyramid Lake Fisheries 2000). Similar efforts directed at endangered suckers have been expended by tribes in the Klamath basin of California and Oregon (NRC 2004b). The White Mountain Apache Tribe has been instrumental in recovery efforts, including captive propagation, of federally Threatened Apache trout, *Oncorhynchus gilae apache* (www.gf.state.az.us/w_c/apache_recovery.shtml).

Tribal hatcheries have been established on numerous rivers in Alaska to supplement wild salmon runs (figure 14.12). Salmon are captured while moving upstream during spawning migrations, stripped of eggs or milt, and the offspring raised in ponds for later release. Many of the criticisms of massive hatchery programs apply less here: Fish that would have spawned in their natal river are used, and "broodstock" are kept for only one spawning event, which minimizes problems associated with translocation of individuals and domestication selection. On the downside, however, is the potential for artificial selection in the rearing ponds and possible competition between massive numbers of artificially selected smolts and naturally spawned wild fish, especially during years of low natural survival. Of mixed value are tribal hatcheries in Washington State, where 20 tribes and coalitions have annually released 30–50 million salmon of five species. The conservation value of such efforts will depend on the balance between deleterious hatchery practices and reforms that address the genetic and behavioral problems discussed here. Of promise is the stated intent of the Northwest Indian Fisheries Commission "to improve and reform harvest and hatchery

Figure 14.12. Supplementation hatcheries. (A) Facilities at the Deer Mountain Tribal Hatchery, Ketchikan, Alaska. (B) A 54-pound chinook salmon that was reared at the Deer Mountain facility and returned to its natal stream alongside the hatchery to spawn.

management . . . [and] minimize the impacts of artificial propagation on wild salmon" (www.nwifc.wa.gov).

Among actively propagated, nonsalmonid fishes are sturgeons. Endangered sturgeon species are cultured in several countries for restoration and stocking, and other sturgeon species are propagated for commercial and recreational harvest (e.g., Logan et al. 1995; T. I. J. Smith and Clugston 1997; Williot et al. 1997). Sturgeon stocks of the Caspian Sea would probably be extinguished from overfishing if it weren't for hatchery production, especially in Iran (Vecsei 2005). Kynard (1997) pointed out that sturgeon culture and stocking for conservation should be undertaken only where extirpation is imminent. Enthusiasm for stocking should be tempered by recognition of possible negative genetic consequences and the greater need for habitat restoration and reduced exploitation. But even "last resort" attempts to save species in imminent threat of extirpation have significant drawbacks because of the genetic bottlenecks associated with small population size (e.g., Anders 1998).

Is conservation aquaculture really more justifiable than stock enhancement for fishing? Imperilment does not guarantee that introductions will have minimal impacts in their new settings; witness the imperiled status of sea lamprey, brown trout, and common carp in their native habitats (see chapter 9). Moyle and Light (1996a) maintained, with supporting data, that pending physiological limitations, most fishes are capable of being invaders and most assemblages are invadable. Limited evidence does however suggest a minimal impact from introducing imperiled fishes, assuming they are introduced into habitats similar to their natural requirements and close to their native range. Five species of endangered omnivorous cyprinids were introduced into a nearby drainage in Sri Lanka, with no apparent effect on the established fish assemblages (Wikramanayake 1990). Translocation has been used as a management tool with some success and apparent minimal impact for fishes of the U.S. Southwest (e.g., Minckley 1995). Imperiled species are often characterized by attributes unlike those possessed by successful invasives (table 4.4 and box 9.1), suggesting that it is generally safe to assume minimal impact. Any translocation effort that carries a species outside its native range should obviously be approached with care, after thorough background research on potential impacts. Regardless, as Philippart pointed out, "Captive breeding should remain a temporary safeguard measure, while awaiting the implementation of measures for protecting species in their restored original habitat . . ." (1995, p. 281).

Is it not also hypocritical to view stocking as admirable for conservation but damnable when the intent is to enhance fisheries? Perhaps, but some important distinctions exist between the two situations. Supplementation of exploited species occurs to maximize yield, usually in response to declining catches; population sizes are perceived as low in terms of angler or fisher satisfaction or economic return. Too often, large numbers of fish have been pumped or dumped into systems with little poststocking assessment of recruitment and even less of ecological impact. As long as catches increase, goals are said to be met. Habitat improvement and harvest regulations are pursued as actions of last resort because of their political unpopularity. Although many past questionable practices are being corrected (e.g., Bowles 1995), the descriptions remain valid for many ongoing operations.

In contrast, most conservation enhancement programs do not start until rare fishes are in danger of extirpation or extinction. Underlying causes of decline are researched, and corrective measures to reverse them are initiated before or while captive propagation and release are attempted (Modde et al. 1995). Typically, small numbers of genetically screened individuals, chosen for maximal genetic variation, are released and monitored closely (Vrijenhoek 1998): "The aim is to produce fish with all the morphological, behavioural and genetic characteristics of the taxa to be conserved" (Philippart 1995, p. 281). Success is defined as a population returning to a size that can sustain itself. Conservation aquaculture is ideally "one component of multifaceted fish restoration and recovery programs . . . designed to be implemented simultaneously with habitat improvement and watershed or ecosystem restoration activities" (Anders 1998, p. 29). Major differences therefore exist in the goals, initial conditions, philosophy, sequence and nature of actions, ecological impact, and outcome of the two approaches.

CONCLUSION

Protein for the Masses

Tropical aquaculture depends heavily on carp and other freshwater herbivores or omnivores grown in ponds fertilized with cheap, locally available agricultural by-products. It is usually extensive in nature, occurs on a small scale, involves native fishes low in the food chain, and often includes some form of polyculture or integrated farming that maximizes efficiency and minimizes waste (see Tacon and DeSilva 1997). Minimal use is made of animal-derived feeds, and protein yields exceed protein investments. Hence carp production in China and India and tilapia production in much of Africa are seldom singled out for environmentally destructive impacts (catfish farming in North America has similar positive energy balances and is generally viewed as a minimally destructive activity, although water quantity and quality issues emerge in large farm situations). The methods of freshwater aquaculture are well studied, and the technology is actively shared (see www.fao.org for an extensive bibliography). Also, many products of freshwater aquaculture are consumed locally and not intended to supply external markets. Hence the least environmental harm comes from simple culture activities that appear to provide the greatest good for humans in the greatest need of protein.

In contrast, some tropical and most temperate aquaculture focuses on diadromous and marine fishes, farmed in net pens and fed nutritionally complete fish meal and fish oil-based diets. Much marine aquaculture involves carnivorous species that are intensively farmed on a large scale, involving transportation of exotic fishes and heavy supplementation with feed and chemicals, with low energetic efficiencies (33% net loss of available fish protein) and considerable production of organic and inorganic wastes. Profitability depends on large-scale production, and the scale of production requires an equally large scale of distribution, which means most marine aquaculture products are exported and marketed to people seldom described as protein deficient. For example, 99% of cultured shrimp are raised in the third world, but virtually all of these are exported to industrial countries (Stonich and Bailey 2000). Technical and logistic complications multiply, technologies are complex and closely guarded, profitability is often low and requires government subsidy, and only recently have methodologies and regulations been developed to minimize environmental impacts.

The upshot of this comparison is that claims that advanced aquaculture will feed starving humans—such as the Global Aquaculture Alliance's motto of "Feeding the World Through Responsible Aquaculture"—ring hollow when current practices are scrutinized. The calculated 33% net loss of protein ignores additional energy costs, distribution problems, and socioeconomic factors, including the multiple transport and processing costs of producing fish meal. Nor can marine stock enhancement programs fill the enormous demand for protein because the programs focus primarily on "luxury species" (Masuda and Tsukamoto 1998, p. 353). From these comparisons, one can easily and correctly conclude that arguments advocating marine fish culture for meeting the world's protein demands, including augmentation of stocks via genetic modification, are specious. Protein-starved children in interior sub-Saharan Africa are unlikely to eat sea-ranched, genetically modified Atlantic salmon.

Misapplication of the Agriculture Analogy

It is often stated that fish will be increasingly obtained from aquaculture rather than capture fisheries, and that this economic transition is analogous to the transition from subsistence hunting to farming and ranching (Goldburg and Triplett 1997). Aquaculture is consequently viewed as the aquatic counterpart of agriculture (see Courtenay and Williams 1992; Courtenay and Moyle 1996; Kautsky et al. 2001). This view contributes substantially to many of the problems summarized in this chapter. Courtenay and Moyle (1996) referred to the introduction paradigm, in which humans feel they have improved on nature by importing, introducing, and eventually domesticating animals and plants. By extension, domesticating fishes will produce an aquaculture product as beneficial as agricultural species. "It is an old, enticing dream: to farm the seas as we farm the land" (Weber 1997, p. 1).

The flaw in this analogy is that agriculture today involves "mostly plants and animals that are so far removed genetically from their wild ancestors that they require care and husbandry to survive [e.g., cattle, sheep, chickens, turkeys], with few persisting in a feral state . . . [whereas] species employed in aquaculture are mostly feral stocks being reared artificially, and most have the capability to return to a feral state if released" (Courtenay and Williams 1992, p. 51). Hence, released and escaped aquaculture species become introduced pests more often than not. Of added significance is that, despite domestication, several feral farm species are capable of inflicting substantial environmental damage when accidentally or deliberately released (e.g., goats, pigs, donkeys, horses). In this respect, the agriculture analogy is valid.

In sum, although optimism about our ability to domesticate fish species may be justified, domestication will come at a cost of lost biodiversity, just as our conquest and domestication of large terrestrial animals has reduced them to a fraction of their former abundance and diversity in most regions of the world (Mowat 1996). Do we need or want to repeat those mistakes in developing culturable fish species?

Ultimately, the benefits must be assessed along with the ecological and other costs of aquaculture, and some balance must be calculated, evaluated, and struck (Waples 1999). Basic questions, such as whether hatcheries are, in fact, cost effective in the production of fish, should first be answered. Also, do hatchery practices actually increase the number of fishes produced, or do they diminish wild fish in the process of trading them for hatchery fish? Determining whether benefits outweigh costs will be difficult, given the variety of currencies involved (dollars earned, jobs created, lives saved, fishing opportunities provided vs. fitness and species lost, habitats destroyed). Increased aquaculture may be an inevitable solution to problems of inadequate protein and income, but this seeming inevitability should be weighed against the environmental costs incurred. Alternative solutions that incorporate habitat restoration and harvest regulation should not be ignored.

PART V Asking Hard Questions, Sorting Out Answers

15. The Ethics of Exploitation and Intervention: Do We Have the Right?

A good rule of angling philosophy is not to interfere with any fisherman's ways of being happy, unless you want to be hated.

—Zane Grey, 1919

In a book purporting to be an objective analysis of fish conservation issues, inclusion of subjective and introspective material could be considered unwarranted; nevertheless, many conclusions reached in this chapter are as much opinions as rational decisions. Ethical decisions can be based on systematic and rational arguments, but they are often also legitimately personal. We must make decisions about the welfare of fishes and how our individual and collective actions influence their future. My chief objective here is to provide information for making those decisions.

Ethical issues and decisions influence our actions toward fishes in numerous arenas. I've chosen to give added scrutiny to questions concerned with commercial and subsistence exploitation, sport fishing, management and conservation intervention, museum collections, and conducting research on fishes. I hope some of the questions posed and answers considered will be unanticipated, or at least thought provoking.

WHAT ARE ETHICS AND WHY DOES THE TOPIC MAKE US SQUIRM?

Ethical questions deal with moral constructs of right and wrong, questions that we must ask ourselves despite a reluctance to do so. The topic of ethics makes many scientists uncomfortable. It deals with the most profound of human dilemmas and requires answers not necessarily based on objective findings. Environmental ethics is an active discipline advanced by academics housed in the philosophy and science departments of colleges and universities (see Rolston 1988; Hargrove 1989; the journal *Environmental Ethics*; http://ethics.sandiego.edu). Distinguishing between right and wrong is seldom easy, and many people, including many scientists, are happy to leave it to philosophers and professional ethicists (see Dallmeyer 2005). However, the accelerated deterioration of the natural world increasingly forces all of us to make decisions about our activities and their consequences based not only on scientific merit, but also on whether they are right or wrong.

Fishes present extraordinary ethical problems because of the direct nature and extent of our impact on them. In numbers, weight, and species, wild fishes are exploited by humans more than any other animal group. As their numbers and diversity decline, we are increasingly forced to decide whether we have the right to continue exploitation at current levels. For answers, we can turn to experts on the subject, or to our own concerns and conscience.

Granted, many people have little trouble justifying taking the life of other organisms for consumption or personal gratification. This justification can come from beliefs that define the place of humans in the universe as ordained by a higher power. It can also come from a cognitive declaration of inalienable rights, regardless of religious conviction. Decisions for such people may be relatively easy because they live more in a world of absolutes. However, knowledge of the principles of ecology and familiarity with the complexities of the natural world make such a simplistic view unsatisfactory. Aldo Leopold's statement (1953, p. 197) that an ecological education makes us live "in a world of wounds" places ethical dilemmas in the path of any who embrace a broader view.

One might ask how these questions were answered in the human past. Many cultures likely never gave them a second thought because of the legendary historical abundance of so many fishes. I have heard that fishermen in some developing nations question the need for conservation measures or even for management regulations. They cannot imagine that God would allow fishes—their source of sustenance—to decline to the point of insufficiency (C. Jennings, pers. comm.). Historically, however, many cultures recognized a potential for scarcity and insufficiency, evidenced by the "first fish" ceremonies that celebrated seasonal abundance brought on by spawning migrations, or by the practice of giving thanks for a successful catch. Such ceremonial recognition of the influence of higher powers on availability of food items could reflect deep religiosity, arising perhaps from painful knowledge of the unpredictability of the resources on which people were critically dependent.

THE ETHICS OF EXPLOITATION—AN EVOLUTIONARY VIEW

From a biological standpoint, it is easy to argue that humans evolved as hunters and that our diet requires animal protein. Our anatomy and physiology reflect natural selection for an ability to perform mental and physical tasks necessary for tracking, luring, outsmarting, pursuing, and subduing animal prey. Little wonder that we gain no small degree of satisfaction if not triumph and exhilaration upon landing a large and powerful fish. Because many activities that give us great pleasure have a direct, positive effect on our fitness, natural selection can explain both the ability to catch prey and any positive feelings surrounding the effort. A free-market philosophy extends this argument to commercial fishing, whereby someone engaged in fishing as a livelihood is providing sustenance directly as food, and indirectly as cash or barter, for self and kin.

Subsistence and Commercial Fishing

Unless you are a strict vegetarian, you are unlikely to have a problem with sustainable fishing, especially subsistence fishing. Commercial fishing—with its basis in market exchange, energy consumption, and exported products—is for many people more problematic. Some improvements have occurred via consumer action, such as seafood watch, ecocertification, and sustainable aquarium collecting practices, but ethical complications arise beyond the issue of sustainably practiced commercial fishing. If we think at the level of impacts on ecosystems, a new set of issues must be considered. For example, what should be our relationship to species with which we compete for fish prey? These predators reach high abundances and remove significant numbers of fishes, including species that are in decline from overfishing.

Corkeron (2004, p. 848) put the question in a specific ecosystem context, "How much of the reduced productivity of the oceans and coasts should remain available to whales?" By some estimates, whales take three to five times more fishes than commercial fisheries (www.iwmc.org). This impact is growing as many whale populations rebound following the general moratorium on whaling instituted in the mid-1980s. Because many people believe that "fewer predators must equal more fish" (Corkeron 2004, p. 848), culling whales has been rationalized in terms of global food security, as a means of increasing the supply of fishes to a growing and hungry human population. Whaling can even be portrayed as part of integrated ecosystem management, justified on ecological as well as ethical grounds (www.icrwhale.org).

Although cetacean control has been promoted as a means of boosting fisheries yields (Simmonds and Hutchinson 1996), global fish stocks plummeted throughout the 1950s and 1960s despite the catastrophic depletion of most large (competing?) whale species. Killing whales does not necessarily lead to more fish. Most large whales eat small crustaceans or clupeoids—both immensely productive, abundant, and short lived—not the higher-level predatory fishes that are targeted by most fisheries.

Whales, however, have advocates, including those who profit from the whale-watching industry. Many of our other competitors for edible fish are less charismatic, and their elimination has been promoted and attempted in marine systems. Researchers have designed studies to estimate the changes in fish abundance that might result from predator removal programs (e.g., Schweder et al. 2000). Winter skates and spiny dogfish have apparently replaced groundfishes on Georges Bank, prompting suggestions that an interesting experiment in ecosystem control would be to target skate and dogfish. Such experimental fishing "would result in an increase in more valuable biomass of groundfish," according to Rothschild (1991, p. 92).

Similar proposals have been offered, and in some cases laws passed, encouraging seabird and fur seal control in South Africa and New Zealand as a means of controlling

Figure 15.1. Recreational fishing is fraught with controversy. Reprinted with kind permission of Jim Borgman and Jerry Scott, ©ZITS PARTNERSHIP, KING FEATURES SYNDICATE

competitors with humans ("culling of pests") and thereby augmenting fish stocks (Crawford et al. 1992; Lalas and Bradshaw 2001). Legal and illegal killing of seals, sea lions, dolphins, sea turtles, and seabirds is practiced around the world in this context. In the northwest Atlantic, management plans designed to recover cod stocks have justified competitor control by invoking trophic cascades (Sugihara et al. 1984). Seal control has been rewarded with a bounty system in eastern Canada, purportedly to protect valuable fishes such as cod (Mowat 1996).

Results of efforts to eliminate species that compete with humans are as inconclusive as the assumptions on which they were constructed are tenuous. Cod in the North Atlantic remain depleted. In southwest Africa, where sardine stocks declined tenfold after the 1960s, overfishing and unfavorable oceanographic conditions overwhelmed any possible effect of seal predation (e.g., D. C. Boyer et al. 2001). Whether any of this human aggression has had an impact on competitor populations or fish stocks remains largely unknown (e.g., Wickens 1996).

It is probably correct to assert that predation by benthic fishes, seabirds, and marine mammals depresses fish populations when those populations are already at historical lows. Predation by cormorants on a number of depleted fish stocks—including lake trout, European grayling, salmons, yellow and Eurasian perch, walleye, eels, and flatfishes—has impaired recovery (Cowx 2002; Link 2002a). Sea lions and orcas take large numbers of migrating Pacific salmon, often in view of fishers and boaters. A key factor here is that many of these fish stocks are already in trouble. The same predators coexisted successfully with their prey alongside preindustrial human societies that were equally dependent on the resource. The predators are an unlikely initial cause of prey population declines. Ethically, do we have the right to eliminate our competitors when their impacts, in the absence of our own overuse and ecosystem disruption, are natural and minimal?

The Ethics of Fishing for Sport

Recreational fishing accounts for at least 4% of the world's fish catch. However, the activity and its impacts on habitat and bycatch are largely unmonitored, and its true extent is unknown (Coleman et al. 2004; Cooke and Cowx 2004). In the U.S., recreational fishing is one of the most popular leisure activities, engaged in by 35–60 million adults who spend \$38–\$41 billion annually (USFWS 1999; S. Waters 2002); multiplier effects indicate that the overall value of sport fishing is \$116 billion (e.g., ASA 2002). A significant fraction of the catch is eaten, but many recreational anglers fish purely for the sport, releasing their catch. As 35 million is about half the number of people who voted in U.S. federal elections in 1998, "anglers" are a significant political force, one that few elected officials want to displease.

Few topics engender more lively discussion in my ichthyology and conservation biology classes than the ethics of sport fishing (figure 15.1). My classes draw students from fisheries management and ecology undergraduate programs, making it easy to find extreme pro and con views, pitting the hook-and-bullet types against tree-hugging vegans. The challenge has been to steer the discussion away from emotional polemics and toward rational, science- and ethics-based decisions. A good starting place is an environmental ethics analysis of the human relationship with animals (e.g., Singer 1993; S. Wilson 2005).

Ethical Arguments against Sport Fishing

Fishing organizations increasingly emphasize an ethical approach to the activity, one that focuses on sustainable resource use and codes of conduct respectful to fish, other anglers, and the environment (e.g., www.fedflyfishers.org, www.nmfs.noaa.gov/ocs/ethics.html). Missing, however, is discussion of the ethics of the sport itself, of whether it is an ethical pursuit. Even in Canada, where provinces have extensive regulations or recommendations on barbless hooks, tackle size, catch-and-release fishing, playing and handling fish, and use of bait, "the ethics of causing of pain and stress are rarely addressed" (Schuppli 2000).

An important, some would say galvanizing, treatment of this issue was published by Eugene Balon of Guelph University, first in 1987, then in expanded form in 2000 (the expanded version lacked some purportedly inflammatory accusations about sadistic behavior). The essay, "Defending Fishes against Recreational Fishing," is an unabashed indictment of sport fishing. Balon's thesis is that sport fishing is "a recreational activity that uses other organisms' lives as gambling dice and their death as a source of enjoyment" (2000, p. 6). His chief arguments are the following:

- A primary allure of angling is its gambling aspect: Angling, like gambling or tobacco smoking, is an unhealthy addiction.
- Angling entails "deliberate disregard of life": "Taking the life of another organism should be done with humility and without the automatic assumption that it is our right." Advertising, profiting from, or boasting about such actions "is plainly unethical, cynical and undignified for reasoning humans."
- The contention that sport fishing groups are involved in conservation is an unfounded myth. Although anglers' associations contribute to preservation of habitats by opposing pollution and development, their ultimate goals are to guarantee their fishing rights and maximize the quality of their fishing experience. Any positive actions are outweighed by the mismanagement of natural resources for the sole benefit of recreational fishing, elimination of species devalued by anglers, haphazard and deliberate introduction of alien species, trammeling of otherwise pristine habitats, injury to nontarget species from discarded line and hooks, poisoning from lead sinkers, and pollution from gasoline engines.
- The chief intent of sport fishing is less the kill or catch, but "in fooling the fish often by a weird type of bait, or in prolonged, 'skillful' landing by means of the finest line possible . . . under the pretense of offering food . . . for little more than our pleasure only, in order to satisfy some craving for torture or to feel superior by fooling the other organism."
- A point implied but not explored by Balon is the impact of sport fishing on natural, evolutionary processes. Whereas other predators target the young, old, or infirm, angling preferentially removes the largest, hardiest, and most fit individuals, thus selecting for smaller sizes and earlier ages at reproduction (see chapter 11).

Balon has no problem with sustainable commercial and subsistence fishing, where quick and efficient capture is preferable and prolonging the kill is undesirable (although death in trawls, in gill nets, and on longlines is characteristically anything but quick). Nor does he denounce collecting for scientific purposes, for much the same reasons (some might accuse him of self-serving bias here). His complaint is that recreational anglers are chiefly interested in the enjoyment derived from matching wits with an intellectually and technologically inferior creature that is fighting for its life and suffering in the process.

Balon is far from alone in his critique of angling. Many authors have criticized "sport" fishing because of the apparent pain, suffering, and fear it induces in hooked fish, and because the struggles of the fish are the main source of enjoyment for the angler. Lord Byron called angling "the cruelest, the coldest, and the stupidest of pretended sports" (*Don Juan*, canto 13, cvi, footnote; he went on to say, "No angler can be a good man"). Descriptions of a fish's attempts to escape often include terms such as "desperation." Hemingway in *The Old Man and the Sea*, described a hooked dolphinfish, *Coryphaena hippurus*, jumping "again and again in the acrobatics of its fear" (Hemingway 1952, p. 71). John McPhee (2002, p. 321), an avid angler, wrestled with the connotations of "playing" a fish: "Never say playing. You are at best torturing and at worst killing a creature you may or may not eat. Playing at one end, dying at the other—if playing is what it is, it is sadism." De Leeuw (1996, p. 386), in a harsher indictment than Balon (2000), concluded, "It is . . . the degree to which hooked fish express their pain and suffering, for which sporting fish are valued. The erratic and rapid swimming, the twisting of the body, the jumping out of the water . . . are all behaviors of fish associated with fear, pain, and suffering." People for the Ethical Treatment of Animals (www.fishinghurts.com) called fishing "the cruelest form of hunting" (see also PISCES: The Anti-Angling Group, www.pisces.demon.co.uk).

Arguments against fishing often compare it to hunting. More than twice as many people in the U.S. fish as hunt (USFWS 1999), and fishing is often seen as less deplorable than hunting because the prey of the hunter is usually a warm-blooded mammal or bird, toward which humans feel more empathy and compassion. In the mind of the ethicist, just the opposite holds. De Leeuw argued, "While hunters make every effort to reduce pain and suffering in their game animals, anglers purposefully inflict these conditions on fish" (1996, p. 373). He continued, "The enjoyment of catching fish for sport . . . consists of purposely inflicting fear, pain, and suffering on fish by forcing them to violently express their interest to stay alive . . . it is precisely the activity of hooking and catching fish, the core of recreational fishing, for which anglers are increasingly being called cruel and immoral" (pp. 387, 380). De Leeuw consequently views fishing as "a morally unjust and wrong act" (p. 380). Yolanda Whitman, in McPhee (2002, p. 325) suggested that fishing is more like bullfighting than hunting. "The intention is to kill, and to extend the act of killing. Fishing is crueler than hunting, in that your goal is to have the fish fight for its life. That's the 'fun.' Hunting, you're trying to kill a creature outright; fishing, you want to 'play' with it."

To bring the distinction home, Balon (2000, p. 4) explored the metaphor of "sport birding": "Putting an earthworm on the hook and casting it away from the hide, the worm and hook are swallowed by a robin. . . . The shrieking and flopping bird is proudly wheeled toward the hide, the hook removed and the bird drowned in the water." McPhee (2002, p. 323) replaced worm and robin with cheese and chipmunk; thus, the chipmunk "runs flat out toward a line of trees, the reel drag clicking. Steady and attentive, the fisherman plays the chipmunk, keeping tension on the line."

Arguments for Sport Fishing

Efforts to justify sport fishing focus on benefits to the anglers, benefits to habitats, proximate and ultimate motivation of those who fish, and scriptural proclamation or reference ("dominion over the fish of the sea," Genesis 1:26; see also LaChat 1996). Many fishing advocates maintain that no real justification is necessary; fishing is something they enjoy doing, it hurts no one (no people), it is often a family and cross-generational activity, it reduces the stress of a hectic lifestyle, and it furthers appreciation of nature (e.g., Redmond and Redmond 1987). Socioeconomic arguments are also raised, such as the ethics of putting someone out of business in the course of restricting or discouraging sport fishing.

No one can dispute the complex enjoyment that an angler derives from the act of fishing, well beyond the catching of fish. As a form of thrill seeking, fishing is relatively safe and environmentally benign (hooked anglers, drownings, and overpowered bass boats excepted). Pleasure and thrills may be proximate or immediate aspects of a positive fishing experience. A deeper answer is often sought to the question of why people like to fish. An evolutionarily logical explanation is that natural selection has favored predatory success. Fishing is therefore a residual trait, selected long ago, related to a more general ability to learn the complex skill set necessary to fish or hunt successfully. "Thomas McGuane has called it 'an act of racial memory,' evoking the atavistic mission of the hunter-gatherer" (McPhee 2002, p. 310). Fishing can then be "justified" as an aspect of our predatory nature. Proximate enjoyment or thrills from the act are part of the positive reinforcement that follows its successful performance (analogously, sex would succeed as a procreative activity without any pleasure sensations, but pleasure makes us seek it further ad infinitum). Viewed thus, we can fairly ask if predation by humans is any less natural than predation in other animal forms. Recreational fishing may have its natural analogs in the "surplus killing" practiced by many predators, including foxes, coyotes, river otters, and bluefish (*Pomatomus saltatrix*) (e.g., Short et al. 2002).

Fishing is also credited with increasing one's appreciation of, and willingness to protect, nature. This argument is difficult to refute and certainly holds for many advocates of the sport, although Balon's (2000) criticisms must still be acknowledged. Similarly, fishing reduces the degree to which we are removed from the acquisition of our food. Whether or not we eat the fish we catch, we usually realize that the fish we eat was caught by someone. This fact is often advertised if not romanticized (the same is seldom the case for other meat products). Fishing therefore makes us participants in and perhaps more appreciative of the process by which our food is obtained. So do hunting, gardening, and clamming. A major difference is that the deer, the tomato, and the cherrystone have little opportunity to resist our actions, nor do we derive much pleasure from prolonging their efforts.

Catch-and-Release Fishing

Probably the most widely invoked response to criticisms of sport fishing is catch-and-release (C&R), formerly called "fishing for fun" (Barnhart 1989). The fish struggles to remain alive, the angler enjoys the fight. Dispense with the dockside photos of joyful fishers and joyless fish. Release the fish.

To Live To Fight Another Day

C&R has both ethical and management benefits, and its history reflects a shift in primary justification. A practice that was initially based on personal, moral actions has been increasingly embraced because of management considerations (Barnhart 1989; Policansky 2002). It has also been increasingly criticized on moral grounds. Back in the "good old days" when fish were much more abundant, releasing your catch was largely a voluntary, individual act, "a point of honour and sportsmanship" (Policansky 2002, p. 78). It was not until late in the 20th century that regulatory C&R came to replace voluntary C&R as a practical management tool. As numbers of anglers and their access to remote and desirable areas increased, as stocks decreased and desirable fishing areas became crowded, the capacity of good fishing areas to sustain large numbers of anglers dwindled. Agencies now managed not just the resource but also its access. Releasing fish became a management tool: More people could fish successfully despite fewer catchable fish (Policansky 2002). Individual cutthroat trout, *Oncorhynchus clarki*, in the Yellowstone River are caught an estimated 9.7 times during the summer fishing season (Schill et al. 1986). "Total C&R" (release-only fishing) is more effective than most other regulatory practices, allowing higher fishing effort without the need for stocking (Policansky 2002). From a management standpoint, little argument can be made against C&R except that its very success puts more pressure on habitats where it is practiced. Overfishing and retention, if allowed, would deter later anglers.

To Be Tortured Again

The ethical case is more tenuous. What are the ethical implications of releasing a fish that has fought to exhaustion at the end of a line? Even fishing enthusiasts are divided on this issue. Is C&R "cruelty masquerading as political correctness" (McPhee 2002, p. 319)? Is sport fishing "a game of dominance followed by cathartic pardons, which, as a nonfishing friend remarked, is one of the hallmarks of an abusive relationship" (Kerasote 1997 in McPhee 2002)? Can hard data contribute to the discussion? For example, what are the mortality rates associated with C&R, and what if any nonlethal stress is placed on a hooked or a repeatedly hooked fish? Short-term hooking mortality is generally and surprisingly low, usually less than 5%, especially when water temperatures are relatively low and capture depths are shallow. Exceptions include striped bass (9%–67%, depending on locale and temperature) (e.g., Muoneke and Childress 1994). The general conclusion is that C&R results in low enough hooking mortality to serve as an effective management tool in most circumstances (Policansky 2002).

Two requirements often instituted to minimize hooking mortality are (1) artificial lures (less likely to be swallowed than baited hooks), and (2) barbless hooks (to make removal less injurious). Few data are apparently available on the bait versus lure issue; most anglers would probably concur that deep hooking is less likely with artificial lures, but controlled studies are needed. The advantages of barbless hooks are surprisingly unsupported by available data, at least as far as injury to fishes is concerned (e.g., Schill and Scarpella 1997). If stress is the major detrimental impact of being hooked and is proportional to how long a fish is played, barbless hooks could not be expected to help greatly. The benefit to anglers after hooking themselves is fairly obvious.

Sublethal effects of angling on released fishes include altered social behavior and impaired reproduction and growth (Lewynsky and Bjornn 1987; Muoneke and Childress 1994). More immediate and easily measured are various physiological indicators of stress, which is known to occur during handling of almost any sort (e.g., Schreck et al. 1997). Stress indicators include elevated cortisol and blood lactate levels, decreased blood pH, and altered serum electrolytes associated with increased anaerobic activity and muscle fatigue. The available evidence indicates that fishing with light tackle, which "gives the fish a fighting chance," increases stress by prolonging the bout.

Working with marine pelagic species—tunas, billfishes, blue sharks (*Prionace glaucus*)—Skomal and Chase (2002) found deleterious changes in physiological factors that were directly related to the length of time a fish was fought (figure 15.2). Cortisol levels in an angled white marlin, *Tetrapturus albidus*, were among the highest reported for any fish. Significant alteration in measured factors occurred after only five minutes of fighting in tunas. Fish released and

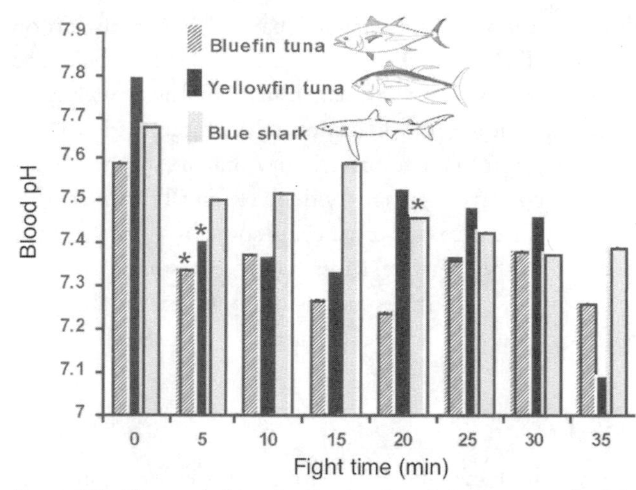

Figure 15.2. Change in blood pH related to fighting time of a fish. Lowered blood pH is associated with higher lactate levels and acidemia and results from physiological changes brought about by exhaustive exercise. Blood pH dropped after only 5 minutes of fight time in tunas and 20 minutes in blue sharks. Values are mean pH; asterisks indicate a statistically significant change. Modified from Skomal and Chase 2002; used with permission. Fish drawings from Bigelow and Schroeder (1953).

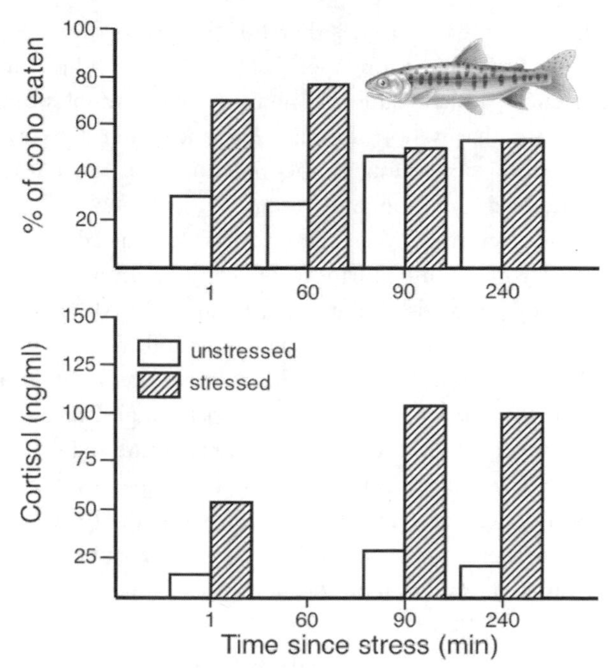

Figure 15.3. Effects of exposure to stressful situations on cortisol levels and ability to avoid predators. "Stressed" coho salmon smolts held in a dry bucket for one minute had elevated plasma cortisol levels for four hours (lower panel). They were also more likely to be captured by a predatory lingcod, *Ophiodon elongatus*, compared with control ("unstressed") fish that were merely transferred between tanks (upper panel). Redrawn from Schreck et al. (1997) based on Olla et al. (1992); coho drawing inset by K. Spencer, www.katespencer.com, in Augerot (2005); used with permission.

tracked after fighting showed limited activity for up to two hours after release, suggesting a mandatory recovery period. In flowing water, the effect might be an inability to hold position against a current. Recovery in marine pelagics was enhanced by resuscitation; the one bluefin tuna, *Thunnus thynnus*, that was not resuscitated died upon release. Skomal and Chase recommended that resuscitation efforts last at least five to seven minutes for bluefin tuna to promote carbon dioxide offloading and oxygen delivery. They postulated that fish that are not sufficiently resuscitated after prolonged angling bouts may be vulnerable to predators.

Experimental as well as anecdotal evidence supports the contention that handling leads to stress and increased vulnerability to predation (figure 15.3). Importantly, although resuscitation may improve physiological indicators, even resuscitated fish are vulnerable during the post-release recovery period. Bonefish, *Albula vulpes*, released by anglers in the Bahamas are frequently seized by lemon sharks, *Negaprion brevirostris*, that follow the action (J. Maynard, S. Gruber, pers. comm.). Sharks may attack over 40% of released bonefish in some locales (Cooke and Philipp 2004). Mortality among bonefish caught and released repeatedly where sharks were absent was less than 5% (Crabtree et al. 1996). If other species suffer similarly from predation after angling, "hooking mortality" data, which are usually based on fish held for observation rather than released, would underestimate such losses.

Socioeconomic and cultural differences lead to different views of C&R. Some native peoples in North America view sport fishing as disrespectful to the fish if it is not eaten (Policansky 2002). C&R is also largely a luxury of well-fed individuals in industrialized nations. And even in the industrialized world, practices differ greatly. For example, in Germany, sport fishing is discouraged and catch-and-release is illegal. "The animal protection law considers it an offense to injure fish for no good reason, and sport fishing in itself is no good reason" (Uwe Schuller quoted in McPhee 2002, p. 327).

Do Fish Feel Pain Like Humans?

Central to the ethical debate over fishing, and critical to the issue of recreational fishing regardless of whether fish are kept or released, is whether fish feel pain. J. D. Rose (2002, p. 29) asked more specifically "whether fishes can experience pain and suffering in a way that resembles our experience, that is, do fishes have a capacity to suffer that meaningfully approximates the psychological impact of pain-induced suffering in humans?" A definitive answer has proven elusive: "Veterinary authors of guidelines for assessing pain in animals more or less give up in despair when they get to fish" (Bateson 1991, p. 832). More general is whether fish suffer at our hands, and if so, what if anything we should do to minimize this suffering. If fish felt pain or suffering, would you stop fishing? The topic also has relevance to the conduct of aquaculture and scientific research, with serious economic consequences. Major scientific societies have dealt at length with the topic, although their conclusions, or avoidance of same, reflect the ambiguity and contentiousness of the issue (e.g., AFS 1999; FSBI 2002).

Much of the discussion hinges on terminology, definitions, and, to a degree, semantics. *Pain, discomfort, suffering, stress, distress, fear, anguish,* and *anxiety* are all relevant terms and constitute a thesaurus of unpleasant reactions and emotions. Dictionaries focus on general descriptions, such as sensations of acute physical hurt caused by injuries. The International Association for the Study of Pain (IASP) is more specific, calling pain "an unpleasant sensory or emotional experience associated with actual or potential tissue damage" (Merskey and Bogduk 1994). IASP suggests that when conducting research on pain in nonhuman animals, "the investigator should try the pain stimulus on himself."

Most definitions make a distinction between pain and *nociception* (but see Broom 2001). Nociception involves only the detection of a noxious or tissue-damaging stimulus, followed by a reflex response. Pain perception involves nonreflexive behavior by an animal that has been adversely affected by a potentially painful stimulus (Sneddon 2003). Some add that pain requires that an animal learn from the experience and thus avoid or minimize future occurrences (Broom 2001; Sneddon et al. 2003). Whereas Broom and Sneddon discuss no more than "perception" of pain, more stringent definitions require "awareness" (e.g., J. D. Rose 2002).

Do fish feel pain? Perhaps. Like humans? What difference does it make? Is pain as we know it the only pain possible? Do you experience pain the same as I do? My wife bore both our children without anesthesia; I'm immobilized by a hangnail. Pain is obviously a relative and subjective phenomenon (e.g., Bateson 1991) and the answer to the question of whether fish feel pain will hinge in part on the definition applied. Some would argue that pain and suffering are uniquely human traits. For those willing to entertain a more open discussion of the topic, evidence can be gathered from anatomical, behavioral, neurophysiological, and cognitive research. Evidence from all categories has been marshaled to answer the question in both the affirmative and the negative. Most wisely, Bateson wrote, "No single criterion provides an all-or-none test for the existence of a subjective sense of pain" (1991, p. 834).

Fish Do Not Feel Pain

The scientific evidence against pain and suffering in fishes was reviewed and synthesized by Rose (2002) in a mechanistic, neurophysiologically based paper that should reassure both anglers and physiologists (see also LaChat 1996; Bermond 1997; Macphail 1998). Rose concluded that many observations of apparent suffering in nonhuman animals resulted from anthropomorphizing—that is, interpreting "reactions by a fish to stimuli that probably would cause pain in a human" (p. 21). Rose maintained that (1) awareness of pain requires consciousness; (2) consciousness resides in the complex neocortex of the cerebrum; (3) only mammals have a neocortex; (4) therefore, fishes cannot experience pain. Fishes do react to injurious or noxious stimuli, but these stimuli are sensed peripherally by nociceptors, generally free nerve endings in the body surface and head with afferent fibers running to the spinal cord and brain stem. Sharks and rays, Rose argued, lack the neural structures needed to process nociceptive information, much less sense pain, "which may explain their relative behavioral unresponsiveness to injury" (p. 32; Sneddon [2003] cited studies that identified nociceptors in elasmobranchs). Pain, which "is always subjective," is "a psychological experience that is separate from behavioral reactions to injurious stimuli"; it "is both a sensory and emotional experience (that requires conscious awareness) . . . [whereas] nociception does not result in pain unless the neural activity associated with it reaches consciousness" (Rose 2002, pp. 15–16). Fishes display nociceptive reactions, but such behaviors should not be confused with an experience of pain: "Working together with the spinal cord, the brain generates responses that cause the organism to 'escape' or 'avoid' these stimuli. These responses are produced by innate neural programs and include withdrawal of the stimulated body part, struggling, locomotion, and in some animals vocalizations" (p. 16).

Rose extended the argument to cover more than injury and more than fishes. He stated, "The brain systems known to be necessary for the experience of fear or other emotional experiences are not present in fishes" (2000, p. 26). In conclusions that will surprise pet owners, animal trainers, and animal behaviorists, he said (p. 25), "because conscious awareness of pain depends on extensively developed frontal lobe neocortex, few (if any) mammals besides humans possess an adequate cortical substrate for pain experience."

Rose's arguments set the bar high: Fishes lack structural homologs to the cerebral cortex, where pain sensation must be housed. Convergent evolution of analogous structures or even pathways is apparently disallowed. "If some form of pain awareness were possible in the brain of a fish . . . its properties would necessarily be so different as to not be comparable to human-like experiences of pain and suffering" (2000, p. 32). Insisting on "human-like experiences" of pain seems a little stringent (see also Braithwaite and Huntingford 2004), analogous to insisting that human-like color perception is the only true form of color vision. Such a line of thinking might conclude that insects can't fly because they lack the wings of birds. Sneddon (2003) maintained that Rose "conveniently ignored the bird and amphibian literature that has proven these animals are capable of pain and yet they do not have a neocortex" (p. 154). Recent findings on sleep in animals show that brain recordings of resting crayfish are characterized by so-called slow waves, previously known only from the mammalian cortex during sleep; crayfishes lack a cortex, and the source of their waves is presently unknown (Ramon et al. 2004). Does this therefore mean they do not sleep? Even the Fisheries Society of the British Isles, otherwise hesitant to take a definitive stand on the ethics of exploiting fishes, concluded that "lack of a neocortex does not mean that fish cannot experience some kind of suffering" (FSBI 2002, p. 2).

Rose sells fishes short when it comes to recognizing their capacity for flexible behavioral responses; he stated, "In spite of the diversity and complexity among species, the behaviors of fishes are nonetheless highly stereotyped and invariant for a given species. . . . The basic behaviors involved in reproduction, feeding and drinking, escape or defense, and reactions to noxious stimuli are controlled by motor patterning mechanisms that are located mainly in the brainstem and spinal cord" (2002, p. 9).

I have spent the majority of my professional career observing fishes underwater and trying to describe and understand their behavior, especially during predator-prey encounters. Along with many colleagues, I have been continually frustrated by variability within species and even within individuals and have concluded that variation in the responses of fishes to stimuli and ecological conditions is extensive and, in fact, adaptive. Fishes are able to assess nuances of difference in the behavior of predators and prey and respond accordingly (Helfman 1990). During predator-prey encounters, this variation can be explained in part by a Threat Sensitivity Hypothesis (Helfman 1989; Helfman and Winkelman 1997): Fish assess the degree of threat imposed by a particular predator at a particular moment and adjust their response accordingly. Strong threat elicits a strong and effective avoidance response, whereas minimal threat elicits less caution. Rose's view of fish behavior, which eschews terms such as *fear* and *flight*, implies a much more simplistic, invariant—and less adaptive—catalog of responses. In my view, fish may not experience fear emotionally in human terms, but their reactions to threatening stimuli are sufficiently analogous to fear reactions that interpretations positing fear and flight are descriptive, instructive, and warranted.

Fish Do Feel Pain

Rose (2002) argued that pain requires consciousness, which is known to reside only in the mammalian (perhaps only the human) neocortex. There are, however, precursors to these higher-level sensations. Initial detection of painful stimuli requires receptors specifically sensitive to harmful or injurious events. The existence of such nociceptors in fishes has recently been demonstrated in the head of rainbow trout, *O. mykiss* (Sneddon et al. 2003; figure 15.4), which is, not coincidentally, where most fish are hooked. Thresholds of response for some nociceptors were six times more sensitive than similar skin receptors in humans (0.6 g in humans vs. 0.1 g in trout), with mechanical stimulation thresholds comparable to those of mammalian eyes. Other criteria for a pain sense include appropriate neurological and physiological components, behaviors that can be viewed as nonreflexive, and a learning capability that would protect an animal from future painful events.

Additional anatomical evidence for pain perception in fishes includes the existence of neuroanatomical components and pathways such as neurotransmitters and neuromodulators (e.g., opioid neuropeptides) similar to those present in the nociception systems of mammals (Rose 2002). Physiologically, fishes respond to injurious and

Figure 15.4. Pain receptors on the lips, snout, and gill cover of a rainbow trout. Such receptors—nerve endings that react to various injurious substances such as strong mechanical stimuli, harsh chemicals, or high temperatures—are a key element in a general "sense" of pain. Triangles = polymodal nociceptors; diamonds = mechanothermal nociceptors; hexagons = mechanochemical receptors. From Sneddon et al. (2003); used with permission.

stressful stimuli as do mammals and many other vertebrates. Nociceptor signals activate hypothalamic neurons, initiating release of cortisol and catecholamine. Morphine blocks reactivity by fishes to injurious stimuli such as electric shocks, although the level of action of morphine is subcortical or peripheral, not central as it is in mammals.

Sneddon (2003) and Sneddon et al. (2003) tested specifically for behavioral responses in rainbow trout indicative of pain perception. These researchers injected acetic acid into the upper and lower lips of trout that had also received intramuscular injections of morphine, with controls for handling and treatment. They found a variety of delayed (i.e., nonreflexive), anomalous reactions in acid-injected individuals that included rubbing their lips on the gravel, resting on the bottom and rocking back and forth, elevated opercular beat rate, and delayed resumption of feeding. A time lag existed between treatment and performance of these apparent "pain-coping" tactics. This lag suggested that the animals had been stressed and that their reactions were not reflexive but were instead "similar to some of the pain-related responses of higher vertebrates and man . . . [suggesting] higher processing is involved" (Sneddon 2003, p. 160). Sneddon et al. (2003) concluded that their evidence on anatomy, physiology, and behavior "fulfils the criteria for animal pain" (p. 1120).

Behaviorally, some fish also exhibit one-trial or similarly rapid avoidance learning when subjected to experimental angling (Beukema 1970a, 1970b; Lewynsky and Bjornn 1987). Also, many fishes when grasped or speared emit vocalizations and other "distress calls," which are at least analogous to the screams of injured birds and mammals. Hence evidence from anatomy, physiology, and behavior suggests that fishes exhibit responses to aversive stimuli and capabilities that are analogous if not homologous to those of higher animals.

Evolutionarily, it seems reasonable that even the lowest or simplest animals should be able to detect and respond to harmful stimuli in an adaptive manner. Pain or its analog(s) provide a framework around which rapid, adaptive learning will develop. Natural selection would quickly eliminate individuals incapable of protecting themselves from potentially damaging conditions or avoiding such conditions in the future (e.g., Bateson 1991; Broom 2001). That many prey fishes have sharp spines reinforced with toxic secretions is strong, evolutionary (and thus circumstantial) evidence that their predators can learn from "painful" experiences.

If fish do in fact feel pain, will you stop fishing?

Where I Stand

I have fished professionally and recreationally, commercially, scientifically, for subsistence, and for pleasure, using just about every method except dynamite. As a devout biological determinist, I rationalize the enjoyment I experience in hooking and fighting a fish as something deep in my genetic makeup, basically ascribing to the atavistic predator scenario. Being able to subdue active, healthy prey—those richest in energy and least likely to carry disease—required skill, cunning, athleticism, and some bravery among our ancestors. Chicks dig long-ball hitters. The rest is human history.

I disagree with Lord Byron's contention that a fisherman cannot be a good man, but I concur that arguments against fishing purely for sport should give any thinking person, especially one concerned about conservation, pause. Along with Thoreau, I often find "I cannot fish without falling a little in self-respect. . . . With every year I am less a fisherman" (1971, p. 192).

INTERVENTION, SCIENCE, AND ETHICS

Exploitation carries a connotation of calculated disregard for the welfare of the fishes, calculation of benefits to us against (discounted) costs to them. But all too often, our efforts at managing and conserving fishes also conflict with their welfare and interests, requiring us to reconsider the rightfulness of our actions.

Mismanagement through Ignorance

Natural systems are complex and the gaps in our knowledge immense, leading frequently to counterproductive and destructive mistakes. When we intervene in nature, even with the best intentions, we risk doing more harm than good. Deserving of criticism in this context are "reclamation" programs that make habitats suitable for often nonindigenous sport fishes via the poisoning of nongame, native trash fishes. One example was the near extermination of the Miller Lake lamprey, *Lampetra minima*, in Oregon in the process of improving trout fishing (R. R. Miller et al. 1989; see chapter 2). Another horror story was the incidental poisoning in 1962 of downstream reaches of the Green River, Wyoming. A reclamation project attempted to eliminate nongame fishes and make the newly built Flaming Gorge Reservoir more trout friendly. More than 21,000 gallons of rotenone were dripped into the Green River; but insufficient neutralizing chemicals were used, and the poison moved far downstream and into Dinosaur National Monument. Four now federally Endangered species of minnow and sucker were among the collateral casualties (see P. B. Holden 1991).

Such mistakes are not limited to management actions intended to augment sport fisheries. In the Virgin River, Utah, rotenone was applied to eradicate introduced red shiner, *Cyprinella lutrensis*, an invasive species often implicated in native fish declines. Miscalculations extended the fish kill an additional 50 km beyond the intended 34 km, into habitat of the endangered Virgin River chub, *Gila seminuda*. Rotenone toxicity was sufficient "to kill a substantial proportion if not most of the fish fauna" throughout the river corridor (Demarais et al. 1993, p. 338). Recolonization of chubs was surprisingly slow from an upstream area that was expected to serve as a source population. Genetic sampling 2.5 years later indicated that very few, closely related adults were breeding with one another. Loss of genetic variability and inbreeding depression were probable results of this "management accident" that had been intended to improve conditions for a species in decline.

Other unanticipated outcomes can arise because of natural complexity. Many endemic desert fishes live in isolated springs, systems disrupted by even small perturbations. Management agencies work hard to minimize human impacts, including fencing out large, nonnative herbivores that trample vegetation and drink or pollute the water. However, some desert springs and their biota apparently persisted because of herbivore disturbance. Ash Meadows National Wildlife Refuge in Nevada and Dalhousie Springs in Witjira National Park, South Australia, experienced a succession of large mammals. First were megafaunal herbivores such as giant marsupials in Australia and various ungulates in North America. When extinguished, these herbivores were replaced by introduced feral livestock (camels, donkeys, and horses in Australia; donkeys, horses, and cattle in Nevada), which grazed down and trampled emergent vegetation. Fencing allowed vegetation to proliferate, choking and drying up springs, creating anoxic conditions, thereby reducing and eliminating endangered endemic fish species (Kodric-Brown and Brown 2004 and pers. comm.). Protecting such systems requires an adaptive approach that recognizes the importance of disturbance in maintaining intact ecosystems.

Electrofishing: First Do No Harm

Electrofishing, stunning fish with electricity, has multiple advantages over nets, traps, explosives, and chemicals for collecting fishes. It employs portable backpack shockers or large boat-mounted and shore-based generators and is most effective in flowing water and over irregular bottoms, where nets are difficult to deploy. In addition, fish can be released in apparent good condition (see J. B. Reynolds 1996 and especially Snyder 2003). Electrofishing was for almost five decades promoted as minimally destructive and basically harmless, the method of choice. Minimal impact was assumed because immediate mortality was minimal and electrofished individuals recovered quickly and swam away. Fishes held for hours and even days showed little obvious damage. This efficient and nondestructive methodology was applied widely, especially in streams containing valuable and relatively fragile salmonid species.

Such was the situation until people began dissecting electrofished individuals and, more important, x-raying them (occasional injuries had been documented as early as the 1940s—e.g., Hauck 1949—but were considered unimportant for many years). Dissections revealed that significant numbers of some species suffered high rates of arterial and aortic hemorrhaging, as well as spinal breakage and deformities (figure 15.5). Injury rates of 20%–40% were

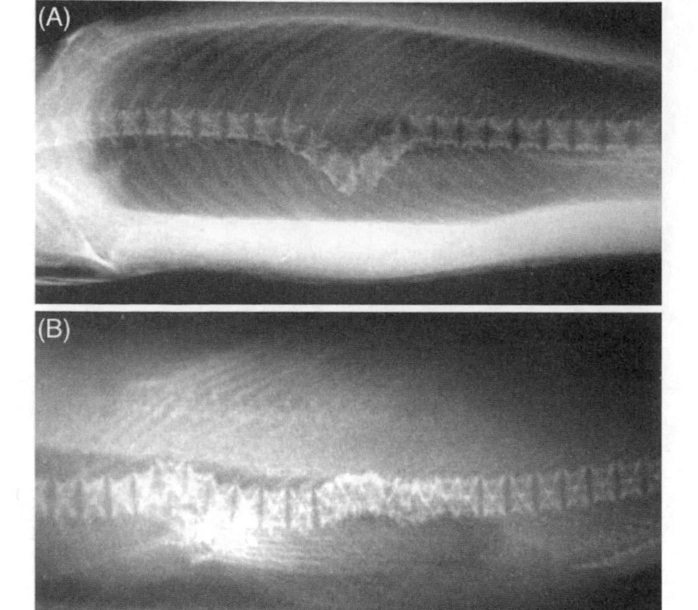

Figure 15.5. Spinal deformities caused by electrofishing. (A) Dorsal and (B) lateral views of a rainbow trout, showing dislocation and breakage of vertebrae. Head is to left in each photo. Photos by N. G. Sharber, Flagstaff, Arizona; used with permission.

not unusual, and rates occasionally reached 98% in trout. Delayed mortality rates of 0%–25% were also documented (Schill and Beland 1995; Nielsen 1998a). X-rays of fish with no obvious external injuries showed injury rates of 6%–32% (Dalbey et al. 1996; Kocovsky et al. 1997). Longer-term studies revealed reduced growth and reproduction due to injuries, stress, and secondary infection (e.g., VanderKooi et al. 2001). One year after electrofishing, rainbow trout that suffered moderate or severe spinal injuries showed little or no increase in length or mass, whereas control fish, including some with minor injuries, grew an average 2 cm and gained 40 g in mass (Dalbey et al. 1996; see also K. G. Thompson et al. 1997). Eggs, embryos, and larvae are very susceptible to electrofishing. Embryonic development, hatching success, and larval growth of endangered razorback sucker, *Xyrauchen texanus*, were all impaired by it (Muth and Ruppert 1997). Besides salmonids, injury-sensitive groups include paddlefish, goldeye, suckers, catfish, sculpins, largemouth bass, and walleye (Snyder 2003).

Simplified, in electrofishing, a strong electric field causes a fish's muscles to contract, forcing it to swim involuntarily toward the anode (galvano- or electrotaxis), basically following the path of electrons. The field becomes more dense (stronger) the closer the fish is to the anode. Normal swimming involves alternating contraction of muscles on either side of the body; in an electric field, muscles on both sides contract simultaneously. At very strong field densities, the muscles become tetanic (completely contracted); the fish is immobilized and floats to the surface or sinks to the bottom. The reaction is physically and physiologically equated to an epileptic seizure (Snyder 2003). Injury occurs when muscular contractions are so intense ("myoclonic jerking") that vertebrae are fractured or displaced; muscles hemorrhage from overcontraction; and neural tissues, including those of the brain, are overstimulated by electricity (e.g., Sharber et al. 1994). Bleeding at the gills, asphyxiation, cardiac arrest, and stress are added concerns (Snyder 2003).

If it were simply a matter of field strength, a balance could be struck between capture efficiency and injury. The degree to which fish are affected, however, is a function of many factors besides output strength and type (DC vs. AC), including current density (a measure of field strength equal to voltage gradient or resistivity); electrode shape and size; and electric wave form, pulse frequency, and pulse width. (For technological improvements, see www.smith-root.com. Smith-Root, Inc., which claims to be "the world leader in electrofishing equipment," provides no information or links on its Web site related to electrofishing-induced injury to fishes. It does offer a two-day course in electrofishing that includes techniques to minimize threat of injury, for $350 per person.)

Injury frequency also varies with length of sampling period; water conductivity and temperature and streambed conductivity; fish proximity to the anode; fish size and species (flesh density, conductivity, body form, swim bladder configuration, behavior); and even conditions under which individuals were reared. Larger fish are more susceptible because a stronger field gradient is established along their length, creating higher current density. In low-conductivity water, capture efficiency is lower but mortality higher because the field is limited to the immediate vicinity of the anode, where it is lethally strong. Hatchery-reared fish are less susceptible to injury than wild fish, perhaps because crowded confinement leads to reduced activity, lower bone and muscle density, and hence weaker muscle contractions and more compressible bone (Hollender and Carline 1994). The lowered sensitivity of hatchery-raised fish has additional conservation implications, namely the contentious issue of whether hatchery-raised salmon should be counted along with wild fish when determining population status of imperiled stocks. Electrofishing-based enumerations would differentially affect the two groups, injuring wild fish more and further tipping the balance in favor of hatchery fish.

At issue from a conservation and ethical standpoint is the trade-off between information gained and fish injured. In the mid-1990s, the growing evidence of injury from electrofishing led to a backlash against its use (e.g., Snyder 1995; Schill and Beland 1995). Schill and Beland pointed out, however, that in a typical fisheries survey program, only a fraction of a stream is actually surveyed (e.g., 500 m of a 10-km reach), fewer than 5% of the fishes are exposed to electrofishing, and in a "worst case" scenario 3%–4% of the shocked fishes die or are injured. Given natural annual mortality rates for stream salmonids of 30%–60%, "many of those few fish 'saved' by an electrofishing ban . . . would soon die of natural causes" (1995, p. 28).

Such mortalities seem a small price for critical information. Fisheries and conservation management alike require sampling of population size and change, age structure, individual growth, and reproductive status and season. In fisheries programs, some degree of injury and mortality is an accepted and unavoidable condition. Poisons, explosives, trawls, and to a lesser extent gill or trammel nets are efficient and relatively unselective of size and species (which is desirable) but have high mortality rates. Seines, block nets, and traps are more selective and less effective, despite a lower incidence of damage and death. Electrofishing is at least as effective as poisons and explosives and less injurious than those and seines, block nets, and traps. Injury rates of 20%–40% and mortality rates of less than 5%–25% are more desirable than 100% mortality. "We may just have to accept that DC-electrofishing injures more fish than we previously thought" (Schill and Beland 1995, p. 29).

Therefore, at the population level, electrofishing has minimal apparent impact, especially among nonsalmonid sport and game species. At the individual level, different rules apply. Age and growth data obtained from repeated, annual electrofishing of the same individuals would be invalid. Causing death, injury, immunosuppression, impaired growth and reproduction, or stress of federally listed species is prohibited as a "taking" under the U.S. Endangered Species Act. It is in the context of assessing the status of listed species—some with population sizes numbering only in the dozens—that alternatives to electrofishing should be considered (Nielsen 1998a; Snyder 2003). Collateral damage to nontarget species should also be considered when sampling in habitats containing imperiled fishes (and other taxa).

Among the alternative methods available are snorkel surveys (Nielsen 1998a). Snorkeling, commonly used for population enumeration in marine surveys, is almost as effective as electrofishing in surveys of stream fishes (e.g., Thurow and Schill 1996). It has obvious limitations in very shallow, swift, or turbid conditions but is useful for surveying the large fishes that are most susceptible to electrofishing damage, as well as enumerating the small fishes that are passed up by electrofishing gear. Snorkeling certainly has to be the least intrusive of any enumeration method, which makes its disadvantages seem minor where endangered species or habitats are involved.

Museum Collections: Necessity or Anachronism?

In chapter 13, collecting for aquaria was challenged on ethical grounds. What about collecting fishes to preserve them on the shelves of museums? Few informed people, aside from the most ardent animal rights advocates, question the essential role that museums play in documenting and protecting biodiversity. However, several colleagues who work in museums have pointed out to me that the public at large is often unappreciative or uninformed about this role, and even many scientists view museum collections as holdovers from 19th-century biology. In fact, museum collections and the collecting activities of taxonomists in such institutions form the framework upon which our catalog of biodiversity is built. Museum collections are also the reference point we will use to assess our success in conserving biodiversity (for a history of fish collections worldwide, see Pietsch and Anderson 1997).

However, museum collectors are not blameless in the arena of abuse of fishes. Professional ichthyologists and fish researchers collect fishes from the wild using a variety of methods, some more humane than others. In recent times, the number of fishes retained during sampling has decreased, "terminal sampling" has grown more difficult to justify, and standardized means of euthanizing fishes (as opposed to tossing them live into a jar of formalin) have been developed, promoted, and increasingly practiced (e.g., Robb and Kestin 2002).

Attitudes have also changed. Thirty years ago, I knew of museum collectors who removed large numbers of known endangered fish species "before they are all gone." A colleague referred to one such researcher as "second only to the Corps of Engineers as a threat to southern fishes." Such practices are extremely rare today, although not all parties adhere to ethical guidelines. I regularly visited

Tellico Creek, a tributary to the Little Tennessee River, for both teaching and research purposes. We kept few if any fish. Tellico is known to contain federally Threatened spotfin chub, *Erimonax monachus*; it is now off limits to me and others because an educator from another institution took classes there repeatedly and preserved large numbers of fish. Researchers and educators alike have to embrace a minimal-impact, leave-no-trace approach while conducting their work. Otherwise, our credibility as practicing conservation biologists will be undermined by justified accusations of hypocrisy (e.g., Spear 2004).

The Ethics of Conducting Research on Fishes

Fishes serve as research subjects and "models" for studies of general physiological phenomena or in human medical research. Growing concern over animal rights and animal welfare issues has prompted the development of guidelines and enforcement of regulations (animal rights and animal welfare proponents have different goals, the former focusing on an end to all human exploitation of animals, the latter emphasizing prevention of suffering and cruelty; see www.sover.net/~lsudlow/ARvsAW.htm). The regulations vary in stringency and inconvenience, depending on your viewpoint. Animal care guidelines and requirements were first applied to mammalian research but have worked their way "down" the zoological hierarchy and now apply to vertebrates in general (e.g., ILAR 1996). Requirements for care, handling, and methods of termination affect the use of fishes in governmental and institutional research programs. Some regulations are direct responses to careless and even cruel practices by researchers in the past. Among those most affected are general physiologists working in nonmedical fields whose manipulations—sometimes referred to as the science of slowly dying animals—must now be justified to an increasingly skeptical public, as well as to funding agencies. To such workers, confirmation of awareness, pain, suffering, fear, distress, discomfort, and other responses in fishes is a major complication.

It is therefore not surprising that evidence refuting suffering in fishes comes from physiologists and turns largely on physiological and neurobiological findings, as interpreted by Rose (2002) in his arguments on the lack of pain awareness in fishes. Rose (2002) concluded his discussion with a nod to animal welfare issues, allowing that good science demands healthy, well-cared-for subjects: "When fishes are used in research, their physical well-being is of great importance for the collection of sound scientific information" (p. 30). That such conclusions would find a receptive audience among anglers is evidenced by the appearance of earlier versions of Rose's arguments in the January 2000 issue of *In-Fisherman* magazine and in an address given in October 2000 to the Wild Trout VII conference. In the latter presentation, Rose equated the actions of anglers with those of a natural predator by concluding, "the struggles of a fish don't signify suffering when the fish is seized in the talons of an osprey, when it is devoured while still alive by a Kodiak bear, or when it is caught by an angler" (p. 13).

CONCLUSION

I have in my lifetime probably killed more fishes—for sport, for food, while making a living, or in the course of conducting research—than the most ardent angler. And for the longest time, I did not question whether my actions were right, whether they could be or even need be justified. But times and conditions have changed. I ask only that each individual at least ask himself or herself, "Is this act ethically justifiable? Could it be modified to promote rather than harm the welfare of fishes?"

Aldo Leopold said that a thing is right "when it tends to preserve the integrity, stability, and beauty of the biotic community. It is wrong when it tends otherwise" (1949, pp. 224–25). His arguments are compelling because we respond to them at both objective and emotional levels. His appeal to emotion has increased acceptance of his ideas among adherents of a conservation ethic. Ethical questions often turn on emotional reactions. No one has surpassed Leopold's eloquence, but we can add to his argument by invoking ecological and evolutionary justification for our actions. We need to think not as a fish, tree, or mountain, but as an ecosystem with its evolved components. Our actions should be governed and directed by how well they approximate the evolutionary process. Humans play a role in ecosystems, and our impacts should be sensitive to the solutions evolved by other ecosystem constituents. We should exploit numbers, age classes, and sizes of animals as if we were a natural predator. We should not remove any and all, disregarding the mortality levels a species has come to expect evolutionarily. Nor should we disregard our impact on other ecosystem components, whether for commerce, pleasure, or research. It is therefore right to act and manage with natural selection in mind. It is wrong to do otherwise.

16. Future Perspectives: Beyond Gloom and Doom?

The realization that the biosphere is driven by an interdependence among biotic and abiotic components challenges both small minds and big governments.

—Judith L. Meyer and Gene S. Helfman, 1993

What does the future hold? What do the examples and lessons discussed in the preceding chapters tell us about the future of biodiversity and of human interactions with fishes and their habitats? What are the role and responsibility of the environmental scientist or the concerned citizen in pointing out problems and offering and participating in solutions? An abundance of thoughtful, creative, workable solutions exists for almost all the problems and situations chronicled in this book; many were presented at the end of individual chapters. In this chapter, I will try to generalize and synthesize, focusing on the positive and constructive, while attempting to avoid sloganeering and preaching. The selected topics have been drawn from literature dealing with fishes but unquestionably reflect issues and solutions that extend beyond fishes.

AN OVERVIEW OF SOLUTIONS

The literature on declining biodiversity among fishes is blessed with almost as many answers as questions. A number of authors have reviewed the general topic and presented a series of solutions (box 16.1). Despite differences in the problems addressed—as well as in focus, background, and motivation—investigators have arrived at a number of similar solutions. Admittedly, much cross-pollination and cross-referencing has occurred among these efforts, but in each case individual investigators and groups have concluded that a particular approach is as applicable and promising for their circumstances as it is for other situations. Equally interesting is the consensus among groups, such as fisheries managers and conservation biologists, that have historically focused on the purportedly conflicting goals of maintaining diversity and maximizing yield.

Features common to these compilations are detailed below. Some are more obvious and elementary than others and require minimal elaboration. Most people concerned about fishes would agree that depleted resources would benefit from legislation based on sound science (hence more research), along with better enforcement of existing regulations. Remaining solutions can be generalized as (1) modifying prevailing attitudes and approaches, and (2) taking specific actions or employing specific tactics.

APPROACHES AND PREVAILING ATTITUDES

As discussed in the first chapter of this book, a major challenge in the effort to conserve biodiversity involves convincing people of the need to value nonutilitarian diversity, not just commodity. It is critically important to go beyond economic arguments for protecting species and ecosystems, because it is difficult if not impossible to place a cash value on many environmental phenomena, not the least of which is biodiversity. One approach is to appeal to the biophilic sense of wonder that many people feel toward natural, evolved phenomena. A strong rationale for protecting nature also exists in religious scripture. Leidy and Moyle (1997) maintained that most of the world's religions and ethical systems have a built-in conservation ethos (e.g., Meffe and Carroll 1997), but it is often not

BOX 16.1. A digest of general and specific solutions to environmental problems involving fishes, as offered by selected authors and agencies.

The authors, their issues, and regions of concern are (1) D. Ludwig et al. 1993 (global fisheries resources), (2) Biber-Klemm 1995 (declining north Mediterranean freshwater diversity), 3. FAO 1995a, 1997b (global fisheries management), (4) Leidy and Moyle 1997 (declining, mostly freshwater, aquatic biodiversity), (5) Winter and Hughes 1997 (declining global and North American aquatic biodiversity), (6) G. M. Mace and Hudson 1999 (sustainability vs. extinction of marine fisheries), (7) NRC 1999b (global fisheries management, with U.S. focus), (8) Montgomery 2003 (salmonid management, with general applicability), and (9) Hilborn 2005 (sustainable fisheries).

SOLUTIONS SUGGESTED BY MULTIPLE AUTHORS

- Legislate and enforce (2, 3, 4, 5, 7, 8, 9).
- Create reserves, as large as possible (1, 4, 5, 7, 8).
- Promote ecosystem-based management (2, 4, 5, 6, 7).
- Be precautionary: Act despite uncertainty and without waiting for scientific consensus (1, 3, 6, 7).
- Monitor results and manage adaptively (1, 5, 6, 7, 9).
- Increase research and monitoring (2, 4, 5, 7).
- Emphasize sustainability (3, 4, 6, 7; opposed by 1).
- Avoid technoarrogance, e.g., technological fixes that treat symptoms rather than causes (1, 5, 8).
- Educate (3, 5).
- Include all stakeholders at all stages (3, 7, 9).
- Reduce fishing effort (7, 8, 9).

SOLUTIONS EMPHASIZED BY INDIVIDUAL COMPILATIONS

- Ludwig et al.: Include human motivation as part of the system studied and monitored; rely on scientists for information but not necessarily for solutions.
- Biber-Klemm: Integrate economics and ecology; publicize regulations and their violation.
- FAO: Use best scientific evidence available, as well as traditional ecological knowledge; maintain biodiversity; conserve population structure; protect critical fisheries habitats and aquatic ecosystems.
- Leidy and Moyle: Integrate conservation into accepted ethical systems; internalize costs of environmental damage; promote water conservation.
- Winter and Hughes: Employ best management practices; promote systematics.
- Mace and Hudson: Encourage communication between resource managers and conservationists.
- NRC: Encourage community-based approaches.
- Montgomery: Learn from history and avoid repeating past mistakes; seek consensus but don't require it; keep in mind that local involvement is essential but local control is easily corrupted; restore habitat; act despite uncertainty and without waiting for scientific consensus.
- Hilborn: Provide management agencies with a clear mandate and recognized authority; protect habitat; diversify fisheries to survive fluctuations in yield.

emphasized. The Evangelical Environmental Network, Creation Care, and similar movements in the U.S. stand as testimony to the growing emphasis on stewardship among religions.

Moyle and Moyle (1995) stressed a need to appeal to intergenerational values, to the responsibility that most people feel toward future generations. If we focus on the economic value of natural goods and services, people tend to discount the future, "to view decisions that provide for the welfare of future generations as sacrifices" (Moyle and Moyle 1995, p. 34). If we recognize that resources belong not only to present generations but also to future generations, we are more likely to use resources in a sustainable manner and make fewer irreversible changes to the environment. Endangered species legislation, codified in most industrialized nations as well as in many developing nations, reflects societal recognition that species are "worth preserving for future generations, even though they currently have little economic value" (Moyle and Moyle 1995, p. 34). Intergenerational obligations are the basis of the concept of sustainable use, which recognizes our moral responsibility to guarantee that the natural wealth we use today remains equally available to and able to support future generations.

Angermeier (2000) argued that we should respect and strive to preserve "naturalness," where nature and natural entities—ecosystem components, levels of diversity, and rates of evolution and extinction—contrast with anthropogenic modifications, especially those brought about by technology and unconstrained by genetic evolution. To achieve conservation goals, "respect for nature must supplant the prevailing world view of human superiority" (p. 373). Such an ideal has great appeal to biologists and conservationists, as witness Aldo Leopold's (1953) articulation of the argument for preserving nonutilitarian biodiversity: "The last word in ignorance is the man who says of an animal or plant: 'What good is it?'" (p. 190). The challenge is finding a course of action that will convince the general public to similarly value nature.

The Precautionary Approach

Many of the policy issues and decisions discussed in this book revolve around a "precautionary" approach to conservation challenges, which states that preventive, anticipatory, and proactive measures should be taken when strong (but not necessarily conclusive) evidence exists that a threat is imminent. It's the environmental equivalent of an anticipatory counterattack. A precautionary philosophy justifies action before a definitive cause-and-effect relationship can be scientifically established, maintaining that incomplete knowledge is no excuse for inaction. Action is justified because if current conditions are allowed to continue, things are going to get worse. If action is delayed, any corrective measures that are still feasible may be more costly and less effective.

A precautionary approach underscores several controversial issues in conservation biology related to scientific uncertainty and consensus, the value of data versus knowledge, qualitative versus quantitative verification, and valid research in conservation (see Ruhl 2004 for a discussion of how a precautionary approach can be misapplied in the face of scientific uncertainty). Environmental problems require field study, but field study is largely observational, hence difficult to manipulate and lacking experimental controls. Field experiments are plagued by limited replication, small sample sizes, high data variances, complex interactions, imprecise measurements, and uncontrolled confounding variables. When subjected to statistical testing, they have a low likelihood of detecting an effect when one exists. As a consequence, "only relatively large effects can be detected with confidence" (Angermeier 1995, p. 151). Bad things may be happening, but we can't detect them because we demand adherence to statistical requirements that are chiefly applicable to a controllable laboratory setting.

Statistical versus Environmental Significance

Coral reef fisheries are a good jumping-off point to explore the relevance of a precautionary approach as it applies to traditional differences between basic ecology and conservation ecology, or what Pister (1999) referred to as the scholarship of discovery, in contrast to the scholarship of application. Let's assume by example that a colorful but rare species of reef seahorse is apparently declining in numbers over a large area because of commercial exploitation for both medicinal and aquarium trades. Conservationists propose an all-out ban on collecting; commercial interests maintain that collecting should continue because a ban would create economic hardship. A resource management agency must decide which course of action to take. The only relevant data come from multiple population censuses made by divers inside and outside a marine reserve. The fish appear to be less abundant outside the reserve, but because of nonstandardized sampling, general rarity, and effective camouflage, the counts at both sites are variable.

A trained population ecologist would analyze the numbers, run appropriate statistical tests, and likely conclude that although the mean population size outside the reserve appeared smaller, the variance in numbers made it difficult to determine whether the difference was real by accepted statistical conventions. Since no statistical difference could be demonstrated, no sound scientific reason existed to support a ban on collecting. Commercial interests would point to this uncertainty and claim that a ban was not supported by science. The conservationist would say that the numbers suggested a smaller population where collecting was permitted, that the species is rare and thought to be in decline globally, and that sufficient justification therefore existed for management agencies to ban all collecting, despite a lack of statistical rigor. At the heart of this discussion are the different goals pursued by traditional researchers and conservation biologists; the former are concerned about the correctness of hypotheses, and the latter about minimizing risk to species. Decisions hinge on dealing with uncertainty in both data and the outcome of a proposed action.

In research, this uncertainty is accepted as an underlying premise, one for which statistical tests provide guidelines. Scientists try to avoid claiming an effect has occurred or a difference exists when no such effect or difference has been shown to occur. Statistical tests are designed to minimize the risk of this so-called Type I error ("the error of rejecting an hypothesis of no difference when it is in fact correct"). Any claim of a difference between outcomes backed by statistical tests has a high degree of certainty, even though such certainty may only validate an observation with minimal ecological significance. An alternative that emphasizes conservation goals is to accept a slightly increased risk (conventionally 10% rather than 5%) of Type I error—that is, accept a greater risk that observed differences are the result of chance, not causative factors. Raising the rejection level in conservation biology is a direct application of the precautionary principle: The costs of being wrong about an effect outweigh the risks of making a statistical blunder.

Fiddling with rejection probabilities is a conservative way to advance scientific knowledge, but it is not the ideal way to preserve biodiversity (NRC 1995). Minimizing Type I error increases the chance that a species will not receive protection when it is needed. In the parlance of statistics, the conservationist is more worried about committing a Type II error: saying that an action has had no effect when an effect has occurred. By minimizing Type II errors, resource managers concerned with protecting species will tend to err on the safe side, at the risk of providing protection that may not be necessary. Statistics can provide guidance under such conditions via power analysis, a means of testing for the probability of a Type II error or "the probability of correctly detecting an effect that is present in nature" (Peterman and M'Gonigle 1992, p. 232). Power analysis can provide vital information on whether effects could be detected given a particular sample size and variability in measurements—that is, on whether the study design was capable of detecting a result had one existed (Peterman 1990; Peterman and M'Gonigle 1992).

Erring on the safe side is the basis of the precautionary principle, which has become as much a tenet of conservation as statistics is a tenet of hypothesis testing. The precautionary principle shifts the burden of proof off the shoulders of conservationists and resource managers onto the shoulders of those who exploit resources (Meffe and Carroll 1997). Faced with uncertainty about the potential effects of some action, the wisest course is to require that those proposing the action demonstrate that it will cause no harm, rather than requiring that those concerned with the resource prove that harm is likely to occur. Analogously, requiring minimization of Type II statistical errors shifts the burden of proof from management agencies to resource users: Fishers would have to demonstrate that their activity has no detrimental effect, rather than management having to show that it does (Peterman 1990). For example, the conservationist can insist that commercial interests must first show that fishing would not affect seahorse populations. The hypothetical data presented above were as inconclusive on this point as they were on the hypothesis that average population size differed between sites. But the precautionary conclusion is that continued fishing could not be justified.

The precautionary approach is relevant to much more than debates among academics. It has become the guiding principle in the realm of environmental decision making at the international level, forming the basis of policy as developed in the Rio Declaration on Environment and Development, the Ozone Layer Protocol, the Climate Change Convention, and the FAO Precautionary Approach to Fisheries, among others (see WWF 1999 and Weeks and Berkeley 2000).

Community Control, Traditional Knowledge, and Best Available Evidence

Interwoven in overviews of how we can improve both biodiversity conservation and fisheries management are recommendations that we must increase our knowledge base, use the best available scientific evidence, and include all stakeholders in decision making. In a related vein, the literature on coral reef fisheries indicates that local community control, especially where limited access to resources is enforced by local involvement, has a strong track record of success. These recommendations and approaches intersect on the issue of the value of traditional ecological knowledge (TEK), information possessed by indigenous people about the natural world and its management, developed and refined over centuries of empirical observation (Berkes et al. 2000; Stripien and DeWeerdt 2002). Our scientific knowledge base has been limited by the number of researchers that can be committed to a particular resource issue, but legions of fishers have spent generations gathering information that is essential for acquiring food in a sustainable manner, information without which they and their families would starve. Quality control affects fitness, and misinformation can be fatal.

We can broaden our knowledge base immensely and promote sustainable resource use if we seek out TEK. We should use not only the best available scientific evidence but also the best available TEK. Social anthropologists and ethnobiologists have developed means of testing the validity of TEK by treating it as testable hypotheses, just as statistical methods can test the validity of conclusions reached from data entered into spreadsheets. Fishers accumulate knowledge, test it, and pass it on over centuries. They pride themselves in their knowledge of their quarry, knowledge that can be verified. Confronted by skeptical anthropologists about the veracity of such information, Johannes emphasized that it differs from unverifiable information about family ties and sexual practices that anthropologists often collect. He interviewed Palauan fisherman who told him where and when reef fishes spawned (Johannes 1978a, 1981). He then went to the places at the assigned times and confirmed that information. His findings doubled our knowledge of reef fish reproductive patterns (Johannes 1978a). This information was later applied directly to the development of fishing regulations in Palau, which has instituted some of the most complete and effective reef conservation measures anywhere. Because the information originated with the stakeholders and was used in locally developed regulations, it possessed a credibility that has facilitated both implementation and enforcement (D. Sumang, pers. comm.; see Bhattacharjee 2005).

Much was made in chapter 3 of disagreement between international and regional lists of imperiled fish species. Some of this disagreement was attributed to insistence that benchmark criteria of decline be met with hard data on population sizes and distributions. For compiling threatened-species lists and natural histories, knowledge may be just as valuable as data. Dismissing traditional knowledge as anecdotal, biased, and inaccurate is making a prejudicial assumption about the intellectual capabilities of the people who have acquired that knowledge. Ideally, scientifically collected data can be combined with TEK to create a complete picture. In reality, throughout much of the world, the only information we have is informal. As demonstrated convincingly by findings such as those of Johannes (1981), Sadovy and Cheung (2003), and Sadovy et al. (2003), informal knowledge can be crucial to the understanding and solution of conservation problems involving fishes (see also Huntington 2000). For many species, collecting informal knowledge judiciously and verifying it wherever possible "is probably one of the only ways whereby sufficient data can be collected, quickly enough, for timely management and conservation action" (Sadovy et al. 2003, p. 361).

ACTIONS I: CONSENSUS IDEAS

Among the actions that can be promoted or taken are those that are more "programmatic" or of relevance to agencies, institutions, and policy formulation, and those that can be undertaken by individuals, regardless of their professional inclination or affiliation.

Adaptive, Ecosystem-Based Management

With little question, the emerging viewpoint is that both conservation and management actions will be more effective if we (1) treat imperiled taxa in the context of the ecosystems in which they occur, and (2) view any plans and actions as experiments that should be modified as conditions change and new information becomes available. "Only intact, healthy ecosystems can provide the complete range of benefits that humans want and need over long periods of time" (McLeod 2005, p. 4). Targeting single species with an unmodifiable blueprint has failed, to the detriment of the target species and the organisms and habitats with which they interact. Adaptive management in an ecosystem context, also called the integrated ecosystem approach or simply ecosystem-based management (EBM), is the starting point for conservation action in the 21st century. It too has a strong basis in TEK and is the solution to numerous problems, as concluded in earlier chapters.

Coastal resources are influenced by terrestrial activities and in turn affect offshore communities. Lake fishes are dependent on events in the surrounding air and watershed. Migratory fishes bring energy from the ocean up rivers and affect and are affected by the surrounding forest and its fauna. EBM recognizes this level of interconnectedness and is more effective than conventional approaches because species did not evolve and do not currently exist in a static vacuum. EBM places value on habitat, takes a multispecies perspective that includes humans, takes into account cumulative and interconnected impacts, and recognizes the need to understand the dynamic and complex nature of ecosystem processes (NRC 1999b; McLeod et al. 2005). An emphasis on adaptiveness—that plans may need modification—is explicit recognition that nature is variable, our knowledge is imperfect, and circumstances change.

Reserves as Ecosystems

A direct result of using EBM as a guiding principle in conservation action is the increased attention given to reserves. Although reserves and parks predate EBM as a formal concept, it is in the context of interconnected and dynamic ecosystems that their true and potential value has come to be appreciated.

Examples of success stories and discussion of the values of reserves, MPAs, fishing refuges, and similar set-aside areas were given earlier (chapters 2, 5, 11, and 12). The consensus is that large no-take reserves, especially when established as networks connected by migration or larval transport, have often been successful in protecting biodiversity and increasing biomass. They achieve this because (1) they provide refuges for imperiled species; (2) they reduce harassment of spawning fishes; (3) they accumulate large fish that produce disproportionately abundant and large larvae; (4) larvae are exported to downstream areas; and (5) emigration (spillover) of larger fish occurs from reserves into nonreserve areas. Reserves are more necessary now than in the past because technology has eliminated many natural refuges previously available to exploited fishes. Because of their effectiveness, reserves provide important refuges in the event other management tactics fail, while serving as realistic, minimally disturbed baselines against which we can compare the effectiveness of many management actions. Not all refuges provide all these benefits, not all are equally successful, and disagreement exists on the universality of benefits. However, successes are frequent and obvious enough that scientists, managers, and stakeholders such as fishers are enthusiastic about the results of reserve creation.

Much remains to be accomplished in this area. Marine reserves are common in nearshore areas but come nowhere close to the theoretically desirable 20% of the oceans that is often recommended. Reserves or even no-fishing zones are woefully lacking on the high seas. The number of freshwater reserves, especially integrated reserves in freshwater systems—those that take a watershed perspective and protect a lake, its inflowing streams, and the surrounding landscape—also remain much too uncommon (e.g., Moyle and Leidy 1992; Leidy and Moyle 1997).

Humans Should Behave like Fishes

One conclusion from the assessment of failed fisheries in chapter 11 was that human impacts would be lessened if fishers acted more like natural predators within the ecosystem, that "the intensity, extent, and frequency of human alterations [should] mimic those of natural regimes" (Angermeier 2000, p. 379). Behaving more naturally means targeting younger rather than older age-classes of prey, exploiting at levels more typical of other predators, sharing resources with those predators, and curtailing fishing or switching to alternative prey types when prey populations decline. This goal of acting more naturally is intertwined with the general concepts of ecosystem-based management, sustainable use, and learning from and valuing natural phenomena. Managing with the entire ecosystem rather than just target species in mind means including humans in the equation. To use ecosystem components sustainably, we must modify human behavior to fit more naturally into the evolved processes and interactions of the ecosystem. This recognizes a fact of utmost importance to overexploited resources—namely, that "people have the freedom to change their behavior, whereas fish do not" (Montgomery 2003, p. 245). But to accomplish a goal of mimicking natural processes, we have to first increase our understanding of how ecosystems function.

Sustainable Development and Use

An ultimate goal of conservation is sustainability. We need to sustain imperiled species and regulate exploitation today to ensure future availability and integrity of the resource—for itself and future human generations. Sustainability in theory is simple. Take no more of a species than it can replace through normal reproduction. As evidenced by the length and depth of the chapters in this book on exploited fishes, most commercial species are exploited at unsustainable rates. Otherwise, they would not be declining. Where have we gone wrong?

The largest causes of unsustainable use are human overpopulation, overconsumption, and excess fishing capacity, especially in industrialized economies. Too many people are killing too many fishes in the name of profit. Historically, human societies generally did not overexploit resources. In large part, this was because of low human population density, but many cultures also had an ethic of sustainability, viewing resources as spirit-imbued gifts from higher powers that were to be respected (Callicott 1989; Lichatowich 1999; Angermeier 2000). People took what they needed and avoided wasteful killing. Granted, preservation methods were often inefficient, which discouraged surplus killing, especially in warmer climates. Still, little evidence exists to suggest that people commonly killed more than they could eat.

The changes that led to our current general condition of overtaxed, unsustainably exploited resources are often traced to the spread of Europeans, with their profit-based, technology-driven, industrial economic system (Mowat 1996). Typically, Europeans engaged in exploitation and domination of ecosystems through unrestrained use of technology (Callicott 1989; Angermeier 2000). Lichatowich (1999, p. 45), talking about the demise of Pacific northwest salmon stocks, captured the process:

> When the maximization of profits becomes the dominant goal, diverse and labor-intensive fisheries controlled by local communities of fishers are usually replaced by expensive technology operated by a few people and governed by distant, centralized bureaucracies. This change weakens the negative feedback loops between fish and fishers, making it virtually impossible for a sustainable relationship to coevolve between the fish and the industrial economy.

Resources became commodities, and profits were reinvested in additional economic ventures, buoyed up during bad times by government subsidies. Overcapitalization and overcapacity ensued. Minimal costs were incurred in the process of driving a common resource into economic extinction because losses were externalized, and profits and effort were simply invested in alternative, more profitable, resources. In the case of Pacific salmon, canneries were even built on barges in the expectation of wiping out a run in one river and moving the operation to another river. Fishery practices basically guaranteed overexploitation: "Fishing fleets so overwhelmed the capacity of the canneries that piles of salmon would accumulate until they began to rot. Cannery workers would then shovel the carcasses back into the river" (Lichatowich 1999, p. 41).

In addition, technological improvements in industrialized fisheries made exploitation of depleted resources possible by increasing the rate at which resources could be extracted. Technological improvements in capture, preservation, and transportation also combined to create export economies. Politics played no small part. Wealth, or its promise, creates access to political power that can be mobilized to promote unlimited resource exploitation (Ludwig et al. 1993). Today, in a global economy, all these forces interact to deplete fishery and other resources. A local resource can sustain a local but not a global culture. Whereas primitive economies fed themselves, free-market economies (try to) feed the world.

Market Force-Free Fishes; Ultimately Unsustainable Fisheries

Conventional fisheries wisdom maintains, erroneously, that when populations decline, economic constraints will put an end to fishing before commercial extinction can occur. Part of the error is that profiteers can take earnings from a resource-based, declining economic activity and invest the money in another, more profitable activity, even as the first resource declines toward unprofitability (e.g., hunt whales into extinction, invest the profits in petrochemical companies). By such reinvesting, they can avoid normal market force limitations on resource use, but at some point the fishes become too rare to bother pursuing.

An even more insidious practice, unrestrained by even extremely low abundances, targets species that are "worth too much not to catch." Sturgeon, giant yellow croaker, coelacanths, bluefin tuna, whale sharks, golden dragon fish, and humphead wrasse can be worth tens and even hundreds of thousands of dollars each. They are, therefore, market force-free. For many of these, economic value escalates with increasing rarity. In other words, "declines in catch per unit of effort will not stop fishing if cash per unit of effort increases with rarity" (Sadovy and Cheung 2003, p. 92). Alarms should go off when a species' commercial worth crosses the threshold into the realm of extreme value.

Mammals, birds, and reptiles have been hunted to extinction. Some authors have maintained that overfishing cannot lead to ecological or biological extinction because fishes are not worth as much and are harder to catch.

Hence, economic extinction will occur first and will "result in reduced fishing pressure for many fish species, thus staving off biological extinction" (Jennings and Kaiser 1998, p. 238). While this scenario may apply to the majority of fish species, those worth as much as an elephant or rhinoceros (bluefin tuna, whale sharks, sturgeons) or as easy to catch as a dodo or a Stellers sea cow (groupers, croakers, and wrasses in spawning aggregations) may suffer a fate similar to that of their nonfish analogs. These fishes should be intensely protected.

The general solution to these problems is commonly referred to as *sustainable development*, which has become a watchword of international conservation efforts directed at exploited resources, such as the those of the UN Commission for Sustainable Development, the IUCN Policy Statement on Sustainable Use of Wild Living Resources, and the FAO Indicators for Sustainable Development of Marine Capture Fisheries (see www.iisd.org and www.iied.org/ for links).

Ludwig et al. (1993) were skeptical of programs, science-based or otherwise, that promoted sustainability as a solution to major environmental problems, citing numerous failed attempts in fisheries management. They pointed out that few proposals were able to present clear blueprints for how sustainability could be achieved, and that reductionist science had proven itself inadequate to deal with the complexity of natural systems, not to mention human greed. If anything, they said, "the scientific community has helped to perpetuate the illusion of sustainable development through scientific and technological progress" (p. 36). Meyer and Helfman (1993) defended sustainability over the alternative of business as usual, which has resulted in diminished resources, failed fisheries, and imperiled species. They wrote that sustainability "is embraced not because of its proven record, but because it offers a viable alternative to uncontrolled exploitation" (p. 569). The commonly prescribed features of box 16.1 can be viewed as the broad outline of the required blueprint.

Legislate, Implement, Enforce, Litigate

It is generally acknowledged that most resource and environmental calamities have sociological, not biological, roots—that "resource problems are not really environmental problems: They are human problems" (Ludwig et al. 1993, p. 36). Fisheries management manages fishers not fishes. To improve conditions, human behavior has to change—never a simple task. While voluntary restraint in resource exploitation is an ideal, restraint requires sacrifice, also difficult to accomplish. Legal frameworks, at local, regional, national, and international levels, are consequently necessary.

National and international agreements and regulations were discussed in chapter 3. Implementation and enforcement of such legislation and regulation tend to vary with scale. Many local and regional species protection laws, such as at the state level in the U.S., carry minimal, if any, penalties. Habitat protection tends to be more effective, even at local and state levels, because administrative agencies have the authority to grant or deny permits based on pollutant discharge and erosion and sedimentation issues. National laws tend to be better armed (e.g., the U.S. ESA; the National Environmental Management: Biodiversity Act of South Africa; the Habitats and Water Framework Directives of the EU). Stiff penalties and jail time are provided, although effectiveness of enforcement varies as a function of resources and commitment. Even where initial enforcement by a constabulary may be practiced, the system can break down at higher levels, such as a judiciary that does not take environmental issues seriously. This is a reality in many tropical countries in regard to marine issues (Y. Sadovy, pers. comm.). The teeth in many national laws are probably sharpest in their capacity to deter harm before it happens, rather than in jailing miscreants after the fact.

International agreements (IUCN, CITES, RAMSAR, Berne, OSPAR) are surprisingly effective, in that compliance is largely voluntary yet widespread. International peer pressure among signatory parties often leads to compliance in the form of national legislation that carries the force of law.

A growing positive trend that maximizes participation in species and habitat protection could be called *preemptive cooperation*. Individuals or organizations participate in agreements to avoid litigation or restrictive legislation. Two examples in the U.S. are habitat conservation plans and safe harbor agreements, both of which balance private development and land use with species protection. Both involve formal agreements between federal agencies and private citizens and emerge from provisions in section 10 of the ESA.

Habitat conservation plans (HCPs) are designed to allow landowners to engage in otherwise lawful activities that may harm listed species or their habitats. Before the activity is approved, however, the landowner must obtain an incidental take permit, which requires submitting an HCP that outlines what the landowner will do to minimize and mitigate the impact of the permitted take on the listed species. The main idea is that individuals of a species or portions of habitat may be expendable in the short term, as long as long-term recovery is ensured (see www.audubon.org/campaign/esa/hcp-guide.html). Although most HCPs involve individual landowners, some bring together people and organizations in a beneficial, streamlining process.

The Etowah Regional HCP of north Georgia is a model plan for managing growth and involving stakeholders. Developed for a river basin just north of Atlanta that is experiencing some of the most rapid population growth and urban development in the U.S., it involves 8 counties, 11 cities, and 18 governmental and nongovernmental agencies and organizations (including environmentalists and developers). The stated, common goal is to enhance "the Etowah watershed through protection of aquatic species and water resources, ensuring continued economic prosperity and quality of life for future generations" (www.etowahhcp.org).

The Etowah River basin is among the most diverse temperate systems in the world, with 91 native fish species, 4 of which are endemic; 9 have already been extirpated (Burkhead et al. 1997). Much of the impetus behind creation of Etowah HCP was the occurrence of at least 9 protected fish species, including the federally Endangered Etowah darter (*Etheostoma etowahae*), Endangered amber darter (*Percina antesella*), and Threatened Cherokee darter (*E. scotti*); 6 other fishes listed by the state of Georgia are candidates for federal recognition (5 federally listed bivalves have apparently been extirpated). Further development could be greatly delayed, if not completely curtailed, because developers are required by law to formulate individual HCPs to assess the impact of their activities and ensure that no harm ("take") befalls listed species or their habitats. Developers have agreed to participate in a regional HCP that includes plans, ordinances, and policies to minimize the impact of development on protected species.

As a result, development can proceed without activating restrictive provisions of the ESA, including the need for individual HCPs. The Etowah HCP includes guidelines to protect riparian buffers, improve erosion and sedimentation control, protect greenspace, and improve stormwater and water resource management. Some critical habitat may be lost in the process; but other areas of habitat will be better protected, and more people will be involved in efforts to protect species and habitats. Harsh reality is that some developers ignore all provisions of the ESA, relying on lax enforcement. This becomes less likely with more eyes watching and less onerous demands. Developers that follow the regulations would be granted incidental take permits, exempting them from litigation in the event protected species or habitats were disturbed. In this manner, "darters and darter habitat will be protected from the most deleterious construction practices, while developers will be protected from costly and time-consuming litigation" (C. Gervich, pers. comm.).

Safe harbor agreements are similar, voluntary, section 10 ESA arrangements between federal agencies (USFWS, NOAA Fisheries) and private landowners that encourage and facilitate species protection on private land. More than half of the species listed under the ESA occur on private land. A major stumbling block in species protection has been concern among private landowners that if listed species are found on their land, or if they improve habitat and attract such species, they will lose use of their property. In a safe harbor agreement, an "enhancement of survival" permit exempts the landowner in the event of future incidental take of a species—that is, landowners will be subjected to "no surprises" resulting from their efforts to improve conditions for a listed species. Before a safe harbor agreement is finalized, the landowner has to show that the species in question will receive a net conservation benefit from the agreement's management provisions (see http://endangered.fws.gov/recovery/harborqa.pdf).

Via a safe harbor agreement, a private landowner in Arizona petitioned to improve a stream running through his property that housed federally threatened Little Colorado spinedace, *Lepidomeda vittata* (www.ecbarranch.com). By restoring riparian vegetation and keeping grazing livestock out of the stream, he will improve water clarity, stream flow, channel stability, food availability, and water temperature, benefiting the spinedace (and helping endangered Southwestern Willow Flycatchers). The landowner is exempt from incidental harm that might befall the listed species as a result of his actions and will not lose use of the property if additional listed species are attracted to the improved habitat. Another relevant safe harbor arrangement has been made between FWS and the Arizona Department of Transportation. Public and private water bodies are being stocked with native endangered fish species such as Gila topminnow,

Poeciliopsis occidentalis, and desert pupfish, *Cyprinodon macularius*, in part to aid in a mosquito eradication program designed to reduce the spread of West Nile virus. Previous plans called for stocking nonnative mosquitofish, *Gambusia affinis*, which are known to prey on and hybridize with native, threatened species. Under conventional provisions of the ESA, citizens and government employees cannot handle endangered species, let alone use them in captive propagation programs (see K. Wilson 2003; www.environmentaldefense.org).

Both HCPs and safe harbor agreements have critics, who focus on compromised species protection and concerns over verifying compliance (e.g., Pedersen 2002; www.audubon.org). In the long run, we have to recognize that species protection efforts will be advanced most if we can gain the participation of individuals who may have been previously hostile to the notion.

Expand Our Knowledge Base

We need to know a lot more about ecosystems and how they work, about the ecology and natural history of imperiled organisms, and especially about the sociological and economic factors that lead to unsustainable use and habitat destruction. Excellent, detailed, and justified catalogs of the broad areas where our knowledge is weakest have appeared repeatedly (e.g., Lubchenco et al. 1991; Naiman et al. 1995). Warren et al. (1997), focusing on fishes, organized our shortcomings into two general categories: accounting information and ecological information. *Accounting information* refers to inventories of locales and ecoregions that tell us what is where and what is missing. *Ecological information* clarifies the factors that determine the generation and maintenance of fish diversity; it includes TEK. Combining the two categories allows us to focus our policy formation, management, protection, and recovery efforts.

Related to accounting, and deserving special mention, is the need to support, train, and appreciate taxonomists, the people who identify and name organisms. Most pleas for revival of taxonomy come from the taxonomists themselves and consequently fall on deaf ears because of the perception of self-promotion. As I have attempted to compile and understand the complex and multifaceted aspects of fish conservation, it has been made repeatedly clear to me that many of the problems arise from insufficient taxonomic study. It is increasingly obvious that conservation begins with taxonomy: We can't begin to manage endangered species without being able to identify them. While taxonomy remains underappreciated, we ironically are calling on taxonomists more and more to be "both Adam and Noah—charged with naming the world's various forms of life as well as saving them from disaster" (DeWeerdt 2002, p. 11).

Avoid Technoarrogance

In box 16.1, a repeated theme is that many conservation problems remain unresolved because of a misplaced abiding faith in technological solutions when, in fact, technology may have caused the problems. Meffe (1992), reviewing the history of salmonid declines, referred to this as "technoarrogance" and said, "Technology can take us only partway, and often down the wrong path at that" (p. 354). Technoarrogance treats symptoms rather than causes: Hatcheries have become a technologically complex solution to the demise of salmon, although the true causes are overfishing and habitat modification. By asking the wrong questions, we get the wrong answers. Reliance on technology obscures the underlying causes because we focus on and become infatuated with the novel gadgetry of the technology. Unwarranted faith in technological solutions to ecological problems arrests the development of more creative and complete solutions (Orr 1994). To paraphrase Meffe (1992) on the philosophy behind promoting agriculture in arid landscapes: If it's a desert, irrigate it; if it then floods, channelize it; and if it then grows too many mosquitoes, spray pesticides or introduce alien fishes. But the desert wasn't the problem. The problem was expecting nature to acquiesce to our desire to make a desert do something unnatural. We too often use technology to bend nature to our will and to sustain overpopulation and overconsumption. In misapplying technology, we alter ecosystems at a rate to which other biota are unable to adapt.

ACTIONS II: PERSONAL ACTIONS

Some of the most significant solutions are the positive actions that individuals— scientists and nonscientists—can take to reverse the losses and promote the conservation of fish species and their ecosystems. The six general categories of action are advocate, communicate, litigate, educate, donate, and participate (regrettably, no catchy acronym comes to mind). These are, of course, in addition to the litany of 101-plus things we can all do to save the earth, including promoting sustainable fisheries through ecocertification and informed consumerism. The main point is to get involved.

Advocate

Let's deal with the difficult stuff first. *Advocacy* can be generally defined as actions taken in support of a cause or issue, with the intent of influencing policies, positions, and programs. It directly entails making recommendations, which makes scientists uncomfortable because they recognize the cloud of uncertainty that surrounds any conclusion derived from scientific study. As discussed earlier, scientists most fear claiming an effect when none has occurred. Credibility is at stake.

Conservation biology is a scientific discipline but, by definition, it is unabashedly a value-laden science (Meffe and Carroll 1997). By choosing their field, "conservation biologists have willingly accepted a moral commitment to protection of the biotic community" (Dallmeyer 2005, p. 418). Conservation biology begins with the premises that the natural world is in trouble (a value statement) and that something needs to be done (i.e., advocacy). The scientist in us cautions against making decisions, while the conservationist badgers us to take action.

Many scientists prefer to provide information and allow others to act on it. But it isn't enough to obtain data on declines in a population of endangered fish or deterioration of a habitat type. Before that information will be translated into positive action to protect the species or habitat, it has to be communicated effectively to the right people, those who will influence if not make a decision. Meffe (2002) emphasized that science alone is not enough to address environmental threats. For conservation to be effective, we need a combination of rigorous science, informed policy in the form of legislation, and implementation of legislation in the form of management. But, Meffe said, "to policy makers and managers the information scientists so carefully provide does not exist unless a special effort is made to move it through policy circles" (p. 366). Hence advocacy. For scientists, advocacy most often entails ensuring that accurate information—the best science—is used in making a decision.

Every scientific society has debated the advocacy issue. Professionals fear accusations of losing their objectivity, an accusation easily countered by remaining objective. Many, if not most, scientific societies have concluded that failing to advocate a position based on reliable, rigorously obtained information is tantamount to advocating the status quo. And the status quo with respect to the environment is unacceptable.

One other impediment to scientific advocacy and environmental activism, at least among academics, is that the scholarship of discovery has traditionally been valued above the scholarship of application. This too is changing, because academic institutions are increasingly required to justify their budgets based on relevance to societal concerns. Few administrators will frown on the positive publicity generated by the application of their school's academic expertise to local, regional, national, and global environmental issues. Yes, some feathers will get ruffled when politicians, economic interests, and financial contributors feel threatened by environmental actions that may affect their bottom line. But these are the interests that have caused the most problems. Remember, the public at large is behind us, and few university administrators want to see headlines such as "University professor silenced for speaking out against pollution." The same should—*should*—apply to government employees.

Advocate, but base your advocacy on knowledge, not opinion.

Communicate

Scientists need to actively communicate their findings to the public and to politicians. Environmental research is relevant; it addresses problems that the public sees as important, which in turn motivates politicians (e.g., Tichacek 1996). But the public does not read scientific journals. Talk to school groups and classes, civic action groups, business groups, environmental interest groups, anyone who is willing to listen. Such interactions make justifying public expenditure on conservation research much easier.

Talk to the press. Seek it out. Many newspapers have environmental writers who are looking for good stories or reliable sources of expertise to confirm the accuracy of what they've written. Write articles yourself for newspapers and popular science or environmentally oriented magazines. Disregard those in the academic community who view such popularization with disdain. They are not solving problems.

Publicize both the good and the bad. Write letters to the editor commending local officials, individuals, and organizations involved in environmental action. Letter writing, to politicians and editorial pages, educates and alerts policy makers and profiteers that someone is paying attention. Importantly, publicize the bad actors. Blow the whistle. Notify governmental agencies of violations of environmental ordinances (waste discharge, sedimentation and erosion violations, fish kills). Riverkeeper organizations often have a hotline to call to report an apparent violation (see www.waterkeeper.org for a list of organizations).

Litigate

Gain an understanding of the legal protection afforded the environment at state and national levels. Better yet, become an environmental lawyer. My interests and abilities led me along a career path in animal behavior research. Events and age if not wisdom directed me toward conservation. Anyone straddling the fence between a career in conservation biology and environmental law should realize that lawyers change things, whereas biologists help lawyers change things. Best yet, get training in both.

Whether or not you have formal training in environmental law, you have the right to sue. Many states have citizen suit provisions, by which individuals can sue other individuals, corporations, and agencies for violation of environmental laws. These provisions in essence deputize citizens to bring actions to protect the environment and biological diversity (www.serconline.org/citizensuits/faq.html, www.defenders.org/states/publications/publicinaction.pdf). Even states that lack such specific legislation have citizen suit provisions written into regulations that address environmental issues, such as water quality regulations. Most federal laws provide for citizen suits. And just the threat of a suit can be effective, without incurring the legal costs of filing a suit and going to court. Often a notice of intent to sue an agency or corporation can lead to negotiations that result in corrective action.

Off the record but widely known, conscientious individuals in many state and federal environmental agencies welcome such lawsuits. Getting sued for failing to enforce environmental regulations is often the only means by which an agency can initiate enforcement. Organizations that lack the personnel and funds to pursue all infractions can do only what the current plan and budget earmark. A lawsuit changes priorities and mobilizes resources. Fines levied against violators can be directly applied to correcting the situation.

Educate

A small amount of information on a specific subject can be critical to saving a species or a habitat. And not just scientists engage in meaningful research. Governmental and academic researchers often lack the time, money, or inclination to gather long-term data on local habitats, but adopt-a-stream and similar environmental action and monitoring groups collect data that can be immensely valuable. When the big box store decides to fill in a local wetland or pave over your favorite stream, a long-term data set and species inventory from that site can be critical to preventing development or minimizing its impacts. Such data sets are often used by environmental experts allied with local citizen action groups to challenge destructive development.

Donate

Give what money you can afford. But how much time can you or should you donate to environmental causes? Given the extent of the problem, all your time could be spent addressing important issues. Fear of overcommitment is probably the major cause of inaction. As a starting point, consider the conventional amount allocated to tithing. I propose that professionals donate, or give back, 10% of their time to environmental causes.

This number is not as arbitrary as it might seem. The concept of the tithe has direct application to the time allocation decision because of the importance ascribed to the living world by most biologists. Perhaps human nature makes us need to believe in something bigger than ourselves. To many biologists, it is the evolutionary process and its end product—biodiversity—that we hold in awe. If adherents of an organized religion can be expected to donate 10% of their worldly goods to promote the activities of that religion, an advocate of biodiversity, professional or not, can rightfully be expected to do the same. Creation deserves no less.

Participate

Get involved personally. Attend public meetings and hearings that deal with the enactment or enforcement of environmental legislation. Get involved in local environmental groups and let them know of your expertise and your willingness to help. Among the "small" actions anyone can engage in is to support nongame conservation programs in your state, even if it's no more than buying a "support wildlife" license plate for your car. These efforts have been highly successful. In Georgia, sales of nongame license plates, at a one-time cost of $20 each, have contributed on average $2 million annually to conservation efforts (www.georgiawildlife.com). The money goes into research, monitoring, habitat restoration, education, and land acquisition. It can also be used as matching funds in proposals to federal agencies, from which sizable sums can be obtained.

Get involved professionally. Many or most professional scientific organizations have subsections or committees devoted to conservation of the organisms, habitats, or processes that are its focus (e.g., the American Fisheries Society has an Endangered Species Committee and a special section on introduced species; the American Society of Ichthyologists and Herpetologists has a Conservation Committee; the Australian Society for Fish Biology has a Threatened Fish Committee). These groups have considerable credibility with the public and with policy makers. If professionals won't provide the information necessary to understand the issues, who can or will?

Many career paths can contribute. Traditional academic pursuits play an obvious role: Academicians can serve in an educational, research, or advisory capacity or help in forming policy. Academicians concerned about conservation should focus their research wherever possible on issues with conservation relevance and encourage their students to do likewise. Government agencies and conservation NGOs offer many of the same opportunities. Also worth consideration is working for a consulting firm dedicated to providing accurate information in background studies or environmental impact assessments (as opposed to the small number of firms that provide biostitutes for corporations and developers).

There are many things we can all do. However, a wealth of possibilities can be paralyzing. Recognize that you can't do it all, and do what you do best. Not everyone is an engaging public speaker anymore than everyone is patient and meticulous enough to conduct tedious, accurate field research. But everyone can make a difference. Most environmental organizations are successful because of the actions of one or a few energetic and dedicated individuals.

To all those concerned citizens who devote so much time and effort and funds to private and nonprofit and nongovernmental organizations dedicated to conservation: thank you. To the rest: Just do something.

CONCLUSION

Gloom and Doom versus a Brighter Tomorrow

Two fairly recent books encapsulate the extremes of outlook and approach in conservation biology. Both have lessons for the future of fishes. They are John Terborgh's (1999) *Requiem for Nature* and (versus) Michael Rosenzweig's (2003) *Win-Win Ecology*. Terborgh's focus is on tropical, terrestrial ecosystems, the impacts of deforestation and globalization, and the failure of conventional conservation programs focused on creating reserves, especially those that include indigenous peoples. As populations grow and technology spreads—two forces that appear inevitable—biodiversity will continue to decline and disappear, he says. His answers lie in top-down, international control that keeps people out, forcibly if necessary.

Rosenzweig agrees that reserves have failed—not because we failed to keep people out, but because we failed to recognize that humans are now too much a part of the landscape to be kept out. Strict reserve policies and manipulative attempts at restoration (a hoped-for return to "natural" conditions) are unlikely to succeed because few if any areas remain unchanged by humanity, and most have undergone irreversible change. Rosenzweig proposes that reservation and restoration ecology be replaced with reconciliation ecology, humans living with nature and vice versa. Instead of approaching conservation as a zero-sum game of either economics or the environment, we should seek precedence in the many win-win instances where human economic, extractive, and even military activity have—often inadvertently—benefited nature. We should capitalize on and promote the happy accidents that occur as a by product of human activity. Despite diametric worldviews, both books emphasize the conservation of large chunks of diverse habitat, ecosystems rather than pieces.

What do the case histories of imperiled fishes tell us about the utility of these two very different approaches? I see promise in applying two tactics that lie at the intersection of the two viewpoints. Terborgh puts little faith in local and community control because he believes small-scale human endeavors are more easily corrupted and overpowered by large economic forces. However, some of the most heartening success stories in fish conservation come at the level of community-based actions, such as local riverkeeper and adopt-a-stream efforts. Similar success has been demonstrated in coral reef fisheries and the aquarium trade. Many analyses point to the likely improvement of conditions if we return to limited-access, locally controlled (and enforced) exploitation, with an emphasis on subsistence rather than large-scale export and profit.

Admittedly, high-seas fisheries and vast habitat expanses are unlikely to benefit from a local approach; in the larger arenas a second tactic at the intersection of the two approaches may have application. Rosenzweig sees little promise in absolute reserves, but the data emerging from no-take MPAs points to the resilience of nature, and to a return to more natural biomass and biodiversity when extractive activities are minimized. Absolute no-take reserves work. To meet human needs, a compromise here mandates a system of rotating no-take reserves, where large networked tracts are allowed to remain fallow for biologically meaningful periods as they recover. During recovery, larvae are exported and fishes spill over into nearby and downstream regions, improving biomass and biodiversity. With rotation, some areas can be exploited while others are allowed to recover.

A Closing Thought

My former teacher, George Barlow, who is far wiser and more experienced than I, wrote: "I have lost my optimism for the future of freshwater fishes (or humans, for that matter)" (2000, p. 268). Perhaps if I reach his age, wisdom will win out over optimism. For now, I remain hopeful if not optimistic. When I give talks on fish conservation issues, I am commonly asked if I expect conditions to improve. Without too much waffling, and when I feel I can be honest, I generally answer that I try to be publicly optimistic whereas I am privately depressed (when I'm dishonest, I just go with optimism, for the fourth reason given below). My depression is a direct result of my professional exposure to the wealth of unhappy stories and projections, many of which appear in these pages. Daily news reports on the declining condition of the world's environments, combined with climate change projections, do little to brighten my mood. But my public optimism is borne of four important facts. First, the projections may prove incorrect. Second, nature is amazingly resilient. Third, given current rates of human population growth and resource exploitation, there's good reason to suspect that our future as a species, at least as a numerically dominant species, is fairly limited (Dr. Barlow's influence perhaps). After tens of millennia or longer, Earth will recover from our activities. New species will arise in place of us and the species we've extinguished. Life will go on.

Fourth, more important and immediately, people are turned off by negativism. We are a hopeful species. A message of gloom and doom, repeated too frequently, will ultimately be ignored. If we want others, like minded or not, to act in defense of the environment, we can't project an image of unrelenting pessimism. For these reasons, I have tried to end each chapter and this book by emphasizing what can and should be done to protect fishes, their habitats, and our future.

The Terborgh-Rosenzweig debate provides just such a concluding theme. It is nature's resilience and potential for restoration, witnessed repeatedly in reserves and protected areas, that gives me hope. Perhaps in some reserves, recovery will be so dramatic that such areas will be left alone in perpetuity, appreciated for their tenacity, and visited only by those who wish to see nature as it might have been before we altered things, there simply and only to teach and enthrall us.

Appendix: Probable Fish Extinctions

Included here is any species for which a fairly definitive statement of extinction has been made by a recognized authority. The conservative approach to fish extinctions recommended by Harrison and Stiassny (1999) would demote many of these species to CR or EN (or even DD) for reasons listed there (see their tables 7, 8, 11, 13).

Scientific name	*Common name*	*Status*	*Former locale*	*Source*
Petromyzontidae				
Eudontomyzon sp.		EX	Byelorussia, Ukraine	Kottelat
Charcharhinidae				
Glyphis gangeticus	Ganges shark	EX	India	H&S
Acipenseridae				
Acipenser nudiventris	ship sturgeon	EX	Black Sea	P. Vecsei, pers. comm.
Huso huso	beluga	EX	Mediterranean	P. Vecsei, pers. comm.
Pseudoscaphirhynchus fedtschenkoi	Syr-Dar shovelnose sturgeon	EX[a]	Kazakhstan	H&S
P. hermanni	small Amu-Dar shovelnose sturgeon	EX[a]	Uzbekistan	H&S
Clupeidae				
Alosa vistinoca		EX	Greece	H&S
Cyprinidae				
Acanthobrama hulensis		EX	Israel	
Acheliognathus elongatus		EX[b]	China	H&S
Algansea barbata	Lerma barb	EX	Mexico	Lyons
Anabarilius alburnops		EX	China	H&S
A. polylepis		EX[b]	China	H&S
Barbus microbarbis		EX	Rwanda	H&S
Cephalokompsus pachycheilus		EX[a]	Philippines	H&S
Chondrostoma genei		EX	Yugoslavia	Povz 1990
C. scodrensis		EX[a]	Albania, Yugoslavia	Kottelat

Scientific name	*Common name*	*Status*	*Former locale*	*Source*
C. soetta		EX	Yugoslavia	Povz 1990
Cyprinella carpio chilia	Maravillas red shiner	EX[f]	China	Kottelat & Chu 1988
C. lutrensis blairi		EX[f]	USA (Mexico?)	Miller
C. micristius		EX[b,f]	China	Kottelat & Chu 1988
C. yilongensis		EX	China	
Epalzeorhynchos bicolor		EW[g]	Thailand	
Evarra bustamantei	Mexican dace	EX	Mexico	
E. eigenmanni	Mexican dace	EX	Mexico	
E. tlahuacensis	Mexican dace	EX	Mexico	
Gila bicolor isolata	Independence Valley tui chub	EX[f]	USA	Miller
G. crassicauda	thicktail chub	EX	USA	
Gnathopogon elongatus suwae	Suwa gudgeon	EX[f]	Japan	Yuma et al. 1998
Gymnocypris biswasi		EX	India	H&S
Lepidomeda altivelis	Pahranagat spinedace	EX	USA	
Leuciscus turskyi		EX	Croatia	
L. ukliva		EX[a]	Croatia	H&S
Mandibularca resinus	bagangan	EX[a]	Philippines	H&S
Notropis amecae	Ameca shiner	EX	Mexico	
N. aulidion	Durango shiner	EX	Mexico	
N. orca	phantom shiner	EX	Mexico, USA	
N. saladonis	sardinita de Salado	EX	Mexico	
N. simus simus	Rio Grande bluntnose shiner	EX[b,f]	USA	Miller
Ospatulus palaemophagus	bitungu	EX[b]	Philippines	H&S
O. truncululatus		EX[a]	Philippines	H&S
Osteobrama belangeri		EW[g]	India	H&S
Pararasbora moltrechtii		EX	Vietnam	H&S
Phoxinellus egridiri	yag baligi	EX[a]	Turkey	H&S
P. handlirschi	cicek	EX[a]	Turkey	H&S
Pogonichthys ciscoides	Clear Lake splittail	EX	USA	
Puntius amarus		EX[a]	Philippines	H&S
P. baoulan	baolan	EX[a]	Philippines	H&S
P. clemensi	bagangan	EX[a]	Philippines	H&S
P. disa	disa	EX[a]	Philippines	H&S
P. flavifuscus	katapa-tapa	EX[a]	Philippines	H&S
P. herrei		EX[a]	Philippines	H&S
P. lanaoensis	kandar	EX[a]	Philippines	H&S
P. manalak	manalak	EX[a]	Philippines	H&S
P. sirang	sirang	EX[c]	Philippines	H&S
P. tras	tras	EX[a]	Philippines	H&S
Rhinichthys cataractae smithi	Banff longnose dace	EX[f]	Canada	Miller
R. deaconi	Las Vegas dace	EX	USA	

Scientific name	Common name	Status	Former locale	Source
R. osculus reliquus	Grass Valley speckled dace	EX[f]	USA	
Sinocylocheilus g. grahami		EX[f]	China	Kottelat & Chu 1988
Sprattlicypris palata	palata	EX[a]	Philippines	H&S
Stypodon signifer	stumptooth minnow	EX	Mexico	
Catostomidae				
Chasmistes liorus liorus	June sucker	EX[e,f]	USA	Miller
C. muriei	Snake River sucker	EX	USA	
Lagochila (Catostomus) lacera	harelip sucker	EX	USA	
Balitoridae				
Sphaerophysa dianchiensis		EX	China	H&S
Characidae				
Brycon acuminatus		EX[d]	Brazil	H&S
Henochilus wheatlandi		EX	Brazil	H&S
Hyphessobrycon parvellus		EX	Brazil	H&S
Ictaluridae				
Noturus trautmani	Scioto madtom	EX[a]	USA	Etnier 1997
Bagridae				
Leiocassis medianalis		EX	China	H&S
Siluridae				
Silurus mento		EX[b]	China	H&S
Schilbeidae				
Irveneia voltae		EX	Ghana	H&S
Platytropius siamensis		EX	Thailand	H&S
Amblycipitidae				
Liobagrus kingi		EX	China	H&S
L. nigricauda		EX	China	H&S
Trichomycteridae				
Rhizosomichthys totae	greasefish	EX	Colombia	
Retropinnidae				
Prototroctes oxyrhynchus	New Zealand grayling	EX	New Zealand	
Salmonidae				
Coregonus alpenae	longjaw cisco	EX	Canada, USA	
C. confusus		EX	Europe	Kottelat
C. fera	fera	EX	Europe	Kottelat
C. gutturosus	gilch	EX	Switzerland	Kottelat
C. hiemalis	gravenche	EX	Europe	Kottelat
C. johannae	deepwater cisco	EX	Canada, USA	
C. kiyi orientalis	Lake Ontario kiyi	EX[f]	Canada, USA	Miller
C. nigripinnis	blackfin cisco	EX	Canada, USA	
C. restrictus	ferit	EX	Switzerland	Kottelat
Oncorhynchus clarki macdonaldi	yellowfin cutthroat trout	EX[f]	USA	Miller
O. clarki ssp.	Alvord cutthroat trout	EX[f]	USA	Miller
O. nerka kawamurai	kuni-masu, black kokanee	EX[f]	Japan	Yuma et al. 1998

Scientific name	*Common name*	*Status*	*Former locale*	*Source*
Salmo montenegrinus		EX	Montenegro	H&S
S. schiefermuelleri		EX	Austria	Kottelat
Salvelinus agassizi	silver trout	EX	USA	
S. inframundus		EX	Europe	Kottelat
S. profundus	tiefsesaibling	EX	Switzerland	Kottelat
S. scharffi	Scharff's char	EX	Ireland	Kottelat
Bedotiidae				
Rheocles sikorae	zona	EX	Madagascar	
Atherinidae				
Chirostoma charari	least silverside	EX	Mexico	Lyons
C. compressum		EX	Mexico	Lyons
Teramulus waterloti		EX	Madagascar	H&S
Adrianichthyidae				
Adrianichthys kruyti	duckbilled buntingi	EX[a]	Sulawesi	H&S
Xenopoecilus poptae	Popta's butingi	EX[a]	Sulawesi	H&S
X. sarasinorum		EX[b]	Sulawesi	H&S
Fundulidae				
Fundulus albolineatus	whiteline topminnow	EX	USA	
Poeciliidae				
Gambusia amistadensis	Amistad gambusia	EX	USA	
G. georgei	San Marcos gambusia	EX[e]	USA	
G. heterochir	Clear Creek gambusia	EX[e]	USA	Robins et al. 1991
Pantanodon madagascarensis		EX[b]	Madagascar	Benstead
Phalloptychus eigenmanni		EX	Brazil	H&S
Priapella bonita	Guayacon ojiazul	EX	Mexico	
Goodeidae				
Allotoca maculata	opal goodeid	CR[a]	Mexico	Miller
Ameca splendens	goodeid	EW[g]	Mexico	
Chapalichthys peraticus	alien splitfin	EX	Mexico	H&S
Characodon garmani	parras characodon	EX	Mexico	
Empetrichthys latos concavus	Raycraft Ranch poolfish	EX[f]	USA	Miller
E. latos pahrump	Pahrump Ranch poolfish	EX[f]	USA	Miller
E. merriami	Ash Meadows killifish	EX	USA	
Skiffia francesae	golden skiffia	EW[g]	Mexico	
Cyprinodontidae				
Cualac tessellatus	cachorrito de Media Luna	EX[b]	Mexico	H&S
Cyprinodon alvarezi	perrito de Potosi	EW[g]	Mexico	
C. ceciliae	cachorrito de la Presa	EX	Mexico	
C. inmemoriam	cachorrito de la Trinidad	EX	Mexico	
C. latifasciatus	perrito de Parras	EX	Mexico	
C. longidorsalis	cachorrito de Charco Palmal	EW	Mexico	
C. nevadensis calidae	Tecopa pupfish	EX[f]	USA	Miller
C. sp.	monkey spring pupfish	EX	USA	Miller

Scientific name	Common name	Status	Former locale	Source
C. spp.	perritos de Sandia	EX	Mexico	
Leptolebias marmoratus	ginger pearlfish	EX[c]	Brazil	H&S
Megupsilon aporus	cachorrito enano de Potosi	EW[g]	Mexico	
Orestias cuvieri	Lake Titicaca orestias	EX[d]	Lake Titicaca	H&S
O. incae		EX	Lake Titicaca	H&S
O. taquiri		EX	Lake Titicaca	H&S
O. tutini		EX	Lake Titicaca	H&S
O. uruni		EX	Lake Titicaca	H&S
Gasterosteidae				
Gasterosteus crenobiontus		EX	Europe	Kottelat
Pungitius kaibarae	Kyoto 9-spine stickleback	EX	Japan	Yuma et al. 1998
Syngnathidae				
Syngnathus affinis	Texas pipefish	EX	USA	R&H
Cottidae				
Cottus echinatus	Utah Lake sculpin	EX	USA	
Percidae				
Etheostoma sellare	Maryland darter	EX	USA	
Sander vitreum glaucum	blue pike	EX[f]	USA	Miller
Apogonidae				
Pterapogon kauderni	Banggai cardinalfish	EW	Indonesia[h]	R&H
Cichlidae				
Astatotilapia martini		EX	Lake Victoria	
A. megalops		EX	Lake Victoria	
Gaurochromis obtusidens		EX	Lake Victoria	
Haplochromis altigenis		EX	Lake Victoria	H&S
H. apogonoides		EX	Lake Victoria	H&S
H. barbarae		EX[b]	Lake Victoria	H&S
H. bareli		EX	Lake Victoria	H&S
H. boops		EX	Lake Victoria	
H. cavifrons		EX	Lake Victoria	
H. cinctus		EX	Lake Victoria	H&S
H. cnester		EX	Lake Victoria	H&S
H. diplotaenia		EX	Lake Victoria	H&S
H. guiarti		EX	Lake Victoria	H&S
H. heusinkveldi		EX	Lake Victoria	H&S
H. hiatus		EX	Lake Victoria	H&S
H. iris		EX	Lake Victoria	H&S
H. lividus		EW[g]	Lake Victoria	
H. maculipinna		EX	Lake Victoria	
H. michaeli		EX	Lake Victoria	
H. nyanzae		EX	Lake Victoria	
H. pachycephalus		EX	Lake Victoria	
H. paraplagiostoma		EX	Lake Victoria	
H. pectoralis		EW[g]	Lake Victoria	
H. pyrrhopteryx		EX	Lake Victoria	

Scientific name	*Common name*	*Status*	*Former locale*	*Source*
H. spekii		EX	Lake Victoria	
H. thuragnathus		EX	Lake Victoria	
H. tridens		EX	Lake Victoria	
H. victorianus		EX	Lake Victoria	
H. spp.		EX[i]	Lake Victoria	
Harpagochromis artaxerxes		EX	Lake Victoria	
Hoplotilapia retrodens		EX	Lake Victoria	
Labrochromis ishmaeli		EW[g]	Lake Victoria	
L. mylergates		EX	Lake Victoria	
L. pharyngomylus		EX	Lake Victoria	
L. teegelaari		EX	Lake Victoria	
Lipochromis microdon		EX	Lake Victoria	
Paratilapia sp. nov		EX	Madagascar	Benstead
Paretroplus menarambo		EW[g]	Madagascar	Benstead
Platytaeniodus degeni		EW[g]	Lake Victoria	
Prognathochromis arcanus		EX	Lake Victoria	
P. argenteus		EX	Lake Victoria	
P. bartoni		EX	Lake Victoria	
P. decticostoma		EX	Lake Victoria	
P. dentex		EX	Lake Victoria	
P. estor		EX	Lake Victoria	
P. flavipinnis		EX	Lake Victoria	
P. gilberti		EX	Lake Victoria	
P. gowersi		EX	Lake Victoria	
P. longirostris		EX	Lake Victoria	
P. macrognathus		EX	Lake Victoria	
P. mandibularis		EX	Lake Victoria	
P. nanoserranus		EX	Lake Victoria	
P. nigrescens		EX	Lake Victoria	
P. paraguiarti		EX	Lake Victoria	
P. percoides		EX	Lake Victoria	
P. perrieri		EW[g]	Lake Victoria	
P. prognathus		EX	Lake Victoria	
P. pseudopellegrini		EX	Lake Victoria	
P. xenostoma		EX	Lake Victoria	
Psammochromis cryptogramma		EX	Lake Victoria	H&S
Pyxichromis parorthostoma		EX	Lake Victoria	
Stomatotepia mongo		EX	Cameroon	H&S
Tristramella magdalenae		EX	Israel	H&S
Xystichromis bayoni		EX	Lake Victoria	
Yssichromis argens		EW[g]	Lake Victoria	
Pomacentridae				
Azurina eupalama	Galapagos damselfish	EX	Galapagos, Cocos Is.	R&H
Eleotridae				
Ratsirakia sp.		EX	Madagascar	Benstead
Gobiidae				
Sicyopterus nigrescens		EX	Hawaiian Islands	H&S
Weberogobius amadi	poso bungu	EX[a]	Sulawesi	H&S

Scientific name	*Common name*	*Status*	*Former locale*	*Source*
Belontiidae				
Betta cf. *tomi*		EX[c]	Singapore	H&S

Source: 1996 IUCN Red List of Threatened Animals, www.wcmc.org.UK/species/animals/animal_redlist.html, unless otherwise noted.
Notes: CR = Critically Endangered, EN = Endangered, VU = Vulnerable, DD = Data Dependent, EX = extinct; EW = extinct in the wild but some individuals maintained in captivity; H&S = Harrison and Stiassny 1999; Miller = R. R. Miller et al. 1989; Kottelat = Kottelat 1998; Lyons = Lyons et al. 1998; Benstead = Benstead et al. 2000; R&H = Roberts and Hawkins 1999. If no source is cited, the reference is IUCN 1996.
[a]Listed as CR by IUCN 1996.
[b]Listed as EN by IUCN 1996.
[c]Listed as VU by IUCN 1996.
[d]Listed as DD by IUCN 1996.
[e]Listed as EN by U.S. Fish & Wildlife Service 1999.
[f]Disqualified by Harrison and Stiassny (1999) as a subspecies of which other subspecies are extant.
[g]Disqualified by Harrison and Stiassny (1999), who do not recognize Extinct in the Wild as a valid extinction.
[h]Banggai cardinalfish have been seen in the wild in North Sulawesi, Indonesia, as the apparent result of aquarium releases (Tackett and Tackett 2001*, Vagelli and Erdmann 2002).
[i]Harrison and Stiassny (1999, tables 9, 11) list 75 other undescribed *Haplochromis* spp. from Lake Victoria that are thought to have gone extinct in recent years, but whose taxonomy and extinction date are unsubstantiated.

Glossary

Actinopterygian. Most higher, bony fishes, in which the fin rays are bony. See *Sarcopterygian.*

Adaptive management. An approach that emphasizes changing management tactics in response to changing conditions (which are continually monitored), rather than "sticking to the plan."

Adipose fin. A small, fleshy fin without supporting spines or rays, set far back on the dorsal surface of many catfishes, characins, salmons, and other groups.

AFS. American Fisheries Society, North America's largest professional organization of scientists, educators, managers, and policy makers concerned with fishes; interests include biology, conservation, and fisheries management (www.fisheries.org).

Allochthonous. Food and materials in a stream that originate from outside a system, such as leaves or insects that drop into a stream.

Allopatric. Species that do not occur (naturally) in the same locale.

Anadromous. Fishes that hatch in fresh water, move to the ocean to grow, and return to fresh water to spawn, such as salmons, sturgeons, and lampreys.

Ancestral. Taxon from which descendant species are derived, often synonymized as primitive or generalized; ancestral traits or conditions are those that appear in an ancestor.

Anthropogenic. Human made or of human origin.

Armored. Hardened and coarsened, as of a stream bottom lacking small sediment particles; occurs when the supply of sediment is less than the stream's ability to move sediment.

Artisanal fisheries. Usually small-scale fisheries practiced by local or native human populations, often involving traditional methods.

ASIH. American Society of Ichthyologists and Herpetologists, an international professional organization dedicated to the study of fishes, amphibians, and reptiles (www.asih.org).

Assemblage. The mix of fish species in a habitat or locale (as opposed to "community," which includes nonfish taxa).

Autochthonous. Produced from within a system, such as algae growing on rocks in a stream. (Winning word in the 2004 National Spelling Bee.)

Backcrossing. Mating of young with the same species as that of their parents, an issue in hybridization.

Benthic. On the bottom; see also *Demersal.*

Bioaccumulation. Increase in amount of a substance in the body of an individual as it ages; of ecological importance because many toxins are not degraded, metabolized, transformed, nor excreted and hence accumulate.

Biogenic. Made by living organisms; e.g., coralline rock is produced by living corals.

Biomagnification. The tendency of substances, often toxins, to increase in concentration as one moves up a food chain; why swordfish and sharks contain more mercury than herring.

Biomarker. A measurable biological response to environmental factors or contaminants.

Biotransformation. A chemical alteration that occurs within an individual or in the environment; e.g., elemental mercury is biotransformed to methylmercury by the action of sulfate-reducing bacteria.

BRDs. Bycatch reduction devices, designed to limit the kill of nontarget organisms captured in fish trawls.

Bycatch. Incidental capture of nontargeted organisms; sometimes "bykill."

CALSC. Catchability-led stock collapse, also called hyperaggregation, the tendency for small schools to aggregate to form large schools, thus keeping fishing pressure high even though stock size has declined.

Captive propagation (= cultured). Reproduction that occurs in a lab or hatchery for commercial or conservation purposes, often followed by release of young in the wild.

Catadromous. Fishes that hatch in the ocean, move to fresh water to grow, and return to the ocean to spawn, such as anguillid eels and mullets.

Cetaceans. Whales, dolphins, porpoises.

Cf.. In taxonomy, "compare with" or "referable to." For example, Nile perch, *Lates* cf. *niloticus*, means a species in the genus *Lates* that is very similar to and may actually be *L. niloticus*, but we're not sure.

Chondrichthyes. The class that contains sharks, skates, rays, and chimaeras.

Closure. The technical term for completion of a dam and start-up of its designed activities.

Commercially extinct. Too rare to be fished profitably.

Compensatory phenomena. Positively density dependent events, i.e., the improvement of survival conditions as populations decline. See *Depensatory*.

Condition, condition factor. An index of the health of individual fish, calculated by dividing weight by length; skinny fish have low condition factors.

Conspecifics. Members of the same species.

CPUE. Catch per unit of effort.

CR. Critically Endangered according to IUCN.

Crepuscular. At twilight, either dusk or dawn.

Demersal. Associated with the bottom; see also *Benthic*.

Depauperate. Of low diversity, lacking in species. See *Speciose*.

Depensatory phenomena. Negatively density dependent events that increase mortality as numbers decrease. See *Compensatory phenomena*.

Derived. Later-appearing taxa within a lineage, often synonymized as advanced or specialized; derived traits appear in a descendant species and are changed from the ancestral condition.

Described. Given a scientific name.

Diadromous. Fishes that move between fresh and salt water during different phases of their lives. See *Anadromous*, *Catadromous*.

Dichromatic. Describes a species whose males and females are different colors; a form of dimorphism.

Dimorphic. Usually refers to anatomical or behavioral differences between males and females (e.g., "sexually dimorphic").

DPS. Distinct population segment, a taxonomic level lower than species, describing a population with distinct traits

Ecosystem (-based) management. An approach that considers interacting rather than solitary components, as in developing fishing guidelines that account for impacts to most species and habitats rather than just a target species.

Elasmobranchs. Sharks, skates, and rays

El Niño. A periodic, large-scale climate fluctuation involving warmer surface waters in the tropical Pacific that persists for much longer than the normal three or four months; also El Niño–Southern Oscillation (ENSO).

Embeddedness. A measure of sediment depth on the bottom of a stream, etc., usually indicating how much hard substrate, such as pebbles and rocks, is covered by sediment.

EN. Endangered according to IUCN.

Endemic. Restricted or native to a geographically defined area.

EPA. U.S. Environmental Protection Agency (www.epa.gov).

ESA. U.S. Endangered Species Act.

Escapement. In managed salmon fisheries, refers to individuals returning to spawn in a given system (i.e., escaping the many sources of mortality on their way back to their natal spawning river or stream).

Estivation. Torpid state entered into by some aquatic organisms during dry periods, often involves burying in the bottom; the aquatic equivalent of hibernation.

ESU. Evolutionarily significant unit, a taxonomic level at or lower than the species level, describing a population with distinct traits that contributes significantly to the genetic diversity and, potentially, the evolution of a species.

Euryhaline. Describes organisms capable of surviving at a wide range of salinity concentrations.

Eutrophication. "The process of increased organic enrichment of an ecosystem, generally through increased nutrient inputs" (Nixon 1995, p. 95).

Extant. Living, opposite of extinct.

Extinct. Globally eliminated; organisms do not go locally extinct. See *Extirpated.*

Extirpated. Locally or regionally eliminated; multiple extirpations lead to extinction.

F1. First filial generation, i.e., the immediate offspring of a mating.

FAO. Food and Agricultural Organization of the UN (www.fao.org); UN agency chiefly involved in fisheries issues.

Fish die-off. General term for large numbers of dead fish, where the causative agent is unknown or involves natural factors.

Fishery-independent sampling. Gathering data on exploited species by means other than sampling the commercial catch, as when a management agency conducts its own trawl surveys.

Fish, fishes. *Fish* refers to one species, *fishes* to more than one species.

Fish kill. Large numbers of dead fish resulting from an identified cause involving a known "responsible party," such as a toxic waste spill.

FWS. U.S. Fish and Wildlife Service (www.fws.gov).

Gonadosomatic index. Ratio of the mass of (usually) a fish's eggs, or sperm, to its body mass; also gonosomatic index.

Gondwana. A supercontinent in what is now the Southern Hemisphere that separated during the Mesozoic, forming the modern continents of Australia, Antarctica, South America, and Africa.

Gradient. The physical slope of a stream; water moves faster in high-gradient streams.

Guild. Species in a community that perform many of the same ecological functions, e.g., the zooplanktivore guild on a reef comprises the many fish species that feed on zooplankton.

HABs. Harmful algal blooms.

High at-risk IUCN ranks. Designations of Critically Endangered (CR), Endangered (EN), and Vulnerable (VU) by IUCN (www.redlist.org).

Hydroperiod. The cycle of water rising and falling, as in a stream; specifically, the length of time an area is flooded by water.

Hypolimnion. The lower layer of water in a stratified water body such as a lake or reservoir, usually at a lower temperature than the upper, epilimnetic region (adjective. hypolimnetic).

Ichthyocide. Chemical used to kill fish (e.g., rotenone, derived from the root of the *Derris* plant); also piscicide.

Ichthyofauna. Species composition of fishes in a defined area or region.

-idae, -id. In taxonomy, usual endings for the name of a family of organisms, e.g., "Salmonidae" and "salmonids" refer to the family that includes salmon, trout, and char.

-iformes, -iforms. In taxonomy, usual endings for the name of an order of organisms, e.g., "Perciformes" and "perciforms" refer to the order that includes families such as perches, sea basses, wrasses, grunts, and croakers.

Imperiled. In this book, generally applies to species with official recognition by the ESA (as Endangered, Threatened, Species of Concern), IUCN (as Critically Endangered, Threatened, Vulnerable), or other national or NGO entity that uses established criteria in designating status; "at risk," and less often "threatened," used similarly.

Impervious surface. Ground cover that does not allow water to percolate into the ground, e.g., asphalt.

Indeterminate growth. The typical lifetime growth pattern in fishes, whereby individuals continue to grow throughout their lives, although the rate may slow considerably in later years.

Intersexuality. Intermediate gender.

Interstitial. In the spaces, between the grains, as in small animals that live among sand grains or in the gravel of a streambed.

Introgression. In hybridization, the injection of alien genes into the genome of the original species via backcrossing.

Iteroparity. Capability of breeding more than once. Contrast with *Semelparity*.

ITQ (individually transferable quota). The quantity of fish a fishing license entitles its holder to catch. The license holder can sell all or part of the quota to another individual.

IUCN. International Union for the Conservation of Nature; premier source for expert-compiled lists of imperiled species, their distribution, and causes of imperilment; now the World Conservation Union (WCU, www.redlist.org).

IUU. Illegal, unreported, and unregulated catches that contribute unknown levels of depletion to commercial species.

K-selected. In ecological terminology, describes species that are relatively long lived, are slow to mature, reproduce infrequently, and produce relatively few offspring; see *R-selected*.

Lacustrine. Living in or pertaining to lakes.

Landlocked. Refers to (usually lacustrine) populations of anadromous species that have been cut off from the ocean; most undertake spawning migrations into tributary streams.

Lentic. A still water body such as a lake or pond. See *Lotic*.

Limnetic. In open water, usually in a lake; similar to pelagic.

Limnology. Study of freshwater.

Lineage. An evolutionary grouping of taxa, referring to ancestral and descendant/derived taxa.

Lotic. Flowing water habitat such as a stream or river. See *Lentic.*

Macroinvertebrates. Aquatic insects, crustaceans, and other invertebrate taxa large enough to see without a microscope.

Macrophytes. Rooted aquatic plants ("weeds" to the unenlightened).

Mainstem. Main channel of a river, as contrasted with tributary streams and smaller rivers that flow into it.

Mesic. Relatively wet habitats, such as marshes and swamps.

Metapopulation. Populations of a species in an area that exchange individuals, and presumable genes, via migration.

MMT. Million metric tons.

Monotypic. A genus or family (occasionally an order) with only one species, e.g., the bowfin, *Amia calva*, is the only living member of its genus (*Amia*), family (Amiidae), and order (Amiiformes).

MPA. Marine protected area, an area set aside as a reserve to protect its resident organisms from human disturbance.

MSY. Maximum sustainable yield.

MT. Metric tons; 1 MT = 1,000 kg.

Natal. Refers to birth place; salmon return to their natal streams to spawn.

Nest associates. Species that spawn in nests made by other species, often taking advantage of the parental care provided by the nest builder.

NGO. Nongovernmental organization, such as Trout Unlimited.

NIS. Nonindigenous species; one that has been moved beyond its natural range or natural zone of potential dispersal; a general term for introduced species.

NMFS. U.S. National Marine Fisheries Service, the federal agency responsible for marine and anadromous fisheries management and conservation; also NOAA Fisheries (www.nmfs.noaa.gov).

NOAA Fisheries. See *NMFS.*

Nociception. Detection of and reflexive response to a noxious or tissue-damaging stimulus.

Nonindigenous. Not native to an area.

Offal. Leftover, often internal parts of a butchered animal, e.g., discarded guts.

Ontogeny. The developmental history of an individual as it grows.

Outbreeding depression. Loss of genetic variation or swamping of local genotypes when large numbers of genetically different individuals are introduced into and breed with a small population; a major concern with escaped, sea-ranched Atlantic salmon.

Overcapacity. Too many boats chasing too few fish.

Overcapitalization. Too much money invested in a nonsustainable or failing enterprise, creating overcapacity

Parr. Early, stream-dwelling life history stage of a salmon or trout, between the alevin or fry and smolt stages, usually characterized by vertical, black "parr marks" on sides.

Pelagics. Open-water fishes that swim near the surface; large pelagics are species such as tunas, billfishes, and sharks; small pelagics are schooling species such as sardines, herrings, anchovies, and small mackerels.

pH. A log-based measure of the acidity or alkalinity of a solution (scales from 0 to 14; 7 is neutral, >7 is more basic, and <7 is more acidic).

Phylogenetic. Relating to the evolutionary history of a group of organisms; taxa listed in "phylogenetic order" go from most ancestral or primitive to most derived or advanced.

Piabas. Amazonian term for ornamental fishes captured for the aquarium trade.

Pinnipeds. Seals, sea lions, walruses.

Piscivore. A fish-eating organism.

Polyculture. Aquaculture with multiple steps, inputs, and products; e.g., runoff of waste from chicken pens fertilizes a pond, promoting growth of algae, which are then eaten by herbivorous fishes.

ppb. Parts per billion; 1 ppb = 1 part per billion or 1 μg/kg body weight or 1 ng/g body weight.

ppm. Parts per million. 1 ppm = 1 part per million or 1 mg/kg body weight or 1 mg/l.

Propagule pressure. the magnitude of introduction effort, or the number of individuals introduced into a system in nonindigenous species introductions.

Propagules. Life history stages, usually larvae, that disperse.

Redds. Gravel nests dug in stream bottoms by female trout and salmon, in which eggs and fry develop.

RfD. Reference dose, or the daily exposure level to noncarcinogens that is unlikely to have deleterious effects over an individual's lifetime.

Riparian. The terrestrial zone immediately bordering a water body, e.g., riparian vegetation along a stream.

Riprap. Large rock, broken concrete, stones, or similar hard items used to stabilize streambanks, etc., and prevent erosion.

River Continuum Concept. A descriptive hypothesis about rivers that explains physical and biological succession and processes from upstream to downstream.

Riverine. Referring to or living in flowing water, as in riverine habitats or organisms.

R-selected. Fast-growing, maturing species that produce many young. See *K-selected.*

Rugosity. Roughness; a measure of the vertical complexity of a surface or habitat; high rugosity means lots of ups and downs.

Sarcopterygian. Relatively primitive, fleshy-finned fishes such as lungfishes and coelacanths. See *Actinopterygian.*

Sea-run. Refers to ecological variants of migratory fishes that spend part of their life cycle at sea, such as steelhead salmon, which are the sea-run variant of rainbow trout. See *Diadromous*, *Anadramous.*

Semelparity. Big-bang spawners that breed once and die, such as Pacific salmon. Contrast with *Iteroparity.*

Smolt. A juvenile salmonid migrating from fresh water to the sea and undergoing a series of physiological changes that make life in seawater possible.

Smoltification. The process of becoming a smolt.

Species flock. A large group of closely related, endemic species in a small area, all of whom are descended from a single ancestor.

Speciose. Of high diversity, having many species in a group or area.

Spp.. More than one species of a genus, but names not specified, e.g., "*Salmo* spp." Means more than one species in the genus *Salmo.*

Squamation. The scales or scale patterns on a fish.

Ssp.. Subspecies of a genus.

Stock. In fisheries management, part of a fish population characterized by a particular migration pattern, area of occupancy, or specific spawning grounds, and targeted by a distinct fishery.

Stream order. A system for identifying progressively larger streams in a watershed; at its simplest, first-order streams have no tributaries, second-order streams are created when two first-order streams come together, etc.

Stunting. The tendency of a population to increase in density and breed at individual sizes so small as to become recreationally and commercially useless.

Sustainability. Using a resource in a way that does not compromise its ability to persist nor its availability to future resource users.

Sympatric. Occurring together.

Taxon. A group of evolutionarily related species (plural: taxa).

T&E. Threatened and endangered, or Threatened and Endangered; used generally to refer to imperiled species but sometimes more specifically, as in the U.S., to refer to species protected by the ESA, which gives protection only to species designated at the rank of Threatened or Endangered.

Thermocline. Vertical zone in a water body where temperature changes rapidly with increasing depth.

Translocated. Moved from one locale to another; often refers to human transport of a species within a defined geographical area such as a continent where it is native to some part.

Undescribed. Refers to a species that has not yet been given an officially recognized scientific name.

Varzea. The gallery forests associated with white-water rivers in Amazonia (Brazilian term in widespread use).

VU. Vulnerable according to IUCN.

Water column. The open-water portion of a body of water, above the bottom.

WCU. World Conservation Union. See *IUCN.*

WHO. World Health Organization; the UN agency with the task of promoting human health.

Xenobiotics. Organic compounds foreign to an organism, naturally occurring compounds distributed via novel pathways, or natural compounds present in unnatural concentrations.

Xenoestrogens. Human-produced estrogen mimics.

Xeric. Relatively dry (= arid) habitats, e.g., deserts; not many fishes are truly xerophilic.

YOY. Young of the year; fish that are not yet a year old in a given year.

References

Aarts, B. G. W., F. W. B. Van den Brink, and P. H. Nienhuis. 2004. Habitat loss as the main cause of the slow recovery of fish faunas of regulated large rivers in Europe: The transversal floodplain gradient. *River Res Appl* 20:3–23.

Abbey, E. 1975. *The monkey wrench gang*. New York: HarperCollins.

ABC (American Bird Conservancy). 2004. 6,000 caspian terns to be shot? www.surfbirds.com/abcbirdsnews.html.

Abila, R. O., and E. G. Jansen. 1997. Socio-economics of the Nile perch fishery on Lake Victoria. www.iucn.org/themes/wetlands/docs/Report2.doc.

Abrahams, M. V., and A. Sutterlin. 1999. The foraging and antipredator behaviour of growth-enhanced transgenic Atlantic salmon. *Anim Behav* 58:933–42.

Adams, S. M., ed. 1990. Biological indicators of stress in fish. *Am Fish Soc Sym* 8. Bethesda, MD: American Fisheries Society.

———, ed. 2002. *Biological indicators of aquatic ecosystem stress*. Bethesda, MD: American Fisheries Society.

Adey, W. H., and K. Loveland. 1998. *Dynamic aquaria: Building living ecosystems*. 2nd ed. San Diego: Academic Press.

Afonso, L. O. B., J. L. Smith, M. G. Ikonomou, et al. 2002. Y-chromosomal DNA markers for discrimination of chemical substance and effluent effects on sexual differentiation in salmon. *Environ Health Persp* 110:881–87.

AFS (American Fisheries Society). 1999. AFS Position Statement. Responsible use of fish and other aquatic organisms. *Fisheries* 24(1): 30–35.

AFS Idaho Chapter. 1995. Why isn't science saving salmon? *Fisheries* 20(9): 4–5.

Agardy, T. 2000. Effects of fisheries on marine ecosystems: A conservationist's perspective. *ICES J Mar Sci* 57:761–65.

Aggarwal, S. G., K. Chandrawanshi, R. M. Patel, et al. 2001. Acidification of surface water in central India. *Water Air Soil Poll* 130:855–62.

Agostinho, C. S., A. A. Agostinho, E. E. Marques, et al. 1997. Abiotic factors influencing piranha attacks on netted fish in the upper Parana River, Brazil. *North Am J Fish Mana* 17:712–718.

AJC (*Atlanta Journal-Constitution*). 2003. Jim Beam will pay for fish killed after warehouse fire. Sept. 10, 2003. www.ajc.com.

Alabaster, J. S., and R. Lloyd. 1982. *Water quality criteria for freshwater fish*. London: Butterworth.

Aladin, N. V., and W. T. Potts. 1992. Changes in the Aral Sea ecosystem during the period 1960–1990. *Hydrobiologia* 237:67–79.

Alam, M. K., and O. E. Maughan. 1995. Acute toxicity of heavy metals to common carp (*Cyprinus carpio*). *J Environ Sci Heal* A 30:1807–1816.

Alava, M. N. R., A. A. Yaptinchay, E. R. Z. Dolumbal, et al. 1997. Fishery and trade of whale shark (*Rhincodon typus*) in the Philippines. Abstract, 77th annual meeting, American Society of Ichthyologists and Herpetologists, Seattle, WA, 57–58. see www.sharktrust.org.

Albaladejo, V. D., and V. T. Corpuz. 1984. *Marine aquarium fish research and resource management*. Philippines: Bureau of Fisheries Aquatic Resources. Cited in Pyle 1993.

Albanese, B., P. L. Angermeier, and S. Dorai-Raj. 2004. Ecological correlates of fish movement in a network of Virginia streams. *Can J Fish Aquat Sci* 61:857–69.

Alcala, A. C., and E. D. Gomez. 1987. Dynamiting coral reefs: A resource destructive fishing method. In *Human impacts on coral reefs: Facts and recommendations*, ed. B. Salvat, 51–60. French Polynesia: Antenne Museum. Cited in Jennings and Kaiser 1998.

Alcala, A. C., and G. R. Russ. 1990. A direct test of the effects of protective management on the abundance and yield of tropical marine resources. *J Conseil International pour l'Exploration de la Mer* 46:40–47.

Alderman, D. J., and T. S. Hastings. 1998. Antibiotic use in aquaculture: Development of antibiotic resistance-potential for consumer health risks. *Int J Food Sci Tech* 33:139–55.

Alexander, G. R., and E. A. Hensen. 1986. Sand bed load in a brook trout stream. *North Am J Fish Mana* 6:9–23.

Alexandratos, N., ed. 1995. *World agriculture: Towards 2010.* Rome: FAO.

Alford, R. 2002. Ky. streams fouled by coal sludge make a comeback; populations recover from spill. *Washington Post*, Dec. 8, A16.

Allan, J. D., R. Abell, Z. Hogan, et al. 2005. Overfishing of inland waters. *BioScience* 55:1041–51.

Allan, J. D., and A. S. Flecker. 1993. Biodiversity conservation in running waters. *BioScience* 43:332–43.

Allen, G. R. 2000. Threatened fishes of the world: *Pterapogon kauderni* Koumans, 1933 (Apogonidae). *Environ Biol Fish* 57:142.

Allen, G. R., S. H. Midgley, and M. Allen. 2002. *Field guide to the freshwater fishes of Australia.* Perth: Western Australian Museum.

Allen, G. R., and R. C. Steene. 1995. Notes on the behavior of the Indonesian cardinalfish (Apogonidae) *Pterapogon kauderni* Koumans. *Rev Fr Aquar* 22:7–9.

Allen, S. K., and R. J. Wattendorf. 1987. Triploid grass carp: Status and management implications. *Fisheries* 12(4): 20–24.

Allendorf, F. W., D. Bayles, D. L. Bottom, et al. 1997. Prioritizing Pacific salmon stocks for conservation. *Conserv Biol* 11:140–52.

Allendorf, F. W., and N. Ryman. 1987. Genetic management of hatchery stocks. In *Population genetics and fishery management*, ed. N. Ryman and F. M. Utter, 141–59. Seattle: Univ. Washington Press.

Allendorf, F. W., N. Ryman, and F. M. Utter. 1987. Genetics and fishery management: Past, present, and future. In Ryman and Utter 1987, 1–19.

Allendorf, F. W., P. Spruell, and F. M. Utter. 2001. Whirling disease and wild trout: Darwinian fisheries management. *Fisheries* 26(5): 27–29.

Allibone, R. M. 1999. Impoundment and introductions: Their impacts on native fish of the upper Waipori River, New Zealand. *J Roy Soc New Zeal* 29:291–99.

Almada-Villela, P. C. 1990. Status of threatened Mexican fishes. *J Fish Biol* 37(suppl. A): 197–99.

Almenara-Roldan, S. 2001. The current status of the aquarium industry in Mexico. *OFI J* 35:1–3. www.ornamental-fish-int.org.

Almenara-Roldan, S., and J. T. Ketchum. 1994. Forgotten islands of the Mexican Pacific. *OFI J* 9:12–14.

Almodóvar, A., G. G. Nicola, and J. Suárez. 2002. Effects of fishery management on populations of brown trout, *Salmo trutta*, in central Spain. In Collares-Pereira, Coelho, and Cowx 2002, 337–45.

Alò, D., and T. F. Turner. 2004. Effects of habitat fragmentation on effective population size in the endangered Rio Grande silvery minnow. *Conserv Biol* 19:1138–48.

Alverson, D. L. 1997. Global assessment of fisheries bycatch and discards: A summary overview. *Am Fish Soc Sym* 20:115–25.

Alverson, D. L., M. H. Freeberg, J. G. Pope, and S. A. Murawski. 1994. *A global assessment of fisheries bycatch and discards.* FAO Fisheries Tech. Paper 339. Rome: FAO.

American Rivers. 1998. *America's most endangered rivers from 1988 to 1998.* www.americanrivers.org/pastendangered.html.

———. 2003. *Maryland success combines restoration with watershed management.* www.americanrivers.org.

American Society for Microbiology. 1995. *Report of the ASM Task Force on Antibiotic Resistance.* Washington, DC: American Society for Microbiology.

Amundsen, P. A., F. J. Staldvik, Y. S. Reshetnikov, et al. 1999. Invasion of vendace *Coregonus albula* in a subarctic watercourse. *Biol Conserv* 88:405–13.

Anders, P. J. 1998. Conservation aquaculture and endangered species: Can objective science prevail over risk anxiety? *Fisheries* 23(11): 28–31.

Andersen, M. E., and J. E. Deacon. 2001. Population size of Devils Hole pupfish (*Cyprinodon diabolis*) correlates with water level. *Copeia* 2001:224–28.

Anderson, D. M., P. M. Gilbert, and J. M. Burkholder. 2002. Harmful algal blooms and eutrophication: Nutrient sources, composition, and consequences. *Estuaries* 25:704–26.

Anderson, E. D. 1990. Fisheries models as applied to elasmobranch fisheries. In Pratt, Gruber, and Taniuchi 1990, 473–84.

Anderson, E. D., J. F. Kocik, and G. R. Shepherd. 1999. Atlantic anadromous fisheries. Unit 3 in NMFS 1999.

Anderson, E. D., R. K. Mayo, K. Sosebee, M. Terceiro, and S. E. Wigley. 1999. Northeast demersal fisheries. Unit 1 in NMFS 1999.

Anderson, R. C., and A. Waheed. 2001. The economics of shark and ray watching in the Maldives. *Shark News* no. 13, IUCN Shark Specialist Group, www.sharktrust.org.

Anderson, S., R. Zepp, J. Machula, et al. 2001. Indicators of UV exposure in corals and their relevance to global climate change and coral bleaching. *Hum Ecol Risk Assess* 7:1271–82.

Andrews, C. 1990. The ornamental fish trade and fish conservation. *J Fish Biol* 37(A): 53–59.

Angermeier, P. L. 1995. Ecological attributes of extinction-prone species: Loss of freshwater fishes of Virginia. *Conserv Biol* 9:143–58.

———. 2000. The natural imperative for Biological Conservervation. *Conserv Biol* 14:373–81.

Angermeier, P. L., and M. R. Winston. 1997. Assessing conservation value of stream communities: A comparison of approaches based on centres of density and species richness. *Freshwater Biol* 37:699–710.

Annett, C. A., R. Pierotti, and J. R. Baylis. 1999. Male and female parental roles in the monogamous cichlid *Tilapia mariae*, introduced in Florida. *Environ Biol Fish* 54:283–93.

Antipa, G. 1909. *Fauna ichtiologica Românìei* (Ichthyological fauna of Romania). Bucharest: Carol Gobl.

AP (Associated Press). 2001. Giant tuna sells for record $173,600. Jan. 5. www.flmnh.ufl.edu/fish/InNews/gianttuna.htm.

Appeldoorn, R. S. 1996. Model and method in reef fishery assessment. In Polunin and Roberts 1996, 219–24.

APPMA (American Pet Products Manufacturing Association). 2002. *2001–2002 National Pet Owners Survey*. Greenwich, CT: Author.

Araujo-Lima, C., and M. Goulding. 1997. *So fruitful a fish: Ecology, conservation, and aquaculture of the Amazon's Tambaqui*. New York: Columbia Univ. Press.

Arcand-Hoy, L. S., and W. H. Benson. 1998. Fish reproduction: An ecologically relevant indicator of endocrine disruption. *Environ Toxicol Chem* 17:49–57.

Armour, C. L., D. A. Duff, and W. Elmore. 1991. The effects of livestock grazing on riparian and stream ecosystems. *Fisheries* 16(1): 7–11.

Aron, W. I., and S. H. Smith. 1971. Ship canals and aquatic ecosystems. *Science* 174:13–20.

Aronson, R. B., W. F. Precht, M. A. Toscano, et al. 2002. The 1998 bleaching event and its aftermath on a coral reef in Belize. *Mar Biol* 141:435–47.

Arthington, A. H., and B. J. Pusey. 2003. Flow restoration and protection in Australian rivers. *River Res Appl* 19:377–95.

Arthington, A. H., J. L. Rall, M. J. Kennard, et al. 2003. Environmental flow requirements of fish in Lesotho rivers using the drift methodology. *River Res Appl* 19:641–66.

Arthington, A. H., and M. J. Zalucki. 1998. *Comparative evaluation of environmental flow assessment techniques: Review of methods*. Canberra: Land and Water Resources Research and Development Corporation.

Arukwe, A. 2001. Cellular and molecular responses to endocrine-modulators and the impact on fish reproduction. *Mar Pollut Bull* 42:643–55.

Arukwe, A., and A. Goksøyr. 1998. Xenobiotics, xenoestrogens and reproduction disturbances in fish. *Sarsia* 83:225–41.

ASA (American Sportfishing Association). 2002. *Sportfishing in America: Values of our traditional pastime*. Alexandria, VA: Author.

ASF (Atlantic Salmon Federation). 2001. Catastrophic salmon escape prompts calls for moratorium on the aquaculture industry. Feb. 22. www.asf.ca/communications/2001/feb/catastrophe.html.

ASFB (Australian Society for Fish Biology). 2001. *Conservation status of Australian fishes, 2001*. www.asfb.org.au/research/tscr/tf_constat2001.htm.

Ashley, K. W., and B. Buff. 1988. Food habits of flathead catfish in the Cape Fear River, North Carolina. *P Annu Conf SE Assoc Fish and Wildlife Agencies* 41:93–99.

ASIH (American Society of Ichthyologists and Herpetologists). 2000. ASIH resolution on transgenic salmon. *Copeia* 2000:1161.

ASMFC (Atlantic States Marine Fisheries Commission). 1981. *Interstate fisheries management plan for the striped bass.* ASMFC Fishery Management Report 1. Washington, DC: Author.

———. 2001. *Amendment 1 to the Interstate Fishery Management Plan for Atlantic Menhaden.* ASMFC Fishery Management Report 37. Washington, DC: Author.

Aspen Institute. 2002. *Dam removal: A new option for a new century.* Washington, DC: Author.

Atkinson, E. 1996. *The feasibility of flushing sediment from reservoirs.* World Commission on Dams, www.wca-infonet.org/id/3524.

Attrill, M. J., and M. H. Depledge. 1997. Community and population indicators of ecosystem health: Targeting links between levels of biological organization. *Aquat Toxicol* 38:183–97.

Attwood, C. G., and B. A. Bennett. 1994. Variation in dispersal of galjoen (*Coracinus capensis*) (Teleostei, Coracinidae) from a marine reserve. *Can J Fish Aquat Sci* 51:1247–57.

Augerot, X. 2005. *Atlas of Pacific salmon: The first map-based status assessment of salmon in the North Pacific.* Berkeley: Univ. California Press.

Auster, P. J. 1998. A conceptual model of the impacts of fishing gear on the integrity of fish habitats. *Conserv Biol* 12:1198–203.

Auster, P., and R. W. Langton. 1998. The effects of fishing on fish habitat. *Am Fish Soc Sym* 22:150–87.

Auster, P. J., R. J. Malatesta, and C. L. S. Donaldson. 1997. Distributional responses to small scale habitat variability by early juvenile silver hake *Merluccius bilinearis. Environ Biol Fish* 50:195–200.

Auster, P. J., R. J. Malatesta, R. W. Langton, et al. 1996. The impact of mobile fishing gear on seafloor habitats in the Gulf of Maine (Northwest Atlantic): Implications for the conservation of fish populations. *Rev Fish Sci* 4:185–202.

Awaluddin, A. 1999. Framework for conservation and management of marine genetic resources. In *Marine genetic resources and sustainable fisheries management*, ed. S. Mustafa and R. A. Rahman, 75–98. Malden, MA: Blackwell Science.

Axelrod, H. R. 2001. Discovery of the cardinal tetra and beyond. In Chao, Petry, et al., 2001, 17–25.

Bacher, D. 2000. *RFA NORCAL chair questions provisions of "Freedom to Fish Act."* www.fishsniffer.com/dbachere/111700freedom.html.

Bailey, C. 1988. Optimal development of third world fisheries. In *North-South perspective on marine policy*, ed. M. A. Morris, 105–28. Boulder, CO: Westview Press. Cited in Vincent and Sadovy 1998.

———. 1997a. Aquaculture and basic human needs. *World Aquaculture* 28:28–31.

———. 1997b. Lessons from Indonesia's 1980 trawler ban. *Mar Policy* 21:225–35.

Baillie, J. E. M., C. Hilton-Taylor, and S. N. Stuart, eds. 2004. *2004 IUCN red list of threatened species: A global species assessment.* Gland, Switzerland: IUCN.

Bain, M. B. 1993. Assessing impacts of introduced aquatic species: Grass carp in large systems. *Environ Manage* 17:211–24.

Baker, I. J., I. I. Solar, and E. M. Donaldson. 1988. Masculinization of chinook salmon (*Oncorhynchus tshawytscha*) by immersion treatments using 17-alpha-methyltestosterone around the time of hatching. *Aquaculture* 72:359–67.

Baker, J. P., and S. W. Christensen. 1991. Effects of acidification on biological communities in aquatic ecosystems. In *Acidic deposition and aquatic ecosystems: Regional case studies*, ed. D. G. Charles, 83–106. New York: Springer-Verlag.

Baker, J. P., W. J. Warren-Hicks, J. Gallagher, et al. 1993. Fish population losses from Adirondack lakes: The role of surface water acidity and acidification. *Water Resour Res* 29:861–74.

Baker, T. T., A. C. Wertheimer, R. D. Burkett, et al. 1996. Status of Pacific salmon and steelhead escapements in southeastern Alaska. *Fisheries* 21(10): 6–18.

Bakke, T. A., P. A. Jansen, and L. P. Hansen. 1990. Differences in the host resistance of Atlantic salmon, *Salmo salar* L., stocks to the monogenean *Gyrodactylus salaris* Malmberg, 1957. *J Fish Biol* 37:577–87.

Bakun, A. 1990. Global climate change and intensification of coastal ocean upwelling. *Science* 247:198–201.

Balon, E. K. 1974. Domestication of the carp *Cyprinus carpio.* Toronto: Royal Ontario Museum Life Sciences Miscellaneous Publication.

———. 1995. Origin and domestication of the wild carp *Cyprinus carpio*: From Roman gourmets to the swimming flowers. *Aquaculture* 129:3–48.

———. 2000. Defending fishes against recreational fishing: An old problem to be solved in the new millennium. *Environ Biol Fish* 57:1–8.

———. 2006. The oldest domesticated fishes, and the consequences of an epigenetic dichotomy in fish culture. *Aqua: J Ichthyol Aquat Biol* 11:47–86.

Balon, E. K., and J. Holcik. 1999. Gabcikovo river barrage system: The ecological disaster and economic calamity for the inland delta of the middle Danube. *Environ Biol Fish* 54:1–17.

Baltz, D. M. 1991. Introduced fishes in marine systems and inland seas. *Biol Conserv* 56:151–77.

Baltz, D. M., and P. B. Moyle. 1993. Invasion resistance to introduced species by a native assemblage of California stream fishes. *Ecol Appl* 3:246–55.

Banarescu, P. 2002. *Romanichthys valsanicola*, a highly threatened species from the Danube River basin in Romania. In Collares-Pereira, Coelho, and Cowx 2002, 174–77.

Banister, K. 1989. Shoals of fish. *Naturopa* (Strasbourg) 62:20–21.

Banner, A., P. Helfrich, and T. Piyakarnchana. 1966. Retention of ciguatera toxin by red snapper, *Lutjanus bohar*. *Copeia* 1966:297–301.

Baquero, J. 1996. The stressful journey of ornamental marine fish. *SPC Live Reef Fish Inform Bull* 1:27–28.

———. 2001. The trade of ornamental fish from the Philippines. In Chao, Petry, et al., 2001, 75–82.

Barber, C. V., and V. R. Pratt. 1997. *Sullied seas: Strategies for combating cyanide fishing in southeast Asia and beyond.* Washington, DC: World Resources Institute.

———. 1998. Poison and profits: Cyanide fishing in the Indo-Pacific. *Environment* 40(8): 5–34.

Barber, R. T., A. K. Hilting, and M. L. Hayes. 2001. The changing health of coral reefs. *Hum Ecol Risk Assess* 7:1255–70.

Barber, W. E., L. L. McDonald, W. P. Erickson, et al. 1995. Effect of the *Exxon Valdez* oil spill on intertidal fish: A field study. *T Am Fish Soc* 124:461–76.

Barbosa, A. C., A. A. Boischio, G. A. East, et al. 1995. Mercury contamination in the Brazilian Amazon: Environmental and occupational aspects. *Water Air Soil Poll* 80:109–21.

Barbour, M. T., J. Gerritsen, B. D. Snyder, et al. 1999. *Rapid bioassessment protocols for use in streams and wadeable rivers: Periphyton, benthic macroinvertebrates and fish*, 2nd ed. EPA 841-B-99-002. Washington, DC: U.S. Environmental Protection Agency.

Bardach, J. E., J. H. Ryther, and W. O. McLarney. 1972. *Aquaculture: The farming and husbandry of freshwater and marine organisms.* New York: Wiley Interscience.

Barker, R. 1997. *And the waters turned to blood.* New York: Simon & Schuster.

Barlow, G. W. 1981. Patterns of parental investment, dispersal and size among coral reef fishes. *Environ Biol Fish* 6:65–85.

———. 2000. *The cichlid fishes: Nature's grand experiment in evolution.* Cambridge, MA: Perseus.

Barnhart, R. A. 1989. Symposium review. Catch-and-release fishing: A decade of experience. *N Am J Fish Manage* 9:74–80.

Barrett, J. C., G. D. Grossman, and J. Rosenfeld. 1992. Turbidity-induced changes in reactive distance of rainbow trout. *T Am Fish Soc* 121:437–43.

Barrett, S. 2000. *FDA and FTC attack shark cartilage.* June 9. www.quackwatch.com/04ConsumerEducation/news/shark.html.

Barse, A. M., and D. H. Secor. 1999. An exotic nematode parasite of the American eel. *Fisheries* 24(2): 6–10.

Barthem, R., and M. Goulding. 1997. *The catfish connection: Ecology, migration, and conservation of Amazon predators.* New York: Columbia Univ. Press.

Bartholomew, J. L. 1998. Host resistance to infection by the myxosporean parasite *Ceratomyxa shasta*: A review. *J Aquat Anim Health* 10.112–20.

Bartley, D. M. 2000. *Responsible ornamental fisheries.* FAO Aquatics Newsletter 24:10–14. www.fao.org/docrep/005/x4933e/x4933e10.htm.

Barton, B. A. 2002. Stress in fishes: A diversity of responses with particular reference to changes in circulating corticosteroids. *Integr Comp Biol* 42:517–25.

Barton, M. 2006. *Bond's Biology of Fishes*, 3rd ed. Clifton Park, NY: Thomson/Brooks Cole.

Bash, J. S., and C. M. Ryan. 2002. Stream restoration and enhancement projects: Is anyone monitoring? *Environ Manage* 29:877–85.

Bassleer, G. 1994. The international trade in aquarium/ornamental fish. *Infofish International* (5): 15–17. Cited in Young 1997.

Bateson, P. P. G. 1991. Assessment of pain in animals. *Anim Behav* 42:827–39.

Baum, J. K., R. A. Myers, D. Kehler, et al. 2003. Collapse and conservation of shark populations in the northwest Atlantic. *Science* 299:389–92.

Baxter, R. M. 1977. Environmental effects of dams and impoundments. *Annu Rev Ecol Syst* 8:255–83.

Bayer, R. D. 2000. *Attraction of birds to hatchery releases of juvenile salmonids at Yaquina Bay and Coos Bay, Oregon*. www.orednet.org/salmon/yaq-coos.htm.

Bayley, P. B. 1995. Understanding large river floodplain ecosystems. *BioScience* 45:153–59.

Bayley, P. B., and M. Petrere Jr. 1989. Amazon fisheries: Assessment methods, current status and management options. *Can Spec Pub Fish Aquat Sci* 106:385–98.

Beacham, T. D. 1983a. Variation in median size and age at sexual maturity of Atlantic cod, *Gadus morhua*, on the Scotian shelf in the northwest Atlantic Ocean. *Fish Bull U.S.* 81:303–21.

———. 1983b. Variation in size and age at sexual maturity of American plaice and yellowtail flounder in the Canadian Maritime Region of the northwest Atlantic Ocean. *Can Tech Rep Fish Aquat Sci* 1196:1–75.

Beamish, R. J., ed. 1995. Climate change and northern fish populations. *Can Spec Pub Fish Aquat Sci* 121.

Beamish, R. J., A. J. Benson, R. M. Sweeting, et al. 2004. Regimes and the history of the major fisheries off Canada's west coast. *Prog Oceanogr* 60:355–85.

Beamish, R. J., and H. H. Harvey. 1972. Acidification of the La Cloche Mountain lakes, Ontario, and resulting fish mortalities. *J Fish Res Board Can* 29:1131–43.

Beck, M. W., K. L. Heck Jr., K. W. Able, et al. 2003. The role of nearshore ecosystems as fish and shellfish nurseries. *Issues Ecol* 11:1–11.

Becker, G. C. 1983. *Fishes of Wisconsin*. Madison: University of Wisconsin Press.

Bednarek, A. T. 2001. Undamming rivers: A review of the ecological impacts of dam removal. *Environ Manage* 27:803–14.

Beechie, T., E. Beamer, B. Collins, et al. 1996. Restoration of habitat-forming processes in Pacific Northwest watersheds: A locally adaptable approach to salmonid habitat restoration. In *The role of restoration in ecosystem management*, ed. D. L. Peterson and C. V. Klimas, 48–67. Madison, WI: Society for Ecological Restoration.

Beets, J. 1997. Effects of a predatory fish on the recruitment and abundance of Caribbean coral reef fishes. *Mar Ecol—Prog Ser* 148:11–21.

Beets, J., and A. Friedlander. 1999. Evaluation of a conservation strategy: A spawning aggregation closure for red hind, *Epinephelus guttatus*, in the U.S. Virgin Islands. *Environ Biol Fish* 55:91–98.

Behnke, R. J. 1992. *Native trout of western North America*. American Fisheries Society Monograph 6. Bethesda, MD: American Fisheries Society.

Belknap, J. 1792. *History of New Hampshire*. Cited in NMFS 1993.

Bell, F. H. 1978. *The Pacific halibut: The resource and the fishery*. Anchorage: Alaska Northwest Publishing.

Bell, J. D., and R. Galzin. 1984. Influence of live coral cover on coral-reef fish communities. *Mar Ecol—Prog Ser* 15:265–74.

Bell, M. A., and S. A. Foster, eds. 1994. *The evolutionary biology of the threespine stickleback*. Oxford: Oxford Univ. Press.

Benaka, L., ed. 1999. Fish habitat: Essential fish habitat and rehabilitation. *Am Fish Soc Sym* 22.

Bender, M. T. 2003. *Can the tuna: FDA's failure to protect children from exposure to mercury in albacore "white" canned tuna*. Montpelier, VT: Mercury Policy Project.

Benke, A. C., R. L. Henry III, D. M. Gillespie, et al. 1985. Importance of snag habitat for animal production in southeastern streams. *Fisheries* 10(5): 8–13.

Bennett, W. A., R. J. Currie, P. F. Wagner, et al. 1997. Cold tolerance and potential overwintering of the red-bellied piranha *Pygocentrus nattereri* in the United States. *T Am Fish Soc* 126:841–49.

Benstead, J. P., P. H. DeRham, J. Gattollait, et al. 2003. Conserving Madagascar's freshwater biodiversity. *BioScience* 53:1101–11.

Benstead, J. P., M. L. J. Stiassny, P. V. Loiselle, et al. 2000. River conservation in Madagascar. In *Global perspectives on river conservation: Science, policy and practice*, ed. J. P. Boon et al., 205–31. New York: Wiley.

Berbari, P., A. Thibodeau, L. Germain, et al. 1999. Antiangiogenic effects of the oral administration of liquid cartilage extract in humans. *J Surg Res* 87:108–13.

Berg, H., M. C. Ohman, S. Troeng, et al. 1998. Environmental economics of coral reef destruction in Sri Lanka. *Ambio* 27:627–34.

Bergersen, E. P., and D. E. Anderson. 1997. The distribution and spread of *Myxobolus cerebralis* in the

United States. *Fisheries* 22(8): 6–7.

Berghahn, R., and M. Purps. 1998. Impact of discard mortality in *Crangon* fisheries on year-class strength of North Sea flatfish species. *J Sea Res* 40:83–91.

Berghahn, R., M. Waltemath, and A. D. Rijnsdorp. 1992. Mortality of fish from the by-catch of shrimp vessels in the North Sea. *J Appl Ichthyol* 8:293–306.

Bergmann, A., H. Hinz, R. Blyth, et al. 2004. Using knowledge from fishers and fisheries scientists to identify possible groundfish "Essential Fish Habitats." *Fish Res* 66:373–79.

Bergstedt, L. C., and E. P. Bergersen. 1997. Health and movements of fish in response to sediment sluicing in the Wind River, Wyoming. *Can J Fish Aquat Sci* 54:312–19.

Berkes, F., J. Colding, and C. Folke. 2000. Rediscovery of traditional ecological knowledge as adaptive management. *Ecol Appl* 10:1251–62.

Berkman, H. E., and C. F. Rabeni. 1987. Effect of siltation on stream fish communities. *Environ Biol Fish* 18:285–94.

Bermond, B. 1997. The myth of animal suffering. In *Animal consciousness and animal ethics*, ed. M. Dol et al., 125–33. Assen, Netherlands: Van Gorcum.

Bernhardt, E. S., M. A. Palmer, J. D. Allan, et al. 2005. Synthesizing U.S. river restoration efforts. *Science* 308:636–37.

Bernoth, E.-M., A. Ellis, P. Midtlyng, et al., eds. 1997. *Furunculosis: Multidisciplinary fish disease research.* New York: Academic Press.

Berra, T. M. 1998. *A natural history of Australia.* San Diego: Academic Press.

———. 2001. *Freshwater fish distribution.* San Diego: Academic Press.

Berra, T. M., and B. J. Pusey. 1997. Threatened fishes of the world: *Lepidogalaxias salamandroides* Mees, 1961 (Lepidogalaxiidae). *Environ Biol Fish* 50:201–2.

Bertram, I. 1996. The aquarium fishery in the Cook Islands: "Is there a need for management?" *SPC Live Reef Fish Inform Bull* 1:10–12.

Best, B. A. 2002. *Trade in coral reef animals, algae and products: An overview.* www.usaid.gov/our_work/environment/water/tech_pubs/coral.reef.report.sections/08.coral_reef_report.2002.trade&mgmt.pdf.

Bestgen, K. R., and S. P. Platania. 1991. Status and conservation of the Rio Grande silvery minnow, *Hybognathus amarus. Southwest Nat* 36:225–32.

Beukema, J. J. 1970a. Acquired hook avoidance in the pike *Esox lucius* L., fished with artificial and natural baits. *J Fish Biol* 2:155–60.

———. 1970b. Angling experiments with carp (*Cyprinus carpio* L.). Part 2: Decreased catchability through one trial learning. *Neth J Zool* 19:81–92.

Beveridge, M. C. M. 1996. *Cage aquaculture*, 2nd ed. Edinburgh: Fishing News Books.

Beverton, R. J. H. 1990. Small marine pelagic fish and the threat of fishing: Are they endangered? *J Fish Biol* 37(suppl. A): 3–16.

Beyer, W. N., G. H. Heinz, and A. W. Redmon-Norwood, eds. 1996. *Environmental contaminants in wildlife: Interpreting tissue concentrations.* Boca Raton, FL: Lewis.

Bhargava, P., J. L. Marshall, W. Bahut, et al. 2001. A phase I and pharmacokinetic study of squalamine, a novel antiangiogenic agent, in patients with advanced cancers. *Clin Cancer Res* 7:3912–19.

Bhattacharjee, Y. 2005. Uncovering the hidden paths of Maine's threatened cod. *Science* 310:1110–11.

Bianco, P. G. 1990. Vanishing freshwater fishes in Italy. *J Fish Biol* 37(suppl. A): 235–37.

Biber-Klemm, S. 1995. Legal aspects of the conservation of endemic freshwater fish in the northern Mediterranean region. *Biol Conserv* 72:321–34.

Bigelow, H. B., and W. C. Schroeder. 1953. Fishes of the Gulf of Maine. *Fish Bull FWS* 74(53).

Bigler, B. S., D. H. Welch, and J. H. Helle. 1996. A review of size trends among North Pacific salmon. *Can J Fish Aquat Sci* 53:455–65.

Bilby, R. E., and P. A. Bisson. 1998. Function and distribution of large woody debris. In *River ecology and management*, ed. R. J. Naiman and R. E. Bilby, 324–46. New York: Springer-Verlag.

Bilby, R. E., B. R. Fransen, and P. A. Bisson. 1996. Incorporation of nitrogen and carbon from spawning coho salmon into the trophic system of small streams: Evidence from stable isotopes. *Can J Fish Aquat Sci* 53:164–73.

———, et al. 1998. Response of juvenile coho salmon (*Oncorhynchus kisutch*) and steelhead (*O. mykiss*) to the addition of salmon carcasses to two streams in southwestern Washington, USA. *Can J Fish Aquat Sci* 55:1909–18.

Birkeland, C. 1997a. Implications for resource management. In Birkeland 1997c, 411–35.

———. 1997b. Introduction. In Birkeland 1997c, 1–12.

———, ed. 1997c. *The life and death of coral reefs*. New York: Chapman and Hall.

———. 1997d. Symbiosis, fisheries and economic development on coral reefs. *Trends Ecol Evol* 12:364–67.

Birstein, V. J., J. R. Waldman, and W. E. Bemis, eds. 1997a. Sturgeon biodiversity and conservation. *Environ Biol Fish* 48:1–438.

Birstein, V. J., W. E. Bemis, and J. R. Waldman. 1997b. The threatened status of acipenseriform species: A summary. *Environ Biol Fish* 48:427–35.

Blaber, S. J. M., and T. G. Blaber. 1980. Factors affecting the distribution of juvenile estuarine and inshore fish. *J Fish Biol* 17:143–62.

Blankenship, H. L., and K. M. Leber. 1995. A responsible approach to marine stock enhancement. *Am Fish Soc Sym* 15:167–75.

Blazer, V. S., D. L. Fabacher, E. E. Little, et al. 1997. Effects of ultraviolet-B radiation on fish: Histologic comparison of a UVB-sensitive and a UVB-tolerant species. *J Aquat Anim Health* 9:132–43.

Blazer, V. S., W. K. Vogelbein, C. L. Densmore, et al. 1999. *Aphanomyces* as a cause of ulcerative skin lesions of menhaden from Chesapeake Bay tributaries. *J Aquat Anim Health* 11:340–49.

Blindheim, J., and H. R. Skjoldal. 1993. Effects of climate changes on the biomass yield of the Barents Sea, Norwegian Sea, and West Greenlans large marine ecosystems. In *Large marine ecosystems: stress, mitigation, and sustainability*. ed. K. Sherman, L. M. Alexander, and B. D. Gold, 185–189. Washington, DC: American Association for the Advancement of Science.

Blinn, D. W., C. Runck, D. A. Clark, et al. 1993. Effects of rainbow trout predation on Little Colorado spinedace. *T Am Fish Soc* 122:139–43.

Block, B. A., H. Dewar, S. B. Blackwell, et al. 2001. Migratory movements, depth preferences, and thermal biology of Atlantic bluefin tuna. *Science* 293:1310–14.

Bock, R. 2002. Why keep natives? *Am Currents* 28(2): 35.

Bodaly, R. A., T. W. D. Johnson, R. J. P. Fudge, et al. 1984. Collapse of the lake whitefish (*Coregonus clupeaformis*) fishery in Southern Indian Lake, Manitoba, following lake impoundment and river diversion. *Can J Fish Aquat Sci* 41:692–700.

Bodaly, R. A., V. L. St. Louis, M. J. Paterson, et al. 1997. Bioaccumulation of mercury in the aquatic food chain in newly flooded areas. In *Mercury and its effects on environment and biology*, ed. H. Sigel and A. Sigel, 259–87. New York: Marcel Dekker.

Boehlert, G. W., and J. B. Morgan. 1985. Turbidity enhances feeding abilities of larval Pacific herring, *Clupea harengus pallasi*. *Hydrobiologia* 123:161–70.

Boersma, P. D. 1997. Magellanic penguins decline in south Atlantic. *Penguin Conserv* 10:2–5.

Boersma, P. D., and J. K. Parrish. 1999. Limiting abuse: Marine protected areas, a limited solution. *Ecol Econ* 31:287–304.

Boet, P., J. Belliard, R. Berrebi-dit-Thomas, et al. 1999. Multiple human impacts by the City of Paris on fish communities in the Seine River basin, France. *Hydrobiologia* 410:59–68.

Bogutskaya, N. G., and A. M. Naseka. 2002a. *Freshwater fishes of Russia*. St. Petersburg, Russia: Zoological Institute of the Russian Academy of Sciences. www.zin.ru/animalia/pisces/eng/taxbase_e/taxbase_index_e.htm.

———. 2002b. Web site and database for freshwater fishes of Russia. A source of information on the current state of the fauna. St. Petersburg, Russia: Zoological Institute of the Russian Academy of Sciences. www.zin.ru/animalia/pisces/eng/taxbase_e/taxbase_index_e.htm.

Bohnsack, J. A. 1993. Marine reserves: They enhance fisheries, reduce conflicts, and protect resources. *Oceanus* 36(3): 63–71.

———. 1994. How marine fishery reserves can improve reef fisheries. *P Gulf Caribb Fish Inst* 43:217–41.

———. 1996. Maintenance and recovery of reef fishery productivity. In Polunin and Roberts 1996, 283–313.

Bohnsack, J. A., and J. S. Ault. 1996. Management strategies to conserve marine biodiversity. *Oceanography* 9:73–82.

Bohnsack, J., B. Causey, M. P. Crosby, et al. 2000. *A rationale for minimum 20–30% no-take protection*. Ninth International Coral Reef Symposium, Bali, Indonesia, 615–19.

Bohnsack, J. A., A.-M. Ecklund, and A. M. Szmant. 1997. Artificial reef research: Is there more than the attraction-production issue? *Fisheries* 22(4): 14–16.

Bolker, B. M., C. M. St. Mary, C. W. Osenberg, et al. 2002. Management at a different scale: Marine ornamentals and local processes. *B Mar Sci* 70:733–48.

Bond, C. E. 1996. *Biology of Fishes*, 2nd ed. Philadelphia: Saunders College Publishing.

Bond, C. E., and T. T. Kan. 1973. *Lampetra* (*Entosphenus*) *minima* n. sp., a dwarfed parasitic lamprey from Oregon. *Copeia* 1973:568–74.

Bonfil, R. 1994. *Overview of world elasmobranch fisheries*. FAO Fisheries Tech. Paper 341. Rome: FAO.

Booth, D. B., and C. R. Jackson. 1997. Urbanization of aquatic systems: Degradation thresholds, stormwater detention, and the limits of mitigation. *J Am Water Resources Assoc* 33:1077–90.

Bortone, S. A., and W. P. Davis. 1994. Fish intersexuality as indicators of environmental stress. *BioScience* 44:165–72.

Borucinska, J. D., J. C. Harshbarger, R. Reimschuessel, et al. 2004. Gingival neoplasms in a captive sand tiger shark, *Carcharias taurus* (Rafinesque), and a wild-caught blue shark, *Prionace glauca* (L.). *J Fish Dis* 27:185–91.

Bosch, J. M., and J. D. Hewlett. 1982. A review of catchment experiments to determine the effect of vegetation changes on water yield and evapotranspiration. *J Hydrol* 55:3–23.

Botsford, L. W., J. C. Castilla, and C. H. Peterson. 1997. The management of fisheries and marine ecosystems. *Science* 277:509–15.

Bottom, D. L., J. A. Lichatowich, and B. E. Riddell. 2006. The estuary, plume, and marine environments. In R. N. Williams 2006, 507–69.

Bouchard, D. A., K. Brockway, C. Giray, et al. 2001. First report of infectious salmon anemia (ISA) in the United States. *B Eur Assoc Fish Pathol* 21:86–88.

Bourdelais, A. J. , C. R. Tomas, J. Naar, et al. 2002. New fish-killing alga in coastal Delaware produces neurotoxins. *Environ Health Persp* 110:465–70.

Bovee, K. D., B. L. Lamb, J. M. Bartholow, et al. 1998. *Stream habitat analysis using the instream flow incremental methodology*. Fort Collins, CO: U.S. Geological Survey-BRD/ITR-1998-0004.

Bowen, B. W., and J. Roman. 2005. Gaia's handmaidens: The Orlog model for Conservation Biology. *Conserv Biol* 19:1037–43.

Bowles, E. C. 1995. Supplementation: Panacea or curse for the recovery of declining fish stocks? *Am Fish Soc Sym* 15:277–83.

Boyce, M. 1992. Population viability analysis. *Annu Rev Ecol Syst* 23:481–506.

Boyce, N. 2000. Survival of the meanest: Galapagos fishermen are winning the struggle for supremacy. *New Scientist* 168(2266): 6.

Boyd, C. E., and C. S. Tucker. 1995. Sustainability of channel catfish farming. *World Aquaculture* 26:45–53.

Boyer, D. C., H. J. Boyer, I. Fossen, et al. 2001. Changes in abundance of the northern Benguela sardine stock during the decade 1990–2000, with comments on the relative importance of fishing and the environment. *S Afr J Marine Sci* 23:67–84.

Boyle, R. H. 1996. Life or death for the Salton Sea. *Smithsonian* (June): 86–97.

Bradshaw, C., A. R. Brand, and L. O. Veale. 2001. The effect of scallop dredging on Irish Sea benthos: Experiments using a closed area. *Hydrobiologia* 465:129–38.

Bradstock, M., and D. P. Gordon. 1983. Coral-like bryozoan growths in Tasman Bay, and their protection to conserve commercial fish stocks. *New Zeal J Mar Fresh* 17:159–63.

Braithwaite, V. A., and F. A. Huntingford. 2004. Fish and welfare: Do fish have the capacity for pain perception and suffering? *Anim Welfare* 13(suppl. S): S87–S92.

Brannon, E. L. 1987. Mechanisms stabilizing salmonid fry emergence timing. In *Sockeye salmon (*Oncorhynchus nerka*) population biology and future management*, ed. H. D. Smith et al., 120–24. Canadian Special Publication, Fisheries and Aquatic Science 96.

Brannon, E. L., D. F. Amend, M. A. Cronin, et al. 2004. The controversy about salmon hatcheries. *Fisheries* 29(9): 12–31.

Branstetter, S. 1999. *The management of the United States Atlantic shark fishery*. FAO Fisheries Tech. Paper 378, vol. 1, 109–48.

Bray, K. 2000. *A global review of IUU fishing*. www.afffa.gov.au.

Bremner, J., and J. Perez. 2002. A case study of human migration and the sea cucumber crisis in the Galapagos Islands. *Ambio* 31:306–10.

Briggs, J. C. 1969a. Panama's sea-level canal. *Science* 162:511–13.

———. 1969b. The sea-level Panama Canal: Potential biological catastrophe. *BioScience* 19:44–47.

———. 1995. *Global biogeography*. New York: Elsevier.

Briggs, M. K., and S. Cornelius. 1998. Opportunities for ecological improvement along the lower Colorado River and delta. *Wetlands* 18:513–29.

Bright, C. 1998. *Life out of bounds: Bioinvasion in a borderless world*. New York: Norton.

Britton, J. C., and B. Morton. 1994. Marine carrion and scavengers. *Oceanogr Mar Biol* 32:369–434.

Broadhurst, M. K. 2000. Modifications to reduce bycatch in prawn trawls: A review and framework for development. *Rev Fish Biol Fisher* 10:27–60.

Brodie, J. 1995. The problems of nutrients and eutrophication in the Australian marine environment. In *The state of the marine environment report for Australia*, ed. L. P. Zann and D. C. Sutton, tech. annex 2, 1–29. Townsville: Great Barrier Reef Marine Park Authority. Cited in Birkeland 1997c.

Brodie, J., C. Christie, M. Devlin, et al. 2001. Catchment management and the Great Barrier Reef. *Water Sci Technol* 43:203–11.

Brooks, B. W., C. M. Foran, S. M. Richards, et al. 2003. Aquatic ecotoxicology of fluoxetine. *Toxicol Lett* 142:169–83.

Brooks, B. W., P. K. Turner, J. K. Stanley, et al. 2003. Waterborne and sediment toxicity of fluoxetine to selected organisms. *Chemosphere* 52:135–42.

Broom, D. M. 2001. The evolution of pain. *Vlaams Diergen Tijds* 70:17–21.

Brouder, M. J. 1999. Relationship between length of roundtail chub and infection intensity of Asian fish tapeworm *Bothriocephalus acheilognathi*. *J Aquat Anim Health* 11:302–4.

Browman, H. I. 2000. Application of evolutionary theory to fisheries science and stock assessment-management. *Mar Ecol—Prog Ser* 208:299.

Brown, B. 1995. *Mountain in the clouds: A search for the wild salmon*, rep. ed. Seattle: Univ. Washington Press.

———. 1997. Disturbances to reefs in recent times. In Birkeland 1997c, 354–79.

Brown, B. E., and R. P. Dunne. 1988. The impact of coral mining on coral reefs in the Maldives. *Environ Conserv* 15:159–65.

Brown, L. R., and L. Starke, eds. 1998. *State of the world 1998: A Worldwatch Institute report on progress toward a sustainable society*. Washington, DC: Norton/Worldwatch Institute.

Brown, S. B., J. D. Fitzsimons, D. C. Honeyfield, et al. 2005. Implications of thiamine deficiency in Great Lakes salmonines. *J Aquat Anim Health* 17:113–24.

Brown, S. K., P. J. Auster, L. Lauck, et al. 1998. *Ecological effects of fishing*. NOAA's State of the Coast Report. Silver Spring, MD: NOAA. state-of-coast.noaa.gov/bulletins/html/ief_03/ief.html.

Brown, S. T. 1999. Trends in the commercial and recreational shark fisheries in Florida, 1980–1992, with implications for management. *N Am J Fish Manage* 19:28–41.

Bruce, B. D., M. A. Green, and P. R. Last. 1998. Threatened fishes of the world: *Brachionichthys hirsutus* (Lacepede, 1804) (Brachionichthyidae). *Environ Biol Fish* 52:418.

Bruton, M. N. 1985. The effects of suspensoids on fish. *Hydrobiologia* 125:221–41.

———. 1995. Have fishes had their chips? The dilemma of threatened fishes. *Environ Biol Fish* 43:1–27.

Bryant, M. D. 1995. Pulsed monitoring for watershed and stream restoration. *Fisheries* 20(11): 6–13.

Brylinsky, M. J., J. Gibson, and D. C. Gordon. 1994. Impacts of flounder trawls on the intertidal habitat and community of the Minas Basin, Bay of Fundy. *Can J Fish Aquat Sci* 51:650–61.

Buckworth, R. C. 1998. World fisheries are in crisis? We must respond! In Pitcher, Hart, and Pauly 1998, 3–17.

Bulger, A. J., B. J. Cosby, and J. R. Webb. 2000. Current, reconstructed past, and projected future status of brook trout (*Salvelinus fontinalis*) streams in Virginia. *Can J Fish Aquat Sci* 57:1515–23.

Bulger, A. J., C. A. Dolloff, B. J. Cosby, et al. 1995. The Shenandoah National Park: Fish in sensitive habitats (SNP: fish) project. An integrated assessment of fish community responses to stream acidification. *Water Air Soil Poll* 85:309–14.

Bull-Tornøe, A. O. 1992. Ornamental fish trade from Lake Malawi. MS thesis, Agricultural Univ. Norway.

Bundy, A. 1998. The red light and adaptive management. In Pitcher, Hart, and Pauly 1998, 361–68.

Bunn, S. E., and A. H. Arthington. 2002. Basic principles and ecological consequences of altered flow regimes for aquatic biodiversity. *Environ Manage* 30:492–507.

Burger, J. 1997. *Oil spills*. Rutgers, NJ: Rutgers Univ. Press.

Burgess, G. H., L. R. Beerkircher, G. M. Cailliet, et al. 2005. Is the collapse of shark populations in the northwest Atlantic Ocean and Gulf of Mexico real? *Fisheries* 30(10): 19–26.

Burkhead, N. M., and H. Jelks. 2001. Effects of suspended sediment on the reproductive success of the tricolor shiner, a crevice-spawning minnow. *T Am Fish Soc* 130:959–68.

Burkhead, N. M., S. J. Walsh, B. J. Freeman, et al. 1997. Status and restoration of the Etowah River, an imperiled southern Appalachian ecosystem. In *Aquatic fauna in peril: The southeastern perspective*, ed. G. W. Benz and D. E. Collins, 375–444. Southeast Aquatic Research Institute Special Publication 1. Decatur, GA: Lenz Design and Communications.

Burkholder, J. M. 1998. Implications of harmful microalgae and heterotrophic dinoflagellates in management of sustainable marine fisheries. *Ecol Appl* 8(suppl.): S37–S62.

———. 2002. *Pfiesteria*: The toxic *Pfiesteria* complex. In *Encyclopedia of Environmental Microbiology*, ed. G. Bitton, 2431–47. New York: Wiley.

Burkholder, J. M., and H. B. Glasgow Jr. 1997. *Pfiesteria piscicida* and other *Pfiesteria*-like dinoflagellates: Behavior, impacts, and environmental controls. *Limnol Oceanogr* 42:1052–75.

———. 2001. History of toxic *Pfiesteria* in North Carolina estuaries from 1991 to the present. *BioScience* 51:827–41.

Burkholder, J. M., A. S. Gordon, P. D. Moeller, et al. 2005. Demonstration of toxicity to fish and to mammalian cells by *Pfiesteria* species: Comparison of assay methods and strains. *P Natl Acad Sci USA* 102:3471–76.

Burkholder, J. M., M. A. Mallin, H. B. Glasgow Jr., et al. 1997. Impacts to a coastal river and estuary from rupture of a large swine waste holding lagoon. *J Environ Qual* 26:1451–66.

Burns, W. C. 1995. *Shark in the Galapagos*. www.ee/lists/ infoterra/1995/03/0062.html.

Bury, N. R., P. A. Walker, and C. N. Glover. 2003. Nutritive metal uptake in teleost fish. *J Exp Biol* 206:11–23.

Busack, C. A., and K. P. Currens. 1995. Genetic risks and hazards in hatchery operations: Fundamental concepts and issues. *Am Fish Soc Sym* 15:71–80.

Butler, J. N., J. Burnett-Herkes, J. A. Barnes, et al. 1993. The Bermuda fisheries: A tragedy of the commons averted? *Environment* 35(1): 7–33.

Butterworth, D. S., K. O. Cochrane, and J. A. A. De Oliveira. 1997. Management procedures: A better way to manage fisheries? The South African experience. *Am Fish Soc Sym* 20:83–90.

Buxton, C. D. 1993. Life-history changes in exploited reef fishes on the east coast of South Africa. *Environ Biol Fish* 36:47–63.

Caddy, J. F., and L. Garibaldi. 2000. Apparent changes in the trophic composition of world marine harvests: The perspective from the FAO capture database. *Ocean Coast Manage* 43:615–55.

Caddy, J. F., and R. Mahon. 1995. *Reference points for fisheries management*. FAO Fisheries Tech. Paper 347. Rome: FAO.

Caddy, J. F., and P. G. Rodhouse. 1998. Cephalopod and groundfish landings: Evidence for ecological change in global fisheries. *Rev Fish Biol Fisher* 8:431–44.

Callicott, J. B. 1989. *In defense of the land ethic: Essays in environmental philosophy*. Albany, NY: SUNY Press.

Cambray, J. A. 1999. Threatened fishes of the world: *Barbus capensis* Smith, 1841 (Cyprinidae). *Environ Biol Fish* 54:150.

Camhi, M. 1996. Costa Rica's shark fishery and cartilage industry. Newsletter of the IUCN Shark Specialist Group. *Shark News* 8.

———. 1998. *Sharks on the line: A state-by-state analysis of sharks and their fisheries*. Islip, NY: Audubon Living Oceans Program.

———. 1999. *Sharks on the line II: An analysis of Pacific state sharks fisheries*. Islip, NY: Audubon Living Oceans Program.

Campana, S. E. 1999. Chemistry and composition of fish otoliths: Pathways, mechanisms and applications. *Mar Ecol—Prog Ser* 188:263–97.

Campana, S. E., W. Joyce, L. Marks, et al. 2002. Population dynamics of the porbeagle in the northwest Atlantic Ocean. *N Am J Fish Manage* 22:106–21.

Campbell, I. C., and T. I. Doeg. 1989. Impact of timber harvesting and production on streams: A review. *Aust J Mar Fresh Res* 40:519–39.

Campbell, P. M., and T. H. Hutchinson. 1998. Wildlife and endocrine disrupters: Requirements for hazard identification. *Environ Toxicol Chem* 17:127–35.

Campbell, R. R. 1997. Rare and endangered fishes and marine mammals of Canada: COSEWIC Fish and Marine Mammal Subcommittee status reports: 11. *Can Field Nat* 111:249–57.

———. 1998. Rare and endangered fishes and marine mammals of Canada: COSEWIC Fish and Marine Mammal Subcommittee status reports: 12. *Can Field Nat* 112:94–98.

Camphuysen, C. J. 1993. *Seabirds feeding on discards in winter in the North Sea*. Den Burg, Texel: Netherlands Institute for Sea Research. Cited in Jennings and Kaiser 1998.

Campos, H., G. Dazarola, B. Dyer, et al. 1998. Categorías de conservación de peces nativos de aguas continentales de Chile. *B Museu Nacional Hist Nat* 47:101–22.

Cargnelli, L. M., S. J. Griesbach, and W. W. Morse. 1999. *Essential fish habitat source document: Atlantic halibut,* Hippoglossus hippoglossus, *life history and habitat characteristics*. NOAA Tech. Memo. NMFS-NE-125.

Carlsson, J., and J. Nilsson. 2001. Effects of geomorphological structures on genetic differentiation among brown trout populations in a northern boreal river drainage. *T Am Fish Soc* 130:36–45.

Carlton, J. T. 1999. The scale and ecological consequences of biological invasions in the world's oceans. In *Invasive species and biodiversity management*, ed. O. T. Sandlund et al., 195–212. Dordrecht, Netherlands: Kluwer Academic.

———. 2000. Global change and biological invasions in the oceans. In *Invasive species in a changing world*, ed. H. A. Mooney and R. J. Hobbs, 31–53. Washington, DC: Island Press.

Carlton, J. T., and J. B. Geller. 1993. Ecological roulette: The global transport of nonindigenous marine organisms. *Science* 261:78–82.

Caro, T. M., ed. 1998. *Behavioral ecology and conservation biology*. Oxford: Oxford Univ. Press.

Carpenter, K. E., and A. C. Alcala. 1977. Philippine coral reef fisheries resources. Part 2: Muro-ami and kayakas reef fisheries, benefit or bane? *Philipp J Fish* 15:217–235. Cited in Jennings and Kaiser 1998.

Carpenter, R. C. 1997. Invertebrate predators and grazers. In Birkeland 1997c, 198–229.

Carpenter, S. R., and J. F. Kitchell, eds. 1993. *The trophic cascade in lakes*. New York: Cambridge Univ. Press.

Carr, J. W., J. M. Anderson, F. G. Whoriskey, et al. 1997. The occurrence and spawning of cultured Atlantic salmon (*Salmo salar*) in a Canadian river. *ICES J Mar Sci* 54:1064–73.

Carr, M. H., and M. A. Hixon. 1997. Artificial reefs: The importance of comparisons with natural reefs. *Fisheries* 22(4): 28–33.

Casey, J. M., and R. A. Myers. 1998. Near extinction of a large, widely distributed fish. *Science* 281:690–92.

Cassani, J. R. 1995. Problems and prospects for grass carp as a management tool. *Am Fish Soc Sym* 15:407–12.

Castro, J. I., C. M. Woodley, and R. L. Brudek. 1999. *A preliminary evaluation of the status of shark species*. FAO Fisheries Tech. Paper 380. Rome: FAO.

Caviedes, C., and T. Fik. 1990. Variability in the Peruvian and Chilean fisheries. In *Climate variability, climate change and fisheries*, ed. M. H. Glantz and L. E. Feingold, 95–102. Boulder, CO: National Center for Atmospheric Research.

CCDR (Canada Communicable Disease Report). 1997. *Ciguatera fish poisoning linked to the ingestion of barracuda in a Montreal restaurant, Quebec*, vol. 23-20. www.phac-aspc.gc.ca/publicat/ccdr-rmtc.

CCME (Canadian Council of Ministers of the Environment). 1999. Canadian water quality guidelines for the protection of aquatic life. In *Canadian environmental quality guidelines*. CCME pub. 1299. Winnepeg: Author.

Cesar, H., K. Warren, Y. Sadovy, et al. 2000. Marine market transformation of the live reef fish food trade in southeast Asia. In *Collected essays on the economics of coral reefs*, ed. H. Cesar, 137–57. Kalmar, Sweden: CORDIO, Kalmar University.

Chan, T. T. C., and Y. Sadovy. 1998. Profile of the marine aquarium fish trade in Hong Kong. *Aquar Sci Conserv* 2:1–17.

Chao, N. H., and I. C. Liao. 2001. Cryopreservation of finfish and shellfish gametes and embryos. *Aquaculture* 197:161–89.

Chao, N. L., P. Petry, G. Prang, et al., eds. 2001. *Conservation and management of ornamental fish resources of the Rio Negro basin, Amazonia, Brazil: Project Piaba*. Manaus: Editora da Univ. Amazonas.

Chao, N. L., and G. Prang. 2002. Decade of Project Piaba: Reflections and prospects. *Ornamental Fish Int J* 39:1–4.

Chao, N. L., G. Prang, and P. Petry. 2001. Project Piaba: Maintenance and sustainable development of ornamental fisheries in the Rio Negro basin, Amazonas, Brazil. In Chao, Petry, et al., 2001, 3–14.

Chapin, F. S., E. S. Zavaleta, V. T. Eviner, et al. 2000. Consequences of changing biodiversity. *Nature* 405:234–42.

Chapman, C. A., and L. J. Chapman. 2003. Deforestation in tropical Africa: Impacts on aquatic ecosystems. In Crisman et al. 2003, 229–46.

Chapman, D. W. 1988. Critical review of variables used to define effects of fines in redds of large salmonids. *T Am Fish Soc* 117:1–21.

Chapman, F. A., S. A. Fitz-Coy, E. M. Thunberg, et al. 1997. United States of America trade in ornamental fish. *J World Aquacult Soc* 28:1–10.

Chapman, L. J., and C. A. Chapman. 2003. Fishes of African rain forests: Emerging and potential threats to a little-known fauna. In Crisman et al. 2003, 176–209.

Charles, A. T. 2001. *Sustainable fishery systems*. Oxford: Blackwell Science.

Chereshnev, I. A. 1996. Threatened fishes of the world: *Salvethymus svetovidovi* Chereshnev & Scopets, 1990 (Salmonidae). *Environ Biol Fish* 46:166.

Chesapeake Bay Program. 2001. *Atlantic menhaden*. wwww.chesapeakebay.net/info/atlantic_menhaden.cfm.

Che-tsung, C., L. Kwang-Ming, and J. Shoou-Jeng. 1997. Preliminary report on Taiwan's whale shark fishery. *TRAFFIC Bull* 17(1). Cambridge, UK: Traffic International.

Chick, J. H., and M. A. Pegg. 2001. Invasive carp in the Mississippi River basin. *Science* 292:2250–51.

Chilcote, M. W., S. A. Leider, and J. J. Loch. 1986. Differential reproductive success of hatchery and wild summer-run steelhead under natural conditions. *T Am Fish Soc* 115:726–35.

Cho, J., and Y. Kim. 2002. Sharks: A potential source of antiangiogenic factors and tumor treatments. *Mar Biotechnol* 4:521–25.

Choat, J. H., and L. Axe. 1996. Growth and longevity in acanthurid fishes: An analysis of otolith increments. *Mar Ecol—Prog Ser* 134:15–26.

Chotkowski, M. A., and J. E. Marsden. 1999. Round goby and mottled sculpin predation on lake trout eggs and fry: Field predictions from laboratory experiments. *J Great Lakes Res* 25:26–35.

Christensen, V. 1996. Managing fisheries involving predator and prey species. *Rev Fish Biol Fisher* 6:417–42. Cited in Jennings and Kaiser 1998.

Christie, P., B. J. McCay, M. L. Miller, et al. 2003. Toward developing a complete understanding: A social science research agenda for marine protected areas. *Fisheries* 28(12): 22–26.

Chu, K. H., P. F. Tam, C. H. Fung, et al. 1997. A biological survey of ballast water in container ships entering Hong Kong. *Hydrobiologia* 352:201–6.

Churchill, J. H. 1989. The effect of commercial trawling on sediment resuspension and transport over the Middle Atlantic Bight continental shelf. *Cont Shelf Res* 9:841–64.

Cisneros-Mata, M. A., L. W. Botsford, and J. F. Quinn. 1997. Projecting viability of *Totoaba macdonaldi*, a population with unknown age-dependent variability. *Ecol Appl* 7: 968–80.

Cisneros-Mata, M. A., G. Montemayor-Lopez, and M. J. Roman-Rodriguez. 1995. Life history and conservation of *Totoaba macdonaldi*. *Conserv Biol* 9:806–14.

CITES (Convention on International Trade in Endangered Species). 2001. *CITES identification guide: Sturgeons and paddlefish*. Ottawa, Canada: Minister of Supply and Services. www.cws-scf.ec.gc.ca.

Claire, E., and R. Storch. 1983. Streamside management and livestock grazing: An objective look at the situation. In *Workshop on livestock and wildlife-fisheries relationships in the Great Basin*, ed. J. Menke, 111–28. Berkeley, CA: U.S. Forest Service.

Clark, C. W. 1985. *Bioeconomic modeling and fisheries*. New York: Wiley.

———. 1996. Marine reserves and the precautionary management of fisheries. *Ecol Appl* 6:369–70.

Clark, J. A., and R. M. May. 2002. Taxonomic bias in conservation research. *Science* 297:191–92.

Clark, M. 2001. Are deepwater fisheries sustainable? The example of orange roughy (*Hoplostethis atlanticus*) in New Zealand. *Fish Res* 51:123–35.

Clark, S., and A. J. Edwards. 1994. The use of artificial reef structures to rehabilitate reef flats degraded by coral mining in the Maldives. *B Mar Sci* 55:726–46.

———. 1995. Coral transplantation as an aid to reef rehabilitation: Evaluation of a case study in the Maldive Islands. *Coral Reefs* 14:201–13.

Clarkson, R. W., and J. R. Wilson. 1995. Trout biomass and stream habitat relationships in the White Mountains area, east-central Arizona. *T Am Fish Soc* 124:599–612.

Clay, C. H. 1995. *Design of fishways and other fish facilities*, 2nd ed. Boca Raton, FL: Lewis.

Clearwater, S. J., A. M. Farag, and J. S. Meyer. 2002. Bioavailability and toxicity of dietborne copper and zinc to fish. *Comp Biochem Phys C* 132:269–313.

Clemmons, J. R., and R. Buchholz, eds. 1997. *Behavioral approaches to conservation in the wild.* Cambridge, UK: Cambridge Univ. Press.

Closs, G. P., and P. S. Lake. 1996. Drought, differential mortality and the coexistence of a native and an introduced fish species in a southeast Australian intermittent stream. *Environ Biol Fish* 47:17–26.

Cocheret de la Moriniere, E. B., J. A. Pollux, I. Nagelkerken, et al. 2002. Post-settlement life cycle migration patterns and habitat preference of coral reef fish that use seagrass and mangrove habitats as nurseries. *Estuar Coast Shelf S* 55:309–21.

Cohen, A. 1997. Sturgeon poaching and black market caviar: A case study. *Environ Biol Fish* 48:423–26.

Colborn, T., D. Dumanoski, and J. P. Myers. 1996. *Our stolen future.* New York: Dutton.

Colclough, S. R., G. Gray, A. Bark, et al. 2002. Fish and fisheries of the tidal Thames: Management of the modern resource, research aims and future pressures. *J Fish Biol* 61(suppl. A): 64–73.

Coleman, F. C., W. F. Figueira, J. S. Ueland, et al. 2004. The impact of United States recreational fisheries on marine fish populations. *Science* 305:1958–60.

Coleman, F. C., C. C. Koenig, and L. A. Collins. 1996. Reproductive styles of shallow-water groupers (Pisces: Serranidae) in the eastern Gulf of Mexico and the consequences of fishing spawning aggregations. *Environ Biol Fish* 47:129–41.

Collares-Pereira, M. J., M. M. Coelho, and I. G. Cowx, eds. 2002. *Conservation of freshwater fishes: Options for the future.* Oxford: Fishing News Books.

Collares-Pereira, M. J., and I. G. Cowx. 2004. The role of catchment scale environmental management in freshwater fish conservation. *Fisheries management and Ecology* 11:303–312.

Collares-Pereira, M. J., I. G. Cowx, J. A. Rodrigues, et al. 2002. A conservation strategy for *Anaecypris hispanica*: A picture of LIFE for a highly endangered Iberian fish. In Collares-Pereira, Coelho, and Cowx 2002, 186–97.

Collier, M. P., R. H. Webb, and E. D. Andrews. 1997. Experimental flooding in Grand Canyon. *Sci Am* (January): 82–89.

Colligan, M., P. Nickerson, and D. Kimball. 1999. Proposed rule on Atlantic salmon. *Federal Register* 64(221): 62627–41, November 17. http://wais.access.gpo.gov.

Conabio (Comision nacional para el conocimiento y uso de la biodiversidad). 2001 (updated 2004). Norma Oficial Mexicana, Nom-059-SEMARNAT-2001 lista 185 especies de peces. www.conabio.gob.mx/conocimiento/ise/fichas/doctos/peces.html.

Concepcion, G. B., and S. G. Nelson. 1999. Effects of a dam and reservoir on the distributions and densities of macrofauna in tropical streams of Guam (Mariana islands). *J Freshwater Ecol* 14:447–54.

Conover, D. O. 1998. Local adaptation in marine fishes: Evidence and implications for stock enhancement. *B Mar Sci* 62:305–11.

———. 2000. Darwinian fishery science. *Mar Ecol—Prog Ser* 208:303–7.

Conover, D. O., and S. B. Munch. 2002. Sustaining fisheries yields over evolutionary time scales. *Science* 297:94–96.

Contreras, S. 1969. Perspectivas de la ictiofauna en las zonas aridas del norte de Mexico. *Mem. Primer Simposio Internacional de Aumento de Producción de Alimentos en Zonas Aridas.* ICASALS, Texas Tech. Publications 3:293–304.

Contreras-Balderas, S., P. Almada-Villela, M. D. Lozano-Vilano, et al. 2002. Freshwater fish at risk or extinct in Mexico: A checklist and review. *Rev Fish Biol Fisher* 12:241–51.

Contreras-Balderas, S., and M. A. Escalante. 1984. Distribution and known impacts of exotic fishes in Mexico. In *Distribution, biology, and management of exotic fishes*, ed. W. R. Courtenay Jr. and J. R. Stauffer Jr., 102–30. Baltimore: Johns Hopkins Univ. Press.

Contreras-Balderas, S., and M. D. Lozano-Vilano. 1994. Water, endangered fishes, and development perspectives in arid lands of Mexico. *Conserv Biol* 8:379–87.

Contreras-MacBeath, T., H. M. Mojica, and R. C. Wilson. 1998. Negative impact on the aquatic ecosystems of the state of Morelos, Mexico, from introduced aquarium and other commercial fish. *Aquar Sci Conserv* 2:67–78.

Converse, Y. K., C. P. Hawkins, and R. A. Valdez. 1998. Habitat relationships of subadult humpback chub in the Colorado River through Grand Canyon: Spatial variability and implications of flow regulation. *Regul River* 14:267–84.

Cooke, S. J., and I. G. Cowx. 2004. The role of recreational fishing in global fish crises. *BioScience* 54:857–59.

Cooke, S. J., and D. P. Philipp. 2004. Behavior and mortality of caught-and-released bonefish (*Albula* spp.) in Bahamian waters with implications for a sustainable recreational fishery. *Biol Conserv* 118:599–607.

Cooper, E. L., ed. 1987. *Carp in North America*. Bethesda, MD: American Fisheries Society.

Copper, P. 1994. Ancient reef ecosystem expansion and collapse. *Coral Reefs* 16:S101–S113.

Cordell, J. 1991. Negotiating sea rights. *Cult Survival Q* 15(2): 5–10.

Corkeron, P. J. 2004. Whale watching, iconography, and marine conservation. *Conserv Biol* 18:847–49.

Corporación Nacional Forestal. 1993. *Libro Rojo de los Vertebrados Terrestres de Chile. Especies Chilenas según Categoría de Conservación (Vertebrados Terrestres)*. www.ssvsa.cl/especies.htm.

Correll, D. L. 2000. The current status of our knowledge of riparian buffer water quality functions. In *Riparian ecology and management in multi-land use watersheds*, ed. P. J. Wigington and R. L. Beschta, 5–10. Middleburg, VA: American Water Resources Association.

Courtenay, W. R. Jr. 1993. Biological pollution through fish introductions. In *Biological pollution: The control and impact of invasive exotic species*, ed. B. N. McKnight, 35–61. Indianapolis: Indiana Univ., Purdue Univ., Indiana Academy of Science.

———. 1997. Nonindigenous fishes. In *Strangers in paradise: Impact and management of nonindigenous species in Florida*, ed. D. Simberloff et al., 109–22. Washington, DC: Island Press.

Courtenay, W. R. Jr., D. A. Hensley, J. N. Taylor, et al. 1984. Distribution of exotic fishes in the continental United States. In *Distribution, biology, and management of exotic fishes*, ed. W. R. Courtenay Jr. and J. R. Stauffer Jr., 41–77. Baltimore: Johns Hopkins Univ. Press.

Courtenay, W. R. Jr., and P. B. Moyle. 1996. Biodiversity, fishes, and the introduction paradigm. In *Biodiversity in managed landscapes: Theory and practice*, ed. R. C. Szaro and D. W. Johnston, 239–52. Oxford: Oxford Univ. Press.

Courtenay, W. R. Jr., H. F. Sahlman, W. W. Miley II, et al. 1974. Exotic fishes in fresh and brackish waters of Florida. *Biol Conserv* 6:292–302.

Courtenay, W. R. Jr., and J. R. Stauffer. 1990. The introduced fish problem and the aquarium fish industry. *J World Aquacult Soc* 21:145–59.

Courtenay, W. R. Jr., and J. N. Taylor. 1984. The exotic ichthyofauna of the contiguous United States with preliminary observations on intranational transplants. In *Symposium on stock enhancement in the management of freshwater fish*, vol. 2, 466–87. Rome: FAO. Cited in Ludwig and Leitch 1996.

Courtenay, W. R. Jr., and J. D. Williams. 1992. Dispersal of exotic species from aquaculture sources, with emphasis on freshwater fishes. In *Dispersal of living organisms into aquatic ecosystems*, ed. A. Rosenfield and R. Mann, 49–81. College Park: Univ. Maryland Sea Grant Program.

Coutant, C. C. 1985. Striped bass, temperature, and dissolved oxygen: A speculative hypothesis for environmental risk. *T Am Fish Soc* 114:31–61.

———. 1990. Temperature-oxygen habitat for freshwater and coastal striped bass in a changing climate. *T Am Fish Soc* 119:240–53.

Coutant, C. C., and R. R. Whitney. 2000. Fish behavior in relation to passage through hydropower turbines: A review. *T Am Fish Soc* 129: 351–80.

———. 2006. Hydroelectric systems and migration. In R. N. Williams 2006, 249–324.

Cowen, R. K. 2002. Larval dispersal and retention and consequences for population connectivity. In Sale 2002a, 149–70.

Cowx, I. 1997. Introduction of fish species into European fresh waters: Economic successes or ecological disasters. *B Fr Peche Piscic* 344–45:57–77 (in French, translation provided by I. Cowx).

Cowx, I. G. 2002a. *Interactions between fish and birds: Implications for management*. Oxford: Blackwell Science.

———. 2002b. Analysis of threats to freshwater fish conservation: Past and present challenges. In Collares-Pereira, Coelho, and Cowx 2002, 201–20. Oxford: Fishing News Books.

Cowx, I. G., and R. L. Welcomme. 1998. *Rehabilitation of rivers for fish*. Oxford: Fishing News Books.

Cox, J. A., T. G. Quinn, and H. H. Boyeter Jr. 1997. Management by Florida's Game and Fresh Water Fish Commission. In *Strangers in paradise: Impact and management of nonindigenous species in Florida*, ed. D. Simberloff et al., 297–316. Washington, DC: Island Press.

Crabtree, R. E., C. W. Harnden, D. Snodgrass, et al. 1996. Age, growth and mortality of bonefish, *Albula vulpes*, from the waters of the Florida Keys. *Fish Bull U.S.* 94: 442–51.

Crawford, R. J. M. 2001. African penguins as predators and prey: Coping (or not) with change. *S Afr J Marine Sci* 23:435–47.

Crawford, R. J. M., L. G. Underhill, C. M. Raubenheimer, et al. 1992. Top predators in the Benguela ecosystem: Implications of their trophic position. *S Afr J Marine Sci* 12:675–87.

Crisman, T. L., L. J. Chapman, C. A. Chapman, et al., eds. 2003. *Conservation, ecology, and management of African fresh waters*. Gainesville: Univ. Press of Florida.

Crivelli, A. J. 1995. Are fish introductions a threat to endemic freshwater fishes in the northern Mediterranean region? *Biol Conserv* 72:311–19.

Crivelli, A., G. Pizat, P. Berrebi, et al. 2000. Conservation biology applied to fish: The example of a project for rehabilitating the marble trout (*Salmo marmoratus*) in Slovenia. *Cybium* 24:211–30.

Cronin, J., and R. F. Kennedy Jr. 1999. *The Riverkeepers: Two activists fight to reclaim our environment as a basic human right*. New York: Simon & Schuster.

Crook, D. A., and A. C. Sanger. 1998a. Threatened fishes of the world: *Galaxias fontanus* Fulton, 1978 (Galaxiidae). *Environ Biol Fish* 53:32.

———. 1998b. Threatened fishes of the world: *Galaxias johnstoni* Scott, 1936 (Galaxiidae). *Environ Biol Fish* 53:154.

Cross, F. B., and R. E. Moss. 1987. Historic changes in fish communities and aquatic habitats in plains streams of Kansas. In *Community and evolutionary ecology of North American stream fishes*, ed. W. J. Matthews and D. C. Heins, 155–65. Norman: Univ. Oklahoma Press.

Crowder, L. B. 1984. Character displacement and habitat shift in a native cisco in southeastern Lake Michigan: Evidence for competition? *Copeia* 1984:878–83.

Crowder, L. B., and E. A. Norse. 2005. The greatest threat: Fisheries. In Norse and Crowder 2005, 183–84.

Crowl, T. A., C. R. Townsend, and A. R. McIntosh. 1992. The impact of introduced brown and rainbow trout on native fish: The case of Australasia. *Rev Fish Biol Fisher* 2:217–41.

Crozier, W. W. 2000. Escaped farmed salmon, *Salmo salar* L., in the Glenarm River, Northern Ireland: Genetic status of the wild population 7 years on. *Fisheries Manage Ecol* 7:437–46.

CSIRO. 2000. *Orange roughy*. Hobart, Tasmania: CSIRO Marine Research. www.marine.csiro.au/leafletsfolder/pdfsheets/38oroughy.pdf.

Cullen, A. P., and C. A. Monteith-McMaster. 1993. Damage to the rainbow trout (*Oncorhynchus mykiss*) lens following an acute dose of UV-B. *Curr Eye Res* 12:97–106.

Daily, G. C., ed. 1997. *Nature's services: Societal dependence on natural ecosystems*. Washington, DC: Island Press.

Dalbey, S. R., T. E. McMahon, and W. Fredenberg. 1996. Effects of electrofishing pulse shape and electrofishing-induced spinal injury on long-term growth and survival of wild rainbow trout. *N Am J Fish Manage* 16:560–69.

Daley, P. 2002. Outbreak of ciguatera poisoning in Calgary, Alberta. Abstract, presentations, Univ. Calgary Faculty of Medicine. www.calgaryhealthregion.ca/postgradmededuc.

Dallmeyer, D. 2005. Toward a sea ethic. In Norse and Crowder 2005, 410–21.

Dalzell, P. 1996. Catch rates, selectivity and yields of reef fishing. In Polunin and Roberts 1996, 161–92.

Damkaer, D. M., and D. B. Dey. 1989. Evidence of fluoride effects on salmon passage at John Day Dam, Columbia River, 1982–1986. *N Am J Fish Manage* 9:154–62.

Daniels, R. A. 2001. Untested assumptions: The role of canals in the dispersal of sea lamprey, alewife, and other fishes in the eastern United States. *Environ Biol Fish* 60:309–29.

Darwin, C. 1871. *Journal of researches into the natural history and geology of the countries visited during the voyage of H. M. S. Beagle round the world, under the command of Capt. Fitz Roy, R.N.* New York: D. Appleton.

Davenport, K. E. 1996. Characteristics of the current international trade in ornamental fish, with special reference to the European Union. *Rev Sci Tech OIE* 15:435–43.

Daves, N. K., and M. F. Nammack. 1998. U.S. and international mechanisms for protecting and managing shark resources. *Fish Res* 39:223–28.

David, B. 2002. Threatened fishes of the world: *Galaxias argenteus* (Gmelin, 1789) (Galaxiidae). *Environ Biol Fish* 63:416.

Davin, W. T. Jr., C. C. Kohler, and D. R. Tindall. 1988. Ciguatera toxins adversely affect piscivorous fishes. *T Am Fish Soc* 117:374–84.

Davis, M. W., and B. L. Olla. 2001. Stress and delayed mortality induced in Pacific halibut by exposure to hooking, net towing, elevated seawater temperature and air: Implications for management of bycatch. *N Am J Fish Manage* 21:725–32.

Dawes, J. 1999. International experience in ornamental marine species management. Part 1: Perspectives. *Ornamental Fish Int J* 26:1–3.

Dayton, P. K., S. F. Thrush, M. T. Agardy, et al. 1995. Environmental effects of marine fishing. *Aquat Conserv* 5:205–32.

Deacon, J. E. 1988. The endangered woundfin and water management in the Virgin River, Utah, Arizona, Nevada. *Fisheries* 13(1): 18–29.

Deacon, J. E., G. Kobetich, J. D. Williams, et al. 1979. Fishes of North America endangered, threatened, or of special concern: 1979. *Fisheries* 4(2): 329–44.

DEFRA (Department of Environment, Food and Rural Affairs). 1997. Water Resources Act 1991 Section 83. www.defra.gov.uk/environment/water/quality/fwfish/98f-noti.pdf.

———. 2003. Frequently asked questions (FAQs) on water framework directive. www.defra.gov.uk/environment/water/wfd/faq.htm#10.

———. 2005. *Salmon and freshwater fisheries review*. www.defra.gov.uk/fish/freshwater/pdf/sffrev.pdf). 17 Jan 2005.

de la Fuente, J., I. Guillen, R. Martinez, et al. 1999. Growth regulation and enhancement in tilapia: Basic research findings and their applications. *Genet Anal—Biomol E* 15:85–90.

de Leeuw, A. D. 1996. Contemplating the interests of fish: The angler's challenge. *Environ Ethics* 18:373–90.

Demarais, B. D., T. E. Dowling, and W. L. Minckley. 1993. Post-perturbation genetic changes in populations of endangered Virgin River chubs. *Conserv Biol* 7:334–41.

DeMartini, E. E. 1993. Modelling the potential of fishery reserves for managing Pacific coral reef fishes. *Fish Bull U.S.* 91:414–27.

———. 1998. How might recruitment research on coral-reef fishes help manage tropical reef fisheries? *Aust J Ecol* 23:305–10.

DeMartini, E. E., and D. A. Roberts. 1990. Effects of giant kelp (*Macrocystis*) on the density and abundance of fishes in a cobble-bottom kelp forest. *B Mar Sci* 46:287–300.

De Meulenaer, T., and C. Raymakers. 1996. *Sturgeon of the Caspian Sea and the international trade in caviar*. Species in Danger Report. Cambridge, UK: TRAFFIC International.

de Moor, I. J., and M. N. Bruton. 1988. Atlas of alien and translocated indigenous aquatic animals in southern Africa. *South African National Scientific Programmes Report* 144:1–310.

Denkhaus, R. 2002. Audubon shot birds, I keep fish: Exploring the naturalist-aquarist connection. *Am Currents* 28(2): 7–8.

Dennis, T. E., S. E. MacAvoy, M. B. Steg, et al. 1995. The association of water chemistry variables and fish condition in streams of Shenandoah National Park (USA). *Water Air Soil Poll* 85:365–70.

Derr, M. 1992. Raiders of the reef. *Audubon* 94:48–54.

Desbonnet, A., P. Pogue, V. Lee, et al. 1994. *Vegetated buffers in the coastal zone: A summary review and bibliography*. Providence: Coastal Resources Center, Univ. Rhode Island.

Detenbeck, N. E., P. W. DeVore, G. J. Niemi, et al. 1992. Recovery of temperate-stream fish communities from disturbance: A review of case studies and synthesis of theory. *Environ Manage* 16:33–53.

Devick, W. S., J. M. Fitzsimons, and R. T. Nishimoto. 1995. Threatened fishes of the world: *Lentipes concolor* Gill, 1860 (Gobiidae). *Environ Biol Fish* 44:325–26.

Devine, J. A., C. E. Adams, and P. S. Maitland. 2000. Changes in reproductive strategy in the ruffe during a period of establishment in a new habitat. *J Fish Biol* 56:1488–96.

DeVivo, J. C. 1996. Fish assemblages as indicators of water quality within the Apalachicola-Chattahoochee-Flint (ACF) river basin. Master's thesis, Univ. Georgia.

Devlin, R. H., J. I. Johnsson, D. E. Smailus, et al. 1999. Increased ability to compete for food by growth hormone-transgenic coho salmon *Oncorhynchus kisutch* (Walbuam). *Aquac Res* 30:479–82.

Devlin, R. H., and Y. Nagahama. 2002. Sex determination and sex differentiation in fish: An overview of genetic, physiological, and environmental influences. *Aquaculture* 208:191–364.

Devlin, R. H., T. Y. Yesaki, E. M. Donaldson, et al. 1995. Transmission and phenotypic effects of an antifreeze GH gene construct in coho salmon, *Oncorhynchus kisutch*. *Aquaculture* 137:161–69.

De Vos, L., and D. Oyugi. 2002. First capture of a coelacanth, *Latimeria chalumnae* Smith, 1939 (Pisces: Latimeriidae), off Kenya. *S Afr J Sci* 98:345–47.

DeWeerdt, S. 2002. What really is an evolutionarily significant unit? *Conserv Biol Pract* 3(1): 10–17.

DeWoody, J. A., and J. C. Avise. 2000. Microsatellite variation in marine, freshwater and anadromous fishes compared with other animals. *J Fish Biol* 56:461–73.

Diamant, A., A. Banet, M. Ucko, et al. 2000. *Mycobacteriosis* in wild rabbitfish *Siganus rivulatus* associated with cage farming in the Gulf of Eilat, Red Sea. *Dis Aquat Organ* 39:211–19.

Diamond, J. 1999. *Guns, germs, and steel: The fates of human societies.* New York: Norton.

———. 2004. *Collapse: How societies choose to fail or succeed.* New York: Viking.

Diamond, J. M. 1987. Extant unless proven extinct? Or, extinct unless proven extant? *Conserv Biol* 1:77–79.

Dickerson, H., and T. Clark. 1998. *Ichthyophthirius multifiliis*: A model of cutaneous infection and immunity in fishes. *Immunological Review* 166:377–84.

Dierking, L. D., K. Burtnyk, K. S. Buchner, et al. 2001. *Visitor learning in zoos and aquariums.* American Aquarium and Zoo Association. www.aza.org/ConEd/VisitorLearning.

Dill, W. A., and A. J. Cordone. 1997. History and status of introduced fishes in California, 1871–1996: Conclusions. *Fisheries* 22(10): 15–18, 35.

Dixon, J. A., L. F. Scura, and T. van't Hof. 1993. Meeting ecological and economic goals: Marine parks in the Caribbean. *Ambio* 22:117–25.

Doadrio, I., and A. Perdices. 1998. Threatened fishes of the world: *Barbus comiza* Steindachner, 1865 (Cyprinidae). *Environ Biol Fish* 51:52.

Dobbins, D. A., R. L. Cailteux, and J. J. Nordhaus. 1999. Flathead catfish abundance and movement in Apalachicola River, Florida. *Am Fish Soc Sym* 24:199–201.

DOC (New Zealand Department of Conservation). 2002a. *Freshwater fish.* www.doc.govt.nz/Conservation/001~Plants-and-Animals/006~Threatened-species/Freshwater-fish.asp.

———. 2002b. *Marine fish.* www.doc.govt.nz/Conservation/001~Plants-and-Animals/006~Threatened-species/Marine-fish.asp.

———. 2004. New Zealand non-migratory galaxiid fishes recovery plan 2003–13. Threatened Species Recovery Plan 53. Wellington: Biodiversity Recovery Unit, Dept. of Conservation.

Doeringer, P. B., P. I. Moss, and D. G. Terkla. 1986. *The New England fishing economy: Jobs, income, and kinship.* Amherst: Univ. Massachusetts Press.

Doherty, P. J. 2002. Variable replenishment and the dynamics of reef fish populations. In Sale 2002a, 327–58.

Domeier, M. L., and P. L. Colin. 1997. Tropical reef fish spawning aggregations: Defined and reviewed. *B Mar Sci* 60:698–726.

Donaldson, T. J., and Y. Sadovy. 2001. Threatened fishes of the world: *Cheilinus undulatus* Rüppell, 1835 (Labridae). *Environ Biol Fish* 62:428.

Done, T. J. 1992. Phase shifts in coral reef communities and their ecological significance. *Hydrobiologia* 247:121–32.

Donnelly, T. W. 1993. Impoundment of rivers: Sediment regime and its effects on benthos. *Aquat Conserv* 3:331–42.

Doppelt, B., M. Scurlock, C. Frissell, et al. 1993. *Entering the watershed: A new approach to save America's river ecosystems.* Washington, DC: Island Press.

Dorea, J. G. 2003. Fish are central in the diet of Amazonian riparians: Should we worry about their mercury concentrations? *Environ Res* 92:232–44.

Douglas, M. E., P. C. Marsh, and W. L. Minckley. 1994. Indigenous fishes of western North America and the hypothesis of competitive displacement: *Meda fulgida* (Cyprinidae) as a case study. *Copeia* 1994:9–19.

Dowd, S. 2001. An aquarist's experience of Amazon River ecotour-expeditions and hobbyist involvement in conservation. In Chao, Petry, et al., 2001, 37–42.

Dowling, T. E, and M. R. Childs. 1992. Impact of hybridization on a threatened trout of the southwestern United States. *Conserv Biol* 6:355–64.

Downs, P. W., and G. M. Kondolf. 2002. Post-project appraisals in adaptive management of river channel restoration. *Environ Manage* 29:477–96.

Drake, M. T., J. E. Claussen, D. P. Phillip, et al. 1997. A comparison of bluegill reproductive strategies and growth among lakes with different fishing intensities. *N Am J Fish Manage* 17:496–507.

Drevnick, P. E., and M. B. Sandheinrich. 2003. Effects of dietary methylmercury on reproductive endocrinology of fathead minnows. *Environ Sci Technol* 37:4390–96.

Driscoll, C. T., G. B. Lawrence, A. T. Bulger, et al. 2001. Acid deposition in the northeastern U.S.: Sources and inputs, ecosystem effects, and management strategies. *BioScience* 51:180–98.

Driscoll, C. T., R. M. Newton, C. P. Gubala, et al. 1991. Adirondack Mountains. In *Acidic deposition and aquatic ecosystems: Regional case studies*, ed. D. F. Charles, 132–202. New York: Springer-Verlag.

Druffel, E. R. M., and B. H. Robison. 1999. Oceanography: Is the deep sea on a diet? *Science* 284:1139–40.

Dubs, D. O., and L. D. Corkum. 1996. Behavioral interactions between round gobies (*Neogobius melanostomus*) and mottled sculpins (*Cottus bairdi*). *Great Lakes Res* 22:838–44.

Ducroty, J. P. 1999. Protection, conservation and biological diversity in the north east Atlantic. *Aquat Conserv* 9:313–25.

Dudgeon, D. 2000a. The ecology of tropical Asian rivers and streams in relation to biodiversity conservation. *Annu Rev Ecol Syst* 31:239–63.

———. 2000b. Large-scale hydrological changes in tropical Asia: Prospects for riverine biodiversity. *BioScience* 50:793–806.

Dudley, S. F. J., and N. A. Gribble. 1999. Management of shark control programmes. In *Case studies of the management of elasmobranch fisheries*, ed. R. Shotton, 819–59. FAO Fisheries Tech. Paper 378, vol. 2. Rome: FAO.

Duff, D. A., L. H. Wullstein, M. Nackowski, et al. 1995. *Indexed bibliography on stream habitat improvement.* Salt Lake City, UT: USDA Forest Service.

Dufour, V. 1997. Pacific island countries and the aquarium fish market. *SPC Live Reef Fish Inform Bull* 2:6–11.

Dugan, J. E., and G. E. Davis. 1993. Applications of marine refugia to coastal fisheries management. *Canadian Journal of Fisheries and Aquat Sci* 50:2029–42.

Dulvy, N. K., S. J. Jennings, N. B. Goodwin, et al. 2005. Comparison of threat and exploitation status in northeast Atlantic marine populations. *J Appl Ecol* 42:883–91.

Dulvy, N. K., J. D. Metcalfe, J. Glanville, et al. 2000. Fishery stability, local extinctions, and shifts in community structure in skates. *Conserv Biol* 14:283–93.

Dulvy, N. K., and J. D. Reynolds. 2002. Predicting extinction vulnerability in skates. *Conserv Biol* 16:440–50.

Dulvy, N. K., D. Stanwell-Smith, W. R. T. Darwall, et al. 1995. Coral mining at Mafia Island, Tanzania: A management dilemma. *Ambio* 24:358–65.

Duncan, J. R., and M. L. McKinney. 2001. Spatial homogenization of the aquatic fauna of Tennessee: Extinction and invasion following land use change and habitat alterations. In Lockwood and McKinney 2001, 245–57.

Dunham, J. B., S. B. Adams, R. E. Schroeter, et al. 2002. Alien invasions in aquatic ecosystems: Toward an understanding of brook trout invasions and potential impacts on inland cutthroat trout in western North America. *Rev Fish Biol Fisher* 12:373–91.

Dunham, R. A. 1996. Fifth international conference for productivity enhancement of the coastal waters. FOID's '96, Pusan, Korea: 5–6. Cited in Hew and Fletcher 1997.

———. 1999. Utilization of transgenic fish in developing countries: Potential benefits and risks. *J World Aquacul Soc* 30:1–11.

Dunham, R. A., C. Chitminat, A. Nichols, et al. 1999. Predator avoidance of transgenic channel catfish containing salmonid growth hormone genes. *Mar Biotechnol* 1:545–51.

Durborow, R. M. 1999. Health and safety concerns in fisheries and aquaculture. *Occup Med* 14:373–406.

Dutta, B. 2001. *Whale sharks: The gentle giants of the sea.* TRAFFIC Dispatches 17. www.TRAFFIC.org/dispatches/wshark.html.

Dybas, C. L. 2002. In Hawaii, the age of aquariums raises concern: Collectors clash with islanders over tropical fish. *Washington Post*, June 3, A07.

Dynesius, M., and C. Nilsson. 1994. Fragmentation and flow regulation of river systems in the northern third of the world. *Science* 266:753–62.

Easton, M. D. L., D. Luszniak, and E. Von der Geest. 2002. Preliminary examination of contaminant loadings in farmed salmon, wild salmon and commercial salmon feed. *Chemosphere* 46:1053–74.

Echelle, A. A. 1991. Conservation genetics and genic diversity in freshwater fishes of western North America. In Minckley and Deacon 1991, 141–53, 477.

Echelle, A. A., and P. J. Connor. 1989. Rapid geographically extensive genetic introgression after secondary contact between two pupfish species (*Cyprinodon*, Cyprinodontidae). *Evolution* 43:717–27.

Echelle, A. A., and A. F. Echelle. 1997. Genetic introgression of endemic taxa by non-natives: A case study with Leon Springs pupfish and sheepshead minnow. *Conserv Biol* 11:153–61.

Echelle, A. A., G. R. Luttrell, R. D. Larson, et al. 1995. Decline of native prairie fishes. In *Our living resources: A report to the nation on the distribution, abundance, and health of U.S. plants, animals, and ecosystems*, ed. E. T. LaRoe et al., 303–5. Washington, DC: National Biological Service.

Ecological Applications. 2003. The science of marine reserves. *Ecol Appl* 13(1) Supplement.

Economidis, P. S., E. Dimitriou, R. Pagoni, et al. 2000. Introduced and translocated fish species in the inland waters of Greece. *Fisheries Manage Ecol* 7:239–50.

Edgar, G. J., P. A. Marshall, P. Mooney. 2003. The effect of the *Jessica* grounding on Galapagos reef fish communities adjacent to the wreck site. *Mar Pollut Bull* 47:296–302.

Edmunds, J. S. G., R. A. McCarthy, and J. S. Ramsdell. 1999. Ciguatoxin reduces larval survivability in finfish. *Toxicon* 37:1827–32.

Edwards, A. J., and A. D. Shepherd. 1992. Environmental implications of aquarium-fish collection in the Maldives, with proposals for regulation. *Environ Conserv* 19:61–72.

Ehrenfeld, D. W. 1970. *Biological Conservervation*. New York: Holt, Rinehart & Winston.

Eleventh Circuit Court. 2007. Alabama Tombigbee Rivers Coalition v. Kempthorne, 477 F. 3d 1250. (11th Cir. Feb. 8, 2007).

Elliott, J. M. 1989. The critical-period concept for juvenile survival and its relevance for population regulation in young sea trout, *Salmo trutta*. *J Fish Biol* 35:91–98.

———. 1994. *Quantitative ecology and the brown trout*. Oxford: Oxford Univ. Press.

Ellis, A. E. 1989. The immunology of teleosts. In *Fish pathology*, 2nd ed., ed. R. J. Roberts, 135–52. Philadelphia: Bailliere Tindall.

Ellis, D., and Associates. 1996. *Net loss: The salmon netcage industry in British Columbia*. Vancouver, BC: David Suzuki Foundation.

Ellis, S. 1999. Aquafarmer information sheet: Farming soft corals for the marine aquarium trade. Center for Tropical and Subtropical Aquaculture pub. 140. Waimanalo, HI. http://library.kcc.hawaii.edu.

Elvira, B. 1995. Native and exotic freshwater fishes in Spanish river basins. *Freshwater Biol* 33:103–8.

———. 1996. Endangered freshwater fish of Spain. In Kirchhofer and Hefti 1996a, 55–61.

Elwood, J. W., and T. F. Waters. 1969. Effects of floods on food consumption and production rates of a stream brook trout population. *T Am Fish Soc* 98:253–62.

Emerson, P., and F. Cox. 2000. Imperfect laws add to danger of perfect storms. *Dallas Morning News*, July 25.

Engas, A., T. Jorgensen, and C. W. West. 1998. A species-selective trawl for demersal gadoid fisheries. *ICES J Mar Sci* 55:835–45.

English, D. B. K., W. Kriesel, V. Leeworthy, et al. 1996. Economic contribution of recreating visitors to the Florida Keys/Key West. In *Linking the economy and environment of Florida Keys/Florida Bay: Visitor nonmarket economic user values*. http://marineeconomics.noaa.gov.

Eno, N. C., D. S. MacDonald, J. A. M. Kinnear, et al. 2001. Effects of crustacean traps on benthic fauna. *ICES J Mar Sci* 58:11–20.

Ensign, W. E., K. N. Leftwich, P. L. Angermeier, et al. 1997. Factors influencing stream fish recovery following a large-scale disturbance. *T Am Fish Soc* 126:895–907.

Epifanio, J., and J. Nielsen. 2000. The role of hybridization in the distribution, conservation and management of aquatic species. *Rev Fish Biol Fisher* 10:245–51.

Epple, R. 2000. *Dam decommissioning: French pilot experiences and the European context*. www.rivernet.org/decom3_e.htm#4.

Erdmann, M.V. 1999. Clove oil: An "eco-friendly" alternative to cyanide use in the live reef fish industry? *SPC Live Reef Fish Inform Bull* 5:4–7.

Erdmann, M. V., and A. Vagelli. 2001. Banggai cardinalfish invade Lembeh Straight. *Coral Reefs* 20:252–53.

Ernst, W., P. Jackman, K. Doe, et al. 2001. Dispersion and toxicity to non-target aquatic organisms of pesticides used to treat sea lice on salmon in net pen enclosures. *Mar Pollut Bull* 42:433–44.

Ervik, A., B. Thorsen, V. Eriksen, et al. 1994. Impact of administering antibacterial agents on wild fish and blue mussels *Mytilus edulis* in the vicinity of fish farms. *Dis Aquat Organ* 18:45–51.

ESA (Ecological Society of America). 1999. *Acid deposition: The ecological response*. A Workshop Report. Washington, DC: Author. www.esa.org.

———. 2000. *Ecosystem services: A primer*. www.actionbioscience.org/environment/esa.html.

Estes, J. A., M. T. Tinker, T. M. Williams, et al. 1998. Killer whale predation on sea otters linking oceanic and nearshore ecosystems. *Science* 282:473–76.

Etnier, D. A. 1972. The effect of annual rechanneling on a stream fish population. *T Am Fish Soc* 101:372–75.

———. 1976. *Percina (Imostoma) tanasi*, a new percid fish from the Little Tennessee River, Tennessee. *P Biol Soc Wash* 88:469–88.

———. 1997. Jeopardized southeastern freshwater fishes: A search for causes. In *Aquatic fauna in peril: The southeastern perspective*, ed. G. W. Benz and D. E. Collins, 87–104. Southeast Aquatic Research Institute Special Publication 1. Decatur, GA: Lenz Design and Communications.

Etnier, D. A., and W. C. Starnes. 1993. *The fishes of Tennessee*. Knoxville: Univ. Tennessee Press.

EU (European Union). 2003. *Health and environment: Use of nonylphenol, nonylphenol ethoxylate and cement*, 26th amend. direct. 76/769/EEC. wwwdb.europarl.eu.int/oeil/oeil_ViewDNL.ProcedureView ?lang=2&procid=6385.

Evans, B. B., and R. J. G. Lester. 2001. Parasites of ornamental fish imported into Australia. *B Eur Assoc Fish Path* 21:51–55.

Ewald, G., P. Larsson, H. Linge , et al. 1998. Biotransport of organic pollutants to an inland Alaska lake by migrating sockeye salmon (*Oncorhynchus nerka*). *Arctic* 51:40–47.

Fahrenthold, D. A. 2004. Abnormal fish found closer to Washington; waste suspected in egg-bearing males. *Washington Post*, Dec. 19, C01.

Fairey, B., K. Taberski, S. Lamerdin, et al. 1997. Organochlorines and other environmental contaminants in muscle tissues of sportfish collected from San Franicsco Bay. *Mar Pollut Bull* 34:1058–71.

Falconer, D. S., and T. F. C. Mackay. 1996. *Introduction to quantitative genetics*, 4th ed. Harlow, UK: Longmans.

Falk, J. H., and L. M. Adelman. 2003. Investigating the impact of prior knowledge and interest on aquarium visitor learning. *J Res Sci Teach* 40:163–76.

Falls, W. W., P. Stinnette, and J. N. Ehringer. 2000. *HCC interdisciplinary live rock project*. Year II Annual Report, Hillsborough Community College. www.nbizz.com/tritonmarine/upload/falls.pdf.

FAO (Food and Agriculture Organization of the United Nations). 1981. *Conservation of the genetic resources of fish: Problems and recommendations*. FAO Fisheries Tech. Reports Paper 217.

———. 1995a. Aquaculture development. Article 9, *FAO Code of Conduct for Responsible Fisheries*. Rome: FAO. www.fao.org/fi/agreem/codecond/ficonde.asp#9.

———. 1995b. *FAO yearbook. Fishery statistics: Catches and landings 1993*, vol. 76. Rome: FAO.

———. 1995c. *Precautionary approach to fisheries: Guidelines on the precautionary approach to capture fisheries and species introductions*. Tech. Paper 350, part 1. Rome: FAO Fisheries.

———. 1997a. *Aquaculture development*. FAO Tech. Guidelines for Responsible Fisheries 5. Rome: FAO.

———. 1997b. *Review of the state of world fishery resources: Marine fisheries*. 3. Western Central Atlantic. FAO Fisheries circular 920 FIRM/c920. Rome: FAO.

———. 1999a. *Indicators for sustainable development of marine capture fisheries*. FAO Tech. Guidelines for Responsible Fisheries 8. Rome: FAO.

———. 1999b. Ornamental aquatic life: What's FAO got to do with it? *FAO News & Highlights*. www.fao.org/news/1999/990901-e.htm.

———. 2000a. *Live ornamental fish*. www.fao.org.

———. 2000b. *The state of world fisheries and aquaculture*. Rome: FAO. www.fao.org/sof/sofia/index_en.htm

———. 2001. *Genetically modified organisms, consumers, food safety and the environment*. Rome: FAO. www.fao.org/DOCREP/003/X9602E/X9602E00.HTM.

———. 2002. *UN atlas of the oceans*. Rome: FAO. www.oceanatlas.org/html.

———. 2003. *FAO yearbook. Fishery statistics: Capture production 2003*, vol. 96/1. www.fao.org/fi/statist/fisoft/fishplus.asp.

———. 2004. *The state of world fisheries and aquaculture (SOFIA) 2004*. Rome: FAO.

Farrell, A. P., W. Bennett, and R. H. Devlin. 1997. Growth-enhanced transgenic salmon can be inferior swimmers. *Can J Zool* 75:335–37.

Farrell, A. P., P. E. Gallaugher, J. Fraser, et al. 2001. Successful recovery of the physiological status of coho salmon on board a commercial gillnet vessel by means of a newly designed revival box. *Can J Fish Aquat Sci* 58:1932–46.

Fausch, K. D. 1988. Tests of competition between native and introduced salmonids in streams: What have we learned? *Can J Fish Aquat Sci* 45:2238–46.

Fausch, K. D., J. Lyons, J. R. Karr, et al. 1990. Fish communities as indicators of environmental degradation. *Am Fish Soc Sym* 8:123–44.

Fausch, K. D., Y. Taniguchi, S. Nakano, et al. 2001. Flood disturbance regimes influence rainbow trout invasion success among five holarctic regions. *Ecol Appl* 11:1438–55.

Fausch, K. D., C. E. Torgersen, C. V. Baxter, et al. 2002. Landscapes to riverscapes: Bridging the gap between research and conservation of stream fishes. *BioScience* 52:483–98.

Fausch, K. D., and R. J. White. 1981. Competition between brook trout (*Salvelinus fontinalis*) and brown trout (*Salmo trutta*) for positions in a Michigan stream. *Can J Fish Aquat Sci* 38:1220–27.

———. 1986. Competition among juveniles of coho salmon, brook trout, and brown trout in a laboratory stream, and implications for Great Lakes tributaries. *T Am Fish Soc* 115:363–81.

FEAP (Federation of European Aquaculture Producers). 2000. *A code of conduct for European aquaculture.* www.feap.org/code.html.

FERC (Federal Energy Regulatory Commission). 1992. *Hydroelectric power resources of the United States: Developed and undeveloped.* Eleventh Publication on Hydropower Resources. Washington, DC: Author.

Ferguson, A., and F. M. Mason. 1981. Allozyme evidence for reproductively isolated sympatric populations of brown trout, *Salmo trutta* L. in Lough Melvin, Ireland. *J Fish Biol* 18:629–42.

Fernando, C. H., J. J. S. Gurgel, and N. A. G. Moyou. 1998. A global view of reservoir fisheries. *Int Rev Hydrobiol* 83 (special issue): 31–42.

Ferreri, C. P., J. R. Stauffer, and T. D. Stecko. 2004. Evaluating impacts of mountain top removal/valley fill coal mining on stream fish populations. In *American Society of Mining and Reclamation Proceedings*, ed. R. I. Barnhisel, 576–92. Lexington, KY: ASMR.

Fewings, D. G., and L. C. Squire. 1999. Notes on reproduction in the estuarine stonefish *Synanceia horrida. SPC Live Reef Fish Inform Bull* 5:31–33.

FGFC (Florida Game and Fresh Water Fish Commission). 1997–98. Aquaculture rules and regulations summary. *Florida Wildlife Code Title 39.* Tallahassee: Author.

Findlay, C. S., D. G. Bert, and L. G. Zheng. 2000. Effect of introduced piscivores on native minnow communities in Adirondack lakes. *Can J Fish Aquat Sci* 57:570–80.

Fine, J. C. 1990. Groupers in love: Spawning aggregations off Honduras. *Sea Frontiers* 36:42–45.

Fisher, J. A. D., and K. T. Frank. 2002. Changes in finfish community structure associated with an offshore fishery closed area on the Scotian Shelf. *Mar Ecol—Prog Ser* 240:249–65.

Fisher, W. L., A. F. Surmont, and C. D. Martin. 1998. Warmwater stream and river fisheries in the southeastern United States: Are we managing them in proportion to their values? *Fisheries* 23(12): 16–24.

Fitzpatrick, J. 1995. Technology and fisheries legislation. Unpublished paper presented at the International Technical Consultation on the Precautionary Approach to Capture Fisheries, Lysekel, Sweden, June 1995. Cited in NRC 1999b.

Fitzsimons, J. M., H. L. Schoenfuss, and T. C. Schoenfuss. 1997. Significance of unimpeded flows in limiting the transmission of parasites from exotics to Hawaiian stream fishes. *Micronesica* 30:117–25.

FKNMS (Florida Keys National Marine Sanctuary). 2000. *Rogue batfish placed on exhibit; New England Aquarium cares for alien species.* Aug. 15. www.fknms.nos.noaa.gov/news/press_release/batfish.html.

Flagg, T. A., F. W. Waknitz, D. J. Maynard, et al. 1995. The effect of hatcheries on native coho salmon populations in the lower Columbia River. *Am Fish Soc Sym* 15:366–75.

Flagg, T. A., B. A. Berejekian, J. E. Colt, et al. 2000. *Ecological and behavioral impacts of artificial production strategies on the abundance of wild salmon populations.* U.S. Dept. Commerce, NOAA Tech. Memo. NMFS-NWFSC-41.

Flecker, A. S., and C. R. Townsend. 1994. Community-wide consequences of trout introduction in New Zealand streams. *Ecol Appl* 4:798–807.

Fleischman, J. 1996. Counting darters. *Audubon* 98(4): 84–89.

———. 1997. Muddying the waters. *Audubon* 99(2): 64–69.

Fleischner, T. L. 1994. Ecological costs of livestock grazing in western North America. *Conserv Biol* 8:629–44.

Fleming, I. A. 1996. Reproductive strategies of Atlantic salmon: Ecology and evolution. *Rev Fish Biol Fisher* 6:379–416.

Fleming, I. A., K. Hindar, I. B. Mjolnerod, et al. 2000. Lifetime success and interactions of farm salmon invading a native population. *P Roy Soc Lond B Bio* 267:1517–23.

Fletcher, G. L., S. V. Goddard, and Y. Wu. 1999. Antifreeze proteins and their genes: From basic research to business opportunity. *Chemtech* 30(6): 17–28.

Flinkman, J., E. Aro, I. Vuorinen, and M. Viitasalo. 1998. Changes in northern Baltic zooplankton and herring nutrition from 1980s to 1990s: Top down and bottom up processes at work. *Mar Ecol—Prog Ser* 165:127–36.

Fobes, N. 1994. *Reaching home: Pacific salmon, Pacific people.* Anchorage: Alaska Northwest Books.

Fogarty, M. J., and S. A. Murawski. 1998. Large-scale disturbance and the structure of marine systems: Fishery impacts on Georges Bank. *Ecol Appl* 8:S6–S22.

Folke, C., N. Kautsky, H. Berg, et al. 1998. The ecological footprint concept for sustainable seafood production: A review. *Ecol Appl* 8:S63-S71.

Fong, Q. S. W., and J. L. Anderson. 2002. International shark fin markets and shark management: An integrated market preference-cohort analysis of the blacktip shark (*Carcharinus limbatus*). *Ecol Econ* 40:117–30.

Font, W. E. 1998. Parasites in paradise: Patterns of helminth distribution in Hawaiian stream fishes. *J Helminthol* 72:307–11.

Forshage, A., and N. E. Carter. 1973. Effects of gravel dredging on the Brazos River. *P Annu Conf SE Assoc Game Fish Comm* 27:695–709.

Foulkes, R. G., and A. C. Anderson. 1994. Impact of artificial fluoridation on salmon species in the northwest USA and British Columbia, Canada. *Fluoride* 27:220–26.

Fowler, C. W. 1999. Management of multi-species fisheries: From overfishing to sustainability. *ICES J Mar Sci* 56:927–32.

Fowler, S., and R. Cavanagh. 2001. CITES update. *Shark News* 13. Newsletter of the IUCN Shark Specialist Group.

Fox, H. E., and M. V. Erdmann. 2000. Fish yields from blast fishing in Indonesia. *Coral Reefs* 19:114.

Frank, K. T., and W. C. Leggett. 1994. Fisheries ecology in the context of ecological and evolutionary theory. *Annu Rev Ecol Syst* 25:401–22.

Franklin, H. B. 2001. The most important fish in the sea. *Discover* 22(9): 44–50.

Freeman, M. C., Z. H. Bowen, K. D. Bovee, et al. 2001. Flow and habitat effects on juvenile fish abundance in natural and altered flow regimes. *Ecol Appl* 11:179–90.

Freyhof, J. 2002. Freshwater fish diversity in Germany: Threats and species extinction. In Collares-Pereira, Coelho, and Cowx 2002, 3–22.

Fricke, H. 2001. Coelacanths: A human responsibility. *J Fish Biol* 59(suppl. A): 332–38.

Fricke, H., K. Hissmann, J. Schauer, et al. 1991. Habitat and population size of the coelacanth *Latimeria chalumnae* at Grand Comoro. *Environ Biol Fish* 32:287–300.

Friedl, G., and A. Wuest. 2002. Disrupting biogeochemical cycles: Consequences of damming. *Aquat Sci* 64:55–65.

Friedlander, A. M., J. D. Parrish, and R. C. DeFelice. 2002. Ecology of the introduced snapper *Lutjanus kasmira* (Forsskal) in the reef fish assemblage of a Hawaiian bay. *J Fish Biol* 60:28–48.

Friedmann, A. S., E. K. Costain, D. L. MacLatchy, et al. 2002. Effect of mercury on general and reproductive health of largemouth bass (*Micropterus salmoides*) from three lakes in New Jersey. *Ecotox Environ Safe* 52:117–22.

Friend, M. 2002. Avian disease at the Salton Sea. *Hydrobiologia* 473:293–306.

Frissell, C. A. 1993. Topology of extinction and endangerment of native fishes in the Pacific Northwest and California (U.S.A.). *Conserv Biol* 7:342–54.

Frissell, C. A., and R. K. Nawa. 1992. Incidence and causes of physical failure of artificial habitat structures in streams of western Oregon and Washington. *N Am J Fish Manage* 12:182–97.

Fritz, A. W. 1987. Commercial fishing for carp. In Cooper 1987, 17–30.

Froese, R., and M. L. D. Palomares. 2000. Growth, natural mortality, length-weight relationship, maximum length and length at first maturity of the coelacanth *Latimeria chalumnae*. *Environ Biol Fish* 58:45–52.

Froese, R., and D. Pauly. 2006. *FishBase*. www.fishbase.org.

FSBI (Fisheries Society of the British Isles). 2002. *Fish welfare*. Briefing Paper 2, Fisheries Society of the British Isles. www.le.ac.uk/biology/fsbi/welfare.pdf.

Fu, C., J. Wu, J. Chen, et al. 2003. Freshwater fish biodiversity in the Yangtze River basin of China: Patterns, threats and conservation. *Biodivers Conserv* 12:1649–85.

Fuller, P. L., L. G. Nico, and J. D. Williams. 1999. Nonindigenous fishes introduced into inland waters of the United States. *Am Fish Soc Spec Pub* 27.

Furness, R. W., and C. J. Camphuysen. 1997. Seabirds as monitors of the marine environment. *ICES J Mar Sci* 54:726–37.

Fyt, S. 1997. Transgenic fish. *Rev Fish Biol Fisher* 7:417–41.

Galat, D. L., L. H. Fredrickson, D. D. Humburg, et al. 1998. Flooding to restore connectivity of regulated large river wetlands. *BioScience* 48:721–33.

Galat, D. L., and B. Robertson. 1992. Response of endangered *Poeciliopsis occidentalis sonoriensis* in the Rio Yaqui drainage, Arizona, to introduced *Gambusia affinis*. *Environ Biol Fish* 33:249–64.

Galbreath, P. F., and G. H. Thorgaard. 1995. Sexual maturation and fertility of diploid and triploid Atlantic salmon x brown trout hybrids. *Aquaculture* 137:299–311.

Galloway, J. N. 2001. Acidification of the world: Natural and anthropogenic. *Water Air Soil Poll* 130:17–24.

Garcia, S. M. 1994. *World review of highly migratory species and straddling stocks*. FAO Fisheries Tech. Paper 337. Rome: FAO.

Garcia, S. M., and C. Newton. 1997. Current situation, trends, and prospects in world capture fisheries. *Am Fish Soc Sym* 20:3–27.

Garman, G. C., and L. A. Nielsen. 1982. Piscivory by stocked brown trout (*Salmo trutta*) and its impact on the nongame fish community of Bottom Creek, Virginia. *Can J Fish Aquat Sci* 39:862–69.

Garthe, S., C. J. Camphuysen, and R. W. Furness. 1996. Amounts of discards by commercial fisheries and their significance as food for seabirds in the North Sea. *Mar Ecol—Prog Ser* 136:1–11.

Gatz, A. J., M. J. Sale, and J. M. Loar. 1987. Habitat shifts in rainbow trout: Competitive influences of brown trout. *Oecologia* 74:7–19.

Gausen, D., and V. Moen. 1991. Large-scale escapes of farmed Atlantic salmon (*Salmo salar*) into Norwegian rivers threaten natural populations. *Can J Fish Aquat Sci* 48:426–28.

Gehrke, P. C., D. M. Gilligan, and M. Barwick. 2002. Changes in fish communities of the Shoalhaven River 20 years after construction of Tallowa Dam, Australia. *River Res Appl* 18:265–86.

Gell, F. R., and C. M. Roberts. 2003. Benefits beyond boundaries: The fishery effects of marine reserves. *Trends Ecol Evol* 18:448–55.

Gende, S. M., R. T. Edwards, M. F. Willson, et al. 2002. Pacific salmon in aquatic and terrestrial ecosystems. *BioScience* 52:917–28.

Gengerke, T. A. 1986. Distribution and abundance of paddlefish in the United States. In *The paddlefish: Status, management and propagation*, ed. J. G. Dillard et al., 22–35. Special Publication 7. Columbia, MD: North Central Division, American Fisheries Society.

Gensemer, R. W., and R. C. Playle. 1999. The bioavailability and toxicity of aluminum in aquatic environments. *Crit Rev Env Sci Tec* 29:315–450.

Gery, J. 1977. *Characoids of the world*. Neptune City, NJ: TFH Publications.

GESAMP (Group of Experts on Scientific Aspects of Marine Environmental Protection). 1996. *Monitoring the ecological effects of coastal aquaculture wastes*. GESAMP Report and Study 57. Rome: FAO.

———. 1997. *Towards safe and effective use of chemicals in coastal aquaculture*. GESAMP Report and Study 65. Rome: FAO.

Gibbs, P. J., A. J. Collins, and L. C. Collett. 1980. Effect of otter prawn trawling on the macrobenthos of a sandy substratum in a New South Wales estuary. *Aust J Mar Fresh Res* 31:509–16.

Gibson, A. J. F., and R. A. Myers. 2002. Effectiveness of a high-frequency-sound fish diversion system at the Annapolis Tidal Hydroelectric Generating Station, Nova Scotia. *N Am J Fish Manage* 22:770–84.

Gibson, C. D. 1998. *The broadbill swordfishery of the northwest Atlantic: An economic and natural history*. Camden, ME: Ensign Press.

Gibson, K. 2003. *Biologist disputes Fish and Game report*. Jan. 5. http://rogueimc.org/2003/05/580.shtml.

Gido, K. B., and J. H. Brown. 1999. Invasion of North American drainages by alien fish species. *Freshwater Biol* 42:387–99.

Gido, K. B., and D. L. Propst. 1999. Habitat use and association of native and nonnative fishes in the San Juan River, New Mexico and Utah. *Copeia* 1999:321–32.

Gido, K. B., and W. J. Matthews. 2001. Ecosystem effects of water column minnows in experimental streams. *Oecologia* 126:247–53.

Gillanders, B. M., and M. J. Kingsford. 2002. Impact of changes in flow of freshwater on estuarine and open coastal habitats and associated organisms. *Oceanogr Mar Biol* 40:233–309.

Given, D. R., and D. A. Norton. 1993. A multivariate approach to assessing threat and for priority setting in threatened species conservation. *Biol Conserv* 64:57–66.

Glynn, P. W. 1997. Bioerosion and coral-reef growth: A dynamic balance. In Birkeland 1997c, 68–95.

Golani, D. 1993. The biology of the Red Sea migrant *Saurida undosquamis* in the Mediterranean and comparison with the indigenous confamilial *Synodus saurus* (Teleostei: Synodontidae). *Hydrobiologia* 271:109–17.

———. 1999. The Gulf of Suez ichthyofauna: Assemblage pool for Lessepsian migration into the Mediterranean. *Israel J Zool* 45:79–90.

Golani, D., L. Orsi-Relini, E. Massutí, et al. 2002. *CIESM atlas of exotic species in the Mediterranean. Vol. 1: Fishes.* Monaco: CIESM The Mediterranean Science Commission. www.ciesm.org/atlas/appendix1.html.

Goldburg, R., M. Elliott, and R. Naylor. 2001. *Marine aquaculture in the United States: Environmental impacts and policy options.* Washington, D.C.: Pew Oceans Commission. www.pewoceans.org.

Goldburg, R., and T. Triplett. 1997. *Murky waters: Environmental effects of aquaculture in the United States.* New York, NY: Environmental Defense Fund. www.edf.org/pubs/reports/aquaculture/.

Goldsmith, E., and N. Hildyard. 1992. *The social and environmental effects of large dams. Vol. 3: A review of literature.* Camelford, UK: Wadebridge Ecological Centre.

Goldstein, R. J., R. W. Harper, and R. Edwards. 2000. *American aquarium fishes.* College Station: Texas A&M Univ.

Golladay, S. W., J. R. Webster, and E. F. Benfield. 1987. Changes in stream morphology and storm transport of seston following watershed disturbance. *J NA Benthol Soc* 6:1–11.

Goñi, R. 1998. Ecosystem effects of marine fisheries: An overview. *Ocean Coast Manage* 40:37–64.

Gonzalez, R. P., A. Leyva, and M. O. Moraes. 2001. Shark cartilage as a source of antiangiogenic compounds: From basic to clinical research. *Biol Pharm Bull* 24:1097–1101.

Goodyear, C. P. 1985. Toxic materials, fishing, and environmental variation: Simulated effects on striped bass population trends. *T Am Fish Soc* 114:107–13.

Gordon, N. D., T. A. McMahon, and B. L. Finlayson. 1992. *Stream hydrology: An introduction for ecologists.* West Sussex, England: Wiley.

Gore, A. 1992. *Earth in the balance: Ecology and the human spirit.* Boston: Houghton Mifflin.

Goreau, T. J., J. Cervino, M. Goreau, et al. 1998. Rapid spread of diseases in Caribbean coral reefs. *Rev Biol Trop* 46(suppl.): 157–71.

Goreau, T., T. McClanahan, R. Hayes, et al. 2000. Conservation of coral reefs after the 1998 global bleaching event. *Conserv Biol* 14:5–15.

Gorman, J. 2003. Did PCBs save the stripers? A fish story. *New York Times*, March 25, F3.

Gorman, O. T., and D. M. Stone. 1999. Ecology of spawning humpback chub, *Gila cypha*, in the Little Colorado River near Grand Canyon, Arizona. *Environ Biol Fish* 55:115–33.

Gosling, L. M., and W. J. Sutherland, eds. 2000. *Behaviour and conservation.* Cambridge, UK: Cambridge Univ. Press.

Goudswaard, K. P. C., and F. Witte. 1997. The catfish fauna of Lake Victoria after the Nile perch upsurge. *Environ Biol Fish* 49:21–43.

Goulding, M. 1980. *The fishes and the forest: Explorations in Amazonian natural history.* Berkeley: Univ. California Press.

Goulding, M., N. J. H. Smith, and D. J. Mahar. 1996. *Floods of fortune: Ecology and economy along the Amazon.* New York: Columbia Univ. Press.

Gradall, K. S., and W. A. Swenson. 1982. Responses of brook trout and creek chubs to turbidity. *T Am Fish Soc* 111:392–95.

Graham, K. 1997. Contemporary status of the North American paddlefish, *Polyodon spathula. Environ Biol Fish* 48:279–89.

Graham, M. 1943. *The fish gate.* London: Faber.

Graham, T. 1996. Managing Palau's aquarium life fishery. *SPC Live Reef Fish Inform Bull* 1:13–18.

Graham, T., and N. Idechong. 1998. Reconciling customary and constitutional law: Managing marine resources in Palau, Micronesia. *Ocean Coast Manage* 40:143–64.

Grant, S. M. 1995. *African rift lakes.* Presentation at Aquarama '95, Singapore, May 1995. www.lakemalawi.com.

Grant, W. S., J. L. Garcia-Marin, and F. M. Utter. 1999. Defining population boundaries for fishery management. In Mustafa 1999a, 27–72.

Graves, J. E. 1996. Conservation genetics of fishes in the pelagic marine realm. In *Conservation genetics*, ed. J. C. Avise and J. L. Hamrick, 335–66. New York: Chapman and Hall.

Greenberg, L. A., and R. A. Stiles. 1993. A descriptive and experimental study of microhabitat use by young-of-the-year benthic stream fishes. *Ecol Freshw Fish* 2:40–49.

Greenlaw, L. 1999. *The hungry ocean: A swordboat captain's journey*. New York: Hyperion.

Greenwood, P. H. 1992. Are the major fish faunas well-known? *Neth J Zool* 42:131–38.

Gregory, J., and P. Clabburn. 2003. Avoidance behaviour of *Alosa fallax fallax* to pulsed ultrasound and its potential as a technique for monitoring clupeid spawning migration in a shallow river. *Aquat Living Resour* 16:313–16.

Gregory, S. V. 1997. Riparian management in the 21st century. In *Creating a forestry for the 21st century: The science of ecosystem management*, ed. K. A. Kohm and J. F. Franklin, 69–85. Washington, DC: Island Press.

Gregory, S., K. Boyer, and A. Gurnell, eds. 2003. The ecology and management of wood in world rivers. *Am Fish Soc Sym* 37.

Gregory, S., H. Li, and J. Li. 2002. The conceptual basis for ecological responses to dam removal. *BioScience* 52:713–23.

Grey, Z. 1919. *Tales of fishes*. New York: Harper & Brothers.

Griffith, J. S., and D. A. Andrews. 1981. Effects of a small suction dredge on fishes and aquatic invertebrates in Idaho streams. *N Am J Fish Manage* 1:21–28.

Grigg, R. W., J. J. Polovina, and M. J. Atkinson. 1984. Model of a coral reef ecosystem. III: Resource limitation, community regulation, fisheries yield and resource management. *Coral Reefs* 3:23–27.

Grimes, C. B. 1995. Marine fisheries section [position statement]. *Am Fish Soc Sym* 15:593–94.

Grippo, M. A., and A. G. Heath. 2003. The effect of mercury on the feeding behavior of fathead minnows (*Pimephales promelas*). *Ecotox Environ Safe* 55:187–98.

Groenewold, S., and M. Fonds. 2000. Effects on benthic scavengers of discards and damaged benthos produced by the beam-trawl fishery in the southern North Sea. *ICES J Mar Sci* 57:1395–1406.

Groom, M. J., G. K. Meffe, and C. R. Carroll, eds. 2006. *Principles of conservation biology*, 3rd ed. Sunderland, MA: Sinauer Associates.

Groom, M. J., and N. Schumaker. 1993. Evaluating landscape change: Patterns of worldwide deforestation and local fragmentation. In *Biotic interactions and global change*, ed. P. M. Kareiva et al., 24–44. Sunderland, MA: Sinauer Associates.

Groombridge, B., ed. 1994. *Biodiversity data sourcebook*. World Conservation Monitoring Centre, WCMC Biodiversity Series 1. Cambridge, UK: World Conservation Press.

Groot, C., and L. Margolis, eds. 1991. *Pacific salmon life histories*. Vancouver, BC: Univ. British Columbia Press.

Gross, M. R. 1984. Sunfish, salmon, and the evolution of alternative reproductive strategies and tactics in fishes. In *Fish reproduction: Strategies and tactics*, ed. G. W. Potts and R. J. Wootton, 55–75. London: Academic Press.

———. 1991. Salmon breeding behavior and life history evolution in changing environments. *Ecology* 72:1180–86.

———. 1998. One species with two biologies: Atlantic salmon (*Salmo salar*) in the wild and in aquaculture. *Can J Fish Aquat Sci* 55(suppl. 1): 131–44.

Gruson, L. 1997. Throwing back undersize fish is said to encourage smaller fry. In Wade 1997, 152–56.

Guier, C. R., L. E. Nichols, and R. T. Rachels. 1984. Biological investigation of flathead catfish in the Cape Fear River. *P Annu Conf SE Assoc Fish Wildlife Agen* 35:607–21.

Guillen, I., J. Berlanga, C. M. Valenzuela, et al. 1999. Safety evaluation of transgenic *Tilapia* with accelerated growth. *Mar Biotechnol* 1:2–14.

Gundlach, E. R., P. D. Boehm, M. Marchand, et al. 1983. The fate of *Amoco Cadiz* oil. *Science* 22:122–29.

Gunn, J. M., and W. Keller. 1990. Biological recovery of an acid lake after reductions in industrial emissions of sulfur. *Nature* 345:431–33.

Gunn, J. M., and S. Sandøy. 2003. Introduction to the *Ambio* special issue on biological recovery from acidification: Northern Lakes Recovery Study. *Ambio* 32:162–64.

Gurnell, A. M., K. J. Gregory, and G. E. Petts. 1995. The role of coarse woody debris in forest aquatic habitats: Implications for management. *Aquat Conserv* 5:143–66.

Gutierrez-Estrada, J. C., J. Prenda, F. Oliva, et al. 1998. Distribution and habitat preferences of the introduced mummichog *Fundulus heteroclitus* (Linneaus) in southwestern Spain. *Estuar Coast Shelf S* 46:827–35.

Gyllensten, U. 1985. The genetic structure of fish: Differences in the intraspecific distribution of biochemical genetic variation between marine, anadromous, and freshwater species. *J Fish Biol* 26:691–99.

Haas, R., and R. Pal. 1984. Mosquito larvivorous fishes. *B Entomol Soc Am* 30:17–25.

Häkkinen, J., E. Vehniäinen, O. Ylönen, et al. 2002. The effects of increasing UV-B radiation on pigmentation, growth and survival of coregonid embryos and larvae. *Environ Biol Fish* 64:451–59.

Hale, M. M., J. E. Crumpton, and R. J. Schuler Jr. 1995. From sportfishing bust to commercial fishing boon: A history of the blue tilapia in Florida. *Am Fish Soc Sym* 15: 425–30.

Hale, R. C., J. Greaves, J. L. Gundersen, et al. 1991. Occurrence of organochlorine contaminants in tissues of the coelacanth *Latimeria chalumnae*. *Environ Biol Fish* 32:361–67.

Hall, K. C., and D. R. Bellwood. 1995. Histological effects of cyanide, stress and starvation on the intestinal mucosa of *Pomacentrus coelestis*, a marine aquarium fish species. *J Fish Biol* 47:438–54.

Hall, M. A. 1998. An ecological view of the tuna-dolphin problem: Impacts and trade-offs. *Rev Fish Biol Fisher* 8:1–34.

Hall, M. A., D. L. Alverson, and K. I. Metuzals. 2000. By-catch: Problems and solutions. *Mar Pollut Bull* 41:204–19.

Hall, S. J. 1999. *The effects of fishing on marine ecosystems and communities*. Malden, MA: Blackwell.

Hallegraeff, G. M. 1998. Transport of toxic dinoflagellates via ships' ballast water: Bioeconomic risk assessment and efficacy of possible ballast water management strategies. *Mar Ecol—Prog Ser* 168:297–309.

Halpern, B. S. 2003. The impact of marine reserves: Do reserves work and does reserve size matter? *Ecol Appl* 13(1): S117–S137.

Ham, K. D., and T. N. Pearsons. 2001. A practical approach for containing ecological risks associated with fish stocking programs. *Fisheries* 26(4): 15–22.

Hameed, F. 2002. First national report to the Conference of the Parties to the Convention on Biological Diversity. Malé: Republic of Maldives, Ministry of Home Affairs, Housing and Environment. www.biodiv.org/doc/world/mv/mv-nr-01-en.pdf.

Hamer, P. A., G. P. Jenkins, and B. M. Gillanders. 2003. Otolith chemistry of juvenile snapper *Pagrus auratus* in Victorian waters: Natural chemical tags and their temporal variation. *Mar Ecol—Prog Ser* 263:261–73.

Hamilton, J. B. 1989. Response of juvenile steelhead to instream deflectors in a high gradient stream. In *Practical approaches to riparian resources management*, ed. R. E. Gresswell et al., 149–57. Bethesda, MD: American Fisheries Society, Montana Chapter.

Hamilton, J. D. 1961. The effect of sand-pit washings on a stream fauna. *Int Assoc Theor App Limnol* 14:435–39.

Hamon, T. R., C. J. Foote, R. Hilborn, et al. 2000. Selection on morphology of spawning wild sockeye salmon by a gill-net fishery. *T Am Fish Soc* 129:1300–15.

Handford, P., G. Bell, and T. Reimchen. 1977. A gillnet fishery considered as an experiment in artificial selection. *J Fish Res Board Can* 34:954–61.

Hannesson, R. 1996. *Fisheries mismanagement: The case of the North Atlantic cod*. Oxford: Fishing News Books.

———. 1998. The role of economic tools in redefining fisheries management. In Pitcher, Hart, and Pauly 1998, 251–60.

Hansen, M. J. 1999. Lake trout in the Great Lakes: Basin-wide stock collapse and binational restoration. In *Great Lakes fishery policy and management: A binational perspective*, ed. W. W. Taylor and C. P. Ferreri, 417–53. East Lansing: Michigan State Univ. Press.

Hanson, L. H., and W. D. Swink. 1989. Downstream migration of recently metamorphosed sea lampreys in the Ocqueoc River, Michigan, before and after treatment with lampricides. *N Am J Fish Manage* 9:327–31.

Harcourt, A. H. 1995. Population viability estimates: Theory and practice for a wild gorilla population. *Conserv Biol* 9:134–42.

Hardin, G. 1968. The tragedy of the commons. *Science* 162:1243–48.

Harding, J. S., E. F. Benfield, P. V. Bolstad, et al. 1998. Stream biodiversity: The ghost of land-use past. *P Natl Acad Sci USA* 95:14843–47.

Hardisty, M. W., and I. C. Potter. 1971. The general biology of adult lampreys. In *The biology of lampreys*, vol. 1, ed. M. W. Hardisty and I. C. Potter, 127–206. San Diego: Academic Press.

Hare, S. R., N. J. Mantua, and R. C. Francis. 1999. Inverse production regimes: Alaska and West Coast Pacific salmon. *Fisheries* 24(1): 6–14.

Hargrove, E. C. 1989. *Foundations of environmental ethics*. Englewood Cliffs, NJ: Prentice Hall.

Harmon, M. E., J. F. Franklin, F. J. Swanson, et al. 1986. Ecology of coarse woody debris in temperate ecosystems. *Adv Ecol Res* 15:133–302.

Haro, A., W. Richkus, K. Whalen, et al. 2000. Population decline of the American eel: Implications for research and management. *Fisheries* 25(9): 7–16.

Harris, A. N., and I. R. Poiner. 1991. Changes in species composition of demersal fish fauna of Southeast Gulf of Carpenteria, Australia, after 20 years of fishing. *Mar Biol* 111:503–19.

Harrison, I. J., and M. L. J. Stiassny. 1999. The quiet crisis: A preliminary listing of the freshwater fishes of the world that are extinct or "missing in action." In *Extinctions in near time*, ed. R. D. E. MacPhee, 271–331. New York: Kluwer Academic/Plenum.

Hart, J. L. 1973. *Pacific fishes of Canada*. Fisheries Research Board of Canada Bull. 180.

Hart, P. J. B., and J. D. Reynolds, eds. 2002. *Handbook of fish biology and fisheries. Vol. 2: Fisheries*. Oxford: Blackwell Science.

Hastein, T., and T. Lindstad. 1991. Diseases in wild and cultured salmon: Possible interaction. *Aquaculture* 98:277–88.

Hatcher, B. G. 1997. Organic production and decomposition. In Birkeland 1997c, 140–74.

Hauck, F. R. 1949. Some harmful effects of the electroshocker on large rainbow trout. *T Am Fish Soc* 77:61–64.

Haugen, T. O., and L. A. Vollestad. 2001. A century of life-history evolution in grayling. *Genetica* 112:475–91.

Hauser, L., G. J. Adcock, P. J. Smith, et al. 2002. Loss of microsatellite diversity and low effective population size in an overexploited population of New Zealand snapper (*Pagrus auratus*). *P Natl Acad Sci USA* 99:11742–47.

Havas, M., T. C. Hutchinson, and G. E. Likens. 1984. Red herrings in acid rain research. *Environ Sci Technol* 18:176A–86A.

Hawkins, C. P., M. L. Murphy, N. H. Anderson, et al. 1983. Density of fish and salamanders in relation to riparian canopy and physical habitat in streams of the northwestern United States. *Can J Fish Aquat Sci*. 40:1173–85.

Hawkins, J. P., and C. M. Roberts. 1993. Effects of recreational SCUBA diving on coral reefs: Trampling on reef-flat communities. *J Appl Ecol* 30:25–30.

Hawkins, J. P., C. M. Roberts, and V. Clark. 2000. The threatened status of restricted-range coral reef fish species. *Anim Conserv* 3:81–88.

Hay, M. E. 1984. Patterns of fish and urchin grazing on Caribbean coral reefs: Are previous results typical? *Ecology* 65:446–54.

Haya, K., L. E. Burridge, and B. D. Chang. 2001. Environmental impact of chemical wastes produced by the salmon aquaculture industry. *ICES J Mar Sci* 58:492–96.

Hayes, E. A. 1997. *A review of the southern bluefin tuna fishery: Implications for ecologically sustainable management*. TRAFFIC Oceania Report. www.TRAFFIC.org.

Haynes, D., J. Muller, and S. Carter. 2000. Pesticide and herbicide residues in sediments and seagrasses from the Great Barrier Reef world heritage area and Queensland coast. *Mar Pollut Bull* 41:279–87.

Hayward, C. J., M. Iwashita, J. S. Crane, et al. 2001. First report of the invasive eel pest *Pseudodactylogyrus bini* in North America and in wild American eels. *Dis Aquat Organ* 44:53–60.

Hearn, W. E., and B. E. Kynard. 1986. Habitat utilization and behavior interactions of juvenile Atlantic salmon (*Salmo salar*) and rainbow trout (*S. gairdneri*) in tributaries of the White River of Vermont. *Can J Fish Aquat Sci* 43:1988–98.

Hecky, R. E. 1993. The eutrophication of Lake Victoria. In Peter Kilham Memorial Lecture. *Int Assoc Theor App Limnol* 25:39–48.

Hedrick, R. P. 1996. Movements of pathogens with the international trade of live fish: Problems and solutions. *Rev Sci Tech OIE* 15:523–31.

Hedrick, R. P., T. S. McDowell, K. Mukkatira, et al. 1999. Susceptibility of selected inland salmonids to experimentally induced infections with *Myxobolus cerebralis*, the causative agent of whirling disease. *J Aquat Anim Health* 11:330–39.

Heil, C. A., P. M. Glibert, M. A. Al-Sarawl, et al. 2001. First record of a fish-killing *Gymnodinium* sp. bloom in Kuwait Bay, Arabian Sea: Chronology and potential causes. *Mar Ecol—Prog Ser* 214:15–23.

Heinz Center. 2002. *Dam removal: Science and decision making.* Washington, DC: Author.

Helfield, J. M., and R. J. Naiman. 2001. Effects of salmon-derived nitrogen on riparian forest growth and implications for stream productivity. *Ecology* 82:2403–9.

———. 2003. Effects of salmon-derived nitrogen on riparian forest growth and implications for stream productivity: Reply. *Ecology* 84:3399–401.

Helfman, G. S. 1989. Threat-sensitive predator avoidance in damselfish-trumpetfish interactions. *Behav Ecol Sociobiol* 24:47–58.

———. 1990. Mode-switching and mode selection in foraging animals. *Adv Stud Behav* 19:249–98.

———. 1993. Fish behaviour by day, night and twilight. In *The behaviour of teleost fishes*, 2nd ed., ed. T. J. Pitcher, 479–512. London: Chapman and Hall.

———, ed. 1999a. Behavior and fish conservation: Case studies and applications. *Environ Biol Fish* 55:7–201.

———. 1999b. Behavior and fish conservation: Introduction, motivation, and overview. *Environ Biol Fish* 55:7–12.

———. 2001. Biodiversity of fishes. In *Encyclopedia of biodiversity*, ed. S. Levin, 755–82. New York: Academic Press.

Helfman, G. S., B. B. Collette, and D. C. Facey. 1997. *The diversity of fishes.* Malden, MA: Blackwell Science.

Helfman, G. S., and E. T. Schultz. 1984. Social transmission of behavioural traditions in a coral reef fish. *Anim Behav* 32:379–84.

Helfman, G. S., and D. L Winkelman. 1997. Threat-sensitivity in bicolor damselfish: Effects of sociality and body size. *Ethology* 103:369–83.

Hellberg, M. E., R. S. Burton, J. E. Neigel, et al. 2002. Genetic assessment of connectivity among marine populations. *B Mar Sci* 70(suppl.): 273–90.

Hemdal, J. 1995. Lake Victoria cichlid species survival plan update. *Aquat Surv* 4(1): 21–22.

Hemingway, E. 1952. *The old man and the sea.* New York: Charles Scribner's Sons.

Hendry, A. P. 2001. Adaptive divergence and the evolution of reproductive isolation in the wild: An empirical demonstration using introduced sockeye salmon. *Genetica* 112:515–34.

Hendry, A. P., J. K. Wenburg, P. Bentzen, et al. 2000. Rapid evolution of reproductive isolation in the wild: Evidence from introduced salmon. *Science* 290:516–18.

Henley, W. R., M. A. Patterson, R. J. Neves, et al. 2000. Effects of sedimentation and turbidity on lotic food webs: A concise review for natural resource managers. *Rev Fish Sci* 8:125–39.

Henry, T. B., and M. C. Black. 2003. Acute and chronic toxicity of the selective serotonin reuptake inhibitor fluoxetine to western mosquitofish *Gambusia affinis.* Abstract, 24th Annual Meeting, Society of Environmental Toxicology and Chemistry, Nov., Austin, TX.

Herbert, M. E., and F. P. Gelwick. 2003. Spatial variation of headwater fish assemblages explained by hydrologic variability and upstream effects of impoundment. *Copeia* 2003:273–84.

Herbold, B., and P. B. Moyle. 1986. Introduced species and vacant niches. *Am Nat* 128:751–60.

Herwig, R. P., J. P. Gray, and D. P. Weston. 1997. Antibacterial resistant bacteria in surficial sediments near salmon net-cage farms in Puget Sound, Washington. *Aquaculture* 149:263–83.

Hesse, L. W., and B. A. Newcomb. 1982. Effects of flushing Spencer Hydro on water quality, fish, and insect fauna in the Niobrara River, Nebraska. *N Am J Fish Manage* 2:45–52.

Hesthagen, T., T. Forseth, R. Saksgård, et al. 2001. Recovery of young brown trout in some acidified streams in southwestern and western Norway. *Water Air Soil Poll* 130:1355–60.

Hew, C. L., and G. L. Fletcher, eds. 1992. *Transgenic fish.* Singapore: World Scientific Publishing.

———. 1997. Transgenic fish for aquaculture. *Chem Ind* 8:311–14.

Hew, C. L., G. L. Fletcher, and P. L. Davies. 1995. Transgenic salmon: Tailoring the genome for food production. *J Fish Biol* 47(suppl. A): 1–19.

Hightower, J. M., and D. Moore. 2003. Mercury levels in high-end consumers of fish. *Environ Health Persp* 111:604–8.

Hilborn, R. 1998. The economic performance of marine stock enhancement projects. *B Mar Sci* 62:661–74.

———. 2005. Are sustainable fisheries achievable? In Norse and Crowder 2005, 247–59.

Hilborn, R., and C. J. Walters. 1992. *Quantitative fisheries stock assessment: Choice, dynamics, and uncertainty.* New York: Chapman and Hall.

Hill, J. A., A. Kiessling, and R. H. Devlin. 2000. Coho salmon (*Oncorhynchus kisutch*) transgenic for a growth hormone gene construct exhibit increased rates of muscle hyperplasia and detectable levels of differential gene expression. *Can J Fish Aquat Sci* 57:939–50.

Hilton-Taylor, C. 2000. *2000 IUCN red list of threatened species*. Gland, Switzerland: IUCN. www.redlist.org.

Hindar, K., N. Ryman, and F. Utter. 1991. Genetic effects of cultured fish on natural fish populations. *Can J Fish Aquat Sci* 48:945–57.

Hitchmough, R. 2002. *New Zealand threat classification system lists 2002*. Wellington, NZ: Department of Conservation.

Hites, R. A., J. A. Foran, D. O. Carpenter, et al. 2004. Global assessment of organic contaminants in farmed salmon. *Science* 303:226–29.

Hixon, M. A. 1991. Predation as a process structuring coral reef fish communities. In Sale 1991, 475–508.

———. 1997. Effects of reef fishes on corals and algae. In Birkeland 1997c, 230–48.

———. 1998. Population dynamics of coral-reef fishes: Controversial concepts and hypotheses. *Aust J Ecol* 23:192–201.

Hoagland, P., D. M. Anderson, Y. Kaoru, et al. 2002. The economic effects of harmful algal blooms in the United States: Estimates, assessment issues, and information needs. *Estuaries* 25:819–37.

Hobson, E. S. 1979. Interactions between piscivorous fishes and their prey. In *Predator-prey systems in fisheries management*, ed. H. Clepper, 231–42. Washington, DC: Sport Fishing Institute.

Hockenberry, J. 2001. *A river out of Eden*. New York: Anchor Books.

Hodgson, G., and J. Liebeler. 2002. *The global coral reef crisis: Trends and solutions*. Reef Check 5-Year Report, Reef Check Foundation, UCLA. www.reefcheck.org.

Hoffman, G. L., and G. Schubert. 1984. Some parasites of exotic fishes. In *Distribution, biology, and management of exotic fishes*, ed. W. R. Courtenay Jr. and J. R. Stauffer Jr., 233–61. Baltimore: Johns Hopkins Univ. Press.

Hogan, Z. S., P. B. Moyle, B. May, et al. 2004. The imperiled giants of the Mekong. *Am Sci* 92:228–37.

Hogan, Z., N. Pengbun, and N. van Zalinge. 2001. Status and conservation of two endangered fish species, the Mekong giant catfish *Pangasianodon gigas* and the giant carp *Catlocarpio siamensis*, in Cambodia's Tonle Sap River. *Nat Hist B Siam Soc* 49:26–282.

Holden, C. 1996. Fish fans beware. *Science* 273:1049.

Holden, P. B. 1991. Ghosts of the Green River: Impacts of Green River on management of native fishes. In Minckley and Deacon 1991, 43–54.

Holder, M. T., M. V. Erdmann, T. P. Wilcox, et al. 1999. Two living species of coelacanths? *P Natl Acad Sci USA* 96:12616–20.

Hollender, B. A., and R. F. Carline. 1994. Injury to wild brook trout by backpack electrofishing. *N Am J Fish Manage* 14:643–49.

Holmlund, C. M., and M. Hammer. 1999. Ecosystem services generated by fish populations. *Ecol Econ* 29:253–68.

Holtby, L. B. 1988. Effects of logging on stream temperature in Carnation Creek, British Columbia, and associated impacts on the coho salmon (*Oncorhynchus kisutch*). *Can J Fish Aquat Sci* 45:502–15.

Holthus, P. 2001. The role of certification for the marine aquarium trade. *InterCoast* 40:22–23 (www.crc.uri.edu).

Honey, M. S. 1999. Treading lightly: Ecotourism's impact on the environment. *Environment* 41:9–32.

Hontela, A., P. Dumont, D. Duclos, et al. 1995. Endocrine and metabolic dysfunction in yellow perch, *Perca flavescens*, exposed to organic contaminants and heavy metals in the St. Lawrence River. *Environ Toxicol Chem* 14:725–31.

Hooke, R. L. 1999. Spatial distribution of human geomorphic activity in the United States: Comparison with rivers. *Earth Surf Proc Land* 24:687–92.

Hopkins, W. A., C. P. Tatara, H. A. Brant, et al. 2003. Relationships between mercury body concentrations, standard metabolic rate, and body mass in eastern mosquitofish (*Gambusia holbrooki*) from three experimental populations. *Environ Toxicol Chem* 22:586–90.

Houde, E. D. 1997. Patterns and consequences of selective processes in teleost early life histories. In *Early life history and recruitment in fish populations*, ed. R. C. Chambers and E. A. Trippel, 173–96. London: Chapman and Hall.

Hrabik, T. R., M. P. Carey, and M. S. Webster. 2001. Interactions between young-of-the-year exotic rainbow smelt and native yellow perch in a northern temperate lake. *T Am Fish Soc* 130:568–82.

Hrabik, T. R., J. J. Magnuson, and A. S. McLain. 1998. Predicting the effects of rainbow smelt on native fishes in small lakes: Evidence from long-term research on two lakes. *Can J Fish Aquat Sci* 55:1364–71.
Hrabik, T. R., and C. J. Watras. 2002. Recent declines in mercury concentration in a freshwater fishery: Isolating the effects of de-acidification and decreased atmospheric mercury deposition in Little Rock Lake. *Sci Total Environ* 297:229–37.
Hubbs, C. L. 1955. Hybridization between fish species in nature. *Syst Zool* 4:1–20.
Hubbs, C. 2001. Environmental correlates to the abundance of spring-adapted versus stream-adapted fishes. *Tex J Sci* 53:299–326.
Hudson, C. 1976. *The southeastern Indians*. Knoxville: Univ. Tennessee Press.
Hughes, R. M., P. R. Kaufmann, A. T. Herlihy, et al. 1998. A process for developing and evaluating indices of fish assemblage integrity. *Can J Fish Aquat Sci* 55:1618–31.
Hughes, R. M., T. R. Whittier, C. M. Rohm, et al. 1990. A regional framework for establishing recovery criteria. *Environ Manage* 14:673–83.
Hughes, T. P. 1994. Catastrophes, phase shifts, and large-scale degradation of a Caribbean coral reef. *Science* 264:1547–51.
Hughes, T. P., A. H. Baird, D. R. Bellwood, et al. 2003. Climate change, human impacts, and the resilience of coral reefs. *Science* 301:929–33.
Humphries, P., and P. S. Lake. 2000. Fish larvae and the management of regulated rivers. *Regul River* 16:421–32.
Hunt, B. P.V., E. A. Pakhomov, and C. D. McQuaid. 2001. Short-term variation and long-term changes in the oceanographic environment and zooplankton community in the vicinity of a sub-Antarctic archipelago. *Mar Biol* 138:369–81.
Hunt, R. L. 1979. *Removal of woody streambank vegetation to improve trout habitat*. Tech. Bull. 115. Madison: Wisconsin Dept. Natural Resources.
Huntington, C., W. Nehlsen, and J. Bowers. 1996. A survey of healthy native stocks of anadromous salmonids in the Pacific Northwest and California. *Fisheries* 21(3): 6–14.
Huntington, H. P. 2000. Using traditional ecological knowledge in science: Methods and applications. *Ecol Appl* 10:1270–74.
Hutchings, J. A. 2000a. Collapse and recovery of marine fishes. *Nature* 406:882–85.
———. 2000b. Numerical assessment in the front seat, ecology and evolution in the back seat: Time to change drivers in fisheries and aquatic sciences. *Mar Ecol—Prog Ser* 208:299–303.
———. 2001a. Conservation biology of marine fishes: Perceptions and caveats regarding assignment of extinction risk. *Can J Fish Aquat Sci* 58:108–21.
———. 2001b. Influence of population decline, fishing, and spawner variability on the recovery of marine fishes. *J Fish Biol* 59(suppl. A): 306–22.
Hutchings, J. A., and J. D. Reynolds. 2004. Marine fish population collapses: Consequences for recovery and extinction risk. *BioScience* 54:297–309.
Hutchings, P. 1990. Review of the effects of trawling on macrobenthic epifaunal communities. *Aust J Mar Fresh Res* 41:111–20.
Huxley, T. H. 1884. Inaugural address. *Fisheries exhibition literature* 4:1–22. Cited in T. D. Smith 1994.
Hylland, K., and C. Haux. 1997. Effects of environmental oestrogens on marine fish species. *Trend Anal Chem* 16:606–12.
IASP (International Association for the Study of Pain). 1979. Pain terms: A list with definitions and notes on usage. *Pain* 6:249–52.
ICCAT (International Commission for the Conservation of Atlantic Tunas). 2001a. *Executive summary of Species Status Reports*. 7.5. BFT-Atlantic bluefin tuna. www.iccat.es, Oct. 2001.
———. 2001b. *Executive summary of Species Status Reports*. 7.9. SWO-ATL-Atlantic swordfish. www.iccat.es, Oct. 2001.
ICES (International Council for Exploration of the Sea). 1995. *Report of the Study Group on unaccounted mortality in fisheries*. ICES CM 1995/B:1. Copenhagen: Author. Cited in NRC 1999b.
ICOLD (International Commission on Large Dams). 2003. *2003 World Register of Dams*. www.icold-cigb.org/PDF/RegisterUserGuide.pdf.
Ikuta, K., A. Munakata, K. Aida, et al. 2001. Effects of low pH on upstream migratory behavior in land-locked sockeye salmon *Oncorhynchus nerka*. *Water Air Soil Poll* 130:99–106.
ILAR (Institute of Laboratory Animal Research). 1996. *Guide for the care and use of laboratory animals*. Washington, DC: Author, Commission on Life Sciences, National Research Council.

IMA (International Maritime Association). 2003. *Special areas and particularly sensitive sea areas.* www.imo.org/Environment/mainframe.asp?topic_id=760.

Immerwahr, J. 1999. *Waiting for a signal: Public attitudes toward global warming, the environment and geophysical research.* Public Agenda/American Geophysical Union Report. April 15. www.agu.org/sci_soc/attitude_study.html.

Impson, D. 1997. Threatened fishes of the world: *Labeo seeberi* Gilchrist & Thompson, 1911 (Cyprinidae). *Environ Biol Fish* 49:480.

———. 2001. Threatened fishes of the world: *Pseudobarbus burgi* Peters, 1864 (Cyprinidae). *Environ Biol Fish* 54:44.

Incerpi, A. 1996. Hatchery-bashing: A useless pastime. *Fisheries* 21(5): 28.

Intrafish. 2001. *Cubans have no qualms eating GM tilapia.* www.intrafish.com/articlea.php?articleID=12777. 29 May 2001.

IPCC (Intergovernmental Panel on Climate Change). 2001. *Climate change 2001: Impacts, adaptations, and vulnerability.* Document from Working Group II of the Intergovernmental Panel on Climate Change. Tech. Summary. www.ipcc.ch/pub/wg2TARtechsum.pdf.

IPHC (International Pacific Halibut Commission). 1998. *The Pacific halibut: Biology, fishery, and management.* IPHC Tech. Report 40. Seattle, WA: Author. www.iphc.washington.edu.

Irish, K. E., and E. A. Norse. 1996. Scant emphasis on marine biodiversity. *Conserv Biol* 10:680.

Irwin, E. R., and M. C. Freeman. 2002. Proposal for adaptive management to conserve biotic integrity in a regulated segment of the Tallapoosa River, Alabama, USA. *Conserv Biol* 16:1212–22.

Isaksen, B., and J. W. Valdemarsen. 1994. Bycatch reduction in trawls by utilizing behavioural differences. In *Marine fish behaviour,* ed. A. Ferno and S. Olsen, 69–83. London: Fishing News Books.

ISOFISH. 2002. *Patagonian toothfish.* www.isofish.org/au/backg/index.htm.

IUCN (International Union for the Conservation of Nature). 1994. *IUCN Red List categories.* Gland, Switzerland: IUCN Species Survival Commission. www.redlist.org.

———. 1996. *1996 IUCN Red List of Threatened Animals.* Gland, Switzerland: Author.

———. 1997. *Threatened fish? Initial guidelines for applying the IUCN Red List criteria to marine fishes.* Gland, Switzerland: Author.

———. 2002. *The IUCN Red List of Threatened Species.* Gland, Switzerland: Author.

———. 2003. *Guidelines for application of the IUCN Red List criteria at regional levels: Version 3.0.* Gland, Switzerland: IUCN Species Survival Commission. www.iucn.org/themes/ssc/redlists/regionalguidelines.htm.

———. 2004. *2004 IUCN Red List of Threatened Species.* www.redlist.org.

Iudicello, S., M. Weber, and R. Wieland. 1999. *Fish, markets, and fishermen: The economics of overfishing.* Washington, DC: Island Press.

Jackson, D. A., and N. E. Mandrak. 2002. Changing fish biodiversity: Predicting the loss of cyprinid biodiversity due to global climate change. *Am Fish Soc Sym* 32:89–98.

Jackson, D. C. 1999. Flathead catfish: Biology, fisheries, and management. *Am Fish Soc Sym* 24:23–35.

Jackson, D. C., and G. Marmulla. 2001. The influence of dams on river fisheries. In *Dams, fish and fisheries: Opportunities, challenges and conflict resolution,* ed. G. Marmulla. FAO Fisheries Tech. Paper 419. Rome: FAO. www.fao.org/DOCREP/004/Y2785E/y2785e02a.htm.

Jackson, J. B. C. 1997. Reefs since Columbus. *Coral Reefs* 16:523–32.

Jackson, J. B. C., M. X. Kirby, W. H. Berger, et al. 2001. Historical overfishing and the recent collapse of coastal ecosystems. *Science* 293:629–38.

Jacobs, M., J. Ferrario, and C. Byrne. 2002. Investigation of polychlorinated dibenzo-p-dioxins, dibenzo-p-furans and selected coplanar biphenyls in Scottish farmed Atlantic salmon (*Salmo salar*). *Chemosphere* 47:183–91.

Jacobson, L. D., F. C. Funk, and G. J. Goiney. 1999. Pacific coast and Alaska pelagic fisheries. Unit 14 in NMFS 1999.

Jagoe, C. H., A. Faivre, and M. C. Newman. 1996. Morphological and morphometric changes in the gills of mosquitofish (*Gambusia holbrooki*) after exposure to mercury (II). *Aquat Toxicol* 34:163–83.

Jara, F., D. Soto, and R. Palma. 1995. Reproduction in captivity of the endangered killifish *Orestias ascontanensis* (Teleostei, Cyprinodontidae). *Copeia* 1995:226–28.

Jeffrey, C. F. G. 2000. Annual, coastal and seasonal variation in Grenadian demersal fisheries (1986–1993) and implications for management. *B Mar Sci* 66:305–19.

Jeffries, D. S., T. A. Clair, S. Couture, et al. 2003. Assessing the recovery of lakes in southeastern Canada from the effects of acidic deposition. *Ambio* 32:176–82.

Jenkins, R. E., and N. M. Burkhead. 1994. *Freshwater fishes of Virginia*. Bethesda, MD: American Fisheries Society.

Jenkins, S. R., B. D. Beukers-Stewart, and A. R. Brand. 2001. Impact of scallop dredging on benthic megafauna: A comparison of damage levels in captured and non-captured organisms. *Mar Ecol—Prog Ser* 215:297–301.

Jennings, S. 2005. Indicators to support an ecosystem approach to fisheries. *Fish and Fisheries* 6:212–32.

Jennings, S., and M. J. Kaiser. 1998. The effects of fishing on marine ecosystems. *Adv Mar Biol* 34:201–352.

Jennings, S., M. J. Kaiser, and J. D. Reynolds. 2001. *Marine fisheries ecology*. Oxford: Blackwell Science.

Jennings, S., and J. M. Lock. 1996. Population and ecosystem effects of fishing. In Polunin and Roberts 1996, 193–218.

Jennings, S., and N.V. C. Polunin. 1996. Impacts of fishing on tropical reef ecosystems. *Ambio* 25:44–49.

Jennings, S., J. D. Reynolds, and S. C. Mills. 1998. Life history correlates of responses to fisheries exploitation. *P Roy Soc Lond B Bio* 265:333–39.

Jennings, S., J. D. Reynolds, and N.V. C. Polunin. 1999. Predicting the vulnerability of tropical reef fishes to exploitation with phylogenies and life histories. *Conserv Biol* 13:1466–75.

Jentoft, S. 1998. Social science in fisheries management: A risk assessment. In Pitcher, Hart, and Pauly 1998, 177–84.

Jewett, S. C., T. A. Dean, B. R. Woodin, et al. 2002. Exposure to hydrocarbons 10 years after the *Exxon Valdez* oil spill: Evidence from cytochrome P4501A expression and biliary FACs in nearshore demersal fishes. *Mar Environ Res* 54:21–48.

Jingran, V. G., and K. L. Sehgal. 1978. *The cold water fisheries of India*. Barrackpore, West Bengal: Inland Fisheries Society of India. Cited in Welcomme 1988.

Jobling, S., M. Nolan, C. R. Tyler, et al. 1998. Widespread sexual disruption in wild fish. *Environ Sci Technol* 32:2498–2506.

Johannes, R. E. 1977. Traditional law of the sea in Micronesia. *Micronesica* 13:121–27.

———. 1978a. Reproductive strategies of coastal marine fishes in the tropics. *Environ Biol Fish* 3:65–84.

———. 1978b. Traditional marine conservation methods in Oceania and their demise. *Annu Rev Ecol Syst* 9:349–64.

———. 1981. *Words of the lagoon: Fishing and marine lore in the Palau District of Micronesia*. Berkeley: Univ, California Press.

———. 1997. Traditional coral-reef fisheries management. In Birkeland 1997c, 380–85.

———. 1998. The case for data-less marine resource management: Examples from tropical nearshore finfisheries. *Trends Ecol Evol* 13:243–46.

———. 2002. The renaissance of community-based marine resource management in Oceania. *Annu Rev Ecol Syst* 33:317–40.

Johannes, R. E., J. Alberts, C. D'Elia, et al. 1972. The metabolism of some coral reef communities: A team study of nutrient and energy flux at Eniwetok. *BioScience* 22:541–43.

Johannes, R. E., and M. Lam. 1999. The live reef food fish trade in the Solomon Islands. *SPC Live Reef Fish Inform Bull* 5:8–15.

Johannes, R. E., and N. J. Ogburn. 1999. Collecting grouper seed for aquaculture in the Philippines. *SPC Live Reef Fish Inform Bull* 6:35–48.

Johannes, R. E., and M. Riepen. 1995. *Environmental, economic, and social implications of the live reef fish trade in Asia and the Western Pacific*. Arlington, VA: Report to the Nature Conservancy and the Forum Fisheries Agency.

Johannes, R. E., K. Ruddle, and E. Hviding. 1993. People, society, and Pacific islands fisheries development and management. *Inshore Fisheries Researc* 5:107. Noumea: South Pacific Commission.

Johannes, R. E., L. Squire, T. Graham, et al. 1999. *Spawning aggregations of groupers (Serranidae) in Palau*. Marine Conservation Research Series Publication 1. Arlington, VA: The Nature Conservancy. www.tnc.org/asiapacific.

Johnsen, B. O., and A. J. Jensen. 1991. The *Gyrodactylus* story in Norway. *Aquaculture* 98:289–302.

Johnson, D. R., N. A. Funicelli, and J. A. Bohnsack. 1999. Effectiveness of an existing estuarine no-take fish sanctuary within the Kennedy Space Center, Florida. *N Am J Fish Manage* 19:436–53.

Johnson, J. A. 1987. *Protected fishes of the United States and Canada*. Bethesda, MD: American Fisheries Society.

Johnson, J. E., and R. T. Hines. 1999. Effects of suspended sediment on vulnerability of young razorback suckers to predation. *T Am Fish Soc* 128:648–55.

Johnson, J. E., and B. L. Jensen. 1991. History and operation of endangered species hatcheries. In Minckley and Deacon 1991, 197–217.

Johnson, K. A. 2002. *A review of national and international literature on the effects of fishing on benthic habitats*. NOAA Tech. Memo. NMFS-F/SPO-57. www.nmfs.noaa.gov.

Johnston, C. E. 1999. The relationship of spawning mode to conservation of North American minnows (Cyprinidae). *Environ Biol Fish* 55:21–30.

Johnstone, R. 1996. Experience with salmonid sex-reversal and triploidisation technology in the United Kingdom. *B Aquacul Assoc Can* 96(2): 9–13.

Jones, E. B. D. III, G. S. Helfman, J. O. Harper, et al. 1999. The effects of riparian deforestation on fish assemblages in southern Appalachian streams. *Conserv Biol* 13:1454–65.

Jones, J. B. 1992. Environmental impact of trawling on the seabed: A review. *New Zeal J Mar Fresh* 26:59–67.

Jones, J. C., and J. D. Reynolds. 1997. Effects of pollution on reproductive behaviour of fishes. *Rev Fish Biol Fisher* 7:463–91.

Jones, R. J., and O. Hoegh-Guldberg. 1999. Effects of cyanide on coral photosynthesis: Implications for identifying the cause of coral bleaching and for assessing the environmental effects of cyanide fishing. *Mar Ecol—Prog Ser* 177:83–91.

Jones, R. J., and A. L. Steven. 1997. Effects of cyanide on corals in relation to cyanide fishing on reefs. *Mar Freshwater Res* 48:517–22.

Jonsson, B. 1997. Review of ecological and behavioural interactions between cultured and wild Atlantic salmon. *ICES J Mar Sci* 54:1031–39.

Jonsson, E., J. I. Johnsson, and B. T. Bjornsson. 1996. Growth hormone increases predation exposure of rainbow trout. *P Roy Soc Lond B Bio* 262(1370): 647–51.

Judy, R. D. Jr., P. N. Seely, T. M. Murray, et al. 1984. *1982 national fisheries survey. Vol. 1: Tech Report: Initial findings*. FWS/OBS-84/06. Washington, DC.

Junger, S. 1997. *The perfect storm: A true story of men against the sea*. New York: Norton

Jungwirth, M., S. Muhar, and S. Schmutz. 1995. The effects of recreated instream and ecotone structures on the fish fauna of an epipotamal river. *Hydrobiologia* 303:195–206.

Jutila, E. 1985. Dredging of rapids for timber-floating in Finland and its effects on river-spawning fish stock. In *Habitat modification and freshwater fisheries*, ed. J. S. Alabaster, 104–8. Rome: FAO.

Kägi, J. H. R., and A. Schäffer. 1988. Biochemistry of metallotheionein. *Biochemistry* 27:8509–15.

Kaiser, J. 1999. Battle over a dying sea. *Science* 284:28–30.

———. 2000. Bringing the Salton Sea back to life. *Science* 287:565.

———. 2002. The science of *Pfiesteria*: Elusive, subtle, and toxic. *Science* 298:346–49.

Kaiser, M. J. 1998. Significance of bottom-fishing disturbance. *Conserv Biol* 12:1230–1235.

Kaiser, M. J., and S. J. de Groot. 2000. *Effects of fishing on non-target species and habitats: Biological, conservation and socioeconomic issues.* Oxford: Blackwell Science.

Kaiser, M. J., and B. E. Spencer. 1995. Survival of by-catch from a beam trawl. *Mar Ecol—Prog Ser* 126:31–38.

Kanehl, P., and J. Lyons. 1992. *Impacts of in-stream sand and gravel mining on stream habitat and fish communities, including a survey on the Big Rib River, Marathon County, Wisconsin*. Wisconsin Dept. Natural Resources Research Report 155.

Kaplan, E. H. 1999. A center for the preservation of fish species: Application of aquaculture to species survival. *Fresh Mar Aquar* (Jan.): 2–5.

Kappel, C. V. 2005. Losing pieces of the puzzle: Threats to marine, estuarine, and diadromous species. *Front Ecol Environ* 3:275–82.

Kapuscinski, A. R., and E. M. Hallerman. 1990a. AFS position statement on transgenic fishes. *Fisheries* 15(4): 2–5.

———. 1990b. Transgenic fish and public policy: Anticipating environmental impacts of transgenic fish. *Fisheries* 15(1): 2–11.

Karr, J. R. 1981. Assessment of biotic integrity using fish communities. *Fisheries* 6(6): 21–27.

———. 1991. Biological integrity: A long-neglected aspect of water resource management. *Ecol Appl* 1:66–84.

Karr, J. R., and E. W. Chu. 1999. *Restoring life in running waters: Better biological monitoring.* Washington, DC: Island Press.

Karr, J. R., K. D. Fausch, P. L. Angermeier, et al. 1986. *Assessing biological integrity in running waters: A method and its rational.* Illinois Natural History Survey Special Publication 5. Champaign, IL.

Katcher, A. H., E. Friedmann, A. M. Beck, et al. 1983. Looking, talking, and blood pressure: The physiological consequences of interaction with the living environment. In *New perspectives on our lives with companion animals*, ed. A. H. Katcher and A. M. Beck, 351–59. Philadelphia, PA: Univ. Pennsylvania Press.

Kauffman, J. B., R. L. Beschta, N. Otting, et al. 1997. An ecological perspective of riparian and stream restoration in the western United States. *Fisheries* 22(5): 12–24.

Kauffman, J. B., and W. C. Krueger. 1984. Livestock impacts on riparian ecosystems and stream management implications: A review. *J Range Manage* 37:430–37.

Kaufman, L. 1992. Catastrophic change in species-rich freshwater ecosystems: The lessons of Lake Victoria. *BioScience* 42:846–58.

Kautsky, N., H. Berg, C. Folke, et al. 1997. Ecological footprint as a means for the assessment of resource use and development limitations in shrimp and tilapia aquaculture. *Aquac Res* 28:753–66.

Kautsky, N., C. Folke, P. Rönnbäck, et al. 2001. Aquaculture. In *Encyclopedia of biodiversity*, vol. 1, 185–98. San Diego: Academic Press.

Keene, J. L., D. G. Noakes, R. D. Moccia, et al. 1998. The efficacy of clove oil as an anaesthetic for rainbow trout, *Oncorhynchus mykiss* (Walbaum). *Aquac Res* 29:89–101.

Keith, P., and J. Allardi. 1996. Endangered freshwater fish: The situation in France. In Kirchhofer and Hefti 1996a, 35–54.

Kelleher, G., C. Bleakely, and S. Wells, eds. 1995. *A global representative system of marine protected areas.* Washington, DC: Environment Department, World Bank.

Kelley, A. M., C. C. Kohler, and D. R. Tindall. 1992. Are crustaceans linked to the ciguatera food chain? *Environ Biol Fish* 33:275–86.

Kelly, D. J., and M. L. Bothwell. 2002. Avoidance of solar ultraviolet radiation by juvenile coho salmon (*Oncorhynchus kisutch*). *Can J Fish Aquat Sci* 59:474–82.

Kelly, D. J., M. L. Bothwell, D. W. Schindler. 2003. Effects of solar ultraviolet radiation on stream benthic communities: An intersite comparison. *Ecology* 84:2724–40.

Kempinger, J. J., K. J. Otis, and J. R. Ball. 1998. Fish kills in the Fox River, Wisconsin, attributable to carbon monoxide from marine engines. *T Am Fish Soc* 127:669–72.

Kenchington, R. A. 1990. *Managing marine environments.* New York: Taylor and Francis.

Kennedy, C. R. 1993. Introductions, spread and colonization of new localities by fish helminth and crustacean parasites in the British Isles: A perspective and appraisal. *J Fish Biol* 43:287–301.

Kennedy, C., and D. MacKinlay, eds. 2000. *Fish toxicology.* Symposium proceedings. Bethesda, MD: American Fisheries Society.

Kennelly, S. J. 1995. The issue of bycatch in Australia's demersal trawl fisheries. *Rev Fish Biol Fisher* 5:213–34.

Kerasote, T. 1997. Catch and deny. *Orion* (Winter 1997): 24–27.

Kerr, R. A., and C. T. McElroy. 1993. Evidence for large upward trends of ultraviolet-b radiation linked to ozone depletion. *Science* 262:1032–34.

Ketola, H. G., P. R. Bowser, G. A. Wooster, et al. 2000. Effects of thiamine on reproduction of Atlantic salmon and a new hypothesis for their extinction in Lake Ontario. *T Am Fish Soc* 129:607–12.

Kidd, A. H., and R. M. Kidd. 1997. Aquarium visitors' perceptions and attitudes toward the importance of marine biodiversity. *Psychol Rep* 81:1083–88.

———. 1999. Benefits, problems and characteristics of home aquarium owners. *Psychol Rep* 84:998–1004.

Kime, D. E. 1993. Classical and nonclassical reproductive steroids in fish. *Rev Fish Biol Fisher* 3:160–80.

———. 1995. The effects of pollution on reproduction in fish. *Rev Fish Biol Fisher* 5:52–95.

———. 1999. A strategy for assessing the effects of xenobiotics on fish reproduction. *Sci Total Environ* 225:3–11.

Kimmel, B. L., and A. W. Groeger. 1980. Limnological and ecological changes associated with reservoir aging. In *Reservoir fisheries: Management strategies for the 80's*, ed. G. E. Hall and M. J. Van Den Avyle, 103–9. Bethesda, MD: American Fisheries Society.

King, F. W. 1987. Thirteen milestones on the road to extinction. In *The road to extinction*, ed. R. Fitter and M. Fitter, 7–18. Cambridge, UK: IUCN/UNEP.

King, M., and U. Faasili. 1999. Community-based management of subsistence fisheries in Samoa. *Fisheries Manage Ecol* 6:133–44. Cited in Johannes 2002.

Kipling, R. 1899. *Captains courageous: A story of the Grand Banks.* New York: Century.

Kirchhofer, A. 1996. Fish conservation in Switzerland: Three case studies. In Kirchhofer and Hefti 1996a, 135–46.

Kirchhofer, A., and D. Hefti, eds. 1996a. *Conservation of endangered freshwater fish in Europe*. Basel, Switzerland: Birkhauser Verlag.

———. 1996b. Introduction. In Kirchhofer and Hefti 1996a, ix–xii.

Kirk, J. T. O. 1983. *Light and photosynthesis in aquatic ecosystems*. Cambridge, UK: Cambridge Univ. Press.

Kiryu, Y., J. D. Shields, W. K. Vogelbein, et al. 2002. Induction of skin ulcers in Atlantic menhaden by injection and aqueous exposure to the zoospores of *Aphanomyces invadans*. *J Aquat Anim Health* 14:11–24.

Kitamura, S., and K. Ikuta. 2001. Effects of acidification on salmonid spawning behavior. *Water Air Soil Poll* 130:875–80.

Kitchell, J. F., ed. 1992. *Food web management: A case study of Lake Mendota*. New York: Springer-Verlag.

Kitchell, J. F., D. E. Schindler, R. Ogutu-Ohwayo, et al. 1997. The Nile perch in Lake Victoria: Interaction between predation and fisheries. *Ecol Appl* 7:653–64.

Knutsen, G. M., and S. Tilseth. 1985. Growth, development and feeding success of Atlantic cod *Gadus morhua* larvae related to egg size. *T Am Fish Soc* 114:507–11.

Kock, K. H. 2000. A brief description of the main species exploited in the Southern Ocean, annex 1 to *Understanding CCAMLR's approach to management*. www.ccamlr.org.

Kocovsky, P. M., C. Gowan, K. D. Fausch, et al. 1997. Spinal injury rates in three wild trout populations in Colorado after eight years of backpack electrofishing. *N Am J Fish Manage* 17:308–13.

Kodric-Brown, A., and J. Brown. 2004. Disturbance is essential for the conservation of desert springs. Abstract, Biennial Conference on the Ecological and Evolutionary Ethology of Fishes, Aug., Saudarkrokur, Iceland.

Koebel, J. W. Jr. 1995. An historical perspective on the Kissimmee River restoration project. *Restor Ecol* 3:149–59.

Koenig, C. C., and P. L. Colin. 1999. Distribution, abundance and survival of juvenile gag grouper, *Mycteroperca microlepis* (Pisces: Serranidae), in seagrass beds of the northeastern Gulf of Mexico. *P Gulf Caribb Fish Inst* 45:37–54.

Kohler, C. C., and J. C. Stanley. 1984. A suggested protocol for evaluating proposed exotic fish introductions in the United States. In *Distribution, biology, and management of exotic fishes*, ed. W. R. Courtenay Jr. and J. R. Stauffer Jr., 387–406. Baltimore: Johns Hopkins Univ. Press.

Kohler, S. T., and C. C. Kohler. 1992. Dead bleached coral provides new surfaces for dinoflagellates implicated in ciguatera fish poisonings. *Environ Biol Fish* 35:413–16.

Kolar, C. S., and D. M. Lodge. 2001. Progress in invasion biology: Predicting invaders. *Trends Ecol Evol* 16:199–204.

———. 2002. Ecological predictions and risk assessment for alien fishes in North America. *Science* 298:1233–36.

Kondolf, G. M. 1994. Livestock grazing and habitat for a threatened species: Land-use decisions under scientific uncertainty in the White Mountains, California, USA. *Environ Manage* 18:501–9.

———. 1995. Five elements for effective evaluation of stream restoration. *Restor Ecol* 3:133–36.

———. 1997. Hungry water: Effects of dams and gravel mining on river channels. *Environ Manage* 21:533–51.

Kornfield, I., and K. E. Carpenter. 1984. Cyprinids of Lake Lanao, Philippines: Taxonomic validity, evolutionary rates and speciation scenarios. In *Evolution of fish species flocks*, ed. A. E. Echelle and I. Kornfield, 69–84. Orono: Univ. Maine Press.

Koslow, J. A., G. W. Boehlert, J. D. M. Gordon, et al. 2000. Continental slope and deep-sea fisheries: Implications for a fragile ecosystem. *ICES J Mar Sci* 57:548–57.

Koslow, J. A., F. Hanley, and R. Wicklund. 1988. Effects of fishing on reef fish communities at Pedro Bank and Port Royal cays, Jamaica. *Mar Ecol—Prog Ser* 43:201–12.

Kottcamp, G. M., and P. B. Moyle. 1972. Use of disposable beverage cans by fish in the San Joaquin Valley. *T Am Fish Soc* 101:566.

Kottelat, M. 1997. European freshwater fishes. *Biologia* 52(suppl. 5): 1–271.

———. 1998. Systematics, species concepts and the conservation of freshwater fish diversity in Europe. *Ital J Zool* 65(suppl.): 65–72.

———. 2000. Overview of the conservation status of European freshwater fishes. Abstract, International Symposium on Freshwater Fish Conservation: Options for the Future, Albufeira, Portugal, 20.

Kovacs, T., S. Gibbons, B. O'Connor, et al. 2004. Summary of case studies investigating the causes of pulp and paper mill effluent regulatory toxicity. *Water Qual Res J Can* 39:93–102.

Kozyreva, A. 1995. Caviar business: If measures are not taken, the king of fishes will disappear. *Rossiiskaya Gazeta*, Moscow, Russia, June 26. Reported in *The Sturgeon Quarterly* 3(3): 12.

Kramer, D. E., and V. M. O'Connell. 1995. *Guide to northeast Pacific rockfishes.* Alaska Sea Grant Marine Advisory Bull. 25, Fairbanks.

Kramer, D. L., and M. R. Chapman. 1999. Implications of fish home range size and relocation for marine reserve function. *Environ Biol Fish* 55:65–79.

Krimsky, S. 2000. *Hormonal chaos: The scientific and social origins of the environmental endocrine hypothesis.* Baltimore: Johns Hopkins Univ. Press.

Krom, M. D., S. Ellner, J. Vanrihn, et al. 1995. Nitrogen and phosphorus cycling and transformations in a prototype nonpolluting integrated mariculture system, Eilat, Israel. *Mar Ecol—Prog Ser* 118:25–36.

Krümmel, E. M., R. W. Macdonald, L. E. Kimpe, et al. 2003. Delivery of pollutants by spawning salmon: Fish dump toxic industrial compounds in Alaskan lakes on their return from the ocean. *Nature* 425:255–56.

Kuang, H. K. 1999. Hong Kong. Appendix IV.1, pp. 295–326. In Vannuccini 1999.

Kulik, G. 1985. Dams, fish, and farmers: The defense of public rights in eighteenth century Rhode Island. In *The countryside in the age of capitalist transformation: Essays on the social history of rural America*, ed. S. Hahn and J. Prude, 25–50. Chapel Hill: Univ. North Carolina Press.

Kurlansky, M. 1997. *Cod: A biography of the fish that changed the world.* New York: Walker.

Kvist, L. P., S. Gram, C. Cacares, et al. 2001. Socio-economy of flood plain households in the Peruvian Amazon. *Forest Ecol Manage* 150:175–86.

KWUA (Klamath Water Users Association). 2003. "Fish kill" vs. "Fish die-off." *KWUA Weekly Update*, July 25. www.klamathbasincrisis.org/articles/KWUA-Newsletter/NL-2003/kwuan1072503.htm.

Kynard, B. 1997. Life history, latitudinal patterns, and status of the shortnose sturgeon, *Acipenser brevirostrum*. *Environ Biol Fish* 48:319–34.

LaChat, M. R. 1996. An argument in defense of fishing. *Fisheries* 21(7): 20–21.

Lachner, E. A., C. R. Robins, and W. R. Courtenay Jr. 1970. *Exotic fishes and other aquatic organisms introduced into North America.* Washington, DC: Smithsonian Institution Press.

Lack, M., and G. Sant. 2001. Patagonian toothfish: Are conservation and trade measures working? *TRAFFIC Bull* 19(1): 1–21.

Lackey, R. T. 2003. Pacific Northwest salmon: Forecasting their status in 2100. *Rev Fish Sci* 11:35–88.

Laegdsgaard, P., and C. R. Johnson. 2001. Why juvenile fish utilise mangrove habitats? *J Exp Mar Biol Ecol* 257:229–53.

Laevastu, T. 1993. *Marine climate, weather and fisheries.* Oxford: Fishing News Books.

Lafferty, K. D., and A. M. Kuris. 1999. How environmental stress affects the impacts of parasites. *Limnol Oceanogr* 44:925–31.

Laikre, L., A. Antunes, P. Alexandrino, et al. 1999. *Conservation genetic management of brown trout (*Salmo trutta*) in Europe.* TROUTCONCERT, EU FAIR CT97-3882. Silkeborg, Denmark: Dept. Inland Fisheries.

Lalas, C., and C. J. A. Bradshaw. 2001. Folklore and chimerical numbers: Reviews of a millennium of interaction between fur seals and humans in the New Zealand region. *New Zeal J Mar Fresh* 35:477–97.

Landsberg, J. H. 1995. Tropical reef-fish disease outbreaks and mass mortalities in Florida, USA: What is the role of dietary biological toxins? *Dis Aquat Organ* 22:83–100.

———. 2002. The effects of harmful algal blooms on aquatic organisms. *Rev Fish Sci* 10:113–390.

Langdon, J. S. 1990. Disease risks of fish introductions and translocations. In *Introduced and translocated fishes and their ecological effects*, ed. D. A. Pollard, 98–107. Australian Bureau of Rural Resources Proceedings 8. Canberra: Australian Government Publishing Service.

Lappalainen, A., and L. Pesonen. 2000. Changes in fish community structure after cessation of waste water discharge in a coastal bay area west of Helsinki, northern Baltic Sea. *Arch Fish Marine Res* 48:226–41.

Larinier, M., F. Travade, and J. P. Porcher. 2002. *Fishways: Biological basis, design criteria, and monitoring.* Rome: FAO.

Larkin, P. A. 1977. An epitaph for the concept of maximum sustained yield. *T Am Fish Soc* 106:1–11.

———. 1979. Predator-prey relations in fishes: An overview of the theory. In *Predator-prey systems in fisheries management*, ed. H. Clepper, 13–20. Washington, DC: Sport Fishing Institute.

Laroche, J., and J. D. Durand. 2004. Genetic structure of fragmented populations of a threatened endemic percid of the Rhone river: *Zingel asper. Heredity* 92:329–34.

Lassuy, D. R. 1995. Introduced species as a factor in extinction and endangerment of native fish species. *Am Fish Soc Sym* 15:391–96.

Last, P. R., E. O. G. Scott, and F. H. Talbot. 1983. *Fishes of Tasmania.* Hobart: Tasmanian Fisheries Development Authority.

Lau, P. P. F., and R. Parry-Jones. 1999. *The Hong Kong trade in live reef fish for food.* Hong Kong: TRAFFIC East Asia and World Wide Fund for Nature.

Law, R. 2000. Fishing, selection, and phenotypic evolution. *ICES J Mar Sci* 57:659–68.

Law, R., and K. Stokes. 2005. Evolutionary impacts of fishing on target populations. In Norse and Crowder 2005, 232–46.

Lawler, J. J., J. E. Aukema, J. B. Grant, et al. 2006. Conservation science: A 20-year report card. *Front Ecol Environ* 4:473–80.

Lea, R. N., R. D. McAllister, and D. A. VenTresca. 1999. Biological aspects of nearshore rockfishes of the genus *Sebastes* from central California. *Cal Dept Fish Game Fish B* 177.

Leary, R. F., F. W. Allendorf, and G. K. Sage. 1995. Hybridization and introgression between introduced and native fish. *Am Fish Soc Sym* 14:91–101.

Leber, K. M., N. P. Brennan, and S. M. Arce. 1995. Marine enhancement with striped mullet: Are hatchery releases replenishing or displacing wild stocks? *Am Fish Soc Sym* 15:376–87.

Lee, A., and R. Langer. 1983. Shark cartilage contains inhibitors of tumor angiogenesis. *Science* 221:1185–87.

Lee, C., and Y. Sadovy. 1998. A taste for live fish: Hong Kong's live reef fish market. *Naga ICLARM Q* (Apr.-June): 38–42.

Lee, C. S., and A. C. Ostrowski. 2001. Current status of marine finfish larviculture in the United States. *Aquaculture* 200:89–109.

Lee, D. S., C. R. Gilbert, C. H. Hocutt, et al. 1980. *Atlas of North American freshwater fishes.* Raleigh: NC State Museum of Natural History.

Lee, R. F., and D. S. Page. 1997. Petroleum hydrocarbons and their effects in subtidal regions after major oil spills. *Mar Pollut Bull* 34:928–40.

Leet, W. S., C. M. Dewees, and C. W. Haugen, eds. 1992. *California's living marine resources and their utilization.* California Sea Grant Report UCSGEP-92–12. Cited in Speer et al. 1997.

Leggett, W. C., and J. E. Carscadden. 1978. Latitudinal variation in reproductive characteristics of American shad (*Alosa sapidissima*): Evidence for population specific life history strategies in fish. *J Fish Res Board Can* 35:1469–78.

Lehane, L., and G. T. Rawlin. 2000. Topically acquired bacterial zoonoses from fish: A review. *Med J Australia* 173:256–59.

Leidy, R. A., and P. B. Moyle. 1997. Conservation status of the world's fish fauna: An overview. In *Conservation biology for the coming decade*, ed. P. L. Fiedler and P. M. Kareiva, 187–227. New York: Chapman and Hall.

Leis, J. M., and M. I. McCormick. 2002. The biology, behavior and ecology of the pelagic, larval stages of coral reef fishes. In Sale 2002a, 171–99.

Lelek, A. 1987. *The freshwater fishes of Europe. Vol. 9: Threatened fishes of Europe.* Wiesbaden, Germany: AULA-Verlag.

Lemly, A. D. 1985. Suppression of native fish population by green sunfish in first order streams of piedmont North Carolina. *T Am Fish Soc* 114:705–12.

Lenat, D. R., and J. K. Crawford. 1994. Effects of land use on water quality and aquatic biota of three North Carolina piedmont streams. *Hydrobiologia* 294:185–99.

Leopold, A. 1949. *A Sand County almanac and sketches here and there.* New York: Oxford Univ. Press.
———. 1953. *Round River: From the journals of Aldo Leopold.* New York: Oxford Univ. Press.
Lessa, R. F., M. Santana, and R. Paglerani. 1999. Age, growth, and stock structure of the oceanic white tip shark, *Carcharhinus longimanus*, from the southwestern equatorial Atlantic. *Fish Res* 42:21–30.
Lever, C. 1992. *They dined on eland: The story of the acclimatization societies.* London: Quiller Press.
———. 1996. *Naturalized fishes of the world.* San Diego: Academic Press.
Levin, P. S., S. Achord, B. E. Feist, et al. 2002. Non-indigenous brook trout and the demise of Pacific salmon: A forgotten threat? *P Roy Soc Lond B Bio* 269:1663–70.
Levin, P. S., and C. B. Grimes. 2002. Reef fish ecology and grouper conservation and management. In Sale 2002a, 377–89.
Levin, P. S., R. W. Zabel, and J. G. Williams. 2001. The road to extinction is paved with good intentions: Negative association of fish hatcheries with threatened salmon. *P Roy Soc Lond B Bio* 268:1153–58.
Levitan, D. R. 1992. Community structure in times past: Influence of human fishing pressure on algal-urchin interactions. *Ecology* 73:1597–1605.
Levitan, D. R., and T. M. McGovern. 2005. The Allee effect in the sea. In Norse and Crowder 2005, 47–57.
Lewis, N. D. 1986. Epidemiology and impact of ciguatera in the Pacific: A review. *Mar Fish Rev* 48(4): 6–13.
Lewis, R. J. 2001. The changing face of ciguatera. *Toxicon* 39:97–106.
Lewis, R. J., and T. A. Ruff. 1993. Ciguatera: Ecological, clinical, and socioeconomic perspectives. *Crit Rev Env Sci Tec* 23:137–56.
Lewynsky, V. A., and T. C. Bjornn. 1987. Response of cutthroat and rainbow trout to experimental catch-and-release fishing. In *Catch-and-release fishing: A decade of experience*, ed. R. Barnhart and T. Roelofs, 16–32. Arcata: California Cooperative Fish Research Unit.
Lichatowich, J. 1999. *Salmon without rivers: A history of the Pacific salmon crisis.* Washington, DC: Island Press.
Lichatowich, J. A., M. A. Powell, and R. N. Williams. 2006. Artificial production and the effects of fish culture on native salmonids. In R. N. Williams 2006, 417–63.
Liddell, W. D., and S. O. Ohlhorst. 1993. Ten years of disturbance and change on a Jamaican fringing reef. *P 7th Int Coral Reef Sym* 1:144–50.
Ligtvoet, W., and F. Witte. 1991. Perturbation through predator introduction: Effects on the food web and fish yields in Lake Victoria (East Africa). In *Perturbation and recovery of terrestrial and aquatic ecosystems*, ed. O. Ravera, 263–68. Chichester, England: Elliss Horwood.
Likens, G. E., F. H. Bormann, and N. M. Johnson. 1972. Acid rain. *Environment* 14(2): 33–40.
Likens, G. E., F. H. Bormann, R. S. Pierce, et al. 1977. *Biogeochemistry of a forested ecosystem.* New York: Springer-Verlag.
Likens, G. E., R. F. Wright, J. N. Galloway, et al. 1979. Acid rain. *Sci Am* 241(4): 44–51.
Lima, S. L., and L. M. Dill. 1990. Behavioral decisions made under the risk of predation: A review and prospectus. *Can J Zool* 68:619–40.
Lindberg, W. J. 1997. Can science resolve the attraction-production issue? *Fisheries* 22(4): 10–13.
Lindeboom, H. J., and S. J. de Groot. 1998. *The effect of different types of fisheries on the North Sea and Irish Sea benthic ecosystems.* Report 1998-1. Den Burg, Texel: Netherlands Institute of Sea Research. Cited in Jennings and Kaiser 1998.
Linhart, O., M. Rodina, and J. Cosson. 2000. Cryopreservation of sperm in common carp *Cyprinus carpio*: Sperm motility and hatching success of embryos. *Cryobiology* 41:241–50.
Link, J. S. 2002a. Ecological considerations in fisheries management: When does it matter? *Fisheries* 27(4): 10–17.
———. 2002b. What does ecosystem-based fisheries management mean? *Fisheries* 27(4): 18–21.
Lintermans, M. 2000. Recolonization by the mountain galaxias *Galaxias olidus* of a montane stream after eradication of rainbow trout *Oncorhynchus mykiss*. *Mar Freshwater Res* 51:799–804.
Lipcius, R. N., L. B. Crowder, and L. E. Morgan. 2005. Metapopulation structure and marine reserves. In Norse and Crowder 2005, 328–45.
Liss, W. J., J. A. Stanford, J. A. Lichatowich, et al. 2006. A foundation for restoration. In R. N. Williams 2006, 51–98. New York: Elsevier Academic Press.
Litvak, M. K., and N. E. Mandrak. 1993. Ecology of freshwater baitfish use in Canada and the United States. *Fisheries* 18(12): 6–13.

Lloyd, D. S. 1987. Turbidity as a water quality standard for salmonid habitats in Alaska. *N Am J Fish Manage* 7:34–45.

Lloyd, L. N., A. H. Arthington, and D. A. Milton. 1986. The mosquitofish: A valuable mosquito-control agent or a pest. In *The ecology of exotic animals and plants: Some Australian case histories*, ed. R. L. Kitching, 7–25. Queensland: Wiley.

Lockwood, J. L., and M. L. McKinney, eds. 2001. *Biotic homogenization: The loss of diversity through invasion and extinction*. New York: Kluwer.

Lodge, D. M. 1993. Biological invasions: Lessons from ecology. *Trends Ecol Evol* 8:133–37.

Lodge, D. M., C. A. Taylor, D. M. Holdich, et al. 2000. Reducing impacts of exotic crayfish introductions: New policies needed. *Fisheries* 25(8): 121–23.

LOE (Living on Earth). 2003. Whistleblower faces the ax. National Public Radio, Nov. 14. www.loe.org.

Loftus, W. F. 1989. Distribution and ecology of exotic fishes in Everglades National Park. In *Proceedings of the 1986 Conference on Science in National Parks. Vol. 5: Management of exotic species in natural communities*, ed. L. K. Thomas, 24–34. Washington, DC: U.S. National Park Service.

Logan, S. H., W. E. Johnston, and S. I. Doroshov. 1995. Economics of joint production of sturgeon (*Acipenser transmontanus* Richardson) and roe for caviar. *Aquaculture* 130: 299–316.

Loiselle, P. V. 2003. Captive breeding for the freshwater fishes of Madagascar. In *The natural history of Madagascar*, ed. S. M. Goodman and J. P. Benstead, 159–74. Chicago: Univ. Chicago Press.

Loomis, J. B. 1996. Measuring the economic benefits of removing dams and restoring the Elwha River: Results of a contingent valuation survey. *Water Resour Res* 32:441–47.

Lorenz, J. J. 1999. The response of fishes to physicochemical changes in the mangroves of northeast Florida Bay. *Estuaries* 22(2B): 500–17.

Lorion, C. M., D. F. Markle, S. B. Reid, et al. 2000. Re-description of the presumed-extinct Miller Lake lamprey, *Lampetra minima*. *Copeia* 2000:1019–28.

Love, M. S., M. Yoklavich, and L. K. Thorsteinson. 2002. *The rockfishes of the northeast Pacific*. Berkeley: Univ. California Press.

Lovely, J. E., B. H. Dannevig, K. Falk, et al. 1999. First identification of infectious salmon anaemia virus in North America with haemorrhagic kidney syndrome. *Dis Aquat Organ* 35:145–48.

Low, L.-L., J. N. Ianelli, and S. A. Lowe. 1999. Alaska groundfish fisheries. Unit 19 in NMFS 1999.

Lowe, C. 2002. Who is to blame? Logics of responsibility in the live reef food fish trade in Sulawesi, Indonesia. *SPC Live Reef Fish Inform Bull* 10:7–16.

Lowe-McConnell, R. M. 1987. *Ecological studies in tropical fish communities*. Cambridge: Cambridge Univ. Press.

———. 1997. EAFRO and after: A guide to key events affecting fish communities in Lake Victoria (East Africa). *S Afr J Sci* 93:570–74.

Lozano-Vilano, M. de L., and S. Contreras-Balderas. 1993. Four new species of *Cyprinodon* from southern Nuevo León, Mexico, with a key to the *C. eximius* complex (Teleostei: Cyprinodontidae). *Ichthyol Explor Fresh* 4:295–308.

Lubbock, H. R., and N. V. C. Polunin. 1975. Conservation and the tropical marine aquarium trade. *Environ Conserv* 2:229–32.

Lubchenco, J., A. M. Olson, L. B. Brubaker, et al. 1991. The Sustainable Biosphere Initiative: An ecological research agenda. *Ecology* 72:371–412.

Lucas, M. C., and E. Baras. 2001. *Migration of freshwater fishes*. Malden, MA: Blackwell Science.

Luck, G. W., G. C. Daily, and P. R. Ehrlich. 2003. Population diversity and ecosystem services. *Trends Ecol Evol* 18:331–36.

Ludwig, D., R. Hilborn, and C. Walters. 1993. Uncertainty, resource exploitation, and conservation: Lessons from history. *Science* 260:17, 36.

Ludwig, H. R. Jr., and J. A. Leitch. 1996. Interbasin transfer of aquatic biota via anglers' bait buckets. *Fisheries* 21(7): 14–18.

Lundberg, J. G., M. Kottelat, G. R. Smith, et al. 2000. So many fishes, so little time: An overview of recent ichthyological discoveries in fresh waters. *Ann Missouri Bot Gard* 87:26–62.

Lunn, K. E., and M.-A. Moreau. 2002. Conservation of Banggai cardinalfish populations in Sulawesi, Indonesia: An integrated research and education project. *SPC Live Reef Fish Inform Bull* 10:33–34.

Lusk, S. 1996. The status of the fish fauna in the Czech Republic. In Kirchhofer and Hefti 1996, 89–98.

Lusk, S., V. Luskova, K. Halacka, et al. 2002. Status and protection of species and intraspecific diversity of the ichthyofauna in the Czech Republic. In Collares-Pereira, Coelho, and Cowx 2002, 24–33.

Lussen, A., T. M. Falk, and W. Villwock. 2003. Phylogenetic patterns in populations of Chilean species of the genus *Orestias* (Teleostei: Cyprinodontidae): Results of mitochondrial DNA analysis. *Mol Phylogenet Evol* 29:151–60.

Lutcavage, M. E., R. W. Brill, G. B. Skomal, et al. 1999. Results of pop-up satellite tagging of spawning size class fish in the Gulf of Maine: Do North Atlantic bluefin tuna spawn in the mid-Atlantic? *Can J Fish Aquat Sci* 56:173–77.

Lutz, D. S. 1995. Gas supersaturation and gas bubble trauma in fish downstream from a midwestern reservoir. *T Am Fish Soc* 124:423–36.

Lydeard, C., and R. L. Mayden. 1995. A diverse and endangered aquatic ecosystem of the southeast United States. *Conserv Biol* 9:800–5.

Lyons, J., G. Gonzalez-Hernandez, E. Soto-Galera, et al. 1998. Decline of freshwater fishes and fisheries in selected drainages of west-central Mexico. *Fish Manage* 23:10–18.

Lyons, J., S. Navarro-Perez, P. A. Cochran, et al. 1995. Index of Biotic Integrity based on fish assemblages for the conservation of streams and rivers in west-central Mexico. *Conserv Biol* 9:569–84.

Lyons, J., L. Wang, and T. D. Simonson. 1996. Development and validation of an Index of Biotic Integrity for coldwater streams in Wisconsin. *N Am J Fish Manage* 16:241–256.

Lytle, D. A., and N. L. Poff. 2004. Adaptation to natural flow regimes. *Trends Ecol Evol* 19:94–100.

MacAvoy, S. E., and A. J. Bulger. 1995. Survival of brook trout (*Salvelinus fontinalis*) embryos and fry in streams of different acid sensitivity in Shenandoah National Park, USA. *Water Air Soil Poll* 85:445–50.

MacCrimmon, H. R., and T. L. Marshall. 1968. World distribution of brown trout, *Salmo trutta*. *J Fish Res Board Can* 25:2527–48.

Mace, G., ed. 1999. *The IUCN criteria review: Report of the marine workshop*. Gland, Switzerland: IUCN.

Mace, G. M., and E. J. Hudson. 1999. Attitudes toward sustainability and extinction. *Conserv Biol* 13:242–46.

Mace, P. M. 1997. The status of ICCAT species relative to optimum yield and overfishing criteria recently proposed in the United States, also with consideration of the precautionary approach. ICCAT working document SCRS/97/74. Cited in Safina 2001b.

Machordom, A., J. L. Garcia-Marine, N. Sanz, et al. 1999. Allozyme diversity in brown trout (*Salmo trutta*) from central Spain: Genetic consequences of restocking. *Freshwater Biol* 41:707–17.

MacIntyre, F., K. W. Estep, and T. T. Noji. 1995. Is it deforestation or desertification when we do it to the oceans? *Naga ICLARM Q* 18(3): 4–6. Cited in Pauly 1995.

Mackinson, S. 2001. Integrating local and scientific knowledge: An example in fisheries science. *Environ Manage* 27:533–45.

Macphail, E. M. 1998. *The evolution of consciousness*. New York: Oxford Univ. Press.

Maes, J., A. W. H. Turnpenny, D. R. Lambert, et al. 2004. Field evaluation of a sound system to reduce estuarine fish intake rates at a power plant cooling water inlet. *J Fish Biol* 64:938–46.

Maggio, T. 2000. *Mattanza: Love and death in the Sea of Sicily*. Cambridge, MA: Perseus.

Magilligan, F. J., and P. F. McDowell. 1997. Stream channel adjustments following elimination of cattle grazing. *J Am Water Resour Assoc* 33:867–78.

Magnien, R. E. 2001. The dynamics of science, perception, and policy during the outbreak of *Pfiesteria* in the Chesapeake Bay. *BioScience* 51:843–52.

Magnuson, J. J. 2002. Signals from ice cover trends and variability. *Am Fish Soc Sym* 32:3–13.

Magnuson, J. J., and B. T. DeStasio. 1997. Thermal niche of fishes and global warming. In *Global warming: Implications for freshwater and marine fish*, ed. C. M. Wood and D. G. McDonald, 377–408. SEB Seminar Series. Cambridge: Cambridge Univ. Press.

Magnuson, J. J., J. D. Meisner, and D. K. Hill. 1990. Potential changes in the thermal habitat of Great Lakes fish after global climate warming. *T Am Fish Soc* 119:254–64.

Magnuson, J. J., C. Safina, and M. P. Sissenwine. 2001. Whose fish are they anyway? *Science* 293:1267–68.

Main, J., and G. I. Sangster. 1985. Trawling experiments with a 2-level net to minimize the undersized gadoid by-catch in a *Nephrops* fishery. *Fish Res* 3:131–45.

Maitland, P. S. 1995. The conservation of freshwater fish: Past and present experience. *Biol Conserv* 72:259–70.

Maitland, P. S., and A. A. Lyle. 1991. Conservation of freshwater fish in the British Isles: The current status and biology of threatened species. *Aquat Conserv* 1:25–54.

———. 1996. Threatened freshwater fishes of Great Britain. In Kirchhofer and Hefti 1996, 9–21.

Majewski, A. R., P. J. Blanchfield, V. P. Palace, et al. 2002. Waterborne 17alpha-ethynylestradiol affects aggressive behaviour of male fathead minnows (*Pimephales promelas*) under artificial spawning conditions. *Water Qual Res J Can* 37:697–710.

Mallin, M. A. 2000. Impacts of industrial animal production on rivers and estuaries. *Am Sci* 88: 26–37.

Mallin, M. A., and L. B. Cahoon. 2003. Industrialized animal production: A major source of nutrient and microbial pollution to aquatic ecosystems. *Popul Environ* 24:369–85.

Manchester, S. J., and J. M. Bullock. 2000. The impacts of non-native species on UK biodiversity and the effectiveness of control. *J Appl Ecol* 37:845–64.

Mandrak, N. E. 1989. Potential invasions of the Great Lakes by fish species associated with climate warming. *J Great Lakes Res* 15:306–16.

Manigold, D. B., and J. A. Schulze. 1969. Pesticides in selected western streams, a progress report. *Pestic Monit J* 3:124–35.

Manire, C. A., and S. H. Gruber. 1990. Many sharks may be headed toward extinction. *Conserv Biol* 4:10–11.

Mann, D. A., D. M. Higgs, W. N. Tavolga, et al. 2001. Ultrasound detection by clupeiform fishes. *J Acoust Soc Am* 109:3048–54.

Manning, N. J., and D. E. Kime. 1984. Temperature regulation of ovarian steroid production in the common carp, *Cyprinus carpio* L., in vivo and in vitro. *Gen Comp Endocr* 56:376–88.

———. 1985. The effect of temperature on testicular steroid production in the rainbow trout, *Salmo gairdneri*, in vitro and in vivo. *Gen Comp Endocr* 57:377–82.

Mantua, N. J., S. R. Hare, Y. Qhang, et al. 1997. A Pacific interdecadal climate oscillation with impacts on salmon production. *B Am Meteorol Soc* 78:1069–79.

March, J. G., J. P. Benstead, C. M. Pringle, et al. 2003. Damming tropical island streams: Problems, solutions, and alternatives. *BioScience* 53:1069–78.

Marchetti, M. P. 1999. An experimental study of competition between the native Sacramento perch (*Archoplites interruptus*) and introduced bluegill (*Lepomis macrochirus*). *Biol Invasions* 1:55–65.

Marchetti, M. P., T. Light, J. Feliciano, et al. 2001. Homogenization of California's fish fauna through abiotic change. In Lockwood and McKinney 2001, 259–78.

Marchetti, M. P., and P. B. Moyle. 2001. Effects of flow regime on fish assemblages in a regulated California stream. *Ecol Appl* 11:530–39.

Mareth, M., N. Bonheur, and B. D. Lane. 2001. *Biodiversity conservation and social justice in the Tonle Sap watershed: The Tonle Sap Biosphere Reserve*. International Conference on Biodiversity and Society, UNESCO/Columbia University Earth Institute, May 22–25. www.earthscape.org/r1/cbs01/cbs01a13aa.html.

Mari, S., J. Labonne, and P. Gaudin. 2002. A conservation strategy for *Zingel asper*, a threatened endemic percid of the Rhone Basin. In Collares-Pereira, Coelho, and Cowx 2002, 149–56.

Marini, F. C. 1996. My notes and observations on raising and breeding the Banggai cardinalfish. *J MaquaCulture* 4(4). http://saltaquarium.about.com.

Marsh, P. C., and J. E. Brooks. 1989. Predation by ictalurid catfishes as a deterrent to re-establishment of hatchery-reared razorback suckers. *Southwest Nat* 34:188–95.

Marsh, P. C., and M. E. Douglas. 1997. Predation by introduced fishes on endangered humpback chub and other native species in the Little Colorado River, Arizona. *T Am Fish Soc* 126:343–46.

Martin, C. 1995. The collapse of the northern cod stocks: Whatever happened to 86/25? *Fisheries* 20(5): 6–8.

Martinez, R., J. Juncal, C. Zaldívar, et al. 2000. Growth efficiency in transgenic tilapia (*Oreochromis* sp.) carrying a single copy of an homologous cDNA growth hormone. *Biochem Bioph Res Co* 267:466–72.

Martin-Smith, K. M. 1998. Effects of disturbance caused by selective timber extraction on fish communities in Sabah, Malaysia. *Environ Biol Fish* 53:155–67.

Marty, G. D., A. Hoffmann, M. S. Okihiro, et al. 2003. Retrospective analysis: Bile hydrocarbons and histopathology of demersal rockfish in Prince William Sound, Alaska, after the *Exxon Valdez* oil spill. *Mar Environ Res* 56:569–84.

Maruyama, T., and J. Hiratsuka. 1992. *Vanishing fishes of Japan*. Osaka: Freshwater Fish Protection Association of Japan. Cited in Bruton 1995.

Maser, C., and J. Sedell. 1994. *From the forest to the sea: The ecology of wood in streams, rivers, estuaries, and oceans*. Delray Beach, FL: St. Lucie Press.

Master, L. 1990. The imperiled status of North American aquatic animals. *Biodivers Net News* 3:1–2, 7–8.

Masuda, R., and K. Tsukamoto. 1998. Stock enhancement in Japan: Review and perspective. *B Mar Sci* 62:337–58.

Matern, S. A. 2001. Using temperature and salinity tolerances to predict the success of the shimofuri goby, a recent invader into California. *T Am Fish Soc* 130:592–99.

Matern, S. A., and K. J. Fleming. 1995. Invasion of a third Asian goby, *Tridentiger bifasciatus*, into California. *Calif Fish Game* 81:71–76.

Mathias, J. A., A. T. Charles, and B. Hu. 1997. *Integrated fish farming*. Boca Raton, FL: Lewis.

Matthews, W. J. 1997. *Patterns in freshwater fish ecology*. New York: Chapman and Hall.

Matthews, W. J., and E. G. Zimmerman. 1990. Potential effects of global warming on native fishes of the southern Great Plains and the Southwest. *Fisheries* 15(6): 26–32.

Mattingly, R. L., E. E. Herricks, and D. M. Johnston. 1993. Channelization and levee construction in Illinois: Review and implications for management. *Environ Manage* 17:781–95.

Mattson, N. S., K. Buakhamvongsa, N. Sukumasavin, et al. 2002. Management and preservation of the giant fish species of the Mekong. In *Cold water fisheries in the Trans-Himalayan countries*, ed. T. Petr and D. B. Swar. Fisheries Tech. Paper 431. Rome: FAO.

Mayden, R. L., and B. R. Kuhajda. 1996. Systematics, taxonomy, and conservation status of the endangered Alabama sturgeon, *Scaphirhynchus suttkusi* Williams and Clemmer (Actinopterygii, Acipenseridae). *Copeia* 1996:241–73.

McAllister, D. E. 1969. Introduction of tropical fishes into a hotspring near Banff, Alberta. *Can Field Nat* 83:31–35.

———. 1988. Environmental, economic and social costs of coral reef destruction in the Philippines. *Galaxea* 7:161–78.

McAllister, D. E., A. L. Hamilton, and B. Harvey. 1997. Global freshwater biodiversity: Striving for the integrity of freshwater ecosystems. *Sea Wind* 11(3): 1–139.

McAllister, D. E., B. J. Parker, and P. M. McKee. 1985. *Rare, endangered and extinct fishes in Canada*. Syllogeus 54. Ottawa: National Museums of Natural Sciences.

McCallister, M. K., R. M. Peterman, and D. M. Gillis. 1992. Statistical evaluation of a large-scale fishing experiment designed to test for a genetic effect of size-selective fishing on British Columbia pink salmon (*Oncorhynchus gorbuscha*). *Can J Fish Aquat Sci* 49:1294–1304.

McClanahan, T. R. 1995. Fish predators and scavengers of the sea urchin *Echinometra mathaei* in Kenyan coral-reef marine parks. *Environ Biol Fish* 43:187–93.

McClanahan, T. R., and S. Mangi. 2000. Spillover of exploitable fishes from a marine park and its effect on the adjacent fishery. *Ecol Appl* 10:1792–1805.

McClanahan, T. R., and N. A. Muthiga. 1988. Changes in Kenyan coral reef community structure and function due to exploitation. *Hydrobiologia* 166:269–76.

McCully, P. 1996. *Silenced rivers: The ecology and politics of large dams*. Atlantic Highlands, NY: Zed Books.

McDowall, R. M. 1990a. *New Zealand freshwater fishes: A natural history and guide*. Auckland: Heinemann Reed.

———. 1990b. When galaxiid and salmonid fishes meet: A family reunion in New Zealand. *J Fish Biol* 37(suppl. A): 35–43.

———. 1994. *Gamekeepers for the nation: The story of New Zealand's acclimatisation societies, 1861–1990*. Christchurch: Canterbury Univ. Press.

———. 1999. Different kinds of diadromy: Different kinds of conservation problems. *ICES J Mar Sci* 56:410–13.

———. 2003. Impacts of introduced salmonids on native galaxiids in New Zealand upland streams: A new look at an old problem. *T Am Fish Soc* 132:229–38.

McDowall, R. M., and G. A. Eldon. 1996. Threatened fishes of the world: *Neochanna burrowsius* (Phillipps, 1926) (Galaxiidae). *Environ Biol Fish* 47:190.

McDowall, R. M., and D. K. Rowe. 1996. Threatened fishes of the world: *Galaxias gracilis* McDowall, 1967 (Galaxiidae). *Environ Biol Fish* 46:280.

McElhany, P., M. H. Ruckelshaus, M. J. Ford, et al. 2000. *Viable salmonid populations and the recovery of evolutionarily significant units*. U.S. Dept. Commerce, NOAA Tech. Memo. NMFS-NWFSC-42.

McFarland, W. N. 1979. Observations on recruitment of haemulid fishes. *P Gulf Caribb Fish Inst* 32:132–38.

McFarland, W. N., J. C. Ogden, and J. N. Lythgoe. 1979. The influence of light on the twilight migrations of grunts. *Environ Biol Fish* 4:9–22.

McGinn, N. A., ed. 2002. *Fisheries in a changing climate*. Bethesda, MD: American Fisheries Society.

McGinnity, P., C. Stone, J. B. Taggart, et al. 1997. Genetic impact of escaped farmed Atlantic salmon (*Salmo salar* L.) on native populations: Use of DNA profiling to assess freshwater performance of wild, farmed, and hybrid progeny in a natural river environment. *ICES J Mar Sci* 54:998–1008.

McGoodwin, J. R. 1990. *Crisis in the world's fisheries*. Stanford, CA: Stanford Univ. Press.

McGrath, D. G., F. DeCastro, C. Futemma, et al. 1993. Fisheries and the evolution of resource management on the lower Amazon floodplain. *Hum Ecol* 21:167–95.

McIntosh, A. R. 2000. Habitat- and size-related variations in exotic trout impacts on native galaxiid fishes in New Zealand streams. *Can J Fish Aquat Sci* 57:2140–51.

McKay, R. J. 1984. Introduction of fishes in Australia. In *Distribution, biology, and management of exotic fishes*, ed. W. R. Courtenay Jr. and J. R. Stauffer Jr., 177–99. Baltimore: Johns Hopkins Univ. Press.

McKenzie, W. D., D. Crews, K. D. Kallman, et al. 1983. Age, weight and the genetics of sexual maturation in the platyfish, *Xiphophorus maculatus*. *Copeia* 1983:770–74.

McKinnell, S., and A. J. Thomson. 1997. Recent events concerning Atlantic salmon escapees in the Pacific. *ICES J Mar Sci* 54:1221–25.

McKinney, M. L. 1997. Extinction vulnerability and selectivity: Combining ecological and paleontological views. *Annu Rev Ecol Syst* 28:495–516.

McLean, E., R. H. Devlin, J. C. Byatt, et al. 1997. Impact of a controlled release formulation of recombinant bovine growth hormone upon growth and seawater adaptation in coho (*Oncorhynchus kisutch*) and chinook (*Oncorhynchus tshawytscha*) salmon. *Aquaculture* 156:113–28.

McLeod, K. L., J. Lubchenco, S. R. Palumbi, et al. 2005. Scientific consensus statement on marine ecosystem-based management. http://compassonline.org.

McManus, J. W. 1996. Social and economic aspects of reef fisheries and their management. In Polunin and Roberts 1996, 249–281.

McPhee, J. 2002. *The founding fish*. New York: Farrar, Straus and Giroux.

McVicar, A. H. 1997. Disease and parasite implications of the coexistence of wild and cultured Atlantic salmon populations. *ICES J Mar Sci* 54:1093–1103.

Meador, M. R. 1996. Water transfer projects and the role of fisheries biologists. *Fisheries* 21(9): 18–23.

Meadows, R. 2001. Turning a radical idea into reality: Removing Edwards Dam in Augusta, Maine. *Conserv Biol in Practice* 2(1): 32–35.

Medley, P. A., G. Gaudian, and S. Wells. 1993. Coral reef fisheries stock assessment. *Rev Fish Biol Fisher* 3:242–85.

Meffe, G. K. 1984. Effects of abiotic disturbance on coexistence of predator-prey fish species. *Ecology* 65:1525–34.

———. 1985. Predation and species replacement in American southwestern fishes: A case study. *Southwest Nat* 30:173–87.

———. 1986. Conservation genetics and the management of endangered fishes. *Fisheries* 11(1): 14–23.

———. 1990. Genetic approaches to conservation of rare fishes: Examples from North American desert species. *J Fish Biol* 37:105–12.

———. 1991. Failed invasion of a southeastern blackwater stream by bluegills: Implications for conservation of native communities. *T Am Fish Soc* 120:333–38.

———. 1992. Techno-arrogance and halfway technologies: Salmon hatcheries on the Pacific Coast of North America. *Conserv Biol* 6:350–54.

———. 2002. Connecting science to management and policy in freshwater fish conservation. In Collares-Pereira, Coelho, and Cowx 2002, 363–72.

Meffe, G. K., and C. R. Carroll. 1997. *Principles of conservation biology*, 2nd ed. Sunderland, MA: Sinauer Associates.

Meffe, G. K., and F. F. Snelson Jr., eds. 1989. *Ecology and evolution of livebearing fishes (Poeciliidae)*. Englewood Cliffs, NJ: Prentice Hall.

Melfi, L. 2003. *Arapaima gigas*. Animal Diversity Web. http://animaldiversity.ummz.umich.edu.

Melvin, E. F., and J. K. Parrish, eds. 2001. *Seabird bycatch: Trends, roadblocks, and solutions*. Fairbanks: Univ. Alaska Sea Grant.

Meng, L., P. B. Moyle, and B. Herbold. 1994. Changes in abundance and distribution of native and introduced fishes of Suisun Marsh. *T Am Fish Soc* 123:498–507.

Merigoux, S., and D. Ponton. 1999. Spatio-temporal distribution of young fish in tributaries of natural and flow-regulated sections of a neotropical river in French Guiana. *Freshwater Biol* 42:177–98.

Meronek, T. G., F. A. Copes, and D. W. Coble. 1995. A summary of bait regulations in the north central United States. *Fisheries* 20(11): 16–23.

Merrill, M., M. C. Freeman, B. J. Freeman, et al. 2001. Stream loss and fragmentation due to impoundments in the upper Oconee watershed. In *Proceedings of the 2001 Georgia Water Resources Conference*, ed. K. J. Hatcher, 66–69. Athens: Univ. Georgia.

Merritt, N. R., and R. L. Haedrich. 1997. *Deepsea demersal fish and fisheries*. London: Chapman and Hall.

Merskey, H., and N. Bogduk, eds. 1994. *Classification of chronic pain*, 2nd ed. Seattle: IASP Press.

Mesnil, B. 1996. When discards survive: Accounting for survival of discards in fisheries assessments. *Aquat Living Resour* 9:209–15.

Messieh, S., T. W. Rowell, D. L. Peer, et al. 1991. The effects of trawling, dredging and ocean dumping on the eastern Canadian continental shelf. *Cont Shelf Res* 11:1237–63.

Meyer, J. L., and G. S. Helfman. 1993. The ecological basis of sustainability. *Ecol Appl* 3:569–71.

Meyer, J. L., L. A. Kaplan, D. Newbold, et al. 2001. *Where rivers are born: The scientific imperative for defending small streams and wetlands*. American Rivers and Sierra Club. www.amrivers.org/doc_repository/WhereRiversAreBorn1.pdf.

Meyer, J. L., E. T. Schultz, and G. S. Helfman. 1983. Fish schools: An asset to corals. *Science* 220:1047–49.

Meyer, J. L., A. Sutherland, K. Barnes, et al. 1999. A scientific basis for erosion and sedimentation standards in the Blue Ridge Physiographic Province. In *Proceedings of the 1999 Georgia Water Resources Conference*, ed. K. J. Hatcher, 321–24. Athens: Univ. Georgia.

Meyers, M. 2001. The pet industry view. In Chao, Petry, et al., 2001, 87 108.

Michael, S. 1996. The Banggai cardinalfish. *Aquarium Fish Magazine* 8(8): 86–87.

Micklin, P. P. 1988. Desiccation of the Aral Sea: A water management disaster in the Soviet Union. *Science* 241:1170–76.

Mieczkowski, Z. 1995. *Environmental issues of tourism and recreation*. Lanham, MD: Univ. Press of America.

Milhous, R. T., M. A. Updike, and D. M. Schneider. 1989. Physical habitat simulation system reference manual: Version II. *U.S. Fish and Wildlife Service Biological Report* 89(16). Washington, DC: USFWS.

Miller, D. J. 1989. Introductions and extinctions of fish in the African great lakes. *Trends Ecol Evol* 4:56–59.

Miller, R. R. 1957. Have the genetic patterns of fishes been altered by introductions or selective fishing? *J Fish Res Board Can* 14:797–806.

———. 1968. Records of some native freshwater fishes transplanted into various waters of California, Baja California, and Nevada. *Calif Fish Game* 54:170–79.

Miller, R. R., J. D. Williams, and J. E. Williams. 1989. Extinctions of North American fishes during the past century. *Fisheries* 14(6): 22–38.

Minckley, W. L. 1973. *Fishes of Arizona*. Phoenix: Arizona Game and Fish Dept.

———. 1991. Native fishes of the Grand Canyon region: An obituary? In *Colorado River ecology and dam management*, ed. Committee to Review the Glen Canyon Environmental Studies, 124–77. Washington, DC: National Academy Press.

———. 1995. Translocation as a tool for conserving imperiled fishes: Experiences in western United States. *Biol Conserv* 2:297–309.

Minckley, W. L., and J. E. Deacon. 1968. Southwestern fishes and the enigma of "endangered species." *Science* 159:1424–32.

———, eds. 1991. *Battle against extinction: Native fish management in the American West*. Tucson: Univ. Arizona Press.

Minckley, W. L., G. K. Meffe, and D. L. Soltz. 1991. Conservation and management of short-lived species: The cyprinodontoids. In Minckley and Deacon 1991, 247–82.

Minns, C. K. 1990. Patterns of distribution and association of fresh-water fish in New Zealand. *New Zeal J Mar Fresh* 24:31–44.

Miranda, L. E., and D. R. DeVries, eds. 1996. Multidimensional approaches to reservoir fisheries management. *Am Fish Soc Sym* 16.

Moberg, F., and C. Folke. 1999. Ecological goods and services of coral reef ecosystems. *Ecol Econ* 29:215–33.

Mobrand, L. E., J. Barr, L. Blankenship, et al. 2005. Hatchery reform in Washington state: Principles and emerging issues. *Fisheries* 30(6): 11–24.

Modde, T., A. T. Scholz, J. H. Williamson, et al. 1995. An augmentation plan for razorback sucker in the upper Colorado River basin. *Am Fish Soc Sym* 15:102–11.

Moe, M. A. 1999. *Marine ornamental aquaculture*. First International Conference on Marine Ornamentals, Honolulu, HI. Cited in E. M. Wood 2001.

Moe, M. A. Jr. 2002. *Culture of marine ornamentals: For love, for money and for science*. Second International Conference on Marine Ornamentals: Collection, Culture and Conservation. http://conference.ifas.ufl.edu/mo/moe.pdf.

Moffie, K. 2002. *The aquarium trade in Hawai'i*. Pacific Fisheries Commission White Paper, Oct. www.westpacificfisheries.net/wpapers/aquarium.html.

Moksness, E., R. Stole, and G. van der Meeren. 1998. Profitability analysis of sea ranching with Atlantic salmon (*Salmo salar*), Arctic charr (*Salvelinus alpinus*), and European lobster (*Homarus gammarus*) in Norway. *B Mar Sci* 62:689–99.

Mol, J. H., and P. E. Ouboter. 2004. Downstream effects of erosion from small-scale gold mining on the instream habitat and fish community of a small neotropical rainforest stream. *Conserv Biol* 18:201–14.

Molloy, J., B. Bell, M. Clout, et al. 2002. *Classifying species according to threat of extinction: A system for New Zealand*. Threatened Species Occasional Publication 22. Wellington: Biodiversity Recovery Unit, Dept. of Conservation.

Mommsen, T. P., M. M. Vijayan, and T. W. Moon. 1999. Cortisol in teleosts: Dynamics, mechanisms of action, and metabolic regulation. *Rev Fish Biol Fisher* 9:211–68.

Monteiro-Neto, C., F. E. De Andrade Cunha, M. Carvalho Nottingham, et al. 2003. Analysis of the marine ornamental fish trade at Ceara State, northeast Brazil. *Biodivers Conserv* 12:1287–95.

Monteith, D. T., C. D. Evans, and S. Patrick. 2001. Monitoring acid waters in the UK: 1988–1998 trends. *Water Air Soil Poll* 130:1307–12.

Montgomery, D. R. 2003. *King of fish: The thousand year run of salmon*. Boulder, CO: Westview Press.

Montgomery, D. R., J. M. Buffington, N. P. Peterson, et al. 1996. Stream-bed scour, egg burial depths, and the influence of salmonid spawning on bed surface mobility and embryo survival. *Can J Fish Aquat Sci* 53:1061–70.

Mooney, H. A. 1999. The Global Invasive Species Program (GISP). *Biol Invasions* 1:97–98.

Moore, A., A. P. Scott, N. Lower, et al. 2003. The effects of 4-nonylphenol and atrazine on Atlantic salmon (*Salmo salar* L.) smolts. *Aquaculture* 222:253–63.

Moore, A., and C. P. Waring. 1996. Sublethal effects of the pesticide Diazinon on olfactory function in mature male Atlantic salmon parr. *J Fish Biol* 48:758–75.

Moore, G., and S. Jennings, eds. 2000. *Commercial fishing: The wider ecological impacts*. Cambridge: British Ecological Society.

Moore, M., and M. Erdmann. 2002. EcoReefs: A new tool for coral reef restoration. *Conserv Pract* 3(3): 41–44.

Morehardt, J. E. 1986. *Instream flow methodologies*. Report NR EPRIEA-4819. Palo Alto, CA: Electric Power Research Institute.

Morel, F. M. M., A. M. L. Kraepiel, and M. Amyot. 1998. The chemical cycle and bioaccumulation of mercury. *Annu Rev Ecol Syst* 29:543–66.

Morgan, J. D., and G. K. Iwama. 1997. Measurements of stressed states in the field. In *Fish stress and health in aquaculture*, ed. O. K. Iwama et al., 247–68, Seminar Series 62. Cambridge: Society for Experimental Biology.

Morita, K., and S. Yamamoto. 2002. Effects of habitat fragmentation by damming on the persistence of stream-dwelling charr populations. *Conserv Biol* 16:1318–23.

Moritz, C. 1994. Defining evolutionarily-significant-units for conservation. *Trends Ecol Evol* 9:373–75.

———. 2002. Strategies to protect biological diversity and the processes that sustain it. *Syst Biol* 51:238–54.

Morizot, D. C., S. W. Calhoun, L. L. Clepper, et al. 1991. Multispecies hybridization among native and introduced centrarchid basses in central Texas. *T Am Fish Soc* 120:283–89.

Morman, R. H., D. W. Cuddy, and P. C. Rugen. 1980. Factors influencing the distribution of sea lamprey (*Petromyzon marinus*) in the Great Lakes. *Can J Fish Aquat Sci* 37:1811–26.

Morrissey, C. A., L. I. Bendell-Young, and J. E. Elliott. 2004. Linking contaminant profiles to the diet and breeding location of American dippers using stable isotopes. *J Appl Ecol* 41:502–12.

Morrisey, D. J., M. M. Gibbs, S. E. Pickmere, et al. 2000. Predicting impacts and recovery of marine farm sites in Stewart Island, New Zealand, from the Findlay-Watling model. *Aquaculture* 185:257–71.

Moser, M. L., and S. B. Roberts. 1999. Effects of nonindigenous ictalurids and recreational electrofishing on the ictalurid community of the Cape Fear River Drainage, North Carolina. *Am Fish Soc Sym* 24:479–85.

Mosqueira, I., I. Cote, S. Jennings, and J. D. Reynolds. 2000. Conservation benefits of marine reserves for fish populations. *Anim Conserv* 3:321–32.

Moss, S. A. 1984. *Sharks: A guide for the amateur naturalist.* Englewood Cliffs, NJ: Prentice Hall.

Mous, P. J., L. Pet-Soede, M. Erdmann, et al. 2000. Cyanide fishing on Indonesian coral reefs for the live food fish market: What is the problem? *SPC Live Reef Fish Inform Bull* 7:20–26.

Mouton, A., L. Basson, and D. Impson. 2001. Health status of ornamental freshwater fishes imported to South Africa: A pilot study. *Aquar Sci Conserv* 3:313–19.

Mowat, F. 1996. *Sea of slaughter: A chronicle of the destruction of animal life in the North Atlantic.* Shelburne, VT: Chapters.

Moyle, P. B. 1984. America's carp. *Nat Hist* 9(84): 42–51.

———. 1991. Ballast water introductions. *Fisheries* 16(1): 4–6.

———. 1994. Biodiversity, biomonitoring, and the structure of stream fish communities. In *Biological monitoring of aquatic systems*, ed. S. L. Loeb and A. Spacie, 171–86. Boca Raton, FL: Lewis.

———. 1995. Conservation of native freshwater fishes in the Mediterranean-type climate of California, USA: A review. *Biol Conserv* 72:271–79.

———. 1996. Status of aquatic habitat types. *Sierra Nevada Ecosystem Project: Final Report to Congress* 2:945–52.

———. 2002. *Inland fishes of California*, rev. exp. ed. Berkeley: Univ. California Press.

Moyle, P., and J. Cech. 2004. *Fishes: An introduction to ichthyology*, 5th ed. Upper Saddle River, NJ: Prentice Hall.

Moyle, P. B., and J. P. Ellison. 1991. A conservation-oriented classification system for the inland waters of California. *Calif Fish Game* 77:161–80.

Moyle, P. B., B. Herbold, D. E. Stevens, et al. 1992. Life history and status of delta smelt in the Sacramento-San Juaquin estuary, California. *T Am Fish Soc* 121:67–77.

Moyle, P. B., and R. A. Leidy. 1992. Loss of biodiversity in aquatic ecosystems: Evidence from fish faunas. In *Conservation biology: The theory and practice of nature conservation, preservation and management*, ed. P. L. Fiedler and S. K. Jain, 127–69. New York: Chapman and Hall.

Moyle, P. B., H. W. Li, and B. A. Barton. 1986. The Frankenstein effect: Impact of introduced fishes on native fishes in North America. In *Fish culture in fisheries management*, ed. R. H. Stroud, 416–26. Bethesda, MD: American Fisheries Society.

Moyle, P. B., and T. Light. 1996a. Biological invasions of fresh water: Empirical rules and assembly theory. *Biol Conserv* 78:149–61.

———. 1996b. Fish invasions in California: Do abiotic factors determine success. *Ecology* 77:1666–70.

Moyle, P. B., and P. R. Moyle. 1995. Endangered fishes and economics: Intergenerational obligations. *Environ Biol Fish* 43:29–37.

Moyle, P. B., and D. Sweetnam. 1994. Delta smelt (*Hypomesus transpacificus*). In *Life on the edge: A guide to California's endangered natural resources*, ed. C. G. Thelander and M. Crabtree, 342–44. Santa Cruz, CA: Biosystems Books.

Moyle, P. B., and J. E. Williams. 1990. Biodiversity loss in the temperate zone: Decline of the native fish fauna of California. *Conserv Biol* 4:275–84.

Moyle, P. B., and R. M. Yoshiyama. 1994. Protection of aquatic biodiversity in California: A five-tiered approach. *Fisheries* 19(2): 6–18.

MSHA (Mine Safety and Health Administration). 2001. Report of investigation. Noninjury Impoundment Failure/Mine Inundation Accident, October 11, 2000. www.msha.gov/impoundments/martincounty/martincountya.pdf.

Muir, W. D., S. G. Smith, J. G. Williams, et al. 2001. Survival of juvenile salmonids passing through bypass systems, turbines, and spillways with and without flow deflectors at Snake River dams. *N Am J Fish Manage* 21:135–46.

Muir, W. M., and R. D. Howard. 1999. Possible ecological risks of transgenic organism release when transgenes affect mating success: Sexual selection and the Trojan gene hypothesis. *P Natl Acad Sci USA* 96:13853–56.

———. 2001. Fitness components and ecological risk of transgenic release: A model using Japanese medaka (*Oryzias latipes*). *Am Nat* 158:1–16.

Muluk, C., and C. Bailey. 1996. Social and environmental impacts of coastal aquaculture in Indonesia. In *Aquacultural development: Social dimensions of an emerging industry*, ed. C. Bailey et al., 193–209. Boulder, CO: Westview Press.

Mumby, P. J., A. J. Edwards, J. E. Arias-González, et al. 2004. Mangroves enhance the biomass of coral reef fish communities in the Caribbean. *Nature* 427:533–36.

Mumby, P. J., E. P. Green, C. D. Clark, et al. 2001. Unprecedented bleaching-induced mortality in *Porites* spp. at Rangiroa Atoll, French Polynesia. *Mar Biol* 139:183–89.

Muncy, R. J., G. J. Atchison, R. V. Bulkley, et al. 1979. Effects of suspended solids and sediment on reproduction and early life of warmwater fishes: A review. EPA Report 600/3-79-042. Washington, DC: USEPA.

Munday, P. L., and S. K. Wilson. 1997. Comparative efficacy of clove oil and other chemicals in anaesthetization of *Pomacentrus amboinensis*, a coral reef fish. *J Fish Biol* 51:931–38.

Munro, J. L. 1996. The scope of tropical reef fisheries and their management. In Polunin and Roberts 1996, 1–14.

Munro, J. L., and I. R. Smith. 1984. Management strategies for multi-species complexes in artisanal fisheries. *P Gulf Caribb Fish Inst* 36:127–41.

Muoneke, M. I., and W. M. Childress. 1994. Hooking mortality: A review for recreational fisheries. *Rev Fish Sci* 2:123–56.

Murawski, S. 2000. Definitions of overfishing from an ecosystems perspective. *ICES J Mar Sci* 57:649–58.

Murawski, S. A., R. W. Brown, S. X. Cadrin, et al. 1999. New England groundfish. Unit 2 in NMFS 1999.

Murawski, S. A., R. Brown, H. L. Lai, et al. 2000. Large-scale closed areas as a fishery-management tool in temperate marine systems: The Georges Bank experience. *B Mar Sci* 66:775–98.

Murawski, S. A., J. J. Maguire, R. K. Mayo, et al. 1997. Groundfish stocks and the fishing industry. In *Northwest Atlantic groundfish: Perspectives on a fishery collapse*, ed. J. Boreman et al., 27–70. Bethesda, MD: American Fisheries Society.

Murphy, B. R., and W. E. Kelso. 1986. Strategies for evaluating fresh-water stocking programs: Past practices and future needs. In *Fish culture in fisheries management*, ed. R. H. Stroud, 303–16. Bethesda, MD: American Fisheries Society.

Murphy, G. I. 1980. Schooling and the ecology and management of marine fish. In *Fish behavior and its use in the capture and culture of fishes*, ed. J. E. Bardach et al., 400–14. Manila, Philippines: ICLARM.

Murray, S. N., R. F. Ambrose, J. A. Bohnsack, et al. 1999. No-take reserve networks: Sustaining fishery populations and marine ecosystems. *Fisheries* 24(11): 11–25.

Musick, J. A. 1999a. Criteria to define extinction risk in marine fishes. *Fisheries* 24(12): 6–14.

———, ed. 1999b. *Life in the slow lane: Ecology and conservation of long-lived marine animals*. American Fisheries Society Symposium 23, Bethesda, MD.

Musick, J. A., G. Burgess, G. Cailliet, et al. 2000. Management of sharks and their relatives (*Elasmobranchii*). *Fisheries* 25(3): 9–13.

Musick, J. A., M. M. Harbin, S. A. Berkeley, et al. 2001. Marine, estuarine and diadromous fish stocks at risk of extinction in North America (exclusive of Pacific salmonids). *Fisheries* 25(11): 6–30.

Mustafa, A., W. Rankaduwa, and P. Campbell. 2001. Estimating the cost of sea lice to salmon aquaculture in eastern Canada. *Can Vet J* 41:54–56.

Mustafa, S., ed. 1999a. *Genetics in sustainable fisheries management*. Malden, MA: Blackwell Science.

———. 1999b. Introduction. In Mustafa 1999a, 3–23.

Muth, R. T., and J. B. Ruppert. 1997. Effects of electrofishing fields on captive embryos and larvae of razorback sucker. *N Am J Fish Manage* 17:160–66.

MWDTF (Montana Whirling Disease Task Force). 1996. *Final report and action recommendations*. Helena, MT: Author.

Myers, M. A., and G. E. White. 1993. The challenge of the Mississippi flood. *Environment* 10(12): 6–35.

Myers, R. A., and N. J. Barrowman. 1996. Is fish recruitment related to spawner abundance? *Fish Bull U.S.* 94:707–24.

Myers, R. A., J. A. Hutchings, and N. J. Barrowman. 1996. The decline of cod in the North Atlantic. *Mar Ecol—Prog Ser* 138:293–308.

———. 1997. Why do fish stocks collapse? The example of cod in Atlantic Canada. *Ecol Appl* 7:91–106.

Myers, R. A., S. A. Levin, R. Lande, et al. 2004. Hatcheries and endangered salmon. *Science* 303:1980.

Myers, R. A., G. Mertz, and P. S. Fowlow. 1997. The maximum population growth rate and recovery times of Atlantic cod (*Gadus morhua*). *Fish Bull U.S.* 95:762–72.

Myers, R. A., and C. A. Ottensmeyer. 2005. Extinction risk in marine species. In Norse and Crowder 2005, 58–79.

Myers, R. A., and B. Worm. 2003. Rapid worldwide depletion of predatory fish communities. *Nature* 423:280–83.

Naesje, T., B. Jonsson, and J. Skurdal. 1995. Spring flood: A primary cue for hatching of river spawning Coregoninae. *Can J Fish Aquat Sci* 52:2190–96.

Nagler, J. J., J. Bouma, G. H. Thorgaard, et al. 2001. High incidence of a male-specific genetic marker in phenotypic female chinook salmon from the Columbia River. *Environ Health Persp* 109:67–69.

Naiman, R. J., J. J. Magnuson, D. M. McKnight, et al. 1995. *The freshwater imperative: A research agenda.* Washington, DC: Island Press.

Nakamura, I. 1985. *FAO species catalogue. Vol. 5: Billfishes of the world.* FAO Fisheries Synopsis 125(5). Rome: FAO.

Nam, Y. K., J. K. Noh, Y. S. Cho, et al. 2001. Dramatically accelerated growth and extraordinary gigantism of transgenic mud loach *Misgurnus mizolepis*. *Transgenic Res* 10:353–62.

NASCO (North Atlantic Salmon Conservation Organization). 1991. *Guidelines to minimize the threats to wild salmon stocks from salmon aquaculture.* Edinburgh: NASCO.

———. 2000. *Salmon farming in Scotland: The great escape.* May 6. http://asf.ca/nasco/nasco2000/escapes.html.

National Association of Evangelicals. 2004a. *For the health of the nation: An evangelical call to civic responsibility.* Oct. 7. www.nae.net.

———. 2004b. *The Sandy Cove Covenant and Invitation.* June. www.nae.net.

Naylor, R. L., R. J. Goldburg, J. H. Primavera, et al. 2000. Effect of aquaculture on world fish supplies. *Nature* 405:1017–24.

———, et al. 2001. Effects of aquaculture on world fish supplies. *Issues Ecol* 8. Washington, DC: Ecological Society of America.

Neff, J. M. 1985. The use of biochemical measurement to detect pollutant-mediated damage to fish. In *Aquatic toxicology and hazard assessment*, ed. R. D. Cardwell, R. Purdy, and R. C. Bahner, 115–83. Philadelphia: American Society for Testing and Materials.

Nehlsen, W., J. E. Williams, and J. A. Lichatowich. 1991. Pacific salmon at the crossroads: Stocks at risk from California, Oregon, Idaho, and Washington. *Fisheries* 16(2): 4–21.

Nelson, J. S. 1994. *Fishes of the world*, 3rd ed. New York: Wiley.

———. 2006. *Fishes of the world*, 4th ed. New York: Wiley.

Nelson, J. S., E. J. Crossman, H. Espinosa-Pérez, et al. 2004. *Common and scientific names of fishes from the United States, Canada, and Mexico*, 6th ed. Bethesda, MD: American Fisheries Society.

Nelson, S. M., and L. C. Keenan. 1992. Use of an indigenous fish species, *Fundulus zebrinus*, in a mosquito abatement program: A field comparison with the mosquitofish, *Gambusia affinis*. *J Am Mosquito Contr* 8:301–4.

Nesler, T. P., R. T. Muth, and A. F. Wasowicz. 1988. Evidence for baseline flow spikes as spawning cues for Colorado squawfish in the Yampa River, Colorado. *Am Fish Soc Sym* 5:68–79.

Netboy, A. 1974. *Salmon: Their fight for survival.* Boston: Houghton Mifflin.

Newcombe, C. P., and J. O. T. Jenson. 1996. Channel suspended sediment and fisheries: A synthesis for quantitative assessment of risk and impact. *N Am J Fish Manage* 16:693–727.

Newcombe, C. P., and D. D. MacDonald. 1991. Effects of suspended sediments on aquatic ecosystems. *N Am J Fish Manage* 11:72–82.

Newman, M. C., and M. A. Unger. 2002. *Fundamentals of ecotoxicology*, 2nd ed. Boca Raton, FL: Lewis.

Newton, L. H., and C. K. Dillingham. 2002. *Watersheds 3: Ten cases in environmental ethics.* Belmont, CA: Wadsworth/Thomson.

Ng, P. K. L., and H. H. Tan. 1997. Freshwater fishes of southeast Asia: Potential for the aquarium fish trade and conservation issues. *Aquar Sci Conserv* 1:79–90.

Nichols, S., A. Shah, G. J. Pellegrin Jr., et al. 1990. Updated estimates of shrimp fleet by-catch in the offshore waters of the Gulf of Mexico 1972–1989. Pascagoula, MS: National Marine Fisheries Service.

Nico, L. G., and P. L. Fuller. 1999. Spatial and temporal patterns of nonindigenous fish introductions in the United States. *Fisheries* 24(1): 16–27.

Nico, L. G., and J. D. Williams. 1996. *Risk assessment on black carp (Pisces: Cyprinidae)*. Final report to the Risk Assessment and Management Committee of the Aquatic Nuisance Species Task Force. Gainesville, FL: U.S. Geological Survey, Biological Resources Division.

Nielsen, J. L. 1998a. Electrofishing California's endangered fish populations. *Fisheries* 23(12): 6–12.

———. 1998b. Threatened fishes of the world: *Oncorhynchus mykiss nelsoni* Evermann 1908 (Salmonidae). *Environ Biol Fish* 51:376.

Niemi, G. J., P. DeVore, N. Detenbeck, et al. 1990. Overview of case studies on recovery of aquatic systems from disturbance. *Environ Manage* 14:571–87.

Nightingale, J., M. Dickens, and D. Vincent. 2001. Aquariums: Some of the reasons why they work so well. *Mar Technol Soc J* 35:18–29.

Nixon, S. W. 1982. Nutrient dynamics, primary production and fisheries yields of lagoons. *Oceanol Acta Sym P*, 357–71. Cited in Birkeland 1997a.

———. 1995. Coastal marine eutrophication: A definition, social causes, and future concerns. *Ophelia* 41:199–219.

NMFS (National Marine Fisheries Service). 1993. Spotlight: Atlantic striped bass. In *Our living oceans: Report on the status of U.S. living marine resources*, 25–32. NOAA Tech. Memo. NMFS-F/SPO-15.

———. 1997. *Status of fisheries of the United States*. Report to Congress, Sept.

———. 1998. *Atlantic swordfish overview*. www.nmfs.noaa.gov/sword.html.

———. 1999. *Our living oceans*. NOAA Tech. Memo. NMFS-F/SPO-41.

———. 2003. *Marine debris removal*. www.pifsc.noaa.gov/crd/marine_debris.html.

———. 2004. *Revised biological opinion for Columbia, Lower Snake River dams*. News release, Nov. 30.

NMFS 1997/NRDC 1998. Draft Amendment 1 to the Fishery Management Plan for Atlantic swordfish, including an environmental assessment and regulatory impact review. Cited in NRDC 1998.

NMFS Southwest. 1999. Pacific highly migratory pelagic fisheries. Unit 18 in NMFS 1999.

NOAA Fisheries. 2004. *Revised biological opinion for Columbia, Lower Snake River dams*. News release, Nov. 30.

NOAA Fisheries Office of Protected Resources. 2005. *Marine and anadromous fish*. www.nmfs.noaa.gov/pr/species/fish.

Noakes, D. J., R. J. Beamish, and M. L. Kent. 2000. On the decline of Pacific salmon and speculative links to salmon farming in British Columbia. *Aquaculture* 183:363–86.

Noaksson, E., M. Linderoth, A. T. C. Bosveld, et al. 2003. Endocrine disruption in brook trout (*Salvelinus fontinalis*) exposed to leachate from a public refuse dump. *Sci Total Environ* 305: 87–103.

Noble, R. L. 1980. Predator-prey interactions in reservoir communities. In *Reservoir fisheries management: Strategies for the 80's*, ed. G. E. Hall and M. J. Van Den Avyle, 137–43. Bethesda, MD: American Fisheries Society.

Noga, E. J. 2000. *Fish disease: Diagnosis and treatment*. Ames: Iowa State Univ. Press.

Nol, P., T. W. Rocke, K. Gross, et al. 2004. Prevalence of neurotoxic *Clostridium botulinum* type C in the gastrointestinal tracts of tilapia (*Oreochromis mossambicus*) in the Salton Sea. *J Wildlife Dis* 40:414–19.

Nordlee, J. A., S. L. Taylor, J. A. Townsend, et al. 1996. Identification of a Brazil-nut allergen in transgenic soybeans. *New Engl J Med* 334:688–92.

Norris, S., and N. L. Chao. 2002. Buy a fish, save a tree? Safeguarding sustainability in an Amazonian ornamental fishery. *Conserv Pract* 3(3): 30–35.

Norse, E. A., and L. B. Crowder, eds. 2005. *Marine conservation biology: The science of maintaining the sea's biodiversity*. Washington, DC: Island Press.

NPCC (Northwest Power and Conservation Council). 2003. *Columbia River basin fish and wildlife program: Twenty years of progress*. Northwest Power and Conservation Council Report 2003-20. Portland, OR.

NRC (National Research Council). 1990. *Decline of the sea turtles: Causes and prevention*. Washington, DC: National Academy Press.

———. 1992a. *Dolphins and the tuna industry*. Washington, DC: National Academy Press.

———. 1992b. *Marine aquaculture: Opportunities for growth.* Washington, DC: National Academy Press.

———. 1994. *An assessment of Atlantic bluefin tuna.* Washington, DC: National Academy Press.

———. 1995. *Science and the Endangered Species Act.* Washington, DC: National Academy Press.

———. 1996a. *Stemming the tide: Controlling introductions of nonindigenous species by ships' ballast water.* Washington, DC: National Academy Press.

———. 1996b. *Upstream: Salmon and society in the Pacific Northwest.* Washington, DC: National Academy Press.

———. 1999a. *Hormonally active agents in the environment.* Washington, DC: National Academy Press.

———. 1999b. *Sustaining marine fisheries.* Washington, DC: National Academy Press.

———. 2000a. *Clean coastal waters: Understanding and reducing the effects of nutrient pollution.* Washington, DC: National Academy Press.

———. 2000b. *Watershed management for potable water supply: Assessing the New York City strategy.* Washington, DC: National Academy Press.

———. 2001. *Marine protected areas: Tools for sustaining ocean ecosystems.* Washington, DC: National Academies Press.

———. 2002a. *Coal waste impoundments: Risks, responses, and alternatives.* Washington, DC: National Academies Press.

———. 2002b. *Effects of trawling and dredging on seafloor habitat.* Washington, DC: National Academies Press.

———. 2002c. *Riparian areas: Functions and strategies for management.* Washington, DC: National Academies Press.

———. 2003. *Ocean noise and marine mammals.* Washington, DC: National Academies Press.

———. 2004a. *Atlantic salmon in Maine.* Washington, DC: National Academies Press.

———. 2004b. *Endangered and threatened fishes in the Klamath River basin: Causes of decline and strategies for recovery.* Washington, DC: National Academies Press.

———. 2005. *Oil spill dispersants: Efficacy and effects.* Washington, DC: National Academies Press.

NRCS (Natural Resources Conservation Service). 1998. *Stream visual assessment protocol.* NWCC-TN-99-1. Portland, OR: National Water and Climate Center.

NRDC (Natural Resource Defense Council). 1998. *Swordfish in the North Atlantic: The case for conservation.* www.nrdc.org.

Nyberg, K., J. Vuorenmaa, M. Rask, et al. 2001. Patterns in water quality and fish status of some acidified lakes in southern Finland during a decade: Recovery proceeding. *Water Air Soil Poll* 130:1373–78.

Oberdörster, E. 2004. Manufactured nanomaterials (Fullerenes, C60) induce oxidative stress in the brain of juvenile largemouth bass. *Environ Health Persp* 112:1058–62.

O'Brien, L., J. Burnett, and R. K. Mayo. 1993. *Maturation of nineteen species of finfish off the northeast coast of the United States, 1985–1990.* NOAA Tech. Report NMFS 113:1–63.

O'Brien, M. H. 1993. Being a scientist means taking sides. *BioScience* 43:706–9.

Odeh, M., ed. 2000. *Advances in fish passage technology: Engineering design and biological evaluation.* Bethesda, MD: American Fisheries Society.

Odenkirk, J., E. Steinkoenig, and F. Spuchesi. 1999. Response of a brown bullhead population to flathead catfish introduction in a small Virginia impoundment. *Am Fish Soc Sym* 24:475–77.

Odum, H. T., and E. P. Odum. 1955. Trophic structure and productivity of a windward coral reef community on Eniwetok Atoll. *Ecol Monogr* 25:291–320.

OECD (Organization for Economic Cooperation and Development). 1997. *Towards sustainable fisheries: Economic aspects of the management of living marine resources.* Paris: Author.

Ogden, J. C. 1997. Ecosystem interactions in the tropical coastal seascape. In Birkeland 1997c, 288–97.

———. 2002. Artificial reefs. *Conserv Pract* 3(4): 46–47.

Ogutu-Ohwayo, R. 1990. The decline of the native fishes of lakes Victoria and Kyoga (East Africa) and the impact of introduced species, especially the Nile perch, *Lates niloticus*, and the Nile tilapia, *Oreochromis niloticus*. *Environ Biol Fish* 27:81–96.

———. 1999. Deterioration in length-weight relationships of Nile perch, *Lates niloticus* L. in lakes Victoria, Kyoga and Nabugabo. *Hydrobiologia* 403:81–86.

Ogutu-Ohwayo, R., R. E. Hecky, A. S. Cohen, et al. 1997. Human impacts on the African Great Lakes. *Environ Biol Fish* 50:117–31.

OHC-NMFS (National Marine Fisheries Service Office of Habitat Conservation). 2000. *Management of Caspian tern and cormorant predation on salmonid smolts in the Columbia River estuary in 2000.* www.nwr.noaa.gov/1habcon/habweb/caspiantern.html.

Ohman, M. C., A. Rajasuriya, and S. Svensson. 1998. The use of butterflyfishes (Chaetodontidae) as bio-indicators of habitat structure and human disturbance. *Ambio* 27:708–16.

Olem, H. 1991. *Liming acidic surface waters.* Chelsea, MI: Lewis.

Oliver, J., and M. Noordeloos, eds. 2002. *ReefBase: A global information system on coral reefs.* www.reefbase.org.

Olla, B. L., M. W. Davis, and C. B. Schreck. 1997. Effects of simulated trawling on sablefish and walleye pollock: The role of light intensity, net velocity and towing duration. *J Fish Biol* 50:1181–94.

———. 1998. Temperature magnified postcapture mortality in adult sablefish after simulated trawling. *J Fish Biol* 53:743–51.

Olsen, D. A., and J. A. LaPlace. 1978. A study of a Virgin Islands grouper fishery based on a breeding aggregation. *P Gulf Caribb Fish Inst* 31:130–44.

Ono, R. D., J. D. Williams, and A. Wagner. 1983. *Vanishing fishes of North America.* Washington, DC: Stone Wall Press.

Ormerod, S. J. 2003. Current issues with fish and fisheries: Editor's overview and introduction. *J Appl Ecol* 40:204–13.

Oro, D., L. Jover, and X. Ruiz. 1996. Influence of trawling activity on the breeding ecology of a threatened seabird, Audouin's gull, *Larus audouinii. Mar Ecol—Prog Ser* 139:19–29.

Orr, D. W. 1994. *Earth in mind: On education and the human prospect.* Washington, DC: Island Press.

Orth, D. J., and O. E. Maughan. 1981. Evaluation of the "Montana Method" for recommending instream flows in Oklahoma. *P Oklahoma Acad Sci* 61:62–66.

O'Scannell, P. O. 1988. Effects of elevated sediment levels from placer mining on survival and behavior of immature Arctic grayling. MSc thesis, Univ. Alaska. Cited in P. A. Ryan 1991.

OSPAR. 2004. *2004 initial OSPAR list of threatened and/or declining species and habitats.* Agreement reference number: 2004–06. www.ospar.org.

Ostenfeld, T., H. E. McLean, and R. H. Devlin. 1998. Transgenesis changes body and head shape in Pacific salmon. *J Fish Biol* 52:850–54.

Ostrander, G. K., K. C. Cheng, J. C. Wolf, et al. 2004. Shark cartilage, cancer and the growing threat of pseudoscience. *Cancer Res* 64:8485–91.

OTA (Office of Technology Assessment). 1993. *Harmful nonindigenous species in the United States.* OTA-F-565. Washington, DC: U.S. Government Printing Office.

———. 1995. *Fish passage technologies: Protection at hydropower facilities.* OTA-ENV-641 GPO stock #052-003-01450-5. Washington, DC: U.S. Government Printing Office.

Owen, J. 2004. Search is on for world's biggest freshwater fish. *National Geographic News*, Dec 14. http://news.nationalgeographic.com.

Owen, R., and M. Depledge. 2005. Nanotechnology and the environment: Risks and rewards. *Mar Pollut Bull* 50:609–12.

Paddack, M. J., and J. A. Estes. 2000. Kelp forest fish populations in marine reserves and adjacent exploited areas of central California. *Ecol Appl* 10:855–70.

Page, L. M., and B. M. Burr. 1991. *A field guide to freshwater fishes.* Boston: Houghton Mifflin.

Pait, A. S., and J. O. Nelson. 2003. Vitellogenesis in male *Fundulus heteroclitus* (killifish) induced by selected estrogenic compounds. *Aquat Toxicol* 64:331–42.

Palau Division of Marine Resources. 1998. *Palau domestic fishing laws 1998.* Koror, Palau: Division of Marine Resources, Republic of Palau.

Palmer, M. A., D. D. Hart, J. D. Allan, et al. 2003. Bridging engineering, ecological and geomorphic science to enhance riverine restoration: Local and national efforts. *Proceedings, National Symposium on Urban and Rural Stream Protection and Restoration*, Philadelphia, PA, June. Reston, VA: American Society of Civil Engineers.

Palumbi, S. R. 2001. Humans as the world's greatest evolutionary force. *Science* 293:1786–90.

———. 2003. *Marine reserves: A tool for ecosystem management and conservation.* Arlington, VA: Pew Oceans Commission. www.pewoceans.org.

Pandian, T. J., and S. G. Sheela. 1995. Hormonal induction of sex reversal in fish. *Aquaculture* 138:1–22.

Paperna, I., M. Vilenkin, and A. P. A. de Matos. 2001. Iridovirus infections in farm-reared tropical ornamental fish. *Dis Aquat Organ* 48:17–25.

Parent, S., and L. M. Schriml. 1995. A model for the determination of fish species at risk based upon life-history traits and ecological data. *Can J Fish Aquat Sci* 52:1768–81.

Park, Y.-S., J. Chang, S. Lek, et al. 2003. Conservation strategies for endemic fish species threatened by the Three Gorges Dam. *Conserv Biol* 17:1748–58.

Parker, R. O. 2002. *Aquaculture science*, 2nd ed. Clifton Park, NY: Thomson Delmar Learning.

Parrish, J. D. 1989. Fish communities of interacting shallow-water habitats in tropical oceanic regions. *Mar Ecol—Prog Ser* 58:143–60.

Parrish, J. K. 1999. Using behavior and ecology to exploit schooling fishes. *Environ Biol Fish* 55:157–81.

———. 2005. Behavioral approaches to marine conservation. In Norse and Crowder 2005, 80–104.

Patten, D. T. 1998. Riparian ecosystems of semi-arid North America: Diversity and human impacts. *Wetlands* 18:498–512.

Patten, D. T., D. A. Harpman, M. I. Voita, et al. 2001. A managed flood on the Colorado River: Background, objectives, design, and implementation. *Ecol Appl* 11:635–43.

Patterson, B. D., and S. M. Goodman, eds. 1997. *Natural change and human impacts in Madagascar.* Washington, DC: Smithsonian Press.

Patterson, K. L., J. W. Porter, K. B. Ritchie, et al. 2002. The etiology of white pox, a lethal disease of the Caribbean elkhorn coral, *Acropora palmata. P Natl Acad Sci USA* 99:8725–30.

Paulay, G. 1997. Diversity and distribution of reef organisms. In Birkeland 1997c, 298–353.

Pauly, D. P. 1979. *Theory and practice of overfishing: A southeast Asian perspective.* Indo-Pacific Fishery Commission, RAPA Report 1987/10. Cited in S. J. Hall 1999.

———. 1995. Anecdotes and the shifting baseline syndrome of fisheries. *Trends Ecol Evol* 10:430.

Pauly, D., and V. Christensen. 1995. Primary production required to sustain global fisheries. *Nature* (London) 374:255–57.

Pauly, D., V. Christensen, J. Dalsgaard, et al. 1998. Fishing down marine food webs. *Science* 279:860–63.

Pauly, D., P. J. B. Hart, and T.J. Pitcher. 1998b. Speaking for themselves: new acts, new actors and a New Deal in a reinvented fisheries management. In *Reinventing fisheries management.* Edited by T.J. Pitcher, P. J. B.Hart, and D. Pauly, pp. 409–415. London: Kluwer Academic Publishers.

Pauly, D., and R. Froese. 2001. Fish stocks. *Encyclopedia of biodiversity*, vol. 2, 801–14. San Diego, CA: Academic Press.

Pauly, D., and J. Maclean. 2003. *In a perfect ocean: The state of fishes and ecosystems in the north Atlantic Ocean.* Washington, DC: Island Press.

Pauly, D. P., G. Silvestre, and I. R. Smith. 1989. On development, fisheries, and dynamite: A brief review of tropical fisheries management. *Nat Res Modeling* 3:307–29.

Pavlov, D. S., Y. S. Reshetnikov, M. I. Shatunovskiy, et al. 1985. Rare and disappearing fishes in the USSR and the principles of their inclusion in the "Red Book." *J Ichthyol* 25:88–99.

Pawson, M. 2001. International shark conservation and management initiatives. *Shark News* 13. Newsletter of the IUCN Shark Specialist Group. www.flmnh.ufl.edu.

Paxton, J. R., and W. N. Eschmeyer, eds. 1998. *Encyclopedia of fishes.* San Diego: Academic Press.

Payne, I. 1987. A lake perched on piscine peril. *New Scientist* 115(1575): 50–54.

Payne, P. M., D. N. Wiley, S. B. Young, et al. 1990. Recent fluctuations in the abundance of baleen whales in the southern Gulf of Maine in relation to their selected prey. *Fish Bull U.S.* 88:687–96.

Pedersen, A. 2002. EU report reveals holes in US Safe Harbor Agreement. *Privacy Laws & Business, International Newsletter.* www.privacyexchange.org.

Perez, J. E., J. Mayz, K. Rylander, et al. 2001. The impact of the transgenic revolution on aquaculture and biodiversity: A review. *Revista Cientifica* 11:101–8.

Perra, P. 1992. By-catch reduction devices as a conservation measure. *Fisheries* 17(1): 28–29.

Perry, A. L., P. J. Low, J. R. Ellis, et al. 2005. Climate change and distribution shifts in marine fishes. *Science* 308:1912–15.

Peterken, C. 1996. Australian threatened fishes: 1996 supplement. *Ninth revision conservation status classification of Australian fishes.* Brisbane: Australian Society for Fish Biology.

Peterman, R. M. 1990. Statistical power analysis can improve fisheries research management. *Can J Fish Aquat Sci* 47:2–15.

———. 2002. Ecocertification: An incentive for dealing effectively with uncertainty, risk, and burden of proof in fisheries. *B Mar Sci* 70:669–81.

Peterman, R. M., and M. M'Gonigle. 1992. Statistical power analysis and the precautionary principle. *Mar Pollut Bull* 24:231–34.

Peterson, C. H. 2001. The "Exxon Valdez" oil spill in Alaska: Acute, indirect and chronic effects on the ecosystem. *Adv Mar Biol* 39:1–103.

Petr, T. 1987. Fish, fisheries, aquatic macrophytes and water quality in inland fisheries. *Water Qual B* 12(3): pages vary.

———, ed. 1995. *Inland fisheries under the impact of irrigated agriculture: Central Asia.* FAO Fisheries Circular 894. Rome: FAO.

Pet-Soede, C., H. S. J. Cesar, and J. S. Pet. 1999. An economic analysis of blast fishing on Indonesian coral reefs. *Environ Conserv* 26:83–93.

Pfeffer, R. A., and G. W. Tribble. 1985. Hurricane effects on an aquarium fish fishery in the Hawaiian Islands. *Proceedings 5th International Coral Reef Symposium* (Tahiti) 3:331–36.

Philipp, D. P. 1992. Comments: Stocking Florida largemouth bass outside its native range. *T Am Fish Soc* 121:686–91.

Philippart, J.-C. 1995. Is captive breeding an effective solution for the preservation of endemic species? *Biol Conserv* 72:281–95.

———. 2000. The demographic explosion of *Silurus glanis* in the Belgian river Meuse: Origin and biological significance. Abstract, International Symposium on Freshwater Fish Conservation: Options for the Future, Albufeira, Portugal, 68.

Phillips, M. J., M. C. C. Beveridge, and R. M. Clarke. 1991. Impact of aquaculture on water resources. In *Aquaculture and water quality*, vol. 3, 568–89. Baton Rouge, LA: World Aquaculture Society. ed. D.E. Brune and J.R. Tomasso.

Pickering, H., and D. Whitmarsh. 1997. Artificial reefs and fisheries exploitation: A review of the "attraction versus production" debate, the influence of design and its significance for policy. *Fish Res* 31:39–59.

Pietsch, T. W., and W. D. Anderson Jr., eds. 1997. *Collection building in ichthyology and herpetology*. American Society of Ichthyologists and Herpetologists Special Publication 3. Lawrence, KS: Allen Press.

PIJAC (Pet Industry Joint Advisory Council). 1996. *U.S. ornamental aquarium industry*. www.petsforum .com. Cited in R. W. Ward 1998.

Pike, A. W., and S. L. Wadsworth. 2000. Sea lice on salmonids: Their biology and control. *Adv Parasitol* 44:233–337.

Pikitch, E. L., D. D. Huppert, and M. P. Sissenwine, eds. 1997. *Global trends: Fisheries management.* American Fisheries Society Symposium 20, Bethesda, MD.

Pillay, T. V. R., ed. 1992. *Aquaculture and the environment.* Oxford: Fishing News Books.

Piller, K. R., and B. M. Burr. 1999. Reproductive biology and spawning habitat supplementation of the relict darter *Etheostoma chienense*, a federally endangered species. *Environ Biol Fish* 55:145–55.

Pilskaln, C. H., J. H. Churchill, and L. M. Mayer. 1998. Resuspension of sediment by bottom trawling in the Gulf of Maine and potential geochemical consequences. *Conserv Biol* 12:1223–29.

Pimentel, D., C. Harvey, P. Resosudarmo, et al. 1995. Environmental and economic costs of soil erosion and conservation benefits. *Science* 267:1117–23.

Pipitone, C., F. Badalamenti, G. D'Anna, et al. 1996. Divieto di pesca a strascico nel Golfo di Castellammare (Sicilia Nord-Occidentale): Alcune considerazioni. *Biologia Marina Mediterranea* 3:200–4.

PIRSA (Primary Industries and Resources South Australia). 1999. *Code of conduct for Australian aquaculture.* www.pir.sa.gov.au.

Pister, E. P. 1991. The Desert Fishes Council: Catalyst for change. In Minckley and Deacon 1991, 55–68.

———. 1999. Professional obligations in the conservation of fishes. *Environ Biol Fish* 55:13–20.

Pitcher, T. J. 1995. The impact of pelagic fish behaviour on fisheries. *Scientia Marina* 59:295–306.

Pitcher, T. J., and D. Pauly. 1998. Rebuilding ecosystems, not sustainability, as the proper goal of fishery management. In Pitcher, Hart, and Pauly 1998, 311–29.

Pitcher, T. J., and P. J. B. Hart, eds. 1995. *The impact of species changes in African lakes.* New York: Chapman and Hall.

Pitcher, T. J., P. J. B. Hart, and D. Pauly, eds. 1998. *Reinventing fisheries management.* Fish and Fisheries Series 23. London: Kluwer Academic.

Pittman, S. J., and C. A. McAlpine. 2003. Movements of marine fish and decapod crustaceans: Process, theory and application. *Adv Mar Biol* 44:205–94.

Plachta, D. T. T., and A. N. Popper. 2003. Evasive responses of American shad (*Alosa sapidissima*) to ultrasonic stimuli. *Acoustics Research Letters Online* 4:25–30.

Plan Development Team. 1990. *The potential of marine fishery reserves for reef fish management in the U.S. southern Atlantic.* NOAA Tech. Memo. NMFS-SEFC-261.

Planes, S. 2002. Biogeography and larval dispersal inferred from population genetic analysis. In Sale 2002a, 201–20.

Planes, S., and G. Lecaillon. 1998. Consequences of the founder effect in the genetic structure of introduced island coral reef fish populations. *Biol J Linn Soc* 63:537–52.

Platania, S. P., and C. Altenbach. 1998. Reproductive strategies and egg types in seven Rio Grande basin cyprinids. *Copeia* 1998:559–69.

Plathong, S., G. J. Inglis, and M. E. Huber. 2000. Effects of self-guided snorkeling trails on corals in a tropical marine park. *Conserv Biol* 14:1821–30.

PMCC (Pacific Marine Conservation Council). 1999. *Rockfish report: The status of west coast rockfish.* www.pmcc.org.

Poach, M. E., P. G. Hunt, E. J. Sadler, et al. 2002. Ammonia volatilization from constructed wetlands that treat swine wastewater. *Trans ASAE* 45:619–27.

Poff, N. L., J. D. Allan, M. B. Bain, et al. 1997. The natural flow regime. *BioScience* 47:769–84.

Poff, N. L., and D. D. Hart. 2002. How dams vary and why it matters for the emerging science of dam removal. *BioScience* 52:659–68.

Poffenberger, J. 1999. Atlantic shark fisheries. Unit 6 in NMFS 1999, 1–4.

Pogonoski, J. J., D. A. Pollard, and J. R. Paxton. 2002. *Conservation overview and action plan for Australian threatened and potentially threatened marine and estuarine fishes.* Environment Australia. www.deh.gov.au.

Policansky, D. 1993a. Evolution and management of exploited fish populations. In *Management strategies for exploited fish populations*, ed. G. Kruse et al., 651–64. Alaska Sea Grant College Program, AK-SG 93–02. Fairbanks, AK: Alaska Sea Grant.

———. 1993b. Fishing as a cause of evolution in fishes. In Stokes, McGlade, and Law, 2–18.

———. 2002. Catch-and release recreational fishing: A historical perspective. In *Recreational fisheries: Ecological, economic and social evaluations*, ed. T. J. Pitcher and C. E. Hollingworth, 74–94. Oxford: Blackwell Science.

Polovina, J. J. 1991. Fisheries applications and biological impacts of artificial reefs. In *Artificial habitats for marine and freshwater fisheries*, ed. W. Seaman Jr. and L. M. Sprague, 153–76. San Diego: Academic Press.

Polunin, N. V. C., and C. M. Roberts, eds. 1996. *Reef fisheries.* London: Chapman and Hall.

Pomeroy, R., R. Agbayani, J. Toledo, et al. 2002. The status of grouper culture in southeast Asia. *SPC Live Reef Fish Inform Bull* 10:22–26.

Pomeroy, R. S., and M. B. Carlos. 1997. Community-based coastal resource management in the Philippines: A review and evaluation of programs and projects, 1984–1994. *Mar Policy* 21:445–64.

Pomeroy, R. S., and R. Rivera-Guieb. 2006. *Fishery co-management: A practical handbook.* Ottawa: International Development Research Centre.

Popper, A. N. 2003. Effects of anthropogenic sounds on fishes. *Fisheries* 28(10): 24–31.

Popper, A. N., and T. J. Carlson. 1998. Application of sound and other stimuli to control fish behavior. *T Am Fish Soc* 127:673–707.

Popper, A. N., and R. R. Fay. 1999. The auditory periphery in fishes. In *Comparative hearing: Fish and amphibians*, ed. R. R. Fay and A. N. Popper, 43–100. New York: Springer-Verlag.

Popper, A. N., R. R. Fay, C. Platt, et al. 2003. Sound detection mechanisms and capabilities of teleost fishes. In *Sensory processing in aquatic environments*, ed. S. P. Collin and N. J. Marshall, 3–38. New York: Springer Verlag.

Portz, D., and H. Tyus. 2004. Fish humps in two Colorado River fishes: A morphological response to cyprinid predation? *Environ Biol Fish* 71:233–45.

Post, J. R., M. Sullivan, S. P. Cox, et al. 2002. Canada's recreational fisheries: The invisible collapse. *Fisheries* 27(1): 6–17.

Postel, S., and S. Carpenter. 1997. Freshwater ecosystem services. In Daily 1997, 195–213.

Postel, S., and B. Richter. 2003. *Rivers for life: Managing water for people and nature.* Washington, DC: Island Press.

Povz, M. 1996. The red data list of freshwater lampreys (Cyclostomata) and fishes (Pisces) of Slovenia. In Kirchhofer and Hefti 1996, 63–72.

Power, G., R. S. Brown, and J. G. Imhof. 1999. Groundwater and fish: Insights from northern North America. *Hydrological Processes* 13:401–22.

Prang, G. 1996. Pursuing the sustainable development of wild caught ornamental fishes in the Middle Rio Negro, Amazonas, Brazil. *Aquatic Survival* 5(1): 1–3.

———. 2001. Aviamento and the ornamental fishery of the Rio Negro, Brazil: Implications for sustainable resource use. In Chao, Petry, et al., 43–73.

Pratt, H. L., and J. I. Castro. 1991. Shark reproduction: Parental investment and limited fisheries—An overview. In *Discovering sharks*, ed. S. H. Gruber, 56–60. Highlands, NJ: American Littoral Society.

Pratt, H. L. Jr., S. H. Gruber, and T. Taniuchi, eds. 1990. *Elasmobranchs as living resources: Advances in the biology, ecology, systematics, and the status of fisheries*. NOAA Tech. Report 90.

Precht, W. F. 1998. The art and science of reef restoration. *Geotimes* 43(1): 16–20.

Pretty, J. L., S. S. C. Harrison, D. J. Shepherd, et al. 2003. River rehabilitation and fish populations: Assessing the benefit of instream structures. *J Appl Ecol* 40:251–65.

Price, K. S., D. A. Flemer, J. L. Taft, et al. 1985. Nutrient enrichment of Chesapeake Bay and its impact on the habitat of striped bass: A speculative hypothesis. *T Am Fish Soc* 114:97–106.

Primavera, J. H. 1998. Tropical shrimp farming and its sustainability. In *Tropical mariculture*, ed. S. deSilva, 257–89. London: Academic Press.

Pringle, C. M., M. C. Freeman, and B. J. Freeman. 2000. Regional effects of hydrologic alterations on riverine macrobiota in the New World: Tropical-temperate comparisons. *BioScience* 50:807–23.

Proudlove, G. S. 1997a. The conservation status of hypogean fishes. *Proceedings 12th International Congress of Speleology. Vol. 3, symposium 9: Biospeleology:* 355–358.

———. *1997b. A synopsis of the hypogean fishes of the world. Proceedings 12th International Congress of Speleology, Vol. 3, Symposium 9: Biospeleology*, 351–54.

———. 2005. *Subterranean fishes of the world: An account of the subterranean (hypogean) fishes described up to 2003 with a bibliography 1541–2004.* Moulis, France: International Society for Subterranean Biology.

Puhlmann, G. 2000. Restoration of oxbow lakes in biosphere reserve mittlere Elbe in Sachsen-Anhalt, Germany. Abstract, International Symposium on Freshwater Fish Conservation: Options for the Future, Albufeira, Portugal.

Pullis, G., and A. Laughland. 1999. Black bass fishing in the U.S. *1996 National survey of fishing, hunting and wildlife-associated recreation*. Report 96-3. Washington, DC: U.S. Fish and Wildlife Service.

Pusey, B. J., and A. H. Arthington. 2003. Importance of the riparian zone to the conservation and management of freshwater fish: A review. *Mar Freshwater Res* 54:1–16.

Pusey, B., M. Kennard, and A. Arthington. 2004. *Freshwater fishes of north-eastern Australia*. Collingwood, Australia: CSIRO.

Pustelnik, G., and O. Guerri. 2002. From the protection of sturgeon *Acipenser sturio*, to river management in general. In Collares-Pereira, Coelho, and Cowx 2002, 143–48.

Pyle, R. L. 1993. Marine aquarium fish. In *Nearshore marine resources of the South Pacific: Information for fisheries development and management*, ed. A. Wright and L. Hill, 135–76. Suva, Fiji: Institute for Pacific Studies.

———. 1996a. How much coral reef biodiversity are we missing? *Global Biodiversity* 6(1): 3–7.

———. 1996b. The twilight zone. *Nat Hist* 105(11): 59–62.

———. 2000. Assessing undiscovered fish biodiversity on deep coral reefs using advanced self-contained diving technology. *Mar Technol Soc J* 34:82–91.

———. In press. How many reef fishes are we missing? Patterns of new species discovery on deep coral reefs in the Indo-Pacific. In *Proceedings of the Deep Reef Fish Symposium*, ed. K. Sulak. La Paz, Mexico: Universidad Autonoma De Baja California Sur.

Pyramid Lake Fisheries. 2000. *Dave Koch Cui-ui Hatchery*. www.pyramidlakefisheries.org/cui-ui.htm.

Quigley, D. T. G., and K. Flannery. 1996. Endangered freshwater fish in Ireland. In Kirchhofer and Hefti 1996, 27–34.

Quinn, J. R. 1994. *The fascinating freshwater fish book: How to catch, keep, and observe your own native fish.* New York: Wiley.

Quirolo, D. 1994. Reef mooring buoys and reef conservation in the Florida Keys: A community and NGO approach. In A. T. White et al., 1994, 80–90.

Raadik, T. A., S. R. Saddlier, and J. D. Koehn. 1996. Threatened fishes of the world: *Galaxias fuscus* Mack, 1936 (Galaxiidae). *Environ Biol Fish* 47:108.

Rabalais, N. N., R. E. Turner, and W. J. Wiseman. 2002. Gulf of Mexico hypoxia, aka "The dead zone." *Annu Rev Ecol Syst* 33:235–63.

Rabeni, C. F., and M. A. Smale. 1995. Effects of siltation on stream fishes and the potential mitigating role of the buffering riparian zone. *Hydrobiologia* 303:211–19.

Radomski, P. J., and T. J. Goeman. 1995. The homogenizing of Minnesota lake fish assemblages. *Fisheries* 20(7): 20–23.

Radonski, G. C., and A. J. Loftus. 1995. Fish genetics, fish hatcheries, wild fish, and other fables. *Am Fish Soc Sym* 15:1–4.

Radtke, R. L. 1995. Forensic biological pursuits of exotic fish origins: Piranha in Hawaii. *Environ Biol Fish* 43:393–99.

Rahel, F. J. 1997. From Johnny Appleseed to Dr. Frankenstein: Changing values and the legacy of fisheries management. *Fisheries* 22(8): 8–9.

———. 2000. Homogenization of fish faunas across the United States. *Science* 288:854–56.

———. 2002. Homogenization of freshwater faunas. *Annu Rev Ecol Syst* 33:291–315.

Rainboth, W. J. 1996. *FAO species identification field guide for fishery purposes: Fishes of the Cambodian Mekong.* FAO: Rome.

Rakes, P. L., J. R. Shute, and P. W. Shute. 1999. Reproductive behaviour, captive breeding, and restoration ecology of endangered fishes. *Environ Biol Fish* 55:31–42.

Raloff, J. 1996. Fishing for answers: Deep trawls leave destruction in their wake—but for how long? *Science News* 150:268–71.

Ramon, F., J. Hernandez-Falcon, B. Nguyen, et al. 2004. Slow wave sleep in crayfish. *P Natl Acad Sci USA* 101:11857–61.

Rand, G. M., and S. R. Petrocelli, eds. 1985. *Fundamentals of aquatic toxicology.* Washington, DC: Hemisphere.

Randall, J. E. 1987a. Collecting reef fishes for aquaria. In *Human impacts on coral reefs: Facts and recommendations*, ed. B. Salvat, 29–39. French Polynesia: Antenne Museum. Cited in Pyle 1993.

———. 1987b. Introductions of marine fishes to the Hawaiian Islands. *B Mar Sci* 41:490–502.

Rask, M., H. Pöysä, P. Nummi, et al. 2001. Recovery of the perch (*Perca fluviatilis*) in an acidified lake and subsequent responses in macroinvertebrates and the goldeneye (*Bucephala clangula*). *Water Air Soil Poll* 130:1367–72.

Ratikin, A., and D. L. Kramer. 1996. Effect of a marine reserve on the distribution of coral reef fishes in Barbados. *Mar Ecol—Prog Ser* 131:97–113.

Ray, J. C., and J. R. Ginsberg. 1999. Endangered species legislation beyond the borders of the United States. *Conserv Biol* 13:956–58.

RBSAPC (University of Georgia River Basin Science and Policy Center). 2001. *Reservoirs in Georgia: Meeting water supply needs while minimizing impacts.* www.rivercenter.uga.edu.

Reade, L. S., and N. K. Waran. 1996. The modern zoo: How do people perceive zoo animals? *App Anim Behav Sci* 47:109–18.

Reaser, J. K., R. Pomerance, and P. O. Thomas. 2000. Coral bleaching and global climate change: Scientific findings and policy recommendations. *Conserv Biol* 14:1500–11.

Red Book of Russia. 2000. *Red book of Russia: Law acts*, official ed. Moscow: State Committee of Russian Federation for Protection of Nature.

Redding, J. M., C. B. Schreck, and F. H. Everest. 1987. Physiological effects on coho salmon and steelhead of exposure to suspended sediments. *T Am Fish Soc* 116:737–44.

Redmond, L. C., and S. L. Redmond. 1987. 50 reasons to fish. *Fisheries* 12(3): 32.

Reebs, S. 2001. *Fish behavior in the aquarium and in the wild.* Ithaca, NY: Cornell Univ. Press.

Reed, J. K. 2002. Deep-water *Oculina* coral reefs of Florida: Biology, impacts, and management. *Hydrobiologia* 471:43–55.

Reed, J. M., and A. P. Dobson. 1993. Behavioural constraints and conservation biology: Conspecific attraction and recruitment. *Trends Ecol Evol* 8:253–56.

REEF (Reef Environmental Education Foundation). 2000. Invasive and exotic species sighted. *REEF Notes* (Spring/Summer): 16. www.reef.org.

ReefBase. 2002. *State of globe's coral reefs chronicled on new, ReefBase, net site shows 2002 to be another worrying year for world's corals.* www.reefbase.org.

Reeves, G. H., J. D. Hall, T. D. Roelofs, et al. 1991. Rehabilitating and modifying stream habitats. *Am Fish Soc Spec Pub* 19:519–58.

Reice, S. R., R. C. Wissmar, and R. J. Naiman. 1990. Disturbance regimes, resilience, and recovery of animal communities and habitats in lotic ecosystems. *Environ Manage* 14:647–59.

Reid, G. M. 1980. "Explosive speciation" of carps in Lake Lanao (Philippines)—Fact or fancy? *Syst Zool* 29:314–16.

Reidmiller, S. 1994. Lake Victoria fisheries: The Kenyan reality and environmental implications. *Environ Biol Fish* 39:29–338.

Reinthal, P. N., K. J. Riseng, and J. S. Parks. 2003. Water management issues in Madagascar: Biodiversity, conservation, and deforestation. In Crisman et al. 2003, 124–42.

Reinthal, P. N., and M. L. J. Stiassny. 1991. The freshwater fishes of Madagascar: A study of an endangered fauna with recommendations for a conservation strategy. *Conserv Biol* 5:231–43.

Reisenbichler, R. R., and S. P. Rubin. 1999. Genetic changes from artificial propagation of Pacific salmon affect the productivity and viability of supplemented populations. *ICES J Mar Sci* 56:459–66.

Reisner, M. 1993. *Cadillac desert: The American West and its disappearing water.* New York: Penguin Books.

Revkin, A. C. 2001. Virus is killing thousands of salmon. *New York Times*, Sept. 7.

Revord, M. E., J. Goldfarb, and S. B. Shurin. 1988. *Aeromonas hydrophila* wound infection in a patient with cyclic neutropenia following a piranha bite. *Pediatr Infect Dis J* 7:70–71.

Reynolds, J. B. 1996. Electrofishing. In *Fisheries techniques*, 2nd ed., ed. B. R. Murphy and D. W. Willis, 221–53. Bethesda, MD: American Fisheries Society.

Reynolds, J. D., and S. Jennings. 2000. The role of animal behaviour in marine conservation. In Gosling and Sutherland 2000, 238–57.

Reynolds, J. E., and D. F. Greboval. 1988. *Socio-economic effects of the evolution of Nile perch fisheries in Lake Victoria: A review.* CIFA Tech. Paper 17. Rome: FAO.

Reznick, D. N. 1993. Norms of reaction in fishes. In Stokes, McGlade, and Law 1993, 72–90.

Rhymer, J. M., and D. Simberloff. 1996. Extinction by hybridization and introgression. *Annu Rev Ecol Syst* 27:83–109.

Ribbink, A. J. 1987. African lakes and their fishes: Conservation scenarios and suggestions. *Environ Biol Fish* 19:3–26.

Ribeiro, C. A. D., L. Belger, E. Pelletier, et al. 2002. Histopathological evidence of inorganic mercury and methyl mercury toxicity in the arctic charr (*Salvelinus alpinus*). *Environ Res* 90:217–25.

Ricciardi, A., and J. B. Rasmussen. 1999. Extinction rates of North American freshwater fauna. *Conserv Biol* 13:1220–22.

Rice, J. 2005. Implementation of the ecosystem approach to fisheries management: Asynchronous co-evolution at the interface between science and policy. *Mar Ecol—Prog Ser* 300:265–70.

Richards, L. J., and J. J. Maguire. 1998. Recent international agreements and the precautionary approach: New directions for fisheries management science. *Can J Fish Aquat Sci* 55:1545–52.

Richards, R. A., and D. G. Deuel. 1987. Atlantic striped bass: Stock status and the recreational fishery. *Mar Fish Rev* 49:58–66.

Richards, W. J., and R. E. Edwards. 1986. Stocking to restore or enhance marine fisheries. In *Fish culture in fisheries management*, ed. R. H. Stroud, 75–80. Bethesda, MD: American Fisheries Society.

Richardson, J. 2005. *Atlas of New Zealand freshwater fishes.* Auckland: New Zealand National Institute of Water & Atmospheric Research. www.niwa.co.nz.

Richardson, M. J., F. G. Whoriskey, and L. H. Roy. 1995. Turbidity generation and biological impacts of an exotic fish *Carassius auratus* introduced into shallow seasonally anoxic ponds. *J Fish Biol* 47:576–85.

Richkus, W. A., and K. Whalen. 2000. Evidence for a decline in the abundance of the American eel, *Anguilla rostrata* (Lesuer), in North America since the early 1980s. *Dana* 12:83–97.

Richmond, R. H. 2005. Recovering populations and restoring ecosystems: Restoration of coral reefs and related marine communities. In Norse and Crowder 2005, 393–409.

Richter, B. D., J. V. Baumgartner, R. Wiggington, et al. 1997. How much water does a river need? *Freshwater Biol* 37:231–49.

Richter, B. D., D. P. Braun, M. A. Mendelson, et al. 1997. Threats to imperiled freshwater fauna. *Conserv Biol* 11:1081–93.

Richter, B. D., A. T. Warren, J. L. Meyer, et al. 2006. A collaborative and adaptive process for developing environmental flow recommendations. *River Res Appl* 22:297–318.

Ricker, W. E. 1972. Hereditary and environmental factors affecting certain salmonid populations. In *The stock concept of Pacific salmon*, ed. R. C. Simon and P. A. Larkin, 19–160. Vancouver: Univ. British Columbia.

———. 1981. Changes in the average size and average age of Pacific salmon. *Can J Fish Aquat Sci* 38:1636–56.

Riegl, B. 2001. Degradation of reef structure, coral and fish communities in the Red Sea by ship groundings and dynamite fisheries. *B Mar Sci* 69:595–611.

———. 2002. Effects of the 1996 and 1998 positive sea-surface temperature anomalies on corals, coral diseases and fish in the Arabian Gulf (Dubai, UAE). *Mar Biol* 140:29–40.

Riegl, B., and K. E. Luke. 1998. Ecological parameters of dynamited reefs in the northern Red Sea and their relevance to reef rehabilitation. *Mar Pollut Bull* 37:488–98.

Riemann, B., and E. Hoffman. 1991. Ecological consequences of dredging and bottom trawling in the Limfjord, Denmark. *Mar Ecol—Prog Ser* 69:171–78.

Rieser, A., C. G. Hudson, and S. E. Roady. 2005. The role of legal regimes in marine conservation. In Norse and Crowder 2005, 362–74.

Rigby, G. R., G. M. Hallegraeff, and C. Sutton. 1999. Novel ballast water heating technique offers cost-effective treatment to reduce the risk of global transport of harmful marine organisms. *Mar Ecol—Prog Ser* 191:289–93.

Rijnsdorp, A. D. 1993. Fisheries as a large-scale experiment on life-history evolution: Disentangling phenotypic and genetic effects in changes in maturation and reproduction of North Sea plaice, *Pleuronectes platessa* L. *Oecologia* 96:391–401.

Riley, A. L. 1998. *Restoring streams in cities.* Washington, DC: Island Press.

Rincon, P. A., J. C. Velasco, N. Gonzales-Sanchez, et al. 1990. Fish assemblages in small streams in western Spain: The influence of an introduced predator. *Arch Hydrobiol* 118:81–91.

Rinne, J. N., J. E. Johnson, B. L. Jensen, et al. 1986. The role of hatcheries in the management and recovery of threatened and endangered fishes. In *Fish culture in fisheries management*, ed. R. H. Stroud, 271–85. Bethesda, MD: American Fisheries Society.

Ritchie, J. C. 1972. Sediment, fish, and fish habitat. *J Soil Water Conserv* 27:124–25.

Rivier, B., and J. Seguier. 1985. Physical and biological effects of gravel extraction in river beds. In *Habitat modification and freshwater fisheries*, ed. J. S. Alabaster, 131–46. Rome: FAO.

Robb, D. H. F., and S. C. Kestin. 2002. Methods used to kill fish: Field observations and literature reviewed. *Anim Welfare* 11:269–82.

Robbins, W. H., and H. R. MacCrimmon. 1974. *The black bass in America and overseas.* Ontario, Canada: Biomanagement Research Enterprises.

Roberts, C. M. 1995. Rapid build-up of fish biomass in a Caribbean marine reserve. *Conserv Biol* 9:815–26.

———. 1996. Settlement and beyond: Population regulation and community structure of reef fishes. In Polunin and Roberts 1996, 85–112.

———. 1997. Ecological advice for the global fisheries crisis. *Trends Ecol Evol* 12:35–38.

———. 2005. Marine protected areas and biodiversity conservation. In Norse and Crowder 2005, 265–79.

Roberts, C. M., J. A. Bohnsack, F. Gell, et al. 2001. Effects of marine reserves on adjacent fisheries. *Science* 294:1920–23.

Roberts, C. M., B. Halpern, S. R. Palumbi, et al. 2001. Designing marine reserve networks. *Conserv Biol in Practice* 2(3): 10–17.

Roberts, C. M., and J. P. Hawkins. 1999. Extinction risk in the sea. *Trends Ecol Evol* 14:241–46.

———. 2000. *Fully-protected marine reserves: A guide.* Washington, DC: WWF Endangered Seas Campaign.

Roberts, C. M., C. J. McClean, J. E. N. Veron, et al. 2002. Marine biodiversity hotspots and conservation priorities for tropical reefs. *Science* 295:1280–84.

Roberts, C. M., N. Quinn, W. Tucker Jr., et al. 1995. Introduction of hatchery-reared Nassau groupers to a coral reef environment. *N Am J Fish Manage* 15:159–64.

Robertson, L. 1997. *Water operations on the Pecos River, New Mexico and the Pecos bluntnose shiner, a federally-listed minnow.* U.S. Conference on Irrigation and Drainage Symposium. Cited in Poff et al. 1997.

Robinet, T. T., and E. E. Feunteun. 2002. Sublethal effects of exposure to chemical compounds: A cause for the decline in Atlantic eels? *Ecotoxicology* 11:265–77.

Robins, C. R., R. M. Bailey, C. E. Bond, et al. 1991. *Common and scientific names of fishes from the United States and Canada*, 5th ed. Bethesda, MD: American Fisheries Society.

Robinson, S. 1984. Collecting tropical marines: Return to reason and common sense. *Freshr Mar Aquar* 7(7): 15–18, 68–73. Cited in Rubec 1986.

Rochet, M.-J. 1998. Short-term effects of fishing on life history traits of fishes. *ICES J Mar Sci* 55:371–91.

Rodhe, H., J. Galloway, and Z. Dianwu. 1992. Acidification in Southeast Asia: Prospects for the coming decades. *Ambio* 21:148–50.

Rodriguez, J. P., and F. Rojas-Suarez. 1996. *Lista roja de la fauna Venezolana, 1995.* www.fpolar.org.ve/librorojo/listaroja.htm.

Roemmich, D., and J. McGowan. 1995. Climatic warming and the decline of zooplankton in the California current. *Science* 267:1324–26.

Rogers, J. B., and Y. L. Builder. 1999. Pacific Coast groundfish fisheries. Unit 15 in NMFS 1999.

Rohter, L. 2001. A collector's item costs Brazilian divers dearly. *New York Times*, Nov. 5. www.nytimes.com/2001/11/05.

Rolston, H. III. 1988. *Environmental ethics: Duties to and values in the natural world.* Philadelphia: Temple University Press.

Romero, A. 1998. Threatened fishes of the world: *Typhlichthys subterraneus* (Girard, 1860) (Amblyopsidae). *Environ Biol Fish* 53:74.

Romero, A., and L. Bennis. 1998. Threatened fishes of the world: *Amblyopsis spelaea* DeKay, 1842 (Amblyopsidae). *Environ Biol Fish* 51:420.

Roni, P., T. J. Beechie, R. E. Bilby, et al. 2002. A review of stream restoration techniques and a hierarchical strategy for prioritizing restoration in Pacific Northwest watersheds. *N Am J Fish Manage* 22:1–20.

Roper Starch Worldwide. 1995. Public attitudes toward zoos, aquariums, and animal theme parks. Cited in Dierking et al. 2001.

Rose, D. A. 1996. *An overview of world trade in sharks and other cartilaginous fishes*. Cambridge, UK: TRAFFIC International. www.TRAFFIC.org.

———. 1998. *Shark fisheries and trade in the Americas. Vol. 1: North America.* Cambridge, UK: TRAFFIC North America. www.TRAFFIC.org/factfile/us-shark_trade.html.

Rose, J. D. 2002. The neurobehavioral nature of fishes and the question of awareness and pain. *Rev Fish Sci* 10:1–38.

Rose, J. D., G. S. Marrs, C. Lewis, et al. 2000. Whirling disease behavior and its relation to pathology of brain stem and spinal cord in rainbow trout. *J Aquat Anim Health* 12:107–18.

Rosenberg, D. M., F. Berkes, R. A. Bodaly, et al. 1997. Large-scale impacts of hydroelectric development. *Environ Rev* 5:27–54.

Rosenberg, D. M., P. McCully, and C. M. Pringle. 2000. Global-scale environmental effects of hydrological alterations: Introduction. *BioScience* 50:746–51.

Rosenberger, A., and P. L. Angermeier. 2003. Ontogenetic shifts in habitat use by the endangered Roanoke logperch (*Percina rex*). *Freshwater Biol* 48:1563–77.

Rosenberger, A. E., and L. J. Chapman. 2000. Respiratory characters of three species of haplochromine cichlids: Implications for use of wetland refugia. *J Fish Biol* 57:483–501.

Rosenfeld, J. 2003. Assessing the habitat requirements of stream fishes: An overview and evaluation of different approaches. *T Am Fish Soc* 132:953–68.

Rosenfield, J. A., and A. Kodric-Brown. 2003. Sexual selection promotes hybridization between Pecos pupfish, *Cyprinodon pecosensis*, and sheepshead minnow, *C. variegatus*. *J Evolution Biol* 16:595–606.

Rosenfield, J. A., S. Nolasco, S. Lindauer, et al. 2004. The role of hybrid vigor in the replacement of Pecos pupfish by its hybrids with sheepshead minnow. *Conserv Biol* 18:1590–98.

Rosenthal, H. 1980. Implications of transplantations to aquaculture and ecosystems. *Mar Fish Rev* 1980:1–14.

Rosenzweig, M. L. 2003. *Win-win ecology: How the earth's species can survive in the midst of human enterprise*. Oxford: Oxford Univ. Press.

Rosgen, D. 1996. *Applied river morphology*. Pagosa Springs, CO: Wildland Hydrology.

Ross, M. R. 1997. *Fisheries conservation and management.* Upper Saddle River, NJ: Prentice Hall.

Ross, R. M., W. A. Lellis, R. M. Bennett, et al. 2001. Landscape determinants of nonindigenous fish invasions. *Biol Invasions* 3:347–61.

Ross, S. T. 1991. Mechanisms structuring stream fish assemblages: Are there lessons from introduced species? *Environ Biol Fish* 30:359–68.

Rosseland, B. O., F. Kroglund, M. Staurnes, et al. 2001. Tolerance to acid water among strains and life stages of Atlantic salmon (*Salmo salar* L.). *Water Air Soil Poll* 130:899–904.

Roth, E., H. Rosenthal, and P. Burbridge. 2000. A discussion of the use of the sustainability index: "Ecological footprint" for aquaculture production. *Aquat Living Resour* 13:461–69.

Roth, N. E., J. D. Allan, and D. L. Erickson. 1996. Landscape influences on stream biotic integrity assessed at multiple spatial scales. *Landscape Ecol* 11:141–56.

Rothschild, B. J. 1986. *Dynamics of marine fish populations*. Cambridge: Harvard Univ. Press.

———. 1991. Multispecies interactions on Georges Bank. *ICES Mar Sci Sym* 193:86–92.

Rowell, C. A. 1993. The effects of fishing on the timing of maturity in North Sea cod (*Gadus morhua* L.). In Stokes, McGlade, and Law 1993, 44–61.

Royce, W. F. 1996. *Introduction to the practice of fishery science*, rev. ed. San Diego: Academic Press.

Rubec, C. D. A. 1994. Canada's federal policy on wetland conservation: A global model. In *Global wetlands: Old world and new*, ed. W. J. Mitsch, 909–17. Amsterdam: Elsevier.

Rubec, P. J. 1986. The effects of sodium cyanide on coral reefs and marine fish in the Philippines. In *The first Asian fisheries forum*, ed. J. L. McLean et al., 297–802. Manila, Philippines: Asian Fisheries Society. www.actwin.com.

———. 1988. The need for conservation and management of Philippine coral reefs. *Environ Biol Fish* 23:141–54.

Rubec, P. J., F. Cruz, V. Pratt, et al. 2000. Cyanide-free, net-caught fish for the marine aquarium trade. *SPC Live Reef Fish Inform Bull* 7:28–34.

Rubin, D. A. 1985. Effect of pH on sex ratio in cichlids and a poeciliid (Teleostei). *Copeia* 1985:233–35.

Ruddle, K. 1996. Geography and human ecology of reef fisheries. In Polunin and Roberts 1996, 137–60.

Ruddle, K., and R. E. Johannes, eds. 1985. *The traditional knowledge and management of coastal systems in Asia and the Pacific*. Jakarta: UNESCO.

Rudstam, L., G. Aneer, and M. Hilden. 1994. Top-down control in the pelagic Baltic ecosystem. *Dana* 10:105–29. Cited in S. J. Hall 1999.

Ruesink, J. L. 2005. Global analysis of factors affecting the outcome of freshwater fish introductions. *Conserv Biol* 19:1883–93.

Ruhl, J. B. 2004. The battle over Endangered Species Act methodology. *Environ Law* 34:555–603.

Ruiz, G. M., J. T. Carlton, E. D. Grosholz, et al. 1997. Global invasions of marine and estuarine habitats by non-indigenous species: Mechanisms, extent, and consequences. *Am Zool* 37:619–30.

Ruiz, G. M., P. W. Fofonoff, J. T. Carlton, et al. 2000. Invasion of coastal marine communities in North America: Apparent patterns, processes, and biases. *Annu Rev Ecol Syst* 31:481–531.

Russ, G. R. 1984. A review of coral reef fisheries. *UNESCO Rep Mar Sci* 27:74–92. Cited in Birkeland 1997b.

———. 1991. Coral reef fisheries: Effects and yields. In Sale 1991, 601–35. San Diego: Academic Press.

———. 2002. Yet another review of marine reserves as reef fishery management tools. In Sale 2002a, 421–33. San Diego: Academic Press.

Russ, G. R., and A. C. Alcala. 1996. Marine reserves: Rates and patterns of recovery and decline of large predatory fish. *Ecol Appl* 6:947–61.

Russ, G. R., A. C. Alcala, A. P. Maypa, et al. 2004. Marine reserve benefits local fisheries. *Ecol Appl* 14:597–606.

Ryan, G. J. 1996. The shipping industry's role in slowing ruffe expansion throughout the Great Lakes. *Fisheries* 21(5): 22–23.

Ryan, P. A. 1991. Environmental effects of sediment on New Zealand streams: A review. *New Zeal J Mar Fresh* 25:207–21.

Ryder, O. A. 1986. Species conservation and systematics: The dilemma of subspecies. *Trends Ecol Evol* 1:9–10.

Ryman, N., and F. Utter, eds. 1987. *Population genetics and fishery management*. Seattle: Univ. Washington Press.

Ryman, N., F. Utter, and L. Laikre. 1995. Protection of intraspecific biodiversity of exploited fishes. *Rev Fish Biol Fisher* 5:417–46.

Ryther, J. H. 1969. Photosynthesis and fish production in the sea. *Science* 166:72–76.

Sadovy, Y. 1992. A preliminary assessment of the marine aquarium export trade in Puerto Rico. *P 7th Int Coral Reef Sym* 2:1014–22.

———. 1994. Grouper stocks of the western central Atlantic: The need for management and management needs. *P Gulf Caribb Fish Inst* 43:43–63.

———. 1998. The live reef fish trade: A role for importers in combating destructive fishing practices—Example of Hong Kong, China. *Proceedings APEC Workshop on the Impacts of Destructive Fishing Practices on the Marine Environment*, 200–7.

———. 2001a. The live reef food fish trade in Hong Kong: Problems and prospects. In *Marketing and shipping live aquatic products*. Univ. Alaska Sea Grant Publication AK-SG-01-03:183-192. Fairbanks, AK: Alaska Sea Grant.

———. 2001b. The threat of fishing to highly fecund fishes. *J Fish Biol* 59(suppl. A): 90–108.

———. 2002. Death in the live fish reef trades. *SPC Live Fish Inform B* 10:3–5.

Sadovy, Y., M. Kulbicki, P. Labrosse, et al. 2003. The humphead wrasse, *Cheilinus undulatus*: Synopsis of a threatened and poorly known giant coral reef fish. *Rev Fish Biol Fisher* 13:327–64.

Sadovy, Y., and W.-L. Cheung. 2003. Near extinction of a highly fecund fish: The one that nearly got away. *Fish and Fisheries* 4:86–99.

Sadovy, Y., and A. S. Cornish. 2000. *Reef fishes of Hong Kong*. Hong Kong: Hong Kong Univ. Press.

Sadovy, Y. J., and J. Pet. 1998. Wild collection of juveniles for grouper mariculture: Just another capture fishery? *SPC Live Reef Fish Inform B* 4:36–39.

Sadovy, Y. J., and A. C. J. Vincent. 2002. The trades in live reef fishes for food and aquaria: Issues and impacts. In Sale 2002a, 391–420.

Safina, C. 1993. Bluefin tuna in the west Atlantic: Negligent management and the making of an endangered species. *Conserv Biol* 7:229–34.

———. 1995. The world's imperiled fish. *Sci Am* 275(5): 46–53.

———. 1997. *Song for the blue ocean*. New York: Henry Holt.

———. 2001a. Fish conservation. *Encyclopedia of biodiversity*, vol. 2, 783–99. San Diego: Academic Press.

———. 2001b. Tuna conservation. In *Tuna: Physiology, ecology, and evolution*, ed. B. A. Block and E. D. Stevens, 413–59. San Diego: Academic Press.

———. 2002. *Eye of the albatross: Visions of hope and survival*. New York: Henry Holt.

Sagua, V. O. 1997. *The contribution of fisheries in man-made lakes and irrigation dams to food security in the Sahelian zone*. CIFA/PD:S/97/2. Rome: FAO.

Said, R., G. Volpin, B. Grimberg, et al. 1998. Hand infections due to non-cholera vibrio after injuries from St Peter's fish (*Tilapia zillii*). *J Hand Surgery—British and European* 23B: 808–10.

Saila, S. B., V. L. Kocic, and J. W. McManus. 1993. Modeling the effects of destructive fishing practices on tropical coral reefs. *Mar Ecol—Prog Ser* 94:51–60.

Sainsbury, J. C. 1996. *Commercial fishing methods: An introduction to vessels and gears*, 3rd ed. Oxford: Fishing News Books.

Sainsbury, K. J. 1988. The ecological basis of multispecies fisheries and management of a demersal fishery in tropical Australia. In *Fish population dynamics*, 2nd ed., ed. J. A. Gulland, 349–382. New York: Wiley.

Sainsbury, K. J., R. A. Campbell, R. Lindholm, et al. 1997. Experimental management of an Australian multispecies fishery: Examining the possibility of trawl induced habitat modification. *Am Fish Soc Sym* 20:107–12.

Sainsbury, K. J., R. A. Campbell, and A. W. Whitelaw. 1993. Effects of trawling on the marine habitat on the north west shelf of Australia and implication for sustainable fisheries management. In *Sustainable fisheries through sustaining fish habitat*, ed. D. A. Hancock, 137–45. Canberra: Australia Bureau of Resource Science.

Sala, E., E. Ballesteros, and R. M. Starr. 2001. Rapid decline of Nassau grouper spawning aggregations in Belize: Fishery management and conservation needs. *Fisheries* 26(10): 23–30.

Sale, P. F., ed. 1991. *The ecology of fishes on coral reefs*. San Diego: Academic Press.

———, ed. 2002a. *Coral reef fishes: Dynamics and diversity in a complex ecosystem*. San Diego: Academic Press.

———. 2002b. The science we need to develop for more effective management. In Sale 2002a, 361–76.

Salvat, B. 1992. Coral reefs: A challenging ecosystem for human societies. *Global Environ Chang* 2:12–18.

Sanchez-Jerez, P., and A. Ramos-Espla. 2002. Changes in fish assemblages associated with the deployment of an antitrawling reef in seagrass meadows. *T Am Fish Soc* 129:1150–59.

Sandøy, S., and R. M. Langåker. 2001. Atlantic salmon and acidification in southern Norway: A disaster in the 20th century, but a hope for the future? *Water Air Soil Poll* 130:1343–48.

Sanz, N., J. L. Garcia-Marin, and C. Pla. 2000. Divergence of brown trout (*Salmo trutta*) within glacial refugia. *Can J Fish Aquat Sci* 57:2201–10.

Saunders, D. L., J. J. Meeuwig, and A. C. J. Vincent. 2002. Freshwater protected areas: Strategies for conservation. *Conserv Biol* 16:30–41.

Savino, J. F., D. J. Jude, and M. J. Kostich. 2001. Use of electrical barriers to deter movement of round goby. *Am Fish Soc Sym* 26:171–82.

Sayer, M. D., J. W. Treasurer, and M. J. Costello. 1996. *Wrasse biology and use in aquaculture*. London: Fishing News Books.

Sazima, I., and S. de Guimaraes. 1987. Scavenging on human corpses as a source for stories about man-eating piranhas. *Environ Biol Fish* 20:75–77.

SBI (Sustainable Biosphere Initiative). 1998. Fisheries as experimental systems in ecology. *B Ecol Soc Am* (April): 165–66.

Schaefer, J. F., E. Marsh-Matthews, D. E. Spooner, et al. 2003. Effects of barriers and thermal refugia on local movement of the threatened leopard darter, *Percina pantherina*. *Environ Biol Fish* 66:391–400.

Scharpf, C. 2000. Politics, science, and the fate of the Alabama sturgeon. *Am Currents* 26(3): 6–14. www.nanfa.org.

———. 2001. American treasures: The importance of keeping native fishes in the aquarium. *Trop Fish Hobbyist* 50(3): 90–94.

———. 2003. Made in America. *Trop Fish Hobbyist* 51(11): 68–74.

Scheidegger, K. J., and M. B. Bain. 1995. Larval fish distribution and microhabitat use in free-flowing and regulated rivers. *Copeia* 1995:125–35.

Shick, J. M., M. P. Lesser, and P. L. Jokiel. 1996. Effects of ultraviolet radiation on corals and other coral reef organisms. *Global Change Biol* 2:527–45.

Schiemer, F., and M. Zalewski. 1992. The importance of riparian ecotones for diversity and productivity of riverine fish communities. *Neth J Zool* 42:323–35.

Schiermeier, Q. 2003. Europe dithers as Canada cuts cod fishing. *Nature* 423:212.

Schill, D. J., and K. F. Beland. 1995. Electrofishing injury studies: A call for population perspective. *Fisheries* 20(6): 28–29.

Schill, D. J., J. S. Griffith, and R. E. Gresswell. 1986. Hooking mortality of cutthroat trout in a catch-and-release segment of the Yellowstone River, Yellowstone National Park. *N Am J Fish Manage* 6:226–32.

Schill, D. J., and R. I. Scarpella. 1997. Barbed hook restrictions in catch-and-release fisheries: A social issue. *N Am J Fish Manage* 17:873–81.

Schindler, D. E., J. F. Kitchell, and R. Ogutu-Ohwayo. 1998. Ecological consequences of alternative gill net fisheries for Nile perch in Lake Victoria. *Conserv Biol* 12:56–64.

Schindler, D. E., M. D. Scheuerell, J. W. Moore, et al. 2003. Pacific salmon and the ecology of coastal ecosystems. *Front Ecol Environ* 1:31–37.

Schleser, D. M. 1998. *North American native fishes for the home aquarium*. Hauppage, NY: Barrons.

Schleyer, M. H., and B. J. Tomalin. 2000. Damage on South African coral reefs and an assessment of their sustainable diving capacity using a fisheries approach. *B Mar Sci* 67:1025–42.

Schlosser, I. J. 1991. Stream fish ecology: A landscape perspective. *BioScience* 41:704–12.

Schmidt, A. S., M. S. Bruun, I. Dalsgaard, et al. 2000. Occurrence of antimicrobial resistance in fish-pathogenic and environmental bacteria associated with four Danish rainbow trout farms. *Appl Environ Microb* 66:4908–17.

Schoenherr, A. A. 1981. The role of competition in the replacement of native fishes by introduced species. In *Fishes in North American deserts*, ed. R. J. Naiman and D. L. Soltz, 173–203. New York: Wiley.

Schofield, P. J., and L. J. Chapman. 1999. Interactions between Nile perch, *Lates niloticus*, and other fishes in Lake Nabugabo, Uganda. *Environ Biol Fish* 55:343–58.

———. 2000. Hypoxia tolerance of introduced Nile perch: Implications for survival of indigenous fishes in the Lake Victoria basin. *African Zool* 35:35–42.

Scholik, A. R., and H. Y. Yan. 2001. Effects of underwater noise on auditory sensitivity of a cyprinid fish. *Hearing Res* 152:17–24.

———. 2002. Effects of boat engine noise on the auditory sensitivity of the fathead minnow, *Pimephales promelas*. *Environ Biol Fish* 63:203–9.

Scholin, C. A., F. Gulland, G. J. Doucette, et al. 2000. Mortality of sea lions along the central California coast linked to a toxic diatom bloom. *Nature* 403:80–84.

Scholz, T. 1999. Parasites in cultured and feral fish. *Vet Parasitol* 84:317–35.

Schramm, H. L. 1996. Resurrecting hatchery bashing: A useless pastime. *Fisheries* 21(11): 26.

Schramm, H. L. Jr., and R. G. Piper, eds. 1995. Uses and effects of cultured fishes in aquatic ecosystems. *Am Fish Soc Sym* 15.

Schreck, C. B. 1990. Physiological, behavioral, and performance indicators of stress. In Biological indicators of stress in fish. *Am Fish Soc Sym* 8:29–37.

———. 2000. Accumulation and long-term effects of stress in fish. In *The biology of animal stress*, ed. G. P. Moberg and J. A. Mench, 147–58. Wallingford, UK: CABI.

Schreck, C. B., B. L. Olla, and M. W. Davis. 1997. Behavioural responses to stress. In *Fish stress and health in aquaculture*, ed. O. K. Iwama et al., 145–70. Seminar series 62. Cambridge: Society for Experimental Biology.

Schueler, T. 1994. The importance of imperviousness. *Watershed Protection Techniques* 1:100–11.

Schultz, I. R., A. Skillman, J.-M. Nicolas, et al. 2003. Fertility of sexually maturing male rainbow trout (*Oncorhynchus mykiss*). *Environ Toxicol Chem* 22:1272–80.

Schuppli, C. A. 2000. *A review of the 1998/99 sport fishing regulations in Canada with specific reference to the humane considerations*. Ottawa: Animal Welfare Foundation of Canada. www.awfc.ca.

Schwartz, M. F., and C. E. Boyd. 1994. Effluent quality during harvest of channel catfish from watershed ponds. *Progressive Fish Culturist* 56:25–32.

Schweder, T., G. S. Hagen, and E. Hatlebakk. 2000. Direct and indirect effects of minke whale abundance on cod and herring fisheries: A scenario experiment for the Greater Barents Sea. *North Atlantic Marine Mammal Commission Scientific Publications* 2:120–32. Cited in Corkeron 2004.

Schwinghamer, P., J. Y. Guigné, and W. C. Siu. 1996. Quantifying the impact of trawling on benthic habitat structure using high resolution acoustics and chaos theory. *Can J Fish Aquat Sci* 53:288–96.

Scientists' Working Group on Biosafety. 1998. *Manual for assessing ecological and human health effects of genetically engineered organisms*. Edmonds, WA: Edmonds Institute.

Scoppettone, G. G., and G. Vinyard. 1991. Life history and management of four endangered lacustrine suckers. In Minckley and Deacon 1991, 359–77.

Scott, M. C. 2001. Integrating the stream and its valley: Land use change, aquatic habitat, and fish assemblages. PhD dissertation, Univ. Georgia.

Scott, M. C., and G. S. Helfman. 2001. Native invasions, homogenization, and the mismeasure of integrity of fish assemblages. *Fisheries* 26(11): 6–15.

Scott, W. B., and E. J. Crossman. 1973. Freshwater fishes of Canada. *Fish Res Board Can B* 184.

Scribner, K. T., K. S. Page, and M. L. Bartron. 2000. Hybridization in freshwater fishes: A review of case studies and cytonuclear methods of biological inference. *Rev Fish Biol Fisher* 10:293–323.

Scudder, G. G. W. 1999. Endangered species protection in Canada. *Conserv Biol* 13:963–65.

SEAFDEC (Southeast Asian Fisheries Development Center). 2001. *Regionalization of the Code of Conduct for Responsible Fisheries*. www.seafdec.org/asean/fcg04.html.

Sear, D. A. 1994. River restoration and geomorphology. *Aquat Conserv* 4:169–77.

Sebens, K. P. 1994. Biodiversity of coral reefs: What are we losing and why? *Am Zool* 34:115–34.

Secor, D. H. 2000. Longevity and resilience of Chesapeake Bay striped bass. *ICES J Mar Sci* 57:808–15.

Secretaría de Solidaridad. 1994. Norma Oficial Mexicana NOM-059-ECOL-1994. *Diaro Oficial* 488 (May 16): 2–60.

Sedell, J. R., G. H. Reeves, and P. A. Bisson. 1997. Habitat policy for salmon in the Pacific Northwest. In Stouder, Bisson, and Naiman 1997, 375–87.

Seehausen, O., J. J. M. van Alphen, and F. Witte. 1997a. Cichlid fish diversity threatened by eutrophication that curbs sexual selection. *Science* 277:1808–11.

Seehausen, O., F. Witte, E. F. Katunzi, et al. 1997b. Patterns of the remnant cichlid fauna in southern Lake Victoria. *Conserv Biol* 11:890–904.

Seehorn, M. E. 1992. *Stream habitat improvement handbook*. Tech. Pub. R8-TP 16. Atlanta, GA: U.S. Forest Service, Southern Region.

Segner, H., K. Caroll, M. Fenske, et al. 2003. Identification of endocrine-disrupting effects in aquatic vertebrates and invertebrates: Report from the European IDEA project. *Ecotox Environ Safe* 54:302–14.

Semmens, B. X., E. R. Buhle, A. K. Salomon, et al. 2004. A hotspot of non-native marine fishes: Evidence for the aquarium trade as an invasion pathway. *Mar Ecol—Prog Ser* 266:239–44.

Serafy, J. E. 1997. Effects of freshwater canal discharge on fish assemblages in a subtropical bay: Field and laboratory observations. *Mar Ecol—Prog Ser* 160:161–72.

Serrano, R., A. Simal-Julián, E. Pitarch, et al. 2003. Biomagnification study on organochlorine compounds in marine aquaculture: The sea bass (*Dicentrarchus labrax*) as a model. *Environ Sci Technol* 37:3375–81.

Servizi, J. A., and D. W. Martens. 1992. Sublethal responses of coho salmon (*Oncorhynchus kisutch*) to suspended sediments. *Can J Fish Aquat Sci* 49:1389–95.

Seymour, E. A., and A. Bergheim. 1991. Towards a reduction of pollution from intensive aquaculture with reference to the farming of salmonids in Norway. *Aquacult Eng* 10:73–88.

Shafland, P. L. 1995. Introduction and establishment of a successful butterfly peacock fishery in southeast Florida canals. *Am Fish Soc Sym* 15:443–51.

———. 1996a. Exotic fish assessments: An alternative view. *Rev Fish Sci* 4:123–32.

———. 1996b. Exotic fishes of Florida, 1994. *Rev Fish Sci* 4:101–22.

Shafland, P. L., and K. J. Foote. 1979. A reproducing population of *Serrasalmus humeralis* Valenciennes in southern Florida. *Florida Sci* 42:206–14.

Shaklee, J. B., and P. Bentzen. 1998. Genetic identification of stocks of marine fish and shellfish. *B Mar Sci* 62:589–621.

Sharber, N. G., S. W. Carothers, J. P. Sharber, et al. 1994. Reducing electrofishing-induced injury of rainbow trout. *N Am J Fish Manage* 14:340–46.

Sharp, G. D., and D. R. McLain. 1995. Fisheries, El Nino–Southern Oscillation and upper temperature records: An eastern Pacific example. *Oceanography* 6:13–22.

Sheldon, A. L. 1987. Rarity: Patterns and consequences for stream fishes. In *Community and evolutionary ecology of North American stream fishes*, ed. W. J. Matthews and D. C. Heins, 203–9. Norman: Univ. Oklahoma Press.

Shenker, J. M., E. D. Maddox, E. Wishinski, et al. 1993. Onshore transport of settlement-stage Nassau grouper *Epinephelus striatus* and other fishes in Exuma Sound, Bahamas. *Mar Ecol—Prog Ser* 98:31–43.

Shepherd, A. R. D., R. M. Warwick, K. R. Clarke, et al. 1992. An analysis of fish community responses to coral mining in the Maldives. *Environ Biol Fish* 33:367–80.

Sheppard, C. R. C., M. Spalding, C. Bradshaw, et al. 2002. Erosion vs. recovery of coral reefs after 1998 El Niño: Chagos reefs, Indian Ocean. *Ambio* 31:40–48.

Sherman, K., N. A. Jaworski, and T. J. Smayda. 1996. *The northeast shelf ecosystem: Assessment, sustainability, and management.* Cambridge, MA: Blackwell Science.

Shiganova, T. A., and Y. V. Bulgakova. 2000. Effects of gelatinous plankton on Black Sea and Sea of Azov fish and their food resources. *ICES J Mar Sci* 57:641–48.

Shipley, J. B., ed. 2004. Aquatic protected areas as fisheries management tools. *Am Fish Soc Sym* 42.

Shoemaker, C. A., P. H. Klesius, and J. J. Evans. 2001. Prevalence of *Streptococcus iniae* in tilapia, hybrid striped bass, and channel catfish on commercial fish farms in the United States. *Am J Vet Res* 62:174–77.

Short, J., J. E. Kinnear, and A. Robley. 2002. Surplus killing by introduced predators in Australia: Evidence for ineffective anti-predator adaptations in native prey species? *Biol Conserv* 103:283–301.

Short J. W., S. D. Rice, R. A. Heintz, et al. 2003. Long-term effects of crude oil on developing fish: Lessons from the *Exxon Valdez* oil spill. *Energy Sources* 25:509–17.

Shulman, M. J. 1985. Recruitment of coral reef fishes: Effects of distribution of predators and shelter. *Ecology* 66:1056–66.

Shulman, M. J., and J. C. Ogden. 1987. What controls tropical reef fish populations: Recruitment or benthic mortality? An example in the Caribbean reef fish *Haemulon flavolineatum*. *Mar Ecol—Prog Ser* 39:233–42.

Shumway, C. A. 1999. A neglected science: Applying behavior to aquatic conservation. *Environ Biol Fish* 55:183–201.

Silsbee, D. G., and G. L. Larson. 1983. A comparison of streams in logged and unlogged areas of Great Smoky Mountains National Park. *Hydrobiologia* 102:99–111.

Silvano, R. A. M., and A. Begossi. 2005. Local knowledge on a cosmopolitan fish: Ethnoecology of *Pomatomus saltatrix* (Pomatomidae) in Brazil and Australia. *Fish Res* 71:43–59.

Simmonds, M. P., and J. D. Hutchinson, eds. 1996. *The conservation of whales and dolphins*. Chichester, UK: Wiley.

Simon, T. P., ed. 1999. *Assessing the sustainability and biological integrity of water resources using fish communities*. Boca Raton, FL: CRC Press.

Simon, T. P., and J. Lyons. 1995. Application of the Index of Biotic Integrity to evaluate water resource integrity in freshwater ecosystems. In *Biological assessment and criteria: Tools for water resource planning and decision making*, ed. W. S. Davis and T. P. Simon, 245–62. Boca Raton, FL: CRC Press.

Simons, D. B., and R. Li. 1984. *Final report for analysis of channel degradation and bank erosion in the lower Kansas River*. MRD Sediment Series 35. Kansas City, MO: U.S. Army Corps of Engineers.

Simpfendorfer, C. A. 2000. Predicting population recovery rates for endangered western Atlantic sawfishes using demographic analysis. *Environ Biol Fish* 58:371–77.

Simpson, P. W., J. R. Newman, A. Keirn, et al. 1982. *Manual of stream channelization impacts on fish and wildlife*. FWS/OBS-82/24. Washington, DC: U.S. Fish and Wildlife Service.

Simpson, S. 2001. Fishy business. *Sci Am* 285(1): 83–89.

Simpson, S. D., M. Meekan, J. Montgomery, et al. 2005. Homeward sound. *Science* 308:221.

Sims, D. W., and P. C. Reid. 2002. Congruent trends in long-term zooplankton decline in the north-east Atlantic and basking shark (*Cetorhinus maximus*) fishery catches off west Ireland. *Fish Oceanogr* 11:59–63.

Sinclair, A. F., and S. A. Murawski. 1999. Why have groundfish stocks declined? In *Northwest Atlantic groundfish: Perspectives on a fishery collapse*, ed. J. Boreman et al., 71–93. Bethesda, MD: American Fisheries Society.

Sindermann, C. J. 1986. Strategies for reducing risks from introductions of aquatic organisms: A marine perspective. *Fisheries* 11(2): 10–15.

Singer, P. 1993. *Practical ethics*, 2nd ed. Cambridge: Cambridge Univ. Press.

Sissenwine, M. P., P. M. Mace, J. E. Powers, et al. 1998. A commentary on western Atlantic bluefin tuna assessments. *T Am Fish Soc* 127:838–55.

Skelton, P. H. 1977. *South African red data book: Fishes*. South African National Scientific Programmes Report 14. Pretoria: Council for Scientific and Industrial Research.

———. 1987. *South African red data book: Fishes*. South African National Scientific Programmes Report 137. Pretoria: Council for Scientific and Industrial Research.

———. 1990. The conservation and status of threatened fishes in southern Africa. *J Fish Biol* 37(suppl. A): 87–95.

———. 2002. An overview of the challenges of conserving freshwater fishes in South Africa. In Collares-Pereira, Coelho, and Cowx 2002, 221–36.

Skelton, P. H., J. A. Cambray, A. Lombard, et al. 1995. Patterns of distribution and conservation status of freshwater fishes in South Africa. *South African J Zool* 30:71–81.

Skjelkvåle, B. L., J. L. Stoddard, and T. Andersen. 2001. Trends in surface water acidification in Europe and North America (1989–1998). *Water Air Soil Poll* 130:787–92.

Skomal, G. B., and B. C. Chase. 2002. The physiological effects of angling on post-release survivorship in tunas, sharks, and marlin. *Am Fish Soc Sym* 30:135–38.

Slaney, T. L., K. D. Hyatt, T. G. Northcote, et al. 1996. Status of anadromous salmon and trout in British Columbia and Yukon. *Fisheries* 21(10): 20–35.

Smith, B. R., and J. J. Tibbles. 1980. Sea lamprey (*Petromyzon marinus*) in Lakes Huron, Michigan, and Superior: History of invasion and control, 1936–1978. *Can J Fish Aquat Sci* 37:1780–1801.

Smith, C. L. 1972. A spawning aggregation of Nassau grouper *Epinephelus striatus* (Bloch). *T Am Fish Soc* 101:257–61.

———. 1985. *Inland fishes of New York*. Albany: New York State Dept. of Environmental Conservation.

Smith, C. T., R. J. Nelson, S. Pollard, et al. 2002. Population genetic analysis of white sturgeon (*Acipenser transmontanus*) in the Fraser River. *J Appl Ichthyol* 18:307–12.

Smith, L. D., M. J. Wonham, L. D. McCann, et al. 1999. Invasion pressure to a ballast-flooded estuary and an assessment of innoculant survival. *Biol Invasions* 1:67–87.

Smith, N. 1972. *A history of dams*. Secaucus, NJ: Citadel Press.

Smith, P., M. Hiney, and O. Samuelsen. 1994. Bacterial resistance to antimicrobial agents used in fish farming: A critical evaluation of method and meaning. *Ann Rev Fish Dis* 4: 273–313. Cited in Goldburg and Triplett 1997.

Smith, P. J. 1994. *Genetic diversity of marine fisheries resources: Possible impacts of fishing.* FAO Fisheries Tech. Paper 334. Rome: FAO.

Smith, P. J., and P. G. Benson. 1997. Genetic diversity in orange roughy from the east of New Zealand. *Fish Res* 31:197–213.

Smith, P. J., F. Ricc, and M. McVeagh. 1991. Loss of genetic diversity due to fishing pressure. *Fish Res* 10:309–16.

Smith, P. W. 1968. An assessment of changes in the fish fauna in two Illinois rivers and its behavior on their future. *T Illinois State Acad Sci* 61:31–45.

Smith, S. H. 1995. *Early changes in the fish community of Lake Ontario.* Tech. Report 60. Ann Arbor, MI: Great Lakes Fishery Commission.

Smith, S. V. 1978. Coral-reef area and the contributions of reefs to processes and resources in the world's oceans. *Nature* 273:225–26.

Smith, T. D. 1983. Changes in size of three dolphin (*Stenella* spp.) populations in the eastern tropical Pacific. *Fish Bull U.S.* 81:1–13.

———. 1994. *Scaling fisheries: The science of measuring the effects of fishing, 1855–1955.* Cambridge: Cambridge Univ. Press.

Smith, T. I. J., and J. P. Clugston. 1997. Status and management of Atlantic sturgeon, *Acipenser oxyrinchus*, in North America. *Environ Biol Fish* 48:335–46.

Sneddon, L. U. 2003. The evidence for pain in fish: The use of morphine as an analgesic. *App Anim Behav Sci* 83:153–62.

Sneddon, L. U., V. A. Braithwaite, and M. J. Gentle. 2003. Do fishes have nociceptors? Evidence for the evolution of a vertebrate sensory system. *P Roy Soc Lond B Bio* 270:1115–21.

Snucins, E., and J. M. Gunn 2003. Use of rehabilitation experiments to understand the recovery dynamics of acid-stressed fish populations. *Ambio* 32:240–43.

Snyder, D. E. 1975. Passage of fish eggs and young through a pumped storage generating station. *J Fish Res Board Can* 32:1259–66.

———. 1995. Impacts of electrofishing on fish. *Fisheries* 20(1): 26–27.

———. 2003. *Electrofishing and its harmful effects on fish.* USGS Info. and Tech. Report USGS/BRD/ITR-2003-0002. www.fort.usgs.gov/products/Publications/21226/21226.asp.

Soballe, D. M., B. L. Kimmel, R. H. Kennedy, et al. 1992. Reservoirs. In *Biodiversity of the southeastern United States: Aquatic communities*, ed. C. T. Hackney et al., 421–74. New York: Wiley.

Soselisa, H. L. 1998. Marine *sasi* in Maluka, Indonesia. *Out of the Shell* 6(3): 1–7. www.dal.ca/~corr.

Soulé, M. E. 1980. Thresholds for survival: Maintaining fitness and evolutionary potential. In *Conservation: An evolutionary-ecological perspective*, ed. M. E. Soulé and B. A. Wilcox, 151–70. Sunderland, MA: Sinauer Associates.

———. 1985. What is conservation biology? *BioScience* 35:727–34.

Spalding, M., and A. Grenfell. 1997. New estimates of global and regional coral reef areas. *Coral Reefs* 16:225–30.

Sparks, J. S., and M. L. J. Stiassny. 2003. Introduction to the freshwater fishes. In *The natural history of Madagascar*, ed. S. M. Goodman and J. P. Benstead, 849–63. Chicago: Univ. Chicago Press.

Sparks, R. E. 1995. Need for ecosystem management of large rivers and their floodplains. *BioScience* 45:168–82.

Spear, J. R. 2004. Minimum-impact research. *Conserv Biol* 18:861.

Speer, L., K. Garrison, K. Lonergan, et al. 1997. *Hook, line, and sinking: The crisis in marine fisheries.* New York: Natural Resources Defense Council.

Speer, L., L. Lauck, E. Pikitch, et al. 2000. *Roe to ruin: The decline of the Caspian Sea sturgeon and the road to recovery.* Caviar Emptor. www.caviaremptor.org/report.html.

Spencer, C. N., B. R. McClelland, and J. A. Stanford. 1991. Shrimp stocking, salmon collapse, and eagle displacement. *BioScience* 41:14–21.

Spurgeon, J. P. G. 1992. The economic valuation of coral reefs. *Mar Pollut Bull* 24:529–36.

SSTC (State Science and Technology Commission). 1991. *The Three Gorges Project and Ecology/Environment*, 8 vols. Beijing: Sciences Press. www.yangtze.com/rsc/tgp/chp1.html.

Stanford, J. A., and J. V. Ward. 1991. Limnology of Lake Powell and the chemistry of the Colorado River. In *Colorado River ecology and dam management*, ed. Committee to Review the Glen Canyon Environmental Studies, 75–101. Washington, DC: National Academy Press.

Stanford, J. A., J. V. Ward, W. J. Liss, et al. 1996. A general protocol for restoration of regulated rivers. *Regul Riv* 12:391–413.

Staniford, D. 2002. *A Big Fish in a Small Pond: The Global Environmental and Public Health Threat of Sea Cage Fish Farming.* Coquitlam, BC: Watershed Watch Salmon Society.

Stanley, E. H., and M. W. Doyle. 2002. A geomorphic perspective on nutrient retention following dam removal. *BioScience* 52:693–701.

Starmans, G. A. N. 1970. *Soil erosion of selected African and Asian catchments.* International Water Erosion Symposium, Prague.

Starnes, L. B. 1983. Effects of surface mining on aquatic resources in North America. *Fisheries* 8(6): 2–4.

Starnes, L. B., G. Compean-Jiminez, D. Dodge, et al. 1996. North American fisheries policy. *Fisheries* 21:26–29.

Starnes, L. B., and D. C. Gasper. 1995. Effects of surface mining on aquatic resources in North America. *Fisheries* 20(5): 20–23.

Stauffer, J. R. Jr., M. E. Arnegard, M. Cetron, et al. 1997. Controlling vectors and hosts of parasitic diseases using fishes. *BioScience* 47:41–49.

Steedman, R. J. 1988. Modification and assessment of an Index of Biotic Integrity to quantify stream quality in southern Ontario. *Can J Fish Aquat Sci* 45:492–501.

Steele, J. H. 1998. Regime shifts in marine systems. *Ecol Appl* 8:S33–S36.

Steelhead Society of B.C. 1999. *Escaped salmon are quickly becoming common place in BC's coastal waters.* Sept 21. www.steelheadsociety.com.

Stefanni, S., and J. L. Thorley. 2003. Mitochondrial DNA phylogeography reveals the existence of an evolutionarily significant unit of the sand goby *Pomatoschistus minutus* in the Adriatic (Eastern Mediterranean). *Mol Phylogenet Evol* 28:601–9.

Steinbeck, J. 1945. *Cannery row.* New York: Bantam Books.

Stergiou, K. I. 2002. Overfishing, tropicalization of fish stocks, uncertainty and ecosystem management: Resharpening Ockham's razor. *Fish Res* 55:1–9.

Stevens, E. D., and R. H. Devlin. 2000a. Gill morphometry in growth hormone transgenic Pacific coho salmon, *Oncorhynchus kisutch*, differs markedly from that in GH transgenic Atlantic salmon. *Environ Biol Fish* 58:113–17.

———. 2000b. Intestinal morphology in growth hormone transgenic coho salmon. *J Fish Biol* 56:191–95.

Stevens, W. K. 1997. Appetite for sushi threatens the giant tuna. In Wade 1997, 180–85.

Stewart, A. 1999. Ban on reef fish urged as 30 fall ill. *South China Morning Post*, March 9. Reprinted in *SPC Live Reef Fish Inform Bull* 5:55–56.

Stewart, J., G. Edmunds, R. A. McCarthy, et al. 2000. Permanent and functional male-to-female sex reversal in d-rR strain medaka (*Oryzias latipes*) following egg microinjection of o,-DDT. *Environ Health Persp* 108:219–24.

Stiassny, M. L. J. 1996. An overview of freshwater biodiversity: With some lessons from African fishes. *Fisheries* 21(9): 7–13.

———. 1999. The medium is the message: Freshwater biodiversity in peril. In *The living planet in crisis: Biodiversity science and policy*, ed. J. Cracraft and F. Griffo, 53–71. New York: Columbia Univ. Press.

Stiassny, M. L. J., and M. C. C. de Pinna. 1994. Basal taxa and the role of cladistic patterns in the evaluation of conservation priorities: A view from freshwater. In *Systematics and conservation evaluation*, ed. P. L. Forey et al., 235–49. Oxford, UK: Clarendon Press.

Stiassny, M. L. J., and N. Raminosoa. 1994. The fishes of the inland waters of Madagascar. *Ann Mus Roy Afrique Centrale Zool* 275:133–49.

Stickney, R. R. 1994. *Principles of aquaculture.* New York: Wiley.

———. 2005. *Aquaculture: An introductory text.* Cambridge, MA: CABI.

St. Louis, V. L., C. A. Kelly, E. Duchemin, et al. 2000. Reservoir surfaces as sources of greenhouse gases to the atmosphere: A global estimate. *BioScience* 50:766–75.

Stobutzki, I., M. Miller, and D. Brewer. 2001. Sustainability of fishery bycatch: A process for assessing highly diverse and numerous bycatch. *Environ Conserv* 28:167–81.

Stokes, K., and R. Law. 2000. Fishing as an evolutionary force. *Mar Ecol—Prog Ser* 208:307–8.

Stokes, T. K., J. M. McGlade, and R. Law, eds. 1993. *The exploitation of evolving resources. Lecture notes in biomathematics*, vol. 99. Berlin: Springer-Verlag.

Stone, R. 1999. Coming to grips with the Aral Sea's grim legacy. *Science* 284:30–33.

Stonich, S. C., and C. Bailey. 2000. Resisting the blue revolution: Contending coalitions surrounding industrial shrimp farming. *Human Organization* 59:23–36.

Stouder, D. J., P. A. Bisson, and R. J. Naiman, eds. 1997. *Pacific salmon and their ecosystems: Status and future options*. New York: Chapman and Hall.

Streater, S. 2003. Drug found in area fish stirs concern. *Fort Worth Star Telegram*, Oct. 17. www.dfw.com.

Stripien, C., and S. DeWeerdt. 2002. Old science, new science: Incorporating traditional ecological knowledge into contemporary management. *Conserv Pract* 3(3): 20–27.

Strüssmann, C. A., T. Saito, and F. Takashima. 1998. Heat-induced germ cell deficiency in the teleosts *Odontesthes bonariensis* and *Patagonina hatcheri*. *Comp Biochem Physiol* 119:637–44.

Stuart, S. N., J. S. Chanson, N. A. Cox, et al. 2004. Status and trends of amphibian declines and extinctions worldwide. *Science* 306:1783–86.

Stubbs, K., and K. Cathey. 1999. CAFOs feed a growing problem. *Endangered Species B* 24(1): 14–16.

Sudekum, A. E., J. D. Parrish, R. L. Radtke, et al. 1991. Life history and ecology of large jacks in undisturbed, shallow, oceanic communities. *Fish Bull U.S.* 89:493–513.

Sugihara, G., S. Garcia, J. A. Gulland, et al. 1984. Ecosystems dynamics. In *Exploitation of marine communities*, ed. R. M. May, 131–53. Berlin: Springer-Verlag.

Sullivan, T. J. 2000. *Aquatic effects of acidic deposition*. Boca Raton, FL: Lewis.

Sumaila, U. R. 1998. Protected marine reserves as hedges against uncertainty: An economist's perspective. In Pitcher, Hart, and Pauly 1998, 303–9.

Sumpter, J. P. 1997. The endocrinology of stress. In *Fish stress and health in aquaculture*, ed. O. K. Iwama et al., 95–118. Seminar Series 62. Cambridge, UK: Society for Experimental Biology.

Sumpter, J. P., and S. Jobling. 1995. Vitellogenesis as a biomarker for estrogenic contamination of the aquatic environment. *Environ Health Persp* 103(suppl.): 173–78.

Sutherland, A. B. 1998. Effects of land-use change on sediment regime and fish assemblages in the Upper Little Tennessee River. MS thesis, Univ. Georgia.

———. 2005. Effects of excessive sediment on stress, growth and reproduction of two southern Appalachian minnows, *Erimonax monachus* and *Cyprinella galactura*. PhD dissertation, Univ. Georgia.

Sutherland, A. B., J. L. Meyer, and E. T. Gardiner. 2002. Effects of land cover on sediment regime and fish assemblage structure in four southern Appalachian streams. *Freshwater Biol* 47:1791–1805.

Sutherland, W. J. 1990. Evolution and fisheries. *Nature* 344:814–15.

Swaby, S. E., and G. W. Potts. 1990. Rare British marine fishes: Identification and conservation. *J Fish Biol* 37(suppl. A): 133–43.

Swanagan, J. S. 2000. Factors influencing zoo visitors' conservation attitudes and behavior. *J Environ Educ* 4:26–31.

Swanson, C., T. Reid, P. S. Young, et al. 2000. Comparative environmental tolerances of threatened delta smelt (*Hypomesus transpacificus*) and introduced wakasagi (*H. nipponensis*) in an altered California estuary. *Oecologia* 123:384–90.

Swartzmann, G. L., and T. M. Zaret. 1983. Modeling fish species introduction and prey extermination: The invasion of *Cichla ocellaris* to Gatun Lake, Panama. In *Developments in environmental modeling. Analysis of ecological systems: State of the art in ecological modeling*, ed. W. K. Lauenroth et al., 361–71. New York: Elsevier Scientific.

Swearer, S. E., G. E. Forrester, M. A. Steele, et al. 2003. Spatio-temporal and interspecific variation in otolith trace-elemental fingerprints in a temperate estuarine fish assemblage. *Estuar Coast Shelf S* 56:1111–23.

Sweatman, H. P. A. 1995. A field study of fish predation on juvenile crown-of-thorns starfish. *Coral Reefs* 14:47–53.

Sweetman, K. E., P. S. Maitland, and A. A. Lyle. 1996. Scottish natural heritage and fish conservation in Scotland. In Kirchhofer and Hefti 1996, 23–26.

Sweka, J. A., and K. J. Hartman. 2003. Reduction of reactive distance and foraging success in smallmouth bass, *Micropterus dolomieu*, exposed to elevated turbidity levels. *Environ Biol Fish* 67:341–47.

Tackett, D. N., and L. Tackett. 2001. Banggai cardinalfish, *Pterapogon kauderni*. *Skin Diver* 50(6): 44.

Tacon, A. G. J. 1996. Feeding tomorrow's fish. *World Aquacul* (Sept.): 20–32.

Tacon, A. G. J., and S. S. DeSilva. 1997. Feed preparation and feed management strategies within semi-intensive fish farming systems in the tropics. *Aquaculture* 151:379–404.

Tamburri, M. N., K. Wasson, and M. Matsuda. 2002. Ballast water deoxygenation can prevent aquatic introductions while reducing ship corrosion. *Biol Conserv* 103:331–41.

Tang, J. Y. M., and D. W. T. Au. 2004. Osmotic distress: A probable cause of fish kills on exposure to a subbloom concentration of the toxic alga *Chattonella marina*. *Environ Toxicol Chem* 23:2727–36.

Taniguchi, Y., Y. Miyake, T. Saito, et al. 2000. Redd superimposition by introduced rainbow trout, *Oncorhynchus mykiss*, on native charrs in a Japanese stream. *Ichthyol Res* 47:149–56.

Tasker, M. L., C. J. Camphuysen, J. Cooper, et al. 2000. The impacts of fishing on marine birds. *ICES J Mar Sci* 57:531–47.

Tatara, C. P., M. Mulvey, and M. C. Newman. 2002. Genetic and demographic responses of mercury-exposed mosquitofish (*Gambusia holbrooki*) populations: Temporal stability and reproductive components of fitness. *Environ Toxicol Chem* 21:2191–97.

Taylor, E. B. 1991. A review of local adaptation in Salmonidae, with particular reference to Pacific and Atlantic salmon. *Aquaculture* 98:185–207.

Taylor, J. N., W. R. Courtenay Jr., and J. A. McCann. 1984. Known impacts of exotic fishes in the continental United States. In *Distribution, biology, and management of exotic fishes*, ed. W. R. Courtenay Jr. and J. R. Stauffer Jr., 322–73. Baltimore: Johns Hopkins Univ. Press.

Taylor, L. 1993. *Aquariums: Windows to nature*. New York: Prentice Hall.

———. 1998. New trends in public aquariums. *Fisheries* 23(2): 16–18.

Taylor, L. R. 1978. Tropical reef fish management: Issues and opinions. In *Papers and comments on tropical reef fish*, ed. L. Taylor and R. Nolan, 3–5. Sea Grant College Program Working Paper 34. Honolulu: Univ. Hawaii.

Taylor, M. S., and M. E. Hellberg. 2003. Genetic evidence for local retention of pelagic larvae in a Caribbean reef fish. *Science* 299:107–9.

Tennant, D. L. 1976. Instream flow regimens for fish, wildlife, recreation and related environmental resources. *Fisheries* 1(4): 6–10.

Terborgh, J. 1974. Preservation of natural diversity: The problem of extinction prone species. *BioScience* 24:715–22.

———. 1999. *Requiem for nature*. Washington, DC: Island Press.

TESS (Threatened and Endangered Species Database System). 2004. U.S. Fish and Wildlife Service. http://endangered.fws.gov/wildlife.html#Species.

Teuscher, D., and C. Luecke. 1996. Competition between kokanees and Utah chub in Flaming Gorge Reservoir, Utah-Wyoming. *T Am Fish Soc* 125:505–11.

Tharme, R. E. 2003. A global perspective on environmental flow assessment: Emerging trends in the development and application of environmental flow methodologies for rivers. *River Res Appl* 19:397–441.

Therriault, T. W., and D. Schneider. 1998. Predicting change in fish mercury concentrations following reservoir impoundment. *Environ Pollut* 101:33–42.

Thomas, M. E. 1995. Monitoring the effects of introduced flathead catfish on sportfish populations in the Altamaha River, Georgia. *P Annu Conf SE Assoc Fish Wildlife Agen* 47:531–38.

Thomas, P. 1990. Molecular and biochemical responses of fish to stressors and their potential use in environmental monitoring. *Am Fish Soc Sym* 8:9–28.

Thompson, D. M. 2002. Long-term effect of instream habitat-improvement structures on channel morphology along the Blackledge and Salmon rivers, Connecticut, USA. *Environ Manage* 29:250–65.

Thompson, K. G., E. P. Bergersen, R. B. Nehring, et al. 1997. Long-term effects of electrofishing on growth and body condition of brown trout and rainbow trout. *N Am J Fish Manage* 17:154–59.

Thompson, W. F. 1916. Statistics of the halibut fishing in the Pacific. In *Report of the British Columbia Provincial Fisheries Commission for 1915*, 65–126. Victoria, BC: Provincial Fisheries Commission.

Thompson, W. F., and F. H. Bell. 1934. Biological statistics of the Pacific halibut fishery. 2: Effect of changes in intensity upon total yield and yield per unit of gear. Report of the International Fisheries Commission 8. Seattle, WA: International Pacific Halibut Commission.

Thoreau, H. D. 1971. *Walden*. Princeton, NJ: Princeton Univ. Press.

Thorpe, J., G. Gall, J. Lannan, et al. 1995. *Conservation of fish and shellfish resources: Managing diversity*. London: Academic Press.

Thorrold, S. R., R. S. Burton, G. P. Jones, et al. 2002. Quantifying larval retention and connectivity in marine populations with artificial and natural markers. *B Mar Sci* 70(suppl.): 291–308.

Thresher, R. E. 1980. *Reef fish: Behavior and ecology on the reef and in the aquarium*. St. Petersburg, FL: Palmetto.

———. 1984. *Reproduction in reef fishes*. Neptune City, NJ: TFH.

Thresher, R. E., and P. L. Colin. 1986. Trophic structure, diversity, and abundance of fishes of the deep reef (30–300 m) at Enewetak, Marshall Islands. *B Mar Sci* 38:253–72.

Thurow, R. F., and D. J. Schill. 1996. Comparison of day snorkeling, night snorkeling, and electrofishing to estimate bull trout abundance and size structure in a second-order stream. *N Am J Fish Manage* 16:314–23.

Tichacek, G. 1996. How to work with your legislators. *Fisheries* 21(5): 35.

Tidwell, J. 2001. Sturgeon fishes and caviar dreams. *Zoogoer*, May/June. http://nationalzoo.si.edu.

Tiemann, J. S., D. P. Gillette, M. L. Wildhaber, et al. 2004. Effects of lowhead dams on riffle-dwelling fishes and macroinvertebrates in a midwestern river. *T Am Fish Soc* 133:705–17.

Tisdall, C. 1994. *Setting priorities for the conservation of New Zealand's threatened plants and animals*, 2nd ed. Wellington: Dept. of Conservation.

Tissot, B. N., and L. E. Hallacher. 2003. Effects of aquarium collectors on coral reef fishes in Kona, Hawaii. *Conserv Biol* 17:1759–68.

Tlusty, M. 2002. The benefits and risks of aquacultural production for the aquarium trade. *Aquaculture* 205:203–19.

Todd, K. 2000. *Tinkering with Eden: A natural history of exotics in America*. New York: Norton.

Tomascik, T. 1991. Settlement patterns of Caribbean scleractinian corals on artificial substrata along a eutrophication gradient, Barbados, West Indies. *Mar Ecol—Prog Ser* 77:261–69.

Tomey, W. A. 1996. *Review of developments in the world ornamental fish trade: Update, trends and future prospects*. Infofish-Aquatech 1996, Kuala Lumpur, Malaysia, Sept. Cited in Young 1997.

Tovar, A., C. Moreno, M. P. Manuel-Vez, et al. 2000. Environmental impacts of intensive aquaculture in marine waters. *Water Res* 34:334–42.

Townsend, C. R. 1996. Invasion biology and ecological impacts of brown trout *Salmo trutta* in New Zealand. *Biol Conserv* 78:13–22.

Townsend, C. R., and T. A. Crowl. 1991. Fragmented population structure in a native New Zealand fish: An effect of introduced brown trout. *Oikos* 61:347–54.

Townsend, C. R., and M. J. Winterbourn. 1992. Assessment of the environmental risk posed by an exotic fish: The proposed introduction of channel catfish (*Ictalurus punctatus*) to New Zealand. *Conserv Biol* 6:273–82.

Tracey, D. M., and P. L. Horn. 1999. Background and review of ageing orange roughy (*Hoplostethis atlanticus*, Trachichthyidae) from New Zealand and elsewhere. *New Zeal J Mar Fresh* 33:67–86.

TRAFFIC (Trade Record Analysis of Flora and Fauna in Commerce) South America. 2000. *Evaluation of the trade of sea cucumber* Isostichopus fuscus *(Echinodermata: Holothuroidea) in the Galapagos Islands during 1999*. www.TRAFFIC.org.

Tratalos, J. A., and T. J. Austin. 2001. Impacts of recreational SCUBA diving on coral communities of the Caribbean island of Grand Cayman. *Biol Conserv* 102:67–75.

Trautman, M. N. B. 1981. *The fishes of Ohio*, rev. ed. Columbus: Ohio State Univ. Press.

Travis, J., F. C. Coleman, C. B. Grimes, et al. 1998. Critically assessing stock enhancement: An introduction to the Mote symposium. *B Mar Sci* 62:305–11.

Traxler, G. S. , J. Richard, and T. E. McDonald. 1998. *Ichthyophthirius multifiliis* (Ich) epizootics in spawning sockeye salmon in British Columbia, Canada. *J Aquat Anim Health* 10:143–51.

Treasurer, J. W., and L. A. Laidler. 1994. *Aeromonas salmonicida* infection in wrasse (Labridae), used as cleaner fish, on an Atlantic salmon, *Salmo salar* L., farm. *J Fish Dis* 17:155–61.

Trippel, E. A. 1995. Age at maturity as a stress indicator in fisheries. *BioScience* 45:759–71.

Troell, M., P. Rohnback, N. Kautsky, et al. 1999. Ecological engineering in aquaculture: The use of seaweeds for removing nutrients from intensive mariculture. *J App Phycol* 11:89–97.

True, C. D., A. S. Loera, and N. C. Castro. 1997. Acquisition of broodstock of *Totoaba macdonaldi*: Field handling, decompression, and prophylaxis of an endangered species. *Progressive Fish-Culturist* 59:246–48.

TSLG (The Social Learning Group). 2001. *Learning to manage global environmental risks. Vol. 1: A comparative history of social responses to climate change, ozone depletion, and acid rain*. Cambridge, MA: MIT Press.

Tuck, G. N., T. Polacheck, J. P. Croxall, et al. 2001. Modeling the impact of fishery by-catches on albatross populations. *J Appl Ecol* 38:1182–96.

Tupper, M., and R. G. Boutilier. 1995. Effects of habitat on settlement, growth, and postsettlement survival of Atlantic cod (*Gadus morhua*). *Can J Fish Aquat Sci* 52:1834–41.

Turgeon, J., and L. Bernatchez. 2003. Reticulate evolution and phenotypic diversity in North American ciscoes, *Coregonus* ssp. (Teleostei: Salmonidae): Implications for the conservation of an evolutionary legacy. *Conserv Genet* 4:67–81.

Turner, S. 1999. Atlantic highly migratory pelagic fisheries. Unit 5 in NMFS 1999.

Turnpenny, A. W. H., and R. Williams. 1980. Effects of sedimentation on the gravels of an industrial river system. *J Fish Biol* 17:681–93.

Tyedmers, P. 2000. Salmon and sustainability: The biophysical cost of producing salmon through the commercial salmon fishery and the intensive salmon culture industry. PhD thesis, Univ. British Columbia.

Tyus, H. M., and J. F. Saunders III. 2000. Nonnative fish control and endangered fish recovery: Lessons from the Colorado River. *Fisheries* 25(9): 17–24.

Ueber, E., and A. MacCall. 1990. The collapse of California's sardine fishery. In *Climate variability, climate change and fisheries*, ed. M. H. Glantz and L. E. Feingold, 17–23. Boulder, CO: National Center for Atmospheric Research.

UICN-South. South American Regional Office of IUCN. 2003. Documentos de Listas y Libros Rojos del Centro de Documentación de la Oficina Regional para América del Sur de la UICN—por País. www.sur.iucn.org/listaroja/publicaciones_esp.htm.

UNEP (United Nations Environmental Programme). 1990. *UNEP environmental management guidelines for fish farming*. Rome: FAO.

———. 1992. *Convention on biological diversity*. www.biodiv.org/convention/articles.asp.

———. 2003. *Global mercury assessment*. UNEP Chemicals Mercury Programme. www.chem.unep.ch.

United States Coral Reef Task Force. 2000. *The national action plan to conserve coral reefs*. Washington, DC: USCRTF. http://coralreef.gov.

Upton, H. F. 1992. Biodiversity and conservation of the marine environment. *Fisheries* 17(3): 20–25.

USACE (U.S. Army Corps of Engineers). 1999. *National inventory of dams*. http://crunch.tec.army.mil/nid/webpages/nid.cfm.

USBR (U.S. Bureau of Reclamation). 2004. *Reclamation begins second year of experimental water releases from Glen Canyon Dam to benefit endangered species*. www.usbr.gov/uc/news/gc_expflows.html.

USEPA (U.S. Environmental Protection Agency). 1990. *The quality of our nation's water: A summary of the 1988 national water quality inventory*. EPA Report 440/4-90-005. Washington, DC: Author.

———. 1994. *Guide to cleaner technologies: Alternative metal finishes*. EPA/625/R-94//007. Washington, DC: Author.

———. 1998. *Compliance assurance implementation plan for concentrated animal feeding operations*. Washington, DC: Author, Office of Enforcement and Compliance Assurance. Cited in Sutherland 1998.

———. 1999. *Progress report on the EPA acid rain program*. EPA 430-R-99-011. Nov. www.epa.gov/acidrain.

———. 2000. *2000 national water quality inventory: Monitoring and assessing water quality*. www.epa.gov/305b/2000report.

———. 2002. *Columbia River basin fish contaminant survey 1996–1998*. EPA 910-R-02-006. Aug. 12. Seattle, WA: USEPA Region 10.

———. 2003. *Update: National listing of fish and wildlife advisories*. USEPA Office of Water EPA-823-F-03-003, May. www.epa.gov/waterscience/fish/advisories/factsheet.pdf.

USFWS (U.S. Fish and Wildlife Service). 1999. *1996 National survey of fishing, hunting, and wildlife-associated recreation*. http://fa.r9.fws.gov/surveys/surveys.html.

———. 2004. *Endangered and threatened wildlife and plants: Interim rule for the beluga sturgeon (*Huso huso*)*. Fish and Wildlife Service 50 CFR Part 17, RIN 1019-AU02. Washington, DC: Author.

Utter, F. 1998. Genetic problems of hatchery-reared progeny released into the wild, and how to deal with them. *B Mar Sci* 62:623–40.

Vagelli, A. 1999. The reproductive biology and early ontogeny of the mouthbrooding Banggai cardinalfish, *Pterapogon kauderni* (Perciformes, Apogonidae). *Environ Biol Fish* 56:79–92.

Vagelli, A. A., and M. V. Erdmann. 2002. First comprehensive survey of the Banggai cardinalfish, *Pterapogon kauderni*. *Environ Biol Fish* 63:1–8.

Valdez, R. A., T. L. Hoffnagle, C. C. McIvor, et al. 2001. Effects of a test flood on fishes of the Colorado River in Grand Canyon, Arizona. *Ecol Appl* 11:686–700.

Valiela, I., J. L. Bowen, and J. K. York. 2001. Mangrove forests: One of the world's threatened major tropical environments. *BioScience* 51:807–15.

Vallejo, B.V. 1997. Survey and review of the Philippine marine aquarium fish industry. *Sea Wind* 11(4): 2–16. Cited in E. M. Wood 2001.

VanderKooi, S. P., A. G. Maule, and C. B. Schreck. 2001. The effects of electroshock on immune function and disease progression in juvenile spring chinook salmon. *T Am Fish Soc* 130: 397–408.

Vanderploeg, H. A., B. J. Eadie, J. R. Liebig, et al. 1987. Contribution of calcite to the particle size spectrum of Lake Michigan seston and its interactions with the plankton. *Can J Fish Aquat Sci* 44:1898–1914.

Van Dolah, F. M., D. Roelke, and R. M. Greene. 2001. Health and ecological impacts of harmful algal blooms: Risk assessment needs. *Hum Ecol Risk Assess* 7:1329–45.

Van Driesche, J., and R. Van Driesche. 2000. *Nature out of place: Biological invasions in the global age.* Washington, DC: Island Press.

———. 2001. Guilty until proven innocent. *Conserv Biol in Practice* 2(1): 8–17.

Van Dyke, F. 2003. *Conservation biology: Foundations, concepts, applications.* Boston: McGraw-Hill.

Vanicek, D. C. 1980. Decline of the Lake Greenhaven Sacramento perch population. *Calif Fish Game* 66:178–83.

Vannote, R. L., G. W. Minshall, K. W. Cummins, et al. 1980. River Continuum Concept. *Can J Fish Aquat Sci* 37:130–37.

Vannuccini, S. 1999. *Shark utilization, marketing and trade.* FAO Fisheries Tech. Paper 389. Rome: FAO.

Van Winkle, W., P. J. Anders, D. H. Secor, et al., eds. 2002. Biology, management, and protection of North American sturgeon. *Am Fish Soc Sym* 28.

Vashro, J. 1995. The "bucket brigade" is ruining our fisheries. *Montana Outdoors* 26(5): 34–37. Cited in Rahel 2002.

Vaughan, D. S., and J. W. Smith. 1999. Southeast menhaden fisheries. Unit 10 in NMFS 1999.

Vecsei, P. 2005. Gastronome 101: How capitalism killed the sturgeons. *Environ Biol Fish* 73:111–16.

Veit, R. R., J. A. McGowan, D. G. Ainley, et al. 1997. Apex marine predator declines ninety percent in association with changing oceanic climate. *Global Change Biol* 3:23–28.

Venter, P., P. Timm, G. Gunn, et al. 2000. Discovery of a viable population of coelacanths (*Latimeria chalumnae* Smith, 1939) at Sodwana Bay, South Africa. *S Afr J Sci* 96:567–68.

Ventura County. 2004. *Matilija Dam ecosystem restoration feasibility study: Final report.* www.matilijadam.org.

Verheye, H. M. 2000. Decadal scale trends across several marine trophic levels in the southern Benguela upwelling system off South Africa. *Ambio* 29:30–34.

Villwock, W. 1993. The Lake Titicaca region in the Peruvian and Bolivian highlands and the consequences of introducing fishes for endemic ichthyofauna and their natural habitat. *Naturwissenschaften* 80:1–8.

Vincent, A. C. J. 1996. *The international trade in seahorses.* Cambridge, UK: Traffic International.

Vincent, A. 2002. Pregnant males set a precedent for CITES. *World Conserv* 3(2002): 26.

Vincent, A. C. J., and H. J. Hall. 1996. The threatened status of marine fishes. *Trends Ecol Evol* 11:360–61.

Vincent, A. C. J., and L. M. Sadler. 1995. Faithful pair bonds in wild seahorse, *Hippocampus whitei. Anim Behav* 50:1557–69.

Vincent, A. C. J., and Y. Sadovy. 1998. Reproductive ecology in the conservation and management of fishes. In Caro 1998, 209–45.

Viney, A. 2004. *SCWF stands firm on the Richard B. Russell Dam.* South Carolina Wildlife Federation. www.scwf.org/articles/index.php?view=84.

Vinyard, G. L., and W. J. O'Brien. 1976. Effects of light and turbidity on the reactive distance of bluegill (*Lepomis macrochirus*). *J Fish Res Board Can* 33:2845–49.

Vitousek, P. M., H. A. Mooney, J. Lubchenco, et al. 1997. Human domination of Earth's ecosystems. *Science* 277:494–99.

Vogelbein, W. K., V. J. Lovko, J. D. Shields, et al. 2002. *Pfiesteria shumwayae* kills fish by micropredation, not exotoxin secretion. *Nature* 418:967–70.

Volpe, J. P., B. R. Anholt, and B. W. Glickman. 2001. Competition among juvenile Atlantic salmon (*Salmo salar*) and steelhead (*Oncorhynchus mykiss*): Relevance to invasion potential in British Columbia. *Can J Fish Aquat Sci* 58:197–207.

Volpe, J. P., E. B. Taylor, D. W. Rimmer, et al. 2000. Evidence of natural reproduction of aquaculture-escaped Atlantic salmon in a coastal British Columbia river. *Conserv Biol* 14:899–903.

Vrijenhoek, R. C. 1989. Population genetics and conservation. In *Conservation for the 21st century*, ed. M. Pearl and D. Western, 89–98. Oxford: Oxford Univ. Press.

———. 1998. Conservation genetics of freshwater fish. *J Fish Biol* 54(suppl. A): 394–412.

Waddington, P. 1997. *Koi kichi*. Stillwater, MN: Voyageur Press.

Wade, N., ed. 1996. *The Science Times book of fish*. New York: Lyons Press.

Wade, P. R. 1995. Revised estimates of incidental kill of dolphins (Delphinidae) by the purse-seine tuna fishery in the eastern tropical Pacific, 1959–1972. *Fish Bull U.S.* 93:345–54.

Wager, R., and P. Jackson. 1993. *The action plan for Australian freshwater fishes*. www.deh.gov.au.

Wahl, D. H., R. A. Stein, and D. R. DeVries. 1995. An ecological framework for evaluating the success and effects of stocked fishes. *Am Fish Soc Sym* 15:176–89.

Waichman, A., N. L. Chao, and P. Petry. 2000. Bio-piracy in Amazonia: Knowing the laws. *OFI Journal* 33. www.ornamental-fish-int.org/biopiracy.htm.

Waichman, A. V., M. Pinheiro, and J. L. Marcon. 2001. Water quality monitoring during the transport of Amazonian ornamental fish. In Chao, Petry, et al., 2001, 279–99.

Wainwright, T. C., and R. G. Kope. 1999. Methods of extinction risk assessment developed for US West Coast salmon. *ICES J Mar Sci* 56:444–48.

Wainwright, T. C., and R. S. Waples. 1998. Prioritizing Pacific salmon stocks for conservation: Response to Allendorf et al. *Conserv Biol* 12:1144–47.

Walker, P. A., and H. J. L. Heessen. 1996. Long-term changes in ray populations in the North Sea. *ICES J Mar Sci* 53:1085–93.

Wallace, J. B., S. L. Eggert, J. L. Meyer, et al. 1997. Multiple trophic levels of a forest stream linked to terrestrial litter inputs. *Science* 277:102–4.

Wallin, J. E. 1992. The symbiotic nest association of yellowfin shiners, *Notropis lutipinnis*, and bluehead chubs, *Nocomis leptocephalus*. *Environ Biol Fish* 33:287–92.

Walseng, B., R. M. Langåker, T. E. Brandrud, et al. 2001. The River Bjerkreim in SW Norway: Successful chemical and biological recovery after liming. *Water Air Soil Poll* 130:1331–36.

Walsh, S. J., N. M. Burkhead, and J. D. Williams. 1995. Southeastern freshwater fishes. In *Our living resources: A report to the nation on the distribution, abundance, and health of U.S. plants, animals and ecosystems*, ed. E. T. LaRoe et al., 144–47. Washington, DC: National Biological Service.

Walters, C. 2000. Natural selection for predation avoidance tactics: Implications for marine population and community dynamics. *Mar Ecol—Prog Ser* 208:309–13.

Walters, C. J., D. Pauly, and V. Christensen. 2000. Ecospace: Prediction of mesoscale spatial patterns in trophic relationships of exploited ecosystems, with emphasis on the impacts of marine protected areas. *Ecosystems* 2:539–54.

Walters, D. M., D. S. Leigh, and A. B. Bearden. 2003. Urbanization, sedimentation, and the homogenization of fish assemblages in the Etowah River basin, USA. *Hydrobiologia* 494:5–10.

Walther, G. R., E. Post, P. Convery, et al. 2002. Ecological responses to recent climate change. *Nature* 416:389–95.

Wang, L., J. Lyons, P. Kanehl, et al. 1997. Influences of watershed land use on habitat quality and biotic integrity in Wisconsin streams. *Fisheries* 22(6): 6–12.

Waples, R. S. 1991a. Genetic interactions between hatchery and wild salmonids: Lessons from the Pacific Northwest. *Can J Fish Aquat Sci* 48(suppl. 1): 124–33.

———. 1991b. Pacific salmon, *Oncorhynchus* spp., and the definition of "species" under the Endangered Species Act. *Mar Fish Rev* 53:11–22.

———. 1995. Evolutionarily significant units and the conservation of biological diversity under the Endangered Species Act. *Am Fish Soc Sym* 17:8–27.

———. 1999. Dispelling some myths about hatcheries. *Fisheries* 24(2): 12–21.

Waples, R. S., G. Gustafson, L. A. Weitkamp, et al. 2001. Characterizing diversity in salmon from the Pacific Northwest. *J Fish Biol* 59(suppl. A): 1–41.

Ward, R. D., M. Woodwark, and D. O. F. Skibinski. 1994. A comparison of genetic diversity levels in marine, freshwater, and anadromous fishes. *J Fish Biol* 44:212–32.

Ward, R. W. 1998. The distribution of exotic and native fishes of the Upper Bullfrog Creek watershed with regard to tropical fish farms in Hillsborough County, Florida. MS thesis, Univ. Georgia.

Warner, R. R., S. E. Swearer, and J. E. Caselle. 2000. Larval accumulation and retention: Implications for the design of marine reserves and essential fish habitat. *B Mar Sci* 66:821–30.

Warren, M. L. Jr., P. L. Angermeier, B. M. Burr, et al. 1997. Decline of a diverse fish fauna: Patterns of imperilment and protection in the southeastern United States. In *Aquatic fauna in peril: The southeast-*

ern perspective, ed. G.W. Benz and D. E. Collins, 105–64. Southeast Aquatic Research Institute Special Publication 1. Decatur, GA: Lenz Design and Communications.

Warren, M. L. Jr., and B. M. Burr. 1994. Status of freshwater fishes of the United States: Overview of an imperiled fauna. *Fisheries* 19(1): 6–18.

Warren, M. L., and M. G. Pardew. 1998. Road crossings as barriers to small-stream fish movement. *T Am Fish Soc* 27:637–44.

Warren, M. L. Jr., H. W. Robison, S. T. Ross, et al. 2000. Diversity, distribution, and conservation status of the native freshwater fishes of the southern United States. *Fisheries* 25(10): 7–31.

Wass, R. C. 1982. The shoreline fishery of American Samoa: Past and present. In *Ecological aspects of coastal zone management*, ed. J. L. Munro, 51–83. New York: UNESCO. Cited in Birkeland 1997b.

Waters, S. 2002. Bassmasters Classic: Super Bowl of bass gets ESPN topspin. *Athens (GA) Banner-Herald*, July 24.

Waters, T. E. 1995. *Sediment in streams: Sources, biological effects, and control.* Bethesda, MD: American Fisheries Society.

Waters, T. F. 1983. Replacement of brook trout by brown trout over 15 years in a Minnesota stream: Production and abundance. *T Am Fish Soc* 12:137–45.

Watling, L. 2005. The global destruction of bottom habitats by mobile fishing gears. In Norse and Crowder 2005, 198–210.

Watling, L., and E. A. Norse. 1998a. Disturbance of the seabed by mobile fishing gear: A comparison to forest clearcutting. *Conserv Biol* 12:1180–97.

———. 1998b. Effects of mobile fishing gear on marine benthos. *Conserv Biol* 12(Special section): 1178–79.

Watson, B. 2003. Something's fishy: Calming aquariums help owners stay in the swim. *Augusta Chronicle*, Your Life Section, April 3, pp. 1, 3.

Watson, C. A., and J. V. Shireman. 1996. *Production of ornamental aquarium fish.* Gainesville: Univ. Florida, Dept. Fisheries and Aquatic Science. http://edis.ifas.ufl.edu.

Watson, R., L. Pang, and D. Pauly. 2001. The marine fisheries of China: Development and reported catches. *Fisheries Centre Research Report* 9(2). Vancouver, British Columbia.

Watson, R., and D. Pauly. 2001. Systematic distortions in world fisheries catch trends. *Nature* 414:534–36.

Watson, R. T., M. C. Zinyowera, and R. H. Moss, eds. 1996. *Climate change 1995. Impacts, adaptations and mitigation of climate change: Scientific technical analyses.* Cambridge: Cambridge Univ. Press. www.ipcc-wg2.org.

Watt, W. D., C. D. Scott, P. J. Zamora, et al. 2000. Acid toxicity levels in Nova Scotian rivers have not declined in synchrony with the decline in sulfate levels. *Water Air Soil Poll* 118:203–29.

Watters, G. T. 1996. Small dams as barriers to freshwater mussels (Bivalvia, Unionoida) and their hosts. *Biol Conserv* 75:79–85.

WCMC (World Conservation Monitoring Centre). 2003. *Checklist of fish and invertebrates listed in the CITES appendices and in EC Regulation 338/97*, 6th ed. Joint Nature Conservation Committee Report 341. www.ukcites.gov.uk.

WDFW (Washington Department of Fish and Wildlife). 2003. *Design of road culverts for fish passage.* Olympia, WA: Author. http://wdfw.wa.gov.

Webb, J. H., D. W. Hay, P. D. Cunningham, et al. 1991. The spawning behavior of escaped farmed and wild Atlantic salmon (*Salmo salar* L.) in a northern Scottish river. *Aquaculture* 98:97–110.

Webb, S. A., and R. R. Miller. 1998. *Zoogoneticus tequila*, a new goodeid fish (Cyprinodontiformes) from the Ameca drainage of Mexico, and a rediagnosis of the genus. *Occasional Papers of the Museum of Zoology, University of Michigan* 725:1–23.

Weber, A., G. S. Proudlove, J. Parzefall, et al. 1998. Pisces (Teleostei). In *Encyclopaedia biospeologica*, ed. C. Juberthie and V. Decu, 1177–213. Moulis-Bucharest: Societe de Biospeologie (Academie Roumaine).

Weber, M.L. 1997. *Farming salmon: a briefing book.* Consultative Group on Biological Diversity. www.seaweb.org/resources/documents/reports_farmingsalmon.

Weber, P. 1994. *Net loss: Fish, jobs, and the marine environment.* Worldwatch Paper 120. Washington, DC: Worldwatch Institute.

———. 1995. Protecting oceanic fisheries and jobs. In *State of the world 1995*, ed. L. R. Brown, 221–37. Washington, DC: Norton/Worldwatch Institute.

Weeks, H., and S. Berkeley. 2000. Uncertainty and precautionary management of marine fisheries: Can the old methods fit the new mandates? *Fisheries* 25(12): 6–15.

Wei, Q., F. Ke, J. Zhang, et al. 1997. Biology, fisheries, and conservation of sturgeons and paddlefish in China. *Environ Biol Fish* 48:241–55.

Weinstein, M. R., M. Litt, D. A. Kertesz, et al. 1997. Invasive infections due to a fish pathogen, *Streptococcus iniae*. *New Engl J Med* 337:589–94.

Weis, J. S., G. M. Smith, and T. Zhou. 1999. Altered predator/prey behavior in polluted environments: Implications for fish conservation. *Environ Biol Fish* 55:43–51.

Weitkamp, D. E., and M. Katz. 1980. A review of dissolved gas supersaturation literature. *T Am Fish Soc* 109:659–702.

Welcomme, R. L. 1984. International transfers of inland fish species. In *Distribution, biology, and management of exotic fishes*, ed. W. R. Courtenay Jr. and J. R. Stauffer Jr., 22–40. Baltimore: Johns Hopkins Univ. Press.

———. 1985. *River fisheries*. FAO Tech. Paper 262. Rome: FAO.

———. 1988. *International introductions of inland aquatic species*. FAO Fisheries Tech. Paper 294. Rome: FAO.

———. 1992. A history of international introductions of inland aquatic species. *ICES Mar Sci Sym* 194:3–14.

———. 1995. Relationships between fisheries and the integrity of river systems. *Regul Riv* 11:121–36.

———. 2001. *Inland fisheries: Ecology and management*. Oxford: Blackwell Science, Fishing News Books.

———. 2003. River fisheries in Africa: Their past, present, and future. In Crisman et al. 2003, 145–76.

Welcomme, R. L., and D. Hagborg. 1977. Towards a model of a floodplain fish population and its fishery. *Environ Biol Fish* 2:7–24.

Wenger, S. 1999. *A review of the scientific literature on riparian buffer width, extent and vegetation*. Athens: Univ. Georgia, Institute of Ecology.

Wetzel, R. G. 2001. *Limnology*, 3rd ed. San Diego: Academic Press.

Weyl, O. L. F., and T. Hecht. 1999. A successful population of largemouth bass, *Micropterus salmoides*, in a subtropical lake in Mozambique. *Environ Biol Fish* 54:53–66.

Whalen, P. J., L. A. Toth, J. W. Koebel, et al. 2002. Kissimmee River restoration: A case study. *Water Sci Technol* 45:55–62.

White, A. T., L. Z. Hale, Y. Renard, et al., eds. 1994. *Collaborative and community-based management of coral reefs: Lessons from experience*. West Hartford, CT: Kumarian Press.

White, R. 2000. *Flushing of sediments from reservoirs*. World Commission on Dams. www.wca-infonet.org/id/117779.

White, R. J., J. R. Karr, and W. Nehlsen. 1995. Better roles for fish stocking in aquatic resource management. *Am Fish Soc Sym* 15:527–47.

Whitfield, A. K., and A. W. Paterson. 1995. Flood-associated mass mortality of fishes in the Sundays estuary. *Water SA* 21:385–89.

Whitfield, P. E., T. Gardner, S. P. Vives, et al. 2002. Biological invasions of the Indo-Pacific lionfish *Pterois volitans* along the Atlantic coast of North America. *Mar Ecol—Prog Ser* 235:289–97.

Whittier, T. R., D. B. Halliwell, and S. G. Paulsen. 1997. Cyprinid distributions in northeast USA lakes: Evidence of regional-scale minnow biodiversity losses. *Can J Fish Aquat Sci* 54:1593–1607.

Wichert, G. A., and D. J. Rapport. 1998. Fish community structure as a measure of degradation and rehabilitation of riparian systems in an agricultural drainage basin. *Environ Manage* 22:425–43.

Wickens, P. 1996. Conflict between Cape (South African) fur seals and line fishing operations. *Wildlife Res* 23:109–17.

Wiebe, W. J., R. E. Johannes, and K. L. Webb. 1975. Nitrogen fixation in a coral reef community. *Science* 188:257–59.

Wijkstrom, U., A. Gumy, and R. Grainger. 2000. Genetically modified organisms and fisheries. In *The state of world fisheries and aquaculture 2000*. Rome: FAO Fisheries Department.

Wikramanayake, E. D. 1990. Conservation of endemic rain forest fishes of Sri Lanka: Results of a translocation experiment. *Conserv Biol* 4:32–38.

Wilkinson, C., O. Linden, H. Cesar, et al. 1999. Ecological and socioeconomic impacts of 1998 coral bleaching in the Indian Ocean: An ENSO impact and a warning of future change? *Ambio* 28:188–96.

Williams, J. D., and G. K. Meffe. 1998. Nonindigenous species. In *Status and trends of the nation's biological resources*, ed. M. J. Mac et al., 117–29. Reston, VA: U.S. Geological Survey.

Williams, J. D., M. L. Warren Jr., K. S. Cummings, et al. 1993. Conservation status of freshwater mussels of the United States and Canada. *Fisheries* 18(9): 2–20.

Williams, J. E., J. E. Johnson, D. A. Hendrickson, et al. 1989. Fishes of North America, endangered, threatened, or of special concern. *Fisheries* 14(6): 2–20.

Williams, J. E., C. A. Wood, and M. P. Dombeck, eds. 1997. *Watershed restoration: Principles and practices*. Bethesda, MD: American Fisheries Society.

Williams, M. 1998. Aquatic resources education for the development of world needs. In Pitcher, Hart, and Pauly, 163–74.

———, ed. 1990. *Wetlands: A threatened landscape*. Oxford: Blackwell.

Williams, R. N., ed. 2006. *Return to the river: Restoring salmon to the Columbia River*. New York: Elsevier Academic Press.

Williams, T. 1992. The Last Bluefin Hunt. *Audubon* 94(4): 14–20.

———. 1994. Killer fish farms. *Audubon* 94(2): 14–22.

———. 2001. Want another carp? *Fly Rod and Reel* (June): 18–24.

Williamson, M. 1996. *Biological invasions*. London: Chapman and Hall.

Williot, P., E. Rochard, G. Castelnaud, et al. 1997. Biological characteristics of European Atlantic sturgeon, *Acipenser sturio*, as the basis for a restoration program in France. *Environ Biol Fish* 48:359–70.

Willis, T. J., R. B. Millar, and R. C. Babcock. 2003. Protection of exploited fish in temperate regions: High density and biomass of snapper *Pagrus auratus* (Sparidae) in northern New Zealand marine reserves. *J Appl Ecol* 40:214–27.

Willson, M. F., S. M. Gende, and B. H. Marston. 1998. Fishes and the forest. *BioScience* 48:455–62.

Willson, M. F., and K. C. Halupka. 1995. Anadromous fish as keystone species in vertebrate communities. *Conserv Biol* 9:489–97.

Wilson, B., and L. M. Dill. 2002. Pacific herring respond to simulated odontocete echolocation sounds. *Can J Fish Aquat Sci* 59:542–53.

Wilson, E. O. 1984. *Biophilia*. Cambridge, MA: Harvard Univ. Press.

Wilson, J., C. W. Osenberg, C. M. St. Mary, et al. 2001. Artificial reefs, the attraction-production issue, and density dependence in marine ornamental fishes. *Aquar Sci Conserv* 3:95–105.

Wilson, K. 2003. Endangered fish restore the balance. *Arizona Daily Wildcat*, online ed. http://wildcat.arizona.edu.

Wilson, S. 2005. Animals and ethics. *The Internet Encyclopedia of Philosophy*. www.iep.utm.edu/a/anim-eth.htm.

Winemiller, K. O. 1990. Caudal eyespots as deterrents against fin predation in the neotropical cichlid *Astronotus ocellatus*. *Copeia* 1990:665–73.

Winemiller, K. O., and D. B. Jepsen. 1998. Effects of seasonality and fish movement on tropical river food webs. *J Fish Biol* 53(suppl. A): 267–96.

Winfield, I., J. M. Fletcher, and P. R. Cubby. 1998. The threat to vendace (*Coregonus albula*) eggs from introduced ruffe (*Gymnocephalus cornuus*) in Bassenthwaite Lake, U.K. *Advances in Limnology* 50:171–77.

Winfield, I. J., J. M. Fletcher, and J. B. James. 2002. Species introductions to two English lake fish communities of high conservation value: A 10-year study. In Collares-Pereira, Coelho, and Cowx 2002, 271–81.

Winston, M. R. 2002. Distribution and abundance of the goldstripe darter (*Etheostoma parvipinne*) in Missouri. *Southwest Nat* 47:187–94.

Winter, B. D., and R. M. Hughes. 1997. AFS draft position statement on biodiversity. *Fisheries* 20(4): 20–26.

Wipfli, M. S., J. P. Hudson, J. P. Caouette, et al. 2003. Marine subsidies in freshwater ecosystems: Salmon carcasses increase the growth rates of stream-resident salmonids. *T Am Fish Soc* 132:371–81.

Wipfli, M. S., J. P. Hudson, D. T. Chaloner, et al. 1999. Influence of salmon spawner densities on stream productivity in southeast Alaska. *Can J Fish Aquat Sci* 56:1600–11.

Witte, F., T. Goldschidt, J. Wanink, et al. 1992. The destruction of an endemic species flock: Quantitative data on the decline of the haplochromine cichlids of Lake Victoria. *Environ Biol Fish* 34:1–28.

Witters, H. E. 1998. Chemical speciation dynamics and toxicity assessment in aquatic systems. *Ecotox Environ Safe* 41:90–95.

Wohl, N. E., and R. F. Carline. 1996. Relations among riparian grazing, sediment loads, macroinvertebrates, and fishes in three central Pennsylvania streams. *Can J Fish Aquat Sci* 53(suppl. 1): 260–66.

Woinarski, J. C. Z., and A. Fisher. 1999. The Australian Endangered Species Protection Act 1992. *Conserv Biol* 13:959–62.

Wolff, N., R. Grober-Dunsmore, C. S. Rogers, et al. 1999. Management implications of fish trap effectiveness in adjacent coral reef and gorgonian habitats. *Environ Biol Fish* 55:83–90.

Wolter, C., and R. Arlinghaus. 2003. Navigation impacts on freshwater fish assemblages: The ecological relevance of swimming performance. *Rev Fish Biol Fisher* 13:63–89.

Wonham, M. J., J. T. Carlont, G. M. Ruiz, et al. 2000. Fish and ships: Relating dispersal frequency to success in biological invasions. *Mar Biol* 136:1111–21.

Wood, E. 1985. *Exploitation of coral reef fishes for the aquarium trade.* Ross-on-Wye, UK: Marine Conservation Society. Cited in E. M. Wood 2001.

———. 2001. *Collection of coral reef fish for aquaria: Global trade, conservation issues and management strategies.* Ross-on-Wye, UK: Marine Conservation Society. www.mcsuk.org.

Wood, E. M., and A. Rajasuriya. 1999. *Sri Lanka marine aquarium fishery conservation and management issues.* Ross-on-Wye, UK: Marine Conservation Society. Cited in E. M. Wood 2001.

Wood, P. J., and P. D. Armitage. 1997. Biological effects of fine sediment in the lotic environment. *Environ Manage* 21:203–17.

Woodward Clyde Consultants. 1980. *Gravel removal studies in arctic and subarctic floodplains in Alaska.* FWS/OBS-80/08. Washington, DC: U.S. Fish and Wildlife Service.

Woodward, D. F., W. G. Brumbaugh, A. J. DeLonay, et al. 1994. Effects on rainbow trout fry of a metals-contaminated diet of benthic invertebrates from the Clark Fork River, Montana. *T Am Fish Soc* 123:51–62.

Woodward, D. F., A. M. Farag, H. L. Bergman, et al. 1995. Metals-contaminated benthic invertebrates in the Clark Fork River, Montana: Effects on age-0 brown trout and rainbow trout. *Can J Fish Aquat Sci* 52:1994–2004.

Wootton, R. J. 1990. *Ecology of teleost fishes.* London: Chapman and Hall.

World Bank. 2000. *Opportunities and constraints of grouper aquaculture in Asia.* Washington, DC: World Bank.

World Commission on Dams. 2000. *Dams and development: A new framework for decision-making.* www.dams.org and www.unep-dams.org.

World Resources Institute. 1996. *World Resources 1996–97.* New York: Oxford Univ. Press.

———. 2002. *Reefs at risk in Southeast Asia.* www.wri.org/b05_live_reef_fish_trade.html.

Worm, B., E. B. Barbier, N. Beaumont, et al. 2006. Impacts of Biodiversity Loss on Ocean Ecosystem Services. *Science* 314: 787–790

Wu, R. S. S. 1995. The environmental impact of marine fish culture: Towards a sustainable future. *Mar Pollut Bull* 31:159–66.

Wunderlich, R. C., B. D. Winter, and J. H. Meyer. 1994. Restoration of the Elwha River ecosystem. *Fisheries* 19(8): 11–19.

WWF (World Wildlife Fund). 1999. The precautionary principle. *Issue brief* (January): 3. Washington, DC: World Wildlife Fund. www.worldwildlife.org.

———. 2001. *The status of wild Atlantic salmon: A river by river assessment.* www.worldwildlife.org/salmon/pubs.cfm.

Wydoski, R. S., and J. Hamill. 1991. Evolution of a cooperative recovery program for endangered fishes in the Upper Colorado River basin. In Minckley and Deacon 1991, 123–35.

Yamamoto, M. N., and A. W. Tagawa. 2000. *Hawai'i's native & exotic freshwater animals.* Honolulu: Mutual Publishing.

Yan, N. D., B. Leung, W. Keller, et al. 2003. Developing conceptual frameworks for the recovery of aquatic biota from acidification. *Ambio* 32:165–69.

Yau, C., M. A. Collins, P. M. Bagley, et al. 2001. Estimating the abundance of Patagonian toothfish *Dissostichus eleginoides* using baited cameras: A preliminary study. *Fish Res* 51:403–12.

Young, L. G. L. 1997. Sustainability issues in the trade for wild and cultured aquarium species. In *Marketing and shipping live aquatic products*, 145–51. Ithaca, NY: Northeastern Regional Agricultural Engineering Service Cooperative Extension.

Yount, J. D., and G. J. Niemi. 1990. Recovery of lotic communities and ecosystems from disturbance: A narrative review of case studies. *Environ Manage* 14:547–69.

Yuma, M., K. Hosoya, and N. Yoshikazu. 1998. Distribution of the freshwater fishes of Japan: An historical overview. *Environ Biol Fish* 52:97–124.

YWRP (Yangtze Valley Water Resources Protection Bureau). 2002. *Questions and answers on environmental issues for the Three Gorges Project.* www.ywrp.gov.cn/english/sxquestion/3.htm.

Zamzow, J. P., and G. S. Losey. 2002. Ultraviolet radiation absorbance by coral reef fish mucus: Photoprotection and visual communication. *Environ Biol Fish* 63:41–47.

Zaret, T. M. 1974. The ecology of introductions: A case study from a central American lake. *Environ Conserv* 1:308–9.

Zaret, T. M., and R. T. Paine. 1973. Species introduction in a tropical lake. *Science* 182:449–55.

Zeller, D., S. L. Stoute, and G. R. Russ. 2003. Movements of reef fishes across marine reserve boundaries: Effects of manipulating a density gradient. *Mar Ecol—Prog Ser* 254:69–280.

Zhang, D. H., F. Cai, X. D. Zhou, et al. 2003. A concise and stereoselective synthesis of squalamine. *Organic Letters* 5:3257–59.

Zholdasova, I. 1997. Sturgeons and the Aral Sea ecological catastrophe. *Environ Biol Fish* 48:373–80.

Zimmermann, T. 1996. If World War III comes, blame fish. *U.S. News and World Report*, Oct. 21, 59–60.

Index

Species Index